Advances in Intelligent Systems and Computing

Volume 558

More information about this series at http://www.springer.com/series/11156

About this series

The series "Advances in Intelligent Systems and Computing" contains publications on theory, applications, and design methods of Intelligent Systems and Intelligent Computing. Virtually all disciplines such as engineering, natural sciences, computer and information science, ICT, economics, business, e-commerce, environment, healthcare, life science are covered. The list of topics spans all the areas of modern intelligent systems and computing such as: computational intelligence, soft computing including neural networks, fuzzy systems, evolutionary computing and the fusion of these paradigms, social intelligence, ambient intelligence, computational neuroscience, artificial life, virtual worlds and society, cognitive science and systems, Perception and Vision, DNA and immune based systems, self-organizing and adaptive systems, e-Learning and teaching, human-centered and human-centric computing, recommender systems, intelligent control, robotics and mechatronics including human-machine teaming, knowledge-based paradigms, learning paradigms, machine ethics, intelligent data analysis, knowledge management, intelligent agents, intelligent decision making and support, intelligent network security, trust management, interactive entertainment, Web intelligence and multimedia. The publications within "Advances in Intelligent Systems and Computing" are primarily proceedings of important conferences, symposia and congresses. They cover significant recent developments in the field, both of a foundational and applicable character. An important characteristic feature of the series is the short publication time and world-wide distribution. This permits a rapid and broad dissemination of research results.

Shahram Latifi
Editor

Information Technology – New Generations

14th International Conference
on Information Technology

 Springer

Editor
Shahram Latifi
Department of Electrical and Computer Engineering
University of Nevada
Las Vegas, Nevada, USA

ISSN 2194-5357   ISSN 2194-5365   (electronic)
Advances in Intelligent Systems and Computing
ISBN 978-3-319-85535-6   ISBN 978-3-319-54978-1   (eBook)
DOI 10.1007/978-3-319-54978-1

Printed on acid-free paper

This Springer imprint is published by Springer Nature
The registered company is Springer International Publishing AG
The registered company address is: Gewerbestrasse 11, 6330 Cham, Switzerland

Contents

Part VII   Software Engineering

ITNG 2017 Organization

General Chair and Editor

Shahram Latifi, UNLV, USA

Vice General Chair

Doina Bein, California State University, Fullerton, USA

Publicity Chair

Kashif Saleem, King Saud University, Saudi Arabia

Track Chairs and Associate Editors

Kohei Arai, Saga University, Japan
Alessio Bucaioni, Mälardalen University, Sweden
Glauco Carneiro, University of Salvador (UNIFACS), Brazil
Narayan Debnath, Winona State University, USA
L. A. Vieira Dias, ITA, Brazil
Ray Hasehmi, Armstrong State University, USA
Si Jung Kim, UNLV, USA
Yenumula Reddy, Grambling state University
Fangyan Shen, New York City College of Technology (CUNY), USA
Ping Wang, Robert Morris University, USA
Christoph Thuemmler, Edinburgh Napier University UK

Track Co-chairs and Session Chairs

Federico Ciccozzi, Mälardalen University, Sweden
Suvamoy Changder, National Institute of Technology, India
Thomas Jell, Siemens, Germany
Saad Mubeen, Mälardalen University, Sweden
Michel Soares, Federal University of Sergipe (UFS), Brazil
Mei Yang, UNLV, USA

Conference Secretary

Mary Roberts, Premier Hall for Advancing Science and Engineering, USA

Industry Partnership

Microsoft, Siemens

Chair Message

Welcome to the 14th International Conference on Information Technology: New Generations – ITNG 2017. It is a pleasure to report that we have another successful year for the ITNG 2017. Gaining popularity and recognition in the IT community around the globe, the conference was able to attract many papers from authors worldwide. The papers were reviewed for their technical soundness, originality, clarity, and relevance to the conference. The conference enjoyed the expert opinion of 112 author and nonauthor scientists who participated in the review process. Each paper was reviewed by at least two independent reviewers. A total of 108 articles were accepted as regular papers, and 13 were accepted as short papers (posters).

The articles in this proceeding address the most recent advances in such areas as wireless communications and networking, software engineering, information security, data mining, informatics, high performance computing architectures, the Internet, and computer vision.

As customary, the conference features two keynote speakers on Monday and Tuesday. There will be a short tutorial session Wednesday morning on deep learning. The presentations for Monday, Tuesday, and Wednesday are organized in two meeting rooms simultaneously, covering a total of 20 technical sessions. Poster presentations are scheduled for the morning and afternoon of these days. The award ceremony, conference reception, and dinner are scheduled for Tuesday evening.

Many people contributed to the success of this year's conference by organizing symposia or technical tracks for the ITNG. We benefited from the professional and timely services of Dr. Doina Bein who, in the capacity of the conference vice chair, was tremendously effective in helping with the website, communicating with authors, and helping with the review and classification of tens of articles. Dr. Yenumula Reddy deserves much credit for spearheading the review process and running a symposium on wireless communications and networking. My sincere thanks go to other symposium and major track organizers and associate editors, namely, Drs. Kohei Arai, Glauco Carneiro, Narayan Debnath, Luiz Alberto Vieira Dias, Ray Hashemi, Fangyan Shen, and Ping Wang. Dr. Kashif Saleem, the conference publicity chair, helped greatly in marketing the ITNG event around the globe.

Others who were responsible for solicitation, review, and handling the papers submitted to their respective tracks/sessions include Drs. Azita Bahrami, Wolfgang Bein, Alessio Bucaioni, Federico Ciccozzi, Si Jung Kim, Saad Mubeen, Armin Schneider, Michel Soares, Christoph Thuemmler, and Mei Yang. Dr. Tanmoy Sarkar of Microsoft was specially instrumental in reviewing the papers submitted to Agile Software track.

The help and support of Springer in preparing the ITNG book of chapters is specially appreciated. Many thanks are due to Michael Luby, the senior editor, and Nicole Lowary, the assistant editor of the Springer supervisor of publications, for the timely handling of our publication order. We also appreciate the hard work by Brinda Megasyamalan, the Springer project coordinator, who looked very closely at every single article to make sure they are formatted correctly according to the publisher guidelines. Finally, the great efforts of the conference secretary, Ms. Mary Roberts, who dealt with the day-to-day conference affairs, including timely handling volumes of e-mails, are acknowledged.

This year, the conference venue is Tuscany Suites Hotel. The hotel, conveniently located within half a mile of the Las Vegas Strip, provides an easy access to other major resorts and recreational centers. I hope and trust that you have an academically and socially fulfilling stay in Las Vegas.

Shahram Latifi
The ITNG General Chair

Reviewers List

Abbas, Haider	Debnath, Narayan	Laskar, John	Qi, Bing
Abreu, Fernando	Di Ruscio, Davide	Li, Shujun	Ratchford, Melva
Abukmail, Ahmed	Dias, L.A. Vieira	Maglaras, Leandros	Reddy, Yenumula
Ahmad, Aftab	Durelli, Rafael	Maia, Marcelo	Rehmani, Mubashir
Ahmed, Adel	Eldefrawy, Mohamed	Malavolta, Ivano	Romano, Breno
Ahmed, Iqbal	El-Ziq, Yacoub	Marques, Johnny	Saadatmand, Mehrdad
Al-Muhtadi, Jalal	Eramo, Romina	Meghanathan, Natarajan	Saleem, Kashif
Anikeev, Maxim	Fernandes, João	Mesit, Jaruwan	Sampaio, Paulo
Araujo, Marco	Figueiredo, Eduardo	Mialaret, Lineu	Sarijari, Mohd
Awan, Fahim	Ford, George	Monteiro, Miguel	Sarkar, Tanmoy
Baguda, Yaqoub	Garcia, Rogério	Montini, Denis	Sbeit, Raed
Bein, Doina	Garuba, Moses	Mubeen, Saad	Schneider, Armin
Bein, Wolfgang	Gawanmeh, Amjad	Mukkamala, Ravi	Shen, Fangyang
Billinghurst, Mark	Hashemi, Ray	Munson, Ethan	Shoraka, Babak
Bucaioni, Alessio	Hsiung, Pao-Ann	Naznin, Mahmuda	Silva, Bruno
Camargo, Valter	Imran, Muhammad	Nguyen, Nikyle	Silva, Paulo
Carlson, Jan	Jell, Thomas	Novais, Renato	Soares, Michel
Carneiro, Glauco	Junior, Methanias	Nwaigwe, Adaeze	Sreekumari, Prasanthi
Chapman, Matthew	Kannan, Sudesh	Oliveira, Adicinéia	Thuemmler, Christoph
Chen, Chun-I	Khan, Zahoor	Oliveira, Marcos	Tolle, Herman
Cheruku, Raghavendar	Khanduja, Vidhi	Orgun, Mehmet	Tsetse, Anthony
Christos Kalloniatis	Kim, Hak	Owen, Richard	Tyler, Alexis
Cicchetti, Antonio	Kim, Si Jung SJ	Pang, Les	Vulgarakis, Aneta
Ciccozzi, Federico	Kinyua, Johnson	Pathan, Al-Sakib	Wang, Jau-Hwang
Clincy, V.	Koch, Fernando	Paulin, Alois	Wang, Yi
Daniels, Jeff	Kohei, Arai	Peiper, Chad	Wang, Ping
Darwish, Marwan	Konduri, Rajeswari	Penteado, Rosângela	Williams, Hank
David, Jose	Kronbauer, Artur	Popa, Vlad	Woo, Woontack

Part I

Networking and Wireless Communications

Performance Enhancement of OMP Algorithm for Compressed Sensing Based Sparse Channel Estimation in OFDM Systems

Vahid Vahidi and Ebrahim Saberinia

Abstract

Long duration of the channel impulse response along with limited number of actual paths in orthogonal frequency division multiplexing (OFDM) vehicular wireless communication systems results in a sparse discrete equivalent channel. Implementing different compressed sensing (CS) algorithms enables channel estimation with lower number of pilot subcarriers compared to conventional channel estimation. In this paper, new methods to enhance the performance of the orthogonal matching pursuit (OMP) for CS channel estimation method is proposed. In particular, in a new algorithm dubbed as linear minimum mean square error-OMP (LMMSE-OMP), the OMP is implemented twice: first using the noisy received pilot data as the input and then using a modified received pilot data processed by the outcome of the first estimator. Simulation results show that LMMSE-OMP improves the performance of the channel estimation using the same number of pilot subcarrier. The added computational complexity is studied and several methods are suggested to keep it minimal while still achieving the performance gain provided by the LMMSE-OMP including using compressive sampling matching pursuit (CoSaMP) CS algorithm for the second round and also changing the way the residue is calculated within the algorithm.

Keywords

Compressed sensing • OMP • OFDM • Channel estimation • Initial • Sparsity

1.1 Introduction

Orthogonal frequency division multiplexing (OFDM) has become a standard scheme for wideband wireless data communications when dealing with channels with large multipath delay spread because of its ability to avoid Inter-Symbol Interference (ISI) and simple one-tap channel equalization for each sub-carrier in frequency domain. This one-tap channel equalization requires the channel parameters to be estimated at the receiver. In a wideband OFDM vehicular wireless communication system, the

discrete channel impulse response can be a sparse vector containing few non-zero elements [1]. While conventional channel estimation methods such as least square (LS) channel estimator and linear minimum mean square error (LMMSE) channel estimator along with new approaches have been studied and implemented in the past decade [2], for sparse channels, compressed sensing (CS) [3] algorithms can be implemented for channel estimation with fraction of the cost in terms of complexity and bandwidth efficiency [4–14].

While there are several approaches to CS, one of the most popular CS greedy algorithms proposed for sparse channel estimation is the orthogonal matching pursuit (OMP) and its variations. OMP was first proposed for sparse image processing in [15] and then studied for general sparse channel estimation in [4] and for OFDM channel estimation in

V. Vahidi (✉) • E. Saberinia
Electrical and Computer Engineering Department, University of Nevada, Las Vegas, Las Vegas, USA
e-mail: vahidi@unlv.nevada.edu; Ebrahim.Saberinia@unlv.edu

© Springer International Publishing AG 2018
S. Latifi (ed.), *Information Technology – New Generations*, Advances in Intelligent
Systems and Computing 558, DOI 10.1007/978-3-319-54978-1_1

particular in [5–14]. Optimum pilot placement design for OMP algorithm was studied in [5–7]. The modified versions of OMP is used for channel estimation in [9–14]. In order to reduce the channel estimation time [9–11], used Compressive Sampling Matching Pursuit (CoSaMP), and [12] used regularized OMP (ROMP). It is indicated in those papers that the accuracy of CoSaMP and ROMP is close to the accuracy of OMP method. On the other hand, in order to increase the accuracy of the channel estimation [13, 14], implemented simultaneous OMP (SOMP) and stagewise OMP (StOMP) respectively.

In this paper, we propose another step to enhance the performance of OMP for sparse OFDM channel estimation using the same number of pilot subcarriers. In the new algorithm, named as LMMSE-OMP, the channel is estimated using CS twice. The first step is exactly the same as the conventional OMP but the estimated channel is not used for OFDM data demodulation. Instead, we use the first channel estimation to create an approximate of the pilot received data with higher signal to noise ratio (SNR). In second step, we rerun the CS channel estimation algorithm with the enhanced pilot data as the input to achieve a better estimate of the channel. Simulation results indicate a considerable improvement in the performance of the channel estimator in comparison to the OMP.

Implementing LMMSE-OMP using OMP in both steps doubles the computational complexity of the channel estimator. The added complexity can be decreased using other CS methods. In particular, if OMP is used at the first step, we can estimate the sparsity of the channel and the variance of the noise along with the first channel estimation and can use CS algorithm such as CoSaMP in second step to decrease the added computational load. Simulation results shows that using CoSaMP instead of OMP in second step decrease the performance slightly while decreasing the added computational complexity significantly.

The remainder of this paper is organized as follows. Section 1.2 describes the system model and elaborates channel estimation problem. Section 1.3 explains the basics of CS based channel estimation. The proposed LMMSE-OMP method is discussed in Sect. 1.4. Simulation results are presented in Sect. 1.5 and finally, Sect. 1.6 concludes the paper.

1.2 System Model

The impulse response of a time varying multipath wireless channel can be presented as [16]:

$$h(\tau, t) = \sum_{k=1}^{K} a_k(t)\delta(\tau - \tau_k(t)) \qquad (1.1)$$

where K is the total number of the paths and a_k and τ_k are the amplitude and delay of the kth path respectively. In an OFDM system, a vector of data symbols (X) in frequency domain is converted to samples of time domain transmitted signal using an inverse fast Fourier transform (IFFT) and, after adding the cyclic prefix (CP), it is converted to analog signal and passed through the channel. The received signal is sampled at the same rate of the transmitter DAC and the digital time domain signal is equal to the transmitted time domain signal passed through the following channel [16]:

$$h(n) = \sum_{l=1}^{L} \alpha_l(n)\delta((n-l)T_s) \qquad (1.2)$$

where $h(n)$ is the discrete equivalent of the channel at time n and T_s is the sampling time of the system. The total number of taps represented by L depends on the maximum delay spread of the channel and T_s and amplitude of each tap is represented by $\alpha_l(n)$. For wideband communications, where T_s can be very small, it is possible that no actual signal reception path falls into some of the taps of digital equivalent channel. Therefore, for some taps the amplitudes $\alpha_l(n)$ are equal to zero. In applications such as wireless vehicular communications, a large number of taps can have zero amplitudes with only a few non-zero amplitude. In this case the channel is called a sparse channel for which $S \ll L$ where S is the number of non-zero taps [5].

In any OFDM system, if the length of cyclic prefix is larger than the channel length ($L \leq C$), we can achieve an approximation of the transmitted data by removing the cyclic prefix from the received data and taking an FFT of the remaining data. In fact, the output of FFT which is called the received signal in frequency domain can be written as [5]:

$$Y = HX + Z \qquad (1.3)$$

where X is the $N \times 1$ vector of transmitted symbols in frequency domain and H is the $N \times N$ channel matrix in the frequency domain. The symbol N indicates the number of subcarriers in OFDM system which also dictates the size of IFFT at the transmitter and FFT at the receiver. The additive white Gaussian noise (AWGN) is represented by $N \times 1$ vector Z. It can be shown that for channels with zero or small Doppler shifts, the channel matrix H is a diagonal matrix whose diagonal elements are equal to N-point FFT of digital equivalent channel $h[n]$ [5]:

$$H = diag\,(Fh) \qquad (1.4)$$

where h is the $N \times 1$ time domain channel vector whose first L elements are equal to α_l, $l = 1, \ldots, L$ and the rest are zeros. The Matrix F is the $N \times N$ Fourier transform matrix.

If the channel H is known, the transmitted symbols X can be estimated from Y by one tap equalization. The minimum mean square error (MMSE) estimate of X is equal to [2]:

$$\widehat{X_k} = \frac{conj\left(\widehat{H}_k\right)}{\left(\widehat{H}_k\right) \times conj\left(\widehat{H}_k\right) + \sigma^2} Y_k, \qquad (1.5)$$

where $\widehat{H}$ is the estimate of the channel and σ^2 is the variance of AWGN. The channel estimation can be performed by sending a pilot signal and applying classical estimation methods such as least square (LS) or linear minimum mean esquire error (LMMSE) schemes. It is indicated in [2] that by sending a $N \times 1$ vector of symbols as pilot and measuring the received $N \times 1$ matrix of received data, the LS estimate of H can be obtained as:

$$\widehat{H}_{ls} = diag\left[\frac{Y_{p_1}}{X_{p_1}} \frac{Y_{p_2}}{X_{p_2}} \cdots \frac{Y_{p_N}}{X_{p_N}}\right]^T \qquad (1.6)$$

where X_{p_i} is the ith transmitted pilot in the frequency domain and Y_{p_i} is the ith received pilot in the frequency domain. A linear minimum mean square estimator (LMMSE) of the channel can be obtained using $\widehat{H}_{ls}$ as [2]:

$$\widehat{H}_{LMMSE} = R_{HH}\left[R_{HH} + \frac{\beta}{\sigma^2}I_N\right]^{-1}\widehat{H}_{ls} \quad , \qquad (1.7)$$

where R_{HH} is the autocorrelation matrix of the frequency domain channel, the number β is defined as $\frac{E\left(X_k^2\right)}{E\left(\frac{1}{X_k^2}\right)}$ when X_k is the transmitted symbol in the frequency domain and σ^2 represents the variance of AWGN. Matrix I_N is the $N \times N$ identity matrix.

1.3   Compressed Sensing for Sparse Channel Estimation

While LS and LMMSE estimates of the channel matrix work for both dense and sparse wireless channels, using CS algorithms for the sparse case can decrease the size of the pilot signal needed and the number of calculations for channel estimation. When a pilot signal is transmitted and we would like to estimate the time domain channel from the received signal Y, one can rewrite the received signal presented in Eq. (1.3) as [5]:

$$Y = X_d F h + Z \qquad (1.8)$$

where X_d is the $N \times N$ diagonal matrix with diagonal elements equal to pilot data X. The time domain channel

vector h is assumed to be sparse with only S non-zero elements where $S \ll L$. The location of non-zero elements of h and their values are to be estimated with CS algorithms. It is indicated in [17, 18] that a good reconstruction of an S sparse data can be obtained by using only a pilot size between 3S and 4S instead of a whole OFDM symbol of size N. If P indicates the number of pilots used, we can separate transmitted and received pilot signals and arrange them in their vectors as [13]:

$$Y^{CS} = X_d^{CS}F^{CS}h + Z^{CS} \qquad (1.9)$$

where X_d^{CS} is a $P \times P$ diagonal matrix of pilots, F^{CS} is a $P \times N$ matrix that its rows are chosen from the rows of F (the rows that corresponds to the pilot places), Z^{CS} is $P \times 1$ AWGN and Y^{CS} is $P \times 1$ receive data corresponding to pilot subcarriers. The estimated channel $\widehat{h}$ would be obtained by solving l_0 minimization problem as [13]:

$$\min\|h\|_0 \ s.t. \left\|Y^{CS} - \varphi h\right\|_2 \leq \sigma \qquad (1.10)$$

where $\varphi = X_d^{CS}F^{CS}$ is the measurement matrix and $\|h\|_0$ defines the number of non-zero elements of h and σ is the variance of AWGN. Since this problem is a NP hard problem, it is indicated in [13] that it can be replaced by a convex optimization problem as follows:

$$\min\|h\|_1 \ s.t. \left\|Y^{CS} - \varphi h\right\|_2 \leq \sigma \qquad (1.11)$$

Where $\|h\|_1$ is the norm 1 of the channel that is obtained by the summation of the absolute values of the channel taps.

While there are several ways to reconstruct h from Y^{CS}, orthogonal matching pursuit (OMP) is a common greedy algorithm for obtaining h. In OMP, we reconstruct Y^{CS} using selected columns of φ. The column of φ that has the largest correlation with Y^{CS} is chosen to initialize the new reconstruction matrix M. Then, we subtract the portion of Y^{CS} that is covered by the new column of M, to calculate a residue which in turn is correlated with columns of φ. In each step a new column is added to matrix M, and the channel of ith iteration is calculated as:

$$\widehat{h}_{OMP_i} = \left(M_i^H M_i\right)^{-1}M_i^H Y^{CS} \qquad (1.12)$$

and a new residue is calculated:

$$R_i = Y^{CS} - \varphi\widehat{h}_{OMP_i} \qquad (1.13)$$

where M_i is the M matrix at ith iteration. The iteration stops when $\frac{\|R_i\|_2}{\|R_{i-1}\|_2} < thresold$. The time domain channel, $\widehat{h}_{omp}$, is obtained through the last iteration of (1.12).

Using $\widehat{h}_{omp}$, one can calculate $\widehat{H}_{ls}$ as:

$$\widehat{H}_{ls} = diag\left(F\widehat{h}_{omp}\right) \qquad (1.14)$$

and $\widehat{H}_{LMMSE}$ using (1.7).

1.4 Enhanced LMMSE-OMP Channel Estimation

The performance of the OMP channel estimation depends on the number of pilot sub carriers and the signal to noise ratio of the received data. We found out that we can enhance the performance of the channel estimation by keeping the same number of pilots but enhancing the SNR by running OMP with noiseless approximate of the received data. In this new method, we first find an approximate of the channel using OMP but instead of using that estimate to demodulate the data subcarrier we create an estimate of the received pilot subcarrier values without noise:

$$Y_{LMMSE}^{CS} = \widehat{H}_{LMMSE} \cdot X_d \qquad (1.15)$$

We use Y_{LMMSE}^{CS} to run another round of OMP and estimate the channel. Simulation results in next section shows that this enhances the performance of the channel estimation significantly.

While we can continue doing a new set of OMP after each estimation, the improvement on performance will decrease while adding to the complexity of channel estimation. It seems that performing OMP twice (one with original noisy received pilot signals to get and initial estimate of the channel and the other using an enhanced version of the received pilot data) gives the most improvement in performance.

The purposed method will have twice the complexity of traditional OMP. If that is a concern we can replace the second run with a lower complexity method such as compressive sampling matching pursuit (CoSaMP) [19]. CoSaMP is an alternative CS greedy algorithm that has lower complexity compared to OMP. Unlike OMP it requires the sparsity of the channel to be known. In CoSaMP algorithm, the matrix M is initiated with $2S$ columns of the measurement matrix φ with the highest correlation with received pilot data. At each iteration, M is updated by adding new columns that describe the reminder. The algorithm runs a fixed number of iteration between $4S$ and $5S$ [20], and produces reasonable channel estimation. In our method, if we first run OMP to find a better approximation of the received data, we can also have an estimate of the sparsity of the channel and use it to run CoSaMP in second round. According to [19] the complexity of OMP is of $O(S.N.P)$ and the complexity of CoSaMP is of $O(N.P)$. Therefore, the complexity of OMP-OMP for enhanced method is of

$O(2S.N.P)$ and the complexity of OMP-CoSaMP is of $O((S + 1).N.P)$.

Since the OMP-CoSaMP method has lower performance than OMP-OMP method, we made a recovery to the CoSaMP round of our channel estimation method. Instead of implementing LS for channel estimation as it is indicated in (12), MMSE method is applied for reconstruction of the sparse channel from matrix M at each iteration of CoSaMP:

$$\widehat{h}_{CoSaMPi} = \left(M_i^H M_i + \frac{1}{SNR}\right)^{-1} * M_i^H Y^{CS} \qquad (1.16)$$

1.5 Simulation Results

In this section, we present the simulation results on the performance of the proposed LMMSE-OMP method and compare it with other estimators. As an example of a sparse channel, we use ITU/Vehicular Type B channel model presented in [1]. The delay spread of the channel is 20 μsec and assuming OFDM signal of 10 MHz bandwidth with $N = 256$ sub-carriers, the digital equivalent channel has $L = 206$ taps. Only $S = 6$ of those taps have a non-zero amplitude. Total number of pilots used for channel estimation is equal to $4S = 24$ pilots. The pilots are placed to generate the best performance according to the method described in [5]. The location of the pilots are as follows:

19,22,23,56,79,107,108,138,153,153,173,177,178,189, 203,210,213,232,235,237,240,254,255,256

Using Monte Carlo simulation, we have calculated the performance of conventional CS channel estimator using OMP and CoSaMP and our purposed method of double running the CS algorithms named OMP-OMP and OMP-COSAMP. We have also evaluated the enhancement when we run the algorithm more than twice. Figure 1.1 shows the plot for the normalized mean square error (NMSE) vs. signal to noise ratio (SNR) for these channel estimators. NMSE of the estimated channel is defined as $\frac{\left\|\widehat{h}-h\right\|_2^2}{\|h\|_2^2}$ [13]. This results show that we can get considerable improvement in channel estimation with the same number of pilots if we process the modified received data for pilots. While it is better to run both algorithms as OMP, if we are concerned about added complexity we can use CoSaMP at the second run and still get a very good result. Looking at the results for rerunning CS algorithm more than two times shows that the improvement is very marginal and may not worth the added complexity. Besides that, just by using MMSE instead of LS, the performance of OMP-CoSaMP would be better than OMP-OMP. This method is indicated by OMP-CoSaMP2 in the simulation results.

Fig. 1.1 The comparison between NMSE performance of different methods

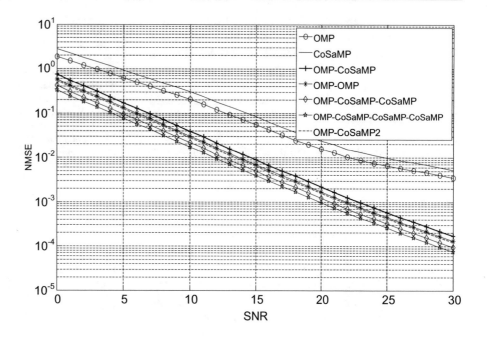

Fig. 1.2 The comparison between BER performance of different methods

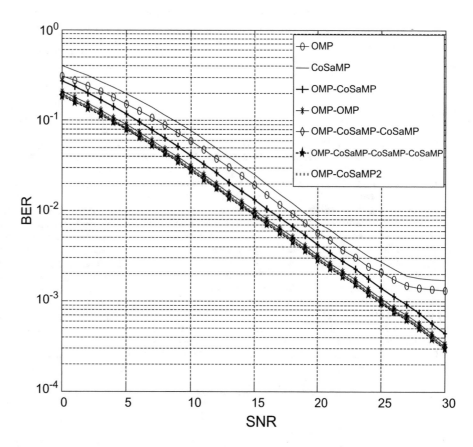

In order to show the effect of enhancing channel estimation performance on the demodulating OFDM signal, we have also simulated a binary phase shift keying (BPSK) OFDM signaling in this channel and demodulated it using various channel estimators described before. Figure 1.2 shows the plot of bit error rate (BER) vs SNR in different channel estimation scenarios. As it is indicated in this figure, a considerable improvement in BER is achieved when the second round of CS method is run with the implementation of more accurate data that where achieved from the first time. It is also shown that there is almost no difference between the performances of OMP-CoSaMP2 method and the method that implements CoSaMP three times after one OMP.

1.6   Conclusion

In this paper, an enhancement to the OMP algorithm for CS estimation of sparse OFDM channel is proposed. The new LMMSE-OMP algorithm uses the same number of pilot sub-carriers as OMP but provide better performance. In LMMSE-OMP the CS channel estimation algorithm is implemented twice. When an initial estimate of the channel, noise variance and sparsity level of the channel are obtained by the first CS algorithm, a new LMMSE estimate of the received pilot data is created and used by second CS algorithm to produce the final estimate of the channel. The simulation results indicate that the proposed LMMSE-OMP method has better performance in comparison to the conventional OMP. Using OMP as the CS algorithm for the first step and CoSaMP as the CS algorithm for the second step provide the best balance between added performance and added computational complexity.

References

1. ITU: International Telecommunication Union, Guidelines for the Evaluation of Radio Transmission Technologies (RTTs) for IMT-2000, ITU-R Recommendation M.1225. (1997).
2. Zou, W., & Lam, W. H. A fast LMMSE channel estimation method for OFDM systems. *EURASIP Journal on Wireless Communications and Networking, 2009*, 752895.
3. Candes, E. J., Romberg, J., & Tao, T. (2006). Robust uncertainty principles: exact signal reconstruction from highly incomplete frequency information. *IEEE Transactions on Information Theory, 52* (2), 489–509.
4. Karabulut, G. Z., & Yongacoglu, A. (2004). Sparse channel estimation using orthogonal matching pursuit algorithm. *Vehicular Technology Conference, 2004. VTC2004-Fall. 2004 I.E. 60th,* 2004, pp. 3880–3884, Vol. 6.
5. He, X., & Song, R. (2010). Pilot pattern optimization for compressed sensing based sparse channel estimation in OFDM systems. *Wireless Communications and Signal Processing (WCSP), 2010 International Conference on, Suzhou*, pp. 1–5.
6. Peng, Y., Alexandropoulos, G. C., Zhao, H., Duan, H. (2013). Performance analysis of OMP-based channel estimation for

OFDM systems with periodical pilots and virtual subcarriers. *Computing, Management and Telecommunications (ComManTel), 2013 International Conference on, Ho Chi Minh City*, Vietnam, pp. 11–16.
7. Qi, C., & Wu, L. (2011). Optimized pilot placement for sparse channel estimation in OFDM systems. *IEEE Signal Processing Letters, 18*(12), 749–752.
8. Yuan, W., Zheng, B., Yue, W., Wang, L. (2011). Two-way relay channel estimation based on compressive sensing. *Wireless Communications and Signal Processing (WCSP), 2011 International Conference on, Nanjing*, pp. 1–5.
9. Gaur, Y., & Chakka, V. (2012). Performance comparison of OMP and CoSaMP based channel estimation in AF-TWRN scenario. *2012 Third International Conference on Computer and Communication Technology.*
10. Li, X., Jing, X., Sun, S., Huang, H., Chen, N., Lu, Y. (2013). An improved reconstruction method for compressive sensing based OFDM channel estimation. *2013 International Conference on Connected Vehicles and Expo (ICCVE), Las Vegas*, pp. 100–105.
11. Pan, H., Xue, Y., Mei, L. Gao, F. (2015). An improved CoSaMP sparse channel estimation algorithm in OFDM system. *Signal Processing, Communications and Computing (ICSPCC), 2015 I.E. International Conference on, Ningbo*, pp. 1–4.
12. Mei, L., Gao, F., Pan, H Xue, Y. (2015). An improved ROMP sparse channel estimation algorithm in OFDM system. *Signal Processing, Communications and Computing (ICSPCC), 2015 I.E. International Conference on, Ningbo*, pp. 1–4.
13. Qi, C., & Wu, L. (2011). A hybrid compressed sensing algorithm for sparse channel estimation in MIMO OFDM systems. *2011 I.E. International Conference on Acoustics, Speech and Signal Processing (ICASSP), Prague*, pp. 3488–3491.
14. Chisaguano, D. J. R., Hou, Y., & Higashino, T. (2015). Minoru Okada ISDB-T diversity receiver using a 4-element ESPAR antenna with periodically alternating directivity. *ITE Transactions on Media Technology and Applications, 3*(4), 268–278.
15. Pati, Y. C., Rezaiifar, R., Krishnaprasad, P. S. (1993). Orthogonal matching pursuit: recursive function approximation with applications to wavelet decomposition. *Signals, Systems and Computers, 1993. 1993 Conference Record of the Twenty-Seventh Asilomar Conference on, Pacific Grove*, pp. 40–44, Vol. 1.
16. Shao, X., Chen, W., Tao, M., Ren, X. (2014). Statistics-based channel estimation and ici mitigation in OFDM system over high mobility channel. *High Mobility Wireless Communications (HMWC), 2014 International Workshop on, Beijing*, pp. 151–155.
17. Candès, E. J., Romberg, J. (2005, January) Practical signal recovery from random projections. In *SPIE International Symposium on Electronic Imaging: Computational Imaging III*, San Jose.
18. Tsaig, Y., Donoho, D. L. (2004). Extensions of compressed sensing. Technical report, Department of Statistics, Stanford University.
19. Tropp, J. A., & Needell, D. (2009). Cosamp:iterative signal recovery from incomplete and inaccurate measurements. *Applied and Computational Harmonic Analysis*, 301–321.
20. Satpathi, S., & Chakraborty, M. (2014). On the number of iterations for convergence of CoSaMP and SP algorithm. arXiv preprint arXiv: 1404.4927.

CARduino: A Semi-automated Vehicle Prototype to Stimulate Cognitive Development in a Learning-Teaching Methodology

Everton Rafael da Silva and Breno Lisi Romano

Abstract

This paper aims to present the development of an semi-automated vehicle prototype using Arduino and sensors that will be controlled by software developed for Android that can simulate the execution of manual and semi-automatic paths according to the user's needs. It will be presented both the physical and the logical development of the proposed vehicle and will be presented a set of experiments to demonstrate the feasibility of its application to different situations, emphasizing cognitive development in a learning-teaching methodology for children, youth and adults. Finally, we carried out a cost analysis in the market, based on some e-commerces available, to design the physical development of the proposed prototype.

Keywords

Component • Arduino • Android • Driver software • Prototype • Cognitive development

2.1 Introduction

This paper aims to show the feasibility of developing a semi-automated vehicle prototype, entitled CARduino, with low cost using the Arduino microcontroller that is controlled by an Android application using Bluetooth communication.

In CARduino locomotion, DC motors powered by batteries are used, in addition to reduction boxes to increase the torque of the motors and circuits used to reverse the direction of rotation.

It is hoped that this research will contribute to the cognitive development of children, youth and adults, complementing existing research in these areas, as will be presented to the course of this paper [1]. Additionally, as a future complementary research, this project will also assist beginners in programming to understand in a practical way, the operation of

programming using blocks (block diagrams) and viewing the operation in practice of the developed program [2].

Another motivation of this work was to unite the practical part of the construction of an automated vehicle prototype with programming for mobile devices with the Android operating system, one of the main market of operating systems. Moreover, as major contributions, will be shown in detail:

- the design of a embedded software in the automated vehicle prototype Arduino;
- the physical construction of this prototype, configuring the Bluetooth communication on at a low level using a specific sensor for this;
- the software design in Android that will communicate with the prototype of the proposed vehicle;
- simulation of the prototype proposed to illustrate its operation to be applied in cognitive development; and
- a cost analysis for design of the proposed prototype.

The following sections will present these outstanding contributions previously.

E.R. da Silva (✉) • B.L. Romano
Federal Institute of Education, Science, and Technology of Sao Paulo, Sao Joao da Boa Vista, SP, Brazil
e-mail: evertonrafael@ifsp.edu.br; blromano@ifsp.edu.br

S. Latifi (ed.), *Information Technology – New Generations*, Advances in Intelligent Systems and Computing 558, DOI 10.1007/978-3-319-54978-1_2

2.2 Related Works

In this section, we introduce some papers related to this research that stood out in recent years in Brazil and in the world: Wirsing (2014) shows the sending and receiving data via Bluetooth, used Arduino and a device with Android operating system [3]; Cardoso (2014) had a home automation system controlled Arduino [4]; Moreira (2012) who presented a weather station model for measuring temperature and humidity of the Amazon [5]; Scriptore (2014) shows the use of a distributed database, which stores information of humidity and temperature measurements using Arduino and Android [6]; Steps (2011) shows the operation of vehicle accessories used Arduino [7]; Lim et al. (2014) discloses a system developed for measuring environmental factors Arduino in a farm field [8]; and Watve (2015) has a verification system for plants using Arduino and Android [9].

In working Wirsing (2014), the Bluetooth device configuration was presented, showing how were made the connections enters the Arduino device and the Bluetooth communication module HC-05 (Fig. 2.1):

As is shown in Fig. 2.1, the Bluetooth module requires 4 connections, which are: the communication ports TX and Arduino RX, GND (Ground or earth) and a 3.3 V connection to power supply. The Android device will connect via Bluetooth with the Arduino.

Some settings are also presented in this paper in relation to communication with Android, where Android has to check if your Bluetooth is active or not, and if it is not active, you have to turn it on. Only after made these checks, Android Arduino can send and receive data via Bluetooth.

In Cardoso's work (2014), part of home automation is presented using the Arduino. This paper presents another form of communication between mobile devices and the Arduino - GSM (Global System for Mobile - Global System for Mobile). With the implementation of GSM as a new type of communication it was also used another communication service - SMS (Short Message Service - Short Message Service) - which could be sent texts with up to 160 characters and a low financial cost.

In this work, has presented a data conversion problem of type float to char, which, to solve the problem, a data conversion table was made. The scanned data had to be converted to ASCII standard for the GSM module could send SMS with the obtained data. The source of this work

was conducted in order to work with other similar sensors, which may add other or work only with the proposed sensor.

In working Scriptore (2014) a database technology was presented distributed, where they are stored read data of humidity and temperature sensors connected to the Arduino. We used a WebService as a solution to systems integration and communication between different applications, thus enabling the storage of data in a MySQL database (Database Management System that uses SQL or Structured Query Language). To achieve the objective of this study also used up an Ethernet Shield - module for Arduino that has the connection function with network and data storage SD Card (Secure Digital Card or no Volatile Memory Card).

Communication and transmission of data and Arduino Ethernet Shield was made by a technology called JSON (JavaScript Object Notation), a lightweight format for the exchange of computer data. The synchronism of the data is done by a WebService PHP (Hypertext Preprocessor - programming language for internet) that makes a request to the Ethernet Shield WebService and returns a query that is stored on the SD Card. With this, a list of measurements is generated and displayed on a software on Android.

Passos (2011) shows in his work a prototype to simulate the control of automotive features. The prototype has some servomotors, which are connected to the Arduino and Bluetooth module. In this work, it is also used a software called Amarino developed for Android that is the prototype of control using the cell accelerometer.

As forgetting a key problem in the vehicle is common, the Passos's research (2011) presents a solution to this problem. With Android software, the car user can connect to the prototype and trigger the opening of locks or car windows, not requiring the intervention of a professional to its opening, so without damaging the car.

Watve (2015) presents an implementation of a system for real time data read from a factory using Arduino and Android. The system is to monitor the humidity and temperature sensors. The software developed for Android uses features such as camera and microphone. The camera is used to log into the factory environment and microphone to receive voice commands within the software. The software also has reporting using graphics, thus presenting the performance of sensors of each environment where they are used. All sensors are used together with the Arduino UNO.

2.3 CARduino Development

In this section, you will be presented as we developed the CARduino - automated vehicle prototype using the Arduino to control sensors / motors and Android to control the interaction of the vehicle with the user. First, it will be presented the tools and components used in vehicle development. It

Fig. 2.1 Connections - bluetooth and Arduino module

will then be shown how to perform the Arduino setup along with Bluetooth HC-05 module and the configuration of the HC-SR04 Ultrasonic Sensor operating in conjunction with the Buzzer (loudspeaker). Finally, development of the complete vehicle prototype will be presented, focusing on both the physical (hardware) and the development of embedded software and the Arduino controller software on Android.

2.3.1 Setting the Bluetooth Module Integrated with Arduino

For a Bluetooth connection to the HC-05 module, there is the need for a hardware-based configuration to enable the administrator mode and a configuration via software to change the name and Bluetooth device password. When Bluetooth modules purchased, they come with a factory default setting, so the configuration is intended to customize the device name, change the password for security measures, and display the MAC address required for connections. The hardware configuration via It is diagrammed in Fig. 2.2.

As shown in Fig. 2.2, a link between 3.3 V Arduino output port directly to pin 34 of the Bluetooth module has been made, aiming to enable its run mode, allowing to amend its factory settings.

Also links to power the Bluetooth module powered directly from the Arduino 5 V output port and GND connected to the VCC and GND of Bluetooth module doors were made.

The TX and RX pins are Bluetooth module communication pins (TX - RX and transmission - reception) and must be configured shape illustrated in Fig. 2.2. The use of resistors

was needed to create a voltage divider for the RX pin, thus preventing damage to the component by excessive electric current. For this reason, we used two 330Ω resistors (ohms) in series reduces the electric current of 5 V to 2.5 V.

In order to set up the Bluetooth Module HC-05 changing your ID, recovering your MAC address and changing your password pairing, it was held the command sequence shown below in Serial Monitor the Arduino IDE:

- **AT + NAME** – Returns the device name;
- **AT + NAME = NOVO_NOME**→Return OK - Sets a new name for the Bluetooth device;
- **AT + ADDR** – Returns the MAC address of the device;
- **AT + PSWD** – Returns the Password (password) to connect to the device;
- **AT + PSWD = NOVA_SENHA**→Returns OK - Sets a new password for the Bluetooth device.

This configuration allows the module to be configured according to user needs, to use and integration with Android.

Table 2.1 shows the source code that allows the Bluetooth module configuration as previously reported. There was the need to use a library called SoftwareSerial.h, for the modification of the serial ports to be used by Bluetooth, in order to avoid conflict between the serial port of the computer with the Bluetooth HC-05 module also works with the serial port.

The Bluetooth Module works with a speed different from the normal serial port (value 9600), so to set this value, we used the command BTserial.begin (38400). This setting should always be within the setup () function code, it will be executed only once, without the need for new configuration later.

Fig. 2.2 Configuration schema of the Bluetooth module

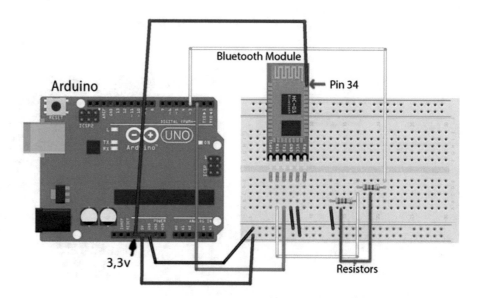

2.3.2 Physical Prototype (Hardware) Automated Vehicle

Figure 2.3 shows the upper and lower parts of CARduino prototype in this research, containing the following components: a Bluetooth module HC-05, an ultrasonic

Table 2.1 Code for the Bluetooth configuration

```
#include<SoftwareSerial.h>
SoftwareSerialBTserial(22,24); // RX | TX
char c = ' ';
void setup(){
   Serial.begin(9600);
   Serial.println("Arduino is ready!");
   BTserial.begin(38400);
}
void loop(){
   if (BTserial.available()){
      c = BTserial.read();
      Serial.write(c);
   }
   if (Serial.available()){
      c = Serial.read();
      BTserial.write(c);
   }
}
```

HC-SR04 sensor, a buzzer, fifteen Jumper cables, two motors with reduction and two modules batteries, one with four rechargeable batteries to power the motors and the other with a high performance battery to power the remaining components.

As can be seen in Fig. 2.3, the top of the breadboard prototype are the HC-SR04 Ultrasonic sensor the buzzer and the batteries. At the bottom of the prototype were coupled Arduino MEGA, the Motor Shield and two engines. All connections necessary to design the CARduino hardware prototype are shown in the schematic of Fig. 2.4.

In this prototype, it could have been eliminated engine power battery but overtax the other battery, having a lot shorter than expected. For this reason, we used four batteries with on / off switch for driving the motors.

Note that we chose to use the Arduino MEGA in CARduino design based on comparative analysis of the different types of Arduino presented on its official website (www.arduino.cc, 2016). This comparison shows that there is a big difference between the amount of communication ports (analog and digital) of the Arduino Uno and Arduino MEGA, most widely used models in the market. The choice for Arduino MEGA justified by this have more analog and digital ports to meet all requirements to be used in this work.

2.3.3 Development of Software in Boarded Arduino

In this section, we will present the development of embedded software on the Arduino to control sensors and automated vehicle engines. The software used for the development was the Arduino IDE 1.65.

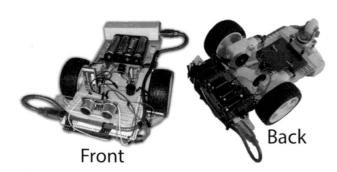

Fig. 2.3 Full prototype of CARduino

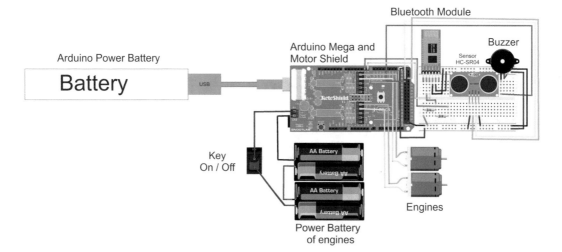

Fig. 2.4 Complete physical schema of CARduino

Because the source code developed and embedded in Arduino was extensive, this section will be presented the operating logic of the embedded software.

The setup () function of the embedded software is the function that must be set all input and output pins of the Arduino and other issues that will be performed only once by the embedded software on the Arduino. The Serial.begin command (9600) means that communication is initiated via the serial port with a 9600 baud rate, which determines the data communication via Bluetooth with the driver software developed on Android.

In the loop () function, it is developed every part of the code that will be repeated forever, following the operation and status changes of embedded software. Initially, defines the instructions for operating the automated vehicle engines. The movement is performed according to the data read by the serial port. This data is received by the Arduino Bluetooth module via the serial communication port. It is worth mentioning that all communication performed by Bluetooth HC-05 module is the serial type.

To illustrate the operation of forward movement of the automated vehicle when the number 8 is read through the serial port, the software will enter a condition, as shown in the last four lines of Table 2.2. First, should stop all motors and then it runs the drive function forward.

Following the same analogy forward movement, when it is read the value 4, the condition executes a piece of code that will move the vehicle to the left. Similarly, when the read value is the number 6, the vehicle will move to the right and, when the value 2 is read, the software performs the backward motion or reverse. Finally, if the value is 5, the software performs the stop command of all motors.

Still on the drive engine, you must determine the maximum speed of each engine so that they have a move in the same speed and in what direction they will turn. For example, in moving forward, the two engines should turn clockwise at the same speed so that the CARduino move straight. The setSpeed () function, available in AFMotor.h library, is responsible for determining the engine speed coupled in Motor Shield used.

It is noteworthy that a function was done to prevent the collision of automated vehicle with obstacles that may be in the path. This collision function works directly with the

ultrasonic sensor HC-SR04 and requires the Ultrassonic.h library for its operation. The value of the distance of obstacles, in centimeters, is returned by ultrassonic.convert function ().

The embedded software on the Arduino for CARduino prototype can be found in the link: https://github.com/EvertonRafaeldaSilva/CARduinoArduinoCode.

2.3.4 Controller Software Development in Android CARduino

In this section, you will be presented the source code of Android software, and explained its operation.

First, one must make a setting within the general configuration file Android (AndroidManifest.xml) to allow the use of the Bluetooth mobile device. For the use of Bluetooth, this setting is mandatory and without it, the Android application can not connect to another Bluetooth device or activate the Bluetooth mobile [10].

Figure 2.5 illustrates the interface developed for CARduino driver software on Android. As can be seen, there are two types of user interactions with the vehicle. The first interaction is manual and there are 5 buttons that can be used to move the CARduino, which are: Front, Ré, right, left and stop. The second interaction is designed to work with custom paths, aiming mainly assist in research related to cognitive development, as previously reported.

As can be seen in Fig. 2.5, to work with this second interaction, one must establish the path to be traveled by CARduino defining a sequence of commands that matches the direction of motion (F - Front T - Back, D - Right, E - Left, P - Stop) and the quantity of centimeters (F or T) or degree (D or E) in which the vehicle should move. For example, if the value entered is F30, the vehicle must move forward and run this command by 30 cm.

Table 2.2 Operation of CARduino engines

```
if(Serial.available() >0)
{
intentrada = Serial.read();
direcaoAT = entrada;
switch (entrada) {
case '8':
pararMotores();
moverParaFrente();
break;
```

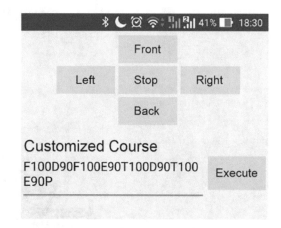

Fig. 2.5 Controller software interface developed on Android

As CARduino was developed using two DC motors, it should be parameterised how long each of the motors is operated by the driver software developed in Android to perform the movements of the vehicle. To meet this goal, we performed a set of ten manual measurements of the prototype developed in this research simulating its motion in a straight line of 100 cm and a 360 °.

In view of the calculations performed, we obtained the result of 0.0218 seconds to travel 1 cm and 0.0051 seconds to go 1. The driver software for Android CARduino can be found on the link:

https://github.com/EvertonRafaeldaSilva/CARduino.

Once finalized the development of Arduino controller software on Android, as reported in this section will be presented in the next section a simulation of its operation.

2.4 Vehicle Operation of Automated Simulation

In this section, we will present the functioning of the automated vehicle in this research through paths sent by the controller software developed on Android. 3 experiments are presented which were used to simulate such an operation, focusing on the functionality of sending a complete path by the controller software to the Arduino, once the manual control has the feature automated direct feedback vehicle.

It is noteworthy that the purpose of this simulation is not only to show the operation of the integrated CARduino with software controlled, but also already provide an initial analysis of the feasibility of using this research to assist in the cognitive development of children, youth and adults. For this reason, the choice of the experiments was based on cognitive exercises ever conducted in other studies, highlighting the research CHIOCHETI (2015).

As already presented in Sect. 2.3.4, the commands to be sent by the controller software on Android for CARduino must be a letter, which identifies the movement (F - Front, T - Back, D - Right, E - Left, P - Stop) and a value in centimeters (front or Rear) or in degrees (Right or Left).

As can be seen in Fig. 2.6, it was defined as Experiment I a route in the shape of a square to treat cognitive issues involving mathematics and logic. The CARduino start the route from the starting point and will go up to 240 cm return to starting position.

Fig. 2.6 Defined experiments in research: Experiment I - Square Path, Experiment II - Triangle Trail and Experiment III - Full Path - Customized Course

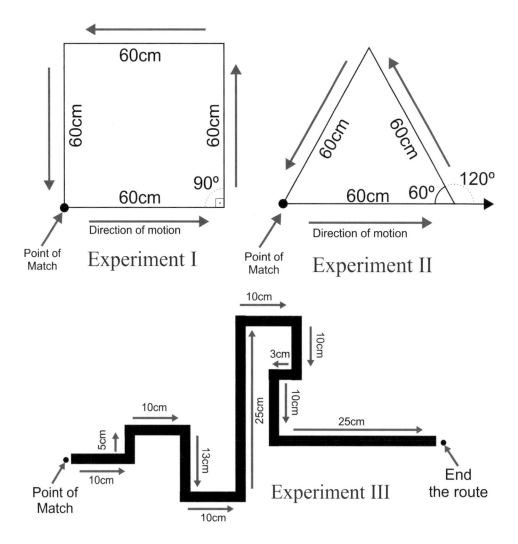

To accomplish this route, the sequence of commands is needed: F60 → E90 → F60 → E90 → F60 → E90 → F60→P.

As Experiment II, we chose to simulate the path of an equilateral triangle with sides measuring 60 centimeters. Thus, this path would treat cognitive issues logic and mathematics somewhat deeper in relation to Experiment I, since one must know the rules that the sum of interior angles of a triangle is 180 degrees and complementary angle an equilateral triangle is worth 180 by subtracting its value.

The route is shown in Fig. 2.6 Experiment II. The vehicle starts moving from the starting point and should follow the direction of the red arrows. The sequence of commands to accomplish this path is: F60 → E120 → F60 → E120 → F60 →P.

To turn the vehicle right triangle in each line join, should apply a value of 120 °, where 60 ° is within the triangle, more than 120 ° outside the triangle, resulting in 180 °.

Finally, as Experiment III, was chosen to simulate a complete path with variations of routes in length and changes in direction, again treating cognitive issues of logic and mathematics. Figure 2.6 illustrates this third experiment.

In this way, we perceive a greater amount of movement of the vehicle. The command to run the route is: F10 → E90 → F5 → D90 → F10 → D90 → F13 → E90 → F10 → E90 → F25 → D90 → F10 → D90 → F10 → D90 → F3 → E90 → F10 → E90 → F25 →P.

2.5 Analysis of Costs CARduino

To meet one of the objectives proposed in this research, this section an overview of the construction costs of CARduino prototype and feasibility of studying this type of technology will be presented. They found 3 sites specializing in trading Arduino components in Brazil: Filipeflop (www.filipeflop.com), Laboratório de Garagem (www.labdegaragem.com.br) and Usina Info (www.usinainfo.com.br).

As can be seen in Table 2.3, the values of the components shown, refer to a query made in January 2016. For this analysis disregarded the value of the resistors.

Given the values compared in Table 2.3, the cost of USD 105.11 to produce a CARduino prototype can be considered low, especially for its variety of applications in use, the gains highlighting that can be obtained for cognitive development children, youth and adults.

Note that this section had only a cost analysis for the development of the physical part (hardware) of CARduino, not including the time effort for the development of both the embedded software on the Arduino as Android in controller software.

Table 2.3 Price comparison component

Component	FilipeFlop	Laboratório de Garagem	Usina Info
Arduino MEGA	USD 31.25	USD 28.12	USD 42.34
Motor Shield	USD 10.92	USD 13.79	USD 14.75
Chassis/ Engines	USD 34,37	USD 18.37	USD 45.00
Sensor HC-SR04	USD 5.91	USD 4.77	USD 5.42
HC-05 Module	USD 15.61	USD 13.26	USD 16.39
Jumpers	USD 6.22	USD 3.73	USD 4.89
Resistors	USD 0.04	USD 0.04	USD 0.11
Total site	**USD 103.38**	**USD 82.09**	**USD 128.86**
Average price	**USD 105.11**		

2.6 Conclusion

This project aimed to design an automated vehicle prototype built with Arduino and controlled with software developed on Android that can perform manual or automatic paths.

Until now research and analyzing the simulation of experiments shown, it is believed that it is feasible to use the prototype designed to cognitive development, for future users can learn to insert custom paths that can process logic issues and more complex mathematics allowing the prototype perform the desired movements.

Analyzing the financial costs of design, it is believed that it is feasible to construct this type of prototype because it presents a low cost of the components used, particularly if they choose in a large scale production. It is worth noting that both the Java programming language as the language for Arduino in development are free, not burdening additional costs for the development of the project, pointing out that this applies also the tools used for development.

An additional contribution of this research is the practical presentation of how to integrate applications developed on Android with circuits produced using Arduino. In addition, not found in the literature any work reportasse systematically, how to enable and use the administrative mode of the Arduino Bluetooth module, this work presented in detail those settings at both the software and the hardware level.

To complement this research, it is suggested as a future work quality analysis of developed code-sources from the point of view of Software Testing to address security issues in embedded software. It is also recommended to apply the automated vehicle prototype developed in conjunction with the controller software for Android in groups of children, adolescents and adults to prove the viability of the same on cognitive issues, since taking the opportunity to identify shortcomings and make the necessary refinements. Finally, one can study the feasibility of adding new features and sensors to increase the range of use of the prototype, for example: rover, access to inhospitable areas, among others.

References

1. Chiocheti, M. S, & Romano, B. L. (2015). Uma abordagem para analisar o desenvolvimento do raciocínio lógico com apoio computacional. Monografia do Curso de Tecnologia em Sistemas para Internet – Instituto Federal de Educação, Ciência e Tecnologia de São Paulo – Campus São João da Boa Vista.
2. Ascencio, A. F. G. (2010). *Fundamentos da programação de computadores: algoritmos, pascal e C/C++* (2nd ed.). São Paulo: Pearson Prentice Hall.
3. Wirsing, B. (2014). *Sending and receiving data via bluetooth with an android device*. White Paper Android Developer.
4. Cardoso, L. F. C. (2014). Sistemas de automação residencial via rede celular usando microcontroladores e sensores. *Revista de Engenharia da Universidade Católica de Petrópolis, 8*(2), 68–83.
5. Moreira, A. S., Portela, A. M., & Silva, R. (2012). Uso da plataforma Arduino no desenvolvimento de soluções tecnológicas para pesquisas de dados atmosféricos na Amazônia. 2012, Revista Científica AMAZÔNICA ISSN 2179 6513, p. 119.
6. Scriptore, D. B., & Junior, José de M. (2014). Banco de dados distribuído para consulta de temperatura e umidade utilizando Arduino e Android. 2014. 6p. Paranavaí, 2014.
7. Passos, B. P. (2011). Sistema Bluetooth para controle de acessórios veiculares utilizando Smartphone com Android. 2011, 83p. Trabalho de Conclusão de Curso (Bacharelado em Engenharia da Computação) Centro Universitário de Brasília UniCEUB, Brasília, 2011.
8. Lim, W., Torres, H. K., Oppus, C. M. (2014). *An agricultural telemetry system implemented using an Arduino-Android interface*. Philippines: IEEE Conference Publications.
9. Watve, O. J. (2015). Implementation of real time factory information system using arduino and android. IEEE Conference Publications.
10. Glauber, N. (2015). *Dominando o Android* (1st ed.). São Paulo: Novatec.

ICI Mitigation for High-Speed OFDM Communications in High-Mobility Vehicular Channels

3

Vahid Vahidi and Ebrahim Saberinia

Abstract

The performance of an Orthogonal Frequency Division Multiplexing (OFDM) system to transmit high bandwidth data from a vehicle to a base station can suffer from Inter-Carrier Interference (ICI) created by high Doppler shifts. In current communication systems, high Doppler shifts can happen because of the high speed of the vehicles such as fixed wings unmanned aircraft vehicles (UAVs) and high speed trains (HST). In next generation wireless systems with high carrier frequency, such as 5G cellular data systems at center frequency between 27.5–71 GHz, even a vehicle moving with moderate speed can cause Doppler shift of several kilohertz. To cancel the ICI, the time variant channel matrix should be estimated in the frequency domain. In this paper, a new channel estimation scheme is presented suitable for high Doppler scenarios. To estimate the channel in the frequency domain, a training sequence in the time domain is transmitted, and both channel amplitudes and Doppler shifts are estimated in time domain. Then, the complete frequency domain channel matrix is constructed from the estimated parameters and used for ICI mitigation. In contrast to conventional methods that only estimate diagonal elements of the frequency domain channel matrix or other partial section of the matrix to reduce the complexity, this new method estimate the complete matrix. Simulation results show significant gain in performance for the complete channel estimation as compared to conventional methods using least square and minimum mean square diagonal elements of the channel estimators in high Doppler scenarios.

Keywords

Channel estimation · Doppler · Equalization · UAV · Payload communication · ICI cancellation

3.1 Introduction

Orthogonal frequency-division multiplexing (OFDM) is a candidate for any wideband wireless data-communication system [1–3]. It makes efficient use of the spectrum, and eliminates inter symbol interference (ISI) in a multipath environment. However, in vehicular wireless communication systems, high speed of the modern vehicles makes use of OFDM challenging because of high Doppler shifts that causes significant inter-carrier interference (ICI). Wireless systems for fixed-wings unmanned aircraft vehicles (UAVs) and high speed trains (HSTs) are examples of high mobility wireless channels. As an example, a low cost supersonic of 1.4-Mach UAV is made in Colorado University [4]. Because of its light weight and small engine, it can fly at low altitudes and it can be used for civilian applications. High Doppler shift can also be the result of high carrier frequency in

V. Vahidi (✉) · E. Saberinia
Department of Electrical and Computer Engineering, University of Nevada, Las Vegas, Las Vegas, NV, USA
e-mail: vahidi@unlv.nevada.edu; ebrahim.saberinia@unlv.edu

normal vehicular speeds. The emerging cellular communication standard 5G is expected to work at center frequency between 27.5-71 GHz in the United States [5, 6]. Thus, even at regular highway car speeds, Doppler shift of several kilohertz affects the received signal.

We have studied the performance of OFDM in high-bandwidth fast-varying channels in [7] for UAV wideband payload communications where simulation results are presented to show that without proper ICI mitigation scheme, the performance of the OFDM system suffers significantly especially when the number of subcarriers becomes larger or the Doppler shift increases. This paper proposes a new ICI mitigation technique that improves the performance of the OFDM system for mobility OFDM systems such as UAV payload communications.

In an ICI free OFDM system, the received signal vector is equal to the multiplication of the diagonal frequency domain channel matrix to the vector of the transmitted symbols [7]. However, the Doppler shift within high mobility channels causes ICI and therefore, the channel matrix becomes non-diagonal. To remove the ICI, the channel matrix should be estimated. Channel estimation (CE) for ICI cancellation has been studied in the literatures [8–21]. In all of these papers, CE is performed by sending pilots in the frequency domain or in the time domain.

Frequency-domain CE techniques have been the subject of several studies [8–12]. Implementing Least Squares (LS) and Minimum Mean Square Errors (MMSE) estimators based on one OFDM symbol as a pilot in frequency domain was proposed in [8]. Using adjacent subcarriers of the frequency domain to mitigate ICI on a specific carrier was proposed in [9]. MinHai et al. [10], assumed that the phase changes in a linear manner with Doppler; therefore, they estimated the Doppler by considering two consecutive channel-transfer functions. Guangxi et al. [11] estimated the channel by zero padding the pilot; they implemented DFT and a high-precision interpolation technique in the frequency domain. Iterative processing of CE in the frequency domain was performed in [12]. Several other authors proposed time-domain CE methods [13–18]. By assuming that the channel impulse response (CIR) varies linearly with time during a block period, Han et al. [13] proposed a time-domain equalization technique. Gupta [14] demonstrated a time-domain ICI-mitigation technique based on DFT estimation of the channel had better performance compared to LS and MMSE estimators. Ahmed et al. [15] assumed that Doppler causes a time-varying attenuation coefficient and a time-varying tap delay; therefore, they transmitted and

cross-correlated a known OFDM reference symbol with the local known symbol. By comparing the difference in pick location, they were able to calculate the Doppler shift. Gupta [16], proposed an iterative time domain Linear Minimum Mean Square Error (TD-LMMSE) that tracked channel variations in the time domain using an LMMSE estimator. For channel estimation, Aggarwal et al. [17] inserted a Pseudo Random Code (PCR) in the time domain during the guard interval. Some studies combined time-domain and frequency-domain techniques for CE. Werner et al. [18], was able to conduct a CE by estimating the channel response in the frequency domain. Afterwards, for the sake of reduction in the interpolation error, a refining step in the time domain was performed. In [19], the Doppler spread was estimated in the frequency domain and complex amplitudes were estimated in the time domain. Most of the schemes reviewed ignore the non-diagonal elements of the channel matrix because of low-mobility channel conditions and to reduce the complexity of channel equalization [8–19]. However, others did take into consideration non-diagonal elements [20, 21]. Ng and Dubey [20] obtained the other elements of the channel matrix with the assumption that the ICI of a subcarrier was just due to its adjacent subcarriers. Nakamura et al. [21] minimized the mean square error, and estimated the Doppler shift and complex amplitudes of all the paths auto regressively. They assumed each time that the received signal occurred because of just one path; therefore, they neglected the ICI of the other paths.

In this paper, we present a new scheme for time-domain CE that uses a complete OFDM symbol as the training sequence instead of using scattered pilots. This approach estimates the channel parameters in the time domain and then constructs a complete channel matrix in the frequency domain. We use MMSE equalization using this estimate for ICI mitigation. To estimate the channel parameters with low complexity, a specific structure is considered for the training sequence and two types of equations were developed for Doppler spread estimation and complex amplitude estimation. Doppler phase shifts are estimated by linear Taylor expansion. Simulation results indicate that this method is very fast, and can mitigate ICI better than classical LS and MMSE approaches. The effectiveness of the proposed method is more obvious in higher Doppler shifts.

The remainder of this paper is organized as follows. Section 3.2 presents the system model. The ICI cancellation method proposed in this study is elaborated in Sect. 3.3. Section 3.4 presents and discusses simulation results, and Sect. 3.5 concludes the paper.

3.2 System Model

The OFDM signal that is transmitted to the channel is represented as:

$$x_n = \begin{cases} \sum_{m=0}^{N-1} X(m)e^{\frac{j2\pi(n+N-N_g)m}{N}} & 0 \leq n \leq N_g - 1 \\ \sum_{m=0}^{N-1} X(m)e^{\frac{j2\pi(n-N_g)m}{N}} & N_g \leq n \leq N - 1 + N_g \end{cases} \quad (3.1)$$

where X is the input signal in the frequency domain, N is the number of subcarriers, n is the index of the transmitted sample, m is the index of the subcarrier, and N_g is the length of the cyclic prefix. The channel model between the transmitter and receiver is modeled as [22]:

$$a_l(t) = \sum_{n=1}^{L} a_n e^{j\theta_n} \cdot e^{j2\pi f_{D_n} kT_{sample}} \cdot g_{total}\left(iT_{sample} - \tau_n\right) \quad (3.2)$$

where L is the number of the non-line-of-sight (NLOS) propagation paths, a_n is the amplitude of the nth path having a Rayleigh distribution developed previously by this research group [7], θ_n is the random delay of the nth path, f_{D_n} is the Doppler of the nth path, τ_n is the delay of the nth path, T_{sample} is the channel resolution for resolvable paths, and g_{total} is the total impulse response of the transmitter and receiver filters.

Based on the model proposed by Haas [22], the received signal after passing through the channel is:

$$y_n = \sum_{i=0}^{L-1} x_{n-i} a_i e^{j2\pi f_i nT_s} + W \quad (3.3)$$

where y_n is the received signal after passing through the channel; a_i and f_i are the amplitude and Doppler frequency of the i th path, respectively; T_s is the sampling time, and W is AWGN. To obtain the received symbols, the DFT of y_n should be calculated. The relation between the received symbols and the transmitted symbols can be expressed as:

$$Y_k = \sum_{m=1}^{N} X_m H_{k,m} + Z \quad (3.4)$$

where Y_k is the k th received symbol, $H_{k,m}$ is the (k,m) element of the channel matrix, and Z is the Fourier transform of AWGN. $H_{k,m}$ was obtained as [21]:

$$H_{k,m} = \sum_{i=0}^{L-1} a_i e^{j2\pi f_i L} e^{\frac{-j2\pi ik}{N}} \frac{1 - e^{jnN\left(2\pi f_i T_s + \frac{2\pi}{N}(m-k)\right)}}{1 - e^{jn\left(2\pi f_i T_s + \frac{2\pi}{N}(m-k)\right)}} \quad (3.5)$$

3.3 Channel Estimation and ICI Cancellation

To construct all the values for $H_{k,m}$, the Doppler shift and the complex amplitude of each path should be estimated. An appropriate training sequence is essential for estimating channels. In the method described in this paper, the structure of the training sequence was designed such that it would be able to estimate the channel amplitude and Doppler shift for all the paths with two sets of linear equations. To estimate these values, a block of pseudorandom codes in the time domain was sent as a pilot. The block's length (L) should be equal or greater than the channel's delay spread. This block was repeated in the same pilot symbol, with a block of zeros of length L inserted between them in order to make sure no interference from the response to the first block was reaching the second block. The use of two blocks became apparent by the following equations, as it enabled the estimation of the Doppler frequency shift from linear phase changes. Approximating $a_i e^{j2\pi f_i nT_s}$ in (3.3) with $a_i(1 + j2\pi f_i nT_s)$ to linearize the equations, the received pilots can be expressed as:

$$\begin{cases} y_1 = P_1 a_1(1 + j2\pi f_1 T_s) + w_1 \\ y_2 = P_2 a_1(1 + j2\pi 2 f_1 T_s) + P_1 a_2(1 + j2\pi 2 f_2 T_s) + w_2 \\ \quad \vdots \\ y_L = P_L a_1(1 + j2\pi L f_1 T_s) + P_{L-1} a_2(1 + j2\pi L f_2 T_s) + \ldots + \\ \quad P_1 a_L(1 + j2\pi L f_L T_s) + w_L \end{cases}$$

$$(3.6)$$

where P_i is the ith transmitted pilot; the other parameters were defined earlier. Now, since the same pilot block was repeated after a zero-padding block, the corresponding received pilots have the following formula:

$$\begin{cases} y_{2L+1} = P_1 a_1(1 + j2\pi(2L+1)f_1 T_s) + w_{2L+1} \\ y_{2L+2} = P_2 a_1(1 + j2\pi(2L+2)f_1 T_s) + \\ \quad P_1 a_2(1 + j2\pi(2l+2)f_2 T_s) + w_{(2L+2)} \\ \quad \vdots \\ y_{2L+L} = P_L a_1(1 + j2\pi(2L+L)f_1 T_s) + \\ \quad P_{L-1} a_2(1 + j2\pi(2L+L)f_2 T_s) + \cdots + \\ \quad P_1 a_L(1 + j2\pi(2L+L)f_L T_s) + w_{2L+L} \quad (3.7) \end{cases}$$

Looking at (3.6) and (3.7), a_i can be calculated from the scaled differnce of y_i and y_{2L+i}. The linear equation would be obtained as:

$$Y_{da\,i} = \left(\frac{2L+i}{2L}\right) \cdot y_i - \frac{i}{2L} \cdot y_{i+2L} = P_i a_1 + P_{i-1} a_2 + \cdots$$
$$+ P_1 a_i + \left(\frac{2L+i}{2L}\right) \cdot w_i - \frac{i}{2L} \cdot w_{i+2L}$$

$$(3.8)$$

Fig. 3.1 Transmitted training sequence

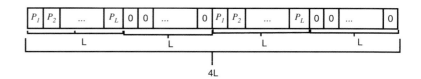

A set of linear equations can be written in matrix form as:

$$Y_{da} = T_1 \times A + w_{diff} \qquad (3.9)$$

where Y_{da} is a $L \times 1$ matrix, $Y_{da}{}^T = [Y_{da\,1}, Y_{da\,2}, \ldots Y_{da\,L}]$, $.^T$ means transpose, A is $L \times 1$ matrix of complex amplitudes $A^T = [a_1, a_2, \cdots, a_L]$, and T_1 is a $L \times L$ lower triangular matrix of training symbols that is presented as:

$$T_1 = \begin{bmatrix} P_1 & 0 & \ldots & 0 \\ P_2 & P_1 & \ldots & 0 \\ \vdots & \vdots & \ddots & \vdots \\ P_L & P_{L-1} & \ldots & P_1 \end{bmatrix} \qquad (3.10)$$

and w_{diff} is a $L \times 1$ matrix, $w_{diff}{}^T = [w_{diff\,1}, w_{diff\,2}, \ldots w_{diff\,L}]$ where $w_{diff\,i} = \left(\frac{2L+i}{2L}\right).w_i - \frac{i}{2L}.w_{i+2L}$.

Since (3.9) is a linear equation, the estimation of A would be obtained as:

$$\hat{A} = (T_1)^{-1} R_a \qquad (3.11)$$

After the calculation of complex amplitudes, the Doppler frequency for each path is calculated based on a similar procedure. Looking at (3.11) and (3.12), f_i can be calculated from the differnce of y_i and y_{2L+i}. The linear equation would be obtained as:

$$Y_{df\,i} = y_{2L+i} - y_i = \hat{a_1} P_i (j2\pi 2Lf_1 T_s) + \hat{a_2} P_{i-1}(j2\pi 2Lf_2 T_s) \\ + \cdots + \hat{a_l} P_1 (j2\pi 2Lf_L T_s) + w_{2L+i} - w_i \qquad (3.12)$$

Where $\hat{a_l}$ s are the estimated complex amplitudes in (3.16). A set of linear equations can be written in a matrix form as:

$$Y_{df} = U_1 \times (D.j2\pi 2LT_s) + w_{diff2}, \qquad (3.13)$$

where Y_{df} is a $L \times 1$ matrix, $Y_{df}{}^T = [Y_{df\,1}, Y_{df\,2}, \ldots Y_{df\,L}]$, D is $L \times 1$ matrix of complex amplitudes $D^T = [f_1, f_2, \cdots, f_L]$ and U_1 is a $L \times L$ lower triangular matrix of training symbols that is presented as:

$$U_1 = \begin{bmatrix} \hat{a_1} P_1 & 0 & \ldots & 0 \\ \hat{a_1} P_2 & \hat{a_2} P_1 & \ldots & 0 \\ \vdots & \vdots & \ddots & \vdots \\ \hat{a_1} P_L & \hat{a_2} P_{L-1} & \ldots & \hat{a_L} P_1 \end{bmatrix} \qquad (3.14)$$

and w_{diff2} is a $L \times 1$ matrix, $w_{diff2}{}^T = [w_{diff2\,1}, w_{diff2\,2}, \ldots w_{diff2\,L}]$ where $w_{diff2\,i} = w_{2L+i} - w_i$.

Since (3.13) is a linear equation, the estimation of D would be obtained as:

$$\hat{D} = \frac{1}{j2\pi 2LT_s}(U_1)^{-1} R_f \qquad (3.15)$$

The structure of the transmitted pilot is presented in Fig. 3.1. For a single OFDM pilot, if the length of the training sequence, N, is more than $4L$ where L is the length of the cyclic prefix, then we can put $\frac{N}{4L}$ of the pilot sequences in Fig. 3.1, but with a different block of pseudorandom codes. As a result, the number of equations would be more than the number of unknowns, which leads to the reduction of the effect of AWGN in the received pilots in the channel estimation procedures discussed as follows.

In the case of $\frac{N}{4L}$ pilot sequences, T_1 in (3.11) and U_1 in (3.15) would be replaced with T and U respectively which are defined as:

$$T^T = \left[T_1, T_2, \ldots T_{\frac{N}{4L}}\right], \qquad (3.16)$$

$$U^T = \left[U_1, U_2, \ldots U_{\frac{N}{4L}}\right], \qquad (3.17)$$

where $T_2 \ldots T_{\frac{N}{4L}}$ and $U_2 \ldots U_{\frac{N}{4L}}$ are defined similar to T_1 and U_1, respectively, but with different pseudorandom codes. Therefore, T and U are $\frac{N}{4L} \times L$ matrixes, which consist of $\frac{N}{4L}$ lower triangular matrixes. Then, (3.11) will be converted to:

$$\hat{A} = \left(T^T T\right)^{-1} T^T Y_{da}, \qquad (3.18)$$

and (3.15) will be converted to:

$$\hat{D} = \frac{1}{j2\pi 2LT_s}\left(U^T U\right)^{-1} U^T Y_{df}, \qquad (3.19)$$

Kay [23] showed that the minimum estimation error would be obtained if $T^T T$ and $U^T U$ were diagonal matrixes. This will be achieved, approximately, by the proposed training sequence shown in Fig. 3.1.

Since P_i s are pseudorandom, the non-diagonal elements of $T^T T$ would be close to zero and the diagonal elements would be obtained as:

$$r_{ai} = \sum_{\substack{n=1 \\ n \neq k(L-l)\ for\ 0<l<i}}^{\frac{N}{4L}} P_n^2 \qquad (3.20)$$

where r_{ai} is the ith diagonal element. If we assume that the linearization is not adding any error to the estimation, the estimated complex amplitudes would be obtained as:

$$\widehat{a}_l = a_i + \frac{1}{r_i} \sum_{\substack{j=1 \\ j \neq k}}^{L} \sum_{k=1}^{L} b\left(P_j a_j P_k + P_k w_{diff\,k}\right) \tag{3.21}$$

where $\widehat{a}_l$ is the estimated complex amplitude and b is a parameter that could be 0 or 1. The other parameters have been defined earlier in the paper. Since training symbols are pseudorandom and noise samples are iid, this approach is very efficient in noise elimination. As a result, most of the estimation error would be because of a linearization error.

Similarly, $U^T U$ can be approximated as a diagonal matrix, and the diagonal elements would be obtained as:

$$r_{f_i} = \sum_{\substack{n=1 \\ n \neq k(L-l) \text{ for } 0<l<i}}^{\frac{N}{4L}} \widehat{a}_i^2 P_n^2 \tag{3.22}$$

where r_{fi} is the ith diagonal element. By the assumption that the linearization does not add any error to the estimation, the estimated Doppler shifts would be obtained as:

$$\widehat{f}_i = f_i + \frac{1}{j2\pi 2LT_s r_{f_i}} \sum_{\substack{j=1 \\ j \neq k}}^{L} \sum_{k=1}^{L} b\widehat{a}_j \widehat{a}_k \left(P_j \widehat{a}_j P_k + P_k w_{diff2\,k}\right) \tag{3.23}$$

Since training symbols are pseudorandom and noise samples are iid, this approach is very efficient in noise elimination. As a result, most of the estimation error would be because of a linearization error.

After obtaining the complex amplitudes and Doppler shifts, the channel matrix of (3.10) can be constructed. While the channel matrix is obtained, the MMSE of the channel matrix should be implemented in order to minimize the error imposed by noise:

$$\widehat{X} = \widehat{H}^* \times \left[\widehat{H} \times \widehat{H}^* + \sigma^2 I\right]^{-1} Y \tag{3.24}$$

where $\widehat{\mathbf{H}}$ is the estimation of the channel matrix, $\mathbf{H}$, σ^2 is the power of noise, Y is the received signal in the frequency domain and $*$ defines conjugate operator.

3.4 Simulation Results

The performance of the proposed method was evaluated using Monte Carlo simulation. A Binary Phase Shift Keying (BPSK) OFDM signal of 10 MHz bandwidth with N = 1024 subcarriers and the cyclic prefix number of 32 was passed through a high mobility UAV communication channel. Specifically, we used the air to ground station channel model Q1 presented in [7]. The time resolution T_{sample} is equal to 100 ns. For the pilot symbol, the value of L is set to 32 and it is constructed using an 8-bipolar Gold sequence with the length of 31 [24] with added bit of "1" to the end. To model the mobility in the channel, two Doppler shifts were considered for the UAV, 800 Hz for low mobility and 8 KHz for high mobility.

Figure 3.2 shows Bit Error Rate (BER) vs. Signal to Noise Ratio (SNR) curves for LS, MMSE, and proposed method for low Doppler shift (800 Hz) and high Doppler shift (8 KHz) scenarios. We can see that for high mobility channels without a ICI mitigation, because of the high BER that is achieved even at 30 dB SNR, the communication is not reliable while our channel estimation and ICI cancellation methods performs much better.

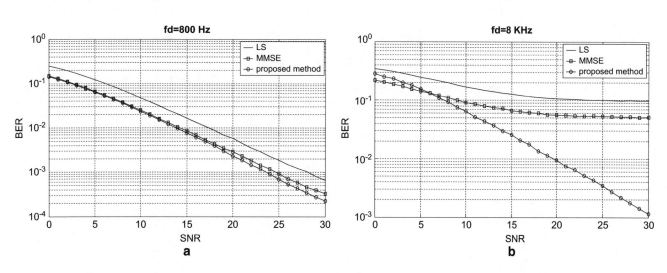

Fig. 3.2 BER vs. SNR for LS, MMSE and the proposed method for the maximum Doppler of (**a**) 800 Hz and (**b**) 8 KHz

3.5   Conclusion

In this paper, a new method for channel estimation and ICI mitigation for high-bandwidth fast-varying vehicular channels was proposed. Using a specific structure for the pilot data and some linearization of the equations, the complex amplitude and the Doppler shift parameters are estimated in time domain. Then the complete frequency domain channel matrix is constructed and used for ICI cancelation. Simulation results show that, this scheme can improve the performance of ICI mitigation compared to the conventional methods in fast-varying channels. The convergence of estimation for the proposed method is very fast in comparison to autoregressive methods used for full channel estimation. In addition, it performs well in the presence of noise because of the use of a pseudorandom training sequence.

References

1. Medbo, J. (1998). Radio wave propagation characteristics at 5 GHz with modeling suggestions for HIPERLAN/2. *Etsi Bran 3eri074A*.
2. ITU-R, Report M.2233. Examples of technical characteristics for unmanned aircraft control and non-payload communications links. 11/2011.
3. ITU: International Telecommunication Union. *Guidelines* for the *evaluation of Radio Transmission Technologies* (RTTs) for *IMT-2000*, ITU-R Recommendation M.1225, 1997.
4. http://www.tecplot.com
5. 4G Americas' 5G Spectrum Recommendations, August 2015.
6. http://www.fiercewireless.com
7. Vahidi, V., & Saberinia, E. (2016) *Orthogonal frequency division multiplexing and channel models for payload communications of unmanned aerial systems*. 2016 international conference on Unmanned Aircraft Systems (ICUAS), Arlington, pp. 1156–1161.
8. van de Beek, J. J., Edfors, O., Sandell, M., Wilson, S. K., & Borjesson, P. O.. *On channel estimation in OFDM systems*. Vehicular Technology conference, 1995 I.E. 45th, Chicago, pp. 815–819.
9. Barhumi, I., Leus, G., & Moonen, M. (2003, September 8–10). *Frequency-Domain equalization for OFDM over doubly-selective channels*. Sixth Baiona workshop on Signal Processing in Communications, Baiona/Spain, pp. 103–107.
10. MinhHai, T., Rie, S., Taisuki, S., & Wada, T. (2015). *A transceiver architecture for ultrasonic OFDM with adaptive Doppler compensation* (pp. 1–6). Washington, DC: OCEANS 2015 – MTS/IEEE Washington.
11. Guangxi, Y., Lunhui, D., & Xiang, C.. (2015). *An efficient channel estimation method for OFDM systems based on comb pilots*. 2015 I.E. Advanced Information Technology, Electronic and Automation Control Conference (IAEAC), Chongqing/China, pp. 1029–1033.
12. Han, J., Zhang, L., & Leus, G. (2016). Partial FFT demodulation for MIMO-OFDM over time-varying underwater acoustic channels. *IEEE Signal Processing Letters, 23*(2), 282–286.
13. Han, K. Y., Ha, K., Sung, K. M., & Lee, C. W. (2000). *Time domain equalization using linear phase interpolation for OFDM in time variant multipath channels with frequency offset*. Vehicular Technology conference proceedings, 2000. VTC 2000-Spring Tokyo. 2000 I.E. 51st, Tokyo, 2000, Vol. 2, pp. 1255–1259.
14. Gupta, A. (2013). Improving channel estimation in OFDM system using time domain channel estimation for time correlated rayleigh fading channel model. *International Journal of Engineering and Science Invention, 2*(8), 45–51.
15. Ahmed, S. (2015). *Estimation and compensation of Doppler scale in UAC OFDM systems* (pp. 1–12). Washington, DC: OCEANS 2015 – MTS/IEEE Washington.
16. Kalakech, A., Darazi, R., Berbineau, M., Simon, E., & Dayoub, I. (2015). *Iterative time domain estimation of rapidly changing channel for OFDM systems in a cognitive radio context*. Technological Advances in Electrical, Electronics and Computer Engineering (TAEECE), 2015 third international conference, Beirut, pp. 268–273.
17. Aggarwal, P., Gupta, A., & Bohara, V. A.. (2015). *A guard interval assisted OFDM symbol-based channel estimation for rapid time-varying scenarios in IEEE 802.lip*. Personal, Indoor, and Mobile Radio Communications (PIMRC), 2015 I.E. 26th annual international symposium, Hong Kong, pp. 100–104.
18. Werner, S., Enescu, M., & Koivunen, V. (2006). *Combined frequency and time domain channel estimation in mobile MIMO-OFDM systems*. Acoustics, Speech and Signal Processing, 2006. ICASSP 2006 proceedings. 2006 I.E. international conference, Toulouse, pp. IV–IV. doi: 10.1109/ICASSP.2006.1660983.
19. Yao, R., Liu, Y., Li, G., & Xu, J. (2015). Channel estimation for orthogonal frequency division multiplexing uplinks in time-varying channels. *IET Communications, 9*(2), 156–166.
20. Ng, W. T., & Dubey, V. K. (2003). *On coded pilot based channel estimation for OFDM in very fast multipath fading channel*. Information, Communications and Signal Processing, 2003 and fourth pacific rim conference on Multimedia. Proceedings of the 2003 joint conference of the fourth international conference, Vol. 2, pp. 859–863.
21. Nakamura, M., Fujii, M., Itami, M., Itoh, K., & Aghvami, A. H.. (2002). *A study on an MMSE ICI canceller for OFDM under Doppler-spread channel. Personal, Indoor and Mobile Radio Communications, 2003. PIMRC 2003. 14th IEEE Proceedings.* 2003, Vol. 1, pp. 236–240.
22. Haas, E. (2002). Aeronautical channel modeling. *IEEE Transactions on Vehicular Technology, 51*(2), 254–264.
23. Kay, S. M. (1993). *Fundamentals of statistical signal processing: Estimation theory* (Vol. I). Englewood Cliffs: Prentice-Hall.
24. Gold, R. (1967). Optimal binary sequences for spread spectrum multiplexing. *IEEE Transactions Information Theory, 13*, 619–621.

Mobile Payment Protocol 3D (MPP 3D) by Using Cloud Messaging

4

Mohammad Vahidalizadehdizaj and Avery Leider

Abstract

Popularity of mobile platform, makes it a proper candidate for electronic payment. However, there are some challenges in this field like privacy protection, security, limitation of mobile networks, and limited capabilities of mobile devices. Traditional e-commerce payment protocols were designed to keep track of traditional flows of data. These protocols are vulnerable to attacks and are not designed for mobile platform. Also, 3D Secure that is an extra security layer of modern payment methods (mainly to prevent card not present fraud), is not proper for mobile platform because of issues like difficulty of viewing authentication pop-up window on a mobile device. In this paper, we propose a new private mobile payment protocol based on client centric model by utilizing symmetric key operations. Our protocol reduces computational cost of Diffie-Hellman key agreement protocol by using algebra of logarithms instead of algebra of exponents, achieves proper privacy protection for payer by involving mobile network operators and generating temporary identities, avoids repudiation attacks by utilizing digital signatures, avoids replay attacks by using random time-stamp generated numbers, and provide better and safer customer experience by utilizing cloud messaging instead of text messaging and pop-up windows in its extra layer of security (3 domain authentication).

Keywords

E-commerce • M-commerce • Mobile commerce • Mobile payment • Privacy protection • Non-repudiation • Replay attack • 3D secure • Verified by visa • MasterCard SecureCode • American express SafeKey • JCB international as J/secure • Diffie-Hellman

4.1 Introduction

E-commerce is any financial transaction over Internet. Most of the time payer uses his credit card in this process. An e-commerce transaction involves purchaser or card holder, merchant, purchaser's credit card issuer (bank), merchant's acquirer (bank), and certification authority for supporting secure transaction execution [14]. Most of these protocols are using a key agreement protocol for establishing a secure connection between the engaging parties [9].

Mobile commerce (m-commerce) is e-commerce activities conducted via mobile platform. Principals of m-commerce are the same as e-commerce plus mobile network operator. Moreover, most of e-commerce protocols are based on public key cryptography that is not efficient in mobile and wireless networks [15]. Some of these protocols are keeping credit card's information on mobile devices or using this information in transactions without proper protection. This is why they are vulnerable to attacks [10].

Mobile devices like smart phones and tablets are becoming very popular among people nowadays [11]. Most of these

M. Vahidalizadehdizaj, PhD (✉) • A. Leider
Computer Science Department, Pace University, New York, NY, USA
e-mail: m.vahidalizadeh@gmail.com; averyleider@gmail.com

© Springer International Publishing AG 2018
S. Latifi (ed.), *Information Technology – New Generations*, Advances in Intelligent
Systems and Computing 558, DOI 10.1007/978-3-319-54978-1_4

devices are light, easy to carry, and convenient to use. These devices are compatible with mobile networks that is widely available in outdoor space. Growth of m-commerce sales continues to be rapid even with the challenges that m-commerce face like slow download times. Forrester predicted 11 percent (of whole e-commerce) growth in m-commerce between 2016 and 2020. Currently m-commerce has 35 percent of e-commerce transactions. Forrester predicted that m-commerce will be 49 percent of e-commerce in 2020. This amount is 252 billion dollars in sales [4].

4.2   Background

In this section, we review existing related protocols. We review Diffie-Hellman as the popular key agreement protocol in payment protocols. We review SET, iKP, KSL, and 3D Secure protocols [7, 15]. Diffie-Hellman is the most famous key agreement protocol. Its calculations are based on algebra of exponents and modulus in arithmetic. Goal of this protocol is to generate a shared session key between two parties [9].

SET defines an open encryption and security specification. This protocol is designed to protect credit card transactions over Internet. Initial version of SET was emerged in a call for security standards by MasterCard and Visa in February 1996 [9]. SET has some problems. In this protocol cardholder is not protected from dishonest merchants that have tendency to charge more than their advertised price or are hackers who put up an illegal website to collect credit card information [15]. Besides, merchant is not protected from dishonest customers who provide invalid credit card numbers or claim refund without any real cause [14]. In this protocol merchant is more vulnerable against fraud, since legislation protects customers in most of the countries in the world [8].

IBM developed iKP ($i = 1, 2, 3$) family of protocols. These protocols are designed to implement credit card transactions between customer and merchant [15]. Each one of the members of the family differs from the others from aspect of level of complexity and security. iKP is direct ancestor of SET. These protocols have been used in Internet since 1996. iKP protocols are unique because of longevity, security, and relative simplicity of the underlying mechanisms. These protocols are based on public key cryptography. They differ from each others from aspect of number of principals who possess public key pairs [14, 15].

KSL is a payment protocol for e-commerce in fixed networks like Internet. This protocol is not proper for mobile platform because of its heavy computation and communication costs [9]. The idea is to reduce the number of people who possess their own key pair in SET protocol. In this protocol all principals except customer should have their own certificates. So, client side computation is lighter. KSL protocol is a nonce based protocol. KSL is an alternative of Kerberos. Drawback of Kerberos is that it uses timestamps. So, it requires at least one loose synchronization between timestamp generator and participants. Principals of KSL protocol are customer, merchant, payment gateway, and financial service provider [11]. All of them except customer should have their own certificates. So, computational cost of customer side is less than other methods [14].

3D Secure is an extra security layer for online credit and debit card transactions in order to prevent card not present fraud. Arcot Systems developed this protocol for the first time. The name 3D came from 3 domains. These domains are acquiring domain, issuer domain, interoperability domain [7]. There are some issues in 3DS. Firstly, there is a pop-up window or inline frame that is coming from a source that is not a familiar domain. It will be very hard for customer to find out if this window is a phishing scam or it is coming from the bank. Also, man in the middle attack and phishing scam is possible in this step. Mobile users may experience some issues to see the pop-up or iframe [7]. Sometimes a 3D Secure confirmation code is required. If the 3DS implementation send it as a text message, customer may be unable to receive it based on the country that he is in. Also, this may cause trouble for the people who change their cellphone number regularly [7].

4.3   Proposed Protocol

In this section, we introduce a new payment protocol that is proper for mobile platform. Our mobile payment protocol is based on client centric model. In this section, we introduce an improved version of Diffie-Hellman key agreement protocol. Our improved key agreement protocol is based on algebra of logarithms and modulus arithmetic. Our intention is to make shared key generation process proper for mobile platform by reducing computational cost. We also try to reduce size of temporary results in our computations. You can see our key agreement protocol in Table 4.1.

You can see the proof of correctness of our key agreement protocol in below. Suppose, there exists two parties A and B. We compare the key that is generated by A to the key that is generated by B. If they are equal, our key agreement protocol is correct. In this proof, we use rules of algebra of logarithms and algebra of modulus arithmetic. We suppose a_i is $a \bmod p_i$ and b_i is $b \bmod p_i$. Integer a is represented by r-tuple $(a_1, \ldots, a_r)$. In this kind of representation, residues should be calculated by multiple divisions. So, $a_i = a \bmod p_i$ ($1 \leqslant i \leqslant r$). Then, based on Theorem 2.1, we have the following [15]:

Table 4.1 Our key agreement protocol

	Action	Description
1	Alice and Bob agree on	q is a large prime number
	three numbers a, p, and q	$a = p^n$ p = 2, 3, ..., n n, u, and v = 1, 2, 3, ..., n
2	Alice picks a secret number u	Alice's secret number = u u mod q is not zero
3	Bob picks a secret number v	Bob's secret number = v v mod q is not zero
4	Alice computes her public number $A = ((u \bmod q) \times \log_p a) \bmod q$	Alice's public number = A $= (\log_p a^{(u \bmod q)}) \bmod q$
5	Bob computes his public number $B = ((v \bmod q) \times \log_p a) \bmod q$	Bob's public number = B $= (\log_p a^{(v \bmod q)}) \bmod q$
6	Alice and Bob exchange their public numbers	Alice knows u, a, p, q, A, and B Bob knows v, a, p, q, A, and B
7	Alice computes $k_a = ((u \bmod q) \times B) \bmod q$	$k_a = ((u \bmod q) \times (\log_p a^{(v \bmod q)})) \bmod q$ $k_a = (\log_p a^{(v \bmod q) \times (u \bmod q)}) \bmod q$ $k_a = (\log_p a^{((u \times v) \bmod q)}) \bmod q$
8	Bob computes $k_b = ((v \bmod q) \times A) \bmod q$	$k_b = ((v \bmod q) \times (\log_p a^{(u \bmod q)})) \bmod q$ $k_b = (\log_p a^{(u \bmod q) \times (v \bmod q)}) \bmod q$ $k_b = (\log_p a^{((v \times u) \bmod q)}) \bmod q$
9	By the law of algebra, Alice's k_a is the same as Bob's k_b,	Alice and Bob both know the secret value k

$$(a_1, \ldots, a_r) \times (b_1, \ldots, b_r) = (a_1 b_1 \bmod p_1, \ldots, a_r b_r \bmod p_r)$$

In our case, we only have one prime number that is q. In addition, based on algebra of logarithm, we have the following [10]:

$$\log_b(m^n) = n \log_b(m)$$

So, let's suppose we have two parties A and B. We follow our protocol to generate a key for each one of them. Then, by the rules that we just mentioned, we prove that the key that is generated for A is equal to the key that is generated for B. Besides, as you see in Table 4.2, results of the calculations in both sides (A and B) are equal.

A:.

$$k_a = ((u \bmod q) \times (\log_p a^{(v \bmod q)})) \bmod q$$
$$k_a = (\log_p a^{(v \bmod q) \times (u \bmod q)}) \bmod q$$
$$k_a = (\log_p a^{((u \times v) \bmod q)}) \bmod q$$

B:.

$$k_b = ((v \bmod q) \times (\log_p a^{(u \bmod q)})) \bmod q$$
$$k_b = (\log_p a^{(u \bmod q) \times (v \bmod q)}) \bmod q$$
$$k_b = (\log_p a^{((v \times u) \bmod q)}) \bmod q$$

As you can see the results in both sides are equal. So, Alice and Bob now have a shared key. They can use this key for symmetric encryption. Note that, this key never traveled via the network during the key generation steps. Also our middle keys never travel via the network. We kept security strength of Diffie-Hellman and reduced its computational cost. We can use our improved key agreement protocol instead of Diffie-Hellman in our mobile payment protocol in order to make it proper for mobile platform.

Our proposed mobile payment protocol's principals are payer, payee, mobile network operator, MPI, ACS, payer's credit card issuer (bank), payee's acquirer (bank), and certification authority for supporting secure transaction execution. Our protocol works with two sets of keys. First set should be shared between payer and his mobile network operator. The second set should be shared between payee and his mobile network operator. Our protocol consists of two-sub protocols that are registration and payment protocols. Payer and payee must register with their own mobile network operators at the beginning. Payer and his mobile network operator should generate a session key by running our improved key agreement protocol. You can see our notations in Table 4.2.

The rest of this section is defining our new mobile payment protocol that we implement in seven steps. At the beginning, payer encrypts registration details such as account information, payer's identity, and his phone number with his shared key. This information should be sent to payer's mobile network operator.

Table 4.2 Notations

Symbols	Descriptions
MNO	Mobile Network Operator
{payer, payee, payerś MNO, payeeś MNO}	A set of engaging parties, which includes Payee, Payer, Payeeś MNO and Payerś MNO
Pay center	Time Stamp and Digital Sign center
PN	Phone Number of Party P
PIN	Party P selected this password identification number
ID	Identity of party P, which identifies party P to MNO
AI	Account information of party P, which including credit limit for each transaction and type of account
R_1	Random and times-tamp generated number by Payer act as Payer's pseudo-ID
R_2	Random and time-stamp generated number to protect against replay attack
K_1	Shared key between payer and his mobile network operator
K_2	Shared key between payer and payee
AMOUNT	Payment transaction amount and currency
DESC	Payment description, which may include delivery address, purchase order details and so on
TID	The identity of transaction
TID_{Req}	Request for TID
PID_{Req}	Request for P identity
Req	Request
MX	The message M encrypted with key X
H(M)	The one way hash function of the message M
i	Used to identify the current session key of X_i and Y_i
$K_{p\text{-}p}$	The secret key shared between Payerś MNO and Payeeś MNO.
Success/Fail	The status of registration, whether success or failed
Yes/No	The status of transaction, whether approved or rejected
Received	Payment receivable update status, which may include the received payment amount
Pr_P	Private key of party P
Pu_P	Public key of party P
CK	client key: a key that is necessary for decoding X_i and Y_i sets on client side
CK_{Req}	Request for client key
T	Current date and time

$$payer \Rightarrow payerś\ MNO :$$
$$\{PN_{Payer}, ID_{Payer}, AI_{payer}\}K1$$

As we mentioned earlier, there are several challenges in designing our payment protocol. One of these challenges is to prevent privacy violation of payer. Most of the current payment protocols are providing identity protection from eavesdroppers. But, they don't provide identity protection from merchant. One of our goals is to avoid possible identity or privacy violation in our payment protocol. In order to overcome the issue of privacy violation, we want to involve mobile network operators in the payment process. Besides, we want to generate temporary identity for our customers in order to provide proper privacy protection for them. We generate this temporary identity based on our customer's phone number and his password identification number. Note that, this ID will be generated after a successful authentication.

During the registration process, payer has to set his password identification number (PIN_{Payer}) in order to access his mobile wallet application. This implementation uses two factor authentication that is an important principle for mobile device access control. Two factor authentication authenticates users in two steps. The first step is authentication with mobile wallet application on his mobile that is something that he has. The second step is password authentication that is something that he knows. Then, ID_{Payer} will be computed by hashing payer's phone number (PN_{Payer}) and password identification number (PIN_{Payer}).

$$ID_{Payer} = PN_{Payer} + \text{Hash}(PN_{Payer}, PIN_{Payer})$$

Then, payer's mobile network operator decodes the message with his shared key (K_1). Payer's mobile network operator stores necessary information into its database.

If registration process is successful, payer's mobile network operator will send confirmation message to inform payer about the result. Confirmation message is also encrypted with the session key (K_1).

Payer's MNO $\Rightarrow$ Payer : {Success/ Failed}Encrypted with K_1

After registration, payer receives mobile wallet application through email or downloads it from his mobile network operator website. Mobile wallet application has symmetric key generation and payment software. After successful installation, a set of symmetric keys ($X = \{ X_1, X_2, \ldots, X_n\}$) will be generated. They will be stored in payer's mobile device and will be sent to his mobile network operator. Payee must go through the similar registration process with his mobile network operator. This enables him to receive the payment amount. Payee generates a set of symmetric keys ($Y = \{Y_1, Y_2, \ldots, Y_n\}$) with his mobile network operator. These keys will be stored into payee's terminal and his mobile network operator's database.

In our protocol, if a person captures details of a payment transaction, he will not be able to use the message again, since all messages are encrypted in our protocol. Besides, these messages include random time-stamp generated numbers in order to protect our protocol from replay attacks. If someone steals the payment device, he can access (X or Y) the shared keys. Therefore, the thief can decode the payment messages and use them for illegal payment. To address this issue, all keys (X and Y) are encrypted in client device (with his key). Note that, this key is only viewable by his mobile network operator. Client does the following steps in order to obtain the client key.

$P \Rightarrow$ P's Mobile Network Operator :
{PN_P, Current Date and Time, $CK_{Request}$}$Pu_{P's MNO}$
P's MNO $\Rightarrow$ P :{CK}Encrypted with Pu_P

Current payment protocols support transaction privacy protection from eavesdroppers. However, they don't support transaction privacy protection from bank. So, it is obvious that who is paying how much to whom for ordering what items in each transaction. Also, some credit card issuers provide categorized spending charts (ex. merchandise, dining, and travel) for their customers. So, the financial institution or bank knows details of the transaction. We want to provide transaction privacy protection in our protocol. For this purpose, we encrypt transaction's details before sending it to pay-center (payment gateway).

The next challenge is to support non-repudiation. We should make sure that after a successful payment, payer or payee can't deny the transaction. For this purpose, we utilize digital signatures. Pay-center is responsible for generating time-stamps and verifying digital signatures in our protocol.

Our proposed payment protocol has seven phases. In our protocol, we verify digital signatures twice. In phase 2, pay-center verifies payer's digital signature and generates our first time-stamp. In phase 6, pay-center generates the second time-stamp and verifies payee's digital signature. Because of these two verifications, we can support non-repudiation in our payment protocol.

It is important to prevent replay attacks in payment protocols. Most of the current payment protocols support this feature. We should support this feature, since it is an essential and fundamental feature of a payment protocol. If we don't have a mechanism to prevent replay attacks, the payment transaction may be used again by an eavesdropper. If an eavesdropper captures one of the transactions, he can manipulate the transaction and use it again for illegal purposes. We also have another restriction about our keys that prevents replay attack. This restriction will be discussed later.

In our protocol, we also have a mechanism to prevent replay attacks. We have two random and time-stamp generated numbers. The first one is payer's pseudo-ID. The second one is to prevent replay attacks. We include this number in our messages in phases 2, 4, 5, 6, and 7 of our payment protocol in order to prevent replay attacks. In this case, a person cannot use a transaction for the second time, since the time-stamp is not matched with the current time. As we mentioned earlier, our proposed payment protocol is composed of seven phases as illustrated in Fig. 4.1. You can see these seven phases with their details in Table 4.3. We designed these steps for mobile platform. These phases should be implemented properly as a payment protocol.

In Summary, payer sends the subtraction request to his mobile network operator. His mobile network operator sends the request to payee's mobile network operator. Payee's mobile network operator sends the request to payee and receives his response. Payee's mobile network operator sends the reply to payer's mobile network operator. If payee accepts the request, payer's mobile network operator will initiate the transaction through the payment gateway (pay-center). If payee rejects the request, payer's mobile network operator will inform the payer about the denial. After a successful transaction pay-center informs mobile network operators about the successful result. Then, they inform their clients about the result of their transaction.

After successful completion of these seven phases, payee will release or deliver the purchased goods or services. As we mentioned earlier, one of the challenges in mobile payment is to prevent replay attacks. To prevent replay attacks, payer's mobile network operator and payee's mobile network operator make sure that symmetric keys (X_i and Y_i) have not been used before processing the current payment transaction. Mobile network operators will keep a list of generated secret keys and expire used symmetric keys from

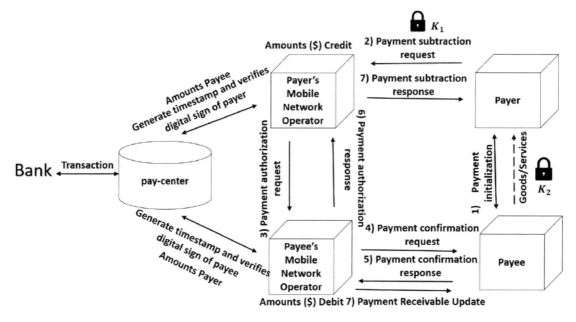

Fig. 4.1 Proposed mobile payment protocol

Table 4.3 Our mobile payment protocol

Phase 1: Payment Initialization:
Payer $\Rightarrow$ Payee: R_1, TID_{Req}, $PayeeID_{Req}$
Payee $\Rightarrow$ Payer: $\{ID_{Payee}, TID, ID_{MNO}\}k_2$
Phase 2: Payment Subtraction Request Payer:
Payer $\Rightarrow$ Payer's MNO: $\{$ ID_{Payee}, ID_MNO, R_1, TID, AMOUNT, DATE, R_2,
H $(ID_{Payee}$, ID_{MNO}, R_1, TID, AMOUNT, DATE,R_2), $\{R_2$, DESC $\}$ $K_2\}$ X_i, i, ID_{Payer}
Payer's MNO $\Rightarrow$ pay-center:
H [$\{ID_{Payee}$, ID_{MNO}, R_1, TID, AMOUNT, DATE, R_2, H $(ID_{Payee}$, ID_{MNO}, R_1, TID, AMOUNT, DATE, R_2),
$\{R_2$, DESC $\}$ $K_2\}X_i$, i, ID_{Payer}]
pay-center $\Rightarrow$ Payer's MNO: generates TimeStamp1 and verifies Payer's digital signature
Phase 3: Payment Authorization Request:
Payer's MNO $\Rightarrow$ Payee's MNO: R_1, ID_{Payee}, TID,
AMOUNT, DATE, $\{R_1$, DESC $\}$ K_2
Phase 4: Payment Confirmation Request:
Payee's MNO $\Rightarrow$ Payee: $\{R_1$,TID, AMOUNT, DATE, $\{R_1$, DESC $\}$ K_2, R_2,
H $(R_1$,TID, AMOUNT, DATE, $\{R_1$, DESC $\}$ K_2, R_2), H $(K_{pp})\}y_i$, i
Phase 5: Payment Confirmation Response:
Payee $\Rightarrow$ Payee's MNO: $\{$ Yes/No, R_2, H $(K_{pp}$, H $(R_1$, TID, AMOUNT, DATE,
$\{R_1$, DESC $\}$ K_2, R_2), $\{$ Yes/No, TID, AMOUNT, DATE $\}$ $K_2\}$ Y_{i+1}
Phase 6: Payment Authorization Response:
Payee's MNO $\Rightarrow$ pay-center: H $(\{$ Yes/No, R_2, H (K_{pp}),
H $(R_1$, TID, AMOUNT, DATE, $\{R_1$, DESC $\}$ K_2, R_2), $\{$ Yes/No, TID, AMOUNT, DATE $\}$ $K_2\}$ $Y_{i+1})$
pay-center $\Rightarrow$ Payee's MNO: generates TimeStamp2 and verifies Payee's digital signature
Payee's MNO $\Rightarrow$ Payer's MNO: Yes/No, TID, AMOUNT, DATE, $\{$ Yes/No, TID, AMOUNT, DATE $\}$ K_2
Phase 7: Payment Subtraction Response:
Payer's MNO $\Rightarrow$ Payer: $\{$ Yes/No, R_2, H $(K_{p\text{-}p})$, H $(ID_{Payee}$, ID_MNO, R_1,
TID, AMOUNT, DATE, R_2), $\{$ Yes/No, TID, AMOUNT, DATE $\}$ $K_2\}$ X_{i+1}
Payee's MNO $\Rightarrow$ Payee: $\{$ Received, R_2, H $(K_{p\text{-}p})$, H $(R_1$, TID, AMOUNT,
DATE, $\{R_1$, DESC $\}$ K_2, $R_2)\}$ Y_{i+1}

Fig. 4.2 Performance evaluation: prime = 7 and 100 iterations

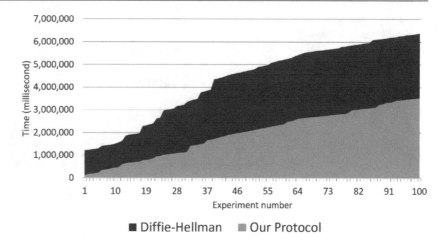

the list. Payer and payee may receive an update notification from their mobile network operators when their key is expired. To update their secret keys, they should connect to their mobile network operator and generate a new session key (K_1) by running our key agreement protocol. Then, they generate a new set of secret keys (X and Y) with a new session key (K_1) in offline mode.

We utilize cloud messaging in order to add an extra layer of security to our payment protocol. Our intention is to prevent card not present fraud in our payment protocol. For this purpose, we try to improve 3D Secure features for authenticating a customer. Note that, 3DS is utilizing behavioral data to understand which transactions are suspicious. This protocol will not prompt all the customers, but it only prompts transactions with risk score shows higher than threshold. There is a behavior model available that assigns each transaction a score based on different factors [17]. If this risk score is less than threshold, the system allows the customer to continue the transaction. But, if this score is higher than threshold, customer will be prompted by a pop-up window or inline frame in the middle of transaction [17].

We borrow all 3DS features for this extra security level, but instead of showing the extra layer as pop-up window or inline frame, we utilize cloud messaging for this purpose [11]. We prevent several issues by utilizing cloud messaging instead of pop-up window (ex. difficulty in viewing pop-up windows in mobile devices, difficulty in receiving one time passwords and confirmation codes via text message, and unable to see the source of pop-up window). We will push a notification to our customer to make sure that he is the legitimate customer. Note that, all mobile operating systems are equipped with cloud messaging tools. The benefit of this approach is that the customer recognizes the source of push notification. Also, if sending a one-time password is required, this approach will be better, since it doesnt have the limitations of text messaging or emails. Note that, texting and sending email from one country to the other one may be

temporarily unavailable or limited, but cloud messaging only require access to Internet [7].

4.4 Performance Evaluation

In this section, we want to Compare execution time of our protocol with DH (Diffie-Hellman). For this experiment, we had a virtual machine with 7 GB of memory, and two CPUs (each CPU had two cores). The operating system was windows seven 64 Bit. We ran our protocol against Diffie-Hellman 100 times to see the difference. In our experiment, we run two protocols against each others 100 times and we kept stacked time of these experiments. You can see the result of these experiments in Fig. 4.2. Horizontal axis is showing number of our experiments and vertical axis is showing the stacked time based on milliseconds. After 100 experiments, average execution time of our protocol was 34,930 milliseconds and average execution time of the original version was 63,116 milliseconds. In the next experiment, we run the protocols against each others 30,000 times. We wanted to see the result for large number of iterations. Our prime number was 7. You can see the result of these experiments in Fig. 4.3. After 30,000 experiments, average execution time of our protocol was 3,899 milliseconds and average execution time of the original version was 6,545 milliseconds.In summary, our protocol was almost twice as fast as Diffie-Hellman. Besides, we compare our proposed payment protocol with existing payment protocols. Table 4.4 shows the result of these comparisons.

4.5 Conclusion

In this paper, we introduced a new payment protocol that is compatible with mobile platform. We decreased computational cost of generating a shared key between two parties. Based on our experiments, our protocol is almost twice as

Fig. 4.3 Performance evaluation: prime = 7 and 30,000 iterations

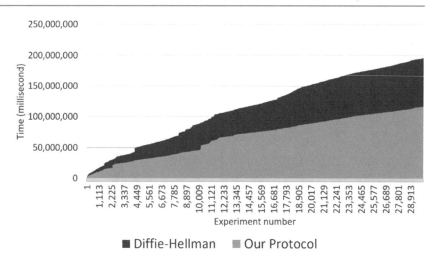

Diffie-Hellman   Our Protocol

Table 4.4 Comparison

Key points	SET	iKP	KSL	MPCP 3D	3DS
Identity protection from merchant	N	N	N	Y	N
Identity protection from eavesdrop	Y	Y	Y	Y	–
Transaction privacy protection from eavesdrop	Y	Y	Y	Y	–
Transaction privacy protection from bank	N	N	N	Y	–
Supporting non-repudiation by digital signature	N	N	N	Y	–
Compatible with mobile devices	N	N	N	Y	–
preventing card not present fraud	N	N	N	Y	–
preventing man in the middle attack and phishing scam	–	–	–	Y	N
No geographical restrictions in the extra security layer	–	–	–	Y	N
customer can see the source of extra security layer	–	–	–	Y	N
Is the implementation of 3D compatible with mobile devices?	–	–	–	Y	N

fast as Diffie-Hellman. We defined two different random and time-stamp generated numbers in order to avoid replay attacks. We utilized digital signature in order to support non-repudiation in our protocol. We defined and extra security layer to prevent card not present fraud. We utilized cloud messaging to implement this extra security layer in order to solve original 3DS issues.

However, there are some parts of this research that can be improved. We will try to reduce the computation and communication costs of key agreement protocol. We try to improve the performance of 3DS behavioral model. So, we should only show our extra security layer for transactions that are more likely a threat. We should extend our protocol in order to support wider range of devices especially popular devices in Internet of things. We will try to utilize other cloud technologies in our payment protocol.

References

1. American Express SafeKey (2016) americanexpress.com. Available https://network.americanexpress.com.
2. Bidgoli, H. (2004). *The internet encyclopedia*. Hoboken, N.J.: Wiley.
3. EMV Migration Forum (2015) near-term solutions to address the growing threat of card-not-present fraud: Card-not-present fraud working committee white paper, April 2015.
4. Forrester: Mobile and Tablet Commerce Forecast 2015 To 2020, Forrester.com, 2016. Available https://www.forrester.com.
5. JCB International J/Secure (2016) global.jcb. Available https://http://www.global.jcb/.
6. MasterCard SecureCode (2016). mastercard.us. Available https://www.mastercard.us.
7. Paul, D., Hongrui, G., Kannan, S., & CA Technologies, Advanced Analytics and Data Science (2014). 3D-secure authentication using advanced models, October 2014.
8. SET Secure Electronic Transaction Specification, Version 1.0, May 31, 1997.
9. Vahidalizadehdizaj, M. (2011). New mobile payment protocol: Mobile pay center protocol 4 (MPCP4) by using new key agreement protocol: VAC2. In *IEEE International Conference on Electronics Computer Technology*, April 8–10, 2011 (Vol. 2, pp. 67–73).
10. Vahidalizadehdizaj, M., & Geranmaye, M. (2011) New mobile payment protocol: Mobile pay center protocol 5 (MPCP5) by using new key agreement protocol: VG1. In *IEEE International Conference on Computer Modeling and Simulation* (Vol. 2, pp. 246–252).
11. Vahidalizadehdizaj, M., & Geranmaye, M. (2011). New mobile payment protocol: Mobile pay center protocol 6 (MPCP6) by using new key agreement protocol: VGC3. In *IEEE International Conference on Computer Modeling and Simulation* (Vol. 2, pp. 253–259).

12. Vahidalizadehdizaj, M., & Tao, L. (2015). A new mobile payment protocol (GMPCP) by using a new key agreement protocol (GC). In *IEEE International Conference on Intelligence and Security Informatics*, May 27–29, 2015 (pp. 169–172).

13. Vahidalizadehdizaj, M., & Tao, L. (2015). A new mobile payment protocol (GMPCP) By using a new group key agreement protocol (VTGKA). In *IEEE International Conference on Computing, Communication and Networking Technologies*.

14. Vahidalizadehdizaj, M., Moghaddam, R. A., & Momenebellah, S. (2011). New mobile payment protocol: Mobile pay center protocol (MPCP). In *International Conference on Electronics Computer Technology*, April 8–10, 2011 (Vol. 2, pp. 74, 78).

15. Vahidalizadehdizaj, M., Moghaddam, R. A., & Momenebellah, S. (2011). New mobile payment protocol: Mobile pay center protocol 2 (MPCP2) by using new key agreement protocol: VAM. In *IEEE Pacific Rim Conference on Computers and Signal Processing*, August 23–26, 2011 (pp. 12–18).

16. Verified by Visa (2016). visaeurope.com. Available https://www.visaeurope.com.

17. Warren, I., Meads, A., Srirama, S., Weerasinghe, T., & Paniagua, C. (2014). Push notification mechanisms for pervasive smartphone applications. *IEEE Pervasive Computing, 13*(2), 61–71.

Event-Based Anomalies in Big Data

Yenumula B. Reddy

Abstract

The data stream generated in a social network or geophysical related or network flow is high speed, continuous, multi-dimensional, and contains massive data. Analytics require the insight behavior of the data stream. The government and business giants want to catch the exceptions to reveal the anomalies and take immediate action. To catch up the exceptions, the analysts need to identify the patterns in a single view of data stream trends, exceptions and catch up anomalies before the system collapses. In this paper, we present a system that detects the variations in the area of interest of data stream. The current research includes the classification of the data stream, detect the event type, commonly used detection methods, and interpret the detected events.

Keywords

Anomaly detection • Big Data • Social network • Modeling • Climate change • False positives • False negatives

5.1 Introduction

Big data refers to a large-scale complex, unstructured information that differs in storage and processing techniques from traditional data. The big data and traditional data differs in three ways: the volume (amount of data), velocity (the rate of data generation), and variety (the types of data– structured and unstructured). The creation of massive data every day (2.5 quintillion bytes per/day) is difficult to organize in traditional database management techniques. This acceleration production of data has created a need for new technologies to analyze data on a real-time basis and detect frauds to save the organizations.

The process of analyzing and mining the complex data (data from internet or sensors) helps to view the affected areas of interest. For example, the interest in environmental data includes many factors. The interests include one of the following or more than one type.

- Temperature and precipitation in a defined geographical area
- Climatological ranking (climate extremes - rain or drought or snow)
- A pattern of life; and social impacts.

Currently, several tools and algorithms are available to create complex queries for analysis of big data in R language. The following are few techniques currently using Big Data analysis.

- Deep learning techniques [1, 2] provided by NVIDIA Caffe and Digits. The model uses convolutional deep learning models and are in developmental state to analyze the complex data.
- The Hadoop platforms including Pig (scripting language platform for complex queries), Hive (managing large data sets in distributed storage using SQL), and Mahout and

Y.B. Reddy (✉)
Department of Computer Science, Grambling State University, Grambling, LA 71245, USA
e-mail: ybreddy@gram.edu

© Springer International Publishing AG 2018
S. Latifi (ed.), *Information Technology – New Generations*, Advances in Intelligent
Systems and Computing 558, DOI 10.1007/978-3-319-54978-1_5

RHadoop (data mining and machine learning algorithms for Hadoop).

Big data preservation has a large number of privacy problems regarding storage, process, retrieve and analyze. The current tools provide access limitations and safe storage. New tools are underway with intellectual property rights, protection of data, access rights, and regulations.

Anomalies in Big Data are the new areas of research. They are the reputations in data deviated from normal or expected data flow. The following examples help us to understand abnormal occurrences in daily life.

- Unusual purchase requests from a department in particular week or month
- Repeated maintenance for the demand for machine or vehicle
- Significant overtime requests from a single person
- Abnormal activities in Facebook

Anomaly detection can identify unexpected activity in the regular data-flow. The activities include the new attacks in networks, frauds in health data (fraud detection), business data, government data, unexpected changes in climate (extremes) data collected from sensors, and social network scams. Some of the malicious events occur in data-flow needs real-time detection, analysis, and remediation. Visual interface will help better in the analysis in the real-time detection of fraud.

5.2   Related Work

Detecting anomalies is a difficult task but valuable in a continuous flow of data. Danger depends on upon the type of abnormalities. Retting et al. [3] introduced anomaly detection model that works batch processing and with data flowing through some nodes. They tested the relative entropy scalability, performance over increased data load as part of the system scalability. The authors concluded that the model works at different scales (geographically and temporally).

Hays and Capretz [4] discussed the contextual anomaly detection in big sensor data. The work presents a technique that uses a well-defined contextual anomaly detection algorithm in real-time analysis. The algorithm is multivariate and used on sensor-generated data. Hao et al. [5] built the technique called Anomaly Marker on cell-based data streams and user-defined thresholds. This method allows users to define the marked area that enables to group interesting points related to anomalies. The model detects the data stream anomalies to a great real-world enterprise server.

Bologa et al. [6] investigated the big data technology and primary methods of analysis. The proposed model applied to insurance fraud detection. The authors presented a case study that relates to the country Romania. Special case includes the analysis for detecting fraud in health insurance. The research explains the related to social network predictive modeling.

Parveen [7] presented malicious activity in large data streams. The thesis discusses the ensemble-based stream mining using supervised, unsupervised learning and graph-based anomaly detection. Parveen concluded that ensemble-based approach is significantly effective compared to traditional methods. The survey paper [8] provides the structured and comprehensive overview of the research on anomaly detection. The authors concluded the effectiveness of particular technique on each category and their advantages.

Kakavand et al. [9] discussed the term frequency and inverse document frequency model for text mining-based anomaly detection. The authors used Mahalanobis Distances Map (MDM) and support vector machine (SVM). They applied the model to discover the hidden correlations between the features and among the packet payloads. Wang et al. [10] discussed the anomaly detection for ultra-wideband radar data. The proposed algorithm helps the signal data passing through the wall. Purdy's white paper [11], the biologically inspired intelligence technology, explains the algorithm called Hierarchical Temporal Memory (HM) works similar to human brain. The application of HTM includes prediction, classification, and anomaly detection.

The remaining part of the paper discusses the proposed method in Sect. 5.3, R Language tools and application for the proposed model, for anomaly detection, and anomaly detection model in Sect. 5.3. Sections 5.4, 5.5 and 5.6 presents the anomaly detection model, event detection and analysis using R discussed, and the conclusions and future research.

5.3   Anomaly Detection Techniques

Social networks are good examples for event detection. Depending on the availability of information events are specified and unspecified. For example, in social networks, it's hard to know the prior information. Therefore, unspecified detection of events technique depends on the temporal signal of the stream of text, images, and videos on a real-time basis. The particular context provides the venue, type of information, time, and description. Some techniques include filtering, clustering, aggregation and query generation helps to analyze the stream of information.

In the text, the similarity is critical. For example, 'breaking news' and 'breakingnews' must be considered as same word counting in the analysis. Grouping similar messages into a single cluster helps in increasing the total weight of the term. Further, new words are included in the panel if they are same or close. Secondly, the number of users interacting with the message and tweets helps in ranking the message. The hashtag terms represent strong indication and can have more weight in the social network messages. Eliminate the stop words and common terms (sleep, work, eat, out, and home) in the analysis. The significant events depend on upon the cross-correlation among the words and number of these words occurs in the message. The peaks are detected using the wavelet maximum and local maxima techniques. Once the event is detected, use latent Dirichlet allocation (LDA) method for topic discovery and activity peaks.

If the specified events are partially or entirely related to the content, we can exploit them using textual analysis, machine learning, and data mining techniques. Reddy et al. discussed the textual analysis using extraction of keywords and their importance [12]. They considered the appropriate weights for the keywords and selection of required document. The relative importance and similarity of words selection is the future research topic.

Sakaki et al. [13] discussed the continuous monitoring of signals to discover new events in near real-time basis in Twitter. They used the filtering techniques and applied to historical data for analysis of past data and monitoring events. Metzler et al. [14] presented the Twitter messages relevant to a user query and a large number of vocabulary match where the message do not contain a query item. Work is in progress regarding comprehensive and structured messages of Twitter events.

For continuously generated text, incremental clustering approaches are suitable. In incremental clustering, the algorithms produce maximum similarity between new tweets and any of the existing clusters. The old and new groups are kept separate or merged depending on the similarity threshold. Summarizing multiple tweets for the same event and providing most common phrase helps in phrase reinforcement and similarity findings. Further, filtering techniques reduce the irrelevant messages and keep the required dataset to a reasonable size.

One of the methods is Seasonal Trend Decomposition using Loess (STL). STL is an algorithm that helps to divide up a time series into the trend, seasonality, and reminder. STL [15] removes an estimated trend component from the time series and splits the data into sub-cycle series. The use of the STL is simple with the following commands in R.

```
>city #displays the data of a year by month
(stored data)
> city.stl = stl (city, s.window="periodic")
> plot(city.stl)
```

Figure 5.1 displays 20 years monthly temperature data of city data. Figure 5.2 shows four graphs of the plot result. The

```
>city
    Jan Feb Mar Apr May Jun Jul Aug Sep Oct Nov Dec
1920 40.6 40.8 44.4 46.7 54.1 58.5 57.7 56.4 54.3 50.5 42.9 39.8
1921 44.2 39.8 45.1 47.0 54.1 58.7 66.3 59.9 57.0 54.2 39.7 42.8
- - - - - -  - - - -
- - - - - - - - - -
1937 40.8 41.0 38.4 47.4 54.1 58.6 61.4 61.8 56.3 50.9 41.4 37.1
1938 42.1 41.2 47.3 46.6 52.4 59.0 59.6 60.4 57.0 50.7 47.8 39.2
1939 39.4 40.9 42.4 47.8 52.4 58.0 60.7 61.8 58.2 46.7 46.6 37.8
```

Fig. 5.1 Data by month for 20 years

first is original data, second seasonal component, third trend component and fourth shows the periodic seasonal pattern extended out from original data. The bar at the right-hand side of each graph shows a relative comparison of the magnitude of each component. The figures show that the change in trend is less than the variation doing to monthly variation (right sidebar helps for comparison). We can replace the current data with any other particular city or place data

```
>city
```

Behavioral change point analysis (BCPA) package in r-language helps to estimate the autocorrelation of the random data. The BCPA package contributes to identifying the changes in animal behaviors obscured by visual inspection or standard techniques (unstructured data). For better understand of data and predict future points, the BCPA uses autoregressive integrated moving average (ARIMA) model. The following code provides usage of the model.

```
>Simp.VT=GetVT(Simp)
>Simp.ws=WindowSweep(Simp.VT, "V*cos(Theta)",
windowsize=50, progress=FALSE,K=2)
>plot(Simp.ws, type="smooth")
>plot(Simp.ws, type="flat")
```

A default "Simp" generates the 200 data sets for x-value and y-value with random time increment. WindowSweep() is the workspace function of BCPA. The sweep function searches most significant change points and identifying the parsimonious model according to an adjusted Bayesian criterion (BIC). The Smaller value for K (of BIC) makes a less sensitive model selection. The Fig. 5.3a, b shows the change point (spike) just after 10, close to 30, and close to 60. Figure 5.3b shows the flat end window.

5.4   Anomaly Detection Model

The proposed model relates to the events at a designated place (global positioning system). The event may be animal or human movement, migration, dispersal, and breeding.

Fig. 5.2 STL Decomposition of city Temperature Time Series

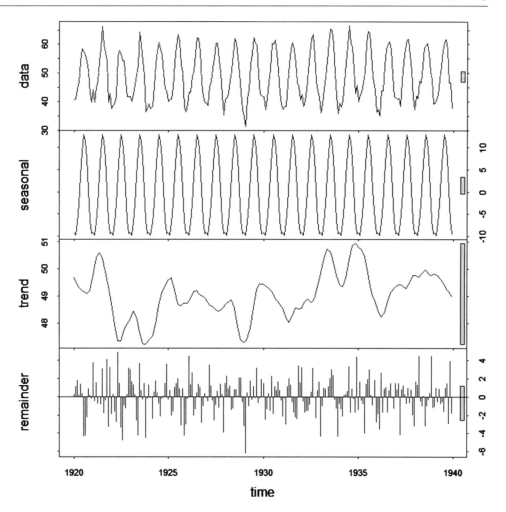

The movements of living beings data are multidimensional and auto-correlated in space and time. The problem is closely related to behavioral change point analysis (BCPA) and is similar to STL algorithm or Simp data as explained in Section 3. The event depends on time ($T = t_1, t_2, ...t_n$)and position P (X, Y). The data decomposition depends on collecting every speed estimate and turning angle into orthogonal components of persistence velocity $V_p(t)$ and turning velocity$V_t(t)$.

$$V_p(T_i) = V(T_i) \cos\left(\psi(T_i)\right) \qquad (5.1)$$

$$V_t(T_i) = V(T_i) \sin\left(\psi(T_i)\right) \qquad (5.2)$$

The persistence velocity captures the tendency and magnitude of a movement in a given direction at a given time interval. The following equation provides the BCPA model of an animal movement at any given time interval and place [16].

$$CorX(t)X(t - \Delta t) = \exp(-\Delta t/\tau) = \rho^{\Delta t} \qquad (5.3)$$

where $X(t)$, the Gaussian time-series depends on mean μ, standard deviation σ, and time scale τ (or autocorrelation coefficient ($0 < \rho < 1$). The likelihood of ρ is calculated using the GetRho function and GetL function.

```
> GetRho <- function (x, t)
+ {
+   getL <- function(rho) {
+   dt <- diff(t)
+   s <- sd(x)
+   mu <- mean(x)
+   n <- length(x)
+   x.plus <- x[-1]
+   x.minus <- x[-length(x)]
+   Likelihood <- dnorm(x.plus, mean = mu +
    (rho^dt) *
+   (x.minus-+   mu), sd = s * sqrt(1 - rho^(2 *
    dt)))
+   logL <- sum(log(Likelihood))
+   if (!is.finite(logL))
+       logL <- -10^10
+       return(-logL)
+ }
+   o <- optimize(getL, lower = 0, upper = 1, tol
    = 1e-04)
+   return(c(o$minimum, o$objective))
+ }
```

Fig. 5.3 (a) Smooth model for
Simp data. (b) flat model for
Simp data

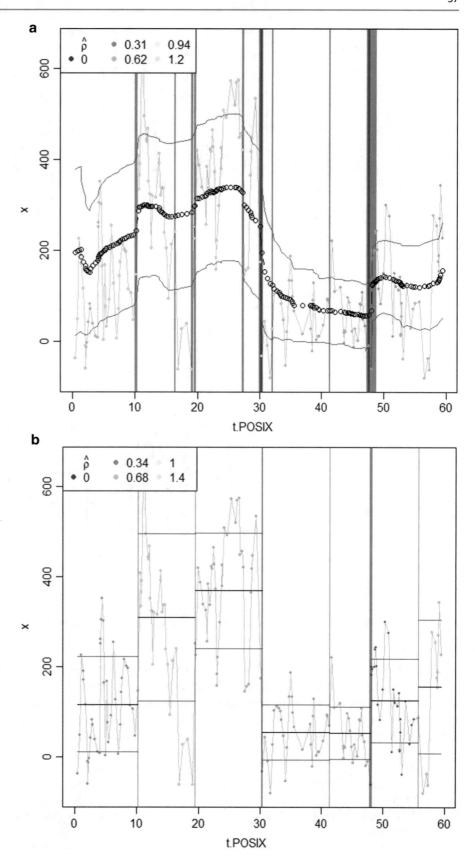

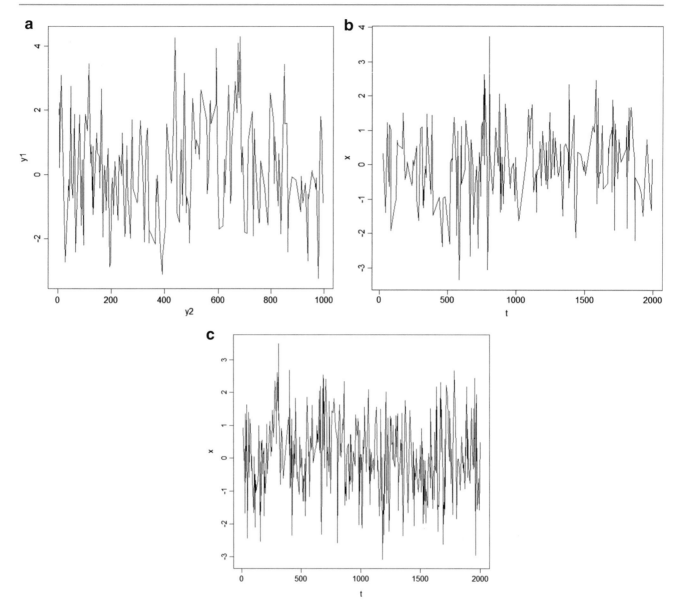

Fig. 5.4 (**a**) The Gaussian time-series for $\rho = 0.8$ and different time units. (**b**) The Gaussian time-series (frequency 200 units) for $\rho = 0.5$ and different time units. (**c**) The Gaussian time-series (frequency 400 units) for $\rho = 0.5$ and different time units

For example, assume ρ (rho) value is set at 0.8 and time t varies 1 to 1000 and data collected for each 200 units of time. The following R code helps to collect the Gaussian time series values for a given likelihood ρ. In the Fig. 5.4a the y2 values are time units and y1 are x values.

```
> rho=0.8
> x.full=arima.sim(1000,model=list(ar=rho))
> t.full=1:1000
> keep=sort(sample(1:1000,200))
> x=x.full[keep]
> t=t.full[keep]
> plot(t,x,type="l")
> GetRho(x, t)
[1]   0.775475      343.329163
```

The estimates of ρ (rho) and likelihood are 0.775475 and 343.329163.

By changing the ρ value and t value, the estimates of ρ (rho) and likelihood are 0.3569982 and 305.9516379 (Fig. 5.4b).

```
> rho=0.5
> x.full=arima.sim(2000,model=list(ar=rho))
> t.full=1:2000
> keep=sort(sample(1:2000,200))
> x=x.full[keep]
> t=t.full[keep]
> plot(t,x,type="l")
> y1=c(x)
> y2=c(t)
```

```
> ybr=c(GetRho(y1,y2))
> ybr
[1]    0.3569982      305.9516379
```

The likelihood calculations shows that likelihood is proportional to ρ. If we change the frequency from 200 to 400 the likelihood of ρ changes to 0.5389303 and 601.8224489 Fig. 5.4c.

Changing the values of ρ , μ , and σ we get different models for likelihood values at different time periods. The BCPA model originally designed to identify the locations where changes are abrupt (assumed to correspond to discrete changes in an animal's behavior). Further, the geo-spatial model in Eq. (5.3) evaluates the movement of species at a particular location. The model can also be used to study the environmental conditions by providing related values to the parameters. Minor modifications may be required depending upon the study of the environment

5.5   Event Detection and Analysis Examples Using R

To work on time series examples, we must type the following commands and make sure all libraries and tools are installed. AnomalyDetectionTs () is the technique that detects the anomalies in seasonal univarient series where the input is series of <timestamp, count> pairs. If the input series is observations, then use the command AnomalyDetectionVec().

```
library(devtools)
install.packages("devtools")
devtools::install_github("twitter/AnomalyDetec-
tion",force=TRUE)
library(AnomalyDetection)
library(wikipediatrend)
install.packages("wikipediatrend")
devtools::install_github("twitter/AnomalyDetec-
tion")
library(AnomalyDetection)
devtools::install_github("petermeissner/wikipe-
diatrend")
install.packages("Rtools")
install.packages("Lock5Data")
install.packages("ggplot2")
require(ggplot2)

res=AnomalyDetectionTs(raw_data,max_anoms=0.02,
direction='both',plot=TRUE)
res$plot
```

The Fig. 5.5a, b shows the report of anomalies with day (September 25th to October 6th) and hour values (24 hours). Figure 5.6 shows the observations as input. Figure 5.5a shows that the observable events are (peaks) on October 1 (lowest and highest) and October 6th (lowest). Figure 5.5b shows (hourly) that the peak event is on October 5th 12th hour. If the time stamp is not available with AnomalyDetectionTs(), use AnomalyDetectionVec(). Figure 5.6 shows for 100 observations with peak at middle and end.

```
res = AnomalyDetectionTs(raw_data, max_anoms=0.05,
direction='both', only_last="day", plot=TRUE)
```

```
res = AnomalyDetectionTs(raw_data, max_anoms=0.05,
direction='both', only_last="hr", plot=TRUE)
```

Create the new_data from raw_data by adding a factor to each data element as 0.01*(1:1000). The new_data is as below. Figure 5.7 shows the observations. Adding the factor, moves the peak position to right (observe Fig. 5.6 and Fig. 5.7).

	index	anoms
1	125	26.5810
2	5320	200.2836
3	6224	154.3940

125	14348	217.4600
126	14358	47.5600
127	14368	257.6600
128	14378	47.7600

```
new_data = raw_data + 0.01*(1:1000)
AnomalyDetectionVec(new_data[,2], max_anoms=0.02,
period=1440, direction='both', plot=TRUE)
```

```
res  =  AnomalyDetectionVec(raw_data[,2],
max_anoms=0.05, period=100, direction='both',
only_last=FALSE,plot=TRUE)
```

5.6   Conclusions

Collecting, organizing, storing, and querying the continuously growing unstructured data is tough due its volume, velocity and size. Further, the unusual or anomalies in the data became a major factor in current day life. The important factor is identifying the unexpected spikes or events in continuous data. For example, the peaks in the Facebook or Twitter data during Thanksgiving, Christmas, and New Year time are very high. Similarly, the tornados or cyclones during June to September are very high. Identifying the spikes at a particular day or hour helps to organize and minimizing the jamming in the communication flow on networks. In the current research, the R-packages are used to study such data and methods provided to detect the events.

The examples were provided using BCPA package and time series package. The Sects. 5.3 and 5.4 provides the events of variation of population, temperature, weather

Fig. 5.5 (**a**) Anomalies seasonal univariate time series. (**b**) Anomalies by hour on last day

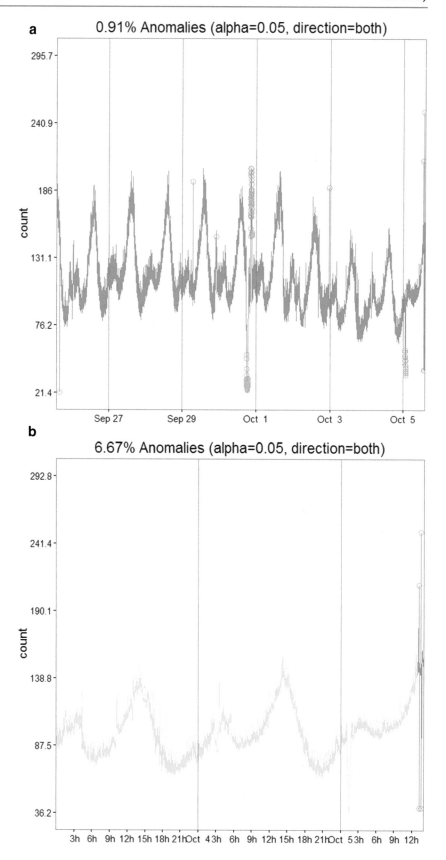

Fig. 5.6 Input as observations

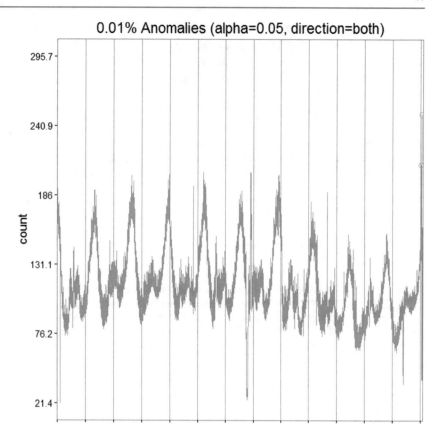

Fig. 5.7 Input as observations
with linear trend

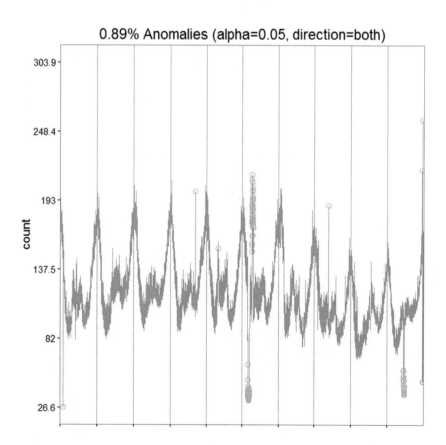

conditions, and movements of animals in the geological area using BCPA package. The document contains the time series examples to detect the events in weather data at different timings (days or hours). This preliminary study helps the table data from stored database. The image analysis is the future study of our research.

Acknowledgments The research work was supported by the AFRL Collaboration Program – Sensors Research, Air Force Contract FA8650-13-C-5800, through subcontract number GRAM 13-S7700-02-C2. The author wishes to express appreciation to Dr. Connie Walton, Director of Sponsored Programs Grambling State University for her continuous support in research

References

1. Berwald, J. (2015, August). *Deep learning for anomaly detection. PIMS-IMA math modeling in industry XIX.*. http://www.mathtube.org/lecture/video/deep-learning-image-anomaly-detection

2. Szegedy, C., Toshev, A., & Erhan, D. (2013). Deep neural networks for object detection. In Burges, C. J. C., Bottou, L., Ghahramani, Z., & Weinberger, K. Q. (Eds.). *NIPS* (pp. 2553–2561). Lake Tahoe: Advances in Neural Information Processing 26.

3. Rettig, L., Khayati, M., Cudre-Mauroux, P., & Piorkowski, M. (2015, November). Online anomaly detection over big data streams. *IEEE International Conference on Big Data*, pp. 1113–1122.

4. Hays, M., & Capretz, M. A. (2014). Contextual anomaly detection in big sensor data. *Journal of Big Data, 1*, 64–71.

5. Hao, M. C., Dayal, U., & Keim, D. A. (2009). Visual analytics of anomaly detection in large data streams. *SPIE 7243, Visualization and Data Analysis 2009, 72430B*, 1–10.

6. Bologa, R., & Florea, A. (2013). Big data and specific analysis methods for insurance fraud detection. *Database Systems Journal, 4* (4), 30–39.

7. Parveen, P. (2013, December). *Evolving insider threat detection using stream analytics and big data.* PhD thesis, University of Texas, Dallas.

8. Chandola, V., Banerjee, A., & Kumar, V. (2009). Anomaly detection: A survey. *ACM Computing Surveys, 41*, 1–72.

9. Kakavand, M., Mustapha, N., Mustapha, A., & Abdullah, M. (2014). A text mining-based an0maly detection model in network security. *Global Journal of Computer Science and Technology, 14*, 21–31.

10. Wang, W., Zhou, X., Zhang, B., & Mu, J. (2013, Febraury). Anamoly detection in big data from UMB radars. *Security Communication Networks, 8*(14), 2469–2475.

11. Purdy, S. *The science of anomaly detection*. White paper. https://numenta.com/assets/pdf/whitepapers/Numenta%20White%20Paper%20-%20Science%20of%20Anomaly%20Detection.pdf

12. Reddy, Y. B. & Hill, D. (2015, November). *Document selection using mapreduce*. IJSPTM.

13. Sakaki, T., Okazaki, M., & Matsuo, Y. (2010). *Earthquake shakes Twitter users: Real-time event detection by social sensors*. Proceedings of the 19th international conference on WWW, 2010, New York, pp. 851–860.

14. Metzler, D.,Cai, C., & Hovy, E. (2012). *Structured event retrieval over microblog archives*. HLT-NAACL, 2012, pp. 646–655.

15. Cleveland, R. B., Cleveland, W. S., McRae, J. E., & Terpenning, I. (1990). STL: a seasonal-trend decomposition procedure based on loess. *Journal of Official Statistics, 6*(1), 3–73.

16. Gurarie, E. *Behavioral change point analysis in R: The bcpa package*. 'bcpa' package is available on CRAN, from 2013 available from install in R.

ACO-Discreet: An Efficient Node Deployment Approach in Wireless Sensor Networks

Tehreem Qasim, Qurrat ul Ain Minhas, Alam Mujahid, Naeem Ali Bhatti, Mubashar Mushtaq, Khalid Saleem, Hasan Mahmood, and M. Shujah Islam Sameem

Abstract

Wireless sensor networks (WSNs) rely on effective deployment of sensing nodes. Efficient sensor deployment with ensured connectivity is a major challenge in WSNs. Several deployment approaches have been proposed in literature to address the connectivity and efficiency of sensor networks. However, most of these works either lack in efficiency or ignore the connectivity issues. In this paper, we propose an efficient and connectivity-based algorithm by modifying the Ant Colony Optimization (ACO) (Liu and He, J Netw Comput Appl 39:310–318, 2014). Traditional ACO algorithms ensure coverage at a high cost and repetitive sensing, which results in resource wastage. Our proposed algorithm reduces the sensing cost with efficient deployment and enhanced connectivity. Simulation results indicate the ability of proposed framework to significantly reduce the coverage cost as well as achieve longer life time for WSNs.

Keywords

Wireless sensor networks (WSNs) • Node deployment • Ant colony optimization (ACO) • Coverage cost reduction • Redundant sensor removal

6.1 Introduction

WSNs require proper deployment of sensors for smooth network operation. The sensing nodes are generally expensive, therefore inefficient and superfluous deployment of

T. Qasim (✉) • Q.A. Minhas • N.A. Bhatti • H. Mahmood
M. Shujah Islam Sameem
Department of Electronics, Quaid i Azam University, Islamabad, Pakistan
e-mail: tehreemqasim@yahoo.com; qminhas@qau.edu.pk; nbhatti@qau.edu.pk; hasan@qau.edu.pk; mshujahislam@gmail.com

A. Mujahid
Informatics Complex, Islamabad, Pakistan
e-mail: mujahidalam@gmail.com

M. Mushtaq
Department of Computer Sciences, Forman Christian College, Lahore, Pakistan
e-mail: mubasharmushtaq@fccollege.edu.pk

K. Saleem
Department of Computer Sciences, Quaid i Azam University, Islamabad, Pakistan
e-mail: ksaleem@qau.edu.pk.com

sensors increases the network cost. Hence an intelligent and effective method is required for this purpose.

Several researchers have explored the sensor deployment techniques in WSNs. In [1] a resource-bounded model is introduced for the optimization of grid coverage problem. In [2] two algorithms are proposed that give average coverage as well as coverage to the grid points that are least covered. In [3] and [4] grid-based placement of sensors is achieved with genetic algorithm, however communication between sensor nodes is not considered [5].

An interesting algorithm proposed recently for deployment issues of WSNs is Ant Colony Optimization (ACO). ACO based algorithms for WSN deployment address the problem of grid-based coverage with low cost and connectivity-guarantee (GCLC). In [6] an ACO based algorithm called EasiDesign is proposed for sensor deployment. This work is further improved in [7], where a similar ACO-based algorithm is proposed with increased number of transitions to reduce the number of deployed sensors. In [8] ACO-Greedy is proposed which uses a greedy migration scheme for ant's locomotion on the graph to further reduce

© Springer International Publishing AG 2018
S. Latifi (ed.), *Information Technology – New Generations*, Advances in Intelligent Systems and Computing 558, DOI 10.1007/978-3-319-54978-1_6

coverage cost. Furthermore, non-uniform coverage radii are used for sensor nodes to increase density of nodes in areas closer to the sink and to handle the large volume of relay traffic in these areas. This approach helps alleviate the energy hole problem. However, in all three ACO based algorithms once a solution is built, a huge number of redundant sensors are present in the solution. This is due to the fact that due to the inherent nature of ACO algorithm, an ant has to continue its tour till every point of interest (PoI) is covered by the deployed sensor network. At some stages during its tour, an ant may have no option but to deploy a redundant sensor/sensors to reach the uncovered points of interest. Removing these redundant sensors can markedly reduce the coverage cost. However identification and removal of these redundant sensors is a major challange and requires diligent formulation to ensure that the operation of the network remains unaffected even after removal of the redundant sensors. In [9], increased search range is proposed for next point selection for an ant but the grid points that are farther away from ant's current location are not prioritized. This scheme uses redundancy check but no proper mechanism is proposed to remove redundancy during execution of the algorithm. Also this scheme sometimes results in serious network connectivity issues if all the redundant sensors are removed [9].

In this paper, an efficient scheme is proposed that works in two phases and has the potential to significantly reduce the coverage cost. The proposed scheme is based on ACO technique and involves two phases. Phase-I of proposed framework is similar to ordinary ACO-based algorithms used for the problem of GCLC [8]. However, a modified heuristic value is used to prioritize those grid points that lie at larger distances from ant's current location. Thus, sensors placed more distantly would result in reduced number of total sensors. Phase-II of the algorithm initiates its operation after a solution is obtained from Phase-I and operates further on the solution to remove the redundant sensors. Hence a proper mechanism has been incorporated into the ACO algorithm to effectively remove unwanted sensors while ensuring connectivity of the network.

Our contributions lie in the following aspects.

1. A two phase ACO based algorithm has been proposed for identification and removal of redundant sensors found in solutions obtained from traditional ACO based algorithms for WSN deployment.
2. The heuristic value used in traditional ACO based algorithms for WSN deployment has been modified to ensure distant placement of sensor nodes and reduce the overall coverage cost of the network.

Rest of the paper is organized as follows. In Sect. 6.2, the problem of GCLC is briefly explained. In Sect. 6.3, the proposed framework is presented. In Sect. 6.4, simulation and results are given. In Sect. 6.5, we conclude the paper.

6.2 The Problem of GCLC

In the problem of GCLC [8], a sensor is to be deployed in such a way that a given set of PoIs as indicated by black dots in Fig. 6.1 are covered by the network with minimum number of sensors. The resultant network has to be fully connected i.e., each node should be able to communicate with other nodes as well as with the sink node through one or more hops.

6.3 Proposed Framework

6.3.1 Phase-I of Proposed Framework

Phase-I of proposed method works like ordinary ACO based algorithms [6–8] for the problem of GCLC. The process is repeated iteratively and during each iteration a fixed number of ants move on a 2-D grid to search for a solution. Each ant starts its tour from the sink and every point that the ant visits is a point of solution (PoS) and a sensor is deployed there. An ant continues its tour until all the PoIs are covered by the deployed sensors. At the end of an iteration the shortest tour comprising the least number of sensors is cached. At the end of all iterations, the smallest of all solutions saved at the end of each iteration is selected. This is where ordinary ACO based algorithms stop. However a lot of redundant sensors are present at this stage. Our formulation aims to remove redundancy while ensuring network connectivity.

In Phase-I, non-uniform coverage radii for sensor nodes and greedy migration scheme similar to [8] have been used. In greedy migration scheme, an ant moves back to a previously visited grid point, for next point selection if more PoIs can be covered after taking the next step from there. For an ant on point i, feasible neighborhood contains all those grid

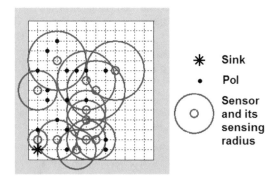

Fig. 6.1 Solution obtained at the end of Phase-I of ACO-Discreet

points that fall within communication radius of sensor on point i. Moreover, effective candidate points (ECPs) are defined as those points in the feasible neighborhood on which sensor placement can cover one or more uncovered PoIs [8]. During execution of the algorithm, for an ant on point i, next point j is selected according to the following formula [8],

$$P_{i,j}(t) = \frac{[\tau_{ij}(t)]^{\alpha}[\eta_{ij}(t)]^{\beta}}{\sum\limits_{r \in S^i_{candidate}} [\tau_{ir}(t)]^{\alpha}[\eta_{ir}(t)]^{\beta}} \qquad (6.1)$$

where, τ_{ij} and η_{ij} represent pheromone value and heuristic desirability for link i, j respectively.

Heuristic desirability [8] modified in this work is given by:

$$\eta_{ij}(t) = L + \sum_{m \in S_{ECP(j)}} l(m) + (0.1 \times Dist_{ij}) \qquad (6.2)$$

The summation term in the above equation represents the total number of ECPs in the communication radius of next possible sensor. Constant L is chosen larger than zero so as to prevent zero from appearing in the denominator of Eq. (6.1). Hence the next possible point where largest number of ECPs fall within sensing radius of deployed sensor is given preference. In this work, the heuristic value used in [8] has been modified and a scaled distance term is included in Eq. (6.2) in the proposed scheme so that if more than one points have equal number of largest ECPs for next possible sensor then the more distant of them may be preferred. This will result in less number of total deployed sensors.

After an ant finishes its tour, the pheromone value for all links evolves as follows [8].

$$\tau_{ij}(t+1) = (1-\rho)\tau_{ij}(t) + \Delta\tau_{ij}(t) \qquad (6.3)$$

where,

$$\Delta\tau = C/total(t) \qquad (6.4)$$

Here, $total(t)$ shows the length of a solution. C is any constant value greater than zero and ρ is the pheromone evaporation value. Hence, links that are part of a short tour receive more new pheromone. And the solution gradually converges to the shortest route. Pheromone constraining is carried out after every T_c iterations to ensure that value of pheromone dosen't increase or decrease beyoned given limits [8].

When Phase-I of the algorithm returns a solution e.g. as shown in Fig. 6.1, Phase-II of the algorithm starts its operation and removes all the sensors that are redundant in nature. Phase-II is described below:

6.3.2   Phase-II of Proposed Framework

The objective of Phase-II of the proposed method is to remove the redundant sensors present in the solution obtained at the completion of phase-I. Redundant sensors (RS) are defined as those sensors which have all the PoIs inside their sensing radii already covered by other sensors [9]. Removal of RSs will not render any PoI uncovered and considerably reduce the coverage cost. Phase-II comprises tours conducted by two ants only. The algorithmic steps of Phase-II are now explained in thorough detail to make its working comprehensible.

The first ant, whose function is identification and removal of redundant sensors, conducts its tour on the solution obtained from Phase-I and visits every PoS. It starts off from the last sensor in the solution and moves in the backwards direction towards the sink. At each step, it checks if the sensor that it visits has all the PoIs in its sensing radius covered more than the desired number of times. This can be done simply by initializing an array *CoveredTimes* with its length equal to the number of PoIs at start of Phase-II. The entry of the array *CoveredTimes* pertaining to a particular PoI is assigned a value equal to, how many sensors it is covered by. If a sensor is found by the first ant with all the PoIs in its sensing radius already covered by other sensors i.e. the values corresponding to all the PoIs in its sensing radius are larger than 1, it is identified as an RS and is not included in the ant's memory; since its removal does not render any critical point uncovered. If for a particular sensor, even a single PoI is found with its corresponding value in *CoveredTimes* equal to 1, it is not removed from the solution since its removal will render the said PoI uncovered. Hence, during its tour, the first ant stores only those visited sensors in its memory that are not redundant.

Algorithm 1 Two-stage ACO-based design for removing redundant sensors in WSNs

1. Phase-I
2. Initialize all parameters;
    // i is counter for iterations.
    // I_{max} is the total number of iterations.
3. **WHILE** (i $\leq I_{max}$) **DO**
4.   **FOR** a = 1 to a = $Ants_{total}$
5.   Move to next grid point using Equation (1);
6.   **END FOR**
7.   Cache the best solution for current iteration;
8.   Update the pheromone values;
9.   **IF** i % T_c = 0
10.   Pheromone Constraining;
11.   **END IF**
12. $i++$

(continued)

Algorithm 1 (continued)

13. **END WHILE**
14. Select the minimum cost solution;
15. Phase-II
    // Execute backwards tour for 1st ant and build a new solution;
16. Visit every PoSs in the solution of Phase-I;
17. **IF** (An RS is found)
18.   Exclude this sensor from the 1st ant's solution;
19. **END IF**
    // Execute tour for 2nd ant;
20. Incrementally add sensors to the 2nd ant's solution from 1st ant's solution that can communicate with each other;
21. **IF** (One or more sensors in 1st ant's solution can't communicate with sensors in 2nd ant's memory)
22.   Save RSs that can communicate with sensors in 2nd ant's memory in a temporary data structure **A**;
23.   Save RSs that can communicate with sensors in 1st ant's solution that aren't included in 2nd ant's memory yet in temporary data structure **B**;
24.   Include an RS common between **A** and **B** in 2nd ant's memory;
25.   Delete **A** and **B**;
26. **END IF**
27. Save the solution;

As shown in Fig. 6.2, at the end of the tour of first ant, a reduced solution is obtained with all the unwanted sensors removed. However, sometimes the connectivity among the nodes left in the first ant's solution is disrupted as could be observed from Fig. 6.2. This is because some of the removed redundant sensors were crucial for connectivity of network. The purpose of second ant's tour is to look for the location of disconnection in first ant's solution and restore the minimum number of removed redundant sensors that are necessary for connectivity. Unlike the first ant, which executes its operation considering the sensing radii, second ant considers the communication radii of the sensor nodes. It starts from the sink and during the first step of the tour looks for all those sensors in the solution of first ant that are connected to the sink via single hop and adds them to its memory. At the second step of its tour, it looks for all those sensors in the solution of first ant that are connected to the nodes present in its own solution it has built so far and again stores them in its memory. Hence, at each step it incrementally builds its solution by adding new sensors from the solution of first ant that can communicate with the nodes present in its own solution. This process continues until it reaches a point where one or more sensors in the solution of the first ant cannot communicate with the nodes present in its own solution. This indicates presence of disconnection in the solution of first ant. At this point, a redundant sensor from the original solution obtained at end of Phase-I needs to be reinstated in order to reestablish communication among nodes. The main challenge lies in searching such sensors. In order to carry out this task the following methodology is adopted.

Upon detection of disconnection the second ant looks for all the sensors in the original solution obtained at the end of Phase-I, which are connected to its own solution built so far and saves them in a temporary data structure. Then it looks for those sensors present in the original solution obtained at end of Phase-I that are connected to the remaining portion of the solution of first ant that cannot communicate with its own solution and saves them in another temporary data structure. The common sensors between these two temporary data structures can be utilized to restore the network connectivity. These are the redundant sensors that were removed by first ant but their presence is crucial for communication between nodes. Hence, one of these common sensors is reinstated to restore the connectivity as shown in Fig. 6.3 and included in the solution of second ant. Now the remaining sensors in the

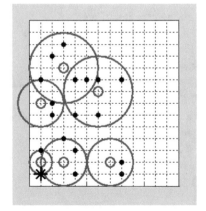

Fig. 6.2 Phase-II of ACO-Discreet, solution built by first ant

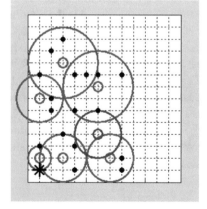

Fig. 6.3 Phase-II of ACO-Discreet, solution built by second ant

solution of first ant that were previously not connected to the solution of second ant can communicate via the sensor restored from Phase-I solution. These sensors are then added to the solution of second ant. This process continues until all the sensors present in first ant's solution are added to the second ant's solution. At the end of Phase-II, a large number of the unwanted sensors are removed in such a way that the functioning of the network remains unaffected. This is explained in Algorithm-I. For reference, we call our proposed algorithm as ACO-Discreet.

6.4 Simulations and Results

Simulations to evaluate the performance of proposed algorithm "ACO-Discreet" are executed in Matlab. Performance of EasiDesign [6], ACO-TCAT [7], ACO-Greedy [8] and ACO-Discreet is compared on a 9×9 grid points in terms of coverage cost (number of deployed sensors), ratio of surviving nodes after 50 rounds of data transmission to the sink node and time overhead. Sensing radius for sensor nodes in EasiDesign [6] and ACO-TCAT [7] remains uniform and is equal to 2 girds. For ACO-Greedy [8] and ACO-Discreet, the sensing radius keeps varying from 1 to 3 grids depending on distance of the sensor from the sink node. Communication radius is considered to be twice the sensing radius. $\alpha = 1$, $\beta = 3$, $\rho = 0.2$ and $T_c = 10$ have been used. Where T_c is the pheromone constraining period [8]. 100 iterations are executed and 10 ants participate in the search for the solution in each iteration.

Figures 6.4 and 6.5 show coverage cost comparison of ACO-Discreet with EasiDesign [6], ACO-TCAT [7] and ACO-Greedy [8]. The simulation results reveal that due to the effectiveness of Phase-II of ACO-Discreet, deployment cost is much less compared to other three algorithms since majority of the redundant sensors are removed.

In order to analyze the ratio of surviving nodes, energy model from [10] is used. To transmit an l-bit packet, the energy consumed is given as,

$$E_{Tx}(l,d) = \begin{cases} lE_{elec} + l\varepsilon_{fs}d^2, & d < d_o \\ lE_{elec} + l\varepsilon_{amp}d^4, & d \geq d_o \end{cases} \quad (6.5)$$

Similarly, to receive an l-bit packet, the energy consumed is given as,

$$E_{Rx}(l) = lE_{elec} \quad (6.6)$$

Figure 6.6 shows ratio of surviving nodes (representing network life) for different algorithms against data transmission to the sink. Each node is assigned 100J energy initially.

It can be seen in Fig. 6.6, that the performance of ACO-Discreet is slightly better than that of ACO-Greedy.

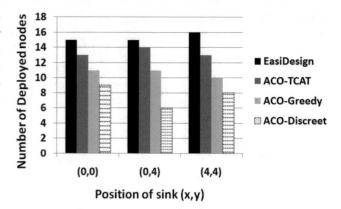

Fig. 6.4 Deployment cost comparison for 9×9 grid points, 20 PoIs

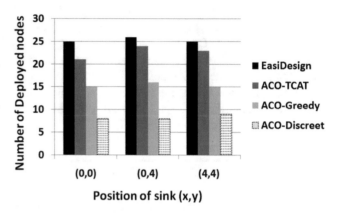

Fig. 6.5 Deployment cost comparison for 9×9 grid points, 40 PoIs

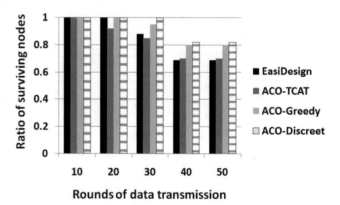

Fig. 6.6 9×9 grid points, 20 PoIs

This is due to reduced number of remaining sensors after Phase-II, which results in less relay traffic.

Table 6.1 shows average time taken by different algorithms for a 9×9 grid, 20 PoIs. The working of ACO-Greedy [8] and ACO-Discreet Phase-I are nearly same, which is why overhead of phase-II is less than two seconds (around 5%). Hence with a very small time overhead, ACO-Discreet can remove the redundant sensors from the solution and considerably reducea the coverage cost.

Table 6.1 Time overhead for different algorithms

Algorithm	Time taken
EasiDesign [6]	14.92 sec
ACO-TCAT [7]	17.12 sec
ACO-Greedy [8]	38.27 sec
ACO-Discreet	40.13 sec

6.5   Conclusion

In this work, a novel ACO based algorithm ACO-Discreet is proposed to solve the problem of GCLC in WSNs which works in two phases. In Phase-I, a solution is built using standard ACO. In Phase-II, further reduction is carried out in the solution obtained from phase-I and a large number of redundant sensors are removed. As a result, an optimal solution is obtained. Due to reduced number of sensors, there is less relay burden on sensor nodes close to the sink. The network lifetime of solutions obtained from the proposed ACO-Discreet is found to be longer than that of other ACO based algorithms. This improved performance is achieved with very little time overhead.

References

1. Dhillon, S. S., Chakrabarty, K., & Iyengar, S. (2002). Sensor placement for grid coverage under imprecise detections. In *Proceedings of the Fifth International Conference on Information Fusion* (Vol. 2, pp. 1581–1587). IEEE.

2. Dhillon, S. S., & Chakrabarty, K. (2003). Sensor placement for effective coverage and surveillance in distributed sensor networks. In *Proceedings of IEEE Wireless Communications and Networking Conference* (Vol. 3). IEEE.

3. Jourdan, D., & de Weck, O. L. (2004). Layout optimization for a wireless sensor network using a multi-objective genetic algorithm. In *2004 I.E. 59th Vehicular Technology Conference, 2004. VTC 2004-Spring* (Vol. 5, pp. 2466–2470). IEEE.

4. Xu, Y., & Yao, X. (2006). A ga approach to the optimal placement of sensors in wireless sensor networks with obstacles and preferences. In *3rd IEEE Consumer Communications and Networking Conference, 2006. CCNC 2006* (Vol. 1, pp. 127–131). IEEE.

5. Ke, W. C., Liu, B. H., & Tsai, M. J. (2007). Constructing a wireless sensor network to fully cover critical grids by deploying minimum sensors on grid points is np-complete. *IEEE Transactions on Computers 56*(5), 710–715.

6. Li, D., Liu, W., & Cui, L. (2010). Easidesign: An improved ant colony algorithm for sensor deployment in real sensor network system. In *2010 I.E. Global Telecommunications Conference (GLOBECOM 2010)* (pp. 1–5). IEEE.

7. Liu, X. (2012). Sensor deployment of wireless sensor networks based on ant colony optimization with three classes of ant transitions. *IEEE Communications Letters, 16*(10), 1604–1607.

8. Liu, X., & He, D. (2014). Ant colony optimization with greedy migration mechanism for node deployment in wireless sensor networks. *Journal of Network and Computer Applications, 39*, 310–318.

9. Singh, S., Chand, S., Kumar, R., & Kumar, B. (2013). Optimal sensor deployment for WSNs in grid environment. *Electronics Letters, 49*(16), 1040–1041.

10. Heinzelman, W. B., Chandrakasan, A. P., & Balakrishnan, H. (2002). An application-specific protocol architecture for wireless microsensor networks. *IEEE Transactions on Wireless Communications, 1*(4), 660–670.

Mohammad Vahidalizadehdizaj and Avery Leider

Abstract

Security in group communication has been a significant field of research for decades. Because of the emerging technologies like cloud computing, mobile computing, ad-hoc networks, and wireless sensor networks, researchers are considering high group dynamics in communication security. Group key agreement protocol is for providing enough security for the group communication. Group key agreement protocol is a way to establish a shared cryptographic key between groups of users over public networks. Group key agreement protocol enables more than two users to agree on a shared secret key for their communications. In this paper we want to introduce a new key agreement protocol (GKA) and a new linear group key agreement protocol (GLGKA). These protocols are proper for the emerging dynamic networks. In GLGKA each member of the group can participate in key generation and management process. Our goal is to provide more efficient key generation approach for group members. This protocol can be used in cloud-based data storage sharing, social network services, smart phone applications, interactive chatting, video conferencing, and etc.

Keywords

Cyber security • Diffie-Hellman • Key agreement protocol • Group key agreement protocol

7.1 Introduction

Security and reliability are two significant factors in modern computing. In this environment most of the services can be provided as shared services. These shared services should have the essential cryptographic factors like data confidentiality, data integrity, authentication and access control to be a secure shared service. It is really important for us to communicate securely over an insecure communication channel. One way to provide this secure channel is using the key agreement protocol method [5].

Note that, digital signature, encryption, and key agreement are three important topics in the field of cryptography. Group key agreement protocol can provide a secure communication channel between more than 2 users over an open network. The idea is having a shared session key between all the members of a group and no one else. Because of the growth of the group-oriented applications like video conferencing, the role of the shared session key between members of a group became critical. A lot of research have been focused on this area so far, but still we need a more efficient and secure solution [3].

Group key agreement protocol is very useful in interactive chatting applications that are really popular these days. Most of the people launch these applications in their cellphones, tablets, notebooks, or their PCs. These applications should provide secure communication channel for their users over an untrusted network. Most of the users

M. Vahidalizadehdizaj (✉) • A. Leider
Department of Computer Science, Pace University, 1 Pace Plaza, New York, NY, USA
e-mail: m.vahidalizadeh@gmail.com; aleider@pace.edu

© Springer International Publishing AG 2018
S. Latifi (ed.), *Information Technology – New Generations*, Advances in Intelligent Systems and Computing 558, DOI 10.1007/978-3-319-54978-1_7

have online profiles and the privacy of these profiles is really important for these users. The application should share a session key between group members. We are going to propose our own group key agreement protocol (GLGKA). In our protocol all users contribute in generation and management of the shared session key [9].

If we want to define a shared session key between two users, we should use a key agreement protocol. One of the most important key agreement protocols is Diffie-Hellman protocol. This protocol was introduced in 1976 by Whitfield Diffie and Martin Hellman. It is a method for defining a shared session key between two users. It uses the public key exchange implementation that was introduced by Ralph Merkle [5].

If we want to define a shared session key for more that two parties, we should use a group key agreement protocol. When the communication channel is not trusted, group key agreement protocols will be useful. In this paper we will introduce a new linear group key agreement protocol named GLGKA. In our protocol, all group members contribute in generation of the shared key. In our protocol there are two sub-protocols for the people who join or leave the group [4].

Defining a shared session key for a group is a complicated task. The challenges are lack of third parties, expensive communication, and limited capabilities of the portable devices. There are also some other challenges for this area that will be discussed later in this paper. For example, in adhoc networks there is not enough trust in the network for communication [2, 4].

In our proposed protocol (GGKA) each member should establish a secure channel with the leader by our proposed secure group communication sub-protocol (SGC). SGC protocol should be used later to transfer the shared session key to the group after a user leaves or joins the group. Our protocol does defines a shared session key for the group with less computation and communication costs. Members may have different devices with different capabilities. Our computations should be fast enough for all kinds of devices. To put it in a nutshell, in our protocol the amount of communication for defining and changing a shared session key for a group is optimized [1, 9].

The rest of the paper is organized as follows. Section 7.2 will review related works. Section 7.3 is about our proposed key agreement and group key agreement protocols. Section 7.4 is our join protocol. Section 7.5 is our leave protocol. Section 7.6 explains about the complexity of our proposed protocols. Section 7.7 is comparison section. Section 7.8 shows our experiments. Finally, Sect. 7.9 concludes this paper.

7.2 Related Works

In this section we are going to review Diffie-Hellman key agreement protocol, which is the most popular protocol in this field. We are also going to review DL08 and KON08

protocols, which are the most efficient group key agreement protocols [8].

7.2.1 Key Agreement Protocol

In this section we review Diffie-Hellmand key agreement protocol, which is the most popular key agreement protocol. Diffie-Hellman is one of the earliest implementations of the public key cryptography, which generates a shared session key between two users. We can use this key later for encrypting or decrypting our information just like a symmetric key. Most of the key agreement protocols use Diffie-Hellman as their basis. RSA also followed Diffie-Hellman method to implement its public key cryptography method [4].

This protocol is based on mathematics. Its fundamental math includes algebra of exponents and modulus arithmetic. To explain how this protocol works, we use Alice and Bob in our example. The goal of this protocol is to agree on a shared secret key between Alice and Bob. Note that, an eavesdropper should not be able to determine the shared session key by observing transferred data between Alice and Bob. Alice and Bob independently generate these keys for themselves in two sides. These symmetric keys will be used to encrypt and decrypt data stream between Alice and Bob. Note that, this key do not travel over the network. The steps of this protocol are shown in Table 7.1. We plan to compare our key agreement protocol (GKA) with this popular key agreement protocol later [8].

Table 7.1 Diffie-Hellman key agreement protocol

Step	Action	Description
1	Alice and Bob agree on two numbers p and g	p is a large prime number and g is called base or generator
2	Alice picks a secret number a	Alice's secret number = a
3	Bob picks a secret number b	Bob's secret number = b
4	Alice computes her public number $X = g^a \bmod p$	Alice's public number = X
5	Bob computes his public number $Y = g^b \bmod p$	Bob's public number = Y
6	Alice and Bob exchange their public numbers	Alice knows p, g, a, X, Y Bob knows p, g, b, X, Y
7	Alice computes $k_a = Y^a \bmod p$	$k_a = (g^b \bmod p)^a \bmod p$ $k_a = (g^b)^a \bmod p$ $k_a = (g^{ba}) \bmod p$
8	Bob computes $k_b = X^b \bmod p$	$k_b = (g^a \bmod p)^b \bmod p$ $k_b = (g^a)^b \bmod p$ $k_b = (g^{ab}) \bmod p$
9	By the law of algebra, Alice's k_a is the same as Bob's k_b, or $k_a = k_b = k$	Alice and Bob both know the secret value k

7.2.2 Group Key Agreement Protocols

The first group key agreement protocol that we are going to review is DL08. Desmedt and Lange proposed a three-round group key agreement protocol in 2007. This protocol is based on pairings and it is proper for groups of parties with different computational capabilities. In this protocol, a balanced group of n parties should have approximately $n/2$ more powerful parties [9].

The number of computations for calculations of signatures and verifications are important factors in computing the complexity of this protocol. We assume that a Digital Signature algorithm is used by the signing scheme. We can assume that a signature generation has the cost of one exponentiation and a signature verification has the cost of two exponentiation [9].

According to this assumptions, the complexity of this protocol includes total number of $(9n/2) + 2n\lg_3[n/3]$ multiplications, $3n/2$ pairings and $3n/2$ exponentiation. The parties will have to transmit $7n/2$ messages and also receive $3n + n\lg_4[n]$ messages. DL08 is a three-round protocol based on the Burmester-Desmedt scheme that achieves the best performance from aspect of cost [9, 10].

The next protocol that we want to review is KON08. It is a cluster-based GKA protocol proposed by Konstantinou in 2010. It is based on Joux's tripartite key agreement Protocol. It has two variants: contributory and non-contributory. This protocol assumes that nodes are clusters with two or three members [9].

In the lower levels nodes belong to only one cluster. In upper levels nodes belong to two clusters. Authentication can be provided by the use of an authenticated version of Joux protocol. Authentication method does not influence the number of rounds or the communication cost of this protocol. In particular, the protocol has $log_2 n/3$ rounds and $4n$ messages should have been transmitted. In the authenticated version, the group has to perform no more than $5n$ scalar multiplications and $11n/2$ pairing computations [2, 6, 9].

7.3 Golden Linear Group Key Agreement Protocol

We divided our main protocol into two sub-protocols. The first one is secure channel sub-protocol. The second one is initiation sub-protocol. Our goal is to define a shared session key for a group of users by these two sub-protocols. In the secure channel sub-protocol, we will use our own key agreement protocol named GKA (Golden Key Agreement Protocol). Based on GKA protocol, all members and leader of the group should share three numbers a, p, and q at the beginning [4, 6, 9].

7.3.1 Golden Key Agreement Protocol (GKA)

This protocol is based on mathematics. The fundamental math includes algebra of logarithms and modulus arithmetic.

Table 7.2 Golden key agreement protocol

Step	Action	Description
1	Alice and Bob agree on three numbers a, p, and q	q is a large prime number $a = p^n$ p = 2, 3, …, n n, u, and v = 1, 2, 3, …, n
2	Alice picks a secret number u	Alice's secret number = u u mod q is not zero
3	Bob picks a secret number v	Bob's secret number = v v mod q is not zero
4	Alice computes her public number $A = ((u \bmod q) \times log_p a) \bmod q$	Alice's public number = A $= (log_p a^{(u \bmod q)}) \bmod q$
5	Bob computes his public number $B = ((v \bmod q) \times log_p a) \bmod q$	Bob's public number = B $= (log_p a^{(v \bmod q)}) \bmod q$
6	Alice and Bob exchange their public numbers	Alice knows u, a, p, q, A, and B Bob knows v, a, p, q, A, and B
7	Alice computes $k_a = ((u \bmod q) \times B) \bmod q$	$k_a = ((u \bmod q) \times (log_p a^{(v \bmod q)})) \bmod q$ $k_a = (log_p a^{(v \bmod q) \times (u \bmod q)}) \bmod q$ $k_a = (log_p a^{((u \times v) \bmod q)}) \bmod q$
8	Bob computes $k_b = ((v \bmod q) \times A) \bmod q$	$k_b = ((v \bmod q) \times (log_p a^{(u \bmod q)})) \bmod q$ $k_b = (log_p a^{(u \bmod q) \times (v \bmod q)}) \bmod q$ $k_b = (log_p a^{((v \times u) \bmod q)}) \bmod q$
9	By the law of algebra, Alice's k_a is the same as Bob's k_b, or $k_a = k_b = k$	Alice and Bob both know the secret value k

For this discussion we will use Alice and Bob as an example. The goal of this process is agreeing on a shared secret session key between Alice and Bob. In this process eavesdroppers should not be able to determine the shared session key. This shared session key will be used by Alice and Bob to independently generate the keys for each side. These keys will be used symmetrically to encrypt and decrypt data stream between Alice and Bob. Note that, our shared session key should not travel over the network. This new key agreement protocol is described in Table 7.2.

Table 7.3 is an example of generating a shared session key between Alice and Bob based on GKA protocol. In this example we choose n = 2, p = 2, and $q = 7 (a = p^n = 4 = 2^2)$.

7.3.2 Secure Channel Sub-protocol

In this part the leader will establish a secure connection to each one of users in the group. Firstly, each user will choose a private key (user i will choose private key p_i). The leader also will choose a private key (p_c). Then, each user will send $A_i = ((p_i \bmod q) \times log_p a) \bmod q$ to the leader. Also the

Table 7.3 GKA example

Step	Action
1	Alice and Bob agree on n = 2, p = 2, a = 4 and q = 7
2	Alice picks a secret number u = 3
3	Bob picks a secret number v = 4
4	Alice computes her public number $A = ((3 \bmod 7) \times log_2 2^2)\, mod\, 7 = 3$
5	Bob computes his public number Bob's public number = B $A = ((4 \bmod 7) \times log_2 2^2)\, mod\, 7 = 1$
6	Alice and Bob exchange their public numbers Alice knows u, a, p, q, A, and B Bob knows v, a, p, q, A, and B
7	Alice computes $k_a = ((u \bmod q) \times (log_p a^{(v\, mod\, q)}))\, mod\, q$ $k_a = ((3 \bmod 7) \times (1))\, mod\, 7 = 3$
8	Bob computes $k_b = ((v \bmod q) \times (log_p a^{(u\, mod\, q)}))\, mod\, q$ $k_a = ((4 \bmod 7) \times (6))\, mod\, 7 = 3$
9	By the law of algebra, Alice's Alice and Bob both know the k_a is the same as Bob's k_b secret value k or $k_a = k_b = k = 3$

leader will broadcast $A_c = ((p_c \bmod q) \times log_p a)\, mod\, q$ to all the group members [6, 7].

Then, each one of the users will calculate $k_i = ((p_i \bmod q) \times A_c)\, mod\, q$. Then, the leader will calculate $k_{c_i} = ((p_c \bmod q) \times A_i)\, mod\, q$. As a result of this sub-protocol, the leader will have a secure line with each one of the members [8].

7.3.3 Initiation Sub-protocol

In this part, the leader will generate a shared session for the users. Firstly, each user chooses his own private key p_i and send it to the leader through the secure channel. After that, the leader will generate a shared session key for the group based on the equation below and will send it to the members through their secure channels.

$$k = (log_p a^{(p_1 p_2 \cdots p_n)})\, mod\, q$$
$$a = p^n \Rightarrow k = (log_p p^{[n(p_1 p_2 \cdots p_n)]})\, mod\, q$$
$$\Rightarrow k = ((p_1 p_2 \ldots p_n) \times log_p p^n)\, mod\, q$$

Note that, this shared session key is only for current users of the group. if anyone leaves or joins the group, the shared group key should be updated. In the next sections we are going to review our join and leave sub-protocols. Join protocol is useful when a new member joins the group and leave protocol is useful when a member leaves the group.

7.4 Join Protocol

When a new user ($user_{n+1}$) want to join a group, he should establish a secure channel with leader of the group. He uses our secure channel protocol (GKA) for this step. Then, he will send his private key with positive sign to the leader through his secure channel. Then, the leader will update his shared session key by multiplying it to the received number ($(p_{n+1} \bmod q)$), since the sign is positive. After that, the leader will broadcast the updated shared session key to the group members. The key will be updated for the leader, members of the group, and the new member Based on the equation below [5].

$$Current\, k = (log_p a^{(p_1 p_2 \cdots p_n)})\, mod\, q$$
$$New\, k = (p_{n+1} \bmod q) \times (log_p a^{(p_1 p_2 \cdots p_n)})\, mod\, q$$
$$= (p_{n+1} \times log_p a^{(p_1 p_2 \cdots p_n)})\, mod\, q$$
$$= (log_p a^{(p_1 p_2 \cdots p_n p_{n+1})})\, mod\, q$$

7.5 Leave Protocol

When an existing user (for example $user_n$) want to leave the group, he should inform the leader before he leaves. He should send his private key with negative sign ($-1 \times p_n$) to the leader through his secure channel. Then, the leader will update his private key by dividing it by the received number. Note that, the leader ignores the negative sign. This negative sign is only symbol of leaving. After that, the leader will broadcast the new shared group key. After that, the leader and the group members will have the same updated key based on the equation below [2, 6, 7].

$$Current\, k = (log_p a^{(p_1 p_2 \cdots p_n)})\, mod\, q$$
$$New\, k = (log_p a^{(p_1 p_2 \cdots p_n)})\, mod\, q\, /\, (p_n \bmod q)$$
$$= (log_p a^{(p_1 p_2 \cdots p_n)/p_n})\, mod\, q$$
$$= (log_p a^{(p_1 p_2 \cdots p_{n-1})})\, mod\, q$$

7.6 Complexity

In this section, we are going to analyze the cost of our protocol. We start from secure channel sub-protocol. In the secure channel sub-protocol, the leader should establish a secure connection with each one of the members based on the private key of the member. The leader will have two calculations and one broadcast. Each member will have two calculations and will send one message and receive one message. In the initiation sub-protocol, the leader will have one calculation and one broadcast. He will also receive n messages from members of the group. Each member will send one message to the leader and receive one message from the leader in this sub-protocol.

In the join protocol, the leader will have three calculations (two for secure channel sub-protocol and one for updating the shared group key), and will send one message to the new user and one broadcast to all members of the group. Each user will only receive one message. In the leave protocol, the leader will have one calculation and one broadcast to members of the group. The leaving member should send one message to the leader and each member except the leaving member should receive one message. In conclusion, our protocol will send $3n$ messages and receive $3n$ messages. Our protocol should do $2n + 2$ calculations for a group of n parties.

7.7 Comparison

In this part we are going to compare our group key agreement protocol with top two group key agreement protocols based on [9]. These protocols are DL08 and KON08. We compare these protocols from aspect of number of rounds, total number of sent messages, total number of received messages, and computations cost. You can see the result of these comparisons in Table 7.4.

Table 7.4 Group key agreement protocols comparison result

Protocol	Rounds	Sent Messages	Received Messages	Computations
DL08	3	$7n/2$	$3n + nlog_4n$	$15n/2 + 2n$ $[log_4n]$
KON08	$log_2n/3$	$4n$	$4n$	$21n/2$
GLGKA	2	$3n$	$3n$	$2n + 2$

As you see in Table 7.4, the comparison result shows that our group key agreement protocol is better than two other protocols from aspect of total number of sent and received messages. Our protocol is also better from them from aspect of computation cost.

7.8 Experiment

In this part we are going to compare Golden Key Agreement Protocol with Diffie-Hellman, which is the most popular key agreement protocol. We implemented these two algorithms and tested them with different parameters. We did all the experiments with a laptop with core-i7 CPU (2670QM 2.20 GHz), 8 GB of DDR3 Ram, and windows seven 64-bit operating system.

The first experiment was with q(GKA) = 15, 485, 863, p(DH) = 15, 485, 863, a = Random(2, 10), p(GKA) = Random(2, 10), g(DH) = Random(2, 10), and n(GKA) = 1. You can see the result of these experiments in Fig. 7.1. We had a loop in this experiment. You can see the number of iterations in each experiment. The second experiment was with q(GKA) = 982, 451, 653, p(DH) = 982, 451, 653, a = Random(2, 10), p(GKA) = Random (2, 10), g(DH) = Random(2, 10), and n(GKA) = 1. You can see the result of these experiments in Fig. 7.2. We had a loop in this experiment. You can see the number of iterations in each experiment.

In the third and last experiment, we ran our protocol against Diffie-Hellman 16 times. Each time we recorded the running time of both protocols. You can see result of these experiments in Fig. 7.3. As you see int Fig. 7.3, our protocol is faster than Diffie-Hellman. In these sixteen experiments the average running time of Diffie-Hellman was 801,268.8125 nanoseconds and Average running time of our protocol in these experiments was 86,662.5625 nanoseconds.

Fig. 7.1 Comparison of golden key agreement protocol with Diffie-Hellman (more than one iteration)

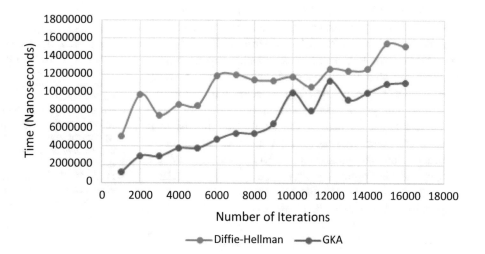

Fig. 7.2 Comparison of golden
key agreement protocol with
Diffie-Hellman (more than one
iteration)

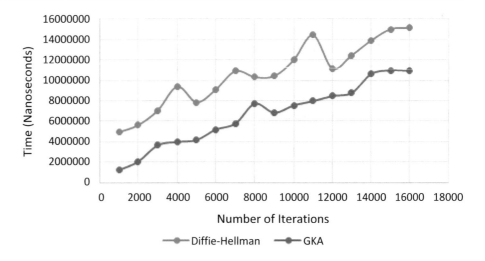

Fig. 7.3 Comparison of GKA
with Diffie-Hellman (one
iteration)

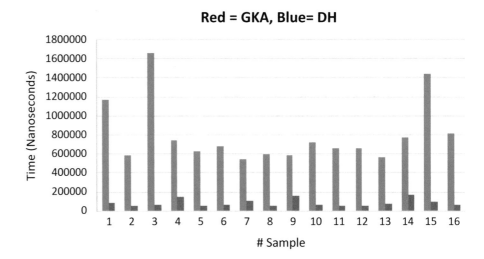

7.9   Conclusion

We recommended two protocols in this paper. The first one
was golden key agreement protocol. We used algebra of
logarithms and modulus arithmetic in this protocol to
improve the performance. We compared this protocol with
Diffie-Hellman (using exponentiation) and we observed that
our protocol is faster. We selected Diffie-Hellman for our
comparison, since it is the most popular key agreement
protocol. We also suggested a group key agreement protocol
named golden linear group key agreement protocol. We
compared our group key agreement protocol with DL08
and KON08 from aspect of the number of rounds, total
number of sent messages, total number of received
messages, and computation cost. DL08 and KON08 are
two of the best group key agreement protocols available
based on. As you seen in Table 7.4, our group key agreement
protocol was better than other protocols.

References

1. Cormen, T. H., Stein, C., Rivest, R. L., & Leiserson, C. E. (2001).
   Introduction to algorithms (2nd ed.). Boston/Burr Ridge/Montréal/
   McGraw-Hill Higher Education.
2. Dizaj, M. V. A. (2011). New mobile payment protocol: Mobile pay
   center protocol 4 (MPCP4) by using new key agreement protocol:
   VAC2. In *IEEE 3rd International Conference on Electronics Computer Technology (ICECT)*, April 8–10, 2011 (Vol. 2, pp. 67–73).
3. Dizaj, M. V. A., & Geranmaye, M. (2011). New mobile payment
   protocol: mobile pay center protocol 5 (MPCP5) by using new key
   agreement protocol: VG1. In *IEEE 3rd International Conference on
   Computer Modeling and Simulation (ICCMS)* (Vol. 2,
   pp. 246–252).
4. Dizaj, M. V. A., & Geranmaye, M. (2011). New mobile payment
   protocol: mobile pay center protocol 6 (MPCP6) by using new key
   agreement protocol: VGC3. In *IEEE 3rd International Conference
   on Computer Modeling and Simulation (ICCMS)* (Vol. 2,
   pp. 253–259).
5. Dizaj, M. V. A., Moghaddam, R. A., & Momenebellah, S. (2011).
   New mobile payment protocol: Mobile pay center protocol (MPCP).
   In *IEEE 3rd International Conference on Electronics Computer
   Technology (ICECT)*, April 8–10, 2011 (Vol. 2, pp. 74–78).

6. Dizaj, M. V. A., Moghaddam, R. A., & Momenebellah, S. (2011). New mobile payment protocol: Mobile pay center protocol 2 (MPCP2) by using new key agreement protocol: VAM. In *IEEE Pacific Rim Conference on Computers and Signal Processing (PacRim)*, August 23–26, 2011 (pp. 12–18).

7. Lee, E. -J., Lee, S. -E., & Yoo, K. -Y. (2008). A certificateless authenticated group key agreement protocol providing forward secrecy. In *2008 International Symposium on Ubiquitous Multimedia Computing* (pp. 124–129).

8. Vahidalizadehdizaj, M., & Tao, L. (2015). A new mobile payment protocol (GMPCP) by using a new key agreement protocol (GC). In *IEEE International Conference on Intelligence and Security Informatics (ISI)*, May 27–29, 2015 (pp. 169–172).

9. Vahidalizadehdizaj, M., & Tao, L. (2015). A new mobile payment protocol (GMPCP) By using a new group key agreement protocol (VTGKA). In *IEEE 6th International Conference on Computing, Communication and Networking Technologies (ICCCNT)*.

10. Yao, G., & Feng, D. (2010). A complete anonymous group key agreement protocol. In: *2010 Second International Conference on Networks Security, Wireless Communications and Trusted Computing* (pp. 717–720).

Implementing an In-Home Sensor Agent in Conjunction with an Elderly Monitoring Network

Katsumi Wasaki, Masaaki Niimura, and Nobuhiro Shimoi

Abstract

In this paper, we present the design and implementation of the in-home sensor agent *MaMoRu-Kun* as an Internet of Things (IoT) smart device, developed via the "Research and development of the regional/solitary elderly life support system using multi-fusion sensors" project. At Akita Prefectural University, the bed and pillow sensors and corresponding monitoring system have been developed to watch elderly individuals at bedtime, in particular those who live alone. As a sensor agent, *MaMoRu-Kun* is connected to the in-home wireless network of the target individuals accommodations and collects trigger information from various switches, motion detection sensors, and a remote controller. This smart device is also able to send its entire set of data along with the status of the sensor to a collection and monitoring server connected via a long-term evolution (LTE) router. We implemented this agent using an Arduino and a Bluetooth-connected Android terminal.

8.1 Introduction

8.1.1 Background and Motivation for Elderly Monitoring Networks

As a consequence of increased life expectancies, healthcare needs faced by older generations are growing. To address this growth, new applications using emerging technologies are needed to contribute to their safety and well-being. More specifically, various research projects are currently focused on responding to the urgent need for networks to watch over the elderly and others in order to realize a community in which individuals can continue to live with a strong level of security [1].

Many application domains exist that involve tracking and monitoring tools for the elderly. One such domains is home monitoring in which the main purpose is to guarantee the safety of elderly individuals when they are alone at home [2–5]. Other proposals focus on a specific part of the house, such as the bedroom [6] or bathroom [7]; not surprisingly, there are numerous privacy issues in using watch-over devices directly, e.g., image sensors or microphones. Alternatively, applications have also been developed with a focus on monitoring health conditions of elderly individuals, such as in [8–10], which evaluate physiological and/or physical activities. These observation methods have a high level of accuracy in terms of detecting various behavior of the target individual through the use of sensors, although there is still the possibility of more invasive watch-over capabilities when other sensors detect high levels of stress.

Key points for watch-over activities include early awareness of something unusual happening with the target individual, automatically contacting specialized institutions, and appropriately handling such situations. Unfortunately, in the case of solitary elderly, it can be difficult to promptly notice something unusual happening through casual observations

K. Wasaki (✉) • M. Niimura
Faculty of Engineering, Shinshu University, Nagano 380–8553, Japan
e-mail: wasaki@cs.shinshu-u.ac.jp; niimura@cs.shinshu-u.ac.jp

N. Shimoi
Faculty of Systems Science and Technology, Akita Prefectural University, Akita 015-0055, Japan
e-mail: shimoi@akita-pu.ac.jp

© Springer International Publishing AG 2018
S. Latifi (ed.), *Information Technology – New Generations*, Advances in Intelligent Systems and Computing 558, DOI 10.1007/978-3-319-54978-1_8

alone (e.g., a porch light remaining on during the daytime, newspapers piling up in a newspaper box, not showing up to meetings, and so on), thus it is necessary to construct an automated mechanism to make it possible to observe the inside of a residence regardless of the time of day or night with a high level of accuracy through a gradual reduction of psychological barriers.

Regarding the observation of an individuals behavior and sleep patterns at night, in [11], Shimoi and Madokoro have conducted preliminary research at Akita Prefectural University involving a smart bed-monitoring system using a bed and pillow embedded with weight and acceleration sensors. Monitoring results show that a newly developed learning function for predicting the need for an ambulance has made it possible to stay abreast of sleeping situations of target individuals on a real-time basis [12]; however, this monitoring system is installed in nursing facilities and the like, thus it is currently not possible to observe an individuals home from a remote location, because the system assumes that a surveillance agent conducts visual observations from inside the same building. Further, during the day, when the target individual is not in bed, it is also not possible to carry out any active observations using the above approach.

Given the above limitations, we observe here that it is important to monitor situations in which the target individual is at home, operating and using electrical appliances and home equipment, i.e., to be aware of hazards to the individual. Solitary elderly are typically unemployed and spend most of their time at home, with their pattern of everyday life remaining the same, often with little communication with the outside world. More specifically, an elderly individual may go out to group activities on Friday afternoons, wake up at approximately six in the morning and go to bed at 10 at night, spend the evening watching television, open

and close the refrigerator to prepare breakfast, lunch, and dinner at home, etc. These examples represent typical patterns of an invariable life that many elderly individuals have.

8.1.2 Research Purpose and Outline

In our present study, we intend to enable the automatic detection of abnormal changes of a target individual at an early stage through a non-invasive collection of triggers that originate from life patterns or cyclic events, also considering reduced psychological barriers, all by installing sensors in the home. To achieve this, as illustrated in Fig. 8.1, our present study proposes the "Solitary Senior Citizens Life Support System of Regional Type," which requires the installation of sensor agents to gather event triggers in the individuals home and learn normal life patterns. Then, in abnormal cases involving changes that widely deviate from these patterns (i.e., an anomaly), our system quickly detects these abnormalities and immediately reports to a predefined individual in charge of the observation network.

In addition to this introduction, the remainder of our paper is organized as follows. First, we review and analyze the requirements for our system in conjunction with the bed monitor and pillow sensors. Based on results of the entire requirements analysis, we then consider the functions required for sensors in the home. Next, we place our sensor agent, *MaMoRu-Kun*, in place within the home as an embedded system, which we examine in terms of its hardware and software specifications. Here, the hardware and software was designed to aggregate connections with various composite sensor devices for trigger collection and processing. We examine the specifications in terms of the sending protocol and format for all data collection servers connected

Fig. 8.1 Overview of our elderly monitoring network, which uses multi-fusion sensors and an in-home agent

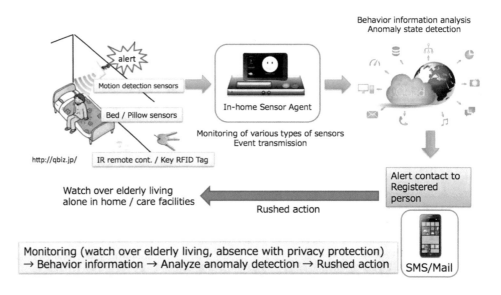

using a wide-area long-term evolution (LTE) wireless network. We also review visualization techniques for displaying the analytical results of monitoring under usual conditions and in case of emergency.

8.2   Requirements Analysis in Conjunction with the Bed Monitor and Pillow Sensors

8.2.1   Requirements Analysis of Our Watch-Over Network

As mentioned above, the most important point for watch-over activities is to be aware of abnormalities regarding the target individual as early as possible. Here, it is necessary to identify a threshold level that requires an emergent response through the combination of exterior and interior observations. Increased observation makes it possible to collect and judge information with higher levels of accuracy, although it is feared that psychological barriers and stress against watch-over systems are heightened and the target individuals are exposed to risks of home invasions.

Given these constraints, we performed a detailed requirements analysis regarding an actual implementation of watch-over network observation items, as well as the feasibility of emergent response actions. Each of these areas are described below and labeled [A] and [B].

[A] *Observation items:* This area governs the observability of the target individuals behavior and living situation, but only under certain conditions, i.e., equipment must be completely non-invasive. Individual observation items are as follows: (A1) going to bed is separate from getting up; (A2) being at home is separate from being away from home; (A3) operation and usage of television(s), air conditioner(s), and other appliances; (A4) frequency of movements within the home; (A5) confirmation of not being bedridden; and (A6) with or without emergent notification.

[B] *Emergent response items:* This area governs the decision and detection of threshold values regarding whether present conditions are within or widely deviated from a normal range, as compared with the ordinary life patterns exhibited by the target individual. Emergent response items are as follows: (B1) data collection and learning during in normal situations; (B2) providing a warning; (B3) automatically notifying emergent responders (e.g., relatives, social welfare workers, administrative officers, and managers for elderly housing); (B4) real-time alert dispatch for the target individual in case of an emergency; (B5) priority of emergent response; and (B6) using gathered data in practice as big data in a anonymized state.

8.2.2   Collaboration with a Bed Monitor and Pillow Sensor

Research is currently underway by Shimoi and Madokoro at Akita Prefectural University regarding bed monitoring and pillow sensors dedicated to watching over solitary elderly while they are sleeping [11]. Their efforts have focused on detecting situations in which the target individual is sleeping, then changes his or her position when waking up; such detection is realized via a piezo load sensor. Further, as shown in Fig. 8.2, in [12], Madokoro et al. are attempting to make multiple observations of an individual while sleeping by installing a three-axis accelerometer within a urethane pillow.

Individually observed data are then transmitted to a monitor terminal via a ZigBee wireless serial line to make it possible to directly gather and store load sensor and acceleration sensor values in real time. Further, monitoring software for initial trials makes it possible to obtain real-time graphical representations of each sensor.

This monitoring system has been set up in nursing facilities and the like; in this system, it is assumed that a

Fig. 8.2 System configuration and sensor layout of the bed monitor and pillow sensors [12]

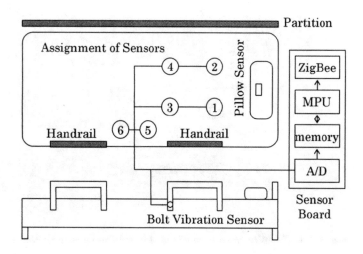

manager visually observes target individuals from within the same building. A radio wave within reach of the target individuals bedroom is transmitted to a main ZigBee receiver, thus transmission are roughly limited to within a house. It is therefore impossible to obtain observation data from outside the individuals home or current location (e.g., a nursing facility).

Technically, it is easy to consolidate observation data of an individual directly in the server in remote locations via a Wi-Fi wireless LAN or wide-area LTE wireless network; however, from the standpoint of psychological barriers and privacy protection, we must avoid directly sending and preserving the target individuals real-time behavior while sleeping to remote places. Counter to this, an emergency report using the functions of the pillow sensors and other such devices around the bed requires a high level of real-time information.

Given the above, we reviewed and developed the specifications below regarding bed monitors and pillow sensors installed in the home within a watch-over network in line with the aforementioned observation items for requirement analysis [A]. The specification items are: (1) the observation data of the bed sensor and pillow sensor are not transmitted in real time; (2) the observation data are received via a sensor agent installed in the house and pretreated to assess the situation during sleep, before awakening, and after leaving bed, i.e., corresponding to requirement (A1); (3) a bedridden state is detected when in bed for a longer period than usual, i.e., corresponding to requirement

(A5); and (4) in case of emergency, such as a sudden illness while sleeping, the target individual can shake the pillow rapidly to immediately generate an emergent notice, i.e., corresponding to requirement (A6).

8.3 Functions and Specifications of a Sensor Agent in the Home

In this section, we describe *MaMoRu-Kun*, a sensor agent for the home. In brief, this agent is a system embedded with a communication function in such a way that it can operate in cooperation with various sensors installed in a watched-over target individual.

8.3.1 Features of the In-Home Agent

As illustrated in Fig. 8.3, the key function of the in-home agent is to detect trigger information from various composite sensors, e.g., physical switches, infrared motion sensors, remote control sensors, and RFID key tags [13], as well as the bed monitor and pillow sensors mentioned above. Detected trigger information is then transmitted to a local server installed separately within the home that learns and stores the target individuals usual life patterns.

During the day, when the target individual is not in bed, the system is unable to perform behavioral observations with only the sensors around the bed, thus we incorporate the

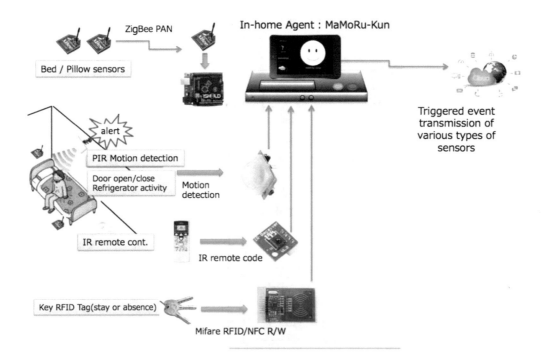

Fig. 8.3 Multi-fusion sensors and detection function using an in-home agent

combined use of infrared motion detection sensors, consumer electronics remote control sensors, and other similar sensors to make it possible to observe other behavioral patterns throughout the day (e.g., opening and closing doors, frequency access of the refrigerator and household electric appliances, etc.). Further, the system determines whether the target individual is at home or not by identifying when an RFID key tag is detected on a tag interface connected with the agent. More specifically, the system automatically determines that the target individual is not at home when this RFID key tag is not detected for a given length of time.

Finally, as noted above, in case of an emergency, such as a sudden illness, the target individual can shake his or her pillow rapidly to cause an emergency notification trigger from the pillow sensor (called *Pillow Shaking*).

As part of an entire watch-over network, the sensor agent has the functionality to send life pattern data under normal conditions and emergency alert notification to all data-gathering and monitoring servers for use on external networks.

The above functions make it possible to implement requirements (A2), (A3), and (A4) for observation items [A] of the watch-over network. The above functions also make it possible to implement requirements (B1) and (B2) of for emergent items [B] through the combined use of the server in the home. In addition, it is possible to implement requirement (B4) via the emergency reporting transmission functionality noted above.

8.3.2 Specifications

Our system is comprised of the following components: (1) a base station (new); (2) an Android terminal for communication (existing); (3) a server to gather data within the home (new); and (4) an LTE wireless router (existing). Here,

(1) and (2) are described in the subsections that follow, while detailed specifications for (3) and (4) are omitted due to space limitations.

1. *Base station:*
   (a) *Exterior appearance:* Exterior appearance: In conformance with Fig. 8.4.
   (b) *Form:* The case footprint is of A4 size, and an Android terminal can be installed on the top board.
   (c) *Specification:* (1-1) Two or more physical switches are provided. (1-2) One or more PIR infrared motion detection sensors are provided; the range of detection must be the space of approximately one eight-tatami room. (1-3) One infrared and remote control data reception module for household use is provided. (1-4) The reader for the RFID tag of the Mifare standard is provided; the reception part of RFID must not be in contact with the top board. (1-5) One or more LEDs to monitor operating conditions of the switch sensors. (1-6) The power supply must be DC7-12V, and power must be fed via the AC adaptor. (1-7) Power consumption must be 5 W or less. (1-8) The case must be of a finless structure. (1-9) The power supply terminal and USB terminal (for firmware rewriting) are provided on the back of the case. (1-10) Connected with the Android terminal using Bluetooth, trigger information from each sensor can be sent via POST by using the HTTP protocol.
2. *Android terminal for communication:*
   (a) *Exterior appearance:* In conformance with Fig. 8.4.
   (b) *Form:* The case footprint is of A5 size, and the display size must be seven to eight inches.
   (c) *Specification:* (2-1) The Android version must be Android 5.0 Lollipop or higher. (2-2) Wi-Fi and Bluetooth communication features must be available. (2-3) The text-to-speech voice synthesis function must be operable by the application, separately

Fig. 8.4 Concept diagram of our in-home agent, *MaMoRu-Kun*s base station

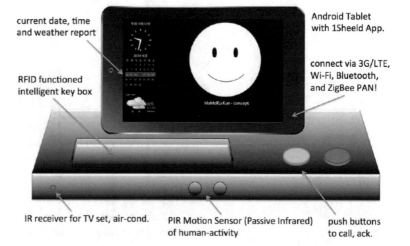

introduced as N2TTS. (2-4) Manager debug mode must be operable for the introduction of the self-build apk. (2-5) The display sleep function can be released at the time of no operation. (2-6) The power supply must be DC5V, and power must be fed via the AC adaptor.

8.4 Design and Implementation of the Home Sensor Agent

We designed and implemented our home sensor agent using peripheral hardware composed of various physical switches, firmware on a microprocessor-embedded Arduino [14], and interface connected via Bluetooth and the 1Sheeld App [15] as a multipurpose connection application introduced to an Android terminal, and a text-to-speech voice synthesis application called N2TTS. Figure 8.5 shows the composition of each functional module (i.e., HW/SW) of the base station.

8.4.1 Hardware Design and Implementation

(H1) Two physical switches and each PIR motion sensor, an RFID-RC522 tag module, an infrared remote control receiver module, and a motion sensor LED are arranged on a peripheral ext. board. These devices are connected using the Arduino (ATMega328) microprocessor via the SPI/PIO interface.

(H2) After processed by the firmware (as described below), trigger information on the processor is transferred to a Bluetooth-connected interface called 1Sheeld via the UART serial.

(H3) After transferred, trigger information is further transmitted to the Bluetooth pairing Android terminal of the other party. The information is designed to be forwarded to the network via the HTTP/TCP/IP stack of the Android terminal.

8.4.2 Software Design and Implementation

(S1) The firmware on the Arduino processor was newly designed and implemented. Multiple Finite State Machine designed the firmware in such a way that it is capable of formatting edge triggers via high/low level detection from various sensors and RFID tag modules, and also conduct tag detection and remote judgment. Two physical switches transmit *"Manual Absent mode"* and *"Manual Home mode"* at right and left, respectively. The PIR motion sensor conducts accumulation counts of approaches and departures, transmitting count values when the trigger disconnects from the sensor. The RFID tag module detects approaches and departures of the Mifare tag given to the key (bunch), and after the module remains in a specific state for more than a specific time period, transmits *"Manual Absent mode"* and *"Manual Home mode,"* respectively. As for infrared remote control data received, results of encoding maker code and a portion of 32-bit data are transmitted in data communication. Motion sensor LEDs blink when physical switches-off and PIR sensor and RFID tag detection of approaches.

Fig. 8.5 Functional block diagram of our in-home agent, *MaMoRu-Kun*

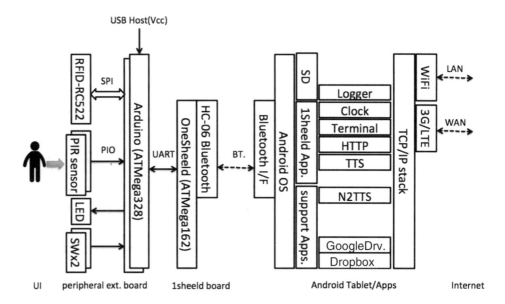

(S2) The multipurpose connection application, i.e., 1Sheeld, implemented on an Android terminal receives trigger information sent from the Arduino firmware and executes processing according to the base stations functional differences (e.g., Logger, Clock, Terminal, HTTP, and TTS).

(S3) We implemented trigger information in such a way that it is sent to the network via Wi-Fi by way of the HTTP/TCP/IP stack. In case of shifting to a home mode, an absent mode manually or automatically, the utterance is implemented via the text-to-speech synthesis application N2TTS for feedback to the target individual.

8.4.3 Prototyping and Load Testing

We produced a prototype based on the above design and specifications. As of September 5, 2016, we manufactured a total of nine prototypes from the breadboard to mass production models. Production costs consisted of a newly developed base station for no more than $100 USD. Further, we purchased an Android terminal, LTE router, and local server for $900 USD, all being commercial off-the-shelf products for a total of approximately $1000 USD per set. We invested approximately 40 person-days in the development of the hardware and firmware, with the size of the firmware slightly less than 500 lines of code (LOC).

To evaluate actual data-collection abilities, we set up the mass production model in a small-scale test model room in Shinshu University. Figure 8.6 shows the installation of the tester in the model room, which consisted of a living room and bedroom in a target individuals home. The model room was furnished with a collapsible cot, an infrared remote control, and various other furniture and fixtures (e.g., refrigerator, electric jar, desk, chair, telephone, etc.).

The continuous operational test period started on May 18, 2016 for all produced testers (i.e., from the breadboard model to the mass production model), with all testers gathering trigger information from various sensors in normal operation. Over the course of testing, an error occurred in which a library or function involving the key-waiting portion of the RFID tag module stopped working due to noise (i.e., an erroneous stop). We therefore devised a method to automatically recover from such a stop in the firmware by monitoring for malfunctions via a Watch Dog Timer (WDT) inserted into the loop-processing portion of the firmware.

8.5 Collection and Analysis of Multi-fusion Sensors and Trigger Information

Trigger information transmitted by home sensor agents, then subsequently collected and stored in a home server, is detailed below, with explanatory notes included in parentheses () representing examples of transmitted data. The information is as follows: (1) two physical switches on a pushdown trigger (e.g., SW1, SW2 = ON); (2) accumulated count values of the PIR motion sensor (e.g., PIR = 1234); (3) edge trigger of the Mifare RFID tag for approaches and departures (e.g., KEY = ON/OFF); and (4) received data from infrared remote control (e.g., RMO = 03:77E1C086).

Agent equipment ID (e.g., ID = 1004), firmware version (e.g., VER = 20160601a), and accumulation of activation frequency (e.g., BOOT = 38) are transmitted by the URL argument along with these POST data. These transmitted data are then inserted and stored in the database installed on

Fig. 8.6 An installation to demonstrate a model room

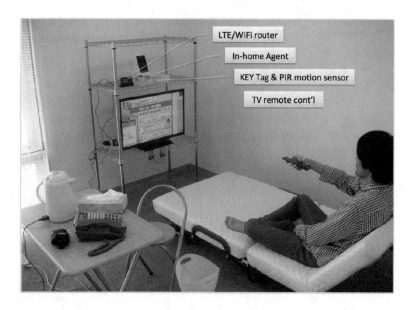

the home server. The transport protocol from the home sensor agent to the home server is based on HTTP and the POST Method standard. Note that data encryption is not active, because communication takes place within the home, but incorporating Secure Socket Layer (SSL) communication is planned for the future.

We attempted to apply visualization to each device. The visualization engine we used was D3.js [16]. Figure 8.7 shows

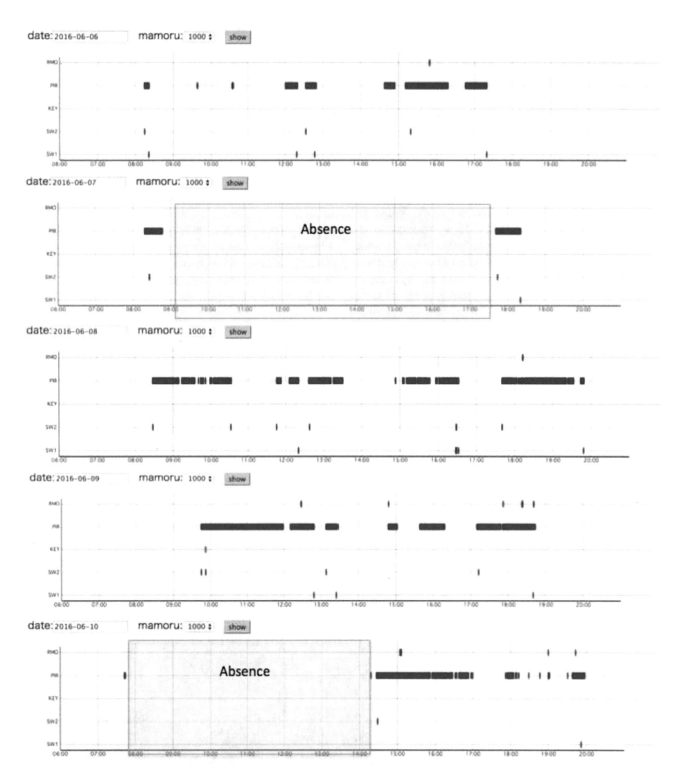

Fig. 8.7 An example trial of device-specific visualization from June 6, 2016 to June 10, 2016 (device ID = 1000)

examples visually displayed with the same equipment ID and various parts of the daily trigger information. The displayed examples were obtained by plotting trigger information (i.e., SW1, SW2, KEY, PIR, and RMO) from a home agent with a certain ID between the hours of 6:00 and 20:00 for five consecutive days. When the PIR motion sensor was continuously detected, it was judged that the target individual was actually *at home*, watching television or moving around the house. In addition, the RMO, a receiver for the infrared remote control system, provided evidence to help determine the target individuals *soundness*, because it switches ON/OFF or changes channel on TV. The KEY and SW triggers, as well as the PIR and RMO triggers, were missing for as long as two days, thus it was determined that the target individual was *not at home* during this time period.

8.6   Conclusions and Future Work

In this paper, we described our home sensor agent, *MaMoRu-Kun*, which was developed under the "Research and development of the regional/solitary elderly life support system using multi-fusion sensors" project; our intent here was to construct a network to watch over the elderly. Based on results of our requirements analysis, we identified specific functions required for the sensor in the house. The hardware andsoftware were designed to establish connections with various composite sensor devices and realize trigger collection and processing. Items added for additional review included visualizing analytical results of the monitoring system under usual conditions, as well as emergency reporting under abnormal circumstances.

With the aim of evaluating the advantages of our proposed system configuration, we replicated a collection site of a houses living room and bedroom, configuring the system to collect actual data in the future by performing a small-scale demonstration experiment. More specifically, trigger information for various switches and motion detection sensors was collected by actually activating prototype sensors in our long-time study. Further, we conducted a load test to send the collected information to all data collection and monitoring servers connected via an LTE router; such information included the information obtained from the bed sensor. And, investigation and prototype making for a system operation at the time of abnormality will continue to achieve the requirements from (B2) to (B6) for emergent response of requirement analysis [B].

Acknowledgements This research was conducted with the support of The Ministry of Internal Affairs and Communications, Strategic Information and Communications R&D Promotion Programme (SCOPE 152302001) in Japan. We are grateful to Prof. Hirokazu Madokoro and A/Prof. Kazuhisa Nakasho at Akita Prefectural University for helpful discussions with this research.

References

1. Mitas, A. W., Rudzki, M., Skotnicka, M., & Lubina, P. (2014). Activity monitoring of the elderly for telecare systems – review. In E. Pietka, J. Kawa, & W. Wieclawek (Eds.), *Information technologies in biomedicine* (Vol. 4, pp. 125–138). Cham: Springer.
2. Jian, Y., Kiong, T. K., & Heng, L. T. (2010). Development of an e-Guardian for the single elderly or the chronically-ill patients. In *Proceedings of Communications and Mobile Computing Conference (CMC)* (pp. 378–382).
3. Gaddam, A., Mukhopadhyay, S. C., & Gupta, G. S. (2011) Trial and experimentation of a smart home monitoring system for elderly. In *Proceedings of Instrumentation and Measurement Technology Conference (I2MTC)* (pp. 1–6).
4. Yan, H., Huo, H., Xu, Y., & Gidlund, M. (2010). Wireless sensor network based E-health system: Implementation and experimental results. In *IEEE Transaction on Consumer Electronics* (Vol. 56, no. 4, pp. 2288–2295).
5. Mayr, H., Franz, B., & Mayr, M. (2010). IHE-compliant mobile application for integrated home healthcare of elderly people. In *Proceedings of Information Technology: New Generations Conference (ITNG)* (pp. 798–803).
6. Schikhof, Y., & Mulder, I. (2008). Under watch and ward at night: Design and evaluation of a remote monitoring system for dementia care. In D. Hutchison, A. Holzinger, T. Kanade, J. Kittler, J. M. Kleinberg, F. Mattern, J. C. Mitchell, M. Naor, O. Nierstrasz, C. Pandu Rangan, B. Steffen, M. Sudan, D. Terzopoulos, D. Tygar, M. Y. Vardi, & G. Weikum (Eds.), *HCI and usability for education and work* (LNCS, Vol. 5298, pp. 475–486). Berlin/Heidelberg: Springer.
7. Chen, J., Kam, A., Zhang, J., Liu, N., & Shue, L. (2005). Bathroom activity monitoring based on sound. In H. W. Gellersen, R. Want, & A. Schmidt (Eds.), *Pervasive computing* (LNCS, Vol. 3468, pp. 65–76). Berlin/New York: Springer.
8. Wtorek, J., Bujnowski, A., Lewandowska, M., Ruminski, J., Polinski, A., & Kaczmarek, M. (2010). Evaluation of physiological and physical activity by means of a wireless multi-sensor. In *Proceedings of Information Technology Conference (ICIT)* (pp. 239–242).
9. Coronato, A., Pietro, G. D., & Sannino, G. (2010) Middleware services for pervasive monitoring elderly and ill people in smart environments. In *Proceedings of Information Technology: New Generations Conference (ITNG)* (pp. 810–815).
10. Arcelus, A., Jones, M. H., Goubran, R., & Knoefel, F. (2007). Integration of smart home technologies in a health monitoring system for the elderly. In *Proceedings of Advanced Information Networking and Applications Workshops (AINAW)* (Vol. 2, pp. 820–825).
11. Shimoi, N., & Madokoro, H. (2013). A study for the bed monitoring system using 3 dimensional accelerometer and piezoelectric weight sensor. *Transactions of SICE (Japanese edition), 49*(12), 1092–1100.
12. Madokoro, H., Shimoi, N., & Sato, K. (2013). Development of non-restraining and QOL sensor systems for bed-leaving prediction. In *IEICE Transactions on Information and Systems (Japanese edition), 96*(12), 3055–3067.
13. MIFARE. *ISO/IEC 14443 Type A 13.56 MHz contactless smart card standard* [Online]. Available http://www.mifare.net/.
14. Arduino, *An open-source prototyping platform for embedded systems* [Online]. Available http://www.arduino.cc/.
15. 1Sheeld, *An Arduino multi-purpose shield with smart-phone* [Online]. Available http://1sheeld.com/.
16. D3, *Data-driven documents* [Online]. Available http://d3js.org/.

How Risk Tolerance Constrains Perceived Risk on Smartphone Users' Risk Behavior

Shwu-Min Horng and Chia-Ling Chao

Abstract

This study focused on smartphone users' intention to install anti-virus software on their devices. In addition to the three tested factors, risk tolerance, perceived, and risk awareness, the models also include several demographic and behavioral variables as controlled factors including tenure of using smartphone, average online time per day, average online time using smartphone per day, gender, age, education level, and monthly expenditure. The results showed that lower risk tolerance, higher perceived risk, and higher risk awareness will lead to higher intention to installation. The constraining effect of risk tolerance on the relationship between perceived risk and intention to installation was also tested significantly. When smartphone users have risk tolerance higher than the threshold, their intention to installation will not be affected by the perceived risk.

Keywords

Risk Tolerance • Perceived Risk • Risk Awareness • Constraining Effect

9.1 Introduction

Smartphones are growing widely popular devices of computing technology. Global sales of smartphones to end users totaled 349 million units in the first quarter of 2016, a 3.9 percent increase over the same period in 2015 [1]. Smartphone sales represented 78 percent of total mobile phone sales in the first quarter of 2016. As smartphones become more popular, and their connectivity and processing power increase, they attract more attention from malware writers around the world.

The danger of smartphone virus has been discussed [2]. Malware describes everything malicious that can do the significant damages to smartphones [3]. Security issue on mobile internet or on smartphones could represent part of the bigger threat, cybersecurity. Significant national attention was focused on promoting individual behaviors that enhance computer and information security [4]. A country level cooperation between Japan and US is underway to build a safer online environment [5]. A study examined the structure of cyber-crimes in developing countries and concluded that the serious cyber-crimes require broadband-intensive applications [6].

The global smartphone security market is expected to grow from $387 million in 2010 to $2,965 million by 2016 with a compound annual growth rate (CAGR) of 44% from 2011 to 2016 [7]. Based on mobile users, typical consumers have at least a 1 in 6 chance of downloading apps that include malware or suspicious URLs, and almost 25 percent of the risky apps that contain malware also contain suspicious URLs [8]. In addition, mobile malware is distributed primarily through infected apps today, unlike the email- and

S.-M. Horng (✉)
Department of Business Administration, National Chengchi University, Taipei City, Taiwan, Republic of China
e-mail: shorng@nccu.edu.tw

C.-L. Chao
Department of Accounting and Information Technology, National Chung Cheng University, Chia-yi, Taiwan, Republic of China
e-mail: cchao2002@netscape.net

S. Latifi (ed.), *Information Technology – New Generations*, Advances in Intelligent Systems and Computing 558, DOI 10.1007/978-3-319-54978-1_9

website-based infections typical of PCs. In summary, the results of this study can provide guidelines for smartphone service providers to develop better strategies and ultimately increase the percent of smartphone users to install the anti-virus software. The objectives of this study include the following:

- Investigate how the users' perceived risk, risk tolerance, and risk awareness affects their intention to install anti-virus software on smartphones.
- Study the constraining effect of risk tolerance on the relationship between perceived risk and the intention to use anti-virus software on smartphones.

9.2   Hypotheses Development

A study discussed the current state of smartphone research and indicated security as one of the important topics for future research [9]. Smartphone applications adopt centralized distribution architectures and are usually available to users from app repositories or app marketplaces. Smartphone operating systems like Android, iOS, and Windows phone expose rich apps to third-party applications. These apps allow applications to use hardware, change phone settings, and read data. Malicious and unscrupulous authors have taken advantages of these resources to the displeasure of users. The threat and damages caused by them have been mentioned previously. However, smartphone users seemed not to consider these problems seriously.

The results from a survey indicated a clear lack of smartphone users' security awareness [10]. They showed that most of smartphone users believe that downloading applications from the app repository is risk-free. They proposed two possible factors misleading users on their behavior. First, distributing an app from an official app repository may mislead the users into believing that the app is secure or that security controls exist in the app repository. Second, smartphone users may be unable to realize that their device is not just a telephone. Therefore, having an anti-virus software installed on smartphones should provide the protection from most malwares, and this research is motivated by this reason.

A study summarized 99 risks on smartphone users and their survey provided the corresponding ranks of those risks [11]. Although the risks on smartphone are mostly personal, it could also cause damage on internet such as the risk of spam. Another survey [12] distinguished the behavioral intention to protect the internet from protect users' own device, and developed the framework based on theory of reasoned action (TRA) [13]. The results suggested that

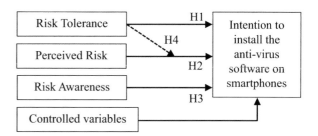

Fig. 9.1 Research Framework

users' intention to perform security-related behavior is influenced by a combination of cognitive, social, and psychological components. Choice architecture can affect smartphone users' willingness to install applications that requested varying permission. Smartphone users are concerned with their privacy and are willing to pay premiums for applications that are less likely to request access to personal information [14].

Smartphone anti-virus software can be considered as an app, and several studies have focused on the intentions to use smartphone applications. A study [15] extended technology acceptance model (TAM) [16] to study users and non-users of three selected mobile applications. Their results suggested that usefulness is a decisive criterion for the use of mobile internet services.

The research framework of this research is illustrated in Fig. 9.1. Note that the risk behavior in this framework is represented by the intention of installing anti-virus software on smartphone, and is conceptually equivalent to risk avoidance.

Risk tolerance has generally been measured presenting subjects with a set of risky events that describe life and death, gambling, or product choice situations. Each subject rates the riskiness of each event. Subjects who choose risky events or rate these as not very risky are defined as risk seekers [17]. A survey studied the risk perception of consumers on internet shopping [18]. The results show that consumers with a higher degree of risk aversion than others tend to perceive internet shopping to be a risky activity. Higher degree of risk aversion is equivalent to lower risk tolerance in which will lead to the behavior of avoiding the risk. Similarly, when considering the use of mobile banking services, users with low risk tolerance will have less intention to use the services, and the tendency to avoid risk. Therefore,

H1: Risk tolerance will have a negative effect on intention to install the software

Perceived risk is commonly considered as the uncertainty regarding possible negative consequences of using a product or service. Perceived risk was defined as "the potential for loss in the pursuit of a desired outcome of using an e-service" at the study of predicting e-service adoption [19]. Perceived risk is negatively related to the intention to adopt a system, to purchase [20], to shop online [21], and to use online applications [22]. However, using anti-virus software is a behavior to avoid risk and users should behave differently compared to the behavior of purchasing online or using online services. When designing a game design framework, perceived threat will have a positive impact on avoidance motivation, which leads to avoidance behavior [23]. Therefore,

H2: Perceived risk will have a positive effect on the intention to install the software

Online users today have serious concerns about their privacy and security while conducting purchases over the Web. They are also concerned about the safety of their email. Many organizations have policies regarding e-mail and Internet use to reduce the risks, and employees' awareness would reduce risks and liability [24]. The awareness of risk online was defined as knowing something about information security [25]. In their study, information security awareness directly and indirectly affect the attitude that leads to the intention to comply with the requirements of the information security policies. When considering personal information security, an anti-virus software installed in smartphones can be viewed as a requirement to prevent the risk. Therefore,

H3: Risk awareness will have a positive effect on the intention to install the software

Based on the summary of several theories, an individual will search for alternatives, when the existing degree of perceived risk is below the general tolerance level [26]. A study investigated how the two factors, risk perception and risk tolerance affects pilot decision-making [27]. The results showed that risk perception and risk tolerance are two separate constructs and negatively related to each other. In another study [28], the intended use of risk-handling activity increases with higher levels of perceived. This relationship is more significant after the level of risk exceeds the individual's acceptance level of risk. Therefore, this study will then hypothesize the existence of threshold levels for perceived risk and risk tolerance, respectively. Users' behavior affected by perceived risk (or risk tolerance) will be constrained, unless their risk tolerance (or perceived risk) measures exceed the threshold levels.

H4: The relationship between perceived risk and risk behavior will be constrained, if risk tolerance is beyond the threshold level

Demographic factors such as age, gender, and experience were found to have effects on several behavioral intentions or activities [29]. They are correlated with the activities of sharing knowledge, and age and gender have been bound to be associated with technology innovativeness [30]. For the users' behavior in mobile internet, experience toward the new technology tested significantly between the intention and behavior [31]. Hence, these factors will be included in the models as controlled variables along with other online behavioral variables such as average time online per day, average smartphone online time per day, education level, cost that willing to pay for the cell phone anti-virus software, and monthly expenditure.

9.3   Data Collection, Analysis and Results

The data were collected from the largest online bulletin board (PPT) in Taiwan. The survey was posted online for approximately one month and the participants were screened by asking whether or not they are smartphone users.

After excluding 34 invalid questionnaires for various reasons, a total of 351 valid samples were available for analysis, yielding an effective rate of 91.2 percent.

The demographic information of the respondents indicate that the top three groups of user are Apple, HTC, and Sony. In general, males outnumber females (54.4%, 45.6%, respectively); most of the respondents have college degree or higher (93%), in the age of 21-30 (68.4%), with more than 3 years of experience using smartphones (57%). A majority of the respondents access the internet with more than three hours per day (85%).

Multi-item, seven-point Liker scale items were used to measure the constructs in the model. Although pre-existing scales and items from the literature were applied wherever appropriate, some items were revised or added to suit the context of this study. Scales were developed based on the review of the most relevant literature as illustrated in Table 9.1. To ensure that face validity, defined as the degree to which respondents judge that items are appropriate to the targeted construct, and is consistent with content validity, the following iterative evaluation process was implemented [32]. A group of college students were asked to assign the questions, developed according to the literature review, to the appropriate constructs including perceived risk, risk

Table 9.1 Content validity

Constructs	Adapted from
RT	Bruner and Hensel (1992); Rippetoe and Rogers (1987); Milne et al. (2000)
PR	Anderson (2010); Ellen and Wiener (1991); Ho (1998); Obermiller (1995)
RA	Beth et al. (2014)
Int	Anderson and Agarwal (2010)

RT Risk Tolerance, *PR* Perceived Risk, *RA* Risk Awareness, *Int* Intention to install the anti-virus software

Table 9.2 Factor analysis and construct

	RA	RT	PR	Int
RT1	0.093	**0.843**	0.056	0.090
RT2	0.144	**0.863**	0.074	0.133
RT3	0.210	**0.852**	0.060	0.168
RT4	0.153	**0.803**	0.148	0.152
RT5	0.223	**0.830**	−0.010	0.078
PR1	0.113	0.02	**0.856**	0.054
PR2	0.086	0.212	**0.845**	0.082
PR3	0.092	−0.038	**0.726**	0.094
PR4	0.065	0.096	**0.863**	0.068
PR5	0.165	0.048	**0.747**	0.117
RA1	**0.851**	0.175	0.119	0.136
RA2	**0.868**	0.218	0.110	0.116
RA3	**0.896**	0.134	0.144	0.153
RA4	**0.892**	0.156	0.069	0.133
RA5	**0.825**	0.170	0.164	0.130
Int1	0.217	0.131	0.150	**0.908**
Int2	0.181	0.211	0.119	**0.913**
Int3	0.156	0.194	0.127	**0.906**

Table 9.3 Reliability indices and correlations among constructs and average variance extracted (AVE)

			AVE	RT	PR	RA	Int
RT	.951	.968	.910	**.954**			
PR	.943	.956	.814	.382	**.902**		
RA	.917	.937	.750	.369	.398	**.866**	
Int	.880	.913	.677	.274	.287	.204	**.823**

Notes:
– The values of second and third columns represent Cronbach alpha and composite reliability, respectively.
– The diagonals represent the square root of average variance extracted (AVE) and off-diagonals are the correlations among constructs.

tolerance, and risk awareness. This procedure was iterated several times until all of the samples reached a total consensus.

Tables 9.2 and 9.3 show the results for reliability and validation tests. The average variance extracted (AVE) estimates the amount of variance captured by a construct's measure relative to random measurement error [33]. Estimates

of composite reliability above 0.7 and AVE above 0.5 are considered supportive of internal consistency [34].

The factor loadings of each item ranging from 0.726 to 0.913 are shown in Table 9.2. According to the suggested guideline [35], all of them are considered excellent. Next, we verified the discriminant validity by examining the square root of the average variance extracted as recommended [33]. The results in Table 9.3 show that the square root of the AVE for each construct is greater than the correlation shared between the construct and other constructs in the model. The average variance extracted for the constructs in the measurement model are all higher than the recommended level of 0.5. Since few correlations are relatively high, we also examined the multicollinearity among constructs. The variance inflation factor (VIF) values for all of the constructs are acceptable.

Factor loadings and cross-loading were calculated and presented in Table 9.3. Each indicator has loading larger than the value of 0.7 and higher than the construct of interest than on any other factor [36]. In sum, the results of content validity, convergent validity, and discriminant validity enable this study to proceed to estimations of the regression models.

Latent constructs for which multiple items are available are combined into one indicator according to the partial disaggregation model [37]. Multiple regression is the method used to study the proposed hypotheses. Intention to install the smartphone anti-virus software represents the dependent variable, and the independent variables include the three tested variables, risk tolerance (RT), perceived risk (PR), and risk awareness (RA), and controlled variables consisting of tenure of using smartphone (Tenn), average online time per day (AOT), average online time using smartphone per day (AOTS), gender, age, education level (Edu), and monthly expenditure (Exp).

The results are shown in Table 9.4. Model I represents the base model used to test hypotheses 1, 2, and 3. To test the constraining effect of risk tolerance on perceived risk, a dummy variable, D_RT was generated and included in model II. D_RT is 0 when it is higher than or equal to the threshold value, and 1 otherwise. The threshold is calculated based on the following procedure. First, the entire sample are ranked based on the value of RT from the highest to the lowest. The correlation coefficient between PR and Int was calculated with the entire sample, and then recalculated by removing the samples with the highest value of RT. This process was reiterated until the final set of sample with the lowest value of RT. A significant different was detected when the threshold value was set to be 4. Out of 324 samples, 22 have RT equal to or larger than 4. The interaction term, D_RT multiplying by PR was included in the second regression model. The significance of the interaction term in the

Table 9.4 Results of the regression models

	Model I (Base)		Model II	
	Beta	T value	Beta	T value
RT	−.222	−4.072***	−.423	−1.630*
PR	.232	4.243***	.144	1.553*
RA	.157	2.970***	.158	2.924***
Tenn	.104	2.008**	.104	2.016**
WTP	.126	2.522**	.123	2.444**
AOT	.046	0.765	.049	0.824
AOTS	−.103	−1.743*	−.100	−1.818*
Gender	.034	0.670	.030	0.631
Age	.140	2.575**	.143	2.576**
Edu	−.060	−1.158	−.057	−1.126
Exp	−.045	−0.844	−.046	−0.365
(D_RT)x(Risk)			.197	1.892*
R^2		.270		
Adj. R^2		.244		
F		10.468***		
F for the R^2 change			3.678**	

***$p < 0.01$, **$p < 0.05$, *$p < 0.1$

second regression model will confirm the constraining effect of RT on PR [38].

In model I, the significant variables indicate that lower risk tolerance, higher perceived risk, higher risk awareness, longer time of using smartphone, willing to pay for higher cost of the software, spending less online time on smartphone, and younger users have higher intention to install the anti-virus software for smartphones. Hypotheses 1 to 3 tested significantly along with five of nine controlled variables. Average online time per day, gender, education and monthly expenditure did not have significant impact on users' intention to install the software. In model II, the constraining effect of risk tolerance is also test significantly. That is, the relationship between perceived risk and intention to installation shows insignificance when risk tolerance is higher than the threshold value.

9.4 Conclusions

Smartphones have been increasingly popular and many users did not know the threat of viruses toward their smartphone. This research studied the factors influencing users' intention to install the anti-virus software at their smartphone. The results showed that lower risk tolerance, higher perceived risk and risk awareness will lead to higher intention to install the anti-virus software at their smartphones, respectively. Anti-virus software vendors could develop strategies increasing users' perceived risk and risk awareness. However, risk tolerance is very likely an internal personal feature and not to be affected by external factors. According to the results of this study, those users with high level of risk

tolerance will not show higher intention to installation, even though they perceive high level of risk.

References

1. Gartner, (2016). Gartner Says Worldwide Smartphone Sales Grew 3.9 Percent in First Quarter of 2016, retrieved on 10/9/2016 from http://www.gartner.com/newsroom/id/2623415
2. Dagon, D., Martin, T., & Starner, T. (2004). Mobile phones as computing devices: The viruses are coming! *IEEE Pervasive Computing, 3*(4), 11–15.
3. Chang, K.S. (2013). Can smartphones get virus/viruses and malware? iphone / android phone only as smart as their users in malware detection, retrieved on 12/24/2013 from http://kschang.hubpages.com/hub/Do-Smartphones-Get-Viruses-Why-iPhones-Android-Phones-and-Others-Are-Only-as-Smart-as-Their-Users
4. Gross, G. (2007). Groups raise concerns about cybersecurity standards, PC World, April 25.
5. Matsubara, M. (2012). A long and winding road for cybersecurity cooperation. *Harvard Asia Quarterly, 14*(1/2), 103–111.
6. Kshetri, N. (2010). Diffusion and effects of cyber-crime in developing economies. *Third World Quarterly, 31*(7), 1057–1079.
7. Marketsandmarkets (2011). global smartphone security market by operating systems, ownership & features (2011–2016), retrieved on 12/28/2013 from: http://www.marketsandmarkets.com/Market-Reports/world-computer-mobile-antivirus-market-156.html
8. Fitton, O., & Prince, D. (2014). The future of mobile devices: security and mobility. Lancaster University, Report ID 73272, pp. 34.
9. Enck W. (2011). Defending users against smartphone apps: Techniques and future directions, ICISS 2011. *LYNC, 7093*, 49–70.
10. Mylonas, A., Kastania, A., & Gritzalis, D. (2013). Delegate the smartphone user? Security awareness in smartphone platforms. *Computers & Security, 34*, 47–66.
11. Felt, A. P., Egelman, S., & Wagner, D. (2012). I've got 99 problems, but vibration ain't one: A survey of smartphone users' concerns. *Proceedings of SPSM', 12*, 33–38.
12. Anderson, C. L., & Agarwal, R. (2010). Practicing safe computing: A multimedia empirical examination of home computer user security behavioral intentions. *MIS Quarterly, 34*(3), 613–643.
13. Ajzen, I., & Fishbein, M. (1980). *Understanding attitudes and predicting social behavior*. Englewood Cliffs, NJ: Prentice-Hall.
14. Egelman, S., Felt, A. P., & Wagner, D. (2013). Choice architecture and smartphone privacy: There's a price for that. In R. Böhme (Ed.), *The economics of information security and privacy* (pp. 211–236). New York: Springer.
15. Verkasalo, H., López-Nicolás, C., Molina-Castillo, F. J., & Bouwman, H. (2010). Analysis of users and non-users of smartphone applications. *Telematics and Informatics, 27*(3), 242–255.
16. Davis, F. D. (1989). Perceived usefulness, perceived ease of use, and user acceptance of information technologies. *MIS Quarterly, 13*(3), 319–340.
17. Dowling, G. R. (1986). Perceived risk: The concept and its measurement. *Psychology and Marketing, 3*(3), 193–210.
18. Tan, S. J. (1999). Strategies for reducing consumers' risk aversion in internet shopping. *Journal of Consumer Marketing, 16*(2), 163–180.
19. Featherman, M. S., & Pavlou, P. A. (2003). Predicting e-services adoption: A perceived risk facets perspective. *International Journal of Human-Computer Studies, 59*(4), 451–474.
20. Pavlou, P. A. (2003). Consumer acceptance of electronic commerce: Integrating trust and risk with the technology acceptance

model. *International Journal of Electronic Commerce, 7*(3), 101–134.

21. Kim, J. (2012). Developing an empirical model of college students' online shopping behavior. *International Journal of Interdisciplinary Social Sciences, 6*(10), 81–109.

22. Lu, H., Hsu, C., & Hsu, H. (1993). An empirical study of the effect of perceived risk upon intention to use online applications. *Information Management & Computer Security, 13*(2), 106–120.

23. Arachchilage, N. A. G., & Love, S. (2013). A game design framework for avoiding phishing attacks. *Computers in Human Behavior, 29*, 706–714.

24. Udo, G. J. (2001). Privacy and security concerns as major barriers for e-commerce: A survey study. *Information Management & Computer Security, 9*(4), 165–174.

25. Bulgurcu, B., Cavusoglu, H., & Benbasat, I. (2010). Information security policy compliance: An empirical study of rationality-based beliefs and information security awareness. *MIS Quarterly, 34*(3), 523–548.

26. Schaninger, C. M. (1976). Perceived risk and personality. *Journal of Consumer Research, 3*(2), 95–100.

27. Hunter, D.R. (2002). Risk perception and risk tolerance in aircraft pilots. Retrieved on 10/12/2012 from http://ntl.bts.gov/lib/19000/19800/19856/PB2003100818.pdf

28. Dowling, G. R., & Staelin, R. (1994). A model of perceived risk and intended risk-handling activity. *Journal of Consumer Research, 21*(1), 119–134.

29. Constant, D., Kiesler, S., & Sproull, L. (1994). What's mine is ours, or is it? A study of attitudes about information sharing. *Information Systems Research, 5*(4), 400–421.

30. Lee, H., Cho, H. J., Xu, W., & Fairhurst, A. (2010). The influence of consumer trait and demographics on intention to use retail self-service checkouts. *Marketing Intelligence & Planning, 28*(1), 46–58.

31. Venkatesh, V., Thong, J. Y. L., & Xu, X. (2012). Consumer acceptance and use of information technology: Extending the unified theory of acceptance and use of technology. *MIS Quarterly, 36*(1), 157–178.

32. Kankanhalli, Q., Tan, B. C. Y., & Wei, K.-K. (2005). Contributing knowledge to electronic knowledge repositories: An empirical investigation. *MIS Quarterly, 29*(1), 113–143.

33. Fornell, C., & Larcker, D. F. (1981). Evaluating structural equation models with unobservable variables and measurement errors. *Journal of Marketing Research, 18*(1), 39–50.

34. Bagozzi, R. P., & Yi, Y. (1988). On the evaluation of structural equation models. *Journal of the Academy of Marketing Science, 16*(1), 74–94.

35. Comrey, A. L., & Lee, H. B. (2013). *A first course in factor analysis.* Hove: Psychology Press.

36. Chin, W. W. (2010). How to write up and report PLS analyses. In V. Esposito Vinzi, W. W. Chin, J. Henseler, & H. Wang (Eds.), *Handbook of partial least squares: Concepts, methods and applications* (pp. 655–690). Berlin: Springer.

37. Bagozzi, R. P., & Edwards, J. R. (1998). A general approach for representing constructs in organizational research. *Organizational Research Methods, 1*(1), 45–87.

38. Siemsen, E., Roth, A. V., & Balasubramanian, S. (2008). How motivation, opportunity, and ability drive knowledge sharing: The constraining-factor model. *Journal of Operations Management, 26*(3), 426–445.

When Asteroids Attack the Moon: Design and Implementation of an STK-Based Satellite Communication Simulation for the NASA-Led Simulation Exploration Experience

10

Bingyang Wei, Amy Knowles, Chris Silva, Christine Mounce, and Anthony Enem

Abstract

The Simulation Exploration Experience (SEE) is an annual, inter-university, distributed simulation challenge led by NASA. A primary objective is to provide a platform to college students to work in highly dispersed teams to design, develop, test, and execute a simulated lunar mission using High Level Architecture. During the SEE in 2016, 19 federates developed by student teams from three continents successfully joined the HLA federation and collaborated to accomplish a lunar mission. The Midwestern State University first participated in SEE and developed a communication satellite federate which broadcasts alert to physical entities on the moon surface about the incoming of an asteroid. This paper describes the design of the communication federate, the federation object model, lessons learned and recommendations for future federate development.

10.1 Introduction

Initiated back in 2011, the annual Simulation Exploration Experience event, formally known as SISO Smackdown, has successfully promoted the awareness of the use of High Level Architecture (HLA) in distributed simulation around the world. During the event, academia, industry and professional associations are joined to collaboratively design, develop, test and demonstrate a simulated lunar mission. SEE provides a excellent platform for college students to learn and practice both M&S and software engineering concepts and principles. More importantly, the opportunity of working closely with M&S professionals in industry and associations is an valuable experience to the students.

During SEE 2016, a lunar mission was simulated. NASA Johnson Space Center (JSC), Kennedy Space Center (KSC) and 12 universities from three continents participated in this year's distributed simulation event. All participants conformed to the IEEE standard for modeling and simulation HLA. As usual, NASA provided basic support to the entire simulation mission, JSC regulated the time of the simulation and KSC provided real-time visualization of the simulation mission. Each university contributed to the lunar simulation by implementing a part of the entire simulation called federate (More HLA terminologies are available in [1]). Here is a brief description of the scenario of SEE 2016: Astronauts are exploring a huge impact crater close to the south pole of the moon called Aitken Basin and several facilities are built: a 3D-printing site by University of Alberta, a supply depot, oxidizer and propellant production facility by University of Bordeaux, an astronaut habitat site by Facens, a cargo rover by Florida Institute of Technology, a fuel rover by University of Nebraska, a lunar buggy and UAV by University of Liverpool.

The peaceful exploration operation is suddenly interrupted by the detection of an incoming asteroid by an asteroid detection system (developed by University of Genoa). The command and communication center [2] developed by University of Calabria alerts all physical entities on the moon surface, and the communication satellites developed by Midwestern State University alert the lunar buggy which is out of the reach of the command and control center.

B. Wei (✉) • A. Knowles • C. Silva • C. Mounce • A. Enem
Department of Computer Science, Midwestern State University, 76308 Wichita Falls, TX, USA
e-mail: bingyang.wei@mwsu.edu

© Springer International Publishing AG 2018
S. Latifi (ed.), *Information Technology – New Generations*, Advances in Intelligent Systems and Computing 558, DOI 10.1007/978-3-319-54978-1_10

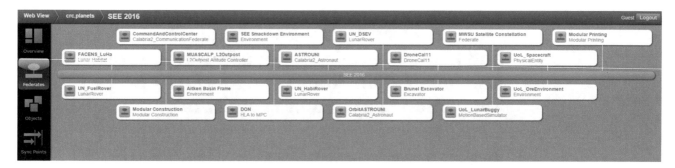

Fig. 10.1 Federates connecting through Pitch RTI

Fig. 10.2 MWSU
communication satellites federate

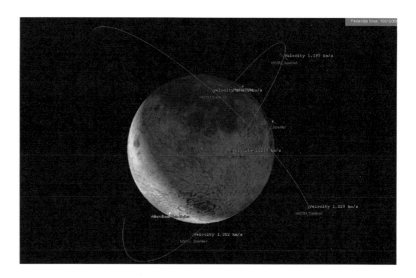

During the simulation, all federates were connected successfully to the runtime infrastructure (RTI) provided by Pitch Technologies and advancing time (see Fig. 10.1). Each rentangle represents a joined federate and the middle gray bar denotes the RTI.

This paper concentrates on one of the federates in Fig. 10.1: the MWSU communication satellites federate developed by the Midwestern State University (see Fig. 10.2). This federate is an aerospace Java application built on top of AGI's proven Systems Tool Kit (STK): the STK Components for Java SDK. In summary, the MWSU communication satellites federate provides the following functions in the lunar mission:

(1) Propagation of constellations of satellites orbiting the moon so that enough coverage of Aitken Basin is achieved;
(2) Inter-communication capabilities between physical entities in the mission;
(3) Visualization of satellites constellation in 3D and all the other physical entities in 2D.

One of the important technical contribution of this work is that MWSU Simulation team developed a HLA interface

implementation for the AGI STK component. The project is available in GitHub (https://github.com/csos95/MWSU-SEE-2016).

The rest of the paper is structured as follows: Sect. 10.2 introduces an important software component that powers our federate, the AGI STK Components; Sect. 10.3 describes the design and development the federate by the Midwestern State University team; Lessons learned from SEE is discussed in Sects. 10.4 and 10.5 concludes the paper and points to some future work.

10.2 AGI STK Components

In order to fulfill the challenging requirements of the MWSU communication satellites federate, the development of the federate needs to rely on robust and free software development kits. By investigating and researching different software, we found that the software applications and development kits offered by Analytical Graphics, Inc. (AGI) is the best choice for modeling, simulating and analyzing the lunar mission. STK Components [3] is a family of low-level class libraries that provide access to specific analytical and 3D visualization capabilities for space exploration.

At the core of the STK Components is the Dynamic Geometry Library (DGL). DGL provides various specialized propagators for modeling the motion of satellites, aircraft, and surface vehicles. Additionally, algorithms like inter-visibility between physical entities given a number of complex, simultaneous constraining conditions in the space are also provided so that the line-of-sight(LOS) information critical in our lunar communication system can be obtained easily. Since DGL adopts the industry recognized time and position which is consistent with that of NASA's environment federate, it makes the positioning of other physical entities in federation accurate. The Insight3D Visualization Library which is a part of DGL allows us to build lightweight yet powerful 3D applications for performance and visualization accuracy of the lunar mission.

Other relevant capabilities provided by DGL:

- Accurate moon modeling;
- Precise platform positioning and orientation modeling for orbit and waypoint propagation;
- Complex numerical and geometry algorithms;
- Manipulate points, axes, vectors, and reference frames.

10.3   MWSU Communication Satellites Federate

The capabilities of STK Components make our federate possible. In the following two subsections, two main modules that constitute the MWSU federate: Satellite Constellation & Lunar Visualization module, and Communication Server module are explained in detail.

10.3.1   Satellite Constellation & Lunar Visualization Module

The Satellite Constellation & Lunar Visualization module first propagates and then renders satellites constellation. STK's numerical propagator is invoked to generate the orbits of different satellites. The constellation is achieved by STK's High Precision Orbit Propagator (HPOP) which is a part of the Orbit Propagation Library (OPL). By numerically integrating the various forces affecting satellites, HPOP brings high fidelity orbit propagation into our MWSU communication satellites.

When the whole federation is running on Pitch RTI, a 6-satellite constellation is designed which could provide a lot of hang time both over Aitken Basin to maximize the time each satellite is in view of the surface entities as shown in Fig. 10.2.

The architecture of the Satellite Constellation & Lunar Visualization module is shown in, Fig. 10.3, a UML class diagram.

An instance of **TheFederate** class that inherits **NullFederateAmbassador** builds up the MWSU communication satellites federate and connects it to the RTI host in JSC by instantiating a **Connection** object. Objects crucial for using HLA services like **RTIambassador**, **EncoderFactory** are created and the whole federate is time constrained during the construction of federate. The **TheFederate** instance is then assigned to **Satellite** objects so that they can invoke the services provided by HLA RTI. **TheFederate** class also contains all the definitions of HLA callback methods: timeConstrainedEnabled, timeAdvanceGrant, removeObjectInstance, provideAttributeValueUpdate, discoverObjectInstance, reflectAttributeValues and receiveInteraction.

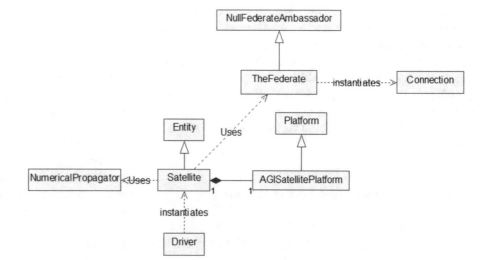

Fig. 10.3 MWSU satellite constellation & lunar visualization module class diagram

Fig. 10.4 Radio message object model template

In this module, **Satellite** is as a subclass of the **PhysicalEntity** object class defined in the SISO_SpaceFOM_entity FOM. Once the MWSU federate was granted time advance by the NASA time regulating federate, each satellite published its logical time, entity name, reference frame and most up-to-date Cartesian coordinate position. Other federates which were interested in obtaining the attributes of satellites could subscribe to and receive the updates sent by the MWSU federate.

Regarding the propagation of orbits, each satellite uses a numerical propagator provided by STK to propagate its orbit. STK Components increases the fidelity and accuracy of its simulation of orbits by providing a variety of environment and force models. In our case, spherical harmonic gravity of the moon and solar radiation force are considered during the propagation of a satellite's orbit. The gravity model of the moon is read in from LunarGravityField_LP100K.grv file. It is then used to construct the immutable field by selecting to the desired fidelity in the degree and order of the represented field, as well as configuring other options such as the inclusion of tidal data. This field is then used to define the force at a given position. In the end, those force models together with six orbital elements which are the parameters required to uniquely identify a specific orbit: semi major axis, eccentricity, inclination, argument of perigee, right ascension of the ascending node (RAAN) and true anomaly are used to propagate the position, orientation, and other attributes of the satellite over time using the numerical propagator which results very good and accurate orbits.

In our effort to visualize satellites and their orbits in STK's Insight3D viewer, we adopted STK Components' **Platform** type which can be used to model satellites, facilities, aircraft, and other "real-world" objects. Simply put, an **AGISatellitePlatform** object is created for each **Satellite** object which stores: the name of the satellite, a time-varying position and orientation calculated by numerical propagator. By adding the **AGISatellitePlatform** into the Insight3D viewer, the visualization of satellites is accomplished.

10.3.2 Communication Server Module

The second important module in the MWSU communication satellites makes communication in the mission possible. The communication server module is used to facilitate lunar communications among different federates. Here is a typical scenario where our communication server might be used: Genoa's asteroid detection system which tracks the approaching asteroid near the moon can broadcast the warning message to all the federates through our communication system.

Messages in the communication system are defined as HLA interactions with transmit (TX) and receive (RX) interactions (Fig. 10.4).

Federates using our communication service should publish the TX interaction and subscribe to RX interaction. The MWSU communication satellites communication server does the reverse order: it publishes RX interaction and subscribes to all the TX interactions.

In order to calculate line-of-sight information between two physical entities in the mission, the MWSU communication satellites federate subscribes to the name, reference frame and position attributes of the **PhysicalEntity** object class and maintains tables of each entity's latest position in the mission during the SEE event. Figure 10.5 is a screenshot of a display listing the discovered entities by the MWSU federate.

The basic work flow of communication server can be illustrated in a sequence diagram (Fig. 10.6):

(1) The communication server receives a TX interaction through the HLA callback method receiveInteraction().
(2) It extracts the source and destination information from the interaction and determines if both the sender and receiver have access to a certain satellite in the satellite constellation.
(3) If so, the communication server builds a RX message and uses the HLA service sendInteraction() to send the RX message, federates which subscribes to the RX message will receive the message.
(4) If not, that TX interaction message will be discarded.

The LOSEvaluate() method invoked by the communication server in the above UML sequence diagram is used to determine the inter-visibility between entities in the lunar scenario. In STK Components, access relation issue between two entities is resolved by specifying access constraint it must satisfy. The simplest and most commonly used access constraint is the Central Body Obstruction Constraint, which

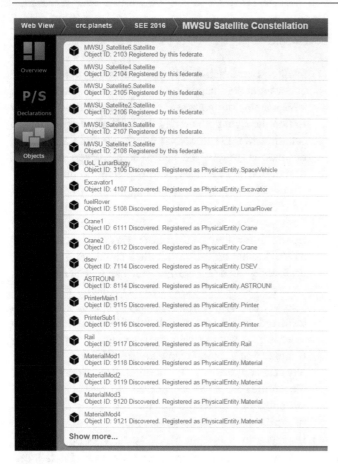

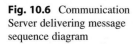

Fig. 10.5 Screenshot of physical entities discovered by MWSU communication satellites federate

requires that the view from one entity to another not be obstructed by a central body such as the moon. Further constraints can be imposed on a single access relation. In our case, we assumed communication capability between a surface entity and satellite is poor near the horizon, so we set an Elevation Angle Constraint with minimum 5° angle on the access relation. This constraint requires that the angle between the link and the plane tangent to the surface of the moon at the location of the entity be between 5° and 175°.

10.4 Lessons Learned and Recommendations for Future Development

During the integration test and the final demo, we have found several problems regarding the entire federation. As MWSU federate is able to locate all the physical entities in the Insight3D viewer, our federate actually verified all the location and motion information of different entities either in the orbit of the moon or on the surface.

In the same time, several problems about our federate were found during the six integration test:

Problem 1: It took too much time to propagate six satellites. Solution: This problem is understandable, since in order to achieve high fidelity, STK numerical propagator calculates a lot of data for one satellite. By using background calculation capability, we paralleled the six propagations instead of propagating six satellites one by one. This greatly reduced the six satellites propagation time.

Fig. 10.6 Communication Server delivering message sequence diagram

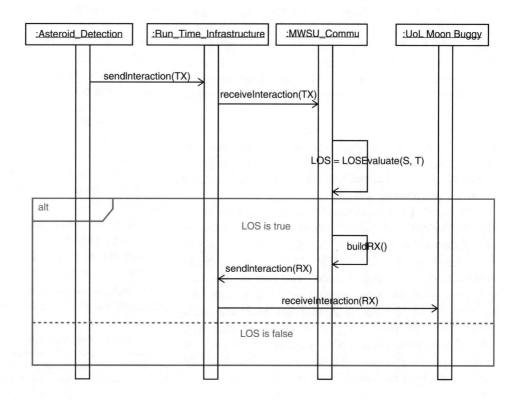

Problem 2: MWSU federate worked well with Pitch RTI but failed to run on MÄK RTI. When testing on MÄK RTI, exception "unsatisfiedLinkError makRtiJava1516e.dll: the specified procedure could not be found" is thrown. Solution: MWSU federate is a Java application built in Eclipse. This is an old problem with MÄK RTI which does not work with Java IDE. We did not solve this problem and have informed engineers from VT MÄK. We hope they can fix this for SEE 2017.

Problem 3: MWSU federate kept throwing NullPointer exceptions due to other teams failing to provide the latest Attribute Handle Value Map or they provide Attribute Handle Value Map that misses certain attribute values. Solution: It is good software engineering principle to not expect federates developed by other teams behave exactly as what you want. Proper and robust exception handling mechanism was added to help solve this problem.

Problem 4: For our Insight3D viewer, we used makers and texts to represent entities on the moon surface and orbiting the moon. Since there are so many of them, the texts and markers could easily bloat the viewer and thus hindered its readability. Solution: We add a distance constraint on the visibility of the marker so that a marker would only become visible when the distance between camera and the entity is less than 1000 km. If we look at the moon at a rather far away distance, we could only see the text representing the name of the entity.

Problem 5: During the testing, we found that some HLA callbacks never get called when they were supposed to be called. Solution: HLA provides more than one different versions of the same callback, for instance, there are three reflectAttributeValues callbacks. We never know which version the RTI will invoke, so instead of just writing one version, it is better to write all the above three in **TheFederate** class and put the same implementation code in three reflectAttributeValues() functions with difference signatures.

Problem 6: When a federate resigned from the federation, the insight3D viewer still kept the marker and text representing the physical entity of that resigned federate. Solution: Clear marker and text when the callback method removeObjectInstance gets called.

Problem 7: There is a synchronization exception during the demonstration in insight3DTimeChanged() method of class **LCANSatManager** due to different threads competing otherEntitiesToBeDrawn list. This would cause the Insight3D viewer to crash. Solution: Our current solution is not good since we only surround the trouble code with try-catch block in order to prevent the entire Insight3D viewer from crashing during the event. Future team may wish to add a mutual exclusion lock on this resource to resolve this. The future team should be aware that this problem has not been solved properly.

Needed improvements and recommendations for future development:

(1) Walker constellation [4] is good enough to provide communication support over the entire moon. Next year's team can explore if STK provides the capability to find the right path among the 14 Walker constellation satellites in order to deliver a message from a source to destination.(Maybe the source is at Aitken Basin and destination is at Hadley Rille). If they do not provide, we could use a graph-related algorithm to implement this feature;

(2) Next year's team could also explore if STK provide the terrain information of Aitken Basin, if so, we our MWSU communication satellites could not only model each entity as a 2D marker, but use real 3D model to simulate the surface of the moon at Aitken Basin. In this year's SEE event, we assumed the moon to be an idealized, spheroidal model; actually it might also consider terrain features such as mountains or ditches;

(3) If next year's team plans to use STK to model the physical entities in 3D model. The orientation of the entities must be considered, so besides the three attributes of PhysicalEntity, our federate also needs to subscribe to the orientation attribute;

(4) Do integration test on MÄK as early as possible. We recommend MÄK set up a web viewer like Pitch does.

10.5 Conclusion

In this paper, we described our experience in designing of the MWSU communication satellites federate which provides communication support to entities involved in a lunar mission. This work has two main contributions. First, the MWSU Sim team developed a HLA interface implementation for the AGI STK component. The other contribution is for educational purpose: during the 4-month period, students in the development team learned HLA standard, programmed using STK Components, used different RTI software implementations and collaborate with students from other universities. Each team in SEE 2016 demonstrated the capabilities that modeling and simulation can bring to solve difficult problems in a challenging environment, from all across the globe, despite language barriers and time zone challenges. By sharing the experience of the Midwestern State University Student Team in participating to the SEE, we hope to guide new teams to participate the coming year's SEE event.

Acknowledgements The authors would like to thank the 2012 UAHuntsville team. The development of MWSU communication satellites federate benefited a lot from their legacy code [5]. Building on top of those nicely written and documented code greatly reduced our developing effort. We would also like to thank James Taylor, for organizing weekly teleconferences, Neil Cameron for hosting the event. Thanks should also be given to our NASA sponsors, Edwin Zack Crues, Dan Dexter, Michael Conroy and Daniel A. O'Neil. They were pulled upon hard by their day jobs and were able to make sure things were running for us to use and provided expertise whenever possible.

We also want to thank the industrial partners, Pitch Technologies and VT MÄK for their support to the SEE 2016.

References

1. HLA tutorial. http://www.pitchtechnologies.com/wp-content/ uploads/2014/04/TheHLAtutorial.pdf. Accessed 15 May 2016.

2. Falcone, A., Garro, A., Longo, F., & Spadafora, F. (2014). Simulation exploration experience: A communication system and a 3D real time visualization for a moon base simulated scenario. In *2014 IEEE/ ACM 18th International Symposium on Distributed Simulation and Real Time Applications (DS-RT)* (pp. 113–120).

3. AGI STK components. http://www.agi.com/products/stk/modules/ default.aspx/id/stk-components. Accessed 15 May 2016.

4. Walker, J. G. (1984). Satellite constellations. *Journal of the British Interplanetary Society, 37*, 559–572.

5. Bulgatz, D., Heater, D., O'Neil, D. A., Norris, B., Schricker, B. C., & Petty, M. D. Design and implementation of lunar communications satellite and server federates for the 2012 SISO Smackdown Federation. In *Proceedings of the Fall 2012 Simulation Interoperability Workshop* (pp. 10–14).

Wireless Body Sensor Network for Monitoring and Evaluating Physical Activity

11

Leonardo Schick, Wanderley Lopes de Souza, and Antonio Francisco do Prado

Abstract

Since the physical inactivity is one of the four main risk factors for the incidence of Non-Communicable Diseases, the World Health Organization has stimulated the creation of actions to promote regular physical activity practices. The Brazilian Ministry of Health established a physical activity program, where people perform physical activities under the supervision of health professionals. In order to real-time monitoring individuals during their physical activity practices we developed an ubiquitous computing environment. This environment is composed of three modules that automatically collect physiological data, and provide indicators which will support public policies for promoting physical activity. This paper presents this environment focusing on the Wireless Body Sensor Network module, and its simulation that was performed using the OMNet++ 5 tool. The simulation results showed a packet loss due to the simultaneous delivery of packets to the coordinator node, which caused a network bottleneck. In order to deal with this problem, we designed a communication protocol to be run at the application layer that allows the host nodes to send packets in turns, avoiding this way the packet loss.

Keywords

Ubiquitous Computing • Wireless Body Sensor Networks • Healthcare Informatics • Physical Activity

11.1 Introduction

The World Health Organization (WHO) has indicated a significant increase in the occurrence of Non-Communicable Diseases (NCDs) in the past ten years [1]. NCDs such as hypertension, obesity, diabetes mellitus, and cancer, were the cause of 38 million of deaths worldwide in 2012, which corresponded to 68% of mortality in that year [2]. WHO has also indicated that the physical inactivity is one of the four main leading risk factors for the development of NCDs [3].

The practice of regular physical activity, besides being fundamental to energy balance and weight control, can lead to: an improvement of the muscular and cardiorespiratory fitness; an improvement of the functional health; and a reduction of the risks of hypertension, coronary heart disease, stroke, diabetes, breast and colon cancer, and depression [4].

According to [5], in Brazil 46% of the adults do not practice at least 150 minutes of physical activity a week, the minimum amount recommended by WHO. To deal with this problem, the Brazilian Ministry of Health established a Physical Activity Program at the Basic Health Units (UBSs), which are small health units integrated to the municipal health system. These UBSs keep Physical Activity Groups (PAGs), where people participate in physical activity sessions under the supervision of health professionals.

L. Schick (✉) • W.L. de Souza • A.F. do Prado
Department of Computing (DC), Federal University of São Carlos (UFSCar), São Carlos, SP, Brazil
e-mail: leonardo.schick@dc.ufscar.br; desouza@dc.ufscar.br; prado@dc.ufscar.br

© Springer International Publishing AG 2018
S. Latifi (ed.), *Information Technology – New Generations*, Advances in Intelligent Systems and Computing 558, DOI 10.1007/978-3-319-54978-1_11

Fig. 11.1 UCEMEPA modules

These sessions generate a large amount of data that needs processing to evaluate the effectiveness of these activities.

In order to manage these physical activities and to evaluate them, the Ubiquitous Computing Group (UCG) of Federal University of São Carlos (UFSCar) developed the Ubiquitous Computing Environment for Monitoring and Evaluating Physical Activity (UCEMEPA) [6]. As illustrated in Fig. 11.1, this environment is composed of three modules: Physical Activity Information System (PAIS); Collective Monitoring Server (CMS); and Wireless Body Sensor Network (WBSN).

This paper focuses on the WBSN module, and is further structured as follows: section II discusses some related work; section III gives an UCEMEPA overview and introduces the WBSN module; section IV details the WBSN module and discusses the PAGs' monitoring approach; section V reports the WBSN simulation and its evaluation; and section VI presents our concluding remarks and gives recommendations for future work.

11.2   Related Work

The literature presents some environments that employ ubiquitous technologies and wireless communication networks to monitor individuals.

A framework is presented in [7], which uses accelerometers to monitor the daily activities of an individual, sending the collected data to a system that has been deployed on a smartphone or a computer. This system calculates the amount of calories consumed by the individual during the day, and allows to visualize the optimal amount of calories still to be consumed that day in order to maintain or reduce weight. This framework aims to prevent obesity, allowing its users to control their daily caloric expenditure.

A system is proposed in [8] for rescuing mainly elderly or disabled individuals when disaster occurs. This system monitors vital sign, location and attitude of these people using multiple sensors attached to their bodies. The collected physical and psychological data are transferred to mobile devices through Bluetooth, and these devices are connected to Internet terminals through WiFi or Wireless LAN. Finally, these data are sent to a health information center.

A platform is presented in [9] for continuous monitoring of the elderly and people with special needs. Sensors are connected to an Arduino board for monitoring the heartbeat, blood pressure, blood oxygen saturation, body temperature,

and the indoor and outdoor location of the individual via GPS. These physiological data are sent to a smartphone, which retransmits them to a server for processing and visualization by health professionals that can issue alerts if any anomaly is detected.

A platform is presented in [10], which is composed of a blood pressure sensor, an analog front-end circuit, and a micro controller. These components are attached to a patient's leg for continuously measuring his blood pressure while he carries out his daily tasks. The collected data is sent to a mobile device, and are analyzed in order to identify chronic venous insufficiency.

All these related works aim to monitor a particular disease and/or a specific group of people. The main distinguishing feature of our work is that WBSN module is part of UCEMEPA environment, which was designed to support the WHO global strategy of preventing NCDs. This environment provides a computational support, not only for monitoring physical activity in the overall population, but also to evaluate the effectiveness of this practice for preventing these diseases. Additionally, the one-to-many approach employed in this module allows to save resources, as opposed to the one-to-one approach employed in these related works.

11.3   Ubiquitous Computing Environment for Monitoring and Evaluating Physical Activity

UCEMEPA is an environment designed to collect data either indoors (e.g., covered sport courts, recreation centers) or outdoors (e.g., walking lanes, open courtyards). Figure 11.2 illustrates a general view of UCEMEPA architecture.

PAIS is a web application designed to register the PAGs and its participants, and to persist the data collected in physical activity sessions. This module allows the health professionals to select the physical activities to be performed in each session, and to visualize by means of tables and graphics the data collected by the other modules. Figure 11.3 shows the PAIS use case diagram with their users and their main functionalities.

CMS is an Android mobile application [11] that runs on a smartphone or tablet, which is carried by the health professional in charge of the physical activity session. This module acts as a gateway, and employs wireless communication interfaces in order to acquire data from the WBSN module,

Fig. 11.2 UCEMEPA
architecture

Fig. 11.3 PAIS use case
diagram

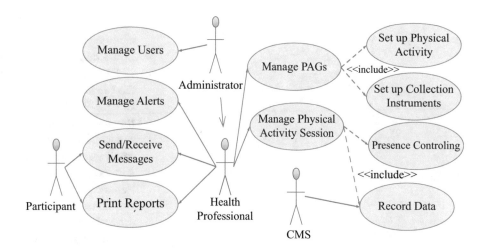

Fig. 11.4 CMS use case diagram

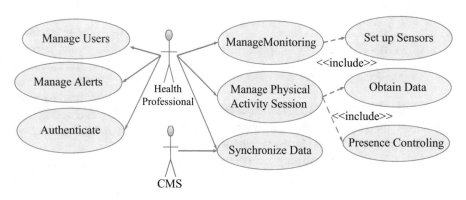

and to send them to the PAIS module. Figure 11.4 shows the CMS use case diagram with their users, and their main functionalities.

WBSN is composed by the interconnection of sensors, and is carried by each PAG participant in order to collect their physiological data, and to send them to the CMS module.

The interaction between the UCEMEPA modules starts with the CMS identifying the active WBSNs. Then, it communicates with these WBSNs in order to collect, to analyze, and to send data to PAIS. Finally, PAIS performs the data persistence, providing these data for visualization. If CMS or PAIS identifies a critical health condition, an alert message is sent to the health professional.

11.4   Wireless Body Sensor Network

Wireless Body Sensor Networks, also referred to as Body Area Networks (BANs), allow monitoring users continuously at real-time, while they are engaged in daily life activities. BANs are mainly composed of wearable or implantable nodes, and one or more wireless networks, which are used to transport collected data from the sensor nodes [12].

Table 11.1 WBSN's sensors

Sensor	Physiological data
ECG	Heartbeat
GSR	Galvanic skin response
SPO2	Pulse and blood oxygen saturation
ACC	Acceleration

Table 11.2 Main features of ZigBee and BLE

	ZigBee	BLE
Radio Frequency	868/915MHz, 2.4GHz	2.4GHz
Data Rate	250Kbps	1Mbps
Distance	10 to 200 meters	10 to 100 meters
Application Throughput	< 0.1Mbps	< 0.2Mbps
Network topology	Star or Mesh	Star only

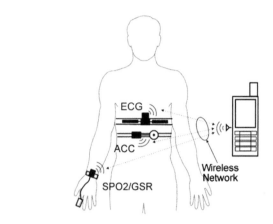

Fig. 11.5 WBSN usage scenario

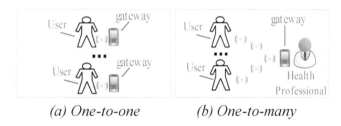

(a) One-to-one   *(b) One-to-many*

Fig. 11.6 Monitoring approaches. (**a**) One-to-one. (**b**) One-to-many

Since the WBSN module must be small and light, so it does not impair the physical activity practices, the open-source platform Arduino [13] was chosen because it allows the construction of Wearable Computing architectures by attaching sensors in pieces of clothing (e.g., gloves, and shirts) [14].

According to [15], some of the most important physiological data to be collected are: heartbeat, blood pressure, body temperature, glycemia, blood oxygen saturation, and respiratory rate. Table 11.1 depicts the sensors that were chosen to compose the WBSN, while Fig. 11.5 illustrates one of its usage scenarios.

Since the presence of a health professional on the PAGs' physical activity sessions is mandatory, the participants do not need to carry mobile devices. Therefore, the one-to-one monitoring approach was replaced by the one-to-many monitoring approach, where the health professional carries the mobile device, which acts as the network coordinator node. These approaches are illustrated in Fig. 11.6.

Besides the economic advantage of using a single device to monitor several individuals, the one-to-many approach brings the following benefits: a less intrusive monitoring,

since the participants will not need to carry other devices apart the sensors; a central management of the sensors; and an easy communication between the mobile device and the server. According to [16], the wireless network technologies recommended to be used with this approach are Bluetooth Low Energy (BLE) [17] and ZigBee [18].

BLE operates in the ISM 2.4 GHz frequency band using the Frequency Hopping Spread Spectrum (FHSS) [16]. The Bluetooth devices are classified into two categories: *Bluetooth Smart*, devices only compatible with BLE that support the Low Energy (LE) controller, and are usually small sized; and *Bluetooth Smart Ready*, devices that support both controllers, the LE and the legacy Basic Rate/Enhanced Data Rate (BR/EDR).

ZigBee operates at 2.4 GHz frequency band, and has a defined rate of 250 Kbps, which is best suited for intermittent data transmission from a sensor or an input device [16]. It has low power consumption, low cost, and ZigBee devices can transmit data over long distances through a mesh network, using intermediate devices to reach more distant ones.

The main features of ZigBee and BLE are well described in [19] and summarized in Table 11.2.

ZigBee was the chosen technology to be employed in the WBSN module, since it has low cost, low battery consumption, and mainly a greater communication range compared with BLE technology.

11.5 WBSN Simulation and Evaluation

In order to have an evaluation of the WBSN module behavior prior to its implementation, the communication between this module and the CMS module was simulated and analyzed using the following tools: OMNet++ 5, an open-source component-based C++ simulation library and framework [20]; and INET 3.4, an OMNet++ framework that provides models for wired, wireless and mobile networks [21].

The WBSN simulation scenario was composed of ten host nodes representing the PAG participants, and one coordinator node representing the health professional. In order to simulate a walking session, the coordinator node was placed in the center of a square area with the host nodes moving around it. Figure 11.7 illustrates this WBSN simulation scenario running in OMNet++, while Table 11.3 depicts the parameters employed in this simulation using User

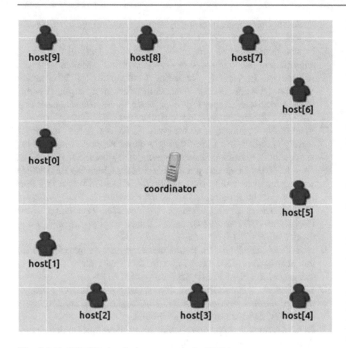

Fig. 11.7 WBSN simulation scenario in OMNet++

Table 11.3 WBSN simulation parameters in OMNet++

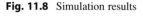

Parameter	Value
Traffic type	UDP
Packets size	100 bytes
Number of nodes	11
Nodes movement speed	1.5 m/s
Total simulation time	600 minutes

Fig. 11.8 Simulation results

Datagram Protocol (UDP) [22] for the communication between the coordinator node and the host nodes.

The simulation was conducted into 10 steps of 60 min each, starting with the square's side set to 10 m and increasing this size by 10 m in the next step, with each node sending a packet to the coordinator node every 30 sec. Figure 11.8 shows the simulation results by means of the packet delivery ratio as function of the square's area.

According to Fig. 11.8(a), there is a packet loss in all simulation steps, attaining almost 61% in the worst case. This loss is due to the simultaneous delivery of packets to the coordinator node, which causes a network bottleneck.

In order to deal with this problem, we designed a communication protocol to be run at the application layer that allows the host nodes to send packets in turns, avoiding this way the packet loss. The coordinator node controls the sending of packets by requesting them in turns for each host node, which must respond before a new request is sent to the next node. This turnaround time was set to 3 sec to be compatible with the send interval.

According to Fig. 11.8(b), this protocol avoids the packet loss in all simulation steps.

11.6 Conclusion

The UCEMEPA is a computational environment for real-time monitoring of individuals during their physical activity practices. This environment was conceived to support the physical activity programs of the Brazilian Ministry of Health. This paper presented the UCEMEPA modules focusing on the WBSN module.

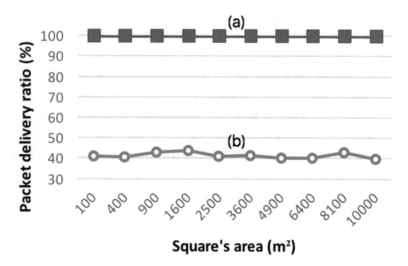

WBSN is composed of sensors and an Arduino board to collect physiological data from PAG participants during a physical activity session. These data are sent to a mobile device carried by the health professional responsible for that session.

Besides the economic advantage of using a single mobile device to monitor several individuals, this one-to-many approach allows a less intrusive monitoring, a central management of the sensors, and an easy communication between the mobile device and the server.

In order to have an evaluation of the WBSN module behavior prior to its implementation, the communication between this module and the CMS module was simulated using the OMNet++ 5 tool.

The simulation results showed a packet loss due to a network bottleneck caused by the simultaneous delivery of packets to the coordinator node. To solve this problem, we designed a communication protocol to be run at the application layer that allows the host nodes to send packets in turns.

As future works, we intend to implement the WBSN module, to deploy it within the UCEMEPA environment, and to evaluate its usability perceived by health professionals and PAGs' participants, during physical activity sessions.

Acknowledgments Our thanks to the Coordination for the Improvement of Higher Education Personnel (CAPES) and to the São Paulo Research Foundation (FAPESP) for sponsoring this research.

References

1. World Health Organization (WHO). (2010). *Global status report on noncommunicable diseases 2010*. ISBN 978924068645-8, 176 p, Geneva.
2. World Health Organization (WHO). *NCD mortality and morbidity*. Available: http://www.who.int/gho/ncd/mortality_morbidity/en. Accessed Sept 2016.
3. World Health Organization (WHO). (2009). *Global health risks – Mortality and burden of disease attributable to selected major risks*, ISBN 9789241563871-8, 27 p. Geneva.
4. World Health Organization (WHO). *Physical activity*. Available: http://www.who.int/ mediacentre/factsheets/fs385/en. Accessed Sept 2016.
5. Brazilian Institute of Geography and Statistics (IBGE). (2014). *Pesquisa nacional de saúde 2013. Percepção do estado de saúde, estilos de vida e doenças crônicas. Brasil, grandes regiões e unidades da federação.* ISBN 978852404334-5, 181 p. Rio de Janeiro.
6. Nunes, D. F. S., et al. (2012). *UCEMEPA: Ubiquitous computing environment for monitoring and evaluating physical activity.* Annals of Eighteenth Americas Conference on Information Systems (AMCIS 2012), Association for Information Systems (AIS) Electronic Library (AISeL), paper 17, http://aisel.aisnet.org/amcis2012/proceedings/AdoptionDiffusionIT/17/, Seattle.
7. Alrajeh, N. A., Lloret, J., & Canovas, A. (2014). A Framework for obesity control using a wireless body sensor network. *International Journal of Distributed Sensor Networks, 10*(7), 06, ISSN 1550-1329.
8. Arai, K. (2014). Rescue system with vital sign, location and attitude sensing together with traffic condition, readiness of helper monitoring in particular for disabled and elderly persons, In: *Proceedings of 11th International Conference on Information Technology: New Generations (ITNG 2014), IEEE Xplore Digital Library, Electronic.* ISBN 978147993188-0, pp. 155–160.
9. Hussain, A. et al. (2015). Health and emergency-care platform for the elderly and disabled people in the Smart City. *Journal of Systems and Software, 110*, 253–263, ISSN 0164-1212.
10. Li, R. et al. Telemedical wearable sensing platform for management of chronic venous disorder, *Annals of Biomedical Engineering*, 44(7), 2282–2291, ISSN 0090-6964.
11. Google. *Develop Apps*. Available: https://developer.android.com/develop. Accessed Sept 2016.
12. Patel, M., & Wang, J. (2010). Applications, challenges, and prospective in emerging body area networking technologies, *IEEE Wireless Communications*, 17, 80–88, ISSN 1536-1284.
13. Arduino. *What is Arduino?*. Available: https://www.arduino.cc/en/Guide/Introduction. Accessed Sept 2016.
14. Bonato, P. (2010). Wearable sensors and systems. *IEEE Engineering in Medicine and Biology Magazine*, 29(3), 25–36, ISSN 0739-5175.
15. Otero, J., Gómez, A. (2007). *Integración de dispositivos biomédicos en sistemas de teleasistencia*, Technical Report CESGA-2007-002, 226.
16. Afonso, J. A.. Maio, A. J. F., & Simoes, R. (2016). Performance evaluation of bluetooth low energy for high data rate body area networks. *Wireless Personal Communications, 90*, 121–141, ISSN 0929-6212.
17. Bluetooth Special Interest Group (SIG). (2014). Bluetooth core specification v4.2, 2722 p.
18. ZigBee Alliance. (2012). *ZigBee specification*. ZigBee Document 053474r20, 604 p.
19. Touati, F. et al. (2015). An experimental performance evaluation and compatibility study of the bluetooth low energy based platform for ECG monitoring in WBANs. *International Journal of Distributed Sensor Networks, 11*(9), 12, ISSN 1550-1329.
20. OMNet++. OMNeT++ Discrete Event Simulator. Available: https://omnetpp.org/ . Accessed Oct 2016.
21. INET Framework. *INET Framework for OMNet++/OMNEST*. Available: https://omnetpp.org/doc/inet/api-current/neddoc/index.html. Accessed Oct 2016.
22. Postel, J. (1980). *User datagram protocol* (p. 3). Marina Del Rey: RFC 768, USC/Information Sciences Institute.

Techniques for Secure Data Transfer

Jeffery Gardner, Jr.

Abstract

The Advanced Encryption Standard better known as the AES algorithm is a symmetric (uses the same key to encrypt and decrypt) cryptographic technique used in most of today's classified and unclassified data transfers. The AES algorithm provides data transfers with layers of security through its mathematical complexity. Alongside this technique of encryption, the act of concealing data within different objects is now becoming an essential component in the art of secure data transfer. This method of hiding secret messages within a file type is known as Steganography. Even though these two elements of secure data transfer differ, they both share the same objective namely to protect the integrity of the data. This paper provides an explanation of Steganography and AES algorithm and how they can be used together to enhance the security of data. The experiments that this paper demonstrates used the AES-256 algorithm.

Keywords

Advanced Encryption Standard • Steganography • Algorithm • S-Box • Cryptography

12.1 Introduction

The Advance Encryption Standard algorithm (AES) is the standard encryption technique for today's classified and unclassified data transfers. The AES algorithm utilizes the Rijndael algorithm [1] that is a symmetric block cipher that operates on blocks of 128 bits giving by the required standard. As a standard, three different key lengths supports AES. They are AES-128, AES-192, and AES-256. The security of the AES algorithm heavily relies on the mathematical complexity of the five different layers of the giving components. The layers of the AES algorithm include the Key Expansion, Substitution Bytes, Shift Rows, Mixed Columns and Add Round Key. These layers will provide the AES algorithm with the necessary protection for secure data transfer.

J. Gardner, Jr. (✉)
Department of Computer Science, Grambling State University, Grambling, LA, USA
e-mail: jgardnerjr14@yahoo.com

Steganography has existed in the "secure transfer" community ever since 440 B.C., where Greek Kings shaved the head of a villager and placed important information on his/her scalp. The villager was then sent back to the village after his/her hair had grown back. With this method in mind, steganography can easily be described as providing security for sensitive data through the means of obscurity.

Steganography compared to cryptography poses a significant difference in objective. Encryption provides communication with security through the use of scramble letters and symbols. With cryptography, any adversary can clearly see that the transmitted message is encrypted and can be passed along for further cryptanalysis or can perform actions that can hinder the transmitted message without the secret key.

This paper will demonstrate how to encrypt data using the AES algorithm and utilizing steganography to embed information in a media file to provide obscurity for elementary data transfer. Section 12.2 explains the design of the AES algorithm and Sect. 12.3 discusses the previously known attacks of AES algorithm. A Sect. 12.4 describes how

© Springer International Publishing AG 2018
S. Latifi (ed.), *Information Technology – New Generations*, Advances in Intelligent Systems and Computing 558, DOI 10.1007/978-3-319-54978-1_12

steganography is used. Section 12.5 discusses related works. Section 12.6 will discuss the application of how the AES encryption and steganography used together for an elementary data transfer.

12.2   Advanced Encryption Standard

The AES algorithm is not only equipped for software purposes but also with providing security to hardware devices. The AES algorithm performs bitwise operations, making encryption for hardware efficient.

Figure 12.1 displays the number of rounds that are generated by each particular key length provided by the AES standard, where

Nk is the number of 32-bit words generated from the key length. These words are known as the sub-keys.
Nb represents the number of 32-bit words, made from the state (the AES standard operates on *Nb* = 4).
Nr is the number of rounds that are utilized for the given key length.

The five components that are responsible for the robustness of the AES algorithm are Key Expansion, Substitution Bytes, Shift Rows, Mix Columns, and Add Key Round. In the AES algorithm, each input value represents in the polynomial form in the $GF(2^8)$.

	NK	Nb	Nr
AES 128	4	4	10
AES 192	6	4	12
AES 256	8	4	14

Fig. 12.1 Standard key combinations of the AES algorithm

Key Expansion is responsible for generating the key schedule and is processed before encrypt and decrypt ciphers. In this process sub-keys are created. For the experiments in this paper, the AES-256 was utilized to test the highest level of security. Because of the AES 256, there were 60 words generated from the key expansion. Each key length of the AES encryption uses the formula in Eq. (12.1) to produce the proper number of sub-words.

$$Nb\,(Nr-1) \qquad (12.1)$$

The substitution bytes are calculated using a pre-computed table known as the S-Box. For testing purposes, the S-Box was developed using C-programming language that was capable of accessing the hardware that derived S-Box values, which sits at the core of the security in the AES algorithm. The process accomplished to obtain a sound understanding of the importance of this component and how values were derived. In developing this individual component, the S-Box is efficient and provides non-linearity and performs a one-for-substitution of an input byte value. To calculate the substitution byte of the data value, the S-Box uses two transformation methods known as the modular inverse and the affine transformation. The modular inverse is responsible for finding the multiplicative inverse of the giving input byte by utilizing the extended Euclidean algorithm.

After the inverse of the giving input value has been obtained, it was then sent through the affine transformation where the S-Box value is the result. Figure 12.2 displays the structure of affine transformation matrix operation. The affine transformation is accomplished by using the formula {out byte} = M {in byte} XOR {v} where *M* is a pre-determined matrix and *v* is the vector of the multiplicative inverse byte.

Figure 12.3 demonstrates the use of the multiplicative inverse and affine transformation to obtain an S-Box value. In Fig. 12.3, the hexadecimal value of {54} was the input byte. The data byte of {54} was processed through the multiplicative inverse transformation finding a value that congruent to:

$$1\ mod\ x^8+x^4+x^3+x+1 \qquad (12.2)$$

In this operation shown in Eq. (12.2), the Galois Field used to represent the polynomial form for the input byte. We represent these values in the $GF(2^8)$ which is the standard field for the AES algorithm. Using the irreducible polynomial $x^8 + x^4 + x^3 + x + 1$ with a degree of 8 that is provided

$$\{out\ byte\} = \begin{bmatrix} 1&0&0&0&1&1&1&1 \\ 1&1&0&0&0&1&1&1 \\ 1&1&1&0&0&0&1&1 \\ 1&1&1&1&0&0&0&1 \\ 1&1&1&1&1&0&0&0 \\ 0&1&1&1&1&1&0&0 \\ 0&0&1&1&1&1&1&0 \\ 0&0&0&1&1&1&1&1 \end{bmatrix} \{in\ byte\} \oplus \begin{bmatrix} 1 \\ 1 \\ 0 \\ 0 \\ 0 \\ 1 \\ 1 \\ 0 \end{bmatrix}$$

Fig. 12.2 Structure of affine transformation matrix operation

Fig. 12.3 Multiplicative inverse and affine transformation

Input Byte (54) → *Inverse GF(2⁸)* → B' (4C) → *Affine Transformation* → S-Box Value (20)

by the norm is a modulo that corresponds to the multiplication of two polynomials. The value B' is the result of the multiplicative inverse. The result of this transformation is the hexadecimal value of {4C}. Finally, the multiplicative inverse result was used to obtain the S-Box value in the affine transformation. In this transformation, a simple matrix multiplication is applied using a pre-determined 4×4 matrix found in Fig. 12.2 and the multiplicative inverse result. The result of these two transformations is the S-Box value.

The inverse substitution byte function operates similar to the forward substitution byte function found in the encryption process providing non-linearity between any two values. However, the inverse substitution method utilizes the affine transformation and multiplicative inverse functions in reverse order. The Affine Transformation is the lead off transformation in this operation and applied first by using the inverse Affine Transformation matrix as in Fig. 12.4.

Figure 12.5 shows the result of this operation is the inverse of the original input value. The value obtained from the affine transformation is then passed to the multiplicative inverse function to obtain the original value.

In this project, experiments took place that tested the speeds between the S-Box and inverse S-Box being accessed by the pre-computed tables as used in majority of AES programs and the hardcode of the various steps in deriving the S-Box and Inverse S-Box values. After conducting ten experiments it was proven that the hardcode of the S-Box and inverse S-Box operations took on average about 7.918 seconds. With the precomputed tables, this same operation took on average about 4.034 seconds. The results of this experiment conveyed that using the per-computed tables is on average 3.839 seconds faster than the hardcoded forward and inverse S-Box operation.

The Shifts Rows component of the AES algorithm is the second procedure of the encryption process and is responsible for shifting each row by a certain number of spaces to the left.

Figure 12.6 displays the basic transformations of the shift rows function. The first row is not shifted at all, the second row is shifted one space to the left, the third row is shifted two spaces to the left, and the fourth row is shifted three spaces to the left. The shift rows operation increases the properties of linearity influences.

Unlike the shift rows operation being the first procedure of the encryption process, the inverse shift rows operation is first of the decryption process. Similar to the shift rows operation in the encrypt process.

Figure 12.7, demonstrates the basis transformations of the inverse shift row operation. The first row is not shifted, the second row is shifted once, the third row is shifted twice and the fourth row is shifted three times. However, the shifts in the decrypt process will be done to the right as.

The Mixed Columns component is responsible for operating on each column of the state. The values of the *state* are polynomials of the $GF(2^8)$. Each column of the *state* undergoes a transformation by multiplying modulo $x^4 + 1$ against an established polynomial resulting in the matrix shown in Fig. 12.8. Each column is multiplied against the 4×4 matrix. This operation is processed until the *Nr-1* round.

$$\{out\ byte\} = \begin{bmatrix} 0 & 0 & 1 & 0 & 0 & 1 & 0 & 1 \\ 1 & 0 & 0 & 1 & 0 & 0 & 1 & 0 \\ 0 & 1 & 0 & 0 & 1 & 0 & 0 & 1 \\ 1 & 0 & 1 & 0 & 0 & 1 & 0 & 0 \\ 0 & 1 & 0 & 1 & 0 & 0 & 1 & 0 \\ 0 & 0 & 1 & 0 & 1 & 0 & 0 & 1 \\ 1 & 0 & 0 & 1 & 0 & 1 & 0 & 0 \\ 0 & 1 & 0 & 0 & 1 & 0 & 1 & 0 \end{bmatrix} \{in\ byte\} \oplus \begin{bmatrix} 1 \\ 0 \\ 1 \\ 0 \\ 0 \\ 0 \\ 0 \\ 0 \end{bmatrix}$$

Fig. 12.4 Inverse affine transformation matrix

Fig. 12.5 Inverse of the input operation

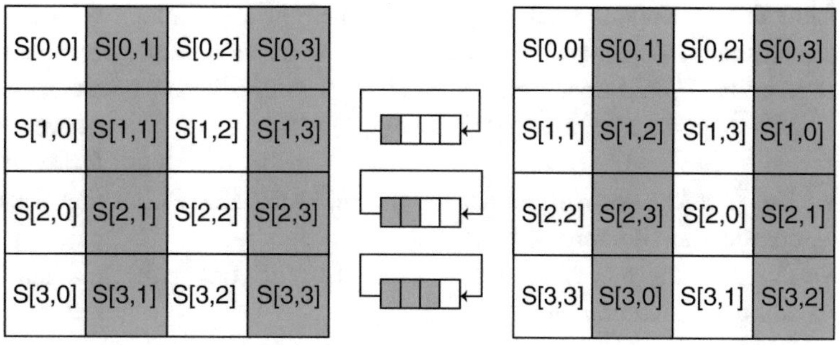

S-Box Value {20} → *Affine Transformation* → B' {4C} → *Inverse GF(2⁸)* → Input Byte {54}

Fig. 12.6 Basic transformations of the shift rows function

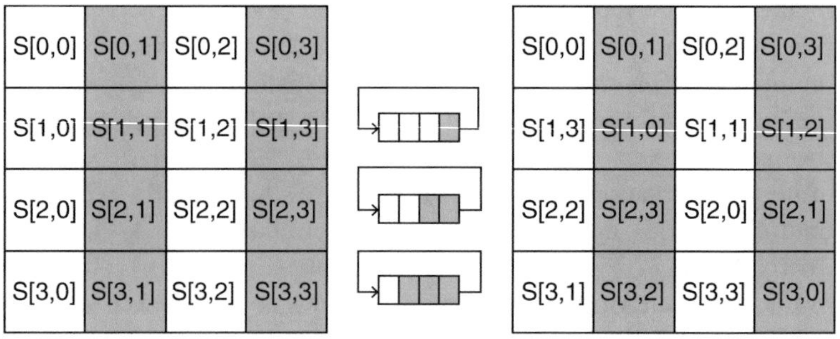

Fig. 12.7 Basis transformations of the inverse shift row operation

$$\begin{bmatrix} S'[0,c] \\ S'[1,c] \\ S'[2,c] \\ S'[3,c] \end{bmatrix} = \begin{bmatrix} 02 & 03 & 01 & 01 \\ 01 & 02 & 03 & 01 \\ 01 & 01 & 02 & 03 \\ 03 & 01 & 01 & 02 \end{bmatrix} \begin{bmatrix} S'[0,c] \\ S'[1,c] \\ S'[2,c] \\ S'[3,c] \end{bmatrix}$$

Fig. 12.8 Transformation matrix

$$\begin{bmatrix} S'[0,c] \\ S'[1,c] \\ S'[2,c] \\ S'[3,c] \end{bmatrix} = \begin{bmatrix} 0e & 0b & 0d & 09 \\ 09 & 0e & 0b & 0d \\ 0d & 09 & 0e & 0b \\ 0b & 0d & 09 & 0e \end{bmatrix} \begin{bmatrix} S'[0,c] \\ S'[1,c] \\ S'[2,c] \\ S'[3,c] \end{bmatrix}$$

Fig. 12.9 4×4 matrix used against the four byte input values of each column

With the inverse mixed columns used in the decryption process, the same logic is used in this operation as it was used in the encryption operation. Fig. 12.9, represents the 4×4 matrix used against the four byte input values of each column. The Add Round Key was responsible for adding the Round Key that was obtained from the key schedule to the *state* utilizing a bitwise XOR operation.

12.3 AES Attacks

The only known attacks on the AES algorithm are side-channel attacks and attacks on weaknesses found in implementation or key management. The AES algorithm must be implemented in the correct strategic way as required by the AES standards. The first key-recovery attacks on full AES were due to Bogdanov et al. [2]. The attack is a biclique attack and is faster than brute force by a factor of about four. It requires 2126.2 operations to recover an AES-128 key. Bogdanov et al. concludes the following results.

- The first key recovery attack on the full AES-128 with computational complexity $2^{126.1}$.
- The first key recovery attack on the full AES-192 with computational complexity $2^{189.7}$.

- The first key recovery attack on the full AES-256 with computational complexity $2^{254.4}$.
- Attacks with lower complexity on the reduced-round versions of AES not considered before, including an attack on 8-round AES-128 with complexity $2^{124.9}$.
- Preimage attacks on compression functions based on the full AES versions

The National Security Agency (NSA) expects that 256 bit AES keys may be cracked by 2018.

12.4 Steganography

Steganography conceals the secret message in plain sight through the use of a cover object. Steganography can be used for concealing and protecting the objects. Concealing is being able to hide the secret message in a cover object without an adversary being aware of any information embedded. Protecting, however, is used in situations where media data has to be protected. Many people use this technique of steganography, such as digital watermarking, in order to protect the original works of authorship to author.

Although steganography has various techniques, the experiments of this paper utilize the least significant bit insertion method for each trial. The two common used methods in steganography are the least significant bit insertion method and the Discrete Cosine Transformation (DCT) method. The least significant bit insertion method is the least complex of the two, being capable of accessing the RGB of each pixel (picture element). The objective of this technique is to swap the least significant bit of each color in every pixel with the bit of the secret data. Because of its low level of complexity the LSB insertion method does not provide efficient robustness. This is vulnerable to transformations such as modifications. In addition to file compressions, a jpeg file format could not withstand this technique, losing information for the extracting process. For the experiments of this paper, a bmp file format was utilized. However, complex mathematical algorithms can be developed to provide scattered positions for the secret data.

Steganography is profoundly a unique process. Unlike other tests (based on speed), the steganography demonstration in this paper is measured by three characteristics. These trials had to satisfy these conditions. First, after embedding the secret data within the cover object is to protect the integrity of the Stego object. Once the Stego object is sent through the data communication medium, a technique is needed to protect the embedded data at the point of retrieval. Next, the stego object must be indistinguishable from the cover object. Once the secret information has been embedded within the cover object, the Stego object should be completely identical to the cover object, displaying no signs of variations. Finally, the extracting process should be accurate. After the receiver has received the stego object and performed the extracting operation, the message should be returned in the original form (sent from source). These characteristics will determine the success of the stenographic operation.

12.5 Related Work

Understanding the basics of cryptography, finite fields, and steganography were important to the development of the AES algorithm and steganography. The reference [3–6] provides the basics. Manoj et al. [7] performed AES based steganography. Manoj utilized steganography with biometrics. The work uses skin tone for the embedding process. The secret data was embedded in one of the high-frequency sub-band of DWT. Data hiding was achieved by cropping an image interactively.

The authors in [8] discussed the secure steganography approach using AES. They used AES-128 and utilized least two significant bits for transformation. The current model uses 256 bits and S-Box for steganography which differs from Ramaiya [8]. The image analysis is characteristic in both cases.

Recently, Arjun et al. [9] discussed steganography based AES algorithm. The authors presented limited literature on AES and usage of LSB function was not seen. The authors used bit plane complexity for steganography technique. This method divided the image into bit planes before using AES algorithm.

12.6 Implementation and Results

In our implementation experiments, a simple 55 byte text file was utilized to demonstrate the operations of enhancing the security of a secret message using cryptography and steganography, which are implemented using the step-down procedures below.

- Because the AES-256 is the desired cryptographic function, the first step is to create a 32-bit key and encrypt the data using the AES-256 algorithm.
- Next, the cover object is loaded for the placement of the encrypted message.
- Once the image is uploaded, the intensity of each pixel is then entered into the access domain. During this process, the Least Significant Bit (LSB) of each required pixel of the Red, Green, Blue color is available for the embedding alteration.
- A stego key is then created in order to conceal the encrypted message. The stego key will be used to grant permission to proceed with extracting process.
- The secret message is then converted into its binary form in so that each bit can be embedded into the Least Significant Bit binary value of each color.
- The message length of the encrypted message is obtained. This will allow the receiver to recover the positions of where the secret message starts and ends in the extracting process.
- Each bit of the encrypted file is then placed at the Least Significant Bit value of each color in the pixels altering the color value slightly.
- The stego object is then sent to the receiver and once received; the receiver will use the stego key in order to perform the extracting operation of the encrypted message.
- After the encrypted message is extracted it is then decrypted using the inverse operation of the AES algorithm.

By using the above procedures, Figs. 12.10a, 12.10b, 12.10c, 12.10d below displays the results of the experiment.

- Figure 12.10a displays the original message (*Hello World, Welcome to the place of Computer Science.*) that was used for the encrypting process using the AES-256 cryptographic function. Since the AES-256 was utilized a key of 32 bytes was created. Note: it is never advised to have the key set in the program but for demonstration purposes it was created inside the program. Once finished, the encrypted message was generated which is found in Fig. 12.10b.
- Figure 12.10c displays two identical images. However, these two images are different. The image on the left is the cover object or the media file that was loaded for the placement of the encrypted message. In this testing, the cover object was a bitmap image file of about 786.5 KB in size. The intensity of each pixel was entered into the access domain for the embedding process.
- The stego key was created in order to allow the program to process the extracting process. After the stego key was

Fig. 12.10 (**a**) The original message for experiment (**b**) Encrypted message for experiment (**c**) Identical images with the encrypted message embedded within the stego object (**d**) The original message is returned after the extracting process

a <u>Original Message</u>

b <u>Encrypted Message</u>

c <u>Medial File</u>

Cover Object   Stego Object

d <u>Decrypted Message</u>

created, the encrypted message was the converted to its binary form in order to be embedded into each color inside of each pixel.

- The Least Significant Bit (LSB) of each required pixel of the Red, Green, Blue color was available for the embedding alteration. Here a Stego key was created in order to conceal the secret message.
- The message length of the secret file was obtained for the positions of where the secret message starts and ends for the steganography decoding process.
- Each bit of the encrypted message was then placed at the Least Significant Bit of the color value of each pixel, altering the color value slightly.
- After the embedding process was completed the resulting image was generated with the encrypted message embedded into it which can be found on the right of Fig. 12.10c. In comparison, the stego object must remain unchanged to the human eye. Capacity of each image was compared to test for any changes in the size. When this procedure was tested there was no different in size, resulting in the image appearing unchanged to the human eye.

- After the stego object reached the receiver's end, the stego key was applied in order to perform the decode process to extract the encrypted message.
- The encrypted message was then decrypted using the inverse operation of the AES algorithm giving the results of Fig. 12.10d. As one can see the decrypted message derived back to the original message resulting in the procedure being a success.

12.7 Conclusion

We conducted the series of experiments to make sure our algorithm works as desired. There is more to explore about the field of Cryptography and Steganography. Because these are the most common foundations of the two components more work will be done to improve effectiveness and efficiency.

Future work will include implementing public key cryptography to enhance the security of the data in transit. Also, experiments and techniques will be developed for processing a steganography program at the cloud level for the improvement of cloud security.

References

1. Daemen, J., Rijmen, V. (1999). AES proposal: Rijndael, AES algorithm submission. September 3. http://www.nist.gov/CryptoToolkit.
2. Bogdanov, A., Khovratovich, D., & Rechberger, C. (2011). *Biclique cryptanalysis of the full AES, Lecture notes in computer science* (Vol. 7073, pp. 344–371). Berlin/New York: Springer.
3. Paar, C., & Pelzl, J. (2010). *Understanding cryptography*. Berlin: Springer.
4. Channalli, S., & Jadhav, A. (2009). Steganography an art of hiding data. *International Journal on Computer Science and Engineering, 1*(3), 137–141.
5. Buchmann, J. (2001). *"5 DES", Introduction to cryptography* (pp. 119–120). New York: Springer.
6. Canright, D. (2005). *A very compact S-Box for AES, Lecture notes in computer science* (Vol. 3659, pp. 441–455). Berlin/New York: Springer.
7. Gowtham, M., et al. (2013). AES based steganography. *IJAIEM, 2*(1), 382–389.
8. Ramaiya, M., et al. (2013). Secured steganography approach using AES. *IJCSEITR, 3*(3), 185–192.
9. Kumthe, A., et al. (2016). Steganography based on AES algorithm and BPCS technique for a securing image. *IJARCSSE, 6*(3), 116–119.

Rate Adjustment Mechanism for Controlling Incast Congestion in Data Center Networks

13

Prasanthi Sreekumari

Abstract

Data Center Transmission Control Protocol (DCTCP) gained more popularity in academic as well as industry areas due to its performance in terms of high throughput and low latency and is widely deployed at data centers nowadays. According to recent research about the performance of DCTCP, the authors found that most of the times the sender's congestion window reduces to one segment which results in timeouts. To address this problem, we modified the calculation of sender's congestion window size for improving the throughput of TCP in data center networks. The results of a series of simulations in a typical data center network topology using Qualnet, the most widely used network simulator demonstrates that the proposed solution can significantly reduce the timeouts and noticeably improves the throughput by more than 10% compare to DCTCP under various network conditions.

Keywords

TCP • Data Centers • Timeouts

13.1 Introduction

In recent years, modern data centers host a variety of services and applications such as web search, social networks and scientific computing for various private, non-profit and government systems [1]. Figure 13.1 shows the conventional Data Center Network (DCN) architecture for data centers adapted from Cisco [2]. The typical data center consists of core, aggregation and access layers. Among that, core layer provides the high-speed packet switching for all incoming and outgoing flows of the data center. In addition, it provides connectivity to multiple aggregation modules and serves as the gateway to the campus core. The aggregation layer defines the Layer 2 domain size and has the responsibility of aggregating the thousands of sessions leaving and entering the data center [3].It also provides value added services such as load balancing, firewalling, offloading to the servers across the access layer switches [3]. The aggregation layer connects to the core layer using Layer-3 10 Gigabit Ethernet links. The traffic in the aggregation layer primarily consists of core layer to access layer and access layer to access layer. Furthermore, the access layer operates in Layer 2 or Layer 3 modes and provides the physical level attachment to the server resources.

For communication between nodes, the vast majority data centers use Transmission Control Protocol (TCP) [4]. However, recent research has shown that the TCP does not work well in the unique data center environment [5]. One of the main reasons for the TCP throughput collapse in DCN is TCP Incast congestion. Incast congestion is a catastrophic loss in throughput that occurs when the number of senders communicates with a single receiver by sending data increases beyond the ability of an Ethernet switch to buffer packets. It leads to severe packet loss and consequently frequent TCP timeouts and thereby reduces the performance of TCP.

P. Sreekumari (✉)
Department of Computer Science, Grambling State University, Grambling, LA, USA
e-mail: s.prasanthy@gmail.com

© Springer International Publishing AG 2018
S. Latifi (ed.), *Information Technology – New Generations*, Advances in Intelligent
Systems and Computing 558, DOI 10.1007/978-3-319-54978-1_13

Fig. 13.1 Conventional data center architecture [2]

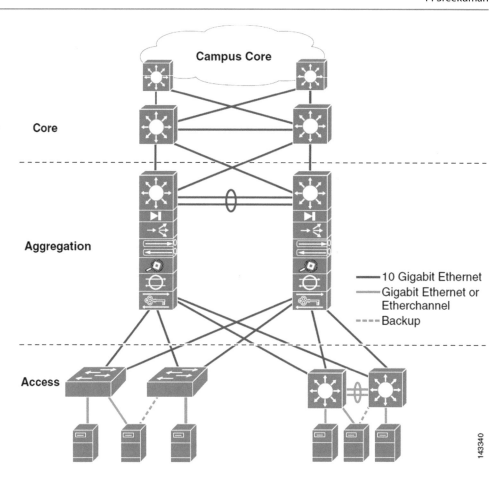

Recently, few attempts have been made to increase the TCP performance in DCN, but still the problem is not completely solved. Among the existing solutions, DCTCP gained more popularity in academic as well as industry areas due to its better performance in terms of throughput and latency. DCTCP uses very small buffer space compared to other existing solutions. As a result, the sender's congestion window size of DCTCP remains very small which leads to TCP timeouts. In this paper, by considering the limitation of DCTCP, we modified the congestion window adjustment scheme of DCTCP and proposed an efficient rate adjustment method which is capable to control Incast congestion and thereby increase the throughput of TCP. The results of a series of simulations in a typical data center network topology using Qualnet demonstrates that the proposed solution can significantly reduce the timeouts and noticeably improves the performance compare to DCTCP in terms of throughput under various network conditions.

The remainder of the paper is organized as follows. In Sect. 13.2, we present the related work. Section 13.3 describes the details of proposed algorithm. In Sect. 13.4, we describe our experimental methodology and present our results. Finally, Sect. 13.5 concludes our work.

13.2   Related Work

The performance degradation of TCP in data center networks mainly due to Incast congestion. This issue has already attracted the attention of many researchers in our research community. In this section, we survey the solutions proposed recently for controlling TCP Incast congestion in data center networks.

Data Center TCP (DCTCP) [6], a variant of TCP designed to operate with very small queue occupancies, without loss of throughput. DCTCP achieves the goals namely high burst tolerance, low latency and high throughput primarily by reacting to congestion in proportion to the extent of congestion. DCTCP propose a simple marking scheme by modifying the Explicit Congestion Notification [ECN] that helps the sources react to network congestion. TCP Fast [7] is a delay based congestion control algorithm proposed for controlling Incast congestion. This main aim of TCP Fast is to maintain a certain queue length at the switch buffer for each flow and keeps the total queue length below the buffer size. In that way, TCP Fast can avoids the droptails over switch buffer and thereby controls the TCP Incast congestion problem.

Incast Congestion Control for TCP (ICTCP) [8] a systematically designed protocol to perform congestion control at the receiver side by adjusting the receive window proactively before packet loss occurs. To perform congestion control at the receiver side since it knows the throughput of all TCP connections and the available bandwidth, ICTCP uses the available bandwidth on the network interface as a quota to co-ordinate the receive window increase of all incoming connections. In addition, ICTCP use the per-flow state to finely tune the receiver window of each connection on the receiver side.

Another protocol proposed for solving the TCP Incast problem is the Incast Avoidance TCP (IA-TCP). IA-TCP is a rate-based congestion control algorithm that controls the total number of packets injected into the network pipe to meet the bandwidth-delay product [9]. IA-TCP was designed to operate at the receive side like ICTCP, which controls both the window size of workers and the delay of ACKs. To control the sending rate of packets, IA-TCP used the measurement of link capacity i.e., the link rate of the interface connected to the top of the rack switch, and it is obtained from the transport layer. In addition, to control the window size of each worker that employs the standard TCP, the receiver exploits the advertisement field in the ACK header.

TCP-FITDC [10] is a delay-based TCP congestion control algorithm proposed for reacting to congestion states of conventional TCP more accurately. TCP-FITDC utilized the modified Congestion Experienced (CE) codepoint of packets defined in DCTCP to categorize the acknowledgments (Acks) into two classes, marked Acks and Unmarked Acks. By analyzing the differences between these two Acks, TCP-FITDC can estimate the network condition more accurately. The main goal of TCP-FITDC is to achieve low latency, high throughput and fast adjustment for TCP when applied to data center networks.

In [11], the authors introduced a new transport protocol which provides Bandwidth Sharing by Allocating Switch Buffer (SAB) to determine the congestion window of each flow. SAB has two main advantages. First, it converges fast. Second, SAB rarely loses packets. This feature can solve the goodput collapse of TCP Incast as well as the unfairness of TCP Outcast since they are both caused by large numbers of packet losses. SAB modifies the sender, switch and the receiver. In addition, SAB proposes a mechanism for adjusting the size of sender's congestion window value which is less than one. Although these algorithms can mitigate the problem of TCP Incast issue, DCTCP gained more popularity in academic as well as industry areas. However, in DCTCP, the size of congestion window reduced to one

frequently which leads to delayed Ack timeouts. For addressing this issue of DCTCP, we propose a new algorithm by adjusting the size of sender's congestion window and thereby control the TCP Incast congestion in data center networks.

13.3 Proposed Algorithm

Sending rate adjustment is an important factor for controlling TCP Incast congestion and thereby improving the performance of TCP in data center networks. In data center networks, the queue length will increase rapidly in a short time due to the concurrent arrival of burst of flows from multiple senders to a single receiver [12]. As a result, switch marks the packets continuously which leads to reduce the sender's congestion window into half of its current size.

In ECN enabled TCP, whenever the sender receives an ECN marked Ack packet, it reduces the size of congestion window into half even if the network is less congested (in the case of single ECN marked Ack packet). This will degrade the performance of TCP. For avoiding the above degradation, DCTCP propose a fine grained reduction function for reducing the size of congestion window based on the value of α. Whenever the sender receives an Ack with congestion experienced code point, the DCTCP sender reduces the size of congestion window using the Eq. (13.1),

$$cwnd \leftarrow cwnd \times (1 - \alpha/2) \qquad (13.1)$$

where cwnd is the size of congestion window and the value of α is calculated from the fraction of marked packets (F) and weight factor (g) according to the Eq. (13.2)

$$\alpha = (1 - g)\,\alpha + g \times F \qquad (13.2)$$

If the value of α is near zero, it indicates that the network is congested lightly. In this case DCTCP reduces the size of congestion window according to Eq. (13.1). However, if the value of α is equal to one, it indicates that the network is highly congested. As a result, DCTCP reduces the sender's congestion window like normal TCP. The above adjustment of congestion window improves the DCTCP sender to control the buffer occupancy at the switch and thereby increases the throughput of data center networks. Recent study [13] shows that one of the main problems in the congestion window estimation of DCTCP is in the choice of α initialization value. If we set zero to α, the sender may suffers from frequent packet losses and retransmission timeouts. On the other hand, if we set one to α, the sender can minimize the queuing delay but the amount of packets to be transferred is much smaller. As a result, the throughput of DCTCP will be reduced.

By considering the above limitations of DCTCP, we modified the DCTCP algorithm particularly the adjustment of sender's congestion window and propose a rate adjustment mechanism for controlling the TCP Incast congestion. When the sender receives the Ack with congestion notification, the sender checks the current network condition based on the outstanding packets and adjust the size of congestion window according to the Eq. (13.3),

$$CWstart = \alpha(CWmax\text{-}CWmin) - Cwcurrent \qquad (13.3)$$

where α is the number of outstanding packets in the network, $CWcurrent$ is the current size of congestion window at the time of receiving the ack packets with congestion notificatio, $CWmax$ and $CWmin$ are the maximum and minimum size of congestion window adjusted before receiving the congestion notification. When the sender receives the Ack of all outstanding packets, sender adjust the sending rate according to Eq. (13.4),

$$CWend = CWcurrent + \beta(CWmax\text{-}CWmin) \qquad (13.4)$$

where the value of β is 02. CWstart and CWend are the starting and ending points of congestion.

13.4 Working Rationale of Proposed Algorithm

We used the DCTCP slow start, fast retransmission and retransmission timeout algorithms. However, we modified only the sending rate when the sender receives the congestion notification via explicit congestion notification algorithm. When the sender receives the congestion notification through the Ack packets with ECE = 1, the sender starts the congestion point and reduces the size of congestion window according to Eq. (13.3). Once the sender receives all the outstanding packets in the network, the sender changed the window size based on Eq. (13.4). Instead of reducing the congestion window into half of the current size, our proposed rate adjustment can utilize the buffer space efficiently.

13.5 Performance Evaluation

In this section, we present the performance of our proposed protocol through comprehensive simulations using Qualnet simulator [14]. We compare the performance of our algorithm with DCTCP as it is the most popular data center transport protocol. We implemented DCTCP in Qualnet using the source code we got from [15]. Our main goal of this work is to increase the performance of TCP by controlling the Incast congestion in data center networks. For achieving our goal, we evaluate the performance of our algorithm in a typical network topology for Partition/Aggregate cloud applications as shown in [16].

As we did in our previous work [16], to simulate the Incast scenario, we used 10 servers connected to a single client via a switch. The link capacity is set to 1 Gbps and link delay is set to 25 μs, RTT 100 μs and RTO min which is equal to 10 ms. The buffer size is set to 64 KB and 256 KB. We vary the SRU size from 10 KB to 128 KB. The marking threshold value 'K' is set according to [6, 17] for 1 Gbps link capacity. The value of the weighted averaging factor 'g' for DCTCP is set to 0.0625 for buffer size 256 KB and 0.15 for 64 KB. An FTP-generic application is run on each source for sending the packets as quickly as possible. We repeated the experiments for 100 times.

To evaluate the performance of our algorithm with DCTCP, we use three important performance metrics namely, goodput, fast retransmissions and flow completion time. First, we calculated the goodput as the ratio of the total data transferred by all the servers to the client and the time required to complete the data transfer. Second, we evaluated the total number of fast retransmission occurred in DCTCP and proposed algorithm and finally we evaluated the flow completion time of both algorithms. We present the results of our evaluation of proposed algorithm with DCTCP in terms of goodput, flow completion time, and fast retransmissions using a single bottleneck TCP Incast scenario. Figure 13.2 shows the performance of our algorithm compared with DCTCP in terms of goodput. The buffer size we set for this simulation is 64 KB with SRU sizes 64 KB and 128 KB. From the result, we observe that even we used a smaller buffer size, the performance of proposed algorithm achieved higher throughput compared to DCTCP. Figure 13.3 presents the comparison of fast retransmissions of 50 senders. From the result, we observed that DCTCP has higher retransmissions than proposed algorithm. One of the main reason of our algorithm gains less fast retransmission is, the efficient utilization of buffer which leads to reduce the

Fig. 13.2 Goodput performance

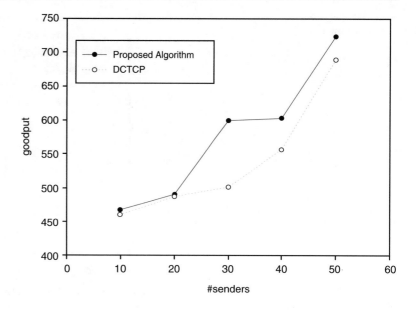

Fig. 13.3 Comparison of fast retransmissions

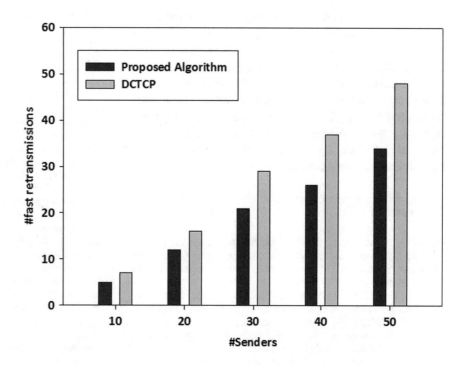

overflow of queue length and thereby reduce the loss of packets from the network.

Figure 13.4 depicts the query completion time of proposed algorithm compared to DCTCP. As we expected, our algorithm has less query completion time due to the efficient sending rate adjustment mechanism.

13.6 Conclusion

In this paper, we have developed a modified DCTCP protocol for improving the performance of TCP by controlling Incast congestion in data center networks. We proposed a

Fig. 13.4 Comparison of query completion time

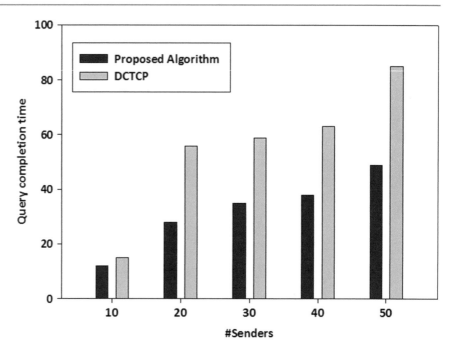

new sending rate adjustment mechanism for avoiding the frequent retransmission timeouts and thereby utilize the buffer space efficiently. We conducted extensive simulation using Qualnet to validate the performance and effectiveness of our algorithm compared to DCTCP in terms of goodput, flow completion time and fast retransmissions. Our experimental results using the typical TCP Incast scenario shows that the proposed algorithm achieves significant improvement in goodput by reducing the number of fast retransmissions as well as query completion time.

References

1. Zhang, Y., & Ansari, N. (First Quarter 2013). On architecture design, congestion notification, TCP incast and power consumption in data centers. *IEEE Communications Surveys and Tutorials, 15*(1), 39–64.
2. Cisco Data Center Infrastructure 2.5 Design Guide. http://www.cisco.com/c/en/us/td/docs/solutions/Enterprise/Data_Center/DC_Infra2_5/DCI_SRND_2_5a_book.pdf.
3. http://www.cisco.com/c/en/us/td/docs/solutions/Enterprise/Data_Center/DC_Infra2_5/DCInfra_1.html.
4. Chen, Y., Griffith, R., Liu, J., Katz, R. H., & Joseph, A. D. (2009). Understanding TCP incast throughput collapse in datacenter networks. In *Proceedings of the 1st ACM workshop on Research on enterprise networking (WREN '09)* (pp. 73–82). New York: ACM.
5. Kant, K. (2009). Data center evolution: A tutorial on state of the art, issues, and challenges. *Computer Networks, 53*(17), 2939–2965. ISSN 1389-1286.
6. Alizadeh, M., Greenberg, A., Maltz, D., Padhye, J., Patel, P., Prabhakar, B., Sengupta, S., Sridharan, M. (2010). Data center TCP (DCTCP). In *Proceedings of the SIGCOMM*, New Delhi, (pp. 63–74).
7. Hwang, J., & Yoo, J. (2014). FaST: Fine-grained and scalable TCP for cloud data center networks. *KSII Transactions on Internet and Information Systems, 8*(3), 762–777. doi:10.3837/tiis.2014.03.003.
8. Wu, H., Feng, Z., Guo, C., Zhang, Y. (2010). ICTCP: Incast Congestion Control for TCP in data center networks. In *Proceedings of ACM CoNEXT*, Philadelphia.
9. Hwang, J., Yoo, J., Choi, N. (2012). IA-TCP: A rate based incast-avoidance algorithm for TCP in data center networks. In *Proceedings of the IEEE ICC*, Ottawa.
10. Zhang, J.,Wen, J.,Wang, J., Zhao, W. (2013). TCP-FITDC: An adaptive approach to TCP incast avoidance for data center applications. *International Conference on Computing, Networking and Communications (ICNC)*, San Diego, (pp. 1048–1052).
11. Zhang, J., Ren, F., Yue, X., Shu, R., & Lin, C. (2014). Sharing bandwidth by allocating switch buffer in data center networks. *IEEE Journal on Selected Areas in Communications, 32*(1), 39–51.
12. Wu, W., & Crawford, M. (2007). Potential performance bottleneck in Linux TCP. *International Journal of Communication Systems, 20*, 1263–1283. doi:10.1002/dac.872.
13. https://eggert.org/students/kato-thesis.pdf.
14. http://web.scalable-networks.com/content/qualnet.
15. http://dev.pyra-handheld.com/index.php.
16. Sreekumari, P., Jung, J., Lee, M. (2015). An early congestion feedback and rate adjustment schemes for many-to-one communication in cloud-based data center networks. *Photonic Network Communications, 31*(1), 23–35.
17. Jiang, C., Li, D., & Xu, M. (2014). LTTP: An LT-code based transport protocol for many-to-one communication in data centers. *IEEE Journal on Selected Areas in Communications, 32*(1), 52–64.

Performance Analysis of Transport Protocols for Multimedia Traffic Over Mobile Wi-Max Network Under Nakagami Fading

14

Fazlullah Khan, Faisal Rahman, Shahzad Khan, and Syed Asif Kamal

Abstract

Due to increased demand in multimedia streaming over Internet, research focus is directed to derive new protocols with added features. These protocols can be best adjusted to multimedia traffic over the Internet. The operations of multimedia networks require fast and high processing communication systems, because sometimes prompt delivery of information is critical. The behavior of different transport protocol can affect the quality of service. The advancement in hardware technologies used at the physical layer and TCP/IP suite has made a great progress, but other layers could not progress with the same pace. This paper argues the challenges posed by multimedia traffic on other layers and their effect on the transport layer services. Numerous papers have illustrated transport protocols for multimedia networks in some preferred and particular scenarios. However, no one have discussed the effect of channel fading on these protocols. Six existing protocols and their performances under Nagagami-m channel are the theme of this paper with respect to suitability for multimedia applications, and the performances of these protocols have been evaluated. This paper concludes that extra effort are essential on the transport protocol for achieving better performance of multimedia networks which will fulfill the various necessities of evolving multimedia applications, specifically under fading channels.

Keywords

Transport protocol • Nakagami fading • Performance comparison • Wi-max IEEE 802-16e • MPEG-4 video

14.1 Introduction

With the advent of new technologies like smart phones, tablets, and personal digital assistance etc., and advancement in the Internet technologies have demanded a large number of multimedia applications. These applications have different needs such as in time delivery, bandwidth utilization, and reliability, for fulfilling quality of service requirements. However, some applications require higher delays and uneven bandwidth instead of high bandwidth and high speed networks, i.e. 3G mobile networks or Wi-Fi.

A significant aspect that has greater influence on the quality of service in multimedia applications is the selection of appropriate transport protocol. The conventional applications used for transferring a file, online transaction, or sending an email make use of relatively slow but reliable transport protocol, e.g. Transmission Control Protocol (TCP) [1]. Whereas for multimedia and game applications a fast and partially reliable transport protocol may be a good choice. However, the diverse nature of multimedia traffic

F. Khan (✉) • S. Khan
Computer Science Department, Abdul Wali Khan University, 23200 Mardan, KPK, Pakistan
e-mail: fazlullah@awkum.edu.pk; mianjan@awkum.edu.pk

F. Rahman • S. Asif Kamal
Electrical Engineering Department, International Islamic University, Islamabad, Pakistan
e-mail: faisalrehman_00@yahoo.com; sakamal@iiu.edu.pk

© Springer International Publishing AG 2018
S. Latifi (ed.), *Information Technology – New Generations*, Advances in Intelligent Systems and Computing 558, DOI 10.1007/978-3-319-54978-1_14

and communication networks poses different challenges to transport protocol. For example, User Datagram Protocol (UDP) [2] and TCP protocol cannot perform well in high performance wireless mobile networks due to the complex service model compared to wired networks. These complex models are necessary for running multimedia applications on new generation application layers which affects the conventional transport layer. In this context UDP and TCP do not perform upto the mark, therefore, a Real-time Transport Protocol (RTP) [3] is a good choice to transmit multimedia traffic on Internet Protocol.

Normally UDP is used for transportation of multimedia applications due to the delay intolerant nature of these applications, where the main hindrance is need of a mechanism for congestion control. To overcome these issues protocols like Multi-path Transport Protocol (MPTCP) [4], Stream Control Transmission Control Protocol (SCTP) [5], and Datagram Congestion Control Protocol (DCCP) [6] are better options to avail for multimedia traffic transportation. Deployment and maintenance of various wireless access technologies used for transportation of multimedia traffic, are easy compare to wired access technologies. These technologies have diverse factors affecting quality of service, e.g. features of Mobile Wi-Max such as high data rates, scalability, security, and mobility. Wi-Max, UMTS, 4G networks are getting popular and have maximized requirements for multimedia applications. UDP is fast, unreliable, and have no congestion control mechanism which puts it into no quality of service category but still a better choice for video streaming. SCTP have multi-homing, multi-streaming features, better bandwidth utilization with ordered and reliable data transportation. DCCP have mechanism of congestion control with un-reliable data transportation.

The de facto standard for Mobile Wi-Max networks is IEEE802.16e [7], which can operate in mobile and static broadband networks. The IEEE 802.16e have data rate up to 63 Mbps, with improved quality of service. It supports end-to-end Internet Protocol based quality of service in different flows. Another feature of this standard is scalability and flexibility based on the offered services as well as network deployment.

The rest of the paper in accordance to the following pattern. In Sect. 14.2, related work is presented followed by transport protocols in Sect. 14.3. Section 14.4 discusses the Nakagami model for channel fading. Section 14.5 describes mobile Wi-Max technology and MPEG-4 Video traffic, and experimental work and evaluations are given in Sect. 14.6. The conclusion of the paper and future research directions and gaps are elaborated in Sect. 14.7.

14.2   Related Work

In wireless multimedia networks transportation is a hard area due to dynamic nature of multimedia applications and routing protocols. Selection of appropriate routing protocol for

Wi-Max technology is considered as a smart task. A good survey on transport protocols in wireless multimedia networks can be found [8], and real-time transport protocols for wireless networks [9].The performance of routing protocols used in mobile Wi-Max is studied [10], quality of service mechanism and routing techniques in a mesh-network of mobile Wi-Max is described [11], whereas [12] analyses statistical performance of Wi-Max and Wi-Fi heterogeneous mesh network.

The performance of Wi-Max technologies has been investigated in [13]. A comparison of transport protocol for voice over IP in Wi-Max network is drawn in [14].The affects of jitter, throughput, delay, and packet loss on video streaming in Wi-Max and ADSL networks is evaluated [15]. Furthermore, packet loss ratio and throughput is evaluated with TCP and its variants using modulation schemes, i.e. BPSK, QPSK, and 16-QAM and64-QAM. Time is very critical factor in transmission of multimedia applications, even more important than reliability. The transmission of MPEG-4 video is analyzed by using UDP, TCP, and DCCP with her variants in [16], with performance parameter similar to [15]. Similarly, [17] have evaluated the same parameters and traffic as in [16] with SCTP protocol over 802.11 wireless networks. Multimedia traffic parameters like throughput and link-capacity using UDP and TCP is analyzed in [18], then comparison among Wi-Max and High-speed Down-link Packet Access is made. The effect of Nakagami fading channel on transport protocol has been studied in [19–23].

In multi-hop wireless networks throughput is measured based on various window sizes in SCTP protocol [24]. Similarly, [25] have analyzed video streaming in CDMA200 wireless networks using SCTP protocol. A comparative study of various transport protocols for video streaming in 802.11 networks is shown in [26] and in fixed Wi-Max in [14]. In short, different performances of transport protocols using different applications can be found in literature. However, the theme of this paper is to use mobile Wi-Max as the wireless access technology with different transport protocols is a good contribution to this field.

14.3   Transport Protocols

In this section different transport protocol are studied briefly. Table 14.1 show names, features, and services offer by these protocols. Following are the details of these protocols;

14.3.1   Multi-path Transport Protocol (MPTCP)

MPTCP is the extended version of TCP. It mainly focuses on efficient utilization of network resources via multi-homing abilities of end users and provides flexibility network connectivity. The efficient utilization is maximized by

Table 14.1 Features and services offered by DCCP, SCTP, RTP, MPTCP, UDP, and TCP

Features and services	DCCP	SCTP	RTP	MPTCP	UDP	TCP
Connection Oriented	✓	✓	✓	✓	✗	✓
Unordered Delivery	✓	✓	✓	✗	✓	✓
Reliability	✗	✓	✗	✓	✗	✓
Congestion Control	✓	✓	✓	✓	✗	✓
Flow Control	✓	✓	✓	✓	✗	✓
Multistreaming	✗	✓	✗	✓	✗	✓
Multihoming	✗	✓	✗	✓	✗	✗

concurrent usage of resources on different network interfaces whereas network connectivity is achieved through multi-paths. MPTCP provides two transport functions; network and application oriented. MPTCP is has better congestion mechanism which is fair with TCP flows and it moves traffic from congested links.

14.3.2 Datagram Congestion Control Protocol (DCCP)

DCCP is unreliable connection oriented protocol used by multimedia applications. It combines unreliable and quick services of UDP with controlling congestion by establish a bidirectional uni-cast connection. DCCP have some versions like Quick-Start for DCCP or faster restart for TFRC. DCCP is a good choice for transferring data in bulks with controlled trade-off between timeliness and reliability.

14.3.3 Stream Control Transmission Protocol (SCTP)

SCTP is reliable session oriented multi-homing. It has the ability to recover lost, corrupt, unordered, and duplicate packets by using re-transmission with selective acknowledgment scheme. It has inherited congestion and flow control mechanism from TCP along with mechanisms of slow start, fast recovery, and congestion avoidance. It is secure against flooding and masquerading attacks.

14.3.4 Real-Time Transport Protocol (RTP)

RTP is designed for real-time applications with end-to-end transportation functionalities. It carries data for uni-cast and multi-cast communication, however, this protocol by itself does guarantee quality-of-service (QoS). RTP is a connectionless protocol but is uses the services of RTP Control Protocol (RTCP) for achieving QoS.

14.3.5 User Datagram Protocol (UDP)

UDP is unreliable connectionless, message-oriented core member of the TCP/IP suite with no error, flow, and congestion control mechanism used in uni-cast/multi-cast networks.

14.3.6 Transmission Control Protocol (TCP)

TCP is the core member of TCP/IP suite, a reliable and connection oriented most widely used transport protocol. It has mechanism for flow control, error control, congestion control, error reporting, and error correction. TCP has many versions with better congestion control mechanism such as FAST TCP, TCP Vegas, and TCP Reno. TCP has extensions for mobile wireless networks, satellite networks and some advancement in original TCP mechanisms.

14.4 Nakagami Fading Channel

In this section we derived the outage probability of Nakagami-m fading channel model. In this model we assume that γ_0 is the required signal-to-noise threshold for decoding the received packet at a node j. The arising of outage event is subject to the condition, where $\gamma(i) < \gamma_0$ within a coherent time of node j. In our model, when node i sends a packet to node j, the outage probability is represented by p_{out_i} as shown in Eq. 14.2.

The Probability Density Function (PDF) of signal-to-noise for Nakagami-m fading channel is given in Eq. 14.1 below,

$$f_\gamma = \frac{1}{\Gamma(m)}\left(\frac{m}{\overline{\gamma}(i)}\right)^m \gamma^{m-1} exp\left(-\frac{m\gamma}{\overline{\gamma}(i)}\right). \qquad (14.1)$$

where m is the shape parameter of the distribution and $\Gamma(.)$ is the gamma function.

$$
\begin{aligned}
p_{out_i} &= P_r(\gamma < \gamma_0) \\
&= \int_0^{\gamma_0} \frac{1}{\Gamma(m)}\left(\frac{m}{\overline{\gamma}(i)}\right)^m \gamma^{m-1} exp\left(-\frac{m\gamma}{\overline{\gamma}(i)}\right) d\gamma \\
p_{out_i} &= 1 - exp\left(-\frac{m\gamma_0}{\overline{\gamma}(i)}\right) \times \sum_{s=0}^{m-1}\frac{\left(\frac{m\gamma_0}{\overline{\gamma}(i)}\right)^s}{s!} \\
p_{out_i} &= 1 - exp\left(-\frac{m\gamma_0 N_0(4\pi)^2 d_i^l}{G_T G_R \lambda^2 P_t}\right) \\
&\quad \times \left(\sum_{s=0}^{m-1}\frac{\left(-\frac{m\gamma_0 N_0(4\pi)^2 d_i^l}{G_T G_R \lambda^2 P_t}\right)^s}{s!}\right).
\end{aligned}
\qquad (14.2)
$$

Fig. 14.1 Mobile Wi-Max system

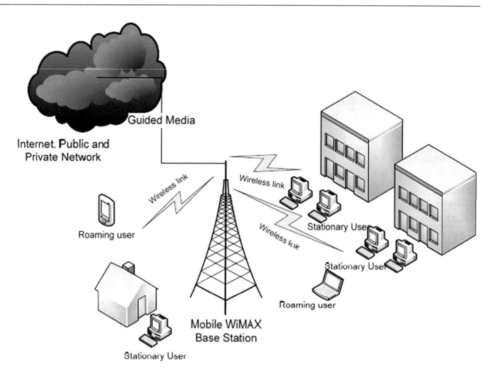

Since packet is divided into M blocks hence, packet error probability will be shown in Eq. 14.3,

$$p_{PHY_i} = 1 - (1 - p_{out_i})^M. \qquad (14.3)$$

$$p_{PHY_i} = 1 - exp\left(-\frac{m\gamma_0 N_0 (4\pi)^2 d_i^l}{G_T G_R \lambda^2 P_t}\right) \times \left(\sum_{s=0}^{m-1} \frac{\left(-\frac{m\gamma_0 N_0 (4\pi)^2 d_i^l}{G_T G_R \lambda^2 P_t}\right)^s}{s!},\right)^M. \qquad (14.4)$$

14.5   Mobile Wi-Max Technology and MPEG-4 Video Traffic

Worldwide Interoperability for Microwave Access (Wi-Max) is wireless metropolitan area network technology. The de facto standard for Wi-Max broadband technology is the IEEE 802.16e (Wi-Max). It offers mobile and fixed broadband networks. It is standardized version of Ethernet intends to provide compatibility and interoperability in broadband wireless access technologies. Theoretically it can transfer data with a speed of 64 Mbps down-link and 28 Mbps up-link on single channel. For mobile networks its bandwidth capacity is 15 Mbps over a range of 3 Km. Wi-Max can operate in different frequency spectrum such as 2.3 GHz, 2.5 GHz, 3.5 GHz, and 5.8 GHz. A mobile

Wi-Max system is depicted in Fig. 14.1. The characteristics that make mobile Wi-Max better than Wi-Fi and other wireless technologies are; high bandwidth, quality of service, scalability, freedom from wires, and security. In the context of mobile Wi-Max mobile stations are connected to a static base station via air interface. Beside connectivity services the base station provides quality of service policies, traffic and radio resource management to mobile stations.

Moving Picture Expert Group version 4 (MPEG-4) is a compression standard for multimedia compression and has reduce bit rate i.e. 4.8 kbps to 64 kbps, with good quality, is a good choice for video streaming over network. It consists of Intra-coded frame, Predicted frame, and Bidirectional frame, where every frame has its own characteristics. We have used MPEG-4 traffic at the application layer and have taken from [27]. The basic object in MPEG-4 is video object plane, a grouping of I frames, P frames, and B frames [26]. MPEG-4 has higher resolution ranging from 176 × 144 to 1920 × 1080 [28]. MPEG-4 is video must be compressed before transferring it over a network [29].

14.6   Simulations

This section discusses only one simulation scenario, performance matrices, and results. We have used network simulator version 2 (NS-2) for simulation with assumption of error free communication and perfect radio conditions. This performance of MCTCP, DCCP, TCP, SCTP, UDP, and RTP have been assessed for video steaming over Wi-Max network shown in Fig. 14.1, and performed extensive experiments on it and the mean results are shown in the paper. Table 14.2 shows list of parameters used in the simulation. The performance matrices used are throughput, packet loss, delay, and jitter. A scenario shown in Fig. 14.2, with multiple mobile stations is simulated for assessing performance of given protocols. The performance matrices are calculated as follow;

- Throughput: It is the number of bits successfully transferred in a particular period of time, i.e., total amount of data received/ total time.
- Packet Loss: It is the number of packets sent but not received correctly, and is calculated as; total number of packet transmitted − total number of packet received.

Table 14.2 The simulation conditions

Parameters	Values
Modulation Schemes	64 QAM
Antenna type	Omni direction
Radio Propagation	Two-ray ground
Queue type	Drop-tail
MAC protocol	IEEE802.11
Routing protocol	AODV
Offered Load	1, 2,1 3,,12 Mbps
Maximum Queue length	100[packet]
Distance between stations	200[m]
Packet size	1024[Byte]
Number of Mobile Stations	8
Speed of Mobile Stations	3Km/h and 5Km/h
Number of Servers	1
Simulation time	50[s]

- Delay: It is the difference in time for a packet being transmitted and received, i.e., Packet Aś receiving time − Packet Aś sending time.
- Jitter: It is the jerk time between two consecutive delays, i.e., Delay(A) − Delay (B).

14.6.1 Multiple Mobile Stations Scenario

The simulation results are based on multiple mobile stations that are connected to the server through base station as depicted in Fig. 14.1. Same simulation parameters are used with different speed of the mobile stations i.e. 3 Km/h and 5 Km/h. The scenario is depicted in Fig. 14.2.

14.6.1.1 Mobile Nodes at a Speed of 3 Km/h

In this scenario video traffic is multi-cast towards the mobile stations and the transport protocol performance with growing number of mobile stations is analyzed. The mobile nodes are moving at a speed of 3 Km/h. The throughput of the network is depicted in Figs. 14.3 and 14.7, where the x-axis shows the number of mobile stations and y-axis represents the throughput in Mbps. It is evident from Figs. 14.3 and 14.7 that the throughput is continuously increasing when the number of mobile nodes is increasing. MPTCP gives us better results in term of throughput because it makes use of multiple paths to avoid congestion and load on a particular node in the network.

Similarly, in Figs. 14.4 and 14.8 the number of packets lost is shown on x-axis with mobile stations on y-axis. The better result in terms of packet lost is given by MPTCP, TCP and DCCP. For MPTCP less packet drops has a reason

Fig. 14.2 Multiple mobile stations downloading video

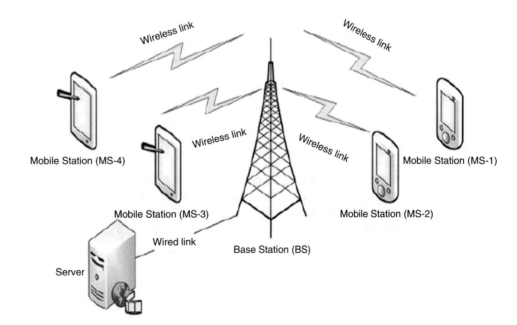

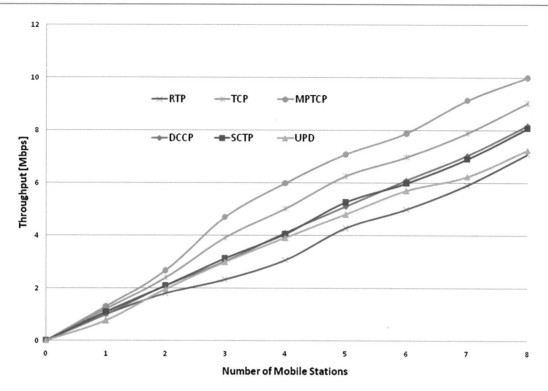

Fig. 14.3 Throughput at a speed of 3 Km/h

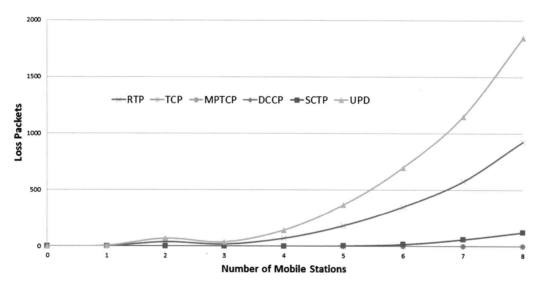

Fig. 14.4 Packet lost at a speed of 3 Km/h

discussed above, whereas TCP and DCCP make connections and have congestion control mechanism with and without reliability and thats why drop less packets compare to other protocols. On the other hand UDP have no congestion con- trol mechanism and due to that it drops highest number of packets. In Figs. 14.5 and 14.9, delay of the network is drawn and MPTCP outperforms all other protocols. In Figs. 14.6 and 14.10, Jitter of the networks is shown where

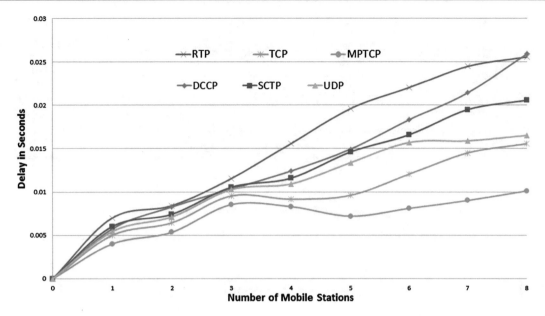

Fig. 14.5 Delay at a speed of 3 Km/h

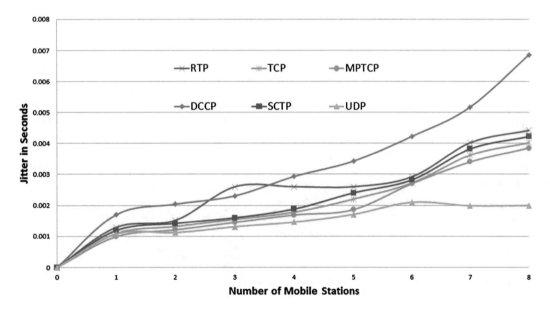

Fig. 14.6 Jitter at a speed of 3 Km/h

UDP gives better results compare to other protocols because the packets arrival rate is constant and continuous.

14.6.1.2 Mobile Nodes at a Speed of 3Km/h

In this scenario video traffic is multi-cast towards the mobile stations and the transport protocol performance with growing number of mobile stations is analyzed. The mobile stations are moving at a speed of five meters per second (Figs. 14.7, 14.8, 14.9, and 14.10).

14.7 Conclusion

In this paper analysis of different transport protocols has been performed by using video traffic under Nakagami channel over mobile Wi-Max network. The results show that MPTCP outperforms all other scheme in terms of throughput, packet loss, jitter, and delay. Similar DCCP and SCTP works better that TCP and UDP in packet loss and throughput but

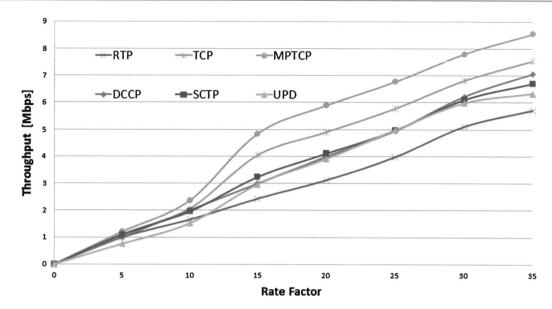

Fig. 14.7 Throughput at a speed of 5 Km/h

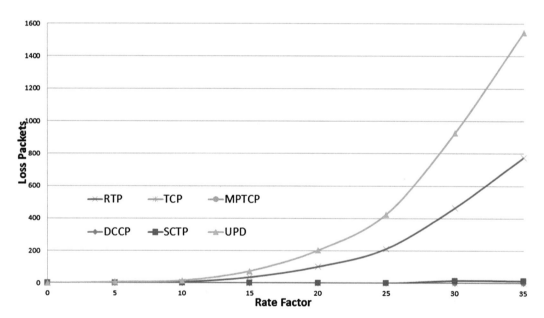

Fig. 14.8 Packet Lost at a speed of 5 Km/h

DCCP and SCTP have more jitter and delay due to which they are not suitable for video traffic. RTP and UDP have higher packet loss which badly affects MPEG transmission. In near future we will extend this research work by considering multiple cells and will analyze transport protocols for transferring voice over Internet protocol and MPEG traffic on Wi-Max and other wireless technologies under different fading channels, such as Suzuki, Weibull, and Riacian. Moreover, the protocols will be analyzed on various complex scenarios and a hybrid approach will be proposed. In short we conclude that after analyzing these six protocols MPTCP provides better QoS for transmission of video traffic over IEEE 802.16e under Nakagami channel, due to the congestion control mechanism it provides.

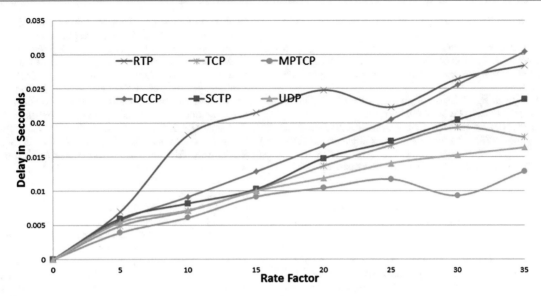

Fig. 14.9 Delay at a speed of 5 Km/h

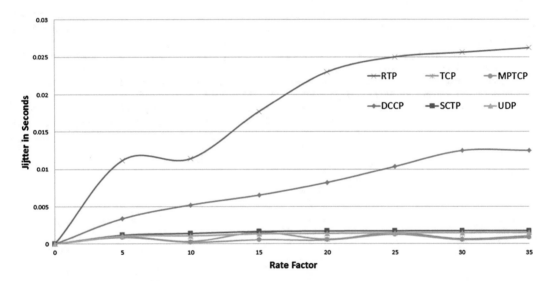

Fig. 14.10 Jitter at a speed of 5 Km/h

References

1. Postel, J. (1981). Transmission control protocol.
2. Postel, J. (1980). User datagram protocol (No. RFC 768).
3. Schulzrinne, H., Casner, S., Frederick, R., & Jacobson, V. (2003). RTP: A transport protocol for real-time applications (No. RFC 3550).
4. Raiciu, C., Handley, M., & Wischik, D. (2011). Coupled congestion control for multi-path transport protocols (No. RFC 6356).
5. Stewart, R. (2007). Stream control transmission protocol (No. RFC 4960).
6. Floyd, S., Handley, M., & Kohler, E. (2006). Datagram congestion control protocol (DCCP).
7. IEEE 802.16e/D5-2004, Part 16: Air interface for fixed and mobile broadband wireless access systems – amendment for physical and medium access control layers for combined fixed and mobile operation in licensed bands, Nov. 2004.
8. Loo, J., Mauri, J. L., & Ortiz, J. H. (Eds.). (2016). Mobile ad hoc networks: current status and future trends. CRC Press.
9. Khan, F., ur Rehman, A., Arif, M., Aftab, M., & Jadoon, B. K. (2016, April). A survey of communication technologies for smart grid connectivity. In *2016 International Conference on Computing, Electronic and Electrical Engineering (ICE Cube)* (pp. 256–261). IEEE.
10. Kaur, G., & Kaur, N. (2016). Dissect the performance of network layer protocols for video streaming over Worldwide Interoperability of Microwaves Access WiMAX). *International Journal of Applied Engineering Research, 11*(10), 7196–7199.
11. Apostolaras, A., Iosifidis, G., Chounos, K., Korakis, T., & Tassiulas, L. (2016). A mechanism for mobile data offloading to wireless mesh networks. *IEEE Transactions on Wireless Communications, 15*(9), 5984–5997.
12. Amin, R., & Martin, J. (2016). Assessing performance gains through global resource control of heterogeneous wireless

networks. *IEEE Transactions on Mobile Computing, 15*(2), 292–305.

13. Abdullah, S. A., & Abduljabbar, R. B. (2014). Streaming video content over NGA (next generation access) network technology. *International Journal of Science, Engineering and Computer Technology, 4*(3), 56.

14. Sarkar, S., Misra, S., Bandyopadhyay, B., Chakraborty, C., & Obaidat, M. S. (2015). Performance analysis of IEEE 802.15. 6 MAC protocol under non-ideal channel conditions and saturated traffic regime. *IEEE Transactions on Computers, 64*(10), 2912–2925.

15. Abdullah, N., & Singh, H. (2014). Performance based comparsion of different routing protocol under different modulation over WiMAX.

16. Kamil, W. A. (2015). Performance evaluation of TCP, UDP and DCCP for video traffics over 4G network (Doctoral dissertation, Universiti Utara Malaysia).

17. Yunus, F., Ariffin, S. H., Syed-Yusof, S. K., Ismail, N. S. N., & Fisal, N. (2014). Transport protocol performance for multi-hop transmission in wireless sensor network (WSN). In *Handbook of research on progressive trends in wireless communications and networking* (pp. 389–409). Hershey: IGI Global.

18. Othman, H. R., Ali, D. M., Yusof, N. A. M., Noh, K. S. S. K. M., & Idris, A. (2014, August). Performance analysis of VoIP over mobile WiMAX (IEEE 802.16 e) best-effort class. In *2014 I.E. 5th Control and System Graduate Research Colloquium (ICSGRC)* (pp. 130–135). Piscataway: IEEE.

19. C. Mukasa., Aalo, V. A., & Efthymoglou, G. (2016, October). On the performance of a dual-hop network with a mobile relay in a Nakagami fading environment. In *2016 I.E. 21st International Workshop on Computer Aided Modelling and Design of Communication Links and Networks (CAMAD)*, Toronto, ON (pp. 43–47). IEEE.

20. Khan, F., ur Rahman, I., Khan, M., Iqbal, N., & Mujahid, A. (2016, September). CoAP-based request-response interaction model for the Internet of things. In *International Conference on Future Intelligent Vehicular Technologies* (pp. 146–156). Cham: Springer.

21. Fida, N., Khan, F., Jan, M. A., & Khan, Z. (2016, September). Performance Analysis of Vehicular Adhoc Network Using Different Highway Traffic Scenarios in Cloud Computing. In International Conference on Future Intelligent Vehicular Technologies (pp. 157–166). Springer, Cham.

22. Younas, N., Asghar, Z., Qayyum, M., & Khan, F. (2016, September). Education and socio economic factors impact on earning for Pakistan-A bigdata analysis. In *International Conference on Future Intelligent Vehicular Technologies* (pp. 215–223). Cham: Springer.

23. Khan, F., Khan, M., Iqbal, Z., ur Rahman, I., & Mujahid, A. (2016, September). Secure and safe surveillance system using sensors networks-internet of things. In *International Conference on Future Intelligent Vehicular Technologies* (pp. 167–174). Cham: Springer.

24. Yunus, F., Ariffin, S. H., Syed-Yusof, S. K., Ismail, N. S. N., & Fisal, N. (2014). Transport protocol performance for multi-hop transmission in wireless sensor network (WSN). In: M. A. Matin (Ed.), *Handbook of research on progressive trends in wireless communications and networking* (Vol. 389). Hershey, PA: IGI Global.

25. Bhebhe, L. (2016). Service Continuity in 3GPP Mobile Networks.

26. Ahmad, S., & Khan, S. A. (2016). Performance evaluation of TCP and DCCP protocols in mobile Ad-Hoc networks (MANETS). *VFAST Transactions on Software Engineering, 10*(1)

27. Tanwir, S., & Perros, H. (2013). A survey of VBR video traffic models. *IEEE Communications Surveys & Tutorials, 15*(4), 1778–1802.

28. http://www.apple.com/quicktime/technologies/h264/

29. http://www.pcguide.com/ref/video/modesColor-c.html

CyberSecurity Education and Training in a Corporate Environment in South Africa Using Gamified Treasure Hunts

Laurie Butgereit

Abstract

Breaches in computer security can have economical, political, and social costs. In a busy corporate IT department, unfortunately, there is often not time enough to train programmers and other technical people on the best practices of computer security and encryption. This paper looks at the use of gamified treasure hunts to encourage participants to follow digital clues in order to find a physical treasure box of chocolates. By following the digital clues, participants in the treasure hunts will learn about such important security tools as Cryptography, GnuPG, SSL, HTTPS, Jasypt, Bouncy Castle, Wireshark, and common attack techniques.

Keywords

Education • Security • Gamification

15.1 Introduction

According to the 2015 Ponemon Institute report on the economic costs of data breaches (or IT security breaks), the average cost of a data breach is $3.79 million and the average cost of an individual record stolen is $154.00. This amounted to a 23% increase in total cost of a data breach and 12% per capital increase in cost since 2013 [1]. The report further summarizes the general root causes of these data breaches with 47% being caused by malicious or criminal activity, 29% because caused by a system glitch, and 25% being caused by human error.

While organizations recognize the importance of IT security to protect digital data, often there is not enough time or money available to send busy employees on appropriate training courses. Programmers need to be kept up-to-date with proper programming techniques and utilities to defend against malicious outside attacks.

This paper describes a research project in using gamified treasure hunts in a corporate IT department in South Africa. These treasure hunts were to encourage programmers and non-programmers (including business analysts, project managers, data base administrators, and testers) to learn more about various tools and utilities such as various types of encryption, various types of hacking utilities, introduction to steganography, etc., in order to encourage the members of the IT department to use these tools, utilities and concepts in their daily work.

15.2 Research Methodology

An Action Research methodology was used in this project. According to Lewin, there are three steps in creating change in an organization or group. These three steps are *unfreezing, moving,* and *refreezing* [2]. The first step of *unfreezing* starts the process by overcoming the existing situation. The second step of *moving* introduces new concepts, ideas, utilities, etc., to the members in the organization. The third and final step of *refreezing* embeds these new concepts and ideas in the organization.

L. Butgereit (✉)
Nelson Mandela Metropolitan University, Port Elizabeth, South Africa

Blue Label Telecoms, Johannesburg, South Africa
e-mail: laurie.butgereit@nmmu.ac.za

© Springer International Publishing AG 2018
S. Latifi (ed.), *Information Technology – New Generations*, Advances in Intelligent
Systems and Computing 558, DOI 10.1007/978-3-319-54978-1_15

15.3 Treasure Hunts in Education

Treasure hunts have been often used in formal education at primary and secondary levels [3]. Often treasure hunts take advantage of the ubiquity of mobile devices [4]. Treasure hunts have been used to encourage outdoor learning [5]. Treasure hunts (and also scavenger hunts) have been used to teach people about facilities or infrastructure which are available to the participant [6].

In most of the reported cases, however, the primary reason to use a Treasure Hunt format was to engage users. In a business environment, however, the adage "time is money" appears to rule and a Treasure Hunt format without a fixed time limit also allows participants to work in their own time and at their own pace. This research, however, looks at Treasure Hunts as a way of dispensing snippets of information to technical people in a busy business environment.

15.4 Gamified Treasure Hunts

Gamified treasure hunts were created which used a combination of digital and tangible artifacts. These gamified treasure hunts were used to present topics in IT security in a way that was fun and exciting.

Johan Huizinga was one of the first people to research the place of games and play in society as opposed to the actual games themselves. Huizinga wrote originally in German and it was later translated into English. In his paper *Homo Ludens: A Study in the Play Element of Culture*, he contrasted the term *Homo Ludens* (Man the Player in the original German) with *Homo sapiens* (Man the Thinker) [7]. In his paper, Huizinga argues that play has five essential elements (1) Play is voluntary (2) Play is outside of ordinary life (3) Play is distinct from ordinary life in both location and time (4) Play requires a certain order (5) Play has no material benefit or profit.

The gamified treasure hunts created in this research project attempted to adhere to these five essential elements with a slight exception to the fifth point. In contrast with the fifth point, a tangible treasure box full of chocolates could be found at the end of the treasure hunts as can be seen in Illustration 15.1.

The treasure hunts were introduced with a poster which included some images (which might be used as clues to the topic) and one or two QR codes to start participants on the treasure hunts. Sample posters can be seen in Illustration 15.2.

Illustration 15.1 Treasure box of chocolates

Illustration 15.2 Sample Posters

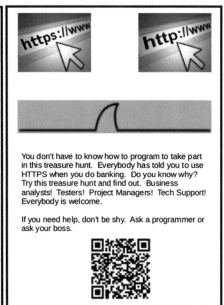

In general, the location of the treasure box of chocolates was virtually digitally hidden somewhere. To encourage competition between the participants, a limited number of chocolates (usually ten) were put in the treasure box. In addition, participants often created an informal "trophy board" of chocolate wrappers on the scrum board. These chocolate wrappers had the participants' names written next to them to brag about their successes at solving the puzzle in the treasure hunt.

15.5 Security Topics Covered

This section covers some of the IT security topics which were covered by the treasure hunts and the types of challenges presented to the participants. The topics included Cryptography, GnuPG, Jasypt, Bouncy Castle, SSL (public key/private key encryption), Wireshark, HTTPS vs HTTP, steganograpy, and various attack strategies such as brute-force attacks and dictionary attacks.

15.5.1 Cryptography

The number of the first treasure hunts were simple introductions to historical cryptography. There were a handful of treasure hunts where the location of the treasure box was encrypted using an historical cypher such as a Caesar Cypher, Vigenère Cipher, Matrix Transformation Cypher and AutoKey Cypher. The QR code linked to a small lesson which explained how brute force attacks can be launched against these cyphers and the participants were invited to program such attacks against these files.

15.5.2 GnuPG

After the topics of historical cryptography, the next treasure hunt was just to learn how to just use symmetric encryption with GnuPG. The QR code on the poster linked to a short explanation extracted from the book *Beautiful Security* by Oram and Viega [8]. The short web page explained that PGP stood for "Pretty Good Security". The web page explained how to download and install GnuPG.

Attached to the web page holding the lesson was a file encrypted with a symmetric key using GnuPG. The web page also gave the key used. The challenge was merely to download the software and decrypt the file. The file held the location of the treasure box.

15.5.3 Jasypt and Bouncy Castle

Jasypt is a simple encryption library for Java. Since this treasure hunt would actually require Java programming, a dual-path treasure hunt was created. The poster starting this treasure hunt had two QR codes. One was clearly labeled for programmers and one was clearly labeled for non-programmers such as business analysts, project managers, database administrators, and testers.

Non-programmers were given a short review of GnuPG and provided with an overview of Jasypt and explained how the programmers were provided with a Java data object which had various fields. Various fields of the object were encrypted with various keys. The non-programmers were to assist the programmers.

The programmers, in the meantime, were provided libraries and documentation about how to use Jasypt. There were also provided with a Java serialized object which had various fields encrypted with various keys.

A programmer and non-programmer needed to pair-up to decrypt the correct fields. Numerous treasure hunts were created using Jasypt and Bouncy Castle as the provider. This gave opportunities to discuss various different algorithms available.

15.5.4 SSL

The treasure hunt informing people how to create public-private keys and use SSL was theoretically a one-path treasure hunt. There was no programming required. However, the treasure hunt did break into two paths depending on whether the user had Linux as an operating system on his or her computer or whether the user had Windows as an operating system.

In both cases, the users were informed how to create an SSL key-pair and shown how the public key was to be distributed and the private key must be kept secret. A specific web site had been developed which allowed users to upload their public key. The public key was then stored in an appropriate location on the server and would allow the participant to then use SFTP to access a specific file.

The file held the location of the treasure box of chocolates. To encourage fun and play, after participants found the treasure box, they were free to move the treasure box to a different physical location in the office and then upload the description of the new location of the treasure box to the server.

15.5.5 Wireshark

Wireshark allows a user to look at the underlying packets of information which are flowing through the user's computer to/from a specific host computer [9].

The goal of this particular treasure hunt was to show both programmers and business analysts how easy it was to view the contents of unencrypted packets of information flowing between the client and server.

The QR code on the poster initiating this treasure hunt provided the participants with instructions on how to download and install Wireshark on their workstations. The treasure hunt challenge also had a small program which simulated an unencrypted client login to the company's server software. In place of valid login credentials, however, the small program had strings indicating the location of the treasure box of chocolates.

15.5.6 HTTPS vs HTTP

A second treasure hunt using Wireshark was held to emphasize the importance of using HTTPS over HTTP. The preparation for this treasure hunt included developing a very simplistic web site which ran using HTTP and did not use HTTPS. The login page of this simplistic web site inserted the location of the treasure box of chocolates inside of HTTP response headers. These HTTP headers were not displayed in the browser that the participant used. The participant would need to use Wireshark to track down these HTTP headers and find the location of the treasure box.

15.5.7 Various Attack Strategies

Four treasure hunts were offered which encouraged the treasure seekers to create various online and/or offline attacks against against simplistic web sites created especially for the treasure hunt and against encrypted strings and files. The philosophy behind these treasure hunts was that unless a programmer had launched any specific attack against data or a web site, he or she would not be able to defend against it.

In the first treasure hunt, an encrypted string containing the location of the treasure box was provided. The string was encrypted using a symmetric key consisting of just a four digit number using a specified algorithm. The programmers were not given the actual number but were given the algorithm. They were encouraged to program a brute force attack trying all possible four digit numbers in order to find the location of the treasure box.

In the second treasure hunt, an encrypted string containing the location of the treasure box was again provided. The string was encrypted with a symmetric key which consisted of just an English word. The programmers were not given the actual word but were told it was a common word. They were encouraged to find an English word list and to program a dictionary attack in order to find the location of the treasure box.

In the third treasure hunt, the encrypted string containing the location of the treasure box was again encrypted with a common English word. The word, however, was in mixed case containing both uppercase letters and lowercase letters.

The programmers needed to again program a dictionary attack but augment it with a recursive algorithm to try all different combinations of uppercase and lowercase letters in the word. Because of the length of time involved in such an attack, the programmers were told that the word contained only five letters or less.

In the fourth treasure hunt, the encrypted string containing the location of the treasure box was again encrypted using a mixed-case common English word. The word, however, had common letters replaced by symbols such as an *a* replaced with *@*, an *e* replaced with *3*, an *l* (el) replaced with *1* (one), an *i* replaced with! , etc. Again, in this case, the programmers were told that the password was five letters or less.

15.5.8 Steganography

Whereas encryption scrambles messages so they can't easily be read, steganography hides messages inside of other objects in such a way that they can not be easily be seen [10]. These treasure hunts required that the programmers learn about bit manipulation. The string containing the location of the treasure box was encoded within images slightly changing color RGB values by no more than 8 (3 bits). These color changes were not visible to the human eye.

In a traditional business computing environment, there is very little opportunity to do bit manipulation. For some programmers, these treasure hunts in steganography were the first time they had ever used bit manipulation. To grab the attention of the participants, popular memes were used as can be seen in Illustration 15.3.

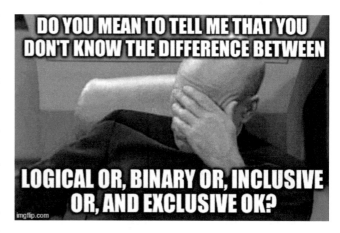

Illustration 15.3 Image introduction a Steganography Treasure Hunt

15.5.9 Going Forward

Additional topics have been scheduled for future security treasure hunts including (1) programmatic use of SSLSockets (2) different encryption algorithms (3) SQL injection (4) parsing SSL certificates.

15.6 Results

Action Research requires that evaluations are done. In this project, the evaluations were done via SurveyMonkey. The survey always first asked if the participant was a programmer, business analyst, project manager, data base administrator or tester.

The results showed that a much higher percentage of the programmers in the office took part in the treasure hunts than the non-programmers in the office. In addition, in the case of dual-path treasure hunts, programmers often did both of the paths of the treasure hunt in order to solve the puzzle and find the treasure box.

Comments on SurveyMonkey included suggestions for more involved security related treasure hunts involving prime numbers, the theory of RSA, homomorphic encryption and cyphers. These suggestions indicate that the treasure hunts, in general, were raising the awareness of security issues among the participants. In addition, management has expressed its satisfaction in this raised awareness and has encouraged additional security related treasure hunts on a number of topics listed in Sec. 5.9.

This raised awareness of security issues indicates that the research project was successful in *unfreezing* and *moving* (as defined by Lewin) the organization with respect to security issues. At this point, despite positive comments from management on these security treasure hunts, it is too early to determine if the research project will be successful in *refreezing* (term also defined by Lewin) the organizational mindset about security issues.

15.7 Conclusion

As the entire world moves to digital storage of information, computer security is becoming more and more important. Data breaches can have astounding economical costs.

Without commenting on the ethics and morality involved, data breaches such as the Ashely Madison breach and the Panama Papers can also have huge social and political fallout. This is true in both the developed world and developing world.

It is critical that members of an IT department have the required training and education to protect the IT infrastructure of their company. Often, however, there is not enough time or money to organize formal training of employees.

This paper reported on an Action Research project which implemented gamified treasure hunts which combined both tangible and virtual artifacts to teach programmers and non-programmers about various security tools such as encryption, certificates, and packet sniffing.

References

1. Ponemon Institute. (2015). 2015 cost of data breach study: Global analysis. Ponemon Institute LLC, Michigan, Tech. Rep. SEW03053-WWEN-03, May 2015.
2. Lewin, K. (1947). Group decision and social change. *Readings in Social Psychology, 3*, 197–211.
3. Rikala, J., & Kankaanranta, M. (2012). *The use of quick response codes in the classroom*. Presented at Mlearn.
4. Wyeth, P., Smith, H., Ng, K. H., Fitzpatrick, G., Luckin, R., Walker, K., Good, J., Underwood, J., & Benford, S. (2008). *Learning through treasure hunting: The role of mobile devices*. Presented at Proceedings of the International Conference on Mobile Learning 2008.
5. Kohen-Vacs, D., Ronen, M., & Cohen, S. (2012). Mobile treasure hunt games for outdoor learning. *Bulletin of the IEEE Technical Committee on Learning Technology, 14*(4), 24–26.
6. McCain, C. (2007). Scavenger hunt assignments in academic libraries: Viewpoints versus reality. *College & Undergraduate Libraries, 14*(1), 19–31.
7. Huizinga, J. *Homo ludens: A study of the play-element in culture* 1944 in German, 1949 in English.
8. Oram, A., & Viega, J. (2009) *Beautiful security: Leading security experts explain how they think*, O'Reilly Media, Sebastopol California, Available: https://books.google.com/books?id=IA8FGmoNWesC&pg=PA263.
9. Baxter, J. H. (2014). *Wireshark essentials*, Packt Publishing, Birmingham, UK, Available: https://books.google.com/books?id=naIbBQAAQBAJ.
10. Johnson, N. F., & Jajodia, S. (1998). Exploring steganography: Seeing the unseen. *Computer, 31*(2), 26–34.

Router Security Penetration Testing in a Virtual Environment

16

Christian Scully and Ping Wang

Abstract

A router is the first line of defense in a typical network. Improper configurations of the router may lead to various security vulnerabilities. Virtualization provides a safe and self-contained environment for network simulation and security testing. This paper uses a virtual penetration testing environment to simulate and analyze the two phases of a typical Advanced Persistent Threat (APT): (1) *incursion* by way of reconnaissance (passive information gathering), and (2) *discovery* by initial compromise to exploit vulnerabilities found in routers linking a corporate network with the untrusted zone of the inherently unsecure World Wide Web.

Keywords

Router • Security • Penetration Testing • Virtualization • NAT

16.1 Introduction

In a typical network infrastructure a router is the gateway connecting the untrusted zone (i.e. the Internet) with the trusted zone of a private internal network. With increasing demand for the Internet of Things (IoT), there are more "things" connected inside a trusted zone, such as home appliances, mobile devices, and vehicles. Gartner forecasted that by 2020, there would be 20.8 billion things connected to the Internet; just this year alone (2016), there are approximately 6.4 billion things connected, with 5.5 million new connections daily [3]. Thus, the router's role in protecting the network is ever more vital. For example, in the commercial sector, critical infrastructures are connected to corporate networks carrying legacy protocols (built without security in mind or with security by obscurity) that remotely control physical generators, water pumps, oil tanks, nuclear reactors, traffic signaling devices, and railways. Should these services be accessible and vulnerable to threats outside the trusted zone, there could be grave damage or widespread destruction [8]. The Aurora Project conducted on a power generator at Department of Homeland Security's Sandia National Labs in 2006 provides an account as to the impact of such vulnerabilities commonly found in Industrial Control Systems [10]. But such network vulnerabilities, compounded with those existing in the things that connect to the Internet, create a variety of challenges that must be addressed to ensure multidimensional hardened protection of critical assets.

Identifying network vulnerabilities through penetration testing is essential to prevention and protection against security threats. This paper reviews major network router attacks and uses VMware virtualization to simulate network routing and penetration testing to identify security vulnerabilities.

C. Scully (✉)
University of Maryland University College, Adelphi, MD, USA
e-mail: scullmchris@gmail.com

P. Wang
Robert Morris University, Pittsburgh, PA, USA
e-mail: wangp@rmu.edu

© Springer International Publishing AG 2018
S. Latifi (ed.), *Information Technology – New Generations*, Advances in Intelligent Systems and Computing 558, DOI 10.1007/978-3-319-54978-1_16

16.2   Review of Literature

16.2.1 Distributed Denial of Service (DDoS)

One of the most common and effective attacks against a network is a DDoS attack, primarily compromising the *availability* of service for the target network. A DDoS attack denies legitimate network user requests by overwhelming data payload (often through an army of rogue zombie devices, or *botnets*) and flooding the inbound connections into the gateway, creating a routing "choke-point." Additionally, public consumers (external to the network) who rely on services are also denied services as a result of packet flooding. Three common types of DDoS attacks are: (1) *ARP (Address Resolution Protocol) poisoning*, where the attacker focuses on ARP request packets on the target network and sends a spoofed MAC address in reply in order to deny service because of improper mapping in the network's ARP table; (2) a *ping of death,* where IP packets containing a payload of over 65,536 bytes flood the system and potentially overwhelm CPU utilization and cause undesirable effects; and (3) a *Smurf attack*, which is similar to a ping of death, but instead overwhelming the target with the processing of ICMP echo replies versus a single, large IP packet [16].

16.2.2 Configuration-Based Vulnerabilities

A primary example of poor router security is the use of default configuration settings. This includes the use of default user names and passwords, IP addresses (e.g. 192.168.1.1), and common protocols (e.g. Telnet, SSH, http, SNMP) used on an operational network. A quick Web search on 'default router passwords' reveals various reference tables useful to adversaries planning to brute force a router; one example of such a table can be found at www.routerpasswords.com, where a dropdown menu allows selection of many common routers, including some Cisco devices.

Additionally, if the router contains wireless technology with outdated default settings like Wired Equivalency Privacy (WEP), this too can provide an open door for an adversary. Stream level encryption, such as WEP and Wi-Fi Protected Access (WPA, a firmware upgrade to WEP), both of which rely on an RC4 stream cipher (only protecting data in transit and not at rest) is easily subject to breach [9].

Finally, routers may not be configured correctly to filter out unnecessary connections to ports, protocols and services. These services may be open temporarily or permanently, using aforementioned mechanisms (e.g. Telnet, SSH, http, SNMP) [7]. This example of poor router security is the

primary focus for penetration testing in this study, though any of the previous examples should be considered.

16.2.3 NAT Vulnerabilities

NAT (network address translation) packets leaving the internal network are usually translated from their internal private and non-routable IP to the external public and routable IP (and the reverse occurs when packets enter the network); a table in the router (or NAT device) provides the translation [6].

Typically, a port scan originating from a source external to the network will not be able to scan the private non-routable addresses, which is why it becomes necessary for an attacker to consider other means of gaining access into the target network (e.g. internal workstation compromise via phishing; or gaining access through an wireless access point with no security or weak encryption); once inside the network, the adversary can guess the private range being used (by a simple IPCONFIG or IFCONFIG command- depending on the target workstation's OS) and learn the size of the network, if the traditional RFC 1918 class structure is followed [14]. However, if the router is vulnerable and the table is accessible, the adversary can learn about the inside network using this approach [6].

16.3   Establish Metrics for Router Testing

As defined in NIST SP 800-115, penetration testing (pen testing) is a security evaluation of applications, systems, or networks conducted by assessors, in which real-world attacks are emulated in order to discover weaknesses that can be leveraged to circumvent security features- using tools and techniques commonly used by adversaries to launch real attacks against live systems and data [12]. The main goal of pen testing is to identify security vulnerabilities. This type of testing is valuable for determining a system's tolerance to attacks, level of attack sophistication, and appropriate countermeasures for detecting and mitigating such threats. In all instances where pen testing would be conducted on live production networks, it is imperative that written permission is obtained from the system owner prior to any pen testing, else the "attack" is considered unethical (e.g. "black hat" hacking) and the pen tester is subject to criminal prosecution [1, 4]. Virtualization, an alternative to live testing on an operational network as demonstrated in this analysis, can provide a safe laboratory-like environment where simulated systems can be configured to operate like live systems, achieving like effects when like vulnerabilities are present. Of course, much would have to be known about the target system, and in some instances, configuration information may be protected (and rightfully so), only to be shared with individuals who possess a valid "need to know."

The primary goal of achieving router security can be best illustrated on a CIA triad (confidentiality, integrity, and availability) [5]. In the simplest of terms, *confidentiality* is to protect secrets; *integrity* is to protect information accuracy and authenticity; and *availability* is to ensure the stability and reliability of a system [9]. The focus of penetration testing is to identify vulnerabilities, risks, as well as gaps in the organization's CIA triad; and even though different sectors may focus primarily on different components of the triad, all components are still vital for complete multidimensional protection of a network.

16.4   Virtualization Methodology

Poor router configuration is the primary focus of study in this paper. The methodology in this study assumes that the attacker already knows the username and password (i.e. *msfadmin*) of the target system; this could easily be replaced by default usernames and passwords (found online) that are not changed upon initial configuration. Knowing both the open ports available to an adversary along with default or supplied login credentials, and the proper exploitation tools (e.g. Metasploit Framework), the adversary has a potential means of gaining initial access into a private network with relative ease.

Prior to conducting the tests contained in this study, and if the IP address is already known, the attacker can conduct a simple access test by opening up a web browser and entering the gateway's externally facing public IP address (assigned by the Internet Service Provider, or ISP). Upon execution, if immediately prompted for username and login credentials (and in some cases the name of the device may be displayed in the prompt), the attacker can try supplied credentials or defaults (found in online router quick reference tables) such as "*admin, password, 1234*" (or in the following methodology: "*msfadmin*"). If this proves successful, the adversary has exclusive access to the router's graphical user interface (GUI, e.g. Netgear or Linksys routers, et al.) or command line interface (CLI, e.g. Cisco routers) and can easily make configuration changes to a router as if he or she was the authorized network/system administrator. Additionally, the adversary can access this interface by taking advantage of the target's Wi-Fi capabilities; in fact, once access is acquired by connecting through the wireless access point (despite whether or not the connection requires authentication), the adversary has successfully gained access to the internal network and can conduct further reconnaissance to map out the entire network behind the router.

Additionally, the pen testing conducted in this study provides additional mechanisms for gaining remote access into a router (e.g. SSH, Telnet, FTP). Many routers, such as Cisco and Netgear, use the SSH or Telnet protocol for remote access. It may be necessary to block these services from inbound connections, which will be further discussed in security countermeasures.

This test was primarily conducted on a Windows 10 laptop with VMware Workstation Player version 12. There were three virtual machines used: one as the adversary workstation with the latest 32-bit version of Kali Linux (with the IP of 192.168.25.134); a target workstation containing "Metasploitable 2," which is an Ubuntu Server 8.04 with added vulnerabilities built for pen-testing (192.168.25.135); and a clean version of Ubuntu Server 8.04 as the normal operating baseline (for the purpose of comparing results with the more vulnerable machine) (192.168.25.136). On the Kali Linux machine, NMap and several auxiliary scanners within the Metasploit Framework console were used to enumerate and attempt to gain access into the target Ubuntu system, as well as the normal operating baseline machine. For router simulation, no actual routers were used (due to cost and virtualization needs); network adapters on each virtual machine were configured for NAT to simulate a dual-network routing configuration. For both hosts being tested, they were assigned a username and password of "*msfadmin*."

16.4.1   Security Benefits of Virtualization

Virtualization can provide a safe, minimum-risk alternative to live pen testing of operational production networks [15]. Creating a notional laboratory environment may even eliminate the requirement of obtaining written permission to conduct a test. Additionally, in virtualization, using an environment such as VMware, the pen tester can create snapshots of normal operating baselines prior to introducing malicious exploits onto the target machine, and revert back to the normal operational baseline snapshot after introducing malware onto the network and analyzing it. This allows for repetitive analysis using different tools to analyze different malware characteristics separately (albeit static and/or dynamic analysis), since it would be ineffective to use all malware analysis tools concurrently in a single exploit test iteration.

16.4.2   Penetration Tests

The tests conducted are reported as follows in the succeeding paragraphs. No written permission was required to perform these tests, as they were conducted in a virtual environment and not on a live production network. All testing was conducted against notional targets.

16.4.2.1 NMap

Using Kali Linux, the nmap –sV –n command was used against both the vulnerable host (.135) and the normal operating baseline (.136). Figure 16.1 shows the open ports discovered in the .135 port scan:

On the operational baseline, the initial NMap scan retrieved 1000 closed ports. Subsequent scans attempted to enumerate the host, including a –A and –O scan. Each revealed the same result of the initial scan, with 1000 closed ports. No more information could be gathered about the normal operating baseline from these tests alone, which may suggest that Linux systems, straight out of the box, are fairly secure by default.

In the scan of the vulnerable machine, there was no need to conduct further (and more aggressive) scanning since the first scan revealed 23 open ports and the type of operating system installed on the target device.

16.4.2.2 MSF Console

Using Metasploit, several vulnerability tests were conducted. First, a TCP scan revealed 26 open ports on the vulnerable host including 3 additional ports that were not discovered with the initial NMap scan. The clean host scan revealed 0 open ports.

Following the initial TCP port scan, several additional tests were conducted against other open ports on the vulnerable host, and the clean host. Figures 16.2 and 16.3 depict a successful FTP and SSH (respectively) brute force attempt against the vulnerable host. On the clean host, all connections were denied, even when supplied with the proper username and password.

16.5 Findings

If the router is configured for security properly, then regardless of whether or not ports, protocols and services are open on the actual devices within the network, such services may remain restricted to the internal network behind a secure router and firewall configuration and likely will not be seen from the outside. Ideally, any scan conducted against a public facing IP address should reveal only those services necessary (e.g. ports 80 and 443 for Internet access), while port scans conducted internally may reveal additional ports, protocols and services not detected in the outside scan. Once an adversary gains access into the target network, further port scanning within will reveal these services as each host is mapped and enumerated using the same techniques discussed in this study.

The findings in this study reveal that if the vulnerable host had been an actual public-facing router connected directly to the untrusted zone of the World Wide Web, the adversary could exploit many open ports, protocols and services that are available. Each enabled service associated with its own open port provides an open door for access, and the

Fig. 16.1 Nmap scan on vulnerable host

```
root@kali2016:~# nmap -sV -n 192.168.25.135

Starting Nmap 7.25BETA1 ( https://nmap.org ) at 2016-10-10 12:01 EDT
Nmap scan report for 192.168.25.135
Host is up (0.00095s latency).
Not shown: 977 closed ports
PORT     STATE SERVICE     VERSION
21/tcp   open  ftp         vsftpd 2.3.4
22/tcp   open  ssh         OpenSSH 4.7p1 Debian 8ubuntu1 (protocol 2.0)
23/tcp   open  telnet      Linux telnetd
25/tcp   open  smtp        Postfix smtpd
53/tcp   open  domain      ISC BIND 9.4.2
80/tcp   open  http        Apache httpd 2.2.8 ((Ubuntu) DAV/2)
111/tcp  open  rpcbind     2 (RPC #100000)
139/tcp  open  netbios-ssn Samba smbd 3.X - 4.X (workgroup: WORKGROUP)
445/tcp  open  netbios-ssn Samba smbd 3.X - 4.X (workgroup: WORKGROUP)
512/tcp  open  exec        netkit-rsh rexecd
513/tcp  open  login?
514/tcp  open  tcpwrapped
1099/tcp open  rmiregistry GNU Classpath grmiregistry
1524/tcp open  shell       Metasploitable root shell
2049/tcp open  nfs         2-4 (RPC #100003)
2121/tcp open  ftp         ProFTPD 1.3.1
3306/tcp open  mysql       MySQL 5.0.51a-3ubuntu5
5432/tcp open  postgresql  PostgreSQL DB 8.3.0 - 8.3.7
5900/tcp open  vnc         VNC (protocol 3.3)
6000/tcp open  X11         (access denied)
6667/tcp open  irc         Unreal ircd
8009/tcp open  ajp13       Apache Jserv (Protocol v1.3)
8180/tcp open  http        Apache Tomcat/Coyote JSP engine 1.1
MAC Address: 00:0C:29:4E:76:8C (VMware)
Service Info: Hosts:  metasploitable.localdomain, localhost, irc.Metasploit
:/o:linux:linux_kernel

Service detection performed. Please report any incorrect results at https:/
Nmap done: 1 IP address (1 host up) scanned in 17.94 seconds
root@kali2016:~#
```

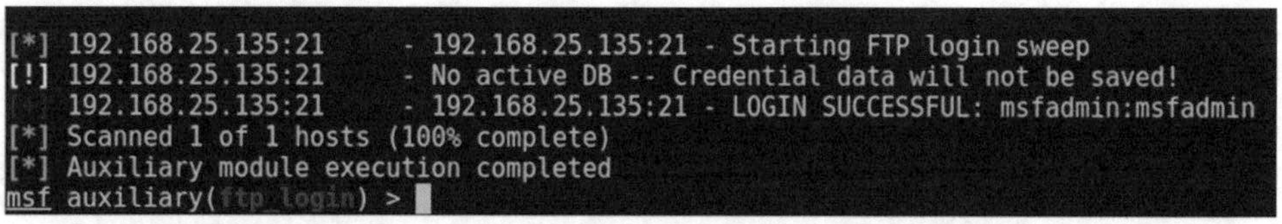

Fig. 16.2 FTP Login scanner against the vulnerable host-successful connection

```
[*] 192.168.25.135:21      - 192.168.25.135:21 - Starting FTP login sweep
[!] 192.168.25.135:21      - No active DB -- Credential data will not be saved!
    192.168.25.135:21      - 192.168.25.135:21 - LOGIN SUCCESSFUL: msfadmin:msfadmin
[*] Scanned 1 of 1 hosts (100% complete)
[*] Auxiliary module execution completed
msf auxiliary(ftp login) > 
```

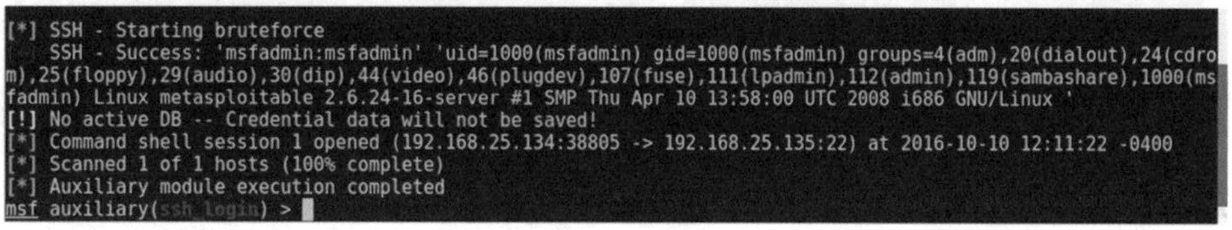

Fig. 16.3 SSH Login scanner against the vulnerable host-successful connection

```
[*] SSH - Starting bruteforce
    SSH - Success: 'msfadmin:msfadmin' 'uid=1000(msfadmin) gid=1000(msfadmin) groups=4(adm),20(dialout),24(cdro
m),25(floppy),29(audio),30(dip),44(video),46(plugdev),107(fuse),111(lpadmin),112(admin),119(sambashare),1000(ms
fadmin) Linux metasploitable 2.6.24-16-server #1 SMP Thu Apr 10 13:58:00 UTC 2008 i686 GNU/Linux '
[!] No active DB -- Credential data will not be saved!
[*] Command shell session 1 opened (192.168.25.134:38805 -> 192.168.25.135:22) at 2016-10-10 12:11:22 -0400
[*] Scanned 1 of 1 hosts (100% complete)
[*] Auxiliary module execution completed
msf auxiliary(ssh login) > 
```

adversary could choose from a wide variety of tools (available on Kali Linux) to craft an exploit that will successfully provide a path into the target network, ultimately allowing the adversary to acquire the means deemed necessary to achieve his or her final objective. With the right amount of information gained in the reconnaissance phase of the APT methodology, these penetration methods are very simple and can take little to no time at all.

16.6   Conclusion and Suggestions

The simulated penetration testing in this study shows that poorly configured network routers are easily vulnerable to security threats and exploitations. The virtualization approach to pen testing proves to be effective and informative with minimal risks. Based on the pen testing results, security countermeasures should be considered. Before routers are physically placed on a live infrastructure, they should be configured properly for optimum security for hardened perimeter protection. Several protection mechanisms prior to introducing a router onto a live production network include: (1) the use of *Access Control Lists (ACL's)* which can allow (legitimate) or deny (rogue) users connection to the internal network, a highly effective method at thwarting DoS attacks [13]; (2) *rate limiting,* which can be used in conjunction with the ACL to throttle back bandwidth on inbound connections, acting as a Quality of Service (QoS) mechanism in order to ensure optimal packet transmission quality and service availability [2], and prevent Permanent DoS (PDoS) attacks [13]; (3) *disabling unnecessary ports,* protocols and services (e.g. ICMP, SSH, Telnet) to prevent scanning [11]; (4) *changing login credentials* from defaults, which will make it difficult for an adversary to consult router reference tables online for brute force attempts; and (5) *securing wireless connections* with WPA-2 encryption (a block level encryption that protects data in transit and at rest) [9]. Additionally, in Wi-Fi security, Service Set Identifiers (SSID) should be changed in such a way that will not reveal the type of router being used (security by obscurity). Finally, a defense-in-depth

strategy, which encompasses a variety of mechanisms necessary to harden a network, is the best approach because it helps to ensure that network defenders are considering, from a multidimensional perspective, all of the possible vectors that an adversary can choose to target a network at any time; and only one vector is required to achieve victory.

References

1. Certified Ethical Hacker, version 8: Ethical Hacking and Countermeasures (Common Body of Knowledge). (2011). *E-commerce council.* Manila: E-C Council Publishing.
2. Cisco IOS Quality of Service Solutions Command Reference, Release 12.2. (n.d.). *Cisco corporation.* Retrieved from: http://www.cisco.com/c/en/us/td/docs/ios/12_2/qos/command/reference/fqos_r/qrfcmd1.html
3. Gartner Press Release: Gartner says 6.4 Billion Connected "Things" Will Be in Use in 2016, Up 30 Percent From 2015. (2015, November 10). *Gartner.* Retrieved from: http://www.gartner.com/newsroom/id/3165317
4. Henry, P. (2016). *SANS Security 502: Perimeter Protection in Depth- Lecture notes. Personal collection of Paul A. Henry.* CISSP, SANS Institute: Bethesda.
5. Huang Y., & Lee, W. (2004). Attack analysis and detection for Ad Hoc routing protocols. *College of Computing: Georgia Institute of Technology.* Retrieved from: http://wenke.gtisc.gatech.edu/papers/raid04.pdf
6. Johnson, D., & Hartpence, B. (2010). *A re-examination of network address translation security.* Rochester Institute of Technology. Retrieved from http://scholarworks.rit.edu/other/761/?utm_source=scholarworks.rit.edu%2Fother%2F761&utm_medium=PDF&utm_campaign=PDFCoverPages
7. Küçüksille, E., YalçŌnkaya, M., & Ganal, S. (2015). Developing a penetration test methodology in ensuring router security and testing it in a virtual laboratory. *Association for Computing Machinery (ACM).* Retrieved from: http://dl.acm.org/citation.cfm?doid=2799979.2799989
8. Koppel, T. (2015). *Lights out: a cyberattack, a nation unprepared, surviving the aftermath.* New York: Crown Publishers.
9. Lee, D. (2014). *(ISC)^2 CISSP Common Body of Knowledge (CBK) Seminar- Linthicum, MD: Lecture notes. Personal collection of Dennis Lee, CISSP, (ISC)^2, Queens.*
10. Meserve, J. (2007). Sources: Staged cyber attack reveals vulnerability in power grid. *CNN.* Retrieved from: http://www.cnn.com/2007/US/09/26/power.at.risk/index.html?iref=topnews

11. Metasploit unleashed: scanner HTTP auxiliary modules. (2016). *Offensive security*. Retrieved from: https://www.offensive-security.com/metasploit-unleashed/scanner-http-auxiliary-modules/

12. NIST Special Publication (SP) 800-115: Technical guide to information security testing and assessment. (2008). *National Institute of Standards and Technology (NIST)*. Retrieved from: http://nvlpubs.nist.gov/nistpubs/Legacy/SP/nistspecialpublication800-115.pdf

13. Rao, S., & Reed M. (2011). Denial of Service attacks and mitigation techniques: Real time implementation with detailed analysis. *SANS Institute: Infosec Reading Room*. Retrieved from: https://www.sans.org/reading-room/whitepapers/detection/denial-service-attacks-mitigation-techniques-real-time-implementation-detailed-analysi-33764

14. RFC 1918: Address Allocation for Private Internets. (1996). *Internet Engineering Task Force (IETF)*. Retrieved from: https://tools.ietf.org/html/rfc1918

15. Sikorski, M., & Honig, A. (2011). *Practical malware analysis: The hands-on guide to dissecting malicious software*. San Francisco: No Starch Press.

16. Waichal, S., & Meshram, B. (2013). Router attacks-detection and defense mechanisms. *International Journal of Scientific & Technology Research, 2*(6). Retrieved from: http://www.ijstr.org/final-print/june2013/Router-Attacks-detection-And-Defense-Mechanisms.pdf

Advanced Machine Language Approach to Detect DDoS Attack Using DBSCAN Clustering Technology with Entropy

Anteneh Girma, Mosses Garuba, and Rajini Goel

Abstract

Service availability is the major and primary security issue in cloud computing environments. Currently existing solutions that address service availability-related issues that can be applied in cloud computing environments are insufficient. In order to ensure the high availability of the offered services, the data centers resources must be protected from DDoS attack threats. DDoS is the major and the most serious security threat that challenges the availability of the data centers resources to the intended clients. The existing solutions that monitor incoming traffic and detect DDoS attacks become ineffective if the attacker's traffic intensity is high. Therefore, it is necessary to devise schemes that will detect DDoS attacks even when the traffic intensity is high; such schemes must deactivate DDoS attackers and serve the legitimate users with available re-sources. This research paper addresses the need to prevent DDoS attacks by defining and demonstrating a hybrid detection model by introducing an advanced and efficient approach to recognize and efficiently discriminate the flood attacks from the flush crowd (legitimate access). Moreover, this paper introduce and discusses, most importantly, the application of multi-variate correlation among the selected and ranked features to significantly reduce the false alarm rate, which is one of the major issue associated with the current available solution.

Keywords

Cloud Security • Clustering • Entropy • DBSCAN • DDoS

17.1 Introduction

Cloud computing uses networks of large groups of servers that typically run low-cost consumer PC technology with specialized connections in order to spread data processing chores across them. This shared IT infrastructure contains large pools of systems that are linked together. The main goal of cloud computing is to apply high performance computing power to perform tens of trillions of computations per second in consumer-oriented applications. While providing these client oriented business applications on shared computing environments using internet service, cloud service providers (CSP) face a huge challenges of security to satisfy their clients' needs.

Information confidentiality, system integrity, and service availability are the three main challenges that every CSP currently face. There are also other major challenges and risks in cloud computing, such as user privacy, data security, data lock-in, disaster recovery, performance, scalability, energy-efficiency, and programmability. There are various ways to look at the security risks for cloud computing, regardless of the service model being examined [1].

A. Girma (✉) • M. Garuba
Computer Science Department, Howard University,
Washington, DC, USA
e-mail: agirma@howard.edu; moses@scs.howard.edu

R. Goel
Department of Information Systems & Decision Sciences, Howard University, Washington, DC, USA
e-mail: rgoel@howard.edu

S. Latifi (ed.), *Information Technology – New Generations*, Advances in Intelligent Systems and Computing 558, DOI 10.1007/978-3-319-54978-1_17

Fig. 17.1 DDoS attack
framework [3]

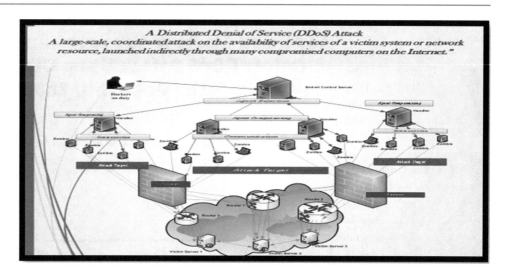

Anteneh G. and et al. [2] discussed and summarized the cloud computing security issues that are associated with all three cloud services. The most common option of using specialized malware to infect the machines of users who are unaware that their machines are compromised or the relatively newer option of amassing a large number of volunteers willing to use DDoS programs in unison. Anteneh G. and et al. [3] explained the organization of DDoS attack as it has been described with Fig. 17.1.

Defensive/detection mechanisms deployed in a distributed/hybrid manner against DDoS attacks have been considered the most promising and are believed to be the potentially effective. The main reasons noted for the success of hybrid detection schemes include: their capability of early DDoS attack detection, their ability to trace back the flooding attack to sources, and their ability to prevent the attack. Different DDoS attacks detection and protection mechanisms are discussed below in the literature review section.

17.2 Literature Review

Researchers have proposed different solution models and have described the impacts of major factors (metrics) used to measure the performance of their different detection schemes. These metrics include defense capability, performance issues, detection response, detection plan, limitations, ease of use, and practicality, as well as how and where we deploy these detection schemes. There are also a number of encouraging and promising analysis made in the area of data mining, which deals with identifying patterns of objects to determine their connectivity (association) or observing a new pattern that can result in a change regarding how the data is organized (classification) or discovering and visually documenting new groups of facts (clustering), or discovering patterns in data that can help in prediction analytics

(forecasting). We have selected one of these four data mining techniques, clustering, for this research to run in parallel with our selected statistical method (Entropy).

Dominca Arila et al. [4] investigate the existing DBSCAN clustering techniques and introduce a new method by using a parallel clustering data mining approach. They exploit the modularity of sequential application and how it can play an important role in redesigning existing software codes, and they present their views on how DBSCAN clustering techniques can enhance the overall performance on a large dimensional dataset. They conclude with the necessity of future research to verify the scalability results for large input sets.

K. Muntaz et al. [5] make a very detailed analysis of the DBSCAN algorithm by explaining its behavior, including how it is organized and its advantages and disadvantages. Moreover, the researchers state the necessary steps involved in the execution of the DBSCAN algorithm, and finally they emphasize that DBSCAN clustering technology plays a very important role in class classification in spatial database environments.

Kamran Khan et al. [6], review and explain different data mining techniques. The researchers survey and explain the advantage and disadvantages of DBSCAN algorithms in the data mining process. After they discover and point out some of the limitations of these DBSCAN model algorithms, they represent their view on how to get these different density-based clustering algorithm to the next level. In another study, Martin Ester et al. [7] evaluate the DBSCAN algorithm using a real data experiment. After their experiment and investigation, they discuss their result, showing that a DBSCAN algorithm is highly effective in determining clusters of different shapes better than the other clustering techniques. The researchers conclude that the application of clustering algorithms is very interesting and produces very good results, and finally they recommend future research that could be done to investigate and enhance the DBSCAN algorithm.

Li Ma et al. [8] discuss another DBSCAN data mining algorithm that has improved features over the normal DBSCAN algorithm. They point out and explain that this improved clustering approach can reduce the number of query objects required and can improve the overall performance of the DBSCAN clustering algorithm. The researchers show how the DBSCAN algorithm can be improved by taking the limitation of the normal DBSCAN algorithm and applying the grid method.

Other researchers, S. Vijayalaksmi and M. Punithavalli [9], present another approach to DBSCAN algorithms using a k-distance algorithm graph model. They also discuss how the EPS and MinPts values can be automatically calculated and applied with DBSCAN clustering technology. They further indicate how this approach enhances the performance and scalability of the clustering function. Finally they demonstrate how their approach can efficiently handle a large data dataset.

In a related analysis, Yanyan Wang et al. [10] present an incremental clustering algorithm for detection purposes - one specifically applied to software vulnerabilities. They investigate the unknown side of software vulnerabilities that can endanger the overall productivity of the software package, and they propose an incremental approach clustering technique to efficiently detect them by building the vulnerability pattern library. Similarly, Jeffrey Erman et al. [11] evaluate two clustering algorithms -- K-Means and DBSCAN and make a comparison with other algorithms previously used for network traffic classification. After making their experiment, they discuss and demonstrate how both the K-Means and the DBSCAN algorithms execute very well with a better performance rate. The experimental result also indicate that DBSCAN clustering capabilities are very attractive even though they produce less accuracy.

Sanjay Chakraborty and N.K. Nagwani [12] investigate the density-based spatial clustering of application with noise (DBSCAN) and describe its incremental behavior in the data mining process. They present the performance of the incremental DBSCAN model by using graphical examples and explain the benefits of this approach. Among the benefits described in their research, they indicate that the performance of their model saves both time and effort; it efficiently improves the time complexity that existed with the traditional DBSCAN algorithm. Another analysis made using a parallel approach while implementing DBSCAN clustering based on MapReduce was conducted by Xiufen Fu et al. [13]. These researchers describe the scalability issue that existed with previous DBSCAN clustering algorithm while handling a large database, and they introduced MapReduce as an ideal platform for parallel programming due to its important and efficient scalability, simplicity, and

fault tolerance. They clearly indicate that it is feasible to run DBSCAN parallel with MapReduce. Likewise, Slava Kisilevich et al. [14] discuss a new DBSCAN clustering technique called P-DBSCAN to explore and analyze selected areas (places) by using collections of pictures. The researchers introduced two new concepts: density threshold and adaptive density. Their main goal is to make a comparison between the traditional DBSCAN clustering algorithm and their newly approached P-DBSCAN model, and a they conclude that the introduction of the density-based approach creates more clusters, while the introduction of adaptive density helps to create small packed clusters with high density.

Zohng Su et al. [15] research to solve a problem associated with retrieving information from the internet based on log data. They introduce another clustering approach using a recursive density based clustering (RDBC) algorithm that is designed based on the DBSCAN algorithm. After experimenting with the new model, they show that the RDBC algorithm is more effective than the DBSCAN in terms of obtaining clustering results.

After reviewing and analyzing the different approaches referenced above, we have selected the DBSCAN clustering algorithm along with Entropy to propose a detection mechanism for DDoS attacks in cloud computing environments. Our approach uses the Entropy application to transform and normalize the large volume of data that could be generated by the DDoS attack, and its uses a DBSCAN clustering algorithm to efficiently discriminate the flood attacks from legitimate access to the cloud computing environment - both at the network and host-based resources.

17.3 Selected Detection Schemes for This Research

We further made a thorough comparative analysis of currently existing detection schemes against their different features affecting their performance factor. The schemes that we investigated and reviewed are Covariance Matrix [16], Entropy [16], Poisson, K-NN Classifier, Kolmogorov Complexity Matrices [17], Dempster Shafer Theory [18], Aggregate Congestion Control (ACC) [19], Hidden Markov Model [20], Decision Tree Model, D-Ward-System [21, 22], and MULTOPS [23]. Among the features that we considered for our analysis include: the detection schemes approach (statistical, knowledge base, machine learning, soft-computing), scalability, complexity, architecture, detection rate, false alarm rate, unknown attack detection, dynamic signature update, real time detection.

As discussed above, we have reviewed different existing DDoS detection schemes having different detection approaches (soft computing, knowledge base, data mining and machine learning, statistical models) for their current strength and weaknesses. After investigating their advantages, disadvantages, and making a through comparative analysis of their features and contributions, we selected two of the most efficient and have better accuracy level in detecting DDoS flooding attacks. Our proposed research model is planning to modulate a hybrid detection model that combines two of these selected schemes at different level of computation to produce a result of high accuracy and detection rate and significantly reduce the false alarm rate. After studying most of the existing detection schemes, we have selected the detection scheme using the machine learning applying the dbscan clustering technology together with Entropy. The main reason for our selection is, we can use entropy which is a statistical method for our data transformation purpose, and dbscan for its efficiency for clustering purpose so that we can easily discriminate the legitimate access from the flood attacks or vice versa. The major functionality of these detecting schemes, their strength, their short comings, and how we are applying them with this research are discussed and explained in detail in Fig. 17.2 below. The major advantage of applying dbscan clustering is its capability to recognize and differentiate the attack from normal access and cluster them separately. It does also help to detect the attacks and the normal access

associated for each selected features correlated multiple unknown attacks at a time. Entropy also has an advantage of using the entire distribution of data to estimate the entropy among selected data features.

17.4 Proposed System

17.4.1 Research Questions and Hypothesis

It is important to provide secure and reliable services in cloud computing environments, and one of the security issues is how to reduce the impact of the distributed denial of services (DDoS) attacks in these environments. The intrusion detection systems (IDS) models that have been developed and distributed by different vendors since the implementation of cloud computing are not well enough to protect the cloud computing environment from DDoS attacks. Instead, DDoS attacks have been a continuous threat to the Internet. The attacks have also increased in sophistication and severity. It has also been found that flooding packages are the most common and effective attack tool among all the attack methods. How to solve this problem is an open question and poses a considerable challenge.

Entropy and DBSCAN Property Matrix

Scheme	Functions	Strength	Shortcoming	How the Scheme is Applied
DBSCAN Clustering	-Efficiently recognize and differentiate the attack from the legitimate users. -Helps to display the attack and normal access for the selected multi-variate strongly correlated features using its density property. -Effectively and efficiently separate (cluster) the attacks from the legitimate access, including also border points.	-Easy to compute -It is effective to detect different flooding attacks. -Detection process is effectively working. -It accurately recognize and differentiate the DDoS attacks and the normal access. -	-Threshold (Epsilon value (EPS)) is arbitrary. Selecting the optimal value is always an issue for big data supposed to have as many clusters as needed. --Has drawback to work with very big data file. -The minimum number of points that make up a cluster (MinPts) is selected by the user, which is sometimes not the correct one While trying to handle an input file having many features.	- We will use dbscan algorithm for clustering purpose. - Try multiple comparison testing limits; - We use Entropy to transform our Input data and create different datasets. - We use Correlational studies to decide the ideal dataset for our research.
Entropy	-Estimate the Entropy among data features for a suspected attack. -If it is beyond a set threshold classify the data as an attack.	- Uses entire distribution of data.	-Requires estimation of the empirical data distribution or reliance on observed (multivariate) data having mass 1/n at each point	- We calculate the entropy value and Min-Max normalization technique together to complete the data transformation.

Fig. 17.2 Selected detecting schemes

Given the accelerated adoption of cloud computing in the industry and the simultaneous growth of DDoS attacks which has increased at an alarming rate, service availability is still considered one of the key challenges in the adoption of cloud computing. Therefore, the need for a better alternative solution model for real-time detection of DDoS attacks remains a central step in avoiding the disruption of cloud services. Moreover, given that flooding packages are the most common considerable challenge to the cloud computing environment and the most effective attack tool among all the attack methods, having a very effective mechanism for determining the behavior patterns of the legitimate users and the attackers will lead to a better way of distinguishing flood attacks from ash events. In this study, we are trying to address the following research questions and formulate two of the corresponding hypothesis with the intention of solving our research questions and improving the current trends in securing cloud computing environments from DDoS attacks.

Research Question 1 How can we effectively discriminate DDoS flooding attacks from sudden increase in legitimate traffic or Flash Events?

Hypothesis 1: Since DDoS attacks could exhaust the victim resources with large amount of resource requests, the application of data mining technology based on the density of objects (Incoming Packets), using clustering technology could discriminate the legitimate access from the flood attacks.

Research Question 2 How can we detect the DDoS attacks as early as possible to keep the availability of the cloud computing services intact?

Hypothesis 2: The multi-variate correlations among the features could identify the patterns behavior of the legitimate user and the flood attacks, and provide additional essential information in ranking and selecting the known parameters as features.

17.4.2 Theoretical Framework and Research Methodologies

17.4.2.1 Introduction
The main idea in our research is that the characteristics of an information system could be described by the correlations among its features. We will rank and select the known parameters as features in order to identify the patterns of all accesses to the cloud computing resources. DDoS flooding attack exhaust both of the network based resources (band width) and host based resources (Servers) very fast in order to shut down the services. Due to the nature to their attack behavior, they deplete the bandwidth of the network by bringing a new traffic behavior to the network. This type

of network can be detected using a stochastic statistical approach so that we can differentiate the normal legitimate access from such an attack. One of the main purpose of our research is to prove the existing correlation among different features of the IP packets and how it could help detecting the security threats and avoid any system degradation or service shut down. The hybrid modelling approach that we select, to prove how these attack behaviors are different from the normal and legitimate accesses; and how we can detect them are, the DBSCAN clustering technique with entropy.

Data mining technology is one of the highly recommended and commonly used data analytics prediction mechanism. These prediction techniques use different clustering algorithms. Even though these algorithms are divided into several groups, there are three major characteristics where clustering algorithms are based and categorized. The three main categories of clustering algorithm include: partitioning, density based clustering, and hierarchical based clustering. Even though each of these clustering technologies have their own contribution in the area of data mining and machine learning technology, we have found out and selected to use the density based clustering techniques for our research as it fits more better for our main concern of how early we could detect anomaly behaviors to detect DDoS attack and how we can separate the legitimate access from the attacks. The Density-Based Spatial Clustering of Applications with Noise (DBSCAN) is designed to identify the clusters of different shapes in a dataset having noises. Though it is primarily considered as an alternative to K-means algorithm for the study of predictive analytics, DBSCAN algorithm doesn't require the number of clusters to be entered in order to run.

17.4.3 The Density-Based Spatial Clustering of Applications with Noise (DBSCAN) Algorithm

The DBSCAN algorithm is designed to determine number of clusters based on the density of a region that is more dependent on the EPS and MinPts parameters which are selected by the user. The EPS parameter represents the reachability maximum distance (radius) which is the distance between the two points considered to be in the same neighborhood. And MinPts represent the reachability of minimum number of points which is the minimum number of data points to be in the same neighborhood and considered to be enough to form a cluster. It is also important to know that it may be very necessary to tune and change the value of the EPS and MinPts parameter to get the best result by using them as threshold parameters. Because of its built-in idea of noise as the name implies, it has been chosen to be one of the most commonly used algorithm for prediction analytics.

DBSCAN is a great tool for outlier and anomaly detection, because most importantly, it finds and labels outliers.

Different researchers approached different anomaly detection problems using the DBSCAN clustering algorithm to come up with the better prediction solutions. Some of the advantages of the DBSCAN algorithm include that number of clusters does not need to be known. The selection of the EPS parameter is very important to come up with a better solution for the required detection prediction.

If we are aware of the possible values of MinPts that represents the minimum number of objects that could form a cluster, we could be able to minimize the unwanted connections of clusters with thin presence of dataset points. DBSCAN algorithm also works with any distance function. But, like the selection of the EPS value, the selection of the distance function should also be carefully made. Because results may be poor for some of the distance function that do not obey standard properties of distance function (Example: Triangular Inequality, Symmetry, Non-Negativity, etc.). We have selected the Euclidean distance function for our research.

Let

$$X_i = [X_{i1}, X_{i2}, X_{i3}...X_{i16}], \text{ and}$$
$$Y_i = [Y_{i1}, Y_{i2}, Y_{i3}...Y_{i16}]$$

The Euclidean Distance:

$$E(X_i, Y_i) = D([X_{i1}, X_{i2}, X_{i3}...X_{i6}, Y_{i1}, Y_{i2}, Y_{i3}...Y_{i6}])$$

$$E(x, y) = \sqrt{\sum_{i=0}^{n} (x_i - y_i)^2}$$

17.4.4 How DBSCAN Is Applied with Our Research

Density-based clustering algorithms partition a cluster which is a region of the data space where there are lots of points, into three types: core points, border points, and Noise. We classify these points as indicators of our performance evaluation matrices (TP, TN, FP, FN) based on the results of dbscan computations. Core points have a large number of other points within a given neighborhood. The parameter MInPts defines how many points counts as a "large number", while the radius parameter defines how large the neighborhoods are around each point.

Let, $Y_i = [Y_{i1}, Y_{i2}, Y_{i3}...Y_{i6}]$ be in the neighborhood of point $X_i = [X_{i1}, X_{i2}, X_{i3}...X_{i6}]$.

If Dist (X, Y) $<=$ radius (EPS value), then it belongs to the cluster. We are running the Euclidean distance function.

According to our assumption we consider these data points as true positive and true negative outcome indicators. Boundary points are points within distance radius of a core point, but there are not enough points Zi such that sufficient neighbors of their own to be considered core. Num (Zi) < MinPts. According to our assumptions these data points are indicators of false negative outcome. If the data point happens to be neither core point nor border point then it is called Noise points. These data points have either too few neighbors or are distance radius from all core points that is greater than the EPS parameter. These are the data points considered to be false alarm rate indicators or false positive outcome.

17.4.5 The Application of Entropy for This Research

Entropy is an information theory that is applied to different scientific researches to determine the uncertainty of a certain object. Many of those researches made in the area of pattern or anomaly detection using a statistical approach, use the concept of entropy to calculate the randomness or uncertainty of the incoming packets X_1; X_2; X_3; ; xn associated with random variable (vector) X, where $X = (X_1; X_2; X_3; ; X_n)$ and each incoming packets has p number of features.

The Entropy of a multivariate discrete random variable is given by:

$$H(x) = -\sum_{x_i \in X} P(x_i)^* Log_2(P(x_i))$$

One of the major process in this research while implementing our hybrid approach is the transformation of our dataset. Entropy calculation is one of the statistical model that we apply strictly for data transformation as a pre-processing data handling model. We have selected six features from the independent incoming IP packets. Each of these packets has the probability of Pi, and we define the entropy function above for normalization of our dataset. We shall compute the probability for each of our selected packet features. The selected features include (Packet windows size, protocol type, source port, destination port, destination IP, and source IP).

17.5 Conclusion

Due to far-below par performance of most of the existing DDoS detection schemes and the very sophisticated nature of these attacks, we have proposed an advanced machine learning approach to detect the distributed denial of services attacks on cloud computing environment with entropy using clustering technology. As we are continuing this research to implement this very effective DDoS hybrid detection schemes, we are looking forward to perform more additional regressive testing to implement this comprehensive approach at both vulnerable side of the cloud computing environment (the network and host level). Counting on our preliminary encouraging and promising experimental results, and the upcoming additional performance testing phases we schedule, this proposed system could be the best alternative lasting solution for cloud computing services availability. We shall present the details of our experimental findings, results, data analysis, and our implementation plan on the next upcoming research paper.

References

1. Vjvek, R.,Vignesh, R., & Hema, V. (2013). ITSI Transactions on Electrical and Electronics Engineering (ITSI-TEEE), *An Innovative Approach to provide Security I Cloud by Prevention of XML and HTTP DDOS Attacks, 1*(1), 2320–8945.
2. Girma, A., Garuba, M., & Li, J. (2015). Analysis of security vulnerabilities of cloud computing environment service models and its main characteristic, ITNG.
3. Girma, A., Garuba, M., Li, J., & Liu, C. (2015). Analysis of dDoS attacks and an introduction of a hybrid statistical model to detect dDoS attacks on cloud computing environment, ITNG.
4. Arlia, D., & Coppola, M. (2001). Experiments in parallel clustering with DBSCAN. *Euro-Par 2001, LNCS, 2150*, 326–331.
5. Mumtaz, K., & Duraiswami, K. (2011). An analysis on density based clustering of multi-dimensional spatial data. *Indian Journal of Computer Science Engineering, 1*(1), 8–12.
6. Khan, K., Rehman, S. U., Aziz, K., Fong, S., Saraavady, S., & Vishwa, A. (2014). *DBSCAN: Past, present, and future*. IEEE, Applications of Digital Information and Web Technologies (ICADIWT). doi: 10.1109/ICADIWT.2014.6814687.
7. Ester, M., PeterKriegel, H., Sander, J., & Xu, X. (1996). A density-based algorithm for discovering clusters in large spatial databases with noise, KDD-96 proceeding.
8. Ma, L., Lei, G., Li, B., Qiao, S., & Wang, J. (2014). MRG-DBSCAN: An improved DBSCAN clustering method based on grid. *Advanced Science and Technology Letters, 8*(2), 119–128.
9. Vijayalaksmi, S., & Punithavalli, M. (2012). A fast approach to clustering datasets using DBSCAN and pruning algorithms. *International Journal of Computer Applications, 60*(14), 1–7.
10. Wnag, Y., Wang, Y., Wu, D., & Ren, J. (2013). An incremental rapid DBSCAN clustering algorithm for detecting software vulnerability. *Journal of Convergence Information Technology, 5*(1), 82–94.
11. Erman, J., Arlitt, M., & Mahanti, A. (2006). *Traffic classification using clustering algorithms*. ACM. SIGCOMM'06 Workshop September 11–15, 2006, Pisa, Italy.
12. SanJah Chakraborty, N. K. (2011). NAGWANI. *International Journal of Enterprise and Business, ISSN (Online), 1*, 2230–8849.
13. Xiufen, F., Hu, S., & Wang, Y. (2014). Research of parallel DBSCAN clustering algorithm based on MapReduce. *International Journal of Database Theory and Application, 7*, 41.
14. Slava kisiele Misslovsl. (2011). P_DBSCAN, A density based clustering algorithms for exploration and analysis of attractive areas using collections of Geo-tagged photos.
15. Su, Z., Yang, Q., Zhang, H., Xu, X., & Hu, Y. (2014). *Correlation-based document clustering using web page*. IEEE Computer Society. OAI: Citesseerx.PSU:10.1.186.4376.
16. Daniel, S. Y., & Wang, X. (2007). Covariance-matrix modeling and detecting various flooding attacks. *IEEE Transactions on Systems, MAN, Cybernetics- Part A: Systems and Humans, 37*(2), 141–142.
17. Kulkarni, A. B., Bush, S. F., & Evans, S. C. (2002). *Detecting distributed denial-of-service attacks using kolmogorov complexity metrics*. GE Research & Development Center, GE Electric Compan. Report Number 2001CRD176, Technical Information Series, February 2002.
18. Lonea, A. M., Popescu, D. E., & Tianfield, H. (2013). Detecting DDoS attacks in cloud computing environment. *International Journal of Computer Communication, 8*(1), 70–78.
19. Mahajan, R., Bellovin, S. M., Floyd, S., Ioannidis, J., Paxson, V., & Shenker, S.. (2002). *Controlling high bandwidth aggregates in the network*, presented at Computer Communication Review (pp. 62–73).
20. Xie, Y., & Yu, S. Z. (2009). A large-scale hidden semi-markov model for anomaly detection on user browsing behaviors. *IEEE/ACM Transactions on Networking (TON), 17*(1), 54–65.
21. Mirkovic, J., Prier, G., & Reiher, P. (2002). Attacking DDoS at the source. In *Proceeding of the 10th IEEE International Conference on Network Protocols (ICNP '02)*. Washington, DC.
22. Mirkovic, J., Prier, G., & Reihe, P. (2003). Source-End DDoS defense. In *Proceedings of 2nd IEEE international symposium on network computing and applications*. Information Science Institute, USC Viterbi- School of Engineering: http://www.isi.edu/~mirkovic/publications/nca03.pdf
23. Abdelsayed, S., Glimsholt, D., Leckie, C., Ryan, S., & Shami, S. (2003). An efficient filter for Denial-of-Service bandwidth attacks. In: *Proceedings of the 46th IEEE Global telecommunications conference (GLOBECOM03)* (pp. 1353–1357). IEEE. doi:10.1109/GLOCOM.2003.1258459. http://ieeexplore.ieee.org/document/1258459/

A Novel Regular Format for X.509 Digital Certificates

18

Alessandro Barenghi, Nicholas Mainardi, and Gerardo Pelosi

Abstract

Digital certificates are one of the key components to ensure secure network communications. The complexity of the certificate standard, ITU-R-X.509, has led to a number of breaches in the TLS protocol security due to certificate misinterpretation by TLS libraries. We argue that the root cause of such an issue is the complexity of the certificate structure, which can be gauged with the framework of formal language theory: the language describing digital certificates is context sensitive. Such a complexity led to handcrafted X.509 parsers, resulting in implementations which are not guaranteed to perform correct language recognition. We highlight the issues in X.509, and propose a new format for digital certificates, designed to be parsed effectively and efficiently, while retaining the same semantic expressiveness. The certificate format can be deployed gradually, is fully specified as a regular language, and is specified as a formal grammar from which a provably correct parser can be automatically derived. We validate the effectiveness of our proposal, and the linear running time provided by the approach, generating an instance of the parser with a production grade lexer/parser generation framework.

Keywords

Digital certificates · X.509 · Language based security · Parsing · Transport layer security

18.1 Introduction

Digital certificates are the key component in providing end-point authentication and data confidentiality in modern transport layer security protocols, among which the prime choice is the Transport Layer Security (TLS) protocol. The proper validation of digital certificates is a crucial point in ensuring that both the origin authentication and transport confidentiality guarantees provided by TLS are indeed in place. Such a validation process is still lacking practical evidence, as reported, among the others, by the authors

of [1]. The causes of this lack of effectiveness can be split in two sets: issues in validating the certificate chains up to trusted anchors, and problems in the correct recognition and interpretation (i.e., parsing) of the certificates. The main cause for the latter problem is the complex structure of the current digital certificate standard format, the ITU-R X.509 [2]. Indeed, the current version of the X.509 standard (ver. 3) is the result of a long history of changes and further amendments, resulting in a format containing some redundancies and a large set of constraints to be enforced. A concrete report on the security critical effects of the complexity of parsing X.509 certificates was reported in [3], where the authors report that the language of the valid X.509 certificates is indeed context sensitive. This implies that it is not possible to prove automatically that an implementation of the X.509 certificate parser is recognizing its fields and enforcing the constraints on their contents. The

A. Barenghi (✉) · N. Mainardi · G. Pelosi
Dipartimento di Elettronica, Informazione e Bioingegneria, Politecnico di Milano, 20133 Milano, Italy
e-mail: alessandro.barenghi@polimi.it; nicholas.mainardi@polimi.it; gerardo.pelosi@polimi.it

© Springer International Publishing AG 2018
S. Latifi (ed.), *Information Technology – New Generations*, Advances in Intelligent Systems and Computing 558, DOI 10.1007/978-3-319-54978-1_18

underlying language complexity continues to create favorable ground for new implementation flaws [6], especially as parsers for digital certificates are handcrafted. To bring into evidence the extent and frequency of TLS parsing implementation flaws, the authors of [7] realized a pseudo-random generator of flawed certificates. Their work reports that issues in the correct validation of certificate chains, or in discarding syntactically invalid certificates, were present in all the most commonly employed TLS libraries.

Contributions. After highlighting the language-theoretic issues affecting the current X.509 certificate format, we propose a new format designed to allow such an automated generation. Our format allows to express the same content of the current X.509 certificate standard, while being specified by a regular language. Such a constraint allows efficient parsers to be automatically generated from the provided grammar specification, providing sound guarantees on the correctness of the implemented recognizer. We designed our format to allow a gradual adoption, as certificates expressed with it will be matched by existing X.509 parsers as v4 certificates, easing the gradual roll-in of the new format. We validate our approach implementing the certificate parser with a classical lexer/parser generation toolchain, employing a deterministic pushdown recognizer to allow richer syntax-error reporting, keeping a linear parsing time.

18.2   An Analysis of X.509

The structure of an X.509 certificate is described in the ITU-R X.509 [2] standard, and in RFC 5280 [8] and its complements. The specification is provided in Abstract Syntax Notation 1 (ASN.1) [9], a semi-formalized language to express Abstract Data Types (ADT) in the form of structures having typed fields. An X.509 certificate is thus described as an ASN.1 ADT throughout its whole structure. ASN.1 allows its users to define ADTs as a composition of either *primitive* or *constructed* data types. Primitive ASN.1 ADTs include arbitrary precision integers, Boolean valued variables, and unique object identifiers (OIDs), represented as sequences of decimal digits separated by dots. In addition to the aforementioned ADTs two primitive types with a significant relevance in the description of the X.509 are the so called BIT STRINGs and OCTET STRINGs: the former one represents a stream of binary digits which may take an arbitrary length, while the latter one forces the length of the bit string to be a multiple of 8. *Constructed* ASN.1 types identify a composition of ADTs, denominated *structure*, and they are specified through the use of appropriate keywords which indicate the constraints on how the composition should be performed. In particular, a SEQUENCE of ADTs indicates that the types in it should be concatenated in the

same order as the one they appear in the description, while a SET allows them to appear in any order, provided that they appear only once. There are also iterative versions of such types, namely SEQUENCE OF and SET OF, in which the structure may be repeated either a bounded or an unbounded number of times. The CHOICE keyword demands that only one out of a list of ADT appears in an instance of the data, while the OPTIONAL keyword states that the ADT may or may not be present in the data instance. It is possible for ADTs typed as BIT STRINGs or OCTET STRINGs to be employed as containers for a structured ADT suffixing their declaration with the CONTAINING keyword, followed by the description of the contents. Lastly ASN.1 provides a syntactic element to allow to disambiguate which ADT out of a sequence of equally typed optional ones is present in an ADT instance in the form of *tags*. Tags are syntactically represented by integers enclosed in square brackets, and can be prefixed to an ADT in a sequence to act as a marker for it.

While ASN.1 allows to detail the structure of an ADT, it does not define how its instances should be encoded. This function is fulfilled by the data encoding documents contained in ITU-R-X690 and following [9], among which the one used to encode X.509 certificates is the Distinguished Encoding Rules (DER) encoding. DER states that all the ASN.1 standard ADTs are encoded as triples of *identifier*, *length* and *content octets*. The identifier octets define the ADT type. In case of ASN.1 tags, such octets are replaced by tag integer value if tag is IMPLICIT (as it can be inferred by ASN.1 tag specification), otherwise the tagged ADT is the content of a constructed type whose identifier octet is the tag integer value. The identifier octets, which also contain the information on whether the ADT is constructed or not, are followed by the length octets, which indicate the length, in bytes, of the subsequent content of the ADT. Given this encoding format, we can now highlight the first language-theoretic issue:

Issue 1 (Matching length-content encoding) *Matching arbitrary lengths with their contents requires a context sensitive recognizer, for which no generic generation algorithm from a specification exists, and no general proof of correctness can be provided* [10].

An X.509 certificate is described by the ASN.1 ADT:

```
Certificate   ::= SEQUENCE   {
tbsCert TBSCertificate,
signatureAlgorithm AlgorithmIdentifier,
signatureValue BIT   STRING   }
```

where the first element of the sequence contains all the relevant information of the certificate (e.g., its subject and issuer, and the subject's public key). The signature-Algorithm element contains the information concerning the cryptographic primitives, together with their parameters, employed by the Certification Authority (CA) to perform the signature on the contents of the `tbsCert` field, and is followed by the actual signature which concludes the certificate. The `tbsCert` element is typed as a `TBSCertificate` ADT, which has the following structure:

```
TBSCertificate ::= SEQUENCE {
  version [ 0 ] EXPLICIT Version DEFAULT v1,
  serialNumber CertSerialNumber,
  signature AlgorithmIdentifier,
  issuer Name,
  validity Validity,
  subject Name,
  subjectPublicKeyInfo SubjectPubKeyInfo,
  issuerUniqueID [ 1 ] IMPLICIT UniqueId OPTIONAL,
  subjectUniqueID [ 2 ] IMPLICIT UniqueId OPTIONAL,
  extensions [ 3 ] EXPLICIT Extensions OPTIONAL }
```

The `TBSCertificate` ADT starts with a mandatory version field (currently containing up to 3), followed by a serial number of the certificate which must be unique among the certificates issued by the same CA. The serial number is followed by a field (`signature`) describing the cryptographic primitives used to perform the signature on the certificate, with the same format as the one contained in the `Certificate` ADT, and required to have the same contents of it by the standard. Such a constraint gives rise to the following:

Issue 2 (Matching AlgorithmIdentifier ADT) *Since this matching is the comparison of an unbounded sequence of bytes in the* `AlgorithmIdentifier` *ADT, such a check entails the need for a context sensitive recognizer* [10], *raising concerns similar to the ones in Issue 1.*

Following the `AlgorithmIdentifier` typed element, a list of valid names for the issuing CA, the validity interval expressed as two dates, and a list of valid names for the subject, i.e., the public key owner, of the certificate are present in the `TBSCertificate` ADT. Lastly, the ADT structure contains an element describing the subject's public key, followed by two optional, and currently deprecated fields, `issuerUniqueID` and `subjectUniqueID`, and by the extensions field, which is mandatory in X.509 v3. The `Extensions` ADT which describes the contents of the extensions field is a SEQUENCE of elements which may include both 17 standardized ones containing additional and security critical pieces of information, such as the allowed use for the public key authenticated by the certificate, and may include custom extensions. The X.509 standard mandates the presence of some extensions only in the case the certificate is self signed, which gives rise to the following issue:

Issue 3 (Validating self signedness) *Syntactically checking if a certificate is self-signed requires to compare the two* Name*typed subject and issuer fields, which contain an arbitrary long list of valid names for the owner of the certificate, and for the issuing CA, respectively. Such a check requires a syntax analyzer able to deal with context sensitive languages* [10].

Furthermore, the X.509 standard allows the complete freedom in defining custom extensions, which may have any content. Such a freedom allows to specify contents which may be ambiguously recognized, providing an example for the following issue:

Issue 4 (Composite ADT ambiguity) *BIT STRING*s and *OCTET STRING*s *declared without* CONTAINING *keyword are always encoded as primitive ADTs, but in some X.509 ADTs (namely* subjectPublicKey *and* extnValue*) they act as constructed ones, resulting in an* ambiguity *. Indeed, an actual primitive bit or octet string may contain any bit stream, including the ones which are valid representation for constructed data types. Whenever an ADT allows an actual primitive string to be present as an alternative to another acting as a container, two different interpretations for the input are possible.*

18.3 Specification of the New Digital Certificate Format

Willing to provide a data format for digital certificates amenable to effective and efficient parsing, we devised it as a regular language over the alphabet Σ constituted of single bytes. Such a design choice allows to turn the formal specification of the data description format into an operative one by means of the algorithm translating a regular grammar into its FSA recognizer [10].

For the sake of clarity, we provide the formal specification of the said language in terms of a syntactic grammar $G = (V_t, V_n, P_s, \text{CERTIFICATE})$, and a lexical one $G_l = (\Sigma, V_t, P_l, \textbf{cert})$, as it is usual for modern technical languages. The lexical grammar describes how the input alphabet is transformed into structured symbols, the *tokens*, which are subsequently used as terminals by the syntax grammar. We will indicate the elements of Σ as two hexadecimal digits, in mono-spaced fonts, e.g., `0x2B`, while the elements of V_t are represented as strings of boldface lowercase alphabetic characters, plus the underscore. The nonterminals of the syntax grammar are reported as angular-brackets encased, uppercase strings, e.g., A. In the following, we employ the Extended Backus Naur Form to define the grammars: transforming them

cert → preamble token_list

preamble → 0x30 0x07 0x30 0x05 0xA0 0x03 0x02 0x01 0x04

token_list → token token_list | token

token → (ctr0|ctr1| ... |ctr127)* token_body term

ctr0 → 0x80   term → 0xFF

token_body → integer | rsa_oid | ascii_string

integer → inttype payload   inttype → 0x02

payload → ([0x00 − 0xFE]|0xFF 0xFF)⁺

rsa_oid → 0x05 0x2A 0x86 0x48 0x86 0xf7 0x0d 0x01 0x01 0x01

ascii_string → 0x06 [0x20 − 0x7F]⁺

Fig. 18.1 Lexical grammar of the proposed certificate format. The lexer yields to the parser the tokens contained in the token list. For some tokens, e.g., OIDs, only a representative sample is reported

⟨CERTIFICATE⟩ → preamble ⟨HASH⟩ ⟨SERIAL⟩ ⟨ISSUER⟩
        ⟨VALIDITY⟩ ⟨SUBJECT⟩ (⟨INTERM_SUF⟩ | ⟨LEAF_SUF⟩)

⟨HASH⟩ → hash_oid any   ⟨VALIDITY⟩ → time time

⟨ISSUER⟩ → ⟨NOREPEATABLE_OID⟩ (oid any)*|(oid any)⁺

⟨SUBJECT⟩ → ⟨ISSUER⟩ ⟨SUBJ_ALT_NAME⟩?
        | ⟨CRITIC_SUBJ_ALT_NAME⟩

⟨INTERM_SUF⟩ → ⟨PK_CERTSIG⟩ (⟨SELF_SIG⟩ ⟨SS_INTERM⟩ |
        ⟨NOSELF_SIG⟩ ⟨NOSS_INTERM⟩)

⟨LEAF_SUF⟩ → ⟨PK_NOCERTSIG⟩ ⟨NOSELF_SIG⟩ ⟨NOSS_LEAF⟩

Fig. 18.2 Portion of the syntactic grammar of a certificate describing the axiom production and the first level of recursion

into plain, left- or right-linear BNF ones to derive the corresponding FSA is straightforward.

Format Lexicon. The lexical grammar of the proposed format, of which the most relevant portions are reported in Fig. 18.1, is designed to provide a strategy for a gradual adoption of the format, and solve the language theoretic issues highlighted in Sect. 18.2. From a lexical standpoint, a certificate is composed of a fixed-length preamble followed by a non empty list of tokens having the 0xFF character as a list separator. The fixed length preamble contains the representation of the prefix of an X.509 certificate up to the version field, included, with the version set to 4. As a consequence any existing library will recognize our format as a new version of X.509 and, if such a format is not yet supported, will point out the error gracefully. Such a strategy allows a gradual roll-out of the new certificates, which is currently considered a crucial factor for adoption [11], as certificates are re-generated by the CAs either upon expiration of the current ones or as a transition policy. The token encoding does not allow a single 0xFF to be present in them; in case such a byte value needs to be represented, a pair of 0xFF characters is used. The use of a unique termination character avoids employing length fields, removing the need for a context sensitive recognizer reported in Issue 1. Each token can be prefixed by one or more *control byte*s, fulfilling the role of ASN.1 tags, i.e., preventing recognition ambiguities between two tokens of the same type. The contents of each token begins with a single byte encoding its type. To avoid the introduction of recognition ambiguities between control bytes and type encoding, the former have their first bit set, while the latter have the first bit clear. Such an encoding allows 127 possible control bytes, and 127 possible primitive types, which we found plentiful to describe the language at hand. We note that the use of control bytes allows us to remove the concept of constructed data types in our format. Indeed, constructed types are redundant delimiters of the structures they repre-

sent (see Sect. 18.2), unless the end of such structure cannot be just inferred from the next token found, a case which can be disambiguated using control bytes. This in turn implicitly solves ambiguities mentioned in Issue 4, since interpretation of a payload as a composition of primitive types is no longer possible.

Statement 1 (Regular Lexicon) *We can now state and prove that the proposed lexical grammar is regular. All the productions of the grammar in Fig.* 18.1 *having only terminal symbols in their right hand side generate a regular sub-language. As a consequence of this, and the fact that regular languages are closed with respect to union and concatenation, also the productions having* **token_list**, **token**, **token_body**, *and* **integer** *generate sub-languages which are regular. Consequentially, for the same reason, the axiom of the lexical grammar* **cert** *generates a regular language.*

Format Syntax. The syntactic grammar for our proposed certificate format, of which the most significant structures are reported in Fig. 18.2, is designed to tackle the issues caused by redundant contents to be checked for equality (Issue 2), and allow consistency checks which required semantic actions (e.g., checking if a certificate is self signed) to be performed syntactically (Issue 3). The CERTIFICATE nonterminal, which is the axiom of the grammar, starts by generating the fixed preamble of the certificate, followed by the hash algorithm employed to compute the signature, so that it is possible to perform the computation of the hash of the message while parsing the certificate contents. Such a fact allows a single-pass certificate validation, yielding good efficiency and limited memory requirements. All the contents of the certificate, save for the initial preamble, are to be hashed and signed by the CA. The certificate structure continues with the certificate serial number, the information concerning the issuer, the certificate validity and the subject of the certificate. We impose a uniqueness constraint on the distinguished names employed in issuer and subject (e.g., common name, organization name), due to impersonation attacks based on duplication of these names [3]. These ones are the set of names defined as mandatory to be supported in [8] (page 21), enforcing the same order of appearance to ease the constraint checking. All data defined up to here must appear in any certificate, regardless of whether it is self signed and whether it may act as an intermediate certificate in a chain or not.

Following the aforementioned pieces of information, the syntactic structure of the certificate is different depending on whether it may be employed as an intermediate in a certificate chain or not, that is its key can be used to verify signatures on other certificates, or not. Such a piece of information must be deduced in X.509 v3 from data which appear in different positions of the certificate (e.g., the appropriate use for the public key specified in the *key usage* extension and the CA flag in *basic constraints* extension). A similar difference in the certificate structure is subsequently introduced to explicitly represent whether a certificate is self-signed or not, resulting in three nonterminals, SS_INTERM, NOSS_INTERM, and NOSS_LEAF being the viable alternatives to generate the suffix of the certificate containing the remaining information. Note that the remaining combination of features is discarded as a self-signed certificate which can be used only as a leaf of a certificate chain is pointless. The case in which a certificate which is both a leaf of its certificate chain and is self signed, i.e. it is a root CA certificate which is the only member of the chain, will have its suffix expanded by the SS_INTERM nonterminal. The suffix of the certificate contains the information pertaining certificate extensions and the signature. An high level overview of its structure is as follows:

$$\langle \text{SUF} \rangle \rightarrow \langle \text{STD\_EXTENS} \rangle \langle \text{CUSTOM\_EXTENS} \rangle \times \langle \text{SIGNATURE} \rangle$$

In X.509 v3 format a number of constraints on either the presence of an extension and a previous field of the certificate, or among extensions exist. To reduce the grammar complexity, we place the information of the extensions depending on previous fields of the certificate right after the field itself. For instance, the *subject alternative names* extension is now placed right after the *subject distinguished name*, while the *extended key usage* and *key usage* ones are placed in this order after the public key algorithm OID and before the public key parameters. These extensions are generated by the PK_CERTSIGN, PK_NOCERTSIGN) nonterminals in the grammar. In contrast with X.509 v3, we require the *key usage* extension to be present, since it is crucial for security. Indeed, X.509 v3 states that if such an extension is missing, a certificate may be used as an intermediate one, which is prone to easy but dangerous misuses. We enforce a fixed order on standard extensions to easily check uniqueness constraint imposed by existing standard.

Such order is designed to minimize the distance between interdependent extensions, leaving extensions with no dependency at all at the end of the list, enforcing appearance ordering in [8].

Besides standard extensions, our format allows to specify custom extensions, maintaining the degree of freedom present in X.509 v3: such extensions appear after standard ones. Finally, the signature field concludes the certificate and contains the signature algorithm OID and the signature itself. Note that there are no algorithm parameters in signature field since we argue that it is sufficient to represent them once alongside the public key. Moreover, note that signature algorithm appears only once effectively solving the issue of context-sensitive checks reported as Issue 2.

We note that, despite our proposed format is missing some fields with respect to the current X.509 v3, we are able to represent with it all of the information contained in the older format, while retaining an easier to recognize, and mechanically generate, format. The only semantic constraint of X.509 v3 which we are not able to check syntactically is the uniqueness of certificate policies extension and custom extensions OIDs. We note that such a shortcoming cannot be coped with using a regular grammar, unless restrictions are imposed on the set of OIDs. However, we note that performing a simple bytewise uniqueness check after the recognition of the entire certificate is quite straightforward, and not likely to be mis-implemented.

Statement 2 (Regular syntax grammar) *Analogously to the procedure employed to prove our lexical grammar regular, we can prove our syntax grammar to be regular as all of its nonterminals either generate finite languages, or are combined together with operations over which regular languages are closed. Thus, the combination of the lexical and syntactic grammars are also regular.*

18.4 Implementation Strategies and Experimental Validation

We realized an implementation of the grammar described in the previous sections, and derived automatically a certificate parser from it. Our purpose was twofold: first of all we validate the absence of ambiguities in it, and then we experimentally validate the fact that our certificate format can be parsed in linear time in the size of the input.

Parser implementation. Our implementation employs the widely consolidated yacc toolchain, composed by a lexical scanner generator, `Flex`, and an LR(k) parser generator, GNU `Bison`. Our choice of a DPDA recognizer was made to provide a more meaningful error reporting, however we note that the implementation and generation of a fully regular recognizer is also possible. In implementing the lexical recognizer, we exploited the multi-state lexer generation feature available in `Flex`, which basically implement the inner product of FSAs. Such a feature allows to describe with ease the distinction between the recognition of payload and type identifiers, partitioning the inner FSA graph. The incoming byte is thus correctly matched depending on the specified lexer state. The multi-state feature is also employed to recognize constraints on public key algorithms and *key usage/extended key usage* extensions. Such constraints are different depending on the public key algorithm recognized which is matched right before them, and a solution relying on lexer states allows a clear specification of their recognition, taking into account the correct set of constraints. We lexically check such constraints as this can be done more easily inspecting the lexical grammar terminals than inspecting whole tokens. Due to the abundance of such constraints, the lexer specification is relatively large, counting 815 lines.

However, the syntax grammar in bison benefits from such an offloading, being only 457 lines long, and counting just 112 rules. Bison manages to generate the parser without conflicts, proving that the proposed grammar is unambiguous as it fulfills the LR(k) condition [10].

Performance Analysis In order to test performances of the generated parser, we need to generate a dataset using a string generator from the same grammar specification. Such generator works in two main phases: first, it parses a yacc-formatted file to build an Abstract Syntax Tree (AST) expanding at least once every grammar production. During generation phase, the algorithm randomly traverses the tree appending strings generated by each terminal symbol found. Although the algorithm is quite straightforward, we needed to implement a new generator from scratch as the current state of the art one, Yagg [12], was not fulfilling our needs. Indeed, while generating all possible strings of given length from a grammar, it has no support for binary alphabet grammars and for lexical grammar multi-states, and it is only able to generate tokens out of a limited set of regular expression operators. Instead, our tool copes with our multi-state employing lexical grammar, deals with a broader set of regular expression operators and, differently from Yagg, it only requires slight modifications to the `flex/bison` specifications, to employ them as input.

We measured the execution times of our parser on a generated dataset made of 90 sample certificates of different lengths. The machine employed for all these tests is a Linux Gentoo 13. 0 amd 64 host based on a six-core Intel Xeon E5-2603v3 endowed with 32 GiB DDR-4 DRAM. The results are reported in Fig. 18.3. The good fit between the experimentally measured timing samples and their linear

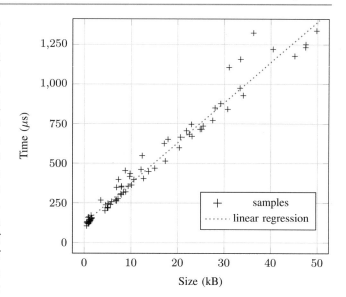

Fig. 18.3 Parsing execution timings as a function of the certificate size

regression clearly shows a linear parsing time. Moreover, we note that our parser recognition times are actually really short, as recognizing 1 kB takes only 25 μs, after a fixed bootstrapping time of 121 μs. To provide a somewhat fair comparison, we checked execution time of the OpenSSL `d2i_X.509_fp` function, which simply reads an ASN.1 DER encoded certificate from a file and stores the information in an internal data structure, leaving the check of syntactic constraints to the validation routine. Such a simpler function is as fast as our complete parsing routine, since OpenSSL employs on average 126 μs, while our parser (stripped and compiled with release-grade optimizations) 127 μs on datasets with an average certificate size of approximately 1500 bytes.

18.5 Conclusion

In this paper, we present a novel format for X.509 certificates. Such format is proven to be unambiguous, regular, and easily deployable onto current infrastructure, as it can be gracefully recognized by X.509 v3 implementations as a v4 certificate. We provide also an implementation of a recognizer employing yacc toolchain, showing linear parsing time.

References

1. Georgiev, M., Iyengar, S., Jana, S., Anubhai, R., Boneh, D., & Shmatikov, V. (2012). The most dangerous code in the World: Validating SSL certificates in non-browser software. In *CCS'12* (pp. 38–49). ACM.

2. International Telecommunication Union. (2012). Recommendation ITU-T X.509: Open systems interconnection – public-key and attribute certificate frameworks. https://www.itu.int/rec/T-REC-X.509-201210-I.

3. Kaminsky, D., Patterson, M. L., & Sassaman, L. (2010). PKI layer cake: New collision attacks against the global X.509 infrastructure. In *FC'10* (LNCS, Vol. 6052, pp. 289–303). Springer.

4. Marlinspike, M. (2002). Internet explorer SSL vulnerability. https://moxie.org/ie-ssl-chain.txt.

5. Marlinspike, M. (2009) Null prefix attacks against SSL/TLS certificates. https://moxie.org/papers/null-prefix-attacks.pdf.

6. Delignat-Lavaud, A. (2014). BERserk vulnerability Part 1 and 2: RSA signature forgery attack due to incorrect parsing of ASN.1 encoded DigestInfo in PKCS1 v1. 5 and certificate forgery in mozilla NSS (Technical report–Intel advanced threat research team). http://www.intelsecurity.com/advanced-threat-research/berserk.html.

7. Brubaker, C., Jana, S., Ray, B., Khurshid, S., & Shmatikov, V. (2014). Using frankencerts for automated adversarial testing of certificate validation in SSL/TLS implementations. In *SP'14* (pp. 114–129). IEEE CS.

8. Cooper, D., Santesson, S., Farrell, S., Boeyen, S., Housley, R., & Polk, W. (2008). Internet X.509 public key infrastructure certificate and CRL. RFC 5280.

9. International Telecommunication Union. (2016). *ITU-T recommendations by series: X series: Data networks, open system communications and security*. https://www.itu.int/rec/T-REC-X.

10. Harrison, M. A. (1978). *Introduction to formal language theory*. Boston, MA: Addison-Wesley Longman Co., Inc.

11. Ben Laurie, E. K., & Langley, A. (2014). Certificate transparency, https://www.certificate-transparency.org/.

12. Coppit, D., & Lian, J. (2005). Yagg: An easy-to-use generator for structured test inputs. In *ASE '05*, Long Beach, CA, November 7–11, 2005 (pp. 356–359). ACM.

A Software-Defined Networking (SDN) Approach to Mitigating DDoS Attacks

19

Hubert D'Cruze, Ping Wang, Raed Omar Sbeit, and Andrew Ray

Abstract

Distributed Denial of Service (DDoS) attacks are a common threat to network security. Traditional mitigation approaches have significant limitations in addressing DDoS attacks. This paper reviews major traditional approaches to DDoS, identifies and discusses their limitations, and proposes a Software-Defined Networking (SDN) model as a more flexible, efficient, effective, and automated mitigation solution. This study focuses on Internet Service Provider (ISP) networks and uses the SDN security implementation at Verizon networks as a case study.

Keywords

DDoS · Mitigation · SDN · ISP · Verizon

19.1 Introduction

Distributed Denial of Service (DDoS) attacks are often launched by a remotely controlled network of Zombies or Botnets with spoofed IP addresses to send malformed packets or service requests to flood target systems and exhaust network bandwidth and server resources intended for legitimate users [35]. DDoS attacks have been on the rise. Report from Akamai shows 126% increase of DDoS attacks in the first quarter compared with the same period a year ago [1].

While some other cyber-attacks aim to compromise confidentiality and integrity of network systems and data, DDoS attacks are a direct and major threat to the availability of

network systems and services. The threat is to interrupt and disrupt online services such as e-commerce and financial operations but also lead to revenue losses and loss of customer trust and confidence for online businesses and Internet Service Providers (ISPs). Therefore, mitigating DDoS attacks in term of preventing, detecting, and responding to the attacks has been a challenging and significant issue to both cybersecurity professionals and researchers. In addition, DDoS mitigation is an important issue not only for enterprise networks but also for ISPs who should play a key role in the DDoS mitigation process as ISPs provide Internet access for online businesses and DDoS attacks can cause increase in bandwidth and even loss of service in the ISP network [24].

There have been wide-ranging research and enterprise solutions on mitigating DDoS flooding attacks on both transport and application layers based on the traditional network infrastructure and assumptions. Each of the research efforts has shown its strengths and limitations in various performance metrics [35]. Enterprise solutions have also been limited in their countermeasures to mitigate large DDoS attacks, especially attacks exploiting vulnerabilities of publicly-accessible network protocols and servers, such as NTP (Network Time Protocol) servers, to achieve massive

H. D'Cruze
University of Maryland University College, Adelphi, MD, USA
e-mail: hubert.dcruze@yahoo.com

P. Wang (✉)
Robert Morris University, Pittsburgh, PA, USA
e-mail: wangp@rmu.edu

R.O. Sbeit · A. Ray
Verizon, Westlake, TX, USA
e-mail: Raed.Sbeit@verizonwireless.com; Andrew.Ray@verizonwireless.com

flooding scale to overwhelm a targeted server [24]. The emerging Software-Defined Networking (SDN) approach has shown increasingly promising improvements in addressing DDoS flooding attacks. The goal of this paper is to seek an optimal mitigation solution to DDoS attacks for ISPs by reviewing and evaluating various solutions using a proposed model of performance metrics. The rest of the paper will examine various mitigation solutions to evaluate their strengths and limitations. The evaluation of the mitigation solutions will be measured by a proposed set of performance metrics. This paper uses the case data and analysis from Verizon ISP network to support the evaluation and conclusions.

19.2   Literature Review

There has been a wide range of research on DDoS mitigation using traditional network protection tactics and heuristics, such as IP trace back, anomaly detection, ingress/egress filtering, ISP collaborative defense, network self-similarity [7, 14, 15, 17]. These approaches, however, have all shown significant limitations as viable solutions to mitigating the DDoS attacks.

The trust management helmet (TMH) model aims to differentiate legitimate users from attackers by registering four types of trusts used to calculate and store as part of a license at clients for session connection with the server securely [30, 31, 33]. But license forgery, replay and deletion attacks could occur and clients may cheat by duplicating or sharing old licenses. A two-tier coordination approach filters distrustful traffic and monitors with a novel RED/Droptail mechanism deployed on network routers [8]. However, spoofed addresses might not be captured by the routers and could open up an opportunity for DDoS attackers. An IP blacklisting approach uses http count filter in addition to CAPTCHA to check the suspected IP address and identify botnets [28]. Similarly, HTTPreject and CAPTCHA mechanisms were proposed to mitigate application-layer botnet-based DDoS attacks while dealing with flash crowd events [2, 26]. But CAPTCHA can be an annoyance to most legitimate users and could negatively impact online business operations.

Entropy-based tactics for addressing DDoS attacks have generated more interest than effective solutions. There was an empirical effort to assess major information metrics, such as Shannon entropy, Hartley entropy, Renyi's entropy to identify low-rate and high-rate DDoS attacks [6]. However, early detection of both low-rate and high-rate DDoS attack remains to be addressed. Entropy vectors of different features from traffic flows were used to form normal patterns using clustering analysis algorithm and identify the deviations from the models that are created [23]. This approach requires effective algorithm to reduce

computational time and memory usage in a high volume and high speed network. To overcome the disadvantages, a fast entropy method using flow-based analysis was proposed [9]. However, it may not be possible to find out the attacker and agents of DDoS attack under this approach.

Use of Cloud computing is not new in defending against DDoS attacks. A dynamic resource distribution approach was proposed based on queueing theory [34]. Nevertheless, cloud hosted servers are still vulnerable to DDoS attacks if they run in the traditional way [3].

Use of honeypots has been a new effort in network defense. A network of virtualized honeypots was proposed to be hosted on physical servers to monitor inbound traffic for malicious activities including flooding packets [10]. But chances are that the routers have already been flooded with malicious requests before the honeypots kick in. An alternative ant-based DDoS detection technique using roaming virtual honeypots was later proposed [27]. But attackers may identify the honeypots and may become an inadvertent launching pad for attacks on the system and worm from honeypot may spread to other networks.

19.3   SDN Model for DDoS Mitigation

"Traditional methods of DDoS detection and mitigation - such as over-provisioning and rate-limiting are simply inadequate against the varied and dynamic methods of DDoS attacks now employed by hackers" [24]. Therefore, intelligent and effective DDoS mitigation solutions are needed for enterprise networks. In addition, businesses often depend on ISP networks for highly available and reliable communication services rather than simple connections [19, 20, 21]. The DDoS mitigation capabilities are even more important for ISP networks, especially mobile networks due to the fast growth and potential in mobile businesses and users and increasing attacks on mobile networks. There have been findings and reports of hackers scanning mobile devices and their capabilities of DDoS attacks to flood mobile networks and effectively deny connections and access for legitimate customers [24]. This paper proposes the Software-Defined Networking (SDN) model as the optimal approach to DDoS mitigation and uses Verizon ISP network as the case for study.

The core performance metrics for measuring DDoS mitigation solutions should be the attributes of being effective, flexible, and programmable or autonomous. Effectiveness means how well the solution can prevent, detect, and stop the attacks with high reliability and accuracy. In addition, traditional DDoS detection and response mechanisms often require costly processing overhead to maintain connection state tables, packet marking, and deployment of additional modules and devices, which lack simple and autonomous management. Therefore, flexibility and automation are desirable features to minimize processing overhead and costs of scalability and operation [25].

SDN is a network architecture well-suited for DDoS attack mitigation. SDN architecture decouples the network control plane and data plane and allows a network to operate and manage network traffic dynamically according to a user need. Features like central control and programmability allow SDN to defend against DDoS attacks and other malicious activities. SDN brings network management to the brim of simplicity, flexibility, and programmability. SDN also allows flexible and programmable configuration and management of the DDoS mitigation schemes with no human intervention and no additional modules and devices [22], [25]. Further studies also confirmed SDN as a simple, flexible, programmable, and cost-effective solution to DDoS mitigation [16, 20]. Thus, implementing SDN will bring immediate benefits of service provisioning with efficiency and effectiveness, lower operating costs, easy configurability and manageability from its unique character of centralized software control and low latency and will improve quality of services as well as end user experience.

The SDN solution to DDoS mitigation is a special fit for ISP networks. An ISP's global view of network traffic is a useful advantage in identifying attack patterns and improving detection. In addition, ISPs are capable of blocking upstream threats at the network edge near the malicious hosts. These advantages may enable ISPs to detect malicious activities preceding a DDoS attack, such as pings and scans, and block these attacks [24]. The following section presents the case of Verizon ISP network and evaluation of their SDN approach to DDoS mitigation.

19.4 Case Study: Verizon Networks

Verizon is a global leader in providing communications, technology solution for our customers to play, work and live. Verizon's mission is to deliver the promise of the connected World. In 2015, Verizon invested more than $11 billion to meet today's surging demand for wireless data and video, but also to get VZ network ready for 5G wireless technology. According to RootMetrics©, the largest independent third-party tester of wireless network performance, Verizon ranks number one in speed, data, reliability and overall network performance in the U.S.

DDoS attacks in general have worsened again this year with reporting partners logging double the number of incidents from last year. This year, Verizon saw a significant jump in the DDoS threat action variety associated with malware. These incident reports came mostly from our partners including US-CERT (computer emergency

readiness team) and Arbor and Akamai networks. The threats mostly involved the repurposing of servers and devices used in amplification and reflection attacks. These attacks rely on improperly secured services, such as Network Time Protocol (NTP), Domain Name System (DNS), and Simple Service Discovery Protocol (SSDP), which make it possible for attackers to spoof source IP addresses, send out a bazillion tiny request packets, and have the services inundate an unwitting target with the equivalent number of much larger payload replies.

There was some indication that there may be two distinct tiers or clusters of DDoS attacks based on bandwidth, velocity, and duration. Here was the worst type of DDoS attack happened to Verizon last year: Instead of a single most common measure, bandwidth has two clusters around 15 and 59 Gbps, while velocity has clusters around 3 and 15 million packets per second. Data on attack duration similarly suggests that clusters around one- and two-day average durations.

ISP networks like Verizon have a "global view of network traffic" that is uniquely useful to DDoS detection [24]. The global view offers very useful insight into attack patterns that are essential to improving detection of attacks. Verizon manages 300,000 endpoints globally, and its DDoS mitigation in terms of detection accuracy and response time has been improved through identification of compromised hosts and knowledge of the network traffic patterns and activities prior to the attack [24].

The DDoS mitigation mechanism at Verizon is featured by a virtualized SDN-enabled infrastructure with a root of trust. Verizon's SDN network also includes resilient applications capable of handling proper discarding of packets in case of a DDOS attack detection [29]. "In case of denial-of-service attacks, virtual anti-DDoS scrubbing functions are dynamically instantiated near the sources. Traffic is automatically routed through these functions. In addition, shaping and prioritization rules are dynamically tuned to give priority to legitimate traffic and minimize the resources consumed by nefarious or suspicious traffic" [29]. The ability of Verizon's virtualized SDN architecture to achieve desired security automation benefits with efficiency and effectiveness is also attributed to its comprehensive security policy and a centralized security controller framework. The centralized security controller framework provides key security benefits of reliable non-disruptive delivery of security solutions, seamless integration with virtualized platforms, flexible protection and remediation scalable to distributed data centers, as well as automatic service management [29]. These benefits are not only critical and effective to the detection and mitigation of DDoS attacks but also simplify deployment of security solutions.

19.5 Conclusions

DDoS attacks are a common threat to enterprise networks and have a potentially serious negative impact on network and service availability for business operations. This paper has reviewed traditional approaches to detecting and mitigating DDoS attacks and identified their limitations and inadequacies to be viable solutions. In evaluation DDoS solutions in terms of their simplicity, flexibility, scalability, and efficiency in delivering and managing detection and mitigation solutions, this study proposes the SDN approach as the optimal solution to defending against DDoS attacks. This study focuses on ISP networks and uses Verizon's virtualized SDN network infrastructure as a case study. Verizon's success in mitigating DDoS threats is attributed to a strong, flexible, automated and efficient SDN security mechanism featuring comprehensive security policy and a centralized security controller.

Network security, including DDoS detection and mitigation, always requires comprehensive efforts from policy to network management to user awareness training. Security measures in addition to a strong SDN network will also contribute DDoS mitigations. For example, service providers may help by securing their services [5]. Blocking access to known botnets servers and patching systems regularly will help to stop malware from turning your nodes into zombies. For larger providers, anti-spoofing filters at the Internet edge can also help to prevent deflection/amplification techniques. To understand how your organization would react to a DDoS attack, conduct regular penetration testing or drills to see where you need to shore up processes and add technology or external mitigation services to help maintain or restore services. Finally, enterprise network administrators and security managers should have a solid understanding of their DDoS mitigation service-level agreements. Make sure that your own DDoS response procedures are built around existing denial of service protections and your operation teams are well trained on how to best engage and leverage these services in cases of DDoS attacks.

References

1. Akamai. (2016). State of the Internet combined executive review. Retrieved from: https://www.akamai.com/us/en/multimedia/documents/state-of-the-internet/state-of-the-internet-report-connectivity-executive-review-q1-2016-akamai.pdf
2. Al-Ali, Z., Al-Duwairi, B., & Al-Hammouri, A.T. (2015). c DDoS attacks and flash crowd events. *2015 I.E. 2nd International Conference on Cyber Security and Cloud Computing.*
3. Alqahtani, S., & Gamble, R. F. (2015). DDoS attacks in service clouds. *2015 48th Hawaii International Conference on System Sciences, 5331–5340.*

4. Behal, S., & Kumar, K. (2016). Trends in validation of DDoS research. *Procedia Computer Science, 85, 7–15.*
5. Benton, K., Camp, L. J., & Small, C. (2013). OpenFlow vulnerability assessment. *Proceedings of the second ACM SIGCOMM workshop on Hot topics in software defined networking – HotSDN '13,* pp. 151–152.
6. Bhuyan, M. H., Bhattacharyya, D., & Kalita, J. (2015). An empirical evaluation of information metrics for low-rate and high-rate DDoS attack detection. *Pattern Recognition Letters, 51, 1–7.*
7. Braga, R., Mota, E., & Passito, A. (2010). Lightweight DDoS flooding attack detection using NOX/OpenFlow. *IEEE Local Computer Network Conference, 408–415.*
8. Chen, C., & Chang, C. (2013). A two-tier coordination system against DDoS attacks. *International Journal of Online Engineering (iJOE), 9(4), 15–21.*
9. David, J., & Thomas, C. (2015). DDoS attack detection using fast entropy approach on flow-based network traffic. *Procedia Computer Science, 50, 30–36.*
10. Deshpande, H. A. (2015). Honey Mesh: Preventing distributed denial of service attacks using virtualized honeypots. *International Journal of Engineering Research & Technology, 4(8), 263–267.*
11. Fichera, S., Galluccio, L., Grancagnolo, S. C., Morabito, G., & Palazzo, S. (2015). OPERETTA: An OPEnflow-based REmedy to mitigate TCP SYNFLOOD attacks against web servers. *Computer Networks, 92, 89–100.*
12. Giotis, K., Argyropoulos, C., Androulidakis, G., Kalogeras, D., & Maglaris, V. (2014). Combining OpenFlow and sFlow for an effective and scalable anomaly detection and mitigation mechanism on SDN environments. *Computer Networks, 62, 122–136.*
13. Jantila, S., & Chaipah, K. (2016). A security analysis of a hybrid mechanism to defend DDoS attacks in SDN. *Procedia Computer Science, 86, 437–440.*
14. Jun, J., Ahn, C., & Kim, S. (2014). DDoS attack detection by using packet sampling and flow features. *Proceedings of the 29th Annual ACM Symposium on Applied Computing - SAC '14,* pp. 185–190.
15. Jyothi, V., Wang, X., Addepalli, S. K., & Karri, R. (2016). BRAIN: Behavior based adaptive intrusion detection in networks: Using hardware performance counters to detect DDoS attacks. *2016 29th International Conference on VLSI Design and 2016 15th International Conference on Embedded Systems (VLSID), 587–588.*
16. Krylov, V., Kravtsov, K., & Sokolova, E. (2016). Fast IP hopping protocol SDI implementation. *Indian Journal of Science and Technology, 8(36), 1–7.*
17. Li, J., Berg, S., Zhang, M., Reiher, P., & Wei, T. (2014). DrawBridge – Software-defined DDoS resistant traffic engineering. *Proceedings of the 2014 ACM conference on SIGCOMM – SIGCOMM '14,* pp. 691–592.
18. Lim, S., Yang, S., Kim, Y., Kim, H., & Yang, S. (2015). Controller scheduling for continued SDN operation under DDoS attacks. *Electronics Letters, 51(16), 1259–1261.*
19. Luo, S., Wu, J., Li, J., & Pei, B. (2015). A defense mechanism for distributed denial of service attack in software-defined networks. *2015 Ninth International Conference on Frontier of Computer Science and Technology, 325–329.*
20. Lu, Y., & Wang, M. (2016). An easy defense mechanism against botnet-based DDoS flooding attack originated in SDN environment using sFlow. *Proceedings of the 11th International Conference on Future Internet Technologies – CFI '16.*
21. Mowla, N. I., Doh, I., & Chae, K. (2014). Multi-defense mechanism against DDoS in SDN based CDNi. *2014 Eighth International Conference on Innovative Mobile and Internet Services in Ubiquitous Computing,* pp. 447–451.
22. Nayana, Y., Gopinath, J., & Girish, L. (2015). DDoS mitigation using Software Defined Network. *International Journal of Engineering Trends and Technology (IJETT), 24(5), 258–264.*

23. Qin, X., Xu, T., & Wang, C. (2015). DDoS attack detection using flow entropy and clustering technique. *2015 11th International Conference on Computational Intelligence and Security*, pp. 412–415.

24. Rodriguez, C. (2015). The expanding role of service providers in DDoS mitigation. *Stratecast Perspectives and Insight for Executives (SPIE), 15*(10), 1–10.

25. Sahay, R., Blanc, G., Zhang, Z., & Debar, H. (2015). Towards autonomic DDoS mitigation using software defined networking. *Proceedings 2015 Workshop on Security of Emerging Networking Technologies*.

26. Schneider, J., & Koch, S. (2010). HTTProject: Handling overload situations without losing the contact to the user. *European Conference on Computer Network Defense, 2010*, 29–34.

27. Selvaraj, R., Marwala, T., & Madhav Kuthadi, V. (2016). Ant-based distributed denial of service detection technique using roaming virtual honeypots. *IET Communications, 10*(8), 929–935.

28. Singh, K. J., & De, T. (2015). DDOS attack detection and mitigation technique based on Http count and verification using CAPTCHA. *2015 International Conference on Computational Intelligence and Networks*, pp. 196–197.

29. Verizon. (2016). Verizon network infrastructure planning: SDN-NFV reference architecture. Retrieved from: http://innovation.verizon.com/

30. Wang, H., Xu, L., & Gu, G. (2015). FloodGuard: A DoS attack prevention extension in software-defined networks. *2015 45th Annual IEEE/IFIP International Conference on Dependable Systems and Networks*.

31. Wang, X., Chen, M., & Xing, C. (2015). SDSNM: A software-defined security networking mechanism to defend against DDoS attacks. *2015 Ninth International Conference on Frontier of Computer Science and Technology*, pp. 115–121.

32. Xiulei, W., Ming, C., Xianglin, W., & Guomin, Z. (2015). Defending DDoS attacks in software defined networking based on improved Shiryaev–Roberts detection algorithm. *Journal of High Speed Networks, 21*(4), 285–298.

33. Yu, J., Fang, C., Lu, L., & Li, Z. (2010). Mitigating application layer distributed denial of service attacks via effective trust management. *IET Communications, 4*(16), 1952–1962.

34. Yu, S., Tian, Y., Guo, S., & Wu, D. O. (2014). Can we beat DDoS attacks in clouds? *IEEE Transactions on Parallel and Distributed Systems, 25*(9), 2245–2254.

35. Zargar, S. T., Joshi, J., & Tipper, D. (2013). A survey of defense mechanisms against distributed denial of service (DDoS) flooding attacks. *IEEE Communications Surveys & Tutorials, 15*(4), 2046–2069.

Integrated Methodology for Information Security Risk Assessment

Ping Wang and Melva Ratchford

Abstract

Information security risk assessment is an important component of information security management. A sound method of risk assessment is critical to accurate evaluation of identified risks and costs associated with information assets. This paper reviews major qualitative and quantitative approaches to assessing information security risks and discusses their strengths and limitations. This paper argues for an optimal method that integrates the strengths of both quantitative calculation and qualitative evaluation for information security risk assessment.

Keywords

Security • Risk • Assessment • Qualitative • Quantitative

20.1 Introduction

Information security risk management is "the process of identifying, assessing, and reducing risks to an acceptable level and implementing the right mechanisms to maintain that level of risk" [15]. Therefore, risk assessment is a critical component in the information security risk management process. Effective risk management is dependent upon a sound risk assessment, also known as risk analysis, which is a process for identifying and evaluating risks [7]. Accurate evaluation of the risks or potential losses associated with the vulnerabilities of information assets is essential to the development of effective risk control process and protection strategies.

A risk is the possibility of an adverse event that would reduce information and business asset [4]. Blakley et al. also pointed out that every information security risk has a cost which can be more or less precisely quantified [4]. Therefore, maximum accuracy in quantifying the cost or loss related to the security risks is certainly an important attribute for any risk analysis methodology. There are two general approaches to risk assessment: quantitative approach and qualitative approach. The quantitative approach uses numeric data, formulas, and calculations to obtain an objective measure of risks. A typical mathematical formulation of risk uses a lower level of granularity of threat and probability to determine an asset's value, exposure, frequency and existing protection measures [6]. The qualitative approach is more subjective and uses expert opinions and perceptions on the probability and impact of a risk to determine the risk level. Both quantitative and qualitative approaches have their own strengths and limitations. For a typical risk assessment, an appropriate approach or methodology should be selected based on the business mission and assessment needs. In addition, critical assets and relevant vulnerabilities and threats need to be identified. Various controls for mitigating the risks need to be identified and evaluated in terms of effectiveness and costs. A cost-benefit analysis (CBA) should be included to support any recommendations of controls.

There are various risk analysis and assessment methodologies currently available. These methodologies are

P. Wang (✉)
Robert Morris University, Pittsburgh, PA, USA
e-mail: wangp@rmu.edu

M. Ratchford
University of Maryland University College, Adelphi, MD, USA
e-mail: melva.ratchford@gmail.com

© Springer International Publishing AG 2018
S. Latifi (ed.), *Information Technology – New Generations*, Advances in Intelligent
Systems and Computing 558, DOI 10.1007/978-3-319-54978-1_20

primarily quantitative or qualitative in nature addressing various dimensions of information security risks, and an organization often faces the challenging task of adopting the optimal or most appropriate methodology. The common goal of risk assessment methodologies is to reach the estimate of overall risk value and the appropriateness of the methodology should fit the needs of the organization [14]. This paper is to briefly review the major approaches to information security risk assessment and propose an optimal and integrated methodology.

20.2   OCTAVE Method

A widely known qualitative methodology for assessing information security risk is the OCTAVE® (Operationally Critical Threat, Asset, and Vulnerability Evaluation^{SM}) method. OCTAVE is a risk-based strategic assessment and planning technique developed by the CERT Coordination Center at Carnegie Mellon University Software Engineering Institute (SEI). OCTAVE method is driven by operational risks and security practices and uses three phases and sub-processes and task activities to build a comprehensive picture of organizational information security needs [1].

The OCTAVE approach has several positive features, including self-direction, flexibility, and comprehensiveness [2]. The method is self-directed, which means that a small internal team can take the lead in analyzing the organizational security needs while incorporating the knowledge of a wide range of employees. It is flexible with different versions and can be customized to fit the needs of different types and sizes of organizations. It is also comprehensive because it focuses on both strategic and organizational risks as well as practical operational security management and technology issues.

However, the end result of OCTAVE risk analysis uses subjective and relative ranking values (high, medium, low) and a descriptive risk-level matrix for risk impact and probability determination. While the relative and categorical rankings may be simple and flexible for individual organizations to use and define, they lack mathematical calculations and quantitative results needed for comparing risk differences [14]. Thus, it would be difficult to use the OCTAVE results as accurate parameters for supporting cost-benefit analysis and decision-making regarding risk control investments and activities.

20.3   CORAS Method

Another qualitative methodology for information security risk assessment is CORAS (Construct a platform for Risk Analysis of Security Critical Systems). CORAS was a framework for model-based risk assessment of security-critical systems developed under the Information Society Technologies program sponsored by the European Union. The CORAS methodology uses UML (Unified Modeling Language) diagrams to represent relationships and dependencies between users and the working environment and the final outcome of risk analysis and risk management decisions are based on the UML class diagrams involving each asset [11].

The major strengths for the CORAS approach include its incorporation of input from and communication among diverse parties and stake-holders as well as improved asset specification and efficiency in the organizational risk analysis process. However, like the OCTAVE methodology, CORAS is a qualitative approach and does not use any precise mathematical calculations but uses an expected value matrix with subjective rankings to determine the expected value of a security risk. The CORAS method is simple and efficient to use but is subjective and lacks accuracy and specificity in risk values.

20.4   IS Risk Analysis Method

Traditional qualitative risk assessment methodologies provide subjective and relative results for risk impact and are not adequate for cost-benefit decision support. To address this limitation of qualitative methods, the IS (information system) Risk Analysis method was proposed based on a business model [13]. The IS Risk Analysis methodology is a systematic quantitative model with four sequential stages which determine the importance levels and valuations of various business functions and IS assets. Mathematical formulas are used to calculate the annual loss expectancy (ALE) for each threat occurrence and organizational disruption.

The ALE calculation in the IS Risk Analysis method is more comprehensive than conventional understanding of loss. The loss of asset due to each threat includes not only the asset replacement cost but also the income loss, the probability of the threat occurrence, and the relative importance of the asset from the viewpoint of the operational continuity. Most importantly, the risk assessment end result is a tangible quantitative monetary value that can be used for making risk management decisions. However, the four stages of the method involve extensive mathematical calculations and may not be simple enough to attract wide participation from management and staff [14].

20.5   ISRAM Method

Another major quantitative method for information security risk analysis is the survey-based ISRAM (Information Security Risk Analysis Method) developed by Karabacak and

Sogukpinar [9]. ISRAM uses a 7-step process including two surveys among managers and staff on the probability and consequence for each of the identified security vulnerabilities. The survey results are numerically represented and used in a formula to calculate the final risk value.

Karabacak and Sogukpinar demonstrated the ISRAM approach with a case study on computer virus infection risks. ISRAM does provide quantitative and objective risk values for supporting risk management decisions. At the same time, the survey instrument used in the ISRAM model includes subjective but numerical evaluations from managers and staff in the operational community. In addition, the survey questions and weight values are customizable with no rigid frames to fit to organizational and business needs. Karabacak and Sogukpinar also claimed the advantage of simplicity and ease-of-use for the method. However, the 7-step process of ISRAM needs extensive preparation and the mathematical formulas are complex and daunting to many potential participants [14].

20.6 Proposal: Integrated Method

The review of the information security risk assessment methodologies presented above reveals strengths and limitations in both qualitative and quantitative approaches. It is important for an organization to adopt an optimal risk analysis method that is accurate in providing end results and customizable according to organizational needs. To facilitate accurate and effective information risk assessment, it is also necessary to identify indirect and hidden risks and costs and compare and analyze various information security risk assessment models and evaluate and prioritize the criteria for selecting effective risk assessment models [3, 8].

Even for qualitative risk evaluation methods, quantifiable and accurate data should be a pre-requisite especially for the area of information security risk analysis [4]. Therefore, an optimal information security risk analysis methodology should integrate the strengths of both qualitative and quantitative methods.

The Risk Management Guide for IT Systems (SP 800–30) published by NIST (National Institute of Standards and Technology) under the U.S. Commerce Department provides an example effort for optimized and integrated direction toward assessing security risks and impacts [10]. The NIST guide uses a qualitative Risk-Level Matrix and subjective and descriptive variables (high, medium, low) to reference risk levels. However, it uses a numeric and quantifiable value to translate the subjective evaluation of the probability of exploitation of a specific vulnerability. The NIST guide also acknowledges the limitations of the

subjective evaluations and emphasizes that risk and impact analysis should consider the advantages and disadvantages of both qualitative and quantitative methods and the additional factors of frequency, cost, and weight of the risk impact for a particular vulnerability [12].

Whitman and Mattord provide an example of a security risk assessment method that integrates both qualitative and quantitative strengths [16]. In the Whitman and Mattord risk assessment model, the equation for determining the risk is: The total risk value *equals* the likelihood of vulnerability occurrence *times* the value of the information asset, *minus* percentage of risk mitigated by current controls, *plus* an element of uncertainty of current knowledge of the vulnerability. It is especially important to include the uncertainty cost factor in the equation since "it is not possible to know everything about every vulnerability" [16]. This method provides numeric and quantifiable end results for risk assessment and management while using the subjective likelihood rating of vulnerability recommended by NIST.

However, the asset loss calculation still reflects the limited traditional concept of annual loss expectancy (ALE). As an improvement for the ALE concept, Bodin, Gordon, and Loeb (2008) introduced the new metrics of expected severe loss and standard deviation of loss in addition to ALE in calculating the total risk value. The expected severe loss is the magnitude of the loss that jeopardizes the organizational survivability [5]. The standard deviation of loss (the square root of the variance of loss) scientifically measures the dispersion of risks and losses. The weighting of each loss parameter can be customized by individual organizations.

20.7 Conclusion

In conclusion, an optimal methodology for information security risk assessment should integrate strengths from both qualitative and quantitative methods to provide accurate and reliable risk data for risk management decision making. The integrated methodology should include the strategic, practical, and customizable phases and processes from the OCTAVE method while incorporating the survey instruments used by the ISRAM method into its processes for diverse and quantifiable input on risk evaluations. In identifying critical assets and their behaviors, the object-oriented UML modeling technique from the CORAS method can be used to improve asset specification. In addition, the risk scores for assets should follow the NIST recommendation that is subjective but numeric and quantifiable. Finally, the total risk value calculation should include ALE, expected severe loss, standard deviation of loss, minus risk mitigation by current controls, and plus uncertainty cost.

References

1. Alberts, C., & Dorofee, A. (2002). *Managing information security risks: The OCTAVE approach.* Boston: Addison Wesley Longman Publishing Co., Inc..

2. Alberts, C., Dorofee, A., Stevens, J., & Woody, C. (2003). Introduction to the OCTAVE approach. Retrieved from http://www.cert.org/octave/pubs.html

3. Anderson, R., & et al. (2013). Measuring the cost of cybercrime. *The Economics of Information Security and Privacy.* Springer.

4. Blakley, B., McDerMott, E., & Geer, D. (2002). *Information security is risk management. NSPW'0I, September 10–13th, 2002, Cioudcroll, New Mexico,* 97–104.

5. Bodin, L. D., Gordon, L. E., & Loeb, M. P. (2008). Information security and risk management. *Communications of the ACM, 51*(4), 64–68.

6. Ghazouani, M., et al. (2014). Information security risk Assessment — A practical approach with a mathematical formulation of risk. *International Journal of Computer Applications, 103*(8), 36–42.

7. Gibson, D. (2015). *Managing risk in information systems* (2nd ed.). Burlington: Jones & Bartlett Learning.

8. Kiran, K. V. D., et al. (2013). A comparative analysis on risk assessment information security models. *International Journal of Computer Applications, 82*(9), 41–47.

9. Karabacak, B., & Sogukpinar, I. (2005). ISRAM: Information security risk analysis method. *Computer & Security, 24*(2005), 147–159.

10. NIST. (2012). *"Guide for Conducting Risk Assessments" (NIST SP800–30 Revision 1) by NIST (2012).* Retrieved from: http://csrc.nist.gov/publications/nistpubs/800-30-rev1/sp800_30_r1.pdf

11. Stolen, K., den Braber, F., Dimitrakos, T., Fredriksen, T., Gran, B. A., Houmb, S., et al. (2002). *Model-based risk assessment – the CORAS approach.* Retrieved from http://www.nik.no/2002/Stolen.pdf

12. Stoneburner, G., Goguen, A., & Feringa, A. (2002). *Risk management guide for information technology systems: Recommendations of NIST.* Retrieved from http://csrc.nist.gov/publications/nistpubs/800-30/sp800-30.pdf

13. Suh, B., & Han, I. (2003). The IS risk analysis based on a business model. *Information & Management, 41*(2003), 149–158.

14. Vorster, A., & Labuschagne, L. (2005). A framework for comparing different information security risk analysis methodologies. Proceedings of SAICSIT 2005, pp. 95–103.

15. Wang, J. A. (2005). Information security models and metrics. *Proceedings of the 43rd ACM Southeast Conference, March 18–20, 2005, Kennesaw, GA.* 178–184.

16. Whitman, M. E., & Mattord, H. J. (2008). *Management of information security* (2nd ed.). Boston: Thomson Course Technology.

Techniques for Detecting and Preventing Denial of Service Attacks (a Systematic Review Approach)

Hossein Zare, Mojgan Azadi, and Peter Olsen

Abstract

This paper analyzes denial of service (DoS) attacks and countermeasures based on a systematic review analysis conducted of papers between 2000 and 2016. The paper is based on three searches. The first was conducted using suitable keywords, the second using references used by selected papers, and, the third considered the most cited English-language articles. We discuss 802.11 along with one of the well-known DoS attacks at physical-level access points. Experts suggest using 802.11w, a "cryptographic client puzzle," and "delaying the effect of request" to provide better protection in this layer. The paper discusses four main network defense systems against network-based attacks—source-end, core-end, victim-end, and distributed techniques—with a focus on two innovative methods, the D-WARD and gossip models. This study also discusses chi-squares and intrusion detection systems (IDSs), two effective models to detect DoS and DDoS attacks.

Keywords

Denial-of-service attacks • IEEE-802.11 • D-WARD • Gossip • Chi-square • Intrusion detection systems • Systematic review

21.1 Introduction

A denial of service (DoS) incident deprives users of access to a service they expect. Goodrich and Tamassia [1] define a DoS attack as any one that impacts the basic functionality of a machine or software and makes it unavailable [1].

H. Zare, MS, PhD (✉)
Department of Health Services Management, University of Maryland University College (UMUC), Adelphi, MD 20783, USA
e-mail: Hossein.Zare@faculty.umuc.edu

M. Azadi, RNC, Informatics MSN, PhD
Johns Hopkins School of Nursing, Baltimore, MD, USA
e-mail: Mazadi2@jhmi.edu

P. Olsen, MSOR, Ae.E
Department of Computer Science and Electrical Engineering, University of Maryland Baltimore County, Catonsville, MD 21250, USA
e-mail: olsen@sigmaxi.net

Distributed denial of service (DDoS) attacks are some of the most frequent disturbance-of-network attacks. This attack takes advantage of hosts on networks with poor or no security. After infecting the first system it goes on to infects others until it attacked the all systems in a particular distribution. This distribution may be all of the systems using particular operating system, all that have a particular application, all on a particular network, or other distributions. Under a DDoS attack, "the arriving packets do not obey end-to-end congestion control algorithms"; they constantly bombard the victim, "causing the well-behaved flows to back off and eventually starve" [2].

21.2 Methodology

This is a review-based paper to address different types of DoS attacks and the most updated preventive methods noted by scholars in recent years. Using a systematic-review

© Springer International Publishing AG 2018
S. Latifi (ed.), *Information Technology – New Generations*, Advances in Intelligent Systems and Computing 558, DOI 10.1007/978-3-319-54978-1_21

approach, we included all kind of studies that provided techniques for detecting and preventing DoS attacks that were published and reported by Scopus [3] from 2000 up to October 2016.

We excluded studies that only reported on DoS attacks without considering techniques for detecting and preventing them. We reviewed publications that covered 802.11, D-WARD and gossip models as well as chi-squares and intrusion detection systems, two effective models to detect DoS and DDoS attacks. Table 21.1 shows our search strategy:

Bibliographies of all selected articles and relevant review articles were reviewed for additional relevant studies. We only included English articles.

21.2.1 Methods of the Review

In step one, we reviewed the titles and abstracts of all 122 articles from the literature search and classified them as "exclude" or "full-review article." All full-review articles were reviewed by all authors, and three articles were selected for this study. Disagreements were resolved through discussion between the original review authors and through a group discussion with all three authors. The following criteria were used to exclude abstracts and full texts:

- non-English language,
- did not address study's question,
- duplicate studies.

Finally, from 122 published documents, 64% were conference papers, 31% were articles, 4% were conference reviews and less than 1% were review papers. After non-articles and non-English papers were eliminated, the number of articles chosen was reduced to 16. From careful consideration of the 16 selected articles, three quality articles were selected for this assignment, and when writing this paper, when required, we looked at the references used by the three selected articles.

21.3 Denial-of-Service-Resistant Intrusion Detection Architecture

A paper published by Mell et al. [4] reported an intrusion detection system (IDS) that is resistant to DoS attacks. The authors claimed the IDS was the only effective model (at the time of the paper's publication) to prevent those attacks [4]. Using this IDS approach, the authors suggested models that hide the hosts and protect the network from flooding DoS attacks, such as:

21.3.1 Active IDS Response

By changing filtering rules in routers and firewalls, IDSs can detect a "flooding" DoS attack. In this technique, IDSs use reconfigured related routers that stop an attack. However, there are several disadvantages to this method; e.g., all routers in an organization must be monitored by the IDS— a potentially expensive option. Filtering on a router level significantly reduces an organization's network speed and legitimate network traffic, often imposing high costs and requiring more IT specialists and other human resources. "IDSs often take time to detect and respond to attacks" [4]. Additionally, if attackers target the IDS response host, they can turn the defensive system off and launch a flooding DoS attack.

21.3.2 Separate Communication Channel and Decentralized Non-hierarchical IDSs

Separating physical wires and creating non-hierarchical IDSs are the two most effective techniques to counter DoS attacks. Yet both are hard to establish and have a high starting cost [4, 5].

Table 21.1 The study search strategies (Search date: October 10, 2016)

TITLE-ABS-KEY	# of papers
(mitigating **dos** OR **distributed dos attacks**)	122 (Conference papers: 78, Articles: 38, Conference Reviews: 5, Reviews: 5)
(**mitigating dos** OR **distributed dos attacks**) AND (LIMIT-TO (DOCTYPE, "**ar**") OR LIMIT-TO (DOCTYPE, "**re**")) AND (LIMIT-TO (LANGUAGE, "**English**"))	37
(**mitigating dos** OR **distributed dos attacks**) AND (LIMIT-TO (DOCTYPE, "**ar**") OR LIMIT-TO (DOCTYPE, "**re**")) AND (LIMIT-TO (LANGUAGE, "**English**")) AND (LIMIT-TO (AFFILCOUNTRY, "**United States**"))	16
Selected papers after reviewing abstracts	3

Fig. 21.1 Critical host's protection against discovery through sniffing [5]

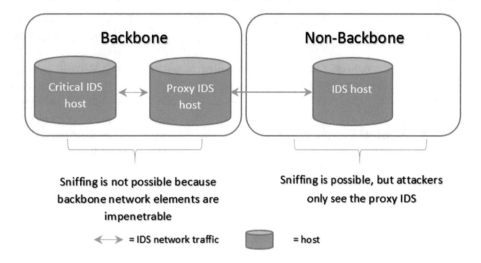

21.3.3 Mobile Recoverable Components

Using IDS components as mobile agents is another effective method. This technique engages "mobile agent critical components [that] randomly move around a network" and make it difficult for attackers to find them [4]. The proposed IDS architecture hides the IDS target and causes hackers to use old attack models—"passive sniffing or active network probing"—to find the IDS [5]. Mell et al.'s paper focuses on this method, which has the following advantages:

(a) The proposed model is able to thwart passive sniffing by placing critical hosts in backbone networks (See Fig. 21.1).
(b) The recommended technique is able to prevent network probing by implementing a filter at the TCP/IP layer while rejecting non-verified packets and without sending any replies.
(c) Finally, the mobile recoverable components technique protects the network/system from DoS attacks by hiding "critical IDS hosts from an attacker's view" and adding multiple backups for critical hosts. If hackers launch a random attack against those critical hosts, the backup hosts cover "the functionality of the halted critical IDS processes" [5].

Despite these mentioned advantages, there are a few limitations for this model. For instance, this model permits internal hackers to attack inter-domain routers. Additionally, if the IDS sensors are placed on insecure hosts, the suggested model is not able to detect their penetration.

21.4 Analyzing Well-Known Countermeasures Against Distributed DoS Attacks

Beitollahi and Deconinck [6] published a study that analyzed several countermeasures against DDoS attacks. They claimed it was "the most complete survey that analyzes the most cited DDoS defense techniques in detail" [6]. These are four main characteristics of DDoS attacks:

21.4.1 Location Where the Attack Is Detected

The "victim" and "routers" as the two most popular locations for detecting attacks [6].

21.4.2 Response Mechanism

The authors believe that "filtering and rate-limiting" are two well-known filtering techniques. The first mechanism uses the attack's signature. The rate-limiting technique focuses on the network traffic; "good traffic is passed through the filter based on good traffic signature," and the rest of the traffic is filtered [6]. Because filtering is an expensive procedure and it is not always possible to generate a comprehensive list of signatures, experts suggest rate-limiting traffic.

21.4.3 Places That a Response Mechanism Applies

"Source-end," "core-end," "victim-end" and distributed are the four main points to apply a response mechanism [6].

21.4.4 Decision Center Place

Similar to response mechanism points, there are three main points to control centers: source-end, core-end and victim-end [6].

21.5 DDoS Defense Techniques

As we described, there are four main points to defend against an attack: *source-end, core-end, victim-end* and *distributed.*

21.5.1 Source-End Defense Technique

Points close to a network's traffic source are called source-end points—such as gateway routes and customer edge routers. This method benefits from filtering, or a rate-limit technique, to detect malicious packets. The technique has little impact on network traffic. Its main challenge is distinguishing malicious traffic and legitimate traffic. One of the well-known end-source defense techniques is "DDoS Network Attack Recognition and Defense," or D-WARD (See Fig. 21.2) [7].

21.5.2 D-WARD

D-WARD is an inline system installed at the exit router of an end network. It works as "source-end components in a distributed defense" or as an "autonomous system" [7]. The system uses continuous control mechanisms for all traffic, incoming and outgoing. The main components of

this system include "observation," "rate-limiting" and "traffic policing." The following is the system's mechanism [7]:

(a) D-WARD uses a set of identifiers or local addresses called the "police address set."
(b) During an observation process, D-WARD collects statistics of communication between a "police address set" and another Internet set, collected at the "aggregate flow."
(c) D-WARD processes the flow of statistics based on this communication between the "police address set" and the Internet and traffic behavior.
(d) D-WARD compares this traffic behavior with a "rate-limiting" component. Then, it will decide whether to impose, adjust or "remove the rate limit on the agflow's sending rate" and define a pattern of communication [7].
(e) D-WARD uses a communication pattern to find suspicious traffic.
(f) This mechanism is useful for TCP, UDP or IMCP protocols. These protocols cover most Internet traffic and potentially could control most DDoS attacks [6].

21.6 Core-end Defense Technique—Gossip Mechanism

In this defense mechanism, "core routers independently detect malicious traffic" by filtering or rate-limiting traffic. In cases of a high volume of traffic, routers are more likely to permit malicious traffic (i.e., a false positive response). Zhang and Parashar [8] developed a unique core-end defense system based on a rate-limiting process, which follows these steps (See Fig. 21.3) [8]:

(a) A local detection node performs different procedures to recognize an attack. In the case of finding suspicious traffic, the local detection node informs the relevant peer group.
(b) If the peer group(s) recognize a suspicious packet, then the local detection makes a connection to them.

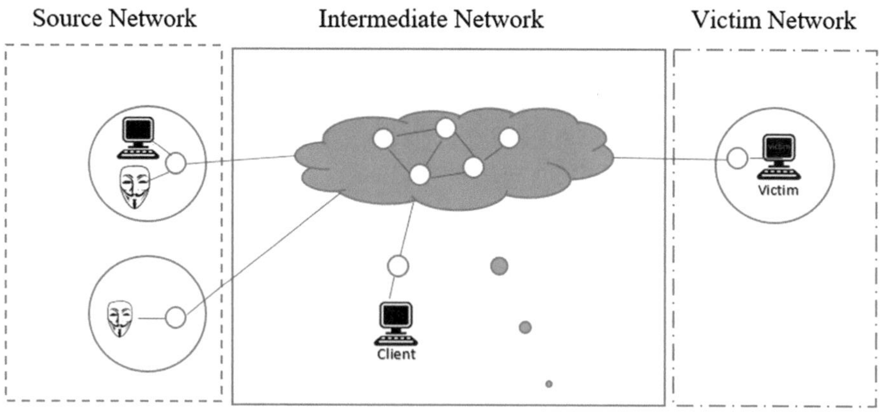

Fig. 21.2 Point of defense (Modified from Mirkovic and Reiher [7])

Source Network   Intermediate Network   Victim Network

```
when ( node n builds a new  (conf, attribute, dest) tuple )
{
        while ( node n believes that not enough of its
        neighbors have received  (const, attribute, dest) tuple)
        {
                m =a neighbor node of p;
                send  (conf, attribute, dest) tuple to m;
        }
}
```

Fig. 21.3 Gossip protocol suggested by Zhang and Parashar [8]

(c) Individual nodes extend countermeasures to prevent "continuance of the attack that it has identified based on the local policies."

(d) Each local defense mode shares the attack's information with other nodes "using a gossip mechanism to improve accuracy of defense"[8].

21.7 Victim-end Defense

This attack could be easily recognized in a victim server because the DDoS generates simulated (or bogus) packets. A victim server can distinguish suspicious packets from legitimate packets. A TCP connection can be established by legitimate users after identifying IP server passes packets. But in a network-layer DDoS, a TCP connection cannot be established by zombie machines and will "simply flood the server with bogus packets" with false IP addresses and meaningless payloads. A victim server can easily detect these bogus packets. The main challenge with any victim-end defense system is that it reverts to a "flash crowd" when a server's workload increases with increased volume usage of legitimate users, making it vulnerable to DDoS attacks [7, 9, 10].

21.7.1 Distributed Defense Technique

This defense system uses mixed methods to detect a DDoS attack, using two or more of the mentioned defense mechanisms. For instance, "source-end & victim-end" or "victim-end & core-end." Performing this system requires infrastructural modification and cooperation between several ISPs. This mechanism is costly and not easy to perform [6].

21.7.2 Well-Known Detection Methods

In addition to the abovementioned defense mechanisms, there are several well-known detection methods, including "sequential change-point detection," "wavelet analysis," "neural networks" and "statistical techniques such as '*chi-square*', '*entropy*' and '*Kolmogorov-Smirnov*'" [11, 12].

21.8 Chi-square Method

The chi-square method uses Pearson's chi-square test to compare the distribution of HTTP, FTP, DNS or other values. After the packet length is bound into a standard range (e.g., 0-64 bytes, 65-128 bytes or 129-256 bytes), a chi-square test can be performed using current traffic and experienced values for a system under attack. If there is a significant difference between current traffic and experienced traffic, an approved hypothesis will show there is an attack. This method also uses the experienced distribution curves to show chi-square values for a source address under DDoS attack. Under normal conditions, when a system is under attack, we expect to see a narrow-long shape, as compared to a bell-shape distribution for a non-attack situation [12].

21.9 DoS Attacks and Countermeasures in 802.11 Wireless Networks

The paper published by Bicakci and Tavil [13] discussed "Denial of Service attacks and countermeasures in IEEE 802.11 wireless networks." The authors provided a comprehensive analysis on IEEE 802.11 access points as some of the most popular access points for most shopping center, airport, and university campus systems. The 802.11 standard includes three main packets: the "management frame, control frame, data frame"—access points are found mainly on a control frame, with the physical layer vulnerable to DoS attacks [13]. The 802.11 permits attackers to interrupt network access completely or selectively by using "few packets and lower power consumption." In a "selective" attack, an individual station is targeted by hacking central basic vulnerabilities. In the next section, we will discuss a few of the most common "selective MAC [Media Access Control] layer attacks" and then talk about the "complete MAC layer attack" [13].

21.10 Detecting MAC Layer Attacks

As mentioned, a main weakness of 802.11 is non-cryptographic communication. The Institute of Electrical and Electronics Engineers (IEEE) introduced 802.11w, providing a sort of cryptographic communication that protects against some DoS attacks [14]. However, "public key cryptography is expensive" [13]. "Cryptographic (client) puzzles" are another solution. This method works on the "puzzle distribution" of a client. The 802.11 could add one more layer as a safeguard, requires verifying a puzzle. This solution is effective for wireless networks but is not able to protect any system from flooding attacks [13].

There are a few non-cryptographic solutions, including "*delaying the effect of request*" and "*decreasing the retry limit.*" The first solution buys more time for an AP to recognize legitimate or false packets, which can potentially eliminate some DoS attacks [15]. The second method is useful for flooding attacks by changing "the retry limit by a smaller value upon detection of DoS attack" [13]. However, researchers report that this solution is difficult to apply "above the firmware level" [16].

21.11 Conclusion

The most common attacks for the network layer are DDoS attacks. They target e-commercial and online banks and any other type of online providers, denying access for legitimate customers [1]. They also threaten the performance of network-based control systems [17].

For this study, we used a systematic review approach to find the most cited papers published in peer-reviewed journals with a main topic of "mitigating DoS or distributed DoS attacks." Here are our main findings:

- At the physical layer, one of the most well-known DoS attacks is an attack on 802.11 access points. Hackers benefit from weaknesses of 802.11 and launch a selective or complete MAC layer attack and knock off the system partially or completely [13].
- Experts suggest using 802.11w, "cryptographic client puzzle" and "delaying the effect of request" techniques and technologies to eliminate DoS attacks [13]. Yet, expensive public cryptographic keys and vulnerabilities to flooding attacks keep doors open for DoS attacks.
- DDoS attacks remain one of the most common attacks at the network layer, using different methods to attack a network—source-end, core-end, victim-end and distributed attacks. Similar techniques defend against these attacks [6].
- The D-WARD is an innovative "source-end & autonomous system" using statistics generated from all incoming and outgoing communication to find communication behavior models, detecting suspicious traffic and protecting the system from DDoS attacks [7].
- The gossip model is a unique core-end defense system developed by Zhang and Parashar [8]. This method improves communication between peer groups and "us [es] a gossip mechanism to improve accuracy of defense" [8].

- The chi-square is one of the effective detection mechanisms for identifying DDoS attacks using chi-square values of networks before and during attacks to recognize a system under attack [12].
- Intrusion detection systems are another detection mechanism [4]. This method hides a critical host from a flooding DoS attack. Separating the communication channel, decentralizing the IDSs and using mobile recoverable components are three main mechanisms to implement IDSs.

Acknowledgements This research received no specific grant from any funding agency in the public, commercial, or not-for-profit sectors. The authors would like to thank Professor Ping Wang for his valuable comments.

Conflict of Interest None declared.

References

1. Goodrich, M., & Tamassia, R. (2010). *Introduction to computer security*. Boston: Addison-Wesley Publishing Company.
2. Ioannidis, J., & Bellovin, S. M. (2002). Implementing pushback: Router-based defense against DDoS attacks. In *Proceedings of Network and Distributed System Security Symposium*. Retrieved April 7, 2017 from http://www.internetsociety.org/doc/implementing-pushback-router-based-defense-against-ddos-attacks
3. Elsevier. SCOPUS Database, www.scopus.com
4. Mell, P., Marks, D., & McLarnon, M. (2000). A denial-of-service resistant intrusion detection architecture. *Computer Networks, 34*, 641–658.
5. Mamun, M. S. I., & Kabir, A. S. (2010). Hierarchical design based intrusion detection system for wireless ad hoc sensor network. *International Journal of Network Security & Its Applications (IJNSA), 2*, 102–117.
6. Beitollahi, H., & Deconinck, G. (2012). Analyzing well-known countermeasures against distributed denial of service attacks. *Computer Communications, 35*, 1312–1332.
7. Mirkovic, J., & Reiher, P. (2005). D-WARD: A source-end defense against flooding denial-of-service attacks. *IEEE Tansactions on Dependable and Secure Computing, 2*, 216–232.
8. Zhang, G., & Parashar, M. (2006). Cooperative defence against ddos attacks. *Journal of Research and Practice in Information Technology, 38*, 69–84.
9. Oikonomou, G. C., Mirkovic, J., Reiher P. L., & Robinson, M. (2006). A framework for a collaborative DDoS defense. In *ACSAC* (pp. 33–42).
10. Anderson, T., Roscoe, T., & Wetherall, D. (2004). Preventing Internet denial-of-service with capabilities. *ACM SIGCOMM Computer Communication Review, 34*, 39–44.
11. Toledo, A. L., & Wang, X. (2008). Robust detection of MAC layer denial-of-service attacks in CSMA/CA wireless networks. *IEEE Transactions on Information Forensics and Security, 3*, 347–358.
12. Feinstein, L., Schnackenberg, D., Balupari, R., & Kindred, D. (2003). Statistical approaches to DDoS attack detection and response. In *DARPA Information Survivability Conference and Exposition, 2003 Proceedings* (pp. 303–314). IEEE. Retrieved April 7, 2017 from http://ieeexplore.ieee.org/abstract/document/1194894/?part=1

13. Bicakci, K., & Tavli, B. (2009). Denial-of-Service attacks and countermeasures in IEEE 802.11 wireless networks. *Computer Standards & Interfaces, 31*, 931–941.

14. IEEE working group. (2010). IEEE standard for information technology–Telecommunications and information exchange between systems–Local and metropolitan area networks–Specific requirements–Part 11: Wireless LAN Medium Access Control (MAC) and Physical Layer (PHY) specifications Amendment 6: Wireless Access in Vehicular Environments. *IEEE standards, 802*, 11p.

15. Bellardo. J, & Savage, S. (2003, August). Denial of service attacks: Real vulnerabilities and practical solutions. In *Proceedings of the 12TH USENIX Security Symposium* (pp. 4–8). Washington, DC.

16. Bernaschi, M., Ferreri, F., & Valcamonici, L. (2008). Access points vulnerabilities to DoS attacks in 802.11 networks. *Wireless Networks, 14*, 159–169.

17. Linda, O., Vollmer, T., & Manic, M. (2009). Neural network based intrusion detection system for critical infrastructures. In *2009 international joint conference on neural networks* (pp. 1827–1834). IEEE. Retrieved April 7, 2017 from http://ieeexplore.ieee.org/abstract/document/5178592/

Open Source Intelligence: Performing Data Mining and Link Analysis to Track Terrorist Activities

Maurice Dawson, Max Lieble, and Adewale Adeboje

Abstract

The increasing rates of terrorism in Africa is a growing concern globally, and the realization of such dreadful circumstances demonstrates the need to disclose who is behind such terrible acts. Terrorists and extremist organizations have been known to use social media, and other forms of Internet-enabled technologies to spread idealism. Analyzing this data could provide valuable information regarding terrorist activity with the use of Open Source Intelligence (OSINT) tools. This study attempts to review the applications and methods that could be used to expose extremist Internet behavior.

Keywords

Social Network Analysis • Twitter • Terrorism • Extremist • Boko Haram • Al Shabaab • Fulani Militants • OSINT

22.1 Introduction

Africa has been plagued with three of the five most dangerous terrorist organizations. Nigeria is home to the world's most dangerous terrorist organization Boko Haram and the Fulani Militants which is the world's fourth deadliest terror group [1–6]. Somali is home to Al-Shabaab which is the world's fifth deadliest terror organization. Analyzing data captured in the Global Terrorism Database (GTD) from the University of Maryland's National Consortium for the Study of Terrorism and Responses to Terrorism (START) from the year 1970–2015 approximately 40,422 incidents have been reported in the Middle East and North Africa [3–7]. The GTD does not include a separate regional analysis of North

Africa. In Sub-Saharan Africa, the number of the incident for the same period is approximately 13,434 [3–7].

Nigeria is Africa's largest economy surpassing South Africa in recent years. However, this country is home to the infamous 419 Scam in which gained international attention due to the number of individuals this scam has taken financial advantage of globally. Somalia is a failed state where terrorist group Al-Shabaab are launching terror attacks in neighboring Kenya and Ethiopia. Both of these countries are strong economies in East Africa. Ethiopia is the home to the African Union (AU) and many other organizations that have a significant presence in Addis Ababa. Kenya is home to one of Africa's best universities, largest military, and growing digital economy.

M. Dawson (✉) • M. Lieble
University of Missouri-St. Louis, 1 University Drive, St. Louis, MO 63121, USA
e-mail: Dawsonmau@umsl.edu; http://www.umsl.edu/

A. Adeboje
Colorado Technical University, 4435 N. Chestnut St, Colorado Springs, CO 80907, USA
e-mail: Adewale.Adeboje@my.cs.coloradotech.edu

22.2 Data Collection Using Open Source Intelligence

Open Source Intelligence (OSINT) is unclassified information or data that is publicly available. OSINT is not to be a substitute for other sources of intelligence but rather complement existing methods to collect information such as

© Springer International Publishing AG 2018
S. Latifi (ed.), *Information Technology – New Generations*, Advances in Intelligent Systems and Computing 558, DOI 10.1007/978-3-319-54978-1_22

Geospatial Intelligence (GEOINT), Signal Intelligence (SIGINT), Human Intelligence (HUMINT), and Measurement Intelligence (MASINT). This data collection method relies on information that is found publicly without the need to request access to it, and it can be used to generate reports [8]. These reports can be behavioral activities, organization operations, political viewpoints, and relationships by simple Twitter retweets, mentions, or Wikipedia that are further analyzed [9, 10]. Technology has made the idea of spying being done by some secret agency has been replaced by analysts who can perform complex analysis with the aid of computer applications in an open society freely sharing data [11].

Data that was captured in this research study was OSINT and analyzed using various software applications. The Metasploit Community Edition (CE), Python, R, RapidMiner, and KNIME were among the applications used in this study. The data was captured through an anonymous account, and further analysis was conducted using the Tor Browser.

22.3 Open Source Software for Intelligence Analysis

Open Source Software (OSS) can enable the search of nefarious activities to be performed on terrorist groups. Since these applications are freely provided, or part of a penetration testing software such as Maltego users have the ability to use these applications. Programs such as Python, R, RapidMiner, KNIME, and others can be utilized for mining data. These applications allow for intelligence to be collected.

Table 22.1 provides details about a small selection of software applications, their description, and potential use. These applications are all open source which allows any individual or organization use them. Many can be loaded onto lightweight Linux distributions which can operate with minimal computing resources to be effective. This will allow for organizations with a limited budget or legacy systems to run these applications as part of their suite of OSINT tools. However, the Internet is a fundamental component to

running these applications as many of these tools crawl the net looking for hashtags, emails, select phases and other digital footprints. The idea behind OSINT is that these items are found freely online through forums, Wikipedia pages, blogs, social media, and more. Thus not having the Internet become an immediate barrier in conducting OSINT on a target.

22.4 Links and Transforms

The research was on the Somalia founded terrorist group Al-Shabaab using the application Maltego which is native in Kali Linux. This was conducted transforms the terrorist entity and labeling it as Al-Shabaab. From there a search was carried out on Twitter accounts that either mentioned the entity or contained tweet about the entity. From there we were able to view specific tweets and retweets that included various elements such as specific key phrases that were deemed to support terrorist activity. From there the tweets were accounts, and tweets were further investigated. The OSINT analysis starts with the terrorist entity Al-Shabaab and from there a transform against Twitter affiliations. From there accounts with tweets to this entity, and-and those that mentioned this entity were captured. From there the social account was mapped to the al_shabaab alias. That alias shows a further connection to a few social media accounts such as an active Instagram account, inactive Bitly link management account, and another. The Instagram account has the profile name al_shabaab with three posts, and under seven followers.

The other entities researched were Boko Haram to include infamous leader Abubakar Shekau. This searched allowed the view of accounts that used his name, and potential sympathizer of the Islamist militant group. A few of the OSINT gathered was followers, tweets, geolocation of tweets, and websites that contained specific phrases. Figure 22.1 shows the analysis conducted on Al-Shabaab and Boko Haram. Figure 22.2 displays the transforms done against the entity Boko Haram, and terrorist leader Shekau. This figure shows news organizations, websites, incoming, and outgoing links to multiple entities.

Table 22.1 Data mining and link analysis applications

Software application	Description & potential use	Source
R language	Language used for statistical computing and graphics.	[12]
Maltego Community Education	Program used to determine the relationships and real world links	[13]
RapidMiner	A program used for all steps of the data mining process including results visualization, validation, and optimization.	[14]
R Studio	IDE for R that allows for the use of R.	[15]
KNIME	Used for enterprise reporting, Business Intelligence (BI), data mining, data analysis, and text mining.	[16]
Python	High level, general purpose programming language.	[17]

Fig. 22.1 Sub-Saharan African
Terrorist Analysis

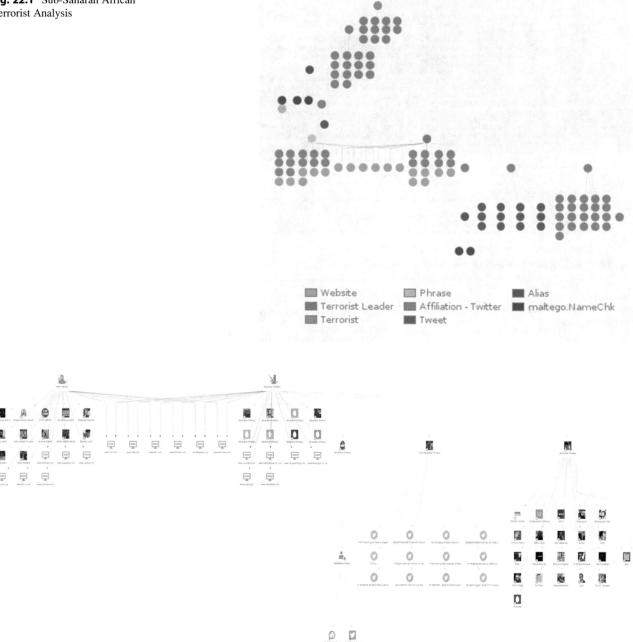

Fig. 22.2 Nigeria–Boko Haram Analysis

22.5   Data Analysis Using R and Python

The R Language is well known for its interactive capabilities as a tool for statistical computing and graphical representation of various statistical procedures. In addition to its analytical capabilities, R is well suited for data mining of all kinds including social network analysis and sentiment analysis. The below image features data collected from Twitter tweets using R. The below information represents details relating to tweets that featured the words "Al-Shabaab," a Somalian militant group. After the relevant tweets were collected, a function that provides all of the usernames of those tweets was utilized as well as a service that provided information about what source was used by those specific users to post those tweets (Fig. 22.3).

Python is a programming language than can be used for various forms of data mining and social media analysis. This tool can be used for sophisticated data analysis as well as simple reports that can be easily understood by a beginning

```
> sapply(tweets, function(x) x$getScreenName())
 [1] "anthonykyal121"  "ferozeahmedboda" "AyahNReza"       "seandw14"       "Joesy642"      "JosamNgoka"      "GeopoliticalJD"  "JohnnieChrome"   "mpendaraha18"
[10] "somali_activist" "alniinawaa4"     "douglaskawuma"   "AbdiAddow1"     "paddymwiine"   "alan_gumisiriza" "LeleiKiplangat"  "ntvuganda"       "john1966olsen"
[19] "PDWilliamsGWU"   "IamRusoke"       "lameck_njeje"    "TerrorFreeSomal" "HambySr"       "younggopp"       "arafrika"
> sapply(tweets, function(x) x$getStatusSource())
 [1] "<a href=\"http://twitter.com/download/android\" rel=\"nofollow\">Twitter for Android</a>"
 [2] "<a href=\"http://twitter.com/download/android\" rel=\"nofollow\">Twitter for Android</a>"
 [3] "<a href=\"http://twitter.com/download/android\" rel=\"nofollow\">Twitter for Android</a>"
 [4] "<a href=\"http://twitter.com\" rel=\"nofollow\">Twitter Web Client</a>"
 [5] "<a href=\"http://twitter.com/download/iphone\" rel=\"nofollow\">Twitter for iPhone</a>"
 [6] "<a href=\"http://twitter.com/download/android\" rel=\"nofollow\">Twitter for Android</a>"
 [7] "<a href=\"http://twitter.com/download/iphone\" rel=\"nofollow\">Twitter for iPhone</a>"
 [8] "<a href=\"http://twitter.com/download/android\" rel=\"nofollow\">Twitter for Android</a>"
 [9] "<a href=\"http://twitter.com/download/iphone\" rel=\"nofollow\">Twitter for iPhone</a>"
[10] "<a href=\"http://twitter.com/download/iphone\" rel=\"nofollow\">Twitter for iPhone</a>"
[11] "<a href=\"http://twitter.com\" rel=\"nofollow\">Twitter Web Client</a>"
[12] "<a href=\"http://twitter.com/download/android\" rel=\"nofollow\">Twitter for Android</a>"
[13] "<a href=\"http://twitter.com/download/iphone\" rel=\"nofollow\">Twitter for iPhone</a>"
[14] "<a href=\"https://mobile.twitter.com\" rel=\"nofollow\">Mobile Web (M2)</a>"
[15] "<a href=\"http://twitter.com/download/android\" rel=\"nofollow\">Twitter for Android</a>"
[16] "<a href=\"http://twitter.com/download/android\" rel=\"nofollow\">Twitter for Android</a>"
[17] "<a href=\"http://snappytv.com\" rel=\"nofollow\">SnappyTV.com</a>"
[18] "<a href=\"http://twitter.com\" rel=\"nofollow\">Twitter Web Client</a>"
[19] "<a href=\"http://twitter.com/#!/download/ipad\" rel=\"nofollow\">Twitter for iPad</a>"
[20] "<a href=\"http://twitter.com/download/android\" rel=\"nofollow\">Twitter for Android</a>"
[21] "<a href=\"http://twitter.com/download/android\" rel=\"nofollow\">Twitter for Android</a>"
[22] "<a href=\"http://twitter.com/download/android\" rel=\"nofollow\">Twitter for Android</a>"
[23] "<a href=\"http://twitter.com\" rel=\"nofollow\">Twitter Web Client</a>"
[24] "<a href=\"http://twitter.com/download/iphone\" rel=\"nofollow\">Twitter for iPhone</a>"
[25] "<a href=\"http://blackberry.com/twitter\" rel=\"nofollow\">Twitter for BlackBerry®</a>"
> |
```

Fig. 22.3 Al-Shabaab Tweet Word Search

Fig. 22.4 Python Jihad Word
Search Program

```
File  Edit  Format  Run  Options  Window  Help
import tweepy
from tweepy import Stream
from tweepy import OAuthHandler
from tweepy.streaming import StreamListener
ckey = 't0KKZp3EI7CylfmfaTFpe6r1K'
csecret = 'fLXY82VFvHc5NF7CBcSLSjGGhfFiOBDVL7Rqa3zOIu5BrI1UNY'
atoken = '706140478878552064-0kCZWapnnIlXoUcKQ2wxUnjxuwKQVck'
asecret = 'yPqbhl8DWGTBwL6eYh5j0vyuW6RL3tZfAyVqP4njm8CBa'

from tweepy import OAuthHandler
class listener(StreamListener):


        def on_data(self, data):
                print (data)
                return True

def on_error(self, status):
        print (status)


auth = OAuthHandler(ckey, csecret)
auth.set_access_token(atoken, asecret)
twitterStream = Stream(auth, listener())
twitterStream.filter(track=["jihad"])
```

user. The coding for Python is relatively similar to that of R but has it very own, unique forms of capabilities that set itself apart from others. Regarding social media analysis, Python is capable of extracting data from Twitter and other social media platforms. An example of the Twitter streaming function through Python is displayed below (Fig. 22.4).

22.6 Findings

Maltego served as a useful tool for link analysis. The platform allowed for detailed mining of particular forms of data. Graphs can be constructed and serve as a sort of an investigation. Furthermore, this program proved useful for cyber threat analysis of many types and ultimately acts as a tool for OSINT. Figure 22.2 displays at social media analysis through Maltego about the search term of "Boko Haram." The output of this function provided specific Twitter accounts that have held some relation to an entity based on each Twitter accounts name or tweeting patterns. Further analysis was conducted on each of these Twitter accounts by performing a function that generated all of the tweets for each Twitter account. In addition to this, sentiment analysis was conducted on each of these tweets to determine a whether or not the tweets for each of these Twitter users was positive, negative, or neutral. Discovered through the research were that many of the tweets have different sentiment values, including positive, negative, and neutral.

22.7   Discussion and Recommendation

Given the results of the data analysis in this study, it is recommended that multiple tools be used for data collection with a human in the loop to check the data collected. The applications were able to grab numerous lines of evidence, but the creation of the links had to be validated by an analyst that was able to verify the credibility of the properties for the entity. In the future entities that have a certain number of links could be weighted at a more trusted level and those with lower links require further analysis. The other recommendation is collect all images and the run an extraction on the Exchangeable Image File Format (EXIF) metadata. This metadata can be mapped on a heat map that correlates with the time stamp found in the metadata. The terrorist's organizations found were not linked to each other or other known groups. In the future research could look at the connections to other bodies such as Non-government Organizations (NGOs), universities, political parties, etc. It is also essential that the types of metrics used to do associations of data elements and associated target behaviors and parametric information. This would be used to define new ways to characterize information to associated objects. A statistical assessment of the accuracy by manual inspection of associations would help intelligence analysts gain confidence in new metrics for the data mining and tracking algorithms.

22.8   Conclusion

The study's objective was to collect OSINT data and conduct a link analysis. Several main conclusions emerged from this research study. The results of the data analysis showed that some OSINT data collected were indeed mapped to terrorist organizations however these links were not all negative. A majority of these links were from news agencies or reporters. Required was an inspection of each entity, and the associated properties to validate if transform conducted was useful.

References

1. Dawson, M., Kisku, D. R., Gupta, P., Sing, J. K., & Li, W. (2016). *Developing next-generation cuntermeasures for homeland security threat prevention* (pp. 1–428). IGI Global: Hershey, PA. doi:10.4018/978-1-5225-0703-1.
2. Dawson, M., & Adeboje, W. (2016). Islamic extremists in Africa: Security spotlight on Kenya and Nigeria. In M. Dawson, D. Kisku, P. Gupta, J. Sing, & W. Li (Eds.), *Developing next-generation countermeasures for homeland security threat prevention* (pp. 93–103). Hershey, PA: Information Science Reference. doi:10.4018/978-1-5225-0703-1.ch005
3. National Consortium for the Study of Terrorism and Responses to Terrorism (START). (2016). Global Terrorism Database [Nigeria]. Retrieved from https://www.start.umd.edu/gtd
4. National Consortium for the Study of Terrorism and Responses to Terrorism (START). (2016). Global Terrorism Database [Somalia]. Retrieved from https://www.start.umd.edu/gtd
5. National Consortium for the Study of Terrorism and Responses to Terrorism (START). (2016). Global Terrorism Database [Kenya]. Retrieved from https://www.start.umd.edu/gtd
6. National Consortium for the Study of Terrorism and Responses to Terrorism (START). (2016). Global Terrorism Database [Ethiopia]. Retrieved from https://www.start.umd.edu/gtd
7. National Consortium for the Study of Terrorism and Responses to Terrorism (START). (2016). Global Terrorism Database [Syria]. Retrieved from https://www.start.umd.edu/gtd
8. Stalder, F., & Hirsh, J. (2002). Open source intelligence. *First Monday, 7*(6), 1–8.
9. Milne, D., & Witten, I. H. (2013). An open-source toolkit for mining Wikipedia. *Artificial Intelligence, 194*, 222–239.
10. Cleveland, S., Jacson, B. C., & Dawson, M. (2016) Microblogging in higher education: Digital natives, knowledge creation, social engineering, and intelligence analysis of educational tweets (pp. 1–19). E-Learning and Digital Media.
11. Steele, R. D. (2000). *On intelligence: Spies and secrecy in an open world*. Washington, D.C: AFCEA International Press.
12. Ihaka, R., & Gentleman, R. (1996). R: A language for data analysis and graphics. *Journal of Computational and Graphical Statistics, 5*(3), 299–314.
13. Bradbury, D. (2011). In plain view: Open source intelligence. *Computer Fraud & Security, 2011*(4), 5–9.
14. Hofmann, M., & Klinkenberg, R. (Eds.). (2013). *RapidMiner: Data mining use cases and business analytics applications*. Boca Raton: CRC Press.
15. Gandrud, C. (2013). *Reproducible research with R and R studio*. Boca Raton: CRC Press.
16. Berthold, M. R., Cebron, N., Dill, F., Gabriel, T. R., Kötter, T., Meinl, T., Ohl, P., Seib, C., Thiel, K., & Wiswedel, B. (2008). *KNIME: The Konstanz information miner* (pp. 319–326). Berlin Heidelberg: Springer.
17. Lutz, M. (1996). *Programming python* (Vol. 8). Sebastapol: O'Reilly.

A Description of External Penetration Testing in Networks

23

Timothy Brown

Abstract

This paper synthesizes multiple sources in order to describe external penetration testing. In a world of ever more sophisticated network attacks this type of knowledge is useful for both security professionals and the laymen. This paper describes some general history of, uses of, and processes of external penetration testing, and communicates the philosophy behind it. This paper concludes by summarizing some potential weaknesses of penetration testing, but also communicates its importance in establishing the security of any system.

Keywords

Penetration testing • Security • Assessment • Scanners • Network

23.1 Introduction

In a world of ever more sophisticated attacks, and ever more internet integration, information security professionals often must take drastic measures to ensure the security of the assets they are charged with. These assets can be numerous: from user information, to passwords, to sensitive organizational files, businesses small and large are increasingly becoming their own data warehouses, storing information of their own and from their customers that could have disastrous effects should they be compromised. But, in the age of online transactions, this information can no longer be locked away. Businesses now rely so fully on the internet that they must necessarily expose some amount of their data and systems, or they could not operate their online services. These public-facing outlets, as well as internal networks, are where the tug-of-war between security and those that would compromise it take place, and go under the microscope during any security and system audit.

External penetration testing, often considered a form of ethical hacking, is a crucially important way that information security professionals probe for the weaknesses and vulnerabilities in a system. This test is one of the oldest methods for assessing the security of a system, and much of computer and network security sprang from it when it was used by the Department of Defense as early as the 1970's to demonstrate the growing need for secure system development programs. [8] The goal of penetration testing, or pen testing, is simple: by approaching the system as an outsider, the security professional can understand it as a would-be hacker, not as someone provided a view of the entire system at once. [5] This change of perspective is designed to prevent the blind spots that could develop when viewing a system only from within or only as a whole. These blind spots of over-familiarity often hide unseen vulnerabilities that a malicious actor could exploit, making it an important part of an IT security assessment. If the IT security assessment was modeled on a spectrum, the penetration test would be at the very bottom, as it takes the finest-grain view possible of the system in an attempt to find any weakness, and examines the system in the greatest detail. This is modeled in Fig. 23.1 [4].

Vulnerabilities within the network structure, or firewall configuration, or server safeguards are only one of the threats penetration testing is designed to assess. If any threat exists, the security expert is going to attempt to identify

T. Brown (✉)
Robert Morris University, Moon, PA, USA
e-mail: trbst106@mail.rmu.edu

© Springer International Publishing AG 2018
S. Latifi (ed.), *Information Technology – New Generations*, Advances in Intelligent
Systems and Computing 558, DOI 10.1007/978-3-319-54978-1_23

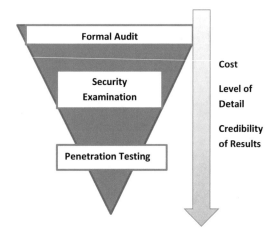

Fig. 23.1 IT security assessment model (Adapted from Ref. [4])

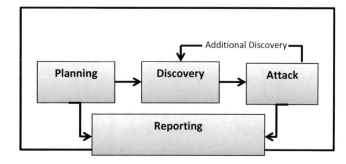

Fig. 23.2 Phases of penetration test (Adapted from Ref. [2])

it. Often, the most overlooked aspects of the system are the physical. The hardware, the users, and the customers are just as much a part of the system as the software. The longest passwords on the most protected servers with the strongest encryption are useless if an employee can be convinced to simply hand them over. The penetration tester is as concerned with these avenues as they are with any other, and they need to be, as these vulnerabilities are some of the most frequently overlooked when performing other kinds of security tests not specifically concerned with them. As outlined in the NIST Technical Guide to Information Security and Assessment [9], pen testing should elucidate how well the system tolerates real world-style attack patterns, the likely level of sophistication an attacker needs to successfully compromise the system, additional countermeasures that could mitigate threats against the system, and the defenders' ability to detect attacks and respond appropriately.

23.2 Process and Technology

When performing a penetration test, setting is particularly important. The goal is to emulate a real attack against a real system, which necessitates creating, or at least mimicking, that situation. This means a useful target for a penetration test is, if not an actual running, production system, something close to it that has been filled with data and brought online in order for it to be tested thoroughly, with all potential uses and behaviors available to be exploited. In most cases, an attacker is not going to have intimate knowledge of the system they are targeting, and this circumstance should be duplicated as much as possible for a "black box" external penetration test [2].

As [1] published, "... the process [of penetration testing] includes gathering information about the target before the

test, identifying possible entry points, attempting to break in (either virtually or for real) and reporting back the findings." These phases are largely compatible across most sources: the process of penetration testing should begin with planning, transition into probing the system and discovery, and then begin attacking the system. Information discovered during the attack should be noted and used to further the scope of the attack, as some aspects of the system will only be discovered during the attack phase but may be crucial to gaining increased access. Once the attack is complete, generating a report that can be understood by management requires care as well. This process is illustrated in Fig. 23.2 [2].

Some parts of the discovery and attack phase can be performed automatically. For instance, network sniffing, port scanning, and service identification are all aspects of network discovery that can be performed simply by letting the appropriate scanner run and accumulate information such as open ports and active hosts or flag potentially vulnerable points of entry, allowing the penetration tester to begin to form a picture of the network and how it operates, and where it can be targeted. [7, 9] describes a few possible scans that can be run automatically, and what kinds of information can be gleaned from them:

> "Information gathered during an open port scan can assist in identifying the target operating system through a process called OS fingerprinting. For example, if a host has TCP ports 135, 139, and 445 open, it is probably a Windows host..."

The penetration tester, by comparing and contrasting the clues from the scanners, can learn what operating system is running on a target server or machine. Knowing the operating environment is essential for the tester to tailor their attempts to that setup, and narrow down which version they may be dealing with.

> "... Some scanners can help identify the application running on a particular port through a process called service identification. Many scanners use a services file that lists common port numbers and typical associated services—for example, a scanner that identifies that TCP port 80 is open on a host may report that a web server is listening at that port—but additional steps are needed before this can be confirmed."

Knowing what kind of applications are operating on the system is one of the best ways to find vulnerabilities. By considering the operating system and applications running, it increases the chances of finding a non-patched piece of software that could present a hole to be exploited.

> "Port scans can also be run independently on entire blocks of IP addresses—here, port scanning performs network discovery by default through identifying the active hosts on the network."

Identifying the active hosts can directly increase the number of potential failure points, gives the tester more knowledge of how the network is operating and what parts are related. This kind of information, served up automatically by scanner programs, serves as a strong starting point for any penetration test.

Armed with this information, the tester moves into the attack phase and must actually begin exploiting these aspects of the system to the best of their ability. [9] enumerates a variety of methods that could penetrate the system based on flaws discovered in the previous phases: misconfigured security systems or missing patches, kernel and other OS flaws, poor input validation and buffer overflow opportunities, applications with too many permissions or with temporarily elevated permissions, and more.

Once the system has been initially penetrated, the tester's goal is to continue to give themselves more and more permissions, which gives them greater and greater access to system features and components. As more permissions are gained, the system can be explored further, creating a feedback loop where the tester learns more and circles back to the discovery phase, broadening their understanding of the system and formulating new avenues of attack (Fig. 23.3).

The final goal of this privilege escalation could vary, from simply gaining enough access to confidential information, to gaining full control of the system, to knocking the entire system off-line or destroying it entirely. While this would not be ideal for a production system undergoing a penetration test, the existence of the capability would still be valuable information for the reporting phase. A tester would not want to deliberately cause this kind of damage, and so must take care to avoid it.

An occasionally overlooked aspect of penetration testing is social engineering. [6] revealed the average user's social engineering vulnerability in 2015 when they conducted a survey questioning employees of a Russian company on their awareness of and resistance to social exploitation by penetration testers, and by extension, real attackers. Only 20% of responders qualified as "secure", with at least half of respondents being rated as "vulnerable"—for reasons ranging from a lack of knowledge of what is and isn't sensitive information, to understanding of browser updates and errors, to the ability of a user to discern between a real website and one designed for phishing. Even those users considered "secure" did not rate highly in their resistance to a simple probing phone call. While a company may be able to defend itself against the technical vulnerabilities revealed from a penetration test, when the tester needs only to make a phone call or send an email to learn a privileged user's password, it becomes elementary to enter even the most secured systems. "Mostly, [users and employees] are not aware of all security threats. There is no tool to prevent social engineering attacks, because those attacks always involve humans… …when employees are educated about security, there is a better chance to prevent a social engineering threat just before it happens." [6] A strong corporate policy and repeated, on-going education of employees is the only real answer to social engineering attacks and should be recommended as part of any report.

23.3 Evaluation

23.3.1 Weaknesses

One of the biggest weaknesses of penetration testing is the inherent risk involved in penetrating any system. The test is most effective when performed on a system that is as close to

Fig. 23.3 Discovery & attack phase steps (Adapted from Ref. [2])

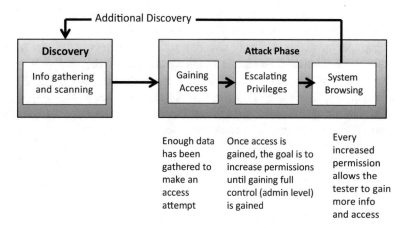

in production and currently running as possible. For best results, it is done on an active, working system. Unintentional consequences are rife in this type of scenario. For instance, it is often difficult to demonstrate a way that a system can be taken down or even damaged without taking the steps that would cause that action and demonstrating that they work. It is entirely possible that penetration testing on a needed system could result in that system being taken down and interrupting business. This risk creates the need to play a balancing act, where you must either trade the best possible penetration report for the safety and security of your business, or trade potential unintended damages for a more full view of the vulnerabilities of your system. In this way, penetration testing can often be self-limiting. The fear of system damage could lead to a penetration tester "holding back" resulting in a penetration test that is not as thorough, but generates a positive testing report [3].

While management may like this positive report, it introduces a second major weakness of penetration testing. Not only is "passing" a penetration test somewhat subjective, it can be outright misleading or can create the wrong mindset for users and managers of a system. [3] elaborates on this:

> "By performing a pen test and coming through with a [passing grade], you might find yourself tempted to think that your network or system is protected and that you can rest easy. This is threatening for two reasons: first, the test [most likely] does not address all important security issues......Unless truly comprehensive, a pen test will only detect possibly entry weaknesses and may miss numerous internal [weaknesses]."

23.3.2 Strengths

The strengths of a well-performed penetration test cannot be understated. From the moment the tester begins gathering information for their attempts, they are exposing with every step the little errors and tweaks of configuration or code that reveal information and leave doors, no matter how small, open. These defects build upon each other, and what starts as a small imperfection can be grown into a massive security hole over the course of testing. This works on two levels. These specific errors are likely to be the path of least resistance for any would-be hacker, so spotting them and fixing them ahead of time is valuable for that reason alone. In addition, a well-constructed report will not mislead or misrepresent the safety of a system as permanent, and impress upon management and those responsible for the system the

importance of constant vigilance and scrutiny. The penetration test must be just as focused on creating a mindset of safety as it is for exposing specific faults [3].

23.4　Conclusion

This kind of external, black box penetration testing is a valuable, powerful tool, but with that power there is risk and responsibility that cannot be underestimated. While different methodologies for penetration testing exist, they all share the same goal: to efficiently find the flaws in a system and expose them with as little disruption as necessary to the workings of a business. When this is executed successfully, a system is explored in the closest possible detail, and vulnerabilities that would not have been found any other way can be secured [10].

References

1. Gupta, A., Kavita, & Kirandeep, K. (2013). Vulnerability assessment and penetration testing. *International Journal of Engineering Trends and Technology, 4*(3), 328.
2. Felderer, M., Büchler, M., Johns, M., Brucker, A. D., Breu, R., & Pretschner, A. (2016). Chapter one-security testing: A survey. *Advances in Computers, 101*, 1–51.
3. Kaur, M. S., & Singh, M. S. (2016). Penetration testing management. *International Journal of Advanced Research in Computer and Communication Engineering, 5*(3), 171–177.
4. Klíma, T. (2016). PETA: Methodology of information systems security penetration testing. *Acta Informatica Pragensia, 5*(2), 98–117.
5. Mattadi, E., & Kumar, K. V. (2015). Evaluation of penetration testing and vulnerability assessments. *International Journal of Electronics Communication and Computer Engineering, 6*(5), 144–148.
6. Michno, B., Nycz, M., & Hajder, P. (2015). Social engineering – Penetration testing. http://sasit.prz.edu.pl/wp-content/uploads/2016/02/SOCIAL-ENGINEERING-PENETRATION-TESTING.pdf
7. Palani, M. U., Vanitha, M. S., & Lakshminarasimman, M. S. (2014). Network security testing using discovery rechniques. *International Journal of Computer Science and Information Technologies, 5*(6), 7146–7153.
8. Said, H. M., Hamdy, M., El Gohary, R., & Salem, A. B. M. (2015). An integrated approach towards a penetration testing for cyberspaces. *European Journal of Computer Science and Information Technology, 3*(1), 108–128.
9. Scarfone, K., Souppaya, M., Cody, A., & Orebaugh, A. (2008). Technical guide to information security testing and assessment. *NIST Special Publication, 800*, 115.
10. Whitman, M. E., & Mattord, H. J. (2003). Principles of information security. Boston, MA: Course Technology.

Fahad Alsolami, Yaqoob Sayid, Nawal Zaied, Isaac Sayid,
and Iman Ismail

Abstract

Cloud storages have recently increased rapidly in many organizations as they provide many benefits and advantages for the users. These advantages include easiness in use, unlimited capacity in storage, scalability and cost effectiveness. This has motivated the concerns regarding the cloud issues including cost, performance and security. Many different schemes have been proposed to address the issues of cloud storage. Each scheme considers these issues from different perspectives without employing an optimal solution that fully satisfies the user requirements. In real world applications, the client's objectives to move to cloud storage vary from single objective to several conflicting objectives. As a result, there is a need to find a tradeoff setup for a multiple cloud storage scheme. This paper proposed Smart Cloud Optimal Resource Selection (SCORS) scheme by applying non-dominated sorting genetic algorithm (NSGA-2) to resolve the user's conflicting requirements. Our optimal solution facilitates setting up multiple cloud storages in an optimized way based on the objectives specified. Since each user has different objectives, all combinations of cloud storage client needs are collected and transformed into different multi-objective optimization problems with constraints and bounds. Experiment results show that SCORS scheme determine the optimal solution from the set of feasible solutions in an efficient way.

Keywords

Cloud storage • Optimal solution • Security

24.1 Introduction

- Nowadays, many people start using cloud storages instead of the traditional way of storage. This is because of the services and features provided by the clouds such as scalability and cost effectiveness [1]. Other services include easiness in use and unlimited capacity in storage. Although these services are beneficial, there are some concerns highlighted from using clouds services [24]. Consequently, there are several cloud storage schemes

suggested by the research community, each has its own strength and weak points centered on the cost, performance and security [1]. Data stored in the cloud are facing issues such as the availability [26]. The availability of the cloud can be forbidden by failure occurrence in internet connections, denial of services attacks or even the regular maintenance that causes service disruption [26]. Moreover, even the stored data is encrypted, the issue of insider attacks remains critical as the insider employee can have illegal copy and access to the data [27, 28]. Also, brute force attacks highlight the issue of the data confidentiality and key management [25]. Other factors such as minimizing the costs [8] and maximizing the performance are considered as needed objectives to be achieved. In real world applications, the client's

F. Alsolami (✉) • Y. Sayid • N. Zaied • I. Sayid • I. Ismail
Department of Information Technology, King Abdulaziz University, Jeddah, Saudi Arabia
e-mail: fjalsolami@kau.edu.sa

objectives to move to cloud storage vary from single objective to several conflicting objectives [10]. As a result, there is a need to find a tradeoff setup for a multiple cloud storage scheme.

- In this paper, we propose an optimal solution called Smart Cloud Optimal Resource Selection (SCORS) that satisfies all the user requirements. SCORS will provide to the user the ability to select the set of criteria he/she only concerned with. If the user interested in more than one criteria that might conflict each other, our proposed framework will satisfy him by using non-dominated sorting genetic algorithm (NSGA-2) [2, 3].
- The remainder of the paper will be as following. Section 24.2 will cover the background of the proposed schemes. In Sect. 24.3, we discuss the research objectives. In Sect. 24.4, we present the research design. Sect. 24.5 covers the research experiments and results. At last, the acknowledgments in Sect. 24.6 and the conclusion will be in the Sect. 24.7.

24.2   Background

- To achieve the security in cloud storage, a number of security schemes have been proposed in the literature. Xiong et al. [4] proposed CloudSeal scheme that provides confidentiality by using two types of encryption methods. Kumbhare et al. proposed Cryptonite [5] which has the so-called strongbox that take the responsibility of protecting and encrypting data. DepSky presented by Bessani et al. [6] which splits the encrypted data into shares and distributes the shares to multiple clouds based on the secret-sharing scheme [7] in order to provide availability and confidentiality. Abu-Libdeh [8] proposed a scheme called Redundant Array of Cloud Storage (RACS) that has advantages in preventing out-of-service problem, economical failures and the issue of the vendor lock-in. Also, a scheme is called High Availability and Integrity Layer (HAIL) is proposed by Bowers [9] that achieve confidentiality by distributing many copies to different clouds and by utilizing the so-called RAID-like technique [29]. Moreover, two other schemes called N-Cloud [10] and CloudStash [7] are proposed by Alsolami et al. Both schemes improve availability and provide confidentially by distributing encrypted data over multiple cloud storages.
- Another scheme is called SPORC proposed by Feldman et al. [11] that encrypts the sent data to the server containing the cloud by utilizing the cryptographic keys of the users. Santos et al. [12] proposed trusted cloud computing platform (TCCP) that makes sure of the

security of the cloud service prior starting the Infrastructure as a Service (IaaS) and the virtual machines. In addition, the scheme proposed by Wang et al. [13] utilizes the erasure coding technique for data distribution in order to have a dependable data. Lei et al. [14] proposed security flexible data storage and sharing scheme (SFDS) that makes the cloud storage secure and private.

- In addition, a scheme is proposed by Sarvabhatla and Vorugunti [15] that is an improved mutual authentication scheme and lightweight in both client and server side. Deshmukh et al. [16] proposed a scheme that focuses on data storage security instead of encrypting the data itself. Mazur et al. [17] proposed a scheme that is fault-tolerant and has protection against cloud security attacks and risks. The Intercloud [18] security scheme uses symmetric key encryption methods. It enhances confidentiality, integrity, reliability, and consistency. Balasaraswathi and Manikandan [19] proposed a scheme for multiple clouds that encrypt the partitioned data and make them secure against hackers attacks since it has a dynamic mechanism for encryption. Depot [21] has fault tolerance feature and provides availability.
- Almost all of previously proposed schemes are meant to serve specific needs or objectives without employing optimal solution. In real world applications, the client's objectives to move to cloud storage [20] vary from single objective to several conflicting objectives. As a result, there is a need to find a tradeoff setup for a multiple cloud storage scheme. This paper suggests an algorithm to automate the selection of optimal setup for general multiple-cloud storage scheme to serve each client according to his needs or objectives.

24.3   SCORS Scheme Objectives

- There are several cloud storage schemes that address the concern of cloud issues. Each scheme has its own strength and weak points centered on the cost, performance and security. Almost all of previously proposed schemes are meant to serve specific needs or objectives without employing an optimal solution. For example, some of these schemes focused on how to provide the security objectives but without concerning an efficient performance. Others are focusing on enhancing the availability and performance but without concerning the overall security in relation. Moreover, further schemes focus on enhanced security which leads to inefficient costs.
- Due to the user prerequisites that might conflict with each other, multiple-objective optimization algorithms need to

be utilized to find the optimal setup for general multiple cloud storage scheme to serve each client according to his needs or objectives. Since each user has different objectives, all combinations of cloud storage client needs are collected and transformed into different multi-objective optimization problems with constraints and bounds. These multi-objective problems are solved using NSGA-2 algorithm [2, 3].

- Generally, clients who want to switch to cloud storage scheme seek for the following objectives: Security, cost and performance. Some of these objectives look too general and need deeper and more precise definition. For example, performance here is meant to be the overall throughput. Security consists of confidentiality, integrity and availability. Consequently, there can be thirty one possible combinations of objectives for a client to choose from as shown in Table 24.1.

24.4 SCORS Scheme Design

- As we mentioned in previous section, we will solve the problem using NSGA-2 [2, 3]. Figure 24.1 illustrates the steps of this algorithm according to our problem in the cloud storage. In the first step, the cloud storages are initialized. After that, the needed objectives will be taken from the client. Then, the processes are continued in Fig. 24.2 which is the applying of the elitist genetic algorithm. In this algorithm, the objective functions are first evaluated. Then criteria considered in the objective functions that include cost, performance, integrity, availability and confidentiality will be ranked according to the user requirements. Next, the child population is created which include the selection of the user first criteria as parent then the process will crossover to the next level of selecting the next criteria. After that, the objective

Table 24.1 All possible combinations of client objectives to use multiple-cloud storage

#	Cost	Performance/throughput	Integrity	Availability	Confidentiality
1	Min				
2		Max			
3			Max		
4				Max	
5					Max
6	Min	Max			
7	Min		Max		
8	Min			Max	
9	Min				Max
10		Max	Max		
11		Max		Max	
12		Max			Max
13			Max	Max	
14			Max		Max
15				Max	Max
16	Min	Max	Max		
17	Min	Max		Max	
18	Min	Max			Max
19	Min		Max	Max	
20	Min		Max		Max
21	Min			Max	Max
22		Max	Max	Max	
23		Max	Max		Max
24		Max		Max	Max
25			Max	Max	Max
26	Min	Max	Max	Max	
27	Min	Max		Max	Max
28	Min		Max	Max	Max
29		Max	Max	Max	Max
30	Min	Max	Max		Max
31	Min	Max	Max	Max	Max

functions are evaluated again to combine the parent and child criteria which then will be ranked again according to the user requirements. The feasible solution set will be generated in the next step and one of them is chosen as an optimal solution. If the selected optimal solution met the user requirements, then the algorithm will stop, otherwise, the child population step is performed again until the user requirements are met.

• The optimization algorithm consists of number of objective functions equal to the objectives needed. For each combination of objectives in Table 24.1, there is a corresponding optimization algorithm to be used. All these objective functions use either one or more of the following decision variables and their bounds:

• Cloud storages to be chosen are expressed by the variable x_i, where $i \in (1,2 \ldots 10)$. Hence, there are 10 decision variables where each one represents specific cloud storage in Table 24.2.

$$0 \leq x_i \leq 1, \; x_i \in (0, 1).$$

• Flag of enabling encryption is expressed by the variable e.

$$0 \leq e \leq 1, e \in (0, 1).$$

• Flag of enabling integrity check is expressed by the variable d.

$$0 \leq d \leq 1, d \in (0, 1).$$

The client objectives are expressed by using objective functions or decision variables. This is due to the nature of these objectives as some of them are measurable while others are just binary values. To unify the expressions, binary objectives are formed as decision variables. The evolutionary NSGA-2 algorithm is going to generate Pareto

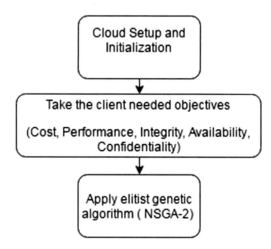

Fig. 24.1 The SCORS flowchart

Fig. 24.2 Elitist genetic algorithm

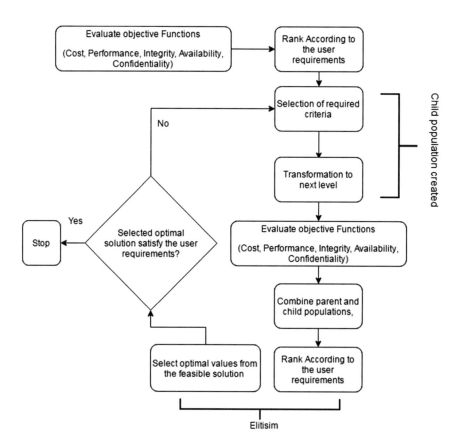

Table 24.2 Storage allocation and traffic prices for amazon S3 standard cloud storages as of 2016-04-25

I	Region name	Region	Endpoint	Avg. latency
1	US East (N. Virginia)	us-east-1	s3.amazonaws.com	188ms
			s3-external-1.amazonaws.com	180ms
2	US West (Oregon)	us-west-2	s3-us-west-2.amazonaws.com	231ms
3	US West (N. California)	us-west-1	s3-us-west-1.amazonaws.com	241ms
4	EU (Ireland)	eu-west-1	s3-eu-west-1.amazonaws.com	290ms
5	EU (Frankfurt)	eu-central-1	s3.eu-central-1.amazonaws.com	91ms
			s3.eu-central-1.amazonaws.com	90ms
6	Asia Pacific (Singapore)	ap-southeast-1	s3-ap-southeast-1.amazonaws.com	171ms
7	Asia Pacific (Tokyo)	ap-northeast-1	s3-ap-northeast-1.amazonaws.com	228ms
8	Asia Pacific (Sydney)	ap-southeast-2	s3-ap-southeast-2.amazonaws.com	344ms
9	Asia Pacific (Seoul)	ap-northeast-2	s3.ap-northeast-2.amazonaws.com	233ms
			s3.ap-northeast-2.amazonaws.com	230ms
10	South America (São Paulo)	sa-east-1	s3-sa-east-1.amazonaws.com	313ms

Note: The price is for clients with data size >1TB and <49TB. Traffic size is of 10TB for both outbound and inbound

optimal solutions to the multi objective optimization problem. The general form of this algorithm:

$$f(x_{i...10}, e, d) = \min(f_1(x_{i...10}, e, d), f_2(x_{i...10}, e, d), \ldots f_n(x_{i...10}, e, d)) \quad (24.1)$$

Where n is number of objective functions to be optimized.

The objective function of minimizing the cost (min c) would be:

$$f_1(x) = \sum_{i=1}^{10} (a_i x_i) + \sum_{i=1}^{10} (b_i x_i) + \sum_{i=1}^{10} (c_i x_i) + \sum_{i=1}^{10} (d_i x_i) + \sum_{i=1}^{10} (e_i x_i) \quad (24.2)$$

Subject to

$$\sum_{i=1}^{10} (x_i) > 0,$$

where a, b, c, d and e are costs of storage allocation, inbound request, outbound request, inbound traffic and outbound traffic.

The objective function of maximizing the performance (throughput) (max P) would be:

$$f_2(x) = \left[\sum_{i=1}^{10} (a_i x_i) + (b \times d) + (c \times e) \right] \quad (24.3)$$

Subject to

$$\sum_{i=1}^{10} (x_i) > 0,$$

where a, b and c are latency of requests to cloud storage, overhead of integrity check and overhead of encryption respectively.

The objective function of maximizing the Availability (max a) would be:

$$f_3(x) = -1 \times \left(\sum_{i=1}^{10} (x_i) \right) \quad (24.4)$$

Subject to

$$2 \leq \sum_{i=1}^{10} (x_i) \leq 8$$

- NSGA-2 algorithm is designed to minimize the objectives. However, the problem to be optimized contains maximization objectives. To solve this, all the maximization objectives were negated.

24.5 Experiments and Results

- The optimization problem in this paper which consists of multiple optimization cases is solved using multi-objective evolutionary algorithm. As mentioned in section III, we can derive thirty one different problems for each case of user needs. Each of these problems is implemented in Java separately with its solvers using MOEA framework. One of the significant features in MOEA framework is that it can deal with mixed-integer multi-objective optimization problems [22].
- The implementation is based on some assumptions as follows. The integrity check and the confidentiality objectives are not measured by numbers. Therefore, they are going to be binary decision variables specified by whether these two objectives are enabled or not. One means enabled, zero means disabled. The performance is measured by throughput value. The throughput is considered by overhead of integrity check, encryption / decryption and total network latency of cloud storages used.

Table 24.3 Average latencies of round trip time gathered from issuing 100 requests per Amazon S3 region.

I	Name	Inbound requests (except deletion requests)		Outbound requests
1	US East (N. Virginia)	$0.005 per 1,000 requests	$0.05 per 10000 requests	$0.004 per 10,000 requests
2	US West (Oregon)	$0.005 per 1,000 requests	$0.05 per 10000 requests	$0.004 per 10,000 requests
3	US West (Northern California)	$0.005 per 1,000 requests	$0.05 per 10000 requests	$0.004 per 10,000 requests
4	EU West (Ireland)	$0.005 per 1,000 requests	$0.05 per 10000 requests	$0.004 per 10,000 requests
5	EU Central (Frankfurt)	$0.0054 per 1,000 requests	$0.054 per 10000 requests	$0.0043 per 10,000 requests
6	Asia Pacific Southeast (Singapore)	$0.0054 per 1,000 requests	$0.054 per 10000 requests	$0.0043 per 10,000 requests
7	Asia Pacific Northeast (Tokyo)	$0.0047 per 1,000 requests	$0.047 per 10000 requests	$0.0037 per 10,000 requests
8	Asia Pacific Southeast (Sydney)	$0.0055 per 1,000 requests	$0.055 per 10000 requests	$0.0044 per 10,000 requests
9	Asia Pacific Northeast (Seoul)	$0.0045 per 1,000 requests	$0.045 per 10000 requests	$0.0035 per 10,000 requests
10	South America (Sao Paulo)	$0.007 per 1,000 requests	$0.07 per 10000 requests	$0.0056 per 10,000 requests

Table 24.4 Prices for amazon S3 standard cloud storages: inbound requests refer to PUT, COPY, POST, or LIST Requests. Outbound requests refer to GET and all other Requests as of 2016-04-25.

I	Name	Allocation price	Price of data transfer out from Amazon S3 to Internet	Price of data transfer in to Amazon S3
1	US East (N. Virginia)	$0.0295 per GB	$0.090 per GB	$0.00 per GB
2	US West (Oregon)	$0.0295 per GB	$0.090 per GB	$0.00 per GB
3	US West (Northern California)	$0.0324 per GB	$0.090 per GB	$0.00 per GB
4	EU West (Ireland)	$0.0295 per GB	$0.090 per GB	$0.00 per GB
5	EU Central (Frankfurt)	$0.0319 per GB	$0.090 per GB	$0.00 per GB
6	Asia Pacific Southeast (Singapore)	$0.0295 per GB	$0.120 per GB	$0.00 per GB
7	Asia Pacific Northeast (Tokyo)	$0.0324 per GB	$0.140 per GB	$0.00 per GB
8	Asia Pacific Southeast (Sydney)	$0.0324 per GB	$0.140 per GB	$0.00 per GB
9	Asia Pacific Northeast (Seoul)	$0.0308 per GB	$0.126 per GB	$0.00 per GB
10	South America (Sao Paulo)	$0.0401 per GB	$0.250 per GB	$0.00 per GB

Other metrics such as file size, and link speed are not considered for two reasons: First, these metrics affect all scenarios evenly since they cannot be changed by other decision variables here in this problem. Second; the purpose here is not to calculate the throughput. Instead, the optimal solution is what is sought for and thus only the decision variables are included in the objectives to either minimize or maximize them. For latencies, all amazon S3 [23] regions are tested and approximate round trip times are recorded as in Table 24.3. Note that the average value is calculated among 100 test cases for the same region. Regions with more than one endpoint are dealt with by relying on the endpoint with lower latency. Regions with more than one endpoint are dealt with by relying on the endpoint with lower latency. The availability is expressed by the number of cloud storages used to store data.

- The cloud storage pricing depends on the total size of the client's files, assuming it is 5TB. The monthly traffic to / from cloud storage to the client is assumed up to 10 TB. The specifications of client device are 12 GB RAM and dual cores Intel I5-2410M overclocked to 2.9 GHz. AES in CFB mode is used to encrypt and decrypt data with key size equals 256 bits. Encryption of 1MB took average of

303 milliseconds for 100 runs. For the integrity check, it is tested using SHA-256 algorithm on a file of 1MB size to generate the checksum and the whole process took an average of 14 milliseconds for 100 runs. The subscribed plan in the Internet service provider (ISP), which is located in Saudi Arabia, includes DSL with 20 Mb/s download speed and 0.5 Mb/s upload speed. Ten cloud storages are used in the experiment. All of them are served by Amazon in different regions [23]. Tables 24.2 and 24.4 contain the index for cloud storages used in the experiments and their details. In the following subsections, three different problems in Table 24.1 are solved.

24.5.1 Solving Problem 31

The optimization problem for satisfying all the five objectives the client can consider in problem 31 includes minimizing cost, maximizing performance, maximizing integrity, maximizing availability and maximizing confidentiality. The objective functions are as follows.

- Min(Cost):

$$f_1(x) = \sum_{i=1}^{10} (a_i x_i) + \sum_{i=1}^{10} (b_i x_i) + \sum_{i=1}^{10} (c_i x_i)$$
$$+ \sum_{i=1}^{10} (d_i x_i) + \sum_{i=1}^{10} (e_i x_i)$$

Variables a, b, c and d are substituted with the corresponding values in Tables 24.2 and 24.4. Therefore, the final form of the objective function is going to be:

$$f_1(x) = 0.1735 \times x_1 + 0.1735 \times x_2 + 0.1764 \times x_3$$
$$+ 0.1735 \times x_4 + 0.1802 \times x_5 + 0.2078 \times x_6$$
$$+ 0.2231 \times x_7 + 0.2318 \times x_8 + 0.2053 \times x_9$$
$$+ 0.3657 \times x_{10}$$

- Max (Performance) = Min (latency + overhead)

$$f_2(x) = \left[\sum_{i=1}^{10} (a_i x_i) + (b \times d) + (c \times e) \right]$$

Variable a is substituted with the corresponding values in Tables 24.2 and 24.4. Variables b and c are already calculated in client machine. Therefore, the final form of the objective function is going to be:

$$f_2(x) = 180 \times x_1 + 231 \times x_2 + 241 \times x_3 + 290 \times x_4 + 90$$
$$\times x_5 + 171 \times x_6 + 228 \times x_7 + 344 \times x_8 + 230$$
$$\times x_9 + 313 \times x_{10} + 14 \times d + 303 \times e$$

- Max(Availability)

$$f_3(x) = -1 \times \left(\sum_{i=1}^{10} (x_i) \right)$$

Applying summation gives the following form:

$$f_3(x) = -1$$
$$\times (x_1 + x_2 + x_3 + x_4 + x_5 + x_6 + x_7 + x_8 + x_9 + x_{10})$$

The confidentiality and integrity objectives have definitive goals. Therefore, constraints are placed to represent these objectives as follows:

$$2 \leq \sum_{i=1}^{10} (x_i) \leq 8$$

$$0 \leq x_i \leq 1, \quad x_i \in (0, 1)$$

$$e = 1, \quad e \in (0, 1)$$

$$d = 1, \quad d \in (0, 1)$$

The optimal solution for problem 31 is obtained from applying NSGA-2 with 10000 generations or evaluations. The results were filtered manually by selecting only three of the best results due to space limitations. Figure 24.3 shows a scattered plot for the optimal solutions in 3D space.

$$x_1 = 1, x_2 = 1, x_3 = 1, x_4 = 1, x_5 = 0, x_6 = 0, x_7 = 0, x_8$$
$$= 0, x_9 = 0, \ x_{10} = 0, e = 1, d = 1$$

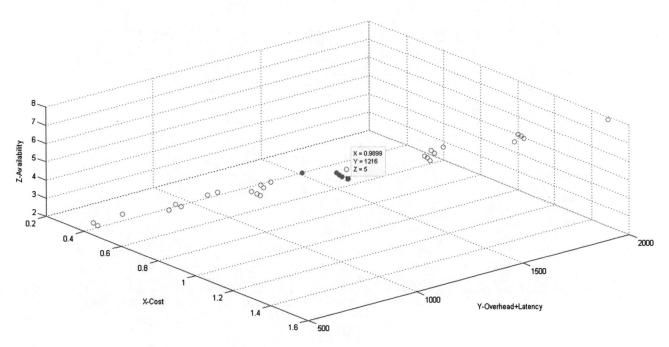

Fig. 24.3 Scattered plot of problem 31 optimal solution in 3D space

$$=> f_1(x) = 0.69690, f_2(x) = 1259.00, f_3(x) = 4.00$$

$$x_1 = 1, x_2 = 1, x_3 = 1, x_4 = 0, x_5 = 1, x_6 = 1, x_7 = 0, x_8$$
$$= 0, x_9 = 0, x_{10} = 0, e = 1, d = 1$$

$$=> f_1(x) = 0.91140, f_2(x) = 1230.00, f_3(x) = 5.00$$

$$x_1 = 1, x_2 = 0, x_3 = 0, x_4 = 0, x_5 = 1, x_6 = 1, x_7 = 1, x_8$$
$$= 0, x_9 = 1, x_{10} = 0, e = 1, d = 1$$

$$=> f_1(x) = 0.98990, f_2(x) = 1216.00, f_3(x) = 5.00$$

24.5.2 Solving Problem 29

The client can choose only four objectives. An example of this case is problem 29 which includes maximizing performance, maximizing integrity, maximizing availability and maximizing confidentiality. The only difference between problem 31 and 29 is that the latter doesn't take minimizing cost into consideration. The optimal solution for problem 29 is obtained from applying NSGA-2 with 10000 generations or evaluations. The results were filtered manually by selecting best three solutions. Figure 24.4 shows scattered plot of the optimal solutions calculated as following:

$$x_1 = 1, x_2 = 0, x_3 = 0, x_4 = 0, x_5 = 1, x_6 = 1, x_7 = 1, x_8$$
$$= 0, x_9 = 0, x_{10} = 0, e = 1, d = 1$$

$$=> f_2(x) = 986.00, f_3(x) = 4.00$$

$$x_1 = 1, x_2 = 0, x_3 = 0, x_4 = 0, x_5 = 1, x_6 = 1, x_7 = 0, x_8$$
$$= 0, x_9 = 0, x_{10} = 0, e = 1, d = 1$$

$$=> f_2(x) = 758.00, f_3(x) = 3.00$$

$$x_1 = 1, x_2 = 0, x_3 = 0, x_4 = 0, x_5 = 1, x_6 = 1, x_7 = 1, x_8$$
$$= 0, x_9 = 1, x_{10} = 0, e = 1, d = 1$$

$$=> f_2(x) = 1216.00, f_3(x) = 5.00$$

24.5.3 Solving Problem 25

An example of choosing only three objectives is problem 25 which includes maximizing integrity, maximizing availability and maximizing confidentiality. The optimal solution

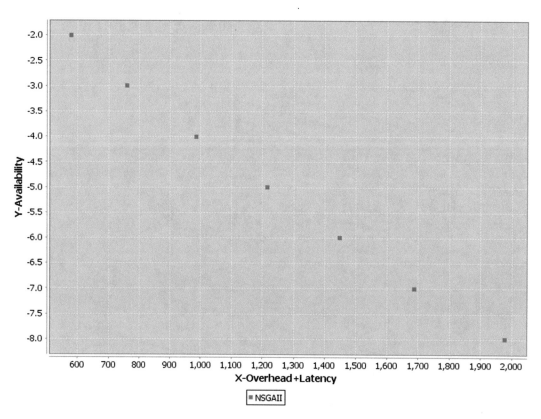

Fig. 24.4 Scattered plot of problem 29 optimal solution

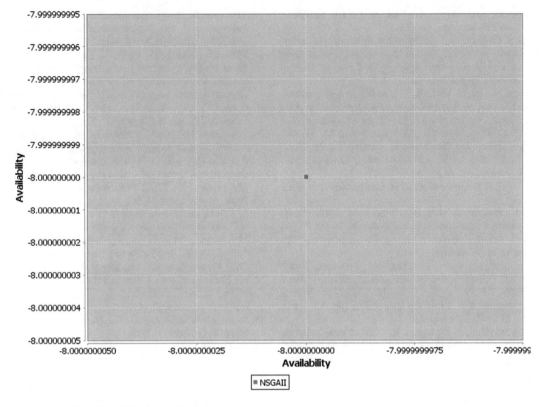

Fig. 24.5 Scattered plot of problem 25 optimal solution

of problem 25 is obtained from applying NSGA-2 with 10000 generations or evaluations. The optimal solution is illustrated in Fig. 24.5.

$$x_1 = 1, x_2 = 1, x_3 = 1, x_4 = 1, x_5 = 1, x_6 = 1, x_7 = 0, x_8 = 1, x_9 = 0, x_{10} = 1, e = 1, d = 1 => f_3(x) = 8.00$$

24.6   Acknowledgments

We would like to thank Professor Edward Chow, Dr. Abdullah Albahdal, and Dr. Seyed M. Buhari for their valuable comments and suggestions.

24.7   Conclusion

Many advantages and benefits have been offered by the cloud storages that led to their advanced and increased deployment for the individuals as well as organizations. This has motivated the concerns regarding the cloud issues including cost, performance and security. Many different schemes have been proposed in the literature without fully satisfy the user requirements. In this paper, we proposed SCORS scheme by applying non-dominated sorting genetic

algorithm (NSGA-2) [2, 3] to resolve a user's conflicting requirements. Finding an optimal solution to set up multiple cloud storages has many benefits for business owners of different objectives. It facilitates setting up multiple cloud storages in an optimized way based on the objectives specified. However, there is a need to build a general scheme that adapts to the proposed optimal solution. Also, some assumptions made in the paper may need to be enhanced to mimic the real life scenarios even better.

References

1. Alsolami, F. J. (2015). *Toward secure sensitive data in the cloud.* Ph.D. dissertation, University of Colorado, Colorado Springs.
2. Deb, K., Pratap, A., Agarwal, S., & Meyarivan, T. (2002). A fast and elitist multiobjective genetic algorithm: NSGA-II. *IEEE Transactions on Evolutionary Computation, 6,* 182–197.
3. Latha, B., Senthilkumar, V. S. (2009). Simulation optimization of process parameters in composite drilling process using multi-objective evolutionary algorithm. In *ARTCom'09. International conference on advances in recent technologies in communication and computing, 2009* (pp. 79–82). IEEE. Kottayam, Kerala.
4. Xiong, H., Zhang, X., Yao, D., Wu, X., Wen, Y. (2012). Towards end-to-end secure content storage and delivery with public cloud. In *Proceedings of the second ACM conference on data and application security and privacy* (pp. 257–266). ACM. San Antonio, TX.
5. Kumbhare, A., & Simmhan, Y., Prasanna, V. (2012). Cryptonite: A secure and performant data repository on public clouds. In *2012 I.E.*

   5th international conference on cloud computing (CLOUD) (pp. 510–517). IEEE. Honolulu, Hawaii.

6. Alsolami, F., & Chow, C. E. (2013). N-cloud: Improving performance and security in cloud storage. In *2013 I.E. 14th international conference on high performance switching and routing (HPSR)* (pp. 221–222) IEEE. Taipei, Taiwan.

7. Alsolami, F., & Boult, T. E. (2014). Cloudstash: Using secret-sharing scheme to secure data, not keys, in multi-clouds. In *2014 11th international conference on information technology: New generations (ITNG)* (pp. 315–320). IEEE. Las Vegas, NV, USA.

8. Bessani, A., Correia, M., Quaresma, B., André, F., & Sousa, P. (2013). Depsky: Dependable and secure storage in a cloud-of-clouds. *ACM Transactions on Storage (TOS), 9*, 12.

9. Shamir, A. (1979). How to share a secret. *Communications of the ACM, 22*, 612–613.

10. AlZain, M. A., Pardede, E., Soh, B., Thom, J. A. (2012). Cloud computing security: From single to multi-clouds. In *2012 45th Hawaii international conference on system science (HICSS)* (pp. 5490–5499). IEEE. Maui, HI.

11. Feldman, A. J., Zeller, W. P., Freedman, M. J., Felten, E. W. (2010). SPORC: Group collaboration using untrusted cloud resources. In *OSDI'10 proceedings of the 9th USENIX conference on operating systems design and implementation* (pp. 337–350). Berkeley, CA: USENIX Association.

12. Santos, N., Gummadi, K. P., Rodrigues, R. (2009). Towards trusted cloud computing. In *HotCloud'09 proceedings of the 2009 conference on hot topics in cloud computing*. Berkeley, CA: USENIX Association

13. Wang, H., Zhao, Z., Li, X., He, C., Qi, Q., Sun, J. (2013). A scheme to ensure data security of cloud storage. In *2013 I.E. international conference on service operations and logistics, and informatics (SOLI)* (pp. 79–82). IEEE. Dongguan.

14. Lei, D., Zhou, K., Jin, H., Liu, J., Wei, R. (2014). SFDS: A security and flexible data sharing scheme in cloud environment. In *2014 international conference on cloud computing and big data (CCBD)*. Wuhan, Hubei. (pp. 101–108).

15. Sarvabhatla, M., Vorugunti, C. S. (2015). A robust mutual authentication scheme for data security in cloud architecture. In *2015 7th international conference on communication systems and networks (COMSNETS)* (pp. 1–6). IEEE. Bangalore.

16. Deshmukh, A., Mihovska, A., Prasad, R. (2012). A cloud computing security schemes: TGOS [threshold group- oriented signature] and TMS [threshold multisignature schemes]. In *2012 world congress on information and communication technologies (WICT)*. (pp. 203–208). IEEE. Trivandrum.

17. Mazur, S., Blasch, E., Chen, Y., Skormin, V. (2011). Mitigating cloud computing security risks using a self-monitoring defensive scheme. In *Proceedings of the 2011 I.E. national aerospace and electronics conference (NAECON)* (pp. 39–45). IEEE. Dayton, OH.

18. Cachin, C., Haas, R., Vukolic, M. (2010) *Dependable storage in the intercloud* (Research Report RZ, 3783).

19. Balasaraswathi, V. R., Manikandan, S. (2014). Enhanced security for multi-cloud storage using cryptographic data splitting with dynamic approach. In *2014 international conference on advanced communication control and computing technologies (ICACCCT)* (pp. 1190–1194). IEEE. Tamilnadu.

20. Popa, R. A., Lorch, J. R., Molnar, D., Wang, H. J., & Zhuang, L. (2011). Enabling security in cloud storage slas with cloudproof. *USENIX Annual Technical Conference, 242*, 31.

21. Mahajan, P., Setty, S., Lee, S., Clement, A., Alvisi, L., Dahlin, M., & Walfish, M. (2011). Depot: Cloud storage with minimal trust. *ACM Transactions on Computer Storage (TOCS), 29*, 12.

22. MOEA Framework (A Free and Open Source Java Framework for Multiobjective Optimization). [Online]. Available: http://moeaframework.org/

23. Amazon Simple Storage Service (Amazon S3). [Online]. Available: http://aws.amazon.com/s3

24. Anthes, G. (2010). Security in the cloud. *Communications of the ACM, 53*(11), 16–18.

25. Choo, K.-K.R. (2010). Cloud computing: Challenges and future directions. *Trends and Issues in Crime and Criminal Justice, 400*, 1.

26. Armbrust, M., Fox, A., Griffith, R., Joseph, A. D., Katz, R., Konwinski, A., Lee, G., Patterson, D., Rabkin, A., Stoica, I., et al. (2010). A view of cloud computing. *Communications of the ACM, 53*(4), 50–58.

27. Cachin, C., Keidar, I., & Shraer, A. (2009). Trusting the cloud. *ACM SIGACT News, 40*(2), 81–86.

28. Rocha, F., & Correia, M. (2011). Lucy in the sky without diamonds: Stealing confidential data in the cloud. In *2011 IEEE/IFIP 41st international conference on dependable systems and networks workshops (DSN-W)* (pp. 129–134). IEEE. Hong Kong.

29. Patterson, D. A., Gibson, G., & Katz, R. H. (1988). A case for redundant arrays of inexpensive disks (RAID). *ACM SIGMOD Record, 17*(3), 109–116.

Towards the Memory Forensics of MS Word Documents

Ziad A. Al-Sharif, Hasan Bagci, Toqa' Abu Zaitoun, and Aseel Asad

Abstract

Memory forensics plays a vital role in digital forensics. It provides important information about user's activities on a digital device. Various techniques can be used to analyze the RAM and locate evidences in support for legal procedures against digital perpetrators in the court of law. This paper investigates digital evidences in relation to MS Word documents. Our approach utilizes the XML representation used internally by MS Office. Different documents are investigated. A memory dump is created while each of these documents is being viewed or edited and after the document is closed. Used documents are decompressed and the resulting folders and XML files are analyzed. Various unique parts of these extracted files are successfully located in the consequent RAM dumps. Results show that several portions of the MS Word document formats and textual data can be successfully located in RAM and these portions would prove that the document is/was viewed or edited by the perpetrator.

Keywords

Digital forensics • Memory forensics • Memory dumps • Microsoft office • Open XML formats • MS word documents

25.1 Introduction

Digital forensics investigates the substances of electronic devices such as computers and smart phones and their permanent and secondary storages. This type of forensics was used to apprehend the criminal that put a collar bomb on a high school student in Australia [1]. Forensics examiners recovered the criminal's name from documents on a USB drive draped around the victim's neck.

Operating systems store most digital contents in files. These files might be used by criminals for various objectives, one of which is to view or edit their contents.

Even though digital forensics tools and techniques can be used to locate these files on the permanent storage of the computer, evidence might be needed to prove that one or more of these files are actually viewed or edited by the perpetrator. Thus, whenever such information is not available directly, evidence can be extracted from the RAM of the used machine.

A RAM dump is a bit-by-bit copy of the physical memory of the system. During RAM analysis, information about the used file can be valuable given that the RAM consists of pages. RAM pages of a process or an opened file are not necessarily to be consecutive; most likely, they are scattered all over the RAM. Thus, a file structure can be helpful in its data recovery, rather than the file system metadata.

Hence, MS Word documents are one of the most commonly used file types [2]. Therefore, extorting textual data from the MS Word document (*docx* file) can be achieved by decompressing the file. Therefore, understanding the structure

Z.A. Al-Sharif (✉) • H. Bagci • T.Abu Zaitoun • A. Asad
Software Engineering Department, Jordan University of Science and Technology, P.O. Box 3030, 22110 Irbid, Jordan
e-mail: zasharif@just.edu.jo; hhasan_2007@hotmail.com; toabuzaitoun@just.edu.jo; aseelasad@yahoo.com

S. Latifi (ed.), *Information Technology – New Generations*, Advances in Intelligent Systems and Computing 558, DOI 10.1007/978-3-319-54978-1_25

of an expanded *docx* file is essential during the analysis process. Accordingly, creating memory dumps during the investigation process is fundamental to the inspection and the evidence extraction process.

A tool [3] is developed so that it allows investigators to automatically decompress (*unzip*) the investigated MS Word document. Additionally, it automatically parses and analyzes the contents of these XML files. The user can extract various parts such as the document's textual data by paragraphs or by textual data blocks. A block is a sequence of textual data (often not a complete paragraph) that is extracted from the XML file. The results of these extracted parts can be saved into simple text files for further analysis, which may incorporate searching for these parts in a RAM dump.

This paper experiments with 10 different MS Word documents. It searches for various parts of their correlated XML files and contents in the consequent RAM dumps. Results show that unique and vital portions of these files are successfully identified.

The rest of this paper is organized as follows. Section 25.2 presents some of our related works and discusses the essential background information that is used during our memory investigation. Section 25.3 presents the used methodology. Section 25.4 presents our experiments and Sect. 25.5 discusses our findings and results. Section 25.6 presents our discussion and future work. Finally, Sect. 25.7 concludes our work.

25.2 Background and Related Work

Digital investigators are tackling memory forensics by various means [4–10]. Most of the time, RAM dumps are created and analyzed in order to find digital evidences in support for legal actions against perpetrators in the court of law. Some researchers put more emphasis on investigating various computer attacks in order to detect malicious activities [11, 12]. Others develop tools, techniques, and frameworks to capture and analyze potential evidences. Ahmad Shosha et al. developed a model prototype of different malicious programs used by perpetrators [13]. Petroni et al. presented a digital forensic tool named FATKit [14] that is dedicated to extract, analyze, and visualize acquired forensic data. Chan Ellick et al. introduced a tool named ForenScope [15] that allows users to disable anti-forensic tools and search for potential evidences in RAM using regular *bash-shell*. This tool keeps the RAM contents intact; it uses only the unused memory space on the investigated machine. Funminiyi Olajide et al. uses RAM dumps to extract user's input information from Windows applications [16]. Johannes et al. proposed a technique to investigate firmware and its major components. He proposed a new technique that improves forensics imaging based on PCI introspection and

page table mapping [17]. Narasimha Shashidhar et al. targeted the *prefetch* folder that is used to speed up the startup time of a program on a Windows machine [18].

On the other hand, this paper targets the memory forensics of MS Word documents. The file of an MS Word document is a compressed (zipped) folder of different XML files and subfolders, each of which is designed to represent a specific feature in the presented document. Simply, you need to replace the *.docx* extension with a *.rar* or *.zip*. This produces a set of subfolders and human readable XML files. During the memory investigation process, these files are employed to identify various parts of the original MS Word document.

For example, the file *_rels/.rels* contains information about the structure of the document. In particular, it contains paths to the metadata information as well as the main XML document that comprises the content of the MS Word document itself. Metadata information is usually stored in the folder named *docProps*. Two or more XML files are stored inside this folder: *app.xml* stores metadata information extracted from the MS Word application and *core.xml* stores metadata from the document such as the author name, last modification time, etc. The *word* folder comprises the actual content of the MS Word document. It contains an XML file named *document.xml* that retains the main text for the document [19–22]. Vital parts extracted from these files can be located in the consequent memory dump.

This paper extracts paragraphs and textual data blocks from the related XML files and searches for them in the RAM of the working machine, not from the hard disk.

25.3 Methodology

Our research model assumes that the perpetrator is viewed or edited an MS Word document that might contain crucial information to the investigation process. This target document might be on any of the machine's secondary storages or it might be downloaded from the Internet using one of the web browsers, ftp servers, or e-mail clients.

The goal is to locate evidence about the potential use (view or edit) of the MS Word document. Thus, our investigation methodology utilizes the internal representation of the *docx* files. Our investigation process extracts and analyzes various components from the XML representation of the target *docx* file and then searches for these contents in the RAM dump of the used machine, see Fig. 25.1.

25.4 Experiments

A RAM memory dump is created while the MS Word document is being viewed/edited and right after it is closed, this dump might include evidences related to that particular file such as parts of its contents or information about its

Fig. 25.1 Investigation model. First, the criminal views or edits an MS Word document. Second, the investigator extorts the document by decompressing it and analyzing its XML files; paragraphs and textual data block are extracted. Finally, these textual data are located in the perpetrator's RAM memory (A memory dump)

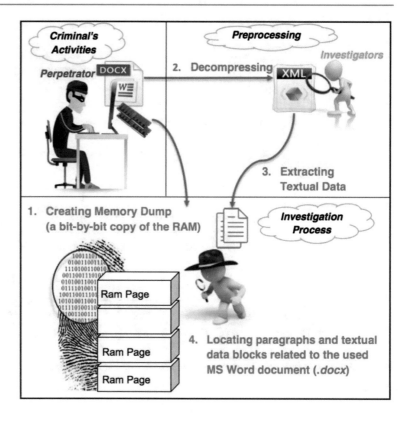

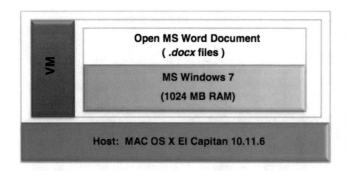

Fig. 25.2 Basic experimental setup: a VM that runs Windows 7 with 1024 MB of RAM

styles and formats. Our experiments locate these substances and validate them against the original contents of the target file. As part of the validation process, we calculate the percentage of the successfully recovered textual instances from the RAM dump.

25.4.1 Experimentation Setup

For these experiments, a Virtual Machine (VM) is created using Oracle's VirtualBox. The VM has 1024 MB of RAM. It runs Windows 7 and uses MS Office 2010 to open and edit these MS Word documents, see Fig. 25.2.

In order to validate our methodology, this paper experiments with 10 different MS Word documents of different sizes and different number of pages and paragraphs. Figure 25.3 presents these files in an ascending order based on their number of pages. It also presents the size of these files in *KB*, their number of paragraphs and the number of textual data blocks that are extracted from the document's internal XML files. However, this figure shows that there is no direct correlation between the size of the MS Word document and the number of paragraphs and textual data blocks that are successfully recovered from a memory dump, see Figs. 25.4 and 25.5.

25.4.2 Experimentation Steps

A RAM memory dump is created for the VM during two different states:

- *State 1*: The RAM is dumped while the MS Word document is being viewed or edited (using MS Word 2010).
- *State 2*: The RAM is dumped after the MS Word is closed.

Each of the 10 different MS Word documents (presented in Fig. 25.3) is decompressed and its internal subfolders and XML files are analyzed. The extracted paragraphs and

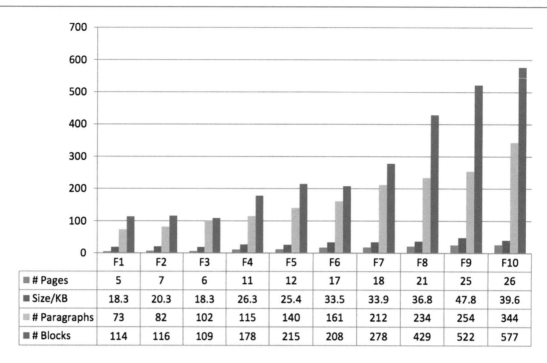

Fig. 25.3 The 10 MS Word documents that are used during our experiments

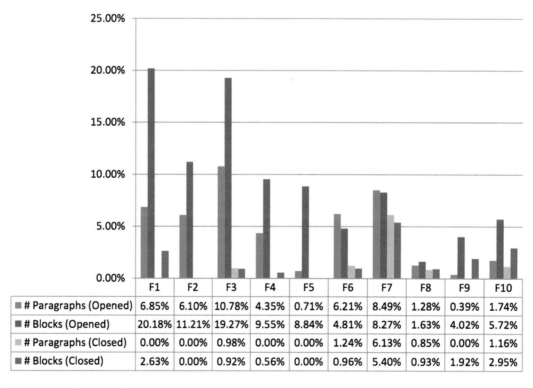

Fig. 25.4 The percentage of successfully recovered paragraphs and textual data blocks

textual data blocks are saved, each in a simple text file. Then, a blind search is applied trying to locate a complete occurrence for each instance in the consequent RAM dump. Following are the steps:

- *Step 1*: The memory dump is searched for complete paragraphs extracted from the *document.xml* file.
- *Step 2*: The memory dump is searched for textual data blocks extracted from the *document.xml* file.

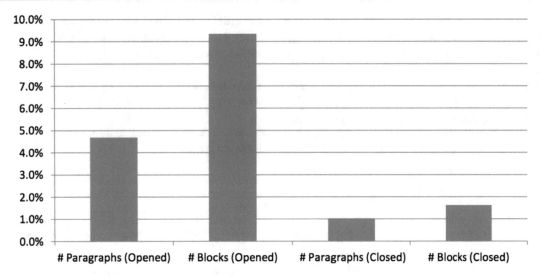

Fig. 25.5 The average percentage of all successfully recovered paragraphs and textual data blocks

25.5 Results

The results show that memory forensics of MS Word documents can be achieved by decompressing the original document and extracting its vital and unique information from the internal XML files that are used to compose the original one. Textual data from the original document are extracted in the form of paragraphs and blocks; a block is a sequence of textual data (not a complete paragraph) that is extracted from the XML representation of the document. A good portion of these textual data are successfully recovered from the consequent RAM dump.

Figure 25.3 shows information about the 10 different MS Word documents that are used during our experiments. The figure shows that these files are of a moderate size that ranges between 18 and 40 KB, and the number of pages in these files ranges from 5 to 26 pages. Finally, the number of successfully extracted textual data (paragraphs and blocks) from the XML files of the decompressed MS Word documents ranges from 73 to 344 paragraphs and from 114 and 577 blocks.

However, Fig. 25.4 presents the percentage of the successfully recovered paragraphs and textual data blocks from the memory dumps that are created during the two states of our experiments. In general, if we compare the two histograms in Figs. 25.3 and 25.4, we can see that the percentage of the recovered paragraphs and blocks decreases whenever the numbers of their corresponding paragraphs and blocks increases in the original document. Furthermore, we can see that the percentage of recovered paragraphs is always less than the corresponding percentage of the recovered blocks; maybe because the size of these blocks is always equal or less than the size of the original paragraph.

For example, when the document is opened, the percentage of the recovered paragraphs ranges from 0.39% to 10.78% whereas the number of recovered blocks ranges from 1.63% to 20.18%. Interestingly, the max and min limits of these percentages of recovered paragraphs and blocks do not match on the same file. This means, it depends totally on the internal representation of the document and its corresponding internal XML representation.

Furthermore, when the document is closed, the percentage of the recovered paragraphs ranges from 0.00% to 6.13% whereas the number of recovered blocks ranges from 0.00% to 5.40%. This means that our chances of locating various paragraphs or textual data blocks of an MS Word document in RAM decreases dramatically after the document is closed. Hence, in some undefined cases the investigator can become lucky enough and might be able to recover a good percentage of the original paragraphs or the correspondent textual data blocks.

Finally, Fig. 25.5 broadly shows that in average the investigator has a good chance locating paragraphs and textual data blocks for an MS Word document in memory dumps whenever this dump is created while the document is open. Furthermore, the figure also shows in average the investigator has a better chance locating textual data blocks than locating complete paragraphs regardless whether the document is opened or closed.

25.6 Discussion and Future Work

Digital investigation has evolved over the years to accommodate the developing technologies and the increasing rates in cyber-threats. Even though most evidences can be reliably located within the permanent storage devices, the RAM memory contains extremely valuable information that sometimes cannot be found elsewhere. Most of the time, a perpetrator activity involves files of various types (whether viewing or editing).

Memory forensic can be used to identify a variety of these activities on a digital device. A memory dump is a bit-by-bit copy of the physical RAM, which consists of pages. These pages belong to the currently or even recently working processes or opened files. A blind investigation of the RAM adds to the challenge. Pages of the same process are not necessarily to be sequentially ordered. Therefore, this may reduce the number of correctly identified and located evidences in the memory dump.

However, the *MS-Office Forensics* tool [3] is built to analyze both of the target file and the RAM dump. It recovers paragraphs and various blocks related to the target file from the RAM dump. This paper experiments with 10 different MS Word documents. In each experiment, only one document is used. This means that located textual data blocks belong to the used document. Though, our future work aims to target multiple documents that are used during a relatively close time. This would make the search process likely to recover parts from multiple documents. Our technique should be able to distinguish blocks by various means such as clustering techniques. Furthermore, our technique experiments only with MS Word documents, however, our future work is aiming to improve this technique to be applicable for all MS Office documents.

25.7 Conclusion

Digital forensics is always in need for the best and easiest methods and techniques to obtain digital evidences that would help build a case against a criminal in the court of law. Memory forensics has been proven effective in digital investigation. Memory investigation of MS Word documents can be achieved by extracting (decompressing) the target document. When a criminal views or edits an MS Word document, we need a technique to prove with concrete evidence that s/he did that on that particular machine. Thus, this paper presented a systematic process that can be used to recover such evidence and to identify criminal's activities by identifying unique parts of the MS Word document in the RAM memory of the used machine. Our technique is based on decompressing the already compressed MS Word document and analyzing the resulting subfolders and XML files. Then, we search for various and unique parts in the RAM dump of the used machine.

Our experimentation shows that our methodology can be valuable to the investigator. It helps establish the evidence that the perpetrator is *actually viewed or edited* the MS word document while performing the crime or to cover the wrongdoing associated with the crime. Various parts related to the MS Word documents are successfully located during various execution states.

Acknowledgements This research was supported in part by Jordan University of Science and Technology.

References

1. McMillan, R. The collar Bomber's explosive tech gaffe. http://www.pcworld.com/article/238353/article.html. Accessed October 28, 2016.
2. Johnson, D. The 8 most popular document formats on the web in 2015. http://duff-johnson.com/2015/02/12. Accessed October 28, 2016.
3. Al-Sharif, Z. A. Ms office forensics. https://sourceforge.net/p/ms-office-forensics/. Accessed January 2, 2017.
4. Al-Saleh, M. I., & Al-Sharif, Z. A. (2012). Utilizing data lifetime of tcp buffers in digital forensics: Empirical study. *Digital Investigation, 9*(2), 119–124
5. Al-Sharif, Z. A., Odeh, D. N., & Al-Saleh, M. I. (2015). Towards carving pdf files in the main memory. In *The International Technology Management Conference (ITMC2015)* (pp. 24–31). The Society of Digital Information and Wireless Communication.
6. Al-Sharif, Z. (2016). Utilizing program's execution data for digital forensics. In *The Third International Conference on Digital Security and Forensics (DigitalSec2016)* (pp. 12–19). The Society of Digital Information and Wireless Communications (SDIWC).
7. Harichandran, V. S., Walnycky, D., Baggili, I., & Breitinger, F. (2016). Cufa: A more formal definition for digital forensic artifacts. *Digital Investigation, 18*, S125–S137.
8. Rafique, M., & Khan, M. (2013). Exploring static and live digital forensics: Methods, practices and tools. *International Journal of Scientific and Engineering Research, 4*(10), 1048–1056.
9. Dezfoli, F. N., Dehghantanha, A., Mahmoud, R., Sani, N. F. B. M., & Daryabar, F. (2013). Digital forensic trends and future. *International Journal of Cyber-Security and Digital Forensics (IJCSDF), 2* (2), 48–76.
10. Cai, L., Sha, J., & Qian, W. (2013). Study on forensic analysis of physical memory. In *Proceedings of the 2nd International Symposium on Computer, Communication, Control and Automation (3CA 2013)*.
11. Ligh, M. H., Case, A., Levy, J., & Walters, A. (2014). *The art of memory forensics: Detecting malware and threats in Windows, Linux, and Mac memory*. Indianapolis: Wiley.
12. Al-Saleh, M., & Al-Sharif, Z. (2013). Ram forensics against cyber crimes involving files. In *The Second International Conference on Cyber Security, Cyber Peacefare and Digital Forensic (CyberSec2013)* (pp. 189–197). The Society of Digital Information and Wireless Communication.
13. Shosha, A. F., Tobin, L., & Gladyshev, P. (2013). Digital forensic reconstruction of a program action. In *Security and Privacy Workshops (SPW), 2013 IEEE* (pp. 119–122). IEEE.
14. Petroni, N. L., Walters, A., Fraser, T., & Arbaugh, W. A. (2006). Fatkit: A framework for the extraction and analysis of digital forensic data from volatile system memory. *Digital Investigation, 3*(4), 197–210.
15. Chan, E., Wan, W., Chaugule, A., & Campbell, R. (2009). A framework for volatile memory forensics. In *Proceedings of the 16th ACM Conference on Computer and Communications Security*.
16. Olajide, F., Savage, N., Akmayeva, G., & Shoniregun, C. (2012). Identifying and finding forensic evidence on windows application.

Journal of Internet Technology and Secured Transactions, ISSN, 2046–3723.

17. Stüttgen, J., Vömel, S., & Denzel, M. (2015). Acquisition and analysis of compromised firmware using memory forensics. *Digital Investigation, 12*, S50–S60.

18. Shashidhar, N. K., & Novak, D. (2015). Digital forensic analysis on prefetch files. *International Journal of Information Security Science, 4*(2), 39–49.

19. Simson, L., & Garfinkel, J. M. (2009). The new xml office document files: Implications for forensics. *IEEE Security and Privacy, 7* (2), 38–44.

20. Park, B., Park, J., & Lee, S. (2009). Data concealment and detection in microsoft office 2007 files. *Digital Investigation, 5*(3), 104–114.

21. Wolpers, M., Najjar, J., Verbert, K., & Duval, E. (2007). Tracking actual usage: the attention metadata approach. *Educational Technology and Society, 10*(3), 106–121.

22. Castiglione, A., De Santis, A., & Soriente, C. (2007). Taking advantages of a disadvantage: Digital forensics and steganography using document metadata. *Journal of Systems and Software, 80*(5), 750–764.

Two Are Better than One: Software Optimizations for AES-GCM over Short Messages

Shay Gueron and Regev Shemy

Abstract

This paper describes some software optimizations for AES-GCM over short messages, applicable for modern processors that have dedicated instructions. By processing two (short) messages in parallel, we achieve better performance than by processing twice, back-to-back, a single (short) message. Additional performance is gained if the using application collects several messages, sorts them by order of length, and the feeds them (in pairs) to the two-message AES-GCM function. For example, our experiments carried out on the latest Intel processor (micro architecture codename Skylake), over a realistic distribution of message lengths, our optimization achieves up to 1.95x speedup, compared to OpenSSL.

Keywords

Component • AES-GCM • TLS • IPSEC • Software optimizations • Skylake

26.1 Introduction

Secure communications on the Internet rely on cryptographic protocols for exchanging a symmetric key between a client and a server, and subsequently using this key with some authenticated encryption mode. The performance of these primitives is crucial for efficient communication [1, 2]. Currently, the most popular authenticated cipher is AES-GCM [2, 3]. On the modern processors, its performance enjoys the AES-NI and PCLMULQDQ instructions for AES and GHASH computations, coupled with extensive software optimization [4–7]. On the latest processors, architecture Skylake. AES-GCM performs at the rate of *0.65* cycles per byte, achieving encryption and authentication at the performance cost of CTR mode encryption alone.

To achieve such performance the processes message needs to be sufficiently long so that algorithmic parallelization can shadow various fixed costs that cannot be parallelized. However, the theoretic performance is not achieved when the messages are short. For example, AES-GCM over an 8KB message performs at *0.65* cycles per byte, but achieves only *2.21* cycles per byte for a *64* bytes message.

Packet size analysis of Internet traffic show that the distribution of these lengths is bimodal: *~44%* of packets are between *40* and *100* bytes, and *~37%* are between *1,400*, *1,500* bytes in size [8]. This motivates the dedicated optimization for AES-GCM on short messages (packets)., which is discussed in the paper.

26.2 Preliminaries

AES-GCM authenticated encryption is defined in [2, 3]. We describe it only briefly here. AES-GCM combines AES-CTR for the encryption, with a polynomial evaluation hash (GHASH) in $GF(2^{128})$ (with the reduction polynomial $x^{128}+x^7+x^4+x+1$) that is turned out to a Message Authentication Code (GMAC).

S. Gueron (✉)
Department of Mathematics, University of Haifa, Haifa, Israel

Intel Corporation, Israel Development Center, Haifa, Israel
e-mail: shay@math.haifa.ac.il

R. Shemy
Intel Corporation, Israel Development Center, Haifa, Israel

© Springer International Publishing AG 2018
S. Latifi (ed.), *Information Technology – New Generations*, Advances in Intelligent
Systems and Computing 558, DOI 10.1007/978-3-319-54978-1_26

The input to the encryption scheme is a nonce (*96* bits), a message M of N_2 bytes, and Additional Authenticated Data (AAD) of length N_1 bytes (for simplicity we consider only M and AAD whose length in bits is divisible by *8*). Denote

$$l_1 = \left\lceil \frac{8N_1}{128} \right\rceil, l_2 = \left\lceil \frac{8N_2}{128} \right\rceil,$$

M (and AAD) is viewed as a sequence of (l_2-*1*) full blocks (*16* bytes) ending with a last block which can be shorter than *16* byes. This block is padded with zeros up to the *16* bytes boundary, and denoted M' (and AAD'). M is first encrypted in CTR mode. Subsequently, GHASH is computed over a string of length $l = l_1 + l_2 + 1$ blocks, obtained by AAD' ‖ M' ‖ LENBLOCK, where LENBLOCK is populated with the encoding (as *8* bytes) of the bit-length of AAD and the encoding (as *8* bytes) of M. The GHASH value is encrypted by XOR-ing is with mask *MASK*. Here, *MASK* is the encryption of nonce, padded with the *4* bytes value 0x00000001.

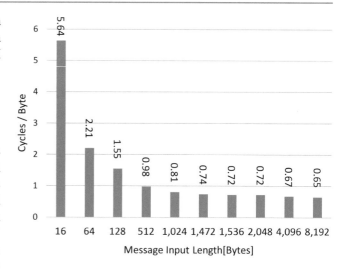

Fig. 26.1 AES-GCM performance (in cycles per byte), based on OpenSSL "speed" utility. The measurements were taken on Intel's micro architecture Codename Skylake. Note that the performance improves with message lengths, and the asymptotic performance is 0.65 C/B for sufficiently long messages.

26.2.1 Efficient AES-GCM Computations on Modern Processors

The computations of AES-GCM consist of $2 + l_2$ AES encryptions (1 for computing the GHASH key (H), *1* for computing *MASK*, and l_2 for encrypting M). In addition, the procedure involves the computation of GHASH over *l* blocks.

Optimized AES-GCM code uses the AES-NI [4, 5] and the PCLMULQDQ instruction [6, 7]. Multiplication in the field $GF(2^{128})$ requires (with the Schoolbook multiplication method) *4* invocations of PCLMULQDQ, and reduction modulo $x^{128}+x^7+x^4+x+1$. A new algorithm for fast reduction [9] carries it out by means of *2* invocations of PCLMULQDQ (and *11* shift/xor operations). We note that by aggregating *8* polynomial products, the reduction can be done only after accumulating *8* products. Hence, the effective cost of the reduction is $1/8 \times 2$ PCLMULQDQ invocations. A detailed account of the optimizations can be found in [6, 7, 9–11]. The optimizations and parallelization are more efficient for long messages. Figure 26.1 shows the performance of AES-GCM (*128* bits key), measured in Cycles/Byte for different message lengths.

Clearly, the performance depends on the length of the message, and there is a big difference between the performance for short and long messages. Indeed, there are some fixed overheads in the computations and they are more pronounced when with short messages. For example, the AES-GCM performance on *8KB* messages is *1.25x* faster than its performance on *1KB* messages.

Remark 1 We point out that the results that OpenSSL[1] [12] speed reports do not take into account some overheads: the

key expansion of the cipher key K, and computing the GHASH key H = AES_K (0), computing some powers of H (in $GF(2^{128})$). In addition, it does not take into account one AES encryption, which is required for preparing *MASK*. While these steps may be negligible for long messages, they are not very small for short message. For example, consider a message of *48* bytes. The CTR mode encryption encrypts *3* blocks, and computing *MASK* requires another encryption, which is *25%* of the overall encryption payload. For an 8KB message, CTR encryption encrypts *512* blocks, and the computation of *MASK* can be indeed neglected. Note also that the performance numbers shown in Fig. 26.1 process messages whose length in bytes is divisible by *16*: these are complete "block" (of *16* bytes) to be processed during the encryption.

Remark 2 [What Can Be Improved?] Note that the performance of AES-GCM on "Skylake", already achieves the fastest possible throughput for long message (see Fig. 26.1, starting from 4KB messages). In fact, if an AES round (using the AESENC instruction) has *1* cycle throughput, then AES-CTR can achieve throughput of at most *10/16=0.625* cycles per byte. From Fig. 26.1, we see that the processor can perform AES-GCM (encryption plus authentication) at (roughly) the same performance that it achieves for the CTR encryption only. This is the highest throughput that the hardware can possibly achieve. This implies that we cannot expect to improve on the AES-GCM performance for large message. Therefore, our optimization target is gaining performance on short messages.

These results illustrate the motivation for speeding up the AES-GCM performance on short messages. Recall the distribution of packet sizes on the Internet [8]: the majority of the packets are short (under 2K bytes), where *44%* of the packets are between *40* to *100* bytes.

26.2.2 Processing Multiple Packets in Parallel

Processing multiple independent messages in parallel, using modern SIMD architectures, improves the resulting performance significantly. Examples with some cryptographic algorithms are shown in [13–15] for hashing. Using SIMD is not the only way to introduce software pipelining in order to turn latency bounded computations to throughput bounded computations. For example, references [4, 5] show how to process multiple messages in parallel for AES-CBC encryption, which is essentially a serial mode, to get the performance of parallelizable modes of operation. Reference [10] shows a technique that gains a *3x* speedup factor for CRC32 computations when a single message is broken (logically) to three chunks, computations are done on these chunks independently so that the processor's pipeline is filled up, and in the end, the 3 results are combined to a single CRC32 value by using some mathematical transformation.

Here, we show how to apply this approach, to AES-GCM. We call this approach "AES-GCM-Multi-Packets". This strategy is attempting to gain performance over short messages, where AES-GCM does not achieve its theoretical performance limit.

For long messages (packets), we point out that the encryption part of AES-GCM can be done in parallel to the GHASH computations. This ensures that the software could be fully pipelined, and the processor could dispatch executing instructions every cycle (this is why the overall performance is better than the sum of the AES-GCM and the GHASH). However, with short messages the pipelining is less efficient, and this is why the performance degrades. By computing AES-GCM on several messages in parallel, the software can be better pipelined, in a way that some of the latencies can be shortened. We give one example. It is known [9, 11]) that throughput of encrypting several blocks together is higher than the throughput obtained from processing each plaintext block separately. Therefore, by processing n packets in parallel, the computations of n masks can be parallelized and sped up.

Remark 3 The Multi-Packets methodology can be applied with a different number of messages (packets) that are processed in parallel. Increasing the number of messages can improve the performance by allowing for better pipelining, but we have to weigh this against the pressure on the availability of registers. We experimented with functions of the type AES-GCM-Multi-Packets-xn (n is the number of messages that the function receives), for n = 2, 4, 8. The improvement gained by n=8, over n=4 was negligible. The improvement gained by n=4 compared to n=2 was in the range of *3%-5%* depending on the message sizes. Nevertheless, we preferred to use n=2 (i.e., use AES-GCM-Multi-Packets-x2) because it gives more flexibility for using it.

Remark 4 The function AES-GCM-Multi-Packets-x2 accepts *2* sets of inputs: (IV$_j$, Message$_j$, AAD$_j$, key schedule$_j$, L$_j$), for *j=1, 2* (L$_i$ is respective message length). While we assume that the AAD is always of *16* bytes length, the message lengths of the two inputs are not necessarily equal. Of course, the function needs to handle different length, in some optimal way. Our solution is to set L = min (L$_1$, L$_2$) and to process, in parallel the first L blocks. The minimum size parallel. The remaining parts of the shorter message are processed serially. For example, if L$_1$ = *128* and L$_2$ = *144*, the function would process the first *128* bytes of both messages in parallel, and then process the remaining 16 bytes of the longer message 2. Of course, when |L$_1$ – L$_2$| is large, the effectiveness of the parallelization is reduced.

Figure 26.2 show the performance results of using our AES-GCM-Multi-Packets-x2 implementation (i.e., processing 2 packets in parallel), compared to using OpenSSL's AES-implementation on the same messages, one-by-one. In this experiment, both implementations processed the same total amount of data (*6,400KB-14,720KB*, each pair has total of *1,000* packets with same message length).

26.3 Further Optimization via Sorting by Packet Sizes

In real applications, the lengths of incoming messages that need to be encrypted, are distributed according to some lengths distribution that is typical to the usage model. We can use the Multi-Packets-x2 as follows: repeated collect two incoming messages, and apply AES-GCM-Multi-Packets-x2 on each

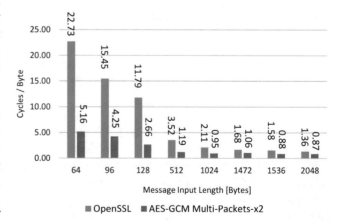

Fig. 26.2 Performance comparison of AES-GCM-Multi-Packets-x2 to OpenSSL. The reported numbers account for the initialization step. The measurements were taken on Intel's micro architecture Codename Skylake. The graph (top) shows the absolute numbers (in cycles per byte), and the table (bottom) shows the relative speedup that the AES-GCM-Multi-Packets-x2 implementation gains. See more details in the text.

Table 2.1

Length	64	96	128	512	1024	1472	1536	2048
Speedup	4.4	3.63	4.43	2.96	2.21	1.58	1.80	1.56

pair. With this approach, Multi-Packets2 will be invoked for pairs of messages that have different lengths.

Better control on the disparity between the lengths of the pairs that are input to AES-GCM-Multi-Packets-x2, is achieved if the application chooses some "window" of size k, collects k messages, sorts them by the message lengths, and then applies AES-GCM-Multi-Packets-x2 on pairs from the *sorted* list. This reduces the length differences of the pairs, compared to what we can expect without sorting.

Table 2.1 shows the results of an experiment that compares the performance with AES-GCM-Multi-Packets-x2 that is fed with unsorted pairs of messages, to the performance of AES-GCM-Multi-Packets-x2 obtained by collecting the messages from a window of k=4 and k=8 messages and sorting them before using AES-GCM-Multi-Packets-x2. In this experiment, message lengths were selected from the following lengths distribution (that simulates the distribution in [8]): *44%* of packets are between 40 and 100 bytes (uniformly distributed over this interval, in granularity of bytes), *37%* are between *1,400, 1,500* bytes (uniformly distributed over this interval, in granularity of bytes), and the remaining lengths (up to *1,536* bytes) are also distributed uniformly on the intervals [*100, 1,400*] bytes.

Figure 26.3 shows the different performance gained by using windows with k=4 or k =8 packets. The reported performance is compared to using OpenSSL's implementation. The performance gain without using a window (i.e., k=2) is *1.59x*. With k=4, the relative speedup is *1.81x*, and with *k=8*, the relative speedup is *1.95x*. The results measured on the micro architecture codename Skylake processor. The results reflect the incremental cost of handling the realistic packet size distribution, where (see Remark **4**) packet byte-lengths are not even divisible by *16* (compare the results to those reported in Fig. 26.2).

Remark 3 The performance of AES-GCM is also affected by the key expansion of the cipher key K, and the computation of the GHASH key H = AES$_K$ (0) together with some powers of H. In AES-GCM-Multi-Packets implementation we added an optimization that parallelizes these overheads (which improved the results by additional ~19%). The results reported in Fig. 26.3 represent the effects of all optimizations.

Other optimizations [16] suggested a different scheduling of packets. Our method is simpler: it only needs to select a window of (even) size k, collect k packets, sort them, and

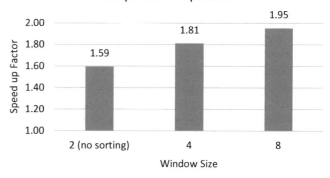

Fig. 26.3 The performance of AES-GCM-Multi-Packets2, using different window sizes (for window size = 2 the application collects pairs of messages and feeds them to the AES-GCM-Multi-Packets2 function without sorting). For window sizes 4 and 8, the application collects 4 (or 8) messages, sorts them first by their lengths and then invokes AES-GCM-Multi-Packets2.

them feed them to the AES-GCM-Multi-Packets-x2 function pair by pair. The (even) number k, is a choice which can be dynamically changed during the application's run time. This allows a great deal of flexibility. Note that even by selecting k=4, we already gain speedup of *1.14x* over the simplest usage of AES-GCM-Multi-Packets-x2 (i.e., k=2). With k=8, we get an incremental speedup of (only) *1.08x*. In other words, small windows ($k \leq 8$) seem to be sufficient in order to approach the optimal performance with a simple scheduling method.

26.4 Conclusions

This paper presented software optimization for AES-GCM, to address the fact that networking applications need to process variable sized messages that are typically short. The optimization builds upon processing multiple packets together, to achieve efficient parallelization. Specifically, we found that AES-GCM-Multi-Packets-x2 (i.e., processing pairs of messages) gave us the best balance between performance, flexibility and complexity. In addition, we showed that using AES-GCM-Multi-Packets-x2 colleting several (more than 2) messages and sorting them by their lengths, improves the efficiency of the implementation. Combining both techniques and applying them to a realistic distribution of message lengths, shows that we can gain a speedup factor of up to *1.95x* (measured on the latest Intel processor, micro architecture codename Skylake).

Acknowledgements This research was supported by the ISRAEL SCIENCE FOUNDATION (grant No. 1018/16).

References

1. Dierks, T., & Rescorla E. (2008). *The Transport Layer Security (TLS) Protocol Version 1.2*. IETF RFC5246, IETF, http://tools.ietf.org/html/rfc5246

2. Viega, J. & McGrew D. *The use of Galois/Counter Mode (GCM) in IPsec Encapsulating Security Payload (ESP)*, IETE RFD 4106. Available at: http://www.rfc-archive.org/

3. Dworkin, M. (2007, November). Recommendation for block cipher modes of operation: Galois/Counter Mode (GCM) for confidentiality and authentication. *Federal Information Processing Standard Publication FIPS 800-38D*, http://csrc.nist.gov/publications/nistpubs/800-38D/SP-800-38D.pdf

4. Gueron, S. (2010). *Intel Advanced Encryption Standard (AES) instructions set (Rev. 3), Intel Software Network*. http://software.intel.com/en-us/articles/advanced-encryption-standard-aes-instructions-set/

5. Gueron, S. (2009). Intel's new AES instructions for enhanced performance and security. Fast Software Encryptiom, 16th International Workshop (FSE 2009). *Lecture Notes in Computer Science, 5656*, 51–66.

6. Gueron, S., & Kounavis M. E. *Intel carry-less multiplication insturction and its usage for computing the GCM mode (Rev. 2.02)*. https://software.intel.com/sites/default/files/managed/72/cc/clmul-wp-rev-2.02-2014-04-20.pdf

7. Gueron, S. (2010, July). Efficient implementation of the Galois Counter Mode using a carry-less multiplier and a fast reduction algorithm. *Information Processing Letters, 110*(14–15), 549–553.

8. Murray D., & Koziniec T. (2012). *The state of enterprise network traffic in 2012*. 18th Asia-Pacific Conference on Communications (APCC), (pp. 179–184). IEEE. http://ieeexplore.ieee.org/document/6388126/?arnumber=6388126

9. Gueron, S. (2013). AES-GCM for efficient authenticated encryption – Ending the reign of HMAC-SHA-1? In *Real-World Cryptography*. https://crypto.stanford.edu/RealWorldCrypto/slides/gueron.pdf

10. Gueron, S. (2012, February). Speeding up crc32c computations with intel crc32 instructions. *Information Processing Letters, 112*(5), 179–185.

11. Gueron, S., & Krasnov V. (2013, March) *Fast implementation of AES-CRT mode for AVX capable x86-64 processors*. http://rt.openssl.org/Ticket/Display.html?id=3021&user=guest&pass=guest

12. OpenSSL: The Open Source toolkit for SSL/TLS, project webpage: http://www.openssl.org

13. Gueron, S., & Krasnov, V. (2012). Simultaneous Hashing of Multiple Messages. *Journal of Information Security, 3*, 319–325.

14. Gueron, S., & Shemy R. (2016, January). [OpenSSL Patch]: Accelerating Multi (MB) CBC SHA256 on architectures that support AVX512 instructions set. http://openssl.6102.n7.nabble.com/openssl-org-4221-PATCH-Accelerating-Multi-Block-MB-CBC-SHA256-on-architectures-that-support-AVX512-it-td62058.html

15. Gueron, S., & Shemy R. (2016, February). [OpenSSL Patch]: Multi Block (MB) SHA 512 for x86_64 Architectures that support AVX2/AVX512 instructions set. http://openssl.6102.n7.nabble.com/openssl-org-4307-PATCH-Multi-Block-MB-SHA512-for-x86-64-Architectures-that-support-AVX2-AVX512-instrt-td63716.html

16. Bogdanov, A., Lauridsen, M. M., & Tischhauser, E. *AES-based authenticated encyrption modes in parallel high-performance software*. https://eprint.iacr.org/2014/186.pdf. Access on 27 July 2014.

BYOD Security Risks and Mitigations

Melva Ratchford, Ping Wang, and Raed Omar Sbeit

Abstract

Adoptions of BYODs (Bring Your Own Devices) have been fast growing in modern organizations, which leads to improvement in convenience and productivity. However, these benefits may be short-lived if companies are not aware of the cost of potential legal implications as well as increases in information security vulnerabilities and risks. This paper discusses the legal and privacy-associated issues that organizations may encounter when adopting BYODs. These include legal considerations on issues regarding US privacy laws, comingled data, device ownership, spoliation of evidence, variety of BYODs, cloud services and mobile solutions. It also proposes some recommendations in order to mitigate these types of risks. In addition, this paper introduces the BYOD policy and management practices at Verizon Wireless as an organizational case study for analysis and recommendations on how to mitigate security risks associated with adoptions of BYODs.

Keywords

BYOD • Privacy • Risk • Policy • Verizon

27.1 Introduction

As the popularity BYOD (Bring Your Own Device) continues to grow, organizations realize that adopting BYOD programs could reduce costs and increase productivity. Most employees would also prefer to use their own devices for both work and personal needs, and adopting BYOD in the work place is a trend that is happening whether the companies like it or not [1]. As early as July 2011, it was reported that 69% of employees were accessing corporate data via their personal mobile devices [2]. The cost benefit for the companies includes a workforce that is able to tele-commute/telework and work after hours as well as reduce hardware cost and maintenance. IT departments perceive cost savings due to less hardware maintenance and overhead, and fewer hardware purchase [3]. In 2013, Gartner predicted that by 2017 half of employers will require their employees to provide their own devices for work [4]. However, all these benefits may be short-lived if companies are not aware of the cost of potential legal implications.

As with any technical implementation that provides access to a company's information, the basic security principles of Confidentiality, Integrity, and Availability (CIA) need to be preserved. This ensures that only authorized personnel have access to corporate data, that the information is not altered with malicious intent, and that it is available when is required. However, due to the nature of BYODs, where the user owns the device and stores personal information mixed with corporate information, the corporations are finding themselves not only in the need to

M. Ratchford
University of Maryland University College, Adelphi, MD, USA
e-mail: melva.ratchford@gmail.com

P. Wang (✉)
Robert Morris University, Pittsburgh, PA, USA
e-mail: wangp@rmu.edu

R.O. Sbeit
Verizon, Westlake, TX, USA
e-mail: raed.sbeit@verizonwireless.com

© Springer International Publishing AG 2018
S. Latifi (ed.), *Information Technology – New Generations*, Advances in Intelligent Systems and Computing 558, DOI 10.1007/978-3-319-54978-1_27

protect corporate data from outside and inside threats, but also they need to protect themselves from legal issues regarding privacy laws when their employees use their own devices [1]. BYOD-related vulnerabilities are exploited by organizational insiders [5]. Disgruntled employees can use the corporate data (i.e. stored in their personally owned device) for nefarious deeds, and the company may need to prove such action in a court of law. The company's lawyers and forensic examiners may find themselves with their hands tide-up when accessing the employee's own device. This is because the employees' personal information is protected by privacy laws, and this consideration if often neglected when adopting BYODs [1].

The purpose of this paper is to discuss legal and privacy-associated issues that organizations may encounter when adopting BYODs. It also proposes some recommendations to mitigate the risks that are introduced when allowing BYODs. This paper also introduces the BYOD policy practices for Verizon Wireless as an organizational case study.

27.2 Legal Considerations

BYOD devices include both smartphones and tablets and newer technologies such as smart watches, smart cars and similar devices, and privacy is a concern when organizations adopt any type of BYOD. Corporations of all sizes readily adopt these technologies allowing their employees to access their corporate systems. Some large corporations adopt expensive management systems for mobile devices (i.e. MDMs – Mobile Device Management and MAMs – Mobile Application Management), and although the data may be protected, there are legal privacy-related issues that may arise when these solutions are implemented. Many small to mid-size companies (maybe with small budgets for security) benefit from the immediate and tangible benefits of BYODs without the knowledge of the legal implications when corporate and personal data reside in the same device and without logical or physical separation. The following sections discuss legal/privacy issues associated with the adoption of BYODs.

27.2.1 U.S. Privacy Laws: The Privacy of an Individual

In order to begin this discussion, it is important to be aware of the privacy laws that protect an individual's personal information. In the United States, the privacy of an individual is protected in by the 4th amendment to the U.S. Constitution [6]. This law directly applies to the property of an employee which resides in the BYODs, and this

amendment is 'the preliminary groundwork to any and all issues related to search and seizure of an individual's device' [3]. Thus, a warrant is required in order for law enforcement to search an item that belongs to an individual. In the context of BYODs, these items are considered 'closed containers' where personal data reside on the device owed by the employee, therefore a warrant is needed in order to search the content of such device [3]. It is interesting to note that BYODs *are not* in the same category of wallets or cars, and this is because this type of items (i.e. wallets or cars) can be searched by police upon an arrest, where the police often find the evidence of the crime in plain sight in the wallet or the car [3].

Besides the 4th amendment, and in lieu of the technological advancement, other laws have been enacted to protect the privacy of cellphones, emails and other electronic data. The laws include the Store Communications Act (SCA), which is part of the Electronic Communication Privacy Act (ECPA) [3]. An organization needs to understand the extent of the meaning of such laws because, based on this knowledge, the BYOD policy needs to include the proper language to protect against privacy-related legal complications when trying to recover (or prove misuse) corporate data from the employee's own device.

27.2.2 Comingled Data

Comingled data refer to mixing personal and company data within the same device location. This can be more prevalent when referring to photo galleries, emails, text messages, web browser history, metadata, and call history where, inherently, text and picture messages reside in the same format and device location [3]. For example, in a case where a disgruntled employee divulges or extracts company's trade secret information, the comingled data may complicate the legal process if the company is trying to prove that proprietary data was extracted by the employee. In a court of law, for example, a digital forensic examiner working on behalf of the company, and in need to extract [potential] company's information from the employee's device, may not be able to do it because the employee's personal data is protected by the SCA/ECP laws [3].

27.2.3 Device Ownership

When BYOD policies are weak or non-existent, device ownership issues can pose problems for an organization trying to protect its data. In a legal defense, for example, an employee can decline the surrendering of his/her device if a company needs the device for analysis [3]. Based on the privacy laws, the employee has the right to refuse to surrender his/her own device claiming it is personal property. However, a company has the right to its proprietary data which reside on the employee's device. The company's BYOD policy needs to be clear in regard to device ownership as it relates to the need to recover or investigate corporate data when fraud is suspected.

27.2.4 Spoliation of Evidence

In legal terms, spoliation of evidence refers to act of negligently or intentionally destroying relevant information that is required during a legal process as part of an investigation [7]. In the BYOD context, a disgruntled employee can make company data 'unavailable' and be free of any wrongdoing [3]. For example, an employee accused by his employer of wrongdoing involving his cellphone turned in his phone to his service provider and exchanged it for another one, and thus, (presumably), discarding the company's data from his possession and avoiding prosecution [3]. Inadvertent spoliation may also occur when a BYOD is heavily used and business data is over-written from memory locations. This may be the case where texts messages or chats (e.g. containing company proprietary information) are rolled over or unintentionally deleted. Therefore, a clear BYOD policy is needed to address such issues and their privacy and legal ramifications.

27.2.5 Variety of BYODs

The variety of smart mobile devices that support network connectivity grows as technology advances. Each type of operating system has different functionality and presents different vulnerabilities. Very often by the time a new device is out, hackers have already figured out a way to break into its security. Smartphones are susceptible to many threats and attacks to include unintentional user downloads that propagate malware capable of controlling resources and collecting sensitive data from the devices [8]. In addition, manufacturers sometimes leave backdoors, not necessarily for malicious intent, but rather for the purpose of troubleshooting. When the main peculiarities of the BYODs operating systems and manufacturers are understood, companies can better prepare and plan for unforeseen expenses they may incur when legal situations arise.

27.2.6 Cloud Services

As with the large variety of mobile device types, the rapid proliferation of cloud services may create legal repercussion when a company is trying to recover corporate data. Knowingly or unknowingly, the employee may be backing up company's data to cloud providers or social media sites. Often these services are freely available or offered for small fees. Sometimes, these 'free' cloud services use the data stored in their sites for data mining or marketing purposes, or the data is exposed to theft, or to outside parties or vendors [3]. All these issues may complicate the job of a company's own investigation when trying to prove that corporate data has been extracted by an employee using his/her own mobile device.

27.2.7 Mobile Solutions: MDMs and MAMs

Many companies adopt mobile solutions to manage BYODs. Mobile Device Management (MDMs) and Mobile Application Management (MAMs) are two of such solutions that separate corporate data from personal data. These solutions incorporate features such as remote wipe that protect and control the corporate data in case of a lost device. Although MDMs and MAMs are solutions that provide company BYOD control and limit the amount of corporate data leakage, some of these solutions are not exempt from 'jail broken' and 'root kits' which are methods use to disable the protection of the device [3]. In addition, these mobile device solutions incorporate certain amount of monitoring that may access sensitive personal data, and the company may be liable for any personal data misuse or loss [1]. In adopting mobile solutions, companies should invest time learning the features and capabilities offered by these systems to ensure they are in compliance with the privacy laws, especially when finding themselves in the need to investigate employees that are suspected of fraud.

27.3 Risk Mitigation

When adopting BYODs, the importance of a strong BYOD policy is paramount in order to mitigate risks associated with litigations and leaks of corporate data. Corporations need to find the method that fits their need and implement the proper safeguards. For example, the U.S. Army considered the use BYODs using an implementation that does not allow any government information to be stored in the device ('zero client'), for personnel 'not directly operating in the field' [1]. Other companies, seeking to increase employee productivity, adopt move invasive implementations that include compliance with data privacy regulations. This is the case of Unisys where, in order to enroll in the program, the employee is required to consent to the installation of a public key infrastructure (PKI) device certificate as well as remote-wipe software [1]. In addition, the employee needs to consent to other policy requirements such as degree of monitoring and possible confiscation of device in case of litigation [1]. The following sections discuss two main considerations aiming at mitigating the risks introduced with BYODs: adoption of a BYOD policy and creation of awareness and training programs.

27.3.1 Define a Strong BYOD Policy

A BYOD policy should address issues regarding the ownership of the data, spoliation of evidence, comingled data, and privacy-related issues that involve compliance with privacy laws. The policy language should specify the company's rights when their data reside in the device. This may include access to comingled spaces within the device such as text messages and photo galleries. Although the formulation of a BYOD policy may require the advice of legal counsel to ensure the policy and terms will hold in a court of law, the authors in [1, 3] suggest the following guidelines:

- In general, the expectation of employee's privacy is limited when company information resides in the device, and the BYOD policy should convey this message.
- When the personal and company's data is comingled, there is no expectation of privacy regarding text, photo messages and other such type of information.
- The employee should waive the right to privacy when device monitoring occurs.
- The company should protect itself against claims associated with the Store Communication Act (SCA). This may include language that indicates that the company does not search and review arbitrarily, but rather under legitimate need.
- The policy should also include an Acceptable Use provision which can include elements such as restriction of applications and cloud service providers, types of websites, duty to protect the company's data stored in the device.
- Include guidelines regarding the use of password and encryption requirements, device loss, trade or sale of the device, or reimbursement provisions.
- The employee must provide 'fully informed and unambiguous' consent to the organization to access and process his/her personal data.
- The organization must take adequate care when handling the employee's personal sensitive data.
- Other policy-related issues involve company's preparation before allowing the use of BYOD. This may include the allocation of IT resources such as help-desk support, procedures regarding employee termination, and other company-related industry considerations (e.g. HIPPA for health industries, Gramm-Leach for banking industries, etc).

27.3.2 Awareness and Training

In mitigating risks of BYOD adoptions, awareness and training cannot be underestimated. Awareness programs teach employers and employees to be mindful of the threats and vulnerabilities (to corporate data) when using BYODs and take proper precautions. Training programs go beyond awareness where skills are taught to employers and employees in order to take specific actions [9]. In the case of BYOD protection, these may refer to activities such as installing virus protection software, use of encryption, use of VPNs when using public Wi-Fi connections, and the understanding privacy-law issues to mention a few.

The awareness and training varies depending on whether the company adds BYOD devices at hoc or through a mobile solution. In either scenario, the employer needs to be aware and trained in order to implement the proper BYOD policy and make informed decisions. For the employees, the awareness and training program addresses the implications of storing company's data in their mobile devices. When reducing the risk associated with litigation and privacy, the employee needs to be aware of the 'legal and investigatory issues that surround the forensic collection of the data in a BYOD environment' [3].

27.4 Case Study: Verizon Wireless

Verizon (VZ) is a global leader in providing communications and technology solution for our customers to play, work and live. Verizon's mission is to deliver the promise of the connected World. In 2015, Verizon invested more than $11 billion to meet today's surging demand for wireless data and video, but also to get Verizon network ready for 5G wireless technology. According to RootMetrics©, the largest independent third-party tester of wireless network performance, Verizon ranks number one in speed, data, reliability and overall network performance in the U.S.

Verizon BYOD Program is a voluntary program designed to enable eligible Verizon employees to use personally-owned electronic devices to work remotely in certain situations. Participation in Verizon's BYOD Program is a privilege. Employee participation in the BYOD Program as well as the BYOD Program itself may be modified or terminated at any time by Verizon for any reason.

The VZ BYOD Mission is to enable employees the option to utilize their personally owned mobile devices to access Verizon eMail and PIM. This includes allowing the employee to replace their corporately owned Personal Digital Assistant (PDA) or tablet device with their personally owned PDA or tablet device of choice. Mobile devices are defined as a smartphone / PDA or an Android or iOS tablet;

BYOD does not include laptops (Windows-based) devices. This initiative is valid ONLY for employees living and working in certain countries. Only "Eligible Employees" can participate in the BYOD Program. An "Eligible Employee" is a Verizon employee who:

1. Is required and authorized to work remotely from time to time; and
2. Was issued, or was offered and declined, a Company Device for working remotely. A Company Device is either:
   A. A Verizon-owned laptop with approved VPN (Virtual Private Network) access (remote Thin Client technology is also valid); or
   B. A Verizon-owned smartphone or tablet on which the employee has been authorized to have access to their Verizon corporate email account;
3. Is not covered by any collective bargaining agreements.

A BYOD Participant may use his/her approved Personal Device to connect to Verizon internal networks and internal systems only to (a) manage email and IM communications, (b) access Verizon contacts (address book) and/or (c) access to certain Verizon Intranet websites webpages. All other access to or use of Verizon internal networks or internal systems via the Personal Device is prohibited. A BYOD Participant must comply with the following BYOD security and technical requirements which are designed to protect the integrity of Verizon's corporate data and are designed to ensure that it remains safe and secure under Verizon control. The corporate container on the BYOD will be managed by Verizon's Mobile Device Management (MDM) enterprise solution to a similar extent that a corporate device is managed; no exceptions. All data classified as VZ Private, Confidential, or Highly Confidential must be inside an IT Security approved encrypted container. The requirements above may be modified at any time by Verizon for any reason. They also include, without limitation: (a) prohibition against circumventing Verizon, carrier or manufacturer technical restrictions or security, and (b) a requirement to cooperate fully with any Verizon efforts to identify and remove any corporate data and/or applications that may be wrongfully transferred to a device other than a Company Device. If a BYOD Participant or the Personal Device is placed on a "legal hold" by Verizon Legal, the requirements to preserve messages and data apply as if the Device was a Company Device.

27.5 Conclusion

The BYOD paradigm may continue for some time as corporations and users find a balanced solution, where corporate-data protection and employees' personal satisfaction can co-exist. However, corporations need to understand, not only the vulnerabilities and security risks introduced when BYODs are allowed, but also the legal situations that may arise when confronted by disgruntled employees. The possibilities of legal implications is sometimes ignored or not realized as companies readily adopt the use of BYODs. The company's perceived savings when adopting BYODs may be overshadowed if they find themselves loosing valuable information or in the middle of a lawsuit due to infringement of privacy, especially when personal and corporate data co-exist within the same device space. Even in the situations where companies adopt mobile solutions that separate corporate and personal data, the issue of privacy regulations needs to be considered for data such as text messages and photo galleries.

However, many of these risks can be mitigated by carefully drafting and implementing a well-balanced BYOD policy that not only addresses the protection of the corporate data but also safeguards against potential litigation related to compliance of privacy laws [10]. These measures coupled with awareness and training programs, after which employees provide consent before enrollment, are part of the safeguards a company needs to undertake in order to protect its more precious asset: its information.

References

1. Absalom, R. (2012). International data privacy legislation review: A guide for BYOD policies. *Ovum Consulting, IT006, 234*, 3–5.
2. Courion Press Info (2011, July 26). 69% of enterprises say employees are connecting personal mobile devices to the corporate network; more than one in five organizations does not have a policy in place to govern this use.
3. Utter, C., & Rea, A. (2015). The 'bring your own device' conundrum form organizations and investigators: An examination of the policy and legal concerns in light of investigatory challenges. *Journal of Digital Forensics Security & Law, V10*, 55.
4. Gartner (2013). Gartner predicts by 2017, half of employers will require employees to supply their own device for work purposes. Retrieved August 31, 2016 from http://www.gartner.com/news room/id/2466615.
5. McCumber, J. (2004). *Assessing and managing security risk in IT systems: A structured methodology*. Boca Raton: CRC Press.
6. U.S. Constitution – Amendment 4. Retrieved August 31, 2016 from http://www.usconstitution.net/xconst_Am4.html.
7. USLegal. (n.d). Spoliation of evidence. Retrieved online August 25 from http://civilprocedure.uslegal.com/discovery/spoliation-of-evidence/.
8. Yong, W., Streff, K., & Raman, S. (2012). Smartphone security challenges. *Computer, 45*, 52–58.
9. Harris, M. A., Patten, K., & Regan, E. (2013). The need for BYOD mobile device security awareness and training. *AMCIS 2013 Proceedings*.
10. Lannon, P. G., & Schreiber, P. M. (2016, Febraury). BYOD policies: Striking the right balance. *HR Magazine* (pp. 71–72).

Evaluation of Cybersecurity Threats on Smart Metering System

Samuel Tweneboah-Koduah, Anthony K. Tsetse, Julius Azasoo, and Barbara Endicott-Popovsky

Abstract

Smart metering has emerged as the next-generation of energy distribution, consumption, and monitoring systems via the convergence of power engineering and information and communication technology (ICT) integration otherwise known as smart grid systems. While the innovation is advancing the future power generation, distribution, consumption monitoring and information delivery, the success of the platform is positively correlated to the thriving integration of technologies upon which the system is built. Nonetheless, the rising trend of cybersecurity attacks on cyber infrastructure and its dependent systems coupled with the system's inherent vulnerabilities present a source of concern not only to the vendors but also the consumers. These security concerns need to be addressed in order to increase consumer confidence so as to ensure greatest adoption and success of smart metering. In this paper, we present a functional communication architecture of the smart metering system. Following that, we demonstrate and discuss the taxonomy of smart metering common vulnerabilities exposure, upon which sophisticated threats can capitalize. We then introduce countermeasure techniques, whose integration is considered pivotal for achieving security protection against existing and future sophisticated attacks on smart metering systems.

Keywords

Smart Metering Infrastructure • Power Engineering • Smart Grid • Cybersecurity • Cyber Threats • Cyber Attack

28.1 Introduction

The modernization of the modern power grid systems otherwise known as the smart grid has been developed for the purpose of enabling bidirectional flows of metering information in order to provide consumers with diverse choices as to how, when, and how much electricity they use. Integrated within the smart grid infrastructure setup is smart metering which core objective is to automate the monitoring of consumers' power consumption, for the purpose of billing and accounting. Smart metering infrastructure otherwise known as Advanced Metering Infrastructure (AMI), is the core component in smart grid infrastructure systems. The functional architecture represents an automated two-way

S. Tweneboah-Koduah (✉)
Centre for Media, Information and Technology, Aalborg University, AC Meyers Vænge 15, DK-2450 Copenhagen, SV, Denmark
e-mail: sat@es.aau.dk

A.K. Tsetse
Department of Computer Science, Northern Kentucky University, Highland Heights, KY 41099, USA

J. Azasoo
Department Computing & Immersive Technologies, The University of Northampton, Northampton, NN2 6JD, UK

B. Endicott-Popovsky, PhD
Center for Information Assurance and Cybersecurity in Education, University of Washington, Bothel, WA 98011-8246, USA

© Springer International Publishing AG 2018
S. Latifi (ed.), *Information Technology – New Generations*, Advances in Intelligent Systems and Computing 558, DOI 10.1007/978-3-319-54978-1_28

communication between a smart utility meter and a utility producer [1]. The metering system monitors consumers' power consumption by collecting information on such consumption and communicating such information back to the utility company for load monitoring and billing [1].

Additionally, smart metering infrastructure provides better monitoring of power consumption and efficient and more transparent billing system to consumers. Thus, the utility providers are able to apply different price models for power consumption based on the time of day and season power is consumed [2]. By design, smart metering enables consumers to access their own real-time use of power consumption information through a web interface and mobile app service. These goals could not have been achieved and realized without the integration of communication technology infrastructure required to gather, assemble, and synthesize data provided by smart meters and other interconnected components.

Smart Metering (SM) has gradually become an interest to both research and industrial communities most importantly to utility companies, energy regulators, energy distribution vendors as well as energy conservation societies [3]. The adoption and use of smart metering are advancing in recent times due to the ability to integrate information and communication technologies with the development of energy infrastructure systems. Notwithstanding, the recent upsurge in cyber attacks against critical infrastructure systems threatens the smooth functioning of smart metering infrastructure development and the electric grid as a whole.

In this paper, we assess some of the major cybersecurity issues in smart metering infrastructure. Our goal is to provide an initial step to classify the system's inherent vulnerabilities and the potential threats capable of exploiting the vulnerabilities. We evaluate this by demonstrating the feasibility and impact of various threat vectors upon a smart metering communication infrastructure network.

This paper is organized as follows. Beginning with this introduction, the next section reviews the state of the art of smart metering system. In Section III, smart metering functional architecture is presented. Section IV explores the evaluation of cybersecurity challenges on smart metering. We discuss the study findings in section V and then conclude the paper in Section VI.

28.2 Related Studies

The concept of smart metering has advanced in recent times due to the integration of information and communication technologies into energy development. Rinaldi classified such integration as cyber interdependency [4]. In a related study, Rinaldi, et al. argued that interdependencies in critical

infrastructure systems give rise to functional and non-functional challenges which do not exist in the single infrastructural system [5]. Accordingly, Li et al, posit that smart metering is part of the smart grid infrastructure system and for that matter, security attacks may take place both in the physical space, as in the conventional power grid, as well as in the cyberspace as in any modern communication infrastructure network [2].

Smart metering infrastructure system is often microprocessor-based which supports wireless connectivity for easy control and monitoring. Li et al. argue that smart meters are massively deployed as access points and are mostly connected to the Internet to engage customers in utility management. These access points, conversely, have become ideal ports for intrusions, penetration and other malicious attacks [2]. Conversely, Li et al, maintain that the openness in the smart metering systems (to the public network) increases vulnerabilities in the grid thereby escalating sophisticated threat attacks on the system. In a related study, Flick and Morehouse claim that cyber threats on critical infrastructure systems in general, and the electricity grid, in particular, has become the subject of increasing research interest both in academia and industry [6]. Contributing to this, Giani, et al., argue, the potential consequences of successful cyber attack on the electric grid is staggering [7]. They stated, smart metering which is part of a Smart Grid infrastructure system, incorporates sensing, communication, and distributed control to accommodate renewable generation, Electronic Vehicle (EV) loads, storage, and many other technologies. These activities substantially increase actionable data transfers making the system more vulnerable to cyber attacks, thus, increasing the urgency of cybersecurity research for electric grids [7].

Prior studies explored various aspects of cyber attacks on smart grid and smart metering systems. For instance, Yan et al, summarize possible vulnerabilities and cybersecurity requirements in smart grid communication systems and surveyed solutions capable of counteracting related cybersecurity threats [8]. In a related study, Wei et al. proposed a framework for protecting power grid automation systems against cyber attacks [9]. The proposal includes integration with the existing legacy systems, desirable performance in terms of modularity, scalability, extendibility, and manageability, alignment to the "Roadmap to Secure Control Systems in the Energy Sector" and future intelligent power delivery systems [9]. Cleveland in [3] argued that while various AMI vendors and customers consider encryption as a security proof solution to the threats of cyber adversaries on AMIs, there are other potential cybersecurity challenges facing AMI systems which require research focus. The challenges Cleveland identified include confidentiality, integrity, data availability and non-repudiation. The issues of privacy, confidentiality, and data availability as cybersecurity threats against smart grid systems have also been discussed in the following studies [2, 10–14].

28.3 Smart Metering Functional Architecture

The future power grid has a tiered architecture to supply energy to consumers [15]. This modern energy infrastructure system starts from power generation through transmission systems to distribution and eventually to the final consumer. A smart grid system strives to use and coordinate various generations, and production as well as the distribution mechanisms of the grid [15]. Smart metering infrastructure is the core component in a smart grid infrastructure system. It functional architecture represents an automated two-way communication between a smart utility meter and a utility producer [1]. Smart meters identify power consumption by collecting information on such consumption and communicate the information back to the utility company for load monitoring and billing for accounting purposes [1]. By generalizing the structures in [1, 2, 15], we present functional smart metering architecture as illustrated in Fig. 28.1. The architecture consists of a micro-load management unit and it hardware subsystem which houses the various hardware components of the system. Each of the structures has its core components and functions explained below.

(i) **Smart Meter:** This is the core of a smart metering infrastructure setup. It acts as the main source of energy-related information or other metrological data and provides interval data for customer energy loads.

(ii) **Smart Metering Communications Network:** Like a traditional communication network, the smart metering network provides a path for information flow within the grid.

(iii) **Customer Gateway:** This acts as the conduit between a smart metering network and the other smart devices in the grid or within the customer facilities, such as a Home Area Network (HAN) or the Neighborhood Area Network (NAN)

Other components within the metering structure include:

(iv) The **Wide Area Network (WAN) Interface:** It collects metering and control information from the Server systems and relays the readings and status of the meter to the server.

(v) The **Home Area Network (HAN):** This serves as the communication medium for device interface sensors,

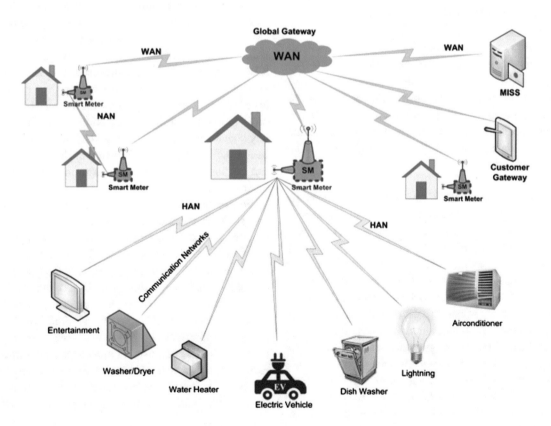

Fig. 28.1 Smart metering communication architecture

actuator/network relays, the In-Home Display (IHD) units, etc. This communication medium can be a single unidirectional or bidirectional or a combination of multiple technologies such as power line carrier (PLC), Ethernet, or wireless communication technologies (e.g. Z-Wave, Bluetooth, ZigBee, WiFi, RF mesh, and WLAN (802.11)).

(vi) The **WAN gateway:** This acts as the link between the metering Unit and the Micro-load metering information system to provide near real-time monitoring and control functions of the metering system and other auxiliary services, by providing access to the electrical utility companies and their consumers.

The utility company gains access to the metering system through a computer interface directly connected to the server. Utility consumers are usually provided access to the metering system through the web and/or mobile application interface, giving consumers the ability to monitor real-time information about energy consumption and billing, as well as performing home automation activities using integrated mobile devices.

(vii) The **Home Area Network (HAN) Gateway**: It provides a communication channel between the main metering unit and the micro-load controllers. As a result, the microcontrollers and load management can be extended to an off-the-shelf micro-load meter for the use of Electric Vehicle (EV) charging systems and other energy consuming loads.

(viii) The **Neighboring Area Network (NAN) Gateway:** Acts as the intermediary tier connecting multiple HANs collectively in the smart grid for the purpose of accumulating energy consumption information from households (the HANs), in a neighborhood and relay the data to the utility company [15] for billing and monitoring.

The Metering Unit **(MU)** is the main control center for the smart metering functional architecture. In the absence of the HAN, the MU monitors the amount of energy being consumed, as well as the ability to curtail power to all consuming devices. Additionally, the Micro-Load Metering Unit (MMU) monitors and reads the consumptions of all the devices and appliances attached to the main meter (including the Electric Vehicle Charging Terminals (EVCT)) by providing granular consumption data for consumption analysis and predicting future energy consumption. The micro-load controller functions to cut-off or connects micro-loads to the main source of electricity via the metering unit. This functionality is linked, to the Direct Load Control (DLC) unit enabling consumers to respond to pricing signals or time-of-use through an application program interface (API) such as Web or App.

28.4 Cybersecurity Challenges in Smart Metering

The conventional metering system is embedded with dedicated power devices, which are mostly integrated with control, monitoring and communication functionalities, using closed networks composed of predictable serial communication links. In contrast, smart metering decouples communication and control functionalities from power devices and is modularized for the purposes of scalability and maintenance [2]. Moreover, smart metering core components are usually commercial off-the-shelf (COTS) products from diverse vendors having unknown incompatibilities.

Cybersecurity challenges of a smart metering system lie in the system's inherent vulnerabilities which expose the infrastructure set to various attacks. Sources of vulnerabilities include firmware, hardware architecture, system applications, as well as a network interface. Besides, the bi-directional communication link between the metering unit and main gateway (and Metering Information Server-MIS) leaves the system open for network-related attacks and protocol failure. Other communication attacks include wireless scrambling, eavesdropping, man-in-the-middle attacks, message modification and injection attacks. For example, IP-based devices are susceptible to IP misconfiguration which does exhibit nondeterministic behavior in times of success attack. IP misconfiguration inevitably decreases system operation and reliability. Besides, smart meters are deployed in smart grid as access points for each customer (in the NAN and HAN), in order to manage utility consumption. These devices are usually connected to the Internet through the metering gateway. In addition to IP spoofing, the gateway (both local and global), can become perfect points for intrusions, DoS attacks, and other Internet-based attacks.

Furthermore, per their design, utility consumers usually interact with the metering system through the web and/or mobile application interfaces. Most of these applications are either web-based or stand-alone. Web-based applications are integrated with the metering system application using Application Programming Interface (API). An outdated API code may be susceptible to various attacks exposing the entire metering system to malicious attacks. Additionally, a poorly configured interface design exposes smart metering system to injection and code execution attacks. Furthermore, in the Home Area Network (HAN), such attacks on a metering device could destabilize the communication system leading to a denial of essential services to interdependent components or devices. In the Neighborhood Area Network (NAN), such an attack could lead to distributed denial of service (DDoS) attacks due to inter-meter communication services.

While many of these systems are designed with security in mind, security misconfiguration could occur at any level or any part of system design exposing the entire setup to software misconfiguration attack. More so, at the firmware level, smart metering components usually have internal memory used for temporary storage and information processing. Like a conventional metering system, power fluctuations in the grid occasionally cause devices to lose memory leading to data loss. Furthermore, intermittent power fluctuations in semiconductor devices may lead to signal loss and system malfunction. Other security challenges in the smart metering infrastructure include component incompatibilities, as well as device-based (physical) attacks, such as natural disasters, illegitimate use of the device (e.g. pilferage), and masquerading. To overcome these challenges will require innovative research and comprehensive system solutions which focus on the architectural redesign, firmware and hardware reconfiguration, network hardening and dynamic system application design.

28.4.1 Smart Metering Cyber Attack

From the above challenges, we present a taxonomy of cybersecurity attack in a smart metering communication system by analyzing system's inherent vulnerabilities vis-à-vis the potential threat actors. In this taxonomy, six types of vulnerabilities are discussed. These are IP Misconfiguration, Injection, DoS, Code Execution, Memory Corruption, and XSS & CSRF. Corresponding threat vectors include physical (device) attack, application (software) attack, network attack, web interface attack, and data attack (see Table 28.1 and Fig. 28.2).

Table 28.1 shows our proposed vulnerability threat matrix. In columns III and IV, threat vectors are matched with their corresponding vulnerabilities.

28.4.2 Attack Vectors

28.4.2.1 Device Attack

This is an attack type capable of compromising smart metering devices. It is the first point of call to compromise the functionality of the entire architecture (depending on the devices involved). In HAN, this type of attack could bring entire network down (especially when the metering unit is the target). Similarly, in a NAN, a device attack may affect the resistance of the network which in the extreme case could lead to DDoS attacks on the entire grid. Device attacks may be caused by IP misconfiguration, memory corruption, and wrongly executed code in the device operating system at the middleware layer.

28.4.2.2 Application Service Attack

This is a type of attack that compromises system applications (Web, Mobile, System, etc) which are run on various components of the system. Smart metering systems run multiple applications both at the local and the server levels. In most cases, these applications are owned by application service providers (ASPs) which are third party vendors. Cyber attacks on these applications will surely compromise the metering system. Common vulnerabilities in this type of attack include SQL injection, code execution, and DoS.

28.4.2.3 Network Attack

This is an attack which aims at compromising intercommunication among devices by either delaying message forwarding or completely failing to deliver. Network attacks may also destruct computational processes within the smart metering system. In HAN, this type of attack aims at destructing the core functionalities of the metering system. Similarly, in a NAN, a network attack may isolate or deny NAN devices from accessing vital information from the neighborhood or addressing messaging request from neighboring devices. Causes of network availability attacks include SQL injection, DoS and code execution in the network infrastructure system.

Table 28.1 Vulnerability-threat matrix

		Vulnerability-threat matrix	
Vulnerabilities (V)	Cyber-attack vectors (AV)	Attack vectors	Vulnerabilities
IP Misconfiguration (**IM**)	Device Attack (**DA**)	DA	IP, MC, CE, D
SQL Injection (**SI**)	Application Attack (**AA**)	AA	SI, D, CE
DoS (**D**)	Network Attack (**NA**)	NA	SI, D, CE
Code Execution (**DE**)	Web Interface Attack (**WiA**)	WiA	SI, D, XC, IP
XSS & CSRF (**XC**)	Data Integrity Attack (**DA**)	DA	SI, CE
Memory Corruption (**MC**)			

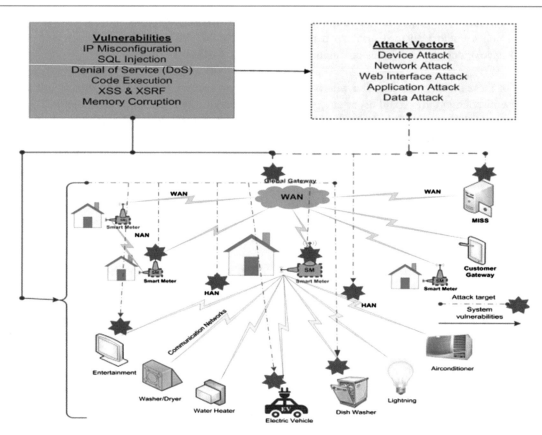

Fig. 28.2 Smart metering and corresponding threat attack

28.4.2.4 Web Interface Attack

This type of attack presents itself as a result of account enumeration, lack of account lockout or weak account credentials. In this case, an attacker may use weak account credentials (either capture plain-text credentials or enumerate accounts) to access the web interface. Web interface attacks may be caused by cross-site scripting (XSS), cross-site reference forgery (CSRF), IP misconfiguration and SQL injection. Other sources include insecure web interface design and weak account credentials. The attack compromises device integrity and could lead to denial of services.

28.4.2.5 Data Integrity Attack

This is an attack whereby the threat agent attempts to compromise system data by inserting, altering or completely deleting data (either stored or in transmission) so as to deceive smart metering to make wrong decisions or compromise its integrity. Data attacks may be caused by SQL injection and code execution which may be executed by a remote attacker.

28.5 Experimental Evaluation of Cyberattacks against Smart Metering

In this section, we demonstrate how SQL injection and DoS attacks could be executed against a smart metering system. These demonstrations were performed on a live server with positive results. In each case, the results show that cyber attack on smart metering systems was successful (Fig. 28.3).

SQL injection attack – Algorithm

```
Print header information
    for URL in target URLs
        for payload in get request payloads
            response = send get request probe to
server
        if response.status code == 500
            print payload and exist for manual attack
        for paylaod in post request payloads
            response = send post request probe to
server
```

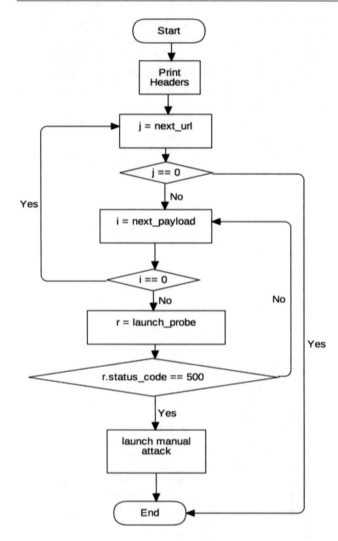

Fig. 28.3 SQL injection attack – Flowchart

SQL injection attack – Python script

This function delivers a payload to the server using the http '*get*' method. To do this, the payload is added to the url. The url sends the request to @params payload {string}. *The request parameters for example*

```
requests.get('http://www.test.com/',
params=payload) will map to http://www.test.com/
?key=value
###
def http_get(url, payload):
    r = requests.get(url, params=payload)
    return process_responds(r)
###
```

This function processes the request to determine if the probe is positive or negative

Probing Get (assuming query is contracted: where id = <defined_param>
('Params ', {'make': "'"})
('Url: ', 'http://metering.grid.com/metering/ meter/topup_history')
   data been sanitised
   data been sanitised
   data been sanitised
   data been sanitised
Probbing Post
   data been sanitised
   data been sanitised
   data been sanitised
   data been sanitised
Vulnerability: *Weakness found (SQL injection)*
Threat: *data sanitised*
Effect: *sensitive information could be disclosed by injection attack*
Impact: *Data confidentiality and integrity could be compromised*
Denial of Service Attack
DoS attack on the Application layer
Attack url:
http://metering.smartmeter.com/metering/server/ dashboard
Tool: loadtest (https://www.npmjs.com/package/ loadtest) requires nodejs to be installed
Test parameter: $ loadtest http://metering. aborsour.com/metering/server/dashboard -t 50 -c 10 --rps 1000.

28.5.1 Discussion

The idea of running both SQLi and DoS attacks on a smart metering system highlights their significant impact on distributed network system, such as smart metering (see Table 28.1). In the case of the former, a payload request was sent to the server to probe the server for vulnerabilities. The server responded with an ACKnowledgement a message header which encourages an attack on the system. This means SQL injection vulnerability in a smart metering system could allow remote authenticated users to execute arbitrary SQL commands via crafted serialized data both on the metering information system server (MISS in Fig. 28.2). For example, SQL injection vulnerability in the login page in the user interface device would allow remote attackers to execute arbitrary SQL commands via a crafted URL.

Per the CVE[1] database, DoS vulnerability remains the most common vulnerability type and can be exploited by

---

[1] Common Vulnerability Exposure.

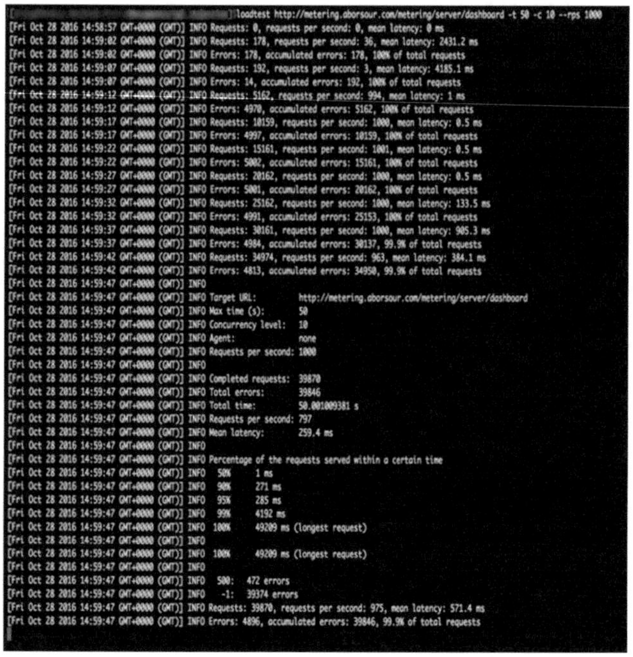

Fig. 28.4 Results (screenshot) of DoS attack

and make case for research effort in this emerging technology. The discussion involved the identification of various vulnerabilities inherent within smart metering components matched with the potential threat vectors capable of exploiting these vulnerabilities. We executed two different attack scenarios (tests) as a proof of concept. Tests results show that vulnerable smart metering system could be abused by various threat actors via crafted vectors.

Finally, it is critical to continue the discussion while at the same time challenging device manufacturers and components' vendors to design, and implement solutions for such mechanisms so as to counteract threats from cyber adversaries of the electrical grid so as to guarantee consumer utmost trust in a smart metering innovation and transformation.

various threat vectors. In the above test, we executed multiple (abnormal) remote requests (1000) to the server from concurrent connections in 50 seconds. The result (Fig. 28.4) shows the server failing or executing arbitrary code (system crushing). For example, a buffer overflow in the Point-to-Point Protocol over the Ethernet (PPPoE) module in the customer gateway when CHAP authentication is configured on the server, could allow remote attackers to cause a denial of service or execute arbitrary code via crafted packets sent during authentication. For example, in CVE-2016-8666, an IP stack in the Linux kernel (before 4.6) allows remote attackers to cause a denial of service (stack consumption and panic) or possibly have an unspecified impact by triggering use of the GRO functions (gro-receive and gro-complete) path for packets with tunnel stacking.

28.6 Conclusion

The core objective of smart grid is to improve efficiency and availability of power by adding more monitoring and control capabilities [16]. This objective is made plausible by the successful integration of a smart metering system which core value is to automate monitoring of consumer power consumption, efficient energy distribution, billing, and accounting. In this paper, an attempt has been made to evaluate the taxonomy of the system inherent vulnerabilities which expose smart metering to various cyber threat vectors

References

1. Yan, Y., Qian, Y., Sharif, H., & Tipper, D. (2013). A survey on smart grid communication infrastructures: Motivations, requirements, and challenges. *Communication Surveys and Tutorials IEEE, 15*(1), 5–20.
2. Li, X., Liang, X., Lu, R., Shen, X., Lin, X., & Zhu, H. (2012). Securing smart grid: cyber attacks, countermeasures, and challenges. *IEEE Communications Magazine, 50*(8), 38–45.
3. Cleveland, F. M. (2008). Cyber security issues for advanced metering infrastructure (ami). In *Power and Energy Society General Meeting-Conversion and Delivery of Electrical Energy in the 21st Century, 2008 IEEE*, pp. 1–5.
4. Rinaldi, S. M. (2004). *Modeling and simulating critical infrastructures and their interdependencies.* System sciences, 2004. Proceedings of the 37th annual Hawaii international conference on, p. 20054.1. 8–pp.
5. Rinaldi, S. M., Peerenboom, J. P., & Kelly, T. K. (2001). Identifying, understanding, and analyzing critical infrastructure interdependencies. *Control Systems IEEE, 21*(6), 11–25.
6. Flick, T., & Morehouse, J. (2010). *Securing the smart grid: Next generation power grid security.* Amsterdam: Elsevier.
7. Giani, A., Bitar, E., Garcia, M., McQueen, M., Khargonekar, P., & Poolla, K. (2011). *Smart grid data integrity attacks: characterizations and countermeasures π.* Smart Grid Communications (SmartGridComm), 2011 I.E. International Conference on, pp. 232–237.
8. Yan, Y., Qian, Y., Sharif, H., & Tipper, D. (2012). A survey on cybersecurity for smart grid communications. *IEEE Communication Surveys and Tutorials, 14*(4), 998–1010.
9. Wei, D., Lu, Y., Jafari, M., Skare, P., & Rohde, K. (2010) *An integrated security system of protecting smart grid against cyber attacks.* Innovative Smart Grid Technologies (ISGT), 2010, pp. 1–7.
10. Liu, J., Xiao, Y., Li, S., Liang, W., & Chen, C. L. (2012). Cyber security and privacy issues in smart grids. *Communication Surveys and Tutorials IEEE, 14*(4), 981–997.
11. Ericsson, G. N. (2010). Cybersecurity and power system communication—essential parts of a smart grid infrastructure. *IEEE Transactions on Power Delivery, 25*(3), 1501–1507.
12. Hahn, A., Ashok, A., Sridhar, S., & Govindarasu, M. (2013). Cyber-physical security testbeds: Architecture, application, and

evaluation for smart grid. *IEEE Transactions on Smart Grid, 4*(2), 847–855.

13. Lu, Z., Lu, X., Wang, W., & Wang, C. (2010). *Review and evaluation of security threats on the communication networks in the smart grid*. Military Communications Conference, 2010-MILCOM 2010, pp. 1830–1835.

14. Metke, A. R., & Ekl, R. L. (2010). Security technology for smart grid networks. *IEEE Transactions on Smart Grid, 1*(1), 99–107.

15. Bou-Harb, E., Fachkha, C., Pourzandi, M., Debbabi, M., & Assi, C. (2013). Communication security for smart grid distribution networks. *IEEE Communications Magazine, 51*(1), 42–49.

16. Clements, S., & Kirkham, H. (2010). *Cybersecurity considerations for the smart grid*. IEEE PES General Meeting, pp. 1–5.

A Layered Model for Understanding and Enforcing Data Privacy

Aftab Ahmad and Ravi Mukkamala

Abstract

In this paper, we propose a layered model for the understanding and enforcing of information privacy. The proposed model consists of three levels. At the lowest level, called the *Read/Write Layer*, privacy is defined as the resistance and resilience to Read or Write violations in the information or information source. At the middle level, *the sharing layer*, a logical privacy connection can be set up between a source and sink based on an *embedded privacy agreement* (EPA). At the highest layer, *the trust layer*, privacy is determined based on the history of sharing between directly connected network entities. We describe how the privacy metrics differ at each layer and how they can be combined to have a three-layer information privacy model. This model can be used to assess privacy in a single-hop network and to design a privacy system for sharing data.

Keywords

Information privacy • Layered model • Metrics • Privacy agreement • Privacy models

29.1 Introduction

Information privacy is a complex phenomenon [1]. The roots of lack of privacy lie in the evolution of the Internet from a data sharing system between a group of trusted researchers, to everybody in the world. Providing privacy is not as simple as making a LAN cable safe from packet sniffers. As a result of years of attacks on our systems and information, we are much more aware and educated now. However, we still struggle with the meaning of privacy. The multi-disciplinary nature of the Internet allows language mixed from medical, financial, legal, and technology sectors. Its multifarious nature demands a solution that applies to data from all sectors, and for information stored or moved *electronically* only. Work in understanding, specifying, quantifying and implementing information privacy is at full throttle. In this paper, we present information privacy provisioning system as a modular concept in order to resolve the complexity of achieving privacy in a public environment. In Sect. 29.2, we summarize the three categories in which information privacy is being investigated, namely, deterministic, probabilistic and framework-based measures. In Sects. 29.3 and 29.4, we elaborate on the proposed model. Conclusions are presented in Sect. 29.5.

29.2 Privacy Metrics and Their Flaws

29.2.1 Deterministic Models

In deterministic models for providing privacy for information stored in a database, records are anonymized through deterministic generalization of identity of the records. A record could have the identity of a person, some sensitive

A. Ahmad
CUNY John Jay College of Criminal Justice, 524 W 59th Street, New York, NY 10019, USA
e-mail: aahmad@jjay.cuny.edu

R. Mukkamala (✉)
Old Dominion University, Norfolk, VA 23529, USA
e-mail: mukka@cs.odu.edu

© Springer International Publishing AG 2018
S. Latifi (ed.), *Information Technology – New Generations*, Advances in Intelligent Systems and Computing 558, DOI 10.1007/978-3-319-54978-1_29

attributes, and some quasi-identifiers. Identities of people can be simply withheld before sharing. Generalizing the quasi-identifiers (QIs) by means such by masking the right three digits of zip codes is an example of this type of privacy enforcement methods. k-anonymity [2], l-diversity [3], t-closeness [4] and their variations are examples of such techniques. Anatomization is yet another technique where separate tables are made for QIs and sensitive attributes (SA), and correlation between the two is removed making it harder to guess SAs [5]. The m-invariance is yet another technique for anonymization [6].

29.2.2 Probabilistic Models

Various probabilistic metrics have been proposed, prominent of which is to employ information theoretic reasoning. Shannon used the concepts of self-information , entropy, and mutual information to define the capacity of a digital system as the maximum average mutual information [7, 8]. However, as pointed out in [9], the average mutual information can't tell the impact of one data record on the ensemble – which is what we want when we talk about privacy preservation. j-measure [10] and i-measure are used as alternatives in this context.

29.2.3 Framework-based Models

Frameworks are guidelines that, if followed strictly, provide a measurement and comparison standard. Some frameworks include the Privacy Framework set forth by APEC [11], the EU Data Privacy Directive [12], the US acts for various sectors, such as HIPPA [13] for healthcare and GLBA for financial sector [14]. Differential privacy based hybrid framework proposed in [15] is proven to have better data sanitization than the deterministic methods discussed above and more efficient than generic secure multi-party computation (SMC). In [16], concepts from another framework, Pufferfish, have been used to come up with a privacy model based on differential privacy. It has been applied to privacy of smart meters [17] as well as on other privacy models [18]. Granularity of the protected data is one of the many concerns in the current solutions. As will be explained in the next sections, the proposed model has a granularity for datasets as small as defined by data owner, including a resolution of one bit (Boolean data).

29.3   Need for a Different Perspective

In the presence of a large number of privacy mechanisms, one would be justified in asking the question, 'why do we need yet another proposal?' Part of the answer lies in the

evolution of various deterministic and non-deterministic measures. It was shown in the l-diversity proposal how k-anonymity can give in to special situations where the mere number in a group is not good enough to be effective, or in the introduction of t-closeness, that generalization can swerve data statistics in a way to leave clues for a data miner. Differential privacy attempts to dampen the impact of a single record, which would impact the utility of the shared data – for example, in case of research on rare traits. As mentioned by some authors of information theoretic measures, the average mutual information has the same effect as differential privacy measure due to the averaging. Besides the individual idiosyncratic issues with the existing proposals, there are no universally accepted goals of privacy preservation.

There are still issues for which solution or even proposals have not been intensely pursued, such as the privacy impact of information published via different unrelated databases. It happens to be such a complex issue that setting up any goals almost guarantees a temporal tag to any proposal. In view of these complexities, we contend that a privacy preservation measures should not only make good use of the available solutions but also provide a unified way of thinking into future. An example of the difference between existing solutions and open issues is permission versus trust. Both impact privacy. While we have a good understanding of how to set up permissions for data usage, measuring the impact of trust on privacy is an open issue. The question then is—can we have a privacy framework in which both can co-exist, namely, the solutions as well as open issue such as trust. In the next section, we discuss the proposed privacy architecture that can be used not only to set goals, like the Pufferfish, but allows research to be integrated in three related yet independent components: setting permissions, leaving room to correlated data identification, relating permission and a data usage agreement to a trust model that can be defined independent of the other two components.

29.4   Proposed Layered Model

An information infrastructure usually has three types of data—user data, control data, and management data. An attack on privacy can happen on either type. An attack on user data can impact the user directly by disclosure to adversary. An attack on control information, such as during connection set up with an access point, can help adversary replace a legitimate data recipient by his own device. An attack on management data, such as domain name server entries, can result in forwarding of user information to rogue servers somewhere else, even in another country, on the Internet.

In our model, which consists of three layers, we mainly consider user data. However, it can be applied to control data as well as the management data. Figure 29.1 illustrates our model.

There are three layers in the user data plane: read/write layer, embedded privacy agreement (EPA) layer and the trust layer. There are two layers in the control plane, the read/write layer and the EPA layer. It is assumed that the control data is only exchanged with a trusted entity. There is only one privacy layer for the management data. It is assumed that a permanent EPA exists for all information and the entity is trusted.

The following example may illustrate the difference of privacy levels for various data types. Suppose an employee connects her tablet to her employer's intranet. It is expected that the intranet device (e.g., Wi-Fi access point) that will connect to the tablet is trusted, by mutual authentication using IEEE 802.1X, for example. The device however offers multiple types of connections, such as a guest connection, user level connection, administrator level connection. Using an EPA, one of these connections can be set up for this user. The data exchanged between the user and the access point is also protected depending on the type of data.

In the next phase of this user connection, the access point communicates with a DHCP server to let it find an IP address for the tablet and assign it. There is no need for an EPA for this phase of connection and the DHCP server is fully trusted. Now that this user has Internet access through access point and the DHCP/NAT server, she can share her personal or business data with her peers on the Internet, say via a social network. During this phase of sharing, not all peers have the same levels of trust, not every data items is to be shared at the same level with all peers, and consequently, subsets of data need to be specified in terms of read or write access given to a peer - requiring all the three layers. At the Read/Write layer, the smallest units of data are specified in

terms of their ability to be read or changed or both during a sharing session. At the EPA layer, a sharing connection is negotiated and implemented for stream(s) of data between a data owner and a data user. At the trust layer, measures of trust are specified against possible violations of EPA(s). We now describe the three layers in detail.

29.4.1 The Read/Write Layer

This is the physical layer of privacy where each atomic unit of information gets coded with a permission value of true or false for one of the three actions, read, write, or both. Figure 29.2 explains this for a data record that contains four attributes, name, age, SSN, and remarks.

In order to find privacy violation, we will need the permission tuple (*Read, Write*) and a probability of violation for each, say ε_R for read violation and ε_w for write violation. Figure 29.3 shows a channel model for the three possible values.

29.4.2 The EPA Layer

The EPA layer specifies protocols between a data owner and a data user. This could be peer-to-peer or through a service provider. The EPA layer specifies at least two types of protocols, a negotiation protocol and a sharing protocol. During the negotiation protocol, two parties decide which attributes are needed and whether they will violate the privacy of an existing such connection. For example, if a life insurance companies collect medical data for their new applicants from a medical information base (MIB) via a subscription service. When a company shares data about its applicants, it may not include information about employer of the applicant. During the negotiation phase, meta data may be exchanged about the information. This metadata may contain types of information to be shared, types of attributes (quasi-identifier and sensitive attribute names), and a set of minimum restrictions on all data to be shared – called Global EPA. Figure 29.4 shows a schematic for the EPA layer protocol for a data warehouse such as proposed in [19].

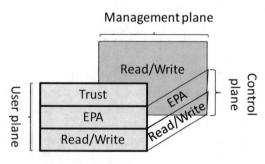

Fig. 29.1 The layered privacy model

Fig. 29.2 The (Read, Write)-specification of attributes. A bar over an action means it is not allowed

Record ID	Name	Age	SSN	Remarks
100100	Michael Joe	38	123-45-6789	Write here
Permissions	(*Read, $\overline{Write}$*)	(*Read, $\overline{Write}$*)	($\overline{Read}, Write$)	(*Read, Write*)

Fig. 29.3 Channel models for
(Read, Write) tuples

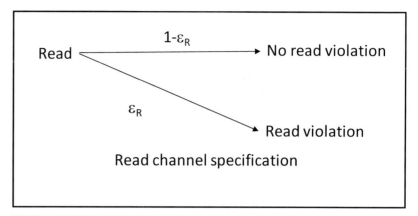

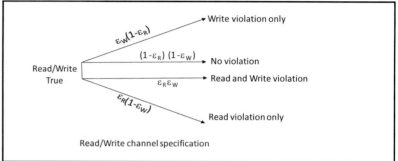

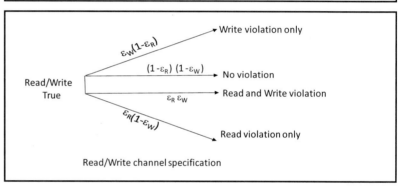

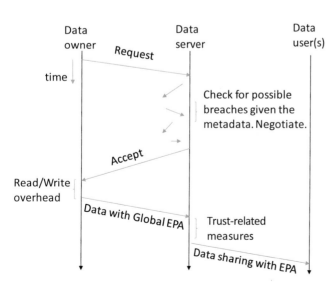

Fig. 29.4 EPA Protocol for data warehouse

29.4.3 Trust Layer

When data is to be released to two or more users, the expectation of the data owner is for all of them to follow the EPA to the letter. However, we learn from experience that all data users are not created equal. The problem with data sharing is that it is an irreversible process. Therefore, a history of trust with the data user must be considered [20]. The trust layer's job is to take measures for enforcement of EPA as close to 100% as possible. Consider the scenario in Fig. 29.5.

This figure shows a data owner connected to five (5) users, each user in turn connected to a bunch and so on. The trust between the data owner and user j is given by τ_j in the figure. A commonsense question is what is the impact of the subsequent users on the EPA between the data owner and user j? This problem has been addressed in network theory, circuit theory and reliability theory. As such, the determination of this impact is highly complex, subjective and hard to generalize. This may be possible on a network route, but not in a social network or business sector. In social networks, such as Twitter, Facebook, Youtube, the trust is much harder to gauge [21] due to human factor. If τ is the probability of trust between all data users and data owner then $\tau = min(\tau_j) \, for \, j = 1, 2, 3, 4, 5$. This poses the question whether same measures against an imperfect trust be applied against all users? It would be a bad engineering practice to apply the same measure against all trust levels. Instead, just like different credit score can get different interest rates, measures should be taken to raise the level of each connection's trust to a minimum threshold. It must be acknowledged here that human factor offers the

biggest challenge. The commonsense solution to that is to assign a price to each data user that reflects the risk. Suppose the risk is an exponential function of $(1-\tau_j)$, then Fig. 29.6 shows a hypothetical price curve for the value of an arbitrarily chosen cost function of $Ae^{\alpha(1-\tau)}$. The rationale for this selection is that it will give a cost of an imperfect trust value (less than 100%) as a function of the cost for a perfect trust (a value of 100%). For example, for a trust value of 0%, the cost is Ae^{α} as compared to one with 100% trust. The graph of Fig. 29.6 is for A and α values of 1, varying the cost between 1 and $e = 2.73$. This is highly arbitrary and the actual cost function should depend on risk for each data sharing instance. This is just for illustrative purpose. The graph uses trust values of $\{\tau_j\} = \{0.9, 0.88, 0.75, 0.95. 0.8\}, j = 1, 2, ..., 5$ from Fig. 29.5.

29.5   Conclusion

In this paper, we have presented a layered model of information privacy. All the proposed layer functions are already implemented in a hodgepodge in most software systems, be it operating systems, database systems and internet applications. Using the proposed approach, the privacy of information can be assured by an embedded privacy agreement (EPA) layer on a link between the data owner and data user. The agreement can be coded at a lower layer by assigning a (Read, Write) tuple to each of the smallest data unit within each data record. Finally, a trust layer provides uniformity in a network of sharing by assigning protection cost according to the risk for the data user having low trust

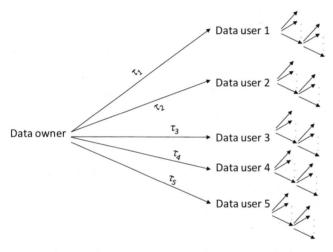

Fig. 29.5 Trust in social networks

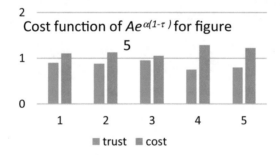

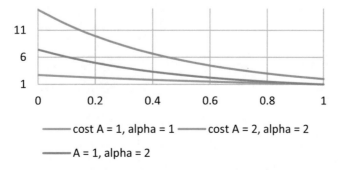

Fig. 29.6 Relating (mis)trust to risk

value. The research on the model is a continuing one and as a next step, we are working on quantification model for the layered privacy model so that systems can be designed for a given privacy level and, conversely, privacy level can be determined for a given system.

References

1. Ahmad, A., et al. (2016) Information privacy domain. *International Journal of Information Privacy, Security and Integrity*. To appear. http://dx.doi.org/10.1504/IJIPSI.2016.082124.
2. Sweeney, L. (2002). Achieving k-anonymity privacy protection using generalization and suppression. *International Journal of Uncertainty, Fuzziness and Knowledge-Based Systems, 10*(05), 571–588.
3. Machanavajjhala, A., Kifer, D., Gehrke, J., & Venkitasubramaniam, M. (2007). l-diversity: Privacy beyond k-anonymity. *ACM Transactions on Knowledge Discovery from Data (TKDD), 1*(1), 3.
4. Li, N., Li, T. and Venkatasubramanian, S. (2007, April). t-closeness: Privacy beyond k-anonymity and l-diversity. In *2007 I.E. 23rd International Conference on Data Engineering* (pp. 106–115). IEEE. Istanbul.
5. Xiao, X. and Tao, Y. (2006, September). Anatomy: Simple and effective privacy preservation. In *Proceedings of the 32nd international conference on Very Large Data Bases* (pp. 139–150). VLDB Endowment. Seoul.
6. Xiao, X. and Tao, Y. (2007, June). M-invariance: towards privacy preserving re-publication of dynamic datasets. In *Proceedings of the 2007 ACM SIGMOD international conference on Management of Data* (pp. 689–700). ACM. Beijing.
7. Rajagopalan, S. R., Sankar, L., Mohajer, S., and Poor, H. V. (2011, October). Smart meter privacy: A utility-privacy framework. In *Smart Grid Communications (SmartGridComm), 2011 I.E. International Conference on* (pp. 190–195). IEEE. Brussel.
8. Makhdoumi, A., Salamatian, S., Fawaz, N., and Médard, M. (2014, November). From the information bottleneck to the privacy funnel. In *Information Theory Workshop (ITW), 2014 IEEE* (pp. 501–505). IEEE. Hobart.
9. Bezzi, M. (2010). An information theoretic approach for privacy metrics. *Transactions on Data Privacy, 3*(3), 199–215.
10. Blanchard, J., Guillet, F., Gras, R., and Briand, H. (2005, November). Using information-theoretic measures to assess association rule interestingness. In *Fifth IEEE international conference on Data Mining (ICDM'05)* (pp. 8–pp). IEEE.
11. Solove, D. J. (2008). *Understanding privacy*. Cambridge: Harvard University Press. Available at understanding-privacy.com.
12. Fromholz, J. M. (2000). The European Union data privacy directive. *Berkeley Technology Law Journal, 15*, 461–484.
13. O'Herrin, J. K., Fost, N., & Kudsk, K. A. (2004). Health Insurance Portability Accountability Act (HIPAA) regulations: Effect on medical record research. *Annals of Surgery, 239*(6), 772–778.
14. Cuaresma, J. C. (2002). The Gramm-Leach-Bliley Act. *Berkeley Technology Law Journal, 17*, 497–517. Boca Raton.
15. Inan, A., Kantarcioglu, M., Ghinita, G., and Bertino, E. (2010, March). Private record matching using differential privacy. *International conference on Extending Database Technology* (pp. 123–134). ACM. Lausanne.
16. He, X., Machanavajjhala, A., and Ding, B. (2014, June). Blowfish privacy: Tuning privacy-utility trade-offs using policies. In *Proceedings of the 2014 ACM SIGMOD international conference on Management of Data* (pp. 1447–1458). ACM. Snowbird.
17. Kessler, S., Buchmann, E., and Böhm, K. (2015). Deploying and evaluating pufferfish privacy for smart meter data. *Karlsruhe Reports in Informatics, 1*, 229–238.
18. Kifer, D., & Machanavajjhala, A. (2014). Pufferfish: A framework for mathematical privacy definitions. *ACM Transactions on Database Systems (TODS), 39*(1), 3.
19. Mukkamala, R., Ahmad, A. and Nvuluri, K. (2016). *Privacy-aware big data warehouse architecture*. IEEE International Congress on Big Data, San Francisco, June 2016.
20. Rutten, W., Rutten, W., Blaas-Franken, J., Blaas-Franken, J., Martin, H., & Martin, H. (2016). The impact of (low) trust on knowledge sharing. *Journal of Knowledge Management, 20*(2), 199–214.
21. Papadopoulos, S., Bontcheva, K., Jaho, E., Lupu, M., & Castillo, C. (2016). Overview of the special issue on trust and veracity of information in social media. *ACM Transactions on Information Systems (TOIS), 34*(3), 14.

BYOD: A Security Policy Evaluation Model

Melva M. Ratchford

Abstract

The rapid increase of personal mobile devices (mainly smartphones and tablets) accessing corporate data has created a phenomenon commonly known as Bring Your Own Device (BYOD). Companies that allow the use of BYODs need to be aware of the risks of exposing their business to inadvertent data leakage or malicious intent posed by inside or outside threats. The adoption of BYOD policies mitigates these types of risks. However, many companies have weak policies, and the problem of exposure of corporate data persists. This paper addresses this problem by proposing a BYOD policy evaluation method to help companies to strengthen their BYOD policies.

This initial research proposes a novel BYOD security policy evaluation model that aims to identify weaknesses in BYOD policies using mathematical comparisons. The results are measurable and provide specific recommendations to strengthen a BYOD policy. Further research is needed in order to demonstrate the viability and effectiveness of this model.

Keywords

BYOD • Policy Evaluation • Evaluation Model • Risk • Security

30.1 Introduction

With the rapid increase of personal mobile devices accessing corporate data (a phenomenon called BYOD – Bring Your Own Device), companies need to be aware of the importance of maintaining corporate data protection in order to ensure the confidentiality, integrity, availability (CIA) of its data [1]. In 2012, a survey conducted by Cisco reported that 95% of the organizations polled permitted the use of employee-owned devices in the workplace [2]. In 2013, another study by Cisco indicated that 9 in 10 Americans used their smartphones for work, where 40% do not password protect them, and 51% connect to unsecured wireless networks using their smartphones [3]. The Gartner Group also predicted that by 2017 half of all companies will actually require employees to use their own mobile device for work [4].

New security risks are introduced with the use of BYODs to include devices easily tampered, lack of security awareness among users, threats and attacks (e.g. spoofing, phishing, data leakage, sniffing, spam, denial-of-service) [5]. BYODs are consumer devices that lack the strict compliance requirements of devices accessing corporate-sensitive data [6]. Security policies are less likely to be enforced in devices the company does not own [7]. Today's workforce expect to be able to access work-related information via their BYOD, and it is up to the company to protect its network and data [8]. A BYOD security policy needs to be in place and enforced. 'A system can be considered secure and trustworthy if the policy enforced by its security administrator is trustworthy too' [9]. However, many companies do not have a BYOD security policy in place, or their policy is weak, lacking technical or organizational considerations, or enforcement mechanisms. This is a problem because their

M.M. Ratchford (✉)
Dakota State University, Madison, SA, USA
e-mail: melva.ratchford@trojans.dsu.edu

© Springer International Publishing AG 2018
S. Latifi (ed.), *Information Technology – New Generations*, Advances in Intelligent
Systems and Computing 558, DOI 10.1007/978-3-319-54978-1_30

corporate data may be exposed to inadvertent data leakage or users with malicious intent (inside/outside threats). Therefore, a company's BYOD security policy should to be evaluated in order to identify the weak strategies that the policy makers can modify and enforce. 'It is possible to evaluate the system security by evaluating its policy' [9].

Current policy evaluation methods involve human intervention to analyze or parse policies written in a natural language (e.g. English) where comparisons are made against published guidelines. This process can produce ambiguous results based on subjective analysis (i.e. the opinion of an individual).

Using design science research methodology, this paper proposes a novel method to evaluate BYOD security policies. The model utilizes an evaluation process based on mathematical analysis that produces quantifiable measurements to provide security levels, identify weak strategies, and provide recommendations. With this information, the company can be in a better position to strengthen the security of its corporate data and mitigate the risks introduced when adopting BYODs.

The organization of the paper is as follows: after this introduction, Sect. 30.2 identifies the literature review, the research gap found in the literature, the requirements needed to fill the gap, and the supporting theories & literature needed to meet the requirements. Section 30.3 presents an overview of the model design. Section 30.4 describes a suitable context to demonstrate the model followed by an evaluation of the problem resolution. Section 30.5 concludes and states future work.

30.2 Literature Review and Underlying Theories

Prior literature provide valuable information in regards to understanding the BYOD paradigm [5, 6, 10, 11]. Several National Institute of Standard and Technology (NIST)'s publications provide further recommendations and guidelines in order to create awareness regarding risks and vulnerabilities to corporate data when BYODs are permitted [1, 12, 13]. However, the literature research finds a gap for a specific and non-ambiguous process to evaluate a company's BYOD security policy.

In order to fill this gap, the following requirements need to be researched and understood:

1. Risks and vulnerabilities associated with BYODs.
2. Methodologies for building security policies.
3. Non-ambiguous evaluating process for policies.

30.2.1 Risks and Vulnerabilities Associated with BYODs

The first requirement (risks and vulnerabilities associated with the adoption of BYODs) involves a thorough understanding of the risks and vulnerabilities introduced to a company when BYODs are allowed. For this purpose, the frameworks proposed by Vorakulpipat et al. [10] and the specific recommendations provided by recognized authorities such as NIST provide the foundation for building a baseline for technical and organizational considerations when mobile devices are allowed. The NIST's 800 publications include the 800-124 Guidelines for Managing the Security of Mobile Devices in the Enterprise [1], 800-114 User's Guide to Telework and Bring Your Own Device (BYOD) Security [12], and 800-46 Guide to Enterprise Telework, Remote Access, and Bring Your Own Device [13]. Main categories such as architecture, authentication, access control, cryptography/encryption, device provisioning, configuration, application requirements, security policy enforcements, auditing, training, and technical support can be expanded to include risks and vulnerabilities at a granular level.

30.2.2 Methodologies for Building Security Policies

The second requirement, (methodologies for building security policies), requires the understanding and developing of a process whereby the BYOD-related risks and vulnerabilities identified above can be addressed in a form of a security policy. Known methodologies for building security policies include McCumber's Cube [14], Peltier's basic concepts for Topic-Specific Policy [15], and Wood's framework for building security policies using the concept of a 'coverage matrix' [16].

The McCumber's Cube methodology is suitable for building a BYOD policy because it ensures that the CIA (Confidentiality, Integrity and Availability) of data is addressed. The data need to be protected while is being transmitted, at rest, and during processing. The security measures to protect the data during its various states need to include the use of technology, creation of policies and the development of training/awareness programs.

Peltier's Topic Specific Policy describes basic components that narrow the topic to one issue [15], making this method appropriate when considering BYOD policy analysis. The policy needs to have thesis statement with clearly identified objectives; it needs to be relevant by specifying to whom/where/how/when does the policy apply; identify roles and responsibilities by position/job/title/job-function; specify terms of compliance to include unacceptable behavior, its consequences and monitoring of compliance; and include additional information providing specific contact information and policy location [15].

In addition, Wood's methodology for writing information security policies describes a coverage matrix [16] that can be further expanded to map the individual provisions (i.e. security requirements) into a data (tree) structure where the main nodes, sub-nodes, and end-nodes delineate the individual policy elements at its most granular levels.

30.2.3 Non-ambiguous Evaluating Process for Policies

The third requirement (a security policy evaluating process), requires the development of a method whereby an ambiguous evaluation process (i.e. based on natural language analyses) becomes a non-ambiguous evaluation method based on mathematical analyses with quantifiable results. This can be achieved using a model proposed by Casola et al. [9]. Their evaluation process consists of a series of algorithms that convert a natural language of a written policy into a binary matrix. It then applies the Euclidean algorithm to calculate the distance between matrices to identify the differences between two security policies (quantifiable data) [9].

The combination of the above underlying theories are used to build the evaluation model proposed in this research.

30.3 Overview of Model Design

The model proposed in this paper, as shown in Fig. 30.1 below, describes a process via which a company's BYOD policy is evaluated against a set of security standards (referred as a reference policy). The evaluation process is a comparison that identifies the differences between the company's BYOD policy and the reference policy. The results of this comparison are non-ambiguous and measurable. The main components of this model are 1) the reference policy and 2) the evaluation process.

30.3.1 The Reference Policy

The natural language of a reference policy proposed in this paper, is built using known and established principles and methodologies defined by the McCumber's Cube methodology, Peltier's Topic-Specific Policy basic concepts, Wood's coverage matrix, and the NIST's recommendations for data

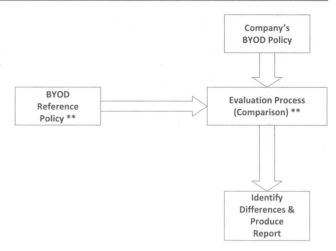

Fig. 30.1 High Level Model

protection when using BYODs. For example, McCumber's Cube methodology is used to ensure that the CIA of data is addressed for each security attribute. For example, data confidentiality during transmission needs to specify measures that include the use of technology (e.g. encryption/VPNs), policy specifics, and human factors that address the need for training/awareness.

Peltier's basic components (as explained in Sect. 2.2) are also included in the construction of the natural language of the reference policy. Then, Wood's coverage matrix is used to organize/format the policy to prepare for the series of transformations required during the policy evaluation process.

30.3.2 The Policy Evaluation Process

The BYOD policy evaluation process presented in this model uses a series of algorithms proposed by Casola et al. [9]. This method performs a number of transformations to convert a natural language of written policy into a binary matrix. It then applies the Euclidean algorithm to calculate the distance between the matrices to identify the differences between two security policies (quantifiable data). In this proposed model, Figs. 30.2, 30.3 and 30.4, show an example comparison between the BYOD reference policy and a company's BYOD policy. These figures show the end-result of this transformation process. The description of the transformation steps themselves (and the weights assigned) are outside the scope of this paper. The examples show the binary matrices created for the reference policy and the policy being evaluated. In this example, matrix R (Fig. 30.2) represents the reference policy, and matrix C (Fig. 30.3) represents a company's policy. The resulting matrix in Fig. 30.4 (after applying the Euclidean algorithm) shows a number value representing the distance between the two matrices.

1. **Let R be the matrix that represents the reference policy**

O	O	O	O
O	O	O	O
O	O	O	O
O	O	O	O
O	O	O	O
O	O	O	O
1	1	1	1
1	1	1	O
1	1	1	1
O	O	O	O
1	1	O	O
1	1	O	O

Fig. 30.2 Example binary representation of a reference policy

2. **Let C be the matrix that represents the company's policy**

0	0	0	0
0	0	0	0
0	0	0	0
0	0	0	0
0	0	0	0
0	0	0	0
1	1	1	0
0	0	0	0
1	1	1	0
0	0	0	0
1	0	0	0
1	0	0	0

Fig. 30.3 Example binary representation of a company's policy

Fig. 30.4 Example Euclidean's distance between matrices

Multiply (R-C)x[(R-C)^T] to find the Trace

⬚	O	O	O	O	O	O	O	O	O	O	O
O	⬚	O	O	O	O	O	O	O	O	O	O
O	O	⬚	O	O	O	O	O	O	O	O	O
O	O	O	⬚	O	O	O	O	O	O	O	O
O	O	O	O	⬚	O	O	O	O	O	O	O
O	O	O	O	O	⬚	1	O	1	O	O	O
O	O	O	O	O	O	⬚	O	1	O	O	O
O	O	O	O	O	O	O	⬚	1	O	O	O
O	O	O	O	O	O	1	O	⬚	O	O	O
O	O	O	O	O	O	O	O	O	⬚	O	O
O	O	O	O	O	O	O	1	O	O	⬚	1
O	O	O	O	O	O	O	1	O	O	1	⬚

So Trace is = Tr(R-C)(R-C)^T)= 1+3+1+1+1 =7

The smaller the value the closest the policies are to each other, indicating that the policy being evaluated is a strong BYOD policy. Likewise, if the distance is represented by a higher value, it indicates the policy being evaluated is weak. In the same manner, each policy provision can be calculated in order to identify the specific weaknesses and thus provide specific recommendations for the weak provisions.

In addition, a visual representation of each policy provision can be provided using a Kiviat's diagram. In Fig. 30.5, one can visually see the weak/strong provisions of a company's policy as they compare to the reference policy. In this example, policy R represents the reference policy and policy C represents the company's policy being evaluated.

Each policy provision (Kn) represents an item of the policy (e.g. provide confidentiality via encryption/VPN, etc). This Kiviat's diagram example shows that the policy being evaluated lacks strength in almost all its policy provisions.

Figure 30.6 below shows the high level model and added steps to reflect more detail. The blue boxes represent the incorporation of the steps for 1) building the reference policy, 2) the steps that involve the transformation from natural language to binary matrix, and 3) the final step of policy comparison (matrix distance calculation) where the results are produced. The creation and transformation of the reference policy is a one-time process. The repetitive process is the evaluation of a company's policy.

30.4 Model Demonstration and Evaluation

30.4.1 Artifact Demonstration

The model can be demonstrated in a case study that uses the natural language of an existing BYOD policy where the process is applied manually.

30.4.2 Artifact Evaluation

Measuring the success or failure of the model can be determined based on the answers to the following questions:

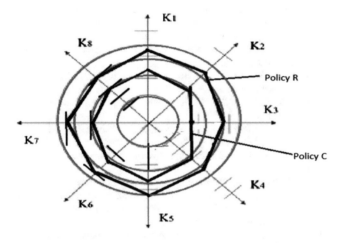

Fig. 30.5 Example Kiviat's representation of a policy comparison

Can this policy evaluation method identify the BYOD risks the company is not addressing in its policy?

Can the level of data exposure be effectively measured?

Does the company find the results of this evaluation useful and clear so that they can implement the necessary changes to their BYOD policy?

Is a reference policy based on the McCumber Cube/Peltier/NIST methodology suitable to create a generic/reference BYOD policy?

Is the reference policy a 'generic' and robust policy to use as an acceptable standard to measure BYOD policies for all size companies of multiple sectors?

Is it feasible/possible to automate the steps proposed by this model?

30.5 Conclusion and Future Work

30.5.1 Conclusion

Corporations need to address the vulnerabilities and security risks introduced when BYODs are allowed. In order to maintain control and mitigate the risks of data leakage/exposure there is a need to have a BYOD security policy in place. However, the policy may be weak. Therefore, an evaluation method that produces measurable results may provide the company with valuable information to strengthen their

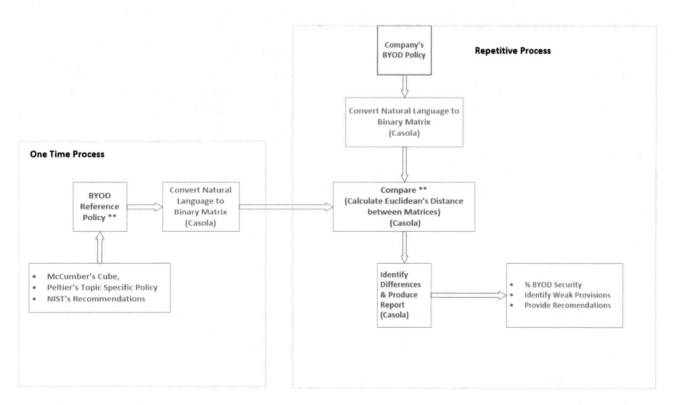

Fig. 30.6 Detailed model

policy hence strengthening the security of its corporate data when BYODs are allowed.

Current literature only provide guidelines to build BYOD policies. Current policy evaluation methods involve human intervention to analyze or parse policies written in a natural language (e.g. English) where comparisons are made against published guidelines. This process can produce ambiguous results based on subjective analysis (i.e. the opinion of an individual). This paper proposes a novel method to evaluate BYOD security policies. The process utilizes an evaluation process based on mathematical analysis that produces quantifiable measurements, security levels and identification of weak strategies.

Using design science research methodology, this work aims to find out if an evaluation model based on a BYOD reference policy built upon established theories such as McCumber Cube methodology, Peltier's basic concepts, and the NIST guidelines for BYOD security coupled with an evaluation process (such as the one presented by Casola) can be used to successfully evaluate a company's BYOD policy.

30.5.2 Future Work

The analysis and description presented thus far represent an initial work into this research. Extensive work is needed in order to build a BYOD reference policy that defines a wide-range of security rules for the reference policy for each main provision/requirement of the natural language of a policy while applying policy concepts based on McCumber's Cube, Peltier and Wood.

At the time of this writing, the intention of this research is to build a 'generic' (but comprehensive) BYOD policy evaluation model independent of the type of organization. However, it is possible to tailor the policy evaluation process by modifying the reference policy in order to meet an industry-specific security requirement(s).

If the model works successfully/effectively, it may be desirable to explore the possibility of automating certain steps of the model by using tools suitable for parsing text, matrix analysis and computation.

References

1. Souppaya, M., & Scarfone, K. (2013). Guidelines for managing the security of mobile devices in the enterprise NIST Special Publication 800-124 Revision 1.
2. Cisco's Technology News Site. (2012). *Cisco study: IT saying yes to BYOD*. Retrieved September 19 from https://newsroom.cisco.com/press-release-content?articleId=854754
3. BYOD Insights. (2013). *A cisco partner network study, report*. Retrieved September 2016 from http://www.ciscomcon.com/sw/swchannel/registration/internet/registration.cfm?SWAPPID=91&RegPageID=350200&SWTHEMEID=12949
4. Gartner. *Gartner predicts by 2017, half of employers will require employees to supply their own device for work purposes*. Retrieved August 31, 2016 from http://www.gartner.com/newsroom/id/2466615
5. Wang, Y., Wei, J., & Vangury, K. (2014). *Bring your own device security issues and challenges*. Consumer Communications and Networking Conference (CCNC), 2014 I.E. 11th, pp. 80–85.
6. Holleran, J. (2014). Building a better BYOD strategy. *Risk Management, 61*, 12–13.
7. Miller, K. W., Voas, J., & Hurlburt, G. F. (2012). BYOD: Security and privacy considerations. *IT Professional, 14*, 53–55.
8. Thompson, G. (2012). BYOD: Enabling the chaos. *Network Security, 2012*, 5.
9. Casola, V., Mazzeo, A., Maxxocca, N., & Vittorini, V. (2007). A policy-based methodology for security evaluation: A security metric for public key infrastructures. *Journal of Computer Security, 15*, 197–229.
10. Vorakulpipat, C., Polprasert, C., & Siwamogsatham, S. (2014). *Managing mobile device security in critical infrastructure sectors. Proceedings of the 7th international conference on Security of Information and Networks*, p. 65.
11. Kumar, R., & Singh, H. (2015). A proactive procedure to mitigate the BYOD risks on the security of an information system. *SIGSOFT Software Engineering Notes, 40*, 1–4.
12. Souppaya, M., & Scarfone K. (2016). *NIST 800-114 Rev 1 user's guide to Telework and Bring Your Own Device (BYOD) security*. Retrieved from http://csrc.nist.gov/publications/drafts/800-114r1/sp800_114r1_draft.pdf
13. Souppaya, M., & Scarfone, K. (2016). *NIST 800-46 Rev 2 guide to enterprise telework, remote access, and Bring Your Own Device (BYOD) security*. Retrieved from http://csrc.nist.gov/publications/drafts/800-46r2/sp800_46r2_draft.pdf
14. McCumber, J. (2004). *Assessing and managing security risk in IT systems: A structured methodology*. CRC Press. Boca Raton.
15. Peltier, T. R. (2016). *Information security policies,procedures, and standards: Guidelines for effective information security management*. Chicago: CRC Press.
16. Wood, C. C. (1995). Writing infosec policies. *Computers & Security, 14*, 667–674.

A Novel Information Privacy Metric

31

Aftab Ahmad and Ravi Mukkamala

Abstract

With the ever-increasing need for sharing data across a wide spectrum of audience, including commercial enterprises, governments, and research organizations, there is an equally growing concern about the privacy of such data. In this paper, we address a specific sharing subdomain where the data is made available only to a selected few and not for public access. Even under this constrained sharing, there are possibilities for privacy violations. The paper addresses issues of quantifying such privacy violations taking into account the degree of trust between the sharing parties. We employ an Embedded Privacy Agreement-based system to evaluate the privacy violation.

Keywords

Privacy agreement • Privacy metrics • Privacy preservation • Trust

31.1 Introduction

The right to privacy of individuals is among the constitutionally granted rights in the USA [1]. With the advent of a new culture based on pervasive information, this right needs an equivalent in the big data world, and curtailing data sharing is not a solution. Instead, the privacy of information and information resources needs to be embedded within data sharing, be it between organizations, individuals, devices, or any combination thereof. We term such an electronic communications system as being *privacy-aware*. For example, a privacy-aware data warehouse server (WS) receives data from one or more clients and disseminates it to one or more users under a sharing agreement, an embedded privacy agreement (EPA) [2].

The issue of data filtering for anonymization prior to publishing becomes complicated when releases of datasets are correlated by other parties. For example, if one data owner releases <I1, I2, S1> attributes of a data where I1 and I2 are quasi-identifiers and S1 is a sensitive attribute, and another publisher releases <I1, I3, S3>. A third-party that has access to both the released data can probabilistically construct <I1, I2, I3, S1, S2> tuples which may compromise individual privacy, since <I1, I2, I3> together may identify an individual with high probability.

In the context of data warehouse, newly released data may compromise the privacy of previously released data. To measure this impact, we need a metric for data privacy. In this paper, we introduce a new privacy metric based on the trust of a third-party with whom data is shared, the attributes shared, and the types of operations (Rwad/Write) allowed on that data. The paper is organized as follows. In Sect. 31.2, we discuss some of the existing privacy-preserving techniques and privacy metrics. In Sect. 31.3, we describe a EPA-based privacy model. Section 31.4 provides details of the proposed privacy metrics. Finally, Sect. 31.5 summarizes the contributions of the paper.

A. Ahmad
CUNY John Jay College of Criminal Justice, 524 W 59th Street, New York, NY 10019, USA
e-mail: aahmad@jjay.cuny.edu

R. Mukkamala (✉)
Old Dominion University, Norfolk, VA 23529, USA
e-mail: mukka@cs.odu.edu

31.2　Privacy Metrics

One can find several proposals in the literature for quantifying information privacy. In this section, we will describe a few popular techniques. Trust and privacy are closely related—a smaller value of trust (between data owner and a third-party) requires a larger extent of privacy measures. Trust determines the expectation component of privacy [3]. The complexity of privacy measure is further enhanced due to the subjectivity of goals in various fields, such as sociology, psychology, business and data sciences. There are algorithms that use clustering analysis for privacy determination [4].

In approaches that follow from information theoretic arguments, such as outlined in [5], the privacy metrics are probabilistic. This work takes the previous work of using average mutual information to the next level, deriving the impact of one sample on mutual information of an ensemble of data. By next level, we mean determining the impact of the disclosure of a single record on the data ensemble called j-measure and the impact of ensemble disclosure on a single attribute called the i-measure [5].

In the absence of a robust theory for measuring information privacy, the route from k-anonymity to t-closeness tells a familiar story, namely, we solve one problem and overlook another. In this case, the generalization of an attribute, say zip code, in k-anonymity does not preclude the possibility of a classification instance with only one dataset for a sensitive attribute. The inclusion of l-diversity guarantees against such an occurrence but may do so at the cost of substantial changes in statistical properties that can lead to statistical inference about individual identity. The t-closeness algorithms added to data generalized already via the k-anonymity and l-diversity algorithms could easily make the published data useless for research and analysis for which it is published. Therefore, privacy of the published data is not a measure that can be studied in isolation without regard to its utility. Just like t-closeness, the differential privacy measures strive to make sure that disclosure of an individual record does not have detrimental effect on the privacy of the rest of the data [6].

The k-anonymity has many extensions. However, (k, l)-diversity reported in [7] is a different graph theory based metric. Another privacy metric, proposed in [8], is based on information theoretic argument of assigning probabilities to each stage of an attack graph. The underlying assumption is that an attacker needs system (private) information. The *mean privacy* metric is defined as the average information needed by an attacker for launching an attack. The main weakness of this work is that privacy does not have to be provided in relation to attacks. As pointed out in the ISACA Whitepaper [9], privacy is highly individual specific and going to be a driving force for the big data. It should be an inherent property for data generated and managed by enterprises. The Pufferfish proposed in [10] seem to assimilate a general framework, in which a data owner can assign privacy metrics to datasets arbitrarily. The Bayesian differential probability model [11] also provides similar insights into what happens when correlated data is shared by independent data owners.

31.3　EPA-Based Privacy Model

In this paper, we look at measuring privacy in an environment where data release is restricted. In other words, data is not published publicly. Instead, data release and access is controlled. This is similar to the way files are accessed in a file system where access control restricts access to files based on the owner's intentions.

In a typical situation, when a data file is shared with a client with certain stipulations on its use and further sharing, there could be violations of the provisions. The probability and degree of sharing-agreement violation will depend on, among other things, the reliability or trust that one can have on that client. Even in day-to-day dealings, we tend to convey confidential information to those that we trust to keep it secret, and have only vacuous conversations with those that we don't trust. So our decision to share or not to share a piece of information with an individual depends on the trust that we have on him/her and the degree of confidentiality required for that piece of information. In this paper, we attempt to quantify privacy of a piece of information based on the trust values of clients that the information is shared with and the operations that they are allowed on the data.

We use EPA or Embedded Privacy Agreement as a means to express a data owner's intentions with respect to sharing of a piece of data. An EPA is a mechanism to express shareability conditions between two parties on a piece of data. While a Global EPA (or data-level EPA) has conditions that are applicable to all fields of the data, a Local EPA (or attribute-level EPA) has field (or attribute) specific conditions. When a data owner shares its data with a client C1 with EPA1, and further C1 shares this data with another client C2 with EPA2, and so on, we may have a tree of EPAs. Obviously, the usability and shareability conditions of a client Cn will not only depend on EPAn, but also on the chain of EPAs from EPA1 to EPAn. We refer to them as EPAs at different levels.

31.3.1　Specification of Embedded Privacy Agreement

Any data in a repository can have two fundamental operations: READ and WRITE. These values can be assigned for each attribute in the form of a tuple (R, W), where R is for READ operation and W for WRITE. R, W can be a 1 or a 0. If read operation is allowed then R = 1, else R = 0. For a data with N fields or attributes, its EPA specifies the type of the attribute—Identifier, quasi-identifier, or sensitive attribute. For identifier j, we denote by $(0,0)_j$ the (R,W) tuple if it should not be shared at all, $(1, 0)$ if it can be read but not be written into, $(0, 1)$ if it can't be read but written into, and $(1, 1)$ if it can be read or written into or both. In the following, we describe some properties of this notation.

31.3.1.1 Fundamental Axiom of Privacy
We can't build a machine with (R, W) = (0, 0) for all data.

Explanation The ability of a machine to protect R and W operations ultimately determines the privacy provided to that data stored and/or processed by it. We can build a machine that does not allow writing to it and use it for one-time writing and one or more reading operations. However, it is no use of having a machine that does not allow reading from it. This axiom restricts the discussion to machines with a utility > 0. A machine that does not allow reading or writing of data can be considered as having a utility of 0. Such a machine is not required to be analyzed as it does not exist in terms of utility.

Corollary Even though we can't have a machine with property (R, W) = (0, 0), we can have an attribute or a dataset with this property. Obviously, the utility of a machine will not suffer if only part of its data can't be read or written.

31.3.1.2 Operation Concatenation
The sum of (R, W)'s for a given attribute in a Global and Local EPA is privacy protective. This implies:

$(0, 0)^{Global} + (x, y)^{Local} = (0, 0)$
$(0, 1)^{Global} + (x, y)^{Local} = (0, y)$
$(1, 0)^{Global} + (x, y)^{Local} = (x, 0)$
$(1, 1)^{Global} + (x, y)^{Local} = (x, y)$

Where a superscript $(.)^{Global}$ implies the tuple is for Global EPA. In other words, if α^{Global} is an R or W operation for an attribute in Global EPA and α^{Local} is the value of the same operation for the same attribute in a local EPA then, the sharing of that attribute is done with a value of $\alpha^{Global}\alpha^{Local}$.

31.3.1.3 Lemma 1
The sharing of an attribute at level k in an EPA tree depends on the sharing tuple of all $(k-1)$ levels before level k.

Proof Starting from level 1, if every two levels are considered as a concatenation just like a global and local EPA, then applying Property 2 follows the Lemma.

31.3.2 Embedded Privacy Agreement (EPA) Matrix

The EPA matrix consists of N columns and 4 rows. The rows are for the (R, W) combinations. An entry of 1 corresponds to the (R, W)-tuple for that attribute being allowed while an entry of 0 means that the corresponding (R, W)-tuple is not allowed. Letting $\{a_n\}$ be the attributes, the EPA matrix A has the following general form:

$$A = \begin{array}{c} \\ (0,0) \\ (0,1) \\ (1,0) \\ (1,0) \end{array} \begin{bmatrix} a_0 & a_1 & \cdots & a_{n-2} & n-1 \\ \alpha_{00} & \alpha_{01} & \cdots & \alpha_{0,n-2} & \alpha_{0,n-1} \\ \alpha_{10} & \alpha_{11} & \cdots & \alpha_{1,n-2} & \alpha_{1,n-1} \\ \alpha_{20} & \alpha_{21} & \cdots & \alpha_{2,n-2} & \alpha_{2,n-1} \\ \alpha_{30} & \alpha_{31} & \cdots & \alpha_{3,n-2} & \alpha_{3,n-1} \end{bmatrix} \quad (31.1)$$

Where $\alpha_{ij} \in (0, 1)$, a 1 meaning that (R, W) tuple is part of EPA and 0 meaning that it is not. This is for the jth attribute.

Example Let a dataset consist of four attributes $\{a_0, a_1, a_2, a_3\}$, with no permission to read or write a_0 *(column 1)*, permission to read but not write a_1 and a_2 (columns 2 and 3), and permission to read and write a_3 (column 4). Then, the EPA matrix A for such an agreement is as follows:

$$A = \begin{bmatrix} 1 & 0 & 0 & 0 \\ 0 & 0 & 0 & 0 \\ 0 & 1 & 1 & 0 \\ 0 & 0 & 0 & 1 \end{bmatrix}$$

In the following, we will describe a theorem about the EPA at level k.

31.3.2.1 Theorem 1
The EPA at level k in the EPA tree is the Hadamard product of the matrices A_i i = 1, 2, ..., k-1, where A_i is the EPA matrix at level i

Proof Using Property 2 of the EPA and the definition of EPA matrix, theorem 1 follows.

Theorem 1 is also the specification of the EPA at level k in the EPA tree.

Interpretation of Theorem 1
Theorem 1 exploits the property of the EPA matrix that an entry is 1 only if an (R, W) tuple is allowed, otherwise the entry is zero. The Hadamard product provides a mechanism to stop propagation of an entry at the next level if it is not allowed at a current level by multiplying the corresponding entry with a zero. The net effect of this specification of EPA is that the original data owner/source is always in control of sharing at any level and when the (R, W) tuple is given a value of zero at any point in an agreement chain, it stops propagating from that point onward.

31.3.3 Privacy Violation

Data is made available to a client along with the respective EPA. Privacy violation may occur after this at the client depending on its trust value. Let us denote trust for a sharing party j by τ_j. Then, for sharing with p parties, the probability of violation $\pi = f(\tau_0, \tau_1, \ldots, \tau_{p-1})$. Following example illustrates this concept: Alice has organized her Facebook friends into n groups based on her relation to each and their relation to other anonymous and non-anonymous users. She is going on a trip and wants to share the news with a group called Close FRiends, but does not want the news to go across her other groups. What is the probability that there will be a violation?

In this scenario, everybody in a group will have the same value for (R, W), likely (1, 1) if private information is shared and the friends can comment on it too. Suppose, the Close Friends group has p members. Also, suppose that friend j in this group has a chain of c_j of friends that leads to one of the Alice's friends in other groups (from which Alice is hiding the news). What is the probability of violation in this case?

In the above case of Alice's travel plans, we see that even if she has the same trust level with all the p friends in a group, violation can take different routes, thus changing the actual value of violation. This illustrates the difficulty in modeling trust in concatenation of data users. In the next section, we describe our approach to measure the privacy in such contexts.

31.4 Privacy and Trust

In this section, we will derive a violation probability for an EPA with a trusted user. Current data sharing systems do not implement a privacy agreement that accompanies the data. Once data is shared, it is not in the control of data owner. Social networks are an example of such systems where a data owner can make trust groups for sharing various types of data, but once data is shared, the owner can't exert any control on its propagation.

Suppose a data owner shares data with p users, each with an EPA and a trust level (expressed as probability). Let τ_j be the trust level between the owner and user j. Let the same EPA be used for sharing data, as shown in Fig. 31.1. In the following, we will derive a privacy violation probability for such case.

31.4.1 Trust in Parallel

In the above example, Alice's connection for sharing is a parallel connection. Violation in any one of the connections is a violation of privacy. If v_i is the probability of violation,

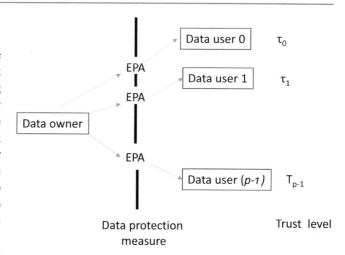

Fig. 31.1 Relation between EPAs and trust

we can say that $v_i = 1 - \tau_j$. The probability of no violation is $1 - \Pi_p (1 - v_i) = 1 - \Pi_p \tau_j$. *We define the log of inverse of the probability of no violation as the amount of privacy inherent in the given connection.*

$$Privacy = -\log\left(1/\left(1 - \prod_p \tau_j\right)\right) \qquad (31.2)$$

From the example of Alice, we see that her information has more inherent privacy if the trust level is high. Ideally, she would like to have a privacy value of infinity. In general, if $(1 - \tau)$ can be thought of as *mistrust* then its inverse is interpreted as Privacy in Eq. (31.2). The use of logarithm provides a unit of measurement. With the log of base 2, the Privacy can be represented as the number of bits.

31.4.2 Trust in Tandem

In the case of a connection to a party that may have multiple parties connected to it in series, we assume that the trust value τ_j ascribed to the immediately connected party describes accurately the complete connection.

31.4.3 Example of Privacy Given Trust

In the social network, suppose, Alice is unable to assign trust values and assumes that $\tau_j = 1$ for everyone. According to Eq. (31.2) the Privacy content of Alice's message is ∞. Alternatively, if, due to lack of a knowledge of trust in her friends Alice assigns a 0 to all trust values, then the Privacy content according to Eq. (31.2) is 0. Lastly, suppose that Alice trusts her friends but is realistic about them being connected to other people and assigns a constant value of $\tau < 1$ to all of them. Then, the amount of privacy according to Eq. (31.2) is given by the following:

Fig. 31.2 A plot of trust versus number of sharers for various values of trust

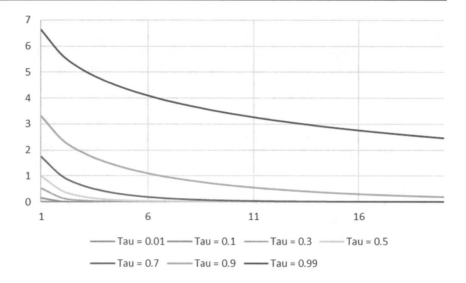

Fig. 31.3 A communications channel like depiction of R/W operations

Assuming an event "Read" occurred in an information system. Let:
ε_R = Probability that this was a violation of privacy, then
Similarly, if an event "Write" occurs, let
ε_W = Probability that this was a violation of privacy, then

$$Privacy = -\log(1/(1 - \tau^p)) \qquad (31.3)$$

Equation (31.3) is plotted for various values of p and τ (Fig. 31.2).

31.4.4 Imperfect Channel

In the expression of Privacy in Equation (31.2), there is the implicit assumption that the assigned values of (R, W) do not change. In practice, the privacy of information may be affected by a channel that is leaking information. Erroneous channel behavior and trust are two different concepts. In terms of digital communications systems, if (R, W) is the transmitted signal, chosen from an EPA, and (r, w) is the received signal then an error occurs if R = 0 but r = 1 (and/or W = 0 and w = 1). Figure 31.3 above shows a communications channel modeling of R and W:

If ε_R is the probability of an R going from 0 to 1 and ε_W be the probability of a W erroneously changing from 0 to 1 than the privacy violation due to channel errors is given by:

$$\varepsilon = 1 - (1 - \varepsilon_R)(1 - \varepsilon_W) = \varepsilon_R + \varepsilon_W - \varepsilon_R \varepsilon_W \qquad (31.4)$$

In a data record with N attributes, if ε_k is the privacy violation for the kth attribute then the privacy violation of the record is $[1 - \Pi_k (1 - \varepsilon_k)]$, where Π_k is the product over all values of k. If ε_l's are small then this expression is approximately $\Sigma_k \varepsilon_k$ and for equal values of ε_k's $= \varepsilon$, it reduces to $N\varepsilon$. In our analysis, we use ε and assume that it can be applied to a single attribute as well as a data record.

Combining ε and τ

Now that we have expressions for privacy violations due to (mis)trust and (erroneous) channel, we propose that *the log sum provides the total privacy inherent* as follows:

$$Privacy = \log\frac{1}{\varepsilon(1 - \Pi\tau)} = -\log(\varepsilon) - \log(1 - \Pi\tau)\text{bits}$$

$$(31.5)$$

Equation (31.5) is a measure of total privacy in an information sharing transaction. It consists of two parts, one in wake of the (privacy) channel errors defined by changes in (R, W) and the other inherent privacy due to the imperfect trust relation. Equation (31.5) does not make any assumptions about the source of evaluation of trust or (R, W). In our work, we assume that (R, W) is provided by data owner to the data dissemination administrator for each attribute to implement the embedded privacy agreement (EPA). The trust values can be provided by a government agency, another agency, such as BBB, the data owner or even the data server.

Privacy measures can be based on the value of Eq. (31.5). In a nut shell, if privacy value is high, there is no need to spend a lot for adding more measures.

Equation (31.5) readily provides a value of zero for public domain where $\varepsilon = 1$ and $\tau = 0$. On the other extreme, for a machine that can be used only for reading provides a privacy that depends on ε_R and τ_I's.

31.5 Conclusion

In this paper, we have provided a way of specifying the embedded privacy agreement and proposed a performance model for information privacy. The specification is done at an attribute level by assigning (R, W)-tuple to each identifier, at the EPA level. We also propose to define privacy in terms of number of bits and divide it into three parts, at the attribute level it is like a channel reliability. At the EPA level, it is like a service level agreement (SLA) and should be taken care through a robust negotiation protocol. At the trust level, it should depend on the trust with the immediate data users. The reason for not considering cascade of trust is that the Internet consists of autonomous systems and there is no rigorous way of ascertaining trust beyond the boundary of two adjacent autonomous systems. We have combined the privacy levels at (R, W) and trust level to come up with a privacy measure in terms of number of bits. In future, we will investigate how this number can be employed in determining optimal privacy controls to be taken for preservation of data privacy when it is shared.

References

1. Turley, J. (2001). Registering publius: The Supreme Court and the right to anonymity. *Cato Supreme Court Review, 57,* 657–683.
2. Mukkamala, R., Ahmad, A. & Nvuluri, K. (2016, June). *Privacy-aware Big Data Warehouse Architecture.* IEEE International Congress on Big Data, San Francisco.
3. Ahmad, A., et al. (2017). Information privacy domain. To appear in the *International Journal of Information Privacy, Security and Integrity.* http://dx.doi.org/10.1504/IJIPSI.2016.082124
4. Sim, K., Gopalkrishnan, V., Zimek, A., & Cong, G. (2013). A survey on enhanced subspace clustering. *Data Mining and Knowledge Discovery, 26*(2), 332–397.
5. Bezzi, M. (2010). An information theoretic approach for privacy metrics. *Transactions on Data Privacy, 3*(3), 199–215.
6. Dwork, C., & Lei, J. (2009, May). Differential privacy and robust statistics. In *Proceedings of the forty-first annual ACM symposium on Theory of Computing* (pp. 371–380). ACM. Bethesda.
7. Trujillo-Rasua, R., & Yero, I. G. k-Metric antidimension: A privacy measure for social graphs. *Information Sciences, 328,* 403–417. 2016.
8. Almasizadeh, J., & Azgomi, M. A. (2014). Mean privacy: A metric for security of computer systems. *Computer Communications, 52,* 47–59.
9. Bojilov, M., Chew, R., Kaitano, F., & Zororo, T. (2013). *ISACA: Privacy & big data* (Whitepaper). Rolling Meadows.
10. Kifer, D., & Machanavajjhala, A. (2014). Pufferfish: A framework for mathematical privacy definitions. *ACM Transactions on Database Systems (TODS), 39*(1), 3.
11. Yang, B., Sato, I., & Nakagawa, H. (2015, May). Bayesian differential privacy on correlated data. In *Proceedings of the 2015 ACM SIGMOD international conference on Management of Data* (pp. 747–762). ACM. Melbourne.

An Integrated Framework for Evaluating the Security Solutions to IP-Based IoT Applications

Gayathri Natesan, Jigang Liu, and Yanjun Zuo

Abstract

As Internet of Things (IoT) applications have taken the center stage of the technology development recently, the security issues and concerns arise naturally and significantly due to IoT connection to the anonymous and untrusted internet. Although the security protocols and technology for the internet applications have been studied for decades, the ubiquity and heterogeneity of IoT applications present unique challenges in handling security issues and problems. In addition to developing new protocols or to upgrade the existing protocols, some research has been done in experimenting security approaches for IoT applications. Since the results of the current research are mainly based on either the related protocols or the applicable approaches, the results of the discussion is often limited to a particular environment or a specific situation. In this paper, based on a thorough study on the existing research accomplishment and published experiment results, an integrated framework is proposed for evaluating the security solutions for IP-based IoT applications with the considerations in hardware constraints, operational constraints and network scenarios. The results of the study shows the potentials in drawing a balanced view in evaluating the security solutions to IP-based IoT applications and laying a step-stone for the further standardization of related IoT protocols and approaches for the security issues.

Keywords

Internet of Things • Device Constraints • Security • DTLS • HIP

32.1 Introduction

The Internet of Things (IoT) is a novel paradigm that is rapidly gaining ground in the scenario of modern wireless telecommunications. The basic idea of this concept is the pervasive presence around us of a variety of things or objects – such as Radio-Frequency IDentification (RFID) tags, sensors, actuators, mobile phones, etc. – which, through unique addressing schemes, are able to interact with each other and cooperate with their neighbors to reach common goals [1]. In coming years, IoT applications would be the foundation of many services and our daily life will depend on its availability and reliable operation. The ability to connect, communicate with, and remotely manage an incalculable number of networked, automated devices via the Internet is becoming pervasive, namely from industrial automation to home automation and smart cities.

The IoT architecture is currently focusing on interoperability, scalability, technology reuse and modularity. Since these focusing pointers have been well studied and developed under the Internet environment, recent building control networks for IoT employ mostly IP-based standards to gain a seamless control over the various nodes. In addition, IP-based IoT devices have the advantage in easily adding

G. Natesan • J. Liu (✉)
Metropolitan State University, St. Paul, Minnesota 55106, USA
e-mail: Jigang.Liu@metrostate.edu

Y. Zuo
University of North Dakota, Grand Forks, North Dakota 58202, USA

© Springer International Publishing AG 2018
S. Latifi (ed.), *Information Technology – New Generations*, Advances in Intelligent
Systems and Computing 558, DOI 10.1007/978-3-319-54978-1_32

and flexibly modifying applications to the devices without changes required on gateways. Moreover, by taking the existing implementation of the proven IP protocols, the development time for IoT applications is reduced, the inter-operability between IoT applications and the internet is established, and the availability of IPv6 with large addressing space to handle ginormous IoT is achieved [19].

While allowing for attractive integration, this shift towards IP-based standards for IoT applications results in new requirements regarding the implementation of IP security protocols on constrained devices as well as the management of the bootstrapping of security keys for devices across multiple manufacturers. Besides, direct usage of Internet protocols for IoT applications leads to either an inefficient operation or an insecure operation as inter-operability becomes a problem when the security solutions are tailored to meet specific IoT application purposes [2]. To support inter-operability while tailoring the existing security solutions, it is obviously needed to establish a framework for standardizing IP-based security protocols for IoT applications.

As the existing studies on security solutions to IP-based IoT applications are more emphasizing on either the application level or on the lower levels of protocols, there is disconnection among the protocols and actual devices along with the security solutions designed for them. In this study, we propose an integrated framework for evaluating the security solutions to IP-based IoT applications. The fundamental idea of this framework is to analyze the security solutions with a three-dimensional perspective, namely the dimensions of protocols, approaches, and IoT devices. In the next section, a background study is provided. The new framework is introduced in Sect. 32.3, while the practical analysis of the framework is discussed in Sect. 32.4, and finally, in Sect. 32.5, the conclusions and the future work are presented.

32.2 Background

Ongoing standardization efforts toward harmonizing Internet protocols for Wireless Sensor Networks (WSN)-based IoT applications have raised hopes of global interoperable solutions at the transport layer and below [9]. Nowadays, there exists a multitude of control protocols for IoT applications. For constructing BAC (Building Automation and Control) systems, the ZigBee standard [ZB], which is an IPv6-based protocol, plays a critical role. Recent trends, however, focus on an all-IP-based-protocol approach for IoT system controls. As a result, IP-based protocols, such as CoAP and 6LoWPAN, have made a leaping progress toward the standardization. The Constrained Application Protocol (CoAP) runs over UDP, not TCP, so that it enables efficient application-level communication for IoT applications, while the 6LoWPAN, which stands for "IPv6 over Low-power Wireless Personal Area Networks," concentrates on the definition of methods and protocols for the efficient transmission and adaptation of IPv6 packets over IEEE 802.15.4 networks. For instance, ZigBee Network, as shown in Fig. 32.1, is a good example in demonstrating how IPv6 protocols are built into IoT devices.

Where in Fig. 32.1, green circles represent ZigBee End Devices (ZED), blue circles indicate ZigBee Routers (ZR), and the red circle shows ZigBee Coordinator (ZC). Each ZED has a limited resource and function, and can communicate with either ZR or ZC, while a ZR can process the information obtained from either ZEDs or other ZRs in terms of receiving data and sending signal message. ZC has the most capability in the network in processing the information received from ZRs or data from ZEDs for it own process or the server requests.

Fig. 32.1 IEEE 802.15.4/ ZigBee Network [6]

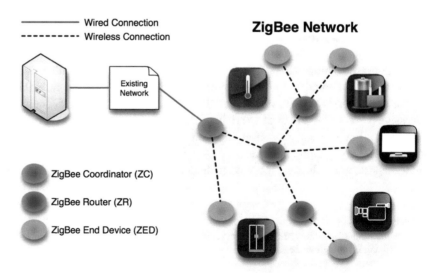

One of its main tasks is the definition of a lightweight version of the HTTP protocol, the Constrained Application Protocol (CoAP) that runs over UDP, not TCP, so it enables efficient application-level communication for IoT as shown in [16]. From the security perspective, proper recommendations on choice of security protocols to the working group can ensure that right security solutions will be in place for IoT in future.

Unquestionably, the main strength of the IoT idea is the high impact it will have on all aspects of everyday-life and behavior of potential users. From the point of view of a private user, the most obvious effects of the IoT applications will be visible in both working and domestic fields. In this context, home automation, assisted living, e-health, and enhanced learning are only a few examples of possible application scenarios in which the new paradigm will play a leading role in the near future. Similarly, from the perspective of business users, the most apparent consequences will be equally visible in fields such as automation for industrial manufacturing, logistics, business/process management, intelligent transportation of people and goods. Obviously, security is a must to make IoT applications acceptable [3] to the users and for that we need a secure architecture in which all device interactions are protected, from joining an IoT network to the secure management of keying materials. To use a simplified OSI network model, Fig. 32.2 below illustrates the comparison of related protocol stacks for Wi-Fi as well as for IoT (6LoWPAN).

Two of the biggest challenges that IoT applications face are the presence of low-powered devices, which need to function for months or years without getting any power recharge, and frequent data exchanges over lousy networks. These unique characteristics and challenges make the as-is use of the existing IP protocols less than ideal and suboptimal [19]. Different standardization bodies and groups are active in creating more interoperable protocol stacks and open standards for the IoT applications. While the IoT is an ambitious paradigm that significantly increases the scale of connected devices, it is highly motivated to consider reusing pre-existing technologies seamlessly with newer and more efficient technologies in managing the complexity and capacity created by IoT applications [21]. Although standardization of IP based IoT protocols has made a leaping progress with the designing of CoAP and 6LoWPAN, there is a stalling in TCP/IP layer for standardization as the IoT applications differ with respect to the security requirements from "no security" to "full blown security" based on the device constraints and operational requirements.

An End-to-End (E2E) protocol provides security even if the underlying network infrastructure is only partially under the user's control. As the infrastructure for Machine-to-Machine (M2M) communication is getting increasingly commoditized, this scenario becomes more likely as the European Telecommunications Standards Institute (ETSI) is currently developing a standard that focuses on providing a "horizontal M2M service platform" [7] to standardize the transport of local device data to a remote data center. For stationary installations security functionality could be provided by the gateway to the higher level network. However, such gateways would present a high-value target for an attacker. If the devices are mobile, for example in a logistics application, there may be no gateway to a provider's network that is under the user's control, similar to how users of smartphones connect directly to their carrier's network. These situations drive the need for end-to-end security solutions.

32.3 An Integrated Framework

In this section we propose an integrated framework in evaluating security solutions to IoT. This integrated framework analyzes the security solutions to IoT from three different dimensions in the IoT devices, the possible approaches, and the existing standards and protocols.

The study of the IoT devices involves the discussion in physical constraints, operational constraints, and network scenarios. As discussed previously, any object in the physical world, a thing, can be connected to the internet by attaching a Micro-Controller Unit (MCU), and works as

Fig. 32.2 Comparison of the related protocol stacks [16]

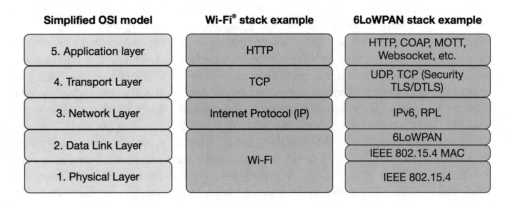

Simplified OSI model	Wi-Fi® stack example	6LoWPAN stack example
5. Application layer	HTTP	HTTP, COAP, MOTT, Websocket, etc.
4. Transport Layer	TCP	UDP, TCP (Security TLS/DTLS)
3. Network Layer	Internet Protocol (IP)	IPv6, RPL
2. Data Link Layer	Wi-Fi	6LoWPAN
		IEEE 802.15.4 MAC
1. Physical Layer		IEEE 802.15.4

either an actuator or a sensor. Those constrained devices exhibit low-power CPUs with few kilo bytes of memory for data and code, and mostly communicate wirelessly, whereas Gateways (GWs) and border routers that connect a Wireless Sensor Network (WSN) to another network, such as the Internet, may communicate over wire [22]. Since IoT devices are very diverse in equipped resources, IETF (Internet Engineering Task Force) proposes a classification of constrained devices based on the size of memory and communication capability [4]. There are three classes defined by the IETF standard. Class 0 defines the IoT devices that have limited memory and cannot communicate directly to the internet while devices classified in class 1 have sufficient memory to hold a tailored IP stack for communicating with the internet. Class 2 is dedicated to the most powerful IoT devices that have large memory for more functionality and capability of using an existing IP stack for their internet connection. Energy is another vital constraint as most of the things might be battery based. This makes the two requirements in memory size and energy capability the most influential factors to evaluate the suitability of the security solutions.

Since constrained devices typically operate in low-power IP networks due to their limited power, memory and computational resources, the reliability of the communication is decreased due to the increase of the loss of packets. Additionally, with increased packet losses, the performance of the system is decreased due to the increase of time and resource spent on the security handshakes in network communication [18]. This is an operational problem caused by constrained devices. So the security solutions to IoT applications should be evaluated for suitability based on those operational constraints in terms of packet loss and handshake frequency.

The approaches used for solving security problems in IoT can be categorized into four groups, hardware-based, delegation-based, centralized, and tailoring-based. In hardware-based approaches, some additional hardware security modules are used. For instance, a Trusted Platform Module (TPM) is a tamper-proof hardware that is used for providing cryptographic computations of public-key-based primitives. A TPM allows for a better key protection, performant disk encryption, and remote attestation of different software modules [5]. A fully implemented two-way authentication security scheme proposed in [13] is a good example in showing how TPMs are used in solving security programs for IoT applications. The system architecture for this proposed scheme is implemented based on a low-power hardware platform suitable to IoT. The authors demonstrated how the TPM enabled publishers can perform a fully authenticated handshake with the subscriber, which acts as the DTLS (Datagram Transport Layer Security) server in this handshake. One of the critical implementation is to allow TPMs to hold keys, such as RSA private keys, in a

protected memory area. Moreover, TPMs have also studied in [10] for the feasibility and applicability in the area of WSNs (Wireless Sensor Networks).

Instead of employing TPMs, intensive tasks in public-key-based session establishment can be delegated from IoT end nodes to some more powerful devices. One such delegation approach is proposed in [11], where a delegation architecture is designed to offload the expensive DTLS connection to a delegation server so that IoT end nodes no longer need to implement expensive public-key cryptography for the connection establishment.

Although the Public-Key-Infrastructure (PKI) is heavily used in IoT security solutions for building the trust between the devices, symmetric-key-based approaches are also implemented in securing E2E (End-to-End) communications for promoting efficiency. While a constrained node must be pre-configured with the shared keys of all entities, a centralized key-distributor is required in sending the shared key to all the constrained nodes for pre-configuration [17]. In addition to DTLS, HIP (Host Identity Protocol) is also implemented and evaluated for key management in [8]. Their experiments demonstrate that although DTLS allows for easier interaction and interoperability with the Internet, HIP performs better due to its smaller memory footprint, less communication overhead and delay and resiliency to packet loss.

Internet of things not only present the constraint over the resources they have, they also challenge the existing security protocols in terms of their needs. To accommodate this situation, some of the existing protocols need to be tailored to meet the requirements. One of the examples is the HIP-DEX (Diet-EXchange) discussed in [12]. As a lightweight protocol extension to HIP, the authors illustrated how HIP-DEX can achieve the security goal by tailoring HIP with a comprehensive session resumption mechanism and an adaptive retransmission mechanism.

In addition to the constraints on IoT operation and devices, the layout of the IoT networks should also be considered in the evaluation of the security solutions to the IoT applications. As discussed in [13], a constrained network interconnected through a gateway to other networks may present four possible communication patters: (1) intra-domain with different nodes; (2) intra-domain with a device; (3) inter-domain to other constrained networks; (4) inter-domain with a host service in the Internet. Since the key distribution varies among different patterns of the network layouts, the reliability and the performance of the IoT applications vary from the choice of the security solutions.

Although there is a close relationship between approaches and protocols, the choice of protocols plays a critical role in evaluating the reliability and performance of a chosen security solution due to the inherent features in each protocol. For instance, DTLS supports both pre-shared keys

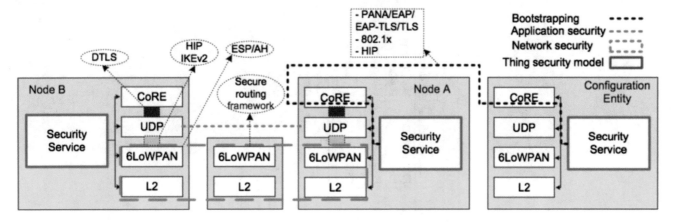

Fig. 32.3 Relationship between IP-based security protocols [9]

and public key based handshakes [20]. The selection of the handshakes in DTLS impacts the resource allocation and time utilization of the security solution. The relationship between various IP-based security protocols can be well illustrated in Fig. 32.3 below.

On the other hand, if the type of the attacks is the main focus, such as DoS (Denial-Of-Services), the selection of the protocols is critical [5, 14].

32.4   Discussion

Based on the previous discussion, the security solutions to IoT applications should be evaluated from three different perspectives in the device constraints, operational constraints, and network scenarios. The summary of the discussion is provided in Table 32.1 below. The comparison is done in such a way in which the most recent security solutions are considered under each category of the approaches discussed and with different protocols under each approach because this helps us to compare how the same protocols under different approaches are catering to different requirements of various IoT devices. Moreover, the results provided in Table 32.1 are based on the suitability for IoT applications with an assumption of a class 1 device with 8 kB of RAM, 64 kB of ROM, and an IEEE 802.15.4 radio interface.

In view of memory consumption, Table 32.1 shows that the delegation architecture (D1) imposes minimum memory requirements because public-key-based operations involved in session establishments are delegated to a more powerful device. Centralized architectures using DTLS (C2) and HIP PSK (C1) are also suitable for memory constrained IoT. The DTLS handshake is more complex and thus it involves the exchange of more messages than in HIP. Further, DTLS runs on the transport layer, i.e. UDP, whereas HIP is carried directly over the network layer, i.e. 6LoWPAN. This directly

increases the overhead due to lower layer per-packet protocol headers. IKEv2 (C3) requires a much higher footprint when compared to HIP and DTLS. The hardware based and tailored protocol approaches may be suitable for case 1 devices but would not be suitable for case 0 devices.

Although the energy consumption can only be compared among the different hardware-based approaches, this attribute is listed in the table to emphasize the importance of the energy consumption in evaluating security solutions to IoT applications. Some observations can be drawn based on the three perspectives discussed previously from Table 32.1. First, it demonstrates the difference between the solutions with various memory foot-prints and different handshaking frequencies with the support facts provided in [15, 20]. Next, from a domain communication point of view, all the approaches support intra-domain communication while all the approaches but centralized support inter-domain communication since centralized approaches do not address the inter-domain communication. Thirdly, due to the requirement for a trusted device, the hardware-based approaches (H1) are the only approaches addressing the mobility requirements for IoT applications. Lastly, due to the vulnerability of IoT applications, all the approaches provide a way to mitigate DoS attacks. Also included in Table 32.1 are other characteristics in session resumption, handshaking type, and trusted entity, which are valuable to be used in evaluating the security solutions to IoT applications.

In general, the results in Table 32.1 indicate that it is not sufficient to consider only either protocol, or approach, or device in evaluating security solutions to IoT applications because all those three areas have impact on the suitability of a particular solution. In other words, an integrated framework that combines the consideration of hardware constraints, operational constraints, and network scenarios should be adopted for evaluating security solutions to IoT applications. For instance, although DTLS has been widely employed in many security solutions, the solutions

Table 32.1 Comparison of security solutions to IoT applications

Approach	Protocol	Device constraints			Operational constraints		Network scenario				Other characteristics		
		RAM (kb)	ROM (kb)	Energy (mJ)	High packet loss	Freq hand-shake	Inter domain	Intra domain	Mobility	DoS support	Session resumption	Hand-shake type	Trusted entity
D1	DTLS	1.45	8.99	NC	✓	✓	✓	✓	NC	✓	✓	PKC certificate based	Domain manager, Device
C1	HIP	1.7	12.9	NC	✓	✓	×	✓	NC	✓	NC	PSK	Domain manager, Device
C2	DTLS	3.9	16.15	NC	✓	✓	×	✓	NC	✓	NC	PSK	Domain manager, Device
C3	IKEv2	9	46.6	NC	×	×	×	✓	NC	✓	NC	PSK	Domain manager, Device
H1	DTLS	18	63	936.4	×	×	✓	✓	✓	✓	NC	PKC	Device
T1	HIP	7	63	NC	×	×	✓	✓	NC	✓	✓	PKC	Device

Suitability for IoT assuming a class 1 device with 8 kB of RAM, 64 kB of ROM, and an IEEE 802.15.4 radio interface

✓ – Suitable , × – Not Suitable, *NC* – Not Considered

C – Centralized, *D* – Delegation, *H* – Hardware Based, *T* – Tailoring protocol

PSK – Pre-shared keys, *PKC* – Public Key Crypto

associated with HIP perform better than DTLS in some cases we studied.

In addition to the facts presented in Table 32.1, some issues and concerns can be drawn from them. In some hardware-based approaches, it is sometime reasonable to use some special-purpose hardware in a normally constrained environment while it is might not be feasible economically if some special-purpose hardware modules are required for some highly constrained and sensitive environment, such as a human-vision enhancement IoT device. Although RSA-based certificates have been widely used in IoT applications, ECC (elliptic Curve Cryptography) based certificates provide the same level of security with considerable smaller key sizes, which means less resource requests and less time and energy consumptions. Therefore, IoT applications that use ECC-based certificates will commonly performance well. The issue related to the capacity of an individual delegation server needs to be considered as well. Although the issue can be mitigated by reducing the number of devices a delegation server can serve, the need for more delegation servers could increase the cost and the complexity of the system.

Centralized approaches require trusting the key distributor so that they can be applicable for only small domains. Public-key based authentication still plays a major role in IoT applications so it is essential for further research using PKI to ascertain that HIP is more efficient than DTLS for memory constrained IoT applications. When a single device is compromised after registering with the Domain Manager, all further communications with other devices are compromised. This is due to the fact that once the credentials are established with the domain manager, the devices can inter-communicate by deriving a fresh unicast TEK.

For tailoring protocol based approaches, there are no additional requirements related to trusted entities as the approaches are tailored without compromising the security, which makes it very interesting. However, merely by tailoring, it cannot cater to low memory requirements which makes it essential to complemented by other approaches.

32.5   Conclusion and Future Work

For a fair comparison, all security approaches used for IoT applications discussed in this study are evaluated under Contiki OS except one which was evaluated with Tiny OS while Tiny OS and Contiki OS render almost similar results for similar configurations. The research approaches chosen employ the light weight protocol versions because they enable the comparison of optimized values. Our goal of the study mainly focusses on the identification of critical factors in assessing IP-based security solutions to IoT applications. The results of our study shows that only analysis on

protocols or approaches is insufficient to evaluate security solutions or recommend protocols for standardization. Based on the existing research and experimental results, an integrated framework is proposed for evaluating IP-based security solutions to IoT applications, which includes all the influencing factors in hardware constraints, operational constraints and network scenarios as identified in this study.

Although the capability of DoS mitigation is provided by most IP-based security solutions, further study on other attack patterns is needed due to the impact of the rapid development of IoT applications. As IoT elements reach to each corner of our daily life, the safety and reliability of the IoT applications are critical to the stability of the society. The ubiquity and heterogeneity of IoT applications serve a stiff challenge in the standardization of the protocols used by IoT devices, operating systems, and software applications. As a result, more research needs to be done in this area before the efficiency and performance of IoT applications can be significantly improved.

References

1. Atzoria, L., Ierab, A., & Morabitoc, G. (2010). The internet of things: a survey. *Computer Networks Magazine, 54*(15), 2787–2805.
2. Bandyopadhyay, D., & Jaydip, S. (2011). Internet of things: applications and challenges in technology and standardization. *Wireless Personal Communications, Springer, 58*(1), 49–69.
3. Bojanova, I., Hulbart, G., & Vaos, J. (2014). Imagineering an internet of anything. *Computer magazine, IEEE, 47*(6), 72–77.
4. Bormann, C., Ersue, M., & Keranen, A. (2013, October). *Terminology for constrained node networks. draft-ietf-lwig-terminology-04 (Work in Progress), IETF.* http://tools.ietf.org/html/draft-ietf-lwig-terminology-04
5. Bui, R. N., Lakkundi, V., Olivereau, A., Serbanati, A., & Rossi, M. (2012). *Secure communication for smart IoT objects: Protocol stacks, use cases and practical examples.* IEEE International Symposium on a World of Wireless, Mobile and Multimedia Networks (WoWMoM'12), 2012, pp. 1–7.
6. Collotta, M., Pau, G., Salerno, V. M., & Scata, G. (2012). *Wireless sensor networks to improve road monitoring.* Wireless Sensor Networks – Technology and Aplications, 2012, pp. 323–346.
7. ETSI TR 102681. (2010). *Machine-to-Machine Communications (M2M); Smart Metering Use Cases.* http://www.etsi.org
8. Garcia Morchon, O. (2013). *Securing the IP-based Internet of Things with HIP and DTLS.* Proceedings of the sixth ACM conference on Security and Privacy in Wireless and Mobile Networks, 2013, pp. 119–124.
9. Heer, T., Morchon, O. G., Hummen, R., Keoh, S. L., Kumar, S. S., & Wehrle, K. (2011). Security challenges in the IP-based Internet of Things. *Wireless Personal Communications, 61*(3), 527–542.
10. Hu, W., Tan, H., Corke, P., Chan Shih, W., & Jha, S. (2012). Toward trusted wireless sensor networks. *Journal ACM Transactions on Sensor Networks (TOSN), 7*(1), 5.
11. Hummen, R., Shafagh, H., Raza S., Voigtzx, T., & WehrleR, K. (2014). *Delegation-based authentication and authorization for the IP-based internet of things. Sensing, Communication, and Networking (SECON), Eleventh Annual IEEE International Conference, IEEE,* 2014, pp. 284–292

12. Hummen, R., Wirtz, H., Henrik Ziegeldorf, J., Hiller, J., & Wehrle, K. (2013). *Tailoring end-to-end IP security protocols to the internet of things*. Network Protocols (ICNP), 21st IEEE International Conference, IEEE, 2013, pp. 1–10.
13. Kothmayr, T., Schmitt, C., Hu, W., Brünig, M., & Carle, G. (2013). DTLS based security and two-way authentication for the Internet of Things. *Ad Hoc Networks, 11*(8), 2710–2723.
14. Moskowitz, R. (2012). *Host Identity Protocol Version 2 (HIPv2)*. IETF: Internet-draft.
15. Moskowitz, R. (2012). HIP Diet EXchange (DEX). Internet Draft draft-moskowitz-hip-rg-dex-06, IETF; Pottie, G. J., Kaiser, W. J. (2000). Wireless integrated network sensors. *Communications of the ACM, 43*(5), 51–58.
16. Olsson, J. (2014, October). *6LoWPAN demystified*. Texas Instruments. Last accessed on 8 Sept 2016 at http://www.ti.com/lit/wp/swry013/swry013.pdf.
17. Perrig, A., Szewczyk, R., Tygar, J. D., Victor, W., & Culler, E. (2002). SPINS: Security protocols for sensor networks. *Journal of Wireless Networks, 8*(5), 521–534.
18. Pottie, G. J., & Kaiser, W. J. (2000). Wireless integrated network sensors. *Communications of the ACM, 43*(5), 51–58.
19. Raza, S. *Lightweight security solutions for the internet of things*. Swedish Institute of Computer Science, SICS Dissertation Series 64. Retrieved on January 27, 2015 from http://www.shahidraza.info/pdf/thesis.pdf
20. Rescorla, E., & Modagugu, N. 2006, April. Datagram transport layer security. RFC 4347. Obsoleted by RFC 6347, updated by RFC 5746.
21. Shafagh, H. *Leveraging public-key-based authentication for the internet of things*. Masters thesis, RWTH Aachen University, Retrieved on January 27, 2015 from http://people.inf.ethz.ch/mshafagh/master_thesis_Hossein_Shafagh_PKC_in_the_IoT.pdf
22. Shelby, Z., Chakrabarti, S., Nordmark, E., & Bormann, C. (2012, November). *Neighbor discovery dptimization for IPv6 over low-Power Wireless Personal Area Networks (6LoWPANs)*. RFC 6775, IETF. http://tools.ietf.org/html/rfc6775

Yu Lu, Fei Hu, and Xin Li

Abstract

Big graph model is frequently used to represent large-scale datasets such as geographical and healthcare data. Deploying these datasets in a third-party public cloud is a common solution for data sharing and processing. Maintaining data security in a public cloud is crucial. Here we target the data integrity issue, which is mainly achieved by hash operations. Existing hash schemes for graphs-structured data are either not suitable to all type of graphs or not computationally efficient for big graph. In this paper, we propose a secure, scalable hash scheme that is applicable to big graphs/trees, and its computation is highly efficient. We use the graph structure information to make our scheme unforgeable. Furthermore, we skillfully tune the scheme to make the graph verification and update processes very efficient. We will prove that our hash scheme is cryptographically secure. Our experimental results show that it has scalable computation performance.

Keywords

Big data • Security • Graphs • Data integrity • Hash function • Merkel hash tree

33.1 Introduction

Big graph, also known as large-scale graph-structured database, is a popular format for representing social network and internet of things (IOT) data[1]. Due to the lack of the capability of storing and processing such size of dataset locally, many researchers and small business owners will utilize the third party cloud service providers such as Amazon EC2 [2] and Turi [3]. If clients do not fully trust the third party, they need a scheme to maintain the integrity of their data. Specifically this scheme should be able to efficiently verify that the data returns form the cloud is exactly what they queried and no data is modified unauthorized. These properties are usually achieved by a hash scheme. Additionally the hash scheme for graph can also be used to check whether two graphs are identical [4].

Hash function is an essential block of modern cryptography. Verification of data integrity, message authentication codes (MAC) and digital signature techniques are all based on hash function. Substantially it is an algorithm to transform any random size bit string to a corresponding unique fixed-length bit string:$H: \{1, 0\}^n \rightarrow \{1, 0\}^l$. For a graph, it seems that we can simply build one vertex table for vertexes and one edge table for edges, then use standard hash functions like SHA-2 [5] to generate hash values for all blocks in these two tables. However, there are several issues. Firstly it is not efficient to store or verify all the hash values if we apply this scheme on some large-scale datasets. Another issue here is this scheme will suffer birthday attack [6]. In a graph-structured database, users always query a part of the graph (a sub-graph), sometime two sub-graphs may

Y. Lu (✉) • F. Hu • X. Li
Department of Electrical and Computer Engineering, University of Alabama, Tuscaloosa, AL, USA
e-mail: ylu56@crimson.ua.edu; fei@eng.ua.edu; xli120@crimson.ua.edu

have same contents but their structures are different regarding to the whole graph. These issues indicate a hash scheme for large graph should be perfect one-way, structure sensitive and efficient to update and verify.

There are several existing hashing schemes for trees and graphs: Merkle-Hash-Tree (MHT) [7] can handle tree-structured graphs under the assumption that any update will not change the graph type from tree to no-tree types. Similar to the MHT method, search-DAG [8] can only handle (Directed Acyclic Graph) DAG. Muhammad et al. [4] proposed perfectly collision-resistant hash functions for graphs (PCHFG) to serve as a universal solution to all types of graphs. PCHFG uses graph traversal schemes to gain labels which can represent the structure of graph (edges), and appends them to the vertexes. In this way the hash values are structure sensitive thus can ensure both the integrity of data contents and structure. Since the hash contains structure information from the whole graph, it is almost unforgeable unless the attacker is omniscient. However, this scheme fails to consider weighted graphs. In a weighted graph implemented with PCHFG, an attacker can change weights on edges without affecting any hash value. Also it didn't support partial share, in some scenario the data owner may want to hide some edges from other users but the labels in vertexes will reveal the existents of edges. Additionally a tiny update requires unacceptable large verification sets to verify.

Motived by these issues we make the following contributions in this paper:

(1) We design and implement a secure hash scheme suitable for all type of graphs. In our scheme we achieve structure sensitive structure integrity assurance through applying graph traversal methods;
(2) We theoretically prove our hash scheme is secure under random oracle;
(3) We further apply our scheme on several real world graphs and analyze the performance based on the simulation results.

33.2 Related Works and Background

Our work is mainly enlightened by two technics TL-MHT [9] and PCHFG[4]. The idea of attaching labels generated during graph traversal methods in PCHFG can make hashes structure sensitive. Utilize TL-MHT to manage hashes can significantly reduce the computational overload and VO size for updating and verifying.

Here we will briefly introduce some basics about MHT and graph hash.

33.2.1 Merkle-Hash-Tree

MHT [7] is a widely used cryptographic technology for large-scale database hash management. The basic idea of MHT is redacting hashes of all data units in the dataset to generate one or multiple root hashes. This allows efficient verifications and updates. We can illustrate the procedure of MHT construction with a simple Binary hash tree (BHT). It works bottom-up. The whole datasets are handled as small data unions. The leaf level of the MHT has the hashes of all data blocks. Each non-leaf-level node contains the hash of the concatenation of all hashes in all its children nodes (in BHT each node has two children). In general BHTs, it computes node hashes as follows: let's denote the leaf-node as: $NH(n) = H(data_u nion(n))$, and non-leaf-node as $NH(m) = H(NH(m1) \mid\mid NH(m2))$ NH denotes node hash, m is the ID of a non-leaf-node while $m1$, $m2$ are IDS of its two children. The BHT is only a special case of MHT, and the form of MHT can be adjusted in different scenarios. For example, the fan-out can be changed. For efficient update and query we can combine balance tree or B+ tree with the MHT.

33.2.2 Graph

The basic components of a graph are vertexes and edges, and it can be represented as $G = (V, E)$, here V is the set of all vertexes and E is the set contains all edges. An edge $e(v_1, v_2, w)$ in E is a link from vertex v_1 to v_2 with the weight w. An edge in an undirected graph can be represented as two inverse directional edges. There are mainly four types of graphs, i.e., the tree, the directed acyclic graph (DAG), the graph with cycles, and the graph with multiple source vertexes (a source vertex is a node without incoming edges).

33.3 Scheme Design and ImplementationM

Refer to our pervious discussion, in our hash scheme we need to reinforce our hash values to make it structure sensitive and robust against birthday attack. Concurrently the whole scheme should be efficient to construct, handle graph verification and update. We will introduce the design and implementation of our scheme step by step in following sections.

33.3.1 Attacker Assumption

We assume the attacker is not omniscient. An attacker can obtain all hash values, some but not all sub-graphs and the security hash methods. Another assumption is that the attacker is a probabilistic polynomial time (PPT) adversary Besides these assumptions we don't limit type of malicious behaviors an attacker may take to compromise the integrity of data.

33.3.2 Review of Standard Hash Scheme

A standard hash scheme contains two probabilistic polynomial time(PPT) algorithms. $\mathcal{H} = (Keygen_{\mathcal{H}}, Hash_{\mathcal{H}})$ [10].

$Keygen_{\mathcal{H}}$ is a probabilistic algorithm takes in a security parameter 1^{λ} and output a key s.

$Hash_{\mathcal{H}}$ takes in a key s and a bit string $\{1, 0\}^{*}$ and output a bit string $Hash_{\mathcal{H}}(x) \in 1, 0^{n}$ here n is implied by s.

In our scheme we will remain this scheme unchanged.

33.3.3 Graph Traversal

The first step of our hash scheme is to generate the labels for all the edges by traveling through the whole graph. A Graph $G(V, E)$ can be traveled either by *Depth-First traversal (DFT)* or *Breadth-First traversal(BFT)*. Here we choose BFT. During the traversal we will assign some labels to the vertexes to indicate their relative positions in the whole graph. For example, as shown in Fig. 33.1a, the vertex v_1 will have the traversal number(TR) label *(1, 0, 1)*. For all of the child vertexes of this vertex, sort them out in the non-decreasing order of their contents. The labels will be $(2, 1, N)(N = 1, 2, 3...)$, here N is the sequence number defined in the sorting. Generally for a vertex, its label is (l, p, N). Here l indicates the level that this vertex is in, and p is the sequence number of its parent node (for a source vertex, p is 0). Table in Fig. 33.1b shows the traversal label of each node, and the BFT result of the graph is in shown in Fig. 33.1a.

Properties of traversal labels: Vertexes in the same level have the same level number l, and p indicates the parent of a node in last level. The following lemma tells that the traversal labels can be used to represent the relative positions of all vertexes in a graph.

33.3.4 Edge Labels

In the shared database, the traversal labels of vertexes may leak some sensitive information. For instance, in the *BFT* of Fig. 33.1a, an internal attacker can use the traversal labels of v_7 and v_6 to deduce the traversal label of v_5. Also attach labels in vertexes cannot secure Therefore, we prefer to use an *Edge-Order Number (EON)* scheme based on the traversal labels to represent the graph structure. The various types of edges in the graph are defined below using the notion of traversal labels. Here we use the BFT result in Fig. 33.1a and its traversal labels in Fig. 33.1b as an example to illustrate our scheme. Edge $e(v_1, v_2)$ is a *tree-edge*, edge $e(v_4, v_5)$ is a *cross-forward-edge*, edge $e(v_6, v_5)$ is a *cross-backward-edge*, edge $e(v_6, v_7)$ is a *forward-edge* and edge $e(v_7, v_3)$ is a *back-edge*. The tree edge is the edge between a parent and its children. Such an edge can be easily captured by the *TN* numbers. However, to capture all types of edges, the relationship between RPI numbers and edges should be defined (Fig. 33.2).

Definition 1. Let T be the BFT result of a graph $G = (V, E)$, and $v_a, v_b \in V$ (l_a, p_a, N_a) and (l_b, p_b, N_b) are the traversal labels corresponding to v_a, v_b. In T, the edge $e(v_a, v_b)$ between v_a, v_b could be one of the following cases:

(1) *tree-edge*: iff $l_a + 1 = l_b$ and $N_a = p_b$;
(2) *forward-edge*: iff $l_a + 1 = l_b$ and $N_a > p_b$;
(3) *backsword-edge*: iff $l_a > l_b$;
(4) *cross-forward-edge*: iff $l_a = l_b$ and $N_a < N_b$;
(5) *cross-backward-edge*: iff $l_a = l_b$ and $N_a > N_b$;
(6) *self-connect-edge*: iff $l_a = l_b, p_a = p_b$ and $N_a > N_b$.

Fig. 33.1 A Graph with BFT and RPI traversal. (**a**) A breadth first tree. (**b**) RPI traversal numbers

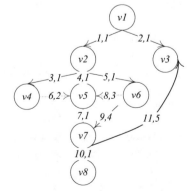

Node	RPI Number
v1	1,0,1
v2	2,1,1
v3	2,1,2
v4	3,1,1
v5	3,1,2
v6	3,1,3
v7	4,2,1
v8	5,1,1

(a) A Breadth First Tree   (b) RPI Traversal Numbers

Fig. 33.2 *BFT* result of DAG
and graphs with cycles. The trees
in (**b**), (**d**) are *BFT* result of
graphs in (**a**), (**c**); each edge in
BFT contains its *EON*

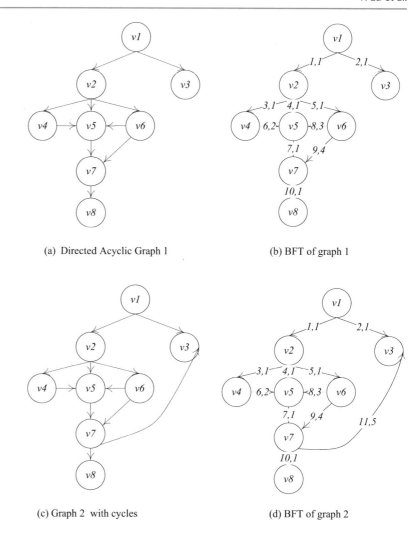

(a) Directed Acyclic Graph 1   (b) BFT of graph 1

(c) Graph 2 with cycles   (d) BFT of graph 2

In this EON scheme each edge has an *EON(O, T)*, here
O indicates the sequence number of the edge among all
edges in no-decreasing order, and *T* represents the type of
this edge (1 means *tree-edge*, 2 *cross-forward-edge*, 3 means
backsword-edge, 4 means emphforward-edge, and 5 means
backsword-edge.).

33.3.5 Procedure of Hashing Graph

Here we define the basic procedure of our hash scheme as
follows.

  $\mathscr{H}$ **and graph traversal:** Input: a graph $G(V, E)$.

1. Find and sort all source nodes in the order of their constant labels.
2. Let x be the first source node in the sorted order. If there
   are no source node in G (V, E), choose the first node as
   the source node.
3. Apply the *BFT* on the graph $G(V, E)$, start from the
   source node x as follows:

(a) If a vertex v_y is visited in *BFS*, assign its *(TN)* to it as
    TN_y.
(b) Prefix the *EON* we defined before to all edges in G.
4. Remove x from the sorted set of source nodes after the
   traversal reach to leaf level.
5. If there are nodes in $G(V, E)$ that are not yet visited, then
   repeat from 2.
6. Compute the hashes of all edges and nodes with a standard hash function.

33.3.6 Birthday Attack Resistance Hash
      for Graph

Data contents in an edge are (EON, Type, source vertex,
destination vertex, weight), which are usually less than
24 bytes [13]. In our scheme all edges from the same source
vertex will have similar *EONs*. Therefore, an attacker can
deduce an unknown edge with its hash value and one of its
adjacent edges within 2^{32} steps. This means that the scheme

is not perfect one-way even if we use a perfect one-way standard hash function. Therefore we should diffuse the relationship between the contents of different edges. Hence, we should add some random redundancy into the edge content. We add one operation into our scheme: appending a unique random number (128 bits) to each edge. Refer to steps in section B, we implement this operation as below.

1. In step 3 (a), it generates a unique random number $R_{e_{(y,z)}}$ for each edge and appends it to the edge.
2. In step 6 we compute the hash of an edge and also computer the hash of the edge appended with a random key as follows:

$$\mathscr{H}(e(y,z)) = Hash((EON)||v_y, v_z||Weight||N_{e_{(y,z)}})$$

The EONs of edges will leak some information of the graph, such as the total number of edges in the graph. Therefore, an order-preserving encryption scheme [11] is required. A randomized traversal number scheme proposed by A. Kundu and E. Bertino in [12] had the feature we need, it can transfer our *EON* into *RandomEdgeOrderNumber* (*REON*) to make it leakage-free while keeping the order unchanged.

33.3.7 MHT Construction Scheme for Graph query

Without a MHT scheme, generating a *VO* needs a time of *O* (*N*), here *N* is the total number of all edges and vertexes. The *VO* size is $O(N - n)$ here *n* is the total number of all edges and vertexes in the update subgraphs. Therefore, when only a small sub graph with M edges (M ≪ E) is queried, the VO size is unacceptably large. Wei et al. [15] proposed a Two-level Merkel Tree (TL-MT) scheme for Big Table integrity protection. In this TL-MT scheme they partitioned the original MHT into two levels, and can thus significantly reduce the construction time and the size of VO. Here we modify this scheme to make it suitable for big graph. In Fig. 33.3 it shows the TL-MT structure for a

graph. The top level is index tree which stores the root hash of the edge table and root hash of the vertex table. As we discussed before, vertexes table can be managed as table-based data, which has been well-studied. Therefore, here we focus on the construction of edge MHT:

1. Store edge hashes in small blocks, combine all hashes (*h1*, *h2*, *hq*) in one block to generate a block hash $BH = \mathscr{H}(h1||h2||hq)$. These block hashes (BH1, BH2, BHn) construct the leaf level of the edge MHT.
2. From the bottom-up we build MHT with a seted fan-out *f*. To generate a *VO* set for subgraph query verification, the server first finds all edge hash blocks spanned by the sub graphs, and redacts the hashes of edges in the blocks but not in the subgraphs. Equation (33.1) below denotes the *VO* set size of subgraphs query, in which *q* denotes the size of each block; *m* is the number of blocks this sub graph query; and s_{sub} denotes the size of sub graphs queried. Since in regular graph query the subgraphs are always connected graph, the subgraphs will span a small number of tables. Therefore the *VO* set size can be reduced. Even in the worst case that subgraphs span many blocks (even equal to its size), it can still perform well. Here we modify this scheme to make it suitable for big graph. In Fig. 33.3 it shows the TL-MT structure for a graph. The top level is index tree which stores the root hash of the edge table and root hash of the vertex table. As we discussed before, vertexes table can be managed as table-based data, which has been well-studied. Therefore, here we focus on the construction of edge MHT:

$$S_{VO_{subquery}} = (q * m - S_{sub}) \tag{33.1}$$

$$H \approx log_f(N) \tag{33.2}$$

$$S_{VO_{subquery}} \approx (f - 1)log_f(N) \tag{33.3}$$

$$S'_{VO_{subquery}} = In(N) * \left(\frac{1}{f * (In(f))^2} + \frac{1}{In(f)} - \frac{1}{(In(f))^2}\right) \tag{33.4}$$

Fig. 33.3 TL-MT For a Graph

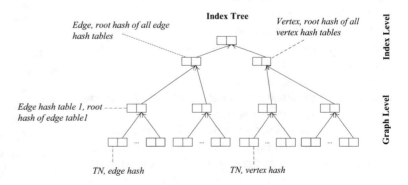

Edge, root hash of all edge hash tables

Index Tree

Vertex, root hash of all vertex hash tables

Edge hash table 1, root hash of edge table1

TN, edge hash   *TN, vertex hash*

Index Level

Graph Level

$$S_{total}(f) = \sum_{n=1}^{N} f^{\log_f(N)} \qquad (33.5)$$

$$S_{total}(f) = \frac{f * N - 1}{N - 1} \qquad (33.6)$$

$$S'_{total}(f) = \frac{1 - N}{(1 - f)^2} \qquad (33.7)$$

Equations (33.5) and (33.6) tell the total size of MHT based on the fan-out f and the number of data blocks N. Equation (33.7) is the differential format of equation (33.6). As we can see, S_{total}' is negative, which means that by increasing the fan-out, the size of MHT can be reduced. However, for a MHT another important feature is update efficiency. In equation (33.3) we calculate the size of VO sets required to update one block hash in MHT. Equation (33.4) is the differential result of equation (33.3) when $f > 3$, $(Inf) < (Inf)$; thus $S'_{VO_{update}}(f) > 0$, which means that the size of VO set increases as fan-out f increases. Since the bottleneck of computer system often exists in bus communications and CPU calculations rather than in memory storage spaces, the fan-out of MHT should be minimized to maintain efficient update.

33.3.8 Data Operations and Corresponding MHT Updates

We start our discussions from regular sub-graph range queries.

Subgraphs query across hash blocks: Only one top hash of all hashes of subgraph will be returned from the server if there is no requirement for redact. Otherwise the VO size will be same as we defined in section G.

Efficient updates: All graph edge updates can be the combination of three types of operations: delete, insert and weight change. In order to achieve efficient updates, we need to limit the influence of updates inside the blocks. And for those updates that will change the BFT results we can use some update operations without changing the BFT structure, as follows:

Weight change operations only involve the edges that need to be updated and will not change the BFT structure. Therefore it can be easily updated.

Deletion operation may change the BFT structure. Thus instead of deleting the edge we can change the type of edge to 0, which denotes that the edge has been deleted. In this way, the structure of BFT can be maintained. Next time

when we update the whole graph, the edges can be completely deleted.

Insertion operation refers to inserting six types of edges we defined in section D. For all edges except the tree edges, the insertion will not change the BFT structure. Inserting a tree edge in leaf level will not change the BFT structure, thus it can be done inside the hash block. For insertion operations that may change the structure of graph, we introduce temporary vertexes and edges to perform an equivalent insertion which will not change the structure.

33.4 Security and Performance Analyze

33.4.1 Correctness

The correctness of our graph hash scheme relies on three facts.

(1) The correctness of the hash function $\mathcal{H}$ we use in our hash scheme;
(2) The edge labels can correctly represent the structure of the graph, as we proved in last section
(3) The top hash we get from MHT is unique for different graphs, this is based on the correctness of MHT scheme.

We use SHA-512[ ]as hash function $\mathcal{H}$ which is pre-proved correct and MHT is a pre-proved scheme. Also as proved in last section the edge labels can correctly represent the structure of the graph we can say that our graph hash scheme is correct.

33.4.2 Security

Similar to correctness problem, the security of our hash scheme depends on three facts:

(1) Under random oracle the hash function $\mathcal{H}$ we use in our hash scheme is secure;
(2) Under random oracle the graph traversal scheme is secure;
(3) Under random oracle the MHT construction scheme is secure.

Due to the page limitation we will not provide the detail prove for (1) and (3) for they are standard schemes which are well proved. Here we focus on the security of part (2). For a hash scheme the security definition is one-way, collision-resistant and leakage-free. The one-way property depends on the hash function and data content (in case of birthday attack). Since the hash function is secure and with the 128 bits random number attached to the edge content, under random oracle it is computational impossible to lunch birthday attack and brute-force attack on our scheme. Also with the *REON* labels graphs with some data content but different structure will have different hash values together with a collision-resistant hash function we can prove that our scheme is collision-resistant. Refer to our discussion in last section with a order-preserving scheme we can make the edge labels leakage-free. With all this three property we can prove our scheme is security under random oracle.

33.4.3 Performance Reslut Analysis

We implement our graph scheme in python 2.7.11 and analysis its performance with some real-world graph structure data. Our experiment was done on a machine with following features: Dell Optiplex 9020, 64 bit windows 7 operating system, core i7-4790 CPU with 16 GB memory size. The real-world graphs we used came from the Stanford Large Network Dataset Collection[13]. In order to get a general trend of performance we also use snap.py[14] to generate several random graph datasets for our experiments. For hash computation we use SHA-512[5]. In Table 33.1 we list the real-world graphs we used together with their sizes (Fig. 33.4).

In the experiment we want to evaluate the computation and communication load of each part in our scheme, the computation complexity is shown by time cost for computation while communication load can be quantized into the size of *VO* set needs to be transmitted through communication channel during the process. There are mainly three parts:

(1) *graph traversal hash computation and MHT construction*: In this part, we perform BFT algorithm on the whole graph to label all edges with *REON* then use SHA-512 hash function to hash all labeled edges appended with a 128 bits random number. Allocate hashes in a fixed range into each block, and the hashes of these blocks will be the leaf nodes of the MHT. As shown in Fig 33.5, the time cost for the whole process increases linearly as the size of dataset increase. Thus the complexity of this part is $O(N)$, here N is the total number of edges and vertexes in a graph-structured dataset. In Fig 33.6 we can see the MHT size also increases linearly with the size of datasets increases and the MHT size is relatively small as for a graph with ten million edges and one million vertexes the MHT size is only 250 KB.

(2) *Hash verification for subgraph query*: In this part, a client receives a sub-graph and a top hash for all hashes of this sub graph (Here the top hash is the hash value of the *concatenatedstring* of all hashes in the sub graph) form the server. In Fig 33.7 we can see the time cost increases linearly as the size of the subgraph increases, the computational complexity is $O(n)$. Only a top hash is required to be send to client, thus the communication load can be ignored.

(3) *Hash verification for graph updates*:

Update involves three type of operations: Insertion, Deletion and Weight update. In our scheme refer to our previous definition, all these there operations have similar computational complexity which is $O(n + q * log_f(N))$ here n is the number of edges required to be updated, N is the number of blocks in the leaf level of MHT and q is the blocks the updates span.

Table 33.1 Graph datasets and their sizes

Graph	V	E
email-EuAll	265,214	420,045
soc-Epinions1	75,879	508,837
Slashdot0902	82,168	948,464
Amazon0302	262,111	1,234,877
Amazon0312	400,727	3,200,440
Wiki-Talk	2,394,385	5,021,410

Fig. 33.4 Time cost for hash computation and MHT construction with different real-world datasets

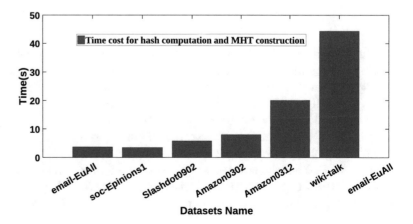

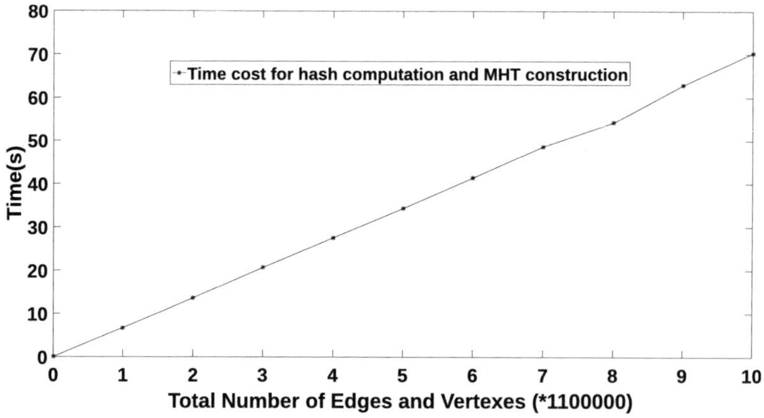

Fig. 33.5 Time cost for hash computation and MHT construction with different random datasets

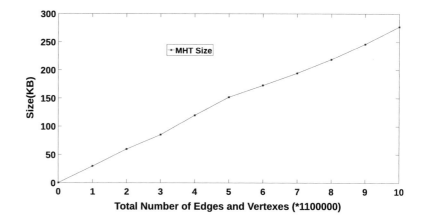

Fig. 33.6 MHT size of different random datasets

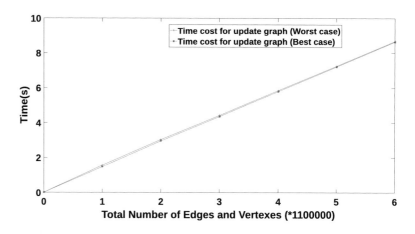

Fig. 33.7 Time cost for verifying different sizes of Sub graphs

We show both the worst case (update spans the most blocks) and the best cases (update spans the least blocks) in Fig. 33.8, in this case since N is only 31259 which is relatively small the difference of time cost is not significant.

The size of VO set requires for updates mainly depends on the number of blocks in the bottom level of MHT this update spans. In Fig. 9 we also show the worst case (update spans the most blocks) and the best cases (update spans the least blocks). For the worst case the size is constant when $n >= N$, and it is $O(q)$ for the best case. Even in the worst cases the

computational complexity and VO sizes is still acceptable. We can say our scheme is efficient for handling graph updating

33.5 Conclusion and Future Work

In this paper we have proposed a hash scheme for big graphs. Our scheme is secure under random oracle and is efficient for graph construction, subgraph query and graph update.

Fig. 33.8 *VO* size for update different sizes of Sub graphs

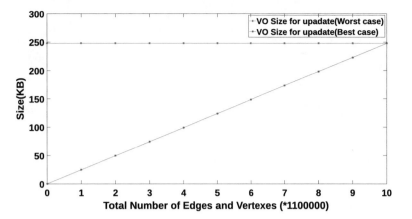

We have implemented our scheme in Python and explored its performance by applying it to some real-world graph-structured datasets as well as some random graph-structured datasets. The experimental results indicate that our scheme is efficient for graph construction, query and update. In the next step we will investigate how to implement our scheme in a distributed big graph application. The challenge is to implement the hash operations in the graph splitting and mergence operations of big graphs.

In next step we will explore how to implement our scheme in a distributed scenario the challenge is to efficiently verify split and merge operations on graphs.

Acknowledgements The authors would like to thank the U.S. NSF (National Science Foundation) for their support through the project DUE-1315328. Any ideas presented here do not necessarily represent NSF's opinions.

References

1. Jagadish, H., & Olken, F. (2004). Database management for life sciences research. *ACM SIGMOD Record, 33*(2), 15–20.
2. Amazon, E. (2010). Amazon elastic compute cloud (Amazon EC2). *Amazon Elastic Compute Cloud (Amazon EC2)*.
3. Low, Y., Gonzalez, J. E., Kyrola, A., Bickson, D., Guestrin, C. E., & Hellerstein, J. (2014). Graphlab: A new framework for parallel machine learning. *arXiv preprint arXiv:1408.2041*.
4. Arshad, M. U., Kundu, A., Bertino, E., Madhavan, K., & Ghafoor, A. (2014). Security of graph data: hashing schemes and definitions. In *Proceedings of the 4th ACM Conference on Data and Application Security and Privacy* (pp. 223–234). ACM.
5. Preneel, B. (1994). Cryptographic hash functions. *European Transactions on Telecommunications, 5*(4), 431–448.
6. Girault, M., Cohen, R., & Campana, M. (1988). A generalized birthday attack. In *Workshop on the Theory and Application of Cryptographic Techniques* (pp. 129–156). Springer.
7. Merkle, R. C. (1989). A certified digital signature. In *Conference on the Theory and Application of Cryptology* (pp. 218–238). Springer.
8. Martel, C., Nuckolls, G., Devanbu, P., Gertz, M., Kwong, A., & Stubblebine, S. G. (2004). A general model for authenticated data structures. *Algorithmica, 39*(1), 21–41.
9. Wei, W., Yu, T., & Xue, R. (2013). ibigtable: Practical data integrity for bigtable in public cloud. In *Proceedings of the Third ACM Conference on Data and Application Security and Privacy* (pp. 341–352). ACM.
10. Katz, J., & Lindell, Y. (2014). *Introduction to modern cryptography*. CRC press.
11. Agrawal, R., Kiernan, J., Srikant, R., & Xu, Y. (2004). Order preserving encryption for numeric data. In *Proceedings of the 2004 ACM SIGMOD International Conference on Management of Data* (pp. 563–574). ACM.
12. Kundu, A., & Bertino, E. (2012). On hashing graphs. *IACR Cryptology ePrint Archive, 2012*, 352.
13. Leskovec, J., & Krevl, A. (2015). {SNAP Datasets}:{Stanford} large network dataset collection.
14. Leskovec, J., & Sosič, R. (2016). Snap: A general-purpose network analysis and graph-mining library. *ACM Transactions on Intelligent Systems and Technology (TIST), 8*(1), 1.

A Cluster Membership Protocol Based on Termination Detection Algorithm in Distributed Systems

SungHoon Park, SuChang Yoo, and BoKyoung Kim

Abstract

In distributed systems, a cluster of computer should continue to do cooperation in order to finish some jobs. In such a system, a cluster membership protocol is especially practical and important elements to provide processes in a cluster of computers with a consistent common knowledge of the membership of the cluster. Whenever a membership change occurs, processes should agree on which of them should do to accomplish an unfinished job or begins a new job. The problem of knowing a stable membership view is very same with the one of agreeing common predicate in a distributed system such as the consensus problem. Based on the termination detection protocol that is traditional one in asynchronous distributed systems, we present the new cluster membership protocol in distributed wired networks.

Keywords

Synchronous Distributed Systems • Cluster membership • Fault Tolerance • Termination Detection

34.1 Introduction

In distributed systems, a cluster of computer should continue to do cooperation in order to finish some jobs. A cluster membership protocol is especially helpful tools to allocate processes in a same cluster with a same view of the membership of the cluster. Whenever a membership change occurs, processes can consent to which of them should do to finish a waiting job or begin a new job. The problem of getting a stable membership view is very same with the one of getting common knowledge in a synchronous distributed system such as the consensus problem [1].

The *Cluster* membership protocol [2] is that every process connected in a network requires getting a stable same cluster membership view if all connected process are belong to just one cluster. The problem was widely discussed at the study community. The reason for this great study is that many distributed systems need a cluster membership protocol [3–7]. In spite of such practically usefulness, to our knowledge there is only a few research that have been committed to this problem in a wired arbitrary connected computing environment.

Depending on process failure and recover, network topologies are changed and process may dynamically connect and disconnect over a wired network. In such wired networks, group membership can be changed so much, making it a special critical module of system software part. In wired arbitrary network systems, a lot of environmental adversities are more common than the static wired network systems such as that can cause loss of messages or data [8]. In particular, a process can easily get to fault by hardware or software problem and disconnect from the wired network. Implementing fault-tolerant distributed applications in such an environment is a complex and difficult behavior [9, 10].

S. Park (✉) • S. Yoo • B. Kim
School of Electrical and Computer Engineering, Chungbuk National University, Cheongju, South Korea
e-mail: spark@cbnu.ac.kr; izibt@nate.com; agiboss210@naver.com

In this paper, we propose a new protocol to the cluster membership protocol in a specific wired distributed computing system. Based on the termination detection protocol that is traditional one in asynchronous distributed systems, we address the new cluster membership protocol. We make up of the rest of this paper as follows. In Sect. 34.2 we address the system model we use. In Sect. 34.3, we describe a specification to the cluster membership problem in a traditional synchronous distributed system. We also address a new protocol to solve the cluster membership problem in a wired arbitrary computing system in Sect. 34.4. In Sect. 34.5, we address conclude.

34.2 Computing System Model, Definition and Assumption

In this section, we describe our models for capturing behavior of distributed systems. We use these models foe reasoning about correctness of our protocol as well as for analysis of distributed computations. Our model for distributed systems is based on notice passing, and all of protocol is around that concept. Many of these kinds of protocol have analogs in the shared memory world but will not be addressed in this paper.

First, we define our system model based on some assumptions and after that we address our goals. We model a distributed system as a loosely coupled messing-passing system without shared memory and a global clock. Our distributed computation model for a wired network is made up of as an undirected graph. That is, the undirected graph is described as $G = (V, E)$, in which vertices V facing each other with set of distributed process $\{1, 2, \ldots, n\}$ ($n > 1$) with unique identifiers and edges E between a pair of process correspond the fact that the two process are in each other's transmission radii. Hence, in our distributed system a channel to directly communicate with each other which changes over time when processes move.

Every process i has a variable N_i, which denotes the neighboring processes, with that i can *directly* communicate the neighboring processes. Every process communicates with a channel that is bidirectional; $j \in N_i$ iff $i \in N_j$. More accurately, in the network $G = (V, E)$, we decide E such that for all $i \in V$, $(i, j) \in E$ if and only if $i \in N_j$. Depending on process's movement, the graph could be disconnected that means that the network is partitioned. Because the processes may alternate their position, N_i position would be unexpectedly changed and therefore G also may be changed accordingly. The assumptions about the processes, wired network and system architecture are followings.

Every process is distinguished by a unique identifier. The unique identifiers are used to distinguish processes during operating the cluster membership search process. Channels and links are bidirectional that means first n first out, i.e. every process receives notices based on the sequence that are delivered over a link between two neighboring processes. Many topology changes may be arbitrary occurred when the process moves in wired networks. That makes a lot of network partitioning and merging. Processes can make a fault to be crash arbitrarily at random and can recover again at any time.

Without network partition, the sender and the receiver do successful notice delivery that means the notice would be successfully delivered only when the two processes remain connected for the all period of notice transfer. Every process has a big receiving buffer enough to avoid buffer overflow all the time in its lifetime. Even though a finite number of topology changes, every process i *eventually* has a same view of cluster membership of the cluster to which i belongs.

34.3 Cluster Luster Membership Specification

We assume that our specification is as followings, it is consist of four properties for a cluster membership protocol.

Safety(1): At any time, all processes in the cluster have a stable consistent view.

Progress: If there are no more changes in the each views of the processes in one cluster, they eventually getting to their stable consistent views.

Validity: If all processes in a event know a view as their local view and they have eventually reached their stable states, then the last process of their sequences of global views are all at same position and must be equal to each other.

Safety(2): When a view is committed as a global view, it cannot be changed.

The first property describes agreement. Consistent history must be an unchanged one for any program that satisfies the specification. The second property shows termination of global view. When the state and event of all processes are unchanged, the processes are eventually getting to close changing their output results. The third property removes trivial solutions where protocols never getting on any new view or always determine on the consistent view.

34.4 Cluster Luster Membership Protocol

At this section, we address a cluster membership protocol that was operated upon the termination detection protocol by scattering computations. After these sections, we will describe in detail the method that this protocol may be accommodated to a distributed system.

34.4.1 Cluster Membership in a Wired Network

We first address our cluster membership protocol in the wired network settings. In which we assume that process and channels have no faults.

The protocol is made up of three phases running at the process that starts the cluster membership protocol.

1. The first phase that is a diffusing phase and it works by first diffusing the "who" notices.
2. The second phase that is a searching phase and it runs by then accumulating the id of every process that is consist of the wired networks. We represent this computation starting processes as the *start process*.
3. The third phase is a closing phase that is managed by deciding the same view and announcing it as a stable new view to all process.

The start process will have the information enough to decide a uniform cluster membership view after taking all process' ids completely and the start process will then broadcast it to the rest of the process in the network. The three kinds of notice, *Who*, *Ack* and *View* are used to manipulate the operations.

As the first phase is diffusing computing phase, *Who* notice is used to make a start of the cluster membership protocol by diffusing the *Who notice*.

1. The first Phase:
   When cluster membership protocol is launched at a start process s, the start process makes a replying queue wl and a accepted queue rl and starts a *scattering computation* by forwarding an *Who notice* to all of its immediate neighboring processes. At the starting point, the replying queue makes up of only its most close neighboring process's ids and the accepted queue has nothing.

   When process i receives a *Who notice* from the neighboring process for the first time, it immediately sends the *Ack* notice to the start process and propagates the *Who notice* to all its neighboring process except the process from which it first accepted an *Who notice*.

   The *Ack* notice sent by process i to the start process contains the ids of all its neighboring process that are needed for the start process to decide the stable view of the process connected with a distributed network. After that, any *Who notice* accepted by other neighboring process will be ignored.

2. The Second Phase:
   Searching phase. When the start process receives the *Ack* notice was taken out from the process j, it takes j out from the replying queue and gets j into the accepted queue and as soon as possible it detects sequentially the each process's id included in the *Ack* notice. If there is the some process in the *Ack* notice which has already been

accepted, i.e. that means it is in the accepted queue, it is dismissed. If it is not in the accepted queue, it is inserted into the replying queue of start process. The start process will be suspends for the *Ack* notice from one.

The replying queue is increasing and decreasing repeatedly when it was accepted based on the accepted *Ack* notices, however the replying queue is continually increasing by accepting the *Ack* notices. But the replying queue at the end could have no element and the replying queue could insert all ids of processes connected to the wired networks whenever the start process accepted the *Ack* notices from all other processes. Therefore the start process eventually has much information enough to decide the stable view of the cluster based on the replying queue. That is because the replying queue could be eventually unoccupied and it means that the start process has accepted the *Ack* notices from all the process.

3. The Third Phase:
   Once the start process has accepted *Ack*s from all other process, it decides the stable view based on the replying queue and forwards a *View* notice to all other process to let know the current view of the cluster. We show some sample running protocol as the protocol execution to explain more specific features. We address the protocol in synchronous setting even though all the behaviors of the protocol are practically asynchronous. We assume that the network shown in Fig. 34.1a is asynchronous. In this shape, and for the all of the paper, thin arrows denote the route of *Who notice*'s move and dotted arrows denotes the way of route of *Ack* notices to the start process.

   As shown in Fig. 34.1, process A is a start process that starts wl_a and rl_b with {B,C} and {A} at each and starts a scattering computation with forwarding out *Who notices* (indicated as "E" in the shape) to its immediate neighbors, viz. process B and C, shown in Fig. 34.1a. As indicated in Fig. 34.1b, process B and C in turn forward the *Who notice* to its most close neighbors only except the start process. It sends the *Ack* notice with close neighboring process queue to the start process A. Hence B and C also send *Who notices* to each other.

But B and C do not acknowledge to the start process about the *Who notices* because process B and C have already accepted *Who notices* from the start process at each. The information of neighboring process is piggybacked upon the *Ack* notice sent by all process.

Upon hearing *Ack* notices from B and C, process A renews $wl_a = \{B,C\}$, $rl_b = \{A\}$ with the close neighboring process information piggybacked at the *Ack* notices.

The *Who notices* is transmitted over the arrows at the edges and the dotted arrows going parallel with the edges denotes *Ack* notices. In Fig. 34.1c, the process D and F also send the *Ack* notices to the starts process at the time they accepted the *Who notice* s from the B and C one by one.

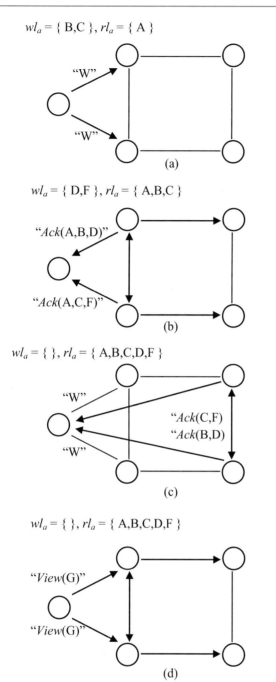

$wl_a = \{ B,C \}, rl_a = \{ A \}$

"W"

"W"

(a)

$wl_a = \{ D,F \}, rl_a = \{ A,B,C \}$

"Ack(A,B,D)"

"Ack(A,C,F)"

(b)

$wl_a = \{ \}, rl_a = \{ A,B,C,D,F \}$

"W"

"Ack(C,F)
"Ack(B,D)

"W"

(c)

$wl_a = \{ \}, rl_a = \{ A,B,C,D,F \}$

"$View$(G)"

"$View$(G)"

(d)

Fig. 34.1 An example of cluster membership protocol execution on the process search protocol

Each of these *Ack* notices includes the ids of the neighbor. All the time, the start *A* accepts all acknowledgments from all of other process except itself in Fig. 34.1d and then determines the stable view between the clusters and forwards it, that is the *View* notice displayed in Fig. 34.1d.

34.5 Conclusion

We have addressed here the study of distributed cluster membership protocol for distributed, wired networks and proved it to be correct based on the symbolic dynamics of

finite state machine obtained by linear probability model. We have also shown that the symbolic formal specification of property in our cluster membership protocol based on linear temporal logic is to reason the protocol correctness.

In real world, the wired network topology is actively and lively changing at random and that dynamic network changed configuration causes frequent connection and disconnection of process over the wired network. In spite of weakness about wired networks, our cluster membership protocol specification guarantees the safety and progress property could be always satisfied.

As mentioned in the introduction, our main goal has been to design cluster membership search protocol and prove decidability of consistent view in as simple a fashion as possible without paying much attention at wired networks to the consistent membership view on every process even though complexity issues of distributed networks.

We are however convinced that process search protocol and linear set theory techniques can considerably relax the strong property of safety in many of our constructions. In particular, a more careful logical design of cluster membership protocol for specific classes of wired networks can be performed using the results from our design of convergence properties of consistent cluster membership view. This could lead to a significant improvement of our protocol from a practical environment point.

Finally, in a practical setting one may generalize the cluster membership protocol to more fit in some distributes systems according to network environments. It will be interesting to explore whether safety or progress could be weakened depending on distributed computing environmental factors.

References

1. Amir, Y., Moser, E., Melliar-Smith, P., Agarwal, D., & Ciarfella, P. (1995). The totem single-ring ordering and membership protocol. *ACM Transactions on Computer Systems, 13*(4), 311–342.
2. Anceaume, E., Charron-Bost, B., Minet, P., Toueg, S. (1995). On the formal specification of group membership services. Technical Report *95*(3), 1534–1559, Computer Science Dept., Cornell Univ., Aug. 1995
3. Anker, T., Chockler, G., Dolev, D., Keidar, I. (1998, August 21–23). *Scalable cluster membership services for novel applications*. Proceedings of workshop on Networks in Distributed Computing (DIMACS 45), 1998, pp. 23–42.
4. Keidar, I., Sussman, J., Marzullo, K., & Dolev, D. (2000, April 15–17). A client server oriented protocol for virtually synchronous cluster membership in WANs. *Proceedings of 20th International Conference Distributed Computing Systems, 3*(1), 234–244.
5. Brunekreef, J., Katoen, J., Koymans, R., & Mauw, S. (1996). Design and analysis of dynamic leader cluster membership protocols in broadcast networks. *Distributed Computing, 9*(4), 157–171.
6. Bottazi, D., Montanari, R., & Rossi, G. (2008). A self-organizing cluster management middleware for distributed ad-hoc networks. *Computer Communications, 31*(13), 3040–3304.
7. Powell, D. (1996). Special section on cluster communication. *Communications of the ACM, 39*(4), 50–97.

8. Pradhan, D., Krishna, P., Vaidya, N. (1996, October 12–15). *Recoverable distributed environments: design and tradeoff analysis.* Proceedings of annual symposium on Fault Tolerant Computing, 1996, pp. 16–25.

9. Briesemeister, L., Hommel, G. (2002, March 5–18). *Localized cluster membership service for wired networks.* Proceedings of international conference on IEEE Parallel Processing Workshops, 2002, pp. 94–100.

10. Hatzis, K., Pentaris, G., Spirakis, P., Tampakas, V., Tan, R.(1999, October 19–21). *Fundamental control protocols in distributed networks.* Proceedings of 11th ACM SPAA, 1999, pp.251–260.

Part III

Information Systems and Internet Technology

An Adaptive Sensor Management System

35

Chyi-Ren Dow, Yaun-Zone Wu, Bonnie Lu, Shiow-Fen Hwang, and Fongray Frank Young

Abstract

This paper proposes an adaptive sensor management system which consists of five main components, including the Sensor Web Interface, Request Dispatcher, Publish/ Subscribe Management, Communication Converter and Message Broker. The Request Dispatcher can automatically allocate the tasks of Sensor Web to each module as the service requested. The Communication Converter let the Message Broker handle the operations of Sensor Web, notify sensors and modify the information of sensors through the Sensor Web Interface. Two sensor control methods are defined for publication and subscription. The experimental results by the designed prototype system show that this system can effectively perform sensor management.

Keywords

IoT • SWE • MQTT • Sensor Management

35.1 Introduction

Sensor Web Enablement (SWE) [1] was proposed by Open Geospatial Consortium (OGC). In the sensor management mechanism, SWE only focuses on describing the information of the sensors and monitoring their schedules. SWE emphasizes on collecting data from the sensors, but does not control the behavior of sensors. Message Queue Telemetry Transport (MQTT) [2] is based on TCP/IP and suitable for IoT environments. It is characterized by its header size fixed on two bytes for various types of packets in MQTT. It transmits and receives information using the publish/subscribe mechanism.

The number of sensors is expected to increase in the future. We need a highly scalable sensor environment. These sensors may be lightweight devices which are powerless and have weaker computing capability. It is necessary to use low-powered protocols for these sensor environments. Different sensors have unique specifications, so the system needs to fuse various data that have collected. It is necessary for a centralized management system to manage the sensors. The mechanism of sensors is usually focused on collecting sensor data, but the control sensor state is seldom discussed.

This paper is organized as follows: Section II discusses the related research and technology. Section III presents our system overview and architecture design. Section IV describes the ideas of Sensor Control Management. Section V shows the system prototype. VI provides the system experimental results. Finally, the conclusions are proposed in Section VII.

C.-R. Dow (✉) • Y.-Z. Wu • B. Lu • S.-F. Hwang • F.F. Young
Department of Information Engineering and Computer Science, Feng Chia University, Taichung, Taiwan
e-mail: crdow@fcu.edu.tw; mobetax81072@gmail.com; bonnielu@mail.fcu.edu.tw; sfhwang@mail.fcu.edu.tw; fryoung@mail.fcu.edu.tw

35.2 Related Work

There are many applications in the sensor network. There are more sensors which are deployed in the environment. The study [3] aimed at sensors integrating. There are many

© Springer International Publishing AG 2018
S. Latifi (ed.), *Information Technology – New Generations*, Advances in Intelligent
Systems and Computing 558, DOI 10.1007/978-3-319-54978-1_35

different formats for the sensors. Therefore, it is difficult to integrate the sensors into the observed system. The research [4] combined SWE with RESTful to transform the message into different formats. In this method, SWE effectively combines with the other protocols. The alternative sensor network was proposed [5]. This research talks about the social network. People who use smartphones are the best observation nodes.

The Internet raises the performance up with the lightweight packet. The MQTT and CoAP [6] are lightweight protocols. The study [7] compare MQTT with CoAP. CoAP, which is based on UDP, is better for small areas than MQTT, which is based on TCP/IP. All things considered, MQTT has better scalability than CoAP in the IoT environments. There is a special characteristic of MQTT that divides messages into three QoS levels: at most once, at least once, and exactly once. The studies [8] propose the different QoS levels. Based on the advantages of MQTT, the system that handles large events was proposed [9].

Management function in the sensor network is required. The Sensor Web Node Management System was proposed [10] to manage the sensor network. We need a system to fuse the information that is collected from different sensors. To solve this problem, they design the Cloud4sens which uses cloud management in the system. This system fuses different data and provides information to users [11].

35.3 System Architecture

This work proposed a system architecture for sensor management. An overview of this system is shown in Fig. 35.1, including the SWE Server, the Management, the Message Broker, sensors and users.

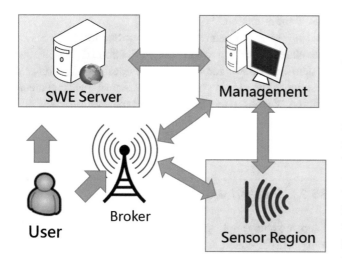

Fig. 35.1 System overview

The broker which relays messages using the MQTT publish/ subscribe mechanism. The SWE Server provides Web Service and standard SWE functions into our platform. The broker transmits messages to the users who in turn subscribes to the topic when the sensors and the management system sends the messages to the topic. The users subscribe to the topic to receive the notification. The management system is responsible for collecting messages and setting the topics. The management system controls the state of all sensors using the publish/subscribe mechanism. Furthermore, the management system can observe the data and control the states of the sensors. The sensors collect information from the environment, traffic and vehicles, and the information is stored in the management system.

The system contains our network management unit, HTTP-to-MQTT broker and converter. In the network management unit, SWE Interface provides the sensor observation service and contains four operational methods, including sensor value insert, sensor value read, service features description and sensor description. It uses XML to exchange the data which collect from the sensors. It also communicates the information and provides the details when a sensor joins the system at the first time. The observation value is stored on the server and accessed by management with this service.

Our system which combines the SWE and MQTT techniques is challenging to provide the real-time service in the sensor network. In the IoT environment, there are many sensors and the system needs scalability. In our system, the traditional SWE standard functions is processed by the SWE interface. We designed a message service that uses the publish/ subscribe mechanism. That is the request dispatcher, publish/ subscribe management, HTTP-MQTT converter and MQTT broker in our system. When the management requests service, the request dispatcher automatically distributes the work to SWE functions. The publish/ subscribe management contains the publish/ subscribe relation between the management and the sensor.

35.4 Sensor Control Management

In the sensor observation and control interface, it has information of the sensors, including the sensor name, the sensor number, latitude and longitude location, grid coordinates, the sensor state, data format, data return frequency, service and communication interface. The sensor has its tag when the sensor registered in the system. We designed a mechanism in which the sensors can exchange the state and control the frequency in the module of the sensor publish/subscribe management, and can also modify the state and frequency with the interface of this mechanism. In the communication interface, we use the platform and have the SWE.

Fig. 35.2 Sensor area example

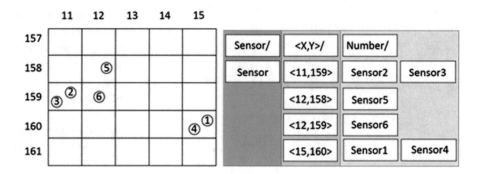

In the control mechanism, there are two parts: control interface and control module. In the control module, all of the sensors in the system subscribe the control topic. To adjust the sensors, the management sends the message to the control topic. The control topic publishes the message to the sensor correctly. The management can decide the control topic and design the change of strategy to modify the sensors. When the change strategy conforms the parameter, the control center will send the message to assign the sensor. We have two methods based on the sensors area and sensor name to control the sensors. Below is an example in Fig. 35.2.

We use the MQTT mechanism in our method and give the topic "Sensor" to all the sensors in our system, and give the topic "<X, Y>" to a grid with latitude and longitude coordinates X and Y. For example, we have the "Sensor5" in grid "<12, 185>". In our method, the topic of this "Sensor5" is defined as "Sensor/<12, 158>/Sensor5". We define all sensors topic in the system with this method. When a new sensor wants to register in the system, the system will give the sensor a number and a topic for this new sensor automatically.

35.5   System Prototype

Our system is implemented through SWE and MQTT. We focused on sensor control, adaptive sensor and sensor management. In order to make these functions easily used by the user, we collected various sensors' information in this platform. We used the tabular form and listed the sensors in the system. It can search and use the sensors more conveniently for users. In the publish/subscribe management, the relation of the manager and the sensors is presented with visualization. We can easily find the publisher and the subscriber through a management system. When we need to control the sensors, it can define and connect the sensors first with visualization.

We accomplished the remote control using the publish/subscribe mechanism with the Sensor Web enablement module. For adaptive sensor control, we made the users control

Fig. 35.3 System prototype

the sensors directly, and the management system accessed the information from the sensor observation module with the time parameter. Moreover, we provided a function to let the user be able to control the frequency for sensors to transmit the data with the MQTT publish/subscribe mechanism. Finally, we combined SWE with the management platform and the observation platform. There is the system prototype shown in the Fig. 35.3.

Our server was implemented on the Ubuntu system and the Mosquitto [12] was used to build the broker. After we built the broker, we defined the IP address and the topic name of the broker. It can use the IP address and the topic name to exchange the information with the MQTT publish/subscribe mechanism. The user can get the information from the control center and the sensors can receive the control message from the management. The control center is implemented on the Window system by using C#. For the communication between the user and the broker, the system can be used to connect to the broker on 4G (LTE). The broker connects to the server with the Ethernet and wireless network.

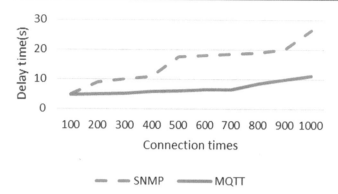

Fig. 35.4 Delay time

35.6 Experimental Results

In order to evaluate the performance of our sensor management system, we design the experimental environment and compare SNMP with MQTT. The management system connects with the sensor by using Wi-Fi is implemented on the Window system. The sensor is a smartphone device.

The management system and sensor use the same device in this experiment. The management system builds on the notebook and the sensor is smartphone. The broker is implemented on the Mosquitto. We use the management to send the control packet with our control method.

Figure 35.4 shows the delay time for SNMP and MQTT. The dash line shows delay time from SNMP. The solid line shows delay time from MQTT. The horizontal axis is the number of times that the management send the packet to sensor. Because the MQTT mechanism only has two-bytes header, we can observe the delay time that MQTT is better than SNMP.

35.7 Conclusions

We proposed a sensor management system which was based on SWE and MQTT. MQTT is the lightweight, low-cost and low-power protocol based on the publish/subscribe mechanism. We used SWE to collect the data from the sensor and provide the sensor control function with MQTT. It can adjust

sensors' state and frequency effectively. With our sensor management system, we not only can control the single sensor, but also control the sensor in the group. For the future work, we will implement a SOS which is based on MQTT. It may increase the performance of sensors for collecting the data from environments.

Acknowledgement The authors would like to thank the Ministry of Science and Technology of Republic of China for financially supporting this research under Contract No. 104-2221-E-035-020-.

References

1. Sensor Web Enablement. http://www.opengeospatial.org/ogc/markets-technologies/swe
2. MQTT. http://mqtt.org/
3. Ala, F., Abdallah, K., Mohsen, G., Ammar, R., & Mehdi, M. (2015, September). Toward better horizontal integration among IoT services. *IEEE Communications Magazine, 53*(9), 72–79.
4. Rouached, M., Baccar, S., & Abid, M. (2012, June). *RESTful sensor web enablement services for wireless sensor networks.* Proceedings of the IEEE Eighth World Congress on Services, pp. 65–72.
5. Al-Fuqaha, A., Guizani, M., Mohammadi, M., Aledhari, M., & Ayyash, M. (2015, November). Internet of Things: A survey on enabling technologies, protocols, and applications. *IEEE Communications Surveys & Tutorials, 17*(4), 2347–2376.
6. CoAP. http://coap.technology/
7. Thangavel, D., Ma, X., Valera, A., Tan, H. X., & Tan, C. K. Y. (2014, April). *Performance evaluation of MQTT and CoAP via a common middleware.* Proceedings of the 9th IEEE International Conference on Intelligent Sensors, Sensor Networks and Information Processing (ISSNIP), pp. 1–6.
8. Jo, H. C., & Jin, H. W. (2015, August). *Adaptive periodic communication over MQTT for large-scale cyber-physical systems.* Proceedings of the IEEE 3rd International Conference on Cyber-Physical Systems, Networks, and Applications, pp. 66–69.
9. Shah, M. A., & Kulkarni, D. B. (2015, May). *Storm pub-sub: High performance, scalable content based event matching system using storm.* Proceedings of the IEEE Parallel and Distributed Processing Symposium Workshop, pp. 585–590.
10. Liu, Q., Mao, S., Li, M., & Li, X. (2014, June). *Release and storage of mine gas monitoring data based on sensor web.* Proceedings of the 22nd international conference on Geo Informatics, 2014, pp. 1–6.
11. Fazio, M., & Puliafito, A. (2015, March). Cloud4sens: A cloud-based architecture for sensor controlling and monitoring. *Communications Magazine, 53*(3), 41–47.
12. Mosquitto. http://mosquitto.org/

A Cloud Service for Graphical User Interfaces Generation and Electronic Health Record Storage

André Magno Costa de Araújo, Valéria Cesário Times, and Marcus Urbano da Silva

Abstract

The development of archetype-based Health Information Systems (HIS) allows the creation of interoperable mechanisms for the Electronic Health Record (EHR) as well as improvements for application maintenance and upgrades. However, we identified a lack of an approach or tool to build dynamic archetype-based data schema in heterogeneous database. This article presents a cloud service able to build Graphical User Interface (GUI) and data schema for use in the healthcare sector. Using data attributes, terminologies and constraints extracted from EHR archetypes, it dynamically generates GUI and data schema in heterogeneous databases. In order to persist EHR data, we used the concept of polyglot persistence to store structured data in a relational database, while non-structured data is stored in a NoSQL database. Finally, we validated the proposed service in a hospital located in northeastern Brazil and demonstrate how health professionals can build GUI for use in the healthcare sector without depending on a software development team.

Keywords

Data management applications and services • Cloud services models and frameworks • Archetypes • Data Schema • Health Information Systems

36.1 Introduction

In the recent years, resources provided by Information Technology (IT) have helped health organizations in developing solutions that improve the quality of services, facilitate access to patient information and increase productivity [1, 2]. Since paper use has been reduced due to EHR, much has been discussed about the use of standards, rules and procedures in the development of HIS. As determined by the best practices of international bodies [3], an HIS must provide an EHR with security mechanisms and uniqueness, preserving the history and evolution of clinical data, which may be reused and shared in other healthcare domains.

Among the main challenges faced in HIS development, the following stand out: (i) the lack of uniformity in modeling EHR data; Many health software manufacturers use their own standards to model EHR attributes and terminologies, thus hindering data exchange between organizations; (ii) the difficulty to store EHR heterogeneous data in a single database; Traditionally, many HIS use a single storage approach to represent the different data types from a health domain, which in this case increases the complexity of health application development; (iii) high maintenance cost caused by ever changing requirements and the natural evolution of an HIS life cycle; Many HIS are designed using traditional development techniques, i.e., changes in data schema and the creation of new features depend on a software development team.

Currently, ISO/EN 13606, Health Level 7 (HL7) and openEHR represent important initiatives to improve health application development [4]. Several research studies

A.M.C. de Araújo (✉) • V.C. Times • M.U. da Silva
Center for Informatics, Federal University of Pernambuco, Recife, Brazil
e-mail: amca@cin.ufpe.br; vct@cin.ufpe.br; mus@cin.ufpe.br

© Springer International Publishing AG 2018
S. Latifi (ed.), *Information Technology – New Generations*, Advances in Intelligent Systems and Computing 558, DOI 10.1007/978-3-319-54978-1_36

indicate the use of openEHR Foundation's archetypes and templates [5, 6] as a viable alternative to standardize EHR data and build extensible applications with better maintainability. An archetype can be defined as a computational expression represented by domain constraints that shape and give a semantic meaning to an EHR, while templates are user interaction interfaces created at runtime from specifications defined in the archetypes [7].

In order to minimize problems caused by ever changing requirements for EHR data, polyglot persistence is seen as an alternative to create more flexible data schema [8]. It consists of a mixed storage approach, combining the consistent characteristics of relational data models with the flexibility of NoSQL data models (i.e., key-value, document and graph). The main idea is to represent structured data using a relational approach, while semi-structured and/or unstructured data are represented in NoSQL data models.

The concept of archetypes and templates has been used to model and create EHR prototypes to support patient clinical care activities [8, 9], represent clinical concepts of legacy systems [10] and model EHR in proprietary database systems [11]. However, although the architecture proposed by openEHR is being discussed and used in the development of health applications [12, 13], it is clear that many organizations still lack the tools and methodologies to support HIS development from archetypes. Difficulties relate mainly to how to build EHR data schema from archetypes in heterogeneous databases and how to provide resources so that health specialists themselves (e.g., physicians, nurses, nutritionists, etc.) can build their own GUI for use in the healthcare sector.

This paper presents a cloud service to assist healthcare professionals in building GUI from existing EHR specifications in archetypes. For this purpose, we developed an approach that extracts from archetypes the data attributes which define the EHR, the terminologies that give a semantic meaning to clinical data and the constraints specified on the attributes. The service then automatically generates a GUI with data handling capabilities and EHR storage in heterogeneous databases using polyglot persistence.

There are three main advantages in using this service. First, the GUI and the data schema are created from a health standard that homogenizes data attributes, terminologies and constraints. Second, data heterogeneity can be represented in different data models; for example, hierarchical data from a laboratory exam may be organized as a collection of documents in a NoSQL database, while a patient's demographics may be stored in a relational database. Finally, it reduces the professional's dependence on a programmer to perform changes in data schema and front-end applications. Here, the idea is to represent EHR clinical data that usually undergo the most changes using flexible data models (i.e., key-value, document and graph), while the

remaining data, which undergo fewer alterations, are stored in relational data schema. Additionally, GUIs are created at runtime by the cloud service, based on archetypes chosen by the domain specialist.

The remaining sections of this paper are organized as follows: Sect. 36.2 describes basic concepts for archetypes and templates and indicates related literature. Section 36.3 presents and discusses the proposed service, while Sect. 36.4 discusses the results obtained. Finally, Sect. 36.5 includes final considerations for this paper.

36.2 Background and Related Work

36.2.1 Archetypes

In order to create GUIs and data schema from archetypes, one must first understand the dual architecture concepts used to model the EHR. For this purpose, we specified a Unified Modeling Language (UML) meta-model that represents the dual modeling concept to characterize a given health domain through archetypes. Figure 36.1 illustrates the meta-model created from openEHR definitions. Each concept definition is given below.

An archetype consists of a computational expression based on a reference model and represented by domain constraints and terminologies [6] (e.g., the data attributes of a blood test). A template is a structure used to group archetypes and allow their use in a particular context of application. It is often associated with a graphical user interface (e.g., a GUI used by a professional to define the elements of a leukogram list, such as leukocytes and neutrophils). Dual modeling is the separation between information and knowledge in health care system architectures. In this approach, the components responsible for modeling the EHR clinical and demographic data are specified through generic data structures, which are composed of data types, constraints and terminologies.

Composition is a main container that groups one or more *Sections*, while a *Section* models a health domain. *Observation* shows information about any events associated with the patient's health or clinical condition, while *Instruction* displays narrative statements about clinical care activities performed on the patient. *Activity* details how to perform clinical care defined in *Instruction*. *Action* contains the record of all actions required during the patient's clinical care. *Evaluation* indicates the diagnosis or assesses the patient's risk.

In an archetype, the specification of attributes is achieved through data entry builders named generic data structures. Such structures allow the representation of EHR data heterogeneity through the following types: ITEM_SINGLE, ITEM_LIST, ITEM_TREE and ITEM_TABLE.

Fig. 36.1 Metamodel to represent openEHR dual architecture concepts

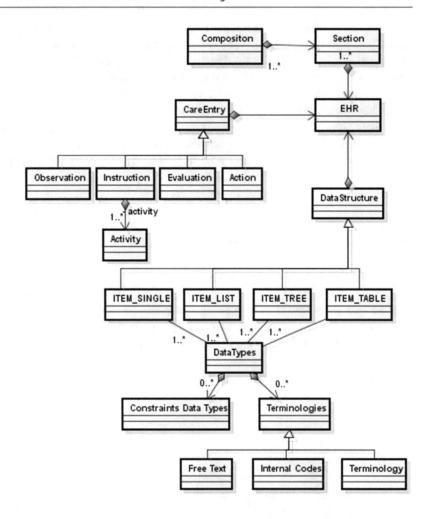

ITEM_SINGLE models a single data attribute such as a patient's weight, height and age. ITEM_LIST groups a set of attributes in a list such as a patient's address containing number, street and zip code. ITEM_TREE specifies a hierarchical data structure that is logically represented as a tree. It can be used, for instance, to model a patient's physical or neurological evaluations. Finally, ITEM_TABLE models data elements by using columns for field definition and rows for field value respectively. Each attribute of a data structure is characterized by a type of data and can have a related set of domain constraints and terminologies. The terminologies give semantic meaning to clinical data and can be represented as a set of health terms defined by a professional.

36.2.2 Related Works

Based on the state-of-the-art works reviewed, we present an analysis of the main related work in the fields of (i) polyglot persistence use in health applications, (ii) generation of templates from archetypes, and (iii) archetype mapping and persistence in databases.

To overcome the rigidity of relational data schema and provide support to the continuous data requirement changes which commonly occur in legacy HIS, Prasad and Sha (2013) [14] specify an architecture and a HIS prototype that allows polyglot persistence and improves health data management in a legacy application. Similarly, Kaur and Rani (2015) [7] specify a polyglot storage architecture to store structured data in a relational database (PostgreSQL), while two NoSQL databases (MongoDB and Neo4j) store semi-structured data, such as laboratory exams and medicine prescriptions. Nevertheless, neither solutions of polyglot persistence make use of archetypes in order to standardize EHR attributes and terminologies.

Späth and Grimson (2010) [12] mapped a set of archetypes for a legacy database and exposed the lack of tools and methodologies that would have helped in modeling archetypes in a database. In this sense, Georg, Judith and Christoph (2013) [15] specified a set of rules to map archetype data attributes in a relational database, generating templates in a specific problem domain. However, the proposed solution could not map archetypes with hierarchical data structures (i.e., ITEM_TREE).

EhrScape (https://www.ehrscape.com) and EHRGen (https://code.google.com/p /open-ehr-gen-framework) frameworks support the health application development process by using the specifications from openEHR and generate templates and data schema from archetype mapping. However, they make use of a single data model to organize EHR data.

The main motivation for the proposed service is to develop a tool capable of building GUI, generating data schema in heterogeneous databases from archetypes and storing EHR data through polyglot persistence.

36.3 Proposed Cloud Service

36.3.1 Architecture and Overview

The service proposed in this paper is designed for the healthcare sector and consists of a computing environment dedicated to the construction of GUI and databases from the specifications existent in EHR archetypes. The core feature of the proposed approach is the extraction of archetypes

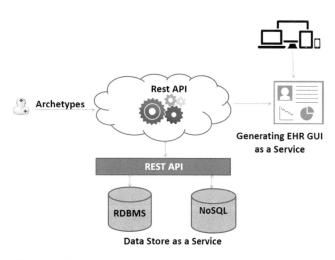

Fig. 36.2 Cloud service architecture

Fig. 36.3 REST API

elements for persistence in different data models, as well as the automatic generation of GUI. Figure 36.2 illustrates the components and architecture relationships designed for the cloud service.

As shown in Fig. 36.2, the cloud service generates data schema and GUI by reading an archetype imported by the user. A REST API extracts from the imported archetype the attributes that define the EHR, the health care terminologies and vocabulary that give a semantic meaning to the clinical data as well as constraints specified on the attributes.

Figure 36.3 shows the data attributes, terminologies in JSON format after the REST API processing. From the extraction of these elements, two operations are performed. First, the data schema are created in a heterogeneous database (i.e., Relational and NoSQL). Second, graphical user interfaces are dynamically created.

The polyglot persistence mechanism in heterogeneous database as well as the generation of GUI proposed in this paper are explained below.

36.3.2 Polyglot Persistence

The polyglot persistence proposed is this work is performed as follows; data of a structured nature such as demographic information is stored in a relational database (i.e., SQL Server), while non-structured data such as a laboratory exam is stored in a NoSQL database. The generation of NoSQL data schema is based on the type of data structures found in archetypes, such as ITEM_SINGLE, ITEM_LIST, ITEM_TREE and ITEM_TABLE.

As the name suggest, ITEM_SINGLE displays a single attribute, while ITEM_LIST groups a set of attributes into a vertical list. Due to the nature of these items, we use the key-value data model to generate data schema. ITEM_TREE displays data attributes in a hierarchical structure containing the several levels of an archetype. In order to maintain such data organization, the cloud service uses the document data model (i.e., collections). Finally, ITEM_TABLE organizes

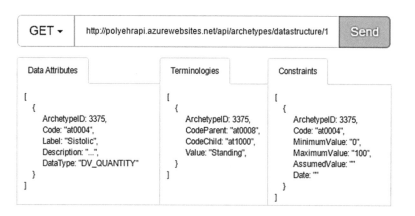

Fig. 36.4 Algorithm for GUI generation

```
Algorithm 1: GUI Generator
Input: DataAttributes[] AS LIST, Constraints[] AS LIST,
       Terminologies[] AS LIST, MetaData[] AS LIST
Output: void()
1    WebForm.Page <- Empty
2    Page.Title <- MetaData['Concept']
3    arrayTextBox[6] AS ARRAY <-
     {'DV_TEXT','DV_COUNT','DV_URI','DV_DATE_TIME','DV_QUANTITY','DV_DURATION'}
4    arrayDropDownList[2] AS ARRAY <- {'DV_CODEDTEXT','DV_ORDINAL'}
5    arrayOptionBoolean[1] AS ARRAY <- {'DV_BOOLEAN'}
6    FOREACH (DataAttribute in DataAttributes[])
7      OBJECT component.Caption <- DataAttribute['DESCRIPTION']
8      IF (DataAttribute['DataType'] EXISTS arrayTextBox[])
9        component.Type <- TEXTBOX
10     ELSE IF (DataAttribute['DataType'] EXISTS arrayDropDownList[])
11       component.Type <- DROPDOWNLIST
12     ELSE IF (DataAttribute['DataType'] EXISTS arrayOptionBoolean[])
13       component.Type <- OPTION
14     END IF
15     FOREACH (Constraint in Constraints[]
             .SELECT(ITEM=> ITEM['CODE'] equals DataAttribute['CODE']))
16       component['CONSTRAINT'][NewIndex].Add(Constraint)
17     LOOP
18     FOREACH (Terminology in Terminologies[]
             .SELECT(ITEM=> ITEM['CODE'] equals DataAttribute['CODE']))
19       component['TERMINOLOGY'][NewIndex].Add(Terminology)
20     LOOP
21     Page.AddComponents(component)
22   LOOP
23   SHOW(Page)
```

the data attributes in a table made of rows and columns. For this type of storage, we use the graph data model, where data found in rows are represented as properties, data from columns as vertices, and attribute description as the relationship between the two.

36.3.3 Generating Graphical User Interfaces

GUI generation is based on data attributes, terminologies and constraints extracted from archetypes. In order to perform this activity, we specified a set of mapping rules that define how to generate and organize GUI components. (i) every GUI is generated from reading an XML archetype; (ii) data attributes extracted from archetypes are used as data entry fields in the GUI; (iii) constraints specified for each data attribute are used with data validation engines in the GUI; (iv) the existing terminologies in each data attribute give a semantic meaning to their respective fields; (v) the GUI data entry field layout is built according to the type of data structures specified in the archetype. Thus, the following rules apply: (a) for the ITEM_SINGLE and ITEM_LIST types, data entry fields are organized directly in the GUI, without the need for tabs. (b) for the ITEM _TREE type, the data attributes grouped into sections are arranged in the GUI through tab(s). (c) the section name defined in the archetype is used to create tab(s) in the GUI. (d) for the ITEM_TABLE type, the number of columns determines the number of tabs

to be created, along with their data attributes organized in the lines. Each tab is named after the column description found in the archetype.

From the definition of mapping rules, we specified an algorithm (i.e., Fig. 36.4) to generate GUI from archetypes. To perform its first task, the algorithm receives data attributes, terminologies and constraints extracted from archetypes as input parameters. With these elements, the algorithm creates an empty GUI (line 1) and gives it the same name as the archetype (line 2). The three vectors shown in lines 3, 4 and 5 define the type of component to be created for each type of data found in the archetype.

In the example above, there are three main components: Text Box, Drop down list (i.e., values drop-down) and Option button (i.e., Boolean choice between two values). In addition to the type to be created, the algorithm validates the information manipulated in each component. For example, the DV_QUANTITY type only accepts integer values, whereas the DV_COUNT type accepts decimal values. The code snippet between lines 6 and 14 scans through data attributes and creates their respective components according to the types found. Then, terminologies and constraints associated to each data attribute are validated. In order to do so, the code snippet between lines 15 and 20 identifies and associates terminologies and constraints to each data attribute. Finally, the created components are added to the GUI (line 21), the loop is finished (line 22) and the form is shown on display (line 23).

36.4   Results and Discussion

In order to evaluate the cloud service proposed in this paper, we have created a set of GUI to manage an EHR in a public hospital located in northeastern Brazil. Here, a physician performs three activities; First, they check and examine the patient and if necessary, accommodate the patient in an observation room. They then perform the clinical tasks and ultimately request medicine from the pharmaceutical department to be administered to the patient. In order to model this activity flow, we relied on the help of health professionals working at the hospital and specified three archetypes: clinical appointment, patient evolution and medicine request. The archetypes built were validated by the physician in charge of the department and are available at polyehr. azurewebsites.net/files/.

We have two main objectives in this evaluation. The first is to assess whether the cloud service correctly generates GUI for the healthcare sector and second, check if the health professionals themselves can generate GUI to use in patient care. Three physicians voluntarily participated in this evaluation.

One physician generated a GUI from the archetypes built and, along with two other professionals, manipulated data from patient care provided at the hospital. Figure 36.5 shows the GUI generated from the medicine request archetype.

In this first phase, we evaluated: (a) XML archetype automatic mapping, (b) individual element selection for each archetype, (c) data schema generation and (d) graphical user interface generation.

For each function carried out in the cloud service, we asked the doctor to indicate if the result was satisfactory, partially satisfactory or unsatisfactory. Each response representing a score of 1, 0.5 and 0, respectively. Table 36.1 shows the results obtained after the physicians' evaluation.

As shown by the results in Table 36.1, all three functions evaluated were satisfactory. That is to say, all terminologies, attributes and constraints specified on the archetypes were correctly mapped by the cloud service, while each user was able to choose which archetype elements to use in order to create data schema. Finally, the GUI were correctly generated using the data manipulation resources.

We then asked for physicians to perform insert, update, delete, and data query operations using the generated GUI. In addition, the physicians answered a set of questions to evaluate the usability of the GUI. In the scale adopted, 5 is fully satisfactory and 1 unsatisfactory.

All data manipulation operations were performed successfully and users were able to log the clinical data in the GUI. The generated GUI obtained a satisfaction level of 93,33%, 86,66 and 80% from the three physicians who participated in this evaluation.

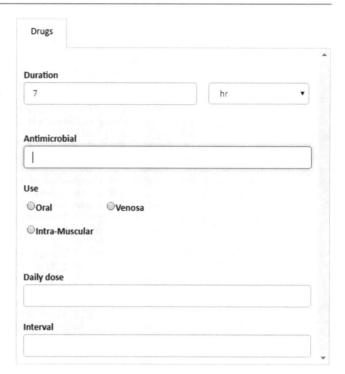

Fig. 36.5 GUI generated for the medicine request archetype

Table 36.1 Cloud service evaluation results

Archetype/Function	A	B	C	D
Clinical appointment	1	1	1	1
Patient's evolution	1	1	1	1
Request Medicines	1	1	1	1

In each GUI created, it is possible to view the stored data, access the editing form and insert new records, as well as print and delete data entries already stored. In this case, the REST API encapsulates the heterogeneity of the manipulation languages used by the adopted data models. The GUI shown in this section and all other resources are available at http://template4ehr.azurewebsites.net/

36.5   Conclusion

In this paper we presented a cloud service able to build graphical user interfaces and store electronic health record using the polyglot persistence concept. In order to do so, we specified a UML metamodel that models and demonstrates how archetype concepts are related. A cloud service architecture was designed and a REST API was developed in order to extract data attributes, terminologies and constraints from archetypes. A polyglot persistence mechanism then creates data schema in heterogeneous databases to store the EHR data. Furthermore, we specified an algorithm able to dynamically generate GUI from the extraction of archetypal elements. Finally, we validated the proposed cloud service in a hospital and demonstrated how health professionals can create and manage a fully functional healthcare application without depending on a software development team.

References

1. Lim, A. K., & Thuemmler, C. (2012). *Opportunities and challenges of internet-based health interventions in the future internet.* 12th International conference on Information Technology – New Generations, 2012.

2. Gurung, S., & Kim, Y. (2015). *Healthcare privacy: How secure are the VOIP/video-conferencing tools for PHI data?.* 12th International conference on Information Technology – New Generations, 2015.

3. Coorevits, P., Sundgren, M., Klein, G. O., Bahr, A., Claerhout, B., Daniel, C., Dugas, M., Dupont, D., Schmidt, A., Singleton, P., De Moor, G., & Kalra, D. (2013). Electronic health records: New opportunities for clinical research. *Journal of Internal Medicine, 274*(6), 547–560.

4. Marco, E., Thomas, A., Jorg, R., Asuman, D., & Gokce, L. (2005). A survey and analysis of electronic healthcare record standards. *ACM Computing Surveys, 37*(4), 277–315.

5. Lezcano, L., Miguel, A. S., & Rodríguez, S. (2010). Integrating reasoning and clinical archetypes using OWL ontologies and SWRL rules. *Journal of Biomedical Informatics, 44*(2), 343–353.

6. Martínez, C. C., Menárguez, T. M., Fernández, B. J. T., & Maldonado, J. A. (2009). A model-driven approach for representing clinical archetypes for Semantic Web environments. *Journal of Biomedical Informatics, 42*(1), 150–164.

7. Lloyd, D., Beale, T., & Heard, S. *openEHR architecture: Architecture overview, 2008.* Available at http://www.openehr.org/releases/ 1.0.2/architecture/overview.pdf. Last accessed Jan 2016.

8. Kaur, K., & Rani, R. (2015). Managing data in healthcare information systems: Many models, one solution. *IEEE Computer Society, 48*(3), 52–59.

9. Buck, J., Garde, S., Kohl, C. D., & Knaup, G. P. (2009). Towards a comprehensive electronic patient record to support an innovative individual care concept for premature infants using the openEHR approach. *International Journal of Medical Informatics, 78*(8), 521–531.

10. Chen, R., Klein, G. O., Sundvall, E., Karlsson, D., & Åhlfeldt, H. (2009). Archetype-based conversion of EHR content models: Pilot experience with a regional EHR system. *BMC Medical Informatics and Decision Making, 9*, 33.

11. Bernstein, K., Bruun, R. M., Vingtoft, S., Andersen, S. K., & Nøhr, C. (2005). Modelling and implementing electronic health records in Denmark. *International Journal of Medical Informatic, 74*(2–4), 213–220.

12. Späth, M. B. & Grimson, J. (2010). Applying the archetype approach to the database of a biobank information management system. *International Journal of Medical Informatics, 80*(3), 205–226.

13. Garde, S., Hovenga, E., Buck, J., & Knaup, P. (2007). Expressing clinical data sets with openEHR archetypes: A solid basis for ubiquitous computing. *International Journal of Medical Informatics, 76*(Suppl 3), 334–341.

14. Prasad, S., & Sha, N. (2013). *NextGen data persistence pattern in healthcare: Polyglot persistence.* Fourth international conference on Computing, Communications and Networking Technologies, 2013.

15. Georg, D., Judith, C., & Christoph, R. (2013). Towards plug-and-play integration of archetypes into legacy electronic health record systems: The ArchiMed experience. *BMC Medical Informatics and Decision Making, 13*(1), 1–12.

Yojiro Harie and Katsumi Wasaki

Abstract

This paper proposes an on-the-fly linear temporal logic model checker using state-space generation based on Petri net models. The hierarchical Petri net simulator (HiPS) tool, developed by our research group at Shinshu University, is a design and verification environment for Place/Transition nets and is capable of generating state-space and trace-process graphs. In combination with external tools, HiPS can perform exhaustive model checking for a state space. However, exhaustive model checking is required for generating a complete state space. On-the-fly model checking is an approach for solving the explosion problem to generate a portion of the overall state space by parallelizing the search and generation processes. In this study, we propose a model checker for a Petri net model to concurrently function with state-space generation using an interprocess communication channel. By utilizing the concept of fluency, we implement automata-based model checking for Petri nets. This implementation achieves high-efficiency for on-the-fly verification, which is independent of the verification and state-space generation processes.

Keywords

Petri net • Model checking • On-the-fly • Linear Temporal Logic (LTL) • Scalability

37.1 Introduction

A test or simulation is used for the verification of system reliability; however, neither of these methods can prove that the system is free of bugs. Formal verification is a method for proving the correctness of system specifications described in a mathematically rigorous framework. Language of temporal ordering specification (LOTOS) [1] and communicating sequential processes (CSP) [2] are examples of formal specification languages. Each language has different features that are related to descriptiveness and abstract level, the behavior of systems must be modeled accurately for verification purposes.

A Petri net is a graphical and mathematical modeling tool that can be applied to many systems. It is effectual to describing and investigation information processing systems that are concurrent, asynchronous, distributed, parallel, non-deterministic, and/or stochastic [3]. Petri nets have the ability to analyze properties, e.g. safety, liveness, boundedness and fairness. In particular, the liveness property on Petri net guarantees that a deadlock never occurs in the system. For modeling with Petri nets, to verify the model is enabled rigorously by analyzing the properties and the behaviors of Petri nets.

The basic class of Petri nets is the Place/Transition net (P/T net). P/T nets contain only the logical relationship of event occurrence. Introducing a temporal concept, two classes of Petri nets are presented, timed Petri nets (TPNs) and stochastic Petri nets (SPNs) [4]. Colored Petri nets (CPNs) are a backward-compatible extension of the concept of Petri nets because its tokens can hold data. CPN Tools [5] is a

Y. Harie (✉) • K. Wasaki
Interdisciplinary Graduate School of Science and Technology, Shinshu University, 4-17-1, Wakasato, Nagano City, Nagano 380-8553, Japan
e-mail: 16st207c@shinshu-u.ac.jp; wasaki@cs.shinshu-u.ac.jp

modeling and analysis tool for CPNs, and platform independent Petri net editor 2 (PIPE2) [6] is a modeling and simulation analysis tool for TPNs and SPNs.

Hierarchical Petri net simulator (HiPS) has been developed and released by our research group as a Petri net design tool [7]. The features of this tool include a common operating method, graphical user interface (GUI), the ability to describe hierarchical Petri nets, and applicability to timed Petri nets. HiPS can simulate a random walk model and observe the behaviors as a firing sequence of a Petri net model. The state-space generator is already implemented in HiPS. The state space consists of the initial state as the initial marking of Petri nets, actions which are defined by firing sequence, and states expressed by the change of markings. This tool has a mechanism for generating a state space and it outputs labeled transition systems (LTS) in the Aldebaran format. Using the construction and analysis of distributed processes (CADP) model checking tools [8] developed by Validation de Systémes/Institut National de Recherche en Informatique et en Automatique (VASY/INRIA), this tool can extract traces that satisfy the conditions in a state space.

Model checking [9] is an automatic verification method using mathematical analysis. Model checking is a useful means of eliminating bugs. However, it can cause state explosion for large complex system models. On-the-fly execution has been proposed to increase the efficiency of model verification. Such execution can perform model checking in parallel with state-space generation.

Model checking can be performed already for Petri net models designed using HiPS together with external tools. However, constructing the overall state space representing the exhaustive behavior of the system model is essential. This is the principal issue with using external tools to verify Petri net models that cause state explosion. Completing on-the-fly verification prior to constructing the overall state space enables HiPS tools to be adapted to large-scale model verification. In the present study, we incorporate an automata-based on-the-fly model checker using specifications written in linear temporal logic (LTL) by converting into fluent automaton.

The remainder of this paper is organized as follows. Section 37.2 presents Petri net and Petri net properties. HiPS, which is Petri net deseign tool, is described in Sect. 37.3. Section 37.4 explains on-the-fly model checking. In particular, model checking for HiPS is described in Sect. 37.5. Techniques for implementing the model checker on HiPS are described in Sect. 37.6. A verification example for ABP is described in Sect. 37.7. Tha last section concludes our paper and suggests the future work.

37.2 Petri Net

A Petri net is a particular kind of bipartite directed graph consisting of two kinds of nodes called places and transitions [3]. Arcs are either from a place to a transition or from a transition to a place. Introducing the concept of conditions and events for modeling, places represent conditions, and transitions represent events. A transition has a specific number of input and output places representing the pre- and postconditions of the event, respectively. The presence of a token in a place is interpreted as indicating the truth of the condition associated with the place. A marking (state) assigns to each place a nonnegative integer. If a marking assigns to place p a nonnegative integer k, we say that p is marked with k tokens. For a marking that represents the states of a system, a Petri net's entire marking of m places is represented by an m-dimensional vector M. In particular, an initial marking state is referred to as the initial marking M_0. The pth component of M, denoted by $M(p)$, is the number of tokens in place p.

Definition 1 (Place/Transition (P/T) net) A Petri net is a 5-tuple, $PN = (P, T, F, W, M_0)$, where $P = \{ p_1, p_2, \cdots, p_m\}$ is a finite set of places,

$T = \{ t_1, t_2, \cdots, t_n\}$ is a finite set of transitions,

$F \subseteq (P \times T) \cup (T \times P)$ is a set of arcs,

$W: F \to \{ 1, 2, 3, \cdots \}$ is a weight function,

$M_0: P \to \{ 0, 1, 2, 3, \cdots \}$ is the initial marking.

A Petri net structure $N = (P, T, F, W)$ without any specific initial marking is denoted by N. A Petri net with the given initial marking M_0 is denoted by (N, M_0).

Definition 2 (Incidence Matrix) Let (P, T, F, W, M_0) be a P/T net and $| P | = m$, $| T | = n$. The incidence matrix $A = [a_{ij}]$ is an $n \times m$ matrix of integers that typical entry is $a_{ij} = a_{ij}^+ - a_{ij}^-$ where $a_{ij}^+ = w(i, j)$ is the weight of the arc from transition i to its output place j and $a_{ij}^- = w(j, i)$ is the weight of the arc to transition i from its input place j.

Definition 3 (Firing Condition) Let (P, T, F, W, M_0) be a P/T net with $^*t = \{ p \mid (p, t) \in F\}$, $t^* = \{ p \mid (t, p) \in F\}$. A transition $t \in T$ is enabled at a marking M, if every place $p \in {}^*t$ satisfies $m(p) \geq w(p, t)$, and every place $p \in t^*$ satisfies $m(p) + w(p, t) \leq k(p)$. The occurrence of t leads to the successor marking m', defined by $m'(p) = m(p)$ if $p \notin {}^*t$ and $p \notin t^*$

or $m(p) - w(p, t)$ if $p \in {}^*t$ and $p \notin t^*$

or $m(p) + w(t, p)$ if $p \notin {}^*t$ and $p \in t^*$

or $m(p) - w(p, t) + w(t, p)$, if $p \in {}^*t$ and $p \in t^*$.

Definition 4 (Reachability) Let (P, T, F, W, M_0) be a P/T net. A marking M_n is reachable from M_0 if there exists a sequence of firings that transforms M_0 to M_n.

By $\sigma = M_0 t_1 M_1 t_2 \ldots t_n M_n$, we denote a firing sequence. In this case, a marking M_n is reachable from M_0 by σ; we denote it by $M_0 [\sigma > M_n$. For a Petri net (N, M_0), the set of all possible markings reachable from the initial marking M_0 is denoted by $R(N, M_0)$ or simply $R(M_0)$. The set of all possible firing sequences from M_0 in a net (N, M_0) is denoted by $L(N, M_0)$ or simply $L(M_0)$.

For a Petri net (N, M_0), in the process of making reachable markings from the initial marking, we can retrieve a tree representation of the reachability markings. If the state space of (N, M_0) is bounded, then the tree representation is called a reachability graph. A reachability graph is defined as follows.

Definition 5 (Reachability Graph) Let there be a Petri net (N, M_0) and a directed graph $G = (V, E)$. V is a set of all states (nodes) in reachability markings of (N, M_0). In other words, the set V is equal to $R(M_0)$. Each pairs $(M_i, M_j) \in E$ is labeled by a transition t_k, where $M_i [t_k > M_j$, and $t_k \in (N, M_0)$.

The exhaustive behavior of systems is called state space. State-space structures in Petri nets are given as reachability graphs.

37.3 The Petri Net Design Tool: HiPS

HiPS is a Petri net design tool developed and released by our research group [7]. This tool can observe the behavior of Petri net models generated by simulating random walks from the initial marking. HiPS tool can work out a hierarchical and timed-net design. HiPS enable a complex hierarchical model to be designed flexibly by considering a subpage described a Petri net model to an object. Figure 37.1 shows a HiPS operating screen and Muller's C-element circuit represented as a P/T-net class [10]. Petri nets have two sets of properties: structural properties, which are related to the structure of the net, and dynamic properties, which depend on the initial marking. HiPS can analyze both sets of properties. Model checking can be performed for Petri net models designed in HiPS in cooperation with external tools. The verification features of HiPS tool are shown below.

(1) State Space Generation: An LTS $L = (S, A, \delta, s_0)$ consists of a non-empty set of states, S, a non-empty set of (transition) labels A, a transition relation $\delta \subseteq S \times A \vee \{\tau\} \times S$, and an initial state $s_0 \in S$. HiPS tool generates the exhaustive behavior of the entire system as a state space, the set of all states reachable from the initial state [10].

State spaces are expressed by the LTS and are output in Aldebaran-Automaton format, which is a file format of the LTS. The LTS labels the transitions between states and describes the behavior of the system based on an event. In addition to the core portion of the tool implemented in C#, HiPS is implemented in C++ for state-space generation from the viewpoint of parallel libraries and execution speed. We use Intel C++ Compiler, an optimizing compiler, and Intel TBB, a C++ parallelization library.

The checking process exhaustively searches the state space to guarantee completely that the properties that must be satisfied in the system are satisfied. In our study, model

Fig. 37.1 Example of principal HiPS GUI design canvas

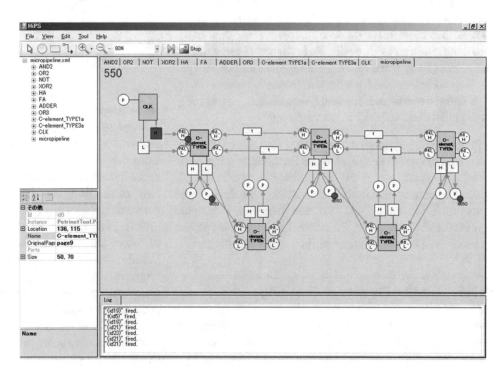

Fig. 37.2 Component structure for generating the reachability graph

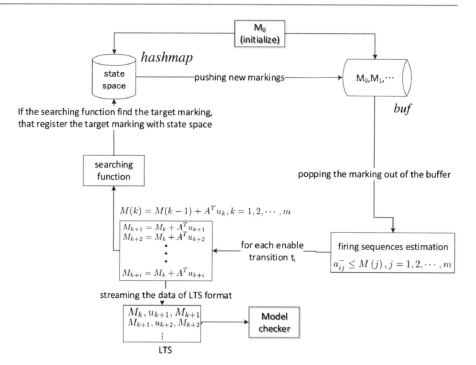

checking requires that the state space be reachable from the initial state via the connection graph, because the model checking is based on automata. Therefore, generating the state space allows only a single thread. This is shown in Fig. 37.2. As the firing assessment from the initial marking of Petri nets is performed sequentially, the generating process adds a marking to the state space if it finds a new marking [10].

(2) Structural Properties Analysis: Structural properties depend on the topological structure of Petri nets. The properties are independent of the initial marking in the sense that these maintain their properties for any initial marking or are concerned with the existence of certain firing sequences from some initial marking. his tool can analyze 7 properties: structurally bounded, (partially) conservative, (partially) repetitive, (partially) consistent, structurally unbounded, unconservative, unconsistent.

(3) Dynamic Properties Analysis: Examples of such dynamic properties analysis include reachability graph analysis, (extended) coverability graph analysis, deadlock analysis, k-bounded analysis, reversibility analysis, and synchronic distance and fairness analysis. Dynamic properties depend on the initial marking, these properties are analyzed by using generated state space actually.

37.4   On-the-Fly Model Checking

Model checking is an automatic method for verifying the correctness properties of systems. The properties of systems are expressed via LTL and computation tree logic (CTL).

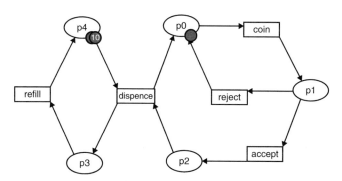

Fig. 37.3 Motivative P/T-net example: vending machine process model (storage capability $= 10$)

Safety properties assert that undesirable events or states never occur under particular conditions. An LTL formula that describes the properties of the system can be converted into an equivalent Büchi and the system automaton. A path from the initial state to an accepting state is a concrete counterexample that the synchronous product does not have an accepting run. Figure 37.3 shows a simplified Petri net model of a vending machine. A desired property of this system is that it should avoid not refilling after a coin is accepted. This vending machine accepts a coin by the *accept* transition and refills the stock by the *refill* transition. The required safety property can be expressed via LTL as **G** (*accept* $\Rightarrow$ **F** *refill*). The problem of state predicates in event-based system descriptions is described in Sect. 37.5.

The state explosion problem increases the processing time and memory space when the target model is enormous. On-the-fly checking is a method for alleviating the state explosion problem [11]. In on-the-fly checking, the searching process operates concurrently with the process of generating the state space. Therefore, on-the-fly checking can terminate searching and generate the state space earlier rather than constructing the overall state space when an acceptable sequence is detected. On-the-fly execution is a useful approach for reducing memory and execution time when model checking has the state explosion problem.

37.4.1 Nested Depth First Search

Nested depth-first search (DFS) is a sequential minimal optimization search algorithm for LTL verification. The acceptance condition of Büchi automata is to frequently pass through an accepting state infinitely. The nested DFS algorithm includes dual DFS. The first DFS begins at the initial state and visits all accepting states. The searching process switches into the second DFS whenever an accepting state is discovered. The second DFS begins to search the cycle graph at the accepting state. The nested DFS algorithm is known to terminate the search process before the generating process finishes constructing the state space.

Figure 37.4 shows an example of exploration procedure of accepting a state in which the Nested DFS for a synchronization product automaton. The search process of the first DFS begins at an initial state S_0 and visits the nodes until an accepting state A_0 is discovered. After detecting such a state, the first DFS process calls the second DFS process and searches an accepting cycle. The accepting path is a concrete counterexample for a given specification when a path reaching an accepting state from an initial state is detected. In Fig. 37.4, the state A_1 is an accepting state. However, the path of A_1 from an initial state is not a candidate for counterexamples. Therefore, there are no accepting cycles of A_1.

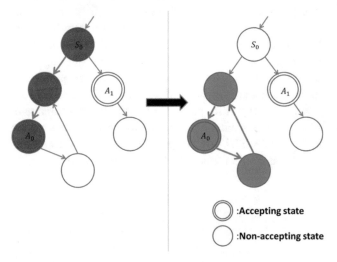

:Accepting state

:Non-accepting state

Fig. 37.4 Exploration procedure of accepting a state in which the Nested DFS for a synchronization product automaton is to be applied

37.5   Model Checking for LTS in HiPS

LTS provides models for describing the sequences of events observed in systems. Therefore, we need to describe a property related to the events in LTL. However, LTL is defined in relation to a Kripke structure. Each relation of a Büchi automaton is labeled with propositions of models. Otherwise, the relations of generated state spaces of LTS are not labeled explicitly by fluent.

Fluent is a truth-value predicate defined by events, and fluent LTL (FLTL) has been proposed [12].

Definition 6 (Fluent) A fluent fl by a pair of sets: $fl = < I_{fl}, T_{fl} >$, I_{fl} is a set of start actions, T_{fl} is a set of end actions, $I_{fl} \wedge T_{fl} \neq \emptyset$. In addition, a fluent fl have the attribute $Initially_{Fl}$. $Initially_{Fl}$ value is true or false at time zero.

A set of labels of Büchi automaton relations consistent with FLTL is different from a set of transition labels in LTS. Therefore, a Büchi automaton that is equivalent to propositions of FLTL cannot synchronize with LTS because both sets of labels are different.

There is a method for converting a Büchi automaton into a tester automaton that is equivalent to LTS with accepting states. A label of transitions as a tester automaton is described by the events that appear in the model. The flow of the verification process is shown in Fig. 37.5. An FLTL formula from the input form is converted to a Büchi automaton is converted to a tester automaton. The searching process constructs a synchronous product with the Büchi automaton and the LTS as its state space and also searches for an acceptable sequence from the synchronous product. The searching process constructs a composed automaton by synchronous products of transition systems with the Büchi automaton and the LTS as its state space and also searches for an acceptable sequence from the composed automaton. The acceptable sequence as a counterexample is shown when that sequence is detected.

37.5.1 Tester Automaton

A tester automaton is generated by the parallel composition of three automata: the Büchi automaton, the Fluent automaton defined by the fluent concept, and the Synchronizer automaton, which operates the connection between the Büchi and the Fluent automata. A tester automaton is defined as follows:

$$Tester_{AM} = (F_{FL_1}||\cdots||F_{FL_n}||Synch_{AM\Phi}||B\ddot{u}chi_{\Phi})\backslash 2^{\Phi}$$

The concept of fluent is proposed as the predicate defined by the event. The fluent automaton is determined by its initial state by the truth-value of the fluent.

Fig. 37.5 Flow diagram of on-
the-fly LTL model checking
in HiPS

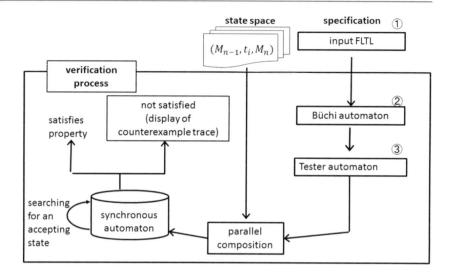

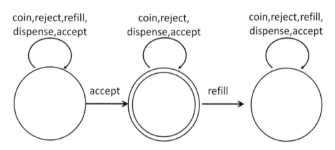

coin,reject,refill,   coin,reject,   coin,reject,refill,
dispense,accept   dispense,accept   dispense,accept

accept   refill

Fig. 37.6 Fluent tester automaton obtained from LTL formula

37.5.2 Graph Contraction

The overall state space subsequent to parallel composition
becomes smaller when conducting graph contraction. The
minimizing technique for the generated tester automaton has
two steps:

STEP 1.   This step contracts internal transitions in the
             tester automaton.
STEP 2.   According to strong bi-simulation equivalence,
             the minimizing technique conducts graph
             contraction.

Contracting formally the tester automaton as the LTS that
has accepting states, we describe the accepting states as self-
loops that are especially labeled. Figure 37.6 shows the

Fluent tester automaton obtained from LTL formula $\{\neg\}\mathbf{G}$
(*accept* $\Rightarrow$ **F***refill*).

37.6 Implements

37.6.1 User Interface

Describing propositions of FLTL formulas on the user interface (UI), each element in the canvas is provided information
as a design entry [13]. The buttons are laid out by each of the
temporal operators ([](Globally), $<>$ (Future), X(Next), U
(Until)) and the logical operators ($\{\neg\}$, $\vee$, $\wedge$) on this form. A
proposition variable in the formula is equivalent to a transition
appeared in the inspected net. The fluent *fluent(a)* consists of
only action a as follows: $Initially_{Fl} = <\{a\}, A -\{a\}>$,
where A is a set of Actions, $Initially_a = false$. Obtained fluents
are converted to equivalent automatons for internal
processing. According to whether the scale of the model
increases, the number of instances of Petri nets increases.
Accordingly, it is difficult to understand what the monitored
transition means. Therefore, the approach of referring to the
instance id of designed models is the cause of the problem of
the increasing cost of FLTL description. We propose FLTL
description support for providing input to the form automatically with the monitored transitions to reduce the description
costs. Figure 37.7 shows an LTL description form that
includes an FLTL formula $\{\neg\}\mathbf{G}$ (*accept* $\Rightarrow$ **F** *refill*).

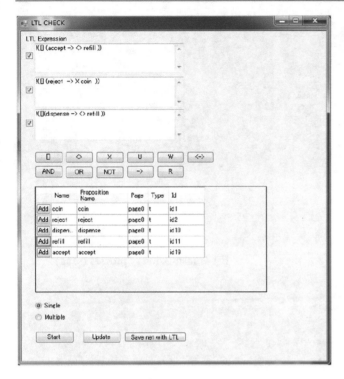

Fig. 37.7 LTL model checking form feature in HiPS

37.6.2 LTL2BA

A Büchi automaton is a type of ω-automaton that extends a finite automaton to infinite inputs [14]. A verification method is proposed, converting a negated LTL formula to an equivalent Büchi automaton. The LTL2BA tool can convert an LTL formula to a never claim. In this study, we export LTL2BA in C++ to implement it in HiPS.

We describe the behavior of the NDFS algorithm. *blue_dfs* and *red_dfs* are searching function. *blue_dfs* can perform accepting states detection, starting from the initial state. *red_dfs* can perform to detect an accepting cycle including the discovered state by *blue_dfs*.

37.6.3 Interprocess Connection Between the Model Checking Process and State-Space Generator

To achieve on-the-fly model checking, we implement the verification process for parallel execution of the state-space generator that has already been implemented in HiPS. The

inter-process communication (IPC) channel is defined as a communication channel for the remote use of the IPC system of the Windows operating system[15]. Through the use of the IPC channel, the state-space generator is expanded to include an LTS transfer function. Operating a remote object to transfer LTS data, the state-space generating process and the verification process can transmit and receive these data.

37.6.4 Verification Example

Alternating bit protocol (ABP) is a network protocol for transferring messages in one direction between connections that makes transfer errors [16]. Transferring a message, the sending side provides an alternative bit for the message. The receiving side compares the expected bit with the alternative bit of the transferred message. If the expected bit differs from the sending one, the receiving side sends an ack message.

The following model of ABP is described by the Petri net in Fig. 37.8. In Fig. 37.8, the sending side, the connection, and the receiving side correspond to the send, medium, and receive modules, respectively. The number of states in this model is 94. Thus, the state space of this model is bounded. Given an LTL fomula ($PUT \Rightarrow \mathbf{F}\ GET$), we obtain a safety sequence as a result. There is a possibility that this model makes endless loops when repeatedly firing the transition sending an error message. Given an LTL formula $\{\neg\}$ ($\mathbf{G}\ (PUT \Rightarrow \mathbf{F}\ GET)$), we obtain a counterexample: $id14$ (PUT) $\Rightarrow$ $^{*}[L1\_id0(send)\_id1(put0) \Rightarrow id1.\ port0(in0) \Rightarrow L1\_id1(medium)\_id11(error0) \Rightarrow id1.\ port/Emphasis>4 (error) \Rightarrow L1\_id3(receive)\_id19(error) \Rightarrow id2.\ port2(in1) \Rightarrow id2.\ port3(out1) \Rightarrow L1\_id0(send)\_id34(resend0)]$.

The erroneous transmission is caused by the medium from the sending side to the receiving side in this counterexample.

37.7 Conclusion

HiPS can perform model checking using the external tool CADP, which is necessary for generating the entire state space. Implementing the on-the-fly execution, capable of the application range of models have been extended. Future work includes support for model correction and simplification of the specification description. SPEC PATTERNS has

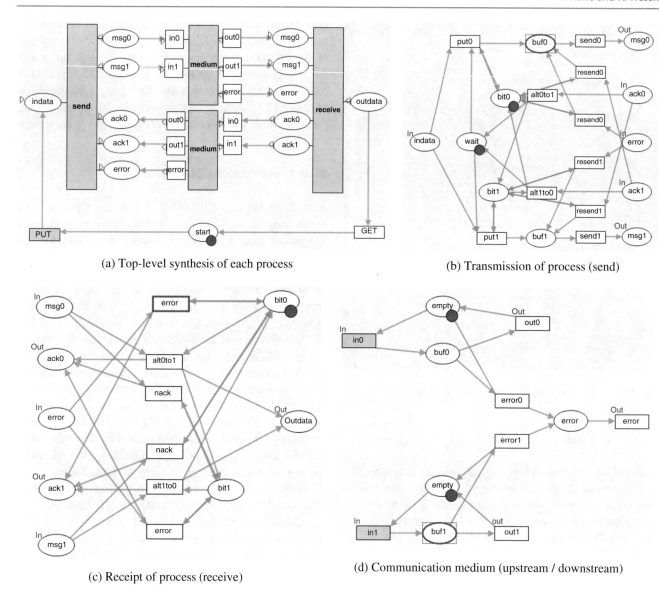

(a) Top-level synthesis of each process

(b) Transmission of process (send)

(c) Receipt of process (receive)

(d) Communication medium (upstream / downstream)

Fig. 37.8 Hierarchical Petri net model example of the ABP. (**a**) Top-level synthesis of each process. (**b**) Transmission of process (send). (**c**) Receipt of process (receive). (**d**) Communication medium (upstream/downstream)

been proposed to describe property specification for finite-state verification [17].

References

1. ISO. (1989). *Information processing systems – Open systems interconnection – LOTOS: A formal description technique based on the temporal ordering of observational behaviour.* Geneve: International Organization for Standardization. Technical Report 8807:1989.
2. Hoare, C. A. R. (1978). Communicating sequential processes. *Communications of the ACM, 21*(8), 666–677. [Online]. Available: http://doi.acm.org/10.1145/359576.359585.
3. Murata, T. (1989). Petri nets: Properties, analysis and applications. *Proceedings of the IEEE 77*, 541–580.
4. Reisig, W. (1985). *Petri nets: An introduction* (EATCS monographs on theoretical computer science, Vol. 4). Berlin/New York: Springer.
5. Westergaard, M., & Verbeek, H. E. CPN tools. http://cpntools.org/.
6. Dingle, N., & Knottenbelt, W. Platform independent petri net editor 2. http://pipe2.sourceforge.net/.
7. University, S. (2016) Hips – Hierarchical petri net simulator. [Online]. Available: http://sourceforge.net/projects/hips-tools/.
8. VASY/INRIA. (2015) CADP toolbox. [Online]. Available: http://cadp.inria.fr/.
9. Clarke, E. M., Grumberg, O., & Peled, D. (2001). *Model checking.* Cambridge: MIT Press.
10. Ohta, I., & Wasaki, K. (2013). Model designing using a Petri net tool and state space generation algorithm for post-verification tool. In *12th Forum on Information Technology, FIT2013* (pp. 171–174).
11. Schwoon, S., & Esparza, J. (2005). A note on on-the-fly verification algorithms. In *Proceedings of the 11th International Conference on*

Tools and Algorithms for the Construction and Analysis of Systems, TACAS2005 (pp. 174–190).

12. Giannakopoulou, D., & Magee, J. (2003). Fluent model checking for event-based systems. In *Proceedings of the 11th ACM SIGSOFT Symposium on Foundations of Software Engineering, 2003.* (pp. 257–266). ACM.

13. Harie, Y., & Wasaki, K. (2015). On-the-fly LTL model checker on the Petri net design tool: HiPS. In *14th Forum on Information Technology, FIT2015* (pp. 139–142).

14. Gastin, P., & Oddoux, D. (2001). Fast LTL to büchi automata translation. In *Proceedings of the 13th International Conference on Computer Aided Verification, CAV2001* (pp. 53–65).

15. Stevens, W. R., Fenner, B., & Rudoff, A. M. (2003). *UNIX network programming* (Vol. 1, 3rd ed.). Delhi: Pearson Education.

16. Tel, G. (2001). *Introduction to distributed algorithms* (2nd ed.). New York: Cambridge University Press.

17. laboratory, S. (2015). Spec patterns. [Online]. Available: http://patterns.projects.cis.ksu.edu/.

Optimally Compressed Digital Content Delivery Using Short Message Service

Muhammad Fahad Khan, Mubashar Mushtaq, and Khalid Saleem

Abstract

Researchers are devising new ways for robust digital content delivery in situations where telecommunication signal strength is very low, especially during natural disasters. In this paper, we present research work targeting two dimensions: (a) We selected the IANA standard for digital content classification, 20 types in 5 categories; applied and compared five different lossless compression schemes (LZW, Huffman coding, PPM, Arithmetic Coding, BWT and LZMA) on these 20 data types; (b) A generic prototype application which encodes (for sending) and decodes (on receiving) the compressed digital content over SMS. Sending digital contents via SMS over satellite communication is achieved by converting digital content into text; apply lossless compression on the text and transmit the compressed text by using SMS. Proposed method does not require Internet Service and also not requires any additional hardware in existing network architecture to transmit digital contents. Results show that overall PPM compression method offers best compression ratio (0.63) among all compression schemes Thus PPM reduces the SMS transmission saving up to 43%, while LZW performs the least with 17.6%.

Keywords

Multimedia Communication • Digital Content Delivery • Satellite communication • Lossless compression schemes

38.1 Introduction

Global System for Mobile Communications (GSM) was launched in early 1990's that has grown exponentially during the last decades. Today, it is the largest telecommunication network technology, with 838 GSM networks, spread in around 234 countries, covering 70% (approximately 4.6 billion) of the world's mobile subscriptions [1, 2].

One of the main disadvantages of GSM network is its infrastructure's disability to withstand natural disasters. The partial or complete damage of telecommunications infrastructure leads to delays and errors in emergency response and disaster relief efforts [3]. There are certain other scenarios where complete telecommunication infrastructure network would not be a feasible option: under-populated regions, war region, ships, planes, mountains etc. [5]. The

M.F. Khan (✉)
Department of Computer Sciences, Quaid-i-Azam University, Islamabad, Pakistan

Department of Software Engineering, Foundation University, Islamabad, Pakistan
e-mail: mfkhan@cs.qau.edu.pk

M. Mushtaq
Department of Computer Sciences, Quaid-i-Azam University, Islamabad, Pakistan

Department of Computer Science, Forman Christian College – A Chartered University, Lahore, Pakistan
e-mail: mubasharmushtaq@fccollege.edu.pk

K. Saleem
Department of Computer Sciences, Quaid-i-Azam University, Islamabad, Pakistan
e-mail: ksaleem@qau.edu.pk

© Springer International Publishing AG 2018
S. Latifi (ed.), *Information Technology – New Generations*, Advances in Intelligent
Systems and Computing 558, DOI 10.1007/978-3-319-54978-1_38

Table 38.1 Sms saving among all content types

PPM	BWT	LZMA	LZW	AC	Huffman
43.28	42.30	42.77	17.59	24.8	28.68

provision of communication facility to these areas is not possible using existing GSM based networks. The only viable solution is satellite communication which is carried through the satellite orbiting around the earth rather than towers. Satellite transmission depends upon atmospheric transport. There are several factors which directly impact the transmission quality and signal strength reduction in satellite based communication, such as climate changes (rain, cloudy), building heights, dense forest, etc. In such a condition, making a voice call is not viable solution because of weak signal strength. It is proposed in [5] that Short Message Service (SMS) could be the best option, rather than voice calls in satellite networks.

Based on the above mentioned method, the work presented in this paper proposes a methodology for sending any type of digital contents over satellite communication through SMS. The proposed method is tested with the help of a prototype application that converts the digital content into ASCII text and then sends it as a payload text of SMS. The proposed method neither changes the existing SMS architecture nor does it require any extra hardware. Rest of the paper is divided into following sections. Sections 38.2, 38.3, and 38.4 present the brief detail of related work, proposed methodology, experiments and evaluations respectively and we conclude the paper in Sect. 38.5.

38.2 Related Work

The related work is presented with respect to digital content delivery over SMS and the digital content compression techniques.

Authors in [6] have presented voice transferring mechanism via SMS over satellite communication. In their proposed method, they converted the voice message into ASCII characters which can be set as payload text of SMS. The major issue with the proposed method is large number of SMS, a problem researched in [5]. The work compares six different audio lossless compression schemes and shows that for proposed method PPM (Prediction by Partial Matching) lossless compression performs well and can reduce audio transmission burden over 20%.

Similarly, [8] presents the approach of sending ECG signal through SMS. In [7], authors discuss the transmission of images using SMS. The work presents a SMS-Image decoder application which converts the color image into gray-scale. Then, it divides the image into smaller blocks

and converts these blocks into binary form along with the header information which contains the block position in the original image. Lastly, the binary data is converted into suitable characters of SMS.

In literature, a number of articles discuss the comparative analysis of compression schemes for specific applications [14, 15]. In [4], authors compared the efficiency of 5 lossless compression schemes for text data and concluded that Shannon Fano algorithm performs well for the large file. [11] discusses the comparison of different finger printing compression techniques. The results show that the modified set partitioning in hierarchical trees (SPIHT) produces better compression ratio for finger print images [11].

Similarly, comparison of compression schemes' impact on finger and face recognition accuracy is presented in [12]. [9] discusses the performance analysis of multimedia compression algorithm.

38.3 Proposed Methodology

SMS service is used for transferring of short messages and it cannot transfer any multimedia content, e.g., voice, image, sound, etc. However, we can overcome this limitation by converting the multimedia content into text, and then it can be sent through SMS. Figure 38.1 gives the overall system architecture of the proposed framework. In the proposed method, first we convert the user selected digital content into byte array output stream. Then, we get the unsigned integer array. Here, we add 255 to all those numbers which fall in the range of 0–31 to move them up to the range of 256 to 287 respectively. The main reason behind this is that, values of 0–31 of ASCII characters may create issue during transference, as such characters are universally reserved for specific functions. For example '0' represents 'null' in ASCII etc. Finally we convert this unsigned integer array into respective ASCII characters. In the next phase six different compression algorithms are applied one by one on these ASCII characters to analyze their performance for the proposed mechanism. The best resultant compressed characters are converted into strings. The index numbers of SMS are flagged and the strings are set as a payload text of SMS (Fig. 38.2).

38.4 Experiments and Evaluations

Each compression algorithm is compared on three terms; Compression Ratio, Compression Factor and Compression Percentage (SMS saving). Measurements are calculated as follow [4].

We selected different types of digital contents categorized by IANA for testing the efficiency of proposed

Fig. 38.1 Proposed system architecture

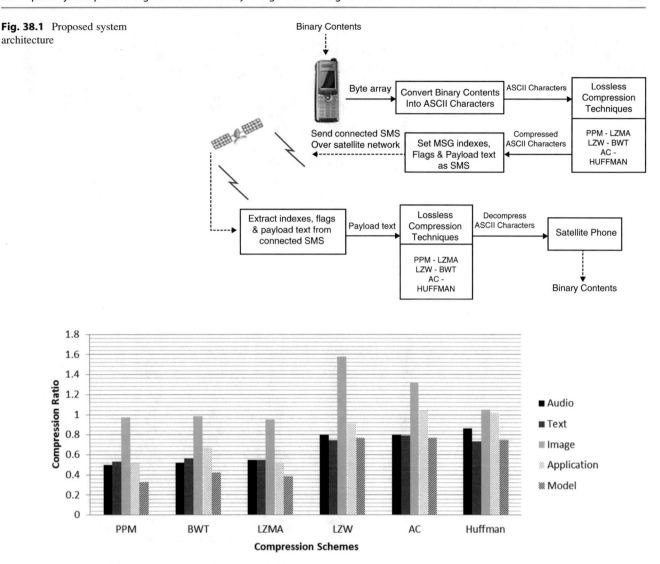

Fig. 38.2 Average compression ratio results for each content category

method [13, 16]. The digital contents are divided into five categories: text, audio, image, application (.trc) and model (.glsl, .dml). We selected different content formats which fall in these categories for the experiments.

We developed an application using J2ME platform for simulating the transference of the digital contents over satellite communication, using SMS. We tested the performance of six lossless compression algorithms (PPM, LZW, LZMA, BWT, AC, Huffman), taken from the IANA classification, on digital contents. These lossless compression algorithms reduce the number of SMS with no loss of quality. According to IANA standards, 20 different formats which are classified into 5 categories are presented in Table I, II and III. The comparison of efficiency of compression algorithms, with the proposed method is based on two factors; number of characters and number of SMS which are directly proportional to each other.

The performance of the six compression algorithms for characters are compared using three factors; compression ratio, compression factor and compression percentage. Figure 38.3 provides the comparison ratio and Fig. 38.4 provides the overall comparison ratio of 6 different compression schemes for all five IANA classes.

The compression ratio of each compression scheme is above 0.90 except for LZW and AC compression algorithm. Smaller compression ratio is the indication that compression technique gain better compression rate and vice versa. But on the other hand, both LZW and AC perform worst for images. Finally, we calculated the overall compression ratio for all digital contents as shown in Fig. 38.6. It clearly depicts that PPM gain smaller compression ratio of 0.63 where LZW have 0.92.

We can see that all compression schemes perform outstandingly, for the presentation category except for LZW

Fig. 38.3 Overall compression ratio for all digital content types with respect to compression techniques

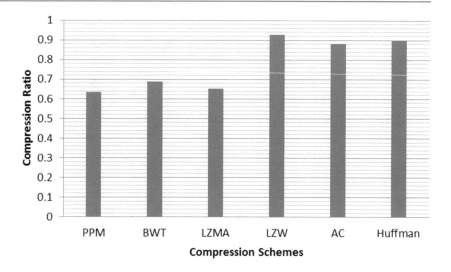

Fig. 38.4 Overall compression factor for all digital content types with respect to compression techniques

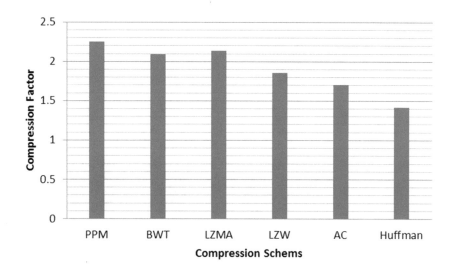

and AC which have 7.04 and 6.12 respectively, as shown in Table II. LZW and AC perform worst for image class having compression factor below 1. Compression factor under 1 means that compression scheme gain very small or negative compression value. Figure 38.5 presents the average compression factor for each class and finally Fig. 38.6 shows the overall compression factor. Results depict that PPM performs better than other compression schemes and it gains compression factor up-to 2.25.

On the other hand all techniques perform worst for image class. Graph in Fig. 38.7 shows the average compression percentage for all classes, and overall performance graph of compression schemes. PPM outperforms other techniques with 43.87% compression percentage whereas LZW shows least performance with 13.12%.

It is clearly depicted from the results trend that almost all compression schemes perform well for four classes; audio, text, model and application but for image class very small compression is achieved because we know that JPEG, GIF

image formats are already compressed by default. It is also shown that Arithmetic Coding (AC) and Huffman Coding gain negative compression for image and application content types, where LZW shows negative compression ratio only for image content type.

Table I, presents the overall performance of compression schemes in terms of transmission/SMS saving. PPM produces 43.28% less SMS for IANA standard whereas LZW performs least among all other compression algorithms with 17.59%. The result of this research will help researchers or R&D professionals to choose better compression algorithm for their respective applications.

38.5   Conclusion

The paper presents a methodology of sending digital contents over satellite through SMS service when signal strength is low, targeting the IANA classification of digital

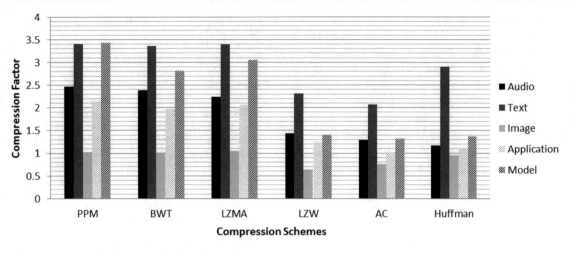

Fig. 38.5   Average compression factor results for each content category

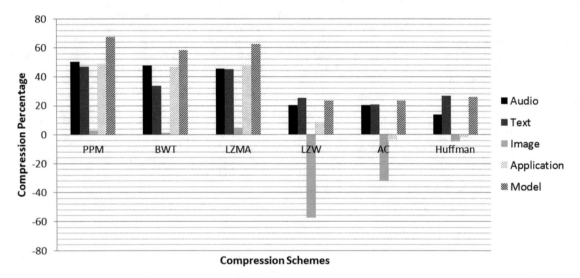

Fig. 38.6   Average compression percentage results for each content category

Fig. 38.7   Overall compression
percentage for all digital content
types with respect to compression
techniques

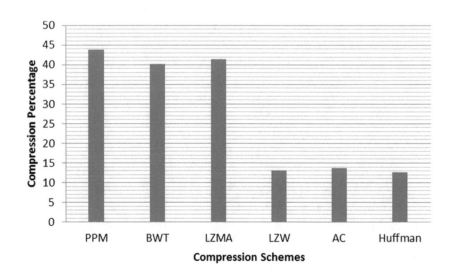

contents. The proposed method has been implemented as an application which converts the content into the text and then set as a payload text of SMS. From the results we can clearly see that PPM compression scheme performs well within the proposed methodology. It reduces the bandwidth usage up-to 43.28% for overall IANA contents. Moreover, proposed method does not change the existing SMS architecture nor does it require any extra hardware. In future, we are planning to improve these results by using filters. We are well aware of the fact that lossless compression does not perform well for such content. Another way to produce good results is by using Vector Quantization (VQ) methods (lossy compression), especially for image content.

References

1. Furletti, B., Gabrielli, L., & Rinzivillo, S. (2012, August 12). Identifying users profiles from mobile calls habits. ACM UrbComp'12, Beijing, China.
2. Measuring the information society. (2012). Online available at: https://www.itu.int/en/ITU-D/Statistics/Documents/publications/mis2012/MIS2012_without_Annex_4.pdf. Last accessed 5 Jan 2017.
3. Townsend, A. M., & Moss M. L. (2005, April). Telecommunications infrastructure in disasters: Preparing cities for crisis communications. Online available at: http://www.nyu.edu/ccpr/pubs/NYU-DisasterCommunications1-Final.pdf.
4. Kodituwakku, S. R., & Amarasinghe (2010). U. S. Comparison of lossless data compression algorithms for text data. *Indian Journal of Computer Science and Engineering, 1*(4), 416–425.
5. Beg, S., Khan, M. F., & Baig, F. (2013). Transference and retrieval of compress voice message over low signal strength in satellite communication. *International Journal of System of Systems Engineering, 4*(2), 174–186.
6. Khan, M. F., & Beg, S. (2012). Transference and retrieval of voice message over low signal strength in satellite communication. *Innovations in Systems and Software Engineering, 8*(4), 293–299.
7. Mahajan, A. A., & Chincholkar, Y. D. (2014, June). Transmission of image using SMS technique. *International Journal of Research in Engineering and Technology, 03*(06), 394–397.
8. Mukhopadhyay, S. K., Mitra, S., & Mitra, M. (2013). ECG signal compression using ASCII character encoding and transmission via SMS. *Biomedical Signal Processing and Control, 8*(4), 354–363.
9. Abdelfattah, E., & Mohiuddin, A. (2010, October). Performance analysis of multimedia compression algorithms. *International journal of computer science & information Technology (IJCSIT), 2*(5), 1–10.
10. Blelloch, G. E. (2010). *Introduction to data compression* (pp. 1–55).
11. Kambli, M., & Bhatia, S. (2010). Comparison of different fingerprint compression techniques. *Signal & Image Processing: An International Journal (SIPIJ), 1*(1), 27–39.
12. Mascher-Kampfer, A., Stögner, H., & Uhl, A. (2007). Comparison of compression algorithms' impact on fingerprint and face recognition accuracy, Electronic Imaging, International Society for Optics and Photonics.
13. Green, R., & Huang, X. Classification of digital content, media, and device types. Online available at: http://www.iskouk.org/sites/default/files/GreenPaper.pdf. Last accessed 5 Jan 2017.
14. Robert, L., & Nadarajan, R. (2006). New algorithms for random access text compression. In *Third International Conference on Information Technology: New Generations (ITNG'06). IEEE.*
15. Alom, B. M. M., Henskens F., & Hannaford M. (2009). "Querying semistructured data with compression in distributed environments" Information technology: New generations, 2009. ITNG'09. In *Sixth International Conference on. IEEE.*
16. Freed, N., Klensin, J., & Hansen, T. (2013). Media type specifications and registration procedures, Request for Comments: 6838, Internet Engineering Task Force (IETF).

The Method of Data Analysis from Social Networks using Apache Hadoop

Askar Boranbayev, Gabit Shuitenov, and Seilkhan Boranbayev

Abstract

This article analyzes data from social networks. The social microblogging system called Twitter is taken as a data source. In the model of distributed computing MapReduce has been used for the implementation of the algorithm for searching the user communities. Apache Hadoop has been chosen as a platform for distributed computing. The program code was developed for retrieving tweets and distributed processing. The analysis of the interests of users of Twitter was conducted.

Keywords

Data • Method • Analysis • Information technology • Social network • Distributed platform • Framework

39.1 Introduction

Development of information technologies stimulates the emergence of new methods and areas of research in different subject areas. In particular, in the world today is marked rapid growth in the number and volume of information, 90% of which was generated in the last 2 years [1]. A large number of social networks, the increasing number of members of these communities on the Internet provides the information base, which of course gives rise to large amounts of data processing problem, analyze the problem of poorly structured data to a qualitatively new level.

Analysis of social networks - one of the most rapidly developing areas of mathematics, sociology, psychology and many other disciplines and is of great interest to researchers. Today, large tracts of personal data became publicly available: biographical facts, correspondence, diaries, photos, video, audio, notes, travel, etc. But unfortunately most of the data presented in unstructured or semi-structured form. Therefore, new mechanisms are needed to effectively address the problem of processing continuously growing volumes of unstructured and semi-structured data from social networks. This paper discusses the method of analysis of data from social networks through the use of Apache Hadoop.

Systematic approach to modeling a wide class of collective social processes in various sectors of society appeared in the 20s of the last century. Psychologist and psychotherapist Jacob Moreno suggested sociogram in which separate individuals presented in the form of points, and the relationship between them - in the form of lines. The idea of using the methods of graph theory to explore the relationships and interactions between people picked up by experts in the field of sociology, psychology, anthropology, political science, economics - so formed direction of Social Network Analysis, to study the structural properties of social relationships, modeled in the

A. Boranbayev (✉)
Nazarbayev University, 53 Kabanbay Batyr Ave, 010000 Astana, Kazakhstan
e-mail: aboranbayev@nu.edu.kz

G. Shuitenov
Department of Information Systems, L.N.Gumilyov Eurasian National University, Astana, Kazakhstan

S. Boranbayev
Department of Information Systems, L.N.Gumilyov Eurasian National University, Astana, Kazakhstan
e-mail: sboranba@yandex.kz

© Springer International Publishing AG 2018
S. Latifi (ed.), *Information Technology – New Generations*, Advances in Intelligent Systems and Computing 558, DOI 10.1007/978-3-319-54978-1_39

form of graphs and networks [2]. An important, but very time-consuming step of this study was to construct a model based on data from various published sources, additional surveys and questionnaires.

39.2 Distributed Platform Apache Hadoop

Currently, social media - social networks, blogs, microblogs, social services, photo hosting, video hosting, etc - are dynamic heterogeneous information sources. They are characterized by the graph (topological) structure formed by various kinds of links between the participants, documents, pages, from which can evolve a variety of information processes.

For example, on Twitter more than 250 million users and over 15 billion connections between them, and on Facebook, more than 500 million users and more than 50 billion relationships. To handle such volumes of data need to bring distributed computing systems and use algorithms that allow parallel processing.

As a platform for distributed computing within the MapReduce model chosen Apache Hadoop platform is free. Hadoop environment enables to analyze unstructured data, is scalable and fault-tolerant, capable of handling a large number of documents.

Apache Hadoop is a freeware set of tools, libraries, and frameworks for the development and execution of distributed programs running on clusters of hundreds or thousands of nodes. Developed in Java within the computing MapReduce paradigm, according to which the application is divided into a large number of identical elementary tasks executable on the cluster nodes and then give the final result [3]. The calculations are organized in the form of composite applications, each of which contains data sources (connected via API Apache Hadoop interface) and compute modules, carrying out their processing. Apache Hadoop is used together with other components as part of an expanding ecosystem Hadoop: means the data in RAM Apache Spark, infrastructure storage Apache Hive and Apache HBase NoSQL data storage system. Apache Hadoop is used mainly with a large number of units (thousands of nodes), usually these sites are not very powerful. It includes several basic modules:

- Hadoop Common – a set of common tools and software for the infrastructure and the entire stack;
- HDFS – distributed file system;
- YARN – system of job scheduling and cluster management;
- MapReduce – platform for the creation and implementation of distributed computing.

A key component is distributed file system HDFS (Hadoop File System), which you save a file, it automatically breaking etsya into blocks of a given size, and each block is duplicated on different nodes in the cluster. replication factor specifies the number of copies of a file in the cluster. Such an implementation of storage allows you to store the system more fault-tolerant and reliable.

The next component is the MapReduce platform - the core of the system of distributed computing. It launches distributed computing tasks, performed in two stages, in which data is divided into blocks, and then grouped the data is processed at each node.

In the model of distributed computing MapReduce has been used for the implementation of the algorithm search the user communities. Computer problem in this model consists of two steps: Map and Reduce. Computational elements of implementing these steps are called, respectively, the mapper (mapper) and redyuser (reducer). At the same time on one task can run a large number of mappers and redyuserov distributed across a cluster of machines. Production data using MapReduce should be represented in key-value format <k; v>. The entire volume of the input data is broken into fragments of a certain size, each such fragment is input to one of the mappers. Incoming pair <kin; vin> is converted to an intermediate pair <kint; vint>. Then, the intermediate data obtained from all mappers, grouped by key kint redyuseram and arrive at the entrance to a <kint; list viint>. Thus, the values corresponding to one key fall into one redyuser. After finishing the output redyusera get a pair of <kout; vout>, are already written to the output file [4].

Figure 39.1 shows the MapReduce operation scheme.

Figure 39.2 provides an overview of the number of words the counting process based on MapReduce platform.

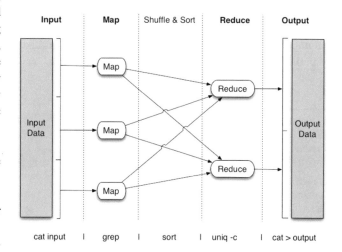

Fig. 39.1 The scheme of MapReduce

Fig. 39.2 Overview of the number of words the counting process based on MapReduce platform

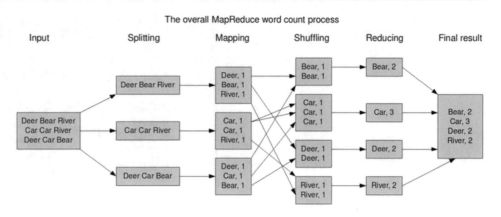

39.3 Data Processing with Social Networks Using Apache Hadoop

In this paper, a distributed system Apache Hadoop 2.7.2 (pseudo-distributed mode - on a single machine all processes run in different processes of the Java-) installed on a virtual machine VMWare, which runs on Ubuntu 14.04 LTS (32-bit). Figure 39.3 shows the installation of VMWare virtual machine.

Virtual Machine Specifications:

- 1 core processor 2.5 GHz
- 2 GB of RAM
- 15 GB of disk space

At the stage of preliminary preparation work with Apache Hadoop Hadoop is necessary to format the file system:

$ bin/hdfs namenode -format
and then start the system (NameNode and DataNode):
$ sbin/start-dfs.sh

Web interface allows you to check the status of Hadoop works through a browser:

http://localhost:50070/

Figure 39.4 shows the Web interface operation status Apache Hadoop.

To solve the problem of extraction of data from a large array of social networks for structured or semi-structured data is used to provide social services API (Application Programming Interface) functions that require different authentication methods and network communication protocols. API - a special set of commands that allows software developers to ask the social networking large amounts of data. The heterogeneity of the ways of access to data complicates the architecture of the data collection system and generally requires solving the problem of unification.

Fig. 39.3 Install VMWare virtual machine

The social system Twitter is taken as a data source, as it has convenient tools for the developer and provides access to the Twitter API functions [5]. After the registration application, the developer gets the API Key and API Secret, which are used for applications access to semi-structured data. As part of working with the Apache Hadoop common programming language has been selected Java. During the development of library Twitter4J it was used. [6]

Twitter API allows you to receive real-time data stream in a format that can filter on geopozitsii (receive tweets only from specified coordinates of the area), as well as content filtering (to receive only the tweets that contain certain words).

Figure 39.5 shows a program code for extracting tweets.

Since the aim was to capture the maximum amount of data that any filters have not been exposed. The application writes random tweets from around the world in a text file. Within 18 hours of the program 3 million tweets were recorded. File size: 343 MB.

Fig. 39.4 Web interface
operation status Apache Hadoop

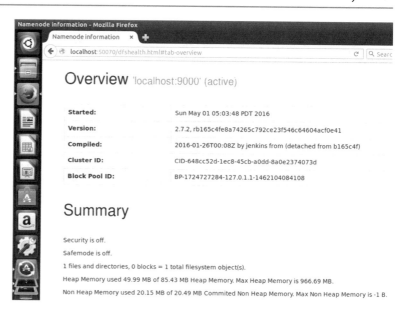

Fig. 39.5 The program code for retrieving tweets

Text data is written in the following format:

- @ <Username> - 1 Tweet
- @ <Username> - 2 Tweet

Figure 39.6 shows the output of the program file.

The task processing analysis of a large data set (tweets) was to determine the most frequently used words, tweets and hash tags, ie search of all words (sequences of characters separated by spaces) in the file and counting the number of mentions for each of them. The specifics of the framework

Apache Hadoop requires some features to consider when developing applications for Hadoop.

The application should be written in Java and includes three separate classes:

- the main class;
- class responsible for the work with Mapper;
- class, responsible for working with the Reducer

Figure 39.7 shows a distributed data processing program code.

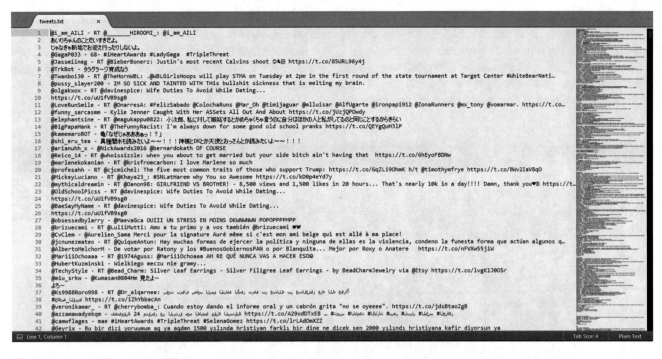

Fig. 39.6 Output of the program file

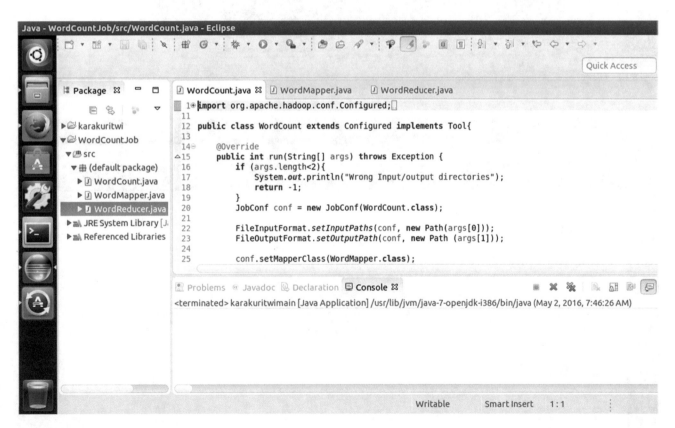

Fig. 39.7 The program code distributed data processing

Annex exported in the form of a JAR file. It we put in the root folder Hadoop in a virtual machine. To execute the program we use the following command:

bin/hadoop jar wordcount.jar WordCount input wordcountoutput (wordcount.jar – the name of the JAR-file; WordCount - the main class of the program;

wordcountoutput - directory for the output file). Figure 39.8 shows a process for performing data processing.

In processing the file it was spent 6 minutes 48 seconds. Figure 39.9 shows the access to the data processing results through the web interface.

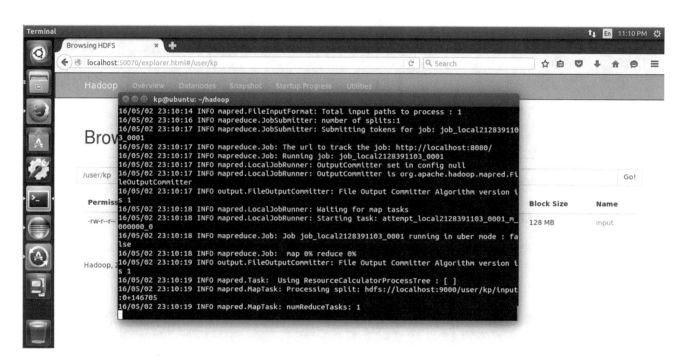

Fig. 39.8 The process of the data processing

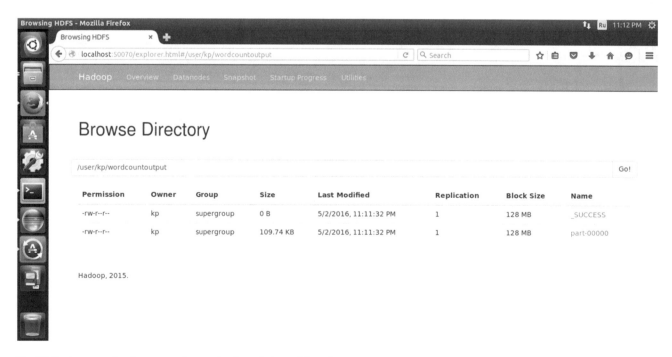

Fig. 39.9 Access to the data processing results through the web interface

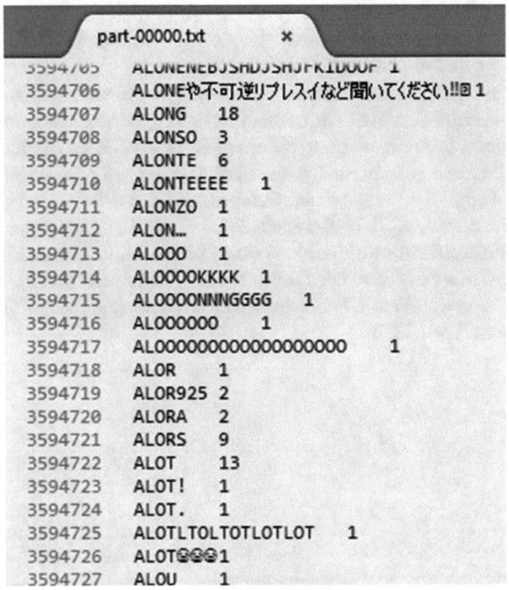

```
part-00000.txt
3594705 ALONENEBJSHDJSHJFKIDOOF 1
3594706 ALONEや不可逆リプレスイなど聞いてください!!▣ 1
3594707 ALONG 18
3594708 ALONSO 3
3594709 ALONTE 6
3594710 ALONTEEEE 1
3594711 ALONZO 1
3594712 ALON… 1
3594713 ALOOO 1
3594714 ALOOOOKKKK 1
3594715 ALOOOONNNGGGG 1
3594716 ALOOOOOO 1
3594717 ALOOOOOOOOOOOOOOOOOO 1
3594718 ALOR 1
3594719 ALOR925 2
3594720 ALORA 2
3594721 ALORS 9
3594722 ALOT 13
3594723 ALOT! 1
3594724 ALOT. 1
3594725 ALOTLTOLTOTLOTLOT 1
3594726 ALOT@@@1
3594727 ALOU 1
```

Fig. 39.10 Output file sorted alphabetically

Output file - part-00000. Recorded in the file format bin. It can be downloaded to the local file system directly from Hadoop web-interface. Figure 39.10 shows the output file sorted alphabetically.

The output file by default sorted alphabetically. To sort the words according to their number use the following command, after creating a text output-sorted.txt Hadoop file in the root directory in a virtual machine:

```
bin / hadoop fs -cat / user / <username> / wordcountoutput /
   part- * | sort -n -k2 -r> output-sorted.txt
```

Figure 39.11 shows the output file sorted by the number of repetitions.

39.4 Conclusion

Social networks are a unique source of data on his personal life and interests of real people. This opens up great opportunities for the solution of research and business problems (many of which previously could not be handled effectively due to lack of data), as well as the creation of support services for users of social networking applications.

Analysis of the company Twitter says that today most people do retweet and repost, if someone else's information reflects their personal interests. Our study confirmed the fact that the predominant way of expressing the interests of users are retweets - more than 42% of processed information by

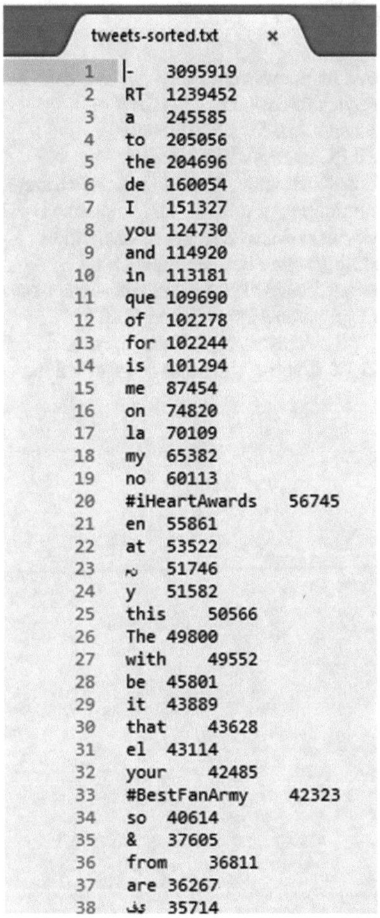

```
tweets-sorted.txt
1 |- 3095919
2 RT 1239452
3 a 245585
4 to 205056
5 the 204696
6 de 160054
7 I 151327
8 you 124730
9 and 114920
10 in 113181
11 que 109690
12 of 102278
13 for 102244
14 is 101294
15 me 87454
16 on 74820
17 la 70109
18 my 65382
19 no 60113
20 #iHeartAwards 56745
21 en 55861
22 at 53522
23 و 51746
24 y 51582
25 this 50566
26 The 49800
27 with 49552
28 be 45801
29 it 43889
30 that 43628
31 el 43114
32 your 42485
33 #BestFanArmy 42323
34 so 40614
35 & 37605
36 from 36811
37 are 36267
38 في 35714
```

Fig. 39.11 Output file sorted by the number of repetitions

us. Thus, according to the analysis of Twitter users post images about 3.5 times more often than video, and photos get 128% more retweets than video. Tweets with text and reference retweets received 86% more in comparison with the photos and video [7]

The above-quoted social media research method can be used to analyze the data based on geolocation data analysis with the search keywords, etc.. For example, the data taken from the popular Instagram network, taking into account factors geopozitsii may give an answer to any points of the city often users take pictures, or to find the most popular hash tags for a specific region or country.

Also, this method can be used for further development and evolvement of research, which was discussed in works [8–10].

The further development of this work may be in line with semantic analysis, the implementation of a distributed information system automated detection in texts motive vocabulary and emotional evaluation of the authors (of opinions) with respect to the objects.

References

1. http://www.vcloudnews.com/every-day-big-data-statistics-2-5-quintillion-bytes-of-data-created-daily/
2. http://www.osp.ru/os/2013/08/13037856
3. https://ru.wikipedia.org/wiki/Hadoop
4. Ryabov, S., & Korshunov, A. (2011). The distributed algorithm for finding communities of users in social networks. – 2011. p. 215.
5. https://dev.twitter.com/overview/documentation
6. Twitter4J: http://twitter4j.org/en/index.html
7. http://texterra.ru/blog/kakoy-tip-kontenta-imeet-samyy-vysokiy-potentsial-rasprostraneniya-v-twitter.html
8. Boranbayev, S., Altayev, S., & Boranbayev, A. (2015). Applying the method of diverse redundancy in cloud based systems for increasing reliability. In *Proceedings of the 12th International Conference on Information Technology: New Generations, ITNG 2015* (pp.796–799) Las Vegas, April 13–15.
9. Boranbayev, S., Boranbayev, A., Altayev, S., & Nurbekov A. (2014). Mathematical model for optimal designing of reliable information systems. In *Proceedings of the 8th IEEE International Conference on Application of Information and Communication Technologies, AICT 2014* (pp.123–127), Astana, October 15–17, 2014.
10. Boranbayev, A.S., & Boranbayev, S.N. (2010). Development and optimization of information systems for health insurance billing. In *Proceedings of the 7th International Conference on Information Technology: New Generations, ITNG 2010* (pp.1282–1284), Las Vegas, April 12–14.

Using Web Crawlers for Feature Extraction of Social Nets for Analysis

Fozia Noor, Asadullah Shah, Waris Gill, and Shoab Ahmad Khan

Abstract

This paper presents a crawler based feature extraction technique for social network analysis. This technique crawl a predefined actor and his associated activities in social network space. From the activities, a set of features are extracted that can be used for a broad spectrum of social network analysis. The utility can act as a middle ware providing a level of abstraction to researchers involved in social network analysis. The tools provide a formatted set of ready features with open APIs that can be easily integrated in any application.

Keywords

Social Network Analysis • Web crawling • Feature extraction • Preprocessing

40.1 Introduction

Social media platforms allow users to share their daily life routine or information in multiple forms i.e. messages, locations, movies and photos etc. with their friends. All these information make data records of an individual disclosing who they are, where they are, what they do, who is in family, what they like, what they plan etc. Nowadays due to high usage of the web, a large amount of social data is created on daily basis. The daily routine activity created by an individual containing enormous amount of information can act as digital Footprints of an individual. The digital footprints of individuals reflect the pictures of digital social structure (up to some extend) hence are of scientific interest for researchers and can be used by different entities to support business, social or commercial purposes.

The effort to study these digital social structures by extracting data from social networks has his roots deep down in previous decades. Different researchers from the field of humanities and social sciences investigated these digital activities to study the society and culture [1]. Different social networks data extraction methods are used by researchers ranging from traditional methods of manual collection to the use of web bots. Existing approaches for social data extraction are: direct database access, API's provided for third party developers, open source software or personalized web crawlers.

Direct database access to social networks data is only reserved for in-house researches, thus not available for public access. API's provides well developed interfaces and data collected is also in clean and well-structured format. But these API's have limit on the quantity and type of data.

F. Noor (✉)
Yanbu University College (YUC), Yanbu, Saudi Arabia

International Islamic University Malaysia (IIUM),
Kuala Lumpur, Malaysia
e-mail: noorf@rcyci.edu.sa

A. Shah
International Islamic University Malaysia (IIUM),
Kuala Lumpur, Malaysia
e-mail: asadullah@kict.iium.edu.my

W. Gill
Lahore University of Management Sciences (LUMS), Lahore, Pakistan
e-mail: waris.gill@outlook.com

S.A. Khan
National University of Sciences and Technology (NUST),
Islamabad, Pakistan
e-mail: shoabak@ceme.nust.edu.pk

© Springer International Publishing AG 2018
S. Latifi (ed.), *Information Technology – New Generations*, Advances in Intelligent
Systems and Computing 558, DOI 10.1007/978-3-319-54978-1_40

Other than these some software application like PowerExchange [2] is also available. Some of this software are quiet successful but need to be paid for full access of features.

Basically an online identity of an individual is comprised of a numerous attributes which can be classified in to generally three different categories; Profile based attributes which are filled by individual at signup, Content based attributes consisting all the activities and Network based attributes which is comprised of social circle of an individual. Major social data extraction tool available in current market has their main focus on monitoring marketing campaigns thus extracts/monitors only specific content of pages. Different general purpose GUI based tools (Nodexl and Pajek) or packages/libraries (Networkx, IGraph) are also available in market. While Tools like NodeXL [3], imports the network based attributes along with some profiles based attributes but do not provides the content based attributes. Thus these tools are basically visualization widget and don't provide a complete data of pages/profiles hence limits the in depth analysis.

Social data extraction can also be done by using crawler who reads HTML documents of the social sites. Different types of crawlers have been designed for different social networks i.e. YouTube, Flickr, Orkut, and Twitter [4]. Crawling data from social nets has many challenges. Some of them are; Complexity of technologies, privacy setting, access restrictions and access rate Limitation through IP banning. In contrast to Visualization tools these crawlers can extracts all the three types of attributes and store them in the data base or file in the format specified.

The wide spectrum of these activities data from different social networks cannot be directly used for synthesis and analysis. Some of these crawled attributes need the cleaning while some of them can be further used to derive attributes. Any analysis, may it be identity resolution, identification of key players or trend analysis requires extraction of a set of features from the social websites like Facebook, tweeter, Instagram, LinkedIn etc. For example the identification of key players requires a gap representation of connectivity among different actors in a network. Similarly finding sub networks or clusters of subnet in a network requires the interconnection and frequency of communication among the actors. The similarity measures in their communication require extraction and tagging of comments and posts. Any analysis of social network depends not only on the availability but also on the quality of extracted and derived features. Hence one of the major challenges in any OSN analysis research is the extraction and crafting of a quality attributes. Once quality attributes will be ready, they can be imported in any of the available visualization or analytical tool for further analyses.

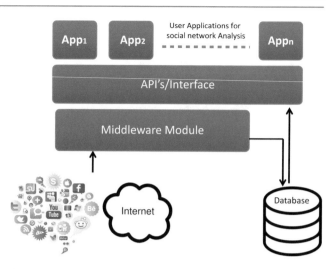

Fig. 40.1 Middleware module high level architecture

The objective of this paper is to discuss the design and implementation of a Middleware Module (shown in Fig. 40.1) that crawls the data from three of the famous social networks, Twitter, Facebook and Instagram. Crawler, given a focal user in a social network quickly obtains nodes in neighbourhood of focal user is discussed in Sect. 40.2. Section 40.3 discusses the pre-processing of the extracted data to make the ready feature to be used for different purposes. Section 40.4 describes the visualization of the structure of the crawled network. The User applications can request the needed attribute from the database for their usage or analysis. Finally, work is summarized in Sect. 40.5.

40.2 Crawling the OSN

The structure of an OSN can be modelled as a graph $G = (V, E)$, with nodes V and edges E. Edges E represents activities/information that is being shared between identities (nodes V). To study these structure broadly two types of analysis strategies used i.e. Complete network analysis and ego network analysis. In case of complete network analysis the social data on internet is too huge to be extracted. It is very hard to scrape complete data of any OSN [5]. Even if extracted and stored offline, its cost will be very high and its analysis by applying different measures will be very difficult. Thus the best way to study any network is to sample the network. Sampling helps to make observations on small set of data and then make inference about complete network. The data can be sampled by following any of the different available approach: based on edges, nodes, walks etc. [6]. Hence improved performance can be attained by limiting the space of study.

Our research is focused on ego network analysis thus the best method of sampling for ego centric network is snowball

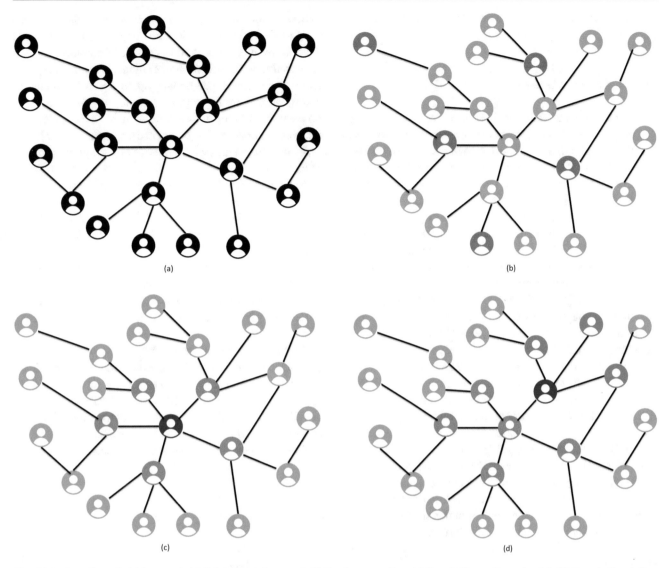

Fig. 40.2 Sampling of social network (**a**) Original graph network (**b**) Random sampling, (**c**) Snowball sampling at level 1; (**d**) Snowball sampling at level 2

sampling using any of the crawling technique (shown in Fig. 40.2). For snowball sampling the crawlers are designed to start the iterative process of crawling from a single focal node "ego". The node can be selected randomly or can be specified specifically. The numbers of nodes extracted are expended by discovering new nodes at every iteration step. Hence we can formally define our problem of data crawling as

"In network G using snowball sampling, the crawler starts with focal user's $f \in V$ and obtains the set of nodes $(v \in V)$ information in the neighbourhood of f at d hops away of f".

As we are aware that in sampling the network, data collected is incomplete due to boundary specification (depth) but there is also a case of getting incomplete data where the data expected to be collected will not be retrieved due to privacy setting of social networks. The contents on OSN's are generally not available to everyone. Social media policy, privacy settings of individual prevent the availability of information. In each OSN user has a right to set their privacy policies. How much information of an individual can be extracted differs for each social media and even for each user depending on the setting of each user profile, even each individual post in addition to network basic privacy setting.

Therefore, we can only access the publically available part of graph G, denoted as $Gp = (Vp, Ep)$, representing publically available nodes Vp and the publically available edges Ep.

40.2.1 Crawler Design

Different architectures for crawler have been proposed and performance bottlenecks have been discussed by researchers [7, 8]. Along with general web crawler's, researcher have also worked to develop focused crawlers like topic crawler for network monitoring [9].Broadly two different techniques are used for crawling; Graph searching and Random walk. In former technique (BFS, DFS) each node in graph is visited only once in the order depending on searching strategy chosen. While in later method, starting point neighbours are selected at random till a termination condition is verified. Random walks allow re-visiting nodes.

The searching technique chosen for traversal is graph searching through Breadth first search (BFS). BFS is a well-known graph traversal algorithm. Major reason for its popularity is that it collects a full view of some particular region in the graph. This particular region in our case is a network of our focal node. BSF is known to give bias samples as it privileges users with high number of social links [10]. However, this effect can be minimized by different techniques [11]. In our case of research as our max depth limit is 3 so this case is not going to arise. The basic flow of work of crawler is almost same for all social networks but the detailed design and operation of a crawler depends on the type of social network and on the types of data to be crawled.

First step of a crawler is to authenticate itself by providing credentials to social network server. Once authen-tication is approved, the page is redirected to the focal seed user f as an initial node and starts exploring the publically available profile, content and network information (friends list) iteratively. In each of the iterations, it visits a node and discovers its direct neighbours up to level 2 (friends of friends) shown in Fig. 40.3 . In this way we can get a complete picture of network around a focal node. The extracted sample nodes connected to make a small sub graph, used in a study can represent any graph on a bigger scale.

Crawlers stores the number of nodes extracted for each level. If some ones friends list is hidden then its network will be hidden. Crawler extracts the full complete profile of each user, visited during crawls. It also collects the content of interaction i.e. Wall posts and photo comments and network i.e. list of friends for each user crawled. The detailed list of information crawler extracts is given in Table 40.1. The most of information extracted is still in raw phase and cannot be used directly as feature for use, thus need pre-processing to be carried out.

The Fig. 40.4 below shows the sample of the text post extracted from the Facebook. For experimentation a professor of university is considered as an ego. Only public number of edges and nodes were extracted. They will vary depending on the individual and publically available information. If the post is the picture or video it is stored with the post Id. The profiles links are deleted and names are bit changed to maintain privacy.

Fig. 40.3 Social network tree of a focal seed node

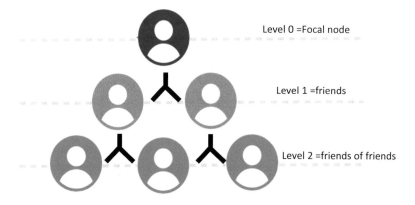

Level 0 =Focal node

Level 1 =friends

Level 2 =friends of friends

Table 40.1 Lists of features extracted

Network	Categories	Extracted attributes
Facebook, Twitter & Instagram	Profile attributes	Name, Date of Birth, Location, Email, Language, Education, Likes, Interest, Profile picture
	Content attributes	Text and photo post/tweets , Date, Time, Comments, Reply, Likes, Shares, Locations
	Network attributes	Followers /Friends list, Groups, Members of groups

Fig. 40.4 Sample post from the Facebook (**a**) Post (if location is tagged) (**b**) post comments and comments likes (**c**) Post likes and shares

Post # 124				
Post	Location	Date and Time	With	With::ProfileLinks
Alumni event	None	March 14, 2015	None	None

(a)

Post # 124				
Likes	Likes::Profile Links		Shares	Shares::Profile Links
Tanzeel	https://www.facebook.com/------------?fref=pb&hc	Arif	https://www.facebook.com/------------?fref-
Saheem	https://www.facebook.com/------------?fref=pb&hc	Rana	https://www.facebook.com/------------?fre
Yasir	https://www.facebook.com/--------------?fref=pb&h	Ibrahim	https://www.facebook.com/------------?fre
Muhammad	https://www.facebook.com/-----------?fref=pb&hc_l	Sajid	https://www.facebook.com/-----------?fref=
		Usman	https://www.facebook.com/------------?fref-

(b)

Post # 124				
Comments	Comments::Profile Links	Comment Date and Time	Comment Text	Comment Likes
Affan	https://www.facebook.com/------------?fref=pb	March , 14,2015 at 4:48pm	Yaa hooooo	None
Siddiq	https://www.facebook.com/------------?fref=pt	March , 14,2015 at 6:15pm	Will be great	None

(c)

Fig. 40.5 Preprocessing layer

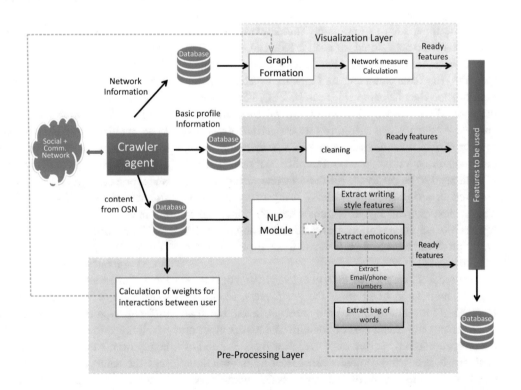

40.3   Preprocessing Layer

Collection of data and its appropriate pre-processing is most critical part of any research. Even if desirable data is collected for any research, that data cannot be used for analysis directly. It needs to be represented in a format that is preferable for analysis. Crawler stores data in raw format and the next major challenge of transforming data into the usable features is carried out in the pre-processing. The pre-processing of data removes or transforms data. The neater, clean and usable the data will be the more easy and handy will be the analysis. Thus data pre-processing plays an important role in analysing any social network.

As explained, the data crawled is categorized in three types, these three types of data are pre-processed as per need and the 4rth type of data; derived data generated from the crawled data during pre-processing as shown in Fig. 40.5.

The first type of data extracted is profiles attributes, which are of interest for many social network researches. Online Identity resolution is one of the most famous research area and different applications developed for this purpose like PeekYou and Pipl are widely used by people now a days. These applications use the basic information like name, age, phone number, email, education as input criteria and search in all possible publically available records for matching entity in real-time. These portals are flexible in their query parameters. All these basic information are stored in the profile information of the user and can be used for different purposes.

Different social networks require different basic information at the sign up stage of the user and shows flexibility in the format of data entry. Thus are not consistent in all networks or even among different users of same network. Like name is available in all networks, but may be defined differently across the networks as name can be one attribute or can be three such as first-name, middle-name and last-name. Even the data stored for each profile can be in different representative format. For example for Location Pakistan, the data value can be Pk, Pak, or Pakistan. For qualification the attribute values can be FSC, FSc, Intermediate or High School all representing up to 12th grade studies. As each social media has its own data format and may be different from other, to make all of them consistent data transformation is needed and conflicting representations of data need to be resolved. All these format difference are resolved in pre-processing stage by using matching rules. The names, interest, favourite quotes etc. are converted in lower case. Duplicate entries are also checked and removed. Descriptive variables of Age, Sex, and Income are converted in numerical.

The second category of data crawled is the content of pages. The content of pages is the post/tweets, Photos or location. Who shares, likes or comments, even the text of comments are also a part of content. The text content of the pages are stored but cannot be used directly in any sort of analysis. It needs the complete process of transformations involving several steps of text processing: white space removal, expanding abbreviation, stemming, stop words removal. All these transformations are carried out in python. Although there are various NLP libraries (i.e. OpenNLP) and NLP web APIs (i.e. AlchemyAPI) that can be integrated within the framework for the analysis of text. There are also many NLP engines available and can be used in the form of web services. NLP web APIs can process tweets, paragraphs and can extract named entities (people, location), topics, and keywords from them. Different researchers used different API to extract named entities from text i.e. tweets or posts. Saif et al. [12] investigated three services Zemanta, OpenCalais, and AlchemyAPI and finally used AlchemyAPI for the Semantic annotation of tweets. Abdel et al. [13] used OpenCalais API to detect named entities in tweets. While Steiner et al. [14] used both AlchemyAPI and OpenCalais to extract entities from Facebook. All these entities are derived data which can be used as features for further text analysis.

The pre-processing of text generates N tokens of interest i.e. words, phrases, symbols, or other meaningful elements. All the data that can be of any interest is extracted from the text and stored in database for later use as features. Stemmed lower case meaning-full words and phrases are stored. Any emoticons used are extracted and stored. In Twitter and Instagram the tweets/shares extracted includes the hash tags and @mention, were not separated at crawling stage. But during pre-processing of tweets @mentions are separated, stored and counted to calculate in frequency of communications. The hashtags are stored in bag of words along with meaning full words and phrases. Emoticons are not easy to be identified. For identifying Emoticons and Slangs, list of emoticons and slangs from Wikipedia is used. Code to extract the slangs and emoticons is written in python that uses regular expression for finding emoticon and Slangs. Likes, shares, comments and reply are also processed and counted to keep track of communication weight.

Figure 40.6 shows the high level of text mining process. Load regular expression; loads the regular expression to extract the useful information from post like phone numbers email address, tags or @mentions depending on the network under processing. After pre-processing the ready features are stored in the Location specified and be accessed by reading out ready files.

The network of user is the friends of user extracted by a crawler. Weight of edges are calculated from the content and stored separately for each kind of communication i.e. shares, likes, @mentions, comments. These can be very useful for multilayer network formation where each layer represent different type of node or edge and multidimensional chain of relationship can be formed. Like network of Facebook can be represented using multiple layers where each layer represent the same actors representing different type of edges between networks i.e. likes , shares, comments etc.

40.4 Visualization of Network

Visualization of network can be used as an explanatory tool to understand network structure and basic network properties. At the last stage of pre-processing network metrics like Degree centrality, Eigenvector centrality,

Fig. 40.6 Text mining process

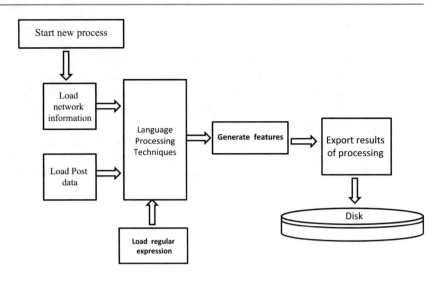

Fig. 40.7 Visualization of Facebook network

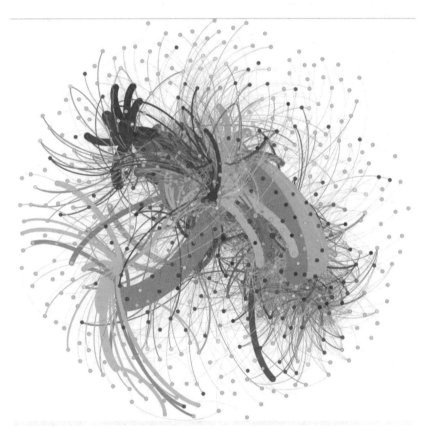

Betweenness, Eccentricity, Number of triangles etc. are calculated for each node and stored in database to be used as ready feature in the analysis. The network measures are calculated by considering each network a single layer network. These Centrality measures are used to find the importance of nodes in the network and are of much interest in case of finding key players in any network of interest.

Figure 40.7 shows the network of Facebook extracted represented in the form of node and edges using different colours. Colours of nodes are based on selected network measure and the thickness of edges represents the communication weights.

All the processed and derived attributes are stored in the form of excel sheets which can be used for further analysis by user application while the visualization is not stored and only available to view the network.

40.5 Conclusion

This study approaches the design of a middleware which can be used to extract the digital foot prints of individual from the selected social networks. It furthers pre-process those unstructured information to turn them into useful data that can be further used for predictive and social analytics. This kind of application can work as middle ware and can make the analytics burden free. As in the current era the extraction of information and its pre-processing consumes the majority of research time.

References

1. Wilson, R. E., Gosling, S. D., & Graham, L. T. (2012). A review of Facebook research in the social sciences. *Perspectives on Psychological Science, 3*, 203–220.
2. https://www.informatica.com/products/data-integration/connectors- powerexchange.html#fbid=UAP0D-8ts7R. Acessed 10 Jan 2017.
3. NODEXL: https://nodexl.codeplex.com/. Acessed 10 Jan 2017.
4. Mislove, A., Marcon, M., Gummadi, K. P., Druschel, P., Bhattacharjee, B. (2007). Measurement and analysis of online social networks. In *Proceedings of the 7th ACM SIGCOMM conference on Internet measurement (IMC'07)*, San Diego.
5. Gjoka, M., Kurant, M., Butts, C., Markopoulou, A. (2010). Walking in facebook: a case study of unbiased sampling of OSNs. INFOCOM, San Diego, Proceedings IEEE, (pp. 1–9)
6. Frank, O. (2011). Survey sampling in networks. In *The SAGE handbook of social network analysis*, London, Sage, (pp. 389–403).
7. Shkapenyuk, V., & Suel, T. (2002). Design and implementation of a high-performance distributed web crawler. In: *Proceedings of the International Conference on Data Engineering*, San Jose, CA, United States, IEEE CS Press (pp. 357–368).
8. Boldi, P., Codenotti, B., Santini, M., & Vigna, S. (2004). Ubicrawler: a scalable fully distributed web crawler. *Software Practice and Experience, 34*, 711–726.
9. Yakushev, A. V., Alexander, V. B., Sloot. P. M. A. *Topic crawler for social networks monitoring Knowledge Engineering and the Semantic Web Volume 394 of the series Communications in Computer and Information Science*, vol 394. Springer, Berlin, Heidelberg, (pp. 214–227).
10. Gjoka, M., Kurant, M., Butts, C., Markopoulou, A.(2010). Walking in facebook: a case study of unbiased sampling of osns In: *Proceedings of the 2010 I.E. Conference on Computer Communications*, (pp. 1–9). San Diego, IEEE.
11. Wilson, C., Boe, B., Sala, A., Puttaswamy, K. P., Zhao, B. Y. (2009). User interactions in social networks and their implications. In *Proceedings of the 4th ACM European conference on Computer systems (EuroSys'09)*, Nuremberg.
12. Saif, H., He, Y., Alani, H. (2012). Alleviating data sparsity for twitter sentiment analysis. In *The 2nd workshop on Making Sense of Microposts*. (#MSM '12): Big Things Come in Small Packages at the 21st International Conference on the World Wide Web (WWW '12), pp. 2–9, Lyon, France.
13. Abel, F., Gao, Q., Houben, G. J., Tao, K. (2011). Semantic enrichment of twitter posts for user profile construction on the social web. In *The Semantic Web: Research and Applications*, (pp. 375–389). Springer-Verlag Berlin, Heidelberg.
14. Steiner, T., Verborgh, R., Vallés, J. G., de Walle, R. (2011). Adding meaning to Facebook microposts via a mash-up API and tracking its data provenance. In *Next Generation Web Services Practices (NWeSP), 7th International Conference on, 2011*, (pp. 342–345).

Detecting Change from Social Networks using Temporal Analysis of Email Data

41

Kajal Nusratullah, Asadullah Shah, Muhammad Usman Akram, and Shoab Ahmad Khan

Abstract

Social network analysis is one of the most recent areas of research which is being used to analyze behavior of a society, person and even to detect malicious activities. The information of time is very important while evaluating a social network and temporal information based analysis is being used in research to have better insight. Theories like similarity proximity, transitive closure and reciprocity are some well-known studies in this regard. Social networks are the representation of social relationships. It is quite natural to have a change in these relations with the passage of time. A longitudinal method is required to observe such changes. This research contributes to explore suitable parameters or features that can reflect the relationships between individual in network. Any foremost change in the values of these parameters can capture the change in network. In this paper we present a framework for extraction of parameters which can be used for temporal analysis of social networks. The proposed feature vector is based on the changes which are highlighted in a network on two consecutive time stamps using the differences in betweenness centrality, clustering coefficient and valued edges. This idea can further be used for detection of any specific change happening in a network.

Keywords

Social network analysis • Temporal analysis • Network measures • Clustering coefficient

Author is currently registered as a PhD student at KICT International Islamic University of Malaysia.

K. Nusratullah (✉)
Yanbu University College, Yanbu, Saudi Arabia

KICT: International Islamic University Malaysia, Gombak, Malaysia
e-mail: Khank@rcyci.edu.sa

A. Shah
KICT: International Islamic University Malaysia, Gombak, Malaysia
e-mail: asadullah@kict.iium.edu.my

M.U. Akram • S.A. Khan
National University of Sciences and Technology (NUST), Rawalpindi, Pakistan
e-mail: usmakram@gmail.com; shoabak@ceme.nust.edu.pk

41.1 Introduction

The concept of change detection in social networks is a research problem that is written several times with different flavors. Different researchers like Katz and Proctor [1] Newcomb [2] performed sociometric test to depict the change in small social groups over time. People in the social groups or communities interact directly or indirectly with each other creating their social networks. These social networks can reveals a true picture of their context i.e. situational information of a person that could be a current job of a person, a present friend circle, a running project, a new entry in the contact list etc. This context can be gathered either from single network or combination of multiple different types of networks like social connection networks, multimedia sharing networks, professional networks or email communication networks.

Social networks are basically dynamic entities that grow and shrink with the passage of time and can be analyzed as static or dynamic topology. Longitudinal research has been used from last few decades to investigate these kinds of networks. Longitudinal studies collects the snapshots of data over a period of time, hence can address the changes in social networks and can answer many queries which are not possible to answer with the single snapshot of data. Longitudinal studies have been used to address different set of problems in social network analysis by many researchers. Like Singer and Spilerman [3], Holland-Leinhardt [4] and Wasserman [5] introduced different longitudinal methods to detect change in social networks. Concept of popularity model, Statistical Process Control (SPC) and transitivity are the extraction of their studies.

In current era of technology email is integrated as a daily activity of different group of people especially in professional environment. According to a survey about 90% of US internet users were involved in the activity of sending and receiving emails [6]. This email network can be a good source to get accurate data about one's professionals and social relations. In research literature different methods are proposed to analyze and visualize email communication for different purposes [7] Moreover different researchers have also focused on analyzing email network on temporal basis; J. S. Hardin detects changes in organizational behavior [8] Peter A. Gloor extracted most active communities and active members of organization [9] while Shahadat Uddin proposed a topological analysis of email network to capture the dynamics of communication [10].The research in this study is also a contribution in the area of longitudinal analysis of email communication networks on the basis of different features extracted or derived from the network. Generally an email network is composed of nodes and edges where nodes represent individuals and edges represents communication among individuals. To perform network analysis multiple features can be extracted from the communication i.e. header, body, set of attachments, bag of words, a subject field, date/time, CC list or can be derived from nodes contribution in networks i.e. actor's network measure. In order to study email over the period of intervals, time can be added as a continuous variable. Addition of time in analysis enables us to explore the process of change and evolvement in network structure in general while nodes or edges in specific.

This paper proposes a framework to capture different segments of an email communication to detect change in structural context of different node in network. To acquire the required result three measures of email network are used as features. Among them two features are network measures; betweenness and clustering coefficient are node specific

i.e. giving the information about node while third feature is valued edge; showing frequency of communication between nodes. The values of these features are determined for each timestamp by using social network analysis. Nodes are ranked on the basis of these values to find the top important contributing nodes in network on different snapshot of time. In last a feature matrix is generated on the basis of presence of top node in consecutive snapshot of time. The values of this feature matrix will detect and categorize the intensity of change in high, low, medium and no change.

The structure of this paper is as follows. Section 41.2 explains the theme of study. Proposed methodology is discussed step by step in Sect. 41.3. As a final point, there is a conclusion and future research direction of proposed research in Sect. 41.4.

41.2 Temporal email Network Analysis

This study contributes in the dynamic analysis of communication network evolved over a period of two years. It's an effort to analyze email communication as a social network where network positions of all nodes are measured. The values of these measures can depict individual's context with in network. A temporal analysis of email communication network, conducted at a number of discrete time moments, is useful in gaining the insight of an individual network for example:

- Addition of new nodes or edges shows the expansion of an individual's networks. It highlights in progress events or new social ties of one's life. The network evolved will be true dynamic if the new nodes are added frequently and associations are created at continuous basis.
- Evolution of nodes in a clusters or cliques can reflect ad hoc groups of individuals formed on temporary basis. These cases arise in organizations during running project for certain time period or can indicate short term relationship in an individual's life.
- Removal of existing nodes can highlight a sudden change in a network. In an organizational context this change can reveal employee's death, job switch, or department change. On other hand removal of edge can reflect cracked association.

Diverse measures have been devised to capture the structural position of nodes. The structural positions of nodes can be measured by finding the quantity of nodes communication or by finding the standing of position of a node depending upon the type of network. The most commonly used measures by researcher to find important position of nodes are degree, eigenvector, betweenness, closeness centrality and clustering coefficient. However this research study only uses three measures whose brief descriptions are given in following

A. *Valued Edges:* Edges represent TO/From feature of email communication. Valued edges or weighted edges can represent the frequent correspondence among nodes in communication [11]. The weight is a number that can indicates either an intensity of relationship, strength of a relationship, or frequency of communication between nodes. There are different ways to represents edge weights in network. One way is to use a numeric figure on edges, second is to draw thick edges as per weight and the third way is to use Edge List.

B. *Clustering Coefficient:* Clustering coefficient is used to measure the connectedness of neighbors. While analyzing an email communication this measure helps to find cliques in communication. Nodes with high clustering coefficient are the persons who use to disseminate the information in network. Mathematically clustering coefficient is called a fraction of possible interconnections. In social network literature clustering coefficient have been used to sightsee the diffusing nodes of network [12, 13]. Figure 41.1 shows a change in clustering coefficient in three different time slots.

C. *Betweenness centrality:* Betweenness centrality is a common network measure to find out the most central node of network. In terms of change detection, with the passage of time a node changes its role. Whenever the betweenness of a node is high, at that time the person is playing an important role in a network. He/she is connected with other group of nodes and they are communicating as influential nodes of network [14]. The fluctuation of this centrality value refers a clear change in position of node in network. Figure 41.2 is showing a scenario representing a network of 10 individuals. At time t_1 Liz is an important node playing a vital role in connecting a sub-network (of two individuals) to main network (of 7 individuals). Role of Liz was basically a hub at time t_1 while in t_2.Liz is not a hub of same sub network.

Before proceeding to proposed methodology let us try to run the concepts discussed above using a sample network of 10 nodes observed at three different time stamps. It will help us in summarizing the idea of proposed research and will make a basis to discuss the proposed framework. Figure 41.3 shows the network observed in three different time stamps i-e t_1, t_2 and t_3. It shows a scenario of an employee (node E), whose job position change in his organization and while

Fig. 41.1 Change in clustering coefficient

t_1: Node b has high clustering coefficient as its neighbor c and are connected to each other. **A clique is found in networks.**

t_2: Value of Clustering coefficient is decreased as its neighbor are no more connected to each other. **A clique is disappeared from networks.**

t_3: The same pattern of communication at t_1 is repeated in t_3. A clique is formed again between same nodes.

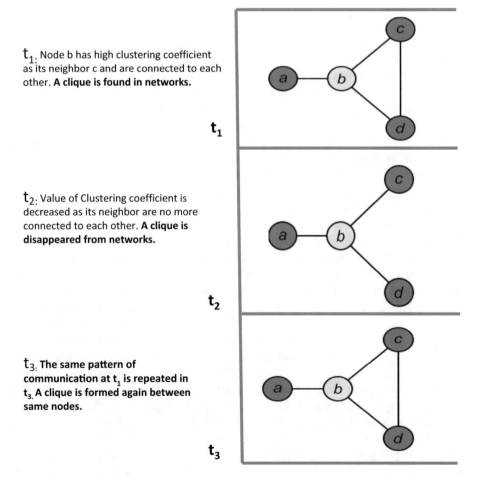

Fig. 41.2 Change in
betweenness centrality

t_1: Between ness of Liz is high in the
above network.

t_1

t_2: With the passage of time Between
ness of Liz is became low as she is not
connecting one group of network to
other.

t_2

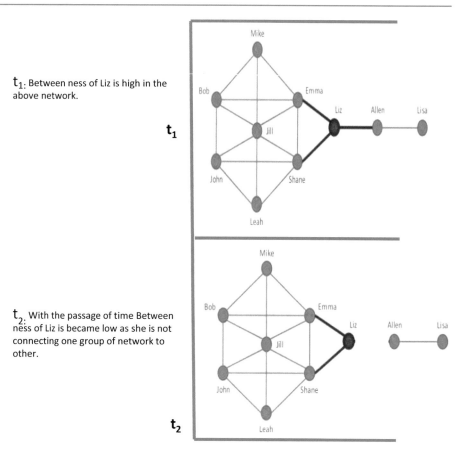

changing the job position how the network of node E grown and changed? In timestamp t_1, it shows that E is a node in network that has only one direct connection with C. Thus he is an isolated person who has no valued edges that reflects less correspondence with colleagues, no clustering coefficient that mirrors few connections with coworkers and no betweenness centrality that shows ineffectual job in organization while A is a main node with high centralities measures at time t1. Time period of t_2 is showing the growing phase of node E. It's getting connected to other nodes. Time stamp t_3 represents the peak time of node E where the weight of its edges became increased, it depicts that his frequency of communication in organization is increased. Betweenness of node E became high in network that reflects an influential position in organization. It's clearly seen that node E is connected with dense group of people in network and he is the best person at time t_3 to disseminate information in organization.

41.3 Proposed Methodology

This section discusses the framework for analysis of social networks using time information. Figure 41.4 shows the flow diagram of proposed method. The main modules are extraction of Network and preparation of data, segmentation of network, network measure extraction and a feature vector formation

In the first module network is broken down in different small nets using time stamp information. Second module is carried to perform analysis of sub nets in order to determine different three network measures. The last module is the formation of feature vector; uses the extracted network measures of each node to identify the change which happened between two consecutive time stamps.

41.3.1 Network Extraction and Data Preparation

The dataset that is observed while experiment, is email communication of a private organization. The structure of organization is well defined according to different level of jobs. Communication that is being observed is related to one department with strength of thirteen workers. Nodes are renamed to maintain privacy of data. The job levels, which involve in communication, are head of the department, coordinators and subordinates. To evaluate the proposed method, email communication between the said job levels is observed for twenty four months. During these units of time, changes are manually annotated to verify the later on results. One of the changes in network is about the change of employee's position. This change in role of any employee causes to add new responsibilities or to get in new assignments. The change can easily reflected in network the frequency of communication of

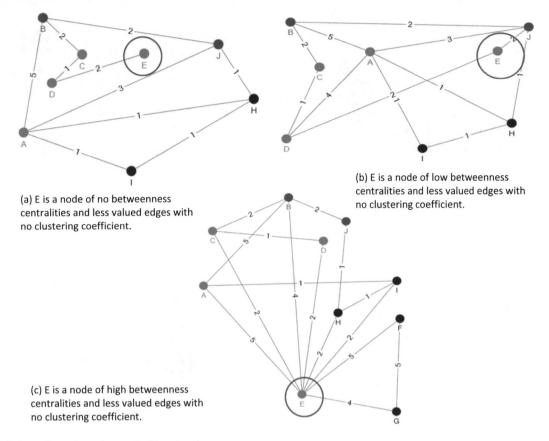

(a) E is a node of no betweenness centralities and less valued edges with no clustering coefficient.

(b) E is a node of low betweenness centralities and less valued edges with no clustering coefficient.

(c) E is a node of high betweenness centralities and less valued edges with no clustering coefficient.

Fig. 41.3 Sub graphs to show changes in E's network

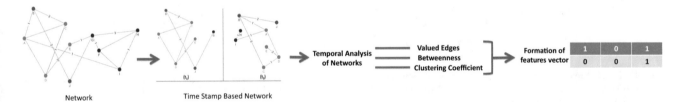

Fig. 41.4 Flow diagram of proposed methodology

that specific employee with other employees will change. Second change in network is expected when head of the department will change. Reflection of this change appears in network by observing the network measures like betweenness etc. related to head of the department's node. From data analysis perspective an email contains several important data elements, for better understanding we can classify email data into Header, Body and optionally Attachments. The Header section of the email contains the primary information about the communication instance such as who sent the email (Source), who was the recipients (Targets), what was it about (Subject) and when did it happen (Timestamp). In order to build a social network we focused on the Header section of the email as it contained all the compulsory information to shape the social network. Each email is a communication instance in the email network making up an edge in the network between Sender

and Receiver. Sender (From) acts as the source node of the network while Recipient(s) (To, CC, BCC) are the target nodes and in this way information from the email header is developed to build up a social network representation of the email data.

41.3.2 Segmentation of Network

Once emails were transformed to social network representation, Gephi is used as a tool to generate the graphs of networks. Graphs are produced for each timestamp i.e.one month. Twenty four graphs are generated. For the reference; only selected graphs are shown in Fig. 41.5. These selected graphs are added to show the change in the role of two nodes U10 and U7.

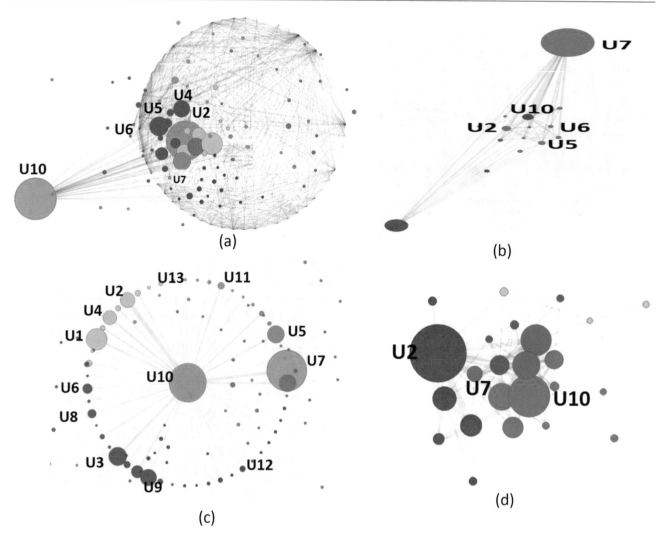

Fig. 41.5 (**a**) Depicts the main graph of last seventeen months of observation. (**b**) Sub graph that shows the initial role of two nodes U10 and U7 during the first seven months. (**c**) Sub graph of last seventeen months to show the peak time of node U10. (**d**) Sub graph to show that U7's role is changed

41.3.3 Network Measure Extraction

After segmentation, social network analysis is performed to extract network measures of each subnet. The changes are concluded on the basis of network measures values extracted for each node in each subnet. Figure 41.5a depicts the main graph of last seventeen months of observation. Figure 41.5b is a sub graph that shows the initial role of two nodes U10 and U7 during the first seven months of observation. It shows that node U10 has a same communication flow as other employee's nodes. This is the time period when U10 is working as a subordinate. At the same time node U7 is communicating as head of the department in networks, Betweenness centrality of U7 is high in the network. In Fig. 41.5c, next seventeen month's communication is shown.

It is better to mention that the graph in Fig. 41.5c is captured at the peak time of node U10. At this time role of U10 is changed to coordinator in organization. His network became more connected. Valued Edges of U10 is increased; more nodes are added in its network and U10 is a communicating as diffusing node. Moreover betweenness centrality became high reflects his importance of job.

41.3.4 Feature Vector Generation

The last step in proposed framework is the generation of feature vector to highlight temporal changes in the networks. As features were extracted for each node, thus each node has its extracted features for all twenty four sub networks. Each Node is ranked based on their feature set values. Table 41.1 highlights top nodes in each time stamp.

To generate feature vector the code in python is written which analyzes the top nodes based on network measure values (given in Table 41.1) and if a top node is not repeated in consecutive time stamps then "1" is added in new feature vector to show change else "0" is added to highlight no change. Table 41.2 shows the final generated feature vector

Table 41.1 Top nodes of twenty four months on the basis of network measures

Timestamp	Valued edges	Betweenness centrality	Clustering coefficient
M1	U1	U1	U4,U6
M2	U2	U7	U7
M3	U2,U5,U6	U6	U3
M4	U6	U7,U3	U3,U7
M5	U6	U5,U7	U3,U7
M6	U5,U6,U7	U5,U7	U4,U7
M7	U1	U1	U4,U6
M8	U9,U10	U9	U7
M9	U2,U10	U7,U9	U7
M10	U2, U10	U2	U7
M11	U2	U10	U2
M12	U1, U3	U2	U10
M13	U1,U3, U7	U2, U10	U2, U10
M14	U5	U5	U10
M15	U4, U10	U5, U2	U3, U7,
M16	U2, U10	U2	U7
M17	U2, U10	U2	U7
M18	U2, U10	U2	U7
M19	U1	U1	U4, U6
M20	U2, U10	U7, U9	U7
M21	U6	U2	U10
M22	U6, U2	U2	U10
M23	U2, U10	U2	U7
M24	U2, U10	U2	U7

Table 41.2 Rate of change on the basis of calculated values of network measures

Timestamp	Valued edges	Betweenness centrality	Clustering coefficient
M1-M2	1	1	1
M2-M3	0	1	1
M3-M4	0	1	0
M4-M5	0	1	0
M5-M6	0	0	1
M6-M7	0	0	1
M7-M8	1	1	1
M8-M9	1	0	0
M9-M10	0	1	0
M10-M11	1	1	1
M11-M12	1	1	1
M12-M13	0	0	0
M13-M14	1	1	1
M14-M15	0	0	0
M15-M16	1	1	1
M16-M17	0	0	0
M17-M18	0	0	0
M18-M19	1	1	1
M19-M20	1	1	1
M20–21	1	1	1
M21-M22	0	0	0
M22-M23	0	0	0
M23-M24	0	0	0

based upon this idea. Now if a vector contains all 0's then it means no change has happened in two time stamps else there will be a change and the intensity of change can be further detected by looking at number of 1's present in the feature vector.

The rate of change that is obtained from feature vector is subjectively verified from the annotated changes. Result shows that proposed framework is capable of capturing all major changes.

41.4 Conclusion

Social network analysis is important to find or highlight any specific information related to that network. A number of applications are being made these days based upon social network analysis. One of important factor in this analysis is to consider temporal information which gives idea about the changes being happened in the network based on any time stamps. This paper presented a framework to highlight temporal changes in a social network. An email dataset of two years has been used to support the idea and three network measures i.e. betweenness centrality, clustering coefficient and valued edge are used to check the properties of a sub network against a time stamp. These findings are then further evaluated to formulate a novel feature vector to highlight temporal changes. In future, machine learning algorithms can be applied on these finding to detect and grade the temporal changes as different labels like low, medium and high.

References

1. Katz, L., & Proctor, C. H. (1959). The configuration of interpersonal relations in a group as a time-dependent stochastic process. *Psychometrika, 24*, 317–327.
2. Newcomb, T. M. (1962). Student peer-group influence. In N. Sanford (Ed.), *The American College : A psychological and social interpretation of the higher learning*. New York: Wiley.
3. Singer, B., & Spilerman, S. (1976). The representation of social processes by Markov models. *The American Journal of Sociology, 82*(1), 1–54.
4. Holland, P. W., & Leinhardt, S. (1977). A dynamic model for social networks. *Journal of Mathematical Sociology, 5*, 5–20.
5. Wasserman, S. (1980). Analyzing social networks as stochastic processes. *Journal of the American Statistical Association, 75*, 280–294.
6. Hansen, D., Shneiderman, B., Smith, M. A. (2010) *Analyzing social media networks with NodeXL: Insights from a connected world*. Morgan Kaufmann. Elsevier Publication.
7. Xiaoyan, F., Hong, S. H., Nikola, S. N., Shen, X., Wu, Y., Xu, K. (2007) Visualization and analysis of email networks, 2007 I.E. APVIS. 329302.
8. Hardin, J. S., Sarkis, G. (2015) Network analysis with the enron email corpus. *Journal of Statistics Education, 23*(2), 3arXiv: 1410.2759 [stat.OT] 4 Aug 2015.
9. Gloor, P. A., Laubacher, R., Zhao, Y., Dynes, S. Temporal visualization and analysis of social networks, NAACSOS Conference, June 27–29, Pittsburgh PA, North American Association for Computational Social and Organizational Science.
10. Uddin, S., Piraveenan, M., Chung, K. K. S., Hossain, L. (2013) Topological analysis of longitudinal networks. Annual Hawaii International Conference on system sciences Proceedings IEEE Computer Society 2013.
11. Andresen, E., Bergman, A., Hallen, L. The role of email communication in strategic networks:patterns observed over time, 19th Annual IMP Conference.
12. Watts, D. J., & Strogatz, S. H. (1998). Collective dynamics of 'small-world' networks. *Nature, 393*(6684), 440–442.
13. Leskovec, J., Kleinberg, J., Faloutsos, C. (2005) Graphs over time: Densification laws, shrinking diameters and possible explanations. In *In KDD* (pp. 177–187).
14. Joseph J. Pfeiffer, Jennifer Neville (2011). Methods to determine node centrality and clustering in graphs with uncertain structure. Report Number: 11–010 Pfeiffer.

The Organisational Constraints of Blending E-Learning Tools in Education: Lecturers' Perceptions

42

Sibusisiwe Dube and Elsje Scott

Abstract

This study investigated and identified the organizational factors that contribute to the poor adoption of technology such as the Sakai Learning Management System in education. Qualitative data were collected through semi-structured interview questions guided by Giddens' structuration model. The participant lecturers were from one of the nineteen universities in Zimbabwe, a developing country where a sluggard uptake of Information and Communication Technologies is currently experienced. The situation not only prejudices the students from enjoying the affordances their counterparts in developed nations enjoy. It has also led to the emergence of the second order digital divide, a problem of concern to the researchers, ICT policy makers and the learning institution management, robbed of the anticipated returns on the costly technological investment. The paper contributes to the limited literature relating to the developing country lecturers' perceptions of e-learning tools in teaching. The findings show that organizational factors play a major role in influencing either the lecturers' positive or negative perceptions of e-learning system tools in education. In addition to the documented individual and technological factors, policy, budget, training, decision making, implementation and consultation techniques have been found to inhibit the successful integration of e-learning tools in the traditional teaching methods.

Keywords

e-learning • Organisational constraints • Lecturers • Education • Sakai learning management systems

42.1 Introduction

The Catastrophe theory posits that at some point in time, a very small increase in pressures to change higher education may produce a dramatic revolution in the delivery of instruction [4]. One such major force that has produced much revolution to the delivery of education has been the technological developments currently evidenced in higher education. Internet based courses have found their way into education through such technological developments as information and communication technologies (ICT). Most common is he electronic learning systems, facilitated by the learning management systems (LMS), which complement the restrictive traditional teaching and learning methods in higher education institutions (HEI). In this regard [14], assert that the emergence of ICTs particularly LMS are causing the brick-and-mortar institutions to blend online instructions with the face to face (f2f) teaching and learning delivery methods. This approach has been necessitated by the affordability of technological devices such as computers and smartphones coupled with the widespread availability of

S. Dube (✉) • E. Scott
Department of Information Systems, University of Cape Town, Cape Town, South Africa
e-mail: sibusisiwenkonkoni@gmail.com; Elsje.Scott@uct.ac.za

S. Latifi (ed.), *Information Technology – New Generations*, Advances in Intelligent Systems and Computing 558, DOI 10.1007/978-3-319-54978-1_42

the internet has provided universities with opportunities to enhance effective teaching and learning online [13] Due to technological changes, education can now move from teacher-centered to learner centered teaching and learning approaches. In addition [3, 9, 20], note that online courses facilitate access to timely course materials independent of time and space. Despite the affordability, availability as well as the affordances of these online environments, their uptake is still at its infancy in developing countries' HEIs a problem discussed in the next section.

42.2   The Problem Statement

HEIs are investing in ICTs due to their affordances in teaching and learning [2, 19]. Nevertheless, little returns are realized from these costly investments, which are underutilized. For instance, survey results from a study by [10] show that, only 20% of the academics at one of the sixteen universities in Zimbabwe, the National University of Science and Technology (NUST) use Sakai LMS tools in teaching since its implementation in 2012. This is contradictory to the Science, Technology, Engineering and Mathematics (STEM), initiative and has led to the emergence of the second order digital divide, a gap between the access and use of ICTs. The current techno savvy students affectionately known as the "Net Generation," or "Millennials," [19] are dissatisfied with the underutilization of both their technical skills and their digital technologies [6]. contend that these students prefer learning techniques that integrate technology. In agreement [15], posit that these students are accustomed with a variety of information and quickly become bored if subjected to traditional teaching methods only. Nevertheless, educators fail these millennials when they conduct lectures in a step-by-step traditional instruction and assessments [7] Studies attribute this situation to socio, economic and political settings in developing African nations [1, 17]. Studies also show that the second order digital divide prejudices the digital native students from keeping abreast with the 1st century living standards [5]. Despite the existence of literature, little is known about the role of the organization in constraining or enabling use of LMS tools in education. Existing studies have a bias towards the user and technological factors [8, 21] and ignore the contextual issues as told by lecturers [18]. This study expands factors in [11], by establishing he organizational factors through answering this key question; what do educators perceive to be the organizational factors constraining the successful integration of LMS tools in teaching?

An answer to this question could inform HEI decision makers on what and how ICT policies should be enacted if a digital divide is to be reduced. It is on this premise that [19],

argues that HEIs need to engage in technologically informed teaching and learning. The results of the study contribute to the limited literature relating to the organization factors of ICT implementation from the educators' view point [17].

42.3   Related Work

Literature exists that discusses the impediments of adopting and using LMS tools in teaching and learning. There are common hindrances to e-learning growth identified by researchers across the academic world. Studies such as [16] have revealed that inadequate ICT and e-learning infrastructure is one of the major challenges hindering the implementation of e-learning in public universities. Infrastructure such as computers, network and internet connectivity, and computer labs are inadequate in most public universities and cannot support the high numbers of students' enrolments. Regarding Zimbabwean universities [15], shows that these institutions lack affordable and reliable Internet bandwidth.

Financial constraints are also some of the challenges hindering the implementation of e-learning in developing countries' public universities. Implementation of e-learning is generally expensive for an average university. For instance, a study by [20] shows that inadequate financing of e-learning is therefore a major barrier to the successful implementation and adoption of e-learning systems such as LMS at universities.

The use of e-learning systems is also hindered by the skills to use the technology. For example, Wanyembi cited in [20] did a survey and found that most of the academics in universities have low ICT and e-learning skills because most of them were trained in the absence of ICT environment. E-learning skills and relevant e-content are critical components necessary for successful implementation and adoption of e-learning in teaching and learning.

Studies have also shown that lack of interest and commitment to use e-learning technologies by the teaching staff in public universities is another challenge to the successful implementation and adoption of e-learning. As such [16], found out that for successfully use technology in class, educators should have a positive attitude.

Finally, studies attribute underutilization of e-learning technologies such as LMS to lack of management commitment. Lack of management commitment at inception and during the project execution stages has a detrimental effect to the successful implementation and utilization of the e-learning at any institution. In this respect [21]. Notes that ICT projects and associated procurements take place in an environment characterized by lack of management continuity and an incentive system that encourages an overly optimistic estimates of the benefits that can be attained from doing the project.

The reviewed literature has identified LMS utilization hindrances with a bias towards human and technological characteristics. Few studies attribute the LMS use inhibitors contextual aspects. One such is the issue of lack of operational e-learning policies, which is yet another challenge hindering the implementation and adoption of e-learning at universities. It has been observed that some universities do not have an e-learning policy and in a case where the policy exists, it is never adhered to. In this regard [20]. Note that most universities are unable to implement their e-learning policies due to budgetary constraints and lack of the necessary e-learning infrastructure. Another being the infrastructural issues, hence the need to investigate such organizational factors from the educators' perspectives operating at universities in a developing African country context. This study is therefore based on the qualitative data that were collected from the educators as discussed in the next section.

42.4   Methodology

Qualitative data were collected from twenty-four purposively sampled lecturers working at NUST a university in Zimbabwe that was established in the year 1991 towards the production of technologically savvy graduates suitable for the current labor market. Fourteen of the twenty-four interviews were recorded in an audio device and averaged twenty minutes while the other ten interviews averaging 10 minutes were recorded as field notes in compliance with the interviewees' request not to be voice recorded. A 90% success rate was achieved against the thirty-three target lecturers. Fig. 42.1 is a representation of the distribution of interviewees by faculty. In-depth interviews were conducted were open ended and semi structured questions were asked each lecturer. The interview questions

were guided by the concepts of Giddens' structuration model elaborated in the next section.

42.5   Theoretical Model

Unlike models commonly used in user perception studies such as the Technology Acceptance Model, Behavioral intention and many similar ones, which are deterministic, people and techno-centric, structuration theory incorporates the social or contextual issues. Despite structuration theory's drawbacks that include being a social theory that is too general and does not directly refer to information technology, the theory's duality of structure and the structuration model were appropriate for explaining the organizational inhibitors to the use of LMS tools in teaching as reported by the lecturers. The major component of Giddens structuration theory is the structuration model. The model is divided into the three dimensions of structure, signification, domination and legitimation as well as the three dimensions of interaction namely: communication, power and sanction. The dimensions of structure and interaction are interlinked by the modalities of interpretive scheme, facility and norm.

The model was useful for understanding the relationship between the agents (Lecturers) and the structure (the university context) with regards to their action and perceptions of LMS tools in teaching. The lecturers draw on interpretive schemes to make sense of the LMS tools in the teaching practice. These meanings are communicated and shared among the lecturers to further produce new interpretive schemes and meanings (signification), which this study sought to establish. More so the university context has a role to play in influencing the meanings the lecturers share. For instance, the university management's use of power in allocating ICT devices compatible with the LMS software could impact negatively or positively on the integration of LMS tools in the teaching and learning process. This kind of domination could either enable or hinder the use of these tools in teaching and learning The final section of the model is the structure of legitimation, which means the lecturers' view of the LMS tools could be influenced by the current teaching culture and norm of the institution. The lecturers could be encouraged or sanctioned by the norm such that they legitimately use the LMS tools as planned or resist them respectively, hence the reproduction of structure. Guided by the structuration model concepts depicted in Fig. 42.2, we collected data through both the audio recorder and field notes for those lecturers not comfortable being recorded. The collected data were transcribed and analyzed using ATLAS ti, a computer software for analyzing qualitative data. Our findings revealed the themes discussed in the subsequent section for findings.

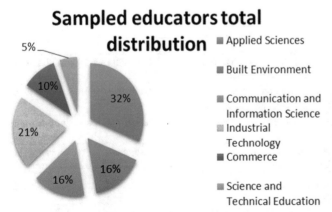

Fig. 42.1 Interview respondents faculty distribution

42.6 Findings

Generally, the lecturers value e-learning systems tools in enhancing teaching and learning. They perceive e-leaning tools such as the LMS to be handy in the provision of effective and efficient teaching and learning. Table 42.1 is a representation of the lecturers' sentiments concerning ICT in education. The Lecturers' comments on Table 42.2 are indicative of the positive views the lecturers have concerning technology in teaching and learning. Nevertheless, our findings also show that there are concerns that need to be addressed if the full potential of technology in education is to be realized. In addition to the known user and technological issues other themes of organizational nature emerged that are containing the use of ICTs in education. These include the policy, budget, training, ICT resources, management support and academic freedom.

42.6.1 Top-Down Approach

The lecturers perceive the university management as too dominating such that they never consult on the technological

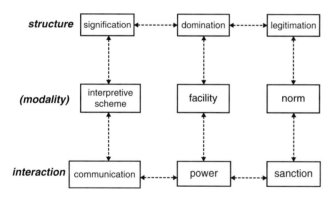

Fig. 42.2 The structuration model (Adopted from Faqih [12])

choices of lecturers before implementation. They argued that there are other e-learning tools worth exploring like e-mails, Facebook, WhatsApp Moodle etc. than forcing the Sakai LMS on them as it violets their academic freedom. They need freedom to choose the ICTs that efficiently and effectively work and suit their teaching needs unlike Sakai that cannot handle large multimedia files and multi users concurrently.

42.6.2 User Support

There was a widespread complaint that the ICTS staff members are not readily available to support users when they encounter problems while interacting with the Sakai LMS as shown in Table 42.2.

42.6.3 Lack of a ICT Budget

The lecturers were also concerned about the inadequate training on how to use the ICTs in the form of the Sakai LMS. They were of the opinion that the ICTs are not financially prioritized hence the inadequate resources for a comprehensive training. One interviewee indicated that the ICT budget has not only been dwindling over the years but has reached an extent were the university management declared via email that there shall be no budget allocation for the training and maintenance of the Sakai LMS in the 2016 onwards, a cost cutting measure. The interviewees argued that the lack of the provision of the necessities such as teas and lunch during training sessions is another reason that explains the low training attendance as depicted in Table 42.3.

Table 42.1 Lecturers' technological perceptions

Lecturers perceptions of technology based teaching	
Interviewee	Views towards technology
A	Technology is advancing we need to get onboard than to be seen lagging behind, only technophobes will resist learning. Use of Sakai should be made mandatory at NUST, it is crucial for students to experience both face to face and e-learning methods
B	I don't fear being invaded by technology, I support use of LMS but I don't have time to implement, will learn when the need arises
C	The benefits of Sakai outweigh the drawbacks; Sakai will reduce students'/lecturer conflicts
D	Students learn differently i.e. visually, through reading, listening and some through hands on experience so the use of e-learning should not overlook that
E	The role of a teacher is to interact with students and gauge their reactions, motivation and concentration levels which cannot be done by the computer. Technology eliminates that bond between students and lecturers and cannot easily share experiences
F	It is acceptable to use e-learning in class
G	It is convenient to use technology for hard copy books are scarce
H	E-learning compliments traditional teaching methods, only technophobes fear being exposed by the technology that they are not efficient
I	Use of technology is the best way to go
J	E-learning enhances learning amongst students due to the vast resources that can be sharable

Table 42.2 User support concerns

User support issues	
Interviewee	User support views
A	ICTS is not forthcoming, but also the chain of command is too long for our problems to be escalated from the departmental to faculty technicians and finally to the ICTS department long chain. Give our departmental technicians enough system access rights to do all the necessary duties
B	The ICTS department is overwhelmed
C	Our technicians need to be capacitated enough (access rights wise) in order to help us since we are closer to them
D	The available support does not go long enough for you to understand
E	Our reported problems either take too long to be solved or are never solved at all
F	There is a need for dedicated e-learning staff to offer us satisfactory user support
G	User support is available but is heavily dependent on how aggressive and persistent you are: there is no immediacy in resolving issues
H	We have reported problems but to no avail
I	We receive few user support problems

Table 42.3 Inadequate training concerns

Training attendance statics				
Faculties	Dates trained	Number attended	Faculty number	Attendance % rate
FAS	16-Aug-16	18	67	27
FOBE	1-Aug-16	17	55	31
CIS	4-Aug-16	9	70	13
FIT	8-Aug-16	5	44	11
Commerce	Not trained	–	–	–
FSTE	21-Jul-16	9	40	23
Total		**58**	**276**	**21**

42.6.4 Lack of ICT Policy

From Table 42.3 has emerged that the NUST institution has no policy in place that can guide the use and accountability of the ICTs like the Sakai LMS. The absence of the documented policy has resulted in a haphazard use of the institutional ICTs. The lecturers are concerned that simply jumping onto the technological wagon without the guiding policy violates the standardization of ICT use in teaching and learning. Furthermore, the institution has failed to recognize the efforts of lecturers by not giving them incentives for using and let alone reprimanding the non-usage of the costly investments, issues only a policy can address. In addition to this is the concern that the lecturers were not consulted prior to choosing an implementing the Sakai LMS, which has proved to be an inhibitor to the use of this technology in education

42.6.5 Lackof Consultation

The lecturers were also concerned about the management's tendency to decide on the choice and implementation of the Sakai LMS without consultation, an academic freedom violation.

42.6.6 Academic Freedom Violation

The interviewees argued that forcing Sakai LMS on them violets their freedom to choose a teaching approach that suits their needs as indicated in Table 42.4. Furthermore, the institutional environment is not conducive for the successful implementation of the e-learning tools due to the inadequate up-to-date infrastructure. The lecturers cited outdated infrastructure as the major cause of regular Sakai system crashes, disrupting the teaching and assessment process, hence its abandonment.

42.7   Conclusion

Although the lecturers who participated in the interviews unanimously value technology such as LMS as tools for teaching, the institution was found to constrain their effort of integrating the ICTs in teaching. The identified constraints included the ICT top-down implementation approach, lack of consultation, inadequate training, lack of ICT budget and policy, inadequate user support and the violation of academic freedom. Therefore, according to Giddens' structuration theory it is not only the individual or technological characteristics but also the organizational context that influences the lecturers' negative or positive perceptions of e-learning tools in education.

Table 42.4 Academic freedom concerns

Interviewee	Views on academic freedom
A	Lecturers should have the academic freedom to use whatever tool best suits them to deliver their work; nothing should be imposed on them
B	The question of whether to use or not use Sakai, should be left to the lecturer to choose, it must be left optional
C	Leave academics to see the benefits, then they won't have to be forced. If you force them they will resist and only do the minimum requirements
D	If lecturers can do their work with whatever tools at their disposal and convenience let them alone, the end results matter
E	Let there be that academic freedom, you cannot put all lecturers in a one size fits all basket. Enforcing Sakai on everyone is a violation of our academic freedom

References

1. Adewumi, S. E. (2011). The eLearning fellowship program at the University of Jos, Nigeria. *Distance Education Aquatic, 32*(2), 261–268.
2. African Development Fund. (2011). Multinational: African virtual university support project phase II. Appraisal Report. African Development Fund.
3. Al-Busaidi, K. A. (2013). An empirical investigation linking learners' adoption of blended learning to their intention of full e-learning. *Behaviour and Information Technology, 32*(11), 1168–1176.
4. Alvarez-Trujillo, H. (2008). Benefits and challenges for the online learner. Lecture. Thomas Jefferson University.
5. Baack, D., Jordan, D. R., & Baack, D. W. (2016). Evolution and revolution in higher education: catastrophe theory's applications and implications. *Midwest Quarterly, 57*(2), 105–125.
6. Barbara, N., & Linda, B. (2014). Changing academic teaching with Web 2.0 technologies. *Innovations in Education and Teaching International, 51*(3), 315–325.
7. Barnes, K., Raymond, C. M., & Pixy, F. S. (2007). Teaching and learning with the net generation. *Innovate: Journal of Online Education, 3*(4), 1–10.
8. Black, A. (2010). Gen Y; Who they are and how they learn. *Educational Horizons, 88*, 92–102.
9. Conde, M. A. (2012). An evolving learning management system for new educational environments using 2.0 tools. *Interactive Learning Environments, 22*(2), 188–204.
10. Dargham, J., Saeed, D., Mcheik, H. (2012). E-learning at school level: challenges and benefits. In *The 13th International Arab Conference on Information Technology*. Jordan: ACIT.
11. Dube, S. & Scott, E. (2014). An empirical study on the use of the sakai Learning Management System (LMS): case of NUST, Zimbabwe. In *Proceedings of the e-Skills for Knowledge Production and Innovation Conference*. Cape Town.
12. Faqih, K.M.S. (2016). Which is more important in e-learning adoption, perceived value or perceived usefulness? Examining the moderating influence of perceived compatibility. In *4th Global Summit on Education GSE*. Kuala Lumpur. World Conferences.
13. Giddens, A. (1984). *The constitution of society*. Cambridge: Polity Press.
14. Hope, K. (2015). Distance education and the evolution of online learning in the United States. *Curriculum and Teaching Dialogue, 17*(1/2), 21–34.
15. Junginger, C. (2008). Who is training whom? The effect of the millennial generation. *FBI Law Enforcement Bulletin, 77*, 19–23.
16. Kabweza, L. (2015). TechZim. [Online] Available at: http://www.techzim.co.zw/2015/02/numbers-6-interesting-facts-telecoms-internet-zimbabwe. Accessed 26 Aug 2016.
17. Khan, S., Hasan, M., & Clement, C. (2012). Barriers to the introduction of ICT into education in developing countries: the example of Bangladesh. *International Journal of Instruction, 5*(2), 61–80.
18. Maina, M. K., & Nzuki, D. M. (2015). Adoption determinants of E-learning management system in institutions of higher learning in Kenya: a case of selected universities in Nairobi Metropolitan. *International Journal of Business and Social Science, 6*(2), 233–248.
19. Mbengo, P. (2014). E-learning adoption by lecturers in selected Zimbabwe State Universities : an application of technology acceptance model. *Journal of Business Administration and Education, 6*(1), 15–33.
20. Ramírez-Correa, P., Arenas-Gaitán, J., & Rondán-Cataluña, F. J. (2015). Gender and acceptance of E-learning: a multi-group analysis based on a structural equation model among college students in Chile and Spain. *PloS One, 10*(10), 1–17.
21. Tarus, J. K., Gichoya, D., & Muumbo, A. (2015). Challenges of implementing E-Learning in Kenya: a case of Kenya Public Universities. *International Review of Research in Open and Distributed Learning, 16*(1), 120–141.

A Rule-Based Relational XML Access Control Model in the Presence of Authorization Conflicts

43

Ali Alwehaibi and Mustafa Atay

Abstract

There is considerable amount of sensitive XML data stored in relational databases. It is a challenge to enforce node level fine-grained authorization policies for XML data stored in relational databases which typically support table and column level access control. Moreover, it is common to have conflicting authorization policies over the hierarchical nested structure of XML data. There are a couple of XML access control models for relational XML databases proposed in the literature. However, to our best knowledge, none of them discussed handling authorization conflicts with conditions in the domain of relational XML databases. Therefore, we believe that there is a need to define and incorporate effective fine-grained XML authorization models with conflict handling mechanisms in the presence of conditions into relational XML databases. We address this issue in this study.

Keywords

XML • RDBMS • Relational • Access Control • Authorization • Conflict Resolution • Security

43.1 Introduction

XML documents over the Web and across corporate networks include critical governmental, financial, medical and scientific information with sensitive data. It is crucial to protect the sensitive data from the access of unauthorized users. Enforcing access restrictions on XML data has become critical in order to have efficient mechanisms to securely store and query XML.

In general, access restriction mechanisms are defined by a set of authorization policies which typically grant or deny access to objects of a data set for specific users. An authorization policy set is likely to include conflicting policies such as grant and deny, grant and partial deny (conditional deny), partial grant (conditional grant) and partial deny, defined on the same object. These authorization conflicts need to be recognized and handled effectively in a secure and reliable access restriction mechanism.

Several researchers have proposed to use the mature relational database technology to store, secure and query XML data [1–5]. Although relational databases support table level and tuple level access control mechanism, they do not provide node-level restriction mechanisms for the hierarchical XML data. Moreover, a relational XML storage needs to be equipped with a fine-grained conflict resolution support.

We propose a rule-based access control model for relational XML databases in this paper. In the proposed access control model, XML authorization rules are converted into relational tuples to be stored in a relational table. Overall security check is conducted inside the relational database.

A. Alwehaibi
Department of Computational Science, NC A&T State University, Greensboro, NC 27405, USA
e-mail: asalweha@aggies.ncat.edu

M. Atay (✉)
Department of Computer Science, Winston-Salem State University, Winston-Salem, NC 27110, USA
e-mail: ataymu@wssu.edu

© Springer International Publishing AG 2018
S. Latifi (ed.), *Information Technology – New Generations*, Advances in Intelligent Systems and Computing 558, DOI 10.1007/978-3-319-54978-1_43

The authorization conflicts are dealt with effectively using a centralized authorization table in the database.

The main contributions of this paper include the followings:

(i) We augment our test bed relational XML Database (XML2REL) [1] with an access control model which enables document authors to define and enforce XML authorization rules which include conflicting grant and deny privileges over XML elements and subtrees stored in a relational database. The XAR2RAR algorithm proposed in this paper takes an XML authorization document and translates it into a relational authorization table to be used in secure query processing.

(ii) We define the concepts of *absolute conflicts* and *partial conflicts* to deal with conflicting authorization policies in a fine-grained access control model.

43.2 Motivation

The authorization conflicts occur in various cases. The conflicts in defining authorization policies may arise due to different reasons. If the authorization policies are not issued by a central division in an enterprise, it is likely to have redundant or conflicting policies proposed by different departments. Even if the authorization policies are issued by a central authority, these policies will not be static. Existing policies may need to be modified as the business rules change or the enterprise evolves. Hence, the existing security rules and the newly proposed rules may cause authorization conflicts. In either case, those conflicts need to be resolved correctly and properly.

A fine-grained access control model:

(i) should allow authorized users to access every single part of the data set that they are permitted.

(ii) should prevent authorized users from accessing any part of the data set that they are not permitted.

In a fine grained access control model, an effective and correct conflict resolution varies from case to case. Therefore, identifying various conflicting situations and dealing with them accordingly is a challenging important task in a fine-grained access control model. The following list includes examples of various conflicting authorization cases based on the security rules defined for accessing student GPA's (Grade Point Average) in an educational domain:

1. When having grant and deny authorizations on the same object with only grant having a condition such as "staff is granted access to GPA > 2.0" and "staff is denied access to GPA".

2. When having grant and deny authorizations on the same object with only deny having a condition such as "staff is granted access to GPA" and "staff is denied access to GPA < 2.0".

3. When having grant and deny authorizations on the same object with both having conditions and the condition of grant is the subset of the condition of deny such as "staff is granted access to GPA > 3.0" and "staff is denied access to GPA > 2.0".

4. When having grant and deny authorizations on the same object with both having conditions and the condition of deny is the subset of the condition of grant such as "staff is granted access to GPA < 3.0" and "staff is denied access to GPA < 2.0".

A fine-grained access control model deals with each one of the various conflict cases differently. For example, while our proposed fine-grained access control model eliminates the grant rule in the first case, in the second case, it does not eliminate but revise the grant rule with the reverse of the deny condition as follows: "staff is granted access to GPA >= 2.0". In case #4, grant rule is not eliminated but revised with the difference of ranges of grant and deny conditions as follows: "staff is granted access to 2.0 <= GPA < 3.0".

Each case needs to be handled properly and differently so that the access control system is ensured to generate and retain fine-grained authorization policies which is a challenging task. We believe that there is need to define and incorporate fine-grained XML authorization models with effective conflict handling mechanisms in the presence of conditional security policies into relational XML databases. We address this issue in this study.

43.3 Related Work

Although a number of access control models were proposed for native XML databases in the literature [6–13], there are lesser access control models proposed by researchers for relational XML databases [2–5, 10]. In [2, 3], authors proposed schema-oblivious XML access control techniques using relational databases. In [4, 5], authors proposed schema-based XML access control approaches for relational databases. However, handling authorization conflicts in the presence of conditions is not elaborated in these studies.

The authorization policies in XML security models are mainly defined either as external subset or internal subset [2]. In external subset approach, the policies are defined in an external file based on the content and structure of the underlying XML document which is denoted by a DTD or XML Schema. Internal subset approach defines and inlines additional tags within the target XML document to incorporate the security information. Internal subset overrides the external subset if they occur together. We choose external subset approach in our study.

Various conflict resolution policies proposed in the literature solve conflicts with general policies such as "denials take precedence or deny overrides" [7, 10]. However, these policies do not consider the cases when the subject is conditionally denied accessing part of the object that was previously granted access. We propose a fine-grained access control model for relational XML databases which address all aspects of handling authorization conflicts including conditional ones in this paper.

43.4   Proposed Authorization Model

In this section, we propose a fine-grained authorization model for relational XML databases. We adopted our previously defined deny-only access control model [4] and enhanced it with the inclusion of grant rules along with deny rules and conflict resolution policies. We also changed the default semantics from *grant* to *deny* in this enhanced authorization model.

In our authorization model, an access control policy contains a set of authorization rules. In this access control model, an authorization rule is a tuple of the form *(subject, object, condition, action, type, mode)* where

- *subject* denotes to a group of users
- *object* refers to the group of data that the subject is concerned with
- *condition* denotes the optional predicate applied to the *object*
- *action* refers to the type of action (select, update, delete, etc.) that the subject is denied or allowed
- *type* indicates whether the rule affects only the object or propagates to the descendants of the object
- *mode* indicates whether the *action* is grant or deny

In our enhanced access control model, we consider both *grant* and *deny* rules. We introduced the *mode* parameter to enable defining opposite authorization rules with *grant* and *deny* modes unlike our previous model where only deny rules are allowed. We introduce conditional authorization rules with the use of *condition* parameter in authorization rules. Therefore, various authorization conflicts may frequently occur in the proposed access control model.

We only consider *select* action for simplicity in this paper. The default semantic of the proposed access control model is *deny*. Therefore, no one can access any part of the XML document unless there is a rule which allows a user to access a specific part, element or attribute of the XML document.

Authorization conflicts occur if two authorization rules with opposite operation modes, namely grant and deny, are defined on the same object. We categorize authorization conflicts as *absolute conflicts* and *partial conflicts* and deal with them accordingly. We introduce the notions of absolute conflicts and partial conflicts in the following definitions.

Definition 4.1 (Absolute Authorization Conflict). An authorization rule can be denoted as $Rule_{sbj}(Obj[predicate])$ where *Rule indicates Grant or Deny, sbj is the subject of the Rule, Obj is the object concerned by the Rule and predicate is the condition defined for the Rule. If the rules* $Grant_s(O[predicate_g])$ *and* $Deny_s(O[predicate_d])$ *are given for the same subject S, object O and if* $predicate_g \subseteq predicate_d$, *then, Absolute Conflict is said to occur between these rules.*

Definition 4.2 (Partial Authorization Conflict). An authorization rule can be denoted as $Rule_{sbj}(Obj[predicate])$ where *Rule indicates Grant or Deny, sbj is the subject of the Rule, Obj is the object concerned by the Rule and predicate is the condition defined for the Rule. If the rules* $Grant_s(O[predicate_g])$ *and* $Deny_s(O[predicate_d])$ *are given for the same subject S, object O and if one of the following conditions holds: (i)* $predicate_d \subset predicate_g$, *(ii)* $predicate_d \cap predicate_g \neq \phi$ *AND* $predicate_g - predicate_d \neq \phi$ *AND* $predicate_d - predicate_g \neq \phi$ *then, Partial Conflict is said to occur between these rules.*

Our general conflict resolution policy is "*the latter rule overrides*" in the presence of *absolute conflicts*. In addition, we fine-tune our conflict resolution policy with the consideration of *partial conflicts* which occur due to the predicates. If a deny rule conflicts with a grant rule and if at least one of them is defined with a predicate, our algorithm analyzes the predicate(s) and partially override the former rule if there exists a *partial conflict*. Therefore, in the case of partial conflicts, conflict resolution policy becomes "*the latter rule partially overrides*".

We consider simple XPath predicates to test target text nodes or attributes within XPath expressions of the rule tuples. We do not consider twig pattern matching and predicates incurred by twig pattern matching in this study. We use a subset of XPath axes, namely *child, descendant* and *attribute* axes, to identify the objects in a rule tuple.

The propagation of an authorization rule to the descendants of the XML object node is not enforced automatically. The *type* parameter in a rule tuple determines the scope of an authorization rule. While a local rule only affects the object node itself, a recursive rule is applied to the object node as well as the descendants of the object node.

A sample XML authorization rule is shown in Fig. 43.1. The *condition* parameter of a rule tuple is combined with the XPath expression of the object *parameter*. The rule tuple

```
<rule>
    <subject>staff</subject>
    <object>//zip[.>48000]</object>
    <action>select</action>
    <type>L</type>
    <mode>grant</mode>
</rule>
```

Fig. 43.1 Sample authorization policy rule

(staff, //zip, [. > 48,000], select, L, grant) indicates that any *zip* element which is greater than 48,000 regardless of its ancestors is granted *select* access for *staff*. The *type L* denotes that this rule is local and does not affect the descendants of *zip* node if there is any descendant node.

43.5 XML-To-Relational Security Policy Conversion

The Security Policy Conversion module translates policy rules and users' information from XML to relational tables. The XML document for users information may include *userID, password, first* and *last name* and *role* of each user. It is straightforward to translate XML subtrees of users' information to relational rows in a table.

XML policy subtrees are mapped into relational tuples to be stored in XML Authorization Table (XAT) in the target relational database. The XAT table is the central place of authorization control in the proposed access control model. The XAT table stores only the grant rules since the default semantic of the proposed access control model is *deny*. Thus, when a query is given against the underlying database, the security system allows the query if there is a policy rule in the XAT table granting the user to execute the requested query against the specified object. Deny policy rules are not stored in the XAT table but dealt with accordingly.

During the mapping process of XML Authorization Rules (XAR) to Relational Authorization Rules (RAR), if the *mode* parameter of the XML authorization rule is *grant,* then a relational rule tuple is inserted into the XAT table. A relational authorization rule (RAR) is deleted from the XAT table if there is a corresponding XAR rule denoted with *deny* mode in the XML authorization document which causes an *absolute conflict*. If an XAR deny rule partially conflicts with a RAR grant rule in the XAT table, then, the RAR rule is updated with a relevant predicate.

We do not map *mode* and *type* parameters of an XAR rule to a RAR rule tuple. The *mode* parameter is not mapped since all the RAR rules in XAT table are grant rules. The *type* parameter is not mapped to a RAR rule tuple since all RAR rules in XAT table are local as the recursive rules are expanded to local ones during the mapping process.

It is a challenge to translate an XML policy subtree to one or more relational tuples in XML Authorization Table (XAT). While it is straightforward to map *subject* and *action* parameters of an XML policy rule to relational columns in the XAT table, there are several issues included in translating an *object* parameter along with its *condition* parameter to its relational equivalents. The issues involved in translating *object* parameters of XAR rules to RAR rules are elaborated in [4]. These issues are primarily due to the '//' axis, recursive scope of the policy and manipulation of the conditions. Conditions can get complex with multiple logical operators. In addition, conditions can be introduced to other nodes besides the context node which increases the complexity of mapping.

We only consider simple conditions for the context node of a path expression for an *object*. The *condition* parameter plays an important role in detecting various conflicts. There are several issues involved in manipulating *condition* parameters (predicates) during the XAR-to-RAR mapping process. These issues include the followings:

- If the operation mode for the XAR rule is *grant,* condition should be extracted and inserted into the XAT table along with other XAR parameters
- If the operation mode for the XAR rule is *deny,* the type of conflict should be detected analyzing the *condition* parameters of both XAR deny rule and the corresponding RAR grant rule
- If there is a partial conflict then a modified condition (new Predicate) needs to be calculated to update the existing RAR rule tuple

We propose XAR2RAR algorithm to convert XML access control rules (XAR) into relational access control rules (RAR) as well as to deal with authorization conflicts. This algorithm detects and resolves authorization conflicts as it translates the XAR rules in the input Authorization.xml document into the RAR rules in the output XML Authorization Table (XAT). The RAR rules in the XAT table are used to enforce the access control policies for XML data stored in a RDBMS. XAR2RAR algorithm is given in Fig. 43.2.

XAR2RAR algorithm processes each XAR rules defined in Authorization.xml within a loop. Firstly, it checks for the *type* parameter. The authorization rule is applied at the node level when *type* is local (L). When type is recursive (R), then the rule is applied at the node level as well as at the subtree rooted at that node. If the path expression of the *Object* has descendant axes then the XAR2RAR algorithm creates an ObjectPattern to find the set of all matching paths for that *Object*. This set of matching paths is called ObjectSet and obtained from a global path table (*AllPaths*) which includes all existing paths in a given XML document. We adapt the

Fig. 43.2 XAR2RAR algorithm

```
00 Algorithm XAR2RAR
01 Input: XAR_Rules Authorization.xml
02 Output: RAR_Rules XAT
03 Begin
04 For each Rule_i in Authorization.xml
05   Predicate, newPredicate = null;
06   If (Object contains Predicate) then Predicate=Object.Predicate End If
07   If (Type='L') then
08     If (Object contains '//') then
09       ObjectPattern = Object.replace ("//", "%/")
10       ObjectSet = "Select path From AllPaths Where path like $ObjectPattern"
11     Else
12         ObjectSet = Object
13     End If
14     For each object_i in objectSet
15       If @Mode='Grant' then
16               Insert into XAT (Subject, object_i, Predicate, Action)
17       Else  /* Mode = 'Deny' */
18         If there exists absolute conflict then
19   Delete From XAT Where (XAT.Subject=Subject AND XAT.Object=object_iAND
                            XAT.Action=Action)
20         Else If there exists partial conflict then
21   Update XAT Set XAT.Predicate=newPredicate Where (XAT.Subject=Subject
                    AND XAT.Object=object_i AND XAT.Action=Action)
22         End If
23       End If
24     End If
25   End For
26   Else  /* Type = 'R'  */
27     If (Object contains '//') then
28       ObjectPattern = Object.replace ("//", "%/")
29       ObjectSet = "Select Path From AllPaths
                       Where (path like $ObjectPattern  OR path like $ObjectPattern/%)"
30     Else
31         ObjectSet = "Select path From AllPaths
                        Where (path like $Object OR path like $Object/%)"
32     End If
33     For each object_i in objectSet
34       If @Mode='Grant' then
35               Insert into XAT (Subject, object_i, Predicate, Action)
36       Else  /* Mode = 'Deny' */
37         If there exists absolute conflict then
38       Delete from XAT Where (XAT.Subject=Subject AND XAT.Object=object_i
                                AND XAT.Action=Action)
39         Else
40           If there exists partial conflict then
41       Update XAT Set XAT.Predicate=newPredicate Where (XAT.Subject=Subject
                        AND XAT.Object=object_i AND XAT.Action=Action)
42           End If
43         End If
44       End If
45     End For
46   End If
47 End For
48 End
```

idea of introducing a global path table to our mapping scheme from XRel approach [14]. If the *type* is recursive, XAR2RAR algorithm places a path expression for the *Object* as well as the path expressions for all the descendants of the *Object* into the *ObjectSet*. Each path expression in the *ObjectSet* is processed as follows:

- If the mode of the XAR rule is *grant* then a RAR rule tuple is inserted into the XAT table (lines 16 and 35)
- If the mode of the XAR rule is *deny* and the rule causes an *absolute conflict* then the conflicting RAR rule tuple is deleted from the XAT table (lines 19 and 38)

- If the mode of the XAR rule is *deny* and the rule causes a *partial conflict* then the conflicting RAR rule tuple is updated in the XAT table (lines 21 and 41)

43.6   Illustrative Case Study

We illustrate the proposed authorization model with a case study in this section. The *department* benchmark data set is used as our sample data set in the case study. The *department* data set's schema (DTD) is shown in Fig. 43.3.

The first set of XAR rules given against the *department* data set are described in the XML document *Authorization1.xml* which is shown in Fig. 43.4.

Our XAR2RAR algorithm translates XML access control rules into relational tuple(s) in XAT table and deals with authorization conflicts as it parses XML authorization documents.

When the first XAR rule is parsed, the object's path expression *//gpa* is extended to its corresponding path expressions. Then, two RAR tuples with the below objects are inserted into XAT table to grant *select* access to all *gpa* nodes for *staff*.

- */department/gradstudent/gpa*
- */department/undergradstudent/gpa*

When the second XAR rule is parsed, firstly, the below expanded path expressions of */department/undergradstudent/address* are extracted from *AllPaths* table:

- */department/undergradstudent/address*
- */department/undergradstudent/address/city*
- */department/undergradstudent/address/state*
- */department/undergradstudent/address/zip*

Then, four RAR tuples with the above objects are inserted into XAT table which grant *select* access to all *address* subtrees of undergraduate students for *staff*.

The third XAR rule shows an example of a conditional authorization rule. The following target object path is granted access for staff subjects for zip codes less than 60,000.

- */department/gradstudent/address/zip[. < 60,000]*

The fourth and fifth rules are processed similar to the second and third rules except they are defined for faculty subjects. The XAT table populated by the XAR2RAR algorithm for the above Authorization1.xml document is given in Table 43.1.

The Authorization2.xml document given in Fig. 43.5 includes additional set of authorization rules which conflict with the ones introduced in Authorization1.xml shown in Fig. 43.4.

First rule in Authorization2.xml denies access to subtrees of address nodes of undergraduate students for *staff*. Since we have grant rules defined for the same target objects, this raises *absolute conflicts* and all of the four tuples corresponding to those objects are deleted from the XAT table.

The second rule denies access to zip codes of graduate students if the zip code is less than 70000 for *staff*. The XAT

Fig. 43.3 DTD of department data set

```
<!ELEMENT department (deptname, gradstudent* , staff*, faculty, undergradstudent*)>
<!ELEMENT gradstudent (name, phone, email, address, office?, url?, gpa)>
<!ELEMENT staff (name, phone, email, office?)>
<!ELEMENT faculty (name, phone, email, office)>
<!ELEMENT undergradstudent (name, phone, email, address, gpa)>
<!ELEMENT name (lastname?,firstname)>
<!ELEMENT address (city, state, zip)>
<!ELEMENT deptname (#PCDATA)>
<!ELEMENT city (#PCDATA)>
<!ELEMENT state (#PCDATA)>
<!ELEMENT zip (#PCDATA)>
<!ELEMENT office (#PCDATA)>
<!ELEMENT phone (#PCDATA)>
<!ELEMENT lastname (#PCDATA)>
<!ELEMENT firstname (#PCDATA)>
<!ELEMENT url (#PCDATA)>
<!ELEMENT gpa (#PCDATA)>
<!ELEMENT email (#PCDATA)>
```

```
<?xml version="1.0" encoding="utf-8"?>
<rules>
      <rule>
            <subject>staff</subject>
            <object>//gpa></object>
            <action>Select</action>
            <type>L</type>
            <mode>Grant</mode>
      </rule>
      <rule>
            <subject>staff</subject>
            <object>/department/undergradstudent/address</object>
            <action>Select</action>
            <type>R</type>
            <mode>Grant</mode>
      </rule>
      <rule>
            <subject>staff</subject>
            <object>//gradstudent//zip[.<60000]</object>
            <action>Select</action>
            <type>L</type>
            <mode>Grant</mode>
      </rule>
      <rule>
            <subject>faculty</subject>
            <object>/department/undergradstudent/address</object>
            <action>Select</action>
            <type>R</type>
            <mode>Grant</mode>
      </rule>
      <rule>
            <subject>faculty</subject>
            <object>//gradstudent//zip[.<70000]</object>
            <action>Select</action>
            <type>L</type>
            <mode>Grant</mode>
      </rule>
</rules>
```

Fig. 43.4 Authorization1.xml document

Table 43.1 XAT table after authorization1.xml parsed

XAT Table

Subject	Object	Predicate	Action
staff	/department/gradstudent/gpa	–	Select
staff	/department/undergradstudent/gpa	–	Select
staff	/department/undergradstudent/address	–	Select
staff	/department/undergradstudent/address/city	–	Select
staff	/department/undergradstudent/address/state	–	Select
staff	/department/undergradstudent/address/zip	–	Select
staff	/department/gradstudent/address/zip	. < 60,000	Select
faculty	/department/undergradstudent/address	–	Select
faculty	/department/undergradstudent/address/city	–	Select
faculty	/department/undergradstudent/address/state	–	Select
faculty	/department/undergradstudent/address/zip	–	Select
faculty	/department/gradstudent/address/zip	. < 70,000	Select

table includes grant rule on zip code for staff subjects if the graduate student's zip code is less than 60000. Since $(zip < 60000) \subseteq (zip < 70000)$, this situation raises an *absolute conflict* and the corresponding tuple is deleted from the XAT table.

The third rule in Authorization2.xml denies access to GPA nodes which are less than 2.0 for *staff*. This rule conflicts with the RAR grant rule for GPA nodes. However the grant rule is unconditional and the deny rule is conditional. Since $(GPA < 2.0) \subset U$, this situation raises a *partial*

```
<?xml version="1.0" encoding="utf-8"?>
<rules>
      <rule>
            <subject>staff</subject>
            <object>//undergradstudent//address</object>
            <action>Select</action>
            <type>R</type>
            <mode>Deny</mode>
      </rule>
      <rule>
            <subject>staff</subject>
            <object>//gradstudent//zip[.<70000]</object>
            <action>Select</action>
            <type>L</type>
            <mode>Deny</mode>
      </rule>
      <rule>
            <subject>staff</subject>
            <object>//gpa[.<2.0]></object>
            <action>Select</action>
            <type>L</type>
            <mode>Deny</mode>
      </rule>
      <rule>
            <subject>faculty</subject>
            <object>//gradstudent//zip[.>60000]</object>
            <action>Select</action>
            <type>L</type>
            <mode>Deny</mode>
      </rule>
</rules>
```

Fig. 43.5 Authorization2.xml document

Table 43.2 XAT table after authorization2.Xml parsed

XAT Table			
Subject	Object	Predicate	Action
staff	/department/gradstudent/gpa	. >= 2.0	Select
staff	/department/undergradstudent/gpa	. >= 2.0	Select
faculty	/department/gradstudent/address	–	Select
faculty	/department/gradstudent/address/city	–	Select
faculty	/department/gradstudent/address/state	–	Select
faculty	/department/gradstudent/address/zip	–	Select
faculty	/department/undergradstudent/address/zip	. <= 60,000	Select

conflict. Thus, we do not delete the corresponding grant tuples in XAT table but update them with the new predicate (GPA $>= 2.0$).

The last rule in Authorization2.xml denies access to zip codes of graduate students if the zip code is greater than 60000 for *faculty* subjects. The XAT table includes grant rule for *faculty* subjects if the graduate student's zip code is less than 70000. Since (zip > 60000) $\cap$ (zip < 70000) $\neq \emptyset$ and differences are not empty set either, this situation raises a *partial conflict* and the corresponding RAR tuple in XAT table is updated with the revised predicate (zip $<= 60000$).

The XAT table populated by the XAR2RAR algorithm after the Authorization2.xml parsed is shown in Table 43.2.

XAT table is central in the process of secure translation of XML queries into SQL queries in our relational XML database system. First, an input XML query is mapped into an SQL query. Then it is rewritten based on the security information obtained from the XAT table. The rewritten secure query returns the permitted query results.

43.7 Conclusions

We proposed a fine-grained access control model with conflict handling mechanisms in the presence of conditions for relational XML databases. We define the concepts of *absolute conflicts* and *partial conflicts* which are instrumental in evaluating and resolving authorization conflicts.

Our proposed algorithm XAR2RAR translates XML authorization rules (XAR) into relational authorization rules (RAR) and stores them in a relational table. XAR2RAR algorithm resolves conflicts due to conditions effectively and deals with overall authorization conflicts elegantly using a centralized authorization table (XAT) in the target relational database.

We introduced conflicts due to grant-deny authorization policies defined on common objects. We did not look into grant-grant or deny-deny conflicts. In case of *partial conflicts*, the underlying RAR rule needs to be updated with a relevant new predicate. The new predicate is produced differently for various situations. We consider elaborating other types of authorization conflicts and introducing a new predicate production algorithm for different cases of *partial conflicts* as potential future work.

References

1. Atay, M., & Lu, S. (2009). *Storing and querying XML: an efficient approach using relational databases*. ISBN 3639115813, VDM Verlag. Saarbrücken.
2. Lee, D., Lee, W., & Liu, P. (2003). Supporting XML security models using relational databases: a vision. In *Lecture notein computer science* (Vol. 2824, pp. 267–281). Berlin/Heidelberg: Springer.
3. Luo, B., Lee, D., Liu, P. (2007). Pragmatic XML access control using off-the-shelf RDBMS. In *Proceedings of the 12th ESORISC (Dresden, September 24–26, 2007)*, (pp. 55–71).
4. Patel, J., & Atay, M. (2011). An efficient access control model for schema-based relational storage of XML documents. In *Proceedings of the 49th ACM Southeast Conference, (Georgia, USA, March 2011)*, (pp. 97–102).
5. Tan, K.L., Lee, M.L., Wang, Y. (2001). Access control of XML documents in relational database systems. In *Proceedings of the International Conference on Internet Computing (IC) (Las Vegas, NV, Jun. 2001)*, (pp. 185–191).
6. Kundu, A., & Bertino, E. (2008). A new model for secure dissemination of XML content. *IEEE Transactions on Systems, Man, and Cybernetics, Part C, 38*(3), 292–301.
7. Damiani, E., Vimercati, S., Paraboschi, S., & Samarati, P. (2002). A fine-grained access control system for XML documents. *IEEE Transactions on Information and System Security (TISSEC), 5*(2), 169–202.
8. Damiani, E., Fansi, M., Gabillon, & Marrara, S. (2008). A general approach to securely querying XML. *Computer Standards and Interfaces, 30*(6), 379–389.
9. Jo, S., & Chung, K. (2015). Design of access control system for telemedicine secure XML docs. *Multimedia Tools and Applications, 74*(7), 2257–2271.
10. Koromilas, L., Chinis, G., Fundulaki, I., & Ioannidis, S. (2009). Controlling access to XML documents over XML native and relational databases. *Secure Data Management LNCS, 5776*, 122–141.
11. Mahfoud, H., & Imine, A. (2012). Secure querying of recursive XML views: a standard XPath-based technique. *Proceedings of the 21st International Conference on World Wide Web*, pp. 575–576.
12. Mirabi, M., Ibrahim, H., Fathi, L., Udzir, N., & Mamat, A. (2011). An access control model for supporting XML document updating. *Networked Digital Technologies., 136*, 37–46.
13. Zhu, H., Lü, K., & Jin, R. (2009). A practical mandatory access control model for XML databases. *Information Sciences, 179*(8), 1116–1133.
14. Yoshikawa, M., Amagasa, T., Shimura, T., & Uemura, S. (2001). XRel: a path-based approach to storage and retrieval of XML docs. Using rel. databases. *ACM Transaction on Internet Technology (TOIT), 1*(2), 110–141.

A Practical Approach to Analyze Logical Thinking Development with Computer Aid

44

Breno Lisi Romano and Adilson Marques da Cunha

Abstract

Logical thinking is essential to one's development. It lays a foundation to acquire knowledge and skills that are used to solve many problems, not only in school, but also to execute tasks and make decisions. Nowadays, interactive technologies have a potential to be applied in education and are presented as a facilitator in both learning and teaching activities. This paper aims to present an application of an approach to analyze the logical thinking development with computer aid. To do so, we developed and applied a test environment to two student groups to assess any improvement in learning and, consequently, in logical thinking. The results pointed out to a visible development of students' logical thinking, showing that they did not have any difficulties to execute the proposed tasks. This allowed us to build an illustrated view of thought trainings that help them to achieve their goals and make decisions. This paper also highlights the importance of using computational tools to support teachers in classrooms and to stimulate the development of logical thinking.

Keywords

Logical thinking • Computers in education • Computational tools

44.1 Introduction

Children today are more technologically skilled than those of the same age group who lived in the past century. Among all the available information, logical thinking is essential to one's development. It lays a foundation to acquire knowledge and mathematical abilities, to analyze information, to solve problems, and to develop oneself both creative and intellectually [1].

The development of logical thinking may help students not only in school, but also to better execute daily tasks, improving their concentration and performance in games and cultural, collective, and social activities. It helps them to better organize how they think, providing a clearer vision of how things happen and in what order.

Specialists say that, as it happens with any muscle to remain in shape, the brain needs to be regularly exercised. Developing children's logical thinking is important because it makes them think more critically, which in turn makes them more prone to present arguments based on a number of criteria and in logically valid principles [2].

Computers can be useful tools when it comes to stimulate the development of logical thinking. When it is used to pass on information to a student, the computer takes the role of a teaching machine and the pedagogical approach is a computer-aid instruction [3].

B.L. Romano (✉)
Federal Institute of Education, Science, and Technology of Sao Paulo, Sao Joao da Boa Vista, SP, Brazil
e-mail: blromano@ifsp.edu.br

A.M. da Cunha
Brazilian Aeronautics Institute of Technology, Sao Jose dos Campos, SP, Brazil
e-mail: cunha@ita.br

© Springer International Publishing AG 2018
S. Latifi (ed.), *Information Technology – New Generations*, Advances in Intelligent Systems and Computing 558, DOI 10.1007/978-3-319-54978-1_44

According to [3], the most noble and irrefutable use of a computer in education is to develop logical thinking or enable problem-solving situations [4].

Given the advances in Information and Communication Technologies (ICT), some computational tools emerged aiming to stimulate and develop logical thinking. These tools go from simple memory games to complex environments in which it is possible to manipulate objects using the command line. Some examples of these tools are Logo [5], Scratch [6], and Alice [7].

Given this scenario, it is possible to notice the relevance of stimulating the development of logical thinking in children, teenagers, and adults with the aid of computers. In this context, the goal of this paper is to present an application of an approach to analyze two specific student groups, by using computers and identifying if it is possible that these groups present an enhanced performance in learning activities and, consequently, more efficiency in executing daily tasks.

44.2 Related Works

Due to the importance of this paper's theme, there are many researches in Brazil and also in the world about the use of computational tools to develop logical thinking. This section presents a summary of some of these researches.

Vasconcelos (2002) did a participatory research on creativity and logical thinking development on the Elementary School students (8th graders – thirteen years old), focusing on heuristic problem solving. This research was based on the internationalist theory of Piaget and presented alternative ways of creating conditions in Math classes to flourish and develop creativity by solving problems that demanded logical thinking [8].

Reis (2006), seeking to help students' education and contribute to changes in educational processes, presented Logo as a tool to build knowledge and also to present linguistic and pedagogic aspects, interactions, interdisciplinarity and teachers' performance assessments in the Logo environment [9].

The research developed by Gorman et al. reports results of 15 third-grade students at a private school in Dallas who learned Logo during a school year. To one-third of the students were given 1 h/week of individual computer time (separated from in-class instructions), and the remaining students received 0.5 h/week of individual computer time. At the end of this school year, the 1-h group performed significantly better than the 0.5-h group on a conditional ruled-learning task. Future research should compare gains from structured languages like Logo with unstructured languages and should use additional assessment techniques [10].

Calder (2010) described how the Scratch tool could be used to design games and develop mathematical concepts. He examined the ways mathematical thinking emerges when children work with Scratch. He described how a class of 6 students used Scratch to design an activity and considered how this facilitated an authentic problem-solving process. The ways their mathematical thinking emerged through this process were outlined, along with further suggestions of using Scratch in classroom situations [11].

Finally, Dim et al. (2011) proposed a discussion about the difficulties of learning in the first years of Computer Science in courses which logic was involved. Considering that the adequate pedagogic tools could improve students' performance, they developed a tool called *Agência Planetária de Inteligência* (APIN) to build a logical-mathematical foundation in high school students and also to support knowledge acquisition through the theory of assimilation [12].

44.3 Methodology

To accomplish the goals of this research, the strategy for applying two computational tools within the defined groups is presented in this section, as well as students' performance assessments and how we assessed applications' data.

44.3.1 Tools

The tools used in this experiment were Super Logo (a Logo language interpreter) and Scratch. The choice of these tools was made due to their computational capacities and ease of use, regarding the fact that they are free and do not require any buyable licenses.

Both tools have educational potential and one of the objectives of this paper was to verify which one of them would perform better as a tool to help the logical thinking of students who partook in the experiments.

44.3.2 Target Audience

Based on the literature and knowing that logical thinking must be stimulated and exercised not only in children, but also in all age groups, a group of teenagers from 14 to 18 years was chosen for the experiment in the proposed computational environment.

The experiments were applied into two different classes of the technical course named, in Portuguese, *Técnico em Informática Integrado ao Ensino Médio,* from the Federal Institute of Education, Science, and Technology of Sao Paulo, Sao Joao da Boa Vista, SP, Brazil. The first class

(**Group 01**) was comprised of students in the high school freshman year (14 years old) and the second class (**Group 02**) was comprised of students in the high school senior year (18 years old). We aimed to assess how much of logical thinking the students were able to develop by having logic and programming courses throughout four years until graduation.

44.3.3 Strategy

Experiments were devised, in order to assess the contribution of the computational tools in the formation/stimulation of logical thinking. They were performed using the same tools and a set of problems and the only difference was the learning stage of each group.

First, the tools (Super Logo and Scratch) were presented to the students. It was also briefly presented how the computer interprets a command line. This explanation aimed to arouse interest of the students on how to use the proposed tools.

To help with the tools introduction, two scripts were created as a support material with instructions regarding each tool. These scripts were distributed to the students to serve as a reference guide to execute the proposed tasks. We also distributed a questionnaire to every student, before the assessment began, to collect personal data that would later be used in the analysis stage.

Each group that partook in the experiments was divided into two parts: half of the students used Super Logo and the other half used Scratch.

After introducing the tools, dividing the groups, and collecting the answered questionnaires, we began to show the students the problems proposed in this research.

To compare the results of the experiments, we used the following criteria:

- Compare the evolution of the high school students' logical thinking by taking into account the four learning years of logic and programming; and
- Compare if any of the tools proved to be efficient in helping teachers.

The way the comparisons were made in our experiments is shown in Fig. 44.1. As it can be seen, the classes were divided into two groups: one that used Super Logo and other that used Scratch. The comparisons relationships are also shown.

44.3.4 Activities

The activities/problems in the experiment and how they were proposed are described in this section. It is worth

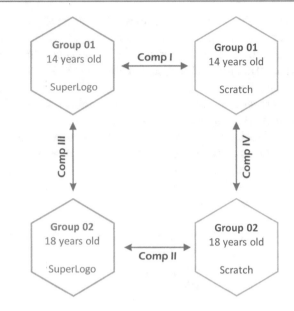

Fig. 44.1 The comparisons between tools and students

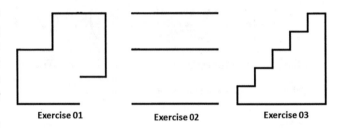

Fig. 44.2 The proposed exercises for Activity I

mentioning that the proposed activities were applied to all groups shown in Fig. 44.1 and that each activity was prepared with an increased level of difficulty as the students progressed.

Activity I consisted of building "drawings" with simple command lines such as "forward" or "to the right". This activity was comprised of 3 exercises. Each exercise had a maximum execution time of 10 minutes. Given these time constraints, each student had up to 3 chances of executing each exercise. The exercises were applied one at a time, with the same time constraints, regardless of group. Upon finishing a try, each student signaled a teacher, who then wrote down the results in that student's questionnaire. When the time for each activity was over, all students stopped their activities and the teachers wrote down the performance of each one of the students in the given activity.

Figure 44.2 presents the 3 proposed exercises for Activity I, called "Take the cursor from point A to point B".

Activity II consisted of creating images that represented regular shapes as geometric forms. In this activity, students could use a few more commands such as repetition control statements ("Repeat a given command X times") and some arithmetic operations to calculate angles to create geometric drawings. This activity was composed of 4 exercises, with an increased level of difficulty as the students progressed. Similar to Activity I, each exercise had a maximum duration time of 10 minutes and each student had up to 3 tries to finish each exercise.

Figure 44.3 shows the 4 proposed exercises for Activity II, called "Drawing geometric drawings".

Due to the difficulty of Activity III, it was composed of only one exercise in which the students should create a more complex drawing. In this activity, the students could use simple commands, repetition statements, and so on, in addition to performing arithmetic operations and using logical thinking to achieve the goal of the exercise. To finish this activity, the students had an unlimited number of tries with a maximum duration time of 60 minutes.

Figure 44.4 shows the proposed exercise for Activity III, called "Drawing a castle".

Exercise 01   **Exercise 02**   **Exercise 03**   **Exercise 04**

Fig. 44.3 The proposed exercises for Activity II

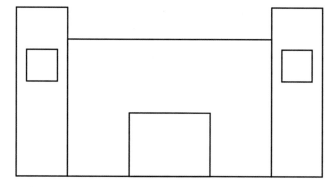

Fig. 44.4 The proposed exercise for Activity III

It is worth to mention that in Activities I and II each student had 3 tries to finish each exercise of each activity, besides the 10-minute duration time. As students finished their exercises or wasted a try due to errors, they called one of the teachers so they could assess the current exercise and write down the results in students' individual sheets.

Figure 44.5 shows a small part of the student sheet template used to report the 3 Attempts of Exercise 01 (10 minutes) for Activity I.

Data gathered from all experiments were input to an electronic sheet system in which graphics were generated with comparisons among classes of experiments and their respective tools. The results are described in the next section.

44.4 Result

In this section, we present the main results from data gathered when using both tools (Super Logo and Scratch) in both students groups (Group 01 and 02) and their comparisons. By doing so, we aim to fulfill the goal of checking if using these tools do indeed help to develop students' logical thinking.

In Group 01, there were 33 students at the time of this experiment. To our surprise, teachers were already using the Scratch in the classroom, as a support tool for teaching logic and introduction to programming.

In Group 02, there were 27 students at the time of this experiment and, only 2 of them knew or had used one of the two computational tools before. However, these 27 students have had 4 years of classes based on logic and programming when compared to Group 01.

Given the questionnaires answered by the students from both groups, we got the following data, presented in Table 44.1.

Fig. 44.5 A small part of the student sheet template used to report the 3 Attempts of Exercise 01 for Activity I

				Start Time	End Time		Not concluded (0%)	25%	50%	75%	Completed Correctly (100%)
								Performance			
Activity I	Exercise 01 (10 min)	Attempt	1	00:00							
				Used method:							
			2								
				Used method:							
			3								
				Used method:							

Table 44.1 The data gathered from students' questionnaires

		Group 01 (33 students)		Group 02 (27 students)	
		Nº	%	Nº	%
Gender	Male	22	67%	18	67%
	Female	11	33%	9	33%
Age	15 a 20	33	100%	27	100%
	21 a 25	0	0%	0	0%
	26 a 30	0	0%	0	0%
	31 a 35	0	0%	0	0%
	>= 36	0	0%	0	0%
Have you done the course of logic or programming?	Yes	0	0%	0	0%
	No	33	100%	27	100%
Have you used the Super Logo or Scratch tools?	Yes	33	100%	2	7%
	No	0	0%	25	93%
If so, what?	SuperLogo	0	0%	1	4%
	Scratch	33	100%	1	4%
Tool that students use during the experiment	SuperLogo	17	52%	14	52%
	Scratch	16	48%	13	48%

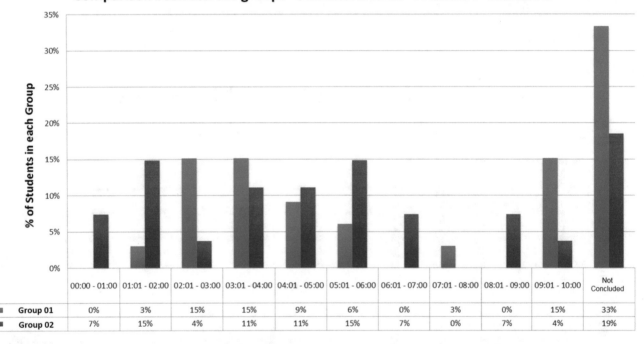

Fig. 44.6 Comparative results for both Groups on Exercise 1, Activity I

44.4.1 Comparative Analyses Between Classes

This section presents a comparison between the two classes of students for each exercise proposed in Activities I, II, and III, regardless of which tool was used.

Figures 44.6, 44.7, and 44.8 show comparisons among Exercises 1, 2, and 3, respectively, from Activity I. As it can be noticed, these figures show the exercises completion percentage of Groups 01 and 02, highlighting the time spent by students to complete these exercises.

Activity I consisted of using the proposed computational tools to create drawings using simple command instructions (forward, to the right, back, and so on). After completing Activity I and analyzing its data, we observed that:

(i) Even though some students could not complete the exercises of this activity, most of them were able to finish them in both groups;

(ii) Most students did not present difficulties to finish the exercises, though some did not finish. This may have

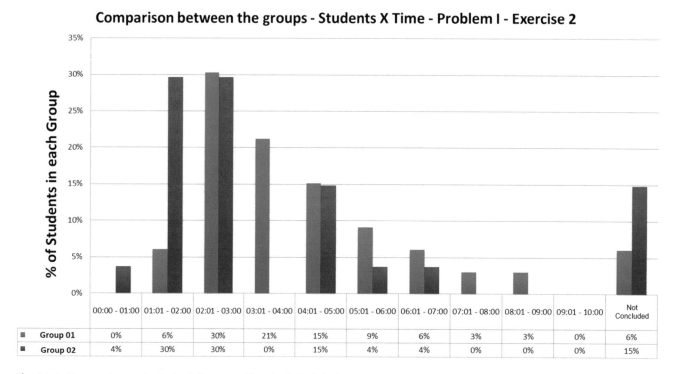

Fig. 44.7 Comparative results for both Groups on Exercise 2, Activity I

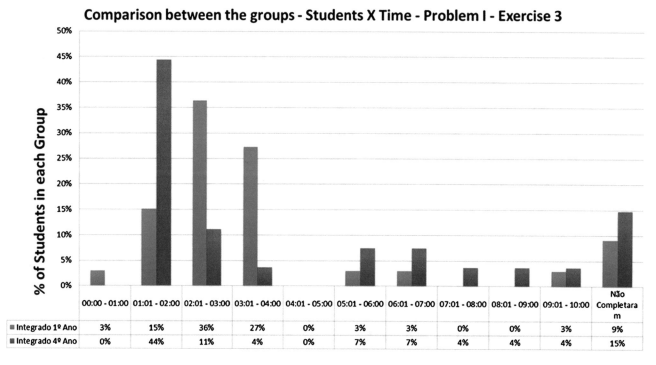

Fig. 44.8 Comparative results for both Groups on Exercise 3, Activity I

happened due to the learning curve of the proposed tools, even though students from Group 01 already knew how to use the Scratch tool;

(iii) A fact that drew our attention was that students from Group 01 were already familiar with Scratch. Because of that, they used repetition statements in Exercise 3. This shows that this particular tool can indeed be helpful to clarify how logical structures and a sequence of commands work.

Figures 44.9, 44.10, 44.11, and 44.12 show a comparison among Exercises 1, 2, 3, and 4, respectively, from Activity II. Activity II consisted of creating regular geometric shapes using simple commands (forward, to the right, back, and so

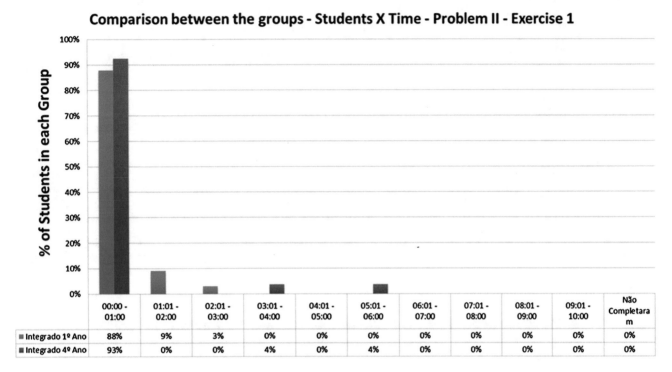

Comparison between the groups - Students X Time - Problem II - Exercise 1

	00:00 - 01:00	01:01 - 02:00	02:01 - 03:00	03:01 - 04:00	04:01 - 05:00	05:01 - 06:00	06:01 - 07:00	07:01 - 08:00	08:01 - 09:00	09:01 - 10:00	Não Completaram
Integrado 1º Ano	88%	9%	3%	0%	0%	0%	0%	0%	0%	0%	0%
Integrado 4º Ano	93%	0%	0%	4%	0%	4%	0%	0%	0%	0%	0%

Fig. 44.9 Comparative results for both Groups on Exercise 1, Activity II

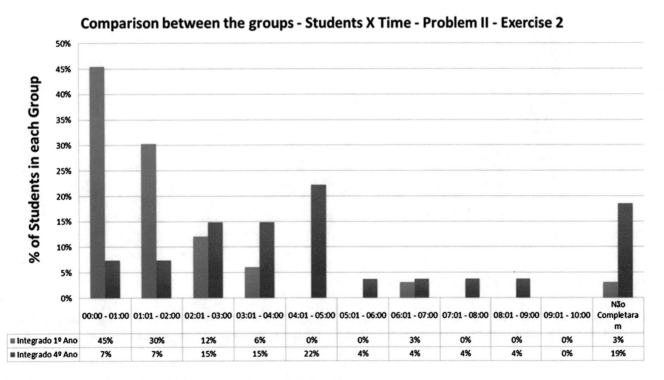

Comparison between the groups - Students X Time - Problem II - Exercise 2

	00:00 - 01:00	01:01 - 02:00	02:01 - 03:00	03:01 - 04:00	04:01 - 05:00	05:01 - 06:00	06:01 - 07:00	07:01 - 08:00	08:01 - 09:00	09:01 - 10:00	Não Completaram
Integrado 1º Ano	45%	30%	12%	6%	0%	0%	3%	0%	0%	0%	3%
Integrado 4º Ano	7%	7%	15%	15%	22%	4%	4%	4%	4%	0%	19%

Fig. 44.10 Comparative results for both Groups on Exercise 2, Activity II

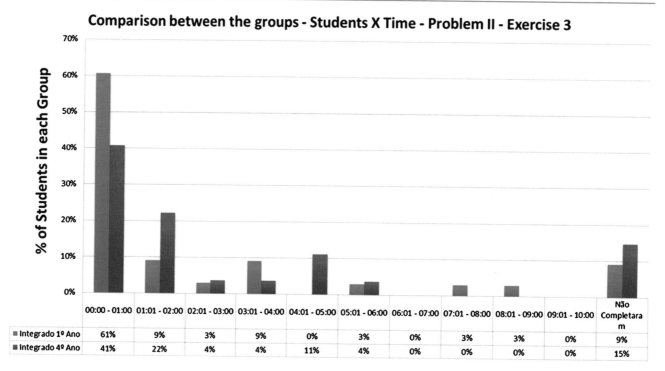

Fig. 44.11 Comparative results for both Groups on Exercise 3, Activity II

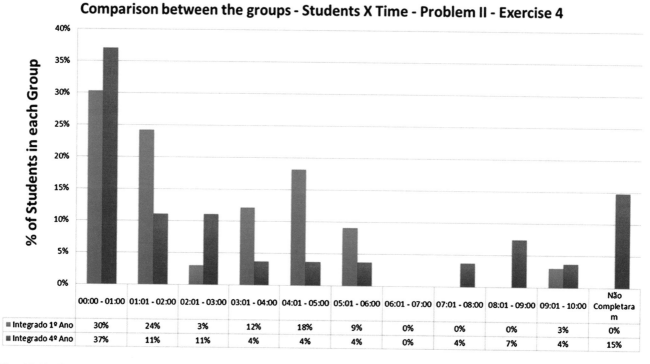

Fig. 44.12 Comparative results for both Groups on Exercise 4, Activity II

on) or repetition structures (repeat N times [commands]). To finish the exercises, some arithmetic operations to calculate angles were also needed. After ending Activity II and analyzing its data, we observed that:

(i) Most of the students from both groups were able to finish the exercises, reaching the goal and showing that their logical thinking was being developed;

(ii) Most of the students did not present difficulties in finishing the exercises, even though some of them did not finish;

(iii) We noticed that in many cases students showed mathematical difficulties, while trying to solve exercises, which may have affected the non-completion of exercises by some of them;

(iv) It was also possible to notice that most of the students from Group 01 did not need much time to solve the proposed exercises. When asked about this, they reported that have had classes about solving similar problems in the Scratch tool, few days before the experiment; and

(v) We also noticed that students from Group 01 did not have difficulties in adapting their knowledge from Scratch to Super Logo. This shows that when one learns something using logical thinking, as a cornerstone, the learned topic can be shared and/or reused whenever necessary, even when means to do it are different.

Activity III consisted of using the proposed tools to create the drawing of a castle. The students could use simple commands (forward, to the right, back, and so on), repetition structures (repeat N time [commands]), and perform some arithmetic operations to calculate some measurements in the drawing. After ending Activity III and analyzing its data, we observed that:

(i) Most of students were able to finish the exercise, reaching the goal and showing refined logical thinking; and

(ii) Since the complexity of Activity III was much higher than the other two activities, it was possible to notice that students from Group 01 had more difficulty to complete it, whereas most students from Group 02 were able to finish it in less time due to their logic study background.

Figure 44.13 shows a comparison for the exercise proposed for Activity III.

44.4.2 Comparative Analysis Between Tools

In this section we present data gathered from this experiment regarding the comparisons of Sect. 44.3.3 – Strategy. These results aim to compare the effectiveness of the proposed tools as supporting tools for teachers in classroom.

A comparison between Groups 01 and 02 and their used tools (Super Logo and Scratch), the percentage of completion, the shortest and longest completion time, and the average completion time are shown in Table 44.2, which presents comparisons defined from Fig. 44.1.

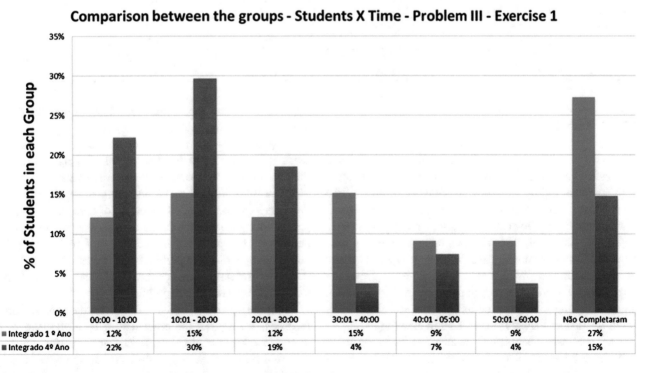

Comparison between the groups - Students X Time - Problem III - Exercise 1

	00:00 - 10:00	10:01 - 20:00	20:01 - 30:00	30:01 - 40:00	40:01 - 05:00	50:01 - 60:00	Não Completaram
■ Integrado 1 º Ano	12%	15%	12%	15%	9%	9%	27%
■ Integrado 4º Ano	22%	30%	19%	4%	7%	4%	15%

Fig. 44.13 Comparative results for both Groups on the Activity III Exercise

Table 44.2 Comparison between the tools used by Group 01 and Group 02

		Group 01		Group 02	
		SuperLogo	Scratch	SuperLogo	Scratch
Activity I: Exercise 01	% Complete	*82%*	*50%*	*100%*	*62%*
	Less Time	00:01:28	00:02:56	00:00:58	00:04:08
	Greater Time	00:09:19	00:09:30	00:08:00	00:09:26
	Average Time	00:04:17	00:06:22	00:03:09	00:06:08
Activity I: Exercise 02	% Complete	*94%*	*94%*	*100%*	*69%*
	Less Time	00:01:56	00:01:56	00:00:47	00:01:34
	Greater Time	00:08:28	00:07:52	00:04:32	00:06:19
	Average Time	00:03:54	00:03:52	00:02:06	00:03:42
Activity I: Exercise 03	% Complete	*94%*	*88%*	*100%*	*69%*
	Less Time	00:01:27	00:00:45	00:01:16	00:01:27
	Greater Time	00:09:56	00:06:39	00:08:44	00:09:53
	Average Time	00:03:19	00:02:46	00:02:15	00:05:13
Activity II: Exercise 01	% Complete	*100%*	*100%*	*100%*	*100%*
	Less Time	00:00:12	00:00:10	00:00:08	00:00:14
	Greater Time	00:01:08	00:02:15	00:00:50	00:05:25
	Average Time	00:00:33	00:00:34	00:00:31	00:01:09
Activity II: Exercise 02	% Complete	*94%*	*100%*	*86%*	*77%*
	Less Time	00:00:11	00:00:20	00:01:49	00:00:19
	Greater Time	00:06:27	00:02:40	00:08:33	00:07:14
	Average Time	00:01:35	00:01:11	00:04:09	00:03:21
Activity II: Exercise 03	% Complete	*100%*	*81%*	*100%*	*69%*
	Less Time	00:00:05	00:00:10	00:00:17	00:00:28
	Greater Time	00:03:04	00:08:14	00:05:12	00:04:37
	Average Time	00:00:42	00:02:23	00:01:44	00:01:46
Activity II: Exercise 04	% Complete	*100%*	*100%*	*100%*	*69%*
	Less Time	00:00:13	00:00:21	00:00:08	00:00:18
	Greater Time	00:09:58	00:05:20	00:07:52	00:09:40
	Average Time	00:02:54	00:02:17	00:01:37	00:04:28
Activity III: Exercise 01	% Complete	*94%*	*50%*	*93%*	*77%*
	Less Time	00:06:18	00:19:52	00:03:22	00:11:05
	Greater Time	00:44:18	00:59:58	00:24:25	00:59:00
	Average Time	00:22:35	00:41:17	00:11:03	00:30:21
Total Average per Tool		*95%*	*83%*	*97%*	*74%*

44.5 Conclusion

The main goal of this research was to present a practical approach to analyze the development of logical thinking with computer aid by providing computational tools to two groups and by assessing if they presented improvements in learning and also in logical thinking due to the tools usage.

To reach this goal, we devised a strategy so the computational tools could be used by two different groups of students in three activities. With the results gathered from the activities, we were able to compare the evolution of the logical thinking in a given period of time and if the tools could be used as a support to teachers in classrooms.

The goal was successfully reached in an experiment performed by students from the technical course named *Tecnico em Informática Integrado ao Ensino Médio*. The main results have shown students' logical thinking

improvements and the importance of using computational tools for supporting teachers in classrooms.

Most students did not present difficulties to complete exercises being able to develop logical thinking that helped them to make decisions towards their objectives.

Regarding the used tools, one thing to highlight, is the previous knowledge each student had of logic and computer programming. Even though the used tools were not known beforehand by some of the students, they used their prior knowledge to solve the problems. So we can say that some cognitive processes were activated when they tried to finish the exercises during the experiments.

From data gathered during the experiments and students feedback, we noticed the efficiency of both computational tools. However, a clarification regarding the performance of each tool is necessary. Firstly, the students who were already familiar with Super Logo showed a better performance when solving the exercises. This is due to the fact that Super Logo commands resemble lines of code of other programming tools. Secondly, some students had noticeable difficulty to develop their logical thinking and also to use repetition statements.

From the students who used Scratch, we noticed an ease of use, regarding repetition statements, due to the fact that this tool offers a graphic option to concatenate commands, making it easier to visualize composed structures. Nevertheless, when it came to simple commands we noticed that some students had difficulties to sort their thinking and to create an ordered and continuous sequence of commands.

Therefore, we suggest that both tools should be used in the classroom for supporting teachers on logic and computer programming.

As future work, we recommend looking for other tools that have potential to help develop logical thinking, since it is likely that new tools will be created to this end.

Another future work proposal is to use the approach of this paper on kids to assess how logical thinking behaves in early stages. Besides that, developing a new dynamic and preferably mobile-friendly tool that could be used the way Super Logo and Scratch were used in this paper would be of tremendous value.

References

1. Oliveira, P. A., & Rocha, A. J. O. (2011). *Raciocínio lógico, conceitos e estabelecimentos de parâmetros para a aprendizagem matemática, In: Revista do Acadêmico de Matemática*. Taguatinga: FACITEC – Faculdade de Ciências Sociais e Tecnológicase.

2. Santana, R. (2013). Método usa exercícios para estimular o cérebro e, assim, manter o raciocínio lógico em dia. Guia Araraquara – Araraquara.

3. Valente, J. A. (1993). Por quê o computador na educação?. In: Valente, J. A. Computadores e Conhecimento: repensando a educação. Campinas, SP: Gráfica da UNICAMP.

4. Valente, J. A. *O computador na sociedade do conhecimento* (pp. 31–43). Brasília: MEC, s/d.

5. Prado, M. E. B. B. (2000). *LOGO – Linguagem de Programação e as Implicações Pedagógicas*. Campinas: NIED – UNICAMP.

6. Resnick, M., et al. (2009). Scratch: programming for all. Magazine - Communications of the ACM - *Scratch Programming for All CACM Homepage archive, 52*(11), 60–67. ACM New York, NY, USA.

7. Alice (2013). Alice project – an educational software that teaches students computer programming in a 3D environment. Disponível em: http://www.alice.org. Acesso em 08 Sep 2016.

8. Vasconcelos, M. C. (2002). Um estudo sobre o incentivo e desenvolvimento do raciocínio lógico dos alunos, através da estratégia de resolução de problemas. Dissertação (Mestrado em Engenharia de Produção) – Universidade Federal de Santa Catarina, Centro Tecnológico. Florianópolis.

9. Reis, M. P. (2006). Brincando com a Lógica. Guaratinguetá: UNESP.

10. Gorman Jr., H., & Bourne Jr., L. E. (1983). Learning to think by learning LOGO: rule learning in third-grade computer programmers. *Bulletin of the Psychonomic Society, 21*(3), 165–167.

11. Calder N. (2010). Using scratch: An integrated problem-solving approach to mathematical thinking. *Australian Primary Mathematics Classroom, 15*(4), 9–14.

12. Dim, C. A., Da Rocha, F. E. L. (2011). "APIN: Uma ferramenta para aprendizagem de lógicas e estímulo do raciocínio e da habilidade de resolução de problemas em um contexto computacional no Ensino Médio". In: XIX Workshop sobre Educação em Computação, Natal. Anais do XXXI CSBC.

A Transversal Sustainability Analysis of the Teaching-Learning Process in Computer Students and Groups

45

Fernanda Carla de Oliveira Prado and Luciel Henrique de Oliveira

Abstract

Since the first UN conference on the environment, in 1972, many agreements seeking to establish goals to balance economic and social growth with environment preservation have been made. Regarding Information Technology (IT), the Green IT concept comes up. Such concept can contribute to a more sustainable environment and ensure economic benefits. In this context, we conducted a transversal field research using the survey method on a sample of 150 students of a technical IT course from five campuses of the Sao Paulo Federal Institute of Education, Science and Technology in order to identify their competences (knowledge, abilities and actions) regarding sustainability in its broad aspect and applied to IT. As a result, there is an opportunity to work on the sustainability concept and use it to turn the students into a collective transformational agent. It was also identified the need to further develop their abilities related to Green IT and its importance to the IT field.

Keywords

Environmental education • Green IT • Professional education

45.1 Introduction

The United Nations Decade of Education for Sustainable Development: 2005–2014 (DESD), coordinated by UNESCO [1], states that sustainable development is comprised by three areas – society, environment and economy – that indicate a long term and continuous changing process supported by culture (values, diversity, knowledge, language, world perspective, way of being, relating, believing and acting). The DESD sustainable development view is "a world where everyone has the opportunity to benefit from quality education and can learn the values, behaviors and life styles required to sustainable development and positive social transformation" [1]. For this reason, the Education for Sustainable Development (ESD) main theme is the respect and concern about a high quality education.

Sachs [2] corroborates this concept by defining that the sustainability tripod integrates social relevance, ecological prudence and economic viability. However, adding another four dimensions enhances this definition: cultural sustainability, balanced territorial distribution, political sustainability (by pairing development and biodiversity preservation) and international political sustainability (by keeping peace and preserving humanity's legacy).

In Brazil, the National Environment Policy (NEP) states that every level and modality of the teaching process must address the sustainability theme and clarifies that environment covers the "interdependency between the natural, socioeconomic and cultural environments". Operational guidelines for professional education in Brazil establish that all technical courses should preferably include three dimensions: "theoretical and practical skills of the

F.C. de Oliveira Prado (✉)
Federal Institute of Education, Science and Technology of Sao Paulo, Sao Joao da Boa Vista, SP, Brazil
e-mail: fernanda.prado@ifsp.edu.br

L.H. de Oliveira
University Center of Associate Colleges, Sao Joao da Boa Vista, SP, Brazil
e-mail: luciel@fae.br

S. Latifi (ed.), *Information Technology – New Generations*, Advances in Intelligent Systems and Computing 558, DOI 10.1007/978-3-319-54978-1_45

profession; general knowledge related to the profession; common attitudes and competences related to a professional area and working in general" [3]. In an organizational context, these dimensions are defined as 'knowledge', 'abilities' and 'actions'. Knowledge means 'know-how' and includes general, theoretical, operational and environmental expertise; abilities are the 'how to do' and relate to one's professional experience; finally, actions mean the 'how to be/act' and refer to personal and professional attributes [4].

During 2008, the Federal Institutes of Education, Science and Technology (FIs) were first created in Brazil. Such institutes are specialized in offering professional and technological education; they seek to strengthen the productive, social and cultural arrangements and to encourage environment preservation researches [5]. By doing so, the FIs cover the economic, social, environmental and cultural dimensions stated by Sachs [2] and DESD [1].

'Information and Communication' is one of the guidelines of the courses offered by the FIs. In turn, Information Technology (IT) can contribute to a more sustainable environment and ensure economic benefits, characterizing a concept known as Green IT [6–8]. Finally, it is of common assumption that talking about Green IT in the classroom contributes to the appropriate use of IT and to the training of ecologically conscious citizens [9–11]. Considering those facts described above, made tried to identify - among senior students of the IT course from five campuses of the São Paulo Federal Institute of Education, Science and Technology (FISP) - their knowledge, abilities and actions regarding sustainability in its broad aspect and applied to IT through Green IT practices and conscious consumption. Given this context, this paper contributes to the sustainability transversality in the teaching-learning process by listing themes that can be discussed in IT courses and, consequently, be part of the students' education on citizenship concepts and work qualifications.

45.2 Information Technology and Sustainability

The increasing consciousness about global warming, electronic garbage disposal and institutions' images are turning the attention to environment protection. Regarding IT, the Green IT concept was conceived and its role is to provide a more sustainable environment and offer economic benefits at the same time [7].

Harmon, Demirkan, Auseklis and Reinoso [8] argue that the main IT sustainability goal is to enable companies to use computational resources in a more efficient way and to improve their global performance. Viotto [12] adds that IT

sustainability is not restricted to being concern about the environment and the reputation of organizations, since it can offer economic benefits, as pointed by Murugesan and Laplante [7].

Murugesan [13] suggests a few ways of how IT can contribute to sustainable development, such as: reducing electric power consumption of datacenters, computers and other computational systems; reconditioning and reusing old equipment, recycling obsolete ones; projecting computers, servers, cooling systems and datacenters focused on energy efficiency and environmental security; and manufacturing electronic components, computers and other subsystems with minimal environmental impact.

Besides that, some simple actions make a difference, like: preferring documents in digital format; reusing printed pages as draft; refilling used printer cartridges; configuring desktops to turn off their screens when idle or setting them to sleep when user is away; turning off any equipment not being used; acquiring products with energy efficiency compliance and/or manufactured by companies concerned about the environment [14–16].

Academic papers about the environmental issue are scarce and localized, but they are more common in the corporate world and specialized media. In 2009, Symantec interviewed IT executives from 1052 companies and found out that Green IT became a priority since it had been receiving bigger funding and more priority. The results also indicated that IT is now the center of such initiatives and professionals seek for a working environment that leans on sustainable practices [17].

In this sense, Tres [18] states that due to the market's lack of professionals with practical knowledge on sustainability, Green IT certification programs have emerged. Some of them are Green IT Citizen and Green IT Foundation [19], CompTIA Green IT [20] and Foundation Certificate in Green IT [21].

From the academic point of view, the importance of working on this subject at all levels of education is evident. However, the lack of formal definitions regarding Green IT, the small number of institutions that offer content related to the theme and the scarcity of related programmatic content and guidelines in schools go beyond Brazilian borders [9, 10]. Evangelou and Pagge [9] propose themes and strategies grouped by education level in order to guide the production of academic material. Robila [10] proposed an introductory module that addresses sustainability in every IT course and reinforces that the development of abilities in Green IT is a competitive edge for these professionals.

Therefore, adding the concept of Green IT to IT courses may help to raise better citizens and professionals that meet the market's demand on this subject.

45.3 Method

To identify the knowledge, abilities and actions regarding sustainability in its broad aspect and applied to IT through Green IT practices and conscious consumption of our sample, was conducted a transversal field research in November 2013, with exploratory and descriptive objectives using the survey method.

The subjects of this research were 150 senior students of the Computer Technician course (offered concurrently with high school) from FISP's campuses. The main differences between the Computer Technician concurrent with high school course and the Computer Technician alone are that the former lasts for 3 or 4 years, has subjects related to high school curriculum and the technical course and an internship period is optional. On the other hand, the latter is modular (lasts for three or four months), and the subjects are all related strictly to the technical scope, having a mandatory internship period.

Only senior students were interviewed due to their experience and their imminent job market insertion. The random probabilistic sample approach was chosen, stratified by the proportion of students by campus and course, thus ensuring the participation of all campuses. This method was applied to the whole population, adopting a confidence level of 90%, considering an error chance of 5% and having an amount of 98 students divided by campuses.

In the absence of a certified instrument to assess the competences on this matter, we created a survey – available at https://pt.surveymonkey.com/r/D2ZC3LC – with opinion measurement questions, grouped by matter and scope as shown in Table 45.1.

Except for the statements regarding the abilities which were focused on assessing the students' understanding of IT sustainability [14–16], the rest of them covered areas such as citizenship and conscious consumption [22]. On the latter, guidelines from the Akatu Institute [23] presented in Table 45.2 were followed.

Finally, at the end of each statement block, students were asked for the source of their knowledge, abilities and actions in order to better analyze the results.

After data gathering through the online survey, a quantitative analysis of this data was performed following three steps:

1. Convert the data to a Microsoft Excel format;
2. Perform a statistical analysis to synthesize and correlate the data using the IBM SPSS software; and
3. Analyze and interpret the correlations and create charts using Microsoft Excel.

The variables used in the statistical analysis came from the survey statements and were used to calculate Person's correlation coefficient. The analysis was focused on the positive and negative correlations with a 95% or 99% certainty.

45.4 Results

The statistical analysis did not show any meaningful correlations between each campus. It is explainable by the fact that all campuses follow the same didactic-pedagogic guidelines, even though they offer different courses. Besides that, the school was pointed as the main source of knowledge, abilities and actions, followed by life experience, family and work (Fig. 45.1).

An interesting data was that 50% of the students said that the school was their only source of knowledge, abilities and actions. When analyzing each course individually, results pointed that the proportion of the Computer Technician course was 76%, whereas the proportion of Computer Technician concurrent with high school course was 74%. This confirms that the course itself does not interfere in this matter.

Regarding the knowledge matter (Fig. 45.2), 62.5% of the students answered all statements correctly, and each statement by itself had a correct answer rate greater than or equal to 50%. Regardless of course, 83% of the students were aware of the environmental impact caused by the world economic and production activities; they agreed that knowing the products' lifespan can help to reduce the amount of electronic garbage. An opportunity to work on the conscious consumption concept was identified: to only buy what one really need; to know and understand one's rights as a consumer; to collect information about companies and their products; and how to correctly discard them are, according to Akatu Institute [22]. These actions that can turn a consumer into a collective transforming agent – given that the act of consuming has an impact on society and on the environment.

The most significant correlations indicated that:

1. 37% of the students said that conscious consumption is related to income;
2. Those who said they were aware about the impact of economic and production activities (89%) agreed that it is important to know more about the products' lifespan (93%), but they do not always donate devices that are no longer being used (36%); and
3. The 74% who acknowledged that individual activities have an impact on everyone else also agreed that it is the government's sole responsibility to eradicate poverty and to protect the environment (84%), and also stated that they would appeal to a consumer protection agency (51%).

Table 45.1 Statements by matter

Matter	Statements
Abilities	I set my computer to turn off the screen when idle
	I set my mobile device to power saving mode
	I set my computer to sleep whenever I am away
	I divide my computer components into ones that can be recycled or reused from those that must be discarded
Knowledge	The world economic and productive activities are modifying the Earth's climate and that can cause severe damages that must be prevented or avoided
	The origin of the products we use is very important since its production may have caused damages to the environment and to society
	Though we live on the same planet, it is an overstatement to say that what one person does affect all others
	Conscious consumption is only possible for people above a certain paygrade because the poor cannot afford to "consciously choose things"
	Plastic, glass, metals and boards are the main components of a computer that can be recycled
	When equipment turns into electronic waste, its best destiny is a chop shop in order to remove its toxic components and recycle its materials
	Only the government has the power to ensure society's balance and address issues like poverty eradication and environment protection
	Replacing meetings by videoconferences contributes to reduce carbon emissions
Actions	When printing, I set the printer to use both sides of the paper sheet
	Instead of throwing away my electronic equipment, I donate them
	I usually do not buy products from certain companies as a way of punishing them for harming society, the environment or local communities
	I usually replace my electronics by newer models
	I shut down equipment not in use
	I prefer to buy equipment with energy efficiency certifications
	I take batteries to a collection station
	I avoid printing
	I already purchased something weighing its environmental pros and cons
	I take into account the proximity and ease of access between my home, work and school/university and try not to go around town
	Whenever possible, I tell people that for them to be healthy it is important to balance eating, physical activities, family time, recreation time and work, among other things
	I use printed pages as draft
	I reuse printer cartridges
	If an advertisement bothers me by being inappropriate or awkward, I speak out and encourage others to do the same
	If I have problems with a product or service and am not able to find a solution with its respective company, I appeal to a consumer protection agency
	I treasure cultural diversity and individual characteristics of each person and stimulate everyone to find and follow their own opinions and feelings

Table 45.2 Guideline on conscious consumption from the Akatu Institute

Theme	Orientation
Consumer's power	Every purchase is an act of power because it has an effect on society and on the environment. By using this power for good, the self-aware consumer becomes a collective transforming agent
Why buy It	Buying only what is necessary is a great lesson on conscious consumption
What to buy	To consciously choose between one product and the other is to compare their effects on social relationships, the economy and the environment, from its production to its disposal
From whom to buy	By choosing products or services from socially and environmentally responsible companies, the consumer contributes to these companies' success
How to use it	Using a product until the end of its lifespan avoids waste, helps to sustainably use nature's and society's resources and avoids garbage hoarding
How to buy	By choosing payment conditions, having the possibility of doing exchanges and getting informed about companies, products and services, the consumer changes himself and society
How to discard	Recycling all materials that can no longer be used is essential to a sustainable use of nature's and society's resources and to avoid garbage hoarding

Fig. 45.1 Sources of competence regarding sustainability

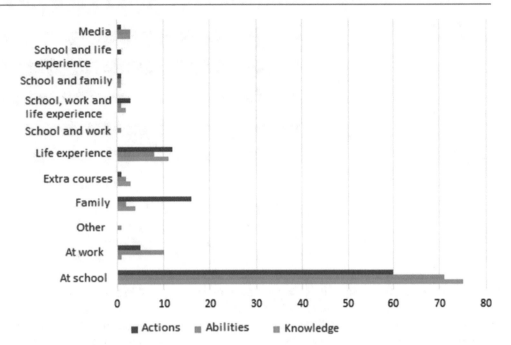

Fig. 45.2 Results for the knowledge matter

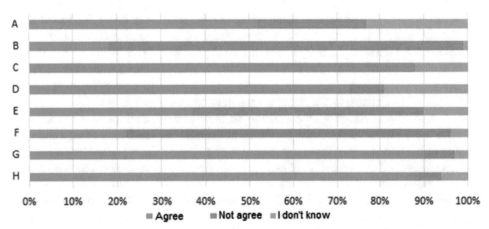

A. Replacing meetings by videoconferences contributes to carbon emissions reduction.
B. Only the government has the power to ensure society's balance and address issues like poverty extinction and environment protection.
C. When equipment turns into electronic waste, its best destiny is a chop shop in order to remove its toxic components and recycle its materials.
D. Plastic, glass, metals and boards are the main components of a computer that can be recycled.
E. Conscious consumption is only possible for people above a certain paygrade because the poor can't afford to "consciously choose things".
F. Though we live on the same planet, it is an overstatement to say that what one person does affect all others.
G. The origin of the products we use is very important since its production may have caused damages to the environment and to society.
H. The world economic and productive activities are modifying the Earth's climate and that can cause severe damages that must be prevented or avoided.

Regarding the abilities (Fig. 45.3), it was noted that students who said they know how to do the procedures mentioned in the statements also had a positive response regarding Green IT. However, when dealing with a more complex activity, most of them (70%) did not know how to do it, which indicates they were not familiar with the subject or that it was poorly developed during hardware classes. On the correlations:

1. Those who knew how to set the computer to turn off the screen when idle (73%) also knew how to set sleep mode when they are away (75%);

Fig. 45.3 Results for the abilities matter

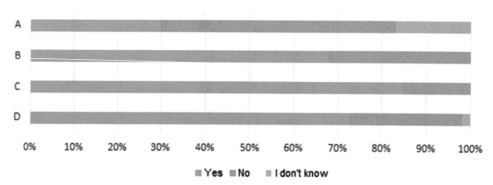

A. I divide my computer components into ones that can be recycled or reused from those that must be discarded.
B. I set my computer to sleep whenever I am away.
C. I set my mobile device to power saving mode.
D. I set my computer to turn off the screen when idle.

2. Those who knew how to turn on the power saving mode in mobile devices (85%) also knew how to set the sleep mode (72%) and usually preferred buying products with an energy efficiency certification (89%); and

3. Those who knew how to distinguish recyclable components from disposable components (30%) had already thought about the effects of a purchase on the environment (53%) and usually do not buy some products as a way to punish companies (73%).

On the actions related to the conscious consumption and citizenship themes (Fig. 45.4), the absolute majority of the students were aware that material goods are not related to quality of life and that they value and respect cultural diversity and individual characteristics of each person. Nevertheless, students still need to be aware that an act of consumption must be bounded by real needs and that their rights as consumers must be respected.

On the Green IT practices (Fig. 45.5), given the ease of access to information and the development of abilities in school, there is a great probability that these practices will start to be used more often and change behaviors as long as they increase awareness.

45.5   Conclusions

Even though there is a political will on Brazil's professional education area to present and work on the sustainability theme, it is still not discussed in the classroom. Nonetheless, the field research showed that the school is considered the main source of knowledge, abilities and actions. These findings confirms that it is important to work on sustainability in its broad aspect and applied to IT in the school environment, as stated by Evangelou and Pagge [9], Robila [10] and Ramalho, Costa, Lopes and Young [11].

When reviewing the state of the art, we noted that most researches about IT and Green IT comes from private organizations. On Green IT practices, the papers prioritize the definition and implementation of techniques to reduce power consumption and decrease carbon emissions. Regarding the method used in this paper, we had no difficulties in getting access to the sample and conduct the survey. However, we had to devote some time creating the survey because there is no other validated instrument available to measure the competences on this paper's theme. We consider the survey presented here as one of the main contributions of this research. It was able to fulfill its goal, and even though the presented results are restricted to our sample, we believe that our method can be applied in other scenarios where such competences must be evaluated.

On the knowledge matter, the students obtained a hit rate of 62.5%, which is considered average in a school environment. On the more complex abilities matter, 70% of the students did not know how to execute them. This might indicate that such matter is not discussed in the classroom. In light of this, we believe that the sustainability theme is discussed in an indirect and maybe unaware manner, covering only citizenship exercise, respect (to others, to the environment and to diversity) and waste reduction.

On the IT contributions to a sustainable development, this paper showed that even though technological evolution is constantly demanding infrastructure and power and contributing to environmental pollution and electronic waste, it is possible to decrease the environmental impact and provide economic benefits through Green IT. We showed the importance of discussing this theme in the IT area in an economic, environmental and social sense, both to organizations and to professionals.

Finally, we hope that this research – available at https://pt.surveymonkey.com/r/D2ZC3LC – can serve as a guideline for other education institutions when including

Fig. 45.4 Results for the action: conscious consumption and citizenship matter

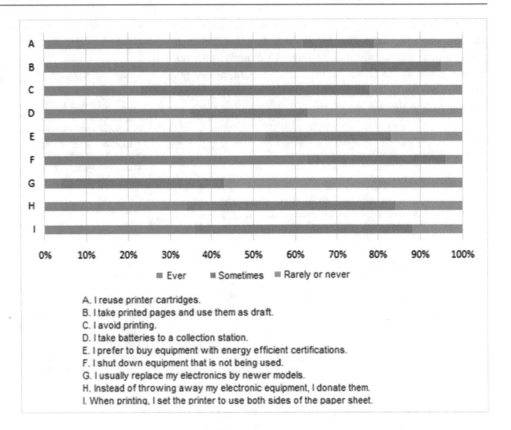

A. I reuse printer cartridges.
B. I take printed pages and use them as draft.
C. I avoid printing.
D. I take batteries to a collection station.
E. I prefer to buy equipment with energy efficient certifications.
F. I shut down equipment that is not being used.
G. I usually replace my electronics by newer models.
H. Instead of throwing away my electronic equipment, I donate them.
I. When printing, I set the printer to use both sides of the paper sheet.

Fig. 45.5 Results for the action: Green IT practices matter

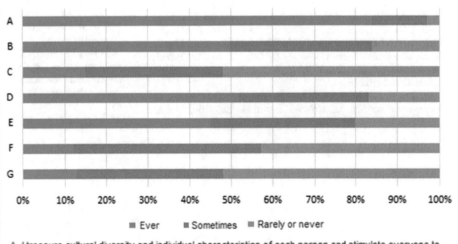

A. I treasure cultural diversity and individual characteristics of each person and stimulate everyone to find and follow their own opinions and feelings.
B. If I have problems with a product or service and am not able to find a solution with its respective company, I appeal to a consumer protection agency.
C. If an advertisement bothers me by being inappropriate or awkward, I speak out and encourage others to do the same.
D. Whenever possible, I tell people that for them to be healthy it is important to balance eating, physical activities, family time, recreation and work, among other things.
E. I take into account the proximity and ease of access between my home, work and school/university and try not to go around town.
F. I already purchased something weighing its environmental pros and cons.
G. I usually don't buy products from certain companies as a way of punishing them for harming society, the environment or local communities.

environmental education in a transversal way for their educational projects. Regarding IT courses, we hope that their curriculums will be updated in order to comply with the market demand for Green IT professionals and, above all, to train students to be empathetic with this matter and aware of their role in the world.

References

1. UNESCO. (2005). Década da Educação das Nações Unidas para um Desenvolvimento Sustentável, 2005–2014: documento final do esquema internacional de implementação, Brasília. http://unesdoc. UNESCO.org/images/0013/001399/139937por.pdf. Accessed 12 ago 2013.
2. Sachs, I. (2002). *Caminhos para o desenvolvimento sustentável.* Rio de Janeiro: Garamond.
3. Ministério da Educação. (1997). Parecer CNE/CEB n. 17, de 3 de dezembro de 1997, Brasília. http://portal.mec.gov.br/setec/ arquivos/pdf_legislacao/tecnico/legisla_tecnico_parecer1797.pdf. Accessed 02 Jul 2013.
4. Roberto, R., Antonello, C. S., & Boff, L. H. (2005). *Os novos horizontes de gestão: aprendizagem organizacional e competências.* Porto Alegre: Bookman Editora.
5. Brasil. (2008). Lei n. 11.892, de 29 de dezembro de 2008. Brasília. http://www.planalto.gov.br/ccivil_03/_ato2007-2010/2008/lei/ l11892.htm. Accessed 08 out 2012.
6. Ozturk, A., Umit, K., Medeni, I. T., Ucuncu, B., Caylan, M., Akba, F., & Medeni, T. D. (2011). Green ICT (information and communication technologies): A review of academic and practitioner perspectives. *International Journal of eBusiness and eGovernment Studies, 3*(1), 1–16.
7. Murugesan, S., & Laplante, P. A. (2011). IT for a greener planet [guest editors' introduction]. *IT professional, 13*(1), 16–18.
8. Harmon R., Demirkan H., Auseklis N., & Reinoso, M. (2010). From green computing to sustainable IT: Developing a sustainable service orientation. System Sciences (HICSS), 2010 43rd Hawaii International Conference on. IEEE. 1–10 January 2010. doi 10. 1109/HICSS.2010.214.
9. Evangelou E. K., & Pagge J. (2015). The urge for GREEN IT courses at universities and technical institutes. Interactive Mobile

Communication Technologies and Learning (IMCL), 2015 International Conference on. IEEE. 19–20 November 2015. doi: 10.1109/ IMCTL.2015.7359627.
10. Robila S. A. (2012). A sustainability component for a first-year course for information technology students. Advanced Learning Technologies (ICALT), 2012 I.E. 12th International Conference on. IEEE. 4–6 July 2012. doi: 10.1109/ICALT.2012.56.
11. Ramalho, A. B., Costa, R. E. G., Lopes, A. F. N., & Young, R. (2010). TI Verde: a tecnologia da informação no campo da sustentabilidade. *Revista da FA7, 1*(8), 107–120.
12. Viotto J. (2009). Informação cada vez mais verde. In: Deloitte Mundo Corporativo. http://www.4sc.com.br/blog/downloads/ mc24.pdf. Accessed 09 Jan 2014.
13. Murugesan, S. (2010). Making IT green. *IT Professional, 2*(12), 4–5. doi:10.1109/MITP.2010.60.
14. Itautec. (2011). Guia do Gestor de TI Sustentável. http://www. itautec.com.br/media/652021/af_guia_gestor_sustentabilidade.pdf. Accessed 27 ago 2013.
15. Porto Digital. (2011). Guia de Boas Práticas para uma TI mais Sustentável. http://www2.fiescnet.com.br/web/recursos/ VUVSR01qa3lOZz09. Accessed 27 ago 2013.
16. Maurer, E. B., & Lanes, L. B. F. (2012). Práticas sustentáveis em TI. *Unoesc & Ciência-ACET, 3*(2), 187–194.
17. Symantec. (2009). Green IT Report – Global: May 2009. http:// www.symantec.com/content/en/us/about/media/GreenIT09_ Report.pdf. Accessed 13 Jan 2013.
18. Tres C. H. (2013). Carreira: Certificação Profissional em TI Verde. http://www.profissionaisti.com.br/2013/10/carreira-certificacao-profissional-em-ti-verde/. Accessed 13 Jan 2014.
19. EXIN. (2014). EXIN Green IT. https://www.exin.com/NL/en/ exams/&fw=green-it. Accessed 13 Jan 2014.
20. CompTia. (2014). CompTIA Certifications. http://certification. comptia.org/getCertified/certifications.aspx. Accessed 13 Jan 2014.
21. BCS. (2014). Green and Sustainable IT. http://certifications.bcs. org/category/15610. Accessed 13 Jan 2014.
22. Instituto Akatu e Ethos. (2010). Pesquisa 2010 – O consumidor brasileiro e a sustentabilidade. http://www.akatu.org.br/Content/ Akatu/Arquivos/file/10_12_13_RSEpesquisa2010_pdf.pdf. Accessed 15 ago 2013.
23. Instituto Akatu. (2004). Teste do consumo consciente: Você é um consumidor consciente? http://centro.akatu.org.br/cr/index.jsp. Accessed 15 ago 2013.

Cloud Computing: A Paradigm Shift for Central University of Technology

Dina Moloja and Tlale Moretlo

Abstract

Education is key in today's generation. It helps the mind to think critically and shape it to produce innovations every day. Academics and Universities of technology are currently exploring new technologies to advance the methods used in teaching and learning. One of the technologies that has recently emerged is Cloud Computing. Cloud Computing is a disseminated computing that allow software and hardware as a service via the internet instead of you having a hardware or software sitting on your desktop or somewhere inside your company's network. Cloud Computing is a type of computing that depends on distributing computing resources instead of having a local servers or personal devices to handle applications. Cloud Computing, as a new paradigm, can offer institutions quality education by providing the latest infrastructure in terms of hardware and software. This paper focuses on the introduction of Cloud Computing at Central University of Technology to improve teaching and learning methods.

Keywords

Cloud computing • Education • Information • Services

46.1 Introduction

With the advantages brought by processing and storage technologies through the internet instead of computer's hard drive, computing resources have become more reasonable, dominant and accessible [5, 8]. The latest technological trend called cloud computing provide resources like CPU and storage that can be used by users through the internet anytime, anywhere and on-demand. In simple terms we can say cloud computing provides shared resources, software as a service, information and other constructive resources through the internet [6, 10].

Sultan (2010) [10] discovered that the development of cloud computing has made a remarkable impact on Information Technology (IT) industry and education. Furthermore, education is depending on Information technology to bring new ways of solving old problems and improving teaching and learning in institutions, specifically Central University of Technology (CUT) [1]. Cloud Computing can be welcomed at Central University of Technology and in education as a tool to bring new innovations whilst saving costs [3, 7]. Currently, CUT use what we call "a black board" to share resources with the students, the problem with the black board is that the server can be down anytime and it has limited computational resource which restrict the staff from sharing whatever they want to share, the way they want, whenever they want. Another problem that CUT is faced with is using a notice boards, whereby they paste information/notices as a way of communicating with the students. The problem with a notice board is that it cannot be accessed outside the campus, not to mention it is expensive and time consuming.

D. Moloja (✉) • T. Moretlo
Department of Information Technology, Central University of Technology, Welkom, South Africa
e-mail: mmoloja@cut.ac.za; mtlale@cut.ac.za

The rate of IT technologies is constantly changing which place more extra financial burden on the institutions [13]. Moreover, to continuously upgrade the hardware and software can be a challenge and consume lot of time, leading to high cost in maintenance, resulting in poor service deliver to the staff and students [1, 4].

Cloud Computing provides the solution to these problems. With cloud computing adopted at CUT, the students and staff can use the platform and applications whether they are on-campus or off-campus depending on their needs. Cloud Computing provides services at the least costs to users like the students and staff members anytime, anywhere as they need the services [7, 14]. Cloud Computing, if implemented at CUT, can be a solution to many problems the institution is faced with.

46.2 What Is Cloud Computing?

Cloud Computing is a disseminated computing that allow software and hardware as a service via the internet instead of you having a hardware or software sitting on your desktop or somewhere inside your company's network [1, 9]. According to Sultan (2010) [10], Cloud Computing is a type of computing that depends on distributing computing resources instead of having a local servers or personal devices to handle applications. There are three commonly used layers in cloud computing, see Fig. 46.1.

- *Infrastructure as a Service (IaaS):* offers infrastructure as per the requirement of students, teachers, researchers with hardware configuration that includes memory, operating system, storage, network capacity globally or locally for a specific work.
- *Platform as a Service (PaaS):* provides applications for deploying platform to allow the researchers/developers to deploy their individual applications without paying for the underlying hardware and software layers.
- *Software as a Service (SaaS):* It offers applications over the network. There is no need of installation and buying the licensed version software. It makes the users free from the maintenance and support.

For the sake of security or personal use or expansion of computing usage there are mainly four types of Cloud Computing deployment models which are as follows [1, 6, 9, 12, 14]

- *Public Cloud:* Resources are common to everyone.
- *Private Cloud:* Owned or leased by an enterprise or company.
- *Hybrid Cloud:* Combination of both Private and Public.
- *Community Cloud:* Shared but for a particular community.

Fig. 46.1 Cloud computing layers and examples [10]

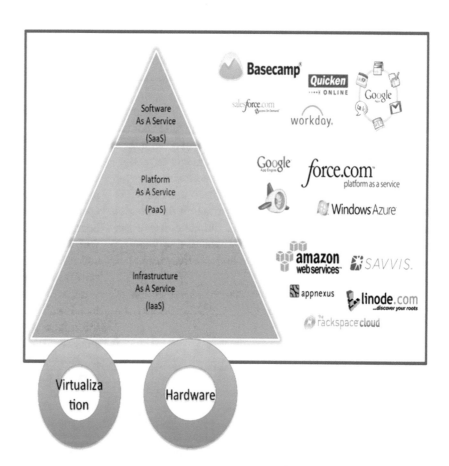

46.3   Literature Review

The potential of cloud computing for improving efficiency, cost and convenience for higher education is being acknowledged by number of US educational and official institutions [2, 14]. The University of California (UC) at Berkeley, for example, found cloud computing to be smart to use in one of their courses that was focused totally on developing and deploying SaaS applications and they were helped by a donation from Amazon Web Services (AWS) [8]. UC was able to transfer its course from locally owned infrastructure to the cloud. One of the main reasons they transferred the course to the cloud was mentioned as having the capacity to attain a huge amount of servers needed for the course in a short period of time [11]. For some universities around the world, the convenience of computing power through cloud computing for research reasons was welcomed.

According to Zissis and Lekkas (2012) [14] one of the key challenges for many laboratories setting up proteomics programs has been the need to attain and sustain a computational infrastructure essential for analysing a huge flow of proteomics data produced by mass spectrometry instruments utilized in determining the elemental composition as well as chemical structure of a molecule. With cloud computing making the analysis less expensive and more accessible, it meant that many more users can set up and customize their own systems and investigators can analyse their data in greater depth than was previously attainable, thus making it possible for them to learn more about the systems they are studying [4].

Major Cloud Computing providers such as IBM and Google are actively promoting cloud computing as tools for research. In 2007 Google and IBM announced a cloud computing university initiative designed to improve computer science students' knowledge of highly parallel computing practices in order to address the emerging paradigm of large-scale distributed computing. And there is also an increasing number of educational establishments that are adopting cloud computing for economic reasons. One of those is the Washington State University's School of Electrical Engineering and Computer Science (EECS). They are faced with budget cuts relating to current economic conditions, as well as continual pressure to do more with less, the EECS selected a cloud platform (namely vSphere 4) from VMware (a leading provider of virtualization technology) as it searched for the best platform to support a move to cloud computing [7, 13].

A number of African educational institutions have adopted cloud computing, fundamentally due to their insufficient IT infrastructures and their incapability to cope with the limitless cycle of hardware and software upgrades. Google has been very successful in targeting the East African educational market. For example, the giant cloud provider has partnered with a number of East African educational establishments (e.g., the National University of Rwanda, the Kigali Institute for Education, the Kigali Institute for Science and Technology, the University of Nairobi, the United States International University, the Kenyan Methodist University and the University of Mauritius) in order to provide Google cloud services (e.g., Gmail, Google Calendar, Google Talk and Google Docs and Spreadsheets) to their students. These universities were also helped by current World Bank grant that supports bandwidth subsidy in universities [3, 5, 13]. Not far away, Microsoft is also helping Ethiopia rolling out 250,000 laptops to its school teachers, all running on Microsoft's Azure cloud platform. The laptops will enable teachers to download curriculum, keep track of academic records and securely transfer student data throughout the education system, without the extra cost of having to build a support system of hardware and software to connect them [11].

The ability of cloud computing to help African education, not only by reducing IT costs but also by making education more efficient than before, is likely to be a very powerful and empowering tool for the advancement of education in this under-developed continent.

Why should CUT be left aside? The implementation of cloud computing at CUT will definitely improve teaching and learning.

46.4   Current Situation at Cut

Education system is always based on marks, figures, notifications, submissions and material sharing. And CUT uses a paper based system to do all these things. Furthermore, CUT sometimes uses what we call a black board "ethuto" which is a learning management system that is managed locally. Ethuto is currently used for submissions, notifications and material sharing but the problem with ethuto is that it is continually upgraded resulting in poor service delivery. Not only that but ethuto has limited space when comes to file sharing and video uploading unlike cloud computing that provide a great deal of storage to use anytime.

Moreover, as part of communication with the students, CUT uses a notice board to paste information and communicate with the students especially when ethuto is not available due to maintenance. Good communication is an essential tool in achieving productivity and maintaining strong relationships at all levels of any institution.

Staff who invest time and energy into delivering clear lines of communication will rapidly build up levels of trust amongst students, leading to increases in productivity, output and morale in general.

Poor communication will inevitably lead to unmotivated students that may begin to question their own confidence in their abilities and inevitably in the institution.

The key challenge with a notice board used by CUT is its inaccessibility. Meaning that it cannot be accessed outside the campus.

There is indeed a need to find a solution to these problems that CUT is faced with and the solution is Cloud Computing services. CUT can overcome these problems by implementing Infrastructure as a service from any cloud service provider on the bases of pay as you go. Also, CUT like any institution is dependent heavily on content management system and it can hire a service to store the content on the cloud and any student or staff or any academia can use that content from anywhere, at any time, on any device.

46.5 Benefits of Cloud Computing for Cut Staff and Students

Cloud Computing has the potential for improving the efficiency, cost and convenience for the universities and higher education at large by providing the following benefits [4, 6, 12]: Ease of use, Storage, User satisfaction, data portability, high availability, reliability, time saver, Low power consumption, resource pooling and scalability.

46.6 Implementation of Cloud Computing in Cut

The possible users of Cloud Computing at CUT are students, support staff and academicians. Each user has their credentials to access a particular cloud service. Adopting SaaS for CUT allows the academics to maintain the attendance of the students without having a lot of attendance registers in the office that can easily be misplaced or lost. SaaS services can allow the academics to conduct online quiz that automatically mark the students and save their grades, saving more time of marking.

Adopting PaaS, CUT can organize practical sessions as and when needed from the Cloud services e.g., developing projects like mobile apps, web apps, etc. Adopting IaaS, staff can upload their study materials or any related content on the cloud and student can access these materials and content anytime, anywhere, whenever they need. In this paper, we suggest the implementation of Infrastructure as a service in public cloud for CUT. The figure below depicts what Cloud Computing can bring to CUT as a higher learning institute if adopted Fig. 46.2.

Fig. 46.2 Various services of cloud for education in higher learning [1]

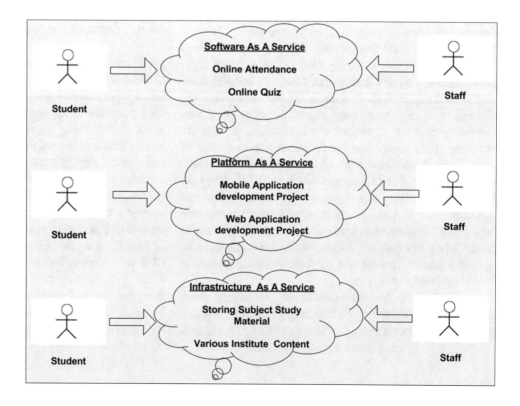

46.7 Conclusion

Cloud computing is a developing computing paradigm that has the potential to deliver opportunities and also have a wide scope in higher education and will benefit CUT if adopted. Also, students are looking more interested in electronic devices as a great source of communiqué. Cloud Computing will be of a good benefit in staff and student's life's if it can be implemented at CUT. Instead of keeping textbooks and heavy bags with them, students will prefer a world without textbooks having much more efficient way to store and access the lectures, assignments, research papers and virtual classes etc.

This paper demonstrates how educational sector has already taken advantage of the benefits which this technology [Cloud Computing] is bringing, not only in terms of cost but also efficiency brought by this paradigm. Hybrid cloud and Infrastructure as a service was suggested in this paper because they best suits CUT needs and can act as a weapon that can be used to solve CUT problems.

The main aim of this paper was to highlight and suggest the implementation of cloud computing, using Infrastructure as a Service in public cloud at CUT.

46.8 Recommendations

A thorough empirical research and investigation is needed to measure the effectiveness of cloud computing if it was to be implemented at CUT.

References

1. Desai, T., & Patel, R. (2016). Cloud computing in educational sector. *International Journal of Innovative Research in Science & Technology, 2*, 191–194.
2. Ercan, T. (2010). Effective use of cloud computing in educational institutions. *Procedia-Social and Behavioral Sciences, 2*, 938–942. Elsevier.
3. Gonzalez-Martinez, B. L. G. S. (2015). Cloud computing education: A state-of-art-survey. *Elsevier, 80*, 132–151.
4. Kuyoro, S. O., Ibikunle F. & Awodele O. (2011). Cloud computing security issues and challenges. *International Journal of Computer Networks, 3*(5), 247–555.
5. Marston, S., Li, Z., Ghalsasi, A., & Zhang, J. (2011). *Cloud computing- the business perspective* (pp. 176–189). Elsevier. Science direct in US.
6. Modi, C., Patel, D., Borisaniya, B., Patel, A., & Rajarajan, M. (2013). A survey on security issues and solution at different layers of cloud computing. *The Journal of Supercomputing, 63*, 561–592.
7. Al Noor, S., Mustafa, G., Chowdhury, S. A., & Hossain, Z. (2010). A proposed architecture of cloud computing for education system in Bangladesh and the impact on current education system. *International Journal of Computer Science and Network Security, 10*, 7–13.
8. Razak, S.F. (2009). Cloud computing in Malaysia universities. *Innovative technologies in intelligent systems and industrial applications* (pp. 101–106). IEEE Xplore in US.
9. Sclater, N. (2010). Cloud computing in education. Policy Brief, Unesco Institute for Information Technology in Education.
10. Sultan, Nabil. (2015). The implications of cloud computing for global enterprise management. In *Global enterprise management* (pp. 39–56). Springer, Palgrave Macmillan US.
11. Sultan, N. (2010). Cloud computing for education: A new dawn. *International Journal of Information Management, 30*, 109–116.
12. Thomas, P. (2011). Cloud computing: A potential paradigm for practising the scholarship of teaching and learning. *The Electronic Library, 29*, 214–224.
13. Zhang, Q., Cheng, L., & Boutaba R. (2010). *Cloud computing: State-of-the-art and research challenges* (pp. 7–18). Springer.
14. Zissis, D., & Lekkas, D. (2012). Addressing cloud computing security issues. *Elsevier, 28*, 583–592.

Peer Music Education for Social Sounds in a CLIL Classroom

Della Ventura Michele

Abstract

Motivation is an important factor in a CLIL (Content and Language Integrated Learning) classroom and it is the key to success in the learning process. ICT (Information and Communication Technology) represents an important tool to improve the students' motivation and the it offers opportunities to develop both academic knowledge and language skills. The aim of this paper is to analyze the impact of the ICT in a CLIL Course and the connected effects relating to the use of the foreign language in a musical field. We present the findings of a case study scenario showed that OPEN SoundS, which is a musical environment designed and developed as a virtual studio where students and teachers (from all over the world) can create collaborative musical projects, is a very usable tool and is highly appreciated by teachers and students. The study highlights, *on the one hand*, a substantial improvement of the learning process of the students thanks to the impact that Open SoundS had on their motivation. *On the other hand*, the importance for teachers to use new tools.

Keywords

CLI • ICT • Motivation • Open SoundS

47.1 Introduction

CLIL can now be considered a real teaching method. CLIL is an acronym for content and language integrated learning. It consists of teaching a curricular subject through the medium of a language other than that which is normally used. In CLIL courses, learners gain knowledge of the curriculum subject while simultaneously learning and using the foreign language. It is important to highlight that "content" is the first word in CLIL because curricular content leads language learning.

The capacity of ICT to facilitate and enhance learning is well documented. According to Vlachos, ICT has a multimodal and vital role to play in CLIL, since it caters for the media and the resources that can enhance multidisciplinary learning, and provides the means that stimulate, guide, and facilitate students in their efforts to express themselves adequately and effectively in the target language [1]. In wider

This paragraph of the first footnote will contain the date on which you submitted your paper for review. It will also contain support information, including sponsor and financial support acknowledgment. For example, "This work was supported in part by the U.S. Department of Commerce under Grant BS123456".
The next few paragraphs should contain the authors' current affiliations, including current address and e-mail. For example, F. A. Author is with the National Institute of Standards and Technology, Boulder, CO 80305 USA (e-mail: author@ boulder.nist.gov).
Michele Della Ventura, is with Music Academy "Studio Musica", Via Andrea Gritti 25, 31,100 Treviso (Italy). (e-mail: dellaventura. michele@tin.it).

D.V. Michele (✉)
Music Academy "Studio Musica", 31100 Treviso, Italy
e-mail: dellaventura.michele@tin.it

© Springer International Publishing AG 2018
S. Latifi (ed.), *Information Technology – New Generations*, Advances in Intelligent Systems and Computing 558, DOI 10.1007/978-3-319-54978-1_47

terms, ICT can transform teaching and learning processes from being highly teacher-dominated to student-centered [2], and that this transformation will result in increased learning gains for students, creating and allowing for opportunities for learners to develop their creativity, problem-solving abilities, informational reasoning skills, communication skills, and other higher-order thinking skills [3].

On the base of the above considerations, it is easy to see that ICT can represent an important tool in a CLIL classroom. Teachers can use ICT for different activity, such as listening, writing, reading and speackig; at the same time, they can use ICT also for practical activity.

In all these cases, the choice of the ICT depends on:

– the group class: different knowledge levels among the students and presence of dyslexic students;
– the teacher: knowledge of the ICT and ability in the use of the ICT.

This article presents a case study referred to a Theory, Ananlysis and Composition (TAC) teaching project developed with a CLIL approach, in Senior High Schools specializing in Music. TAC is a complex discipline because of the countless music grammar rules and the countless specific words that it deals with.

The main objective of this project was to check on and assess the impact of the use of ICT in the students' learning process: increase the students' motivation in order to see if it corresponds to an improvement of his/her academic results. For this reason it has been decided to introduce the use of the on-line platform Open SoundS in the classroom.

The paper is structured as follows. Sect. 47.2 describes the motivation to learning. Sect. 47.3 describes the computer environment Open Sounds. Sect. 47.4 shows an experimental test that illustrate the effectiveness of the proposed method and finally, conclusions are drawn in Sect. 47.5.

47.2 Motivation to Learning

Motivating students is one the most difficult parts of being a high school educator.

Motivation is defined by psychologists as an internal process that activates, guides, and maintains behavior over time. In other words motivation gets you going, keeps you going, and determines where you're trying to go [4].

According to Brophy, motivation to learn is [5] "*a student tendency to find academic activities meaningful and worthwhile and to try to derive the intended academic benefits from them*".

The constructs can be viewed as belonging to three categories [6]:

1. constructs that refer to students' traits and states, such as interest and curiosity: students in general education programs who are interested or curious about topics are oriented toward inquiry and discovery, both of which are instructionally desirable;
2. constructs that refer to students' beliefs, such as self-determination (the ability to make choices and have some degree of control in what we do and how we do it [7]), goal orientation (an objective or outcome that students pursue, is a goal why students pursue it is referred to as their goal orientation, and the result is goal-directed behavior [8]), self-regulation (students who are self-regulating know what they want to accomplish when they learn: they bring appropriate strategies to bear and continually monitor their progress toward their goals [9]), and self efficacy ("*beliefs in one's capabilities to organize and execute the courses of action required to produce given attainments*" [10]);
3. constructs that refer to students' expectations: expectations of students, and the strategies based on these expectations, play an important role in increasing or reducing students' motivation in general education programs [11].

Motivation is an important factor in the learning of foreign languages but it is particularly important in the learning of a curricular subject through a foreign languages (CLIL). Motivation influences learners to choose a task, get energized about it, and persist until they accomplish it successfully, regardless of whether it brings an immediate reward [12]. In this regard, ICT represent a motivational tool to learn. A relationship seems to exist between the opportunities that ICT present and motivation for students: when technology use aligns with authentic or "real-world" applications, motivation can be enhanced [13].

Here are some effects that it is possible to obtain using ICT on the learning process, according to the conceptpt of motivation:

- creation of a pleasant and supportive atmosphere;
- make learning stimulating and enjoyable;
- increase of the student motivation by promoting cooperation among the learners;
- increase of the students' expectation of success;
- improvement of poor handwriting and languages skills;
- development of communication skills;
- encouragement of self-pacing with increased capacities to deal with individual learning styles as students can work at the pace and intensity suitable to their needs;
- change teacher practices, planning tools and assessment grids;
- allow students to learn independently.

47.3 Why Open SoundS

This section briefly reviews the individual features and design aspects of Open SoundS. This will provide a good understanding as to why Open SoundS can be considered an educational tool to assist in learning.

OPEN SoundS is an on-line platform that offers a new dimension in training on the Net: the possibility to produce and share music remotely within communities.

In particular, the goal is to test the extension of a model of informal learning, and its integration in a creative key, into educational paths/processes meeting the expectations of both the society of knowledge and information, and the individual educational and vocational needs of students.

The OPEN SoundS Community is a meeting point for students and the various actors involved in the creation of music in a collaborative, remote and transnational dimension. The community involves:

- students from schools, music academies, universities, research centres and relevant networks (school, producers and/or artists networks);
- teachers from the various music education areas;
- musicians and other operators from the communication industry (cinema, TV, advertising);
- community of music ICT professionals and software developers;
- companies involved in the multimedia supply chain and, in particular, in electronic music (specialized manufactures of music computers and software).

OPEN SoundS is a virtual learning environment dedicated to the creation of music. The networks of interest in this area are related to:

- the development of creative musical products;
- skills around the operation of technological equipment (software and hardware) for the production and creation of music;
- the acquisition of more general skills related to collaborative learning environments.

Through OPEN SoundS students and teachers can:

- access a virtual learning environment dedicated to collaborative music production which functions remotely and transnationally;
- work with students from all over the world to create and share music remotely within the education system;
- access training and information resources for the conscious and strategic use of digital music technology and the Net.

The group starts to create a content type "project" (*Create Project*), that means, a collaborative music production in a dedicated environment that allows to:

- describe the musical project in every aspect, cultural and technical;

- upload any type of music files necessary for its implementation;
- view the tracking of the contribution to the final production provided by each student team;
- post and view comments about the creative productions.

OPEN SoundS, through a highly innovative and creative practice in fact wants to be a means to stimulate and support for the development of key competencies for initial and continuing training.

With Open SoundS students can collaborate in groups. They should be allowed to talk freely to one another [14]. Encourage learners to talk in English as much as possible because this will give them practice in using the vocabulary of the discipline (TAC), and also topic-specific vocabulary [1]. Discussion also helps with the understanding of what they are doing and why [15, 16] (i.e. musical harmonization, musical modulation, . . .). More able learners develop their communication skills by clarifying their ideas as they explain them to others [1, 16–18] (i.e. realization of specific instrument staff). Less able students are usually supported by other group members, and feel more confident to contribute ideas.

These are the reasons why the teacher must ask themselves (before, during and after the development of the activity) which task motivate students, which task involve interaction, which task develop thinking skills and which task need language support.

To summarize, in Table 47.1 there are some indicators that the teacher could consider to evaluate a practical activity on Open SoundS.

47.4 Application and Analysis /Research Method

The research presented in this article refers to a pilot project that analyzes the effects on learning and on teaching brought by the implementation of the on-line platform Open SoundS in the classroom lesson. The discipline forming the object of the project is Theory, Analysis and Composition (TAC). TAC is a complex discipline because of the countless music grammar rules and the countless specific words that it deals with. At the same time it is a (main) discipline that characterize the entire course of musical studies (curricula discipline).

The research was conducted for a time period of 6 months (from November 2015 to April 2016) and it engaged the fourth grade of the Music High School, with a total of 24 students (11 girls and 13 boys).

In the first period of work (from November 2015 to January 2016) students participated in the lessons in the classroom listening to the explanations of the teacher, taking notes and

Table 47.1 Indicators to evaluate practical activity

Assessment criteria	Good	Adequate	Lacking
Defining the musical problem and providing a solution	The student formulates the problem and provides a solution	The student formulates the problem but does not provide a solution	The student does not formulate the problem and does not provide a solution
Making observations	The student provides procedures requiring the use of specific musical grammar rules	The student Provides basic procedures requiring the use of general musical grammar rules	The student does not provide any procedures
Making corrections	The student identifies mistakes and provides help	The student identifies some mistakes and has difficulties in providing help	The student does not identify mistakes
Providing solutions	The student produces solutions with examples	The student produces a solution with an example	The student does not produces a solution
Evaluating the procedure	The student produces a complex solution that involves several musical grammar rules	The student produces a solution that involves some general musical grammar rules	The student does not produce a solution
Explaning the solution	The student makes correct explanation using an appropriate musical language	The student makes correct explanation using a range of familiar musical grammar words	The student does not make explanation

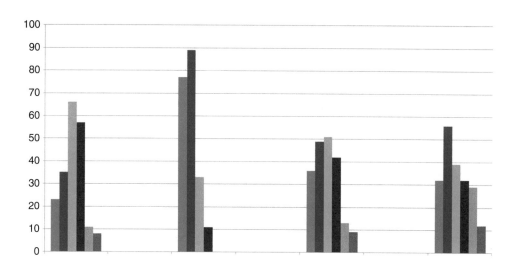

Fig. 47.1 Results of the examinations

studying the lecture notes were given to them by the teachers (of the curricula discipline and of the foreign language).

In the second period (from February 2016 to April 2016) students participated in the lessons using the platform Open SoundS.

At the end of each period, an examination was passed in the classroom: students had to realize a musical piece for a video (supplied by the teachers and different for each period) explaining the choices and the procedures.

Fig. 47.1 shows the results of the two examinations, on the base of the following parameters (Table 47.2).

Besides the numeric results that can be read in the diagrams, one of the most important things that emerged refers to the increase, in the second period, of the number of students that improved the specific language and the range of musical grammar world. At the same time there was an improvement also in the use of the musical grammar rules.

The learning improvement appeared for students who already drew a high profit: the process was positive for them as well, inasmuch as they learned to select the information they found based on the group members (particularly with reference to the less able students).

47.5 Discussion and Conclusions

This paper has presented an analysis regarding how ICT can have a positive impact on student motivation and on learning outcomes, in a CLIL Clasroom. The introduction of Open SoundS was truly satisfying: there was a positive and significant impact both on the learning and on the teaching which was subsequently mirrored by the results reached at a didactic level.

Motivation is one of the characteristics which influence how students approach their learning and how it is important

Table 47.2 Results of theexaminations

	1st period	2nd period	1st period	2nd period	1st period	2nd period
	Good		Adequate		Lacking	
Use of specific musical grammar rules	23	35	66	57	11	8
Use of general musical grammar rules	77	89	33	11	0	0
Pertinence of musical language	36	49	51	42	13	9
Use of a range of familiar musical grammar words	32	56	39	32	29	12

in academic achievement. The results show that the use of ICT in teaching and learning had a significant positive impact on student motivation, such as increasing students' attitude to learn, improving classroom behavior, and providing better performance of learning outcome.

Further investigation is needed to find out if it is possible to obtain better results in different disciplines in the field of music education.

References

1. Vlachos, K. (2009). The potential of information communication technologies (ICT) in Content and Language Integrated Learning (CLIL): The case of english as a second/Foreign language. In D. Marsh, P. Mehisto, D. Wolff, R. Aliaga, T. Asikainen, M. J. Frigols-Martin, S. Hughes & G. Langé. (Ed.), *CLIL practice: perspectives from the field (University of Jyväskylä, Finland)* (pp. 189–198).
2. Unesco. (2003). Manual for pilot testing the use of indicators to assess impact of ICT use in education [online]. http://www.unescobkk.org/education/ict/resource.
3. Youssef, B., Mounir, D. (2008). The impact of ICT on student performance in higher education: Direct effects, indirect effects and organizational change. In *The economics of e-learning [online monograph]*. Revista de Universidad y Sociedad del Conocimiento (RUSC). Vol. 5, no. 1. UOC.
4. Slavin, R. E. (2000). *Educational psychology: Theory and practice*. Needham Heights: Allyn and Bacon.
5. Brophy, J. E. (1988). *On motivating students* (pp. 201–245). New York: Random House.
6. Glynn Shawn, M., Aultman, L. P., & Owens, A. (2005). Motivation to learn in general education programs. *The Journal of General Education, 54*(2), 150–170.
7. Deci, E. L. (1996). Making room for self-regulation: some thoughts on the link between emo tion and behavior. *Psychological Inquiry, 7*, 220–223.
8. Pintrich, P. R., & Schunk, D. H. (2000). *Motivation in education: Theory, research, and appli cations*. Columbus: Merrill. Pribbenow, D. A., & Golde, C. M.
9. Tuckman, B. W. (2003). The effect of learning and motivation strategies training on college students' achievement. *Journal of College Student Development, 44*, 430–437.
10. Bandura, A. (1997). *Self-efficacy: The exercise of control*. New York: Freeman.
11. Rosenthal, R., & Jacobson, L. (1968). *Pygmalion in the classroom*. New York: Holt, Rinehart & Winston.
12. Della Ventura, M. (2015). iPAD in music education: A case study of collaborative learning with Dyslexic learners. In *Proceedings of the international conference on interdisciplinary social science studies*. Oxford: IOS Press.
13. Brian, C., & Housand, A. M. (2012). The role of technology in gifted Students' motivation. *Psychology in the Schools, 49*(7), 706–715.
14. Mehisto, P., et al. (2008). *Uncovering CLIL* (p. 29). Oxford: Macmillan Educ..
15. Hayward, D. (2003). *Teaching and assessing practical skills in science*. Cambridge: Cambridge University Press.
16. Barbero, T., & Maggi, F. (2011). Assessment and evaluation in CLIL. In D. Marsh & O. Meyer (Eds.), *Quality interfaces*. Eichstätt: EAP, Eichstaett Academic Press UG.
17. Barbero T. (2012). Innovative assessment for an innovative approach. Perspectives. *A Journal of TESOL Italy,* Special Issue on CLIL, Vol. XXXVII, n. 2, Fall 2010.
18. Serragiotto, G. (2007). Assessment and evaluation in CLIL. In D. Marsh & D. Wolff (Eds.), *Diverse contexts*. Frankfurt am Main: Converging Goals, Peter Lang.

Teaching Distributed Systems Using Hadoop

48

Ronaldo C.M. Correia, Gabriel Spadon, Danilo M. Eler,
Celso Olivete Jr., and Rogério E. Garcia

Abstract

Databases and Distributed Systems have a fundamental relevance in Computer Science; they are usually presented in courses where the high-level of abstraction characterizes the teaching and learning processes. Consequently, the teaching method needs to evolve to fulfill the present requirements. Therefore, grounded in these concepts, the main goal of this paper is to introduce a teaching methodology via benchmark tests. Our methodology was conducted using the Hadoop framework, and it is innovative and proved effective. Our methods allow students to be exposed to complex data, system architecture, network infrastructure, trending technologies and algorithms. During the courses, students analyzed the performance of some computational architectures through benchmark tests on local and on the cloud. Along with this scenario, they evaluate the processing time of each architecture. As a result, our methodology proved to be a support learning method, which allows students to have contact with trending tools.

Keywords

Learning methodology • Distributed systems • Hadoop • Student-centered learning

48.1 Introduction

The Big Data concept emerged from data scalability problems. Such term describes methods and paradigms of data generation, collection, and analysis in a methodological framework. Which is responsible for dealing with high scalability, reliability and fault tolerance on data [1]. Big Data is still challenged by network infrastructure and data collection. Its frameworks are supported by tools as Hadoop *MapReduce* and Hadoop *Distributed File System* (HDFS), that appears either in commercial solutions and in researches [2]. The MapReduce provides support to the analytical processing of the data while the HDFS performs a distributed file-block storage [3].

An examination of the factors that affect the distribution of the data is warranted. In this sense, Khalid, Afzal, and Aftab [4] and Souza et al. [5] formalized some of these arguments, but none of them focused on computational models as a student-centered teaching methodology. As a consequence, this paper introduces a teaching and learning methodology that focuses on assisting students in the understanding of Databases, Distributed Systems, and other related topics. Our approach has the aim of clarifying theoretical concepts. It was conducted using the Hadoop framework as a learning tool; and, though it, students are exposed to complex data, system architectures, network infrastructure, trending technologies and algorithms.

Our results are based on explaining the structure of our course as well as the description of our methods. In order to validate such methods, we show a case study, where one of our students performed a series of tests with the purpose of

R.C.M. Correia • G. Spadon (✉) • D.M. Eler • C. Olivete Jr.
R.E. Garcia
Sao Paulo State University, Sao Paulo, Brazil
e-mail: ronaldo@fct.unesp.br; spadon@fct.unesp.br; daniloeler@fct.unesp.br; olivete@fct.unesp.br; rogerio@fct.unesp.br

© Springer International Publishing AG 2018
S. Latifi (ed.), *Information Technology – New Generations*, Advances in Intelligent
Systems and Computing 558, DOI 10.1007/978-3-319-54978-1_48

building a cluster. We address this methodology to understand how much the incomprehension of basic concepts can negatively influence students that are exposed to trending technologies and its abstractions.

Accordingly, this paper is organized as follows: Sect. 48.2 presents some related works; Sect. 48.3 presents the course architecture, a brief description of relevant server layouts, and an overview of Hadoop; Sect. 48.4 presents our methodology and a case study that shows how it behaves in a real scenario; at last, conclusions come in Sect. 48.5.

48.2 Related Works

From the usage of Hadoop as a learning tool, we can derive two approaches to engaging it in the learning process. The first one describes the search for meaning to the data, in other words, data mining processes. The other one focus on the architecture that describes the hardware where Big Data tools are employed; through it, students are encouraged to search for better configurations in the framework, as well as to improve its architecture while seeking for better performance. Given these approaches, our assumption is that these two can turn Hadoop into a learning tool.

The literature introduces former works, which focused on teaching-learning methodologies for different types computer science courses. This is the case of Souza et al. [6] which introduces a novel methodology for courses of Formal Languages and Automata Theory, while Souza et al. [7] introduced a continuous methodology that covers all the Theoretical Computing courses. Also, there is Correia et al. [8] which focused on new approaches for teaching data structures and algorithms for undergraduate students. However, when comes to methods for introducing trending technologies like Hadoop, there is a lack of a clear and precise methodology.

Consequently, the aimed result of our method is to teach how to introduce the process of evaluating the performance of a system, by guiding students in choosing the most suitable model for each particular cluster-planning case. In this sense, the first challenge is to transform Hadoop into a learning tool, which will enable the students' formation as better and up-to-date professionals.

48.3 Course Architecture

The relevance of Big Data led us to suit our course in a teaching model that can introduce trending topics to students. As it is known, there are a variety of computational models, from centralized to distributed. Many of them can be used to deploy the Hadoop tool. Consequently, it is important to review the literature and understand how each one contrasts the others. Thus, we divided our course into topics of: *Scale-up architecture*; *Scale-out architecture*; *Client–Server model*; *Client–Server with Master-Slave model*; and *Hadoop server-roles*. In the following, we present a summarized view of our course.

48.3.1 Course Topics

Coulouris et al. [9] defined a distributed system as the one where computers are independent, but they work for a common result; such computers are attached to a network where each one communicates with each other, coordinating actions by exchanging messages. Contrariwise, centralized computing is not divided at all, and every resource is located on a single server, which can be a server or a client to others. The centralized model does not face any network problems like latency and delayed packages because they do not need to be connected to a network to complete a task. Given these concepts, models, and architectures, Big Data can be generalized on *scale-up* and *scale-out*, which are stated in the following according to Henrique and Kaster [10].

48.3.1.1 Scale-Up Architecture
This architecture describes the server that centralizes hardware, which increases the processing performance, memory capability, and storage management. Specifically, scale-up servers make use of high throughput, avoiding latency and network delays. Such architecture implies in a limit where a single server cannot centralize and manage all resources. The closer to its limit higher is the monetary cost, and at this point, the distribution is the answer to improve the performance [4].

48.3.1.2 Scale-Out Architecture
The scale-out architecture focuses on model expansion, performing a distributed processing and loading balance. In this architecture, each computer manages their memory, clock, and storage. However, they work for a single purpose, and they can share their resources to solve common problems. That is because of the distribution, which makes the processing capacity higher. Most of the Big Data frameworks, such as Hadoop, are deployed through this architecture. The meaning is the fault tolerance factor is increased by the high availability of the data. Before introducing Hadoop and its roles, we introduce to students two models derived from the scale-out one. The first one is the Client-Server and the other one Client-Server with Master-Slave. These concepts enable the understanding of how works the interaction of the computers within a given architecture.

Client–Server Model

A server is a software on looping that is running on a computer. Such software is waiting to share their resources. A client, on the other hand, is an intermittent process that needs to be intentionally started on a host computer. This architecture is depicted by Fig. 48.1.

Client–Server with Master-Slave Model

As well as the previous model, the idea of servers and clients are the same. The difference between them is how the processes communicate with each other and how they are distributed. This model is unidirectional, and it usually has more slaves than masters. The master is the only process that receives requests from clients; slaves share information neither to external clients nor to each other. The slave works as a server process, but it just replies the master's requests, and the master works as a client and as a server at the same time. As a client, the master makes job requests to the slaves, and as a server, the master receives external requests and coordinate all slaves. This architecture is depicted by Fig. 48.2.

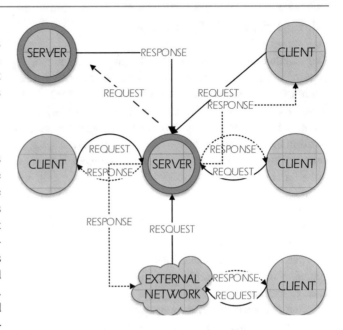

Fig. 48.1 Representation of the Client-Server model

48.3.2 Hadoop Framework

Hadoop is a framework that focuses on processing a large amount of data. It was designed to solve problems of data scalability and complex analysis, which depends on a stable model for significant processing tasks that cannot be entirely performed on database systems [10]. When Hadoop is deployed on a grid or cluster, its information has more than one copy stored on different servers, increasing the data availability. In this scenario, when a server is down, the first server that has the requested information reply to the master's request. Its architecture was planned to be independent and to solve their integrity and scalability problems. The idea behind it focuses on the problem solution and not on the environment planning.

Each computer on a cluster/grid hosts a standalone framework that is managed by the master, and each one controls a distinct group of assigned roles. That is, the master hosts the NameNode and the JobTracker/MapReduce, as well as the slaves host the DataNode/HDFS and the TaskTracker/MapReduce. Hadoop can work with more then one DataNode and TaskTracker on different slaves, as well as SecondaryNameNode, but just one NameNode and JobTracker on the master server. These roles are briefly described in the following:

(A) *JobTracker* – divide and distribute problems;
(B) *TaskTracker* – solves small problems;
(C) *DataNode* – stores the file blocks;
(D) *Secondary NameNode* – perform assistance;

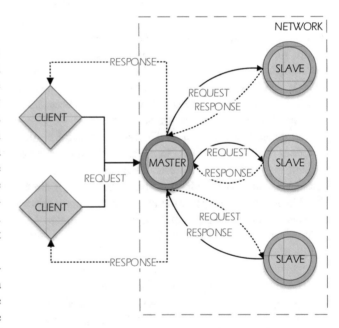

Fig. 48.2 Representation of the Client-Server with Master-Slave model

(E) *NameNode* – manage connections and requests, stores the file block list that contains the paths for each stored file.

Following, we introduce our methodology, presenting the most important points that should be considered by teachers in order to obtain good results.

48.4   Proposed Methodology

Our methodology consists of allowing the students to acquire the practical knowledge from the theoretical one. Such practical knowledge is defined as an approach to identify a better deployment model to Hadoop via benchmark. The benchmark has the aim to enhance the system performance, and on our methodology, we used it as a way to demonstrate to students the difference between computational architectures. The motivation of our work is the trending topics, which are not entirely explored by the students. Consequently, with this approach, we can present a single topic with new teaching strategies, contributing to the robustness of knowledge.

There are several conditions to be considered before choosing the cluster architecture. To this end, there is the need to evaluate the performance of each possible scenario before making a choice. As a consequence, we need to cause the same load and stress in each architecture and perform metrical analysis though benchmarks.

As a first step, we instruct our students to dismantle all the available computers and to build groups of computers with similar hardware. In the face of more than one similar group, they evaluated each one singularly. Subsequently, for each group, we instructed them to perform a stress test, seeking for information about overheat and overhead of the processor. To this end, we used *TeraSort* method. Such method allows us to take advantage of all cores of the processors in the task of sorting a terabyte file of numbers.

The students were encouraged to remove from the cluster, machines with less than two terabytes of storage space or without gigabyte Ethernet driver. This step is particular to each case, but it is an important one; without it, a cluster could have an inferior performance when the computers try to balance its hardware to match an inferior computer.

When the stress test is complete, it is important to retest different distributions. The students used Open Source, Enterprise, and Professional ones. At this point, it is important to have a group discussion with the students to analyze each group of machines. Once they have selected the ones that will be in the cluster, they worked to guarantee the lowest CPU time, which are the ones that may achieve a better platform for data analysis. In this sense, they tested different computing architectures among scale-up, scale-out and geographically distributed server.

The others distributions were tested the same as the first, using a benchmark. Once we already tested processing time, core capability and the hard drives, the students checked the processing time considering the network impact. For that, they made network attacks to the cluster computers using Distributed Denial of Service – DDoS, making some of them fail in the middle of the test. With the computers failing, the

Hadoop should reach the next computer, with the requested information, and that will influence in the processing time by causing the abortion of the job.

At the end of the test, there should be sufficient information to make a choice about the machines to be used, the distribution of Hadoop and its computing architecture. To validate the obtained results, the collected information should be compared to other clusters of similar hardware.

To illustrate the usage of our methodology and its steps, we demonstrate at next it appliance and the decisions made by a student that with our guidance was able to build a cluster.

48.4.1   Case Study

To perform this case study, it was used 30 tower computers. The hardware that was available were not uniform, and this was the first problem faced. To overcome it, the computers were disassembled in order to separate the common hardware and to allocate it to new machines. As a result, it was achieved different processing architectures, half AMD, and the other half Intel. By using a stress test, more specifically the *TeraSort*, it was decided to use the Intel architecture. That was because the Intel hardware does not overheat as easily as the AMD one.

Subsequently, it was proceeded with investigations about the distribution of Cloudera,[1] Hortonworks[2] and MapR[3] of the Hadoop. In this sense, it was required to differentiate each one from each other by testing them. Such test was made by using Pi Estimator. This method requires two input, the number of mappings and another one for the samples. For each hundred looping, it was increased the exponential factor until it reaches five hundred and twelve mappings and one billion of samples.

The test was estimated to be completed in 36 hours for each distribution. For each one, the output was not considered. This is because the tests focused just on the processing time. Also, time is crucial to fit performance to the cluster.

As so, the first test was on the Hortonworks; the test took 34 hours and did not present any error. The second test was on the MapR. The test took 53 hours and returned one error message, omitted almost 24 results, kept on looping on two experiments and aborted the Pi Estimator at the last step of the test. The last test was on Cloudera where the test took 36 hours and did not present any error. The benchmark results were divided by collection cycle, and they are depicted in Fig. 48.3a, b, c.

---

[1] www.cloudera.com

[2] www.hortonworks.com

[3] www.mapr.com

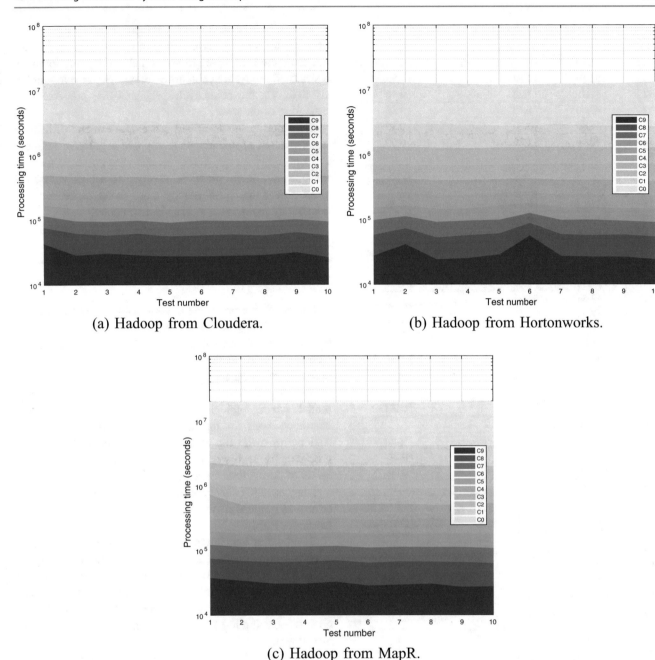

Fig. 48.3 Benchmark of the Hadoop distributions. (**a**) Hadoop from Cloudera. (**b**) Hadoop from Hortonworks. (**c**) Hadoop from MapR

For a complete analysis, it was performed the same test in the *Apache Hadoop*, which results are detailed in the Table 48.1. Analyzing the results, it was concluded that among the different platforms of data management, the usage of *Apache Hadoop* was faster than others. That is because it took less than 30 hours to complete the tasks without a single error. In a superficial analysis, it is possible to list some possibilities of why this happened: the network, task starvation, deadlock, and others.

Lastly, to quantify the cloud server performance on benchmark tests, it was deployed a few servers on a VPS to compare them together with a geographically distributed one. It is important to note that most of the cloud servers are centralized on different datacenters. Consequently, computers might be faster and the results not straightly comparable.

However, for each cloud benchmark it was plotted the time spent on each Pi Estimator cycle on dispersion graphs, which are depicted on Fig. 48.4a, b, c. The cycles of each analysis showed that the distributed and the centralized model, obtained similar values, alternating between which is faster. Further, the geographically distributed model

Table 48.1 Apache benchmark results (values in seconds)

Apache Hadoop benchmark results			
Cycle	Mean (μ)	Standard deviation	Variance (σ^2)
T0	51.60	1.60	1.70
T1	51.70	1.20	0.90
T2	49.60	0.90	0.80
T3	48.60	1.30	0.30
T4	50.40	1.20	0.00
T5	56.50	1.20	0.00
T6	73.20	2.20	0.60
T7	85.40	0.50	0.20
T8	110.30	2.10	0.08
T9	1098.70	6.30	0.70

Distributed cloud server results			
Cycle	Mean (μ)	Standard deviation (σ)	Variance (σ^2)
T0	16.86	1.64	2.71
T1	17.57	1.24	1.55
T2	17.58	0.90	0.81
T3	17.94	1.31	1.72
T4	20.88	1.29	1.67
T5	26.82	1.27	1.63
T6	41.97	2.24	5.05
T7	293.73	5.48	30.00
T8	568.81	9.91	98.28
T9	6316.61	97.29	9465.80

Centralized cloud server results			
Cycle	Mean (μ)	Standard deviation (σ)	Variance (σ^2)
T0	14.49	0.47	0.22
T1	15.46	0.30	0.09
T2	16.46	0.48	0.23
T3	16.47	1.60	2.57
T4	26.50	0.35	0.12
T5	46.54	2.51	6.30
T6	79.68	1.77	3.15
T7	109.20	2.74	7.49
T8	203.97	3.10	9.58
T9	1766.34	16.59	275.25

Geographically distributed cloud server results			
Cycle	Mean (μ)	Standard deviation (σ)	Variance (σ^2)
T0	39.23	4.93	24.32
T1	291.50	78.44	6152.84
T2	125.43	47.62	2267.97
T3	986.29	434.80	189055.37
T4	601.16	243.69	59389.66
T5	2260.01	790.81	625395.16
T6	4390.56	1070.37	1145706.50

provided worst times in all cycles. The centralized results show an almost continuous processing time in eighty seconds. The distributed one shows an interesting time difference in all cycles. Taking into account this time variation, the highest processing time of it is almost half of the one from the centralized model. Finally, the geographically distributed model was able to be tested up to cycle number six (T6). On the other cycles, the exit rate prevented the execution of the test.

Considering the described process, at the end of tests, it was collected sufficient data to suppose about investment in local or cloud server. The described case points to the fact that the investment could be reverted to a highly scalable cloud server that could deploy more slaves as needed, not demanding investment in local infrastructure.

48.5 Conclusions

In this paper, we presented a methodology based on benchmark analysis to guide the learning on courses of Distributed Systems, Data Bases and Computers Network for undergraduate students, by acquiring knowledge of the most suitable architecture for Hadoop.

We theorize the process of choosing a computational model, which implements Hadoop as teaching approach, where we seek for the knowledge construction to allow students to make the concept concrete. According to our proposal, for each step of benchmark analysis, the student needs to define time metrics across a reliable test method, and they need to collect and analyze statistically all results to be able to suppose about the best deployment model for each particular case.

Besides the benchmark result, the student can interact directly with the system and the hardware, and this is always important to understand a real system and how it works. The method helps to connect the student with different distributions and architecture models of an application, directly contributing to a decision about local and cloud investment.

This approach can open possibilities to make questions about what is the best platform, the best system or software architecture, encouraging new proposals, like our previous works [5, 11]. We developed our methodology while performing two similar studies in different applications, applying it in all research in development at our research group. Our evaluation consisted in confronting the results of each common application in various scaling methods at local and in the cloud. We decided to perform it as a learning methodology because it has a strong influence on past researches made by us and by others in our group.

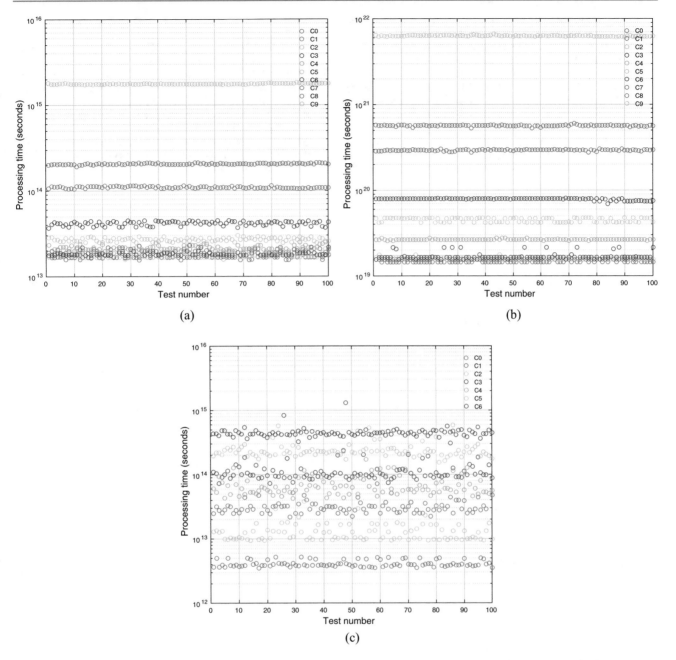

Fig. 48.4 Benchmark of the computational architectures. (**a**) Distributed Hadoop. (**b**) Centralized Hadoop. (**c**) Geographically distributed Hadoop

Acknowledgements We are grateful to the *Department of Mathematics and Computer Science* (DMC) at *Faculty of Science and Technology* (FCT) in *Sao Paulo State University* (UNESP) by the resources granted to this research.

References

1. Sagiroglu, S., & Sinanc, D. (2013). Big data: A review. In *2013 International Conference on Collaboration Technologies and Systems (CTS)* (pp. 42–47). IEEE.

2. Müller, M. U., Rosenbach, M., & Schulz, T. (2013). Living by the numbers. Big data knows what your future holds. In *SPIEGEL ONLINE*, vol. 17.

3. Shvachko, K., Kuang, H., Radia, S., & Chansler, R. (2010). The Hadoop distributed file system. In *2010 I.E. 26th Symposium on Mass Storage Systems and Technologies (MSST)* (pp. 1–10). IEEE.

4. Appuswamy, R., Gkantsidis, C., Narayanan, D., Hodson, O., & Rowstron, A. (2013). Scale-up vs scale-out for Hadoop: Time to rethink? In *Proceedings of the 4th Annual Symposium on Cloud Computing* (p. 20). ACM.

5. Souza, G. S., Messias Correia, R. C., Garcia, R. E., & Olivete, C. (2015). Simulation and analysis applied on virtualization to build

Hadoop clusters. In *2015 10th Iberian Conference on Information Systems and Technologies (CISTI)* (pp. 1–7). IEEE.

6. Souza, G. S., Olivete, C., Correia, R. C. M., & Garcia, R. E. (2016). Combined methodology for theoretical computing. In *Frontiers in Education Conference (FIE)* (pp. 1–7). IEEE.

7. Souza, G. S., Olivete, C., Correia, R. C. M., & Garcia, R. E. (2015). Teaching-learning methodology for formal languages and automata theory. In *Frontiers in Education Conference (FIE)* (pp. 1–7). IEEE.

8. Messias Correia, R. C., Garcia, R. E., Olivete, C., Costacurta Brandi, A., & Cardim, G. P. (2014). A methodological approach to use technological support on teaching and learning data

structures. In *2014 I.E. Frontiers in Education Conference (FIE)* (pp. 1–8). IEEE.

9. Coulouris, G. F., Dollimore, J., & Kindberg, T. (2005). *Distributed systems: Concepts and design*. Boston: Pearson Education.

10. Henrique, G. J., & Kaster, D. d. S. (2013). Consultas por similaridade em big data: Alternativas e soluções. Faculdade Estadual de Londrina, Technical Report.

11. Santana, V. J., Spadon de Souza, G., Messias Correia, R. C., Garcia, R. E., Medeiros Eler, D., & Olivete, C. (2015) Scalable information system using event oriented programming and NoSQL. In *2015 10th Iberian Conference on Information Systems and Technologies (CISTI)* (pp. 1–6). IEEE.

MUSE: A Music Conducting Recognition System

Chase D. Carthen, Richard Kelley, Cris Ruggieri, Sergiu M. Dascalu,
Justice Colby, and Frederick C. Harris Jr.

Abstract

In this paper, we introduce Music in a Universal Sound Environment(MUSE), a system for gesture recognition in the domain of musical conducting. Our system captures conductors' musical gestures to drive a MIDI-based music generation system allowing a human user to conduct a fully synthetic orchestra. Moreover, our system also aims to further improve a conductor's technique in a fun and interactive environment. We describe how our system facilitates learning through a intuitive graphical interface, and describe how we utilized techniques from machine learning and Conga, a finite state machine, to process inputs from a low cost Leap Motion sensor in which estimates the beats patterns that a conductor is suggesting through interpreting hand motions. To explore other beat detection algorithms, we also include a machine learning module that utilizes Hidden Markov Models (HMM) in order to detect the beat patterns of a conductor. An additional experiment was also conducted for future expansion of the machine learning module with Recurrent Neural Networks (rnn) and the results prove to be better than a set of HMMs. MUSE allows users to control the tempo of a virtual orchestra through basic conducting patterns used by conductors in real time. Finally, we discuss a number of ways in which our system can be used for educational and professional purposes.

Keywords

Pattern recognition • Machine learning • Music • Hidden Markov models • Education

49.1 Introduction

Although music is fundamentally an aural phenomenon, much of the communication required for an ensemble to produce music is visual in nature. In particular, musical conductors use hand gestures to communicate with musicians. Musical conductors only form of feedback for improving their conducting is mainly received from a band

and is very limited without the aid of a band. Beginning conductors lack an efficient way to practice conducting without a band present to help them hone their skills and lack the skill of experienced conductors that have worked many hours with bands. There are currently many gesture recognition software that have been built for the purpose of making music easier and enjoyable.

Recent advances in sensor technology are making such gesture recognition feasible and economical; cameras such as Microsoft's Kinect and the Leap Motion Controller are able to estimate the kinematics and dynamics of arms and hands in real time, meaning that conductors' gestures can now be accurately and cheaply recorded by the computer. There have been a few projects that have investigated and implemented different methods of capturing conducting.

C.D. Carthen (✉) • R. Kelley • C. Ruggieri • S.M. Dascalu
J. Colby • F.C. Harris Jr.
Department of Computer Science and Engineering,
University of Nevada, Reno, Nevada, USA 89557,
e-mail: chase@nevada.unr.edu; richard.kelley@gmail.com;
cris.ruggieri@gmail.com; dascalus@cse.unr.edu;
organicjustice@gmail.com; fred.harris@cse.unr.edu

© Springer International Publishing AG 2018
S. Latifi (ed.), *Information Technology – New Generations*, Advances in Intelligent
Systems and Computing 558, DOI 10.1007/978-3-319-54978-1_49

These methods have utilized the Leap Motion and Kinect in order to control music with gestures in different ways. Projects such as those from the developers WhiteVoid that utilizes a single Leap Motion, baton, and speakers while allowing users to control the tempo and individual instruments [1]. Another project [2], utilizes the Kinect in order to capture conducting gestures for manipulating tempo and the volume of individual instruments. There have been some old systems in the past that have been used outside of conducting but more for creating a coordinated performance such as [3] where many cellphones were synchronized for playing a symphony. These systems can provide a way for experienced or new users to learn conducting with feedback in a new way.

MUSE has been designed for beginning conductors to learn and enjoy conducting. MUSE allows users to learn simple conducting patterns and utilizes the Leap motion sensor's API [4], and algorithms such as: a finite state machine recognizer called Conga [5] and HMMs. These algorithms allow MUSE to effectively detect a user's conducting basic conducting patterns and control the tempo of a given song. An additional experiment was also conducted with rnn to determine if it is better than a HMM and to be later Incorporated into MUSE. The rnn experiment is included since not many papers seem experiment with using an rnn. MUSE provides a graphical user interface (GUI) for users to control and select different songs and even record previous performances of songs. Unlike other previous conducting projects MUSE has been designed as an architecture that is expandable.

In this paper we first describe the related work. Secondly, we then outline the architecture of our system and some of its functionality. We then give details on the algorithms used to capture conducting by the system and their accuracy. In the next section, We discuss our system's applications more throughly. Finally, we conclude with a discussion of further applications and future work. Throughout the rest of this paper we will refer to the Leap Motion as the Leap.

49.2   Background

The task of building software catered to beginner conductors requires an understanding of basic conducting patterns. As mentioned earlier there have been other software designed to recognize conducting with different degrees of complexity. Some software simply captures the change in velocity and alters the tempo, while other software may attempt to capture a gesture formation in conjunction with the change in velocity. MUSE is able to capture both the change in velocity and a gesture pattern. The gesture patterns of conducting is described in the Conducting Patterning and Ictus section.

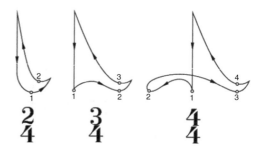

Fig. 49.1 Three basic conducting patterns that are taught are shown above. On *the left* is a two beat pattern, in *the middle* is a right three beat, and on *the right* is a four beat pattern. The beats or ictus's are located at the ends of *the arrows* [6]

49.2.1 Conducting Patterns and the Ictus

MUSE has been designed to recognize basic conducting patterns that are taught to beginning conductors. Figure 49.1 gives an example of these basic patterns. These conducting patterns consist of ictus's or where the beat occurs typically when the conductor's hand switches in a major direction. Figure 49.1 demonstrates three conducting patterns where the left most has two ictus's, the middle has three ictus's, and the rightmost has four ictus's. Basic conducting patterns are typically used to help beginning conductors to gain a grasp on keeping tempo with a band and to lead into more complex conducting styles.

49.2.2 Sensors for Gesture Recognition

In order to perform gesture recognition, the first obvious requirement is a system of sensors to record raw data from a person's motions. Previous systems for capturing gestures can be broken into two broad approaches: camera- based systems and controller based systems. A camera-based system utilizes camera's to capture a person's hand or baton motions, while a controller based system utilizes some kind of controller to capture a user's hand motion. Both types of recognition make use of gesture recognition algorithms to detect the beats of a conducting pattern.

An example of a controller used for gesture recognition, in particular, is the controller for Nintendo's Wii known as the Wiimote. The Wiimote has been successfully used to recognize a number of gestures with a fairly small training set [7]. The Wiimote has accelerometers that make it possible to keep track of orientation and makes use of a infrared sensor bar to get position information. The WiiMote is a great tool for creating a conducting application. Despite the WiiMote and other controller based systems being useful for capturing gestures for the use of conducting, they introduce a new interaction that may seem foreign to a conductor.

There are primarily three types of camera based systems: traditional intensity cameras, stereo camera systems, and depth cameras. In this paper our system uses depth based system camera like the Kinect. The Kinect has been used to build a commercially-successful pose recognition system for humans in indoor settings [8]. In this paper, we use a system similar to the Kinect to track hand motions specifically.

There has been some work on gesture recognition in conducting including work that was previously discussed in the Introduction section. In particular, the Conga framework [5] that uses finite state machines to recognize conducting, dynamic time warping (DTW) which has been used in [9, 10] to improve the overall accuracy of detecting conducting patterns, and HMMs [11]. The systems in previous apporaches with HMMs, DTW, and others are very accurate and have 95% or above accuracy. However, finite state machines such as Conga are constructed by hand and manual construction obviates the need to perform a lengthy training process, the use of a non-statistical approach can lack the robustness required to deal with noisy sensors and inexperienced conductors. Despite the accuracy of these apporaches, our system uses the Conga frame to recognize different conducting patterns and extends the Conga framework by using HMMs to perform the necessary sequence analysis to deal with noise. However, beats are still captured by the Conga framework exclusively.

49.3 System Architecture and Functionality

MUSE system architecture has been designed with a culmination of multiple dependencies which are: QT to be used as a Graphical User Interface (GUI) [12], rtmidi to handle communication to other midi input and output devices [13], libjdksmidi to parse midi files and handle midi events [14], the Leap Motion API to capture motion information from the Leap Motion device [4], Nick Gillian's Gesture Recognition Toolkit (GRT) to create a Hidden Markov HMM for recognizing conducting gestures, implementing the Conga framework in C++ to capture a majority of conducting gestures, and OpenGL to visualize the conductor's hand [15], and zlib for compressing records generated by user's performances [16]. The rnn is currently not part of MUSE and will be added later. These dependencies have made it possible to build MUSE.

49.3.1 Overview of the Architecture

The overall picture of MUSE's architecture can be seen in Fig. 49.2. MUSE can be broken done into four simple layers being its presentation layer, midi processor layer, hardware layer, and pattern recognition layer. The presentation layer consists of the GUI and everything necessary to create a

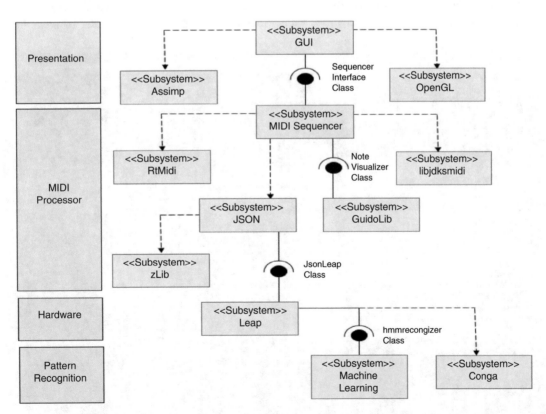

Fig. 49.2 The system architecture is demonstrated in this figure. The system consists of a presentation layer that shows information to the user, a midi processor for manipulating the musical output of the system, a pattern recognition layer where all code for detecting conducting patterns is placed. All interfaces within MUSE demonstrated in this figure as well

virtual view of the user's hand. The hardware layer of MUSE is simply the Leap Motion itself and the Leap Motion API required to use the Leap. The Pattern Recognition consists of two different beat pattern recognizers being the Conga finite state machine and the use of GRT to create a HMM. The Midi Processor layer handles processing midi files by controlling midi playback speed in the case of changing conductor input and has the capability of serializing and recording a user's hand motions for future playback. In the next following sections the presentation layer and MIDI processor layer are explained in detail. The pattern recognition layer is explained later in Sect. 49.4.

49.3.1.1 Presentation Layer

The GUI of MUSE is designed in a similar fashion to any music player in the sense that it allows the user to select and play songs at their leisure. As the song plays, the user may provide gestural input via a Leap Motion controller to guide musical playback. Also, they can alter the play-style or even the volume of the instruments which are provided as controls in the GUI. MUSE has components for choosing specific models of conducting and visualization as well as opening MIDI files. A screen shot of MUSE's GUI can be seen in Fig. 49.3 and the individual components of the screen are detailed as follows:

- OpenGL Visualizer: The OpenGL visualizer displays the motions of a user's hand and allows the user to see what

the Leap is detecting. This interface is provided so that the user may be able to determine what is the best area to conduct over the Leap.

- Channel Manipulation: The channel manipulation which is the sixteen slider bars next located next to the OpenGL Visualizer. In this section of the GUI the user is able to manipulate different 16 different midi channels of MIDI files that are playing. Each channel correlates to a different instrument that is playing on the MIDI file. The volume sliders will also generate specific volumes based on the configuration of each MIDI file that is opened. The sliders operate and provide information in real-time because the variables for volume are changing constantly.

- Tempo Manipulation: A tempo manipulation toggle allows users to change the tempo of each song (for the song to play faster or slower). When a MIDI file is loaded the tempo within the midi file is automatically loaded and can be changed with the plus and minus below the tempo next to the OpenGL visualizer. It will also change in real-time as a user moves their hand or conducting instrument over the Leap depending on the velocities of their movements.

- Beat Pattern Detection Menu: Located above the tempo selection and to the left of the OpenGL visualizer are the options to choose a 2-pattern, 3-pattern, and 4-pattern. Above these options are the displays for the desired beat pattern and the current detected beat for that pattern. We introduced this feature in order to make detecting beat

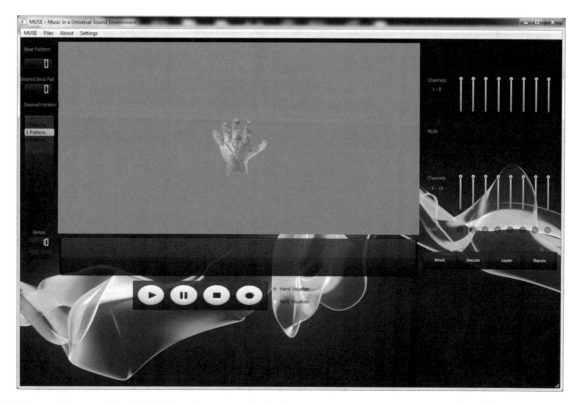

Fig. 49.3 An example screenshot of MUSE that demonstrates MUSE's GUI

Fig. 49.4 A close up of the Note Visualizer. The visualizer updates in real time, according to the gestures of the user and displays notes of a selected instrument to the user

patterns easier and more accurate. We moving towards a system where it expects certain beats patterns and to grade the user.

- Note Manipulation Section: Located next to the next below the channel manipulation is the note manipulation section. This section is dedicated to manipulating different play styles of the midi which we have denoted as attack, staccato, legato, and marcato. Only one of these states can be on at any given time. This functionality was added as a precursor for detecting the type of play attached to different types of conducting that are usually captured in a conductors motion.
- Files Menu: The user interface has drop down menus that will be utilized in order to open files, exit the program, record conducting, and to provide information on the program itself. The recording functionality will be explained later in Sect. 49.3.1.2.
- Musical Note Visualizer: Demonstrated below the below the OpenGL Visualizer is small view of the musical note visualizer that the user can see what notes are currently playing. An example of the note visualizer output can be seen in Fig. 49.4.

49.3.1.2 Midi Processor Layer

The midi processor consists of a midi sequencer, an rtmidi module, libjdksmidi module, Guidolib module, and a JSON module. Each of these modules has a unique independent/dependent responsibility to the midi processor. The libjdks module is responsible for reading and parsing a midi file for all midi events existent in the midi. The sequencer is responsible for determining what midi events to feed to the Rtmidi module for playback dependant upon input given from the pattern recognition level. The rtmidi module allows users to have the option to send midi signals to other midi output devices that could be connected to other software or synthesizers.

Also, MUSE allows a user to record their performance. This requires the midi processor layer to keep track of all motion created by the user made during the playback of a song. Every frame of information that is acquired from the Leap is recorded and timestamped. When a user has finally finished recording a song and their gesture motions, the motions are serialized into a JSON format with the necessary information. The JSON string generated is then compressed with zlib and saved at a location specified by the user. When a user goes to select this recorded file again for playback, the user's recording will be played back in the same fashion as before. This functionality could be used a feed back to give the user insight on how well they are doing with their conducting. We choose to serialize our data as JSON in order to account for future expandability for other user data that may be beneficial.

49.4 Machine Learning Module Theory and Results

In this section the rnn and HMMs used for predicting beat patterns are discussed as a part of the Pattern Recognition layer. Both of these classifiers were strictly made to classify the patterns themselves, but not the individual beats inside of the pattern. Also the HMMs have so far only been implemented for the machine learning module. The machine learning module was built to automatically determine which pattern is being detected without the user selecting a preselected pattern. Right now MUSE primarily relies upon the user's selection and uses Conga to determine the change in tempo based on Conga primarily.

Based on the details from the Conga paper [5], we have implemented a finite state machine to be used with MUSE. We have spent time analyzing the Leap's output when someone is conducting in order to figure out what is needed to capture an ictus or down beat in a conducting pattern. Our finite state machine keeps track of the x and y position, velocity, and acceleration. Our finite state machine effectively looks for zero crossings in the velocity and acceleration that denotes a major change in direction, which is an ictus in conducting. A unique finite state machine was created for a two beat pattern, three beat pattern where the second beat is on the left, three beat right pattern where the second beat is on the right, and four beat pattern. These four patterns, that are recognized by Conga, are used for MUSE and for training both the HMMs and rnn.

The HMMs in the Pattern Recognition layer uses any velocity measurements received and computed by the Leap Motion API as input. Inside the Pattern Recognition layer of MUSE, the velocity measurements are preprocessed into a discrete alphabet and then passed into a HMM for each pattern supported by MUSE. There are four unique HMMs representing each beat pattern supported by MUSE. Predicting the beat pattern is then selected based on the model with the greatest likelihood. This classification is not accepted as the ground truth until the pattern is completed. Conga is used to determine if a pattern has truly been completed. In order to capture the ordering and complexity of beat patterns a left to right Hidden Markov model structure was chosen to model the data captured from the Leap. The left to right Hidden Markov model was chosen because a beat pattern is a cyclic in nature. An alternative model type is an ergodic HMM, which does not model a beat pattern well because it does not have a strict path for state sequences, which makes the ergodic model a bad choice.

In creating the discrete alphabet for the HMM only the x and y components of the velocity data are used.

The velocity vector captured from the Leap is encoded with the following function:

$$D = \left\lfloor vec.roll + \frac{1}{2} \right\rfloor \mod 16 \qquad (49.1)$$

Where vec.roll is the roll of the current velocity with respect to the z axis, which can be computed by the Leap Motion API. The roll was chosen for the decoding function because it points in the direction that the hand is currently going. The above formula in Equation 49.1 that uses the API was inspired by another paper [17] and the equation encodes acceleration vectors into 16 different quadrants to be passed into the HMMs and rnn.

An rnn was built with keras [18] to test how well it would work in comparison to the HMMs. The rnn's architecture is a single long short term memory (lstm) layer follow by a regular feed-forward layer and the output is squashed with a softmax layer. The rnn's objective function is a categorical cross entropy function. The rnn has four outputs were each output indicates the most probable beat pattern.

The MUSE system utilizes data collected from the Leap Motion API to train HMMs and a RNN. The Leap Motion API provides data at rates of 100 FPS to 200 FPS enough to predict conducting gestures. A dataset constructed from 140 data samples from six different people who conducted for 30 seconds was used to train both HMMs and the RNN. The conga finite machine was used to label this dataset and it was split into 80% training and 20% testing. Another dataset was labeled manually by pressing every time a beat occurs from one person with 40 datasets that were 10 seconds each. Table 49.1 demonstrates the overall results of the rnn and

HMMs over both datasets. The rnn performs better than the HMMs on both datasets with 98.1% on the Conga labeled and 87.3% on the manually labeled dataset.

The difference in performance between the Conga and Manually labeled datasets can be explained by the way the two datasets were labelled. In processing both these datasets the individual patterns in each recording were split apart and labeled. For example, if one recording had four patterns within it, then those four patterns would be broken done into four individual sequences and labelled. In labeling the Conga finite state machine will have some delay in determining the end of a beat pattern and is limited by the design of the finite state machine created for the pattern. In the manually labelled dataset the end of a pattern is limited by the skill and delay of the key press by the one labeling the dataset. The difference in the results demonstrates that the end of patterns were very different between the two datasets and actually impacts the two classifiers.

49.5 Discussion

MUSE has been designed for students and teachers to use this system for learning basic conducting patterns. A lot of the features have been tailored for constructing future games and assignments for teachers and students. However, MUSE is need of improvements overall, especially in more robust beat detection algorithms and the inclusion of other sensors. A lot of other projects have focused on building systems that are able to effectively capture conducting accurately. Unlike these systems, MUSE's overarching goal is to capture the motions of a conductor and provides users feedback in a scalable system. MUSE has incorporated a way to save past performances that is independent of other sensors that could allow for the addition of new sensors.

MUSE has several different features that makes it easily adaptable for teaching purposes and scalable for other sensors that may be added further on. MUSE is capable of taking in midi and producing output to different output sources. Users are able to save and record performances, alter the playing style of music, and choose different conducting patterns that they wish to be recorded. Having the capability to select different beat patterns allows for different conducting scenarios to be constructed in the future. Having the ability to select different beat styles allows for MUSE to have the extensibility for algorithms that detect the style of conducting. Using midi as an output source is more flexible then audio as it can be controlled robustly. The user is also able to view how notes are changing through time, effectively allowing a user to examine a visual representation of the music. Despite these features, MUSE has several improvements that could be made.

Table 49.1 Comparison of the rnn versus the HMMs on the Conga labeled dataset and Manually labeled dataset

	Conga labeled	Manually labeled
rnn	**98.1%**	**87.3%**
HMMs	78.9%	50.7%

MUSE's HMMs were created for the purposes of capturing four basic conducting patterns. These algorithms are accurate enough to perceive the conducting patterns and even give the user feedback. The feedback from MUSE can be severely impacted by any inaccuracies from the limited range and sensitivity of the sensor. Despite these inaccuracies, the rnn demonstrated in the previous may be beneficial for improving the overall accuracy of MUSE's current use of HMMs.

The rnn discussed in the previous section was found to be more accurate that the hmms. There have been very few experiments with music conducting and determining a beat pattern. It was found that the rnn outperforms classifying individual beat pattern segments. However, it has not been added to MUSE yet. These results require further investigation and a real time classifier needs to be made for different types of rnns.

49.6 Conclusions and Future Work

The MUSE application coalesces a variety of different tools and concepts to provide a versatile set of functionality behind an intuitive interface. Having incorporated a large amount of research as to the conventions of conducting as well as the needs of the music community, MUSE offers a solution to a ubiquitous problem in a unique way. Through the inclusion of a finite state machine beat recognizer using the Conga framework, the application provides an accurate method to not only keep track of a user's conducting patterns but also record their movements. There are some future plans to improve the accuracy of the conducting recognition system by using techniques such as DTW mentioned in the introduction and the rnn. We also have plans to incorporate different sensors such as the Kinect. Another goal is to implement the system as a web application and to incorporate MUSE into a virtual reality application, such as a six-sided cave or an Oculus Rift. Besides tracking movements, the machine learning within MUSE demonstrates the ability to be adaptable for other algorithms. With a dynamic visualization component incorporated into the GUI as well as a robust manipulation of sound output, the technical back-end is well hidden to the user as they utilize MUSE to its full potential.

Acknowledgements This material is based in part upon work supported by: The National Science Foundation under grant number (s) IIA-1329469. Any opinions, findings, and conclusions or recommendations expressed in this material are those of the author (s) and do not necessarily reflect the views of the National Science Foundation.

References

1. Stinson, L. (2014). Conduct a virtual symphony with touchscreens and an interactive baton. [Online]. Available: http://www.wired.com/2014/05/interactive-baton/.
2. Toh, L.-W., Chao, W., & Chen, Y.-S. (2013). An interactive conducting system using kinect. In *2013 I.E. International Conference on Multimedia and Expo (ICME)* (pp. 1–6).
3. Levin, G., Shakar, G., Gibbons, S., Sohrawardy, Y., Gruber, J., Semlak, E., Schmidl, G., Lehner, J., & Feinberg, J. (2001). Dialtones (a telesymphony). [Online]. Available: http://www.flong.com/projects/telesymphony/.
4. Api reference. [Online]. Available: https://developer.leapmotion.com/documentation/index.html.
5. Lee, E., Grüll, I., Kiel, H., & Borchers, J. (2006). Conga: A framework for adaptive conducting gesture analysis. In *Proceedings of the 2006 Conference on New interfaces for Musical Expression* (pp. 260–265). IRCAM – Centre Pompidou.
6. Smus, B. (2013). Gestural music direction. [Online]. Available: http://smus.com/gestural-music-direction/.
7. Schlömer, T., Poppinga, B., Henze, N., & Boll, S. (2008). Gesture recognition with a Wii controller. In *Proceedings of the 2nd International Conference on Tangible and Embedded Interaction*, ser. TEI'08 (pp. 11–14). New York: ACM. [Online]. Available: http://doi.acm.org/10.1145/1347390.1347395.
8. Shotton, J., Sharp, T., Kipman, A., Fitzgibbon, A., Finocchio, M., Blake, A., Cook, M., & Moore, R. (2013). Real-time human pose recognition in parts from single depth images. *Communication ACM, 56*(1), 116–124. [Online]. Available: http://doi.acm.org/10.1145/2398356.2398381.
9. Schramm, R., Rosito Jung, C., & Reck Miranda, E. (2015). Dynamic time warping for music conducting gestures evaluation. *IEEE Transactions on Multimedia, 17*(2), 243–255.
10. Schramm, R., Jung, C. R., & Miranda, E. R. (2015). Dynamic time warping for music conducting gestures evaluation. *IEEE Transactions on Multimedia, 17*(2), 243–255.
11. Rabiner, L. (1989). A tutorial on hidden Markov models and selected applications in speech recognition. *Proceedings of the IEEE, 77*(2), 257–286.
12. Qt – home. (2016). [Online]. Available: http://www.qt.io/.
13. Scavone, G. P. Introduction. [Online]. Available: https://www.music.mcgill.ca/~gary/rtmidi/.
14. Koftinoff, J. jdksmidi. [Online]. Available: https://github.com/jdkoftinoff/jdksmidi.
15. The industry's foundation for high performance graphics. [Online]. Available: https://www.opengl.org/.
16. Roelofs, G., & Adler, M. (2013). zlib home site. [Online]. Available: http://www.zlib.net/.
17. Schmidt, D. (2007). Acceleration-based gesture recognition for conducting with hidden markov models. Ph.D. dissertation, Ludwig-Maximilians-Universitaet.
18. Chollet, F. (2016). keras. https://github.com/fchollet/keras.

Using Big Data, Internet of Things, and Agile for Crises Management

50

James de Castro Martins, Adriano Fonseca Mancilha Pinto, Edizon Eduardo Basseto Junior, Gildarcio Sousa Goncalves, Henrique Duarte Borges Louro, Jose Marcos Gomes, Lineu Alves Lima Filho, Luiz Henrique Ribeiro Coura da Silva, Romulo Alceu Rodrigues, Wilson Cristoni Neto, Adilson Marques da Cunha, and Luiz Alberto Vieira Dias

Abstract

This paper describes the use of Scrum Agile Method in a collaborative software project named Big Data, Internet of Things, and Agile for Accidents and Crises (BD-ITAC). It applies the Scrum agile method and its best practices, the Hadoop ecosystem, and cloud computing for the management of emergencies, involving monitoring, warning, and prevention. It reports the experience of students from three different courses on the graduate program in Electronics and Computer Engineering at the Brazilian Aeronautics Institute of Technology (Instituto Tecnologico de Aeronautica – ITA), during the first semester of 2016. The major contribution of this work is the application of an interdisciplinary Problem-Based Learning, where students have worked asynchronously and geographically dispersed to deliver valuable increments. This work was performed during four project sprints on just sixteen academic weeks. The main project output was a working, developed, and tested software. During all project, a big data environment was used, as a transparent way to fulfill the needs for alerts and crises management.

Keywords

Alerts and Crisis • Hadoop Ecosystem • Cloud Computing • Big Data • Internet of Things

50.1 Introduction

The World Conference on Natural Disaster Reduction held in Kobe, Hyogo, Japan [1], adopted, among other strategic goals: "The more effective integration of disaster risk considerations into sustainable development policies, planning and programming at all levels, with a special emphasis on disaster prevention, mitigation, preparedness, and vulnerability reduction".

According to Zschau and Küppers [2], the early warning systems proved to be essential in the 90's, as part of an ecosystem to help identify, predict, and communicate risks to the community. The authors support the importance of dealing with these systems as being of interest to scientific, public, and private sectors. For the success of the risk prevention, as mentioned by the authors, these systems should necessarily be considered as part of a partnership tied with the communication of different levels of activity.

According to Smith and Petley [3], there are two types of risk perception: the objective perception, which is derived from empirical data; and the subjective perception, perceived from a group of individuals. According to the authors, these points of view should not be considered antagonistic, but only complement the decision-making and the communication given to a crisis or disaster. In this context, Zschau

J. de Castro Martins (✉) • A.F. Mancilha Pinto • E.E.B. Junior • G.S. Goncalves • H.D.B. Louro • J.M. Gomes • L.A.L. Filho • L.H.R.C. da Silva • R.A. Rodrigues • W.C. Neto • A.M. da Cunha • L.A.V. Dias
Computer Science Department, Brazilian Aeronautics Institute of Technology (Instituto Tecnologico de Aeronautica – ITA), Sao Jose dos Campos, SP, Brazil
e-mail: james76cm@gmail.com; amancilha@gmail.com; edizon.basseto.junior@gmail.com; gildarciosousa@gmail.com; henrique.dblouro@gmail.com; jose.marcos.gomes@gmail.com; lineulimasjc@gmail.com; luizcoura@gmail.com; romuloadmr@gmail.com; wcristoni@gmail.com; cunha@ita.br; vdias@ita.br

and Küppers [2] state that there is the following defined process for early warning systems:

- Evaluation and forecasting - when objective and perceptive input observations should be considered;
- Alerts and communication - when forecasting should be converted into messages, which are sent to agencies and appropriate groups; and
- Response to alerts - when alerts are converted into actions.

In the last decades, the main technological challenges involved web-based technologies for emergency responses. They provided resource management for social networks, as a priority tool to spread information during emergencies. They were understood as a collaborative platform and as a primary information source for data discovery [4].

In this context, during the first semester of 2016, graduate students of three different courses from the Electronic and Computing Engineering Program at the Brazilian Aeronautics Institute of Technology (*Instituto Tecnologico de Aeronautica - ITA*) have developed, within just 16 weeks, a software prototype. This prototype addressed emergency events involving monitoring, warning, and prevention, and also applied an interdisciplinary Problem-Based Learning (PBL) model.

50.2 Background

The proposed interdisciplinary PBL was applied for the following three courses: CE-240 - Database Systems Project, CE-245 - Information Technologies, and CE-229 - Software Testing. Its objective was to develop an academic system prototype, using the Scrum agile method and its best practices. It involved: big data; Internet of things; the Hadoop ecosystem; cloud computing and security; and emergency management, involving monitoring, warning, and prevention.

50.2.1 Problem-Based Learning (PBL)

According to Ertmer et al. [5], Problem-Based Learning is an instructional learner-centered approach that empowers learners to conduct research, integrate, theory and practice, and apply knowledge and skills to develop a viable solution to a defined problem.

50.2.2 The Scrum Agile Method and Practices

According to Rubin [6], Scrum is a refreshingly simple and people-centric framework based upon the values of honesty,

openness, courage, respect, focus, trust, empowerment, and collaboration. In Scrum, work is performed in iterations or cycles of up to a calendar month called sprints.

It involves only roles, activities, and artifacts. The roles are: Product Owner (PO), Scrum Master (SM), and Team Developers (TD). The activities are: Sprint, Sprint Planning Meeting, Daily Meeting, Sprint Review, and Sprint Retrospective. And the artifacts are: Product Backlog (PB), Sprint Backlogs, and potentially shippable Product Increments [6].

50.2.3 Big Data

Big data are large pools of data that can be captured, communicated, aggregated, stored, and analyzed. This concept is now part of every sector and function of the global economy. The definition of big data can vary by sector, depending on the software tools that are commonly available. Large volumes of datasets are common in most industries. Moreover, big data in many sectors today will range from a few dozens of terabytes to multiple petabytes [7, 8].

It consists of increasing: volume (i.e., amount of data); velocity (i.e., speed of data in and out); variety (i.e., range of data types and sources); veracity (i.e., the quality of captured data that can vary greatly, affecting accurate analysis); and value (i.e., referring to the ability to turn data into value) [9].

50.2.4 Internet of Things (IoT)

According to Gubbi [10], IoT is a new technological context, involving computational and communication environments. In this context, the big challenge is on autonomous interaction of devices and technologies, making this technology invisible to its final user.

50.2.5 The Hadoop Ecosystem

According to Verner [11], the Hadoop is a framework that has several other sub-frameworks, each composed of a module or library that receive a denomination. This framework is a largely used platform for big data, developed by the Apache Software Foundation, and mostly written in the Java language. This powerful open-source tool enables the processing of large volumes of data.

50.2.6 Cloud Computing and Security

According to the NIST [12], cloud computing is a model for enabling ubiquitous, convenient, on-demand network access to a shared pool of configurable computing resources (e.g., networks, servers, storage, applications, and services) that can be rapidly provisioned and released with minimal management effort or service provider interaction.

According to Singh and Malhotra [13], in a cloud-computing environment there must be concern with information security, especially when it comes to networks, virtual machines security, user interface security, compliance, and privacy.

50.2.7 The Management of Emergency Monitoring, Warning, and Prevention in Big Events Situations

The Big Data, Internet of Things, and Agile for Accidents and Crises (BD-ITAC) project proposed the creation of a central facility of information processing named the Situation Room, which should be able to manage: a set of subjective pieces of information sent by remote users, via mobile devices; a set of social network data; and empirical information sent from IoT sensors. In order to analyze and evaluate this data, the BD-ITAC project considered the existence of an Operator. Its main role was to determine whether the information received in the Situation Room was to be transformed into an alert. If so, it would be transmitted to the competent authorities to help the process of decision making, or sent to the inhabitants of the regions under risk.

50.3 The Project Overview

In this project, several artifacts were used in the development phase. The vision of the project has limited its scope, while the architecture defined the role of each Scrum Team. Trello [14] was used to control User Stories (US) and their tasks. Burndown Charts were used to show the progress of each Scrum Team to complement their assigned tasks. In addition, Google Docs and Google Sites were used to support documentation and communication.

50.3.1 The Project Vision Artifact

According to the Scrum agile method, a project must have a Vision Artifact to help to accomplish its main objective. In this project, its Vision Artifact was a collaborative effort from all the participants as follows:

> For *public and private organizations involved with monitoring, alerts, and small and big emergency situations,* **who** *need to deal with large volume of data, internet of things, agile methods, and other emergent technologies,* **the** *BD-ITAC project* **is** *a system based upon social networks which runs on cloud computing.* **Differently** *from the existing products of Universities, Research*

Institutes, or Enterprises, **this product** *was developed in just 16 weeks, within an academic environment, using the best practices of the Scrum agile method and following software quality, reliability, safety, security, and testability requirements.*

50.3.2 The BD-ITAC Project Architecture

After researching and several discussions, the project participants built the final architecture, as shown in Fig. 50.1.

50.4 The Project Development

All students from the three courses were divided into 6 Scrum Teams. Each Scrum Team had an assigned role, as shown in Table 50.1.

The BD-ITAC project had four executed sprints, lasting one month each. They are described in the following sections, according to the main values delivered from each ST based on the Product Backlog.

50.4.1 Sprint 0

The Sprint 0 main objective was the training of all students in the technologies involved in the project and in the fundamentals of agile method and its best practices. Also, the participants had to be familiarized with the use of tools such as Google Hangouts, Trello, and Burndown Charts. They took courses on Big Data Fundamentals, Hadoop fundamentals, Data Science Methodology, Hadoop Hive, Introduction to NoSQL, Introduction to Data Analysis Using R, Moving Data into Hadoop, and Spark.

The Scrum was studied during the first lectures and practiced through a group exercise called "LEGO® 4 Scrum" [15]. Also, during Sprint 0, the six ST were consolidated, as well as the definition of each role as follows: Scrum Master, Product Owner, and Team Developer (TD) for each team.

The LEGO® 4 Scrum dynamic exercise provided a very realistic experience in the use of the Scrum through the use of LEGO® pieces to build a fictional college town. The product development was incremented in each Sprint, with value-based prioritization, collaboration, and cooperation between Scrum Teams (ST). This aimed to accomplish a clear goal and the iterative improvement process provided by the lessons learned at the end of each Sprint. It helped the students in truly learning and practicing the content that had been presented during classes.

At the end of Sprint 0, the following value artifacts were produced:

Fig. 50.1 The BD-ITAC project architecture

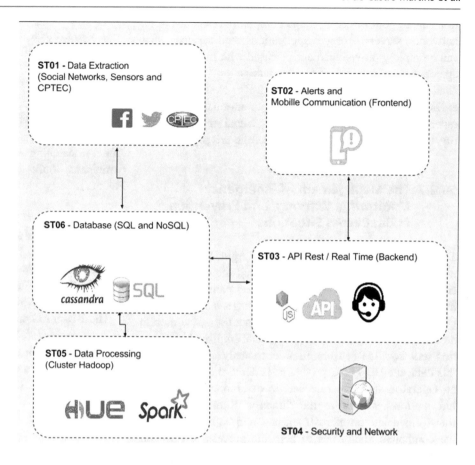

Table 50.1 The BD-ITAC project macro-functions

Scrum teams IDs	Macro-functions
TS#01	Data extraction (Via social networks/sensors)
TS#02	Alerts and mobile Communication
TS#03	Rest API/Real time
TS#04	Network and security
TS#05	Data processing (Hadoop cluster)
TS#06	Database (Relation and NoSQL)

- Training in required technologies and understanding the main concepts for agile development; and
- Project's vision.

50.4.2 Sprint 1

The Sprint Backlog for Sprint 1 had a total of 20 User Stories. Figure 50.2 shows the Sprint 1 Burndown Chart, where the planning effort (IDEAL, on blue line) of all 6 TS has been reached by their performed effort (REAL, on red line), during 4 weeks.

These US were mainly based on specific individual tasks from each ST, since integration was not a requirement in this first Sprint.

In order to show registered events, some criteria were defined for events' selection, for crises registration, and also for the use of data crisis indicators. At this point, it was also created and implemented a relational data model, using MySQL and a non-relational data model using Cassandra [16] databases.

The features implemented during Sprint 1 focused on providing alert registration and the creation of the logical infrastructure for the BD-ITAC project. Therefore, an event registration feature was implemented in a mobile app, as well as a web service simulator.

The project infrastructure was implemented on a private cloud, initially providing three servers protected by a firewall, as well as all the tools and technologies required for the beginning of the project, such as Apache Hadoop, the Database Systems and the Web Server. Figure 50.3 shows the infrastructure cloud computing and Fig. 50.4 shows the access test from the TS 06.

In parallel, the ST responsible by the processing of big data worked on deciding the best techniques for exporting the data from the relational databases into the Hadoop Data File System (HDFS).

Fig. 50.2 The Sprint
1 burndown chart

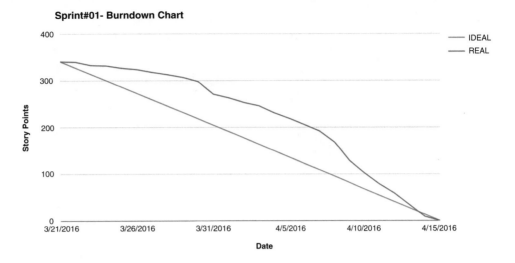

Fig. 50.3 The access screen in
the infrastructure cloud-
computing

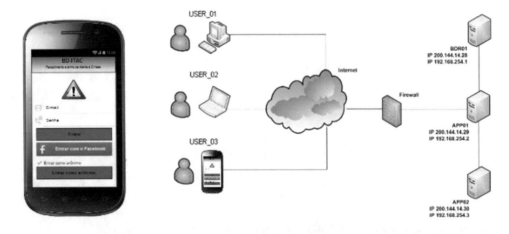

Fig. 50.4 The access test from
the TS 06 (tm06)

50.4.3 Sprint 2

During the Sprint 2, a total of 25 US were implemented. The cumulative results from Sprint 1 and 2 represented 36.88% of the total US from the Product Backlog. Figure 50.5 shows the Sprint 2 Burndown Chart, where the planning effort (IDEAL, on blue line) of all 6 TS has been reached by their performed effort (REAL, on red line), during 4 weeks.

In order to achieve the maximum integration and to effectively use the BD-ITAC project infrastructure, each ST has prioritized its tasks. During this Sprint 2, the integration took place mainly at the Situation Room.

Some events occurred at this time. Crisis alerts and images have been captured on mobile applications from the web interface. Data collection and the mobile application enabled social networks communications through an Application Programming Interfaces (API). Figure 50.6 shows a snapshot of the Control Room homepage and Fig. 50.7

shows mobile snapshots with alert information and crisis indicators by region.

Some communications were enabled with the Center for Weather Forecasting and Climate Studies (Centro de Previsao de Tempo e Estudos Climaticos - CPTEC). Also, some communications were enabled with the National Institute for Space Research (Instituto Nacional de Pesquisas Espaciais - INPE). Both of them were considered as sources of public information.

Through the integration of these technologies, it became possible to receive warnings through social networks, and also through photos sent from the mobile application. It was also implemented an electronic sensors simulator, to get sensor data from areas under potentially risk.

The BD-ITAC project infrastructure was improved through the implementation of a security policy. Figure 50.8 shows the reduction of brute force attacks from 33,041 to 783 in just 5 days.

Fig. 50.5 The Sprint
2 burndown chart

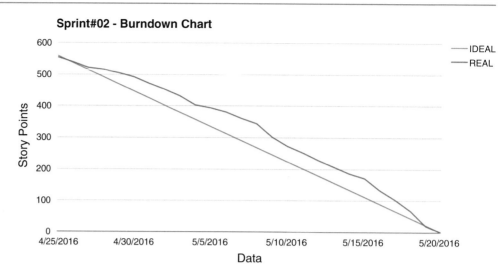

Fig. 50.6 A snapshot of the control room homepage

#ID	Description / Local	Start	Finish	Count Avisos	Is Alerta	Active
1	Monitoramento Crise Alagamento em São José dos Campos Caieiras / SP	01/04/2016	30/06/2016	0	Yes	Yes
3	teste Santa Rosa Do Purus / AC	12/06/2016		2	No	No

Fig. 50.7 Mobile snapshots with
alert information and crisis
indicators by region

Fig. 50.8 The BD-ITAC project showing reduction of Brute force attacks

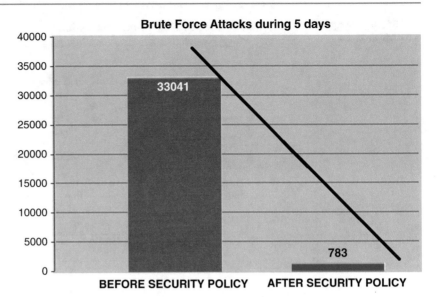

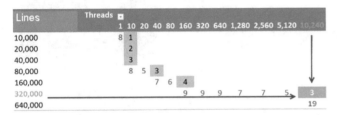

Fig. 50.9 The stress test performed on Cassandra

Tools such as Hue [17] and Spark [18] were used for processing unstructured data from the data collection mechanisms. The MySQL and Cassandra databases were also enhanced, to allow receiving large amounts of data. Figure 50.9 shows the stress test performed on Cassandra, where 320.000 Lines of Code (LoC) were executed using 10.240 threads in just 3 s.

50.4.4 Sprint 3

During this Sprint 3, 27 US were implemented. Figure 50.10 shows this Sprint 3 Burndown Chart, where the planning effort (IDEAL, on blue line) of all 6 TS has been reached by their performed effort (REAL, on red line), by the end of the 4th week of this Sprint.

During Sprint 3, a crisis simulation sequence was established based upon the structure of an Assigned Mission (AM). Figure 50.11 shows the BD-ITAC project architecture used to run the AM.

The AM involved a fictitious crisis to be located at the city of Mariana, on the state of Minas Gerais, Brazil, on the wintertime of 2016. It started with a partial breach of a dam compromised on its main structure. This dam was undergoing renovations and went through emergency reforms, since a previous breaking of another dam located upwards on its feeding river, also located in the city of Mariana.

All ST have concluded their academic project tasks and US, by focusing on the AM, effectively integrating and deploying their Apps making them available through a cloud-computing infrastructure.

Alerts registrations were successfully tested in the Situation Room through mobile Web Apps, by receiving collected data from Facebook, Twitter, and IoT sensors. Figure 50.12 shows a simulation of gathering information from Facebook and Twitter, 104,265 posts from Facebook and 2269 posts from twitter, summing up 39 posts per second in this crisis.

The mobile Web App started, by supplying the Situation Room with data entered by users, simulating residents actions in the city of Mariana, supporting the decision making process with realistic information and competent technicians and authorities, and informing the local population about the impending danger.

At this point, a big data processing system was fully integrated with relational (MySQL) and non-relational (Cassandra) databases. The BD-ITAC system project was tested for usability and performance, and some alerts were successfully processed in Hadoop Ecosystem using Apache Spark, Hive [19], and Hue.

The relational database had acceptable performance tests in the existing infrastructure, as shown in Table 50.2. A data load with 100 threads was simulated with JUnit and it was executed in just 1 s. The same test was performed with 200 threads and it ran in 2 s. As more tests were executed, the results have shown a linear test result. The BD-ITAC system project started to show instability, when the test

Fig. 50.10 The Sprint
3 burndown chart

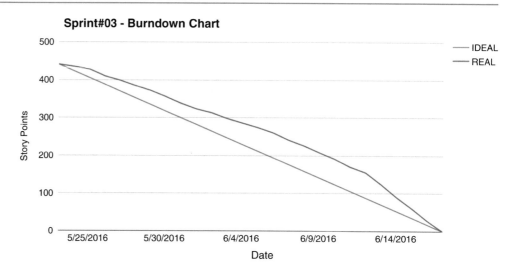

Fig. 50.11 The BD-ITAC
project architecture used to run
the AM

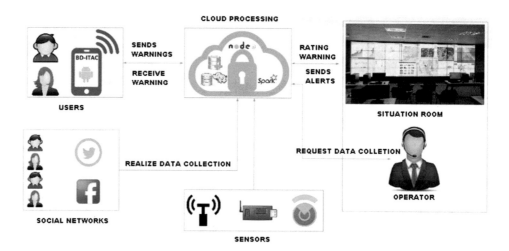

Fig. 50.12 The simulation of gathering information from facebook and twitter

Table 50.2 The MYSQL data load performance test with Junit

Number of threads	Response time (s)	Connection failure (%)
100	1	0
500	5	0
1000	10	10

Table 50.3 The NoSQL data load performance test with Cassandra-stress

Records	Threads	Time (s)
10,000	10	1
20,000	10	2
40,000	10	3
80,000	40	3
160,000	160	4
320,000	10,240	3
640,000	10,240	19

reached 1000 threads, i.e., approximately 10% of threads started to return the "connection failure" message.

The same data load test, performed using MySQL with JUnit, was also executed against Cassandra database system. The tool used was the Cassandra Stress. The results in Table 50.3 show the infrastructure reached its best performed with a threshold of 320,000 records, simulating 10,240 threads simultaneously. This data load test finished in 3 s. Tests executed with greater values of records and threads failed, when considering the cloud infrastructure deployed.

50.5 Conclusion

This paper described an academic developed prototype using the Scrum agile method and interdisciplinary Problem Based Learning, as an opportunity for students to academically exercise the development of a real complex project.

By developing the BD-ITAC project, it was possible to manage the combinatorial explosion of data through the use of new solutions that enabled collecting, storing, and processing large volumes of data, through big data technologies. Managing the development of this BD-ITAC interdisciplinary project with Scrum and its best practices has provided quick, efficient, and continuous results.

The obtained results through the use of a set of open-source tools allowed the conclusion of this agile BD-ITAC academic prototype on time fulfilling a proposed Assigned Mission (AM).

The main contributions of this BD-ITAC academic project was summarized in the application of its final architecture, providing real life experience on implementing a high performance and a high availability system initially designed only for academic purposes but that was able to alert, manage, detect, and prevent possible crisis in real emergency situations.

50.6 Future Works

The authors suggest the continuation of this BD-ITAC project research and development, by using data provided by remote sensors or sensors attached to Unmanned Aerial Vehicles (UAV) in another academic project involving Problem-Based Learning. These types of devices, involving real-time embedded system development and Acceptance Test Driven-Development (ATDD), would broad this project scope, making it possible to investigate and discover new technologies that can advance the big data and IoT fields of research and development.

Acknowledgement The authors would like to thank: the Brazilian Aeronautics Institute of Technology (ITA); the Casimiro Montenegro Filho Foundation (FCMF); and the 2RP Net Enterprise, for their infrastructure and support to the development of the BD-ITAC project, allowing its Proof of Concept (PoC) in an academic and simulated real environment.

References

1. World Conference on Natural Disaster Reduction (2005, Jan 18–22). *Hyogo framework for action 2005–2015- building the resilence of nations and communits to disasters.* Available: http://www.unisdr.org/wcdr. Last access: 10/04/2016.
2. Zschau, J., & Küppers, A. N. (2003). *Early warning systems for natural disaster reduction* (1st ed.). Berlin/Heidelberg/New York: Springer.
3. Smith, K., & Petley, D. N. (2008). *Environmental hazards: Assessing risk and reducing disaster* (5th ed.). New York: Routledge Taylor and Francis Group.
4. Mehrotra, S., Qiu, X., Cao Z., & Tate A. (2013). Technological challenges in emergency response. *IEEE Intelligent Systems, 28*(5).
5. Ertmer, P. A., Hmelo-Silver, C., Walker, A., & Leary, H. (2015). Overview of problem-based learning: Definitions and distinctions: Exploring and extending the legacy of Howard S. Barrows. In *Essential readings in problem-based learning* (p. 5). West Lafayetin: Purdue University Press.
6. Rubin, K. S. (2013). Scrum framework. In *Essential scrum: A practical guide to the most popular agile process.* 13 20 Saddle River: Addison Wesley/Pearson Education Inc.
7. Snijders, C., Matzat, U., & Reips, U. D. (2012). Big data: Big gaps of knowledge in the field of internet. *International Journal of Internet Science, 7*, 1–5.
8. Hashen, I. A. T., Yaqoob I., Anuar N. B., & Khan, S. U. (2016). The rise of big data on cloud computing: Review and open research issues. Available: https://www.researchgate.net/publication/264624667. Last access: 10/04/2016.
9. IBM – Big Data & Analytic Hub. (2015, March 19). Available: http://www.ibmbigdatahub.com/blog/why-only-one-5-vs-big-data really-matters. Last access: 07/04/16.
10. Gubbi, J., Rajkumar, B., Marusic, S., & Marimuthu, P. (2013). Internet of Things (IoT): A vision, architectural elements, and

future directions. *Future Generation Computer Systems, 29,* 1645–1660.

11. Venner, J. (2014). *Pro Hadoop* (2nd ed.). New York: Apress.

12. *Recommendations of the National Institute of Standards and Technology.* (2011). The NIST definition of cloud computing. http://dx. doi.org/10.6028/NIST.SP.800-145.

13. Singh, A., & Malhotra, M. (2015). Security concerns at various levels of cloud computing paradigm: A review. In International journal of computer networks and applications

14. Kniberg, H., & Skarin, M. (2010). Kanban and scrum making the most of both. In *Enterprise software development series INFOQ.*

15. Scrum simulation with lego. Available: https://www.lego4scrum. com/. Last access: 10/04/16.

16. Cassandra. Available: http://cassandra.apache.org/. Last access: 10/04/2016.

17. Hue is a web interface for analyzing data with Apache Hadoop. Available: http://gethue.com/. Last access: 08/23/2016.

18. Apache spark tm lightning-fast cluster computing. Available: http:// spark.apache.org/. Last access: 10/04/2016.

19. Hive. Available: http://hive.apache.org/. Last access: 10/04/2016.

An Agile and Collaborative Model-Driven Development Framework for Web Applications

Breno Lisi Romano and Adilson Marques da Cunha

Abstract

Given the needs to investigate and present new solutions that combine agile modeling practices, MDD, and collaborative development for clients and developers to successfully create web applications, this paper goal is to present an Agile and Collaborative Model-Driven Development framework for web applications (AC-MDD Framework). Such framework aims to increase productivity by generating source code from models and also reducing the waste of resources on the modeling and documenting stages of a web application. To fulfill this goal, we have used new visual constructs from a new UML profile called Agile Modelling Language for Web Applications (WebAgileML) and the Web-ACMDD Method to operate the AC-MDD Framework. The methodology of this paper was successfully applied to an academic project, proving the feasibility of our new framework, method, and profile proposed.

Keywords

Agile • MDD • Development Web Applications • Collaborative

51.1 Introduction

In the last decades, software development became a strategic area to many companies and market sectors, considering its high demands and quick technological advances [1]. It is worth mentioning that in this scenario web applications have become the main business for emerging companies, making the Web a standard implementation platform for Business-to-Business (B2B) applications [2].

Additionally, Liang (2007) emphasizes that the development of these Web applications are under pressure due to the evolution of businesses' needs, short deadlines, and limited resources. To deal with these things, a different approach is required, in order to turn business requirements into software artifacts and deal with the constant evolution of Internet application workflow [3].

In this context, agile methodologies began being used to develop Web applications. These methodologies represent project management approaches that guide the production process of computational systems, being flexible to the environment dynamics that are common in software development while generating value and innovation [4].

A research conducted in 2011 by VersionOne with people from different sectors of the software development industry showed that 60% of companies' projects already used agile methodologies. The most common were Scrum, Kanban, and eXtreme Programming (XP).

The research also presented the following main worries, regarding the adoption of agile methodologies in new projects: lack of documentation, lack of an engineering process for development, difficulty in planning long term activities, among others [5].

B.L. Romano (✉)
Federal Institute of Education, Science, and Technology of Sao Paulo, Sao Joao da Boa Vista, SP, Brazil
e-mail: blromano@ifsp.edu.br

A.M. da Cunha
Brazilian Aeronautics Institute of Technology, Sao Jose dos Campos, SP, Brazil
e-mail: cunha@ita.br

© Springer International Publishing AG 2018
S. Latifi (ed.), *Information Technology – New Generations*, Advances in Intelligent Systems and Computing 558, DOI 10.1007/978-3-319-54978-1_51

Fig. 51.1 The scope of the
proposed approach of this paper

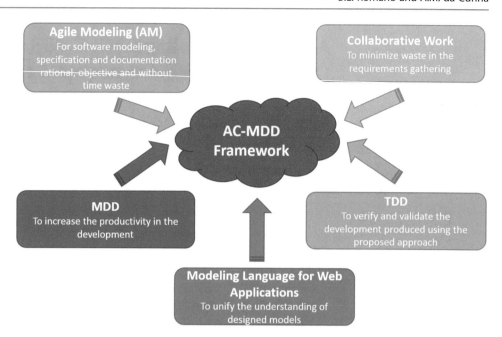

As it can be noticed, the lack of a proper documentation is one of the major concerns in software development companies. In most cases, the documentation is created from the models during the Project and System Analysis stage.

The principles and practices of Agile Modeling (AM) were conceived to circumvent the previously mentioned difficulties. Its idea is to use the modeling efforts in a more rational manner and stay focused on the project. By doing this, project's development is likely to become more efficient. AM suggests that the documentation must be done in a rational manner, maximizing client's investment in the development process [6].

Another interesting remark about agile methodologies practices is their constant search for productivity increase in software development. To do so, AM may also use the Model-Driven Development (MDD) approach, which defines the transformation mechanisms that allow the generation of source-code in a given programming language from high level and abstract models [7].

As mentioned before, the development of Web applications requires agile development methodologies with short iterative cycles due to short deadlines. Ambler (2003) proposed an agile version of MDD called Agile Model Driven Development (AMDD) to take advantage of both techniques based on models and agility [6]. The core principle of AMDD is to create rational models (just barely good enough) that can be modified throughout project's course. This approach implies in smaller iterations, encourages the participation of everyone involved, and allows requirements to evolve along with the project [3].

In the last few years, agile development approaches and MDD practices have been increasingly used in industry to develop Web applications, but not in a combined fashion.

Rumpe (2004) says that this happens because even though both approaches are market trends and there exists a possibility of combining them, they are, in essence, conflictive with one another [8].

Given this scenario, we identified the needs to investigate and present new solutions that combine agile modeling practices, MDD, and collaborative development between clients and developers to successfully create Web applications in a given programming language and to overcome the difficulties mentioned before. The scope of our proposed approach of this paper is shown in Fig. 51.1.

The goal of this paper is, therefore, to present and conceive an Agile and Collaborative Model-Driven Development framework for Web applications named AC-MDD Framework. Such framework should improve the productivity in developing applications, be consistent with the requirements, generate source code from models, and also reduce the waste of efforts involved in modeling and documenting Web applications.

51.2   Related Works

In this section, we present a review of the fundamental aspects and distinguishing features of the existing AMDD processes.

51.2.1 The AMDD High-Level Life Cycle

As the name suggests, AMDD is the agile version of MDD. According to Amber (2007), the goal of MDD is to allow the creation of understandable models. Generating software from these models, however, is not possible for every development team. In this context, AMDD is a more realist approach in a sense that its goal is to describe how developers and stakeholders can work together to create models that are "just barely good enough" [9].

Additionally, according to Matinnejad (2011), AMDD can be better described as a "smart compromise". Due to some contrasting points of view between agile development and MDD, an effective compromise must be reached, in order to obtain the advantages of modeling without suffering from its disadvantages. That is the goal of AMDD [10].

51.2.2 The Sage Process

The Sage process uses MDD for an iterative and incremental approach to develop high confidence reactive multi-agent systems [11]. The main goal of this process is to benefit from the advantages of agile development to implement such systems by improving the documentation of the agile process. Given this intent, an MDD method is proposed that includes four intertwined models that capture developers' decisions and creates the documentation. The four proposed Sage models are: Environmental Model, Behavioral Model, Design Model, and Execution Time Model.

Kirby (2006) highlights that the Sage process takes advantages of the UML class diagrams to represent elements from different models. Besides, the process defines attributes for each class, making it possible for them to interact with one another. Finally, this process does not incorporate an explicit agile process. Instead, it uses a subset of agile practices and principles [11].

51.2.3 The MDD-SLAP Process

The System Level Agile Process (SLAP) is an agile methodology based on Scrum, developed and approved by Motorola. SLAP divides the software life cycle into successive iterations and each iteration consists of three Sprints: Architecture and Requirements; Development; and System Integration Resources Tests. Motorola's MDD is a process based on the V-Model and includes the following development activities: Application and Architecture Requirements; High Level Project Requirements Analysis; Detailed Project; Source Code Generation; and Tests.

This process seeks to increase the development pace, to guarantee a frequent deliver of functionalities and to improve product quality [12]. Besides, this process uses SLAP and MDD, both used by Motorola, and establishes a simple correspondence between Sprints' SLAP and Motorola's MDD process.

51.3   The AC-MDD Framework

Using theoretical concepts and related works, we conceived the **A**gile and **C**ollaborative **M**odel-**D**riven **D**evelopment Framework for Web Applications (AC-MDD Framework). This framework is shown in Fig. 51.2 and corresponds to the agile model driven development of Web applications scenario, based on the life cycle of AMDD and the Scrum and the XP agile development process.

We must initiate the development of a Web application with the agile modeling activity in a business level (high level abstraction) of the application. We must then model both the technical requirements and user stories (sub activity 0.1) and application's architecture and technology (sub activity 0.2).

Since this activity happens only once and aims to better contextualize the main functional and non-functional requirements of Web applications, we say that the business agile modeling is in Sprint 0 (initial phase) of the development.

The parts involved in this first moment are those of system analyst and stakeholders, represented by the client. The involvement of these parts is justified by the fact that they have specific modeling abilities, as well as domain knowledge to be used to develop the application.

The outputs of the Web Application Business Agile Modeling are: documentation of the web application context; documentation of the main stakeholders; use case models and their scenarios; user stories model used for development; and the documentation of the architecture and technology features.

In the first activity, the "Web Application Business Agile Modeling", we must define a set of technical requirements and user stories to be refined, modeled, developed, and tested in a Sprint development. Its definition must be done considering the technical abilities of the development team and their workload. Due to this, there may exist several development Sprints, depending on the complexity of the Web application and the hours needed by the development team.

Once the set of technical requirements and user stories is chosen, the system analysts and stakeholders initiate a refinement of their respective modeling in sub activity 1.1 from the "Web Application Development" activity, applying the principles and practices of AM and focusing only on the current Sprint.

In the next step, and without the participation of the stakeholders, the system analysts must work together with developers to fulfill sub activity 1.2, the "Detailed Agile Modeling of the Web Application Functionalities", which must be done in the Sprint. If some refinement is required at the end of this sub activity, we suggest returning to sub

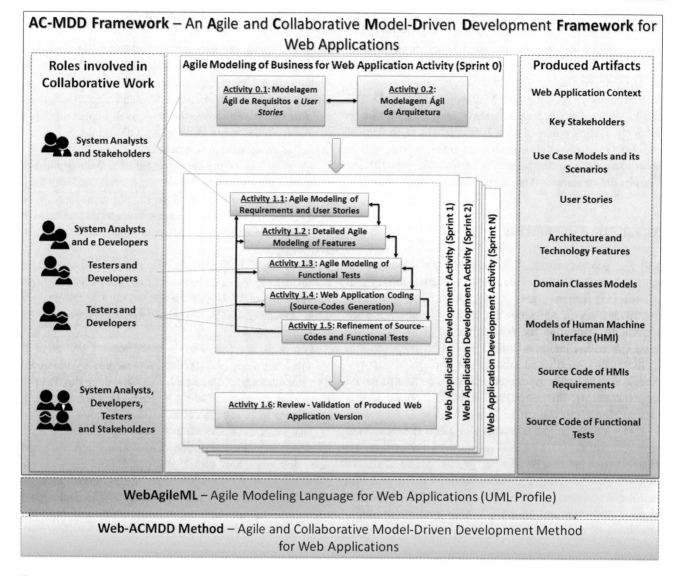

Fig. 51.2 The AC-MDD Framework

activity 1.1, "Refinement of the Business Agile Modeling", to reduce the occurrence of future problems in future stages.

The sub activity "Detailed Agile Modeling of Functionalities" uses the concept of Just in Time (JIT), following the same suggestion from AMDD to model only in time and only enough to solve the problem. The two main outputs from this activity are the main classes model and the Computer Human Interfaces (CHIs), considering the features of the web application.

Testers and developers do the next sub activity and it represents the "Functional Tests Agile Modeling" of the Web application (sub activity 1.3). In this activity they must link to the user stories the agile test modeling to verify system's behavior. They must focus on the functional tests and browsing experience, as well as the successful and alternative scenarios from the use cases. Once again, if

they identify that a refinement is needed, at the end of this activity we suggest them to go back to the "Detailed Agile Modeling of Functionalities" (sub activity 1.2).

Once the test modeling of the Web application is finished, developers and testers execute the next two sub activities defined in the AC-MDD Framework shown in Fig. 51.2. These sub activities are: Implementation of the Web Application (sub activity 1.4) and Web Application Source Code/ Test Refinement (sub activity 1.5).

In sub activity 1.4, developers and testers generate source code both for the CHIs and functional tests with the support of Web-ACMDD. Source-codes are generated for specific technologies and defined in the "Web Application Business Agile Modeling" activity. The algorithms to generate application's source-code use the concept of MDD, but are not presented in this paper because they are not part of its scope.

In sub activity 1.5, developers must refine the source code by complementing functionalities and applying the "layout" and "look & feel" from the CHIs. We stress at this point that the "look & feel" feature is not yet part of the source code generation of the AC-MDD Framework, and for this reason it must be manually done by the developers. In this sub activity, testers must execute the tests and check for the existence of flaws in the generated source-codes.

There exists a flow from the Web application test and source code refinement (sub activity 1.5) to every other previously done sub activity. This is justified by the fact that this flow aims to reduce the occurrence of flaws identified in the development stage and also to correct them in the very same development sprint by returning to the activity responsible for correcting the flaws.

The last proposed sub-activity in the AC-MDD Framework, called "Revision - Validation of the current Web Application Version" (sub activity 1.6), must be done by every part of the development team. This sub-activity consists of presenting the output of the last development sprint to the stakeholders, aiming to validate the requirements identified in the Web Application Business Agile Modeling (Sprint 0). If any refinement is needed, new user stories must be created to be covered in the next Development Sprint.

As shown in Fig. 51.2, visual constructs available in a new UML profile called Agile Modeling Language for Web Applications (WebAgileML) were used throughout the whole AC-MDD Framework. To implement this framework, we conceived the Agile and Collaborative Model-Driven Development Method for Web Applications (Web-AMDD method), in order to make it operational. In other words, we followed the steps of the Web-AMDD method by using different sets of visual constructs from the WebAgileML to enable the execution of the AC-MDD Framework.

51.4 The Web-ACMDD Method

The Web-ACMDD method offers a feasible solution to the following topics: inherent and existing difficulties regarding the productivity while developing Web Applications; difficulties in the requirements gathering stage; and efforts in models and documentation that have no value to clients.

Besides that, we applied the Web-ACMDD method to the whole AC-MDD Framework using the WebAgileML as a cornerstone (Fig. 51.2). In this sense, from the moment we identified that the Web-ACMDD application was not only a set of heuristics and/or good practices, but also a defined method characterized by a set of scientific and technological concepts, we identified the 12 steps of the Web-ACMDD method shown in Fig. 51.3.

The 12 steps of the Web-ACMDD are shown in Fig. 51.3, using the UML Activity Diagram. In each step, we systematically show how to properly go through it and try to document the most relevant elements of the AC-MDD Framework shown in Fig. 51.2.

In each step, we also detail the visual notations from the WebAgileML used in the AC-MDD Framework. They depend on the Web-ACMDD method requirements to make the framework operational.

The new WebAgileML profile was created to cover features that are not yet covered by UML and SysML, but are required to apply the Web-ACMDD to the development of Web applications. This profile was proposed to allow the agile modeling of Web applications and took into account the following features: development of agile processes, AM, UML, SysML, and WebML.

From this point, it was possible to conceive and project a WebAgileML metamodel. The highlights were:

- The new stereotypes and their respective properties to represent new visual constructs;
- UML elements that these stereotypes extend; and
- The relationships between the visual constructs.

A fragment of the WebAgileML profile metamodel for the elements proposed in steps 1 and 2 of the Web-ACMDD method is shown in Fig. 51.4.

A fragment of the WebAgileML profile metamodel for the elements proposed in steps 4, 5, and 6 of the Web-ACMDD method is shown in Fig. 51.5.

A fragment of the WebAgileML profile metamodel for the elements proposed in step 8 of the Web-ACMDD method is shown in Fig. 51.6.

Finally, part of the WebAgileML profile metamodel for the elements proposed in step 9 of the Web-ACMDD method is shown in Fig. 51.7.

It can be noticed that the visual constructs defined in the WebAgileML profile, even though directly related with the Web-ACMDD method proposed in this paper, can be used regardless of their role in the development of web applications. This is due to the fact that the WebAgileML represents an extension of UML and adds features that were not available in the latter, enabling development teams who are already familiar with UML to take advantage of such new features.

51.5 Applying the Web-ACMDD Framework to an Academic Case Study

This section describes an experiment performed using case studies from an academic project to validate the AC-MDD Framework and the WebAgileML proposed in this paper.

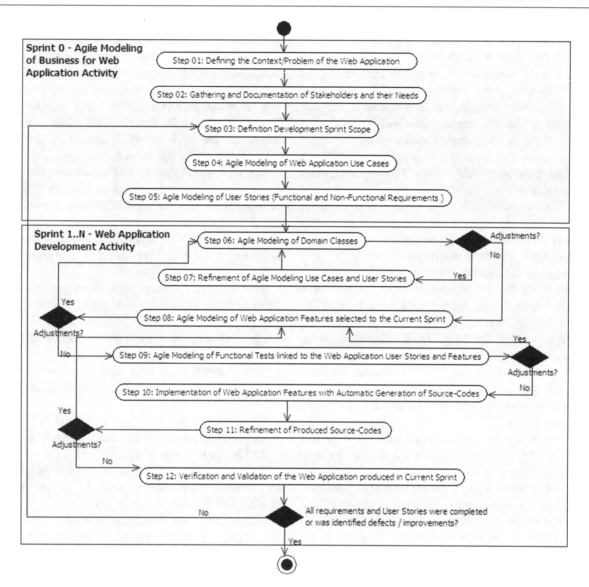

Fig. 51.3 The Web-ACMDD Method

The case study is part of an academic project called "**IF Enchentes-SBV**", the development of a collaborative environment to control and alert about floods and meteorological conditions for the *São João da Boa Vista County*. It was developed by students from an Integrated Technical Course of Informatics, from the *Instituto Federal de Educação, Ciência e Tecnologia de São Paulo* (IFSP), in the São João da Boa Vista Campus, São Paulo, Brazil.

The project IF Enchentes-SBV aims to help people from *São João da Boa Vista*, located in the countryside of São Paulo state, to deal with a real flood problem they've been facing in the last few years.

The idea of the project is to develop a Web application and an Android app to allow people from *São João da Boa Vista* to see real-time information about floods in different parts of the town they live in. The flood data were measured

by meteorological stations (Data Collection Platform – DCP) and their prototypes created during the project.

This project is justified by the fact that, in the last few years, the *São João da Boa Vista* city has been facing serious problems of floods and overflows in different parts of town. To help inhabitants avoid material damages and life risks, we propose the development of a collaborative computational environment (Web application, Android app, and DCP prototypes) to help them get real-time information, both from DCPs and from other inhabitants, about these climate issues.

In this paper, we took the Web application part of the collaborative computational environment of the IF Enchentes-SBV project as the case study. The application covers five functionalities: User/Administrator Management; DCPs Management; DCPs Measurements Management;

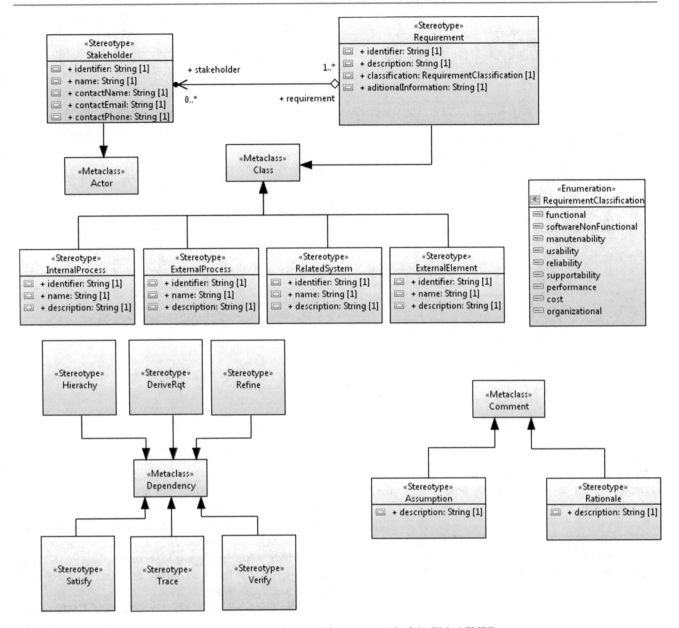

Fig. 51.4 The WebAgileML metamodel for the elements proposed in steps 1 and 2 of the Web-ACMDD

Reports and Charts Management; and DCPs Critical Alerts Management.

Due to scope, this section focus on the DCPs Management functionality using the AC-MDD Framework, the Web-ACMDD method, and the WebAgileML profile. The agile modeling using the visual constructs of WebAgileML is shown in Fig. 51.8.

It is important to notice that Fig. 51.8 represents an agile modeling using WebAgileML of CRUD (Create, Read, Update, and Delete) functions of the DCPs Management.

At first, the user accesses a webpage to see all DCPs in the database. From this point, he/she can insert, edit, or remove a DCP. The dotted arrows in Fig. 51.8 represent the navigation flow between webpages and each link has a different purpose:

- Normal Link – represents a parameter passing situation between any of the model elements and a webpage and allows browsing between two webpages;
- OK Link – represents a successful operation and directs to the next element or webpage to be processed by the

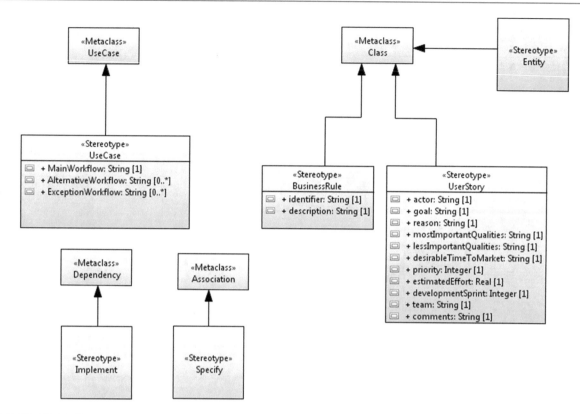

Fig. 51.5 The WebAgileML metamodel for the elements proposed in steps 4, 5, and 6 of the Web-ACMDD

web application; and

- KO Link – represents a failed operation and directs to the next element or webpage to be processed by the Web application.

Given the agile modeling in Fig. 51.8, we execute Step 10 of the Web-ACMDD method, which is responsible for the implementation of the web application functionalities with automatic source-code generation. The CHIs representing the CRUD from the DCP Management are shown in Figs. 51.9 and 51.10.

51.6 Conclusion

This paper's goal was to investigate and conceive an AC-MDD Framework that helps increase the productivity in web applications development by being consistent with system's requirements, generating source code from the models, improving the collaboration between everyone involved, and reducing waste of resources in the modeling and documenting stages of a Web application.

The AC-MDD Framework relates to Web application agile model-driven development and was proposed based on the life cycle of AMDD, the Scrum, and the XP agile development approaches.

From the AC-MDD Framework scenario, we proposed the Web-ACMDD method as a solution to improve the productivity while developing Web applications by reducing efforts spent on their modeling and inconsistencies in requirements gathering.

Additionally, we also conceived a new UML profile called WebAgileML to allow the agile modeling of web applications. We took into account the following characteristics: agile development, AM, UML, SysML, WebML, and IFML.

It is worth mentioning that up to now the AC-MDD Framework, Web-ACMDD, and WebAgileML profile applications are not yet finished. However, these applications are already running and some partial results were presented in this paper. These results focused on the output agile models along with their CHIs on the IF Enchentes-SBV project, which was the case study used in this paper.

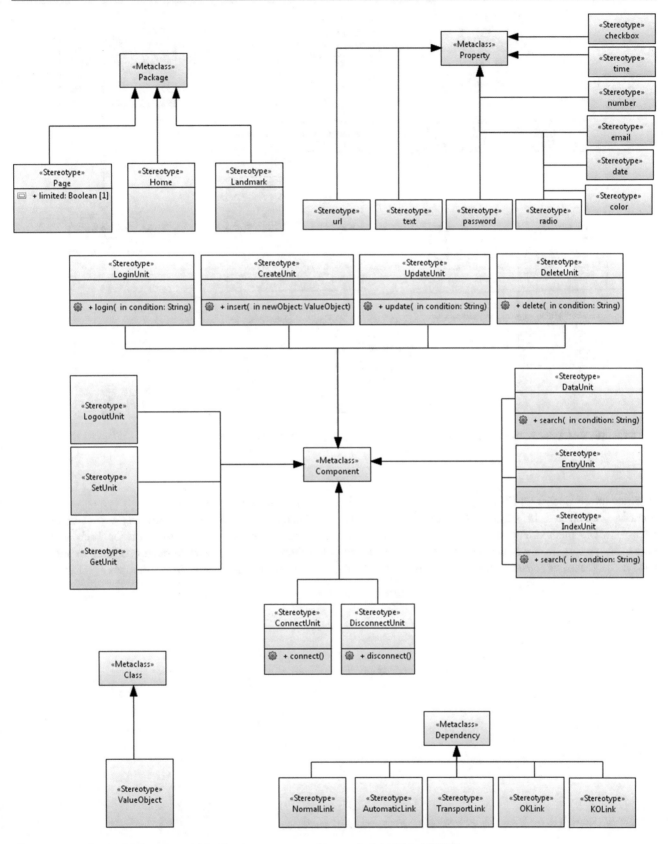

Fig. 51.6 The WebAgileML metamodel for the elements proposed in step 8 of the Web-ACMDD

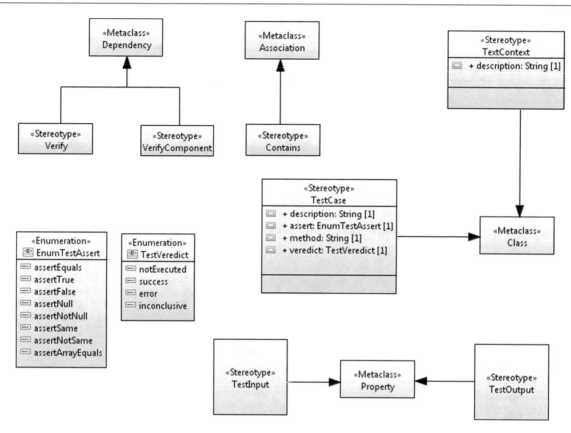

Fig. 51.7 The WebAgileML metamodel for the elements proposed in step 9 of the Web-ACMDD

Since the IF Enchentes-SBV project is still under development, changes in the AC-MDD Framework, Web-ACMDD method, and the WebAgileML may occur. However, we can already notice an increase in productivity and a decrease of efforts involved in modeling and documenting a Web application.

Finally, we believe that the timely aspect of this area demands a constant improvement on the Web application development productivity, reducing efforts to model the system and inconsistencies in requirements gathering stage.

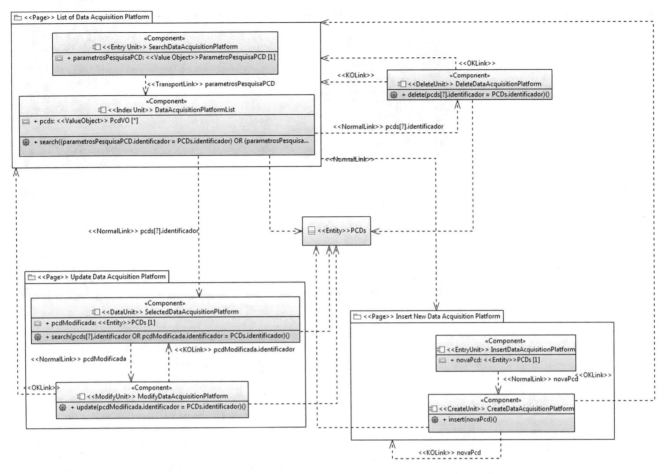

Fig. 51.8 The Agile modeling of the macro functionality of the DCPs Management from the IF Enchents-SBV project using the AC-MDD Framework

Fig. 51.9 The CHI to list the PCDs in the database

Fig. 51.10 The CHI to insert a
new DPC in the database

References

1. Beck, K. (2004). *Programação extrema (XP) explicada: Acolha as mudanças*. Porto Alegre: Bookman.
2. Brambilla, M., Ceri, S., Comai, S., & Fraternali, P. (2006). A case tool for modelling and automatically generating web service-enabled applications. *International Journal of Web Engineering and Technology, Inderscience Publishers, 2*(4), 354–372.
3. Liang, X., Marmaridis, I., Ginige, A. (2007) Facilitating agile model driven development and end-user development for evolvingweb-basedworkflow applications. In: e-Business Engineering, 2007. ICEBE 2007. IEEE International Conference on. [S.l.: s.n.] (pp. 231–238).
4. Berni, J. C. A. (2010). Gestão para o processo de desenvolvimento de software científica, utilizando uma abordagem ágil e adaptativa na micro empresa. 77f. Dissertação (Mestrado no Programa de Pós-Graduação em Engenharia de Produção na Universidade Federal de Santa Maria, Rio Grande do Sul.
5. Versionone (2011). State of agile survey – The state of agile development. [S.l.]. Accessed on October 01, 2017. Available in https://www.versionone.com/.
6. Ambler, S. W. (2003) Agile model driven development is good enough. IEEE Software, 20(5), 71–73. ISSN 0740-7459.
7. Romano, B. L., Silva, G. B. E., Cunha, A. M. D., Mourao, W. I. (2010). Applying mda development approach to a hydrological project. In IEEE. *Information technology: New generations (ITNG)*. 2010 Seventh International Conference on. [S.l.] (pp. 1127–1132).
8. Rumpe, B. (2014). Agile test-based modeling. *International Conference on Software Engineering Research and Practice (SERP06), 1*, 10–15.
9. Ambler, S. W. (2003). Agile model driven development (AMDD). Xootic Magazine. Accessed on October 01, 2017. Available in http://www.agilemodeling.com/essays/amdd.htm.
10. Matinnejad, R. (2011). Agile model driven development: An intelligent compromise. In *IEEE. Software engineering research, management and applications (SERA)*. 2011 9th International Conference on. [S.l.] (pp. 197–202).
11. J. Kirby Jr. (2006). Model-driven agile development of reactive multi-agent systems. In *30th Annual international computer software and applications conference (COMPSAC'06)*. Published in: IEEE Software [S.l.: s.n.]. v. 28(2), March-April 2011, (pp. 297–302). ISSN 0730–3157.
12. Zhang, Y., & Patel, S. (2011). Agile model-driven development in practice. *IEEE software, IEEE Computer Society, 28*(2), 84.

Strauss Cunha Carvalho, Renê Esteves Maria, Leonardo Schmitt, and Luiz Alberto Vieira Dias

Abstract

This work aims to generate White-box Test Cases from a mainframe NATURAL code fragment, using the Control Flow Graph Technique. Basically, it enables a code fragment analysis, generating its control flow represented by a Graph perspective. As a consequence, it provides the automatic generation of White-box Test Cases. Furthermore, this work contributes to attain a lower degree of difficult in the execution of White-box Tests within a Mainframe environment. Also, it brings a significant contribution related to the execution time, inherent to the software testing. At the end, it is added a testing expertise to NATURAL development teams. Usually there is not enough available time to test all possible paths in a algorithm, even in a simple application. Based on it, this work provides automatic generation of test cases for empowerment of team decision, regarding which test cases are more relevant to be performed.

Keywords

Software testing and engineering • Software quality • Control flow technique

52.1 Introduction

According to Molinari [1], there is a relationship between the developed systems quality and their respective testing activities. Therefore, the software testing process becomes essential for product quality assurance or developed service. Consequently, it is possible to observe in public or private organizations, an increasing demand for the software development among the most diverse segments such as commercial, industrial, government, military, and scientific.

Usually, testing software activities increase the development cost. According to Pressman [2], is possible to see a direct relation between time to defect detection with the development cost. In other words, when more early a defect detection occurs, the saving is up to hundred times lesser, if comparing to the final development phase.

Therefore, there is not enough time to test all possible paths in an algorithm, even for simple applications. Due to financial limitations, the spent time for testing activities can be directly added to the development cost. It increases the final product price [2].

Through the years, the software testing area has been developed models, methods, and techniques, that match the ideal cost/benefit [3]. According to Pressman [2], it is a way to identify internal software defects, named White-box Tests

S.C. Carvalho (✉)
Computer Science Division, Aeronautics Institute of Technology – ITA, São José dos Campos, Brazil

Brazilian Federal Service of Data Processing – SERPRO, Rio de Janeiro, Brazil
e-mail: strauss.carvalho@serpro.gov.br; strauss@ita.br

R.E. Maria • L.A.V. Dias
Computer Science Division, Aeronautics Institute of Technology – ITA, São José dos Campos, Brazil
e-mail: rene@ita.br; vdias@ita.br

L. Schmitt,

Brazilian Federal Service of Data Processing – SERPRO
Rio de Janeiro, Brazil
e-mail: leonardo.schmitt@serpro.gov.br

© Springer International Publishing AG 2018
S. Latifi (ed.), *Information Technology – New Generations*, Advances in Intelligent Systems and Computing 558, DOI 10.1007/978-3-319-54978-1_52

(WbT). Within the WbT is possible to use the Control Flow Technique (CFT), it enables a clear code review.

Based on this context, this work has developed a WbT tool, using a variation of the CFT, named Control Flow Graphs (CFG). The developed tool, which works within Mainframe NATURAL code fragments [4], aims to create test cases automation.

Moreover, it generates a graph that represents the code flow. The created graph intends to decrease the code analysis difficulty. At the end, this work tries to improve the software quality, reducing time to create test case, optimizing time to perform tests, and bringing the defect detection to early development stages.

This work is organized in the following structure: Section two is devoted to describe the work background regarding the testing software area; Section three describes the proposed tool. Section four discusses and presents obtained results from two experiments where the proposed WbT tool had been applied. Finally, in section five concluding remarks are presented.

52.2 Software Testing

This section describes relevant concepts that involve this work. Conforming the standard IEEE 610.12-1990 [5], the test activity is defined by the system or component operation process, under specific conditions. Indeed, it requires the registration and observation regarding collected results, enabling the system or component evaluation.

According to Crispin and Gregory [3], the agile software development requires a new concept to the traditional tester role, named agile tester. The agile tester is a team member that comprises an agile development team (also know as cross-functional team). Also, the agile tester has to be involved with the business and technical areas, in order to improve and refine the product requirements. It is fundamental to boost the development in the right direction.

In addition, the main agile tester skill is turning product requirements into automated tests. In most cases, they are experienced exploratory testers trying to understand user wishes and needs. Therefore, the software quality is a common goal to the development team [3].

52.2.1 Agile Testing Quadrants

The agile testing quadrants has been used as a guide to the development team. Initially, the quadrants were created by Brian Marick, presenting a numerated matrix divided into four cells. Each cell number does not imply a required order to be followed [3].

According to Crispin and Gregory [3], the agile testing quadrant is a simple taxonomy that enables a test planning to the agile development team. Therefore, it provides a minimum required self-organization to start the development within each sprint. The Fig. 52.1 is showing the agile testing quadrants. This work is based on the Q1 and Q2.

The Q1 focuses the prime testing techniques used in the agile development, like the Test-driven Development (TDD) [3]. The TDD provides to team members the possibility to develop features without worrying about later changes. Consequently, it improves the development and application quality. Moreover, it helps the team makes better decisions regarding its project design.

As reported by Meszaros [6], this quadrant has two practices: (1) **unit test**, to validate a smaller part of the application, that means objects or methods; (2) **component**

Fig. 52.1 The Agile testing quadrants (Adapted from [3])

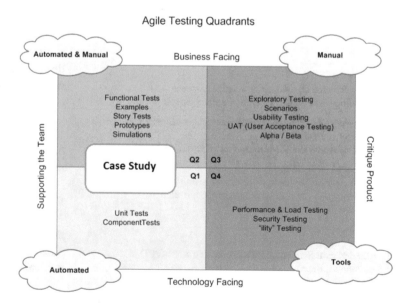

test, to validate the major part of the application as a set of classes that provides the same task. Usually, the Q1 tests are developed with xUnit tools, they aim to measure the internal software quality.

Unit tests are not design by costumers, it will not understand the internal development aspects. They must be performed within a continue integration approach. In fact, it provides the source code quality assurance [8]. Therefore, when the team members are writing their unit tests, their testing functionality occurs before its own existing.

This work is concentrated in the Q1 and Q2. After the algorithm writing, this work contributes to its graphical visualization of possible paths, and to automatic generating its respective test cases. Thus, functional testings are performed focusing on the customer understanding that establishes quality criteria.

At the end, this work provides a friendly language that could be used by all involved resources. In other words, tests are written in such manner that business experts can understand the implemented features through this friendly language.

52.2.2 White-Box Testing

According to Myers [7], the tests based on implementation are named White-box Testing (WbT) and Structural Testing. The test selection is based on information obtained from source code, in order to perform different software parts disregarding its specification.

The WbT is based on the internal software architecture. It engages techniques to identify internal structure defects. According to Beck [8], to perform WbTs is required agile testers with enough knowledge about the technology used. Also, the internal software architecture.

Conform to Pressman [2], to execute the WbT some techniques are used to identify a software internal structure defects, like control flow and data stream.

52.2.3 Control Flow Technique

The Control Flow Technique (CFT) shows the possible paths in an algorithm by symbolic representations. Usually, graphs are used to visualize the software logic controls. Applying the graph usage, this technique can be known as Control Flow Graphs (CFG) [2].

The Fig. 52.2 is showing a CFG sample. Each vertex represents algorithm logic controls, each arc corresponds the possible paths between logic controls. There are five vertexes and seven arcs in the Fig. 52.2.

According to Pressman [2], the CFG creation can be a hard activate, when the analysed algorithm has compound

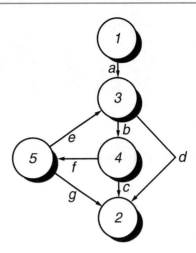

Fig. 52.2 A control flow graph from [2]

conditions. That means, when exist boolean operations as or, and, not-and, and not-or. Therefore, this work is using the CFG and tries to reduce its creation complexity.

52.2.4 Test Cases

Conforming to Craig and Jaskiel [9], test cases are comprised by inputs, outputs, execution restriction rules, and expected results/behaviors. It enables document and test applications, under predefined conditions.

As reported by Hetzel [10], the test case main goal is formally communicate and identify software defects. It allows the software quality evaluation.

However, there is a significant effort for manual test case writing. Especially when it is trying to cover all possible paths. It consumes more development time, consequently it increases the final project cost [3].

Thereby, some approaches were proposed tools for generating test cases, from certain specifications. This work has been investigated the proposed approaches from Santiago and Arantes apud Marinke [11].

Based on those approaches, this work propounds a WbT tool based on CFG. The next section describes the proposed tool. Basically, it uses a NATURAL fragment code for generating its flow graph and test cases.

52.3 Proposed Tool

Initially, this work was looking to source codes written in NATURAL Language for Mainframe environments [4]. It was observing difficulties faced by organizations where trying to perform WbT in this type of source code. Moreover, there is a high cyclomatic complexity inherent

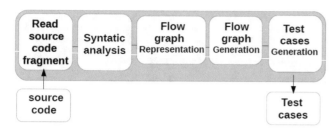

Fig. 52.3 The proposed tool execution path

in its legacy programs. In most cases, this type of language have no object-oriented design that provides this kind of situation.

The code fragments used in this work had been developed at the Brazilian Federal Service of Data Processing (SERPRO) [12]. It is relevant to highlight that keeping classified information, there are no confidential information and/or business aspects. This work is focusing in analysis of algorithms within their Mainframe context.

This proposed tool used two different code fragments as experiments. Both were exposed to the same path. Initially, their NATURAL source codes were read. Then, their flow graphs were automatic generated. Finally, their respective test cases were produced. This path can be divided into five steps, as the Fig. 52.3 illustrates.

The Fig. 52.3, shows a diagram containing the proposed tool five steps. Moreover, The Fig. 52.3 clarifies the proposed tool execution path. Which starts from the source code fragment until the test cases output. Each one of the five steps are described within this section.

The proposed tool was written in the Python programming language [13]. Therefore, using Python native modules, the first step consists in read the input code line by line. It enables the next step the source code syntactic analysis.

The second step performs a syntactic analysis from the last step output. The syntactic analysis (also known as parsing) term refers the translation process between language programs. Based in a formal grammar, it aims to recognize the program syntactic structure.

After parsing and during the execution time, the third step converts the source code into a flow graph. Thus, the source code analyzed is represented through a flow graph. The adjacency matrix is used as a data structure for the graph representation. Also, this matrix has been stored in the main memory.

In the fourth stage, the previous graph representation is converted into a visual mode. Based on adjacency matrix stored in the main memory, this proposed tool is using Python graphs modules. It provides to the development team a quickly visualization of their implemented possible paths.

Using the third step provided matrix, the last step is devoted to generate the respective test cases corresponding

to the input source code. Within this step, each graph arcs is visited, for each visit a test case is generated. It produces a complete path coverage to the analyzed code fragment.

At the end, the development team has to decide which test cases will be performed within the current sprint. In fact, to avoid an enormous amount of test cases, it is recommended to break down the functionality into smaller parts to be tested.

52.4 The Main Results

This section tackles the main results obtained from two conducted experiments. Moreover, it shows their NATURAL code fragments, as well their respective CFGs and Test Cases. There is a cyclomatic complexity crescent between both experiments. That means, the first experiment is more simple then the second experiment.

52.4.1 First Experiment

The Fig. 52.4 is showing the source code fragment used in the first experiment. This fragment has two conditional structures in sequence, the IF and ELSE statements.

Following the proposed tool execution path, the Fig. 52.5 illustrates the extracted CFG from the first code fragment (see Fig. 52.4). Also, their test cases are showing in the Fig. 52.6.

52.4.2 Second Experiment

The second experiment was based in the source code fragment shown in the Fig. 52.7. It is an algorithm with higher cyclomatic complexity then the first experiment.

Looking to its line nineteen (see Fig. 52.7), it is possible to observe multiple conditions using the logical operator AND. Furthermore, there is a dense sequence of IF and ELSE statements. It increases the amount of possible paths.

Following the proposed tool execution path, the Fig. 52.8 illustrates the extracted CFG from the second code fragment (see Fig. 52.7). Also, their test cases are showing in the Fig. 52.9. At the end, the second experiment had been generated eighty-eight test cases (see Fig. 52.9).

52.5 Analyses and Discussions

The experiments performed by this work proved the relation between the source code cyclomatic complexity with the amount of test cases. Therefore, it was observed a significant productivity gain when automate the test cases generation.

Fig. 52.4 The source code fragment used in the first experiment

```
1    IF HIST-VIEW.NU-TERMINAL-IP  NE '                      '
2      MOVE HIST-VIEW.NU-TERMINAL-IP                  TO #NU-TERMINAL-IP(#I)
3    ELSE
4      MOVE HIST-VIEW.TERMINAL-USUARIO               TO #NU-TERMINAL-IP(#I)
5    END-IF
6    MOVE 'PEMP'                             TO #IND-MUDANCA(#I)
7    MOVE HIST-VIEW.DA-EVENTOS(06)           TO #DATA-WORK
8    MOVE #DT-WORK-1                         TO #DATA-EVENTO(#I)
9    IF HIST-VIEW.IT-PORTE-EMPRESA= '01'
10     MOVE 'MICROEMPRESA'                    TO #ALTERACAO(#I)
11   ELSE
12     IF HIST-VIEW.IT-PORTE-EMPRESA= '03'
13       MOVE 'EMPRESA DE PEQUENO PORTE'     TO #ALTERACAO(#I)
14     ELSE
15       IF HIST-VIEW.IT-PORTE-EMPRESA= '05' OR= ' '
16         MOVE 'DEMAIS'                     TO #ALTERACAO(#I)
17       END-IF
18     END-IF
19   END-IF
```

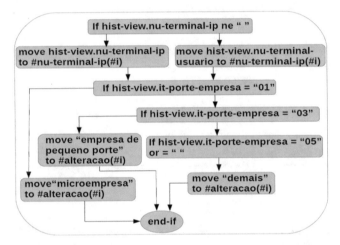

Fig. 52.5 The CFG from the first experiment

	A	B	C	D	E	F
1		if hist-view.nu-terminal-ip ne " "	If hist-view.it-porte-empresa = "01"	If hist-view.it-porte-empresa = "03"	If hist-view.it-porte-empresa = "05"	Result
2	move hist-view.nu-terminal-ip to #nu-terminal-ip(#i)					
3	Hist-view.nu-terminal-usuario to #nu-termi-nal-ip(#i)					
4	move "Microempre-sa" to #alteracao(#i)					
5	move "empresa de pequeno porte" to #alteracao(#i)					
6	mode "demais" to #alteracao(#i)					

Fig. 52.6 The test cases from the first experiment

Also, it was observed that scanning flow graph arcs it is possible to produce a complete set of all possible paths. Indeed, when test cases have an automatic generation, it increases the possibility to detect defects at an early development stage.

According to Pressman [2], there is a strong relationship between the defect detection moment and the final product cost. The final cost trends to stabilize when a defect is found at an early development stage. Moreover, Pressman says that the saving is up to a hundred times when compared to the final development stage.

Another important observed result is the code fragment visual representation trough the CFG. In this representation any team member is able to analyze the algorithm behavior. Also, it was providing malicious codes identification, like hidden IF statements.

It is recommended, to scalable the tool generated from this research, before to create the CFG, to use a lexical analytics from source code. Thus, the beings need not check it. Other recommendation its may compromise the structural coverage of the test cases necessary that programmers manual intervention decide which tests to run.

Therefore, to decrease the complexity between users communication and to scalable this proposed tool, It is suggested: (1) to create a grammar and apply it in the syntactic analysis; (2) improve it to allow the analysis of multiple condition and subroutine calls; and (3) add a Graphical User Interface (GUI).

52.6 Conclusion

This work has investigated the WbT utilization through CFT, within NATURAL source code fragments. As well as generating automatic from the analysed fragment its

Fig. 52.7 The source code
fragment used in the second
experiment

```
 1   IF FCPJ_CO-EVENTOS-N(*) = 101 AND STATUS_CO-STATUS-PEDIDO EQ 83
 2     IF #AX-TEM-EV-INTERESSE-EST EQ 'S'
 3        MOVE #CO-UF-AUX    TO #AX-CONV-PPA
 4        PERFORM PESQUISAR-CNAE-INTERESSE-CONV
 5        IF #AX-CNAE-INTERESSE-CONV-IC = 'S'
 6           PERFORM ADICIONAR-ARRAY-CONV-PPA
 7        END-IF
 8     END-IF
 9     IF #AX-TEM-EV-INTERESSE-MUNIC EQ 'S'
10        MOVE #CO-MUN-AUX TO #AX-CONV-PPA
11        IF #AX-CONV-PPA  = #AX-ARRAY-CONVENENTES (*)
12           PERFORM ADICIONAR-ARRAY-CONV-PPA
13        END-IF
14     END-IF
15     IF #CO-UF-AUX EQ 'MG'
16        MOVE '31  '  TO #AX-CONV-PPA
17        PERFORM ADICIONAR-ARRAY-CONV-PPA
18     END-IF
19     IF  #CO-UF-AUX EQ 'MA' AND #AX-TEM-EV-INTERESSE-JUNTA = 'S'
20        MOVE '21  '  TO #AX-CONV-PPA
21        PERFORM ADICIONAR-ARRAY-CONV-PPA
22     END-IF
23     ESCAPE ROUTINE
24   ELSE
25     IF FCPJ_CO-EVENTOS-N(*) = 246
26        PERFORM SETAR-ARRAY-CONV-PPA-246
27     ELSE
28        PERFORM SETAR-ARRAY-CONV-PPA
29     END-IF
30   END-IF
31   PERFORM SETAR-ARRAY-CONV-INFO
```

corresponding graph, and subsequently its respective test cases.

The proposed tool is based on Q1 and Q2, aiming to identify the possible paths in NATURAL algorithms. Moreover, it enables a friendly visualization of the algorithm paths.

The SERPRO provides the necessary NATURAL codes fragments from its development environment, to validate this work. It should be stressed that the provided fragments have no confidential information and/or business aspects.

The proposed tool was aimed to analyse its algorithm aspects only.

At the end, both carried out experiments have produced the expected result. That means, the CFG generation and respective test cases were generated for each fragment.

Based on experiment results, the proposed tool proved effective in the CFG creation and automated test cases generation. Thus, it provides the defects detection at early development stages, ensuring the software quality and reliability.

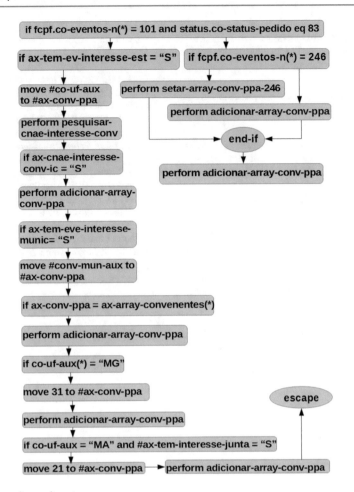

Fig. 52.8 The CFG from the second experiment

Fig. 52.9 The test cases from the first experiment

		if fcpj.co-eventos-n(*) = 101 and status.co-status-pedido eq 83	if ax-tem-ev-inter esse-est = "S"	if #ax-cnae-inter esse-conv-ic = "S"	if ax-tem-ev-interes sse-muni c = "S"	if #ax-conv-ppa = #ax-array-conve nentes (*)	if #co-uf-aux = "MG"	if co-uf-aux eq "MA" and #ax-tem-interess e-junta = "S"	if fcpj.c o-event os-n(*) = 246	Result
2	move #co-uf-aux to #ax-conv-ppa									
3	perform pesqui-sar-cnae-inte-resse-conv									
4	perform adicio-nar-array-conv-ppa									
5	mun-aux to #ax-conv-ppa									
6	perform adicio-nar-array-conv-ppa									
7	move "31 " to #ax-conv-ppa									
8	perform adicio-nar-array-conv-ppa									
9	move "21 " to #ax-conv-ppa'									
10	perform adicio-nar-array-conv-ppa									
11	perform setar-ar-ray-conv-ppa-246									
12	perform adicio-nar-array-conv-ppa									

Acknowledgements The authors thank: to ITA, for its support during this work development; to SERPRO, for providing two NATURAL code fragments and allowing our work; and to 2RP Net enterprise for supporting hardware infrastructure for this work.

References

1. Molinari, L. (2003). *Testes de Software: Produzindo Sistemas Melhores e Mais Confiáveis* (4th ed.). São Paulo: Editora Érica.
2. Pressman, R. S. (2010). *Software engineering: A practitioners approach* (7th ed.). New York: McGraw-Hill.
3. Crispin, L., & Gregory, J. (2009). *Agile testing: A practical guide for testers and Agile teams* (1st ed.). Crawfordsville: Addison-Wesley.
4. Software Ag. *NATURAL and Adabas Database*. [Online]. Available: https://empower.softwareag.com/Products/. Accessed October 10, 2015.
5. Computer Society of the IEEE (1990). *IEEE standard glossary of software engineering terminology*. New York: Institute of Electrical and Electronics Engineers.
6. Meszaros, G. (2007). *xUnit test patterns: Refactoring test code* (1st ed.). Boston: Addison-Wesley.
7. Myers, G. J., Sandler, C., & Badgett, T. (2011). *The art of software* (3rd ed.). Hoboken: Wiley.
8. Beck, K. (2005). *Extreme programming explained: Embrace change* (2nd ed.). Boston: Addison-Wesley.
9. Craig, R. D., & Jaskiel, S. P. (2002). *Systematic software testing* (1st ed.). Boston: Artech House.
10. Hetzel, C. W. (1986). The complete guide to software testing. *Quality And Reliability Engineering International, 2*(4), 274–274.
11. Vijaykumar, M. N. L., & Senne, E. L. F. (2012). *Geração de Testes Caixa Branca para aplicações Multithreads: Abordagem por Statecharts*. INPE – Instituto Nacional de Pesquisas Espaciais.
12. SERPRO. *Brazilian Federal Service of Data Processing*. [Online]. Available: https://www.serpro.gov.br/home. Accessed August 12, 2015.
13. Python Foundation. *Python 2.7.10 documentation*. [Online]. Available: https://docs.python.org/2/. Accessed June 06, 2015.

Requirements Prioritization in Agile: Use of Planning Poker for Maximizing Return on Investment

53

Vaibhav Sachdeva

Abstract

Agile Methodologies are gaining popularity at lightning pace and have provided software industry a way to deliver products incrementally at a rapid pace. With an attempt to welcome changing requirements and incremental delivery, requirements prioritization becomes vital for the success of the product and thereby the organization. However, prioritizing requirements can become a nightmare for product owners and there is no easy way to create a product backlog, 1 to n list of prioritized requirements. For an organization to succeed, it is crucial to work first on requirements that are not only of high value to the customer but also require minimum cost and effort in order to maximize their Return On Investment (ROI). Agile values and principles talk about software craftsmanship and ways to write good quality code and thereby minimize introduction of any new technical debt. However, no solution is described on how to handle existing technical debt for legacy projects. To maintain a sustainable pace, technical debt needs to be managed efficiently so that teams are not bogged down. This paper talks about estimating priority using planning poker, modified Fibonacci series cards, and provides a multi-phase solution to create a product backlog in order to maximize ROI. This paper also provides a method of handling and prioritizing technical debt and the impact of non-functional requirements on technical debt prioritization. The solution proposed is then substantiated using an industrial case study.

Keywords

Agile Methodologies • Prioritization • Return on Investment (ROI) • Feature Business Value (FBV) • Technical Debt Value (TDV)

53.1 Introduction

Agile methodologies [1] have revolutionized the software industry and have provided various frameworks like Scrum and Extreme Programming (XP) to allow rapid delivery of customer requirements [2–4]. The Agile Manifesto [5] provides four values and twelve principles that form the foundation for all the agile frameworks. For the scope of this paper we are going to concentrate on scrum [6]. The term scrum comes from rugby and means tight formation. Scrum is a framework for developing and managing complex projects and is based on transparency, inspection, and adaption [7]. In scrum focus is on working on highest priority items in a sprint, varies from 1 to 4 weeks with preference towards a smaller time period [8], getting customer feedback frequently, and adjusting based on the feedback.

Figure 53.1 shows the project management triangle [9]. Scope, cost, and time are the three variables that determine the quality of a project. In scrum, cost and time are kept

V. Sachdeva (✉)
Department of Computer Science, The University of Texas at Dallas, Richardson, TX 75080, USA
e-mail: vaibhav.sachdeva@utdallas.edu

© Springer International Publishing AG 2018
S. Latifi (ed.), *Information Technology – New Generations*, Advances in Intelligent Systems and Computing 558, DOI 10.1007/978-3-319-54978-1_53

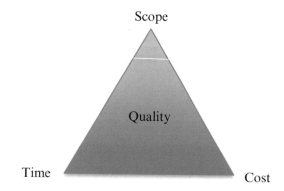

Fig. 53.1 Project management triangle

constant and scope is made variable, hence it is important to deliver requirements to customers that not only deliver maximum value to customer but also maximize ROI.

There are various requirements classification techniques available [10–12], however in literature, requirements are broadly divided into two categories: Functional Requirements (FRs) and Non-Functional Requirements (NFRs). In FRs, focus is on the functionality [13] or behavior [14] of the system. For example, withdrawal from a bank account is a FR. NFRs [15] are soft goals and define the characteristics, quality, constraints of the product and process. For example, the withdrawal transaction should take less than 2 min is a NFR. In industry, there is another way to classify requirements: customer facing requirements and technical debt. Customer facing requirements are the same as FRs. Technical debt is generally the consequence of trade offs made in order to meet time deadlines [16]. For example, build environment improvement is technical debt.

Requirements prioritization is a vital step in any product development and requires immense time and market research and involves major challenges [17]. In scrum, the prioritized requirements are stored in a product backlog [8]. High-level requirements are known as features in scrum. Features are further broken down into user stories [18], which are vertical slices of requirements that are small enough to fit in a sprint and provide incremental customer value. During sprint planning, scrum teams use planning poker to estimate relative effort of each user story and then pull user stories into their sprint backlog based on their capacity and velocity. Once the sprint is over, the user stories completed within the sprint are demoed to the customers to get their feedback. Any feedback is converted into more user stories and is added back into the product backlog by the product owner in a prioritized order. Figure 53.2 shows the flow of user stories in and out of the product backlog.

Let us consider an online shopping portal product backlog as shown in Table 53.1. The product backlog is a combination of customer facing features and technical debt.

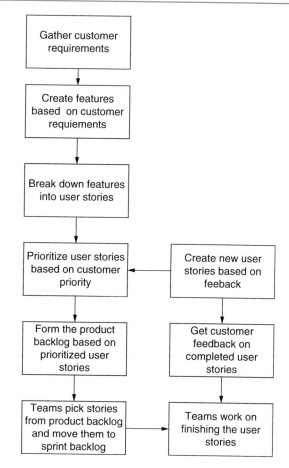

Fig. 53.2 Product backlog flow

Table 53.1 Online shopping portal requirements

Products requirements	Priority
View items	1
Add items to cart	2
View cart	3
Pay for items	4
Track order	5
Login	6
Search for items	7
Sort items	8
Add promo code	9
Add items to wish list	10
Buy from wish list	11
Improve deployment model	12
Enhance database design	13
Automation support for browser	14
Add unit test cases	15
Add system logs	16

Creating the product backlog requires lot of effort and time, as prioritization is a challenging task. Features and user stories in the product backlog are prioritized in a priority order based on customer needs. However, for any successful business, focus needs to be on delivering items with maximum ROI and not only priority. ROI is defined as the return on an investment as compared to its cost, as in

$$ROI = ((Return - Cost) \div Cost) \times 100 \quad (53.1)$$

Therefore, requirements prioritization in agile is very crucial but product owners also need to focus on ROI. This paper utilizes planning poker, a simple technique to help compare items relatively rather than absolutely and uses a modified Fibonacci series, to prioritize features relatively to each other rather than providing them an absolute priority. This paper provides an elegant multi-phase solution to handle and prioritize customer facing requirements and technical debt based on ROI and how NFRs play a vital role in the prioritization process.

This paper is structured as follows. Section 53.2 talks about some existing prioritization techniques and issues with the current approach. Section 53.3 uses planning poker for feature prioritization and proposes a multi-phase approach to prioritize backlog based on ROI. Section 53.4 presents an industrial case study that employs the proposed solution for product backlog prioritization and how it affects priority and business. Section 53.5 talks about some threats to validity. Section 53.6 portrays the future work involved and conclusions are drawn in Section 53.7.

53.2 Existing Techniques and Challenges

Patrick Lencioni rightly said, "If everything is important, then nothing is." In a world where customers find everything important, for a business selling products/solutions/software to these customers, it is vital to prioritize the requirements to provide maximum customer satisfaction and business value. When it comes to prioritization of requirements, there is no formal approach available. Mostly prioritization is followed by gut feelings and hunch of the managers. Scrum talks about prioritizing user stories in the product backlog based on their priority to the customer [8]. However, when multiple customers are involved, priorities clash and organizations need to focus on features that will provide them maximum ROI. Assigning absolute priorities to features can be very tedious. As shown in Table 53.1, requirements are prioritized 1 to n. Some techniques like sample selection [19], cost-value approach [20], group recommendation [21], and client-driven [22] are aimed towards customer priority rather than maximizing ROI and applicability of all these techniques to agile is unknown. The use of these techniques involves tremendous effort and is not worth the precision and accuracy that is required.

Legacy projects that move from other existing software development methodologies like waterfall to agile come with vast technical debt. Also it is impossible to dodge technical debt to projects that emerge directly from agile methodologies. Prioritizing technical debt is really important to maintain a good balance in the organization. Agile frameworks like scrum do not handle technical debt explicitly. Generally it is maintained with customer requirements in a common product backlog and often overlooked. As seen in Table 53.1, a common backlog is used and all technical debt items have been given low priority and are most likely to be disregarded for a long time.

53.3 Multi-phase Solution

The solution suggested in this paper involves maintaining separate product backlogs for customer facing requirements and technical debt, as there is no easy way to compare the priority of the two items. Also maintaining different backlogs will bring visibility to technical debt and serve as a conversation starter for negotiation. For prioritization of requirements, planning poker will be used.

The solution has been divided into two categories. Section 53.3.1 provides a four-phase approach to prioritize customer-facing requirements. Section 53.3.2 provides a solution to prioritize technical debt and discussed the role of NFRs in prioritization.

53.3.1 Prioritizing Customer Facing Requirement

Agile introduced the concept of planning poker to estimate relative effort of user stories. Planning poker is a consensus based estimating technique [23]. It uses planning poker cards that are based on a modified Fibonacci series. The series, not restricted to, generally constitutes of 1, 3, 5, 8, 13, 21, and so on. Planning poker involves relative comparison rather than assigning an absolute value to any item. Planning poker helps in more accurate estimates due to involvement of multiple experts in the estimation and negotiation until consensus is reached. A major portion of planning poker relies on conversation that leads to multiple players agreeing on the same value.

Phase 1 of the solution requires the use of planning poker to estimate the priority of the features in the product backlog by the product owners. This will involve selecting a base feature and finding the relative priority of other features as compared to the base feature. Each player will hold a card and reveal it to everyone at the same time. If the value of the cards is same, that is the priority of the feature being estimated. If the values differ, players with extreme values negotiate and cards are redrawn until consensus is reached. At the end of the game, all features will have a value associated with them that is relative to all other features. The number associated with each feature is known as the Feature Business Value (FBV).

Phase 2 requires the engineering team to use the same planning poker technique to calculate the relative effort involved in completing the features. As features consist of user stories, the effort required to complete a feature is the sum of the effort required for the underlying user stories i.e.

$$Effort\ (F_i) = \sum Effort(User\ Stories_i) \qquad (53.2)$$

The important point to consider is that the effort is relative and the units of measurements need to be same as priority in order to calculate ROI. The effort calculation should be done independent of priority estimate in order to avoid biasing.

Phase 3 requires calculate the ROI of each feature. If F_i denotes the ith feature, calculating ROI can be easily done using Eq. (53.3).

$$ROI\ (F_i) = FBV\ (F_i) \div Effort\ (F_i) \qquad (53.3)$$

Phase 4 requires prioritizing the features in the backlog based on decreasing ROI values calculated in Phase 3. Figure 53.3 describes the process diagram for the same.

53.3.2 Prioritizing Technical Debt

Time pressure on projects generally leads to cutting corners, taking shortcuts, which ultimately leads to addition of technical debt [24]. Legacy projects generally tend to come along with enormous technical debt and managing it is crucial for the success of any business. As technical debt is separate from customer facing requirements, the advised solution is to maintain a separate backlog for technical debt that should be prioritized within it self in order to maximize ROI. This also gives explicit visibility to technical debt so that it is not ignored. NFRs play a vital role in prioritizing technical debt and need to be handled carefully. The solution includes a five-phase approach.

Phase 1 includes determining the priority of the technical debt items relative to each other. As described in Sect. 53.3.1, planning poker is an easy and efficient way to prioritize items relative to each other. As technical debt impacts engineering teams, it is essential to include product owners and engineering team during the planning poker game. One big misconception that product owners have is that technical debt has no impact to the customers or product and therefore does not require their attention. Involving product owners during planning poker will be the conversation starter to understand the impact of technical debt to the overall product and thereby the customers. Technical Debt Value (TDV)

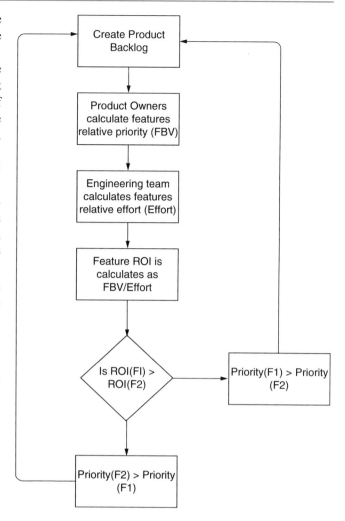

Fig. 53.3 Customer requirements prioritization process diagram

denotes the value assigned to each item during planning poker game.

Phase 2 involves estimating the effort involved in completing the technical debt items. Similar to Sect. 53.3.1, this uses planning poker and only includes the engineering team, as they are the ones executing the tasks. The effort required for a feature would be the cumulative sum of the effort required for the underlying user stories and Eq. (53.2) is used for the same.

An important aspect of requirements is NFRs, which are generally ignored due to the focus on FRs and speedy delivery of product. NFRs like security, performance, maintainability, usability, etc. do not define specific behavior or functionality of the product and are rather used to measure some specific characteristics of the product or process under construction. NFRs are a vital part of any product and along with FRs complete the system. Phase 3, a vital phase of the solution involves understanding the high level NFRs that are affected by each technical debt item and get the total count. This count is denoted as NFRs Count (NFC). More than

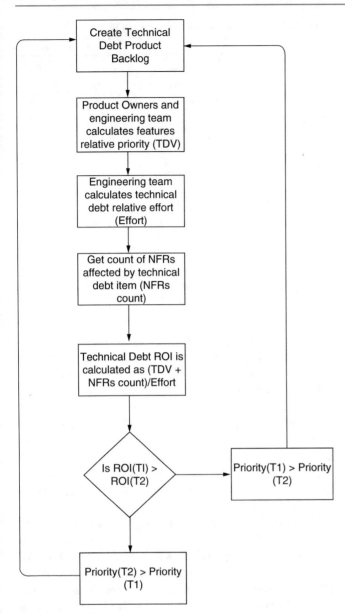

Fig. 53.4 Technical debt prioritization process diagram

Table 53.2 Release MVP

F17869	F17654	F13467	F13461	F21346
US341	US675	US1443	US243	US5436
US342	US13785	US1642	US354	US5555
US346	US13789	US1643	US498	US5614
US451		US1644	US492	US5617
US15673		US1645	US289	US5730
		US1647		US5779
		US1648		US5780
		US1649		

Table 53.3 Product backlog based on priority

Priority	Feature ID
1	F13467
2	F17654
3	F13461
4	F17869
5	F21346
6	T54673
7	T53216
8	T3719
9	T7716

53.4 Industrial Case Study

In this section, I am using a multinational IT organization's software project to describe the use of the solution proposed in this paper to prioritize customer facing requirements and technical debt in two separate backlogs. The software is used to maintain test cases and their results for a call manager that is used by different products. Table 53.2 shows the features that are a part of the product backlog. Each feature has multiple user stories in it. The user stories is yellow constitute the Minimum Viable Product (MVP) for the next release. MVP is a product with the right amount of features to generate revenue and allow better understanding of the market instead of delivering the entire set of features [25].

The product backlog as shown in Table 53.2 just represents customer-facing features. Technical debt, that is the part of the same backlog, before the application of the proposed solution, has been ignored for the release and is the lowest priority item. Such kind of prioritization hurts the product in the long term. The product owners prioritize the product backlog based on customer requirements and come up with a 1 to n list of features in a prioritized order. Table 53.3 shows the product backlog that the product owners came up with based on a priority order.

The first step of the solution is to divide the backlog into two backlogs: one for customer requirements and the other one for technical debt.

defining what each NFR means, it is important that each stakeholder understands the involved NFR and there are no discrepancies. Focus needs to be laid on communication rather than definition.

Phase 4 involves calculating the ROI of each technical debt item. If T_i denotes i^{th} technical debt item, ROI can be calculate using

$$ROI\ (T_i) = (TDV\ (T_i) + NFC\ (T_i)) \div Effort\ (T_i) \quad (53.4)$$

Phase 5 involves prioritizing the technical debt backlog in a descending order of ROI calculated in 53.4.

Figure 53.4 describes the process diagram involved in technical debt prioritization.

Table 53.4 ROI calculation

Feature ID	FBV	Effort	ROI = FBV/Effort
F13467	13	39	0.333
F17654	13	6	2.167
F13461	8	15	0.533
F17869	5	32	0.156
F21346	5	29	0.172

Table 53.5 Product backlog based on ROI

Priority	Feature ID
1	F17654
2	F13461
3	F13467
4	F21346
5	F17869

Table 53.6 Technical dept backlog ROI calculations

Tech Debt ID	TDV	Effort	NFC	ROI = (TDV + NFC)/Effort
T54673	13	18	7	1.111
T53216	8	9	4	1.333
T3719	8	21	2	0.476
T7716	5	13	3	0.615

As proposed in Sect. 53.3 of this paper, planning poker can be used to calculate FBV. To calculate effort involved for each feature, effort of all user stories that are part of the MVP for the release is summed up and the resulting number is the total effort required to complete the required feature for the release. For example, for feature F13461, there are 5 user stories with effort estimate of 3 user story points each. As all user stories are required for the release, the total effort as shown in Table 53.3 is 15 user story points. Effort for all other features is calculated in a similar manner.

Equation (53.2) is then applied to calculate the ROI for each feature. Table 53.4 shows the FBV, effort, and ROI for each feature. Once the ROI has been calculated, the features are prioritized in a descending order of ROI as shown in Table 53.5. The contrast in Tables 53.3 and 53.5 is clearly evident. A feature that has highest customer priority might not provide maximum ROI. For a business to be successful, it is important to work on items with maximum ROI.

The second part of the solution requires prioritizing the technical debt backlog. Product owners and engineering team play planning poker to determine TDV.

Similar to customer backlog, effort is calculated by summing up effort of all user stories in the technical debt feature. For example T3719 has 3 user stories with effort as 13, 3, and 5 user story points respectively. Therefore the total effort required is 21 user story points. Table 53.6 shows the product backlog for technical debt. The high level NFRS

Table 53.7 Technical dept backlog based on ROI

Priority	Tech debt ID
1	T53216
2	T54673
3	T7716
4	T3719

Table 53.8 Revised MVP for release

T53216	F17654	F13467	F13461	T54763	F21346
US5918	US675	US1443	US243	US878	US5436
US5900	US13785	US1642	US354	US888	US5555
US6111	US13789	US1643	US498	US890	US5614
		US1644	US492		US5617
		US1645	US289		US5730
		US1647			US5779
		US1648			US5780

been considered for this project include: security, performance, usability, maintainability, modifiability, reliability, portability, stability, and testability. Each technical debt user story is measured against these NFRS to get NFC. For eg. T53216 impacts security, performance, stability, and reliability and therefore NFC is 4. NFC for all technical debt items is calculated in a similar manner. ROI is calculated using Eq. (53.4) as shown in Table 53.6.

Once the ROI has been calculated the technical debt items are prioritized in a descending order of their ROI as shown in Table 53.7.

So now we have two backlogs prioritized in order to maximize ROI. The next step involves deciding what goes into the sprint backlog during sprint planning meeting. The easiest way to do this is to have a conversation with the product owners. As they were already involved in determining TDV, they technical debt items will not be ignored. In the case study, as shown in Table 53.8, MVP for the release was modified. F17869 was dropped due to its low ROI and T53216 was added.

53.5   Threats to Validity

The solution assumes that FBV, TDV and effort are calculated in vacuum to prevent any bias. However, it is impossible to suggest that no bias was present in the project taken under consideration. Further, the solution has been verified with just one industrial project. The solution needs to be applied to a variety of projects to confirm its validity. There is no easy way to determine the actual impact on revenue by the proposed solution and requires further research.

53.6 Future Work

The solution described in this paper does not handle requirement dependencies. The paper does not describe what a good mix of technical debt and customer-facing items looks like. Once the two backlogs have been prioritized, further research needs to be done to understand what will constitute the sprint backlog, as it will be a combination of the two backlogs. Moreover, the priority calculated based on ROI will be used as a guideline and will require some human intervention. This paper doesn't handle risk explicitly while prioritizing the backlog. Interleaving of risk and priority needs to be carefully handled and requires extra research. Also the solution has been verified with a single project as of now with keeping threats to validity in mind. Applying the solution to more industrial projects will help explore the limitations and improve the solution.

53.7 Conclusion

The focus in agile needs to shift from working on highest priority items to items that provides maximum ROI for maximum business success. Priority and effort need to be relative instead of absolute and planning poker is a powerful tool that can be used for estimation. In order to deliver products quickly to customers, organizations should not disregard the importance of working on technical debt items. Product owners need to apprehend the impact of technical debt items on the product and help prioritize them accordingly. NFRs play a vital role in prioritizing technical debt items. A good balance of technical debt items and customer facing items can help a business achieve maximum success in the long term.

Acknowledgment The author thanks Dr. Lawrence Chung for his constant guidance, valuable ideas, and various references.

References

1. Highsmith, J., & Cockburn, A. (2001). Agile software development: The business of innovation. *Computer, 34*(9), 120–127.
2. Marcal, A. S. C., Furtado Soares, F. S., & Belchior, A. D. (2007). Mapping CMMI project management process areas to SCRUM practices. Software Engineering Workshop, 2007. SEW 2007. 31st IEEE. IEEE.
3. Abrahamsson, P. et al. (2003). New directions on agile methods: a comparative analysis. Software Engineering, 2003. *Proceedings. 25th International Conference on Software Engineering.* IEEE, pp. 244–254.
4. Beck, K. (1999). Embracing change with extreme programming. *Computer, 32*(10), 70–77.
5. Beck, K. et al. (2001). Manifesto for agile software development. Available: http://agilemanifesto.org.
6. Sutherland, J., Schwaber, K. (2007). The scrum papers: nuts, bolts, and origins of an agile process. Citeseer, [Online]. Available: http://scrumfoundation.com/library.
7. Schwaber, K. (2004). *Agile project management with scrum.* Redmond: Microsoft Press.
8. Schwaber, K., & Sutherland, J. (2011). The scrum guide. Scrum Alliance. [Online]. Available: https://www.scrum.org/resources/scrum-guide.
9. Atkinson, R. (1999). Project management: cost, time and quality, two best guesses and a phenomenon, its time to accept other success criteria. *International Journal of Project Management, 17*(6), 337–342.
10. IEEE Computer Society. (1998). Software Engineering Standards Committee, and IEEE-SA Standards Board. IEEE Recommended Practice for Software Requirements Specifications. Institute of Electrical and Electronics Engineers.
11. Kotonya, G., & Sommerville, I. (1998). Requirements engineering: processes and techniques. Wiley Publishing, Chichester.
12. Van Lamsweerde, A. (2001). Goal-oriented requirements engineering: A guided tour. Requirements Engineering, 2001. *Proceedings. Fifth IEEE International Symposium on Requirements Engineering.* IEEE. p. 249, August 27–31, 2001.
13. IEEE Standards Association. (1990). Standard glossary of software engineering terminology. IEEE Std: 610–12.
14. Anton, A. I. (1997). Goal identification and refinement in the specification of software-based information systems.
15. Jarke, M., et al. (1993). Theories underlying requirements engineering: An overview of Nature at genesis. Requirements Engineering, 1993. *Proceedings of IEEE International Symposium on.* IEEE.
16. Lim, E., Taksande, N., & Seaman, C. (2012). A balancing act: what software practitioners have to say about technical debt. *IEEE Software, 29*(6), 22–27.
17. Babar, M. I., Ramzan, M., & Ghayyur, S. A. K. (2011). Challenges and future trends in software requirements prioritization. Computer Networks and Information Technology (ICCNIT), 2011 International Conference on. IEEE. pp. 319–324.
18. Cohn, M. (2004). User stories applied: For agile software development. Addison-Wesley Professional.
19. Shao, P., Sample selection: An algorithm for requirements prioritization, ACM-SE 46 Proceedings of the 46th Annual Southeast Regional Conference on XX (pp. 525–526).
20. Ryan, K., & Karlsson, J. (1997). Prioritizing software requirements in an industrial setting. *Proceedings of the 19th international conference on software engineering.* ACM.
21. Felfernig, A., & Ninaus, G. (2012). Group recommendation algorithms for requirements prioritization. *Proceedings of the third international workshop on recommendation systems for software engineering.* IEEE Press.
22. Racheva, Z., Daneva, M., & Herrmann, A. (2010). A conceptual model of client-driven agile requirements prioritization: Results of a case study. *Proceedings of the 2010 acm-ieee international symposium on empirical software engineering and measurement.* ACM.
23. Grenning, J. (2002). Planning poker or how to avoid analysis paralysis while release planning. Hawthorn Woods: Renaissance Software Consulting 3 (2002).
24. Kruchten, P., Robert L. Nord, & Ozkaya, I. (2012). Technical Debt: From Metaphor to Theory and Practice. Ieee software 29.6 (2012).
25. Ries, E. (2009). Minimum viable product: a guide. Startup Lessons Learned.

EasyTest: An Approach for Automatic Test Cases Generation from UML Activity Diagrams

Fernando Augusto Diniz Teixeira and Glaucia Braga e Silva

Abstract

The test cases generation is one of the great challenges for the Software Test Community because of the development efforts and costs to create, validate and test a large number of test cases. The automation of this process increases testing productivity and reduce labor hours. One technique that has been adopted to automate test cases generation is Model Based Testing (MBT). This paper proposes the EasyTest approach to generate test cases from UML Activity Diagrams aiming to integrate Modeling, Coding and Test stages in a software process and to reduce costs and development efforts. The proposed approach suggests an early detection of defects even in the modeling stage to prevent that unidentified defects are embedded in the coding stage. The work also presents the use of the generated test cases before and after the coding stage. To verify the proposed approach, this work also presents the EasyTest Tool to provide interoperability with the JUnit framework.

Keywords

Test Automation • Model Based Testing • Gray- Box Testing • JUnit • TDD

54.1 Introduction

Empirical studies show that the test activity consumes a significant part of costs and development efforts and this expense becomes often cost-time prohibitive. The test cases generation represents a complex task in the test process because of the development effort and cost to create, validate and test a large number of test cases. To deal with this complexity some tools automate the generation of test cases from produced source code. Despite the benefits with automation, this technique still has problems to verify all system functionalities scenarios [1]. Furthermore, once a software defect is found the code must be fixed and verified again. This process causes rework, delays and increases the development costs.

Assuming that the earlier a defect is found the cheaper it is to fix it, the Model-Based Testing (MBT) technique appears to be promising. MBT consists of using various types of formal models to derive a set of test cases. MBT has a better performance than code-based approaches because it is a mixed approach of source code and requirements specification for testing the software [2]. Furthermore, MBT is more promising in terms of cost because the generated test cases can be used as starting point for the construction of a defect-free code. Unified Modeling Language (UML) models have long been used with the MBT technique because they have a lot of relevant information about system specification for test case design [3].

We propose an approach called "EasyTest" to generate test cases from UML activity diagrams aiming to integrate Modeling, Coding and Test stages in a software process and to reduce costs and development efforts. The EasyTest approach suggests an early detection of defects even in the modeling stage to prevent that unidentified defects are

F.A.D. Teixeira (✉) • G. Braga e Silva
Institute of Technologics and Exact Sciences Federal, University of Vicosa – UFV, Florestal, Minas Gerais, Brazil
e-mail: fernando.augusto@ufv.br; glaucia@ufv.br

© Springer International Publishing AG 2018
S. Latifi (ed.), *Information Technology – New Generations*, Advances in Intelligent Systems and Computing 558, DOI 10.1007/978-3-319-54978-1_54

embedded in the coding stage. The generated test cases can be used in two scenarios: before and after the coding stage. If no code was produced, the test cases generated from the UML activity diagrams can be used by developers to produce defect-free code with Test Driven Development (TDD). Otherwise, after the coding stage, the generated test cases can be executed to verify (according to model specifications) and validate (expected behavior of an operation) the produced code in a test execution tool. To verify and to automate the proposed approach this work also presents the EasyTest Tool that provides interoperability with the JUnit frame-work.

The paper is organized as follows: In Sect. 54.2, some related works are discussed. The EasyTest approach is presented in Sect. 54.3. Section 54.4 addresses the EasyTest tool. Finally, in Sect. 54.5, we discuss some conclusion and future works.

54.2   Related Works

There are different techniques and tools proposed for the automated generation of test cases. Some tools generate test cases from source code, but as mentioned by Shamshiri et al. [1] some tools did not present good test results for all the required functionalities.

Several studies address the use of MBT technique to generate test cases from different types of design and specification models. Nebut et al. [4] propose a test case generation approach from UML Use Case models and emphasize problems with interpretation caused by the use of natural language. Some researches address the generation of test cases from UML Class models and Object Constraint Language (OCL) specifications [5, 6].

Some works present techniques for transforming UML activity diagrams into graphs to generate test cases [7, 8] and provide tools to validate their proposals. However, these studies did not show a way to integrate the technique with other stages of a software process.

Pakinam et al. [3] extend the work of Linzhang et al. [7] to provide an architecture for automated generation of test cases from UML Activity Diagrams although they don't present an implementation of the proposal. Jena et al. [2] propose an approach similar to the work of Pakinam et al. [3] which uses a genetic algorithm to reduce the number of test cases while keeping the same coverage.

Test case generation techniques are commonly evaluated regarding the coverage criterion [9–11]. This criterion is one of the key metrics for evaluating the techniques performance. Other metrics such as time, cost, effort and generation complexity are also considered in the quality evaluation of the generated test cases [12].

The set of generated test cases can be used before the coding stage to produce defect-free code using the Test-Driven Development (TDD) strategy. These test cases provide a meaningful representation of alternative flows and branch conditions, enabling an easier application of the TDD for programmers. Latorre [13] shows how Unit Test-Driven Development (UTDD), a TDD subcategory, has a good performance in learning and application with programmers of different skill levels. Janzen and Saiedian [14] highlight how automated tools for unit tests have assisted improvements of TDD itself.

Based on the work of Linzhang et al. [7], the proposed approach consists of a gray-box test technique. This technique deals with problems which used to be ignored by both black-box and white-box techniques. It extends the logical coverage criteria of white box technique and finds all the possible paths from the model which describes the expected behavior of an operation. Then it generates test cases which can satisfy the path conditions expected by black-box technique.

54.3   EasyTest Approach

This section presents an automatic approach to generate test cases from UML activity diagrams using gray-box technique. The EasyTest approach comprises three phases, as shown in Fig. 54.1: (1) importing activity diagrams in XMI; (2) test cases generation; and (3) applying test cases.

54.3.1   Phase 1: Importing Activity Diagrams in XMI

Phase 1 aims to obtain and extract relevant information, about test case generation process, from XMI activity diagrams and then provide it to Phase 2. These information comprise properties of activities (vertices) and flows (edges). XMI (XML metadata Interchange) standard is defined by OMG (Object Management Group) and is widely used for exchanging metadata between UML modeling tools. Phase 1 uses activity diagrams drawn in different UML modeling tools as long as they have been exported to XMI according to OMG specification.

The EasyTest approach proposes interoperability with different types of UML modeling tools, importing XMI files in order to reuse activity diagrams previously drawn to avoid rework. This strategy is cheaper in terms of effort and time than those applied in other works that require the manual design of activity diagrams in their tools [7, 8].

To illustrate the use of EasyTest approach in a real case of development process, we adopt a case study based on a Control System for *Aedes Aegypti* that is being developed

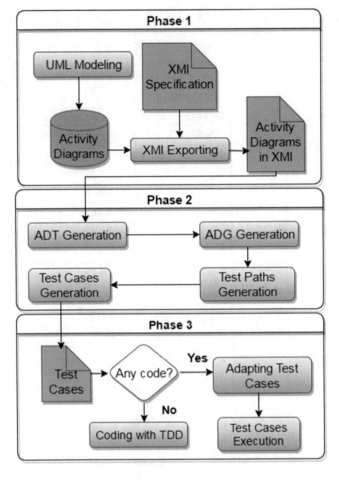

Fig. 54.1 EasyTest Approach

Fig. 54.2 Activity diagram for the operation "Create Complaint"

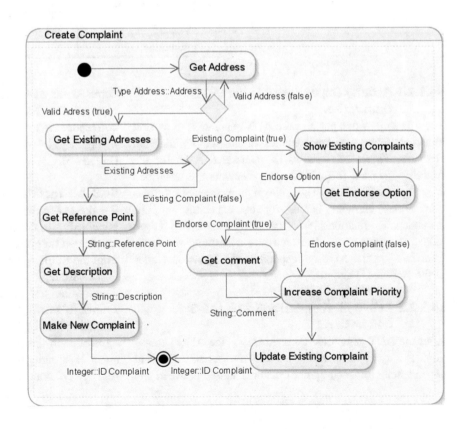

in an academic interdisciplinary project at Federal University of Viçosa – campus Florestal. For Phase 1, an Activity Diagram was produced to represent an use case from this system that refers to an operation called "Create Complaint" used by citizens to report an occurrence of *Aedes Aegypti* vector in an infested area.

The activity diagram mentioned above is illustrated in Fig. 54.2. This diagram was produced in Oracle JDeveloper 12c tool (version 12.1.3.0.0) and exported to XMI format.

It is important to note that some edges in the diagram contain labels with textual stereotypes representing data flow for that edge. For example, the edge "String:Reference Point" contains the variable name "Reference Point" of the type "String" from the activity "Get Reference Point" to "Get Description". Values like that are relevant for the successful generation of test cases in Phase 2.

54.3.2 Phase 2: Test Cases Generation

Phase 2 is responsible for processing the XMI Activity Diagram to find Test Paths and then generate the final test cases. This phase was based on some techniques proposed by [2, 3, 15].

The development of this phase is comprised by four steps that are presented below as well as the applied improvements proposed in this work.

Fig. 54.3 Generated ADT

ID	Element Name	Dependency	Input	Expected Output
0	Initial Node			
1	Get Address	0		Address
		2	Valid Address(false)	Address
2	Validate Address	1	Address	Valid Address(true)
		1	Address	Valid Address(false)
3	Get Existing Addresses	2	Valid Address(true)	
4	Validate Existing Addresses	3		Existing Complaint(true)
		3		Existing Complaint(false)
5	Show Existing Complaints	4	Existing Complaint(true)	
6	Get Reference Point	4	Existing Complaint(false)	Reference Point
7	Get Endorse Option	5		Endorse Option
8	Get Description	6	Reference Point	Description
9	Validate Endorse Option	7	Endorse Option	Endorse Complaint(false)
		7	Endorse Option	Endorse Complaint(true)
10	Make New Complaint	8	Description	ID Complaint
11	Increase Complaint Priority	9	Endorse Complaint(false)	
		12	Comment	
12	Get comment	9	Endorse Complaint(true)	Comment
13	Activity Final Node	14	ID Complaint	
		10	ID Complaint	
14	Update existing complaint	11		ID Complaint

54.3.2.1 Activity Dependency Table (ADT) Generation

The first step of this phase is the ADT generation to present inputs, outputs and dependencies of all activity diagram elements. Figure 54.3 shows the generated ADT for the Activity Diagram (Fig. 54.2). The elements are described at each row of the table. The table columns present information about element name, inputs, expected outputs, dependencies relationships between elements and edges values in some cases. The elements containing the term. "Validate" in the ADT correspond to the Decision Nodes of the Activity Diagram.

54.3.2.2 Activity Dependency Graph (ADG) Generation

After the ADT generation the next step is the ADG generation. The symbols in ADT become vertices in ADG and dependencies of each symbol become edges between the corresponding vertices. The ADG generation is a significant part of the EasyTest approach to gather test paths corresponding to the activity diagram.

Figure 54.4 shows the generated ADG for the ADT (Fig. 54.3).

54.3.2.3 Test Paths Generation

For the generation of Test Paths from the ADG generated previously, we developed an algorithm based on the depth-first search for graphs. The final set of Test Cases are defined from the resulting test paths.

The strategy used to generate these Test Paths impacts directly the test cases quality. The algorithm was built aiming to satisfy the criterion "Transition Coverage". As defined in [9, 16], Transition Coverage intends to verify all transitions in the activity diagram. The transition coverage result is a ratio between the verified transitions and all Activity Diagram transitions. This criterion guarantees that each loop condition is checked and an iteration is executed at least one time. By this way, we also guarantee that other coverage criteria as "Activity Coverage" and "Branch Coverage" are satisfied. The "Activity Coverage" intends to verify all Activity States in the Activity Diagram. The "Branch Coverage" intends to verify all conditions branches in the Activity Diagram so that all edges are checked.

Based on the "Transition Coverage" and using the generated ADG (Fig. 54.4) the following Test Paths were obtained:

Test Path 1: 0, 1, 2, 3, 4, 5, 7, 9, 11, 14, 13
Test Path 2: 0, 1, 2, 3, 4, 5, 7, 9, 12, 11, 14, 13
Test Path 3: 0, 1, 2, 3, 4, 6, 8, 10, 13
Test Path 4: 0, 1, 2, 1, 2, 3, 4, 5, 7, 9, 11, 14, 13

Another criterion that could be considered is "All Activity Path Coverage". This criterion intends to verify all different activity sequences from the Initial Node to the Final Node. Considering this criterion for the ADG shown in Fig. 54.4 we would get six Test Paths. The choice of the "Transition Coverage" was based on the fact that if the loop is executed at least one time and a Final Node is reached, other tests for the same loop condition are not required. Thereby, the "Transition Coverage" can provide the same coverage with fewer test cases.

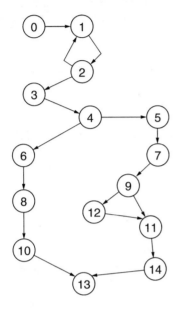

Fig. 54.4 Generated ADG

54.3.2.4 Test Cases Generation

The last step of Phase 2 is the test cases generation from the ADT and Test Paths. According to the number of Test Paths, this step generated four Test Cases for the given Example. Figure 54.5 shows one of these generated test cases in Phase 2 of the EasyTest approach. It can be observed that all input and output values of each path vertice are detailed as well as the input and output values of the test case.

Fig. 54.5 Example of a generated test case

Test Case Number	Test Path Vertice ID	Vertice Input	Vertice Output	Test Case Input	Test Case Expected Output
1	0			Address::Type Address @Address !=null; Endorse Option:: Boolean @Endorse Option ==false;	ID Complaint:: Integer;
	1		Address		
	2	Address	Valid Address(true)		
	3	Valid Address(true)			
	4		Existing Complaint(true)		
	5	Existing Complaint(true)			
	7		Endorse Option		
	9	Endorse Option	Endorse Complaint(false)		
	11	Endorse Complaint(false)			
	14		ID Complaint		
	13	ID Complaint			

54.3.3 Phase 3: Applying Test Cases

Phase 3 has the purpose of applying the generated test cases in the test execution and coding stages. In the EasyTest approach, the resulting test cases can be used to verify or produce code in any programming languages. Different Test Cases Execution tools can be used in this approach as long as a mapping has been designed to adapt the test cases for the code structure used in the selected tool.

Phase 3 comprises two usage scenarios within the software development process: before and after the coding stage. The possibility of application in different scenarios allows a better adherence of the testing activity in the software development process. The two different scenarios are discussed below.

54.3.3.1 Before Coding Stage: Test Driven Development (TDD)

In this scenario, TDD is applied to build defect-free code. Once the test cases have been generated, in addition to a meaningful representa-tion of alternative flows and branch conditions, the process of developing code becomes much simpler because the code structure will be guided by these resources. As shown by [13], When test driven development is applied with unit tests, we have a good performance in learning and application with programmers of different skill levels.

Although in the TDD context models are not usually created, the EasyTest approach can provide advantages in terms of time and effort because of the automatic generation of test cases. In addition, the approach can help programmers to produce defect-free code from the beginning of the development software process because they can use test cases to verify all functionalities and their flows, thus compensating for the time required to develop the activity diagrams.

To verify this scenario, the authors made a quick single experiment to develop a defect-free code applying TDD for the operation "Create Complaint". For the experiment, the code was produced based on test cases generated by the approach and the activity diagram (Fig. 54.2). From these resources, it was possible to quickly identify conditional and iterative blocks as well as the possible values for each flow.

For the experiment, we used EClEmma, a free Java code coverage tool. This tool was used to evaluate the line coverage of the produced code and a coverage of nearly 92% was achieved. This result highlights the quality of the generated test cases and how much the code is compatible with the activity diagram. However, other experiments will be made to check the results for different contexts.

54.3.3.2 After Coding Stage: Exporting Generated Test Cases

In this scenario, we have an activity diagram and a code made for it. The source code is used to verify if it matches with the results of the test cases execution, that is, if the code execution logic is consistent with the activity diagram expected behavior. Otherwise, it is necessary to evaluate the changes required in the code to satisfy the modeling specifications. As long as the code is compatible with the model we have a valid test for all code functionalities and flows.

Also, we have an automatic generation of test code that comprises the generated test cases for execution in any Test Execution Tool. To achieve this, it is required a mapping between the generated test cases and the technical structure of the selected tool.

Once the activity diagram is source code independent, we assign generic names to classes and operations which will later be adjusted by testers/programmers.

54.4 EasyTest Tool

To verify the EasyTest approach, we developed a Java tool called "EasyTest tool" that provides automatic support for the Phases 2 and 3. According to the Phase 1 of the approach, EasyTest Tool uses activity Diagrams represented in XMI files. This tool version provides, in step "Adapting Test Cases", test cases for Java through a mapping for integration with the JUnit framework.

The EasyTest tool provides an interface where for each generated test cases, specific instructions and fields are displayed so the tester/programmer can insert the required test data. In addition, the interface displays a dynamic graph highlighting corresponding Test Paths for each Test Case to support the filling of fields by the user. With these resources the tool can generate a code for JUnit capable to test all functionalities and flows, reducing efforts and time of creating test cases and increasing quality of test cases in spite programmers of different skill levels.

54.5 Conclusion and Future Work

This work presented the EasyTest approach of automatic test cases generation from UML activity diagrams and addressed the use of these test cases in two scenarios: before and after the coding stage. For the first scenario, the approach supplies TDD programmers with relevant information about the functionality logic in order to ease the process of developing a defect-free code. For the second scenario, the approach proposes an automatic generation of test code that comprises the test cases to execute them in any Test Execution Tool. To verify this scenario, the EasyTest tool was developed to provide interoperability with JUnit.

The EasyTest is based on gray-box technique, allowing the identification of problems which used to be ignored by both black-box and white-box techniques as it extends the logical coverage criteria and finds all the paths from the designed model, which describes the expected behavior of an operation.

Based on some experiments and observations, the EasyTest approach can provide some gains in terms of cost, effort and time. This gains can be observed in contexts such as: reuse of activity diagrams previously designed; earlier detection of defects before coding stage; support for the defect-free code development; and increase in the test case quality because the use of a gray-box technique.

For future works, we intend to make more experiments for the two scenarios of the EasyTest Phase 3, involving a set of senior testers/programmers to evaluate and refine the EasyTest approach.

References

1. Shamshiri, S., Just, R., Rojas, J. M., Fraser, G., McMinn, P., & Arcuri, A. (2015). Do automatically generated unit tests find real faults? an empirical study of effectiveness and challenges (t). In *Automated software engineering (ASE), 2015 30th IEEE/ACM international conference on* (pp. 201–211). Lincoln, Nebraska: IEEE.
2. Jena, A. K., Swain, S. K., & Mohapatra, D. P. (2014). A novel approach for test case generation from uml activity diagram. In *Issues and challenges in intelligent computing techniques (ICICT), 2014 international conference on* (pp. 621–629). Ghaziabad: IEEE.
3. Boghdady, P. N., Badr, N. L., Hashem, M., & Tolba, M. F. (2011). A proposed test case generation technique based on activity diagrams. *International Journal of Engineering & Technology IJET-IJENS, 11*(03), 35–52.
4. Nebut, C., Fleurey, F., Le Traon, Y., & Jezequel, J.-M. (2006). Automatic test generation: A use case driven approach. *IEEE Transactions on Software Engineering, 32*(3), 140–155.
5. Chang, C.-K. & Lin, N.-W. (2015). Utgen: A black-box method-level unit-test generator for junit test-platform. In *Trustworthy systems and their applications (TSA), 2015 second international conference on* (pp. 1–7). Hualien: IEEE.
6. Bouquet, F., Grandpierre, C., Legeard, B., Peureux, F., Vacelet, N., & Utting, M. (2007). A subset of precise uml for model-based testing. In *Proceedings of the 3rd international workshop on advances in model-based testing* (pp. 95–104). London: ACM.
7. Linzhang, W., Jiesong, Y., Xiaofeng, Y., Jun, H., Xuandong, L., & Z. Guoliang (2004). Generating test cases from uml activity diagram based on gray-box method. In *Software engineering conference, 2004. 11th Asia-Pacific* (pp. 284–291). Busan: IEEE.
8. Mingsong, C., Xiaokang, Q., & Xuandong, L. (2006). Automatic test case generation for uml activity diagrams. In *Proceedings of the 2006 international workshop on automation of software test* (pp. 2–8). Shanghai: ACM.
9. Chen, M., Mishra, P., & Kalita, D. (2008). Coverage-driven automatic test generation for uml activity diagrams. In *Proceedings of the 18th ACM great lakes symposium on VLSI* (pp.139–142). Orlando: ACM.
10. Chen, M., Qiu, X., Xu, W., Wang, L., Zhao, J., & Li, X. (2009). Uml activity diagram-based automatic test case generation for java programs. *The Computer Journal, 52*(5), 545–556.
11. McQuillan, J. A., & Power, J. F. (2005). A survey of uml-based coverage criteria for software testing. In *Department of computer science*. Kildare: NUI Maynooth Co..
12. Nirpal, P. B., & Kale, K. (2011). A brief overview of software testing metrics. *International Journal on Computer Science and Engineering (IJCSE), 3*(1), 204–211.
13. Latorre, R. (2014). Effects of developer experience on learning and applying unit test-driven development. *IEEE Transactions on Software Engineering, 40*(4), 381–395.
14. Janzen, D. S., & Saiedian, H. (2005). Test-driven development: Concepts, taxonomy, and future direction. *Computer, 38*(9), 43–50.
15. Boghdady, P. N., Badr, N. L., Hashim, M. A., & Tolba, M. F. 2011. An enhanced test case generation technique based on activity diagrams. In *Computer engineering & systems (ICCES), 2011 international conference on* (pp. 289–294). Cairo: IEEE.
16. Swain, R. K., Panthi, V., & Beher, P. K. (2013). Generation of test cases using activity diagram. *International journal of computer science and informatics, 2*(2), 2231–5292.

An Integrated Academic System Prototype Using Accidents and Crises Management as PBL

Lais S. Siles, Mayara V.M. Santos, Romulo A. Rodrigues, Lineu A.L. Filho, João P.T. Siles, Renê Esteves Maria, Johnny C. Marques, Luiz A.V. Dias, and Adilson M. da Cunha

Abstract

This paper aims to describe the agile development of an integrated system for accidents and crises management. This academic project prototype was developed at the Brazilian Aeronautics Institute of Technology, on the second Semester of 2015. The project has involved 80 undergraduate and graduate students at the same time from four different electronic and computer engineering courses. The Scrum Framework was combined with Problem-Based Learning (PBL), to develop a prototype within just 17 academic weeks. The prototype was developed as a Proof of Concept (PoC) and applied within a natural disaster scenario management, involving the four segments of: Civil Defense, Health Care, Fire Department, and Police Department. At the end of the project, it was possible to deliver an integrated academic system project prototype, associating a Control Room with Web Applications connected through Cockpit Display Systems (CDSs). Students were able to work geographically dispersed, using free cloud-based tools, and the Safety Critical Application Development Environment (SCADE), from ®;ANSYS Esterel Technologies, combining multiple types of hardware like Raspberry Pi and Arduino, and different sets of open-source tools.

Keywords

Scrum framework • Problem-based learning • Real-time embedded systems • Safety critical application development environment • Software quality

55.1 Introduction

According to Coombs [1], crisis is an event that represents unpredictable threats. It can have negative effects on organizations, industries, or stakeholders, especially if improperly handled. It can threaten public safety and cause financial and reputation losses.

Large number of people is affected every year by natural disasters, involving human injuries and/or material damages. According to the Annual Disaster Statistical Review [2], from 2004 to 2013, the annual average of people killed by natural disasters reached 99,820 worldwide. In 2014, natural disasters affected over 140.8 million people causing US $99.2 billion on damages.

On this context, an integrated system project prototype was developed and named ACMIS (meaning Accidents and Crises Management Integrated System) project, in Portuguese SIGAC (Sistema Integrado de Gerenciamento de Acidentes e Crises), following the PBL (Problem-Based Learning) model [3], during the second Semester of 2015.

L.S. Siles • M.V.M. Santos • R.A. Rodrigues • L.A.L. Filho
J.P.T. Siles • R.E. Maria • J.C. Marques (✉) • L.A.V. Dias
A.M. da Cunha
Computer Science Division, Aeronautics Institute of Technology –
ITA, São José dos Campos, Brazil, Brazil
e-mail: laiss.siles@gmail.com; mayara.vms@gmail.com;
romuloadmr@gmail.com; lineulimasjc@gmail.com;
joaosiles@gmail.com; rene@ita.br; johnny.marques@gmail.com;
vdias@ita.br; cunha@ita.br

This Project involved the four segments of Civil Defense, Fire Department, Health Care, and Police Department. All of them integrated within a unique and centralized Control Room for managing and monitoring middle proportion accident and/or crises scenarios.

In just one Semester, 80 undergraduate and graduate students from four different courses at the Brazilian Aeronautics Institute of Technology (Instituto Tecnológico de Aeronáutica – ITA) have built an academic system project prototype as a Proof of Concept (PoC). The development has used the ®;ANSYS Esterel Technologies SCADE, a Safety Critical Application Development Environment [4], with Scrum framework [5], within just 17 weeks.

This paper is organized as follows. Section 55.2 describes the background, Sect. 55.3 the ACMIS project overview, Sect. 55.4 its development and main challenges, Sect. 55.5 the ACMIS Project compliance with safety standards, and Sect. 55.6 the concluding remarks and recommendations.

55.2 Background

The ACMIS Project was collaboratively developed at the same time by 80 undergraduate and graduate students from the following four courses: CES-65 Embedded Systems Project; CE-230 Software Quality, Reliability, and Safety; CE-235 Real-time Embedded Systems; and CE-237 Advanced Topics in Software Testing.

Initially, students were divided into 8 interdisciplinary Scrum Teams (STs). At the end of the semester, each ST had to present its built in embedded system with high cohesion and low coupling as a component of the major ACMIS project. Students from STs had to work geographically dispersed, using free cloud-based tools. An ACMIS business project academic partner provided all needed hardware.

At the end of their courses, students had to present and report on homepages their participations on building and integrating the final ACMIS Project prototype process. The assessment of this entire development process was done by Professors playing the roles of the enterprise stakeholders and investors for the construction of the ACMIS Project prototype.

55.2.1 Real-Time Embedded System

A real-time embedded system is a microprocessor system in which the computer is completely encapsulated on the device or system it controls. It performs a set of predefined tasks, normally with specific requirements, and monitoring and/or reacting from hardware behavior [6].

55.2.2 Accidents and Crises Management

Accidents and crises management is a process that can prevent or decrease the damage that a crisis can inflict on organizations and their stakeholders. Public safety must be the primary concern in crisis situations. It can be divided into three phases: pre-crisis, including prevention and preparation; crisis response, involving immediate reaction to a crisis; and post-crisis, involving continuous follow-ups and information [7].

55.2.3 Scrum Framework

The Scrum is an iterative and incremental framework to organize and manage complex problems. It is based on empirical process control and aims to constantly improve its development practices through iterations and feedbacks. It contains predefined roles, artifacts, and ceremonies [8].

55.2.4 SCADE

SCADE is a suite of software tools and programming languages, created through a partnership between Airbus, Schneider Electric, and Verilog enterprises. It provides a development environment for critical applications such as Cockpit Display Systems (CDSs) for aircrafts, railways, nuclear power plants, and other domains through model oriented programming. The SCADE environment can generate certified code by international standards such as DO-178C [9], providing tools for automatically generating, testing, and documenting software [4]. For this ACMIS Project, the ®ANSYS Esterel Technologies has provided ITA with academic licenses for the SCADE Suite 16.2.1.

55.2.5 Arduino and Raspberry Pi Boards

Arduino is an easy to use yet powerful open-source platform that can be used in a great variety of hardware and software projects. It eases interactions between software and many devices such as sensors, motors, LCD displays, among others. An Arduino board can provide processing on inputs turning them into outputs[10]. Raspberry Pi is a small single computer board, released in 2008 by the Raspberry Pi Foundation with the main goal of encouraging young people interaction with the programming and low-end computing world. The Model-B version of the Raspberry has been used on this ACMIS Project. It has enabled the construction of a cheap and versatile platform in which SCADE Applications, Python, and C code could be executed [11].

55.3 The ACMIS Project Setup

During the first week of the ACMIS Project development, an PBL was proposed to the students, involving the knowledge domain of Accidents and Crises Management. In order to apply the Scrum, each student had to choose one of the following Scrum roles to perform: Product Owner (PO), Scrum Master (SM), or Team Developer (TD). A PO, an SM, and about seven TDs comprised each ST.

In order to scale up Scrum, it was adopted the Scrum of Scrums technique [8]. An initial research involved segment selections of accidents and crises scenarios [7]. Naturally, Civil Defense, Health Care, Fire Department, and Police Department were chosen, because they usually should work together on planning and acting during crises. Afterword, each segment was divided into two macro-functions and each one was assigned to a specific ST within a segment. Table 55.1 shows the ACMIS Project distributed functions by Segments, Scrum Teams IDs, and Macro-functions.

Table 55.1 The ACMIS Project Distributed Functions

Segments	Scrum teams IDs	Macro-functions
Civil defense	TS#01	Monitoring room
	TS#05	Coordination
Health care	TS#02	Ambulances
	TS#06	Emergency room
Fire department	TS#03	Search and rescue
	TS#07	Rescue and aftermath
Police department	TS#04	Occurrence report
	TS#08	Preventive security

Figure 55.1 shows the initial ACMIS Project architecture with each segment building one CDS using the SCADE Tool, linked with a Control Room. In this architecture, each CDS sends real-time data for the database to be displayed at the Control Room.

The ACMIS Project architecture was continuously improved, in order to allow the delivery of a single and practical end product from all STs at the end. The next section describes the four developed ACMIS Project sprints.

55.4 The ACMIS Project Sprints

The ACMIS Project had short-time duration, following four Scrum sprints of four time-boxed weeks each. Its development was divided into the following sprints: Sprint 0 provided tutorials in Scrum and SCADE, Sprint 1 provided the first set of functions development on each ST, Sprint 2 the second set of functions development and local integrations, and Sprint 3 provided the final ACMIS Project integration. Each Sprint was performed in four weeks.

55.4.1 Sprint 0

During the Sprint 0, each ST had to face the forming phase [12], which means that each student had to discover their teammates skills. In order to meet with their teammates, students took part in weekly meetings, by using free web conferencing tools like Google Hangout On Air and Skype within four weeks. Through this Sprint 0, each PO had created User Stories (USs) for the Sprint Backlog (SB) discussing each one of them with their teammates, enabling incrementally product enhancements and ST understandings.

Fig. 55.1 The initial ACMIS project architecture

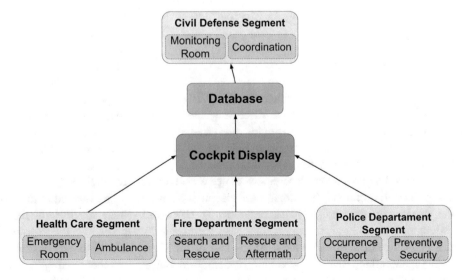

After each student voluntarily has chosen its ST, professors balanced the number of participants on each ST. At the end of this Sprint 0, 8 STs have completed this forming phase. In order to adapt the Scrum of Scrums technique to this Project, two new roles were created and named General Product Owner (GPO) and General Scrum Master (GSM) for keeping project value and the agile method working. Online weekly meetings helped the GPO and the GSM to align STs needs and progresses, facilitating interactions with their POs and SMs. Stakeholders monitored the Project process, by using the common artifacts available from free cloud-based tools like Google Sites, Google Drive, and Google Groups.

After enough iteration, POs collaboratively wrote the ACMIS Project Vision Artifact, as follows:

> To public companies and/or individuals involved in accidents and crises that needs to develop, integrate, and manage computer and embedded systems, the Accidents and Crises Management Integrated System (ACMIS) project is a Real-time Embedded System that provides the academic and interdisciplinary learning skills needed for Computer Engineers. Differently from Universities or Research Institutes as S2ID, Cadenas, WGRA, CGE, Galante, and others, the final product will be academically developed in just 17 weeks, with agility, quality, reliability, safety, and testability, certifiable by the DO-178C.

After four weeks learning the SCADE software tool and interacting with their teammates on weekly meetings, students were ready for collaborating with the proposed project development and implementation. The ACMIS Project source-code management was provided by a free version control named Git. The students chose a web-based Git repository hosting service named GitHub for hosting the Project source-code. Thus, all the developed code became available at the URL address https://github.com/ projetosigac/projetosigac.

At the end of this Sprint 0, it was possible to visualize and identify that students had several background knowledges from the industry such as, application developers, network administrators, database developers, and educators.

55.4.2 Sprint 1

After the initial phase of learning the SCADE software environment, meeting teammates, discussing the Product Backlog, and exploring technologies and methods for accidents and crises management, a reduced set of USs were highly prioritized by the POs within each TS.

The First Planning Meeting ceremony aimed at delivering value through a minimal functional product. Each ST used the Planning Poker technique [4], in order to grant USs effort estimations, creating granular tasks that were then taken by the TDs, according to their skills. At the end of this First Planning Meeting, each ST was able to define its Sprint Backlog (SB).

Since the ACMIS Project students were geographically dispersed and had different time schedules, online platforms usage allowed better communication and collaboration, and all involved artifacts were stored on the GitHub and/or Google Drive.

Every week, students hosted three types of online meetings: **POs Meetings**, to maintain the focus on the product value aligned with professors expectations playing the role of a fictitious enterprise investor; **SMs Meetings**, to keep track of impediments and facilitate the ST integrations; and **ST Meetings**, to monitor each task status, keeping the ST updated with information brought by SMs and POs.

The main goal of this Sprint 1 was the core functionalities defined from ST#01 and ST#05. As components of the Civil Defense Segment, they were responsible for registering, activating, coordinating, and monitoring the crisis and/or accident macro functions, as a starting point, immediately forwarding it to other Segments. The secondary goal of Sprint 1 was to implement the Victims Registration Module, according to the predefined color pattern, from the meeting with the Fire Department Segment representatives on Sprint 0, considering black color assigned to dead people, red color to people needing urgent care with high priority, and yellow to people waiting for some assistance with low priority [14].

The ST#02 and ST#06 were responsible for the Health Care Segment assistance defining its main goal for this Sprint 1 as the implementation of a module to locate vehicles and crews nearby crisis and/or disaster spots. These STs also were responsible to calculate the required resources through a Web App, as shown in Fig. 55.2. In parallel, a CDS for keeping track of available beds in hospitals, and a CDS to be installed on vehicles, showing the capacity of nearby hospitals designed using the SCADE Display.

On this Sprint 1, ST#03 and ST#07 from the Fire Department Segment and ST#04 and ST#08 from the Police Department Segment focused their effort in understanding how SCADE technologies could be integrated with databases through the REST API architecture [15]. At that time, some prototypes using the SCADE environment were developed, according to David White and Gerard Luttgen [16] and the SCADE official documentation [4]. After several conducted tests by all STs from the Fire and Police Department Segments, it was possible to send data from the SCADE software tool to the REST API to persist information (e.g. victims and events information). At that time, it was also possible to create an Entity-Relationship model (ER-model) to support the whole ACMIS Project required data.

By the end of this Sprint 1, all the STs were able to share their results and receive feedbacks from professors and peers, at the Sprint Review Meeting. On this Sprint, 21 USs were placed on the Sprint Backlog, 18 USs have satisfied quality criteria being approved by POs, and 5 USs were assigned to the next Sprint 2 Backlog.

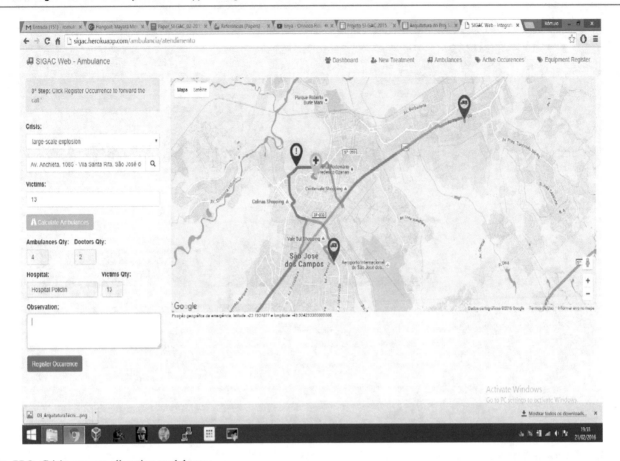

Fig. 55.2 Crisis resource allocation module app

According to Tuckmans four-stage model[12], the STs went through the forming phase, during the Sprint 0. On this Sprint 1, the main roles of the ACMIS Project as well as the needed technologies to be used were still unclear for mostly TDs, who strongly relied on the PO of each ST to clarify the main points.

During the last week of this Sprint 1, students performed the Sprint Retrospective Meeting, in order to show the strengths and drawbacks of their STs at the ongoing ACMIS Project as a whole. At that time, it was possible to identify the following recommendations from this Sprint 1 Retrospective: better time management, need for enhancement effort balance, and more integration improvement for the next Sprint.

55.4.3 Sprint 2

During the Sprint 2 Planning Meeting, some new USs were selected by the STs, based on the obtained results from the Sprint 1 and Professors feedback, and four USs not finished during the previous Sprint 1 had to be included in this Sprint 2.

On this Sprint 2, the STs received the following hardware development kits: Raspberry Pi; Arduino; multiple types of sensors; and some boards for connecting components.

It was also available for the Project development an AR. Drone Parrot 2.0[17], a quadcopter drone with USB ports, Wi-Fi interfaces, and a HD camera. Its open Application Programming Interface (API) allowed great variety of flight operations (e.g. take-offs, path programming, altitude, and speed control), enabling SCADE application tests and future PoC simulations.

On this Sprint 2, most STs overcame its forming maturing level stage and evolved into the storming maturing level stage[12]. At this time, students from the CE-237 Software Testing course had created high level software testing plans [18], considering all Project involved technologies. In order to enable the Test Driven Development (TDD) technique [19] in the ACMIS Project, test cases based on testing plans became available before this Sprint 2 development begins.

On this Sprint 02, ST#01 and ST#05 concentrated their efforts on the REST API communication improvement between the SCADE applications and the Project database. Two CDS prototypes were implemented, allowing sending

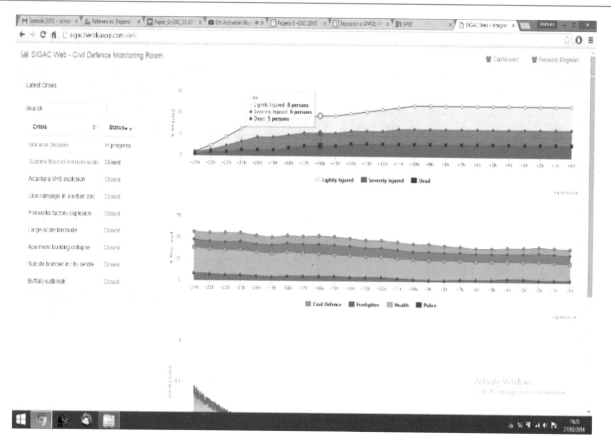

Fig. 55.3 The civil defense monitoring module

data for the database hosted in the cloud (i.e. Amazon Web Services platform). At the same time, the Web App Module was implemented, in order to visualize data extracted from the database in the Monitoring/Control Room, as shown in Fig. 55.3. The data graphically show the amount of people working in a crisis or accident scenario by segments and people affected status, in real-time. At this time, only authorized users from the four segments have access to it.

On this Sprint 2, ST#02 and ST#06 developed an Ambulance Inventory Component for the Web App to enable vehicles equipped with right technical resources to be allocated to specific crisis types. In parallel, both STs initiated a Victim Monitoring System development, by using the available hardware kits and SCADE tools. This system was able to display victims vital signs in real-time from the ambulance CDSs to send the collected data via REST API to the cloud database, enabling real-time monitoring through the Web App and Data Storage.

On this Sprint 2, ST#05 also developed a Multi Sensor Software for the quadcopter drone used for data assimilation in crisis environments, when access by human teams would not be possible due to unacceptable risk levels. The proposed solution used a Raspberry Pi coupled with temperature, moisture, fuel gases, and smoke sensors. Moreover, an SCADE application ran in the Raspberry Pi and interacted

Fig. 55.4 Police segment multi-sensor drone concept

with sensors, sending real-time data over wireless network to the CDS Control Station.

Figure 55.4 shows the multi-sensor quadcopter drone initial implementation, based on the available hardware kits. The ST#05 used basic home devices (i.e. lighter and hair dryer), in order to test hardware functionality and its

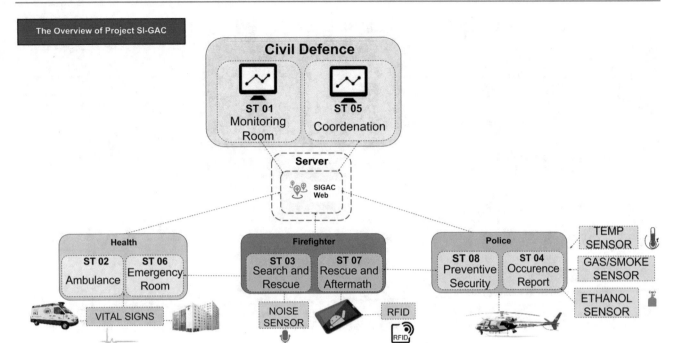

Fig. 55.5 The final project architecture

integration with the ACMIS Project. The Arduino was used to connect sensors and interact with the SCADE application running in a Raspberry Pi. The ST#08 conducted tests with the AR.Drone API, in order to verify if it could be controlled by SCADE applications.

By the end of this Sprint 2, STs presented the main results and had proposed an initial integration solution, 16 USs have achieved the quality criteria and were approved by POs, and 7 USs did not meet the requirements and had to be assigned to the next Sprint 3.

As in the previous Sprint, students on the last week of this Sprint 2 participated on the Sprint Retrospective to identify the main points to be improved. At that time, they post the following recommendations for the next and final Sprint 3: needs for equally workload distribution throughout the Sprint; needs for better communication among STs; and focus on the Project integration.

55.4.4 Sprint 3

The Sprint 3 had focus on integration among technologies and features developed from the previous Sprints. Thus, the Project Architecture was improved based on the results obtained so far, as shown in Fig. 55.5. ST#01 and ST#05 concluded the Web App features hosted in the cloud (i.e. Heroku platform). They optimized the REST API for receiving data sent through SCADE applications and also provided information for other STs.

In order to visually allocate resources for different segments involved in an accident and/or crisis, ST#01 developed the Resource System Management Module, as shown in Fig. 55.6.

ST#02 and ST#06 have developed a module for monitoring heartbeat and corporal temperature. They also used an Arduino to connect the module with sensors, as shown in Fig. 55.7. These data were integrated with the Web App allowing better victims monitoring and optimizing the hospitals bed management.

ST#04 focused in the sensors implementation (i.e. smoke, gas, temperature, and ethanol) integrated and coupled with the AR.Drone. At the same time, ST#08 was able to perform snapshots with the AR.Drone camera showing real-time crisis images in the Web App.

ST#03 and ST#07 accomplished an integration with sensors and Raspberry Pi, based on the ST#04 implementation, developing a Mobile Application Module to hardware readers to receive RFID signals, in order to register victims and send their data to the CDS Monitoring/Control Room, in real-time. This implementation was able to increase victims triage, better allocating then, according to their needs in available environments.

Table 55.2 shows the STs performance during the 3 sprints. In Sprint 3, the last executed, 580 Story Points were delivered from 595 initially estimated Story Points.

At the end of this Sprint 3, most STs evolved from the maturity level stage of storming to the maturity level stage of norming[12]. At that time, they were able to solve their internal differences and better collaborate to each other.

Fig. 55.6 Resource system management module

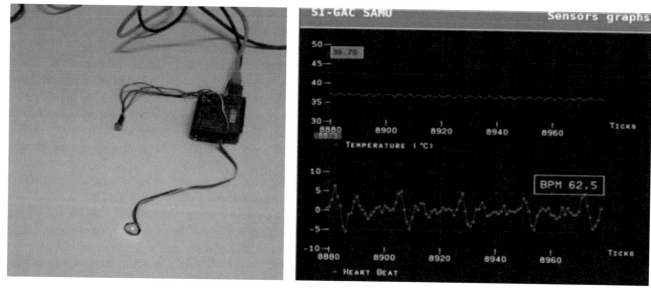

Fig. 55.7 Temperature and heartbeat sensors

Moreover, some TDs were even more committed with the Project goals and became even more proactive.

55.5 The ACMIS Project Compliance with Safety Standards

The DO-178C has 71 safety objectives. Additionally, the DO-331 is a supplement that contains modifications and additions to the DO-178C associated to the use of Model Driven Development (MDD). For the ACMIS Project, the students proposed only a subset of the DO-178C objectives.

The following premises were also used in the ACMIS Project:

(1) USs were considered system and software high-level;
(2) SCADE models were considered software low-level requirements; and
(3) SCADE is a set of system tools that can be qualified, generates source-codes that comply and trace to models, and conforms to standards.

Table 55.2 ST performance during sprints

Sprint	Estimated story points	Delivered story points
1	356	338
2	468	387
3	595	580

The authors of this paper also believe that compliance to other safety standards as ISO 62304[20] or IEC 12207[21] are very similar.

55.6 Conclusion

This paper described a successful academic experience of 80 undergraduate and graduate students, with the integrated system development for accidents and crises management, using embedded and real-time system, with quality, reliability, safety, and testability, supported by Scrum agile method and its best practices.

The ACMIS Project system development involved four integrated segments of Civil Defense, Fire Department, Health Care, and Police Department in a Monitoring/Control Room. It communicated with hardware and CDS developed using the SCADE environment tools. The academic system provided managing and monitoring accidents and crises to minimize its effects on public safety and assets.

At the end of this ACMIS Project, it was possible to successfully perform an assigned mission, using a system prototype in a fictitious natural disaster crisis situation of middle proportions. This assigned mission was evaluated by professors, and demonstrated to entrepreneurs and local authorities, in order to show the Project relevance for the community.

The main contribution of this work was the agile and SCADE software tools environment usage combined with multiple types of hardware, and a set of open-source tools, allowing students geographically dispersed to develop an academic integrated system for a real social problem management.

55.6.1 Future Works

It is suggested to continue this ACMIS Project by using more structured and unstructured data provided by Data Collection Platforms (DCP), other sensors, and social networks, expanding the Project scope and enabling brand new technologies like Big Data and Data Analytics.

Acknowledgements The authors thank: to ITA, for its support during this ACMIS Project development; to ANSYS Esterel Technologies, for providing full academic SCADE software tool license; and to 2RP Net enterprise SPOT Project for supporting some hardware and software infrastructure for this academic proof of concept.

References

1. Coombs, W. T. (2014). *Ongoing crisis communication* (4th ed.). Florida: SAGE Publications, Inc.
2. Guha-sapir, D., Hoyois, P., & Below, R. (2015). *Annual disaster statistical review 2014: The numbers and trends.* Brussels: CRED.
3. Ramos, M. P., Matuck, G. R., Matrigrani, C. F., Mirachi, S., Segeti, E., Leite, M., Da Cunha, A. M., & Dias, L. A. V. (2013). Applying interdisciplinarity and agile methods in the development of a smart grids system. In *2013 10th International Conference on Information Technology: New Generations*, Las Vegas (pp. 103–110).
4. Esterel Technologies. (2015). *SCADE suite.* [Online]. Available: http://www.esterel-technologies.com/products/scade-suite/. Accessed: 10 September, 2015.
5. Moe, N. B., & Dingsyr, T. (2008). In A. Sillitti, O. Hazzan, E. Bache, & X. Albaladejo (Eds.), *Scrum and team effectiveness: Theory and practice* (Lecture notes in business information processing, Vol. 77, pp. 11–20), October 2016. Berlin/Heidelberg: Springer.
6. Shaw, A. C. (2000). *Real-time systems and software* (1st ed.). New York: Wiley.
7. de Oliveira, M. (2009). *Manual de gerenciamento de desastres. Ministério da Integração Nacional, Secretaria Nacional de Defesa Civil, Universidade Federal de Santa Catarina*, Centro Universitário de Estudos e Pesquisas sobre Desastres.
8. Sutherland, J., Viktorov, A., & Blount, J. (2007). Distributed scrum: Agile project management with outsourced development teams. In *System Sciences, 2007. HICSS 2007. 40th Annual Hawaii International Conference*, Hawaii. doi:10.1109/HICSS.2007.180.
9. RTCA. (2011). DO-178C software considerations in airborne systems and equipment certification, Radio Technical Commission for Aeronautics.
10. Arduino. (2015). Available https://www.arduino.cc/. Accessed 10 Dec 2015.
11. Raspberry P. I. (2016). Available https://www.raspberrypi.org/. Accessed 02 Feb 2016.
12. Tuckman, B. (1965). Developmental sequence in small groups. *Psychological Bulletin, 63*(6), 384–399. DOI avaliable http://doi.acm.org/10.1145/161468.16147.
13. Mahni, V., & Hovelja, T. (2012). On using planning poker for estimating user stories. *Journal of Systems and Software, 85*(9), 2086–2095.
14. Oliveira, F. A. G. (2016). *Anlise do mtodo START para triagem em incidentes com mltiplas vtimas: Uma reviso sistemtica.* Available http://repositorio.ufba.br/ri/handle/ri/13977. Accessed 13 May 2016.
15. InfoQ: A Brief Introduction to REST. Available http://www.infoq.com/articles/rest-introduction. Accessed 11 May 2016.
16. White, D., & Lüttgen, G. (2008). *Embedded systems programming Accessing Databases from Esterel. EURASIP Journal on Embedded Systems, 2008*(1), 961036.
17. AR DRONE Parrot 2.0: Available http://ardrone2.parrot.com/. Accessed 03 Feb 2016.
18. Bentley, J. E., Bank, W., & Charlotte, N. C. *Software testing fundamentals—Concepts, Roles, and Terminology.* Available http://www2.sas.com/proceedings/sugi30/141-30.pdf. Accessed 11 May 2016.
19. George, B., & Williams, L. (2003). An initial investigation of test driven development in industry. In *Proceedings of the 2003 ACM Symposium on Applied Computing – SAC'03*, New York. doi:10.1145/952532.952753.
20. International Standardization Organization (2006). ISO 62304 medical device software – Software life cycle processes.
21. International Electro Technical Commission (2008). IEC 12207 systems and software engineering software life cycle processes.

James de Castro Martins, Adriano Fonseca Mancilha Pinto, Gildarcio Sousa Goncalves, Rafael Augusto Lopes Shigemura, Wilson Cristoni Neto, Adilson Marques da Cunha, and Luiz Alberto Vieira Dias

Abstract

This paper describes the use of Agile Testing Quadrants, using Scrum Agile Method in a collaborative software project named Big Data, Internet of Things, and Agile for Accidents and Crises (BD-ITAC). It applies the Scrum agile method and its best practices, the Hadoop ecosystem, and cloud computing for the management of emergencies, involving monitoring, warning, and prevention. It reports the experience of students, during the first semester of 2016, from three different courses on the graduate program in Electronics and Computer Engineering at the Brazilian Aeronautics Institute of Technology (*Instituto Tecnologico de Aeronautica* - ITA). The major contribution of this work is the academic application of Agile Testing Quadrants for Problem-Based Learning. The students worked asynchronous and geographically dispersed to deliver valuable increments. This work was performed during four project Sprints on just sixteen academic weeks. The main project output was a working, developed, and tested software.

Keywords

Scrum Agile Method • Agile Testing Quadrants • Big Data • Cloud Computing • Management of Emergencies

56.1 Introduction

According to Bach [1], "Testing is an infinite process of comparing the invisible to the ambiguous, in order to avoid the unthinkable happening to the anonymous". Also according to Pressman [2], the test process must have the highest probability to find the majority of defects with minimal effort in the shortest possible time.

Jorgensen [3] states "Why do we test? The two main reasons are to make a judgment about quality or

acceptability and to discover problems. We test because we know that we are fallible - this is especially true in the domain of software and software-controlled systems".

Platz [4] ratifies the importance of doing software testing on its Tricentis' newly published report. In this report, the white paper named 'Software Fail Watch: 2015 in Review' presents a strong case for the continued necessity of enterprise software testing and also for the following Top 3 Government Bugs of 2015:

1) The face recognition software of the Department of Motor Vehicles (DMV) denied twins' driver-licenses, because the computer could not tell them apart;
2) After a malware attack, the police department was forced to pay hackers data ransom in bitcoins; and
3) The software failure paralyzed the health care service for over 10 days, denying care to over 10,850 patients.

J. de Castro Martins (✉) • A.F.M. Pinto • G.S. Goncalves • R.A.L. Shigemura • W.C. Neto • A.M. da Cunha • L.A.V. Dias
Computer Science Department, Brazilian Aeronautics Institute of Technology (Instituto Tecnologico de Aeronautica – ITA), Sao Jose dos Campos, Sao Paulo, Brazil
e-mail: james76cm@gmail.com; amancilha@gmail.com; gildarciosousa@gmail.com; rafael.shigemura@gmail.com; wcristoni@gmail.com; cunha@ita.br; vdias@ita.br

© Springer International Publishing AG 2018
S. Latifi (ed.), *Information Technology – New Generations*, Advances in Intelligent Systems and Computing 558, DOI 10.1007/978-3-319-54978-1_56

In this context, during the first semester of 2016, graduate students from three different courses on the Electronic and Computing Engineering Program at the Brazilian Aeronautics Institute of Technology (*Instituto Tecnologico de Aeronautica - ITA*) participated of a project, which aimed to develop software for Problem-Based Learning (PBL) with security and testability, in just 16 academic weeks.

This academic project prototype was named Big Data, Internet of Things, and Agile for Accidents and Crises (BD-ITAC). The main contributions of this project provide real-life experience on implementing a high performance and availability system, initially designed to alert, manage, detect, and prevent possible crises in real emergency situations [5].

This paper aims to present software testing based on the Agile Testing Quadrants, according to Gregory and Crispin [6]. The challenge was to apply it all in an interdisciplinary PBL of alerts and crises management, involving the Scrum agile method, big data, and cloud computing, as the focus of this paper.

This paper aims to present the results of a software testing research based on the Agile Testing Quadrants, according to Gregory and Crispin [6]. The main challenge of this research was its application on an interdisciplinary PBL project management of alerts and crises, involving the Scrum agile method, big data, and cloud computing.

56.2 Background

56.2.1 The BD-ITAC Project

According to Martins et al. [5], the BD-ITAC project was an interdisciplinary PBL developed by students from the following three academic courses: CE-240 - Database Systems Project, CE-245 - Information Technology, and CE-229 - Software Testing.

It has involved Scrum agile method and its best practices, big data; internet of things; Hadoop ecosystem; cloud computing and security; and also emergency management for monitoring, warning, and prevention.

All students from the three courses were divided into 6 Scrum Teams. Each Scrum Team had an assigned role, as shown in Table 56.1.

The adopted architecture is shown in Fig. 56.1.

Table 56.1 The BD-ITAC project macro-functions

Scrum teams IDs	Macro-functions
TS#01	Data Extraction (via Social Networks / Sensors)
TS#02	Alerts and Mobile Communication
TS#03	Rest API / Real Time
TS#04	Network and Security
TS#05	Data Processing (Hadoop Cluster)
TS#06	Database (Relational and NoSQL)

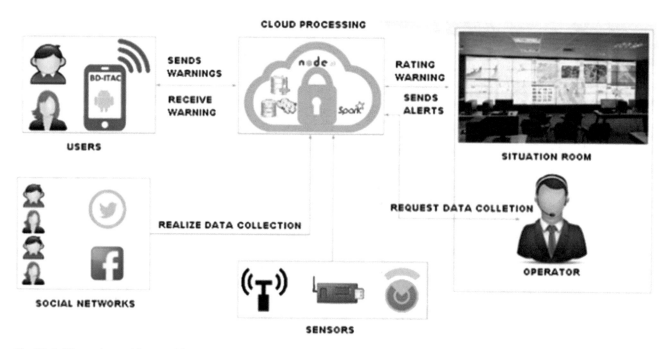

Fig. 56.1 The project architecture [5]

Fig. 56.2 The Agile testing quadrants [6]

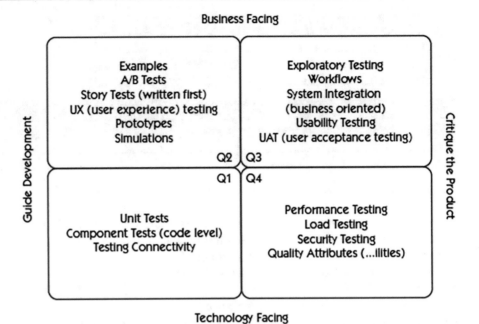

56.2.2 The Problem-Based Learning (PBL)

According to Ertmer et al. [7], "PBL is an instructional approach that has been used successfully for over 30 years and continues to gain acceptance in multiple disciplines. It is an instructional learner-centered approach that empowers learners to conduct research, integrate, theory and practice, and apply knowledge and skills to develop a viable solution to a defined problem."

56.2.3 The Scrum Agile Method and Practices

According to Rubin [8], Scrum is a refreshingly simple and people-centric framework based upon the values of honesty, openness, courage, respect, focus, trust, empowerment, and collaboration. In Scrum, work is performed in iterations or cycles of up to a calendar month called Sprints.

56.2.4 The Agile Testing Quadrants

According to Gregory and Crispin [6], the Agile Testing Quadrants are based on a matrix by Brian Marick. They state that "It's important to understand the purpose behind the Quadrants and the terminology used to convey their concepts. The quadrant numbering system does not imply any order. You don't work through the quadrants from 1 to 4, in a sequential manner. It's an arbitrary numbering system so that when we talk about the Quadrants, we can say 'Q1' instead of 'technology-facing tests that guide development' ".

The quadrants are: Q1 - for technology-facing tests that guide development and it can be units test, components test, and testing connectivity; Q2 - for business-facing tests that guide development and it includes prototypes and simulations in Q2 because they are small experiments to help us understand an idea or concept; Q3 - for business-facing tests that critique (evaluate) the product and includes User Acceptance Testing (UAT); and Q4 - for technology-facing tests that critique (evaluate) the product". Figure 56.2 shows the Agile Testing Quadrants [6].

Gregory and Crispin in [6] also clarified "The Quadrants are merely a taxonomy or model to help teams plan their testing and make sure they have all the resources they need to accomplish it. There are no hard-and-fast rules about what goes in which quadrant. Adapt the Quadrants model to show what tests your team needs to consider. Make the testing visible so that your team thinks about testing first as you do your release, feature, and story planning. This visibility exposes the types of tests that are currently being done and the number of people involved. Use it to provoke discussions about testing and which areas you may want to spend more time on".

56.3 The Agile Software Testing and Development in the Project

On the BD-ITAC project functionalities and values were developed and delivered through the extensive use of software testing. Each Scrum Team has planned the software testing execution through test cases. In all of them, there were 176 cases of planned and implemented tests.

Header Section			
Scrum Team			
Author		Created On	
Reviewed		Reviewed On	

Test Case Section							
ID User Story	Description - User Story	Test Case	Test Steps	What To Expect	Test Data	Actual Result	Status

Fig. 56.3 The spreadsheet model of the used test cases in the BD-ITAC project

According to Jorgensen [3] "the essence of software testing is to determine the set of test cases for the item to be tested. A test case is a recognized work product."

Figure 56.3 shows a spreadsheet model of the used test cases in the BD-ITAC project.

Still according to Jorgensen [3], there are two fundamental approaches to test cases: functional testing (white box) and structural testing (black box). Both approaches have several methods of identification of different test cases and are often just called test methods.

Perry [9] defines functional tests as virtually all validation tests and inspects how the system performs. Examples of this include: (1) Unit testing – to verify that the system functions properly; (2) Integrated testing – where the system runs tasks that involve more than one application or database to verify that it performed the tasks accurately; (3) System testing - to simulate operation of the entire system and to verify that it ran correctly; (4) User acceptance - this real-world test means the most to your business. Once your organization staff begin to interact with your system, they'll verify that it functions properly. Perry [9] also defines structural testing as a dynamic technique in which test data selection and evaluation are driven by the goal of covering various characteristics of the code during testing.

In this context, the various types of software testing were implemented during the Sprints, according to the Agile Testing Quadrants and to the value delivered at the end of each BD-ITAC project Sprint [5].

56.3.1 The Sprint #0

Its main objective was the training of all students in the involved technologies and in the fundamentals of agile method and its best practices, following the produced value of: training in required technologies, understanding the main concepts for agile development, and defining the project vision [5].

The Product Backlog (PB) [8] of the BD-ITAC project was created to direct User Stories (US) and to plan the next

Sprints, assisting in the preparation of test cases to be implemented.

56.3.2 The Sprint #1

Its US focused on: the definition of acceptance criteria for alerts, the implementation of network infrastructure in cloud computing, the event log in a mobile app; and in the web service simulator. Therefore, 33 test cases were developed and the software tests described below were considered the most representative in the Agile Testing Quadrants.

From Quadrant 3, the UAT has been defined in a register of crises, based on events and selection criteria for the use of crisis-indicator data.

From Quadrant 1, the JUnit 4.x [10] tool was used to run the unit tests and from Quadrant 2, the Mock was created in Java language [11], intending to simulate sending and receiving alerts and events' data.

One of the test cases to register an event for a possible system user was composed of the following four steps: (1) In the main system screen, the user must click the "Registering Event" button and wait for the loading screen; (2) Select the Crisis Rating "Earthquake". (3) Inform, in the description field, the following text "Earthquake registration test in App Alert BD-ITAC"; and (4) Click the Save button.

The system should integrate with the Mock (simulating API) and must return to the main application screen. Figure 56.4 shows the JUnit 4.x presentation tool screen in the MockTestRunner class with a score of failure.

The unit and integration tests were used to evaluate functional quality of technical components and the Android app [12], involving the graphical interface and integration processes with Mock.

From the Quadrant 2, the Fig. 56.5 shows the mobile app screen performing integration with Mock.

The Mocha tool [13] was used for unit testing to view a list of indicators of crises and alerts via HTTP / GET (request). This was done in JSON format to access and/or set up a solution development for environment NodeJS.

Fig. 56.4 The snapshot of JUnit 4.x tool application

```
1   package br.ita.bditac.test;
2
3   import org.junit.runner.JUnitCore;
4   import org.junit.runner.Result;
5   import org.junit.runner.notification.Failure;
6
7   public class MockTestRunner {
8     public static void main(String[] args) {
9       Result result = JUnitCore.runClasses(HaversineTest.class);
10      for (Failure failure : result.getFailures()) {
11        System.out.println(failure.toString());
12      }
13    }
14  }
```

Fig. 56.5 The mobile app integrating on Mock

From Quadrant 1, the Fig. 56.6 shows the use of Mocha tool with crisis' description.

From Quadrant 1, the connectivity tests on network infrastructure have been automated through TestInfra tool [14]: to check the installation and the status of the SSH service [15]; to check the installation of iperf3 tool [16]; and to check and verify the connection to MySQL. Figure 56.7 shows the creation of SSH test parameters.

56.3.3 The Sprint #2

The test cases planned for this Sprint mainly aimed at the creation implementation of the social networks data collection environment, the proposed security policy, and the usability of non-relational database. At the end, 112 test cases were performed. The most significant of them are addressed as follows.

From Quadrant 2, the environment of social networks data collection was tested through a story test involving the integration of a Java API and Twitter social network. According to Gregory and Crispin [6], a story test defines expected behavior for the code to be delivered by the story.

The Twitter posts capture routine was written in Java language, using the Twitter4J library. It was responsible for capturing and communicating messages. It was performed in order to check the georeferencing of the post "rain". Figure 56.8 shows a code fragment setting up the search routine with the georeferencing from the Twitter.

From Quadrant 3, the implemented Security Policy was used as UAT to establish the acceptance testing of the network infrastructure. It was the basis for the achievement of connectivity and security tests. From Quadrant 1, Fig. 56.9 shows the test of opened doors, in Web01 server, by TestInfra tool.

From Quadrant 4, Fig. 56.10 presents a rkhunter checking to data files by Lynis tool [17]. Perry [9] states that security testing is designed to evaluate the adequacy of protective procedures.

From Quadrant 4, for the performance of the non-relational database (Cassandra) test, the JMeter [18] and the Cassandra Stress [19] tools were used. Figure 56.11 shows an output of the JMeter tool and Fig. 56.12 presents Cassandra's performance test with 100,000 partitions in just 3 s.

According to Crispin and Gregory [20], performance, load, and scalability testing all fall into Quadrant 4, because of their technology focus.

56.3.4 The Sprint 3

The software test set used during the last Sprint were organized in 31 test cases and focused using the mobile app, the integration with the Situation Room, and the performance testing in the non-relational database.

Fig. 56.6 The snapshot of an unit test using the Mocha tool with crisis' description

```javascript
describe('Crisis business', function() {
  describe('#save()', function() {

    it('should save without error', function(done) {

      var crisis = {
        name: 'Teste',
        email: 'teste@gmail.com',
        phone: 111111,
        place: 'Teste',
        type: 3,
        title: 'Teste'
      };

      var stub = sinon.stub(app.dao.crisis, 'save');
      stub.returns( function(crisis, callback){
        return callback(null, {status: 'ok'});
      });

      crisisBusiness.save(crisis, function(err, data) {
      });

      stub.called.should.be.equal(true);

      done();
```

```python
def test_ssh_service(Service):
    ssh = Service("ssh")
    assert ssh.is_running
    assert ssh.is_enabled
```

Fig. 56.7 The creation of test parameters using the TestInfra tool

```java
FilterQuery filter = new FilterQuery();

double[][] location = new double[2][2];

location[0][0] = Double.parseDouble(coords[0]);
location[0][1] = Double.parseDouble(coords[1]);
location[1][0] = Double.parseDouble(coords[4]);
location[1][1] = Double.parseDouble(coords[5]);

filter.locations(location);

twitterStream.addListener(listener);

twitterStream.filter(filter);
```

Fig. 56.8 The code fragment set up search routine with georeference from Twitter

From Quadrant 1, the FBInfer tool [21] was used for component test. On June 9, 2016, it was conducted static analysis of the developed code for the mobile app. It was founded 32 defects, out of these, 72% came from third-party components and 28% from internal development. The conclusion is depicted in Fig. 56.13 that shows the analysis report after the test.

From Quadrant 1, unit tests were performed in all application modules to consider the mobile app finished. Figure 56.14 shows unit tests from the Android Studio Interface Development Environment (IDE) [22].

From Quadrant 3, Fig. 56.15 shows usability test on the developed mobile app. It represented registration's simulation of a flooding in the city of *Bauru, Sao Paulo*, Brazil.

From Quadrant 3, sources of data collected from social networks, IoT sensors, and information sent from mobile app, were integrated into the web app available at the Situation Room. Figure 56.16 shows the usability of web application at the Situation Room, receiving alert inputs.

From Quadrant 4, performance tests of infrastructure and non-relational databases were carried out with a view from the expectation to meet characteristics of big data proposed by the BD-ITAC project. A performance test was conducted using the SolarWinds tool [23]. Figure 56.17 presents the monitoring shown that, from Jun 5 to 6, 2016, all available infrastructure resources fully met project demands by SolarWinds tool. On average, they used less than 10% of the maximum capacity of all assets (network, processor, and disk memory, space, and speed).

From Quadrant 4, the processing of data traffic was tested by using the TestInfra tool to check data settings and sortings, as quickly as possible, in the big data environment. Figure 56.18 shows, among other information, verification results from the Spark [24] submited to the TestInfra tool.

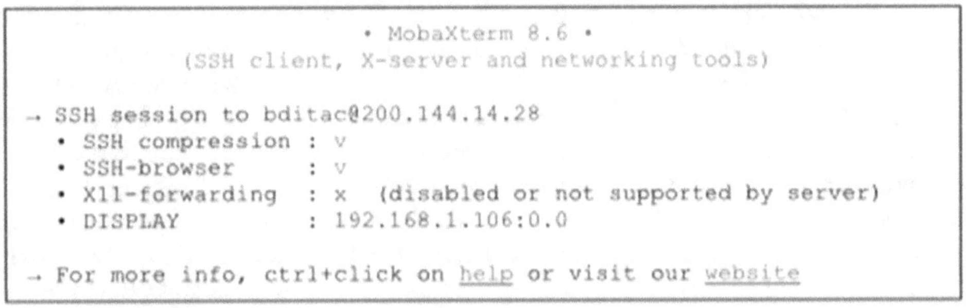

Fig. 56.9 The test of opened doors, in Web01 server, by TestInfra tool

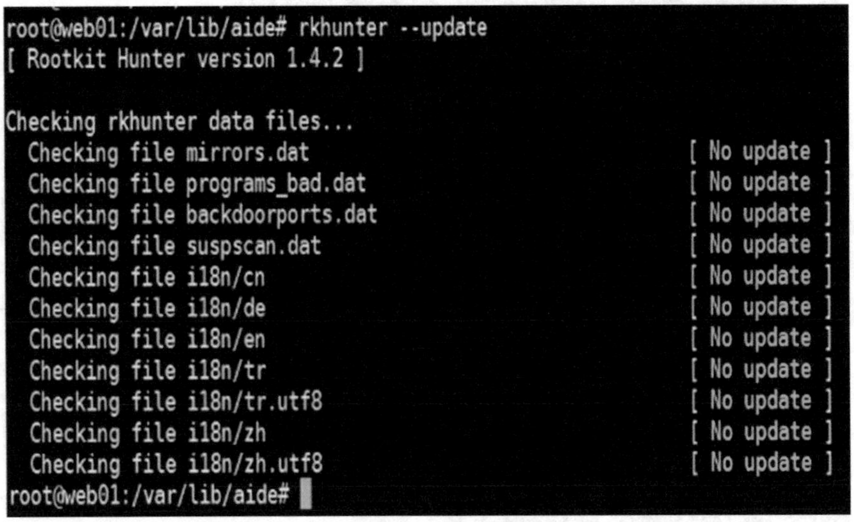

```
                        • MobaXterm 8.6 •
               (SSH client, X-server and networking tools)

→ SSH session to bditac@200.144.14.28
  • SSH compression : v
  • SSH-browser      : v
  • X11-forwarding   : x  (disabled or not supported by server)
  • DISPLAY          : 192.168.1.106:0.0

→ For more info, ctrl+click on help or visit our website
```

Fig. 56.10 The rkhunter checking to data files by Lynis tool

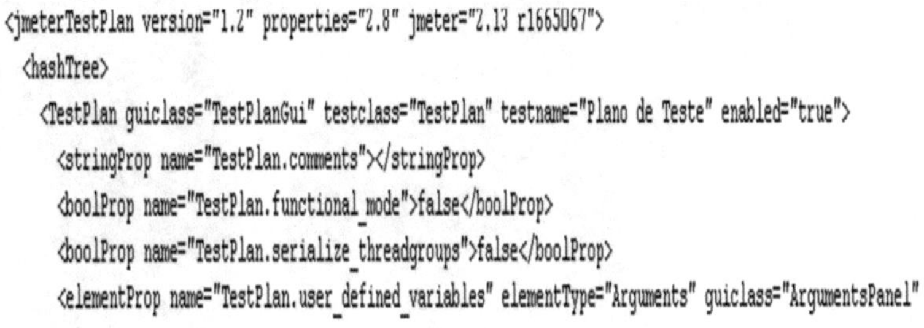

```
root@web01:/var/lib/aide# rkhunter --update
[ Rootkit Hunter version 1.4.2 ]

Checking rkhunter data files...
  Checking file mirrors.dat                         [ No update ]
  Checking file programs_bad.dat                    [ No update ]
  Checking file backdoorports.dat                   [ No update ]
  Checking file suspscan.dat                        [ No update ]
  Checking file i18n/cn                             [ No update ]
  Checking file i18n/de                             [ No update ]
  Checking file i18n/en                             [ No update ]
  Checking file i18n/tr                             [ No update ]
  Checking file i18n/tr.utf8                        [ No update ]
  Checking file i18n/zh                             [ No update ]
  Checking file i18n/zh.utf8                        [ No update ]
root@web01:/var/lib/aide#
```

Fig. 56.11 The output of the JMeter tool

```
<jmeterTestPlan version="1.2" properties="2.8" jmeter="2.13 r1665067">
  <hashTree>
    <TestPlan guiclass="TestPlanGui" testclass="TestPlan" testname="Plano de Teste" enabled="true">
      <stringProp name="TestPlan.comments"></stringProp>
      <boolProp name="TestPlan.functional_mode">false</boolProp>
      <boolProp name="TestPlan.serialize_threadgroups">false</boolProp>
      <elementProp name="TestPlan.user_defined_variables" elementType="Arguments" guiclass="ArgumentsPanel"
```

From Quadrant 4, the performance of Cassandra was analyzed by using the JMeter tool. Figure 56.19 shows the performance test, where 320,000 Lines of Code (LoC) were executed, by using 10,240 threads in just 3 s.

56.4 The Main Results

The analysis of the main obtained results was based on the presentation of the main features developed on the BD-ITAC project, along with the used tools and software testing types. The Agile Quadrants Testing were represented for each testing performed. Figure 56.20 shows the Agile Quadrants Testing for PBL of alerts management and crises involving Scrum agile method, big data, and cloud computing, investigaited in this paper.

During the Sprint 0, expertise building and training of students as agile developers and the learning of key technologies involved in the BD-ITAC project have contributed to the creation of the Product Backlog, a fact that helped the implematation of the next Sprint test cases.

During the Sprint 1, acceptance criteria have been defined for the crisis register - Quadrant 3. The development of the mobile app unit tests were conducted by the JUnit 4.x tool and it has created a Mock in Java language - Quadrants 1 and 2. To show warning indicators, via http protocol, it was used the Mocha tool to perform unit testing - Quadrant 1. The network infrastructure conectivity was tested by the TestInfra tool - Quadrant 1.

During Sprint 2, for research on Twitter, a story test was performed - Quadrant 2. The implemented Security Policy has provided the use of pre-defined acceptance criteria - Quadrant 3. On this occasion, it was used the TestInfra tool as a connectivity test (Quadrant 1) and also the Lynis tool as

a security test (Quadrant 4). At the end, performance testings were implemented on Cassandra using the JMeter and the Cassandra-Stress tools - Quadrant 4.

Finally, aiming to deliver the final product, during the Sprint 3, component tests were applied through the FBInfer tool - Quadrant 1. Unit tests were performed using the Android Studio IDE - Quadrant 1. The usability test was developed for mobile app - Quadrant 3. It tested usability at the Situation Room homepage - Quadrant 3. Final performance tests were conducted by using SolarWinds, TestInfra, and JMeter tools - Quadrant 4.

```
Results:
op rate                   : 33294 [WRITE:33294]
partition rate            : 33294 [WRITE:33294]
row rate                  : 33294 [WRITE:33294]
latency mean              : 6.1 [WRITE:6.1]
latency median            : 4.5 [WRITE:4.5]
latency 95th percentile   : 12.9 [WRITE:12.9]
latency 99th percentile   : 30.3 [WRITE:30.3]
latency 99.9th percentile : 80.0 [WRITE:80.0]
latency max               : 131.1 [WRITE:131.1]
Total partitions          : 100000 [WRITE:100000]
Total errors              : 0 [WRITE:0]
total gc count            : 2
total gc mb               : 1140
total gc time (s)         : 0
avg gc time(ms)           : 83
stdev gc time(ms)         : 22
Total operation time      : 00:00:03

END
```

Fig. 56.12 The Cassandra's performance test

56.5 Conclusion

The objective of this paper was to present the results of a software testing research, based on the Agile Testing Quadrants, according to Gregory and Crispin [6]. The main challenge of this research was its application on an interdisciplinary PBL project management of alerts and crises, involving the Scrum agile method, big data, and cloud computing.

This paper has tackled the integrated Problem-Based Learning (PBL) and the Scrum agile method and its best practices applying the Agile Testing Quadrants into the Big Data, Internet of Things, and Agile for Accidents and Crises (BD-ITAC) project.

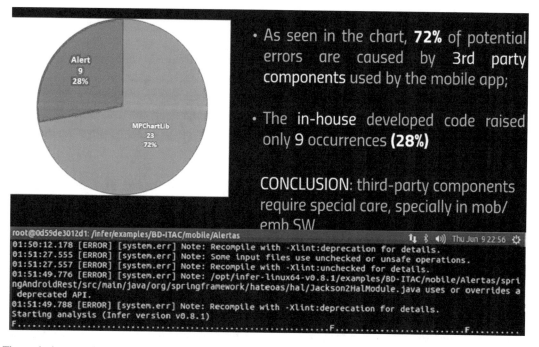

Fig. 56.13 The analysis report from the FBInfer test held on June 9, 2016

br.ita.bditac.ws.client.alerta in Alertas: 11 total, 1 failed, 10 passed 20.17 s

Collapse | Expand

AllTests		20.17 s
EventoTests		1.06 s
EventoTests.test01PostEvento	passed	939 ms
EventoTests.test02GetEventoById	passed	47 ms
EventoTests.test03GetEventoByIdInexistente	passed	76 ms
AlertaTests		19.08 s
AlertaTests.test01PostAlerta	passed	57 ms
AlertaTests.test02GetAlertaById	passed	51 ms
AlertaTests.test03GetAlertaByIdInexistente	passed	47 ms
AlertaTests.test04GetAlertaByRegiao	failed	16.37 s
AlertaTests.test05GetAlertaByRegiaoInexistente	passed	2.50 s
AlertaTests.test06GetRegiaoComAlerta	passed	33 ms
AlertaTests.test07GetRegiaoSemAlerta	passed	15 ms
IndicadoresTests		32 ms
IndicadoresTests.test01GetIndicadores	passed	32 ms

Fig. 56.14 The unit tests from the Android Studio IDE

Fig. 56.15 The image of a flood
sent, via mobile app from the
BD-ITAC project

Home / Crisis / List

#ID	Description / Local	Start	Finish	Count Avisos	Is Alerta	Active			
1	**Monitoring flooding crisis in Sao Jose dos Campos** Caieiras / SP	01/04/2016	30/06/2016	0	Yes	Yes			
3	test Santa Rosa Do Purus / AC	12/06/2016		2	No	No			
4	test Santa Rosa Do Purus / AC	12/06/2016		1	No	Yes			
5	test São José Dos Campos / SP	12/06/2016		1	No	Yes			
6	**River flooding caused by flood** Bauru / SP	16/06/2016		1	No	Yes			

Fig. 56.16 The web application usability at the situation room receiving alert inputs

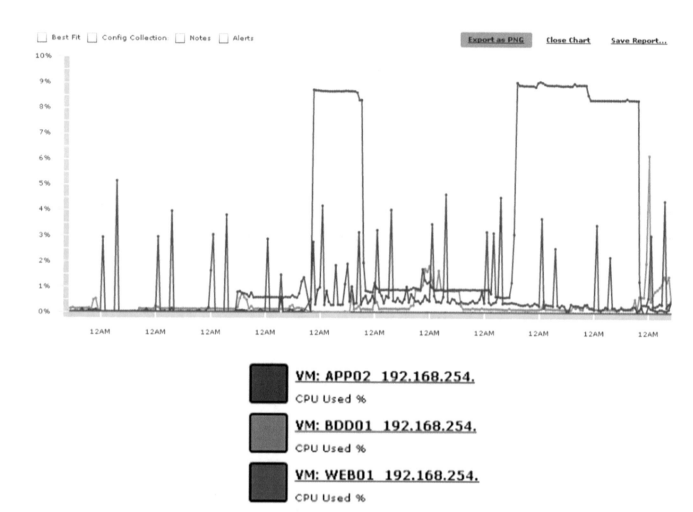

Fig. 56.17 The network monitoring screen from Jun 5 to 6, 2016, by SolarWinds tool

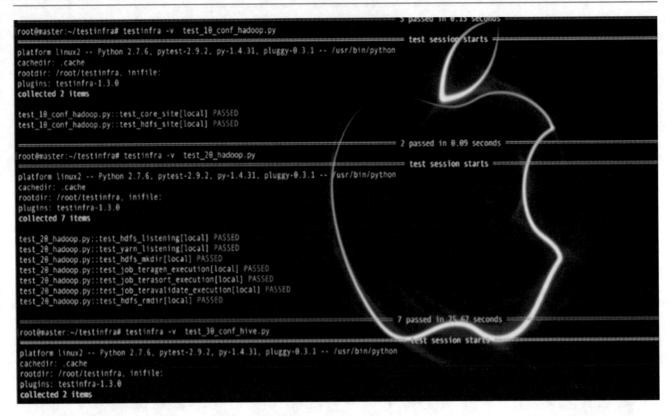

Fig. 56.18 The check results screen from Spark submited to the TestInfra tool

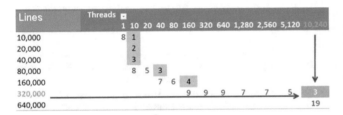

Fig. 56.19 The performance test on Cassandra where 320,000 LoC were executed by using 10,240 threads in just 3 s by the JMeter tool

The application of agile software testing in the BD-ITAC project was developed in four Sprints. Sprint 0 involved the training of all students in technologies and fundamentals of agile method and its best practices. In Sprint 1, 2, and 3, a total of 176 test cases were succesfully planned and implemented.

On this process, a total of 13 tools was used, five from Quadrant 1, three from Quadrant 2, and also five from Quadrant 4. At the end, it was possible to carry out connectivity and performance tests with the same software testing tool named TestInfra.

The main contribution of the research sinthesized in this paper was the BD-ITAC project application of Agile Testing Quadrants.

The authors of this research believe that software testing tools, usability testing, and User Acceptance Testing (UAT) have directly contributed to the main delivery value of each Sprint of the Scrum Agile Method developed in the project.

The authors recommend a natural continuation of this project with the application of Internet of Things (IoT) in another academic project involving Problem-Based Learning.

56.6   Future Works

In addition to the used technologies reported in this paper, the authors suggest the continuation of this research, in different interdisciplinary Problem-Based Learning applications. This can be done, by using computer systems in real-time enviroments for analyses and automatic alerts, mainly for continuous delivery, automated building, and DevOps techniques together.

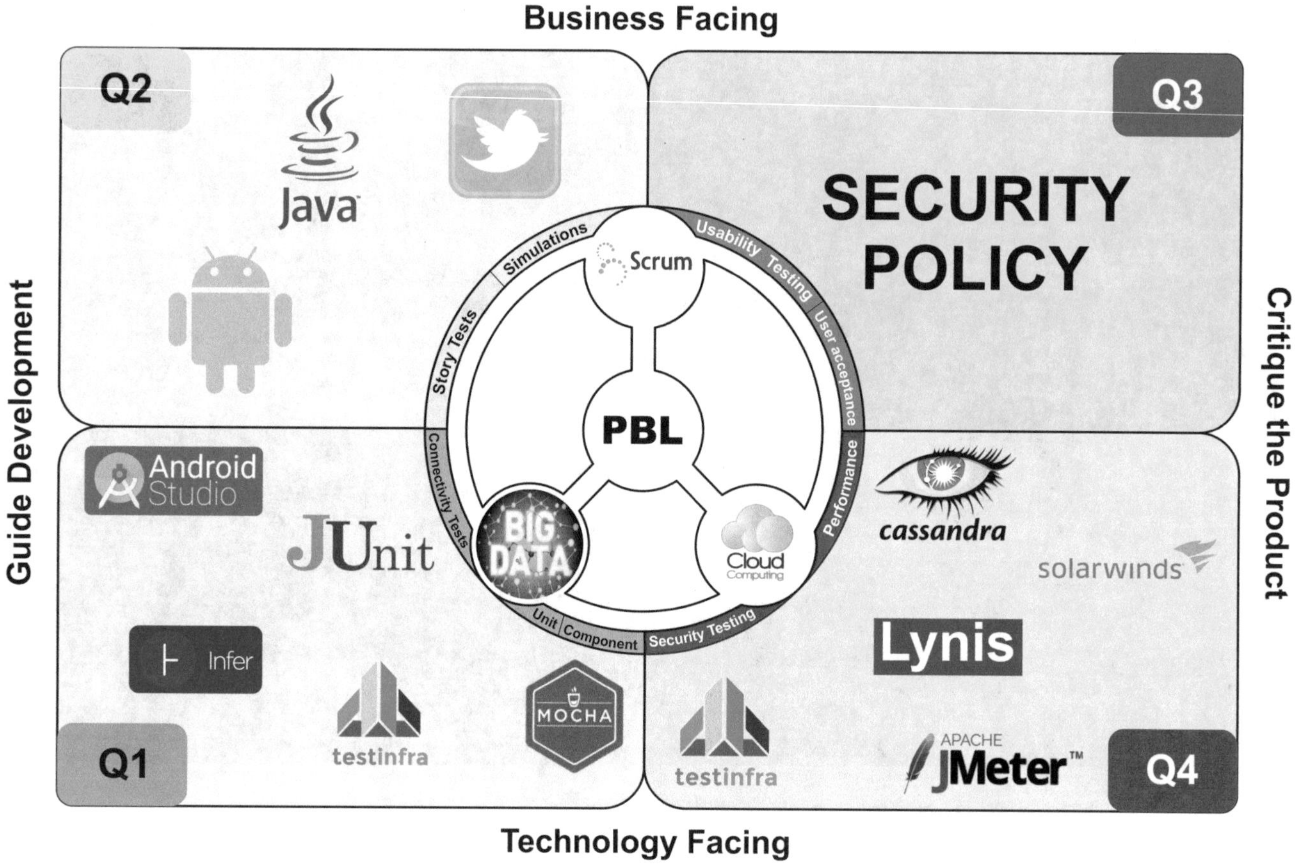

Business Facing

Fig. 56.20 The Agile Testing Quadrants for PBL of alerts and crises management involving Scrum agile method, big data, and cloud computing

Acknowledgment The authors would like to thank: the Brazilian Aeronautics Institute of Technology (ITA); the Casimiro Montenegro Filho Foundation (FCMF); and the 2RP Net Enterprise, for their support and infrastructure to the development of this BD-ITAC project, allowing its Proof of Concept (PoC) in an academic and simulated real environment.

References

1. Bach, J. (2014). Testing Excellence Learn Software: Testing Best Software Testing Quotes. [Online]. Available: http://www. testingexcellence.com/best-software-testing-quotes. Last access: 10/20/2016.
2. Pressman, R. S. (2014). Software testing techniques. In *Software engineering a practitioner's approach* (6th ed., p. 389). New York: McGraw-Hill.
3. Jorgensen, P. C. (2014). A perspective on testing. In B. Raton (Ed.), *Software testing – a Craftsman's approach* (4th ed., p. 3). London, UK/New York: CRC Press.
4. W. Platz. Tricentis. Software fail watch: 2015 in review. [Online]. Available: http://www.tricentis.com/resource-assets/report-software-fail-watch-2015-in-review/. Last access: 10/20/2016.
5. Martins, J. C., Pinto, A. F. M., Junior, E. E. B., Goncalves, G. S., Louro, H. D. B., Gomes, J. M., Filho, L. A. L., Silva, L. H. R. C., Rodrigues, R. A., Neto, W. C., Cunha, A. M., & Dias, L. A. V. (2017). Using big data, internet of things, and agile for crises management. In *ITNG*. Las Vegas.
6. Crispin, L., & Gregory, J. (2015). Using models to help plan. In *More agile testing learning journeys for the whole team* (pp. 101–117). Saddle River: Addison Wesley.
7. Ertmer, P. A., Hmelo-Silver, C., Walker, A., & Leary, H. (2015). Overview of problem-based learning: Definitions and distinctions: Exploring and extending the legacy of Howard S. Barrows. In *Essential readings in problem-based learning* (p. 5). West Lafayette: Purdue University Press.
8. Rubin, K. S. (2013). Scrum framework. In *Essential scrum: A practical guide to the most popular agile process* (pp. 13–20). Saddle River: Addison Wesley: Pearson Education Inc.
9. Perry, W. E. (2007). *Effective methods for software testing: Includes complete guidelines, checklists, and templates* (3rd ed. pp. 70–228). Indianapolis: Wiley Publishing.
10. JUnit 4.x. [Online]. Available: http://junit.org/junit4/. Last access: 10/20/2016.
11. Oracle. Java Software. [Online]. Available: https://www.oracle.com/java/index.html. Last access: 10/20/2016.
12. Android. [Online]. Available: https://www.android.com/intl/en_us/. Last access: 10/20/2016.
13. Mocha. [Online]. Available: https://mochajs.org/. Last access: 10/20/2016.
14. TestInfra.Testinfra test your infrastructure. [Online]. Available: http://testinfra.readthedocs.io/en/latest/. Last access: 10/20/2016.
15. Barrett, D., Silverman, R., & Byrnes, R. (2005). A recommended setup. In *SSH, the secure shell: The definitive guide* (2nd ed., p. 398). Sebastopol: O' Reilly.
16. Buyya, R., & Dastjerdi, A. V. (2016). Experimental results. In *Internet of things: Principles and paradigms* (p. 113). Amsterdam: Elsevier.

17. Lynis. [Online]. Available: https://cisofy.com/lynis/. Last access: 10/25/2016.
18. Jmeter. [Online]. Available: http://jmeter.apache.org/. Last access: 10/20/2016.
19. Vivek, M. (2014). Cassandra performance tuning. In *Beginning apache Cassandra development* (pp. 153–169). New York: Apress.
20. Crispin, L., & Gregory, J. (2009). Performance, load, stress, and scalability testing. In *Agile testing a practical guide for testers and agile teams* (p. 233). Saddle River: Addison Wesley.
21. FBInfer. [Online]. Available: http://fbinfer.com/. Last access: 10/21/2016.
22. Android Studio. [Online]. Available: https://developer.android.com/studio/features.html. Last access: 10/20/2016.
23. Solarwinds Monitor & Test. [Online]. Available: http://www.solarwinds.com/solutions/website-speed-test. Last access: 10/20/2016.
24. Spark. [Online]. Available: http://spark.apache.org/. Last access: 10/20/2016.

Integrating NoSQL, Relational Database, and the Hadoop Ecosystem in an Interdisciplinary Project involving Big Data and Credit Card Transactions

57

Romulo Alceu Rodrigues, Lineu Alves Lima Filho, Gildarcio Sousa Gonçalves, Lineu F.S. Mialaret, Adilson Marques da Cunha, and Luiz Alberto Vieira Dias

Abstract

The project entitled as Big Data, Internet of Things, and Mobile Devices, in Portuguese *Banco de Dados, Internet das Coisas e Dispositivos Moveis* (BDIC-DM) was implemented at the Brazilian Aeronautics Institute of Technology (ITA) on the 1st Semester of 2015. It involved 60 graduate students within just 17 academic weeks. As a starting point for some features of real time Online Transactional Processing (OLTP) system, the Relational Database Management System (RDBMS) MySQL was used along with the NoSQL Cassandra to store transaction data generated from web portal and mobile applications. Considering batch data analysis, the Apache Hadoop Ecosystem was used for Online Analytical Processing (OLAP). The infrastructure based on the Apache Sqoop tool has allowed exporting data from the relational database MySQL to the Hadoop File System (HDFS), while Python scripts were used to export transaction data from the NoSQL database to the HDFS. The main objective of the BDIC-DM project was to implement an e-Commerce prototype system to manage credit card transactions, involving large volumes of data, by using different technologies. The used tools involved generation, storage, and consumption of Big Data. This paper describes the process of integrating NoSQL and relational database with Hadoop Cluster, during an academic project using the Scrum Agile Method. At the end, processing time significantly decreased, by using appropriate tools and available data. For future work, it is suggested the investigation of other tools and datasets.

Keywords

NoSQL · RDBMS · Hadoop ecosystem · Big Data · Credit card transactions · Data Integration

57.1 Introduction

The use of credit cards has been increasing worldwide. According to the World Payments Report [1], the number of non-cash transactions was around 358 billion in 2013. Figure 57.1 shows the global growth rate of 27.7 billion non-cash transactions in 2013, comparing to 23.3 billion in 2012. Since this amount of data rises yearly, it became necessary to use Big Data tools to facilitate the process of storage, consumption, and analysis of huge datasets [2–5].

In Computer Science and Engineering courses, it is clear that there is still an inadequate pedagogical practice in face

R.A. Rodrigues (✉) • L.A.L. Filho • G.S. Gonçalves • L.F.S. Mialaret • A.M. da Cunha • L.A.V. Dias
Computer Science Department, Brazilian Aeronautics Institute of Technology (Instituto Tecnologico de Aeronautica – ITA), Sao Jose dos Campos, SP, Brazil
e-mail: romulo.rodrigues@cptec.inpe.br; lineulimasjc@gmail.com; gildarciosousa@gmail.com; lmialaret@gmail.com; cunha@ita.br; vdias@ita.br

S. Latifi (ed.), *Information Technology – New Generations*, Advances in Intelligent Systems and Computing 558, DOI 10.1007/978-3-319-54978-1_57

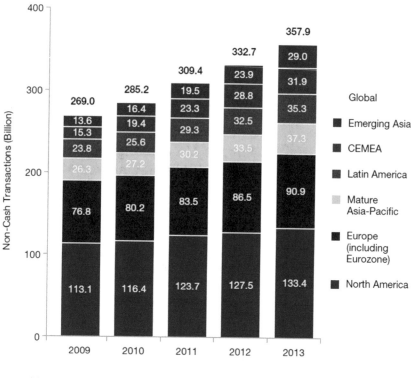

Fig. 57.1 The number of non-cash transactions worldwide [1]

of a reality that has required multidisciplinary, multidimensional, global, and contextualized preparation. The realization of practical activities in database courses for emergent technologies is an important complement to the theoretical part [6, 7].

This paper presents the integration of two database systems with the Hadoop Ecosystem in an academical project. The integration of NoSQL Cassandra and MySQL databases with the Hadoop Ecosystem enabled big data processing. The proposed project main objective was to manage a large amount of financial data to detect credit card frauds in the least amount of time. Frauds involving computer generated financial transactions are responsible for yearly losses over billions of dollars [8].

57.2 Background

Several technological tools were used to deliver this Proof of Concept (PoC). Therefore, it is necessary to provide some background knowledge in order to familiarize with the investigated technologies, tools, and issues.

57.2.1 NoSQL Concept

NoSQL database [9, 10] are alternative to relational database systems. It mitigates the scalability problem in relational database. It emerged in a Big Data context in which large volumes of data required high processing power. Basically,

```
CREATE TABLE employees (
company text,
name    text,
age     int,
role    text,
PRIMARY KEY (company,name)
);
```

Fig. 57.2 The employees CQL script

NoSQL database can be categorized into four different data models: Column Family Storages [11] and Key Value Storages, supported by Cassandra; Document Storages; and Graph Database Storages.

57.2.2 Apache Cassandra

The Apache Cassandra [12] is a scalable open source NoSQL database that provides high availability and high performance. After being developed by Facebook, it became an Apache project in 2010. Cassandra uses CQL (Cassandra Query Language), which is very similar to SQL to manipulate objects and data.

The CQL CREATE TABLE statement allows the creation of composite primary keys. For instance, the CQL shown in Fig. 57.2 creates the table EMPLOYEES, which is keyed on COMPANY and NAME columns, referring to the name of the company and the name of the employee.

When a query retrieves the data table, the result set suggests that one row does exist for each combination of company and employee, as shown in Fig. 57.3.

company	name	age	role
OSC	eric	38	ceo
OSC	john	37	dev
OSC	anya	29	lead
RKG	ben	27	dev
RKG	chad	35	ops

Fig. 57.3 The data table example

	eric:age	eric:role	john:age	john:role		
OSC	38	ceo	37	dev		

	anya:age	anya:role	ben:age	ben:role	chad:age	chad:role
RKG	29	lead	27	dev	35	ops

Fig. 57.4 The column family storage example

However, this sample data is stored in two rows, one with four columns and the other with six columns. The order of the columns in the PRIMARY KEY field defines the storage mode. The COMPANY column is used to create the ROWS and the NAME column defines the COLUMNS of each row, as shown in Fig. 57.4.

57.2.3 Big Data and Hadoop

The term Big Data usually refers to large datasets generated by companies and institutions, which often share some common features as the 5 V's: Volume, Variety, Velocity, Veracity, and Value [13, 14]. The source of this data is heterogeneous, for instance, sensors, supercomputers, security cameras, servers, and social networks. In the context of Big Data, the analysis of these large data volumes can be used to acquire new "insights", such as information on human behavioral patterns, climate change, security alerts, or any type of strategic information that brings added value to an institution [13, 14].

In recent years, the increase in size and variety of datasets, associated with global policy in searching for solutions with good cost benefit, allow companies like Facebook and Yahoo to search for scalable solutions that could analyze terabytes or even petabytes of data.

57.2.4 Hadoop Ecosystem

The Hadoop ecosystem used in the BDIC-DM project was comprised of the following tools: Hadoop Common, Hadoop Distributed File System (HDFS), MapReduce, and Apache Hive.

57.2.4.1 Hadoop Common

Lately, datasets have reached hundreds of gigabytes. The demand for analysis of unstructured data such as logs, social networks, and data sensors increased the needs for horizontal scalable topology solutions. This has provided tools for Data Scientist to extract information from large volumes of data, in parallel manner [11].

The Apache Hadoop is based on the Google MapReduce and Google File System technology originated from the Yahoo company. It is a framework written in Java, optimized for analyzing large volumes of structured and unstructured data, using relatively inexpensive hardware. Hadoop is a powerful tool for sequence analysis and does not replace relational database. It works as complement, allowing the management of different types of data in large volumes. In the paradigm used by Hadoop, the processing power is brought to the data, unlike relational solutions that rely on large data transfers. Therefore, data does not have to be moved through a network from point A to point B, for instance, from a traditional storage system into a cluster [15].

57.2.4.2 HDFS (Hadoop Distributed File System)

The Hadoop most important component is its distributed file system, which provides high fault tolerance through data replication among system nodes. This allows the growth of the system's capacity, in a transparent manner, and high performance in the transmission of the data over clusters' network. The HDFS works with large blocks of data (64 MB default) due to the fact that Apache Hadoop performance is superior when the access to the data sequentially takes place, in other words, the number of "seek" operations is reduced.

Very large files are ideal for Hadoop. It looks for the beginning of each block, keeping blocks sequentially from that point. An important feature of HDFS is the topology awareness, so the system assumes that the available bandwidth is greater when the nodes are located in the same rack. The access cost of such data increases with the distance among racks, since the network traffic must pass through several switches. The system uses its topology awareness in order to create a distribution policy that maximizes reliability and performance while accessing data [16].

57.2.4.3 MapReduce

Hadoop MapReduce [17] is a programming model based on Google's MapReduce. It was created for operations with large amount of data, while hiding the complexity of developing parallel applications, making it accessible to larger number of users.

MapReduce is a core component of Apache Hadoop and it can be used in transparent manner through its ecosystem tools such as Apache Hive.

57.2.4.4 Apache *Hive*

Although the MapReduce model facilitates the creation of applications that benefit from parallel computing environments, implementing this type of software for several users is still complex, especially for non-java developers. In 2007, Facebook has identified the needs to bring familiar database concepts into MapReduce, increasing the range of professionals who could benefit from this technology.

The Apache Hive offers SQL like interface language (Hive Query Language - HQL), abstracting the process of implementing MapReduce. Hive users can write HQL queries and submit them to clusters through its command line or web interface. The application is responsible for creating and distributing all the necessary jobs and returns the results on the screen. The Apache Hive, however, is not a tool designed for real-time analysis, its main use is in high-volume batch analysis [18].

57.3 The Project Overview

The BDIC-DM project was an academic e-Commerce prototype development, integrated to a Big Data processing architecture, supporting the analysis of large volumes of sensitive data. The project used the Scrum method for Agile Development and the Interdisciplinary Problem Based Learning (IPBL) approach, involving 60 graduate students from the following three courses: CE-240 - Database Systems Project, CE- 245 - Information Technology, and CE-229 - Software Testing, at the Brazilian Aeronautics Institute of Technology (*Instituto Tecnologico de Aeronautica* - ITA).

In order to develop this BDIC-DM project, students were divided into the following five Scrum Teams (ST): ST01 - e-Commerce, ST02 - Supporting the Maximum of Transactions per Second, ST03 - Data Analysis, ST04 - Fraud Detections, and ST05 - Supplying Large Amount of Data, as shown in Fig. 57.5.

The project prototype had to be delivered in just 17 academic weeks. Its development was divided into four sprints of four weeks each. A Sprint 0, lasting four weeks, was performed for leveling students' capability. Considering that the Sprint 0 was used to elaborate the project architecture and to acquire knowledge from technologies and tools employed students had three effective sprints to deliver the end product.

The project main objective was to implement and integrate OLAP and OLTP to an e-Commerce system, therefore enabling the extraction of insights such as fraudulent behavior, sales statistics, and credit card use. The focus of this paper is on the OLAP implementation of the BDIC-DM project developed by ST03 and ST05.

Initially, the NoSQL Cassandra was the only database to be used in the BDIC-DM project. However, after analysis by developers, the use of an RDBMS was needed, since the e-Commerce portal required a relational database. The main purpose of the portal was to simulate the sale of some products, considering payments by credit cards. All tables, but the transaction table responsible for the financial transaction, as shown in Fig. 57.6, were allocated into the relational database [19].

It was decided to use the Apache Hadoop, as the OLAP analysis solution, isolating the processing of large volumes of transactions from the OLTP environment, using Cassandra optimized for high performance. The scalability offered by

Fig. 57.5 The BDIC-DM project general architecture

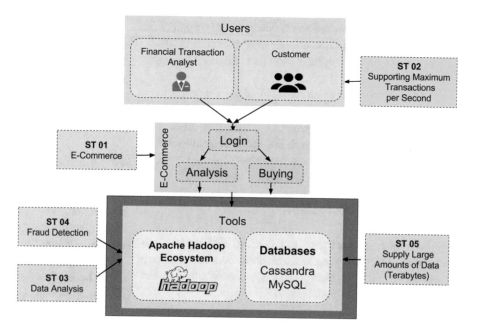

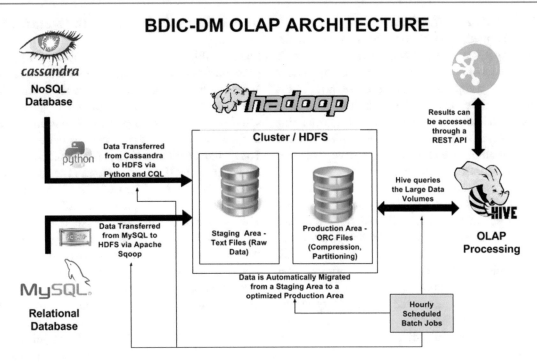

Fig. 57.6 The data transfer complete process

Hadoop combined with ease implementation of jobs needed through the Apache Hive were crucial for the chosen solution [18]. The Apache Hive was identified as the ideal solution for analyses of financial transactions imported from Cassandra, avoiding also the use of the main database for historical analyses of intensive computation resources. With the use of Hive, it was possible to implement high level queries, although the team had no specialists in implementing parallel algorithms or knowledge in low level implementation of MapReduce.

The HDFS provided an environment for storing and processing the assimilated data from the relational and NoSQL database, providing significant gains in disk space and fault tolerance.

The financial transactions involving purchases made by customers through the e-Commerce portal were stored in the NoSQL Cassandra. The main purpose of the proposed solution was to provide separation between financial transactions and relational database. Thus, it was possible to apply fraud detection algorithms to detect credit card frauds faster, increasing efficiency.

57.4   The Project Development

The BDIC-DM project was developed in the following sprints: Sprint 1 - Moving data into Hadoop, involving the dataset denormalization; Sprint 2 - Accessing the dataset; and Sprint 3 - Performance improvements, involving testing architecture.

57.4.1   Sprint 1: Moving Data into Hadoop

Moving large data volumes into HDFS can be challenging due to limitations in corporate networks and storage systems.

In the first Sprint of the BDIC-DM project, it was identified the needs to transfer data between a relational database (MySQL) and a NoSQL database (Cassandra), using the Hadoop Hive from the HDFS for dataset analyses (OLAP), enabling extraction of insights from historical data.

In order to solve this problem, the Apache Sqoop tool was chosen due to its capacity for high performance of data transfers involving databases and HDFS. Sqoop [20] allows transfers to be executed in parallel and enables direct execution of MapReduce jobs in the remote data. For example, this flexibility allows queries to be submitted to the relational database, filtering results from the source, reducing network traffic and storage requirements in HDFS.

In Cassandra, the native command COPY TO CSV (Comma-Separated Values) was used through a combination of Python Scripts and CLASH (Cassandra Language Shell). The data were transferred to HDFS with the use of Hadoop [21] PUT command.

To integrate the data transfer processes, the Operating System Level, more specifically CRON jobs, had a set of batch jobs implemented and scheduled to run hourly. The HDFS stored raw data as it entered the integration system in the staging area. Figure 57.6 shows the complete process.

57.4.1.1 Denormalizing the Dataset

Although the normalization process is a common and recommended procedure in relational databases, this can be a problem for massive data volumes. The JOIN operations are a costly type of computation that can cause drops in the performance of Apache Hive queries, if used in datasets of terabytes and petabytes. In this scenario, denormalizing datasets, which consists of basically gathering information scattered from multiple tables to a larger table, becomes good practice of performance optimization.

One of the queries developed during the BDIC-DM project involved providing sales statistics by location. The Entity Relationship (ER) model used in the MySQL database followed the normalization standard. As data were transferred to the HDFS, the denormalization process occurred, thus creating a single table, combining the original four tables from the MySQL database.

The normalization was performed using a query on the Apache Hive, which transferred the data stored in text file format from a staging (normalized) area to a new production area (denormalized).

The file format chosen for storing files in the production area was the Optimized Row Columnar (ORC), one of the latest advances of Apache Hive. The ORC format [22] proved to be the most effective way to store and access data in HDFS for the BDIC-DM project scenario, bringing several optimizations such as automatic indexing and the ability to identify the type of stored data. In addition, the ORC format allows the use of data compression through ZLIB and SNAPPY libraries.

In order to test the normalization and conversion process, 10 GB of synthetic data were generated using the Generate Data tool [23] and inserted into the BDIC-DM project databases (MySQL e NoSQL). After transferring the dataset from the databases to the staging area (text files), and then to the production area (ORC files with SNAPPY compression), a reduction of 59% in the use of disk space was observed, as shown in Fig. 57.7.

57.4.2 Sprint 2: Accessing the Dataset

Since the flow of data from the databases into the HDFS had been implemented in Sprint 1, so the focus shifted into accessing the data effectively through Apache Hive. In this research, the use of Hive external tables was a suitable solution. This feature consisted of indicating the file paths in the HDFS, during the process of table creation in Hive. The system could then automatically access the data without the need of INSERT operations. The great advantage of external tables is the Data Definition Language (DDL) independence from the data stored in HDFS. For instance, if a table is deleted by mistake, the data remains intact.

The HQL query represented in Fig. 57.8 retrieves the amount of transactions divided by country and grouped by month for the past 365 days based on the system date. For the BDIC-DM project, a total of nineteen queries were created with the aim of classifying transaction volumes, amount of sales in specific cities, holidays, and so forth. Those queries were scheduled in hourly basis batch jobs, as shown in Fig. 57.2. The results were then made accessible to the web portal through a REST API, which is beyond the scope of this paper.

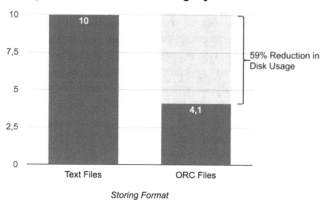

Fig. 57.7 The disk space usage by the testing dataset

Fig. 57.8 The data being accessed through HQL query

```
SELECT
        trans_loc_id, country_name, MONTH(trans_date),
        COUNT(MONTH((trans_date)) as month
FROM
        transactions
JOIN
        locations ON (
                trans_loc_id = loc_id AND trans_date > DATE_SUB
(FROM_UNIXTIME (unix_timestamp ()),365)
GROUP BY
        trans_loc_id, country_name, MONTH(trans_date)
LIMIT 250;
```

57.4.3 Sprint 3: Performance Improvements

During the execution of the first jobs in the BDIC-DM project, it was noticed some queries running for a very long time, although the standard recommendations from the official documentation of the Apache Hive had been followed. For instance, the query shown in Figure 57.9 took 10 minutes and 14 seconds to finish when applied to the data in the staging area, and 8 minutes and 19 seconds when applied to the production area.

The migration of the storing format from text to ORC files brought benefits such as access to new features, automatic indexing, and compression. However, to improve query performance some important good practices had to be implemented.

The most important feature implemented in the production area was the automatic table partitioning. This technique consisted of indicating to Hive one or more fields that should be used to separate the data into HDFS subdirectories. For instance, the staging area version of the transaction table contained the field "data" in "YYYY-MM-DD" format. By adding the "PARTITIONED BY (YEAR, MONTH, DAY)" syntax to the DDL of this table in production, it informed Apache Hive to automatically organize the data into smaller files contained into subdirectories named YEAR, MONTH, and DAY, as the data was migrated to ORC format by the batch jobs. This technique application was very beneficial when querying large datasets, since it became possible to access fractions of the data, for instance, the year of 2012, which consists of a narrow set of the original dataset.

At the end of this sprint 3, an assigned mission was performed and the same query was then executed in the same dataset (production area only), taking one minute and fifteen seconds to finish, which represented a reduction of 78% in processing time, compared with its first execution in the staging area.

57.4.4 Testing Architecture

The HQL language is quite similar to relational SQL, but it has some differences that caused some issues early in the project, since the development team was using the language for the first time. As syntax errors and some unexpected results began to appear, it became necessary to validate the queries through unit tests before they were submitted to the production environment. The development team implemented an architectural test based on the Hive Runner framework, which is based on JUnit, well-known in the java world [24].

The architectural testing consisted of using SQLite and .NET language to create an Oracle to validate a set of expected results through a set of synthetic data generated by the Generate Data tool [23, 25]. A set of automated test cases were then implemented in the Hive Runner to validate the queries used in this project.

The decision of using software testing techniques reduced the time occupied by the team in query refactoring due to syntax mistakes, thus significantly increasing this project productivity. Figure 57.9 contains a diagram of the architectural test described above.

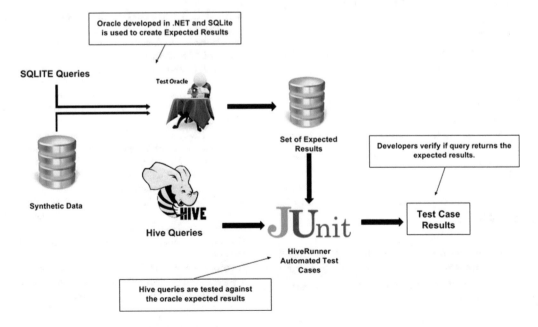

BDIC-DM QUERY TESTING ARCHITECTURE

Fig. 57.9 The architectural test used in the BDIC-DM project

57.5 Results and Discussion

The use of HDFS provided a 59% reduction of disk space usage for storing data by using ORC format. Query performance gains were observed on Apache Hive after the denormalization of dataset and employment of table partitioning. It is believed that in a larger dataset the gains would be even more evident.

During this BDIC-DM project, it was possible to apply the proposed architecture in Fig. 57.6 through a generated data assimilation infrastructure based on Apache Sqoop and features of Cassandra to export data in CSV format. It was also possible to centralize varied data sources in a single environment that combines storage power (HDFS) and processing (MapReduce). By the OLTP and OLAP processing isolation, it was possible to identify significant performance gains in the system, since Cassandra was only used for online transactions.

57.6 Conclusion

This paper discussed the integration of technologies for the generation, storage, and processing of large volumes of data, using the Interdisciplinary Problem Based Learning (IPBL) methodology, within the Electronic Engineering and Computer Science Graduate Program, in the area of Informatics, at the Brazilian Aeronautics Institute of Technology (Instituto Tecnologico de Aeronautica - ITA).

The obtained results came from the implementation of good development practices and also from the commitment of the students that took part in the BDIC-DM project, who spent a great amount of technical effort to make the execution of this project possible.

The proposed architecture at the beginning of the project proved to be capable of fulfilling the assigned mission. The BDIC-DM project has provided a realistic development and research experience of the topics discussed in this document.

Finally, it is concluded that the integration of the Hadoop ecosystem technology, NoSQL, and relational databases enables a wide range of implementations that bring important values such as processing and storage scalability, performance, and flexibility.

57.7 Future Works

It is suggested to continue this project by using the Apache Spark [26], Stack, and Cassandra database in an elastic cloud-based storage environment, enabling the execution of MapReduce jobs in memory.

Acknowledgment The authors would like to thank the Brazilian Aeronautics Institute of Technology (ITA), for the support and contribution during this project, the 2RP Net enterprise, and the SPOT Project for supporting some hardware and software infrastructure to develop this academic proof of concept.

References

1. Lees, A., & King, M. (2015). *World payment report*. Capgemini Consulting and Royal Bank of Scotland (pp. 1–36), Vol. 1.
2. Hey, T., Tansley, S., & Tolle, K. (2009). The fourth paradigm: Data-intensive scientific discovery. In *E-Science and information management*. Berlin/Heidelberg: Springer.
3. Reed, D. A., & Dongarra, J. (2015). Exascale computing and big data. *Communications of the ACM, 58*(7), 56–68.
4. Chen, M., Mao, S., & Liu, Y. (2014). Big data: A survey. *Mobile Networks and Applications, 19*(2), 171–209. Available at: http://dx.doi.org/10.1007/s11036-013-0489-0.
5. Tsai, C. W., et al. (2015, October). Big data analytics: A survey. Journal of Big Data, 2, 1–32. Available at: http://dx.doi.org/10.1186/s40537-015-0030-3.
6. Guerra, V. da C., et al. (2014, April). Interdisciplinarity and agile development: A case study on graduate courses. In ITNG 2014 – Proceedings of the 11th international conference on information technology: New generations (pp. 622–623). Las Vegas: IEEE Computer Society.
7. da Cunha, A. M., et al. (2008). Estudo de Caso abrangendo o Ensino Interdisciplinar de Engenharia de Software. *Fórum de Educação em Engenharia de Software, 43*(8), 80–88. Available at: https://goo.gl/m8JUJc.
8. Carneiro, E. M., et al. (2015, April). Cluster analysis and artificial neural networks: A case study in credit card fraud detection. In *2015 12th international conference on information technology – New generations* (pp. 122–126). Las Vegas.
9. Tiwari, S. (2011). *Professional NoSQL*. Indianapolis: Wiley.
10. Hecht, R., & Jablonski, S. (2011, December). NoSQL evaluation: A use case oriented survey. In *Proceedings – 2011 international conference on cloud and service computing, CSC 2011* (pp. 336–341).
11. Harrison, G. (2015). *Next generation databases*. New York: Apress.
12. Apache-Camel (2011). Apache Cassandra. The Apache Software Foundation. Available at: http://camel.apache.org/index.html. Accessed 17 September 2016.
13. Venner, J. (2009). *Pro Hadoop*. New York: Apress.
14. Ishwarappa, & Anuradha, J. (2015). A brief introduction on big data 5Vs characteristics and hadoop technology. *Procedia Computer Science, 48*(C), 319–324. Available at: http://dx.doi.org/10.1016/j.procs.2015.04.188.
15. Bhosale, H. S., & Gadekar, D. P. (2014). A review paper on big data and Hadoop. *International Journal of Scientific and Research Publications, 4*(10), 2250–3153. Available at: www.ijsrp.org.
16. Shvachko, K., et al. (2010). The Hadoop distributed file system. In *2010 I.E. 26th symposium on mass storage systems and technologies, MSST2010* (pp. 1–10). Incline Village.
17. Dean, J., & Ghemawat, S. (2008). MapReduce: Simplified data processing on large clusters. *Communications of the ACM, 51*(1), 107–113. Available at: http://doi.acm.org/10.1145/1327452.1327492.
18. Thusoo, A., et al. (2010). Hive – A petabyte scale data warehouse using hadoop. In *Proceedings – International conference on data engineering* (pp. 996–1005). Long Beach.

19. Codd, E. F. (1990). *The relational model for database management: Version 2*. Boston: Addison-Wesley Longman Publishing Co.
20. Sqoop. The Apache Software Foundation. Available at: http://sqoop.apache.org/. Accessed 20 Sept 2016.
21. White, T. (2015). *Hadoop: The definitive guide* (4th ed.). Sebastopol: O'Reilly Media, Inc..
22. Huai, Y., et al., (2014). Major technical advancements in apache hive. In *SIGMOD'14*. Snowbird.
23. Generatedata. Available at: http://www.generatedata.com/. Accessed 20 Sept 2016.
24. HiveRunner. Available at: https://github.com/klarna/HiveRunner. Accessed 21 Sept 2016.
25. SQLite. Available at: https://www.sqlite.org/. Accessed 12 Sept 2016.
26. Apache Spark™ – Lightning-fast cluster computing. Available at: http://spark.apache.org/. Accessed 20 Sept 2016.

Enhancing Range Analysis in Software Design Models by Detecting Floating-Point Absorption and Cancellation

58

Marcus Kimura Lopes, Ricardo Bedin França, Luiz Alberto Vieira Dias, and Adilson Marques da Cunha

Abstract

Floating-point subtleties are often overlooked by software developers, but they can have considerable impact over critical systems which make extensive manipulations of real numbers. This work presents a method for detecting floating-point absorption and cancellation when performing variable range analysis in design models. Employing this method as early as in the design phase permits cheaper detection and treatment of such floating-point anomalies. Our method works by analyzing sums and subtractions in a model and verifying if, given two variable ranges, there are sub-ranges of them on which absorption and cancellation can occur. We also provide the number of canceled bits for each sub-range in order to permit better assessment of the impact of each detected cancellation. We implemented our method in a range analysis tool that operates over SCADE models: results are presented in an HTML report and cancellations are depicted in graphs. This method can be used for early design analysis or as a basis for more complex, end-to-end numerical precision verification.

Keywords

Floating-point • Loss of Precision • Static Analysis

58.1 Introduction

The use of floating-point types to represent and manipulate real ($\mathbb{R}$) numbers in computers is widespread: most modern microprocessors comply, at least partially, to the IEEE 754 [1] standard. In addition, popular programming languages usually contain types such as `float`, `single` or `double`, which refer to data types defined in IEEE 754. However, the subtleties of floating-point arithmetic are not always clear to average programmers, who may not know

the situations where floating-point arithmetic and the real, infinitely precise one, may present noticeable differences. Unexpected behavior that arises from misuse of floating-point is especially problematic in safety-critical control systems (such as aircraft flight controls, medical devices and nuclear plants), as these make extensive use of real numbers and their malfunctioning can cause loss of lives and resources. One infamous example of system malfunction as a consequence of floating-point misunderstanding is the failure of the Patriot missile defense battery [2], which caused 28 deaths.

While the development of safety-critical software products usually follows strict processes, based on guidance such as the DO-178C [3] and IEC 61508-3 [4], floating-point arithmetic – with all its complexity – is no more than one single point of concern in an entire product life cycle. Hence, specific guidance related to floating-point is scarce and there

M.K. Lopes • R.B. França
EMBRAER S.A., São José dos Campos, SP, Brazil
e-mail: marcus.lopes@embraer.com.br; ricardo.franca@embraer.com.br

L.A.V. Dias (✉) • A.M. da Cunha
Instituto Tecnológico de Aeronáutica (ITA), São José dos Campos, SP, Brazil
e-mail: vdias@ita.br; cunha@ita.br

© Springer International Publishing AG 2018
S. Latifi (ed.), *Information Technology – New Generations*, Advances in Intelligent
Systems and Computing 558, DOI 10.1007/978-3-319-54978-1_58

is little to no availability of mature, commercial-off-the-shelf (COTS) floating-point analysis solutions.

In this paper, we present a simple method for early detection of floating-point absorption and cancellation in software products, using their design models as inputs, in order to detect code parts where floating-point errors are likely to be amplified. While we obviously assume that software design is performed mostly using model-based development, this assumption makes sense in the development of safety-critical control systems because design models are usually suitable for simulation, automatic code generation or both. Our proposed method is implemented over the RangeAnalyzer tool, developed by Honda and Vieira Dias [5], as its inputs are the ranges of each variable subject to sum or subtraction and their tool provides these ranges for our use. While this paper is not necessarily restricted to safety-critical software, we do assume the use of methods and languages (such as SCADE) that are much more popular in safety-critical embedded systems than anywhere else.

This paper is structured as follows: Sect. 58.2 presents elementary aspects of floating-point. In Sect. 58.3 we assess works that are related to ours. Section 58.4 describes our proposed method for absorption and cancellation detection, Sect. 58.5 presents the implementation of our method – together with an example of analysis in a SCADE model – and, in Sect. 58.6, we draw conclusions and present perspectives of future work.

58.2   Floating-Point Basics

In this section, we limit ourselves to a basic presentation of floating-point representations, absorption and cancellation. Much deeper information is available in the IEEE 754 standard or in other works, such as Goldberg's [6].

A floating-point number, in contrast to an integer or fixed-point one, has a variable resolution – an integer, for example, always has resolution of one. This variable resolution exists due to the presence of an exponent in any floating-point representation: while a 32-bit signed integer has one sign bit and a 31-bit significand, a IEEE 754-compliant 32-bit floating point has one sign bit, 23 significand (also known as mantissa) bits (with a resolution of 2^{-23}) and eight exponent bits. In this paper, we focus on the 32-bit floating-point format, usually known as `single` or `float`: it is the less precise of the binary floating-types prescribed by IEEE 754, thus being more vulnerable to noticeable arithmetic discrepancies with respect to the traditional arithmetics. Moreover, a number of microcontrollers and microprocessors do not have hardware implementation for other floating-point formats.

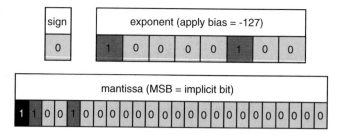

Fig. 58.1   The number 50 in single precision floating-point

As shown in Fig. 58.1, the number 50, is represented in floating-point as $(2^0 + 2^{-1} + 2^{-4}) * 2^5$. The number 6.25 (which is $50 * 2^{-3}$) would be represented as $(2^0 + 2^{-1} + 2^{-4}) * 2^2$. We can notice that these two numbers have identical mantissas and only their exponents vary – this permits an enormous range of numbers (which is proportional to the number of bits used to encode the exponent) but implies a variable resolution, since the least significant mantissa bit is worth $2^{-23} * 2^{exponent}$. In other words, the absolute precision of numbers close to 6.25 is finer than the absolute precision of numbers close to 50. Notice that Fig. 58.1 shows 24 (rather than 23) mantissa bits: since all normalized numbers respect the constraint $1 \leq mantissa < 2$, floating-point implementations "save" one bit.

Since floating-point is a discretized representation of real numbers, with a finite number of bits, it does not always follow traditional arithmetics. Sums and subtractions are particularly tricky because we need to represent both numbers in the exponent of the larger magnitude (in order to sum their mantissas) and, before completing the operation, make sure the result remains a normalized floating-point.

When summing numbers with large magnitude differences, we may have **absorption**, i.e. the smaller magnitude becomes meaningless with respect to the larger one. In single precision, using the IEEE 754 default rounding mode, $5 * 10^6 + 0.1 = 5 * 10^6$ because 0.1 is so small compared to $5 * 10^6$ (which has a resolution of 0.5) that it has no effect in a sum.

When subtracting numbers of similar magnitude, we may have **cancellation**. In cancellation, the most significant bits of each number are similar and cancel themselves out after the subtraction. While this is not a problem if the inputs of the subtraction were exact floating-point numbers (a "benign" cancellation), any rounding errors already present in the inputs would become proportionally larger after cancellation – a "catastrophic" cancellation. If we perform 50-49.99999 in single precision floating-point, our result would be $1.14 * 10^{-5}$, rather than 10^{-5}: since 49.99999 is not an exact floating-point number, its last bit had to be rounded and its actual value had a small error. After the subtraction, the absolute error remained the same but it became relatively large with respect to the result.

For a subtraction $x - y$, where $x > y > 0$, the number of canceled-out bits is given by the "Theorem on Loss of Precision" [7]: if there are positive integers p and q such as $2^{-p} \leq 1 - (y/x) \leq 2^{-q}$, the number of lost bits is between q and p.

58.3 Related Work

Despite the many years of research regarding floating-point arithmetic analysis, there is still a lack of options when looking for mature, ready-to-use methods and tools. Most of the relevant recent work remains either on theory or on tools that are not available for commercial use.

The most important work related to ours is certainly the development of the Fluctuat tool and its use in airborne software programs [8]. This tool performs static analysis on C programs, using the abstract interpretation formal method, and provides floating-point accuracy analysis by comparing the actual C-coded variables with idealized, real ($\mathbb{R}$) ones. Later works such as [9] and [10] criticize this tool as providing too conservative results – which might be too imprecise and give false alarms even if the algorithms are sufficiently precise – but Benz does mention that Fluctuat and static analysis tools in general "work exceptionally well in special domains like embedded systems". However, Fluctuat is not available for commercial use – this impeded its evaluation in our study.

One recent attempt to make static analysis less conservative is [11]. The authors combine abstract interpretation and model checking to analyze floating-point errors in Java programs. Otherwise, the basic concept used is the same: they compare actual values generated (via symbolic execution) by a program with the idealized ones. While the concept does look promising, the authors acknowledge scalability issues – a problem that commonly plagues formal analysis of floating-point computations.

Another possibility for formal analysis of floating-point computations is the use of theorems and proof assistants. Harrison [12] successfully used the HOL Light theorem prover to perform formal verification on several floating-point algorithms; Akbarpour and Tahar [13] used HOL to perform error analysis of digital filters – which are commonly used in flight control software. In order to make use of such an approach in a repeatable fashion, we would need to automate the task of parsing C code into HOL or any other language that contains a formal model for the IEEE 754 - floating-point arithmetic.

Given the limitations and complexity of static analysis, some researchers advocate the use of dynamic, run-time floating-point error evaluation. This approach was used in several works, such as [9] and [14] and can even be used to help developing mixed-precision code, where the precision of a variable is chosen automatically by the tool [10, 15].

In our work, we opted to focus on single precision floating-point and static analysis in order to permit early analysis without any concern about developing test cases. Although we do not intend to give concrete end-to-end evaluation of floating-point precision in algorithms, we consider that our tool can provide easily understandable insights that may be helpful for system and software designers.

58.4 Detecting Absorption and Cancellation in Range Analysis

In this work, we use simple means to detect absorption and cancellation in SCADE models, assuming that all function inputs have their ranges defined and using the algorithms (based on Abstract Interpretation) and tool described in [5] to propagate these ranges along all intermediate variables. Whenever the analyzer encounters sum and subtraction blocks (i.e. those who cause absorption and cancellation), we can use the range of their inputs to check for absorption or cancellation.

In this section, we use interval notation (e.g. $[min_x, max_x]$) to describe a given range x. All the equations are based on 24-bit single precision mantissa: replacing 24 with 53, for example, would represent the double precision numbers.

58.4.1 Absorption

While absorption does not necessarily degrade the precision of algorithm, it may be evidence of poor design, as some variable value is potentially ignored inside a computation. Since absorption is linked to absolute values, we can safely ignore range signs and treat sums and subtractions in the same manner. In order to minimize the number of false alarms, we will consider that an absorption happened whenever the maximum absolute values of two summed ranges differ by a magnitude larger than 2^{24}, as depicted in Equation 58.1. In this way, we will detect absorption whenever a variable whose value can be potentially enormous is summed to another one whose value cannot be nearly as high.

$$\neg \left(2^{-24} < \frac{\max(|min_1|, |max_1|)}{\max(|min_2|, |max_2|)} < 2^{24} \right) \quad (58.1)$$

58.4.2 Cancellation

Our main contribution in this paper is the cancellation detection during model range analysis – other works [9, 10] also emphasize the importance of detecting cancellations. Here, we do not try to separate benign cancellations from catastrophic ones, since the inputs of safety-critical systems

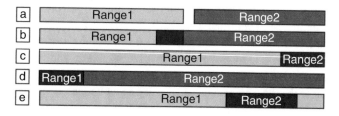

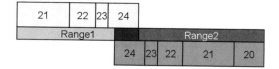

Fig. 58.2 Possibilities for subtracting two non-negative ranges

Fig. 58.3 Graphical view of bits lost in a subtraction

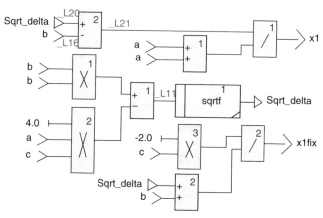

Fig. 58.4 Simplified quadratic equation in SCADE

often come from other systems and were probably rounded or truncated.

In order to render our cancellation detection algorithm easier to understand, we present here a simplified version of it. The first simplification is the use of non-negative ranges: actual ranges can be fully negative or even have a negative lower limit and a positive higher one; then we shall reorder the subtraction (if both operands are negative) and avoid using the Theorem on Loss of Precision for numbers of different signs. Another simplification in this paper is that we sort the ranges in order to ease the algorithm presentation: whenever "Range1" and "Range2" are mentioned, we assume Range1 is the range with lowest lower limit or, if both ranges have the same lower limit, Range1 shall have the lowest higher limit.

With these simplifications, the two ranges to be subtracted can be represented by one of the situations depicted in Fig. 58.2. The first one represents two disjoint ranges, the second one represents two ranges with a partial intersection and, in the other three, one range is a sub-range of the other. Notice that in situations (c) and (d) we have a common limit between the two ranges. As we shall see in this section, these situations, as well as situations (a) and (b), are much simpler than situation (e).

Notice that we may have all 24 bits canceled out if two ranges intersect; the number of canceled out bits gets smaller as we use numbers that are farther away from the intersection. This intuition is presented in Fig. 58.3.

The formalization of this intuition can be done with the Theorem on Loss of Precision: knowing that $1 - (y/x) \leq 2^{-q}$ means that we have surely lost q bits of precision, we can fix q as 24 and y as min_2 (i.e. the lower limit of Range2) in order to find the values in Range1 where all bits are canceled out. We have $1 - (min_2/x) \leq 2^{-24}$, which implies $x \geq min_2 * (1 - 2^{-24})$. Since all values of x shall belong to Range1, the actual sub-range where all bits are lost is the intersection of $[min_2 * (1 - 2^{-24}), \infty)$ and Range1 itself, excluding all

values of Range1 that are larger than max_2 in situation (e) of Fig. 58.2. A similar reasoning can be used to find out the sub-ranges of Range1 where 23, 22 (and so on) bits are lost: n bits ($n > 0$) are lost in the intersection of $[min_2 * (1 - 2^{-n}), min_2 * (1 - 2^{-(n+1)}))$ and Range1. The same method is used for Range2: all bits are lost in $(-\infty, max_1(1 - 2^{-24})] \cap Range2$ and n bits are lost in $(max_1(1 - 2^{-(n+1)}), max_1(1 - 2^{-n})] \cap Range2$.

Situation (e) of Fig. 58.2 is trickier than the others because, while the entire Range2 is subject to loss of all bits, the loss of precision in Range1 shall be computed twice: one excluding the values of Range1 that are larger than max_2 and another one excluding the values of Range1 that are smaller than min_2.

In our method, we opted to compute the sub-ranges for all possible numbers of cancelled bits because the impact of cancellation may vary according to tolerances deemed acceptable by system and software designers. If a relative error of, say, 10^{-4} is deemed acceptable for a computation, we need to preserve at least 14 bits (i.e. have at most 10 canceled out bits) of a mantissa, as $10^{-4} \approx 2^{-13.28}$.

58.5 Results

Since we implemented this method to evaluate single-precision computations, we could safely use double precision when computing ranges. If the model to be analyzed is supposed to be run in larger precision, using extended-precision libraries (such as MPFR[1]) is important in order to make sure that the analysis tool behaves more closely to actual arithmetics than the design model.

For confidentiality and simplicity reasons, we present the results of our implementation using the quadratic equation $ax^2 + bx + c$ as an example. Figure 58.4 shows a SCADE

---

[1] http://www.mpfr.org

Fig. 58.5 Range analysis report

Name	Calculated LR	Calculated UR	Comments
Input List			
a	1.0	1.0	
b	5000.0	5000.0	
c	0.01	1000.0	
Internal List			
Sqrt_delta	4999.599983	4999.999996	
_L11	24996000	24999999.96	WARNING: Possible absorption. Absolute difference value could be greater than 2^24
_L21	-0.400016	-3.999999e-006	WARNING: Cancellation Report: SUB_L21.html

Fig. 58.6 Cancellation graph in the quadratic equation

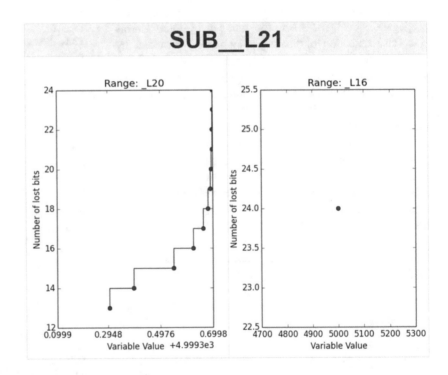

model that finds one of its roots, assuming that b is positive. As shown by Press et al. [16], the direct computation of $-b + \sqrt{b^2 - 4ac}$ (with $b > 0$) would result in a cancellation, while computing the same equation root with $-2c(b + \sqrt{b^2 - 4ac})$ avoids this problem.

We executed our implementation using the model of Fig. 58.4 with the following ranges:

$$a = [1, 1]; b = [5000, 5000]; c = [0.01, 1000] \qquad (58.2)$$

We noticed, as expected, that smaller values of c imply $\sqrt{b^2 - 4ac} \approx b$. While no cancellation was reported in the data flow that leads to $x1fix$, the tool did report a potential absorption when computing $b^2 - 4ac$ (_L11 in Fig. 58.4) and a potential cancellation in $\sqrt{b^2 - 4ac} - b$ (_L21), as shown in Fig. 58.5.[2]

Whenever a cancellation occurs, the tool generates a graph showing the number of bits that were possibly

---

[2] Due to paper size constraints, we show only a part of the original RangeAnalyzer report.

canceled out in both variables involved in the subtraction. As shown in Fig. 58.6, larger values of _L20 may cause larger cancellation, while _L16 (which corresponds to input *b*), being fixed, is responsible for a full cancellation in the worst case (i.e. if _L20 is large).

58.6 Conclusion and Future Work

In this paper, we presented a method that enables early detection of potential floating-point absorption and cancellation phenomena in SCADE models, using Honda's RangeAnalyzer tool as the basis for our implementation of the proposed method. We consider that early floating-point analysis can help in the development of numerically stable operators which help providing less error propagation along a complex software program.

There is ample scope for future work in this subject: while our method helps finding cancellation "hotspots", a more complete analysis should include floating-point error propagation analysis. Also, the number of false alarms (i.e. false detection of floating-point issues) provided by our method depends on the precision of the underlying variable range analyzer: the less imprecision present in the range analysis, the smaller is the number of false floating-point alarms. Hence, it would be interesting to compare the precision of RangeAnalyzer with respect to other available tools (e.g. Frama-C value analysis[3]) in order to minimize the number of false alarms, which are often the major drawback of tools based on abstract interpretation. Another open perspective is the integration of this analysis with tests and other techniques in order to fulfill the objectives required by documents such as the DO-178C.

References

1. IEEE Computer Society (2008). *IEEE standard for floating-point arithmetic*. New York: Institute of Electrical and Electronics Engineers.
2. Carlone, R. V. (1992). Patriot Missile Defense: Software problem led to system failure at Dhahran, Saudi Arabia (GAO/IMTEC-92-26), U.S. Government Accountability Office, Technical report.
3. DO-178C (2011). *Software considerations in airborne systems and equipment certification*. Washington, DC: Radio Technical Commission for Aeronautics (RTCA).
4. International Electrotechnical Commission (2010). *Functional safety of electrical/electronic/programmable electronic safety-related systems – Part 3: Software requirements*. Geneva: International Electrotechnical Commission (IEC).
5. Honda, R. M., & Dias, L. A. V. (2013). Rangeanalyzer: An automatic tool for arithmetic overflow detection in model-based development. In *ITNG 2013*, Las Vegas (pp. 254–259).
6. Goldberg, D. (1991). What every computer scientist should know about floating-point arithmetic. *ACM Computing Surveys, 23*(1), 5–48. [Online] Available: http://doi.acm.org/10.1145/103162.103163
7. Cheney, E. W., & Kincaid, D. (2012). *Numerical mathematics and computing* (7th ed.). Boston: Brooks/Cole Publishing Co.
8. Delmas, D., Goubault, E., Putot, S., Souyris, J., Tekkal, K., & Védrine, F. (2009). Towards an industrial use of FLUCTUAT on safety-critical avionics software. In M. Alpuente, B. Cook, & C. Joubert (Eds.), *Formal methods for industrial critical systems* (Lecture notes in computer science, Vol. 5825, ch. 6, pp. 53–69). Berlin/Heidelberg: Springer. [Online] Available: http://dx.doi.org/10.1007/978-3-642-04570-7_6
9. Benz, F., Hildebrandt, A., & Hack, S. (2012). A dynamic program analysis to find floating-point accuracy problems. In *Proceedings of the 33rd ACM SIGPLAN Conference on Programming Language Design and Implementation, PLDI'12* (pp. 453–462). New York, NY: ACM.
10. Lam, M. O. (2014). *Automated floating-point precision analysis*. Ph.D. dissertation, University of Maryland, College Park.
11. Ramachandran, J., Pasareanu, C. S., & Wahl, T. (2015). Symbolic execution for checking the accuracy of floating-point programs. *ACM SIGSOFT Software Engineering Notes, 40*(1), 1–5.
12. Harrison, J. (2006). Floating-point verification using theorem proving. In M. Bernardo & A. Cimatti (Eds.), *Formal Methods for Hardware Verification, 6th International School on Formal Methods for the Design of Computer, Communication, and Software Systems* (Lecture notes in computer science, Vol. 3965, pp. 211–242). Bertinoro: Springer.
13. Akbarpour, B., & Tahar, S. (2007). Error analysis of digital filters using HOL theorem proving. *Journal of Applied Logic, 5*(4), 651–666.
14. Bao, T., & Zhang, X. (2013). On-the-fly detection of instability problems in floating-point program execution. In *Proceedings of the 2013 ACM SIGPLAN International Conference on Object Oriented Programming Systems Languages & Applications, OOPSLA'13* (pp. 817–832). New York, NY: ACM. [Online] Available: http://doi.acm.org/10.1145/2509136.2509526
15. Rubio-González, C., Nguyen, C., Nguyen, H. D., Demmel, J., Kahan, W., Sen, K., Bailey, D. H., Iancu, C., & Hough, D. (2013). Precimonious: Tuning assistant for floating-point precision. In *Proceedings of the International Conference on High Performance Computing, Networking, Storage and Analysis, SC'13* (pp. 27:1–27:12). New York: ACM. [Online] Available: http://doi.acm.org/10.1145/2503210.2503296
16. Press, W. H., Teukolsky, S. A., Vetterling, W. T., & Flannery, B. P. (1992). *Numerical recipes in C: The art of scientific computing* (2nd ed.). New York: Cambridge University Press.

---

[3] http://frama-c.com/value.html

Requirements Prioritization Using Hierarchical Dependencies

59

Luay Alawneh

Abstract

Software development environments are susceptible to changes in requirements due to the abrupt updates in market needs. This imposes a burden on the software development team to adhere to these new changes by prioritizing them based on their significance to the project. However, prioritization of the list of requirements is not a spontaneous task as it involves several attributes such as requirements importance, complexity, cost, and completion time. Stakeholder requirements are usually interrelated where they may contribute to the same use cases and the same quality attributes that are identified during the specifications phase. Thus, considering the relationships among requirements at the different specification levels should be investigated in the prioritization process. In this paper, we propose a new approach for requirements prioritization using the relationships between the stakeholder requirements and their derived specifications in the form of use cases and non-functional requirements.

Keywords

Derived requirements • Use cases • Functional requirements • Non-functional requirements • Requirements pyramid

59.1 Introduction

Software development is perhaps one of the highly demanding areas where customer satisfaction may change quickly when vital requirements are not accurately delivered. Consequently, the management should carefully address the changes in requirements and the impact of certain requirements on the product's quality [1]. This includes conducting impact analysis on the existing features in order to measure the feasibility and the consequences of applying the new requirements. The management should accurately measure the importance of each new requirement in order to forecast the project timeline in the next releases.

In agile software development environments, requirements should be prioritized to be included in the next releases which may result in rearranging the list of requirements based on new high priority requests [2]. Requirements could be related to the same feature in the software system where one is concerned with the functionality of this feature and the other is responsible for maintaining its quality. Requirements which target the quality of service are classified as non-functional requirements. Non-functional requirements should not be of less importance than their functional counterparts since the user experience is also evaluated based on the quality of the software system such as its performance and usability [7]. For example, the user should be able to access the required features easily and should not experience any major delays in the response coming from the target system. Therefore, it is important to consider all kinds of requirements when preparing the project timeline. It is evident that, in normal situations, the list of requirements could not fit a single release. Hence,

L. Alawneh (✉)
Software Engineering Department, Jordan University
of Science & Technology, Irbid, Jordan
e-mail: lmalawneh@just.edu.jo

© Springer International Publishing AG 2018
S. Latifi (ed.), *Information Technology – New Generations*, Advances in Intelligent
Systems and Computing 558, DOI 10.1007/978-3-319-54978-1_59

requirements prioritization should be considered in order to gain the customer satisfaction and maintain the level of quality in the software system.

In a software project, new requirements may come from different stakeholders such as the customer management, product users, product owners, software architects, and software developers. Usually, requirements from the business side should be of high priority. However, in some cases, requests from the technical side should be taken seriously in order to maintain the quality of service. The rearrangement of the requirements in the project timeline should be handled carefully since delaying important requirements ignorantly may impact the success of the project. Thus, a keen requirements prioritization process should be applied in order to reach an agreement among the stakeholders.

There exist many requirements prioritization techniques that solve this problem [4, 10]. Despite the adoption of these techniques in the literature and the validity of their prioritization results, they have been applied without taking into consideration the hierarchical dependencies (stakeholder needs and their derived requirements) among requirements. Derived requirements are the detailed requirements that were extracted from the stakeholder needs in the form of use cases or non-functional requirements.

In this paper, we propose an approach for prioritizing stakeholder requirements based on the dependencies among their derived requirements. For example, two different requirements introduced by two different stakeholders may share common features, or may contribute to the same non-functional requirements. Moreover, we suggest to use the relativeness among the derived requirements in order to calculate the relative values and costs between two stakeholder requirements [4].

The rest of the paper is organized as follows. Section 59.2 discusses the related work. In Section 59.3, we detail the proposed approach. Section 59.4 presents a running example that explains and discusses the relativeness calculation and the requirements prioritization steps. Section 59.5 concludes the paper with a glimpse on the future directions.

59.2 Related Work

An abundant number of research studies ranked prioritization as one of the very important steps in the requirements engineering process [5].

One of the most popular techniques for decision making based on prioritization is the Analytic Hierarchy Process (AHP) proposed by Saaty [3]. The AHP works by comparing all the requirements according to the relative values between each pair of requirements (a matrix is constructed in this step). Then, a normalized matrix is extracted from the relative values matrix and finally a numerical value is derived for each requirement that shows its importance based on a specified criterion.

AHP could be applied as a multi-criteria method where the relative values from several criteria can be compared to guide in decision making. Even though the AHP method proved to be very reliable, it suffers from the scalability problem as more comparisons are needed when the number of requirements increases [1, 6]. A number of techniques were proposed in the literature that are based on AHP [4, 5]. Perini et al. [1] proposed a case-based driven approach, part of the Case-Based Ranking framework [6], to overcome the scalability problem in AHP by applying machine learning techniques to infer the requirements rankings approximations. Another approach that applied the AHP technique based on two criteria is the Cost-Value approach [9]. To overcome the scalability problem caused by the number of pairwise comparisons, Tonella et al. proposed to benefit from algorithms found in genetics [14].

Babar et al. [8] proposed PHandler, a decision support system that targets the prioritization of requirements in large scale systems. PHandler is a knowledge based system that uses a hybrid approach of three techniques: the Back Propagation Neural Network (BPNN), the Value-Based Intelligent Requirement prioritization (VIRP), and AHP. The authors applied the hybrid approach to large scale requirements with the objective of removing the biases imposed by the different stakeholders.

McZara et al. [10] proposed a new method for requirements prioritization referred to as SNIPR which is based on natural language processing and satisfiability modulo theories solvers. The authors conducted an experiment that involved software engineers and compared their approach to the weighted sum technique. The results show that their approach outperforms in terms of execution time and accuracy. Azar et al. [13] presented a value-oriented prioritization (VOP) framework that uses the stakeholder ratings to establish the links between requirements and business values. This approach derives from the AHP approach since it constructs a comparison matrix based on the requirements' relative values and costs.

Shao et al. [11] proposed DRank, a semi-automatic requirements prioritization method that takes into consideration the dependencies among requirements and the stakeholders' preferences. The approach proposes that integration among subjective prioritization, requirements' contributions, and the business order should be taken into consideration to get accurate rankings of requirements and to eliminate any biasness in the prioritization results. However, the authors state that their approach is still prone to the issue of subjectivity especially in the process of requirements evaluation. Kukreja et al. [12] presented a prioritization tool that can handle prerequisites in the requirements prioritization process. Another interesting study proposed by Tuunanen et al. [15] suggested that the cultural context imposed by the differences of the stakeholders should be considered when conducting requirements prioritization.

In this paper, we use the *cost* and *value* criteria to measure the priorities of requirements. In the following, we present our approach for requirements prioritization based on the relationships between high-level and derived requirements.

59.3 Approach

We present a new approach for requirements prioritization that takes into consideration the relationships among the stakeholders' needs and the derived requirements in the form of use cases and non-functional requirements.

Figure 59.1 depicts the relationships between the stakeholder needs and their derived requirements. This figure is an excerpt of the requirements pyramid [16] where we only present the agreed upon requirements and the derived use cases and non-functional requirements.

This example represents four requirements RQ1-RQ4 and their derived use cases (UC1-UC6) and non-functional requirements (NFR1 and NFR2). It shows that RQ1 is involved in UC1 and UC2. Similarly, RQ2 contributes to UC2, UC3 and NFR1. It is clear that RQ1 and RQ2 are interrelated through UC2. Thus, we use these dependencies as the basis in the calculation of the priorities among requirements.

Figure 59.2 depicts the steps involved in the prioritization process. Usually, new requirements are continuously gathered from the different stakeholders in the project. The newly elicited requirements should be integrated in the project timeline. This task is not straightforward as it should involve the input of the different stakeholders in the project in order to agree on which requirements should have precedence.

The integration step should result in an agreement with the stakeholders on the feasibility of the newly added requirements and their importance to the success of the project. The problem faced at this stage is how to embed the new requirements in the timeline without impacting the overall progress by taking into account the time, budget and resources constraints. Thus, existing requirements may be

delayed or suspended if the newly added requirements should prevail.

A sound prioritization method should be applied to have the best rearrangement of the requirements in order to attain the most value in the project. The integration and the rearrangement steps are iterative until the best schedule is reached. The importance of the requirement is determined using several factors such as the stakeholders' opinions, complexity, time to implement, risks, and the severity of failures. The latter is very important since it may affect the delivery time of a certain requirement but not its priority. This is due to the fact that high severity requirements should adhere to stringent validation in order to ensure the quality of service.

However, subjectivity in evaluating the requirements could interfere negatively to the project success. Our approach works on prioritizing stakeholder requirements (RQ) based on their derived use cases (UC) and non-functional requirements (NFR) and the relationships among them. The advantage of prioritizing the requirements at the RQ level instead of the UC/NFR level is that in the latter, the relationships among the requirements are not directly involved in the prioritization process. This may have an impact on the quality of the prioritization results as they could be biased towards a certain feature. Moreover, the number of comparisons involved at the RQ level is usually less than that in the UC/NFR level. Taking this into consideration could improve the performance and accuracy of prioritization techniques such as AHP [3]. It should be noted that the proposed approach would not bring any value, in terms of performance, to the prioritization process when the ratio between the number of requirements and their derived ones is close to 1.

We follow the cost-value approach to prioritize requirements based on their relative values (degree of importance) to the project and their relative costs (man-hour) to implement. The *relativeness computation* step calculates the relative *values* and the relative *costs* between requirements $(RQ_1 \ldots RQ_m)$ using the *values* and *costs* that were assigned

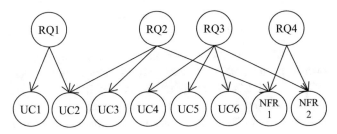

Fig. 59.1 Requirements hierarchy

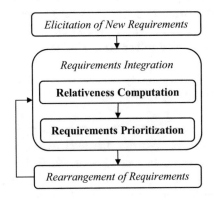

Fig. 59.2 Requirements prioritization approach

to the derived requirements. The assignment of costs and values to the derived requirements is more accurate as they include more specifications.

The *relativeness R* between RQ_i and RQ_j for any criterion is computed as follows:

$$WC(RQ_i) = \sum_n WC(UC_n) + \sum_m WC(NFR_m) \quad (59.1)$$

$$WC_s(RQ_i, RQ_j) = \sum_k WC(RQ_i) \cap WC(RQ_j) \quad (59.2)$$

$$\overline{RQ}_i = WC(RQ_i) - WC_s(RQ_i, RQ_j) \quad (59.3)$$

$$R_{i,j}(RQ_i, RQ_j) = \begin{cases} \dfrac{\overline{RQ}_i}{\overline{RQ}_j} & \\ \dfrac{1}{\overline{RQ}_j} \text{ when } \overline{RQ}_i = 0 \\ \overline{RQ}_i \text{ when } \overline{RQ}_j = 0 \end{cases} \quad (59.4)$$

Equation 59.1 calculates the total weights of the derived requirements covered by RQ_i where WC represents the total weight for RQ_i based on the weights of the derived requirements for a specific criterion C. Eq. 59.2 computes the sum of weights (WC_s) for the shared requirements between RQ_i and RQ_j. In Eq. 59.3, the value in $\overline{RQ}_i$ measures the weight that will be considered in the calculation of R which is the weight of the covered derived requirements by excluding the weights of the shared ones. The reason behind this is that we need to get the weights of the distinct derived requirements in order to reduce the existence of any biasness.

Equation 59.4 calculates the relativeness R based on the weights of the two compared requirements. We apply this step for the value and cost criteria. Thus, the output of this step is the *relative value* R_v and the *relative cost* R_c between each pair of requirements. The output of this step will then be input to the prioritization step.

In this paper, we used the AHP method in order to find the priorities of requirements. After calculating the priorities, the final step will be to rearrange the requirements in the subsequent releases. In the next section, we provide a running example that illustrates our prioritization approach.

59.4 Running Example

We use the example presented in Fig. 59.1 to explain the steps involved in preparing the relative *values* and *costs* between the high level requirements. Table 59.1 shows the four stakeholder requirements and their corresponding derived use cases and non-functional requirements. RQ4 is

Table 59.1 Requirements hierarchy

Requirement	Derived requirements
RQ1	UC1, UC2
RQ2	UC2, UC3, NFR1
RQ3	UC4, UC5, UC6, NFR2
RQ4	NFR1, NFR2

Table 59.2 Weight and man-hour

Requirements	Value	Man-hour (cost)
UC1	24	80
UC2	6	40
UC3	5	200
UC4	4	100
UC5	12	80
UC6	20	220
NFR1	18	400
NFR2	11	90

only concerned with non-functional requirements which are shared with RQ2 and RQ3.

Table 59.2 shows the *values* and *costs* (in terms of man-hour) that were assigned to the derived requirements. These numbers were assigned by the stakeholders in the requirements engineering process (more specifically in the evaluation and agreement phase). Even though these figures could be biased towards one of the stakeholders in the project, they should be a result of a rigorous evaluation.

To calculate the *relative value* between RQ1 and RQ2, we use Eqs. 59.1, 59.2, 59.3 and 59.4 as follows.

$$\overline{RQ}_1 = W(UC1, UC2) - W(UC2) = 24$$

$$\overline{RQ}_2 = W(UC2, UC3, NFR1) - W(UC2) = 23$$
$$\rightarrow R_{1,2} = \frac{24}{23} \approx 1.04$$

Similarly, the *relative value* between RQ2 and RQ3 can be calculated as:

$$\overline{RQ}_2 = W(UC2, UC3, NFR1) - W(\varnothing) = 29$$

$$\overline{RQ}_3 = W(UC4, UC5, UC6, NFR2) - W(\varnothing) = 47$$

$$\rightarrow R_{2,3} = \frac{29}{47} \approx 0.62$$

It should be noted that this calculation should only be done once for each pair of requirements as the relative value between RQ1 and RQ2 is the reciprocal of that of RQ2 and RQ1. Table 59.3 presents the relative values between the four requirements by using the values in the *Value* column from

Table 59.3 Relative values

	RQ1	RQ2	RQ3	RQ4	Relative value
RQ1	1.00	1.04	0.64	1.03	0.2209
RQ2	0.96	1.00	0.62	1.00	0.2126
RQ3	1.57	1.62	1.00	2.00	0.3641
RQ4	0.97	1.00	0.50	1.00	0.2024

Table 59.4 Relative man-hour

	RQ1	RQ2	RQ3	RQ4	Relative cost
RQ1	1.00	0.13	0.24	0.24	0.0972
RQ2	7.50	1.00	1.31	2.67	0.7220
RQ3	4.08	0.77	1.00	1.00	0.4087
RQ4	4.08	0.38	1.00	1.00	0.3878

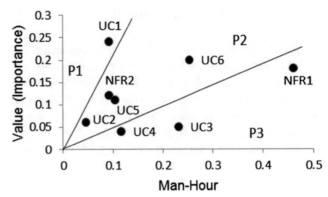

Fig. 59.4 Prioritization using original approach

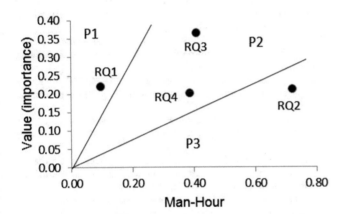

Fig. 59.3 Prioritization using proposed approach

Table 59.5 Original vs. proposed approach

Original approach		Proposed approach		
UC + NFR	Priority (P)	RQ	$\sum P$	$\bar{P}$
UC1	P1	RQ1	P1	P1
UC2	P2	RQ1, RQ2	P3 + P1	P2
UC3	P3	RQ2	P3	P3
UC4	P3	RQ3	P2	P2
UC5	P2	RQ3	P2	P2
UC6	P2	RQ3	P2	P2
NFR1	P3	RQ2 + RQ4	P3 + P2	P3
NFR2	P2	RQ3 + RQ4	P2 + P2	P2

Table 59.2. The *Relative Value* column is calculated using the AHP method.

Similarly, Table 59.4 shows the *relative cost* (in terms of man-hour) by using the values in the Man-hour column from Table 59.2. Then, we use the AHP method to compute the *relative value* and the *relative cost* values [4].

Figure 59.3 depicts the cost-value diagram for the four requirements using our approach. The diagram has three regions P1, P2, and P3 where P3 has the lowest priority and P1 has the highest priority. In the list of stakeholders requirements RQ1 has the highest priority, RQ3 and RQ4 have medium priority and RQ2 has the lowest priority in the list.

Similarly, Fig. 59.4 shows the cost-value diagram for the eight features generated using the original method. In both cases, we used the AHP method to prioritize the requirements in each requirements level. The relative value in the original approach is straightforward. For example, the relative value between UC1 and UC2 is 24/6.

Table 59.5 presents a comparison between the original approach (using AHP only) and our proposed approach. The $\sum P$ column combines the priorities for each derived requirement based on the high level requirement it belongs to. For example, UC2 occurs in RQ1 and RQ2 with priorities P3 and P1 respectively. Therefore, the priority will be the average of P3 and P1, which is equal to P2 as shown in column $\bar{P}$. The only requirement that has a priority level in our approach that is different from the original approach is UC4.

The results, in Table 59.5, show that the prioritization using the proposed approach provided very similar results to those when applying AHP without considering the hierarchical dependencies. The advantage of our approach is that the size of the comparison matrix is reduced to the number of requirements at the higher level. Moreover, the slight difference, caused by UC4, could be due to the fact that UC4 is interrelated with UC5, UC6 and NFR2 which all have P2 level. However, by inspecting the location of UC4 in Fig. 59.4, it is clear that it is very close to the border of P2 level.

59.5 Conclusion and Future Work

We presented a requirements prioritization approach based on the hierarchical dependencies among requirements. We provided a new method to reduce the number of comparisons in a prioritization technique such as the AHP method. To the best of our knowledge, this is the first approach that targets the dependencies among requirements hierarchies. A running example shows that the results of our approach are very accurate with respect to the original AHP method.

In this paper, we applied our approach on a small number of requirements. In the future, we need to test the proposed approach on several examples. More importantly, we intend to apply our approach on the prioritization of a list of requirements from the industry.

References

1. Perini, A., Susi, A., & Avesani, P. (2013). A machine learning approach to software requirements prioritization. *IEEE Transactions on Software Engineering, 39*(4), 445–461.
2. Petersen, K., & Wohlin, C. (2009). A comparison of issues and advantages in agile and incremental development between state of the art and an industrial case. *Journal of Systems and Software, 82* (9), 1479–1490.
3. Saaty, T. L. (2008). Decision making with the analytic hierarchy process. *International Journal of Services Sciences, 1*(1), 83–98.
4. Herrmann, A., & Daneva, M. (2008). Requirements prioritization based on benefit and cost prediction: An agenda for future research. In *Proceedings of the 16th IEEE international requirements engineering conference* (pp. 125–134).
5. Achimugu, P., Selamat, A., Ibrahim, R., & Mahrin, M. (2014). A systematic literature review of software requirements prioritization research. *Information and Software Technology, 56*, 568–585.
6. Avesani, P., Bazzanella, C., Perini, A., & Susi, A. (2005). Facing scalability issues in requirements prioritization with machine learning techniques. In *Proceedings of requirements enginering conference* (pp. 297–306).
7. Chung, L., Leite, P., & Cesar, J. (2009). On non-functional requirements in software engineering. In *Conceptual modeling: Foundations and applications* (Vol. 5600, pp. 363–379). Berlin, Heidelberg: Springer.
8. Babar, M. I., Ghazali, M., Jawawi, D. N., Shamsuddin, S. M., & Ibrahim, N. (2015). PHandler: an expert system for a scalable software requirements prioritization process. *Knowledge-Based Systems, 84*, 179–202.
9. Karlsson, J., & Ryan, K. (1997). cost-value approach for prioritizing requirements. *IEEE Software, 14*(5), 67–74.
10. McZara, J., Sarkani, S., Holzer, T., & Eveleigh, T. (2015). Software requirements prioritization and selection using linguistic tools and constraint solvers—A controlled experiment. *Empirical Software Engineering, 20*(6), 1721–1761.
11. Shao, F., Peng, R., Lai, H., & Wang, B. (2016). DRank: A semi-automated requirements prioritization method based on preferences and dependencies. Journal of Systems and Software. doi:10.1016/j.jss.2016.09.043, (in-press).
12. Kukreja, N., Payyavula, S. S., Boehm, B., & Padmanabhuni, S. (2013). Value-based requirements prioritization: Usage experiences. *Procedia Computer Science, 16*, 806–813.
13. Azar, J., Smith, R. K., & Cordes, D. (2007). Value-oriented requirements prioritization in a small development organization. *IEEE Software, 24*(1), 32–73.
14. Tonella, P., Susi, A., & Palma, F. (2013). Interactive requirements prioritization using a genetic algorithm. *Information and Software Technology, 55*, 173–187.
15. Tuunanen, T., & Kuo, I.-T. (2015). The effect of culture on requirements: A value-based view of prioritization. *European Journal of Inforation Systems, 24*(3), 295–313.
16. Zielczynski, P. (2007). *Requirements management using IBM Rational RequisitePro* (1st ed.). Upper Saddle River: IBM Press.

Part VI

Data Mining

An Optimized Data Mining Method to Support Solar Flare Forecast

Sérgio Luisir Díscola Junior, José Roberto Cecatto, Márcio Merino Fernandes, and Marcela Xavier Ribeiro

Abstract

Historical Solar X-rays time series are employed to track solar activity and solar flares. High level of X-rays released during Solar Flares can interfere in telecommunication equipment operation. In this sense, it is important the development of computational methods to forecast Solar Flares analyzing the X-ray emissions. In this work, historical Solar X-rays time series sequences are employed to predict future Solar Flares using traditional classification algorithms. However, for large data sequences, the classification algorithms face the problem of "dimensionality curse", where the algorithms performance and accuracy degrade with the increase in the sequence size. To deal with this problem, we proposed a method that employs feature selection to determine which time instants of a sequence should be considered by the mining process, reducing the processing time and increasing the accuracy of the mining process. Moreover, the proposed method also determines which are the antecedent time instants that most affect a future Solar Flare.

Keywords

Solar flare • Forecasting • Time series • Data mining • Feature selection

60.1 Introduction

Solar Flares release large amounts of X-ray flux that can potentially damage electronic devices of crewed space crafts and communication devices, as well as destabilize power grid stations on Earth. So, it is imperative the development of works that analyzes X-ray flux levels emitted by Sun and Solar Flare events in order to create mechanisms to warn in advance authorities and organizations of potential Solar Flares.

Generally, works about Solar Flare forecasting can be divided into two trends: (1) purely statistical forecasting

methods, and (2) forecasting methods based on Data Mining (DM). Works belonging to the first trend try to establish a statistical distribution pattern of solar data and further calculate the statistics involved in the dataset to foreseen Solar Flares. The second employs data mining algorithms to build suitable learning models. Basically, the **Data Mining process** is divided in the following steps [10]: (1) establish the DM goals; (2) collect the data; (3) data preprocessing: transform and clean the data, choose the DM task (classification, regression or clustering), and choose a specific DM algorithm; (4) apply the DM algorithm over the preprocessed data; (5) evaluate the results; (6) produce the model in an operational environment. The data mining approaches have the advantage to make the domain specialist knowledge a potential key part of the forecasting process. The preprocessing is a key step of DM process because it impacts in the success of the overall process. In this work, we employ a new preprocessing methodology that maps time series onto

S.L.D. Junior (✉) • M.M. Fernandes • M.X. Ribeiro
Department of Computer Science, Federal University of São Carlos, São Carlos, SP, Brazil
e-mail: sergio.discola@dc.ufscar.br

J.R. Cecatto
National Institute for Space Research, São José dos Campos, SP, Brazil

© Springer International Publishing AG 2018
S. Latifi (ed.), *Information Technology – New Generations*, Advances in Intelligent Systems and Computing 558, DOI 10.1007/978-3-319-54978-1_60

sequences with the aid of the specialist knowledge about Solar Flares behaviour to maximize the method outcomes.

In most literature research regarding Solar Flare forecasting, it can be observed that a very challenging task is to define the most significant features used in the Solar Flare forecasting process. We can find papers that use magnetogram vector [3, 15, 16], sunspot number, sunspot area [5], radio flux, X-ray flux [9], among others, as input to the process of Solar Flare forecasting. A significant feature used in Solar Flare forecasting is the intensity of the X-ray flux emitted by the Sun because it establishes a relation between a Solar Flare Event and its impact on Earth using an event classification range. Furthermore, works on solar flare forecasting usually are based on a dataset that describes solar features in a specific instant of time. This means that most works do not consider the historical evolution of the features in an Active Region in the forecasting method. In this sense, our proposed method employs feature selection methods into the Solar X-ray time series to increase the method's accuracy and computational performance, by selecting the most representative X-ray levels to determine a future Solar Flare. In addition, the historical evolution of the X-ray levels emitted by the Sun is the core data employed to generate the learning model in the proposed method. Through the use of the feature selection methods, we looked for patterns across the X-ray time series by selecting the most appropriate antecedent time instants of the X-ray time series to determine whether a Solar Flare will occur.

60.2 Method Description

Solar Flares are classified according to the strength L of X-ray levels released during the event. It can be classified as $B(L<1E-06), C(1E-06 \leq L<1E-05), M(1E-05 \leq L<1E-04)$ and $X(L \geq 1E-04)$. Flares classified as B are the weakest event and the ones classified as X are the strongest and may cause the major damages on Earth.

In order to forecast Solar Flares, the proposed data mining method encompasses 3 steps: (1) map solar time series onto sequences; (2) perform feature selection; (3) mine the preprocessed solar data.

Mapping solar time series onto sequences

The initial X-ray time series $X(t) = \{ t, x \mid t \in \mathbf{I} \times Sample\_Rate$, where $\mathbf{I}$ is an Integer, $Sample\_Rate$ is the sample rate of X-ray levels, and $x \in \mathbf{R}$ of X-ray values}, was converted in a dataset tuple $XTS(t) = \{ t, x, class \mid t \in \mathbf{I} \times Sample\_Rate, x \in \mathbf{R}$ and $class \in \{ B, C, M, X \}\}$, where $class$ corresponds to the solar flare classification given to the x X-ray value. Examples of $XTS(t)$ tuples are given in Table 60.1.

Table 60.1 Example of dataset tuples- XTS(t)

X-ray time series		
t	x(t)	Solar Flare Class
0	3.65E-7	B
5	3.92E-7	B
10	4.09E-7	B
15	4.04E-7	B
20	3.92E-7	B
25	3.94E-7	B
30	3.84E-7	B
35	3.80E-7	B
40	3.80E-7	B
45	3.83E-6	C
50	3.84E-7	B
55	3.90E-7	B
60	6.47E-7	B
65	6.75E-7	B
70	5.24E-7	B

The proposed method splits the dataset $XTS(t)$ into sequences of X-ray time series composed by:

(1) the first n observations called *Current Window* $C(t) = \{x(t), ss \leq t \leq Current\_Window\_End\}$, where ss is the *Sequence Start* $= Sample\_Rate \times Sequence\_Step$, and *Sequence_Step* is the step that the sequences are slided to build the next sequence; *Current_Window_End* $= ss + Sample\_Rate \times (n - 1)$, and; n is the size of the current window;

(2) the next j observations called jump $J(t) = \{x(t), Current\_Window\_End + Sample\_Rate \leq t \leq Jump\_Window\_End\}$, where: *Jump_Window_Size* $= Current\_Window\_End + Sample\_Rate \times j$, and; j is the size of the jump window. Note that the values within the Jump Window are the observations between the Current Window and the Future Window. Although, this values are not taken into account by the algorithm that produces the final tuple, the number of observations within the Jump Window (the Jump Window size) is used to set the anticipation period of the forecasting.

(3) the last f observations called *Future Window* $F(t) = \{x(t), Jump\_Window\_End + Sample\_Rate \leq t \leq Future\_Window\_End\}$, where: *Future_Window_End* $= Jump\_Window\_End + Sample\_Rate \times f$, and; f is the size of the *Future Window*).

Considering Table 60.1, the resulting sequences is presented in Table 60.2.

Note that the first value of f1 in Table 60.2 (3.65E-7) corresponds to the X-ray level occurred in time instant 0 (x (0)), the second value (3.92E-7), to the level in time instant

Table 60.2 Example of the Sequence Built – (Attributes: *ss* – time instant when each sequence is formed; *f1–f12* – X-ray values at the "ss" time instant)

ss	Current window				Jump				Future window			
	f1	f2	f3	f4	f5	f6	f7	f8	f9	f10	f11	f12
0	3.65E-7	3.92E-7	4.09E-7	4.04E-7	3.92E-7	3.94E-7	3.84E-7	3.80E-7	3.80E-7	3.83E-6	3.84E-7	3.90E-7
5	3.92E-7	4.09E-7	4.04E-7	3.92E-7	3.94E-7	3.84E-7	3.80E-7	3.80E-7	3.83E-6	3.84E-7	3.90E-7	6.47E-7
10	4.09E-7	4.04E-7	3.92E-7	3.94E-7	3.84E-7	3.80E-7	3.80E-7	3.83E-6	3.84E-7	3.90E-7	6.47E-7	6.75E-7
15	4.04E-7	3.92E-7	3.94E-7	3.84E-7	3.80E-7	3.80E-7	3.83E-6	3.84E-7	3.90E-7	6.47E-7	6.75E-7	5.24E-7

Table 60.3 Example of the generalized sequences

ss	Current window				Jump				Future window			
	f1	f2	f3	f4	f5	f6	f7	f8	f9	f10	f11	f12
0	x(0)	x(5)	x(10)	x(15)	x(20)	x(25)	x(30)	x(35)	x(40)	x(45)	x(50)	x(55)
5	x(5)	x(10)	x(15)	x(20)	x(25)	x(30)	x(35)	x(40)	x(45)	x(50)	x(55)	x(60)
10	x(10)	x(15)	x(20)	x(25)	x(30)	x(35)	x(40)	x(45)	x(50)	x(55)	x(60)	x(65)
15	x(15)	x(20)	x(25)	x(30)	x(35)	x(40)	x(45)	x(50)	x(55)	x(60)	x(65)	x(70)

Table 60.4 Final tuple composed by attributes f1–f4 and class

ss	f1	f2	f3	f4	Class
0	3.65E-7	3.92E-7	4.09E-7	4.04E-7	Yes
5	3.92E-7	4.09E-7	4.04E-7	3.92E-7	Yes
10	4.09E-7	4.04E-7	3.92E-7	3.94E-7	No
15	4.04E-7	3.92E-7	3.94E-7	3.84E-7	No

Table 60.5 Generalized tuple: attributes f1–f4, class

ss	f1	f2	f3	f4	Class
0	x(0)	x(5)	x(10)	x(15)	Yes
5	x(5)	x(10)	x(15)	x(20)	Yes
10	x(10)	x(15)	x(20)	x(25)	No
15	x(15)	x(20)	x(25)	x(30)	No

5 (x(5)), and son on. So, we can generalize the sequences shown at Table 60.2 as follows presented in Table 60.3.

The sequence mapping allows traditional classification methods to generate models that predict **future** labels giving **current** values of X-ray levels. Actually, traditional classification methods predicts **current** labels giving **current** values of any nature. In this sense, we build a method that correlates **current** values with **future** labels as follows. Considering the first sequence of Table 60.2, a new tuple is generated using the X-ray levels contained on the *current window*. The tuple will add a new attribute named *Class* so that the new tuple will be composed by attributes *f1 to f4* of Table 60.2 and the new attribute called *class*. The value of this new attribute will be set as **Yes** if the correspondent Solar Flare Classification of the maximum X-ray level located at the future window of the analyzed sequence is C, M or X, and **No**, otherwise.

In the example of Table 60.2, just the class of the first two sequences are set to **Yes** because there are no X-ray levels greater than 1.0E-07 occurring as the future window of the other sequences, that is the maximum of a class *B* Solar Flare. For this example, the final tuples built are presented in Table 60.4.

Table 60.5 shows the general form of Table 60.4.

The size of the *current window*, the *jump window* and the *future window* can be customized as well as the *shift* between two consecutive sequences according to the domain specialist needs. Tables 60.2 and 60.3 also show the values between the Current Window and the Future Window encomprising the Jump Window. Recall that the values within the Jump Window are not employed but the number of observations within this window are employed to generate the final tuple composition of Table 60.4. So, if the specialist wants the forecasting of one day of advance and considering that the sample rate of the observations are five minutes, he should set the jump window size as 288 observations.

Performing Feature Selection in Solar Data Sequences

In real scenario, the preprocessed sequences usually have a large number of features. However, for large data sequences, the classification algorithms face the problem of "dimensionality curse", where the algorithms performance and accuracy degrade with the increase in the sequence size. In order to reduce the number of features and consequently to reduce the data to be submitted to the mining process, the preprocessed sequences are submitted to a feature selection process. The feature selection process also suggests the time instants that are most important to be considered by the forecasting model. Our proposed method employs two supervised feature selection to perform dimensionality reduction in the sequence data Starminer [12] and Relief [8].

Starminer [12] is an statistical association-rule based algorithm, which mines rules associating a feature f_i and a class c_k, if f_i has a uniform and a particular behaviour among

Fig. 60.1 The proposed method

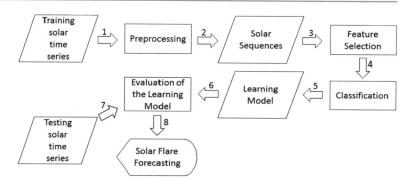

the instances of the c_k class in the training dataset, producing rules of the form $f_i \rightarrow c_k$. A rule is generated if a statistical hypothesis test rejects the hypothesis that feature f_i has the same mean over the instances of class c_k and the instances of the remaining dataset. The Starminer algorithm returns the features that are presented in the mined rules as the selected ones. We chose Starminer because: (1) it produces rules relating features and the classes that they most describe, which can be validate by domain expert; (2) it has a linear computational cost, having a low computational cost when compared with most feature selection methods; (3) it selects the most relevant features and discards the irrelevant ones, not returning a ranking.

Relief [8] is a distance-based feature selection method. An weight w_i is associated to each feature f_i. This weight indicates the relevance of the feature and it is updated at each iteration. After a given number of iterations the weight of each feature composes the relevance vector. An interaction consists of randomly selecting an instance x of a class c_k, the dataset instances are ranked according to the distance that they present from the x, using the Euclidian distance. The closest same-class instance is called 'near-hit', and the closest different-class instance is called 'near-miss': $w_i = w_i(x_i nearHit_i)^2 + (x_i nearMiss_i)^2$. The feature weight increases if its feature values differ less in the nearby instances of the same class than in the nearby instances of the other class. Even though it has a huge computational cost, we chose Relief because it is one of the baseline feature selection algorithm most employed in literature.

Mining the solar data sequences

In this step the data sequences, considering only the relevant features mined by the feature selection approach are submitted to a traditional classification method. A question that may arises is "how traditional classification methods will be capable to predict **future** labels using the new dataset"? As shown in Table 60.4, the X-ray values of the current window (*f1 to f4*) is being related with the maximum Solar Flare occurred at the future window. So that, when a traditional classification method is trained with that preprocessed dataset, it will make a model relating a set of X-ray values with its related future ones. Then, if we

apply X-ray values collected in a snapshot not yet analyzed during the classification training phase to the produced model, this one will predict the future label, namely class of the Solar Flare. In this way, the traditional classification method will be able to predict future labels using current values. It is important to emphasize that the *future window* is shifted from the *current window* by a *jump window*. The size of the *jump window* corresponds to the number of observations that the forecasting will be performed. In the example, the forecasting method will predict labels in advance of 4 observations. If each observation corresponds to 5 minutes, this means that the method will foreseen Solar Flares from 20 minutes of antecedence.

A general scheme of the proposed work in shown in Fig. 60.1:

60.3 Experiments

The experiments performed in this work used Solar data from 2014 and 2015 provided by "U.S. Dept. of Commerce, NOAA, Space Weather Prediction Center" at http://www.swpc.noaa.gov/products/goes-X-ray-flux using X-ray levels captured in a five minute sample rate. X-ray flux emitted by Sun is monitored by two X-ray sensors located at the GOES (Geostationary Operational Environmental Satellite) satellite. The first sensor captures X-ray flux in the 0.5–4.0 Angstrom passband, while the other captures it in the 1–8 Angstrom passband. Solar Flares can be classified according to the intensity of X-ray in the 1 to 8 Angstroms passband as described in Sect. 60.2.

The experiments were performed in three phases. First, we employed the proposed method without feature selection. In the second phase, we used "Starminer" in the method. Finally, in the third phase, we used "Relief Attribute Evaluation". For each phase, we used the X-ray time series from 2014 as training dataset in order to produce a learning model with one day of forecasting advance, and, then, the proposed method were tested employing the data collected during 6 months of 2015.

A set of six experiments, divided in three phases, were performed to validate the proposed method as presented in

Table 60.6 Description of the attributes used in the experiments

Configuration attribute	Description	Possible values
Feature selection	This attribute tells the Feature Selection Method used	Not used Starminer Relief Attribute Evaluation
Current window	This attribute tells the number of days of the window used to compose the sequence that will be labeled with the "future" class	Integer value
Jump	This attribute tells the number of observations that will be considered to build the next sequence	Integer value
Future window	This attribute tells the number of days of the window used to look for the "future" class of the previouly defined "Current window"	Integer value
Data interval/ training	This attribute tells the interval considered to catch the data used during the training phase of the classification method used in the experiments	1 year 2014
Data interval/ testing	This attribute tells the interval considered to catch the data used during the testing phase of the experiments.	6 months 2015
Classification method	This attribute tells the Data Mining Classification Method used during the validation of the proposed forecasting method	J48, IBK, NaiveBayes, OneR SVM (SMO – Weka – Polykernel)

Table 60.7 Experiments planning

Phase. Test Id	Feature selection method	Current window	Jump	Future window	Data interval Training	Testing	Classification method
1.1	Not used	1	1	1	1 year 2014	6 months 2015	J48, IBK, NaiveBayes, OneR, SVM (SMO – Weka – Polykernel)
2.2	Starminer						
3.3	Relief Attribute Evaluation						
1.4	Not used	2	1	1			
2.5	Starminer						
3.6	Relief Attribute Evaluation						

Table 60.7. These experiments were configured using the attributes described in Table 60.6:

Table 60.7 presents the configuration of the experiments performed:

It was considered two main configurations during each phase: (1) one day for the *Current Window*, and; (2) two days for the *current window*. *Jump* and *Future Window* was configured as one day for both phases. This means that we analyzed the influence of X-rays for one day and two days for the *current window* in the production of the forecasting model. The implementation of the classification methods were obtained from Weka [6]. The classification methods employed were: J48, IBK, Naivebayes, OneR, Support Vector Machine [1, 2, 7, 11, 13].

The experiments resulted in very high rates of accuracy in the majority of the classification methods (see Fig. 60.3), but in a deeper analysis of the results, it was observed that the majority of the classifications methods also resulted high differences between the True Positive (TP) and True Negative (TN) rates. This difference is due the unbalanced characteristic of the Solar Flares time series, since high-intensity solar flares are rare. For that reason, we employed the ROC area information [4] to qualify the method efficacy. The biggest ROC Area consists on the best results obtained, suggesting the best and most balanced results of TP and TN rates as well as a good accuracy.

As we can see in Fig. 60.2, Naive Bayes classifier got the highest ROC areas. Particularly, Test 3.6 was the highest as it reached 0.799 of ROC area. In the other hand, we observe that this experiment got an accuracy of 72.7%. This accuracy is pretty high considering the unbalanced dataset, and it got a balanced TP and TN rates, or 70.9% and 79.7%, respectively (as it was already suggested by the ROC Area result).

Test 3.6 results were also very interesting for some reasons: (1) as shown in Table 60.7, Test 3.6 is configured with a "current window" of 2 days and, according to our specialist domain, this period has the most significant pattern that can be observed to forecast a solar flare; (2) if we use "Relief Attribute Selection" and considering two days of "current window", the most significant instants to use in the forecasting model are comprised in the following intervals (given in hours): $1 \leq instant \leq 9$ (start of the

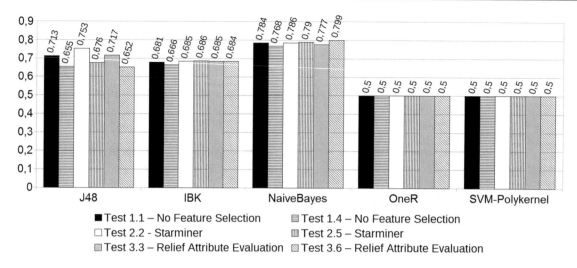

Fig. 60.2 ROC area

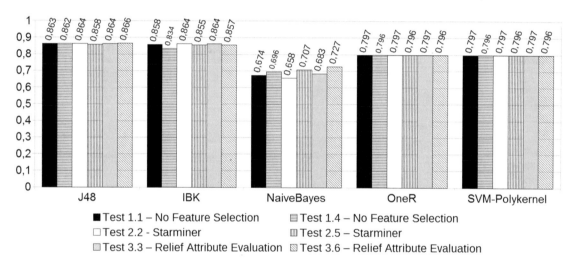

Fig. 60.3 Accuracy

first day), $18 \leq instant \leq 24$ (end of the first day), and $33 \leq instant \leq 48$ (last 16 hours of the second day); (3) this feature selection in addition with Naive Bayes classifier resulted the best accuracy and ROC area among all the tests and further, it enabled a speedup of 3.2 which means a execution time 3.2 faster than if we consider all features of the dataset ($speedup = time_{noRelief}/time_{withRelief} = 61.9/19.18$); (4) the overall accuracy, considering the tests performed with 2 days of "current window", increased from 69.6% (without feature selection) to 72.7% (considering the feature selection) as well as the ROC Area, from 0.768 to 0.799.

The forecasting method results are presented in Figs. 60.2, 60.3 and 60.4.

In order to aid the comprehension of the results obtained, we show the most significant time analysis in Fig. 60.5

60.4 Conclusions

Most works found in literature about Solar Flare forecasting analyzes several features regarding Active Regions on snapshots and, then, use them to produce forecasting models. In this work, we consider the historical evolution of a X-ray level time series to produce the forecasting model. Furthermore, we studied, through the usage of feature selection methods like "Starminer" and "Relief Attribute Evaluation", the most significant instants to consider in the forecasting method. This approach resulted in high accuracy results which balanced TP and TN rates, as well as optimizing the overall forecasting method in terms of performance speedup. The developed method reached an accuracy of 72.7%, with a TP rate of 70.9% and TN rate of 79.7%. Another great

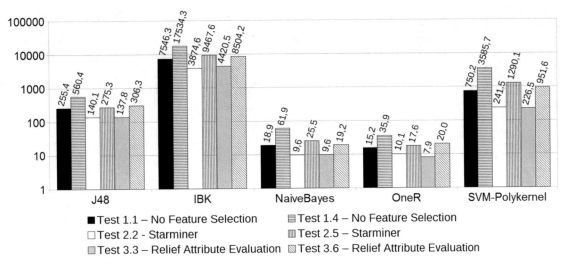

Fig. 60.4 Total time = Training time + Testing time (Time in seconds)

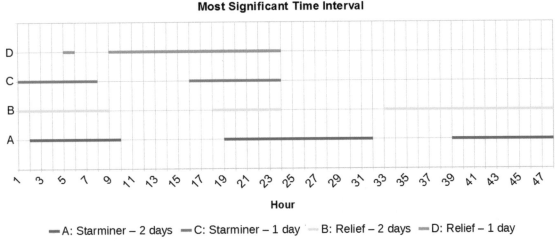

Fig. 60.5 Most significant time interval analysis

contribution of this work was the analysis and results obtained of the most significant interval to consider in the forecasting method. It was achieved that, in the scope of our experiments, the best time interval was within two days for "current window" and comprised the initial and end of the first day, and the last 16 hours of the second day. This result corroborate the theory from domain specialists that the analysis of two days of data can determine possible Solar Flares. For future work, we intend to use this method to forecast the five levels of Solar Flares individually.

References

1. Aha, D. W., Kibler, D., & Albert, M. K. (1991). Instance-based learning algorithms. *Machine Learning, 6*(1), 37–66.
2. Besnard, P., & Hanks, S. (1995). In: *Proceedings of the Eleventh Conference on Uncertainty in artificial intelligence*, Montreal, 18–20 Aug 1995. Morgan Kaufmann Publishers.
3. Bobra, M. G., & Couvidat, S. (2015). Solar flare prediction using SDO/HMI vector magnetic field data with a machine-learning algorithm. *The Astrophysical Journal, 798*(2), 135.
4. Fawcett, T. (2004). *ROC graphs: Notes and practical considerations for researchers*. Palo Alto, CA.: Technical Report HP Laboratories.
5. Gallagher, P. T., Moon, Y.-J., & Wang, H. (2002). Active-region monitoring and flare forecasting – I. Data processing and first results. *Solar Physics, 209*(1), 171–183.
6. Hall, M., Frank, E., Holmes, G., Pfahringer, B., Reutemann, P., & Witten, I. (2009). The WEKA data mining software: An update. *SIGKDD Explorations, 11*(1), 10–18.
7. Holte, R. C. (1993). Very simple classification rules perform well on most commonly used datasets. *Machine Learning, 11*(1), 63–90.
8. Kira, K., & Rendell, L. A. (1992). A practical approach to feature selection. In *Proceedings of the Ninth International Workshop on Machine learning*, College Park (pp. 249–256)
9. Li, R., & Zhu, J. (2013). Solar flare forecasting based on sequential sunspot data. *Research in Astronomy and Astrophysics, 13*(9), 1118–1126.
10. Maimon, O., & Rokach, L. (2010). *Data mining and knowledge discovery handbook* (2nd ed.). New York: Springer.

11. Quinlainn, J. R., & Ross, J. (1993). *C4.5: Programs for machine learning*. San Mateo: Morgan Kaufmann Publishers.
12. Ribeiro, M. X., Balan, A. G. R., Felipe, J. C., Traina, A. J. M., & Traina, C. (2009). Mining statistical association rules to select the most relevant medical image features. In *Mining complex data* (pp. 113–131). Berlin/Heidelberg: Springer.
13. Scholkopf, B., Burges, C. J. C., & Smola, A. J. (1999). *Advances in kernel methods: Support vector learning*. Cambridge: MIT Press.
14. Thijssen, J. M. (2007). *Computational physics* (2nd ed.). Cambridge: Cambridge University Press.
15. Yu, D., Huang, X., Wang, H., & Cui, Y. (2009). Short-term solar flare prediction using a sequential supervised learning method. *Solar Physics, 255*, 91–105.
16. Yu, D., Huang, X., Hu, Q., Zhou, R., Wang, H., & Cui, Y. (2010). Short-term solar flare prediction using multiresolution predictors. *The Astrophysical Journal, 709*(1), 321–326.

A New Approach to Classify Sugarcane Fields Based on Association Rules

61

Rafael S. João, Steve T.A. Mpinda, Ana P.B. Vieira, Renato S. João, Luciana A.S. Romani, and Marcela X. Ribeiro

Abstract

In order to corroborate the acquired knowledge of the human expert with the use of computational systems in the context of agrocomputing, this work presents a novel classification method for mining agrometeorological remote sensing data and its implementation to identify sugarcane fields, by analyzing Normalized Difference Vegetation Index (NDVI) series. The proposed method, called RAMiner (**R**ule-based **A**ssociative classifier **Miner**) creates a learning model from sets of mined association rules and employs the rules to constructs an associative classifier. RAMiner was proposed to deal with low spatial resolution image datasets, provided by two sensors/satellites (AVHRR/NOAA and MODIS/Terra). The proposal employs a two-ways classification step for the new data: Considers the conviction value and the conviction-based probability (a weighted accuracy formulated in this work). The results given were compared with others delivered by well-known classifiers, such as C4.5, zeroR, OneR, Naive Bayes, Random Forest and Support Vector Machine (SVM). RAMiner presented the highest accuracy (83.4%), attesting it is well-suited to mine remote sensing data.

Keywords

Agrometeorological • Data mining • Association rules • Associative classification

61.1 Introduction

Over the recent past, there has been a growing interest in the incorporation of computational technologies to the agricultural sector. It allows improvements and increasing in the productivity, benefiting farmers, investors of agrobusiness and consequently the society in general. Data mining techniques applied to the agriculture context has improved the decision making process, once the analysis of huge volumes of data by human is impossible without the aid of computer methods. Specially when that analysis consider changes over the time or space. Sugarcane cultivation is an excellent scenario to run spatio-temporal computational systems. Due to that kind of plantation has many changes over the time. Changes like the geographical expansion of the cultivation, common reforms in planting, period when another cultivation is managed while the soil nutrients are

R.S. João (✉)
Federal University of São Carlos (UFSCar), São Carlos, Brazil

L3S Research Center, University of Hannover, Hannover, Germany
e-mail: rafael.joao@dc.ufscar.br

S.T.A. Mpinda • A.P.B. Vieira • M.X. Ribeiro
Federal University of São Carlos (UFSCar), São Carlos, Brazil
e-mail: steve.mpinda@dc.ufscar.br; ana.vieira@dc.ufscar.br; marcela@dc.ufscar.br

R.S. João
L3S Research Center, University of Hannover, Hannover, Germany
e-mail: joao@l3s.de

L.A. Romani
Embrapa Agricultural Informatics – Campinas, Campinas, Brazil
e-mail: luciana.romani@embrapa.br

© Springer International Publishing AG 2018
S. Latifi (ed.), *Information Technology – New Generations*, Advances in Intelligent
Systems and Computing 558, DOI 10.1007/978-3-319-54978-1_61

reinstated, and so on. Some changes imply that, in a certain time, a region be considered a sugarcane field and other times not.

Actually, the availability of data is increasing in the sugarcane cultivation as well because this kind of planting is nowadays highly mechanized through vehicles (harvesters, transport trucks and others) equipped with sensors that generate large volumes of data. One important data provider for sugarcane fields are the orbital sensors. However the amount of data available for agriculture, comprising data from different sensors, increases exponentially.

In this work, the analysis was conducted on sets of low spatial and high temporal resolution images. Two well-known sensors widely used because of their confiability are the family of satellites from National Oceanic and Atmospheric Administration (AVHRR/NOAA) and the Moderate Resolution Imaging Spectroradiometer (MODIS/Terra). Among other information, the sensors AVHRR and MODIS, respectively aboard on NOAA and Terra satellites, provide the Normalized Difference Vegetation Index (NDVI), calculated by the difference between the reflectance of the near-infrared and the reflectance of the red; divided by the sum of them. The NDVI values consist in a number in the interval $[-1,1]$ referring to a geographical point (lat/long) in a specific time. The higher the NDVI value, the higher is the biomass of that point in the time. Due to the fact that NDVI is commonly employed by the experts in the data analysis process, we considered that value in this work. For instance, the kind of cultivation in a given geographical area, how long is from harvest, and more, can be estimated by the use of data mining tasks when considering NDVI series. In order to provide a good analysis of data, it is necessary a robust database as well as suitable computational methods.

A novel method, called RAMiner (**R**ule-based **A**ssociative classifier **Miner**), is proposed in this work to identify sugarcane fields based on series of NDVI extracted from low resolution images. RAMiner implements an associative classifier with sets of association rules, named learning model, generated from sets of labeled NDVI series. The learning model is considered for counting the number of rules that can be used to classify new instances as sugarcane regions or not. The classification is given in RAMiner by two measures, the conviction value and the conviction-based probability, described in details later in this paper. To evaluate the obtained results, some stable implementations of well-known classifiers algorithms in the framework Weka (Waikato Environment for Knowledge Analysis) were selected, specifically ZeroR, oneR, C4.5 tree, Naive Bayes, Random Forest and SVM. These algorithms were chosen considering that they are state-of-the-art classifiers.

This work is organized and presented as follow: In Sect. 61.2, there are presented some related works, which were considered as inspiration to conduct this job. In Sect. 61.3 the proposed method is detailed with its execution steps. The Sect. 61.4 describes the conducted experiments and the obtained results when considering a real scenario. Finally, in Sect. 61.5, the conclusion and possibilities of future works are presented.

61.2 Related Work

As can be seen in the literature, there are many works that apply data mining tasks in the context of agriculture [1], the most part of them are related to the task of classification. In this context, association rules were firstly employed in [2] for diagnosing soybean diseases, where information given by the expert joint with association rules were employed to construct a knowledge base to analyze health and disease plants.

Bhatia and Gupta [3] proposed the usage of quantitative multidimensional association rules to analyze agricultural data warehouses. In that work, the continuous data were previously discretized in ranges defined by categorical attributes associated to the quantitative ones. After the discretization process, the processed data were submitted to a traditional association rule mining algorithm, revealing patterns about the agricultural data.

Ramesh and Ramar [4] applied some traditional classifiers (Naive Bayes, C4.5 and Random Forest) to analyze soil data characteristics, such as humidity, temperature and physical-chemistry properties. In [5] a classification approach based on k-means to identify ground types can be seen and the works [6–10] and [11] describe the use of traditional classifications techniques to predict milk smell, sound of birds, forest variation, eggs fertility, nitrogen index of corns and cracks on eggs, respectively.

Over the years, traditional classifiers and association rules have been promising techniques employed in literature to analyze agricultural data. However, to the best of our knowledge, this work is the first one that proposes an associative classifier to predict the existence or absence of sugarcane regions analyzing sensor series data.

61.3 Proposed Method

The method described in this work, called RAMiner, implements an associative classifier to identify sugarcane fields based on NDVI series. Different from previous approaches, RAMiner constructs a model based on association rules to infer a class. This model is high-level comprehensive and can be used by specialists to analyze and validate the data. The model is used to classify new data and tends to bring high accuracy results when compared with the traditional classification methods. Basically, the method RAMiner consists of three main steps:

(1) Association rule mining and generation of the learning model;
(2) Computation of the frequency matches between data and the mined association rules;
(3) Classification employing conviction or conviction-based probability estimation, based on the frequencies computed;

61.3.1 Step 1: Learning Model

In the first step, all frequent patterns are identified and used to generate sets of association rules by the use of the traditional association rule mining algorithm Apriori [12]. In this work, an instance is defined as a NDVI series of a spatial point, attached to its location (latitude and longitude). A set of association rules (C) is generated from the training instances previously known as being from sugarcane regions. On the other hand, the set N is generated from the training instances previously known as not being from sugarcane regions.

Generalizing this step of the method, a set of association rules is generated for each class of the training set. The generated rules are of the form $X \rightarrow Y$, where X and Y are disjoint sets of items, and each item has the form $month_i = NDVI$. Note that for the learning model generation, a set of labeled instances for each possible class should be supplied. In this work, the labeled instances were given in a mask of sugarcane fields provided by Canasat project (http://www.canasat.inpe.br).

61.3.2 Step 2: Rule Matches Counter

Now the number of rules matched by each instance is calculated. This step implements a poll to find what is the set of rules (set C or set N) that each instance S most matches. We define that an instance S matches to a rule R if all itemsets belonging to R occurs in S, i.e. if all the attributes of S occur in R and their values belong to the same interval of the attributes in R. At the end of this step, each instance of the test set (not yet labeled) is associated to its matches and no-matches counter.

61.3.3 Step 3: Classification

Finally, the counters values calculated in Step 2 are used to compute the *conviction* measure. Let A_S be the number of matches of instance S in the set C (sugarcane association rules), and B_S the number of no-matches of S in C. D_S is the number of matches of S in N (non-sugarcane association rules), and F_S the number of no-matches of S in N. The *conviction of Yes* measure (CvY_S) of an instance S is the belief of S belongs to the sugarcane class (Yes class) and it is presented in Equation 61.1.

$$CvY_S = \frac{W * A_S + F_S + 1}{A_S + B_S + F_S + 1} \qquad (61.1)$$

The conviction measure works as a weighted probability, where W is an inductive bias: the higher the W value is, the higher impact of the matches that the analyzed instance S has in the training sugarcane rules set C. Following the same idea, the *conviction of No* measure (CvN_S) of an instance S is the belief that S not belongs to the sugarcane class and it is given by the Equation 61.2.

$$CvN_S = \frac{W * D_S + B_S + 1}{F_S + D_S + B_S + 1} \qquad (61.2)$$

The conviction values CvY_S and CvN_S are employed by RAMiner to provide a new weighted way to infer the class of an new instance S. RAMiner also employs the *conviction-based probability* $P c_S$ (presented in Equation 61.3) to assign a label for a new instance S.

$$Pc_S = \frac{CvY_S}{CvY_S + CvN_S} \qquad (61.3)$$

The conviction measure and conviction-based probability can be employed to classify a new instance in RAMiner:

- using the *conviction* measure: if $CvY_S > CvN_S$ assign S to Yes class (sugarcane), otherwise to No class;
- using the *conviction-based probability*: if $Pc_S > t$, assign S to Yes class (sugarcane), otherwise to No class;

Therefore, there are two techniques that are used to classify a new instance in RAMiner. The user can set the best technique to use. The higher conviction value defines the class of an instance. If the conviction-based probability value is higher than a threshold t, the instance is classified as Yes. RAMiner was implemented almost in Java language, aided by some scripts in PHP. Its method is presented in more details in Algorithm 1.

Algorithm 1 The RAMiner method

1: **procedure** RAMINER($dbY, dbN, dbT, t, W, bp, usePC$)
2:   $set\ C \leftarrow Apriori(dbY)$
3:   $set\ N \leftarrow Apriori(dbN)$
4:   **for** each S $\in$ dbT **do**
5:     $A_s \leftarrow countMatches(C)$
6:     $B_s \leftarrow countNoMatches(C)$
7:     $D_s \leftarrow countMatches(N)$
8:     $F_s \leftarrow countNoMatches(N)$
9:     $CvY_S \leftarrow (W*A_S+F_S+1)/(A_S+B_S+F_S+1)$
10:    $CvN_S \leftarrow (W*D_S+B_S+1)/(F_S+D_S+B_S+1)$
11:    $Pc_S = \leftarrow CvY_S/(CvY_S + CvN_S)$
12:    **if** ($usePcv$) **then**
13:      **if** ($Pc_S > t$) **then**
14:        label S to Yes
15:      **else**
16:        label S to No
17:      **end if**
18:    **else**
19:      **if** (($CvY_S > CvN_S$) OR
20: (($CvY_S == CvN_S$) AND ($bp == true$))) **then**
21:        label S to Yes
22:      **else**
23:        label S to No
24:      **end if**
25:    **end if**
26: **end for**
27: **return** S
28: **end procedure**

RAMiner receives as input: the training sets (instances from the class Yes – sugarcane) *dbY* and *dbN* (from the class No); the test set *dbT*; the values of weight *W* to calculate the conviction values; the *usePC*, a *Boolean* input that informs if the method should use the *conviction-based probability* to classify (*usePC = true*) or just the conviction measure (*usePC = false*); and the assumption of *bp*. The parameter *bp* is a *Boolean* flag as well, which describes if the city being managed is considered a strong or a weak sugarcane producer. The parameter *bp* is useful in cases where a tie happens, i.e., when RAMiner finds an equal conviction value to classify an instance *S* as being or not being a sugarcane field, the value of *bp* decides the result; conducts the classification to the class *No* when *bp* is *false* and to the class *Yes* (sugarcane) otherwise.

In lines 2 and 3, the training datasets are submitted to the algorithm Apriori to generate the sets *C* and *N* of association rules, i.e., the learning model. After the number of matches

are computed for each instance from the test set (lines 5–8) and the measures of conviction and conviction-based probability are then computed (lines 9–11). At the end, the instance *S* is classified either using the conviction measure or the conviction-based probability, according to the *usePC* set by the user (lines from 12 to 25). The conviction-based probability works as an weighted probability. High *w* values indicate high contribution of the rules matches to classify the new instances.

61.4 Experiment and Results

61.4.1 About the Data

To create the learning model, four labeled datasets were considered: (1) two of them containing NDVI series from sugarcane fields collected by the sensors (NOAA and MODIS) and (2) two datasets containing NDVI series referring to non-sugarcane fields. Also six datasets with NDVI series of geographical regions referring to three distinct Brazilian cities (Morro Agudo, Araçatuba e Piracicaba) were employed in the experiments as test sets. Each city is represented by two datasets: one with information given by NOAA sensor and other by MODIS sensor. The three cities were chosen based on the assumption given by the domain human expert that one city is a high level sugarcane producer, one a medium and one a low producer. The main goal of this experiment is to analyze the performance of RAMiner when considering datasets with different behaviors.

Figure 61.1 shows a slice of NDVI series, from one of the datasets considered by RAMiner (*dbY*, *dbN* and *dbT*). In the figure are listed the couple of geographical coordinates (lat/long) followed by the NDVI series, each one referring to one month of the year. Note that the figure shows only five NDVI values for each geographical point, as a sample.

The datasets managed by RAMiner are listed in Table 61.1 with their respectively number of instances. Araçatuba, Morro Agudo and Piracicaba are the test sets, instances of the cities to be classified. Sugarcane Fields and Non-Sugarcane Fields are training datasets, instances previously known as being from sugarcane regions and not.

Due to some peculiarities of each dataset from the satellites MODIS and NOAA, there were slightly differences, such as how many NDVI values each dataset contais, difference between geographical resolutions, order of the data, etc. A pre-processing script was implemented to

Fig. 61.1 A sample of the datasets. From left to right: lat/long points followed by the NDVI series, each NDVI value referring to one month

-223.905.211.983.664	-525.320.396.455.983	0.664740	0.641619	0.632000	0.658915	0.632653
-223.905.211.983.664	-525.221.239.454.160	0.635294	0.664336	0.558824	0.633588	0.493976
-224.104.410.892.768	-526.212.809.472.390	0.500000	0.508571	0.481481	0.473054	0.414634
-224.104.410.892.768	-524.824.611.446.868	0.492823	0.528796	0.490196	0.421053	0.341772
-224.204.010.347.320	-524.725.454.445.045	0.431694	0.436782	0.405714	0.253521	0.177665
-224.303.609.801.872	-524.824.611.446.868	0.487437	0.485714	0.421053	0.255474	0.252525

standardize that datasets. This process concerns in the following aspects:

- Select a one year window of data, regarding to the same time interval (from 2004-01-04 to 2005-30-03);
- Merge the biweekly data of MODIS to monthly data, as presented in NOAA;
- Reduce the precision difference of both sensors (lat/long), increasing the intersection area;
- Sort instances, in ascending order, by their latitude and longitude values;
- Transform NDVI values from numeric to categorical. Following the NDVI ranges given by the expert and presented in Table 61.2.

61.4.2 Describing the Experiment and Obtained Results

As aforementioned, each parameter of RAMiner was idealized to contributes in the improvement of the obtained results, by combining variations of their values. That strategy provides a customization of RAMiner to run in the finest set up. In the experiment, it was proposed to assign the values $\{0.2, 0.3, 0.4, 0.5, 0.6\}$ to the t parameter and the values $\{2, 4, 6, 8, 10, 15, 20, 25, 30\}$ to the W. Each one of that 90 possible combinations (45 when the bp parameter was *true* and 45 when $bp = false$) was considered as a set up to run RAMiner with all the instances – NDVI

Table 61.1 Datasets managed by RAMiner in the proposed experiment and their respectively size, in number of instances

Dataset	MODIS	NOAA
Araçatuba	1205	1034
Morro Agudo	1425	1220
Piracicaba	1427	1222
Sugarcane fields	23,542	22,754
Non-sugarcane fields	177,614	151,402

Table 61.2 NDVI range categories

Category	Means	NDVI interval
High	High	0.8 to 1.0
Mod_High	Moderately high	0.6 to < 0.8
Mod_Low	Moderately low	0.4 to < 0.6
Low	Low	0.2 to < 0.4
Very_Low	Too low	> 0 to < 0.2
Soil_Exposure	Exposed soil	≤ 0

series associated to a lat/long – of the datesets. In each execution RAMiner provides two measures for each instance of the dataset, the conviction and the conviction-based probability. However only three of the total values were considered by RAMiner to the process of classification, they were: the biggest conviction obtained on the instances (convMax), the average of the convictions (convAvg) and the average conviction-probability (probAvg).

The learning model was composed by four sets of association rules: from sugarcane fields, considering the sensors (1) NOAA and (2) MODIS and from non sugarcane fields, considering (3) NOAA and (4) MODIS. Figure 61.2, shows a slice from (1) containing five rules with their respective confidence values. The rule ♯1 could be used, for instance, to infer (with 97% of confidence) that: "*If in May, July and August the value of NDVI were Mod_Low, June may present the same behavior as well, i.e., Mod_Low.*". If a high number of rules in (1) match to the selected instance, then it can be used to classify the region (lat/long of the instance) as a sugarcane field.

In addition that datasets presented in Table 61.2, the learning model creation step provides four sets of data (association rules), which were merged into two main association rules set, C (Yes class rules) and N (No class rules), generated of the training instances. The set C (yes) has 99 rules generated from MODIS Sugarcane Fields, and 75 from NOAA Sugarcane Fields. The set N (No) has 81 rules generated from the mining of MODIS Non-Sugarcane Fields, and 80 from NOAA Non-Sugarcane Fields.

Figures 61.3 and 61.4 show a pictorial representation of the results delivered by RAMiner when dealing with the dataset of Morro Agudo. The figures present the obtained conviction (Fig. 61.3) and conviction-based probability (Fig. 61.4) measures by the variation of the parameters W and t, respectively. In the figures, the axis x refers to the one of the 90 parameter combinations (described previously) whereas the axis y refers to the obtained value of the estimated measure (in the left side) and the parameter value assumption (in the right side).

For the both figures, the crosshatched line represents the parameter value whereas the continuous line represents the value obtained (conviction/conviction-based probability).

As can be seen in Fig. 61.3, the variation of W does not imply in significantly changes to the conviction values. Worse results were obtained when $W = 10$, however the difference between the best accuracy (52.8%) and the

Fig. 61.2 A sample containing five association rules from the learning model with their confidence values

```
1. may=Mod_Low jul=Mod_Low aug=Mod_Low 7822 ==> jun=Mod_Low 7571    conf:(0.97)
2. may=Mod_Low jul=Mod_Low nov=Mod_Low jan=Mod_Low 7091 ==> jun=Mod_Low 6849    conf:(0.97)
3. may=Mod_Low jul=Mod_Low jan=Mod_Low 8273 ==> jun=Mod_Low 7988    conf:(0.97)
4. apr=Mod_Low may=Mod_Low jul=Mod_Low jan=Mod_Low 7331 ==> jun=Mod_Low 7077    conf:(0.97)
5. may=Mod_Low jul=Mod_Low sep=Low jan=Mod_Low 7390 ==> jun=Mod_Low 7127    conf:(0.96)
```

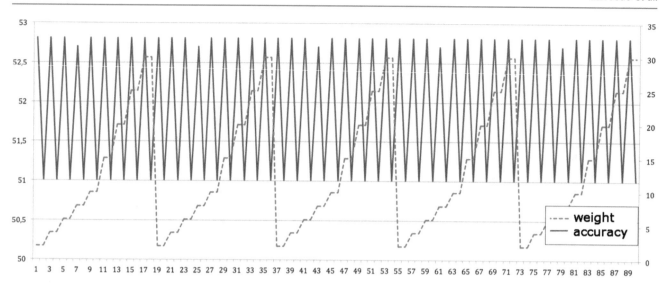

Fig. 61.3 Conviction value obtained in the dataset of Morro Agudo when the parameter W varies

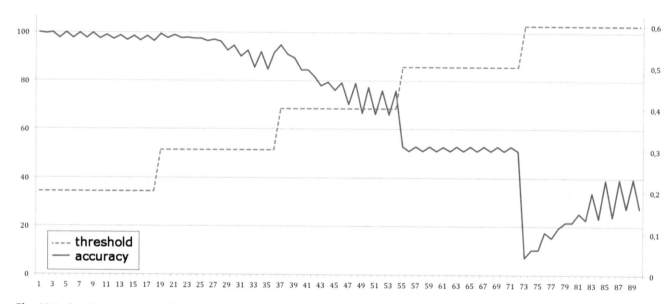

Fig. 61.4 Conviction-based probability value obtained when t is varying in the dataset of city Morro Agudo

worst (51%) is slightly smaller. On the other hands, Fig. 61.4 shows that as long as t increases significant decreases in the results can be observed. The higher values were delivered when a small threshold was considered (less than 0.4), but decreased when the threshold grew.

In total RAMiner delivered 360 results for each instance of each dataset (2 measures by 90 parameter combinations, considering 2 sensors). The best results obtained (the higher ones) were selected to be compared to the results given by the well-known classifiers, which delivered 36 results. The results are summarized in Table 61.3. The first three columns of the table provide a reference name, the approach considered and which sensor provides the dataset. The fourth, fifth and sixth columns show the result obtained in each city (A., M.A. and P. mean Araçatuba, Morro Agudo and Piracicaba, respectively). The last column presents an

average value obtained from the three results of the cities, considering the algorithm in the line.

The approaches references presented in Table 61.3 are:

- C4.5 tree run under MODIS dataset (c4.5-m) and NOAA (c4.5-n)4;
- Naive Bayes run under MODIS (nB - m) and NOAA (nB - n) datasets;
- One Rule run under MODIS (1R-m) and NOAA (1R-n) datasets;
- Zero Rule run under MODIS (0R-m) and NOAA (0R-n) datasets;
- Random Forest run under MODIS dataset (RF-m) and NOAA (RF-n);
- Support Vector Machine (SVM) run under MODIS dataset (SVM-m) and NOAA (SVM-n);

- RAMiner – maximum conviction measure from NOAA dataset (convMax-n) and MODIS (convMax-m);
- RAMiner – average conviction measure from NOAA dataset (convAvg-n) and MODIS (convAvg-m);
- RAMiner – average conviction-based probability from NOAA dataset (probAvg-n) and MODIS (probAvg-m).

The three best and the three worst results are boldfaced in the table. Note that the three best results (see table) were obtained by the use of RAMiner, four of the five highest measures as well. On the other hands, two of three worse results were given by algorithm 0R – see Araçatuba (11.48%) and M. Agudo (24.23%). The results obtained from Piracicaba prove that 0R also provides the worst result for the city (33.16%). However it is not one of the boldfaced values due to is not one of the three lower values. In general, the Random Forest approach was the unique that delivered similar results for both sensors, but RAMiner was better than almost all other classifiers in most of obtained results. Supposing a minimum threshold of 50%, it can be seen in the table that RAMiner delivers almost all of the best results. Morro Agudo has 8 accuracies higher than 50%, 5 of them given by RAMiner, Piracicaba has accuracies higher than 50% only when the instances were classified by RAMiner.

The accuracies presented in Table 61.3 are also presented in bar chart format, in Fig. 61.5. Each bar refers to one approach in one of the three cities. Approaches are organized in groups of three bars, one per city, Araçatuba, Morro Agudo and Piracicaba, respectively. The first 12 groups are related to the well-known classifiers and the last six groups of bars (in the right side) show accuracies given by RAMiner.

Table 61.3 and Fig. 61.5 taken together show that the accuracies obtained in the three cities presented a similar behavior, beside some slightly differences. Only Araçatuba presented different results from the both other. In Araçatuba, just 4 of 9 results above 50% were delivered by RAMiner. Also when considering MODIS sensor the accuracies were

Table 61.3 Accuracies obtained by RAMiner and the considered well-known classifiers in the datasets of the three cities, Araçatuba (A.), Morro Agudo (M.A.) and Piracicaba (P.). The results consider both the sensors, NOAA and MODIS. The three best and the three worst results are boldfaced

Reference	Algorithm	Sensor	Accuracy (%)			
			A.	M.A.	P.	Avg.
c4.5-m	C4.5	MODIS	39.32	32.76	44.90	38.99
c4.5-n	C4.5	NOAA	48.47	46.92	45.92	47.10
nB-m	N. Bayes	MODIS	35.46	29.69	36.76	33.97
nB-n	N. Bayes	NOAA	52.84	45.05	45.41	47.76
1R-m	One Rule	MODIS	39.12	**24.57**	39.07	34.25
1R-n	One Rule	NOAA	59.34	54.77	44.73	52.94
0R-m	Zero Rule	MODIS	69.51	56.31	47.90	57.90
0R-n	Zero Rule	NOAA	**11.48**	**24.23**	33.16	22.95
RF-m	R. Forest	MODIS	64.93	49.31	45.32	53.18
RF-n	R. Forest	NOAA	65.65	50.85	45.92	54.14
SVM-m	SVM	MODIS	46.56	49.73	48.85	48.38
SVM-n	SVM	NOAA	41.73	41.62	41.80	41.71
convMax-n	RAMiner	NOAA	49.38	52.81	54.07	52.08
convMax-m	RAMiner	MODIS	**83.34**	51.01	48.69	61.01
convAvg-n	RAMiner	NOAA	47.29	52.73	52.87	50.96
convAvg-m	RAMiner	MODIS	65.42	47.73	46.36	53.17
probAvg-n	RAMiner	NOAA	66.29	69.66	**70.02**	68.65
probAvg-m	RAMiner	MODIS	**77.41**	68.45	64.97	70.27

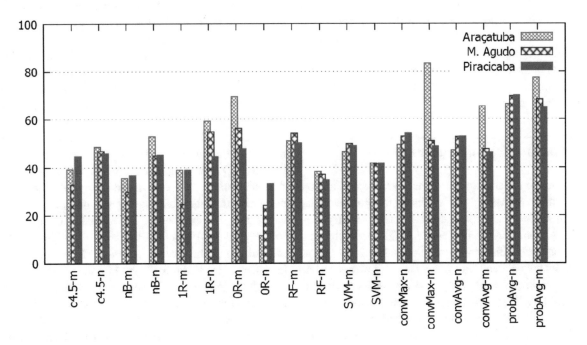

Fig. 61.5 Accuracies comparative

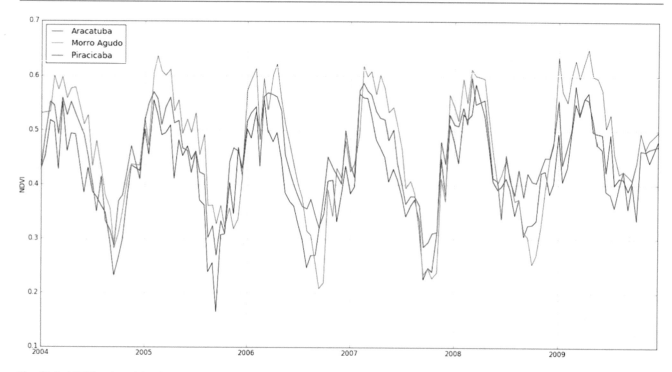

Fig. 61.6 NDVI series of the three cities on time interval from 2004 to 2009

slightly better than when considering NOAA. According to the expert analysis, that slight difference in the results of Araçatuba can be justified observing NDVI series of the three cities on bigger time interval (from 2004 to 2009), presented in Fig. 61.6. There are sharper drops in Araçatuba than in other cities. It may be caused by common reforms in planting, whereas other cultivations are managed as a way to improve the soil quality. Probably there were more changes than in other cities. Another fact given by the expert is the fact that in the analyzed time period the planting magnitude of Araçatuba was significantly lower than nowadays, due to was a region being introduced in the sugarcane scenario. So any small climatological change would result in significantly variations in the graph.

61.5   Conclusions and Future Works

This project was undertaken to design a novel classification method, called RAMiner and evaluate its run to mine agrometeorological data. The main goal was to identify sugarcane fields by analyzing NDVI series. RAMiner employed a majority voting classifier strategy and considered low spatial resolution image datasets. The accuracies of RAMiner were obtained by two measures, the conviction and conviction-based probability. They were compared to the results obtained with other well-known classifiers in the literature.

Despite of the results depending on many factors, like geographical position, time interval analyzed, planting magnitude and others, RAMiner presented almost of all highest accuracies. The incorporation of the conviction measure makes RAMiner more reliable, since it allows a weighted probability calculus by the conviction-based probability measure. Also, the incorporation of the W and t parameters makes RAMiner suitable for many other areas, since it provides an flexible method to set up RAMiner to deal with different datasets, as a way to maximize the performance. In the proposed experiment the best results were observed when considering the threshold parameter t less than 0.4.

As expected, RAMiner proved that is suitable to deal with cities that have different planting magnitude, as had being idealized. The results delivered by RAMiner have raised important questions about the city Araçatuba, pontificating historical particularities of the city. Even fast RAMiner is not linear; The learning model generation step has the complexity of the algorithm Apriori ($O(KQz)$), where K is the instances, Q is the candidates generated of each instance and z is the number of items in the instances. The coinciding points identification step has quadratic complexity, in the worst case and the loop of RAMiner has complexity on $O(CP * (LM + 1))$, where CP is the size of Coinciding Points set and LM is the size of Learning Model set (sum of the instances of the learning datasets for the yes and no classes).

The optimization of its complexity would be an interesting focus for a further work. As well as to extend the set of well known algorithms with more state-of-the-art approaches, such as Classification Based on Associations (CBA) and compare their results.

It is important to notice that due to NDVI series considered in this work refer to a sequence of numbers in a period of one year, there is an implicit temporal aspect associated to the data. This work did not deal directly with the temporal knowledge in the data, it would might be considered in a further work.

References

1. Mucherino, A., Papajorgji, P., & Pardalos, P. M. (2009). *Data mining in agriculture*. New York: Springer.
2. Michalski, R. S., & Chilausky, R. L. (1980). Learning by being told and learning from examples: An experimental comparison of the two methods of knowledge acquisition in the context of developing an expert system for soybean disease diagnosis. *International Journal of Policy Analysis and Information System, 4*, 125–161.
3. Bhatia, J., & Gupta, A. (2014). Mining of quantitative association rules in agricultural data warehouse: A road map. *IJISIS, 3*, 187–198.
4. Ramesh, V., & Ramar, K. (2011). Classification of agricultural land soils: A data mining approach. *Agricultural Journal, 6*, 82–86.
5. Verheyen, K., & Hermy, M. (2001). The relative importance of dispersal limitation of vascular plants in secondary forest succession in Muizen forest. *Journal of Ecology, 89*, 829–840.
6. Brudzewski, K., Osowski, S., & Markiewicz, T. (2004). Classification of milk by means of an electronic nose and SVM neural network. *Sens Actuators, 98*, 291–298.
7. Fagerlund, S. (2007). Bird species recognition using support vector machines. *Signal Processing, 38637*, 8.
8. Brudzewski, K., Osowski, S., & Markiewicz, T. (1998). Satellite remote sensing for forestry planning: A review. *Scandinavian Journal of Forest Research, 13*, 90–110.
9. Das, K. C., & Evans, M. D. (1992). Detecting fertility of hatching eggs using machine vision II: Neural network classifiers. *Transactions of the ASAE. American Society of Agricultural Engineers, 35*, 2035–204.
10. Karimi, Y., Prasher, S. O., Patel, R. M., & Kim, S. H. (2006). Application of support vector machine technology for weed and nitrogen stress detection in corn. *Computer Electronics Agriculture, 51*, 99–109.
11. Patel, V. C., McClendon, R. W., & Goodrum, J. W. (1994). Crack detection in eggs using computer vision and neural networks. *Artificial Intelligence Applications, 8*, 21–31.
12. Agrawal, R., & Srikant, R. (1994). Fast algorithms for mining association rules in large databases. In *Proceedings of the 20th International Conference on Very Large Data Bases, VLDB*, Santiago (pp. 487–499).

Danilo Medeiros Eler, Ives Renê Venturini Pola, Rogério Eduardo Garcia, and Jaqueline Batista Martins Teixeira

Abstract

Text mining is an important step to categorize textual data by using data mining techniques. As most obtained textual data is unstructured, it needs to be processed before applying mining algorithms – that process is known as pre-processing step in overall text mining process. Pre-processing step has important impact on mining. This paper aims at providing detailed analysis of the document pre-processing when employing multidimensional projection techniques to generate graphical representations of vector space models, which are computed from eight combinations of three steps: stemming, term weighting and term elimination based on low frequency cut. Experiments were made to show that the visual approach is useful to perceive the processing effects on document similarities and group formation (i.e., cohesion and separation). Additionally, quality measures were computed from graphical representations and compared with classification rates of a k-Nearest Neighbor and Naive Bayes classifiers, where the results highlights the importance of the pre-processing step in text mining.

Keywords

Text mining • Document pre-processing • Visualization • Document similarity • Multidimensional projection

62.1 Introduction

Many documents have been generated and stored by distinct institutions, organizations and researchers over the years. Considering the amount of data and its complexity, text mining techniques are widely employed to classify, organize and generate useful knowledge from textual data [1]. Usually, this kind of data is unstructured and need to be pre-processed before mining tasks.

Document pre-processing is composed by important and essential steps for text mining tasks. The steps usually filter documents of interest, eliminate irrelevant terms, and assign weights to terms. However, due to its non-automatic nature, where the user needs to define parameters, pre-processing receives less importance in the wholly text mining scenario [2]. In order to analyze the pre-processing effects and parameter configurations in text mining tasks, quality measures are used to verify which pre-processing step leads to the best mining result [3] – for example, classification rate can be employed to evaluate the pre-processing effect in the classification accuracy.

In this paper we highlight the importance of document pre-processing by employing multidimensional projection techniques [4] to generate graphical representations based on text similarities as a form of textual data representation [5, 6]. The visual analysis facilitates the comprehension of pre-processing effects to document similarities, that is, which steps or parameter configuration can improve both

D.M. Eler (✉) • I.R.V. Pola • R.E. Garcia • J.B.M. Teixeira
Faculdade de Ciências e Tecnologia, Departamento de Matemática e Computação, UNESP – Universidade Estadual Paulista, Presidente Prudente-SP, Brazil
e-mail: daniloeler@fct.unesp.br; ivesrene@gmail.com; rogerio@fct.unesp.br; jt.jaque@gmail.com

© Springer International Publishing AG 2018
S. Latifi (ed.), *Information Technology – New Generations*, Advances in Intelligent Systems and Computing 558, DOI 10.1007/978-3-319-54978-1_62

groups cohesion and separation, or the correct classification rate. To illustrate the usefulness of visual analysis, distinct vector space models [7] are computed from document collections by varying the pre-processing steps, such as stemming [8], term weighting based on TF-IDF [9], and reduction of amount of terms based on frequency cut [10]. The combination of pre-processing steps results in eight distinct vector space models that are visualized in 2D space by using multidimensional projection techniques.

The main contribution of this work is the visualization of the pre-processing effects, where the graphical representations show the cohesion or separation of document groups when changing the parameters configuration of pre-processing. In order to validate the best vector space model, Silhouette Coefficient [11] and Neighborhood Hit [5] quality measures were computed to each projection.

This paper is organized as follows. Section 62.2 presents the theoretical foundation of pre-processing steps commonly employed in text mining tasks and a brief foundation about multidimensional projection techniques. Section 62.3 presents the employed methodology used to analyze the document pre-processing effects. Section 62.4 presents the performed experiments with a discussion about visual analysis of vector space models computed from combination of distinct document pre-processing. Section 62.5 concludes the paper, summarizing the main achievements and projecting further works.

62.2 Theoretical Foundation

This section describes the theoretical foundations used in this work, regarding to document pre-processing and multi-dimensional projections.

62.2.1 Document Pre-processing

Most data mining tasks are applied in structured data. However, data from document collections for text mining tasks are unstructured. So, several pre-processing steps must be executed to compute a structured model for a document collection, listed as:

- **Document Selection:** identification of the documents employed in the text mining task;
- **Tokenization:** words or terms identification from the document collection;
- **Stop-word elimination:** elimination of irrelevant words or terms that commonly appear in any document (i.e., articles and prepositions);
- **Stemming:** reduces the word for a minimized form, extracting the root of the word. For instance, the root form of "consignment" is "consign";

- **Luhn Cut:** eliminates terms based on terms frequency. Usually, this process establish a low cut from which only the terms with a frequency count over the cut value are considered;
- **Weighting:** establish a weighting value to reduce or increase the term influence to each document. For instance, the most known approach is TF-IDF (Term Frequency – Inverse Document Frequency).

After the document collection is pre-processed, a vector space model [7] is computed, commonly called as *document × term* matrix, in which each row represents a document, each column represents a term, and each element is the frequency (TF) or the term influence in the respective document (e.g., TF-IDF). Thus, data mining algorithms, such as clustering and classifiers, can be employed to the structured document collection. Additionally, most of information visualization techniques also require structured data, and therefore, those steps are useful for document collection visualization as well.

The next section presents the concepts of the employed visualization technique.

62.2.2 Multidimensional Projection

Multidimensional Projection have been used to aid the exploration of multidimensional spaces computed from distinct domains [6, 12–14]. Multidimensional Projection techniques transform data sets described in a $\mathbb{R}^m$ space (with m attributes) into a $\mathbb{R}^n$ space ($m > n$) [4], preserving in the projected space the structure from the original multidimensional space. Therefore, these techniques can be employed to perform a dimensionality reduction allowing a better data set visualization into the computer screen (2D space). Thus, similar instances are placed near in 2D space and dissimilar ones are placed far, keeping as much as possible the instances relationship from the original multidimensional space.

62.3 Visual Analysis of Document Pre-processing Effects

This paper highlights the importance of the document pre-processing steps in text mining tasks. First, consider the methodology process in Fig. 62.1. Given a document collection, three pre-processing steps can be executed:

- **Stemming:** in this step is decided if Porter's stemming [8] will be employed to get the root of each word;
- **Frequency Cut:** some terms can be discarded in this step [10] by choosing a low frequency cut value (LC_{Thr}); otherwise, if no cut is set (LC_1), all terms

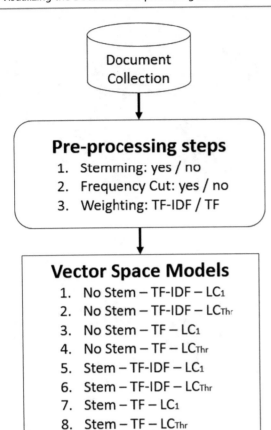

Fig. 62.1 Methodology process

whose frequency is greater than one will be used to compute the vector space model. In this paper, the evaluation of the low frequency cut value (LC_{Thr}) was performed by an automatic approach based on Otsu's Threshold Selection Method [15];

- **Term Weighting:** the term frequency (TF) can be used as coordinate values in vector space model or term weights can be assigned based on TF-IDF [9, 16].

Once all vector space models are computed, a multidimensional visualization approach creates a graphical representation (projection) to represent the document similarities in a 2D space. In order to evaluate which projection represents the better group of collections, two quality measures were employed, the Silhouette Coefficient [11] and the Neighborhood Hit [5], as detailed in the next section.

62.4   Experiments

This section describes two experiments performed to show the document pre-processing effects. We used document collections of scientific papers and news. They are listed in the following:

- **CBR-ILP-IR:** 574 papers from three areas of Artificial Intelligence: Case-Based Reasoning (CBR), Inductive Logic Programming (ILP), Information Retrieval (IR);
- **NEWS-8:** 495 news from Reuters, AP, BBC and CNN, classified in eight classes;

All experiments followed the document pre-processing steps described in Sect. 62.3, resulting in eight distinct vector space models (VSM) from combination of stemming, term weighting and frequency cutting steps. The Least Square Projection (LSP) [5] technique was employed to generate the graphical representations. LSP preserves the instances neighborhood relations from multidimensional space (i.e., vector space model) into a lower dimensional space – in this paper, the projected space is 2D. Additionally, we also present the traditional approach of measuring the vector space model quality based on correct classification rate. For that, we used k-Nearest Neighbor (K-NN) and Naive Bayes classifiers from Weka,[1] with default parameters and k-NN set with 1-NN.

In the first experiment, the CBR-ILP-IR data set was pre-processed and eight VSMs was evaluated, from which LSP technique generated graphical representations, as shown in Fig. 62.2. Each color represents a class of a document and the closer the points the more similar they are. The term frequency cutting value 475 was set by Otsu's threshold method [15]. Its worth to note that stemming did not produced great changes in the clusters, but comparing Fig. 62.2b, f, the use of stemming caused the spread of the red group. The frequency cut improved the green group cohesion as can be seen when comparing Fig. 62.2e, f. The term weighting based on TF-IDF does not improve the group formation and spread some groups, as one might observe in Fig. 62.2h, f. As presented in Table 62.1, we used Silhouette Coefficient to identify the VSM that better preserve the groups cohesion and separation. Silhouette Coefficient shows the best projection for the CBR-ILP-IR data set, as presented in Fig. 62.2b – VSM computed with stemming, TF-IDF and low cut 475. Additionally, Fig. 62.4a shows the Neighborhood Hit quality measure evaluated for each presented projection in Fig. 62.2. Thus, one might note that the projection shown in Fig. 62.2b preserves the document similarities when the number of neighbors is increased.

A similar experiment was executed with NEWS-8 data set: eight VSMs were evaluated by varying the pre-processing steps and LSP technique generated graphical representations for each VSM. Again, the term frequency cutting value 51 was set by Otsu's threshold method. The resulting projections are shown in Fig. 62.3. It can be

---

[1] Weka is a system composed by several data mining algorithms – available in http://www.cs.waikato.ac.nz/ml/weka/.

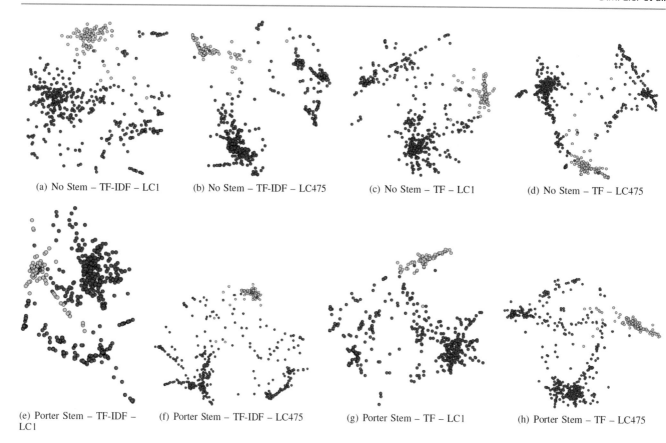

(a) No Stem – TF-IDF – LC1   (b) No Stem – TF-IDF – LC475   (c) No Stem – TF – LC1   (d) No Stem – TF – LC475

(e) Porter Stem – TF-IDF –   (f) Porter Stem – TF-IDF – LC475   (g) Porter Stem – TF – LC1   (h) Porter Stem – TF – LC475
LC1

Fig. 62.2 Graphical representations generated for each vector space model from CBR-ILP-IR data set. The Least Squares Projection (LSP) [5] technique was employed to evaluate the projections. (**a**) No stem – TF-IDF – LC1. (**b**) No stem – TF-IDF – LC475. (**c**) No stem – TF – LC1. (**d**) No stem – TF – LC475. (**e**) Porter stem – TF-IDF – LC1. (**f**) Porter stem – TF-IDF – LC475. (**g**) Porter stem – TF – LC1. (**h**) Porter stem – TF – LC475

Table 62.1 Silhouette Coefficient evaluated for each projection presented in Fig. 62.2 and classification rate from k-NN and Naive Bayes classifiers – Vector Space Models from CBR-ILP-IR data set

Vector space model	Silhouette	k-NN	Naive bayes
No stem – TF-IDF – LC1	0.48	48%	97%
No stem – TF-IDF – LC475	0.67	93%	98%
No stem – TF – LC1	0.57	48%	97%
No stem – TF – LC475	0.58	93%	97%
Stem – TF-IDF – LC1	0.43	57%	97%
Stem – TF-IDF – LC475	0.56	91%	95%
Stem – TF – LC1	0.58	57%	97%
Stem – TF – LC475	0.58	91%	95%

notable that the best result is presented in Fig. 62.3f – VSM computed with stemming, TF-IDF and low cut 51 – showing that stemming can improve the group cohesion, even though it did not occur when comparing Fig. 62.3d, h, in which stemming scattered some group of documents. The frequency cut improved the group cohesion and separation for

all projections, except for those shown in Fig. 62.3g, h, in which some group separation was decreased. Regarding the term weighting, the only improving was noted when comparing Fig. 62.3e, f. As shown in Table 62.2, the Silhouette Coefficient was evaluated for each projection and confirms the visual inspection: the projection presented in Fig. 62.3f represents the best VSM. With a similar result, the projections presented in Fig. 62.3b, d, which does not use stemming in document pre-processing, were evaluated with similar quality based on Silhouette Coefficient. Additionally, Fig. 62.4b shows the Neighborhood Hit quality measure computed for each projection presented in Fig. 62.3, confirming the analysis based on visual inspection and Silhouette Coefficient. Again, one might note the consistency in the quality of the best VSMs as the number of neighbors increases.

Comparing the silhouette values with classification rates of k-NN and Naive Bayes classifiers, presented in Tables 62.1 and 62.2, it is possible to note that silhouette

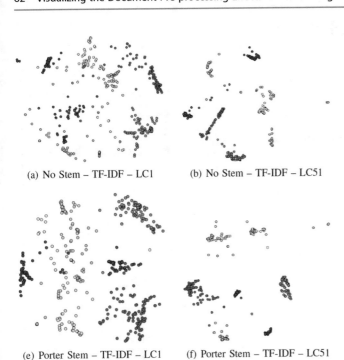

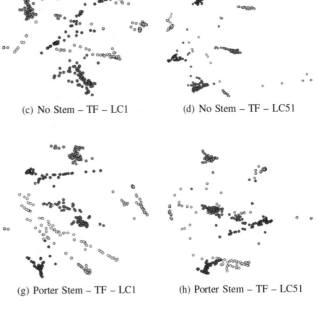

(a) No Stem – TF-IDF – LC1   (b) No Stem – TF-IDF – LC51   (c) No Stem – TF – LC1   (d) No Stem – TF – LC51

(e) Porter Stem – TF-IDF – LC1   (f) Porter Stem – TF-IDF – LC51   (g) Porter Stem – TF – LC1   (h) Porter Stem – TF – LC51

Fig. 62.3 Graphical representations generated for each vector space model from NEWS-8 data set. The Least Squares Projection (LSP) [5] technique was employed to compute the projections. (**a**) No stem – TF-IDF – LC1. (**b**) No stem – TF-IDF – LC51. (**c**) No stem – TF – LC1. (**d**) No stem – TF – LC51. (**e**) Porter stem – TF-IDF – LC1. (**f**) Porter stem – TF-IDF – LC51. (**g**) Porter stem – TF – LC1 (**h**) Porter stem – TF – LC51

Table 62.2 Silhouette Coefficient computed for each projection presented in Fig. 62.3 and classification rate from k-NN and Naive Bayes classifiers – Vector Space Models from NEWS-8 data set

Vector space model	Silhouette	k-NN	Naive bayes
No stem – TF-IDF – LC1	0.48	88%	97%
No stem – TF-IDF – LC51	0.65	97%	98%
No stem – TF – LC1	0.53	88%	97%
No stem – TF – LC51	0.75	96%	98%
Stem – TF-IDF – LC1	0.48	90%	97%
Stem – TF-IDF – LC51	0.77	97%	98%
Stem – TF – LC1	0.56	90%	97%
Stem – TF – LC51	0.53	97%	98%

coefficient can indicate the best vector space models, as well as the Neighborhood Hit graphs presented in Fig. 62.4.

62.5 Conclusions and Future Works

Text mining is a valuable task for distinct research fields. The success of text mining tasks is highly dependent of the document pre-processing in which documents are selected or eliminated, relevant words are chosen, terms are eliminated, and weights are assign to terms. Thus, the raw textual data is transformed in a vector space model (VSM) capable of

discriminate each document. Usually, classifiers are employed to verify the quality of VSMs computed from distinct combination of pre-processing steps or parameter configuration. However, this approach do not show the document similarities neither group cohesion and separation.

In this paper, we presented a visualization-based approach to understand and analyze the document pre-processing effects in text mining. Graphical representations were generated with multidimensional projection techniques to show the document similarities and group formation in 2D space. Thus, it was possible to perceive the pre-processing effects to each document, to group of documents, to each class of documents, and to the wholly collection. This approach was supported by two distinct quality measures employed from the projections computed for each vector space model – Silhouette Coefficient and Neighborhood Hit. We also used k-Nearest Neighbor and Naive Bayes classifiers to show that the traditional approach based on classification rates to choose the best vector space model is also precise. However, our approach played an important role by enabling detailed inspection of the pre-processing effects.

Further works involve to analyze other pre-processing steps, for instance, outliers removal and stop-words elimination. Additionally, we also intend to analyze different stemming algorithms as well as distinct weighting methods.

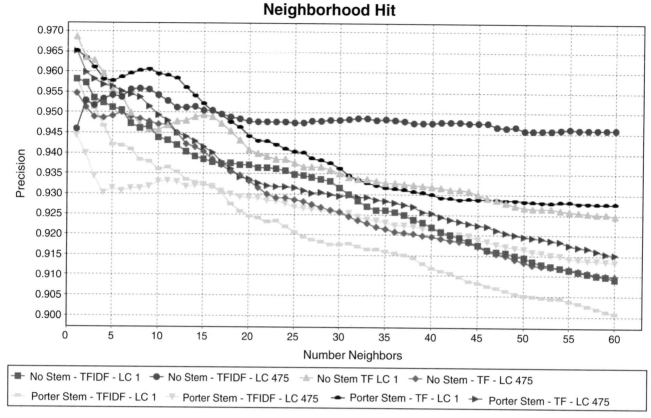

(a) Neighbourhood Hit from CBR-ILP-IR

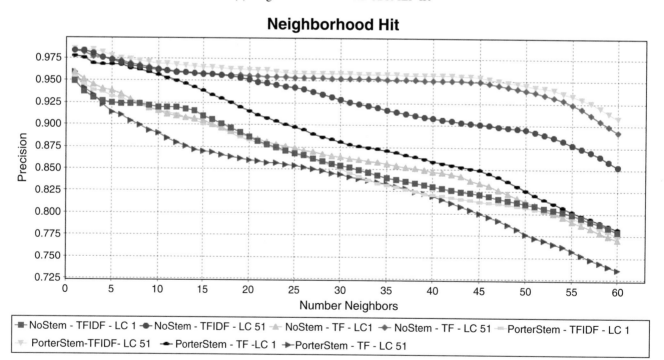

(b) Neighbourhood Hit from NEW58

Fig. 62.4 Neighbourhood Hit graphics for analyzing the quality of graphical representations computed from vector space models generated from CBR-ILP-IR and NEWS-8 data sets. (**a**) Neighbourhood hit from CBR-ILP-IR. (**b**) Neighbourhood hit from NEWS8

Acknowledgements The authors acknowledge the financial support of the Brazilian financial agency São Paulo Research Foundation (FAPESP) – grant # 2013/03452-0.

References

1. Hu, Y., Milios, E. E., & Blustein, J. (2012). Enhancing semi-supervised document clustering with feature supervision. In *Proceedings of the 27th Annual ACM Symposium on Applied Computing, SAC'12* (pp. 929–936). New York: ACM. [Online] Available: http://doi.acm.org/10.1145/2245276.2245457.

2. Nogueira, B. M., Moura, M. F., Conrado, M. S., Rossi, R. G., Marcacini, R. M., & Rezende, S. O. (2008). Winning some of the document preprocessing challenges in a text mining process. In *Anais do IV Workshop em Algoritmos e Aplicações de Mineração de Dados – WAAMD, XXIII Simpósio Brasileiro de Banco de Dados- SBBD* (pp. 10–18). Porto Alegre: SBC. [Online] Available: http://www.lbd.dcc.ufmg.br:8080/colecoes/waamd/2008/002.pdf.

3. Chandrasekar, P., & Qian, K. (2016). Undefined, undefined, undefined, and undefined, the impact of data preprocessing on the performance of a Naive Bayes classifier. In: *2016 I.E. 40th Annual Computer Software and Applications Conference (COMPSAC)*, Atlanta (Vol. 02, pp. 618–619).

4. Tejada, E., Minghim, R., & Nonato, L. G. (2003). On improved projection techniques to support visual exploration of multidimensional data sets. *Information Visualization, 2*(4), 218–231.

5. Paulovich, F. V., Nonato, L. G., Minghim, R., & Levkowitz, H. (2008). Least square projection: A fast high precision multidimensional projection technique and its application to document mapping. *IEEE Transactions on Visualization and Computer Graphics, 14*(3), 564–575.

6. Eler, D. M., Paulovich, F. V., Oliveira, M. C. F. d., & Minghim, R. (2008). Coordinated and multiple views for visualizing text collections. In *Proceedings of the 12th International Conference on Information Visualisation (IV'08)* (pp. 246–251). Washington, DC: IEEE Computer Society.

7. Salton, G., Wong, A., & Yang, C. S. (1975). A vector space model for automatic indexing. *Communications of the ACM, 18*, 613–620.

8. Porter, M. F. (1997). *An algorithm for suffix stripping* (pp. 313–316). San Francisco: Morgan Kaufmann Publishers Inc.

9. Salton, G., & Yang, C. S. (1973). On the specification of term values in automatic indexing. *Journal of Documentation, 29*(4), 351–372.

10. Luhn, H. P. (1958). The automatic creation of literature abstracts. *IBM Journal of Research and Development, 2*(2), 159–165. [Online] Available: http://dx.doi.org/10.1147/rd.22.0159.

11. Tan, P.-N., Steinbach, M., & Kumar, V. (2005). *Introduction to data mining* (1st ed.). Boston: Addison-Wesley Longman Publishing Co., Inc.

12. Eler, D., Nakazaki, M., Paulovich, F., Santos, D., Andery, G., Oliveira, M., Batista, J. E. S., & Minghim, R. (2009). Visual analysis of image collections. *The Visual Computer, 25*(10), 923–937.

13. Paulovich, F. V., Eler, D. M., Poco, J., Botha, C., Minghim, R., & Nonato, L. G. (2011). Piecewise Laplacian-based projection for interactive data exploration and organization. *Computer Graphics Forum, 30*(3), 1091–1100.

14. Bodo, L., de Oliveira, H. C., Breve, F. A., & Eler, D. M. (2016). Performance indicators analysis in software processes using semi-supervised learning with information visualization. In *13th International Conference on Information Technology: New Generations (ITNG 2016)* (Advances in intelligent systems and computing, pp. 555–568). Las Vegas, NV: Springer International Publishing.

15. Eler, D. M., & Garcia, R. E. (2013). Using Otsu's threshold selection method for eliminating terms in vector space model computation. In *International Conference on Information Visualization*, Barcelona (pp. 220–226). IEEE Computer Society.

16. Salton, G., & Buckley, C. (1988). Term-weighting approaches in automatic text retrieval. *Information Processing and Management*, 513–523.

Complex-Network Tools to Understand the Behavior of Criminality in Urban Areas

63

Gabriel Spadon, Lucas C. Scabora, Marcus V. S. Araujo, Paulo H. Oliveir, Bruno B. Machado, Elaine P. M. Sousa, Caetano Traina, and Jose F. Rodrigues

Abstract

Complex networks are nowadays employed in several applications. Modeling urban street networks is one of them, and in particular to analyze criminal aspects of a city. Several research groups have focused on such application, but until now, there is a lack of a well-defined methodology for employing complex networks in a whole crime analysis process, i.e. from data preparation to a deep analysis of criminal communities. Furthermore, the "toolset" available for those works is not complete enough, also lacking techniques to maintain up-to-date, complete crime datasets and proper assessment measures. In this sense, we propose a threefold methodology for employing complex networks in the detection of highly criminal areas within a city. Our methodology comprises three tasks: (i) Mapping of Urban Crimes; (ii) Criminal Community Identification; and (iii) Crime Analysis. Moreover, it provides a proper set of assessment measures for analyzing intrinsic criminality of communities, especially when considering different crime types. We show our methodology by applying it to a real crime dataset from the city of San Francisco—CA, USA. The results confirm its effectiveness to identify and analyze high criminality areas within a city. Hence, our contributions provide a basis for further developments on complex networks applied to crime analysis.

Keywords

Complex networks • Crime analysis • Criminal communities • Pattern analysis

63.1 Introduction

Complex networks have long been used to model social behavior, spatial patterns, spreading of epidemics, and urban structures. They are capable of representing from neural connections to subway networks [1]. In the context of street networks, which are of particular interest for this work, complex networks are able to describe flow relationships and social behavior in an urban zone [2].

G. Spadon (✉) • L.C. Scabora • M.V.S. Araujo • P.H. Oliveir
B.B. Machado • E.P.M. Sousa • C. Traina • J.F. Rodrigues
University of Sao Paulo, Sao Paulo, Brazil
e-mail: spadon@usp.br; lucascsb@usp.br; araujo@usp.br;
pholiveira@usp.br; brandoli@icmc.usp.br; parros@icmc.usp.br;
caetano@icmc.usp.br; junio@icmc.usp.br

Social behavior as violence and social disorder may be related to the urban structure of a given city. For instance, low-flow areas tend to be propitious for crime events, since such areas end up being less surveilled when compared to high-flow areas. For that reason, crime events usually occur within regions that can be detected based on their structural and behavioral properties [3]. In this context, this paper provides a methodology for aiding the analysis of urban criminality—e.g. how specific crimes occur in certain areas and how their types are related to each other by being in neighboring areas, that is, by being similar.

The hypothesis of this work is that *by employing network mapping techniques, allied to distance-based properties of graphs, it is possible to identify and trace the relationship between areas that are highly criminal within a city. Our*

© Springer International Publishing AG 2018
S. Latifi (ed.), *Information Technology – New Generations*, Advances in Intelligent
Systems and Computing 558, DOI 10.1007/978-3-319-54978-1_63

assumption is that similar crimes occur in adjacent regions. Such crimes emerge from, among other factors, the urban organization, and their inter-similarity carries correspondence to the distance from one region to the other. Technically, our proposal is based on: (i) community detection algorithms applied to datasets of isolated crime types; (ii) pattern and similarity analyses, which compare crimes occurred in distinct regions within the same city. Based on those techniques, we propose a threefold methodology consisted of: **(i) Mapping of Urban Crimes** —we describe how to combine a city's georeferenced urban structure with crime records into a complex network with potential for analytical tasks; **(ii) Criminal Community Identification** —we show how to detect criminal communities by exploring geographical as well as structural properties; **(iii) Crime Analysis** —we methodologically compare crime patterns by identifying relationships between their communities, especially in the face the of diverse crime types.

Our methodology contributes to the understanding of criminality in the context of urban organization. We work on issues related to: (i) *an approach to identify highly criminal areas*; (ii) *the characterization of the city space by inferring the homogeneity of its crimes*; (iii) *the identification of city sectors that present behaviors related to the sparsity of their crimes*; and (iv) *the delineation of the city space by identifying regions that are related, acting as crime spots for different crime types*.

The remainder of the paper is structured as follows. Section 63.2 presents the related work, discussing how our methodology stands out from theirs. Section 63.3 describes the dataset used in this work and the methodology proposed. Section 63.4 discusses the results obtained from the application of our methodology to the crimes dataset, which describes a real scenario from the street network of the North American city of San Francisco. Finally, Sect. 63.5 presents the conclusion.

63.2 Related Work

Several works in the literature have dealt with the topics of urban networks and criminality. This section describes some of these works organized into two categories related to the phases of our proposal: **Mapping of Urban Crimes**, which focuses on representing raw crime data with a complex network, and **Crime Analysis**, which refers to identifying criminal behaviors and patterns by analyzing urban street networks.

Mapping of Urban Crimes. Spicer et al. [4] describe a theoretical framework capable of mapping urban crimes into related locations in a city. Their work discusses the pros and cons of mapping methods. To represent a city,

they use street networks and GIS software, describing methods for mapping georeferenced elements into a complex network. However, their methods associate a crime with the nearest node or edge based on address geocoding, not having any relation with graph measures. Moreover, all methods are superficially introduced, not presenting any formal fundaments.

Shinode and Shinode [5] aim at introducing a search-window method to analyze urban crimes in street networks. They represent a search window as a subarea that concentrates a high number of crimes in small regions; besides that, the crimes are used to identify clusters spatially and temporally. The authors validate their methods through an empirical case study. However, their criminal data derive from 911-calls made in 1996; that is, they used a dataset that carries few details and that is outdated, especially when one considers the conclusions claimed by the authors. Such information implies a superficial analysis of the crime behavior and the ways to prevent it.

Crime Analysis. White et al. [6] represent criminal activities via a georeferenced complex network, through which they compare criminal demographics and identify criminal communities. The demographic regions are created by clustering the urban areas using socioeconomic borders, whereas the crime network is formed by linking crime locations that are close according to the Euclidean distance. To identify the criminal communities, the authors employ a label propagation technique. Their results show that the crime network can be constructed without requiring information about individual crimes. However, their crime network does not take the urban topology into account, so their comparison is based only on socioeconomic information.

Rey et al. [7] employ spatiotemporal techniques to represent and analyze urban crimes. Their approach provides a way to quantify neighborhood criminality and is able to identify patterns from the data. Moreover, they discuss spatial crime analyses as a way to determine the influence of a crime on its surroundings, analyzing an entire city and providing an evaluation of its crimes. However, their work is based on outdated data, derived from a police district in the period from 2005 to 2009. Besides, their methods rely on black-box GIS technologies, used to summarize criminal areas by geocoding each crime into a quarter-mile grid cell. Furthermore, they analyze only city cells having residential units.

63.3 Proposed Methodology

This section describes the phases of our methodology according to an example.

63.3.1 Background and Datasets

We refer to an urban street network as a directed georeferenced graph $G = \{V, E\}$ composed of a set E of $|E|$ edges (street segments) and another set V of $|V|$ nodes (streets intersections). We refer to an edge $e \in E$ as an ordered pair $i, \ j, i \in V$ and $j \in V$, in which i is named *source* and j is named *target*. Each node $v \in V$ has coordinates l_{at} and l_{on}, where l_{at} denotes the node's latitude and l_{on} the longitude. We refer to a set $C, C \cap V = \emptyset$, of crimes, each with coordinates l_{at} and l_{on} within the area of the city. Each crime c has a value that belongs to a domain of 39 crime types—the most common types are *assault*, *theft*, and *minor crimes*.

Graph Source. Our approach uses electronic cartographic maps extracted from the *OpenStreetMap* (OSM) [8] platform. OSM provides maps representing spatial elements, e.g. points, lines and polygons, as objects. Such objects, which are abstractions from the real-world geographical space, are represented by a triple of attributes that uniquely identifies them in their set, granting each object an identification number and georeferenced coordinates. These objects allow to represent OSM data as a georeferenced graph.

Crime Source. The crime dataset C comes from the *San Francisco OpenData* (SFO) initiative,[1] a central clearinghouse for public information about the city of San Francisco—CA, USA. SFO provides data concerning public issues. We use a set of events that corresponds to the period between Jan 1st, 2003 through May 5th, 2016. It has 1,916,911 instances of 39 crime types.

63.3.2 Mapping of Urban Crimes

The first phase of our methodology receives the crime dataset as input, denoted as set C, where each element $c \in C$ corresponds to a crime event occurred in a specific latitude and longitude. We proceed by mapping the crime events to the nodes of the urban street network, represented as the graph $G = \{V, E\}$. Given a crime c, its corresponding network node is the closest node $v, v \in V$, as given by the Euclidean distance calculated from the coordinates of c and v. This distance refers to the real length between elements thought the Earth's surface, which is derived from the law of cosines and defined as:

$$d_{ij}^E = \mathcal{R} \times cos^{-1}(sin(l_{at}^i)sin(l_{at}^j) + \\ cos(l_{at}^i)cos(l_{at}^j)cos(\triangle_{ij}^{l_{on}})) \qquad (63.1)$$

where l_{at}^i and l_{at}^j are latitudes, $\triangle_{ij}^{l_{on}}$ is the difference between longitudes l_{on}^i and l_{on}^j, $\mathfrak{R}$ is the earth's radius (6,371 km), and d_{ij}^E denotes the Euclidean distance (E) between points i and j. All values are represented in radians [9]. With respect to the georreferenced network, d_{ij}^E also represents an edge's weight, which refers to the street length (in meters) that connects the streets' intersections (i.e. nodes) i and j.

Next, we enrich the network representation with a new property, making it reflect the geographical distribution of crimes. The result is a new set of nodes V', with each node $v \in V'$ having a set with three properties $prop(v) = \{l_{at}, l_{on}, |C_v|\}$, where l_{at} is the latitude, l_{on} is the longitude, and $|C_v|$ is the quantity of crimes mapped to node v. Formally, the set of crimes mapped to a node v, denoted as C_v, is defined as:

$$C_v = c \in C | d_{vc}^E < d_{wc}^E, \ \forall \ w \in \ V, \ v \neq w \qquad (63.2)$$

Note that this phase is performed separately for each kind of crime. For example, if performing the mapping for crime type *assault*, we get a version of the graph in which the nodes have a property indicating the number of crimes that are categorized as an *assault*. After processing each kind of crime, we get a new property in the nodes. In a mapping perspective, each processing yields a new graph that differs with respect to the properties of the nodes; formally, $? = G_0, G_1, \ldots, G_n$, where each $G = V', E, \ \forall \ G \in ?$.

63.3.3 Criminal Community Identification

Intuitively, a community is understood as a group of nodes that have a probability of connecting to each other that is greater than the probability of connecting to nodes out of the group [10]. Formally, consider $G = \{V', E\}$ as a city graph and $p(i, j)$ as the probability of nodes i and j to define a connection; a community set is understood as a subset of nodes $\mathfrak{C}$. A node i is considered a member of the community $\mathfrak{c} \in \mathfrak{C}$ if its probability of connecting to a node $j \in \mathfrak{c}$ is greater than the probability of connecting to a node j' in another community $\mathfrak{c}' \in \mathfrak{C}$, as follows:

$$p(i,j) \geq p(i,j') \ | i \notin \mathfrak{c}, i \notin \mathfrak{c}', \ \forall \ j \in \mathfrak{c}, \ \forall \ j' \in \mathfrak{c}', \mathfrak{c} \neq \mathfrak{c}'$$

The elements of a community define a connected induced subgraph $G(\mathfrak{c})$ in which each node is reachable from all other nodes of the same community. Formally, $G(\mathfrak{c}) = V'', E'$, $V'' \subseteq V' \Leftrightarrow \forall \ i \in \mathfrak{c}, \ \forall \ j \in \mathfrak{c} \ | \ d_{ij} < \infty$, where $d_{ij} : V' \times V' \to \mathbb{R}$ is an arbitrary distance function which returns the distance from any given pair of nodes.

Next, we proceed with the identification of communities. To identify them concerning each crime type, we opted for

---

[1] Available on "data.sfgov.org".

the tool Nerstrand,[2] which carries out a fast multi-thread detection. This tool, which yields results broadly accepted in the literature, identifies communities without any information about how many of them exist in the city. The input for this phase is an undirected graph, whereas the output is a set of number-labeled communities. The tool allows the user to attach weights to the graph elements; for our application, the edges' weights are the distance between two distinct nodes, and nodes' properties are the number of mapped crimes.

Our dataset comprises 39 crime types, however, processing and analyzing all of them would exceed the available space of this work. Besides, our techniques and results are generalizable for any number of crime types. For those reasons, we worked only with the three most recurrent ones, which comprise almost 40% of our dataset and represent: (i) assault, i.e. to inflict injury on a person intentionally; (ii) theft, i.e. to take a property off a person's possession by stealth, with no brute force; and (iii) minor crimes, such as offenses that do not involve any loss or injury. They represent, respectively, 167,832 (8.76%), 392,338 (20.47%) and 204,451 (10.67%) crimes. As previously explained, for each crime type, we perform a mapping over the graph, each one with a different number of crimes mapped into their nodes. Further on, we discuss that each crime-mapped graph can produce a specific criminality scenario; that is, when considering each crime type, we get a different set of communities, each one characterized by that kind of crime and by the topology of the network.

The next step, after identifying communities, is to filter out the less relevant ones for each type of crime. From the perspective of criminality analysis, relevance here means choosing the highly criminal communities. This approach reduces the excessive number of identified communities, as well as characterizes the most relevant ones. Accordingly, we worked with the *top five highly criminal communities*, which were identified by using two variables: (i) the highest criminal average, i.e. the average number of crimes per node; (ii) community size threshold, which prioritizes communities containing at least a specified number of nodes. For the purpose of this work, the threshold has been set to 100 nodes, since communities with at least 100 nodes are spatially well distributed, and can better represent the regions of the city.

63.3.4 Crime Analysis

In this last phase, with the communities already detected, we proceed to analyze their interrelationship and interaction.

Therefore, here, we consider the communities in pairs of types; for example, we analyze communities of assault and theft to verify whether they occur together or if one type of crime makes the other more intense. The idea is to obtain insights about their geospatial dependency, pointing to regions that share space of more than one crime type and, consequently, are potentially more dangerous. To this end, this phase has two stages. The first stage obtains the geospatial similarity between pairs of crime types. The second stage computes the Homogeneity and Completeness scores.

63.3.4.1 Measuring the Similarity of Communities

A similarity value enables us to characterize the relationship of distinct events based on their spatial behavior, more specifically their spatial distribution. Such similarity is based on their distances and considers the possibility of a non-criminal region containing crimes as well. This is because a non-criminal area may have its behavior affected by a close criminal area.

We proceed with the computation of the similarity measure by considering the previously detected top five highly criminal communities. The similarity of crime events is based on the distance between elements belonging to any two sets of communities $\mathfrak{C}$ and $\mathfrak{F}$, where each one represents a distinct crime type. Note that, at this point, after communities were detected based on the topology of the city, we proceed using only the nodes and their positioning—i.e. we do not consider the edges nor the city topology for this processing. Such additional information is not necessary since we consider only the position of highly criminal communities in order to characterize the geospatial relationship of crime types. Hence, we define the similarity $\mathcal{S}_\mathcal{K}(\mathfrak{C}, \mathfrak{F}) : \mathfrak{C} \quad x \quad \mathfrak{C} \to [0, 1]$; the closer to 0, the more spatially apart are the sets of nodes:

$$\mathcal{S}_\mathcal{K}(\mathfrak{C}, \mathfrak{F}) = 1 - \frac{\sum_{}^{\forall u \in \mathfrak{C}} \sum_{}^{\forall v \in \mathfrak{F}} d_{uv}^E}{|\mathfrak{C}| + |\mathfrak{F}|} \qquad (63.3)$$

Such measure identifies, at the same time, the similarity between intra-and inter-community elements. The result 1 denotes that the elements within community $\mathfrak{C}$ are highly close to each other and to the elements of community $\mathfrak{F}$, intra-and inter-similar respectively, and 0 denotes the opposite.

63.3.4.2 Identifying the Behavior of Crimes

The characterization of communities identified in a city can be complemented by the Homogeneity and Completeness scores. They are derived from an entropy-based cluster evaluation [11], which works by analyzing the presence and

---

[2] Available on "www-users.cs.umn.edu/~lasalle/nerstrand/".

absence of criminality in the network's communities. Such scores quantify how intensely each crime type manifests across a community set $\mathfrak{C}$. To this end, the Homogeneity score *evaluates the criminality* in a community by identifying the criminal distribution among their nodes. Further, the Completeness score is used to *quantify how intrinsic is the criminality* in a community by measuring their distribution among all the communities, both in terms of a specific crime type.

Homogeneity. Regarding the top-five criminal communities, it is desired to know what are their predominant aspects. For the purpose of achieving a better assessment of criminal communities, the Homogeneity score is able to measure how uniform, i.e. homogeneous, a community is. It is defined by $\mathcal{H}(\mathfrak{C}) : \mathfrak{C} \quad x \quad \mathfrak{C} \rightarrow [0, 1]$ as follows:

$$\mathcal{H}(\mathfrak{C}) = 1 - \frac{\sum\limits_{i=1}^{|\mathfrak{C}|}\sum\limits_{j=0}^{|Q|} \frac{|\mathfrak{C}_{i_j}|}{|V'|} \times log_2\left(\frac{|\mathfrak{C}_{i_j}|}{|\mathfrak{C}_i|}\right)}{\sum\limits_{k=0}^{|Q|} \frac{|\mathfrak{C}_k|}{|V'|} \times log_2\left(\frac{|\mathfrak{C}_k|}{|V'|}\right)} \quad (63.4)$$

where $| V' |$ is the number of nodes in all the communities of set $\mathfrak{C}$ (the top five in our example) of a crime type, and $Q = \{ 1, 0\}$, which, in our scenario, indicates the nodes with and without crimes, respectively. We used $\mathfrak{C}$ to denote the set that contains all communities and $\mathfrak{C}_i \in \mathfrak{C}$ to indicate a single community. Finally, $\mathfrak{C}_{i_j}, j \in Q$, represents nodes in community i in which crimes are present or absent, whereas $\mathfrak{C}_k$, $k \in Q$, represents the same idea but among all communities. For instance, $\mathfrak{C}_k$, $k = 0$, represents all nodes in all communities that have no crime, while $\mathfrak{C}_{i_j}$, $i = 1, j = 1$, indicates the nodes from community 1 that have crime(s).

The Homogeneity score equals 0 when the presence and absence of crimes are proportionally similar across all the communities. Contrarily, the value 1 denotes that all communities in $\mathfrak{C}$ have crimes in all their nodes or no crime at all.

It is noteworthy that the Homogeneity score solely is insufficient to provide the needed comprehension about the criminal behavior in a city. This is because a single community can have all its nodes characterizing either crime occurrences or no crime at all, which, in turn, could yield the same Homogeneity score. For this reason, we also employ the Completeness score that, combined with the Homogeneity score, can better describe the distribution of crimes across a city.

Completeness. If crimes do not concentrate in a community, it is likely that there are multiple communities with criminal nodes spread across the city, demanding surveillance at multiple criminal spots. Alternatively, if crimes concentrate, they tend to be grouped within a single community, requiring

police patrols on specific regions. To determine if crimes are more or less concentrated, we employ the Completeness score, which is directly obtained from the whole community set. The Completeness score equals 0 when the presence of crimes is totally scattered, i.e. equally absent across all communities. On the other hand, the higher the value, the more concentrated are the criminal and safe zones. The definition of the Completeness score is symmetric to the Homogeneity (see Equation 63.4) definition—for more details, see [11].

63.4 Results and Discussions

This section presents and discusses the results of the experiments regarding the three phases of our methodology.

Influence of Crime Mapping. The first experiment aimed at analyzing the impact of the urban crime mapping (first phase of our methodology) on the criminal community identification process. This process was carried out twice: once considering the topological graph, i.e. the graph that considers only nodes and edges, and once over the network after the urban crime mapping, regarding the newly-included edges' weights (distances in meters) and nodes' weights (number of crimes). As a result, we obtained the number of 134 communities from the topological graph and, when considering the information introduced by the crime mapping phase, the numbers of identified communities were: 210 for *assault*, 215 for *theft* and 211 for *minor crimes*.

As a first result, we noticed the discrepancy between the numbers of criminal communities for each graph. Such difference is due to the presence/absence of crime-related data derived from the crime mapping phase. The Nerstrand algorithm considers the crime-related data (the nodes' and edges' weights), ensuring that the nodes within each identified community share similar characteristics. This behavior is not the same when considering the topological graph. Since there is no crime information attached to its nodes and edges, the algorithm examines only the connections between nodes. Therefore, considering only the network's topology, there is no guarantee that the crimes will be optimally grouped among the identified communities. These findings show that identifying communities through a topological graph is an inaccurate way to represent crime relationships in a city.

Communities' Design Aspects. The second experiment aimed at comparing the results from the previous analysis by considering their design aspects. Table 63.1 presents the most highly criminal communities obtained directly from the network's topology and from the three graphs generated via the Crime Mapping process. The table shows twelve communities sorted by crime average, among which are the top five criminal ones.

The table describes a higher crime average in communities derived from topology only. Such higher average occurs because, in this context, there is a lower number of communities, each one containing a higher number of criminal nodes.

As shown in the table, the number of crimes per community still stands out when comparing the graph from topology and the "complete" graphs—i.e. the graphs generated by the Crime Mapping process. This is because the complete graph used no attribute that characterizes a crime in the community identification process, so there is no way to assure that they are related just by being in the same community. Therefore, hereafter we discuss only the graphs with crime-related data.

Table 63.1 Communities with the highest crime average, identified from: (i) *topology*, (ii) *assaults*, (iii) *theft* and (iv) *minor crimes*. Column *Avg* denotes the crime average and # denotes the number of nodes in each community

	Communities identified from:							
	Topology		Assault		Theft		Minor Crimes	
	Avg	#	Avg	#	Avg	#	Avg	#
00	1063.00	2	362.00	2	4030.00	3	545.92	12
01	157.67	788	230.59	17	245.00	2	457.00	2
02	116.30	802	55.26	316	201.00	17	356.41	17
03	110.75	4	47.44	226	60.15	281	281.91	67
04	108.80	5	41.00	2	49.74	277	140.13	142
05	77.93	515	34.69	87	47.08	156	110.15	188
06	77.50	2	34.06	149	37.88	66	73.80	5
07	65.86	539	29.53	238	32.25	4	72.99	143
08	44.17	744	27.42	142	25.92	296	71.37	286
09	43.88	1129	22.14	248	25.77	342	68.50	4
10	36.28	635	15.52	281	21.79	307	58.72	32
11	35.88	518	14.20	54	21.10	371	54.43	144

Regarding such graphs, Table 63.1 shows a significant number of communities with few nodes. This is evidence that the crimes are not evenly distributed in the network. Therefore, if someone eventually attempted to eradicate or prevent crimes in the entire city based on characteristics of a local region, such attempt would fail because the crimes are not concentrated in a single criminal community.

San Francisco's Crime Neighborhood. To allow analyzing a group of communities containing a significant number of crimes, we have selected the region from the city of San Francisco that comprises the highest crime indices. This region contains the five communities, as mentioned earlier, of each crime type, which are among the twelve communities shown in Table 63.1. The top five communities of each type also meet the requirement regarding the number of nodes, previously determined as a threshold of 100 nodes (see Sect. 63.3.4). Specifically, considering the table, the top five communities for *assault* are in lines 2, 3, 6, 7 and 8; the top five for *theft* are in lines 3, 4, 5, 8 and 9; and the top five for *minor-crimes* are in lines 4, 5, 7, 8 and 11. Figure 63.1 depicts the entire selected area of San Francisco, as well as highlights each community (and each geospatial intersection of communities).

To quantify the relationship between communities in the city space, we used Equation 63.3 to obtain the similarity between communities. Through this measure, we compared each pair of crime types, computing the similarity between community sets $\mathfrak{C}$ of one type and community sets $\mathfrak{F}$ of another type. By doing so, we achieved a similarity value $\mathcal{S}_{\mathcal{K}}$ $(\mathfrak{C}, \mathfrak{F})$ of 0. 66 for assault *vs.* theft, 0. 65 for assault *vs.* minor crimes and 0. 58 for theft *vs.* minor crimes. Such results suggest that there is a relationship between crimes of different types, especially when comparing assault *vs.* theft and assault *vs.* minor crimes, since the values 0. 65 and 0. 66 can

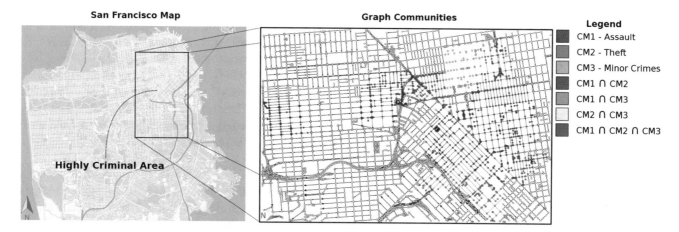

Fig. 63.1 The most highly criminal region from San Francisco—CA, USA, identified by evaluating the crime density in the city from 01/01/2003 to 05/18/2016. Each colored area represents the occurrence of a crime type or the intersection of a crime type with another type of crime. In the legend, each CM (from 1 to 3) denotes a type—*assault*, *theft* and *minor crimes*

Table 63.2 The Homogeneity and Completeness scores used to evaluate the quality of each community set and to quantify how intrinsic is the criminality within them

Crime type	Homogeneity	Completeness
Theft	0.015588	0.004159
Minor Crimes	0.014014	0.003441
Assault	0.013557	0.003234

be considered significant enough if one ponders over what these numbers represent—that the majority of the area of these two communities might be more dangerous due to the occurrence of more than one crime type.

More specifically: **(i)** it is very likely that criminal activities occur at intersections of multiple crime types, which we call *crime hubs*—regions that attract or disseminate criminality; **(ii)** those crimes are spatially related, so it is possible that they have been caused by the same people or gang; **(iii)** places, where more than one crime type occurs, are high-priority regions for public safety improvement.

Communities' Intrinsic Criminality. The previous section discussed the geospatial disposition of different crime types with respect to intersections between criminal regions. This section, in turn, brings results concerning the Homogeneity and Completeness scores for the communities in the most highly criminal area of San Francisco. This experiment aimed at measuring how intrinsic was the criminality within the neighborhoods of San Francisco during the studied period. To understand this measure, recall that the number of crimes of some nodes is zero. For the *assault* type, there are 1,071 nodes, of which 132 (12.33%) contain at least one crime and 939 (87.68%) contain none. For *theft*, there are 903 nodes, of which 133 (14.73%) are criminal and 770 (85.27%) are not criminal. For *minor crimes*, there are 1,352 nodes, of which 177 (13.09%) present crimes and 1,175 (86.91%) do not. These values and percentages refer to the entire community set for each crime type.

The presence and absence of crimes were considered to calculate the scores presented in Table 63.2, which shows that *assault* has the lowest values for both measures. Regarding Homogeneity, the value indicates that this crime type presents a reduced concentration of criminal nodes within each community. This is straightly inferable because, besides the low Homogeneity value indicating an inexpressive occurrence of crimes, the actual number of crimes is very low; a pattern that holds for other types of crime as well.

Since the value for Completeness is also the closest to 0, we can reliably assume the occurrence of crimes to be scattered across all the *assault* communities. The *theft* type has values close to 0 as well, but the highest ones when compared to the values of other crime types. The greater difference corresponded to the Completeness score, whose value suggests a geospatial crime concentration more

elevated. Regarding Completeness (see Sect. 63.3.4), such higher concentration tends to occur in a smaller set of communities rather than across all of them, which is more likely for the *assault* type.

A higher crime concentration means that a particular crime type is less likely to intersect with another crime type. This is because, in such a case, the criminal nodes would occupy a smaller area in the city space. Furthermore, such a crime type could be considered easier to prevent, since its occurrences would be in a more compact area of the city. On the other hand, a lower concentration of a crime type means that it exerts a stronger influence on the network as a whole, being easier to propagate to remote areas and demanding global approaches both to eradicate it and to prevent it.

63.5 Conclusion

In this paper, we propose a threefold methodology for identifying and analyzing criminal spatial patterns in urban street networks. Our methodology is based on the hypothesis that *by employing network mapping techniques, allied to distance-based properties of graphs, it is possible to identify and trace the relationship between areas that are highly criminal within a city.* To demonstrate our methodology, we analyzed criminal communities from real crime data representing *assaults*, *thefts* and *minor crimes* in the city of San Francisco—CA, USA. The methodology comprises the following phases: (i) Mapping of Urban Crimes, (ii) Criminal Community Identification and (iii) Crime Analysis. The latter employs the well-established Homogeneity and Completeness scores to analyze the identified criminal communities more deeply. Our main achievement confirm the hypothesis of our work, allowing us to state that different crime types share common spaces, characterizing areas that lack strategies for crime prevention, and that particular crime types are sparser than others in the city space.

The highlighted contributions are: **(i)** the use of a complex network to represent the real-world space, enabling a complete analysis of the city; **(ii)** independence from socioeconomic information, allowing the analyses of crimes based on solely their spatial disposition; **(iii)** the assessment of the impact of criminal regions, considering the similarity between distinct crime types and their Homogeneity and Completeness scores.

Finally, as future work, we intend to continue expanding our methodology developing tools to analyze crimes and their types considering a temporal perspective. Our aim is to provide insights about the criminal behaviors in a city, allowing to determine how a crime spreads throughout the city and when interventions were made by the governors were effective to prevent them, using time windows and comparing the results with socioeconomic data.

Acknowledgements We are grateful to CNPq (National Counsel of Technological and Scientific Development) by the support (grants 9254-601/M, 444985/2014-0, and 147098/2016-5), to FAPESP (Sao Paulo Research Foundation) by the assistance (grant 2015/15392-7, 2016/02557-0, 2016/17330-1) and to Capes (Brazilian Federal Agency for Support and Evaluation of Graduate Education) by the found to this research.

References

1. Boccaletti, S., Latora, V., Moreno, Y., Chavez, M., & Hwang, D. U. (2006). Complex networks: Structure and dynamics. *Physics Reports, 424*(4–5), 175–308.

2. Porta., S., Strano, E., Iacoviello, V., Messora, R., Latora, V., Cardillo, A., Wang, F., & Scellato, S. (2009). Street centrality and densities of retail and services in Bologna, Italy. *Environment and Planning B: Planning and Design, 36*(3), 450–465.

3. Deryol, R., Wilcox, P., Logan, M., & Wooldredge, J. (2016). Crime places in context: An illustration of the multilevel nature of hot spot development. *Journal of Quantitative Criminology, 32*(2), 305–325.

4. Spicer, V., Song, J., Brantingham, P., Park, A., & Andresen, M. A. (2016). Street profile analysis: A new method for mapping crime on major roadways. *Applied Geography, 69*, 65–74.

5. Shiode, S., & Shiode, N. (2013). Network-based space-time search-window technique for hotspot detection of street-level crime incidents. *International Journal of Geographical Information Science, 27*(5), 866–882.

6. Serrano, H., Oliveira, M., & Menezes, R. (2015). *The spatial structure of crime in urban environments* (Lecture notes in computer science (including its subseries lecture notes in artificial intelligence and lecture notes in bioinformatics), Vol. 8852, pp. 102–111).

7. Rey, S. J., Mack, E. A., & Koschinsky, J. (2012). Exploratory space-time analysis of burglary patterns. *Journal of Quantitative Criminology, 28*(3), 509–531.

8. Haklay, M., & Weber, P. (2008). OpenStreet map: User-generated street maps. *IEEE Pervasive Computing, 7*(4), 12–18.

9. Konstantopoulos, T. (2012). *Introduction to projective geometry.* Mineola: Dover Publications.

10. Estrada, E., Fox, M., Higham, D. J., & Oppo, G. L. (2010). *Network science: Complexity in nature and technology.* Cambridge: Cambridge university press.

11. Rosenberg, A., & Hirschberg, J. (2007). V-measure: A conditional entropy-based external cluster evaluation measure. In *Proceedings of the 2007 Joint Conference on Empirical Methods in Natural Language Processing and Computational. Natural Language Learning*, Prague (Vol. 1, pp. 410–420).

Big Data: A Systematic Review

Antonio Fernando Cruz Santos, Ítalo Pereira Teles,
Otávio Manoel Pereira Siqueira, and Adicinéia Aparecida de Oliveira

Abstract

Big Data has been gathering importance in the last few years, specially through bigger data generation, and, consequently, more accessible storage of said data. They are originated at social media or sensors, for example, and are stored to be transformed in useful information. The use of Big Data is becoming more common in several fields of business, mainly because it is a source of competitive differential, through the analysis of the stored data. This study has the objective of executing a systematic review towards presenting a broad vision of Big Data. Were analyzed 466 publications from 2005 until March 2016.

Keywords

Big Data • Analytics • Dimensions • Tools

64.1 Introduction

Information is 21st century's oil. That can be observed through corporations' constant search for more data, and, consequently, its change into information that may become the differential or an opportunity in business. Corporations are gathering more and more data that may come from different sources, with a big range of variety and complexity. That leads to a change in the data management strategy, called Big Data [1].

According to [2] Big Data is a generic term for data that can't be contained on regular repositories, bulky data that can't be stored in a single server, or that isn't structured so it can fit in a rows and columns database. Gartner Group defines Big Data as the three Vs, or dimensions: volume, velocity and variety. Other authors also add two extra dimensions: value and veracity [3–7].

Big Data can be and is being used in several areas of business, like healthcare, finance and education, just to name a few. With huge amounts of data stored by corporations it's possible to produce useful information to business, creating opportunities and market differentials [8]. This concept is now part of every sector and function of the global economy [9].

This study's main goal is to present the state of the art about Big Data, its characteristics, most used tools, and its main applications. The purpose of the paper is to help, through a systematic review, understanding the current situation of this field of research, also serving as basis to other researches. The systematic review was accomplished using on two research bases, considering articles since 2005.

The rest of this paper is organized as follows: Section 64.1, contains an introduction about the article; Section 64.2 presents second section that presents a theoretical background about Big Data; Section 64.3 explains the results and discussions regarding the found publications; and, at last, the Section 64.4 contains the final considerations about this study.

A.F.C. Santos (✉) • Í.P. Teles • O.M.P. Siqueira • A.A. de Oliveira
Department of Computing, Federal University of Sergipe,
Aracaju, SE, Brazil
e-mail: fernandoafcs@gmail.com; italop.teles@gmail.com;
otaviomps@gmail.com; adicineia@ufs.br

© Springer International Publishing AG 2018
S. Latifi (ed.), *Information Technology – New Generations*, Advances in Intelligent
Systems and Computing 558, DOI 10.1007/978-3-319-54978-1_64

64.2 Big Data

Big data is an abstract concept that includes other characteristics besides the idea of massive data. Despite its well known importance, there are still different definitions for the term. Broadly speaking, Big Data regards to sets of data that could not be perceived, acquired, managed and processed by traditional Information Technology (IT) tools, inside a reasonable time lapse [10, 11].

There are several definitions. However, some similarities are common, like the volume, velocity and variety characteristics, so Big Data can be featured under the 3Vs optics. Those V's, also known as dimensions, are characteristics that define what is Big Data. Some authors that add two other dimensions, respectively, value and veracity (Fig. 64.1).

Volume is the most evident dimension in what concerns to Big Data, precisely because of the huge amount of data made available that needs to be analyzed and processed. Before, the increase of the data volume was mainly because of the increment on the number of transactions and its details concerning granularity. This increase is too small considering the size of the volume on Big Data. It began with the increment of interactions between users online, the so-called weblogs, considering the growth in popularity of the internet. Through these logs, it was possible to understand the behavior of the web users. Then, came social networks, like Twitter and Facebook that, combined with the capacity increment and number of devices compatible with the internet, resulted in a leap of data production. All this data

volume needs to be stored and then analyzed, so useful information can be extracted. In summary, the volume dimension means that with the massive data generation and collection, its scale becomes bigger and bigger [6, 10, 13].

The variety dimension is as diversified as the data sources and formats. Previously, the data to be analyzed came mainly from transactional systems, where data was stored in a higly structured manner. However, as time passed, other kinds of data began to be stored and analyzed. Among these, other kinds of data are the semi-structured, like eXtensible Markup Language (XML) and JavaScript Object Notation (JSON), and unstructured data, like audio, video or text files, for instance. This variety of data makes analysis more challenging, specially combining this dimension – variety – with volume [6, 12].

Velocity is a critical factor when working with Big Data. This dimension is connected to the speed in which data must be stored and analyzed, so it can create maximum commercial value in the shortest time frame possible, or in real time [10, 12].

The value dimension defines the value that the data can add to the organization. That is to say, data, on itself, does not generate useful information, needing identification, transformation and extraction of the information from the data through analysis that add value to the business. [5, 12].

At last, the veracity dimension, which includes two aspects: the consistency of the data (assurance), which can be defined as its statistical reliance, and the reliability of the data, defined through a series of factors, including its source, gathering and processing methods, in addition to an easy and

Fig. 64.1 Big Data's dimensions [12]

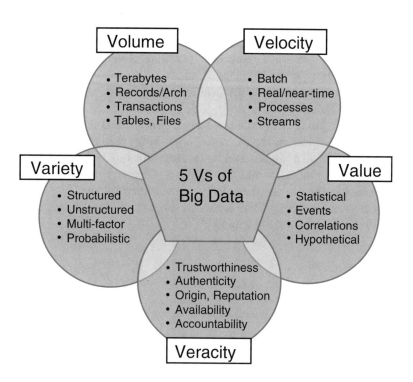

reliable infrastructure. In general, veracity assures that the data used are trustworthy, authentic and protected from unauthorized access [12].

As it appears, Big Data has well defined characteristics to be considered so. According to [14], Big Data solutions are ideal to: analyze not only structured data, but also semi-structured and unstructured, depending on the variety of its sources; analyze all data, or a good part of it, against a sample; and to exploratory and iterative analysis, when the business measures aren't predetermined.

64.3 Methodology

On this study, the research approach was the literary review, as proposed by [15]. To perform the systematic review was important to determine from the beginning the protocol to be followed. At first, the variables of interest of this study were defined, which are:

- The countries of the authors;
- Dimensions (Volume, Velocity, Variety, Veracity and Value);
- Application field (healthcare, financial sector, smart cities, etc.);
- Used tools, their functions and if they are open source;
- The goal of the study (architectural proposal, framework proposal, survey, features, etc.);
- Description of the case study.

This study began in March 2016, researching on bibliographic databases. The search string was created to gather the biggest amount of papers possible about Big Data, focusing mainly in architectural, platforms and technological proposals. Only papers in english were incorporated. No criteria of date of publication was stablished.

The basis for the research were IEEE Explorer and Spring Link, with the search string being: Big Data AND (Architecture OR Platform OR Technologies).

After the initial research, 5311 articles were found on the two consulted databases (see Table 64.1).

On the first step, all the publications that were not articles were removed. Then, the articles that were not available, either because were paid or had pending. On the next step, it were removed the studies that were incomplete or in the form of "short paper". Were also excluded the articles that did not contained the term "Big Data" on its title, abstract or among the keywords, resulting in a significant reduction. The reason for this is that the search engines for each database are different concerning to filtering publications.

On the second step, the publications were removed after reading the title and the abstract, in that order. That was the most labor intensive phase, since it was required a thorough

Table 64.1 Amount of articles selected after executed the protocol

Step	IEEE	Springer link
After the search string query	2051	3260
Not articles	104	101
Restricted access	0	102
Incomplete articles	586	26
Does not contain *Big Data* on the title, abstract or keywords	168	2645
Removed after the title was read	304	130
Removed after abstract was read	522	157
Result	**367**	**99**

reading so the articles would be contemplated or not. Were removed specially the articles that were mainly focused on infrastructure and not Big Data. After this phase, 466 articles remained. On this phase it was also built a software to help analyze the variables on interest of this study, and therefore, store the data extracted in a database. After that was possible to analyze the articles easily and quickly.

64.4 Results and Discussion

For this research, 466 articles were analyzed, all of them published after 2010 (Fig. 64.2).

The figure represents a progressive number of published articles on Big Data over the years. The first article to be analyzed is from 2011. In the following years, the number of publications increased, as we can see in 2015, when two 222 articles were found.

In what concerns to the country of publishing, around 52 different countries were identified (Fig. 64.3).

The figure shows the nine countries that were cited the most, and includes an "other countries" item, for the remaining ones identified. The countries with the most number of publications are, in order, China, United States, India and United Kingdom, respectively with 117, 108, 54 and 30 publications each, composing 57% of all articles. The other countries represent 27% of the publications, with 145 articles.

From the 466 articles shown, the variables determined for this study were analyzed. The dimensions, also called "the V's" of Big Data are: volume, variety, velocity, veracity and value [16]. On this context, the articles were analyzed to understand which dimensions are the most discussed about. It is important to point that in the same article, several or none of the dimensions may be addressed (Fig. 64.4).

The dimensions more discussed were Volume, Velocity and Variety, in 328, 269 and 264 articles, respectively. This can be easily explained, since these are the dimensions that originally described what Big Data was [4].

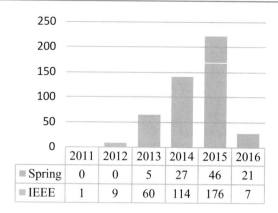

Fig. 64.2 Amount of selected articles per year versus base

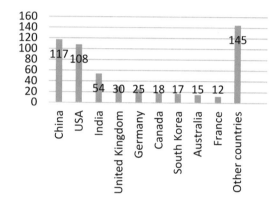

Fig. 64.3 Amount of articles selected for country

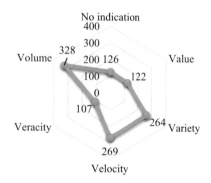

Fig. 64.4 Dimensions presented in the articles

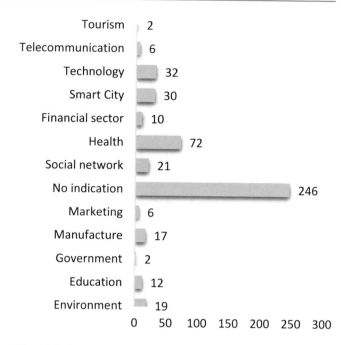

Fig. 64.5 Application presented on the articles

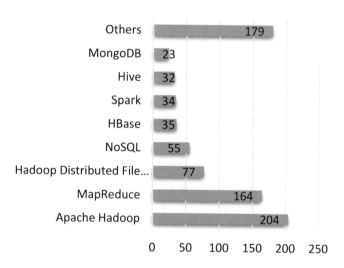

Fig. 64.6 Tools presented in the articles

Another variable that was analyzed regarded was the application field of Big Data (Fig. 64.5).

A total of 13 different application fields were identified. For the articles that wouldn't mention any application, the value "No indication" was assigned. As it can be seen, among the indications of business fields, the one that stands out is the healthcare industry, with 72 articles, followed by technology and smart cities. It is remarkable that the healthcare industry stood out among the others, since the

search string was not directed to any specific application. According to [17] the intense use of Big Data on healthcare happens not only because of the amount of data handled (electronic patient registration, exams, for example), but also for the diversity of kinds of data and the velocity in which they must be managed.

The tools discussed on the articles were also analyzed. A total of 35 tools were discussed on the 466 articles analyzed (Fig. 64.6).

Among the most mentioned tools is Apache Hadoop, because it is a broadly known and used tool when it comes to Big Data [10]. The Apache Hadoop processes great volumes of data using distributed nodes that allow it to

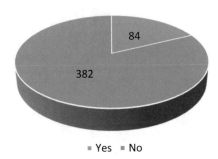

Fig. 64.7 Articles' main objective

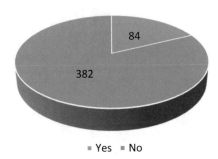

Fig. 64.8 Articles that present a case study or not

process faster a bigger amount of data [18]. Another tool named in 164 articles was MapReduce, which is an open source framework developed to handle massive data, structured and unstructured, and that allows scalable programming [8, 18]. In a different analysis, it was considered if the tools were open source or not. From the 35 tools only 10 are not open source, what can be considered a good sign, since they don't present any limitation of their use, and, doing it so, they help popularize the use of Big Data.

The articles were also analyzed towards their main proposed goal (Fig. 64.7). Several goals were identified among the selected publications.

From the 466 analyzed articles, 135 are conceptual and/or characterizing ones. These articles define Big Data, its dimensions, applications among other things. Coming in second among the applications, architecture and frameworks are also much discussed, with 56 and 49 articles, respectively.

The last chosen variable was the verification if the article described a case study or not (Fig. 64.8).

From the 466 articles analyzed, 84 described case studies, corresponding to 18% of the total. It can be considered a small percentage, but as the study on Big Data is relatively new, these numbers tend to grow.

64.5   Conclusions

The purpose of this study was to bring a broad picture on the state of the art about Big Data. According to the analysis conducted, it was possible to foresee that, despite being a new subject, the number of publications about the theme is increasing, what shows the importance and the interest on the matter. A fact that assures this point is the diversity of countries with publications concerning the focal point of this study.

It was also possible to realize that Big Data is being applied in several areas of business, adding value to it and creating new opportunities. The area that stood out was healthcare, with the larger number of articles found and analyzed.

Another topic regarded the focus of the presented studies. Being a new subject, many of the articles are more related to conceptualization and/or characterization on the theme, with a total of 135 publications.

As a future work, a reference architecture will be proposed, aiming the healthcare field of study, using the information presented on this article as basis.

References

1. Gartner Group. (2011). Gartner says worldwide enterprise IT spending to reach $2.7 Trillion in 2012. [Online]. Available: http://www.gartner.com/newsroom/id/1824919.
2. Davenport, T. (2014). Big data no trabalho. Brazil: Campus.
3. Chandarana, P., & Vijayalakshmi, M. (2014). Big data analytics frameworks. In *Proceedings of the international conference on circuits, systems communication and information technology applications 2014* (pp. 430–434). Mumbai.
4. Laney, D. (2011). 3D data management: Controlling data volume, velocity, and variety. *Information and Software Technology, 51*(4), 769–784.
5. Liu, Z., Yang, P., & Zhang, L. (2013). A sketch of big data technologies. In *2013 seventh international conference on internet computing for engineering and science* (pp. 26–29).
6. Ramesh, B. (2015). *Big data: Architecture* (vol. 11). India: Springer India.
7. Ward, J. S., & Barker, A. (2013). Undefined by data: A survey of big data definitions. *arXiv.org, 1*, 2.
8. Benjelloun, F.-Z., Lahcen, A. A., & Belfkih, S. (2015). An overview of big data opportunities, applications and tools. In *Intelligent systems computer vision (ISCV), 2015* (pp. 1–6).
9. Maria, R. E., et al. (2015). Applying scrum in an interdisciplinary project using big data, internet of things, and credit cards. In *Proceedings of 12th international conference on information technology: New generations ITNG 2015* (pp. 67–72). Linz.
10. Chen, M., Mao, S., & Liu, Y. (2014). Big data: A survey. January, *19*, 171–209.
11. Schneider, R. D. (2012). *Hadoop for dummies* (1st ed.). EUA: John Wiley & Sons.
12. Demchenko, Y., Grosso, P., De Laat, C., & Membrey, P. (2013). Addressing big data issues in scientific data infrastructure. In

Proceedings of 2013 international conference on collaboration technologies and systems (CTS 2013) (pp. 48–55).

13. Seay, C., Agrawal, R., Kadadi, A., & Barel, Y. (2015). Using hadoop on the mainframe: A big solution for the challenges of big data. In *12th international conference on information technology: New generations (ITNG), 2015* (pp. 765–769).

14. Zikopoulos, P., & Eaton, C. (2011). *Understanding big data: Analytics for enterprise class hadoop and streaming data*. New York: McGraw-Hill Education.

15. Kitchenham, B. (2004). *Procedures for performing systematic reviews*. United Kingdom: Keele University Technical Report.

16. Li, S., & Ni, J. (2015). Evolution of big-data-enhanced higher education systems. In *2015 eighth international conference on internet computing for science and engineering* (pp. 253–258).

17. Raghupathi, W., & Raghupathi, V. (2014). Big data analytics in healthcare: Promise and potential. *Health Information Science and Systems, 2*, 1–10.

18. Katal, A., Wazid, M., & Goudar, R. H. (2013). Big data: Issues, challenges, tools and good practices. In *Sixth international conference on contemporary computong, IC3 2013* (pp. 404–409).

Evidences from the Literature on Database Migration to the Cloud

65

Antonio Carlos Marcelino de Paula, Glauco de Figueiredo Carneiro, and Antonio Cesar Brandao Gomes da Silva

Abstract

Context: The cloud computing paradigm has received increasing attention because of its claimed financial and functional benefits. A number of competing providers can support organizations to access computing services without owning the corresponding infrastructure. However, the migration of legacy system from the database perspective is not a trivial task. **Goal**: Characterize reports from the literature addressing the migration of legacy systems to the cloud with emphasis on database issues. **Method**: The characterization followed a four-phase approach having as a start point the selection of papers published in conferences and journals. **Results**: The overall data collected from the papers depicts that there are six main reported strategies to migrate databases to the cloud and twelve reported issues related to this migration. **Conclusion**: We expect that the strategies, approaches and tools reported in the primary papers can contribute to lessons learnt regarding how companies should face the migration of their legacy systems databases to the cloud.

Keywords

Cloud computing · Cloud migration · Databases

65.1 Introduction

Clouds are a large pool of virtual resources such as hardware, development platforms, and services. These resources can be dynamically adjusted to a variable load, allowing for optimal resource utilization [1, 2]. Cloud Computing (CC) has emerged as a model, where essential components of a computer system such as software applications, platforms for software development or physical infrastructure are considered as services and are delivered according to users needs. Moving to the cloud means giving up incumbent information systems practices and facing the initial perception of losing control of data that in a previous scenario were stored in local servers [3, 4]. This scenario highlights the importance to understand not only opportunities but also the challenges regarding the migration and adoption to the cloud computing paradigm [5–7].

Previous papers have investigated the adoption and acceptance of public cloud services at both the individual and organizational levels [8]. From the organizational perspective, several papers have identified factors that influence and affect cloud computing adoption, such as maintenance and improvement of market share and competitiveness [9]. To the best of our knowledge, we did not find studies characterizing reports from the literature focusing on issues related to the database migration of legacy systems to the cloud. This could help disseminating lessons learnt, experience reports, best practices and related issues for both practitioners and researchers [10, 11]. We found the paper [12] that organizes the efforts of legacy systems migration to the cloud as scenarios and connect them with a list of reusable solutions for the application data migration in the form of patterns. However, they do not discuss these

A.C.M. de Paula (✉) · G. de Figueiredo Carneiro · A.C.B.G. da Silva
Salvador University – UNIFACS, Salvador, Brazil
e-mail: antonio.paula@unifacs.br; glauco.carneiro@unifacs.br;
antoniocesar01@gmail.com

© Springer International Publishing AG 2018
S. Latifi (ed.), *Information Technology – New Generations*, Advances in Intelligent
Systems and Computing 558, DOI 10.1007/978-3-319-54978-1_65

scenarios in terms of issues and challenges faced by practitioners neither how to deal with these issues.

The rest of this paper is organized as follows: Sect. 65.2 outlines the research methodology; Sect. 65.3 identify and characterize evidences collected from the literature; Sect. 65.4 presents and discusses the results of this characterization. At last, the concluding remarks as well as limitations and scope for future research were discussed in Sect. 65.5.

65.2 Methodology

Identify the needs for a characterization study. To the best of our knowledge, there is no previous study characterizing the migration of legacy systems from the database perspective based on evidences published in the literature. The analysis of data provided by primary studies and the consequent identification of strategies and related issues could help in establishing knowledge on how the companies should adopt and migrate their legacy systems to the cloud according from the database perspective. These findings can be a useful reference to develop guidelines and to support companies to face challenges related to the use of cloud computing.

Specifying the Research Question. Considering this context, we focused on the following research questions.

Study Research Question 1 (SRQ1): What are the main strategies to migrate legacy systems to the cloud from the database perspective according to evidences provided by papers published in the literature?

Study Research Question 2 (SRQ2): What are the main issues related to this migration according to evidences provided by papers published in the literature?

Study Research Question 3 (SRQ3): Which cloud service providers offer support to migrate databases to the cloud according to evidences provided by papers published in the literature?

Adopted Strategy. The characterization had as a starting point a list of 66 papers selected by a Systematic Literature Review (SLR) [9] that considered aspects related to cloud computing migration. The goal is to identify in these papers evidences of migration from the database perspective. These papers were selected based on quality criteria defined in the SLR. Figure 65.1 conveys the phases of the characterization study to reach the goals SRQ1, SRQ2 and SRQ3.

Phase 1: Applying the Search String. The search string is presented as follows and applied in all the 66 papers from the SLR, as can be seen in Fig. 65.1. The result of this phase was a list of 42 papers containing the string *"database"*. **Phase 2: Selecting the Papers.** The 42 papers from the previous phase were analyzed to identify evidences that could answer SRQ1, SRQ2 and SRQ3. The result of this phase was a list of nine selected papers. **Phase 3: Manual Inclusion.** This phase resulted in the inclusion of four papers suggested by the authors. The result of this phase was a list of thirteen papers. Nine papers from the phase 2 plus four papers included in this phase. **Phase 4: Identifying Evidences from the Papers.** In this phase we identified relevant evidences from the 13 selected papers from Phase 3 to build the mental model presented in Fig. 65.2.

65.3 Evidences from the Literature

In this section, we identify and characterize evidences collected from papers represented and numbered in Fig. 65.2. We followed the instructions described in Fig. 65.1 to select the papers to answer the SRQ1, SRQ2 and SRQ3. The papers are listed in the Reference Section.

65.3.1 Strategies for Database Migration

As a result of the analysis of the selected papers, we identified six strategies to answer the first research question (**SRQ1**): *What are the main strategies to migrate legacy systems to the cloud from the database perspective according to evidences provided by papers published in the literature?* These strategies are presented as follows.

Strategy I: Snapshot (nodes 2.1 in Fig. 65.2 and mentioned by reference P2): The goal of the database snapshot strategy is to move it to a read-only option to the cloud, therefore providing only a static view of the database. The database snapshot is transactionally consistent with the source database at the moment of the snapshot's creation. In [P2], the authors mention the need of an appropriate snapshot plan where no data changes are accepted. The databases are copied from production server to a Virtual Machine (VM). This is the simplest and less flexible strategy.

Fig. 65.1 Phases of the characterization study

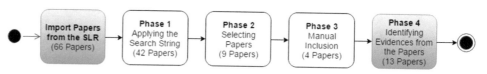

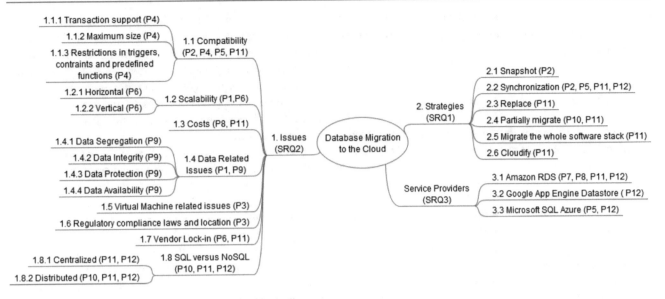

Fig. 65.2 Findings from the selected papers published in the literature

Strategy II: Synchronization (nodes 2.2 in Fig. 65.2 **mentioned by references P2, P5, P11):** According to [P2], the main goal of the synchronization strategy is to first create the database, index and repository servers in the cloud. The second step consists in the synchronization with their physical counterparts. The third step consists in performing load balance and migrating the web servers. It then goes on resync features. It is worth to mention that a real-time synchronization is not simple to be achieved for large database sizes. This task can take several hours to resync which can lead to an repository in the cloud temporarily out of date. To overcome this limitation, the **Strategy I Snapshot** can be used together with **Strategy II**. The SQL Azure Data Sync is an example of **Strategy II**, given that there is a – Based on the Sync framework, its main function is to make database synchronizations between SQL Azure and on-premises databases [P5]. According to [P11], a hybrid solution is an example of solution that would serve the requirements of different legacy system, for example, a Health Insurance Company(HIC) and External Auditing Company(EAC): by creating a separate database in the Cloud which only holds the data required for auditing purposes, and moving to the Cloud also the part of the business layer allowing to execute queries on them by EAC, HIC can operate normally. By anonymizing the personal data and ensuring synchronization between the "local" and "remote" data, HIC fulfills its obligation. The synchronization of data between the two data sets is unidirectional (from the local to the remote).

Strategy III: Replace (nodes 2.3 in Fig. 65.2 **mentioned by P11):** Replace component(s) with Cloud offerings. This is the least invasive type of migration, where one or more (architectural) components are replaced

by Cloud services. The data and/or business logic have to be migrated to the Cloud service. A series of configurations, rewiring and adaptation activities to cope with possible incompatibilities may be triggered as part of this migration. Using Google App Engine Datastore in place of a local MySQL database is an example of this migration type, other example of migration is SQL Server Migration Assistants, it's a tool that can be used to migrate DRS to Microsoft SQL Server model, then, we can conect with Azure SQL. To do this same migration for another Cloud Service like Amazon Relational Database Service (RDS) it will be necessary configure MySQL isntance as a replication source, launch and prepare Amazon RDS to running MYSQL, after this, it's possible use a tool called RDS Provisioned IOPS to do the migration more simple.

Strategy IV: Partial migration (nodes 2.4 in Fig. 65.2 **mentioned by P11):** Partially migrating some of the application functionality to the Cloud entails migrating one or more layers of the full application, or a set of architectural components from one or more layers implementing a particular functionality. Using a combination of Amazon SimpleDB and EC2 instances to host the auditing data and business logic for HIC is an example of such a migration.

Strategy V: Migrate the whole software stack (nodes 2.5 in Fig. 65.2 **mentioned by P11):** Migrate the whole software stack of the application to the Cloud. This is the classic case of migration to the Cloud where the application is encapsulated in VMs to execute on the Cloud.

Strategy VI: Cloudify (nodes 2.6 in Fig. 65.2 **mentioned by P11):** the term *Cloudify* consists in migration the full application to the cloud as a composition of services running on the Cloud. As in the case of component replacement (nodes 2.3 in Fig. 65.2), cloudification requires the migration of data and business logic to the Cloud, in addition to any adaptive actions to address possible incompatibilities.

65.3.2 Issues

We found eight issues to answer the second research question (**SRQ2**): *What are the main issues related to this migration according to evidences provided by papers published in the literature?* These issues will be presented in the next paragraphs of this subsection. These eight typos of issues are related to compability, maximum size, scalability, costs, data related issues, virtual machine related issues, regulatory compliance laws and location, vendor Lock-in and SQL versus NoSQL.

Compatibility (node 1.1 in Fig. 65.2 **represented by P2, P4, P5, P11):** According to [P2], compatibility is a remaining problem in the context of migrating database of legacy systems to the cloud. In the selected paper [P4], the authors mentioned that some features are not supported in the selected database of the Platform as a Service (PaaS) provider, a replacement solution should be found before hand. Considering that PaaS supports only light-weight databases, the implementation and use of *triggers, constraints, and predefined functions* (node 1.1.3 in Fig. 65.2) will probably not be viable. The cloud databases supported by some providers may not have been implemented some predefined functions necessary to migrate the database to the cloud.

The *maximum size* (node 1.1.2 in Fig. 65.2) of the database in cloud is another issue that should be analyzed. Each provider define a specific maximum size to support. In this case, if the database size, that will be migrated to the cloud, exceeds the maximum size, some effort to reduce the database size will be required.

Scalability (node 1.2 in Fig. 65.2 **represented by P1, P6):** According to [P6] and [P1], scalability deals with the ability of the software system to manage increasing complexity when given additional resources. Scalability with large data set operations is a requirement for cloud computing environments. Horizontal scalability refers to the ability of the cloud provider to scale add more machines into the pool of resources. As a result, it implements load balancing on demand to enable an effective application delivery solution. Distributed hash table (DHT), column-orientation, and horizontal partitioning are examples of features that are used in horizontal scalability solutions. Vertical scalability refers to the ability of the cloud provider to scale through the inclusion of more resources such as CPU and memory depending on the identified need to the existing machines. The authors of [P6] state that applications that fail to vertically scale may end up costing more when deployed in the

cloud because of the additional demand on compute resources required as demand increases. In the case of Force.com, according to [P6], it was designed to run a simple business application and it is based on a database centric architecture. This points out a scalability issue as it partitions its database per application. This means a challenge to be faced if an application needs to scale more than what a single database can provide.

Costs (node 1.3 in Fig. 65.2 **represented by P8, P11):** According to [P8], the database license costs varies significantly depending on the choice of the cloud database S/W. For example, in cases where a open-source database is selected, there is no license cost. On the other hand, when a commercial database such are Oracle or Microsoft SQL Server is selected, then the license fee is a function of physical CPU sockets (processors) and core counts. However, calculating the license cost is not a trivial task. Subtle factors such as virtualization, total number of effective users, whether replication is applied or not all should be taken into account in determining the final license cost. As a result of comparing the costs with Cloud databases, the prevalent pricing policy is per-socket (instead of per-core) charging for traditional (non-virtualized) server, this puts the use of commercial database in the cloud at a relative disadvantage. And, at the same time it encourages the use of SaaS database service when using the cloud. This suggests that a reconsideration of S/W licensing structures, particularly as applicable to large-scale parallel machines, may be worthwhile for making cloud-based hosting more appealing. [P11] states the some Cloud providers offer different licensing options to their consumers. For the Amazon RDS for example, consumers can bring their own license for MySQL, Oracle or IBMDB2, get charged for a per-hour license using Oracle DB, or pay a one-time charge per RDS instance to get reduced hourly charging rates. Some companies include licensing fees for free with each account.

Data Related Issues (node 1.4 in Fig. 65.2 **represented by P1, P9):** According to [P1] and [P9], security issues such as ensuring access control, data access, and availability of individual customer resources are requirements for a multi-tenant environment. **Data segregation**: A relevant cloud characteristic is multi-tenancy where several users data might be stored at the same physical location using hypervisor techniques under the concept of virtualization. This means that an organizations data may be mingled in various ways with other users data causing confidential data leakages. However, there are only few occurrences of this scenario when users data storage are subject to data leaks. **Data availability**: [P9] also states that cloud providers must provide on-request and reliable service with highest up-times. If an organizations data gets locked-in and cloud providers fail to provide access, this service disruption could pose potential financial damage to the organisation and its clients. **Data integrity** is a relevant security issue in the cloud. According to [P9], data integrity is measured by the level of secure channels in place for handling transactions. This is why data have to be transferred among servers and databases though secure channels (e.g. SSL), every transaction has to be verified for legitimacy (e.g. checksums), certain level atomicity, isolation and durable. APIs handling the transactions have to be reliable, well recognised and time-tested (e.g. simple object access protocol (SOAP)). **Data protection**: [P9] also discusses the concept of customer data protection as a core concern of security and privacy measures in cloud computing. The authors also state privacy is a moral and legal right of individuals. For that reason, theu mention that data owners need to be ensured that their data is not shared with any third party. Storing data and applications that reside outside the organizations premises poses the potential risk of unauthorised access and processing of the data and application. Customers may lose control over their critical assets. Data confidentiality and privacy risks may be more critical when providers reserve the right to change their terms and conditions. Apart from the data theft from external attackers, data leakage can also be carried out by the employees of the service providers. Therefore, measures such as privacy policy, data subject consent and control, transparency of data, data operations, and assurance of data protection are necessary and should be included in the Service Level Agreement (SLA).

Virtual Machine Related Issues (node 1.5 in Fig. 65.2 **represented by P3):** According to [P3], when using IAAS services to migrate systems to cloud can offer certain risks. The inappropriate management and configuration of VMs can causes several problems, that can be dangerous to the systems stored in it. A VM related security risks assessment in order to obtain an understanding of the security risks and impact should be performed

Regulatory compliance laws and location (node 1.6 in Fig. 65.2 **represented by P3):** are also discussed in [P3], when storing data in the clouds, companies may be hiring services with providers that have data centers distributed across multiple locations. Nonetheless, some companies use sensitive data and this data access or storage cannot be according to regulatory compliance laws. In this case data needs to be moved to another geographical location.

Vendor Lock-in (node 1.7 in Fig. 65.2 **represented by P6, P11):** is discussed by [P6] and [P11]. Considering that a user should be able to move data/programs to another cloud provider, cloud users must be protected against the so called "data lock-in. In fact, there are currently no standardized ways to plug into a cloud, and this challenges the migration to a new cloud provider. This would be a good chance for providers to compete on openness from the outset. Currently, due to the lack of interoperability among cloud platforms and the lack of standardization efforts, cloud

providers cannot guarantee that a cloud user can move its data/programs to another cloud provider on demand. Cloud Computing would become much more appealing if protection against data lock-in is fully implemented. This issue is of paramount importance and, together with the security topics, comprise a key factor that will determine the extent to which Cloud Computing will be widely adopted by the enterprises.

SQL versus NoSQL (node 1.8 in Fig. 65.2 **represented by P10, P11, P12):** On the choice among NoSQL and SQL databases are presented by [P12], but without considering Cloud-related factors like service and deployment model. Two main types of Cloud data hosting solutions can be distinguished with respect to the application interaction with the Cloud data store. The first type allows interaction on a fine granular level, e.g., by using SQL after migrating the database hosted traditionally to an Amazon EC2 instance. The second type provides a service interface to interact with the Cloud data store such as provided by Amazon SimpleDB. The data store becomes a data service, which in turn requires interaction on the level of the service interface that is more coarse grained compared to the interaction when using SQL for instance. By choosing between SQL and NoSQL for example, each solution is targeting a specific application domain and therefore does not come with all features. The offered functionality might be configurable, but not extensible.

65.3.3 Service Providers

In the selected papers we identified three service providers (Amazon Web Services (AWS), Google and Microsoft Azure) to answer the third research question (**SRQ3**). These providers have data centers in different regions of the world. This should be a criteria to be considered in the migration due to possible network restrictions and number of hops. Moreover, comparing what they offer, it is possible to gather information regarding a possible strategy to migrate databases to the cloud. All providers offer ephemeral storage (temporary allocation of data in volatile caching), as well as support for block storage, object storage, relational database, archiving, and NoSQL. Depending on the migration requirements, these services can be contracted either isolated or combined.

According to [P4], is important to verify all features needed by the application are available. if it occurs some features needed could not be supported by database provider. In this case, Microsoft Azure,[1] Amazon AWS[2] and Google

---

[1] https://azure.microsoft.com/en-us/services/sql-database/

[2] https://aws.amazon.com/rds/sqlserver/pricing/

Cloud[3] are potential candidates for this purpose. Tools can be used to convert a database into another. Some options of tools offered by these providers can be seen in the following urls[4,5,6]

Amazon RDS (node 3.1 in Fig. 65.2 mentioned by P7, P8, P11, P12): is a web service that set-up, operates and scales a relational database on AWS. The Amazon RDS provides six well-known database engines to choose from, including Amazon Aurora, Oracle, Microsoft SQL Server, PostgreSQL, MySQL and MariaDB. With this solution it is possible to configure read replicas and multi-availability zone deployment.

Google App Engine Datastore (node 3.2 in Fig. 65.2 mentioned by P12): is a nonrelational replicated data store offered on a Public Cloud as SaaS.

Microsoft SQL Azure (node 3.3 in Fig. 65.2 represented by P5, P12): It offers a wide range of databases options to its users among Oracle, IBM DB2, MariaBD, Postgres.

65.4   Results and Discussions

Strategies Analysis. Based on Fig. 65.2, we identified in the literature evidences of six strategies to support the migration of legacy systems databases to the cloud. As already mentioned in the previous section, the choice of these strategies depends on the migration goals, legacy database characteristics and the compatibility of these characteristics with the cloud database provider. Considering that these strategies present different scenarios, the migration planning is essential for both the appropriate strategy selection and effective database migration to the cloud.

Issues Analysis. As can be seen in Fig. 65.2, eight issues were identified in the literature. These issues represent challenges that can impact the database migration to the cloud. The *compatibility* issues collected from the literature were related to restrictions of the size of the database, support to transactions, as well as triggers and predefined functions. These issues should be previously analyzed to support the migration planning. This analysis should take into account the details of services offered by each provider. To this end, accessing the provider portal is advisable, configuring its free tier with the database services to verify compliance with the database requirements of the legacy system is also an important step. The *scalability* issues

collected from the literature were related to vertical and horizontal scalability. And providers offer different solutions to this end.[7,8,9,10]

Provider Services Analysis. We found evidences reporting the migration of legacy database to three cloud providers. The selection of these providers followed the original database characteristics that had to be adapted and adjusted to the selected provider.

Threats to Validity. There may be bias in data extraction. However, this was addressed through defining a data extraction form to ensure consistent extraction of relevant data to answering the research questions. The findings and implications are based on the extracted data. Another possible threat is the selection bias. We addressed this threat during the selection step as described in Fig. 65.1, i.e. the papers included in the characterization were identified through a selection process which comprises of multiple phases. The papers identified from a preview systematic review conducted by the authors were accumulated from multiple literature databases covering relevant journals and proceedings. One possible threat is bias in the selection of publications. This was addressed through specifying a research protocol that defines the research questions and objectives of the study, inclusion and exclusion criteria, search strings, the search strategy and strategy for data extraction. Another possible bias is the fact that the 66 papers considered as a start point could not represent the appropriate target for strategies and issues to migrate legacy systems from the database perspective. The string used to select these 66 papers focused on migration of legacy systems to the cloud, therefore this start list represent a appropriate sample from which papers to answer the research questions were selected. The set of 13 papers selected in Phase 4 as described in Fig. 65.1 is a potential external validity threat. However, the papers were selected having as a start point the ones published between 2005 and 2015 from a previous systematic review conducted by the authors. For this reason, the set of the 13 papers are considered representative enough as a sampling for this characterization.

65.5   Conclusions

This paper presented a characterization of main strategies reported in the literature addressing issues related to the migration of legacy systems databases to the cloud. To

---

[3] https://cloud.google.com/sql-server/

[4] https://azure.microsoft.com/en-us/campaigns/azure-vs-aws/mapping/

[5] https://aws.amazon.com/dms/

[6] https://cloud.google.com/free-trial/docs/migrate-google-cloud-platform

[7] https://azure.microsoft.com/en-in/documentation/articles/documentdb-partition-data/

[8] https://cloud.google.com/datastore/

[9] https://aws.amazon.com/simpledb/

[10] https://console.ng.bluemix.net/catalog/services/cloudant-nosql-db

Table 65.1 Selected papers

ID	Author, title	Venue	Year
P1	J. Alonso, L. Orue-Echevarria, M. Escalante, J. Gorronogoitia and D. Presenza, Cloud modernization assessment framework: Analyzing the impact of a potential migration to Cloud	MESOCA	2013
P2	J. Wu, P. Teregowda, K. Williams, M. Khabsa, D. Jordan, E. Treece, C. L. Giles, Migrating a Digital Library to a Private Cloud	IC2E	2014
P3	A. Michalas, N. Paladi and C. Gehrmann, Security aspects of e-Health systems migration to the cloud	HEALTHCOM	2014
P4	Q. H. Vu and R. Asal, Legacy Application Migration to the Cloud: Practicability and Methodology	SERVICES	2012
P5	P. J. P. da Costa and A. M. R. da Cruz. Migration to Windows Azure Analysis and Comparison	PROTCY	2012
P6	B. Rimal, A. Jukan, D. Katsaros and Y. Goeleven. Architectural requirements for cloud computing systems: an enterprise cloud approach	JGC	2011
P7	M. Manuja, Moving agile based projects on Cloud	IAdCC	2014
P8	B. C. Tak, B. Urgaonkar and A. Sivasubramaniam, Cloudy with a Chance of Cost Savings	TPDS	2012
P9	H. Mouratidis, S. Islam, C. Kalloniatis and S. Gritzalis. A framework to support selection of cloud providers based on security and privacy requirements	JSS	2013
P10	C. H. Costa, P. H. Maia, N. C. Mendonca, and L. S. Rocha. Supporting Partial Database Migration to the Cloud Using Non-Intrusive Software Adaptations: An Experience Report	CLOUDWAY	2015
P11	V. Andrikopoulos, T. Binz, F. Leymann, and S. Strauch. How to adapt applications for the Cloud environment	COMPUTING	2013
P12	S. Strauch, O. Kopp, F. Leymann, and T. Unger. A taxonomy for cloud data hosting solutions	DASC	2011
P13	M. Vasconcelos, N. C. Mendona, and P. H. M. Maia. Cloud Detours: A Non-intrusive Approach for Automatic Software Adaptation to the Cloud	ESOCC	2015

achieve this goal we followed a four-phase approach having as a start point the selection of papers published in conferences and journals. The evidences identified in the selected papers provide a panoramic view of six strategies reported to migrate databases from legacy systems to the cloud. These strategies have different goals depending on the migration needs, constraints and resources available (Table 65.1).

References

1. Fehling, C., Leymann, F., Retter, R., Schupeck, W., & Arbitter, P. (2014). *Cloud computing patterns: Fundamentals to design, build, and manage cloud applications*. Wien: Springer.
2. Zhao, J.-F. & Zhou, J.-T. (2014). Strategies and methods for cloud migration. *International Journal of Automation and Computing, 11* (2), 143–152. [Online]. Available http://link.springer.com/article/10.1007/s11633-014-0776-7nhttp://link.springer.com/10.1007/s11633-014-0776-7
3. Lee S.-G., Chae, S. H., & Cho, K. M. (2013). Drivers and inhibitors of SaaS adoption in Korea. *International Journal of Information Management, 33*(3), 429–440.
4. Brummett, T., & Galloway, M. (2016). Towards providing resource management in a local iaas cloud architecture. In S. Latifi (ed.), *Information technology: new generations* (pp. 413–423). Switzerland: Springer.
5. Li, Z., O'Brien, L., Cai, R., & Zhang, H. (2012). Towards a taxonomy of performance evaluation of commercial cloud services. In *2012 I.E. 5th International Conference on Cloud Computing (CLOUD)* (pp. 344–351). Piscataway: IEEE.
6. Li, Z., O'Brien, L., Zhang, H., & Cai, R. (2012). On a catalogue of metrics for evaluating commercial cloud services. In *Proceedings of the 2012 ACM/IEEE 13th International Conference on Grid Computing* (pp. 164–173). Piscataway: IEEE Computer Society.
7. Morioka, E., & Sharbaf, M. S. (2015). Cloud computing: Digital forensic solutions. In 2015 12th International Conference on Information Technology-New Generations (ITNG) (pp. 589–594). Las Vegas: IEEE.
8. Chang Y.-W., & Hsu, P.-Y. (2016). Investigating switching intention to cloud enterprise information systems: An analysis at the organizational level. *International Journal of Information Management.*
9. Paula, A. C. M., & Carneiro, G. F. (2016). Cloud computing adoption, cost-benefit relationship and strategies for selecting providers: A systematic review. In *Proceedings of the 11th International Conference on Evaluation of Novel Software Approaches to Software Engineering* (pp. 27–39).
10. Ribeiro, F. M. Jr, da Rocha, T., Santos, J. C., & Moreno, E. D. (2016). A model-driven solution for automatic software deployment in the cloud. In *Information technology: new generations* (pp. 591–601). Switzerland: Springer
11. Khalil, I. M., Khreishah, A., Bouktif, S., & Ahmad, A. (2013). Security concerns in cloud computing. In *2013 Tenth International Conference on Information Technology: New Generations (ITNG)* (pp. 411–416). Piscataway: IEEE.
12. Strauch, S., Andrikopoulos, V., & Bachmann, T. (2013). Migrating application data to the cloud using cloud data. In *e 3rd International Conference on Cloud Computing and Service Science (CLOSER)* (pp. 36–46). Citeseer

Improving Data Quality Through Deep Learning and Statistical Models

66

Wei Dai, Kenji Yoshigoe, and William Parsley

Abstract

Traditional data quality control methods are based on users' experience or previously established business rules, and this limits performance in addition to being a very time consuming process with lower than desirable accuracy. Utilizing deep learning, we can leverage computing resources and advanced techniques to overcome these challenges and provide greater value to users.

In this paper, we, the authors, first review relevant works and discuss machine learning techniques, tools, and statistical quality models. Second, we offer a creative data quality framework based on deep learning and statistical model algorithm for identifying data quality. Third, we use data involving salary levels from an open dataset published by the state of Arkansas to demonstrate how to identify outlier data and how to improve data quality via deep learning. Finally, we discuss future work.

Keywords

Data quality • Data clean • Deep learning • Statistical quality control • Weka

66.1 Introduction

In the era of Big Data, engineers are often challenged to innovate software platforms to store, analyze, and manage vast amounts of data in often diverse datasets. However, without improved data quality at data warehouses or data centers, engineers are not able to properly analyze this data due to garbage in, garbage out (GIGO) computing behavior. Hence, improving and controlling data quality has become one of the most critical business strategies for companies, organizations, governments, and financial service providers.

One of the essential tasks for data quality is to detect data quality problems, especially outlier detection.

Dr. Richard Wang, a professor at MIT and leader in the field of data quality, speaks to a theoretical data quality framework as possessing qualities that are intrinsic, contextual, easily, and representational, accessible [1]. In software industry, popular data quality software tools contain many data profiling modules, but these functions usually are based on simple statistical algorithms [2]. In addition, these tools often do not offer insight into how to improve data quality problems through machine learning algorithms.

Machine learning is a promising subfield of computer science that approaches human-level artificial intelligence (AI), and is already utilized in self-driving cars, data mining, and natural speech recognition software [3–6]. These AI algorithms can either be unsupervised or supervised. In particular, deep learning is a type of machine learning that is one of the supervised methods [7–9]. It is clear that improving data quality is essential to data mining, data analysis, and Big Data processes. [16] mentions data quality

W. Dai (✉) • W. Parsley
Information Science, University of Arkansas at Little Rock, Little Rock, AR, USA
e-mail: wxdai@ualr.edu; wmparsley@ualr.edu

K. Yoshigoe
Computer Science, University of Arkansas at Little Rock, Little Rock, AR, USA
e-mail: kxyoshigoe@ualr.edu

© Springer International Publishing AG 2018
S. Latifi (ed.), *Information Technology – New Generations*, Advances in Intelligent Systems and Computing 558, DOI 10.1007/978-3-319-54978-1_66

assessment architecture via AI, but does not discuss how to improve and how to identify data quality with unique details based on AI and statistical modes. To the best of our knowledge, no literature published speaks to improving data quality through both machine learning and statistical quality control models.

Our first contribution is a data quality architecture based on deep learning and statistics model that enables identify data quality problems, and it also offers improve data quality.

The rest of this paper is organized as follows. Section 66.2 is related work; Sect. 66.3 discusses software architecture and data flow; Sect. 66.4 describes the data source and data structure used in this work; Sect. 66.5 describes data preparation; and Sect. 66.6 discusses machine learning modes, including deep learning and statistical quality controls. Section 66.7 is a summary of our work, and Sect. 66.8 is future discussion.

66.2 Related Work

66.2.1 Outlier Detection Technologies

Outlier data is data that is so different from the rest of our dataset that is can skew models and statistical measures. Hawkins officially defined it as "…an observation which deviates so much from the other observations as to arouse suspicions that it was generated by a different mechanism" [10]. However, outlier data does not necessarily mean the data is erroneous or problematic, but this data is at greater risk for being erroneous or skewing models. Outlier detection can be used in different industries. For example, [11] mentions that intrusion detection systems, credit card fraud monitoring systems, medical diagnosis, law enforcement, and earth science all utilize this technology.

According to [11], there are many outlier detection processes helping people to identify outlier data, including probabilistic and statistical models, linear correlation analysis, proximity-based detection, and supervised outlier detection.

66.2.2 Statistical Quality Control

Probabilistic and statistical models can be used for outlier detection, including normal distribution and Zipf distribution. Moreover, these models work well in one-dimensional and multidimensional data.

Statistical process control (SPC) [12–14] is a procedure of quality control based on statistical methods. A normal distribution is a symmetric, continuous, bell-shaped distribution of a variable. Figure 66.1 displays a normal probability distribution as a model for quality characteristics within the limits at three standard deviations (σ) on either side of the mean (μ).

The base of SPC is the central limit theorem [14]. In addition, a control chart is one of the key techniques of SPC, and this chart contains a center line (CL) and lower and upper control limits (LCL and UCL in formulas (66.1 and 66.2).

$$UCL = \mu + 3\sigma \qquad (66.1)$$

$$LCL = \mu - 3\sigma \qquad (66.2)$$

where μ is population mean, and σ is population standard deviation.

Fig. 66.1 Normal distribution curve

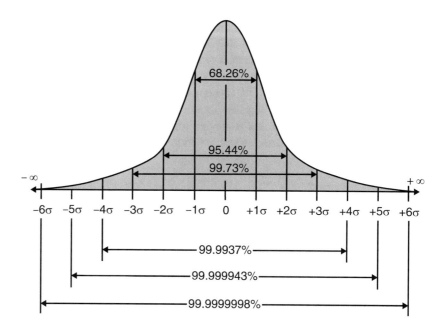

A control chart is a graph or chart with limit lines, or control lines. There are basically three kinds of control lines: UCL, CL, and LCL (see Fig. 66.2).

66.2.3 Deep Learning

Deep learning is a computational model made up of multiple hidden layers of analysis used to understand representations of data so that analysts are able to better understand complex data problems (see Fig. 66.3). Deep learning network includes Convolutional neural networks and Backpropagation networks.

In this paper, we develop algorithms based on KNIME and WEKA [16–19], [20] because these applications are visually interactive tools, and the software supports JAVA programs.

66.3 Overview Architecture

In this paper, we utilized a data quality architecture (see Fig. 66.6) with three parts: data preparation, machine learning, and data visualization. Data preparation focuses on identifying basic problems and cleaning the data; Machine learning contains neutral network and statistical

quality control model for discovering complex outlier data; Data visualization then displays risk data, error data, and correct data.

The architecture also shows the process of data manipulation. At the data preparation stage, the data profiling program reads the data file (CSV file type); particularly for the purpose of transforming string type and date type data values into numeric values; data quality problems should be discovered at this stage as well, such as incorrect sex code or date. At the deep learning network stage, the data is then split into training data and testing data for use by the neural network algorithms. Our deep learning network will offer predictive salaries at this point as well. At the statistical quality control module stage, the program calculates the errors between predictive salary (P) and real salary (T). If the difference rate is out of UCL or LCL, this data will be marked as outlier data. Finally, at the data visualization stage, the program will produce reports through visualization models, which is not the focus of this work and is omitted from this paper.

66.4 Data Source

For our data source, we accessed the state of Arkansas' open data library and selected their public information sheets containing employee salaries. We downloaded it as a csv-format file from the website (transparency.arkansas.gov). The dataset is free use and open data and contains fiscal year, agency, pay class category, pay scale type, position number, position title, employee name, percent of time, annual salary, etc.

In this paper, we assume that most of officers' salaries are accurate and correct because this data comes from Arkansas state human resource department. However, we also believe that this data may contain some errors.

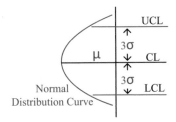

Fig. 66.2 Quality control chart of SPC

Fig. 66.3 Deep learning network

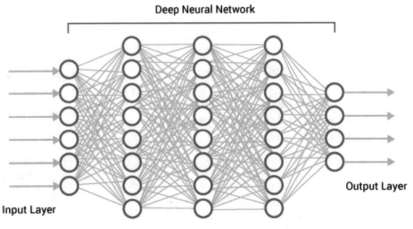

Fig. 66.4 Data preparation program

Data Preparation

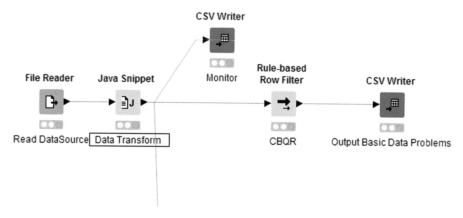

66.5 Data Preparation

The data preparation stage of our project contains Data Transform and Data Clean programs so that basic data quality problems can filtered as shown in Fig. 66.4. The Data Transform program is the process of transferring data from String, Boolean, and Character types into numeric types via a hash code function.

Data Transform program is completed by a program written in JAVA. There are three major benefits after transferring numeric data.

1. Standardization data: The string value i long and takes up excess storage space, but an integer value is simple and short. For instance, a hash code of "DEPT OF PARKS AND TOURISM" is 1971843741, resulting in highly efficient compression.
2. Identify problems: a tiny change of string leads to a big difference in hash code. For example, a hash code of "DEPT OF PARKS AND TOURISM" is 1971843741, but "DEPT OF PARK AND TOURISM" is 1553109818.
3. Machine learning has been shown to run more efficiently with numeric values.

66.5.1 Transferring String Type

If the data was null or empty, the data was marked with a special integrate value for future procedure function. The pseudo-code is here:

```
Read data
IF object is null THEN
   iHashCode = -1;
ELSEIF object is empty THEN
```

```
   iHashCode =0;
ELSE
   iHashCode = hashcode(object)
EndIF
```

66.5.2 Transferring Date Type

Another kind of data being translated is date type, such as date of birth or career service dates. For example, the program transfers date of birth into days using the following pseudo-code:

```
Read data
   iFlag = -1
   IF date value is incorrect THEN
      Mark error data, return error
   ELSE
      Calculate days, return iDays
   ENDIF
```

66.5.3 Data Cleaning

Check Basic Quality Rule (CBQR) means that the program checked the element data quality problems via logical rules. For example, instances where the data is empty but it should not be an empty value. The pseudo-code is here:

```
Read data
   iFlag = -1
   IF  iHashCode == -1 THEN
       IF the data cannot be NULL value THEN
          iFlag =0
       END IF
```

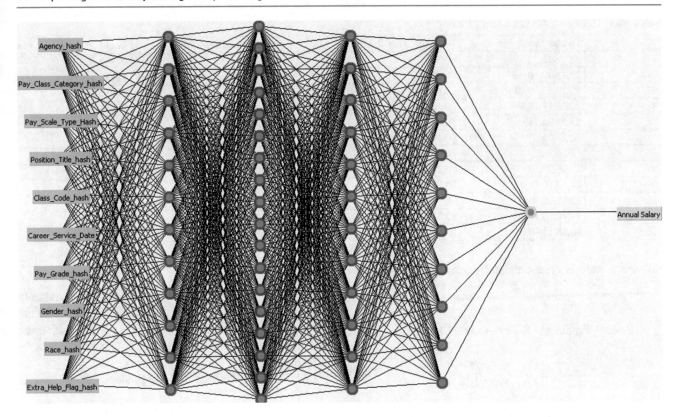

Fig. 66.5 Backpropagation network

```
        ENDIF
      IF iHashCode == 0 THEN
        IF the data cannot be EMPTY value THEN
            iFlag =0
        END IF
      ENDIF

      IF iFlag=0 THEN
        Push the data into element quality problem
        data
      EndIF
Close Fetch
```

66.6 Deep Learning

The *Machine Learning* program is able to learn patterns from data sets. The data was separated into two parts: training data and testing data. The *Machine Learning* program trains itself through training data, and it tests the accuracy of the *Machine Learning* algorithm via testing data. Weka has a test option for choosing different training data sets or test sets.

There are many algorithms for prediction. After comparing Back- propagation(BP) network, Convolutional Neural Network (CNN), Support Vector Machine (SVM), and

Regression models, testing results show that BP network is the best choice.

As shown in Fig. 66.5, there are six Layers BP networks, including 10 input nodes, four hidden layers (each hidden layer has 12, 18, 12, and 10 nodes, respectively), and one output node. Furthermore, input nodes contain *agency* hashcode, *pay-class-category* hashcode, *pay-scale-type* hashcode, etc. Output node is *annual salary*.

After testing different hidden layers of BP algorithms, we compared test results, including correlation coefficient, mean absolute error, and root mean squared error as shown in Table 66.1.

According to the table, when hidden layers are 12, 18, 12, and 10, the correlation coefficient has the maximum value. Additionally, mean absolute error and root mean squared error are at their minimum values.

66.7 Statistical Quality Control Model

Outlier detection is based on a statistical quality control model. Our deep learning model outputs a predicated value when it reads input data. However, the predicated value usually is not equal to actual value because of error.

The error for the dataset equates to the difference between predicted salary and actual salary (Formula 66.3), and the

Table 66.1 Different hidden layers for BP

#	Hidden layers	Correlation coefficient	Mean absolute error	Root mean squared error
1	12,18,12	0.7756	8381.9728	12342.309
2	12,18,12,10	0.7846	8149.966	12133.4712
3	12,18,12,10,10	0.0225	12602.9016	19612.9531
4	12,18,12,10,8	0	12602.9034	19612.9514
5	12,18,24,10	0.7768	8701.5927	12315.6109
6	12,18,24	0.7712	8445.6757	12405.1537
7	12,36,24	0.75	8605.0589	13045.9201
8	12,36,24,10	0.7707	8424.1713	12472.9748
9	24,18,12,10	0.7765	8362.1652	12303.3004
10	12,18,10,10	0.7801	8300.8978	12352.81
11	12,18,16,10	0.7774	8543.8297	12277.3229
12	10,18,12,10	0.7787	8437.5677	12289.3999

difference ratio is error divided by actual salary (Formula 66.4).

$$\text{Error} = \text{Predicated Salary} - \text{Acutal Salary} \quad (66.3)$$

$$\text{Difference Ratio} = \frac{Error}{\text{Acutal Salary}} \quad (66.4)$$

Moreover, quality control charts need to take into account mean and standard deviation. However, we can incorporate the population mean and standard deviation of the population through the central limit theorem. This theorem has four important properties [15]:

1. The mean of the sample ($\mu_{\bar{x}}\mu_{\bar{x}}$) means will be the same as the population mean (μ).

$$\mu_{\bar{x}} = \mu \quad (66.5)$$

2. The standard deviation of the sample ($\sigma_{\bar{x}}$) means will be less than the standard deviation of the population (σ), and it will equal to the population standard deviation divided by the square root of the sample size.

$$\sigma_{\bar{x}} = \frac{\sigma}{\sqrt{n}} \quad (66.6)$$

3. If the original variable is normally distributed, the distribution of the sample means will be normally distributed, for any sample size n.
4. If the distribution of the original variable might not be normal, a sample size of 30 or more is needed to use a normal distribution to approximate the distribution of the sample means. The larger the sample, the better the approximation will be.

Before calculating our quality control chart, we randomly selected 100 rows to calculate the standard deviation of the sample ($\sigma_{\bar{x}}$) and mean of the sample ($\mu_{\bar{x}}$) for use in the Difference-Ratio column as partial sample data (Show in Table 66.2).

After choosing 100 samples, we produced the following:

$$\mu_{\bar{x}} = -0.23 \text{ and } \sigma_{\bar{x}} = 0.22 \quad (66.7)$$

According to formulas 66.5, 66.6, and 66.7, the population mean (μ) and standard deviation of the population (σ) are:

$$\mu = \mu_{\bar{x}} = -0.23 \text{ and } \sigma = \sqrt{n} \times \sigma_{\bar{x}}$$
$$= \sqrt{100} \times 0.22 = 2.2 \quad (66.8)$$

According to formula 66.8, the upper and lower control limits (UCL and LCL) are:

$$\text{UCL} = \mu + 3\sigma = -0.23 + 3 \times 2.2 = 5.37$$

$$\text{LCL} = \mu - 3\sigma = -0.23 - 3 \times 2.2 = -6.83$$

If difference ratio is out of UCL or LCL, this data is out of three sigma (3σ), and will be marked as outlier data, or risk data. In Table 66.3, some data is marked as outlier data because the difference ratio is more than 25. For example, salaries for some full-time employees for the Arkansas state government were only 21 cents per year.

66.8 Conclusion

We have built a unique framework for outlier detection in data profiling, including data preparation, machine learning, and statistical quality control models. After cleaning data in data preparation, machine learning model outputs have been

Table 66.2 Part of sample data

ID	Actual annual salary	Predicated annual salary	ERROR	Difference ratio
14	153588.66	89940.368	-63648.3	-0.414407496
49	102937.54	97946.535	-4991.01	-0.048485761
19	95229.36	75777.512	-19451.8	-0.204263139
86	88607.79	87640.989	-966.801	-0.010911016
93	63733.87	34973.542	-34760.3	-0.498471231
27	69733.87	32259.939	-37473.9	-0.537384932
63	64970.88	45202.44	-19768.4	-0.304266157
42	61614.18	24997.06	-36617.1	-0.594296962
57	61566.96	32533.174	-29033.8	-0.471580634
75	60879.1	35386.174	-25492.9	-0.418746762
81	57852.08	25052.524	-32799.6	-0.566955518
1	56264.64	50359.792	-5904.85	-0.104947761
66	54999.98	21391.136	-33608.8	-0.611070113
4	50678.99	42406.105	-8272.89	-0.163240921
95	50607.23	19280.438	-31326.8	-0.619018113
13	50000.08	52144.907	2144.827	0.042896471

Table 66.3 Outlier data of annual salary

ID	Actual annual salary	Predicated annual salary	ERROR	Difference ratio
14138	0.21	30607.41	30607.20	145748.59
10435	0.21	26211.88	26211.67	124817.48
7238	0.21	26089.28	26089.07	124233.67
1032	0.21	25926.96	25926.75	123460.71
2934	0.21	25596.23	25596.02	121885.82
9102	0.21	23615.14	23614.93	112452.06
2179	0.21	23167.60	23167.39	110320.90
198	2040	91925.23	89885.23	44.06
13131	2040	91925.23	89885.23	44.06
9719	2040	65939.44	63899.44	31.32
11540	2040	53485.19	51445.19	25.22
9998	6240	22113.57	15873 57	2.54
9567	5824	19029.40	13205.40	2.27
17602	5824	18935.52	13111.52	2.25
15575	6760	21978 68	15218.68	2.25
15078	5824	18474.99	12650.99	2.17
2641	5824	18461.80	12637.80	2.17
5988	39399.84	123755.51	84355.67	2.14
2313	6240	19483.77	13243.77	2.12
7521	39399.84	122850.90	83451.06	2.12
15555	39399.64	122850.90	83451.06	2.12
6731	7280	22524.36	15244.36	2.09

able to predict salaries. Then, outlier data was detected by our statistical quality control model.

Our work integrates deep learning networks and statistical quality control models for improving data quality. This idea has a great potential to reduce workload and improve performance as system developers and business users do not need to create excessive data rules to identify data quality problems, as deep learning algorithms automatically learn data patterns through training datasets.

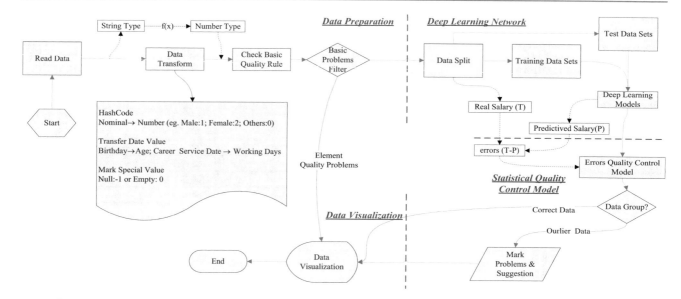

Fig. 66.6 Logical architecture

66.9   Future Work

Data profiling of information quality is very important for the advancement of Big Data. Deep learning is a very promising approach for potentially solving many Big Data challenges. We will continue to explore other neural network models with various datasets, and k-folds validation. We will also conduct these experiments on GPU-based server to expedite the deep learning performance and to seek effective data visualization solutions.

References

1. Strong, D. M., Lee, Y. W., & Wang, R. Y. (1997). Data quality in context. *Communications of the ACM, 40*(5), 103–110.
2. Michalski, R. S., Carbonell, J. G., & Mitchell, T. M. (2013). *Machine learning: An artificial intelligence approach*. Berlin: Springer Science & Business Media.
3. Alpaydin, E. (2014). *Introduction to machine learning*. Cambridge, MA/London: MIT Press.
4. Murphy, K. P. (2012). *Machine learning: A probabilistic perspective*. Cambridge, MA: MIT Press.
5. Natarajan, B. K. (2014). *Machine learning: A theoretical approach*. San Mateo: Morgan Kaufmann.
6. Schmidhuber, J. (2015). Deep learning in neural networks: An overview. *Neural Networks, 61*, 85–117.
7. LeCun, Y., Bengio, Y., & Hinton, G. (2015). Deep learning. *Nature, 521*(7553), 436–444.
8. Deng, L., Hinton, G., & Kingsbury, B. (2013). New types of deep neural network learning for speech recognition and related applications: An overview. In *IEEE international conference on acoustics, speech and signal processing (ICASSP), 2013* (pp. 8599–8603). IEEE.
9. Hawkins, S., He, H., Williams, G., & Baxter, R. (2002). Outlier detection using replicator neural networks. In *Data warehousing and knowledge discovery* (pp. 170–180). Berlin Heidelberg: Springer.
10. Aggarwal, C. C. (2015). Outlier analysis. In *Data mining* (pp. 237–263). Springer International Publishing.
11. Montgomery, D. C. (2009). *Statistical quality control* (Vol. 7). New York: Wiley.
12. Leavenworth, R. S., & Grant, E. L. (2000). *Statistical quality control*. New York: Tata McGraw-Hill Education.
13. DeVor, R. E., Chang, T.-h., & Sutherland, J. W. (2007). *Statistical quality design and control: Contemporary concepts and methods*. Upper Saddle River: Prentice Hall.
14. Bluman, A. G. (2009). *Elementary statistics: A step by step approach*. New York: McGraw-Hill Higher Education.
15. Berthold, M. R., Cebron, N., Dill, F., Gabriel, T. R., Kötter, T., Meinl, T., Ohl, P., Sieb, C., Thiel, K., & Wiswedel, B. (2008). *KNIME: The Konstanz information miner*. Berlin Heidelberg: Springer.
16. O'hagan, S., & Kell, D. B. (2015). Software review: the KNIME workflow environment and its applications in genetic programming and machine learning. *Genetic Programming and Evolvable Machines, 16*(3), 387–391.
17. Hall, M., Frank, E., Holmes, G., Pfahringer, B., Reutemann, P., & Witten, I. H. (2009). The WEKA data mining software: An update. *ACM SIGKDD Explorations Newsletter, 11*(1), 10–18.
18. Mark, H., Frank, E., Holmes, G., Pfahringer, B., Reutemann, P., & Witten, I. H. (2009). The WEKA data mining software: An update. *ACM SIGKDD Explorations Newsletter, 11*(1), 10–18.
19. Fournier-Viger, P., Gomariz, A., Gueniche, T., Soltani, A., Wu, C.-W., & Tseng, V. S. (2014). SPMF: A java open-source pattern mining library. *The Journal of Machine Learning Research, 15*(1), 3389–3393.
20. Fournier-Viger, P., Gomariz, A., Gueniche, T., Soltani, A., Wu, C. W., & Tseng, V. S. (2014). SPMF: A java open-source pattern mining library. *The Journal of Machine Learning Research, 15*(1), 3389–3393.

A Framework for Auditing XBRL Documents Based on the GRI Sustainability Guidelines

67

Daniela Costa Souza and Paulo Caetano da Silva

Abstract

The adoption of XBRL by the Global Report Initiative (GRI) in the disclosure of sustainability reports contributes to the increase of their quality; however, the heterogeneity of enterprise information systems and the adoption of efficient audit processes are potential obstacles to the use of such reports in the corporate setting. Although the adoption of XBRL represents an improvement for the analysis and use of sustainability data, it lacks an integration framework that allows users to benefit from the standardization proposed by the GRI. In an attempt to solve system and data integration issues in this heterogeneous setting and also to improve the efficiency of sustainability report audit, a service framework is proposed. The framework consists of a process model for the generation and analysis of sustainability reports, an architecture to structure the information environment of organizations, operators for analytical processing (OLAP) data sustainability and an analysis of sustainability data.

Keywords

GRI • XBRL • OLAP • LMDQL • Sustainability analysis

67.1 Introduction

The disclosure of non-financial data, especially sustainability reports, has gained great importance in the corporate setting. Many companies choose to disclose such reports; however, the lack of uniformity in the disclosure makes it difficult to analyze the data and information presented. Regarding veracity and sufficiency, *stakeholders* see the reports negatively, when it comes to business practices [17].

This may be due to the fact that it is a voluntary disclosure, which can be drawn from several regulations and does not need to be audited. Moreover, the diversity of information systems in organizations hampers the collection, analysis and the way these data are integrated and disclosed. In this context, the use of Web Services associated with the use

of OLAP tools assists in the integration process and analytical processing for the auditing of sustainability reports, which may provide greater credibility of the information that is provided. Thus, it is necessary to use an environment which is capable of efficiently and effectively audit sustainability reports. In order to solve these problems, this paper proposes a framework for auditing XBRL documents based on the GRI sustainability guidelines [9–11].

After this introduction, this work is structured as follows: Section II presents a literature review; Section III proposes the integration process flow and the auditing of sustainability reports based on the GRI guidelines; the continuous audit architecture for the corporate setting; an analysis of the compliance of sustainability reports with GRI guidelines; GRI operators, whose function is to perform analytical processing of XBRL/GRI data; and a steps for the analysis of sustainability data. Finally, Section IV presents a case study and then the conclusions and future work are presented.

D.C. Souza (✉) • P.C. da Silva
Salvador University (UNIFACS), Salvador 41820-460, Brazil
e-mail: dannyscosta@msn.com; paulo.caetano@pro.unifacs.br

© Springer International Publishing AG 2018
S. Latifi (ed.), *Information Technology – New Generations*, Advances in Intelligent
Systems and Computing 558, DOI 10.1007/978-3-319-54978-1_67

67.2   Literature Review

The e-Financial Process Integration Framework (EFPIF) is a framework which is used for the integration of financial information and investment analysis [12]. The goal of this framework is to provide greater transparency regarding the information available to customers. This transparency is materialized with the disclosure of financial information to help evaluate and improve the understanding of the investment setting, in order to reach a reduction in the risk and loss of investments.

A change from the traditional audit model to continuous audit is suggested to assist in decision-making processes in companies [24]. Continuous audit is a technology innovation of the traditional audit, which is based on automation and, though its concepts have been established for almost two decades, in practice it is still something new [2]. The integrity of the data in financial reports is questionable; however, continuous audit is considered an effective and safe way to facilitate early detection of fraud in financial reports [14]. According to [6], there is growing interest in exploring the continuous audit methodology and auditors recognize that the traditional audit is obsolete. A study on the integration of information systems and auditing reveals several future implications for audits. The authors state that continuous auditing and continuous monitoring of the information that travels through organizations are tools that can assist *stakeholders* in detecting errors and financial fraud [13].

A model for auditing through software systems called Continuous Auditing Web Service (CAWS) is based on Web Service and XML technologies [18]. This model has been developed in order to reduce complexity in the transmission of data and also to aggregate security to systems. To do so, the following technologies have been used: XML, WS-BPEL (used in the composition of new services) and Web Service (used to avoid incompatibility in data access and exchange). A collaborative model based on SOA for audit systems is proposed, which uses XML standards and data transformation applications developed by companies and software vendors [7]. The adoption of SOA involves risks, these risks often manifest themselves during an implementation of SOA solution and arise mainly due to insufficient detail in the SOA project [16].

The adoption of XBRL by GRI provides greater facility in the collection and analysis of sustainability data, besides significantly influencing the improvement of quality of data that composes the report is discussed in Leibs [25].

67.3   A Service Framework for Sustainability Reports

In this section the service framework is shown from the process model, the architecture of the corporate setting, the sustainability operators and steps for the audit of sustainability reports. More details of the architecture and operators can also be seen in the article "OLAP-based Sustainability Report Auditing" [4].

67.3.1 Integration Process Flow

The process model is composed of five steps that are performed following an operational flow. The steps, illustrated in Fig. 67.1, consist of activities developed in the activity flow in order to perform an analysis of the

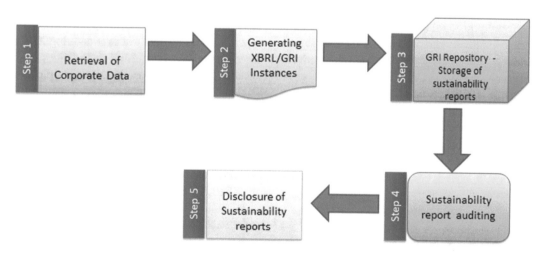

Fig. 67.1 Integration process flow and continuous audit in sustainability reports based on the GRI guidelines

auditing of the organization's sustainability reports, which will assist *stakeholders* in decision making.

The first step is the retrieval of corporate data of the organization, which can be extracted from information systems, business processes and other data sources. In this step, an ETL process is performed, which will allow the obtained data to be extracted, transformed and loaded into a database for further processing, or sent directly to Step 2. This step is responsible for collecting data from the corporate setting so that all the subsequent layers of the model can consume them. Based on the data collected, the flow starts, which will enable the integration and continuous audit of sustainability reports. In the second step, the data obtained in Step 1 are standardized by the use of XBRL taxonomy of the GRI and XBRL instances are generated, representing sustainability reports. The third step is another ETL process, in which the data contained in the XBRL instances are converted and stored in a relational database, since this environment is more efficient for the use of OLAP tools and other data analysis techniques, e.g. data mining. The fourth step is where the sustainability report audit process takes place. This audit may be done through LMDQL-based OLAP operators [20, 22], or the traditional OLAP operators and data mining techniques. In the fifth step is the presentation layer, whose function is to allow the auditor, through an interface, to perform queries which will allow the disclosure of the organization's data regarding sustainability.

67.3.2 Corporate Setting Architecture of Continuous Audit

The framework proposed in this paper has an architecture for the corporate setting, which is service-based. The proposed architecture is divided into two conceptual layers: (1) Integration infrastructure, which is intended to promote the access and retrieval of data within the informational structure of the company. In this layer the corporate setting, extraction services and standardization services are found; and (2) Global Reporting Initiative, whose function is to integrate the architecture proposed by the GRI to the informational scenario of the organization. In this layer, persistence, audit and distribution services are found. The services are materialized in the form of Web Services to meet the necessary integration in order to design the collection environment and the recovery of sustainability data. Thus, all adjacent layers of the model can consume these data, as presented in [3, 4]. Figure 67.2 illustrates the framework proposed using SOA, XBRL, GRI and Continuous Auditing.

It is expected that through this architecture it will be possible to provide the necessary means to mitigate problems of access and standardization of data on the sustainable performance of organizations. In the next section

the analysis of the compliance of sustainability reports is presented.

67.3.3 Analysis of the Compliance of Sustainability Reports

The GRI guidelines are developed through a process that involves a network of *stakeholders*, including company representatives, workers, financial markets, auditors and specialists in several areas.

Based on the guidelines proposed by the GRI, the authors [1, 8] propose two indices for the analysis of compliance of sustainability reports: GEE (Degree of Effective Disclosure) and GAPIE (Degree of Full Compliance) [4].

To calculate GAPIE, the total number of Full Compliance indicators, "APL", i.e. the total indicators that had their content reported according to what is required by the GRI guidelines, is added to the total number of indicators omitted with a reason, "OJ", divided by the total number of Core indicators, which are essential indicators for sustainability reporting, subtracted by the total number of indicators that are Not Applicable, "NA", which are indicators that do not apply to the organization. Figure 67.3 shows the formula for the calculation of this index.

To calculate GEE, the total number of Full Compliance indicators, "APL", is divided by the total number of Core indicators, subtracted by the total number of Not Applicable indicators, "NA", as shown in Fig. 67.4.

Another way of assessing compliance is through the use of the GRIConformity index. This index allows to compare the data reported by organizations to the data required by the GRI guidelines. Through the use of this index, the auditor can analyze and assess whether organizations disclose sustainability data in compliance with the guidelines, as shown in Fig. 67.5.

The next section presents the operators based on these three indices, which will allow to audit sustainability reports based on the guidelines proposed by the GRI and the studies proposed by [1, 8].

67.3.4 Operators for the Assessment of Compliance

The use of tools for the analytical processing of data (OLAP) to perform strategic analyses of an organization assists in identifying trends and patterns in order to better conduct businesses. LMDQL (Link Based Multidimensional Query Language) [20–23] is a language that is intended to conduct analytical processing of multidimensional data expressed in XML documents interconnected by links. LMDQL is derived from Multidimensional Expression

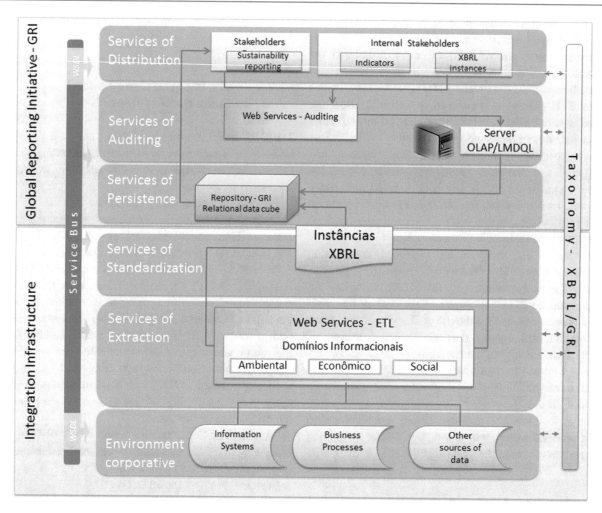

Fig. 67.2 A service framework for sustainability reports

Fig. 67.3 GAPIE formula

$$GAPIE = \frac{TOTAL\ INDICATORS\ ``APL" + TOTAL\ INDICATORS\ ``OJ"}{TOTAL\ INDICATORS\ ESSENTIAL\ -\ TOTAL\ INDICATORS\ ``NA"}$$

Fig. 67.4 GEE formula

$$GEE = \frac{TOTAL\ INDICATORS\ ``APL"}{TOTAL\ INDICATORS\ ESSENTIAL\ -\ TOTAL\ INDICATORS\ ``NA"}$$

Fig. 67.5 GRIConformity formula

$$GRIConformity = \frac{TOTAL\ INDICATORS\ DISCLOSED\ BY\ THE\ ORGANIZATION}{GRI\ TOTAL\ INDICATORS}$$

(MDX) [15, 25, 26] and executes OLAP queries on relational databases and XML-based documents.

This paper proposes the use of Mondrian server [19] to execute OLAP queries through an extension of LMDQL language, in order to manipulate sustainability data issued in GRI/XBRL reports. To make this possible, a Web Service to access the Mondrian server is proposed.

For the analytical processing of sustainability documents, considering the GRI guidelines and GAPIE, GEE and GRIConformity indices, three operators were specified:

GRIConformity, GRIGapie and GRIGee. In Table 67.1, the syntax of these LMDQL operators for sustainability auditing is presented. The three operators (i.e. GRIConformity, GRIGapie and GRIGee) have a MemberSet type as a parameter, which refers to a member set of a data cube, according to the specifications of LMDQL operators.

To perform an LMDQL query based on these operators, illustrated in Fig. 67.6, a parameter is provided referring to the sustainability element (or a set of elements) contained in the sustainability report to be assessed, e.g. Element. A specific element can be used, i.e. [Element].[EconomicImpactUseProductsServicesDescription]). If an analysis of a set of elements is needed, the keyword "children" must be used, referencing all children members of the "Element" dimension (contained in the database), i.e. "[Element].children". In the tests carried out in this paper, the following dimensions were specified: (i) Entity, which refers to the name of the companies that issue sustainability reports; (ii) Document, which corresponds to sustainability documents that the company issues; (iii) Element, which refers to the elements that correspond to the sustainability indicators contained in these documents and (iv) Time, the time period to which the document belongs.

67.3.5 Sustainability Analysis

Figure 67.7 shows the steps proposed for the analysis of sustainability data. To the analysis of sustainability data, a qualitative, exploratory and documental survey was carried out. Therefore, the files disclosed on the GRI website were analyzed, since a detailed report with the data disclosed by organizations is available on the website.

Step 1: This step consists of the retrieval of XBRL instances available on the GRI website [10]. From these instances it is possible to start the analysis of the organization's sustainability data;

Step 2: This step comprises the extraction of XBRL instances into a data sheet. Excel was used here, whose purpose is to prepare the data for the analysis of the organization's sustainability data;

Step 3: At this stage with the retrieved instance, taxonomy indicators are coded according to the GRI guidelines and described according to their content disclosed, and to their profile, i.e. essential or additional.

Step 4: A lack of information that is essential for calculating Gapie and Gee is observed, i.e. GRI code, indicators and classification [1, 8–11]. This information is inserted as columns in the spreadsheet, linking them to the data of XBRL instances;

Step 5: This step includes the coding of indicators. It is very important, in order to separate the type of indicator that is being used in the report, since the GRI taxonomy has profile indicators and economic, social and environmental performance indicators, the latter in turn are separated into essential and additional indicators. As the calculations of GRIGapie and GRIGee operators only use the essential indicators, unlike GRIConformity that makes use of all indicators, it was necessary to carry out this separation [9–11]. To assist the coding of the indicators, the public taxonomy database CoreFiling [5] was used, which identifies the indicators of the organization's analyzed files, separating them and relating them to the codes described in the GRI Taxonomy manual;

Step 6: In this step the information provided by organizations was analyzed, which reports their sustainable practices, and for this, the information was classified according to the information classification model in compliance with what is required by the GRI essential indicators. The criteria used was based on models recommended by [1, 8];

Step 7: In this step, an Excel ETL process for relational DBMS is carried out;

Step 8: This step will allow the execution of operators, described in Section III D, which will evaluate the compliance of the information disclosed by organizations in their sustainability reports;

Table 67.1 LMDQL operators for sustainability auditing

GRIConformity (<*MemberSet*>)
GRIGaple (<*MemberSet*>)
GRIGee (<*MemberSet*>)

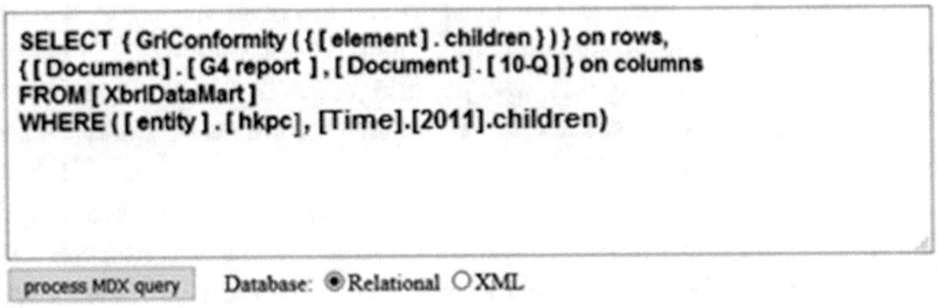

Fig. 67.6 OLAP query using an LMDQL operator for sustainability auditing

```
SELECT { GriConformity ( { { [ element ] . children } } ) } on rows,
{ { [ Document ] . [ G4 report ] , [ Document ] . [ 10-Q ] } } on columns
FROM [ XbrlDataMart ]
WHERE ( [ entity ] . [ hkpc ], [Time].[2011].children)
```

process MDX query   Database: ● Relational ○ XML

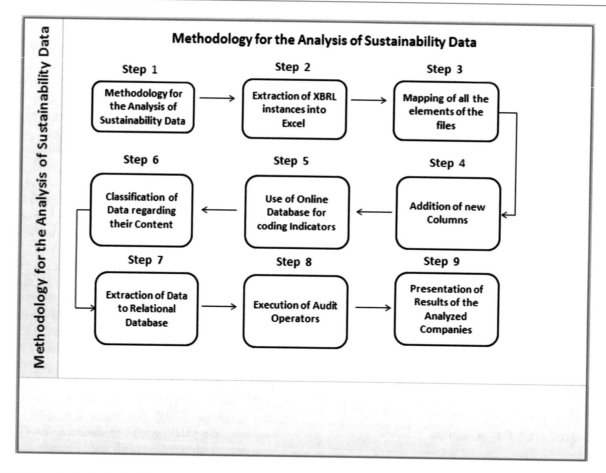

Fig. 67.7 Steps for the Analysis of Sustainability Data

Step 9: This step corresponds to the presentation of results of the analysis of compliance of sustainability reports.

67.4   A Sample of Sustainability Analysis

For the analysis of sustainability data, according to the steps proposed in Section III E, first the files on the GRI website were retrieved. Then, to analyze the data, an ETL process of the files into an Excel spreadsheet was carried out. After that, the following activities were performed: mapping of all elements of the files, addition of new columns, online database use for coding indicators, classification of data concerning their content; extraction of data from Excel spreadsheet into the relational database; execution of audit operators and finally, presentation of results. The auditor may choose the database that will perform the query. A relational database was used, as it has a better performance for OLAP queries than XML databases.

The relational database model used in this work is based on XBRL Data Model [23], which makes it independent of the organization's business model that uses it, facilitating its

application in different contexts. To load the data in the relational database an ETL process (Extract, Transform, Load) was carried out on sustainability reports.

- GRIConformity Operator

Figure 67.8 shows an example of use of the GRIConformity operator. To perform the query, the auditor reports which document he is to analyze. Two types of documents have been specified, G4 Report and 10-Q. The first refers to the XBRL document which contains all the indicators informed by the GRI taxonomy, and the second is the XBRL document issued by the organization under analysis. It is necessary to specify the company that issued the report (e.g. [Entity]. [hkpc]). The options "relational" and "XML" in the Database field are attributes of the Mondrian server extension to the implementation of LMDQL [20–23], with which the analyst chooses on which database paradigm he wants to perform the query. After the query, the operator classifies the elements as "yes" if they are in compliance, and "no", if they are not. Fields containing the character "-" indicate no results for the query executed by the GRIConformity operator.

- GRIGapie operator

```
SELECT { GriConformity ( { [ element ] . children } ) } on rows,
{ [ Document ] . [ G4 report ] , [ Document ] . [ 10-Q ] } on columns
FROM [ XbrlDataMart ]
WHERE ( [ entity ] . [ hkpc], [Time].[2011].children)
```

process MDX query   Database: ● Relational ○ XML

Results:

	[Document.Document] [all] [G4 Report]		[Document.Document] [all] [10-Q]	
	Value	GRI Conformity	Value	GRI Conformity
EmployeesNumberEmploymentContractGenderAdditionalD	-	-	-	-
EmployeesReceivedRegularPerformanceCareerDevelcome	1	Yes	-	-
EmployeeTurnoverNumber	206	Yes	-	-
EmployeeTurnoverRate	0,306	Yes	-	-
EmployeeWagesBenefits	864.600.000	Yes	-	-
EmploymentAspectManagementApproachOverallDescripti	-	-	-	-
EnergyAspectManagementApproachOverallDescription	-	-	-	-
EnergyConsumption	680.039	Yes	-	-

Fig. 67.8 Example of use of the GRIConformity operator

```
SELECT {GRIGapie ( { [ element ] . children } ) } on rows
{ [ Document ] . G4 report essential ] , [ Document ]. [ 10-H ] } on  columns

FROM [ XbrlDataMart ]
WHERE ( [ entity ].[hkpc], [Time].[2011]. children
```

process MDX query   Database: ● Relational ○ XML

Results:

	[Document.Document].[all].[G4 Report essential]		[Document.Document].[all].[10-Q]	
	Value	GRI Gapie	Value	GRI Gapie
AspectBoundaryLimitationOutsideOrganizationDescrip	0	undeclared. Gaple: 2	–	–
AspectMaterialOutsideEntitiesIdentificationLocatio	0	undeclared. Gaple: 4	–	–
AspectMaterialOutsideOrganizationFlag	0	undeclared. Gaple: 6	–	–
AspectMaterialWithinOrganizationFlag	0	undeclared. Gaple: 8	–	–
WhyLaborPracticesGrievanceMechanismsAspectMaterial	–		–	
WhyLocalCommunitiesAspectMaterialDescription	0	undeclared. Gaple: 304	–	
WhyMarketPresenceAspectMaterialDescription	0	undeclared. Gaple: 306	–	
WorkforceInjuryRate	4,4	yes. Gaple: 315	–	

Fig. 67.9 Example of use of GRIGapie operator

This operator has a query structure that is similar to GRIConformity. After the query, the operator checks the essential elements that should be declared and classifies them as "yes" if they have been declared, and "undeclared" if the organization has not disclosed information that is essential to the report. In Fig. 67.9, the result of the query executed by GRIGapie operator is shown, referring to the 432 elements

disclosed by HKPC in 2011. The character "-" indicates no results for the query executed by the GRIGapie operator.

• GriGee operator

Similarly to other operators, after the query, the operator checks the essential elements that should be declared and classifies them as "yes" if they have been declared and "undeclared" if the organization has not disclosed the

```
SELECT {GRIGee ( { [ element ] . children } ) } on rows
{ [ Document ] . G4 report essential ] , [ Document ]. [ 10-H ] } on  columns
FROM [ XbrlDataMart ]
WHERE ( [ entity ].[hkpc], [Time].[2011]. children
```

process MDX query · Database: ⦿ Relational ○ XML

Results:

	[Document.Document].[all].[G4 Report essential]		[Document.Document].[all].[10-Q]	
	Value	GRI Gee	Value	GRI Gapie
AspectBoundaryLimitationOutsideOrganizationDescrip	0	undeclared. Gee: 2	–	–
AspectMaterialOutsideEntitiesIdentificationLocatio	0	undeclared. Gee: 4	–	–
AspectMaterialOutsideOrganizationFlag	0	undeclared. Gee: 6	–	–
AspectMaterialWithinOrganizationFlag	0	undeclared. 8	–	–
AvailabilityProductsServicesLowIncomesDescription	0	undeclared. Gee: 10	–	–
WorkforceInjuryRate	4,4	yes. Gee: 315	–	–
WorkforceNumber	1.348	yes. Gee: 316	–	–

Fig. 67.10 Example of use of the GRIGee operator

information that is essential to the report. In Fig. 67.10 the result of the query performed by the operator GRIGee is shown, referring to the 432 indicators presented by HKPC.

By performing these queries, it was found that HKPC presented 432 indicators, of which only 78 were key indicators. These indicators were disclosed annually, 8 in 2011, 8 in 2012 and 62 in 2013. The GRIGapie, GRIGee and GRIConformity calculations were made per year. Thus, for the year 2011, HKPC presented, respectively, 1% of GRIGapie and GRIGee, also 1% in the year 2012; and in 2013 it presented 8% of GRIGapie and 7% of GRIGee. When calculating GRIConformity, the 432 indicators from 2645 taxonomy indicators were divided, generating a result of 16%. As shown, though HKPC discloses information related to its sustainable practice, it is still in non-compliance.

67.5 Conclusion and Future Work

In this work a service framework was presented to audit sustainability reports based on the GRI guidelines, which consisted of the integration process flow, architecture of the corporate setting, operators for the assessment of compliance of sustainability data and a method for sustainability analysis. This framework may contribute to the development of a model that is able to simplify the collection, analysis, comparison and disclosure of data related to the sustainability performance of organizations. For the analysis of reports, audit operators were presented, and continuous monitoring of organizations' sustainability reports may be performed with them. It is expected that through this work and the use of technology

involving SOA and XBRL, the framework can bring greater reliability and security to *stakeholders* in decision making, with regard to the fundamental dimensions of corporate sustainability that are: social, environmental and economic. This framework is a scale model that grows as the organization establishes excellence in sustainability, internationalization and standardization of data. For future work, it is intended to develop the detail level of the framework presented, creating services for the collection and processing of data, services for OLAP analysis and data mining, in addition to those required for the disclosure of sustainability results.

References

1. Carvalho, Fernanda Medeiros de. e Siqueira, José Ricardo Maia de. Os Indicadores Ambientais nas Normas de Balanço Social. In *Encontro Nacional Sobre Gestão Empresarial e Meio Ambiente*, 8, 2005, Rio de Janeiro 2007.
2. Chan, D. Y., & Vasarhelyi, M. A. (2011). Innovation and practice of continuous auditing. *International Journal of Accounting Information Systems, 12*(2), 152–160.
3. Costa, D., Silva, M. A. P., & Silva, P. C. (2016). A service *framework* to audit sustainability reporting based on GRI rules. In *International conference on information systems and technology management – CONTECSI – 13ª Edição*. São Paulo.
4. Costa, D., Silva, M. A. P, & Silva, P. C. (2016). OLAP-based sustainability report auditing international. In *Conference on internet and web applications and services – ICIW*. Valencia.
5. COREFILING. Public taxonomies. 2016. Available at: <https://bigfoot.corefiling.com/yeti/resources/yeti-gwt/Yeti.jsp#tax~(id~287*v~439)!net~(a~3963*l~1114)!lang~(code~en)!rg~(rg~1*p~1)>. Accessed: 13 Sept 2016.
6. Cruz, E. M., Costa, D., & Silva, P. C. (2014). Sustainability reports based on XBRL through a service-oriented architecture approach.

In *Third international conference on challenges in environmental science and computer engineering (CESCE 2014), 2014.* London.

7. Qiushi, C., Zuoming, H., & Jibing, H. (2013) A collaborative computer auditing system under SOA-based conceptual model.

8. Dias, Lidiane Nazaré da Silva. Análise da Utilização dos Indicadores do Global Reporting Initiative nos Relatórios Sociais em Empresas Brasileiras. 2006. Dissertação (Mestrado em Ciências Contábeis) – FACC/UFRJ, Rio de Janeiro, 2006. FIBRIA. Relatórios de Sustentabilidade. Available at: <http://livros01. livrosgratis.com.br/cp030028.pdf>. Accessed: 13 Oct 2016.

9. GRI, Directrizes para a Elaboração de Relatórios de Sustentabilidade © 2000–2006. Amsterdam, 2006.

10. GRI™ (2013) G4 XBRL schema. [OnLine]. Available at: <http:// xbrl.globalreporting.org/2014-12-01/Forms/AllItems.aspx>. Accessed: 13 Oct 2016.

11. GRI – Global Reporting Iniciative (2014). For the guide lines and standard setting – G4. Available at: <https://www.globalreporting. org/Pages/default.aspx>. Accessed: 13 Oct 2016.

12. Huang, D.-Y., Huang, W.-T., & Tsai, R.-L. (2006). The investment integration framework based on XBRL and web services. *International Journal of Electronic Business Management, 3,* 173–180. Available at: <http://www.doaj.org/doaj?func=abstract&id=2 98446>. Accessed: 13 Oct 2016.

13. Kanellou, A., & Spathis, C. (2011). Auditing in enterprise system environment: A synthesis. *Journal of Enterprise Information Management, 24*(6), 494–519.

14. Lin, C.-C., Lin, F., & Liang, D. (2010). An analysis of using state of the art technologies to implement real-time continuous assurance. *Proceedings – 2010 6th World Congress on Services, Services-1 2010,* art. no. 5577269, pp. 415–422.

15. MDX (2016). Multidimensional expressions (MDX) reference. Available at: <https://msdn.microsoft.com/en-us/library/ms145506. aspx>. Accessed: 13 Oct 2016.

16. Monteiro, E., Silva, P.C. (2015). Risk management lifecycle implementation services in SOA. In *12th international conference on information technology: New generations.* Las Vegas. doi:10.1109/ ITNG.2015.137 [ITNG link] [IEEE link].

17. O'dwyera, B., Unermanb, J. E., & Hessionc, E. (2005). User needs in sustainability reporting: Perspectives of stakeholders in Ireland. *European Accounting Review, 14*(4), 759–787.

18. Murthy, U. S., & Groomer, S. M. (2004). A continuous auditing web services model for XML-based accounting systems. *International Journal of Accounting Information Systems, 5,* 139–163.

19. Pentaho (2011). Logical model. Available at: <http://mondrian. pentaho.com/documentation/schema.php#Cubes_and_ Dimensions>. Accessed: 12 Sept 2016.

20. Silva, P. C., & Times, V. C. (2009). LMDQL: Link-based and multidimensional query language. In: *DOLAP 09 – ACM twelfth international workshop on data warehousing and OLAP, 2009.* Hong Kong.

21. Silva, P. C., Times, V. C., Ciferri, R. R., & Ciferri, C. D. (2012). Analytical processing over XML and XLink. *International Journal of Data Warehousing and Mining (IJDWM), 8*(1), 52–92 . IGI Global.

22. Silva, M. A. P., & Silva, P. C. (2014). Financial forensic analysis. In *13th IADIS international conference WWW/INTERNET (ICWI).* Porto. ISBN: 978-989-8533-24-1. Available at: <http://www. iadisportal.org/digital-library/financial-forensic-analysis>. Accessed: 12 Sept 2016.

23. Silva, M. A. P., & Silva, P. C. (2014). Analytical processing for forensic analysis. In *First international workshop on compliance, evolution and security in cross-organizational processes (CESCOP 2014).* Ulm. doi:10.1109/EDOCW.2014.60.

24. Sikka, P., Filling, S., & Liew, P. (2009). The audit crunch: Reforming auditing. *Managerial Auditing Journal, 24*(2), 135–155.

25. Leibs, S. (2007). Sustainability reporting: Earth in the balance sheet. CFO. [Online]. Available at: <http://www.cfo.com/article. cfm/10234097/>. Accessed: 27 Dec 2016.

26. Spofford, G. (2001). *MDX solutions: With Microsoft SQL server analysis services* (p. 163). New York: Wiley.

Mining Historical Information to Study Bug Fixes

Eduardo C. Campos and Marcelo A. Maia

Abstract

Software is present in almost all economic activity, and is boosting economic growth from many perspectives. At the same time, like any other man-made artifacts, software suffers from various bugs which lead to incorrect results, deadlocks, or even crashes of the entire system. Several approaches have been proposed to aid debugging. An interesting recent research direction is *automatic program repair*, which achieves promising results towards the reduction of costs associated with defect repair in software maintenance. The identification of common bug fix patterns is important to generate program patches automatically. In this paper, we conduct an empirical study with more than 4 million bug fixing commits distributed among 101,471 Java projects hosted on GitHub. We used a domain-specific programming language called *Boa* to analyze ultra-large-scale data efficiently. With *Boa's* support, we automatically detect the prevalence of the 5 most common bug fix patterns (identified in the work of Pan et al.) in those bug fixing commits.

Keywords

Software bugs • Automatic error repair • Bug fix patterns • Human-like patches • Software fault

68.1 Introduction

There are more bugs in real-world programs than human programmers can realistically address [4]. The battle against software bugs exists since software existed. It requires much effort to fix bugs, e.g., Kim and Whitehead [3] report that the median time for fixing a single bug is about 200 days. Program evolution and repair are major components of software maintenance, which consumes a daunting fraction of the total cost of software production. Research in *automatic program repair* has focused on reducing defect repair costs and are therefore especially beneficial. Moreover, research in this direction has already produced promising results. For example, Le Goues et al. [4] reported that their approach was able to automatically fix 55 out of 105 bugs. However, the research community has limited knowledge on the nature of bug fixes [7] and still does not have general consensus on which kinds of software bugs are most common [8].

This paper presents an in-depth investigation on the bug fixing commits of Java programs, taken from several million human-made bug fixes from GitHub (the world's largest collection of open-source software, with more than 23 million public repositories). This software repository contains an enormous collection of software and information about software [1]. We used a domain-specific programming language called *Boa* [1] to analyze ultra-large-scale data efficiently. In particular, we analyzed the characteristics of those bug fixing commits from different perspectives in order to answer the four research questions below:

E.C. Campos (✉) • M.A. Maia
Faculty of Computing, Federal University of Uberlândia, Uberlândia, Brazil
e-mail: eduardocunha11@gmail.com; marcelo.maia@ufu.br

© Springer International Publishing AG 2018
S. Latifi (ed.), *Information Technology – New Generations*, Advances in Intelligent Systems and Computing 558, DOI 10.1007/978-3-319-54978-1_68

RQ₁: *Considering bug fixing commits associated with only one file, which file types are usually changed to fix a bug?*

RQ₂: *Which kinds of statements appear more frequently in bug fixing commits?*

RQ₃: *What is the prevalence of the 5 most common bug fix patterns identified in the work of Pan et al. in the analyzed bug fixing commits?*

RQ₄: *How many distinct developers committed these bug fixing commits? And how many committers are there for each analyzed Java project?*

To sum up, our contributions are as follows:

- We have shown that developers often forget to add `IF` preconditions in the code. One proof of this is that the bug fix pattern that most appeared in the analyzed bug fixing commits of *Boa* dataset was IF-APC (Addition of IF Precondition Check);

- Our findings suggest that mutation based program repair may need to consider multi-language programming and bugs in non-source files (e.g., configuration files). Our results confirm the findings obtained by Zhong and Su [11] (please see the Findings 1, 2, 13, and 14 for more details). This is relevant because current automatic repair approaches have been evaluated on only source files belonging to a limited number of programming languages (such as C and Java).

the nature of bug fixes and automatic program repair. Furthermore, the empirical study focuses on only one aspect of automatic program repair, namely the search space of fixing bugs. They mined repair models from manual fixes, and the mined repair models improve random search. Our study provides findings to better understand and improve these approaches. For example, we notice that many bugs reside in source files of different programming languages or in non-source files (e.g., configuration files). Xuan et al. [10] proposed *Nopol*, an approach to automatic repair of buggy conditional statements (e.g., `if-then-else` statements). In our dataset, the bug fix pattern that appears more frequently is IF-APC (*Addition of IF Precondition Check*), totaling 29.2019% of the analyzed bug fixing commits. Kim et al. [2] proposed an approach called PAR to fix bugs in Java code automatically. PAR is based on repair templates: each of the PAR's ten repair templates represents a common way to fix a common kind of bug. Soto et al. [9] conducted a large-scale study of bug fixing commits in Java projects. Their findings provide useful insights for automatic program repair tools in Java. They created *Boa* programs to detect the PAR's bug fix patterns [2] and provided an informative approximation of their prevalence in the *Boa* dataset. We used the same dataset in our study but we created *Boa* programs to detect the five most common bug fix patterns identified in the work of Pan et al. [8]. Moreover, we do not limit our study to bug fix patterns. We also investigated other aspects related to human-made bug fixes such as the kinds of statements that appear more frequently in bug fixing commits and the kinds of files that are usually changed to fix a bug.

68.2 Related Work

Zhong and Su [11] designed and developed *BugStat*, a tool that extracts and analyzes bug fixes. They conducted an empirical study on more than 9,000 real-world bug fixes from six popular Java projects. Their results provide useful guidance and insights for improving the state-of-the-art of automatic program repair. We study a much larger dataset [1] with 101,471 Java projects. Moreover, we designed *Boa* programs that automatically detect the five most common bug fix patterns identified in the work of Pan et al. [8]. Martinez and Monperrus [6] analyzed the links between

68.3 Dataset and Characteristics

In this paper, we use the *September 2015/GitHub* dataset offered by *Boa* [1], including 554,864 Java projects with 23,226,512 commits. *Boa* identifies 4,590,405 as bug fixing commits distributed among 101,471 Java projects (18.2875%). In other words, 81.7125% of Java projects present in this dataset do not have any bug fixing commit. In this paper, we focus our analysis on these 101,471 Java projects because our goal is to study bug fixes and patterns. Figure 68.1 shows a query written in *Boa* language that

```
1   counts: output sum of int;
2   p: Project = input;
3
4   exists (i: int; match(`^java$`, lowercase(p.programming_languages[i])))
5   foreach (j: int; p.code_repositories[j].kind == RepositoryKind.GIT)
6     foreach (k: int; isfixingrevision(p.code_repositories[j].revisions[k].log))
7       counts << 1;
```

Fig. 68.1 Querying number of bug-fixing commits in Java GitHub projects using *Boa* language

returns the number of bug fixing commits in Java GitHub projects. The built-in function `isfixingrevision` (line 6) uses a list of regular expressions to match against the revision's log (i.e., commit's log message). If there is a match, then the function returns true indicating the log most likely was for a commit fixing a bug.

Figure 68.2 shows the distribution of bug fixing commits among 101,471 Java projects. As we can see in this figure, 81% of these projects have 1 to 15 bug fixing commits. Only 9% of these projects have 51 or more bug fixing commits. This pie chart shows that is not common to see open-source Java projects hosted on GitHub with a large number of bug fixing commits (e.g., more than 50 bug fixing commits).

Programming Language: As returned by *Boa*, the major language of a project is the one with the highest percentage of source code, considering the files in the project. Figure 68.3 shows the distribution of the analyzed projects per number of programming languages. As we can see, 56,414 out of 101,471 Java projects (i.e., 55.5961%) use only one programming language (i.e., Java). However, 10,837 out of 101,471 Java projects (i.e., 10.6798%) use five or more programming languages.

Kinds of changed files: Table 68.1 shows the kinds and descriptions of changed files present in the *Boa* dataset and the number of changed files per file kind. We consider a file *changed* if it is new, modified, or deleted in a commit. In total, 52,052,571 files were changed.

68.4 Results

In this section, we used the described dataset to answer the four research questions listed in the paper's introduction.

RQ$_1$: *Considering bug fixing commits associated with only one file, which file types are usually changed to fix a bug?*

Figure 68.4 shows the number of 1-file fixing commits per File Kind. As shown in Fig. 68.4, the 2 kinds of changed files that appear most frequently in the 1-file fixing commits are: SOURCE_JAVA_JLS3 and UNKNOWN. The number of 1-file fixing commits related to these 2 kinds of changed files are respectively, 776,834 and 857,738. Text and binary files are changed least frequently. This is unsurprising, since such files are often documentation, and binaries should be changed rarely. XML files in Java projects usually represent build files or configuration files (the names of the most found configuration files end with "xml" or "properties" [11]); 7.7363% of analyzed 1-file fixing commits are related to changes in XML files. As these bugs are not related to source files, they could not be fixed by current automatic program

Distribution of Fixing commits among Java projects

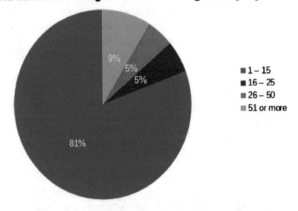

Fig. 68.2 Distribution of bug fixing commits among Java GitHub projects

Fig. 68.3 Number of programming languages in each Java project using GIT

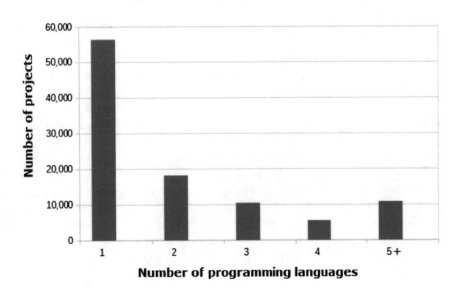

Table 68.1 Kinds of changed files present in the *Boa* dataset (JLS: Java Language Specification)

File Kind	Total	Description
SOURCE_JAVA_JLS4	83,798	The file represents a Java source file that parsed without error as JLS4
TEXT	541,023	The file represents a text file
BINARY	752,945	The file represents a binary file
SOURCE_JAVA_ERROR	2,073,558	The file represents a Java source file that had a parse error
SOURCE_JAVA_JLS2	2,607,413	The file represents a Java source file that parsed without error as JLS2
XML	6,818,299	The file represents an XML file
SOURCE_JAVA_JLS3	15,748,967	The file represents a Java source file that parsed without error as JLS3
UNKNOWN	23,426,568	The file's type was unknown

Fig. 68.4 Number of 1-file fixing commits per File Kind

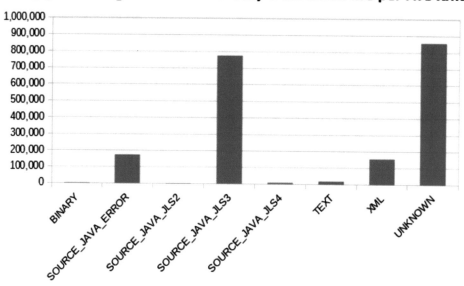

repair techniques [11]. Rather more surprising is how frequently UNKNOWN files are changed. We deepen our analysis in these committed UNKNOWN files and found that they are related to other programming languages like: C++, C, PHP, JavaScript, CoffeeScript, Erlang, Groovy, Scala, Python, Emacs Lisp, etc. Although the analyzed projects were mainly written in Java, 45,057 out of 101,471 Java projects (i.e., 44.4038%) use 2 or more programming languages (see Fig. 68.3 for additional details). This results showed that 42.3352% of fixing commits that have changed 1 file are related to changes in non-Java source files.

Summary of RQ1. We notice that many bugs reside in non-Java source files (e.g., source files of different programming languages like Scala, Groovy, PHP, etc.) or non-source files (e.g., XML files). Many implementations of research techniques that automatically repair software bugs target programs written in C language (e.g., Prophet [5], GenProg [4]). Thus existing approach may be insufficient in fixing certain

bugs. However, it is desirable to understand where such bugs reside, so we could investigate their nature and explore corresponding repair approaches.

RQ2: *Which kinds of statements appear more frequently in bug fixing commits?*

To answer this question, we use *Boa* to compute the number of fixing commits that added a particular statement kind in order to solve the corresponding bug. We investigate the following 13 statement kinds present in *Boa* Programming Guide[1]: ASSERT, BLOCK, BREAK, CATCH, CONTINUE, FOR, IF, RETURN, SYNCHRONIZED, THROW, TRY, SWITCH, and WHILE. The statement kind BLOCK is somewhat different because it was designed by *Boa* inventors to characterize a statement that contains a list of statements within it (e.g., the statements in the method body).

---

[1] http://boa.cs.iastate.edu/docs/dsl-types.php(verifiedon20/07/2016)

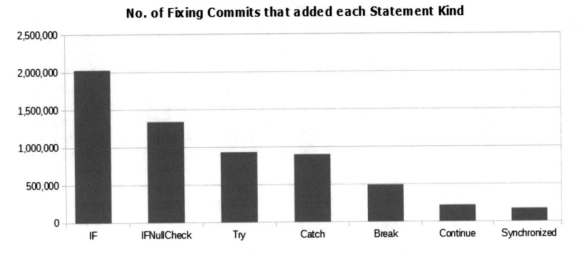

Fig. 68.5 Number of Fixing Commits that added each Statement Kind

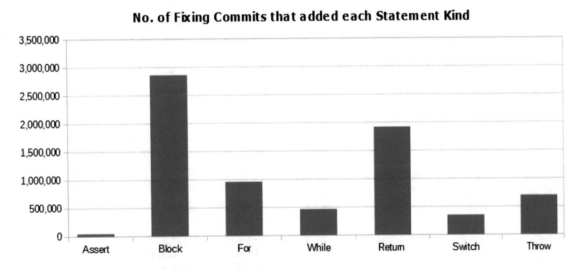

Fig. 68.6 Number of Fixing Commits that added each Statement Kind

Concerning the IF Statement, we investigate how many fixing commits added null checks. The IFNullCheck statement is an IF statement where the boolean condition is of the form: null == expr OR expr == null OR null != expr OR expr != null. Basically, we build a query written in *Boa* language that counts how many null checks were previously in the file (previous version of the file, if exists) and how many null checks are currently in the file (actual version of the file). If there are more null checks than previously, the bug fixing commit is counted. We performed this algorithm for all changed files and fixing commits of our dataset (i.e., 52,052,571 and 4,590,405, respectively) and for all 13 statement kinds aforementioned.

Summary of RQ₂. This research question is very important to investigate the nature of bug fixes in terms of statement kind that is added to fix a particular bug. Moreover, we can identify the prevalence of some statement kinds with respect to others. For instance, Figs. 68.5 and 68.6 show the results we obtained. Figure 68.5 shows the prevalence of IF and IFNullCheck statements with respect to the others (more specifically, 1,340,488 fixing commits added IF null checks). This number corresponds to 29% of all analyzed fixing commits. Moreover, Fig. 68.6 shows the prevalence of BLOCK and RETURN statements with respect to other statement kinds like iteration commands (e.g., WHILE, FOR).

RQ₃: *What is the prevalence of the 5 most common bug fix patterns identified in the work of Pan et al. in the analyzed bug fixing commits?*

Pan et al. [8] identified 27 bug fix patterns through manual inspection of the bug fix change history of seven open-source Java projects. They found that the most common categories of bug fix patterns are Method Call and IF-related. Moreover, the most common individual patterns are: MC-DAP (Method Call with Different Actual Parameter Values), IF-CC (Change in IF conditional), and AS-CE (Change of Assignment Expression). Below we detail each one of these five bug fix patterns.

(1) **Change of IF Condition Expression (IF-CC):** The bug fix changes the condition expression of an `IF` condition [8]. Example:

```
- if (listBox.getSelectedIndex() == 0)
+ if (listBox.getSelectedIndex() > 0)
```

(2) **Method Call with different actual parameter values (MC-DAP):** The bug fix changes the expression passed into one or more parameters of a method call [8]. Example:

```
- String.getBytes("UTF-8");
+ String.getBytes("ISO-8859-1");
```

(3) **Method Call with different number of parameters or different types of parameters (MC-DNP):** The bug fix changes a method call by using different number of parameters, or different parameter types. This may be caused by a change of method interface, or use of an overloaded method [8]. Example:

```
- getSolrQuery(f.getFilter());
+ getSolrQuery(f.getFilter(),analyzer);
```

(4) **Change of Assignment Expression (AS-CE):** The bug fix changes the expression on the right hand side of an assignment statement. The expression on the left-hand side is the same in both the bug and fix versions [8]. Example:

```
- names[0] = person.getName();
+ names[0] = employees[0].getName();
```

(5) **Addition of IF Precondition Check (IF-APC):** This bug fix adds an `IF` predicate to ensure a precondition is met before an object is accessed or an operation is performed. Without the precondition check, there may be a `NullPointerException` error or an invalid operation execution caused by the buggy code [8]. Example:

```
- repo.getFileContent(path);
+ if (repo != null && path != null)
+     repo.getFileContent(path);
```

Pan et al. [8] also discovered that there is a similarity of bug fix patterns across projects. This indicates that developers may have trouble with individual code situations, and that frequencies of bug introduction are independent of application domain [8]. However, the main drawback of the bug fix patterns approach stems from its automation. We therefore automatically detect these five bug fix patterns, estimating their prevalence in the dataset presented in Sect. 68.3.

We use *Boa* language to detect common bug fix patterns in the historical information of the projects. *Boa* provides domain-specific language features for mining source code [1]. *Boa's* capabilities are powerful, but limited in the precision it enables in detection of the aforementioned bug fix patterns. For example, it cannot directly `diff` two files. Rather than finding exact counts of bug fix patterns, we approximate by processing pre- and post-fix files separately. Fortunately, these five patterns can be detected by *Boa*, as we describe below. For each pattern, we create a query written in *Boa* language. In the following paragraphs, we describe in natural language each of the five algorithms designed to detect the five bug fix patterns aforementioned.

(1) **How many bug fixing commits change one or more IF Condition Expressions (IF-CC)?** To answer this question and to detect this pattern, for both pre- and post-fix versions of a buggy file, we count how many `IF` conditions and expressions of these `IF` conditions appear. Then, if the number of `IF` conditions is the same between these two versions of the file (to ensure that it is a modification and not an addition or deletion), we check whether the number of expressions of these `IF` conditions is different between these two versions of the file. If it's true, the pattern was detected and the bug fixing commit is recorded. For more information of what kind of expressions we consider, see the section `ExpressionKind` of this page.[2] *We found that 196,283 out of 4,590,405 (4.2759%) bug fixing commits change one or more IFcondition expressions.*

(2) **How many bug fixing commits change the parameter values of the method calls (MC-DAP)?** To answer this question and to detect this pattern, for both pre- and post-fix versions of a buggy file, we count how many method calls appear and we also built 2 strings (i.e., one string for the pre-version and another string for the post-fix version of these file) containing the parameter values (i.e., string literals) of all method calls. Then, if the

---

[2] http://boa.cs.iastate.edu/docs/dsl-types.php

number of method calls is the same between these two versions of the file, we compare if the two strings are different. If it's true, the pattern was detected and the bug fixing commit is recorded. *We found that 290,818 out of 4,590,405 (6.3353%) bug fixing commits change the parameter values of the method calls.*

(3) **How many bug fixing commits change the number or type of parameters of the method calls (MC-DNP)?** To answer this question and to detect this pattern, for both pre- and post-fix versions of a buggy file, we count how many method calls and method parameters appear. Then, if the number of method calls is the same between these two versions of the file, we check whether the number of parameters are different. If it's true, the pattern was detected and the bug fixing commit is recorded. For this pattern, due *Boa* limitations, it was not possible to identify the types of method parameters present in the method calls. *We found that 192,375 out of 4,590,405 (4.1908%) bug fixing commits change the number of parameters of the method calls.*

(4) **How many bug fixing commits change one or more assignment expressions (AS-CE)?** To answer this question and to detect this pattern, for both pre- and post-fix versions of a buggy file, we count how many assignment expressions and expressions of these assignments appear. Then, if the number of assignment expressions is the same (to ensure that it is a modification and not an addition or deletion), we check whether the number of expressions between these two versions of the file are different. If it's true, the pattern was detected and the bug fixing commit is recorded. For more information of what kind of expressions we consider, see the section `ExpressionKind` of this page.[3] *We found that 511,299 out of 4,590,405 (11.1384%) bug fixing commits change one or more assignment expressions.*

(5) **How many bug fixing commits added a null check precondition (IF-APC)?** To answer this question and detect this pattern, for both pre- and post-fix versions of a buggy file, we count how many null checks appear. Then, if the number of null checks in the current version of the file is greater than in the previous version of these file, the pattern was detected and the bug fixing commit is recorded. The `IFNullCheck` statement is an `IF` statement where the boolean condition is of the form: `null == expr` OR `expr == null` OR `null != expr` OR `expr != null`. *We found that 1,340,488 out of 4,590,405 (29.2019%) bug fixing commits added a null check precondition.*

Summary of RQ$_3$. Figure 68.7 shows a bar chart with the number of bug fixing commits distributed among the five studied bug fix patterns. As shown in this figure, the bug fix pattern that appears more frequently is IF-APC (29.2019% of the analyzed bug fixing commits). Observe that several bug fixing commits match this bug pattern in order to avoid *NullPointerException* errors.

RQ$_4$: How many distinct developers committed these bug fixing commits? And how many committers are there for each analyzed Java project?

To answer this research question, we build a *Boa* program that retrieves the username of the person who committed the revision. As the username is unique per person, it was possible to identify how many bug fixing commits each person committed and how many distinct developers committed these bug fixing commits. Figure 68.8 shows that a large number of open-source projects have only a single committer (33.6933%). Generally, open-source projects are small and have very few committers and thus problems affecting large development teams may not show when analyzing open-source software.

Summary of RQ$_4$: We found that only 130,488 distinct developers distributed among 101,471 Java projects committed 4,590,405 bug fixing commits. This shows how much the bug fixing commits are concentrated in a few people on the project and the need to better share knowledge in software teams for future software maintenance tasks.

68.5 Threats to Validity

Correctness of Boa programs. The correctness of our analysis depends on both our *Boa* programs and its Domain-Specific Types.[4] For example, we rely on *Boa* to identify bug fixing commits. However, precisely accomplishing this is an open problem. To mitigate the risk of implementation errors, we released our *Boa* programs.[5] Because *Boa* does not provide an easy mechanism to identify precise, statement-level diffs between commits, our template matching and analysis of code changes (by counting each statement kind or expression kind) only provide estimates of behavior. We consider our results as informative approximations.

---

[3] http://boa.cs.iastate.edu/docs/dsl-types.php

[4] http://boa.cs.iastate.edu/docs/dsl-types.php#Expression

[5] https://github.com/eduardocunha11/BoaPrograms (verified on 15/09/2016)

Fig. 68.7 Number of bug fixing commits per bug fix pattern

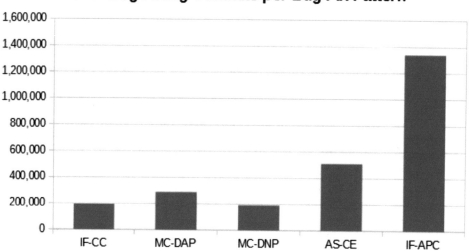

Fig. 68.8 Number of committers in each Java project using GIT

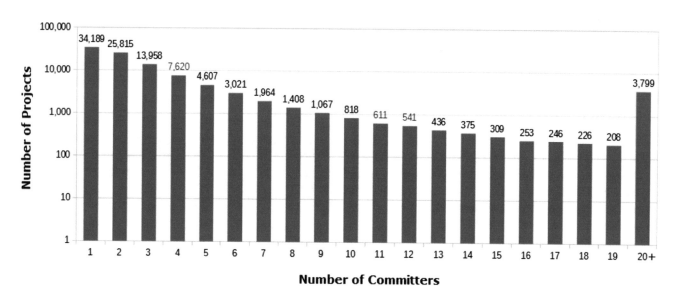

Systems are all open-source. All systems examined in this paper are developed as open-source. Hence they might not be representative of closed source development since different development processes could lead to different bug fix patterns. Despite being open-source, several of the analyzed projects have substantial industrial participation.

68.6 Conclusion

This paper explored the underlying patterns in bug fixes mined from software project change histories. We rely on *Boa* to identify bug fixing commits and to detect the five most common bug fixing patterns identified by Pan et al. [8]. We have conducted a large-scale empirical study with

more than 4 million bug fixing commits distributed among 101,471 Java projects hosted on GitHub to answer four research questions about bug fixes.

The findings of our study provide useful insights for automatic repair tools in Java. For example, future work may explore how to locate bugs in non-source files or source files of different programming languages present in a Java project and how to fix them with advanced techniques. Another example of follow-up work would be to propose an approach to automatic repair of assignment expressions (AS-CE bug fix pattern). Our work also confirms the results of a recent work [6] that showed that IF conditions are among the most error-prone program elements in Java programs (IF-APC bug fix pattern).

Overall, our findings motivate additional study of repair in Java and future research may leverage such knowledge to fix more bugs.

Acknowledgements This work was partially supported by CAPES, CNPQ and FAPEMIG.

References

1. Dyer, R., Nguyen, H. A., Rajan, H., & Nguyen, T. N. (2015). Boa: Ultra-large-scale software repository and source-code mining. *ACM Transactions on Software Engineering and Methodology, 25* (1), 7:1–7:34.
2. Kim, D., Nam, J., Song, J., & Kim, S. (2013). Automatic patch generation learned from human-written patches. In *Proceedings of the ICSE'13* (pp. 802–811). Piscataway: IEEE Press.
3. Kim, S., & Whitehead, E. J, Jr. (2006). How long did it take to fix bugs? In *Proceedings of MSR'06* (pp. 173–174). New York: ACM.
4. Le Goues, C., Dewey-Vogt, M., Forrest, S., & Weimer, W. (2012). A systematic study of automated program repair: Fixing 55 out of 105 bugs for $8 each. In *Proceedings of ICSE'12* (pp. 3–13). Piscataway: IEEE Press.
5. Long, F., & Rinard, M. (2016). Automatic patch generation by learning correct code. In *POPL'16* (pp. 298–312). New York: ACM.
6. Martinez, M., & Monperrus, M. (2015). Mining software repair models for reasoning on the search space of automated program fixing. *Empirical Software Engineering, 20*(1), 176–205.
7. Monperrus, M. (2014). A critical review of "automatic patch generation learned from human-written patches": Essay on the problem statement and the evaluation of automatic software repair. In *Proceedings of the ICSE'2014* (pp. 234–242). New York: ACM.
8. Pan, K., Kim, S., & Whitehead, E. J, Jr. (2009). Toward an understanding of bug fix patterns. *Empirical Software Engineering, 14* (3), 286–315.
9. Soto, M., Thung, F., Wong, C. -P. Le Goues, C., & Lo, D. (2016). A deeper look into bug fixes: Patterns, replacements, deletions, and additions. In *Proceedings of MSR'16* (pp. 512–515). ACM.
10. Xuan, J., Martinez, M., DeMarco, F., Clément, M., Lamelas, S., Durieux, T., Le Berre, D., & Monperrus, M. (2016). Nopol: Automatic repair of conditional statement bugs in Java programs. *IEEE Transactions on Software Engineering.*
11. Zhong, H., & Su, Z. (2015). An empirical study on real bug fixes. In *Proceedings of ICSE'15* (pp. 913–923). Piscataway: IEEE Press.

An Empirical Study of Control Flow Graphs for Unit Testing

Weifeng Xu, Omar El Ariss, and Yunkai Liu

Abstract

This paper conducts an empirical study of a Control Flow Graphs (CFG) visualizer from which various test coverages can be exercised directly. First, we demonstrate how control structures are extracted from bytecode and compound conditions are decomposed into simple multi-level conditions in Java bytecode. Then, we visualize the decomposed compound conditions in CFG. The layout of a CFG is calculated by extending a force-directed drawing algorithm. Each control node of CFG represents a simple condition. The empirical study shows that (1) the tool successfully decomposes a compound condition into simple multi-level conditions and (2) the extended force-based layout algorithm produces the best layout for visualizing CFG.

Keywords

Bytecode • CFG • Force-directed drawing algorithm • Compound condition • Test coverages

69.1 Introduction

Control Flow Graph (CFG) is commonly used by software practitioners for white-box testing [1] to examine if their unit code works as expected. White-box testing relies on control structures, i.e., conditions, in CFG to achieve test coverages, including the decision, condition, and multi-condition coverages. Several graph visualization tools can be used for visualizing CFG, including Graphviz [2] and Eclipse

plug-in [3]. However, the existing tools cannot be directly used for extracting test cases: (1) Graphviz essentially is a graph auto-layout problem. It arranges the positions of each vertex and edge of a graph automatically. It relies on different tools to extract each vertex and edge of a program before drawing the CFG, (2) Eclipse plug-in can generate CFG from source code. However, it treats all conditions, both simple and compound, as a single node. Additional steps are therefore needed to identify simple conditions in compound conditions to cover condition and multi-condition coverages. The Eclipse tool will not work if the source code is not available.

In our previous work [4], we have briefly presented an automated tool to construct and visualize CFG from Java bytecode to demonstrate control structures of Java code. However, the previous paper does not specifically address the challenge of how different test coverages can be directly exercised, mainly, condition and multi-condition coverages. The paper solves two related issues of such a tool: (1) how can control structures be recognized by the tool, especially compound conditions in source code? For example, the

W. Xu (✉)
Department of Computer Science, Bowie State University, Bowie, MD, USA
e-mail: wxu@bowiestat.edu

O. El Ariss
Department of Computer Science, Penn State University, State College, PA, USA
e-mail: oue1@psu.edu

Y. Liu
Department of Computer & Information Science, Gannon University, Erie, PA, USA
e-mail: liu006@gannon.edu

© Springer International Publishing AG 2018
S. Latifi (ed.), *Information Technology – New Generations*, Advances in Intelligent
Systems and Computing 558, DOI 10.1007/978-3-319-54978-1_69

compound condition *(a > b) && (b > c)* contains two simple conditions *a > b* and *b > c*. (2) how nicer does the force-directed approach proposed in our previous work comparing other visualization techniques such as non-force-based algorithms. In this paper, we first demonstrate how predicates (compound conditions) in source code can be decomposed into several simple conditions in bytecode. Then, we discuss the force-directed layout algorithm [5] in details to draw CFG nicer than the non-forced-based drawing algorithms do.

We organize the rest of this paper as follows: Sect. 69.2 shows control structures in bytecode with the old running example in [4]. Sections 69.3 and 69.4 describe how to construct and visualize a bytecode CFG. Section 69.5 shows empirical study results. Sections 69.6 and 69.7 review the related work and conclude the paper.

69.2 Control Structures in Bytecode

69.2.1 Running Example

We use the Zodiac problem [6] as a running example to illustrate how our proposed tool is developed to visualize CFG with decomposed simple conditions for white-box testing. The Zodiac problem finds the zodiac sign, e.g., *Invalid, Aries, Taurus, Gemini* and *Cancer*, for a given date. Each sign type is represented by an integer 0 to 5, respectively. Note that only five types of zodiac signs are shown for the demonstration purpose. The source code of Zodiac program is shown below. All predicates of the nested if-statements in the Zodiac problem source code are compound conditions.

```
int zodiac(int month, int day, int year) {
    int result = 0;
    if (month == 3 && day >= 21 || month == 4 && day <= 19) {
        result = 1;
    } else if (month == 4 || month == 5 && day <= 20) {
        result = 2;
    } else if (month == 5 || month == 6 && day <= 20) {
        result = 3;
    } else {  result = 4; }
    return result;
}
```

The Java source code of the Zodiac problem is compiled into bytecode to be executed. Table 69.1 shows the bytecode of the Zodiac problem. Java bytecode is a stack-based language, which pops data (operand) from the top of the stack and pushes data back to the stack [7]. The stack is commonly referred to as an operand stack For example; the bytecode instruction *if_icmpne n* pops the top two integers off the stack and then to compare them. If the two integers are not

Table 69.1 Zodiac bytecode instructions

------------Block 1	------------Block 10	------------Block 20
[1] iconst_0	[17] goto 44	[32] if_icmpeq 39
[2] istore 4	------------Block 11	------------Block 21
[3] iload 1	[18] iload 1	[33] iload 1
[4] iconst_3	[19] iconst_4	[34] bipush 6
------------Block 2	------------Block 12	------------Block 22
[5] if_icmpne 9	[20] if_icmpeq 27	[35] if_icmpne 42
------------Block 3	------------Block 13	------------Block 23
[6] iload 2	[21] iload 1	[36] iload 2
[7] bipush 21	[22] iconst_5	[37] bipush 20
------------Block 4	------------Block 14	------------Block 24
[8] if_icmpge 15	[23] if_icmpne 30	[38] if_icmpgt 42
------------Block 5	------------	------------Block 25
[9] iload 1	Block 15	[39] iconst_3
[10] iconst_4	[24] iload 2	[40] istore 4
------------Block 6	[25] bipush 20	------------Block 26
[11] if_icmpne 18	------------Block 16	[41] goto 44
------------Block 7	[26] if_icmpgt 30	------------Block 27
[12] iload 2	------------Block 17	[42] iconst_4
[13] bipush 19	[27] iconst_2	[43] istore 4
------------Block 8	[28] istore 4	------------Block 28
[14] if_icmpgt 18	------------Block 18	[44] iload 4
------------Block 9	[29] goto 44	[45] ireturn
[15] iconst_1	------------Block 19	
[16] istore 4	[30] iload 1	
	[31] iconst_5	

equal, execution branches to the instruction line *n*. The two integer values are pre-loaded from a local variable table using *iconst* or *iload* instructions. Table 69.1 shows that compound conditions in Java source code are represented by simple multi-level predicates in bytecode. For example, the compound condition *if (month == 3 && day >= 21)* in the program source code is to test if a Zodiac sign is *Aries* for a given date. The compound condition is decomposed into two simple predicates [5] *if_icmpne 9* and [9] *if_icmpge 15*.

69.2.2 Extracting Control Structures

Bytecode statement instructions include the following four types: (1) load and store (e.g. *aload, istore*), (2) Arithmetic and logic (e.g. *iadd, fcmpl*), (3) Type conversion (e.g. *i2b, d2i*), (3) Object creation and manipulation (*new, putfield*), and (4) Operand stack management (e.g. *swap, dup2*). Bytecode control instructions include (1) Control transfer (e.g. *ifeq, goto*) and (2) Method invocation and return (e.g. *invokespecial, ireturn*). In order to extract control structures for a given method, the instructions of a method need to be divided into program blocks where the control flow may change.

Definition 1: Block Markers A Java method *M* consists of a sequence of bytecode instructions represented by numbered IDs from 1 to *n*. A block marker points to the first instruction of a sequence of instructions (i.e., a block or segment). Block markers of M are represented by $L = \{l \mid l$ is the start of an instruction where a control flow may change, $1 < n\}$. Block markers are often organized in an ascending order. Formally, a block l is defined as:

$$
\begin{cases}
i \text{ and } i+1 & \text{if } i \text{ is a control instruction } \textbf{\textit{if}} XXX, \textbf{\textit{invoke}} XXX & (a) \\
s & \text{if } i \text{ is a control instruction } \textbf{\textit{if}} XXX \text{ s or } \textbf{\textit{goto}} \text{ s} & (b) \\
i & \text{if } i = 1 & (c) \\
i - 1 & \text{if } i \text{ is a control instruction } \textbf{\textit{Xreturn}} & (d)
\end{cases}
$$

Where ifXXX and invokeXXX represent instructions that start with words "if" and "invoke". They are referred to as if and invoke statements in the rest of paper, respectively. Note that formulas (c) and (d) are the starting and end markers. For example, for a given instruction [5] (i.e., if_icmpne 9), the next discovered block markers are $\{5, 6, 9\}$ based on formulas (a) and (b). By applying the definition of block markers, the Zodiac program bytecode instructions are divided into 28 segments as shown in Table 69.1. The dashed lines are the visual representations of block markers which divide instructions into instruction segments. The IDs of instructions under the dashed lines are block markers.

Definition 2: Marker Clear Sequences A marker-clear sequence is a segment of bytecode instructions without containing any other markers except the starting instruction. Marker-clear sequences of the method M is a set, which can be defined as: $S = \{(i, j) \mid l \notin (i, j],$ where i, j are instructions, $l \in L\}$. For example, s $(1, 4)$ is a marker-clear sequence as there are no block markers between instructions 1 and 4 besides instruction 1. S $(1, 5)$ is not a marker-clear sequence as instructions 5 is a marker in addition to instruction 1. There is a special case when a marker-clear sequence only contains one statement, e.g., instruction 5.

Definition 3: Program Blocks A program block b is the longest marker-clear sequence for a given block marker. Program blocks of the method M are defined as a set: $B = \{(i, j) \mid \exists s (i, j) \in S, i \in L, s (i, j)$ is the longest marker-clear sequence for the given $i\}$. For example, s $(1, 3)$ is a marker-clear sequence. However, it is not a program block since there exists another marker-clear s $(1, 4)$ such that $4 > 3$.

Definition 4: Control and statement Blocks A control block o is a special program block that contains one control instruction. Control blocks of the method M are defined as a set: $O = \{(i, j) \mid \exists b (i, j), b$ contains a control instruction$\}$. For example, blocks 2 and 28 are control blocks. It is worth noting that (1) there are two types of blocks in a CFG, i.e., statement blocks and control blocks. Program blocks are

statement blocks if they are not control blocks, such as blocks 1 and 3. (2) *if* or *invoke* control statement is the only instruction in its control block. For example, block 2 only contains one control statement [5].

Definition 5: Bytecode CFG A bytecode CFG of a Java method M is a 5-tuple G (V, E, s, t, e), where G' (V, E) is a simple digraph. The vertex set $V = Vs \cup Vc$ where Vs and Vc represent statement and control blocks in M, respectively. The edge set E represents the flow of controls between statement and control blocks in M, i.e., $E \subseteq \{Vs \rightarrow Vc \cup Vc \, ^d{}_d Vs\}$ where d is a predicate decision with either True or False value. s is a start vertex that represents the entry point of M, and t is a termination vertex that represents the exit point of M. e contains one edge $e1 = s \rightarrow V$ and a set of edges $e2 \subseteq \{v \rightarrow t\}$. It indicates that a program only has one incoming edge and may have a set of e2 if it has multiple return statements.

69.3 Bytecode CFG Construction

Constructing a CFG, G (V, E, s, t, e) is to identify vertices V and identify edges E for a given method M.

69.3.1 Identifying the Vertices

There exists a one-to-one mapping relationship between vertices and program blocks. The type of the vertices is based on whether a marker is a control statement in a program block or not. For example, blocks 2 and 18 are control vertices because their marker are either *if* or *goto* control statements. A vertex v is defined as:

$$
v = \begin{cases}
vc & \text{if } \exists o (i,j) \in O & (e) \\
vs & else & (f)
\end{cases}
$$

Note that control statements, i.e., *if* and *goto*, are associated with destination markers $l \in L$ to where the control should be redirected. To determine the destination, a mapping function is needed to find the block for the given marker.

Definition 7: Block Search Function A block search function is a function f (l) that determines v for a given marker $l \in L$.

69.3.2 Identifying Edges

At runtime, method blocks are executed in a sequential order unless a block is a control block. If a control block is an unconditional control block, such as *goto l*, then it redirects the execution flow to the new block starting. In case the control block contains a condition control block, e.g., *if l* statement, it redirects the execution flow to the new block starting with *l* and the next separating marker in L as well. The latter indicates that the conditional statement can be evaluated either as true or false. The algorithm is described as follows.

```
Algorithm: Identifying Edges E
Inputs: L: a set of markers
Outputs: E: a set of edges
Procedure
  E ←{}
  For each l ∈ L do
  v1 ← f(l)
  If l is a control statement with a new target l'∈ L
       v2 ← f(l')
       E ← E∪e(v1, v2)
  End if
  If l is a conditional control block and l is the last
  marker
       l' ← next l
       v2 ← f(l')
  E ← E∪e(v1, v2)
  End if
  End for
  return E
End procedure
```

Three steps are needed to construct a complete CFG. It includes: (1) adding the source *s* and sink *t* vertices to the CFG, (2) adding an edge (s, f(l)) from the source vertex to the vertex containing the first marker, and (3) add edges *e* from the return statements to the sink. *e* represents an edge which defined as {(f(l), t) | where l is a return statement}. The CFG of Zodiac program is shown in Fig. 69.1. Vertices 0 and 29 represent the source and sink, respectively. e = {(0, 1) (28, 29)}, which represents the edges that connect to the source and sink vertices, respectively.

69.3.3 Compound Conditions in CFG

The compound condition *if (month == 3 && day >= 21)* and its decomposed simple predicates *if_icmpne 9* and *if_icmpge 15* are visualized in the CFG of Zodiac bytecode in Fig. 69.1. Note that (1) the two conditional blocks, i.e., blocks 2 and 4, containing the two predicates must be in the same path to be equivalent to the compound condition with the && operation, e.g., $p = 0 \rightarrow 1 \rightarrow 2 \rightarrow 3 \rightarrow 4 \rightarrow 9 \rightarrow 10 \rightarrow 28 \rightarrow 29$. (2) Each predicate used in bytecode is not necessarily equivalent to each simple condition used in the compound condition. For

example, the simple condition month == 3 often will be replaced with the bytecode predicates *if_icmpeq*, however, JVM may use *if_icmpne* instead due to the bytecode optimization.

To achieve test coverage for a given coverage criterion, such as decision, condition, and multi-condition coverages, we need to trace the desired outcomes for each predicate in the given path. A path containing the outcomes of predicates is called a tagged path. Specifically, a tagged path is a sequence of edges that have at least one tagged edge. A tagged edge is defined as v d >u. The variable *v* is the source block that represents a predicate in a statement, where *d* is a tagged value for *v*. This value represents the expected outcome of *v* (i.e., true or false) for reaching *u*. For example, for the component condition, *month == 3 && day >= 21*, with expected evaluation as True. The tagged path should be achieved as p: $0 \rightarrow 1 \rightarrow 2^F F 3 \rightarrow 4^T T 9 \rightarrow 10 \rightarrow 28 \rightarrow 29$, where F (false) and T (true) are the expected outcomes of the control instructions 2 and 4. Note that (1) the condition coverage is automatically achieved by traversing the bytecode of CFGs. Condition coverage requires each simple predicate to be evaluated to both true and false. It is not difficult to prove, as a compound condition is decomposed into simple predicates and each predicate has two possible outcomes. Table 69.2 shows that the condition coverage for compound condition *month == 3 && day > = 21* can be satisfied by three test cases, i.e., three paths. (2) Multiple condition coverage is not guaranteed. This criterion requires that all combinations of predicate results need to be tested. For example, considering the following code *if (a or b) and c then*, multiple condition coverage needs $2^3 = 8$ test cases. The coverage criterion is not guaranteed by visiting the bytecode CFG due to the "short-circuit" operation. That is, the second predicate will not be evaluated if the result can be deduced solely by evaluating the first operand. For example, if the predicate 2 is not evaluated as True, then predicate 4 will have no chance to be executed.

69.4 An Extended Force-Based CFG Visualizer

We have extended the force-based layout algorithm to visualize CFG that is generated from bytecode. The original force-based layout algorithm is used to calculate layouts of undirected graphs, and it has two drawbacks: (1) it does not distinguish between two special vertices, i.e., source and sink vertices, from other vertices. The force-based algorithm misplaces source and sink; (2) it does not distinguish between two special types of edges, i.e., loop and jump, from other edges. The loops and jumps represent *for/while* and *goto* statements. These special edges often overlap with

Fig. 69.1 The CFG of Zodiac
Bytecode

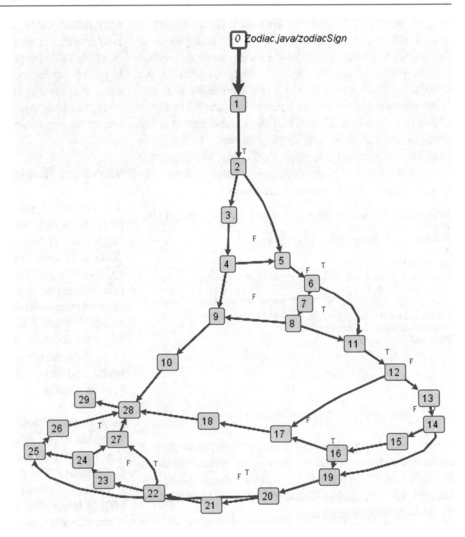

Table 69.2 Tagged paths that achieves condition coverage for predicates (Blocks) 2 and 4

Path ID	Tagged Paths	Covered Predicates
1	$0 \rightarrow 1 \rightarrow 2^F F 3 \rightarrow 4^T T 9 \rightarrow 10 \rightarrow 28 \rightarrow 29$	2 = False, 4 = True
2	$0 \rightarrow 1 \rightarrow 2^T T 5 \rightarrow 6^F F 7 \rightarrow 8^F F$ $9 \rightarrow 10 \rightarrow 28 \rightarrow 29$	2 = True
3	$0 \rightarrow 1 \rightarrow 2^F F 3 \rightarrow 4^F F 5 \rightarrow 6^F F 7 \rightarrow 8^F F$ $9 \rightarrow 10 \rightarrow 28 \rightarrow 29$	2 = False, 4 = False

other edges. The extended force-based algorithm introduces an additional force, i.e., a gravitational force, to arrange the positions of special vertices and edges in a given CFG.

69.4.1 Extended Force-Based CFG Visualization Algorithm

Force-directed algorithms are the most flexible and popular algorithms for calculating layouts of undirected graphs.

They calculate the layout of a graph using graph structure information only. For a given directed graph $G = \{V, E\}$, a force-directed algorithms model edges as springs and vertices as charged particles. Springs are attractive forces based on Hooke's law. Springs are used to attracting vertices to each other. Different from springs, charged particles represent repulsive forces based on Coulomb's law. Charged particles are used to separate all vertices. Force is represented in a vector, which includes magnitude and direction. A force-directed algorithm starts with assigning a random position for each vertex. Then, each vertex calculates the attractive and repulsive forces that were applied to them and moves to a new position. The calculation and movement activities repeat until the graph reaches equilibrium states. In equilibrium states for a given graph, edges tend to have uniform length because of the spring forces, and nodes that are not connected by an edge tend to be drawn further apart because of the electrical repulsion.

The extended force-directed algorithm introduces two additional vertices, source s and sinks t in bytecode CFG, $G = \{V, E, s, t, e\}$. A source vertex is a vertex with in-degree of zero, while a sink vertex is a vertex with out-degree of zero. Traditionally, all vertices of a CFG are arranged in the form of top-to-bottom where s and t are placed on the top and bottom positions, respectively. To rearrange s and t, we introduce a third force, named Earth Gravitational Force, to the original force-based algorithms. The gravity of Earth, which is denoted as $\vec{T}$, refers to the acceleration that the Earth impacts to objects on or near its surface. The Earth Gravitational Force is defined as $\vec{T}(v) = mg$, where m is the mass of the vertex and g is the gravitational content. The extended algorithm is described as follows:

```
Algorithm The extended force-based CFG visualization
Input: G = {V, E, s, t,e}
Output: new location of each v

Procedure
Place vertices of G in random locations
  Repeat M times
    Calculate the force F⃗(v) on each vertex
    Move the vertex based on force on vertex
    Draw graph on screen
End of Procedure
```

The force $\vec{F}(v)$ is defined as:

$$\vec{F}(v) = \sum_{(u,v)\in V\times V} \vec{H}_{uv} + \sum_{(u,v)\in E} \vec{C}_{uv} + \sum_{(v)\in E} \vec{T}_v \quad (69.1)$$

Where $\vec{H}_{uv}$ represents the attractive force between two connected vertices, u and v, which is calculated based on Hooke's law. $\vec{C}_{uv}$ represents the repulsive force among any vertices. This is calculated based on Coulomb's law. Finally, $\vec{T}_v$ is the gravitational force.

69.4.2 Positioning Source and Sink Vertices

Figure 69.2 shows the automated layout calculation with the original force-based algorithm with two forces applied to the CFG of Zodiac problem. Vertices in Fig. 69.2 (a) are assigned to random positions. Figure 69.2 (b) shows the

equilibrium states of the CFG. Figure 69.3 illustrates the layout with the new earth gravitational force. Figure 69.3 (a) (b) (c) (d) shows the CFG layouts at different iterations. Fig. 69.3 (d) indicates that the extended force-based algorithm can position the source and sink vertices correctly when reaching equilibrium state while Fig. 69.2 (b) does not succeed to do that.

69.4.3 Positioning Loops and Jump Edges

There are two types of special edges, loops and jumps (i.e., loop and if-else statements) in CFG. For example, v2 in Fig. 69.1 is a predicate node containing an if-else statement. Without an appropriate positioning algorithm, the edge (v2, v5) will be a straight line and maybe overlaps with (v2, v3), (v3, v4), and (v4, v5). We need to position the special edges to avoid edge overlaps. The idea of avoiding edge overlaps is to attach some invisible vertices to these edges as these edges often cross over several vertices. These invisible vertices will make curves when applying extended forced-based layout algorithm. There are three steps to attach invisible vertices to edges:

Step 1: Identifying dominator relationships. In a CFG, a vertex v dominates another vertex w, denoted as (w^v), if and only if every directed path from s to w in the graph contains v. The dominators of vertex w is defined as dom (w) = {v | v ^ w}. For example, dom (v5) = {v0, v1, v2}.

Step 2: Identifying special edges. The vertex $v_{shortest} = v \in$ dom (w) has the shortest path from itself to w, where $v_{shortest}$ is the starting node and w is the ending vertex, i.e., the special edge is defined as ($v_{shortest}$, w).

Step 3: Attaching invisible vertices to special edges. The number of the invisible vertices equals to the number of vertices from $v_{shortest}$ to w.

Fig. 69.2 The Layout of *Zodiac* CFG with Original force-based Algorithm. (**a**) Initial state. (**b**) Equilibrium state

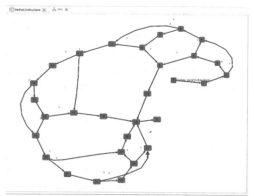
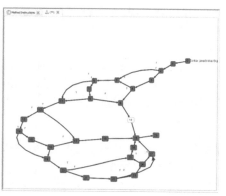

(a) Initial state   (b) Equilibrium state

Fig. 69.3 The Layout of *Zodiac* Program with the Extended Force-based Algorithm. (**a**) Initial state. (**b**) Iteration 10. (**c**) Iteration 20. (**d**) Equilibrium state

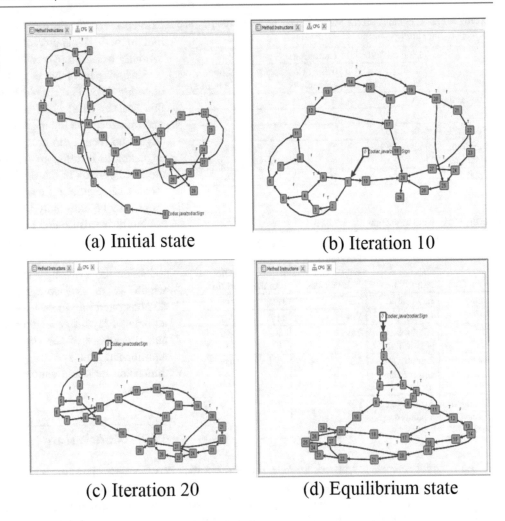

(a) Initial state

(b) Iteration 10

(c) Iteration 20

(d) Equilibrium state

69.5 Empirical Study

The empirical study aims at answering two following questions: **Q1**: How well does the bytecode approach handle compound conditions? **Q2**: To what extent does the extended force-based approach fits the purpose of visualizing CFG for unit testing? In addition to ZodiacSign, three programs are chosen to evaluate our approach, including Triangle program, the NextDate program [6], and the Vending Machine program [8]. The NextDate program computes the next date for a given date. The Vending Machine program is a classical job-interview question for software testing positions. The size of the four programs (in Java and bytecode) and the number of conditions are listed in Table 69.3. We selected these programs since they cover the different number of conditions, which can be categorized into three groups, small (8 and 10), medium (19), and large (41). To evaluate how well the tool interprets compound conditions automatically, we choose four programs as empirical study subjects. These programs contain (1) five simple conditions, including, ==, >, <, <=, and <=, and <>. (2) Various compound conditions containing and ∧, or ||, and ! (not) operations. Table 69.3 shows 36 Java compound conditions are present in the four programs. These compound conditions are successfully decomposed into 78 simple conditions.

To evaluate how the extended force-based approach fit the purpose of visualizing CFG, we compare our approach with other non-force-based algorithms and the original force-based algorithm. The focus of the comparison is to: (1) the ability to position source and sink vertices (2) and the prettiness of the generated graph. The prettiness of the graph is measured by the average number of crossings and the dimension ratio of CFG. The average number of crossings is the number of crossed edges among all edges in the four programs. The lower the number of crossed edges contained in a CFG, the better the layout algorithm is. The dimension ratio of CFG is the percentage of the height and width. We prefer the ratio to be closer to the golden ratio, i.e., 1.6. We set up the same initial distance of two vertices as one unit to make a fair comparison. Observations made from Table 69.4 are described as following:

Table 69.3 Program size, java compound conditions, and corresponding simple conditions

	Line of Code		Conditions	
	Java	Bytecode	# Java Compound Conditions	Simple Conditions
ZodiacSign	13	45	3	10
Triangle	22	27	3	8
Next Date	48	51	9	19
Vending Machine	82	68	21	41
Total			36	78

Table 69.4 CFG algorithm comparison

Approach	Algorithm	Position source and sink vertices	Prettiness	
			Average # crossings	Dimension ratio
Non-force-based	BSF-based	Yes	13%	0.5
	Loop	Yes	11%	3.1
	Hierarchical positioning	Yes	2%	2.9
Force-based	Attractive and repulsive forces	No	3%	1.4
Extended Force-based	Attractive , repulsive gravity	Yes	3%	1.4

1. The original force-based algorithm does not address the issue of positioning source and sink vertices.
2. Among all approaches, the hierarchical positioning algorithm produces layouts with a least average number of crossings (2%). The BSF-based algorithm produces the highest number of crossings. Both force-based approaches produce the same number of crossings, and this figure is similar to what is generated by the hierarchical positioning algorithm.
3. Among all approaches, the force-based approaches produce the best dimension ratio, i.e., 1.4.
4. Overall, the extended force-based approach produces the best layouts. It can correctly position the source and sink vertices.

69.6 Related Work

Extracting control structures from Java bytecode is one of the key activities for static analysis and model checking. Most of the extracting algorithms are based on a set of rules [9]. Sinha et al. [10] propose a control-flow graph extraction algorithm for both Java source and bytecode. Amighi et al. [11] present a sound control-flow graph extraction for Java programs with exceptions using the formal method. The approach attempts to define the rules semi-formally based on bytecode to build control structures.

Several papers have studied the layout algorithm for drawing CFG. The simplest approach is based on breadth-first search (BFS) [12]. BFS-based positioning algorithms first transform the input graphs into trees by ignoring certain edges and then try to compute the locations of vertices and edges for the resulting trees. Loop Positioning Algorithm is a modified version of the BFS-based positioning algorithm. It focuses on separating loops from vertices. However, outer loops may be extremely long due to the nature of the BFS. Hierarchical positioning algorithm [13] has added two more features to improve on the layout: (1) the addition of dummy vertices to improve the smoothness of loops, and (2) the reduction of crossings to paint a good-looking graph, which is an NP-complete problem. This paper has demonstrated a force-based positioning algorithm for automated layout arrangement. It is an iterative approach, and the layout of the graph is determined based on the equilibrium states of a graph. It produces best layout dimension ratio and can correctly position source and sink vertices.

69.7 Conclusion

This paper presents a novel approach to building an automated tool to visualize CFG for exercising unit testing techniques without the need for the source code. The CFG is constructed from bytecode to handle compound conditions for different testing coverages. Our proposed algorithm for visualizing the CFG is an extension to the force-based layout algorithm by introducing the earth gravitational force. A video demonstration of our tool is available at https://www.youtube.com/watch?v=Ey4JfVhhHQg.

References

1. ABET Accreditation. [Online]. Available: http://www.abet.org/. Accessed 24 Sept 2019.
2. Graphviz - Graph Visualization Software. [Online]. Available: http://www.graphviz.org/. Accessed 2 Feb 2016.
3. Eclipse Foundation. [Online]. Available: http://www.eclipse.org/. Accessed 30 Nov 2015.
4. Xu, W., Syed, A. R., Zheng, Q. (2015). A force-directed program graph visualizer for teaching white-box testing techniques. In *ASEE annual conference and exposition*. Seattle.
5. Eades, P. (1984). A heuristic for graph drawing. *Congressus Numerantium, 160*(42), 149.
6. Jorgensen, P. C. (2008). *Software testing: A craftsman's approach* (3rd ed.). Boca Raton: Auerbach Publications.
7. Lindholm, T., Yellin, F., Bracha, G., & Buckley, A. (2013). *The java® virtual machine specification*. Boston: Addison-Wesley Professional.

8. How would you test a vending machine?. (2011). [Online]. Available: http://www.softwaretestingquestions.net/category/testing-in-the-wild/. Accessed 24 Jan 2013.

9. Zhao, J. (1999). Analyzing control flow in ava bytecode. In *16th conference of Japan Society for software science and technology*. Linz.

10. Sinha, S., & Harrold, M. J. (2000). Analysis and testing of programs with exception handling constructs. *IEEE Transactions on Software Engineering, 26*(9), 849–871.

11. Amighi, A., PC, de Gomes., Gurov, D., Huisman, M. (2012). Sound control-flow graph extraction for Java programs with exceptions. In *10th international conference on software engineering and formal methods*. Berlin/Heidelberg.

12. Di Battista, G., Eades, P., Tamassia, R., Tollis, I. G. (1998). Graph drawing: Algorithms for the visualization of graphs. Upper Saddle River: Prentice Hall PTR.

13. Würthinger, T. (2006). *Visualization of java control flow graphs*. Linz, Austria: Johannes Kepler University.

A New Approach to Evaluate the Complexity Function of Algorithms Based on Simulations of Hierarchical Colored Petri Net Models

Clarimundo M. Moraes Júnior, Rita Maria S. Julia, Stéphane Julia, and Luciane de F. Silva

Abstract

This paper proposes a new approach to estimate the complexity function of algorithms based on automatic simulations of formal models. In order to cope with this purpose, it is necessary to count on specification techniques that, in addition to modeling the control flow of algorithms, allow performing complete simulations of the generated models for diverse scenarios. In this way, this work uses Hierarchical Colored Petri Nets to model the algorithms to be analyzed. Further, it uses the resources of CPN tools to create a set of control functions that will make possible to simulate the generated models automatically, in such a way that these simulations correspond to the real execution of the algorithms, even for non uniform data. The complexity function of the algorithms will be retrieved from these simulations. With the aim of validating this new approach, the search algorithm Minimax operating with non uniform data is used as a case study. The technical motivation for this choice is that the dynamic control flow inherent to Minimax (caused by the presence of numerous deviation instructions) is very appropriate to test the correct behavior of the control functions that direct the automatic simulations of the models. Further, the application of the algorithm to non uniform data forces the creation of additional control functions that enable the models to also handle this kind of data. The results obtained confirm the correctness of the approach proposed.

Keywords

Model simulation • CPN tools • Colored petri net • Algorithm analysis • Complexity function

70.1 Introduction

In the computation area, the focus to solving a determined kind of problem by means of software is the proposal of a viable algorithm for treating this problem. In this context, a fundamental parameter to be taken into consideration to evaluate the appropriateness of such solution is the runtime

required by this algorithm to resolve the problem [4]. Various analytical techniques have been proposed to calculate this runtime complexity, such as: asymptotic analysis, which measures to which extent the runtime increases compared to input data increasing; the recurrence techniques used to evaluate recursive algorithms; and probability analysis, which makes use of input distribution probabilities in order to estimate the expected runtime [4]. Such techniques of analysis are essentially mathematics-based and do not require that the algorithms are implemented in order to be used. However, depending on the characteristics of the algorithm under analysis, the task of estimating its runtime complexity through one of these

C.M.M. Júnior (✉) • R.M.S. Julia • S. Julia • L. de F. Silva
Computer Science Faculty, Federal University of Uberlândia, Uberlândia, Brasil
e-mail: clarimundo@iftm.edu.br; rita@ufu.br; stephane@ufu.br; lucianefatsilva@gmail.com

© Springer International Publishing AG 2018
S. Latifi (ed.), *Information Technology – New Generations*, Advances in Intelligent Systems and Computing 558, DOI 10.1007/978-3-319-54978-1_70

techniques can be very arduous – or even impossible – as frequently occurs in parallel computation, for example, whenever the runtime depends on parameter updating provided in real time by message exchanges among the involved processors [2].

Motivated by these arguments, here the authors present a new approach to estimate the complexity function of algorithms without the need to implement them, based on the construction of appropriated formal models whose simulations can be associated to their real execution. Therefore, additionally to represent the control flow of an algorithm, the models produced must be able to be simulated. In order to cope with these requirements, this work uses the CPN Tools, since the proprieties inherent to the Hierarchical Colored Petri Nets (HCPN), as well as the functionalities that can be obtained through the use of operators and through the implementation of appropriate control functions, make this formalism very promising to achieve such purpose. It is interesting to point out that in a previous work the authors proposed an approach to model the control flow of algorithms by means of CPN Tools [10]. Nevertheless, these models could not be simulated in such a way as to represent the real execution of the algorithm, since they did not count on operators and control functions that are essential to represent the dynamics of the real execution of the algorithm for real data. Thereby, this paper extends the approach presented in [10] in the following way: by using the CPN Tools built-in time operator to evaluate the time in the course of the algorithm simulation (the dynamics of values of this operator is inspired on the Assymptotic Notation method [4] and [12]), and by implementing some control functions that direct the scenario simulations in such a way that they follow the same dynamics that would be followed by the real execution of the algorithm. At the end of each simulation, the value calculated by the time operator is plotted in a graphic "runtime X datum". The curve obtained once all the simulations are concluded corresponds to the complexity function of the algorithm. It is important to point out that, depending on the problem to which an algorithm is applied, the calculation of its runtime must take into consideration the handling of non uniform data, that is, of data whose sizes vary greatly throughout the execution of the algorithm (such as search trees whose subtrees have very distinct branching factors). In this case, the theoretical approach-based runtimes calculations establish that the parameters that represent the data must be estimated through their average values. For example, in [17] the author show that, in the complexity function $O(b^m)$ related to the recursive search algorithm Minimax [16] – where the parameters b and m correspond, respectively, to the branching factor and to the depth of the search tree – if the subtrees of this search tree are not uniform (that is, they do not have the same branching factor), then the value of b corresponds to the average value of the branching factors related to these subtrees. It happens, for instance, when this algorithm is used to point out the best move in player agents [14]. Whenever treating these cases, the control functions of the present approach must generate the parameters related to the simulated data by means of uniform stochastic distribution.

The approach presented here can be particularly useful to be applied to distributed algorithms whose complexity functions can not be appropriately calculated by means of traditional analytical methods, a fact that happens, for example, whenever these calculations depend on information retrieved from the real execution of the algorithms that can not be seized in the analytical methods (such as information provided by means of message exchanges). In these cases, the model simulations performed in the present approach will be able to generate this information without the need of really executing the algorithms. As it would be extremely hard to face the technical and theoretical complexity involved in the implementation of the approach proposed in this paper directly in the context of distributed algorithms, here the authors decided to validate this approach primally in the context of challenging serial algorithms. That is why the search algorithm Minimax operating with non uniform data is used in this paper as a case study. This choice is based on technical and practical motivations.

The technical motivation is based on the following defies: first, to cope with the dynamic control flow inherent to Minimax (caused by the presence of numerous deviation instructions) in the algorithm modeling and in the automatic simulations of the models; second, to cope with situations in which the algorithm is applied to problems that deal with non uniform data, a fact that increases the intricacy of estimating the complexity function of algorithms. The practical motivation is the fact that Minimax has its complexity function determined by analytical methods. The known Complexity Function of this algorithm can be then used as a comparative parameter to estimate the acuity of the complexity function generated by the method proposed herein.

This paper is structured as it follows: Sects. 70.2 and 70.3 present the related works and the theoretical foundations, respectively; Sect. 70.4 describes how to model the control flow of the main kind of commands, giving a special emphasis to those that provoke frequent deviations in the control flow of the algorithms; Sect. 70.5, presents how to model the algorithm and describes the use of the time operator and the creation of the control functions that make possible the automatic simulations of the model in such a way as to estimate the runtime of the algorithm; finally, Sect. 70.6 describes the conclusions and the future work proposals.

70.2   Related Works

Some works have proposed the use of Petri Net (PN) to model specific types of commands involved in algorithms. In [7], the focus was exclusively to model assignment and iterative commands by means of Ordinary PN. In [10], the authors extend [7] by using Hierarchical Colored Petri Net (HCPN) to model also commands that provoke intense deviations in the control flow, like conditional, recursive and repetition commands. It is important to point out that distinctly from the approach proposed in the present paper, in [10] the authors do not define the data flow control policies that are fundamental to allow for the execution of complete simulations of distinct scenarios from which the calculation of the complexity function of algorithms becomes feasible. In other words, the functions created in [10] just allow for simulating specific scenarios for testing the appropriateness of the control flow in specific paths of the model. For this, before starting any simulation to test a determined path of the model, the tokens that represent the data in the places that belong to this path must be all instantiated. Differently from this, in the present work the instantiation of the tokens corresponding to the data – and, consequently, the definition of the paths in the course of the simulations – are dynamically defined in real time, according to the data that are being used in the simulated scenario.

In [1] and [3], the authors use scenario simulation-based methodology to analyze the performance of algorithms. Differently from the approach of the present paper, the application of these methodologies requires that the involved algorithms are effectively implemented.

An interesting proposal was suggested in [6], where the authors defend the use of empirical computational complexity to understand the performance and the scalability of programs. The idea of the authors was to present a tool that can generate, from a program with a great quantity of lines of code, a small number of blocks of commands (clusters) and execute these clusters in order to analyze its performance under different workloads. It should be highlighted that in [6], like in other works previously cited and distinctly from the approach in this paper, the algorithm being analyzed must be implemented.

70.3   Theoretical Foundations

70.3.1 Asymptotic Notation

The time complexity function represents the mathematic expression that relates total time spent by the algorithm with the size of its inputs. With the expression at hand, the obtained curve shows behavior which can be compared to the behavior of a function. This comparative method is known as *asymptotic approximation* [4] and [12] of a curve obtained by means of another curve that is used as a reference. It is worth noting that the time complexity function simply provides the growth rate for the algorithm's execution time. In order to find the time complexity function of an iterative algorithm, an unitary temporal cost is associated to each command of the algorithm. Next, its complexity is estimated by summing up the number of executions of each of these commands from the stipulated iterations.

70.3.2 Algorithm Minimax

In the area of games, the Minimax [16] is a serial search algorithm used in automatic game agents that possess perfect information (each player has their turn to play thus allowing that every player knows the current state of game). Through a search tree, the Minimax aims at choosing the appropriate movement (action) from the current state of the game (root-node). The solution is found from the values calculated for the leaf-nodes. Such values are back propagated through the tree's previous levels (that alternately represent the maximizer level for the agent and minimizer for the opponent). The search is made in-depth, always starting at the leftmost nodes of the tree. When a leaf-node is found, an evaluation is made according to a pre-established heuristic. In assessing a leaf-node, the algorithm back propagates or does not evaluate the node from the previous level. The theoretical time complexity function for Minimax is $O(b^m)$, where b is the branching factor of the tree and m is its maximum depth. It is important to note that for some applications, b can vary widely from node to node, for example, in the checkers agent presented in [14].

70.3.3 Hierarchical Colored Petri Net

The Hierarchical Colored Petri Nets (HCPN) are an extension of the Colored Petri Nets (CPN) [9]. CPN is a graphic modeling language that permits the association of the PN [13] formalism (allowing also representation of synchronization and competition) with the programming paradigm of functional languages (authorizing the definition and manipulation of complex data types) [9]. The main idea of CPN is to aggregate complex information (colors) to the tokens, allowing in general for the creation of models with compact structures. In this way, HCPN extends CPN by including a hierarchy concept provided by means of subnets (or pages) that enhance the clarity of the generated models, since these subnets define specialized parts of the modeled system. Further, particularly considering the needs related to the objectives to be reached in the present paper, HCPN permit the use of operators that accounts the time in the course of the simulation and the implementation of control functions that direct the simulations in such a way that they follow the same dynamics that would be followed by the real execution of the system.

70.4 Modeling Command Structures

This section describes how to model the control flow of the main kind of commands, giving a special emphasis to those that provoke frequent deviations in the control flow of the algorithms. This modeling is analogous to that proposed by the authors in [10]. As in [10], in this paper the modeling of these command structures fits the following semantics: each command is modeled by a transition; further, in the models each data structure is represented by a token allocated in a place that corresponds to a specific processing state.

70.4.1 Assignment Command

Figure 70.1a shows a sequence of two attribution commands for which the modeling is represented by the transitions $X: = 2$ and $Y: = X + Z$, respectively, as shown in

Fig. 70.1b. When a successive firing of these transitions occurs, the token initially situated in place p_1 (Fig. 70.1b) moves to places p_2 (Fig. 70.1c) and p_3 (Fig. 70.1d), successively. This fact simulates the sequential execution of the modeled attribution commands. This possibility of following, on the model, the dynamic of the command execution is a gain that is obtained through the use of the PN.

70.4.2 Conditional and Iterative Commands

The modeling and the automatic simulation of commands whose real execution involves deviation in the control flow, such as conditional, iterative and recursive commands, require specific abilities from the modeling tool that is used. That is why the authors of the present paper have been motivated to use the resources of HCPN. Figure 70.2b, c, d show modeling examples of conditional and iterative command. The conditional command control flow shown in Fig. 70.2a (lines 1–4) is illustrated in Fig. 70.2b, 70.2c. Figure 70.2b represents the situation in which the condition $(x > 0)$ of the command is true, and Fig. 70.2c represents the situation in which such a condition is false. The representations of the transition firings t_1 (Fig. 70.2b) and t_2 (Fig. 70.2c) simulate the logic test performed on the

Fig. 70.1 Modeling assignment commands

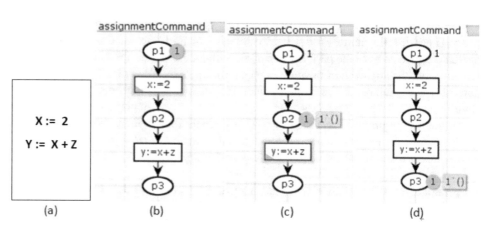

Fig. 70.2 Modeling conditional and iterative commands

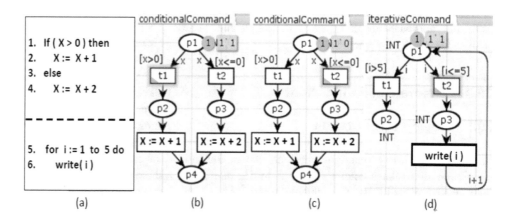

conditional command which deviates in this manner the flow of the token to the left or to the right path of the model, respectively. HCPN allows for modeling this dynamics by counting on the following very own special characteristics: it operates with tokens that can receive specific values (instances) of the modeled problem. These values can be tested by conditions known as Guards [9] that are associated to given transitions. The firing of a transition will only occur if its Guard produces a true value, as shown in Fig. 70.2b, 70.2c.

Another CPN ability lies in the fact that it allows for the association of labels to the arcs of the model, in such a way as to make possible to update the instances of the tokens. This is the case occurring in the output arc related to the transition t_3 in Fig. 70.2d. This figure represents an iterative command shown in Fig. 70.2a (lines 5–6), where the iterations occur with i varying from 1 to 5 (i increased from 1 at each iteration). The token situated in place p_1 possesses instances that are generated through the updating of the variable i. While the Guard of transition t_2 is true, transitions t_2 and t_3 are fired, alternately, thus representing the command iterations. When the Guard associated to transition t_2 becomes false (and that of transition t_1 is true), the end of the iterations occur through the firing t_1.

70.4.3 Recursive Command

Figure 70.3 exemplifies a recursive function named $f1()$. Lines 2, 3 and 4 represent, respectively, the test for the stop criterion of the recursive calls, the recursive returns and the recursive calls for this function. The control flow of a recursive function consists of two stages in the following order: sequence of recursive calls are represented by means of data that are progressively pushed onto a stack – that is initially empty – until a stop criterion is satisfied. In the opposite way, the recursive returns are represented by means of progressive popping of data from the stack, until it becomes empty. The figures of the next subsections show how to model the control flow of the recursive function $f1()$ presented in the example of Fig. 70.3, taking 2 as the initial value for the input parameter "x".

```
1.   function f1 ( int x )
2.       if ( x = 0 ) then
3.           return x
4.       x = x + f1 ( x - 1 )
```

Fig. 70.3 Example of recursive function

70.4.3.1 Modeling Recursive Call
In Fig. 70.4a, a token with value $x = 2$ is allocated in place p_3, enabling the transition "call". This represents the fact that a recursive call is about to happen. When this transition is fired, a token is inserted (stacking) in place *push* that represents a stack and another is inserted in place f_1 with the new decreased value $x = 1$ (Fig. 70.4b). The Guard of transition t_2 being true $(x < > 0)$, t_2 is then fired (Fig. 70.4c). Once again, the transition *call* is ready for firing. Once a new firing occurs, a new token is inserted (stacking) in place *push* (Fig. 70.4d).

70.4.3.2 Modeling Recursive Return
After the second insertion (stacking) of tokens in place *push* (Fig. 70.4d), a new token with value $x = 0$ appears in place f_1. As only the Guard associated with t_1 is true, this transition is fired, producing a token in place *ret* (Fig. 70.5a). This models the occurrence of the stop criterion (line 2 of Fig. 70.3). Noteworthy here is that this self-same token appears automatically in place *ret*1. This fact is only possible because both places *ret* and *ret*1 are fusion places [9] (what happens in *ret* also happens in *ret*1). Therefore, the flow of the token can be deviated from one place to another within the model without the need for an explicit link between them. This allows one to simulate on the model, in an efficient manner, the data flow deviation caused by recursive returns. This resource that exists on CPN Tools was another aspect that motivated the authors of this study to choose it as simulation environment.

Figure 70.5a shows further that the transition t_4 is ready for firing since its Guard is true. When the transition t_4 is fired (Fig. 70.5b), the first token removed (unstacking) from place *push* is the last token inserted in this place during the recursive calls, as would be expected. This control is made possible by means of Guard $x1 = x2$ associated with the transition t_4. In sequence, Fig. 70.5c shows the removal of the second token in place *push*, emptying the stack completely. Finally, as the Guard of transition t_3 is true (Fig. 70.5c), t_3 is fired (Fig. 70.5d), thus simulating the end of the recursion.

70.5 Calculating Complexity Functions Through Model Simulations

This section, presents how to calculate complexity functions through automatic model simulations using the Minimax algorithm to prove the technique. Therefore, this section goes on to include the discussion pertinent to the use and the implementation of the following resources: the time operator and the control functions, which correspond to a relevant extension introduced in [10] with the aim of

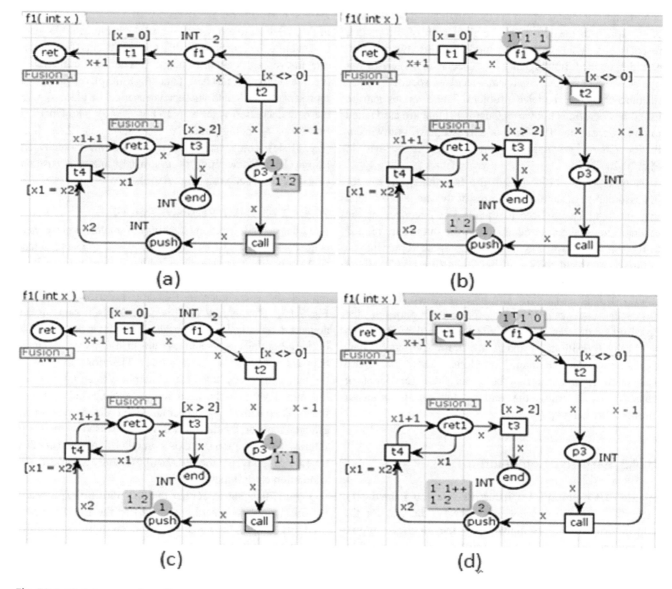

Fig. 70.4 Modeling recursive call

allowing the automatic simulation of the models even for non uniform data. In this way, Sect. 70.5.1.1 shows how the control functions are implemented. Further, Sect. 70.5.2 shows how the time operator is used to calculate the complexity function of the Minimax by means of automatic simulations. In order to validate the experimental results computed through simulation, in the same subsection the complexity function obtained experimentally is compared with that obtained by means of the analytical method.

70.5.1 Modeling the Minimax

It is important to highlight three aspects. (1) The semantics of control flow modeling for the commands that make up the Minimax were the same as those proposed in Sect. 70.4.

(2) The analysis of the runtime complexity of the Minimax algorithm depends only on the depth and the ramification factor of the explored search tree [16]. (3) All declarations for data structures, variables and functions that exist on the model were elaborated using the functional programming language Standard ML, which exists on CPN Tools. In [5] and [10] details were given as to the declaration of such elements through CPN Tools. As stated earlier, the authors of this work modeled the Minimax algorithm using an HCPN and the CPN Tools. This algorithm has a set of commands that can be structured into three functional parts: *VerifyLeaf*, *VerifyMax* and *VerifyMin* as shown in Fig. 70.6.

An overview of the informal modeling of the algorithm, which corresponds to a structure of such parts, is shown in Fig. 70.7. In such a figure, the Min-Max module represents

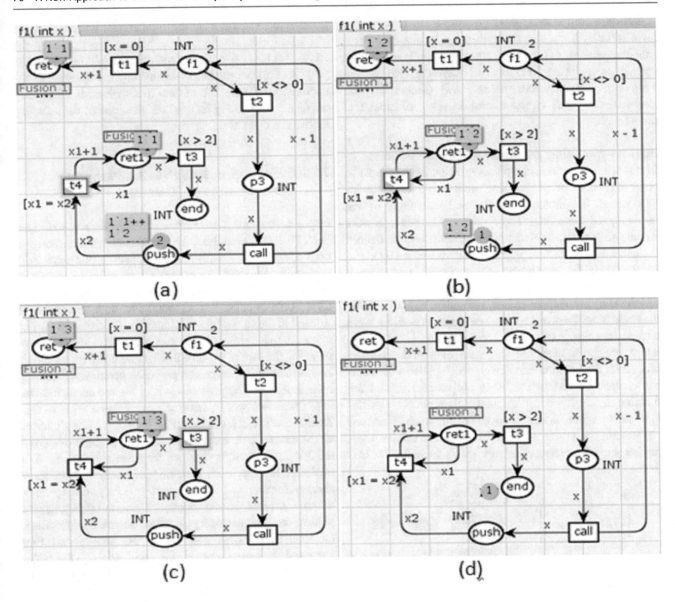

Fig. 70.5 Modeling recursive return

```
1.    fun Min-Max ( n : node , depth : int , bestmove : move) : real =
2.        if leaf ( n )  return  evaluate ( n )         Part 1 - VerifyLeaf
3.        if n is a max node
4.            besteval := - infinity
5.            for each child of n
6.                v := Min-Max ( child , d + 1 , bestmove )
7.                if v > besteval
8.                    besteval := v                     Part 2 - VerifyMax
9.                    thebest = bestmove
10.           bestmove := thebest
11.           return besteval
12.       if n is a min node
13.           besteval := + infinity
14.           for each child of n
15.               v := Min-Max ( child , d + 1 , bestmove )
16.               if v < besteval                       Part 3 - VerifyMin
17.                   besteval := v
18.                   thebest = bestmove
19.           bestmove := thebest
20.           return  besteval
```

Fig. 70.6 Pseudo-code of the Minimax

the main net, thus allocating the most abstract hierarchical level of the model. The remaining modules of Fig. 70.7 correspond to the subnets of the model that represent the substructures *VerifyLeaf*, *VerifyMax* and *VerifyMin*. The filled arrows in Fig. 70.7 represent the communication between the main net and the subnets, whereas the dashed arrows represent the communication existing between the subsets. The need to model the communication between distinct parts of the model in the same hierarchical level, as well as in distinct hierarchical levels, was fundamental to the authors' motivation for the use of the HCPN and the CPN Tools environment. In fact, the hierarchical resource present on the HCPN, as well as fusion places, abstract transitions and sockets resource present on the CPN Tools environment provided robust technical support for the implementation of the present approach. In [10], the authors showed how to connect the most abstract model (main net) to subsets (low level models) using abstract transitions. Beside that, for a subnet to connect to a main net, the use of input and output sockets is necessary. Figure 70.8 shows a part of the *VerifyMin* subnet, where places *Part3* (socket *In*) and *Part4* (socket *Out*) represent, respectively, its input and output places.

70.5.1.1 Modeling the Minimax Control Policies

In this work the control functions must be implemented in order that the modeled algorithm produces all the scenarios that would be produced by the real execution of the implemented algorithm. These control functions, besides dealing with parameters that carry information that defines the current status of the exploration of the tree under simulation (such as: depth, branching factor, node being explored), should also produce values (token instances), which define in an automatic way the path to be taken during the simulation. To achieve this, three control functions were created (see Fig. 70.8): the Boolean function *verifLoop(v)*, which checks if there are still sons of a node *n* to be processed; the function *decsons(v)*, which updates the number of sons of node *n* that were not processed; and the function *defsons(v)*, which stochastically produces a particular ramification factor of node *n*. In such functions *v* represents the structured input parameter that carries information about the current simulation status, including the tree

nodes *n* to be involved in the simulations. Besides that, the last function is composed of a built-in function from the CPN Tools called the *uniform distribution function* [18]. This allows the model to simulate the dynamic of the exploration of non uniform trees (trees whose subtrees present distinct branching factors) which is one of the technical motives of this article.

70.5.2 Simulation Model: Estimating the Runtime of Minimax

Inspired by the Asymptotic Notation method (see Sect. 70.3.1), the authors of the present work associate a time operator $@ + 1$ [9] at each transition that models an algorithm command (for example, transitions t_{13} and t_{27} in Fig. 70.8). Each time that one of these transitions is fired, the CPN Tools calculates its temporal cost on a time indicator situated in the simulation environment. Therefore, at the end of the complete simulation of a scenario, this indicator has as its value, the sum of the time units referent to the simulation. In this manner, the approach proposed herein was validated by means of different simulations of the model, as follows: depth searches were performed using trees with various depths. Each search carried out in these trees was simulated on the model 30 times, thus obtaining an average value over the 30 time indicators obtained for these simulations. As a matter of fact, for each scenario simulated, the number of children for each node of the search tree vary and depend on the uniform distribution function. Because of that, several replications are necessary in order to reach a stabilized medium value when considering the sum of all time operators activated during the simulation. This value corresponds to the runtime of the estimated algorithm by means of the approach proposed herein. Following on, the

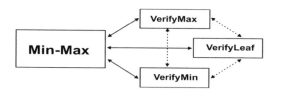

Fig. 70.7 Overview of the Minimax model

Fig. 70.8 Partial model subnet VerifyMin

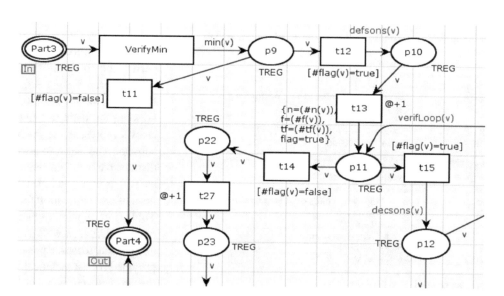

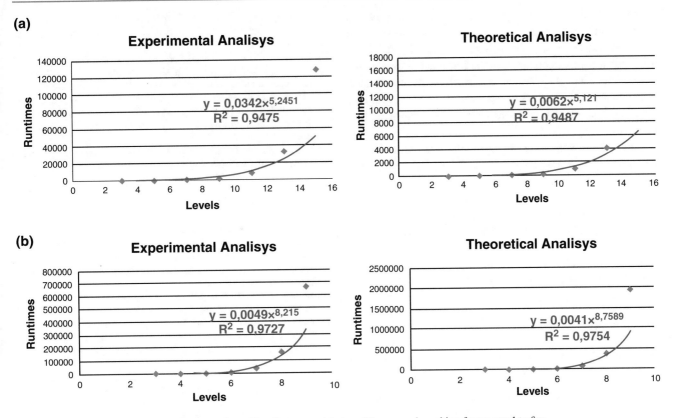

Fig. 70.9 Complexity functions: (**a**) average branching factor equal to two (**b**) average branching factor equal to five

theoretical value of the runtime of the Minimax algorithm was also calculated by means of the traditional analytical method (see Sect. 70.3.2). Finally, the runtimes retrieved from both approaches were plotted on a graphic and compared by means of their respective correlation factors (R^2). Such validation was performed considering two distinct scenarios. In order to keep the coherence, in each scenario, the calculation of the runtime in both approaches was executed taking into account the same values for the average branching factor and for the tree depths. Figure 70.9a illustrates the first validation scenario in which the tree depth values range in the set {3, 5, 7, 11, 13, 15} and the average branching factor is equal to two (taking [1,3] as interval in the uniform distribution). In Fig. 70.9a the left hand regression curve represents the Minimax complexity function estimated by the simulation model proposed herein. The right hand regression curve represents the Minimax runtime complexity obtained by the theoretical analysis. Both are exponential regression curves and both have correlation factors that are extremely close (and tending to the value 1), a fact that proves the correctness of the approach proposed herein. In fact, the regression curve of the present approach has a correlation factor (R^2) equal to 0.9475, whereas the regression curve of the theoretical approach has a correlation factor equal to 0.9487. Figure 70.9b shows the second validation scenario in which the

simulations were performed in trees with an average branching factor equal to five and with maximum depths range in the set {3, 4, 5, 6, 7, 8, 9}. The regression curve obtained experimentally (left-hand side of Fig. 70.9b) presented a correlation factor equal to 0.9727 and the curve obtained theoretically (right-hand side of Fig. 70.9b) presented a factor equal to 0.9754 (again both R^2 are extremely close).

70.6   Conclusions and Future Works

This paper presents a new approach to estimate the complexity function of algorithms by modeling and simulating them in the formal environment of the CPN tools. An important point to be emphasized in the presented approach is that the paths in the models defined by each complete simulation correspond exactly to the sequence of commands that would be executed in the actual processing of the algorithm. These paths are defined in the course of each simulation through control functions that dynamically instantiate the data structures associated to the tokens in the corresponding places that belong to these paths. The duration of each simulation is calculated by a specialized time operator and is plotted in a graphic "runtime X data". At the end of the simulations, the graphic represents then the complexity function. With the aim of validating this new approach, the search algorithm Minimax operating with non uniform data was used as a case study. The authors performed in particular a statistical comparison between the Minimax graphic that is calculated from the present approach with the graphic obtained from theoretical-based analysis. The results confirm that both graphics present an extremely close runtime behavior.

As a future work proposal, the authors intend to use the methodology proposed here to estimate the complexity functions of distributed algorithms for which a pure analytical approach is not feasible. In particular, the modeling and simulation of the parallel version of Minimax called Young Brother Wait Concept is under study at the present time.

Acknowledgements The authors would like to thank CAPES, FAPEMIG and CNPq for financial support.

References

1. Arcuri, A., Iqbal, M. Z., & Briand, L. (2010). Formal analysis of the effectiveness and predictability of random testing. In *ACM ISSTA* (pp. 219–229).
2. Barbosa, V. C. (2003). *An introduction to distributed algorithms.* The MIT Press. ISBN-10:0262514427, ISBN-13:978-0262514422.
3. Borovska, P., & Lazarova, M. (2007). Efficiency of parallel Minimax algorithm for game tree search. In *Proceedings of the International Conference on CompSysTech*, New York (pp. 14:1–14:6).
4. Cormen, T. H., et al. (2009) *Introduction to algorithms* (3rd ed.). MIT Press.
5. CPN Tools. http://cpntools.org.
6. Goldsmith, S. F., Aiken, A. S., & Wilkerson, D. S. (2007). Measuring empirical computational complexity. In *Proceeding ESEC-FSE'07* (pp. 395–404).
7. Hanzalek, Z. (2003). *Parallel algorithms for distributed control – A petri net based approach.* Thesis, Departament of Control Engineering, Czech Technical University in Prague, Czech Republic.
8. Jensen, K., Kristensen, L. M. & Wells L. (2007). Coloured petri net and CPN tools for moedlling and validation of concurrent systems. *Journal on Software Tools Technology Transfer,* 213–254. Springer.
9. Jensen, K., & Kristensen, L. M. (2009). *Coloured petri nets – Modeling and validation of concurrent systems.* Berlin: Springer.
10. Junior, C. M. M., Julia, R. M. S., & Julia, S. (2015). Modeling recursive search algorithms by means of hierarchical colored petri nets and CPN tools. In *New Generations on 12th ITNG*, Las Vegas (pp. 788–791).
11. Kishimoto, A., & Schaeffer, J. (2002). Distributed game-tree search using transposition table driven work scheduling. In *Proceedings OF ICPP*, Beijing (pp. 323–330).
12. Kleinberg, J., & Tardos, E. (2013) *Algorithm design* (1st ed.). Harlow: Pearson.
13. Murata, T. (1989). Petri nets: Properties, analysis and applications. In *Proceedings IEEE* (pp. 541–580).
14. Neto, H. C., & Julia, R. M. S. (2007). LS-Draughts – A Draughts learning system based on genetic algorithms, neural network and temporal differences. In *IEEE Congress on Evolutionary Computation* (pp. 2523–2529).
15. Newborn, M. (1988). Unsynchronized iteratively deepening parallel alpha-beta search. *IEEE TPAMI, 10*(5), 687–694.
16. Norvig, P., & Russel, S. (2009). *Artificial intelligence: A modern approach* (3rd ed.). Prentice Hall.
17. Rocki, K. M. (2011). *Large scale monte carlo tree search on GPU.* Thesis, Departament of Computer Science, School of Information Science and Technology, The University of Tokyo, Japan.
18. Sprent, P., & Smeeton, N. C. (2007). *Applied nonparametric statistical methods* (4th ed.). Boca Raton: Chapman and Hall/CRC.

Detection Strategies for Modularity Anomalies: An Evaluation with Software Product Lines

71

Eduardo Fernandes, Priscila Souza, Kecia Ferreira, Mariza Bigonha, and Eduardo Figueiredo

Abstract

A Software Product Line (SPL) is a configurable set of systems that share common and varying features. SPL requires a satisfactory code modularity for effective use. Therefore, modularity anomalies make software reuse difficult. By detecting and solving an anomaly, we may increase the software quality and ease reuse. Different detection strategies support the identification of modularity anomalies. However, we lack an investigation of their effectiveness in the SPL context. In this paper, after an evaluation of existing strategies, we compared four strategies from the literature for two modularity anomalies that affect SPLs: God Class and God Method. In addition, we proposed two novel detection strategies and compared them with the existing ones, using three SPLs. As a result, existing strategies showed high recall but low precision. In addition, when compared to detection strategies from the literature, our strategies presented comparable or higher recall and precision rates for some SPLs.

Keywords

Detection Strategies • Modularity Anomalies • Software Product Lines

71.1 Introduction

Software reuse consists of using existing code to develop new systems [20]. Reuse requires a satisfactory code modularity for its effective application [4, 20]. Since existing software components may contain problems that should not be propagated to new systems, modularity anomalies may make reuse difficult. The literature also references modularity anomalies as bad smells [13]. Therefore, by detecting and solving an anomaly before reusing a component, we may increase the component quality and decrease time and efforts spent on maintenance, for instance [12, 18].

In this context, we need effective methods to support the detection of modularity anomalies and, consequently, reuse [3]. Different detection strategies have been proposed to support the identification of anomalies [1, 8, 14]. Besides that, although several anomalies may affect reuse [8], we lack an investigation on the effectiveness of the existing strategies for detecting anomalies in Software Product Lines (SPL). An SPL is a set of systems that share common and varying features [4]. By combining features, we generate different SPL products [4]. SPL aims to support reuse with decreasing maintenance efforts to developers [20].

In this paper, we investigate modularity anomalies in SPL, since anomalies affect negatively the SPL modularity and makes reuse difficult. After an *ad hoc* literature review, we compared four detection strategies from the literature for two well-known modularity anomalies that may affect

E. Fernandes (✉) • P. Souza
Federal University of Minas Gerais, Belo Horizonte, Brazil
e-mail: eduardofernandes@dcc.ufmg.br; priscilasouza@dcc.ufmg.br

K. Ferreira
Federal Center for Technological Education of Minas Gerais, Belo Horizonte, Brazil
e-mail: kecia@decom.cefetmg.br

M. Bigonha • E. Figueiredo
Federal University of Minas Gerais, Belo Horizonte, Brazil
e-mail: mariza@dcc.ufmg.br; figueiredo@dcc.ufmg.br

© Springer International Publishing AG 2018
S. Latifi (ed.), *Information Technology – New Generations*, Advances in Intelligent
Systems and Computing 558, DOI 10.1007/978-3-319-54978-1_71

SPLs: *God Class* [13] and *God Method* [14]. In addition, we proposed novel detection strategies to support the identification of both anomalies. We designed the novel strategies given the small amount of strategies from the literature that are based on traditional, well known, and ease to compute software metrics. We analyzed three SPLs.

As a result, we presented a novel detection strategy for each anomaly, i.e., *God Class* and *God Method*. Through the comparison of our strategies with the ones from the literature, we observed positive results with respect to our strategies. For MobileMedia, our strategies obtained comparable recall and the highest precision results for both anomalies. In the case of Berkeley DB, we obtained the best and second best recall when compared to the existing strategies. Finally, our strategy for *God Class* obtained the best recall and precision rates, in comparison with the others, in the case of TankWar.

71.2 Background

Modularity anomalies are symptoms of deeper problems in the modularity of systems [13]. Several types of anomalies may affect the modularity of a system, such as *Lazy Class*, *Feature Envy*, and *Long Parameter List* [13, 14]. In this study, we investigated two types of anomaly: *God Class* [13] and *God Method* [14]. We chose these anomalies because (i) they have different detection strategies in the literature for comparison and (ii) although both are general-purpose anomalies, they can affect negatively the SPL design. *God Class* is a class that contains excessive knowledge of the system and responsibilities [13]. *God Method* is a large method with high complexity and many responsibilities [14].

Two approaches may support the detection of modularity anomalies [18]. Manual detection relies on code inspection. Automated detection counts on the support of detection strategies, i.e., compositions of metric-based rules that define when a specific software component, e.g., class, method, or package, is prone to contain a modularity anomaly [14]. In turn, tools aim to support the automated detection of anomalies. These tools apply some type of detection strategy or equivalent techniques [6, 18, 22].

A Software Product Line (SPL) is a configurable set of systems that share common and varying features [20]. There are four types of features in a product line: mandatory, optional, alternative inclusive (OR), and alternative exclusive (XOR) [20]. Each product from an SPL is composed by general features that define the SPL basis (mandatory features) and specific features that differ a product from others (optional, OR, or XOR features) [4]. Artifacts of an SPL may contain modularity anomalies like in other types of software systems. However, there are few studies to investigate anomalies in this specific context [4].

71.3 Study Settings

Sections 71.3.1, 71.3.2, and 71.3.3 present the study goal and research questions, steps, and artifacts.

71.3.1 Goal and Research Questions

In this study, we were specifically concerned with detection strategies to identify modularity anomalies that hinder reuse in SPL. We then designed new strategies for these anomalies. We also conducted a comparative study of detection strategies in the SPL context. To guide our study, we designed two research questions as follows.

RQ1. *Are the existing detection strategies for modularity anomalies effective in the SPL context?*
RQ2. *Are the novel detection strategies more effective than the existing ones in the SPL context?*

71.3.2 Study Steps

We designed seven study steps discussed as follows. Steps 1–5 composed the study phase called *Selection of Artifacts*. In Step 1, we selected the SPLs for analysis. Step 2 consisted of the selection of modularity anomalies, based on anomalies that we were able to detect in the chosen systems from Step 1. Step 3 encompassed the selection of strategies from the literature for comparison. Step 4 was dedicated to the creation of new strategies for the anomalies chosen from Step 2. Our strategies relied on well-known anomaly definitions [13, 14] and the SPL characteristics [4]. In the same step, we compared such strategies with strategies from the literature provided by Step 3. Step 5 consisted of selecting detection tools for modularity anomalies. This step was essential to support the definition of reference lists of anomalies for SPLs, collected from Step 1, without a previously computed reference list of anomalies. A reference list of anomalies is an itemization of anomalies that occur in a given system. Experts in a system can generate reference lists [19]. Otherwise, such lists may rely on the detection results provided by a detection tool.

The remaining steps, Steps 6 and 7, composed the last phase called *Comparative Study*. Step 6 comprised the comparison of existing detection strategies from the literature to answer RQ1. Finally, Step 7 targeted on RQ2 through the comparison of novel detection strategies with the existing ones.

71.3.3 Selected Artifacts

In Steps 1 and 2, we chose three SPLs extracted from a repository [21]: MobileMedia [10], Berkeley DB, and TankWar. These systems are implemented in AHEAD or FeatureHouse and have from 2 K to 42 K number of lines of code (LOC) [15]. We selected MobileMedia based on the availability of reference lists for *God Class* and *God Method*. We selected Berkeley DB and TankWar because of the three following reasons. First, Berkeley DB is one of the largest systems in the SPL catalog. Second, we are able to import the code of these systems in FeatureIDE for automated anomaly detection and generation of reference lists. Third, there are at least two occurrences of *God Class* and *God Method* in each system. Details in the discussion of Step 5.

A set of 11 pre-computed software metrics are provided by the SPL repository. Coupling between Objects (CBO) [5], Lines of Code (LOC) [15], Number of Attributes (NOA) [15], Number of Constant Refinements (NCR) [1], Number of Methods (NOM) [15], and Weighted Methods per Class (WMC) [5] are class-level metrics. McCabe's Cyclomatic Complexity (Cyclo) [16], Method Lines of Code (MLOC) [15], Number of Method Refinements (NMR) [2], Number of Operations Overrides (NOOr) [17], and Number of Parameters (NP) [15] are method-level metrics.

In Step 3, we selected strategies for each anomaly. After an *ad hoc* literature review, we selected two strategies for *God Class* and two for *God Method*. Although we have found other strategies, our selection relied on the metrics provided by the SPL catalog. To adapt each strategy, we discarded clauses with metrics that are unavailable in the SPL catalog. We derived thresholds for the metrics – i.e., values that support the characterization of a given metric [21] – based on the selected SPL repository and Vale's Method [21]. We chose Vale's method because it provided us sufficient flexibility to define detection strategies. To derive thresholds, we used the R statistical environment[1] and the entire SPL repository [21], with 33 SPLs, as target systems. Table 71.1 presents each strategy adapted from the literature. In Step 4, we proposed a new strategy for each modularity anomaly (details in Sect. 71.5).

In Step 5, we extracted the reference lists of *God Class* and *God Method* for MobileMedia from a previous work [19], with 7 instances per anomaly. Experts from the MobileMedia development team composed these lists. For Berkeley DB and TankWar, we did not find reference lists. Therefore, we generated them based on the results of JSpIRIT [22], a plug-in tool for Eclipse IDE. The automated detection in BerkeleyDB and TankWar supported the composition of reference lists. Based on a SLR [9], we chose

Table 71.1 Detection strategies from the literature

God Class
GC1 [21]: [(LOC > 77) AND (WMC > 17) AND (CBO > 0)] OR (NCR > 1)
GC2 [11]: (WMC > 34) AND (NOM > 14) AND (NOA > 8)
God Method
GM1 [6]: (MLOC >50)
GM2 [14]: (MLOC >13) AND (Cyclo >2)

Table 71.2 Results for existing strategies

SPL	Strat.	TP	FP	TN	FN	Recall	Precision
MobileMedia	GC1	5	10	126	2	71%	33%
	GC2	0	0	7	136	0%	0%
	GM1	0	1	365	7	0%	0%
	GM2	3	30	336	4	43%	9%
Berkeley DB	GC1	17	80	524	0	100%	18%
	GC2	15	21	583	2	88%	88%
	GM1	11	53	5618	1	92%	17%
	GM2	11	441	5230	1	92%	2%
TankWar	GC1	1	11	76	0	100%	8%
	GC2	0	2	85	1	0%	0%
	GM1	2	10	297	0	100%	17%
	GM2	2	66	241	0	100%	3%

JSpIRIT because of its availability for download and sufficient recall and precision. The tool found 35 *God Class* and 37 *God Method* instances for Berkeley DB, and 2 and 3 instances, respectively, for TankWar. After the validation of the results by the paper's authors, we obtained (i) 17 and 12 instances for *God Class* and *God Method*, respectively, for Berkeley DB and (ii) 1 and 2 instances for the same anomalies, for TankWar. The study artifacts are available in the research website.[2]

71.4 Comparison of Existing Strategies

This section aims to answer RQ1. Table 71.2 presents recall and precision, per SPL, for both *God Class* and *God Method*. TP is the number of true positives (correct identification of real anomalies). FP is the number of false positives (incorrect identification of real anomalies). TN is the number of true negatives (correct non-identification of anomalies). Finally, FN is the number of false negatives (incorrect non-identification of anomalies) [7]. The used formula are $Precision = TP / (TP + FP)$ and $Recall = TP / (TP + FN)$ [7].

In general, we observed a mean and median recall of 66% and 90%, respectively, i.e., a significant result. However, regarding precision, we observed mean and median of 16%

---

[1] https://www.r-project.org

[2] http://labsoft.dcc.ufmg.br/doku.php?id = about:itng17-detection-strategies

and 9%, respectively, a low result. Therefore, our data suggests that the existing detection strategies were not sufficiently effective in the SPL context. GC1 presented the highest recall rate for all SPLs and a slight highest precision for two of them. Therefore, our data suggests that GC1 was more effective than GC2. In turn, both GM1 and GM2 performed similarly in terms of recall for two of the three SPLs, although precision was slightly higher for GM1. We provide a discussion per SPL as follows.

MobileMedia Regarding *God Class*, GC1 was the only strategy able to detect code anomaly instances, with recall and precision rates of 71% and 33%, respectively, against 0% for both measures in the case of GC2. The low percentages for GC2 may relate to the high threshold for WMC, computed based on MLOC that is generally low for the analyzed SPLs. Regarding *God Method,* we observed similar behavior, in which GM1 was unable to detect anomalies, although GM2 provided 43% and 9% of recall and precision, respectively. Again, the justification for the non-effectivity of GM1 relies on the high threshold for MLOC.

Berkeley DB Regarding *God Class* detection strategies, GC1 presented 100% of recall (a result 12% higher than GC2), but a low precision of 18% (70% lower than for GC2). Considering the significant difference between precision rates, we observed that GC2 was more effective than GC1. Although such results differ from the previous ones, they reinforce our assumptions that MLOC for the SPLs affected the results, since the methods from classes of Berkeley DB are significantly larger than MobileMedia. With respect to *God Method*, both GM1 and GM2 presented the same recall, but GM1 provided 15% more precision than the other strategy, probably for the same reason discussed in the case of MobileMedia. Therefore, we assume that GM1 was slightly more effective than GM2.

TankWar Regarding *God Class,* GC2 was unable to detect anomalies, while GC1 presented 100% and 8% of recall and precision, respectively. Besides the considerations valid for MobileMedia and Berkeley DB, NCR contributed to the high recall observed for GC1. On the other hand, NRC contributed to low precision, by causing an increase in the number of FP, mainly because the threshold is very low. Note that, in SPL, several methods tend to have more than 1 refinement, due to the modularization of features. We conclude that GC1 was more effective than the other strategy. However, regarding *God Method*, both GM1 and GM2 obtained the same recall. In turn, GM1 presented a precision only 14% higher than the other strategy, due to the higher, stricter threshold for MLOC. We conclude that GM1 was slightly more effective than GM2.

71.5 Comparison with Novel Strategies

This section aims to answer RQ2. Table 71.3 presents the novel strategies for *God Class* (GC3) and *God Method* (GM3). We took into account the findings from Sect. 71.4 to design each strategy aiming better recall and precision. For CG3, we used LOC because a high number of code lines may indicate excessive responsibilities of the class. We also used NOA and NOM because a high number of attributes (NOA) and methods (NOM) suggests excessive knowledge of the class (attributes) and responsibilities (methods). Finally, we used WMC because a high weight of the class indicates that the class is doing more than it should do. For GM3, we used MLOC because a high number of code lines is a symptom of complex method. We also used NP because a large list of parameters may point that the method requires too much knowledge of the current or external classes. Finally, we used Cyclo because it indicates too many responsibilities of the method.

Table 71.4 presents recall and precision for the novel strategies, per SPL. We observed moderate rates of recall, but low precision – we obtained moderate-to-high rates only for MobileMedia. For *God Class*, GC1 presented the highest rates of recall for Berkeley DB and TankWar, and the highest precision for MobileMedia and TankWar. However, with respect to *God Method*, we observed a significant rate of precision for MobileMedia.

MobileMedia Regarding *God Class*, our strategy GC3 presented the highest precision (100%), 67% higher than for the second highest (GC1). This positive result relates to the discard of metrics that generated high FP in the other strategies (e.g. NCR). However, GC1 presented recall of 71%, a result 42% higher than for GC3, probably because GC3 is stricter since it has more metrics to compare. Regarding *God Method*, GM3 presented the highest precision rate

Table 71.3 Novel strategies proposed in this study

God Class
GC3: (LOC > 77) AND (NOA > 4) AND (NOM > 10) AND (WMC > 17)
God Method
GM3: (MLOC >13) AND (NP > 2) AND (Cyclo >3)

Table 71.4 Results for the novel strategies

SPL	Strat.	TP	FP	TN	FN	Recall	Precision
MobileMedia	GC3	2	0	136	5	**29%**	**100%**
	GM3	1	1	365	6	**14%**	**50%**
Berkeley DB	GC3	17	51	553	0	**100%**	**25%**
	GM3	6	139	5532	6	**50%**	**4%**
TankWar	GC3	1	4	83	0	**100%**	**20%**
	GM3	0	10	296	3	**0%**	**0%**

(50%), a result 41% higher than for GM2. This positive result relates to lower threshold for MLOC combined with two metrics that support the identification of *Long Method* even in SPL (i.e. NP and Cyclo). However, the precision for GM2 was 29% higher than for GM3. This result may relate to the very low thresholds for NP and Cyclo, although such metrics tend to be low in SPL. We conclude that GC3 and GM3 were more precise.

Berkeley DB Regarding *God Class*, our strategy GC3 presented the highest recall (100%), similarly to GC1. Note that GC2 obtained 88% of recall, a result only 12% lower than our strategy. Regarding precision, GC3 presented the second highest rate, although it is 63% lower than for GC2. Therefore, our data suggests that GC2 was the most effective strategy, in general. For *God Method*, GM3 obtained the second highest recall, against 92% for both GM1 and GM2. In addition, GM3 had low rates of recall (50%) and precision (4%) when compared to the highest rates obtained by GM1 (92% and 17%, respectively). Therefore, GM2 was the most effective strategy. Overall, our findings for Berkeley DB have similar justification than in the case of MobileMedia, for both GC3 and GM3.

TankWar analysis Regarding *God Class*, we observed the highest recall and precision (100% and 20%, respectively) for our strategy GC3. Although GC1 and GC3 presented the same recall of 100%, our strategy presented a 12% higher precision. The justification is similar to the MobileMedia case. We conclude that GC3 was more effective than the GC1. For *God Method*, GM3 did not find anomalies, and the highest results were obtained by GM1 (100% of recall and 17% of precision). In this case, an anomalously low NP (even for SPLs) for most of the methods affected significantly GM3.

71.6 Threats to Validity

We discuss threats to validity as follows.

Construct and Internal Validity We designed our study with steps for replication. The study settings relied on the literature. We also provided the study artifacts in the research website. These treatments aim to minimize problems with study replication and reliability. Moreover, we conducted a careful data collection to prevent missing data, incorrect selection of metrics, and inappropriate use of threshold derivation methods. We double-checked the collected data. Therefore, we expect that our data collection is reliable for analysis. Regarding the automatically generated reference lists of anomalies, we chose an effective tool from a previous work [9].

Conclusion and External Validity We carefully performed the data analysis to minimized problems with data interpretation. We based the choice of mathematical computation (recall and precision) on previous studies. Finally, some factors may prevent the generalization of our research findings, since we proposed strategies based on the authors' background and subjective perceptions of anomalies. To minimize this problem, we carefully chose metrics from the available metric set, and we defined each strategy in details.

71.7 Related Work

Fontana et al. (2012) [12] conducted a literature review and a comparison of detection tools for modularity anomalies. They concluded that the tools provide significantly different results for a same anomaly, some results are redundant, and the tools' agreement is high for anomalies as *Large Class*. Moha et al. (2010) [18] also evaluated tools. The authors compared a set of tools with a new one proposed by them. By relying on reference lists of anomalies built after manual code inspection, they computed recall and precision of the tools. However, their study did not compare an extensive set of tools, and there was no agreement computation.

In a previous work [9], we conducted a systematic literature review on detection tools for modularity anomalies. We found 84 tools, 29 of them available for download. We also compared tools by computing recall, precision, and agreement. We observed that most of the tools rely on detection strategies based on metrics. We also observed redundant results of the tools, and conclude that new detection strategies may be explored to improve the detection effectiveness. In the present study, we aimed to contribute by filling the gap observed in our previous work [9] focused on the SPL context.

71.8 Conclusion and Future Work

In this paper, we evaluated detection strategies for modularity anomalies in SPL. We compared detection strategies from the literature, for two well-known anomalies that may affect the SPL modularity: *God Class* and *God Method*. This comparison aimed to assess if they are effective in the SPL context. We then proposed novel strategies, one for each anomaly, and compared them with the existing strategies. Our study analyzed three SPLs.

As a result, when comparing existing detection strategies, we have observed high recall, with respective mean and median of 66% and 90%. However, we observed low precision with respective mean and median of 16% and 9%. We concluded that the existing strategies were not effective for SPLs. In turn, when comparing our novel strategies with the existing ones, we observed higher recall for our strategies, but low precision as observed for the other strategies. As future work, we suggest the investigation a larger amount of anomalies, new detection strategies, and the analysis of other SPLs.

Acknowledgements This work was partially supported by CAPES, CNPq (grant 424340/2016-0), and FAPEMIG (grant PPM-00382-14).

References

1. Abilio, R., Padilha, J., Figueiredo, E., Costa, H. (2015). Detecting code smells in software product lines. In *Proceedings of the 12th ITNG* (p. 433–438).

2. Abilio, R., Vale, G., Figueiredo, E., Costa, H. (2016). Metrics for feature-oriented programming. In *Proceedings of the 7th WETSoM* (p. 36–42).

3. Almeida, E., Alvaro, A., Lucrédio, D., Garcia, V., Meira, S. (2004). RiSE Project. In *Proceedings of the 5th IRI* (pp. 48–53).

4. Apel, S., Batory, D., Kästner, C., Saake, G. (2013). Feature-oriented software product lines. Berlin Heidelberg: Springer Science & Business Media.

5. Chidamber, S., & Kemerer, C. (1994). A metrics suite for object oriented design. *Transactions on Software Engineering (TSE), 20* (6), 476–493.

6. Fard, A., & Mesbah, A. (2013). JSNose. In *Proceedings of the 13th SCAM* (pp. 116–125).

7. Fawcett, T. (2006). An introduction to ROC analysis. *Pattern Recognition Letters, 27*(8), 861–874.

8. Fenske, W., Schulze, S. (2015). Code smells revisited. In *Proceedings of the 9th VaMoS* (pp. 3–10).

9. Fernandes, E., Oliveira, J., Vale, G., Paiva, T., & Figueiredo, E. (2016). A review-based comparative study of bad smell detection tools. In *Proceedings of the 20th EASE*.

10. Figueiredo, E., Cacho, N., Sant'Anna, C., Monteiro, M., Kulesza, U., Garcia, A., Soares, S., Ferrari, F., Khan, S., Castor Filho, F., Dantas, F. (2008). Evolving software product lines with aspects. In *Proceedings of the 30th ICSE* (pp. 261–270).

11. Filo, T., Bigonha, M., Ferreira, K. (2015). A catalogue of thresholds for object-oriented software metrics. In *Proceedings of the 1st SOFTENG* (pp. 48–55).

12. Fontana, F., Braione, P., & Zanoni, M. (2012). Automatic detection of bad smells in code. *Journal of Object Technology (JOT), 11*(2), 5–1.

13. Fowler, M. (1999). *Refactoring: Improving the design of existing programs*. Reading: Addison-Wesley Publishing.

14. Lanza, M., & Marinescu, R. (2007). *Object-oriented metrics in practice*. Berlin Heidelberg: Springer Science & Business Media.

15. Lorenz, M., & Kidd, J. (1994). *Object-Oriented software metrics: A practical guide*. Englewood Cliffs: Prentice-Hall.

16. McCabe, T. (1976). A complexity measure. *Transactions on Software Engineering (TSE), SE-2*(4), 308–320.

17. Miller, B., Hsia, P., Kung, C. (1999). Object-oriented architecture measures. In *Proceedings of the 32nd HICSS* (pp. 8069–8086).

18. Moha, N., Gueheneuc, Y.-G., Duchien, L., & Le Meur, A.-F. (2010). DECOR. *Transactions on Software Engineering (TSE), 36* (1), 20–36.

19. Paiva, T., Damasceno, A., Padilha, J., Figueiredo, E.,Sant'Anna, C. (2015). Experimental evaluation of code smell detection tools. In *Proceedings of the 3rd VEM* (pp. 17–24).

20. Pohl, K., Böckle, G., van der Linden, F. (2005). Software product line engineering. Berlin Heidelberg: Springer Science & Business Media.

21. Vale, G., & Figueiredo, E. (2015). A method to derive metric thresholds for software product lines. In *Proceedings of the 29th SBES* (pp. 110–119).

22. Vidal, S., Marcos, C., & Díaz-Pace, J. (2014). An approach to prioritize code smells for refactoring. *Automated Software Engineering (ASE), 23*, 1–32.

Randomized Event Sequence Generation Strategies for Automated Testing of Android Apps

David Adamo, Renée Bryce, and Tariq M. King

Abstract

Mobile apps are often tested with automatically generated sequences of Graphical User Interface (GUI) events. Dynamic GUI testing algorithms construct event sequences by selecting and executing events from GUI states at runtime. The event selection strategy used in a dynamic GUI testing algorithm may directly influence the quality of the test suites it produces. Existing algorithms use a uniform probability distribution to randomly select events from each GUI state and they are often not directly applicable to mobile apps. In this paper, we develop a randomized algorithm to dynamically construct test suites with event sequences for Android apps. We develop two frequency-based event selection strategies as alternatives to uniform random event selection. Our event selection algorithms construct event sequences by dynamically altering event selection probabilities based on the prior selection frequency of events in each GUI state. We compare the frequency-based strategies to uniform random selection across nine Android apps. The results of our experiments show that the frequency-based event selection strategies tend to produce test suites that achieve better code coverage and fault detection than test suites constructed with uniform random event selection.

Keywords

GUI testing • Automated testing • Mobile apps • Android

72.1 Introduction

Mobile devices are increasingly powerful resources that enable individuals to perform computing tasks anywhere at anytime. Google's Android holds the largest share of the mobile Operating System (OS) market worldwide [8]. Mobile apps now provide services to end users in critical domains such as e-commerce, banking and healthcare.

Faulty mobile apps may lead to devastating consequences. Only about 16% of users are likely to try a failing app more than twice [15]. By 2017, the mobile app market will be a $77 billion industry [6]. Therefore, the success of a mobile app may depend on how thoroughly it is tested during its development life cycle.

The services provided by a mobile app are often accessed by executing GUI events such as clicking a button or entering data into text boxes. Mobile apps are Event Driven Systems (EDSs) that are often tested with automatically generated GUI event sequences. Model-based testing techniques use an abstract model of the Application Under Test (AUT) to automatically generate event sequences. Model construction is an expensive process and model-based testing approaches may generate infeasible test cases since the model may not always accurately reflect the

D. Adamo (✉) • R. Bryce
University of North Texas, 3940 North Elm Street, 76205 Denton, TX, USA
e-mail: DavidAdamo@my.unt.edu; Renee.Bryce@unt.edu

T.M. King
Ultimate Software Group, Inc., 2250 North Commerce Parkway, 33326 Weston, FL, USA
e-mail: Tariq_King@ultimatesoftware.com

© Springer International Publishing AG 2018
S. Latifi (ed.), *Information Technology – New Generations*, Advances in Intelligent Systems and Computing 558, DOI 10.1007/978-3-319-54978-1_72

runtime state of the AUT. Dynamic GUI testing techniques are often based on dynamic event extraction and do not require an abstract model of the AUT [3]. Dynamic event extraction-based algorithms generate event sequences through repeated selection and execution of available events from GUI states at runtime. Existing algorithms and techniques for dynamic test suite construction focus on other types of GUI-based software (such as desktop and web applications), and are often not effective for mobile apps because they are not adapted to GUI patterns that are unique to mobile applications. Android devices have "home" and "back" navigation buttons that exhibit domain-specific behavior and are available in every GUI state of an Android app. These navigation events often close the AUT and are disproportionately more likely to be selected during dynamic test suite construction, compared to other events that are only available in specific GUI states. Algorithms that do not compensate for the ubiquity of these events may produce test suites that contain too many short event sequences that do not adequately explore the AUT's GUI. Dynamic GUI testing techniques identify potential events from GUI states at runtime and use an event selection strategy to choose events to execute. The choice of event selection strategy may influence the quality of dynamically constructed test suites. While there is a significant body of work on automated GUI testing tools and model-based techniques [1, 10, 11, 13, 16, 17], prior research gives little attention to event selection strategies for dynamic GUI testing of mobile apps. Uniform random selection with pseudo-random numbers may generate test suites in which certain GUI events are executed often, while some others are executed rarely or not at all. Thus, the test suites may fail to exercise parts of the GUI that cover significant amounts of code or expose faults.

In this work, we develop a randomized dynamic event extraction-based algorithm to automatically construct test suites for Android apps. The algorithm allows specification of fixed probability values to prevent disproportionate selection of navigation events. We also develop two frequency-based event selection strategies and empirically compare them to uniform random selection on nine Android apps. Our frequency-based event selection algorithms maintain a history of event selection frequencies and use the frequency information to dynamically alter event selection probabilities during test suite construction. The objective is to generate test cases that are biased toward infrequently selected events. The first algorithm (frequency weighted) assigns weights to each event in a GUI state such that the weight of an event is inversely proportional to the number of times it has been previously selected. The greater the weight of an event, the more likely it is to be selected. The second algorithm (minimum frequency) always chooses an event that has been selected least frequently in a GUI state.

This work makes the following contributions: (i) we develop a randomized test suite construction algorithm for dynamic event extraction-based testing of Android apps (ii) we develop two frequency-based event selection algorithms as alternatives to uniform random event selection (iii) we empirically evaluate the frequency-based and uniform random event selection strategies on nine Android apps in terms of code coverage and fault detection. The results of our experiments show that the frequency-based strategies tend to produce test suites that achieve better code coverage and fault detection than test suites constructed with uniform random event selection.

The rest of this paper is organized as follows. Section 72.2 discusses related work on dynamic GUI testing. Section 72.3 introduces our test suite construction algorithm for Android apps. We present our frequency-based event selection strategies in Sect. 72.4 and empirically evaluate them in Sect. 72.5. Section 72.6 discusses the results of our empirical study, Sect. 72.7 discusses threats to validity and Sect. 72.8 concludes the paper.

72.2 Related Work

Bae et al. [3] perform an empirical comparison of model-based and dynamic event extraction-based GUI testing techniques. They evaluate dynamic event extraction-based testing with uniform random event selection on Java desktop applications. The study shows that a dynamic event extraction-based approach achieves better code coverage and has a lower tendency to generate nonexecutable event sequences compared to a model-based approach. Adaptive random testing (ART) may improve the effectiveness of random testing by evenly spreading test cases across the input domain. Chen et al. [5] show that ART can reduce the number of test cases needed to find the first fault by up to 50%. However, the majority of work in ART focuses only on programs with numeric inputs. Liu et al. [10] adapt ART to mobile app testing and show that an ART approach is more effective for fault detection than random testing. ART is inefficient even for trivial problems because it repeatedly calculates distances among test cases and generates multiple test case candidates, most of which are discarded [2]. Carino [4] develops algorithms to generate and execute single fixed-length event sequences for Java desktop applications. Dynodroid [11] automatically generates a single sequence of inputs for Android apps. In this paper, we develop an algorithm to generate test suites with multiple distinct event sequences of varying length. We use the algorithm to evaluate two frequency-based event selection strategies that dynamically alter event selection probabilities based on the prior selection frequency of events in each GUI state.

72.3   Dynamic Event Sequence Generation

Figure 72.1 shows an algorithm to automatically construct test suites with GUI event sequences for Android apps. The algorithm takes the following input: (i) the application under test (ii) number of event sequences (test cases) to generate (iii) the probability of selecting the *back* event, and (iv) the probability of selecting the *home* event.

Android devices have "home" and "back" navigation buttons that are available in every GUI state of an Android app. The "home" event always closes the AUT regardless of its GUI state. The "back" event often closes the AUT, but its behavior in an event sequence also depends on the GUI state of the AUT and the set of states explored by preceding events. The presence of these navigation events in every GUI state makes them disproportionately more likely to be selected, compared to other events that are only available in specific GUI states. The algorithm in Fig. 72.1 allows specification of fixed probability values for the "home" and "back" events and uses the specified value to probabilistically terminate test cases. A single test case ends when an event closes the AUT. The algorithm begins with an empty test suite and generates multiple event sequences as test cases. Before construction of each event sequence, the algorithm clears all data created by the previous session and initializes an empty event sequence. After restarting the

AUT, the algorithm identifies the events that are available in the AUT's initial GUI state and uses a selection strategy to choose an event. The algorithm executes the selected event and adds it to the event sequence. Selecting and executing an event leads to a new GUI state. This event selection and execution process is repeated in each GUI state until the algorithm selects an event that closes the AUT. The algorithm also ensures that duplicate event sequences are discarded.

The implementation of the *selectEvent* function call on line 13 determines the strategy used to select events from each GUI state. A uniform random strategy uses a uniform probability distribution over the event sequence space to randomly select an event in each GUI state. Each event in a GUI state is equally likely to be selected and the probability distribution never changes. Uniform random selection is often implemented with pseudorandom number generators that select a random event from the set of available events in each GUI state. We discuss frequency-based alternatives to uniform random selection in Sect. 72.4.

72.4   Frequency-Based Event Selection Strategies

72.4.1 Frequency Weighted Selection

In this strategy, event selection probabilities are determined by the number of times each event has been previously selected. Similar to uniform random selection, each event in a GUI state may be selected. Unlike uniform random selection, every event in a GUI state is *not equally likely* to be selected. In any given GUI state, events that have been previously selected fewer times relative to other events, have a higher likelihood of selection. This selection strategy dynamically changes event selection probabilities during test suite construction by keeping track of the selection frequency of each event.

Figure 72.2 shows the frequency weighted selection algorithm. The algorithm takes the set of available events in a GUI state as input. The weight of each event in a GUI state is given by:

$$weight(e) = \frac{1}{N(e) + 1} \qquad (72.1)$$

Input: application under test, AUT
Input: number of test cases, n
Input: back button probability, p_{back}
Input: home button probability, p_{home}
Output: test suite, T

```
 1:  T ← φ
 2:  testCaseCount ← 0
 3:  while testCaseCount < n do
 4:      clear app data and start AUT
 5:      testCase ← φ
 6:      repeat
 7:          if random(0, 1) ≤ p_back then
 8:              selectedEvent ← createBackEvent()
 9:          else if random(0, 1) ≤ p_home then
10:              selectedEvent ← createHomeEvent()
11:          else
12:              events ← getAvailableEvents()
13:              selectedEvent ← selectEvent(events)
14:              execute selectedEvent
15:              testCase ← testCase ∪ {selectedEvent}
16:          end if
17:      until application exits
18:      if testCase ∉ T then
19:          T ← T ∪ {testCase}
20:          testCaseCount ← testCaseCount + 1
21:      end if
22:  end while
```

Fig. 72.1 Randomized test suite construction algorithm

where e is an event and $N(e)$ is the number of times the event has been previously selected. The algorithm makes a random selection biased by the weight of each available event i.e. events with greater weights are more likely to be selected. Finally, the algorithm updates the selection count for the selected event.

Input: set of available events in GUI state, *events*
Output: selected event, *selectedEvent*
1: **function** FREQWEIGHTEDSELECTION(*events*)
2:     $totalWeight \leftarrow 0.0$
3:     **for** *event* in *events* **do**
4:         $totalWeight \leftarrow totalWeight +$ weight(*event*)
5:     **end for**
6:     $selectedEvent \leftarrow$ *first event in set of events*
7:     $selectionWeight \leftarrow random(0,1) \times totalWeight$
8:     $weightCnt \leftarrow 0.0$
9:     **for** *event* in *events* **do**
10:        $weightCnt \leftarrow weightCnt +$ weight(*event*)
11:        **if** $weightCnt \geq selectionWeight$ **then**
12:           $selectedEvent \leftarrow event$
13:           updateSelectionCount(*selectedEvent*)
14:           **return** *selectedEvent*
15:        **end if**
16:     **end for**
17:     updateSelectionCount(*selectedEvent*)
18:     **return** *selectedEvent*
19: **end function**

Fig. 72.2 Frequency weighted event selection algorithm

Input: set of available events in GUI state, *events*
Output: selected event, *selectedEvent*
1: **function** MINIMUMSELECTION(*events*)
2:     $candidates \leftarrow \phi$
3:     $minCount \leftarrow \infty$
4:     **for** *event* in *events* **do**
5:        $timesSelected \leftarrow$ getSelectionCount(*event*)
6:        **if** $timesSelected < minCount$ **then**
7:           $candidates \leftarrow \phi$
8:           $candidates \leftarrow candidates \cup \{event\}$
9:           $minCount \leftarrow timesSelected$
10:        **else if** $timesSelected == minCount$ **then**
11:           $candidates \leftarrow candidates \cup \{event\}$
12:        **end if**
13:     **end for**
14:     $selectedEvent \leftarrow$ selectRandom(*candidates*)
15:     updateSelectionCount(*selectedEvent*)
16:     **return** *selectedEvent*
17: **end function**

Fig. 72.3 Minimum frequency event selection algorithm

72.4.2 Minimum Frequency Selection

This strategy considers only events that have been selected least frequently in a given GUI state. Unlike uniform random and frequency weighted selection, there are instances in which some events in a GUI state have no chance of selection. This strategy gives exclusive consideration to the least frequently selected events in a GUI state.

Figure 72.3 shows the minimum frequency selection algorithm. The algorithm takes the set of available events in a GUI state as input. It iterates through the set of available

events and identifies the subset of events that have been selected the least number of times. All events that are not in this subset are discarded. If there is more than one event that has been selected the least number of times, the algorithm makes a uniform random selection (i.e. random tie breaking). Finally, the algorithm updates the selection count of the selected event.

72.5 Evaluation

72.5.1 Research Questions

We conduct an empirical study to address the following research questions:

RQ1: Do the frequency-based event selection strategies generate test suites that achieve higher code coverage than those generated with uniform random event selection?

RQ2: Do the frequency-based event selection strategies generate test suites that detect more faults than those generated with uniform random event selection?

72.5.2 Subject Apps

We evaluate each event selection strategy on nine real-world Android apps. Each app is publicly available in the F-droid app repository[1] and/or Google Play Store.[2] We selected each app using the following criteria: (i) the app must be GUI-based i.e. no system services (ii) the app's bytecode can be automatically instrumented to collect code coverage metrics without manually modifying its source code. Many Android apps are designed to prevent bytecode instrumentation. We used techniques described by Zhauniarovich et al. [17] to automatically instrument the bytecode of each app. Table 72.1 shows characteristics of the selected apps. The apps range from 1,026 to 5,736 source lines of code (SLOC), 3,597 to 22,169 blocks of bytecode and 1,000 to over 50,000 downloads from Google Play Store. Download metrics are only available for apps in the Google Play Store.

72.5.3 Implementation

We implemented each event selection strategy as part of a prototype tool called *Autodroid*. The tool takes packaged Android apps (APK files) as input and automatically generates test suites with event sequences. *Autodroid*

---
[1] https://f-droid.org/

[2] https://play.google.com/store/apps

Table 72.1 Characteristics of selected Android apps

App name	Domain	Downloads	# Bytecode blocks	# SLOC
Tomdroid v0.7.2	Productivity	10,000–50,000	22,169	5,736
Droidshows v6.5	Entertainment	50,000–100,000	16,244	3,322
Loaned v1.0.2	Lifestyle	100–500	9,781	2,837
Budget v4.0	Finance	10,000–50,000	9,129	3,159
A time tracker v0.23	Productivity	1,000–5,000	8,351	1,980
Repay v1.6	Finance	1,000–5,000	7,124	2,059
SimpleDo v1.2.0	Productivity	100–500	5,355	1,259
Moneybalance v1.0	Finance	–	4,959	1,460
WhoHasMyStuff v1.0.25	Productivity	1,000–5,000	3,597	1,026

collects log files for each test suite it generates. We identify faults by analyzing the log files for unhandled exceptions. The test suites, log files and coverage metadata used in this paper are publicly available online.[3]

72.5.4 Experimental Setup and Design

For each subject app, we use each event selection strategy to generate 10 test suites on an Android 4.4 emulator. Each test suite contains 200 event sequences. The length of an event sequence is determined by the probability of selecting an event that terminates the AUT. We set the probability of the "back" and "home" events in each GUI state to 5% each. Existing test suites for most apps in the F-droid repository have less than 40% block coverage [9]. Our preliminary experiments with each event selection strategy show that the 5% probability value constructs test suites that achieve greater than 40% block coverage for each subject app.

The independent variable in our evaluation is the choice of event selection strategy. We evaluate the following event selection strategies: (i) *Rand* – uniform random event selection (ii) *FreqWeighted* – frequency weighted event selection and (iii) *MinFrequency* – minimum frequency event selection. To answer our research questions, we use two measures of test suite effectiveness as dependent variables: *block coverage* and *number of unique exceptions found*. A *(basic) block* is a sequence of code statements that always executes as a single unit. Block coverage is a measure of the proportion of code blocks executed by a test suite. Exceptions typically indicate faults in an Android app.

72.5.5 Results

Code Coverage: Table 72.2 shows the mean and median block coverage across ten test suites for each app and each event selection strategy. The *FreqWeighted* test suites achieve higher mean and median block coverage than

Table 72.2 Summary block coverage statistics

Measure	Rand	FreqWeighted	MinFrequency
Tomdroid			
Mean (%)	50.10	**51.68**	**52.11**
Median (%)	48.38	**51.25**	48.39
Droidshows			
Mean (%)	55.78	**56.19**	55.38
Median (%)	54.15	**54.30**	**55.80**
Loaned			
Mean (%)	60.94	58.58	**64.21**
Median (%)	58.79	57.30	**63.71**
Budget			
Mean (%)	73.42	**75.08**	**76.28**
Median (%)	72.99	**74.68**	**76.73**
A time tracker			
Mean (%)	67.55	**71.50**	**73.58**
Median (%)	68.35	**71.93**	**74.08**
Repay			
Mean (%)	**48.71**	47.79	48.35
Median (%)	46.16	45.98	**48.90**
SimpleDo			
Mean (%)	49.68	**51.30**	**50.67**
Median (%)	50.00	**51.56**	**52.38**
Moneybalance			
Mean (%)	**91.24**	87.59	85.43
Median (%)	**91.34**	86.88	86.14
WhoHasMyStuff			
Mean (%)	82.44	**83.39**	**83.90**
Median (%)	83.32	**83.61**	**83.95**

uniform random selection in 6 out of 9 apps. *MinFrequency* generates test suites that achieve higher mean block coverage compared to uniform random event selection in 6 out of 9 apps. The *MinFrequency* test suites also achieve higher median block coverage in 8 out of 9 apps compared to uniform random event selection.

To facilitate comparisons across apps, we used min-max scaling [14] to normalize the block coverage measurements for each app. We combined the rescaled data from all apps and performed a Mann-Whitney U-test [12]. The p-values in Table 72.3 indicate that *MinFrequency* shows statistically significant improvement in code coverage over uniform random event selection at the $p < 0.05$ level.

---

[3] https://github.com/davidadamojr/random_strategy_suites

Table 72.3 Comparison of block coverage measurements for random and frequency-based event selection

Null hypothesis	Alternate hypothesis	p-value
$Cov(FreqWeighted) \leq Cov$ (Rand)	$Cov(FreqWeighted) > Cov$ (Rand)	0.18
$Cov(MinFrequency) \leq Cov$ (Rand)	$Cov(MinFrequency) > Cov$ (Rand)	0.03

Table 72.4 Average number of unique exceptions

Application	Rand	FreqWeighted	MinFrequency
Tomdroid	11.4	**11.5**	**12.6**
Droidshows	2.5	**3.1**	**2.7**
Loaned	0	0	0
Budget	0.3	**0.6**	**0.7**
A time tracker	0.1	**0.2**	0.1
Repay	0.4	**0.5**	**0.8**
SimpleDo	0.2	**0.5**	0.2
Moneybalance	**1**	0.9	0.6
WhoHasMyStuff	0.4	**0.8**	**0.6**

Table 72.5 Comparison of number of unique exceptions found with random and frequency-based event selection

Null hypothesis	Alternate hypothesis	p-value
Faults (FreqWeighted) $\leq$ Faults (Rand)	Faults (FreqWeighted) > Faults (Rand)	0.02
Faults (MinFrequency) $\leq$ Faults (Rand)	Faults (MinFrequency) > Faults (Rand)	0.12

Fault Detection: Table 72.4 shows the average number of unique exceptions found by each event selection strategy for each of the apps. The *FreqWeighted* strategy found a higher average number of faults than uniform random selection in 7 out of 9 apps. The *MinFrequency* strategy found a higher average number of faults than uniform random event selection in 5 out of 9 apps. None of the event selection strategies generated test suites that found any faults in the *Loaned* app.

We rescaled and combined the experiment data from all apps and performed a Mann-Whitney U-test[12]. The p-values in Table 72.5 indicate that *FreqWeighted* shows statistically significant improvement in fault detection over uniform random selection at the $p < 0.05$ level.

72.6 Discussion and Implications

72.6.1 Potential Correlations and Factors Affecting Effectiveness

To gain further insight into the results of our experiments, we examined the test suites. Uniform random selection generates test suites that contain a lower number of unique events than those generated with the frequency-based strategies. This may be a factor in the diminished effectiveness of uniform random selection compared to the frequency-based strategies in terms of code coverage and fault detection.

One or both frequency-based strategies generated test suites that achieved better mean or median code coverage in all subject apps except *Moneybalance*. This suggests that the effectiveness of each event selection strategy may be affected by the nature of the AUT's GUI. Most of Moneybalance's functionality can only be accessed by successfully filling multiple validated text input fields. Since the frequency-based strategies take a more systematic approach to event selection, uniform random selection is more likely to repeat valid text entry sequences and may be more suitable for apps where most of the functionality can only be accessed by repeating particular events.

Frequency weighted selection generated test suites that achieved better code coverage and showed the most significant improvement in fault detection over uniform random selection. This may be attributed to a number of factors. A significant body of work in adaptive random testing (ART) shows that improvements in test case diversity lead to improvements in fault detection ability of test suites [5, 10]. To estimate test case diversity, we calculated the sum of the minimum Hamming distances[7] from each event sequence in a test suite to every other event sequence in the test suite. Frequency weighted event selection generated test suites with equal or greater test case diversity compared to test suites generated with uniform random selection. The frequency weighted test suites also had a tendency to cover a higher number of unique events compared to uniform random event selection. Frequency weighted event selection biases test suite construction toward events that have been previously selected fewer times relative to other events. The increase in test case diversity and bias toward unexplored events may be a factor in the significant improvement in fault detection observed with frequency weighted selection.

Minimum frequency event selection generated test suites that achieved the most significant improvement in block coverage over uniform random selection. However, it was less effective than frequency weighted event selection at finding faults despite achieving better code coverage. This may be due to a number of factors. Unlike uniform random and frequency weighted selection, minimum frequency selection only considers the subset of events that have been selected least frequently in each GUI state. We observed that minimum frequency event selection tends to generate test suites that cover a higher number of unique events compared to uniform random selection and frequency weighted selection. However, the test suites also tend to have lower test case diversity than those generated with uniform random and frequency weighted selection. The consideration of only a subset of events in each GUI state may play a role in the relatively low test case diversity of test suites constructed with minimum frequency selection. The increase in the number of unique events covered with minimum frequency selection may be a factor in its tendency to achieve better code coverage than uniform random and frequency weighted selection. The decrease in test case diversity may contribute to its lower tendency to find faults compared to frequency weighted event selection.

72.6.2 Practical Implications for Testers

Testers need to consider the characteristics of the AUT when choosing an event selection strategy to automatically construct test suites. Uniform random selection is a simple and effective choice for small apps with GUIs that consist primarily of validated text input fields. Test suites generated with frequency weighted event selection are effective at finding faults while those generated with minimum frequency selection tend to maximize code coverage. In future work, we will study the effectiveness of our techniques on additional Android apps.

72.7   Threats to Validity

The principal threat to validity of this study is the generalizability of the results as we use a limited number of subject applications. The size and complexity of the AUT may affect the results obtained with our techniques. We minimized this threat by selecting apps of different sizes and from multiple domains. The randomized nature of the event selection algorithms is also a threat to validity. To minimize this threat, we ran the algorithms 10 times for each app. Our assessment of fault finding effectiveness is limited to faults that are exposed as unhandled exceptions. The techniques presented in this work may lead to different results for other types of faults. This work is limited to GUI events and does not consider system events (e.g., changes in network connectivity) that may affect app behavior.

72.8   Conclusions and Future Work

Dynamic event extraction-based techniques can automatically generate event sequence test cases for Android apps. During test case generation, events are typically selected from the AUT's GUI with a uniform probability distribution. In this work, we develop a randomized test suite construction algorithm and evaluate two frequency-based event selection strategies as alternatives to uniform random selection. The first algorithm assigns weights to each event based on its selection frequency and then makes a frequency

weighted selection. The second algorithm always chooses an event that has been selected the least number of times in a GUI state. Both algorithms dynamically alter event selection probabilities based on the prior selection frequency of events. We compared the frequency-based strategies to uniform random selection across nine Android apps. The results show that the frequency-based strategies tend to generate more effective test suites compared to uniform random selection. The **frequency weighted** strategy achieves the most significant improvement in fault detection while the **minimum frequency** strategy achieves the most significant improvement in code coverage. In future work, we will evaluate the cost of the frequency-based strategies in terms of time. We will develop more sophisticated test suite construction algorithms that consider system events such as changes in network connectivity. We will also adapt our algorithms to other mobile app platforms.

Acknowledgements This work is supported in part by Ultimate Software Group, Inc. Any opinions, findings, and conclusions expressed herein are the authors' and do not reflect those of the sponsors.

References

1. Amalfitano, D., Fasolino, A. R., Tramontana, P., Ta, B. D., & Memon, A. M. (2015). MobiGUITAR: Automated model-based testing of mobile apps. *IEEE Software, 32*(5), 53–59.
2. Arcuri, A., & Briand, L. (2011). Adaptive random testing: An illusion of effectiveness? In *Proceedings of the 2011 International Symposium on Software Testing and Analysis* (pp. 265–275). New York: ACM.
3. Bae, G., Rothermel, G., & Bae D.-H. (2014). Comparing model-based and dynamic event-extraction based GUI testing techniques: An empirical study. *Journal of Systems and Software, 97*, 15–46.
4. Carino, S. (2016). *Dynamically testing graphical user interfaces.* Ph.D. dissertation, The University of Western Ontario.
5. Chen, T. Y., Kuo, F.-C., Merkel, R. G., & Tse, T. (2010). Adaptive random testing: The art of test case diversity. *Journal of Systems and Software, 83*(1), 60–66.
6. Entrepreneur.com. (2014). By 2017, the app market will be a $77 billion industry (infographic). [Online]. Available: https://www.entrepreneur.com/article/236832 (Accessed 10-25-2016).
7. Hamming, R. W. (1950). Error detecting and error correcting codes. *Bell System Technical Journal, 29*(2), 147–160.
8. IDC Research. (2016). Smartphone OS market share, 2016 q2. [Online]. Available: http://www.idc.com/prodserv/smartphone-os-market-share.jsp (Accessed 10-25-2016).
9. Kochhar, P. S., Thung, F., Nagappan, N., Zimmermann, T., & Lo, D. (2015). Understanding the test automation culture of app developers. In *2015 I.E. 8th International Conference on Software Testing, Verification and Validation (ICST)* (pp. 1–10). Graz: IEEE.
10. Liu, Z., Gao, X., & Long, X. (2010). Adaptive random testing of mobile application. In *2010 2nd International Conference on Computer Engineering and Technology* (vol. 2, pp. 297–301). Piscataway: IEEE.
11. Machiry, A., Tahiliani, R., & Naik, M. (2013). Dynodroid: An input generation system for android apps. In *Proceedings of the 2013 9th Joint Meeting on Foundations of Software Engineering* (pp. 224–234). New York: ACM.

12. Mann, H. B., & Whitney, D. R. (1947). On a test of whether one of two random variables is stochastically larger than the other. *The Annals of Mathematical Statistics, 18*1, 50–60.

13. Memon, A. M. (2007). An event-flow model of GUI-based applications for testing. *Software Testing Verification and Reliability, 17*(3), 137–158.

14. Mohamad, I. B., & Usman, D. (2013). Standardization and its effects on k-means clustering algorithm. *Research Journal of Applied Sciences, Engineering and Technology, 6*(17), 3299–3303.

15. TechCrunch.com. (2013). Users have low tolerance for buggy apps only 16% will try a failing app more than twice. [Online]. Available: https://techcrunch.com/2013/03/12/users-have-low-tolerance-for-buggy-apps-only-16-will-try-a-failing-app-more-than-twice/ (Accessed 10-25-2016).

16. Zaeem, R. N., Prasad, M. R., Khurshid, S. (2014). Automated generation of oracles for testing user-interaction features of mobile apps. In *2014 I.E. Seventh International Conference on Software Testing, Verification and Validation* (pp. 183–192). Los Alamitos: IEEE.

17. Zhauniarovich, Y., Philippov, A., Gadyatskaya, O., Crispo, B., & Massacci, F. (2015). Towards black box testing of android apps. In *2015 10th International Conference on Availability, Reliability and Security (ARES)* (pp. 501–510). Piscataway: IEEE

Requirement Verification in SOA Models Based on Interorganizational WorkFlow Nets and Linear Logic

Kênia Santos de Oliveira, Stéphane Julia, and Vinícius Ferreira de Oliveira

Abstract

This paper presents a method for requirement verification in Service-Oriented Architecture (SOA) models based on Interorganizational WorkFlow nets. In SOA Design, a requirement model (public model) only specify tasks which are of interest of all parties involved in the corresponding interorganizational architectural model (a set of interacting private models). Architectural models involve much more tasks: they contain the detailed tasks of all the private processes (individual workflow processes) that interact through asynchronous communication mechanisms in order to produce the services specified in the requirement model. In the proposed approach, services correspond to scenarios of Interorganizational WorkFlow nets. For each scenario of the public and private models, a proof tree of Linear Logic is produced and transform into a precedence graph that specifies task sequence requirements. Precedence graphs of the public and private models are then compared in order to verify if all the existing scenario of the requirement model also exist in the architectural model. The comparison of the models (public to private) is based on the notion of branching bisimilarity that prove behavioral equivalence between distinct finite automatas.

Keywords

Interorganizational WorkFlow net • Petri nets • Service-oriented architecture • Linear logic • Bisimilarity

73.1 Introduction

Workflow management systems have not been increasing only within individual organizations but also between different organizations that distributed theirs services within distinct business offices. It exists then a need to work with workflow processes distributed through complex collaboration mechanisms in order to achieve common goals.

K.S. de Oliveira (✉) • S. Julia • V. Ferreira de Oliveira
Federal University of Uberlândia, Computing Faculty,
Uberlândia, Brazil
e-mail: keniasoli@gmail.com; stephane@ufu.br;
viniciusfdeoliveira@gmail.com

Business processes distributed over different organizations are referred to as interorganizational workflow [1]. In particular, interorganizational workflow approaches allow organizations with complementary skills to perform jobs that are not within the range of a single organization. Interorganizational workflow can be easily represented by Interorganizational WorkFlow nets (IOWF-net) [2]. A IOWF-net is derived from the definition of a WorkFlow net (WF-net [3]) which is essentially a Petri net model used to represent a business process [4].

In this context of distributed service, the concept of Service Oriented Architecture (SOA) has been highlighted. SOA consider services as the primary means in the realization of strategic goals [5]. In the current marketplace, the technology platform most associated with the realization of

© Springer International Publishing AG 2018
S. Latifi (ed.), *Information Technology – New Generations*, Advances in Intelligent Systems and Computing 558, DOI 10.1007/978-3-319-54978-1_73

SOA is Web services [5] which are available in a distributed environment provided by Internet. The organizations can then provide their own services through complex tasks which can be implemented using a combination (or composition) of several Web Services [6]. In [7], a Web service is modeled through the help of a WorkFlow net supplemented by an interface.

An important issue in Software Engineering is to ensure that an architectural model reproduces the behavior of the model obtained through the requirement analysis activity. Verify that the behavior of the requirement model exists in the corresponding architectural model minimizes risks of failure in projects, increase the guarantee of software quality and avoids rework costs [8]. Some works, such as [8] and [9], address this issue. However, specifically in the context of SOA only a few works addressing this problem can be found in scientific literature [10]. It is then of great interest to propose an approach that verifies if the requirements defined in analysis models (in terms of dynamic behavior) are present in SOA models.

This article presents an approach based on a kind of comparative analysis between requirement and architectural models. The SOA models are represented by interorganizational WorkFlows nets. As interorganizational WorkFlows nets is based on Petri net theory the modeling and analysis of workflows process are performed in a formal way. In particular, the analysis is based on Linear Logic proof that are produced from interorganizational workflow processes modeled by IOWF-nets. As a matter of fact, some studies have already shown the relationship between the Petri net theory and Linear Logic, such as [11] and [12], since there is an almost direct translation of the structure of a Petri net into formulas of Linear Logic. To show the equivalence of the two distinct models (requirement and architectural models), a new definition of semantic equivalence of distinct models can be produced, as it has been done, for example, in the context of process algebras with the notion of bisimulation [13]. A new definition of semantic equivalence in the context of Linear Logic compatible with the notion of bisimulation will then be introduce in this work in order to verify if requirements specified in an analysis model also exist in the corresponding architectural model in the context of SOA.

The remainder of the article is organized as follows. In Sect. 73.2 the definition of Interorganizational WorkFlow nets, an overview of Linear Logic and the notion of Branching Bisimilarity is presented. The approach to formally detect requirements present in SOA models is proposed in Sect. 73.3. Finally, the last section concludes this work with a short summary, an assessment about the approach presented and an outlook on future work.

73.2 Theoretical Background

73.2.1 Interorganizational WorkFlow Net

An Interorganizational WorkFlow net (IOWF-net) is a Petri net that models an interorganizational workflow process. According to [4], the formal definition of a IOWF-net is based on a tuple $IOWF - net = \{PN_1, PN_2, \ldots, PN_n, P_{AC}, AC\}$, where:

(1) $n \in \mathbb{N}$ is the number of Local WorkFlow nets (LWF-nets);
(2) For each $k \in \{1, \ldots, n\}$: PN_k is a WF-net with source place i_k and sink place o_k;
(3) P_{AC} is the set of asynchronous communication elements (communication places). Each asynchronous communication element corresponds to a place name in P_{AC};
(4) AC corresponds to the asynchronous communication relation. It specifies a set of input transitions and a set of output transitions for each asynchronous communication element.

To clarify the concepts defined above, consider the synthetic example presented in Fig. 73.1. This IOWF-net has two LWF-nets: A and B. Each one has only one source place (iA for LWF-A and iB for LWF-B) and one sink place (oA for LWF-A and oB for LWF-B). The places PC1 and PC2 are the communication places.

In [4], Unfolded Interorganizational WorkFlow nets are defined. The unfolding of an IOWF-net corresponds to the transformation of an IOWF-net into a simple WF-net. In the unfolded net, i.e. the U(IOWF-net), all the LWF-nets are included into a single workflow process considering a start transition and a termination transition. A global source place i and a global sink place o have then to be added in order to respect the basic structure of a simple WF-net. Considering in Fig. 73.1 the dashed arcs, the start place i,

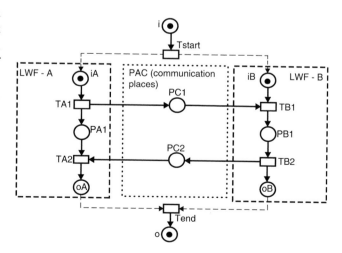

Fig. 73.1 An IOWF-net example

the start transition Tstart, the end place o, and the end transition Tend, the IOWF-net that shows the collaboration between local WorkFlow nets LWF-A and LWF-B can be transformed into the corresponding U(IOWF-net) that respects then the properties of a simple workflow process.

The correctness of a workflow process is associated to the verification of the soundness property [14]. The soundness property will ensure that all the process instances of a WorkFlow net (cases) will be correctly treated and finalized (no deadlock situations and no duplication of the information). Many studies have already considered the analysis of the soundness property in the interorganizational workflow process case as in [4] and [15].

73.2.2 Linear Logic

Linear logic is a refinement of classical and intuitionistic logic and was introduced by Jean-Yves Girard [16]. Instead of emphasizing truth, as in classical logic, or proof, as in intuitionistic logic, Linear Logic emphasizes the role of formulas as resources.

In the Linear Logic, there are several connectives, but in this paper only the *times* connective, denoted by $\otimes$, and the *linear implies* connective, denoted by $\multimap$ are used. The *times* connective represents simultaneous availability of resources; for instance, $A \otimes B$ represents the simultaneous availability of resources A and B. The *linear implies* connective represents a state change; for instance, $A \multimap B$ denotes that consuming A, B is produced; after the production of B, A will not be available anymore.

The following definition presented in [12] show how to translate a Petri net model into Linear Logic formulas:

- A marking M is a monomial in $\otimes$ and is represented by $M = A_1 \otimes A_2 \otimes \ldots \otimes A_k$ where A_i are place names.
- A sequent $M, t_i \vdash M'$ represents a scenario where M and M' are respectively the initial and final markings, and t_i is a list of non-ordered transitions.

A sequent can be proven by applying the rules of the sequent calculus. In this paper, the following rules are used:

- $\multimap_L$ rule – expresses a transition firing and generates two sequents (the right sequent represents the subsequent remaining to be proved and the left sequent represents the consumed tokens by this firing;
- $\otimes_L$ rule – is used to transform a marking in an atoms list;
- $\otimes_R$ rule – transforms a sequent such as $A, B \vdash A \otimes B$ into two identity ones $A \vdash A$ and $B \vdash B$.

Linear Logic proof tree is read from the bottom-up. The proof stops when the identity sequent $o \vdash o$ appears in the proof tree, when there is not any rule that can be applied or when all the leaves of the proof tree are identity sequents.

The Linear Logic proof trees can be transformed into precedence graph, as show in [17]. The precedence graph is obtained by labeling the corresponding proof trees. To label a proof tree, each time the $\multimap_L$ rule is applied, the corresponding transition t_i label the application of the rule, as well as the atoms produced and consumed. Furthermore, the initial event must be labeled by i_i and the final event must be labeled by f_i. Once the labeling is performed, each identity sequence represents the association of two views of the same atom: the left part of an identity sequent is labeled by the event that produced it and the right part of on identity sequent is labeled by the event that consumed it. The labels are shown in the proof tree above the atoms and below the rules $\multimap_L$. In precedence graph, the vertices are events and the arcs are identity sequent, i.e. relation between the event that produced the atom and one that consumed the atom [17]. The Sect. 73.3.2 shows how proof trees and the corresponding labelings are built.

73.2.3 Branching Bisimilarity

The idea of bisimilarity was first introduced in [18] and can be interpreted in the following manner: two processes are equivalent if and only if they can always copy or simulate the actions of each other. Bisimilarity does not make the distinction between external (observable) actions and internal (silent) actions; therefore it is not a suitable equivalence concept for processes with internal behavior.

Branching bisimilarity was first introduced in [19] and is a variant of bisimilarity; however, branching bisimilarity distinguishes external behavior from internal behavior. The distinction between external and internal behavior captures the idea that an environment observing two processes might not be able to see any differences in their behavior while internally the two processes perform different computations [13]. Therefore, to be able to make a distinction between external and internal behavior (hidden events), silent actions can be introduced. Silent actions are actions that cannot be observed. Usually, silent actions are denoted with the action label τ.

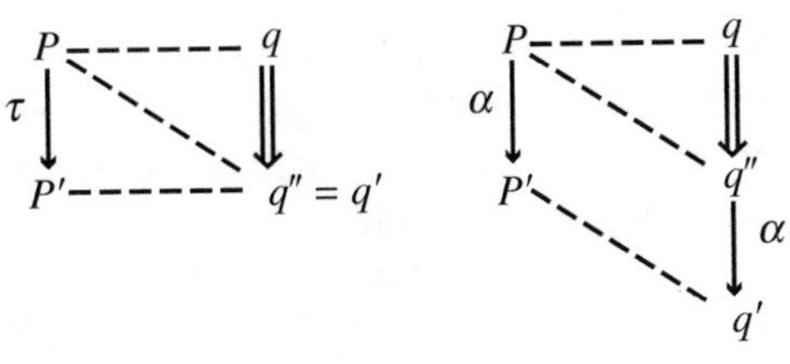

Fig. 73.2 The essence of branching bisimulation

Fig. 73.3 Public model
(requirement)

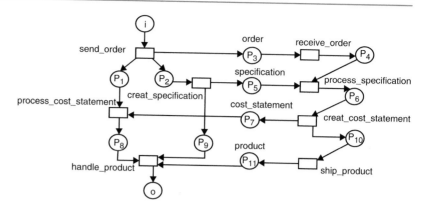

Figure 73.2 presented in [13] show the essence of a branching bisimulation.

In Fig. 73.2, τ represents a silent action, α represents an observable action, and p, q, p', q', q'' represents processes.

On the left side of Fig. 73.2, the process p can evolve into another process p' by executing a silent action (τ) and the process q can evolve into another process q'' by executing a sequence of zero or more τ actions (this is represented by relation '$\Rightarrow$'). Then, it is possible to observe on the left side of the figure that the process p has an equivalence relation with the processes q and q'' and the process p' has an equivalence relation with the process q''. These facts clearly state that two equivalent processes will continue equivalent after the introduction of some additional silent actions in one of the processes or even in both.

On the right side of Fig. 73.2, the process p can evolve into another process p' by executing an observable action (α), the process q can evolve into another process q'' by executing a sequence of zero or more τ actions (relation '$\Rightarrow$') and the process q'' can evolve into another process q' by executing an observable action (α). Then, it is possible to observe on the right side of the figure that the process p has an equivalence relation with the processes q and q'', and the process p' has an equivalence relation with the process q'. These facts clearly state that two equivalent processes will continue equivalent after the introduction of some additional observable actions in one of the processes only if the same observable actions also exist in the other process and respect the same sequence constraints in both processes.

73.3 Requirement Verification in SOA Models

In [20], Aalst defined the P2P-WorkFlow nets, which can be used to model SOA. In such an approach, two views of the system are considered: the public one and the private one. A public WorkFlow net specifies the expected system requirements the parties involved will have to perform. A private WorkFlow net typically contains several tasks which are only of local interest and which do not appear in the public WorkFlow net model.

Figure 73.3 represents a public WorkFlow net example and Fig. 73.4 represents the corresponding private model modeled by an IOWF-net. These examples were presented in [20] and involves two business partners: a contractor (LWF – contractor) and a subcontractor (LWF – subcontractor) as shown in Fig. 73.4. First, the contractor sends an order to the subcontractor. Then, the contractor sends a detailed specification to the subcontractor and the subcontractor sends a cost statement to the contractor. Based on the received specification, the subcontractor manufactures the desired product and sends it to the contractor.

In Fig. 73.4, a global source place i and a global sink place o were added in order to respect the basic structure of a simple WF-net. In fact, p_0 corresponds to the start place of the local WorkFlow net LWF-contractor, p_{26} corresponds to the start place of the local WorkFlow net LWF-subcontractor, p_{27} corresponds to the end place of the local WorkFlow net LWF-contractor and p_{28} corresponds to the end place of the local WorkFlow net LWF-subcontractor.

The P2P approach ensures that the IOWF-net obtained is sound since the private WorkFlow nets (Fig. 73.4) are constructed based on a public WorkFlow net (Fig. 73.3) using specific rules. Such rules restrain the design patterns allowed for the construction of the private model and are based on the concept of branching bissimilarity presented in Sect. 73.2.3. But in practice, organizations build theirs workflow processes the way they want without worrying too much about rules. Considering this fact, Passos and Julia [15] presented an approach that verifies if the main business relationships between involved organizations can be provided safely. The approach identifies deadlock-freeness scenarios in interorganizational workflow processes modeled by IOWF-nets through the use of Linear Logic proof trees and can be applied to interorganizational workflow processes not necessarily sound. In particular, the approach can be applied to models that respect relaxed versions of the soundness criterion.

Therefore, our approach is based on the ideas presented in [20] and [15] whose main purpose is to verify if the behavior defined in the public model (requirement model) is also present in the private model (architectural model)

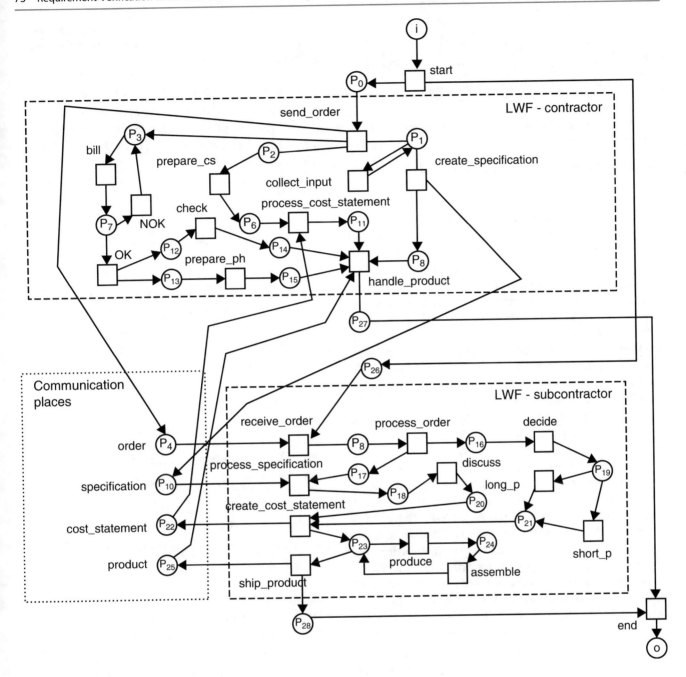

Fig. 73.4 Private model (architecture)

without depending on construction rules that ensure the soundness of the process.

73.3.1 Proposed Method

The approach presented in this work considers that the requirement model corresponds to a public WorkFlow net and the architectural model is the set of private WorkFlow nets that interact through asynchronous communication mechanisms in order to produce the services specified by

the requirement model. Therefore, the architectural model corresponds to an IOWF-net. The services of the SOA model are then the scenarios of the IOWF-net. A scenario in the context of a workflow process corresponds to a well defined route mapped into the corresponding WF-net and, if the WF-net has more than one route (places with two or more output arcs), more than one scenario has to be considered then.

To introduce the proposed approach, the examples presented in Figs. 73.3 and 73.4 are used. In these examples, both the requirement model and the architectural model are

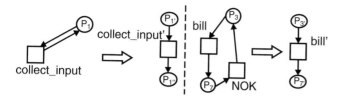

Fig. 73.5 Transformation of iterative route constraints into single tasks

sound and it exists a single scenario described by the public model.

In this work, iterative routes of WorkFlow net models will be replaced by simple global tasks, as it is generally the case of hierarchical approaches based on the notion of well formed blocks [21]. Figure 73.5 shows examples of iterative routes constraints that exist in the WorkFlow net of Fig. 73.4 and theirs transformation into single tasks.

The proposed method in this work respects the following sequence of steps: (1) build the proof trees of Linear Logic for each scenario of the public WorkFlow net and transformed the obtained proof trees into the corresponding precedence graphs; (2) build the proof trees of Linear Logic for each scenario of the private WorkFlow net (unfolded model) and transformed the obtained proof trees into the corresponding precedence graphs; (3) verify the equivalence between the precedence graphs of the public model and the precedence graphs of the private model.

Because the precedence graphs show in a formal way the sequencing constraints of a set of activities performed by a WorkFlow net, they can be seen as a king of operational semantic associated to a workflow process. The equivalence between two WorkFlow nets (the public and the private one) can consequently be verified using the branching bisimilarity concept presented in Sect. 73.2.3 whose purpose is to compare operational semantics of distinct formal behavioral models. In our approach, the activities of the private model that do not appear in the public model will be considered as silent actions. When removing the silent actions of the precedence graphs of the IOWF-net (private models), the obtained reduced graph should be the same as the one of the public WorkFlow net when the behavior of the architectural model reproduces the behavior of the public model (requirement verification of the architectural model).

73.3.2 Method Application

To illustrate the approach, we use the small example of Figs. 73.3 and 73.4. The transitions of the models will be named according to the initial letters of the activities associates to them. For example, *send_order* and *receive_order* will be represented by t_{so} and t_{ro}, respectively.

The transitions (activities) of the requirement model (WorkFlow net of Fig. 73.3) are represented by the following formulas of Linear Logic:

$$t_{so} = i \multimap P_1 \otimes P_2 \otimes P_3, \ t_{ro} = P_3 \multimap P_4,$$
$$t_{cs} = P_2 \multimap P_5 \otimes P_9, \ t_{ps} = P_4 \otimes P_5 \multimap P_6,$$
$$t_{ccs} = P_6 \multimap P_7 \otimes P_{10}, \ t_{pcs} = P_1 \otimes P_7 \multimap P_8,$$
$$t_{sp} = P_{10} \multimap P_{11}, \ t_{hp} = P_8 \otimes P_9 \otimes P_{11} \multimap o.$$

Considering the unique scenario (named Sr) of the requirement model, the following sequent needs to be proven:

$$i, t_{so}, t_{ro}, t_{cs}, t_{ps}, t_{ccs}, t_{pcs}, t_{sp}, t_{hp} \vdash o$$

The proof tree for the requirement model is then the following.

$$
\cfrac{
\cfrac{
\cfrac{
\cfrac{
\cfrac{
\cfrac{
\cfrac{
\cfrac{
\cfrac{
\cfrac{
\cfrac{
\cfrac{P_8 \vdash P_8 \quad P_9 \vdash P_9 \quad P_{11} \vdash P_{11}}{P_8, P_9, P_{11} \vdash P_8 \otimes P_9 \otimes P_{11}} {\otimes_R} \quad o \vdash o }{P_{10} \vdash P_{10} \quad P_9, P_8, P_{11}, P_8 \otimes P_9 \otimes P_{11} \multimap o \vdash o} {\multimap_L}
}{
\cfrac{P_1 \vdash P_1 \quad P_7 \vdash P_7}{P_1, P_7 \vdash P_1 \otimes P_7} {\otimes_R} \quad P_9, P_{10}, P_8, P_{10} \multimap P_{11}, t_{hp} \vdash o} {\multimap_L}
}{P_1, P_9, P7, P_{10}, P_1 \otimes P_7 \multimap P_8, t_{sp}, t_{hp} \vdash o} {\otimes_L}
}{P_6 \vdash P_6 \quad P_1, P_9, P_7 \otimes P_{10}, t_{pcs}, t_{sp}, t_{hp} \vdash o} {\multimap_L}
}{
\cfrac{P_4 \vdash P_4 \quad P_5 \vdash P_5}{P_4, P_5 \vdash P_4 \otimes P_5} {\otimes_R} \quad P_1, P_9, P_6, P_6 \multimap P_7 \otimes P_{10}, t_{pcs}, t_{sp}, t_{hp} \vdash o} {\multimap_L}
}{P_1, P_4, P_5, P_9, P_4 \otimes P_5 \multimap P_6, t_{ccs}, t_{pcs}, t_{sp}, t_{hp} \vdash o} {\otimes_L}
}{P_2 \vdash P_2 \quad P_1, P_4, P_5 \otimes P_9, t_{ps}, t_{ccs}, t_{pcs}, t_{sp}, t_{hp} \vdash o} {\multimap_L}
}{P_3 \vdash P_3 \quad P_1, P_2, P_4, P_2 \multimap P_5 \otimes P_9, t_{ps}, t_{ccs}, t_{pcs}, t_{sp}, t_{hp} \vdash o} {\multimap_L}
}{P_1, P_2, P_3, P_3 \multimap P_4, t_{cs}, t_{ps}, t_{ccs}, t_{pcs}, t_{sp}, t_{hp} \vdash o} {\otimes_L}
}{i \vdash i \quad P_1 \otimes P_2 \otimes P_3, P_3 \multimap P_4, t_{cs}, t_{ps}, t_{ccs}, t_{pcs}, t_{sp}, t_{hp} \vdash o} {\multimap_L}
}{i, i \multimap P_1 \otimes P_2 \otimes P_3, t_{ro}, t_{cs}, t_{ps}, t_{ccs}, t_{pcs}, t_{sp}, t_{hp} \vdash o}
$$

To generate the precedence graph, the proof tree should be labeled as explained in Sect. 73.2.2. For space reasons, only part of the proof trees are labeled as following:

$$
\cfrac{
\cfrac{
\cfrac{\overset{t_{pcs}}{P_8} \vdash \overset{t_{hp}}{P_8} \quad \overset{t_{cs}}{P_9} \vdash \overset{t_{hp}}{P_9} \quad \overset{t_{sp}}{P_{11}} \vdash \overset{t_{hp}}{P_{11}}}{\overset{t_{pcs}}{P_8}, \overset{t_{cs}}{P_9}, \overset{t_{sp}}{P_{11}} \vdash \overset{t_{hp}}{P_8} \otimes \overset{t_{hp}}{P_9} \otimes \overset{t_{hp}}{P_{11}}} {\otimes_R} \quad o \vdash o
}{\vdots}
}{
\cfrac{\overset{i_i}{i} \vdash \overset{t_{so}}{i} \quad \overset{t_{so}}{P_1} \otimes \overset{t_{so}}{P_2} \otimes \overset{t_{so}}{P_3}, P_3 \multimap P_4, t_{cs}, t_{ps}, t_{ccs}, t_{pcs}, t_{sp}, t_{hp} \vdash o}{} {\multimap_L \ t_{hp}}
}{i, i \multimap P_1 \otimes P_2 \otimes P_3, t_{ro}, t_{cs}, t_{ps}, t_{ccs}, t_{pcs}, t_{sp}, t_{hp} \vdash o} {\multimap_L \ t_{so}}
$$

The precedence graph of the requirement model is presented in Fig. 73.6. In this graph, the vertices represented the activities and the arcs the conditions that activate the activities. It can be seen, in particular, as the formal specification of the requirement (in term of behavior) expected

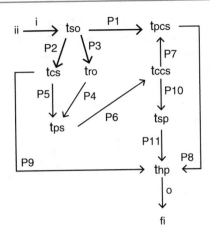

Fig. 73.6 Precedence graph of the requirement model

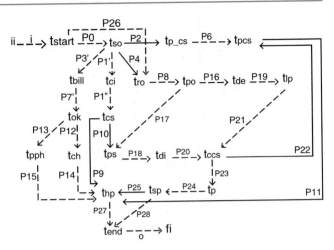

Fig. 73.7 Precedence graph of the architectural model

from the SOA that will implement such a service. It represents too a possible view of the operational semantic associated to the corresponding workflow process.

The transitions of the architectural model (IOWF-net of Fig. 73.4) are represented by the following formulas of Linear Logic:

$$t_{start} = i \multimap P_0 \otimes P_{26}, t_{so} = P_0 \multimap P_1' \otimes P_2 \otimes P_3' \otimes P_4,$$
$$t_{ci} = P_1' \multimap P_1'', t_{pcs} = P_2 \multimap P_6, t_{bill} = P_3' \multimap P_7',$$
$$t_{ro} = P_4 \otimes P_{26} \multimap P_8, t_{cs} = P_1'' \multimap P_9 \otimes P_{10},$$
$$t_{pcs} = P_6 \otimes P_{22} \multimap P_{11}, t_{ok} = P_7' \multimap P_{12} \otimes P_{13},$$
$$t_{po} = P_8 \multimap P_{16} \otimes P_{17}, t_{ps} = P_{10} \otimes P_{17} \multimap P_{18},$$
$$t_{ch} = P_{12} \multimap P_{14}, t_{pph} = P_{13} \multimap P_{15},$$
$$t_{de} = P_{16} \multimap P_{19}, t_{di} = P_{18} \multimap P_{20},$$
$$t_{lp} = P_{19} \multimap P_{21}, t_{ccs} = P_{20} \otimes P_{21} \multimap P_{22} \otimes P_{23},$$
$$t_p = P_{23} \multimap P_{24}, t_{sp} = P_{24} \multimap P_{25} \otimes P_{28},$$
$$t_{hp} = P_9 \otimes P_{11} \otimes P_{14} \otimes P_{15} \otimes P_{25} \multimap P_{27},$$
$$t_{end} = P_{27} \otimes P_{28} \multimap o.$$

One of the scenario of the IOWF-net of Fig. 73.4 (named Sa) corresponds to the following sequent of Linear Logic:

$$i, t_{start}, t_{so}, t_{ci}, t_{p\_cs}, t_{bill}, t_{ro}, t_{cs}, t_{pcs}, t_{ok}, t_{po}, t_{ps}, t_{ch},$$
$$t_{pph}, t_{de}, t_{di}, t_{lp}, t_{ccs}, t_p, t_{sp}, t_{hp}, t_{end} \vdash o$$

For space reasons, just the first and the last lines of the labeled proof trees are shown, as following:

$$\cfrac{\cfrac{\begin{matrix}P_{28}\\t_{sp}\end{matrix} \vdash \begin{matrix}P_{28}\\t_{end}\end{matrix} \quad \cfrac{P_{27}}{t_{hp}} \vdash \cfrac{P_{27}}{t_{end}}}{\begin{matrix}P_{28}\\t_{sp}\end{matrix}, \begin{matrix}P_{27}\\t_{hp}\end{matrix} \vdash \begin{matrix}P_{27}\\t_{end}\end{matrix} \otimes \begin{matrix}P_{28}\\t_{end}\end{matrix}} \otimes R \qquad \cfrac{t_{end} \vdash \overset{o}{f_i} \multimap L}{\vdots}}{}_{t_{hp}}$$

$$\overline{\overset{i}{i_i}, i \multimap P_0 \otimes P_{26}, t_{so}, t_{ci}, t_{p\_cs}, t_{bill}, t_{ro}, \ldots, t_{hp}, t_{end} \vdash o}$$

The precedence graph of the architectural model is then presented in Fig. 73.7. In this graph, all sequences of activities performed in the IOWF-net are clearly specified. In the precedence graph of the architectural model, dashed lines are used when an arc is link to a silent activity (one of its vertices corresponds to a silent activity). As a matter of fact, these activities are of interest only to their respective private WorkFlow net.

By removing the silent actions of a precedence graph, it is necessary to connect the precedent activity that is connected to the silent activity to the successor activity of the silent activity. For example, by removing the activity t_{pcs} in Fig. 73.7 a new directed arc is created between the activities t_{so} and t_{pcs}. Removing all the silent activities of the precedence graph of Fig. 73.7, the precedence graph of Fig. 73.8 is obtained.

Comparing the reduced precedence graph of the architectural model with the precedence graph of the requirement model, it can be observed that both models executed the same activities respecting the same sequential constraints.

The additional arcs iP_{26}, $P_{28}o$, $P_3'P_7'P_{13}P_{15}$, $P_3'P_7'P_{12}P_{14}$ and $P_8P_{16}P_{19}P_{21}$ that exist in the graph of Fig. 73.8 and that do not exist in the graph of Fig. 73.6 are simply redundant constraints that can be removed without modifying the requirement specification. For example, the arc $P_3'P_7'P_{13}P_{15}$ simply state that activity t_{so} has to happen before activity t_{hp}. But this statement already exists through the sequence of arc $P_1'P_1''$, P_9 for example.

Removing the redundant arcs of the graph of Fig. 73.7, the precedence graphs of Figs. 73.6 and 73.8 are exactly the same. Therefore, for this example, it can be concluded that the requirements defined in the public model are present in the architectural model.

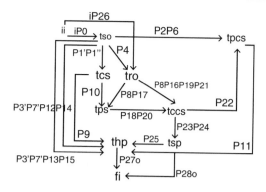

Fig. 73.8 Reduced precedence graph of the architectural model

73.4 Discussion and Conclusion

This paper presented an approach for requirement verification in SOA models based on Interorganizational WorkFlow nets and Linear Logic. Our purpose was to show that, in context of SOA, all scenarios present in the requirement model (public model) was also present in the architectural (private) model. This approach was based in particular on the construction of proof trees of Linear Logic and of precedence graphs that show the operational semantic of distinct models. Building the precedence graph for each scenario of the requirement model and for each scenario of the architecture model, it was possible to compare the graphs and check if they simulate each other's behavior, respecting the notion of branching bisimilarity. The precedence graph is only build for the sequents of Linear Logic syntactically correct, i.e. deadlock-freeness scenarios that reach the ending places of WorkFlow nets.

The advantages of the presented approach is that the organizations do not have to be constrained by external actors to build their private workflow processes, as it is the case of the approach presented in [20]. Therefore, the organizations involved in the interorganizational workflow process can simply verify if the set of requirement scenarios of an analysis model are present in an available SOA model candidate to the implementation of a required service.

One advantage of using Linear Logic's proof trees to qualitative analysis (soundness property) is that it enables the study of sub-processes, considering its partial marking [22]. Linear Logic allows to study the reachability problem of Petri nets considering only partial marking of the net and not necessarily the global marking of the whole model. Another advantage of using Linear Logic is that information related to task execution intervals is directly computed from the proof trees made to prove soundness.

As a future work proposal, the approach will applied to situations that consider more than one scenario in the

requirement model and whose architectural model is not necessarily sound, as it is the case in most of the existing systems that have to deal with possible deadlock situations for example.

References

1. Riempp, G. (1998). *Wide area workflow management: Creating partnerships for the 21st century.* New York: Springer.
2. van der Aalst, W. M. P. (2000). Loosely coupled interorganizational workflows: Modeling and analyzing workflows crossing organizational boundaries. *Information and Management, 37,* 67–75.
3. van der Aalst, W. M. P. (1998). The application of Petri nets to workflow management. *The Journal of Circuits, Systems and Computers, 8,* 21–66.
4. van der Aalst, W. M. P. (1998). Modeling and analyzing interorganizational workflows. In *International Conference on Application of Concurrency to System Design* (pp. 262–272). IEEE Computer Society Press.
5. Erl, T. (2009). *SOA principles of service design.* Upper Saddle River: Prentice Hall.
6. Klai, K., Ochi, H., & Tata, S. (2013). Formal abstraction and compatibility checking of web services. In *20th International Conference on Web Services (ICWS)* (pp. 163–170). IEEE.
7. Passos, L. M. S., & Julia, S. (2015). Deadlock-freeness scenarios detection in web service composition. In *12th International Conference on Information Technology – New Generations* (pp. 780–783).
8. Goknil, A., Kurtev, I., & Berg, K. V. D. (2014). Generation and validation of traces between requirements and architecture based on formal trace semantics. *Journal of Systems and Software, 88*(3), 112–137.
9. Maté, A., & Trujillo, J. (2012). A trace metamodel proposal based on the model driven architecture framework for the traceability of user requirements in data warehouses. *Information Systems, 37*(8), 753–766.
10. Zernadji, T., Tibermacine, C., Cherif, F., & Zouioueche, A. (2015). Integrating quality requirements in engineering web service orchestrations. *Journal of Systems and Software, 122,* 463–483.
11. Girault, F., Pradin-Chezalviel, B., & Valette, R. (1997). A logic for Petri nets. *Journal européen des systèmes automatisés.*
12. Riviere, N., Pradin-Chezalviel, B., & Valette, R. (2001). Reachability and temporal conflicts in t-time Petri nets. In *9th International Workshop on Petri Nets and Performance Models.*
13. Basten, T. (1998). In terms of nets system design with petri nets and process algebra. Ph.D. dissertation, Eindhoven University of Technology, Eindhoven, Netherlands.
14. van der Aalst, W. (1996). Structural characterizations of sound workflow nets. Eindhoven University of Technology, Computing Science Reports/23.
15. Passos, L. M. S., & Julia, S. (2014). Linear logic as a tool for deadlock-freeness scenarios detection in interorganizational workflow processes. In *IEEE 26th International Conference on Tools with Artificial Intelligence* (pp. 316–320).
16. Girard, J.-Y. (1987). Linear logic. *Theoretical Computer Science, 50*(1), 1–102.
17. Diaz, M. (2009). *Petri nets: Fundamental models, verification and applications.* Reading, MA: Wiley-ISTE.
18. Park, D. (1981). Concurrency and automata on infinite sequences. In *5th GI-Conference on Theoretical Computer Science* (pp. 167–183). Berlin: Springer.

19. van Glabbeek, R. J., & Weijland, W. P. (1996). Branching time and abstraction in bisimulation semantics. *Journal of the ACM, 43*(3), 555–600.

20. van der Aalst, W. M. P. (2003). Inheritance of interorganizational workflows: How to agree to disagree without loosing control? *Information Technology and Management, 4*(4), 345–389.

21. Valette, R. (1979). Analysis of Petri nets by stepwise refinements. *Journal of Computer and System Sciences, 18*(1), 35–46.

22. Passos, L. M. S., & Julia, S. (2016). Linear logic as a tool for qualitative and quantitative analysis of workflow processes. *International Journal on Artificial Intelligence Tools*, 1–25. 1 650 008.

Cambuci: A Service-Oriented Reference Architecture for Software Asset Repositories

Márcio Osshiro, Elisa Y. Nakagawa, Débora M.B. Paiva*, Geraldo Landre, Edilson Palma†, and Maria Istela Cagnin

Abstract

Reuse of assets results in faster execution of a software project. Considering the importance of repositories to support reuse of assets, we highlight the benefits of using Reference Architecture (RA) to facilitate the development of repositories. Reference architectures of repositories found in the literature are specific to a particular type of asset or represent only some functionality, and they do not fully meet the expected results of the Reuse Asset Management Process of ISO/IEC 12207. This paper presents a service-oriented reference architecture for software asset repositories, named Cambuci. For this, the systematic process ProSA-RA for supporting definition of reference architectures is used. In order to investigate the quality of the Cambuci's description, we conducted two evaluations. The first based on a checklist and the second based on one instantiation of this RA. From the results obtained, we observed improvement opportunities in the description of Cambuci and the support offered by it in the development of repositories.

Keywords

Reference architecture · Repository · Software asset · ISO/IEC 12207 · SOA

74.1 Introduction

Systematic reuse of software assets leads to a swifter and less costly execution of software projects[7, 15]. To perform the management of software assets reuse [6], support from repositories is needed [1, 2, 4, 9], which are important computational tools to facilitate the practice of reuse by developers.

In parallel, Reference Architectures (RA) have been proposed to ease software development. However, only three RAs were found in the domain of software assets repositories, being two of them for specific types of assets [8, 16] and the other, besides being asset type independent, it mainly deal with software assets sharing between repositories and third party tools [5]. It is noted that none of them fully complies with the results expected by the Software Assets Management Process from the ISO/IEC 12207 standard [6], whose intention is to manage the life cycle of reusable assets from their conception until their deactivation. In addition to this process, other processes of the software life cycle are defined by the ISO/IEC 12207.

With the above-mentioned, it is observed the importance of the definition of a RA that contemplates the management of software assets to support the development of repositories. This RA must also comply with the Service-

*Financial support by Fundect (T.O. n° 219/2014 and 102/2016)
†Financial support by Capes

M. Osshiro
Federal Institute of Mato Grosso do Sul (IFMS) – Campo Grande (MS), Três Lagoas, Brazil
e-mail: marcio.osshiro@ifms.edu.br

E.Y. Nakagawa
University of São Paulo (USP) – São Carlos (SP), São Paulo, Brazil

D.M.B. Paiva (✉) • G. Landre • E. Palma • M.I. Cagnin
Facom – Federal University of Mato Grosso do Sul (UFMS) – Campo Grande (MS), Campo Grande, Brazil
e-mail: istela@facom.ufms.br

© Springer International Publishing AG 2018
S. Latifi (ed.), *Information Technology – New Generations*, Advances in Intelligent Systems and Computing 558, DOI 10.1007/978-3-319-54978-1_74

Oriented Architecture (SOA) [3], due to the loose coupling and interoperability that this type of architecture must provide, as such repositories can be used by many computational tools, such as for modeling, simulation, programming, and version control tools.

This paper presents the definition of a service oriented RA for software assets repositories, named Cambuci. For that, the systematic process to define RAs called ProSA-RA [10] was used. This process was chosen because, aside from being well-documented, it has been already applied in the establishment of many RAs [10]. The evaluation of Cambuci was done by means of a specific questionnaire (checklist FERA [13]) and also by instantiating Cambuci. As a result, it was observed that Cambuci can contribute for the development of repositories and, consequently, increase the practice of reuse by software developers.

This paper is organized as follows: Sect. 74.2 discusses related work, Sect. 74.3 describes the establishment of Cambuci, Sect. 74.4 shows evaluations conducted of Cambuci and results obtained and Sect. 74.5 describes conclusion and future work.

74.2 Related Work

This section presents three RAs of assets repositories [5, 8, 16] that were also considered during the establishment of the Cambuci. These three RAs were identified from a systematic mapping conducted by Osshiro [12], and also updated in May 2016.

The RA from Meland et al. [8] takes only security assets into account, whose main architectural requirements are: communication with the repository from external tools, feedback from the users, usage permissions according to defined policies, and statistics about popularity of assets. The RA from Hongmin et al. [5] uses SOA to facilitate the integration and communication of the repository with other repositories and tools. It has architectural requirements related mainly to the classification of assets, user groups control, and access privileges to the assets. The RA from Yan et al. [16] deals only with assets for business process models and contemplates the main requirements of a traditional repository and specific requirements for business process models repositories, such as the management of life cycle of business processes and configuration of business processes. It was noticed that none of the gathered RAs fully complies with the results expected by the Software Assets Management Process from the ISO/IEC 12207 standard [6], neither were established with the support of a systematic RA definition process. Beyond that, the architectural views from the above-mentioned works lie on high abstraction levels.

74.3 Establishment of Cambuci

The objetive of Cambuci is to support the architectural design phase during the development of repositories of diverse types of software assets, such as requirements, analysis and design models, source code, test data, among others. It is emphasized that the architectural requirements for Cambuci were mainly extracted from ONTO-ResAsset [14] and RefTEST-SOA [11]. The first is an ontology that represents, in one single source, knowledge about specification and management of reusable assets according to ISO/IEC 12207. The second is a service-oriented RA that supports the development of service-oriented testing tools. This RA provides a set of architectural requirements in the SOA context for CASE tools, as it is the case of software asset repositories.

For the establishment of Cambuci, the four steps of ProSA-RA were followed. During the step RA-1 of ProSA-RA, the following information sources were used: (i) ONTO-ResAsset; (ii) RAs of assets repositories [5, 8, 16] (Sect. 74.2); (iii) architectural requirements of the context of services from RefTEST-SOA; and (iv) one expert in the software reuse domain.

In the step RA-2 of ProSA-RA, aided with the information sources selected in the previous step, the elicitation of the architectural requirements for Cambuci was performed, which were divided into two groups. The first group refers to the architectural requirements for the software assets repositories domain (extracted from sources (i) and (ii)) and the second group refers to the architectural requirements for the context of services (extracted from source (iii)). Table 74.1 illustrates the 17 architectural requirements of Cambuci related to the domain of software assets repositories and Table 74.2 shows only part of the 14 architectural requirements of our RA due to limitation of space and related to the context of services.

Each requirement in Table 74.1 is classified in concepts about specification and management of reusable assets and in sublayers from the application layer (primary services, general orthogonal services, orthogonal support services, and orthogonal organizational services) from the overview, which is one the architectural views used in the description of Cambuci. Each requirement from Table 74.2 is classified in SOA concepts and in orchestration, intermediation, presentation, and service layers, also from the overview.

In the step RA-3 of ProSA-RA, the architectural design of Cambuci was made. Four architectural views were developed: overview, module view, runtime view and deployment view. Due to space limitations, only the overview of Cambuci is described in detail.

Table 74.1 Architectural requirements for software assets repositories

ID	Description	Concepts	Sub-layer
AR-SA_1	The RA must enable software assets repositories to include a new asset, that can be composed by various artifacts	Asset specification	Primary service
AR-SA_2	The RA must enable software assets repositories to provide a mechanism for acceptance and certification of assets	Asset management	Primary service
AR-SA_3	The RA must enable software assets repositories to disable assets that will not be used anymore	Asset management	Primary service
AR-SA_4	The RA must enable software assets repositories to allow the classification of one asset and also the record of the usage context of the asset	Asset specification	Primary service
AR-SA_5	The RA must enable software assets repositories to record dependencies between assets	Asset specification	Primary service
AR-SA_6	The RA must enable software assets repositories to notify the concerned about changes that happen in the asset	Communication	Orthogonal organizational services
AR-SA_7	The RA must enable software assets repositories to support the search and recovery of the assets	Asset management	Primary service
AR-SA_8	The RA must enable software assets repositories to support the navigation between assets	Asset management	Primary service
AR-SA_9	The reference architecture must enable software assets repositories to accept assets from multiple sources, with the objective of facilitate the integration between different teams and different repositories	Asset management	Primary asset
AR-SA_10	The reference architecture must enable software assets repositories to create and store multiples versions of the same asset	Configuration management	Orthogonal support services
AR-SA_11	The reference architecture must enable software assets repositories to manage configuration, for example, the definition of items from the asset that are configurable, the control of changes of items from the asset that are configurable	Configuration management	Orthogonal support services
AR-SA_12	The reference architecture must enable software assets repositories to support the record of the users impressions about the version of the asset that they used	Communication	Orthogonal organizational services
AR-SA_13	The reference architecture must enable software assets repositories to record metrics collected about the usage of the asset	Metric	Orthogonal organizational services
AR-SA_14	The reference architecture must enable software assets repositories to provide information related to the reuse, reuse initiatives, more used assets, etc	Communication	Orthogonal organizational services
AR-SA_15	The reference architecture must enable software assets repositories to support access according to the role one user holds	Security	General orthogonal services
AR-SA_16	The reference architecture must enable software assets repositories to ensure the assets integrity, that is, they will not suffer unauthorized modifications	Security	General orthogonal services
AR-SA_17	The RA must enable software assets repositories to be able to manage transactions, ensuring the atomicity, consistency, isolation and durability	Persistence	General orthogonal services

Legend: AR-SA refers to a architectural requirement in the context of software assets repositories

The **Cambuci overview**, showed in Fig. 74.1, presents the services related to the architectural requirements identified in step RA-2. The services from the domain of software assets repositories are allocated in the application layer, and the other services are distributed among the orchestration, intermediation, and presentation layers, according to their roles. Besides that, this view also shows the persistence layer of the software assets.

In the application layer, the services are divided into four groups. Group I (primary services of software assets repositories) refers to the services of specification and management of assets. Group II (orthogonal support services) refers to the support services of the repository such as the dependency control between assets and configuration management. Group III (orthogonal organizational services) refers to the services that support the organizational activities from the repository according to ISO/IEC 12207. Group IV (general orthogonal services) refers to the services with general purposes, such as the planning and management of assets and the feedback from the users of the repository.

The orchestration layer enables creation of new services, performing a composition of already existing services, and the services quality layer verifies the compliance of quality requirements present on other layers of services. This layer guarantees services from other layers are provided with a specified quality. For this, it monitors the provided services and in the case where they will not offer a specified quality, an event is fired to the provider responsible for the service.

Table 74.2 Architectural requirements for the context of services

ID	Requirements	Concepts	Layer
AR-S_1	The RA must enable the developed software assets repositories to persist different types of assets, such that the repositories can be easily integrated	General requirement	Orchestration
AR-S_2	The RA must enable software assets repositories implemented in distinct programming languages and under different platforms to be easily integrated	General requirement	Orchestration
AR-S_3	The RA must provide mechanisms for the services of the software assets repositories to be able to be published and discovered afterwards by client applications	Service publication	Intermediation
AR-S_4	The RA must provide mechanisms for the services of the software assets repositories to be able to be composed by orchestration or used by client applications	Interaction between services, services composition	Orchestration
AR-S_5	The RA must enable the development of software assets repositories that provide information about their characteristics and normative usage directions by means of standardized descriptions	Service description	Presentation
AR-S_6	The reference architecture must enable the development of software assets repositories that provide semantic descriptions, thus allowing their classification in services repositories	Service description, service publication	Presentation, intermediation
AR-S_7	The reference architecture must enable the development of software assets repositories that have at their disposal, information and documents related to their quality characteristics	Service description, service quality	Presentation, service quality
AR-S_8	The reference architecture must provide mechanisms for capture, monitoring, record and signaling of non-compliance of quality requirements established between service providers and service clients	Service quality	Service quality
AR-S_9	The reference architecture must enable the development of scalable software assets repositories, able to evolve incrementally by means of the composition of new functionalities available in the shape of services	Services composition	Business process
AR-S_10	The reference architecture must make possible that software assets repositories services and compositions of these services to be uniformly processed, that is, be able to be published, located and used in the same way	Service description, service publication, interaction between services, services composition	Presentation, intermediation, business process
AR-S_11	The reference architecture must enable services from the software assets repository to be able to interact directly or by means of a service bus	Interaction between services	Business process
…	…	…	…

Legend: AR-S refers to a Architectural Requirement for the context of services

For example, in Cambuci, update of assets can only be done by authorized people, a search for assets must return results before a maximum period of time, disabled assets can not be accessed, users are notified when one asset is modified before a maximum period of time after the occurrence of a modification, among others.

The intermediation layer enables services to be published, discovered, associated, and provided, having the following elements: (i) services registration, which allows publication of a service and search for available services; (ii) service agent, which does the role of mediator between service client and service provider; and (iii) services scheduler, which processes the services requests with dependencies.

Since Cambuci is a service-based RA, its presentation layer has the following elements: (i) service description, which defines how a service can be called and what this

service answers; (ii) service engine, which processes services requests; and (iii) controller, which is responsible for handling services routed by the service engine. The persistence layer stores software assets by means of a database management system or a system of files and directories.

Through the orchestration layer, clients can search functionalities in Service Registration, request functionalities of services directly (using the Service Description) when they know the address of services, or interact through a Service Agent.

The **module view** represents Cambuci through packages, sub-packages, classes, and interfaces. These elements contain functionalities corresponding to the services from the overview of Cambuci. For each service category of each layer of the overview, one corresponding package was identified to compose the module view. Services from each category were defined as sub-packages of the package of the

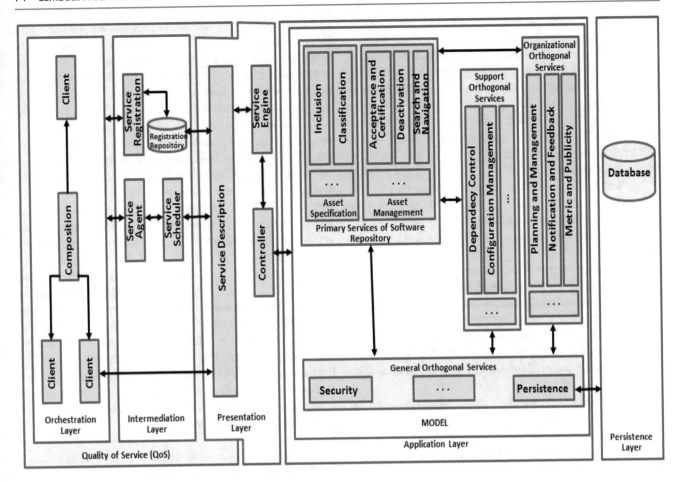

Fig. 74.1 Overview of the Cambuci

corresponding category. For example, services related to software assets repositories, described in the application layer of the overview, are implemented by the packages:

- `PrimaryServicesSoftwareAsset Repository`,
- `OrthogonalSupportServices`,
- `OrthogonalOrganizationalServices`, and
- `GeneralOrthogonalServices`.

The package `PrimaryServicesSoftwareAsse-tRepo- sitory` is sub-divided into two packages: `AssetSpecification` and `AssetsManagement`.

The **runtime view** represents Cambuci by components in different levels of abstraction and their respective interfaces. This view presents the structure of the software assets repository while it is running and shows the interactions between its components.

The **deployment view** represents the hardware structure over which the components of Cambuci are allocated and also shows the network connections that denote interactions between those components. Associated with each element from this view, artifacts that can be generated and

technologies that can be used for the deployment of a software assets repository are suggested; for example, SOAP or REST communication protocols, service description language WSDL, service engine AXIS2 and MySQL database.

74.4 Evaluation of Cambuci

Evaluation of Cambuci was done by means of a checklist, called FERA [13], and also through an instantiation of Cambuci. FERA enables to evaluate the construction and descriptions of RAs and is composed of four steps, containing questions that represent the perspectives of different stakeholders, such as software architects, domain specialists, managers, and developers [13].

The first step verifies the completeness of the general information, construction, and content of RAs. The second step verifies the suitability of RAs to be released for use. The third step makes the conclusion of the general analysis of RAs. The fourth step evaluates specificities of the application domain of RAs. In this work, questions from the first three steps were kept as initially proposed. In the fourth step, four questions were proposed about the domain of software assets repositories, covering concepts from SOA, repositories, and software assets.

Joined the evaluation of Cambuci, 20 graduate students from the course Software Development (class of 2015/2) from the Computer Science Graduate Program from Facom/UFMS and one RA specialist. Training was given to all students to uniform the knowledge about software architecture, RA, software assets, repositories, SOA, and Cambuci. Every doubt from the students was answered during the training.

Beginning the evaluation, the whole documentation of Cambuci was made available to the 21 evaluators. The time spent reading the documentation was one hour and 50 minutes and the time to answer the checklist was two hours.

After ending the checklist, all answers from evaluators were tabulated by question. Each question can have one of the following answers: yes (that is, the documentation fully complies with what is being evaluated by the question), no (that is, the documentation does not comply with what is being evaluated by the question), partially (that is, the documentation partially complies with what is being evaluated by the question).

It was observed that the documentation of Cambuci, on average, fully complies with 55.5%, partially complies with 17% and does not comply with 27.5% of what is evaluated in first step. Based on analysis of results of the first step of FERA, it was concluded that the documentation of Cambuci is satisfactory, containing the main information and complying with the main RA requirements raised by the checklist. The weak point of Cambuci, in relation to the first step, is that it does not present the record of changes that were made. Another point is related to the design rationale, since record of architectural decisions was not made during conception of the RA.

With the results from the second step of FERA, it is observed that the Cambuci's documentation fully complies with 36% of the requirements evaluated in this step, partially complies with 31% and does not comply with 33% of them. From the analysis of answers, it is indicated that some points need to be better specified, such as: description of the stakeholders, identification of variability points, pitfalls, tracking of what can be implemented using components from third parties or open sources, and an implementation schedule.

Since the third and fourth steps of FERA have a reduced number of questions, the percentages from the results of the Cambuci evaluation were calculated by question for those steps. In the third step, it was verified that 48% of the evaluators considered possible to determine probable changes in Cambuci, 62% considered the Cambuci's documentation complete and 76% of the evaluators asserted that the Cambuci's views can be instantiated in concrete architectures of software assets repositories. In the fourth step, it was observed that 52% of the evaluators were able to identify benefits offered by SOA to Cambuci, 76% of the evaluators affirmed that the main requirements from the software assets repositories domain were identified, 43% can glimpse the integration of an instance of Cambuci with other systems/tools, and 81% of the evaluators considered that all important requirements from the domain are documented in the architectural description of Cambuci.

In addition to the evaluation supported by FERA, Cambuci was instantiated as well. For that, Cambuci was used as a base for the development of a service-oriented library, written in Java, named Hidra. This library offers services matching the main functionalities of software assets repositories: 71% of the architectural requirements for repositories (IDs 1, 2, 3, 4, 5, 6, 7, 9, 10, 11, 16, 17 from Table 74.1) and 50% of the architectural requirements for services from Cambuci (IDs 1, 2, 3, 4, 9, 10, 11 from Table 74.2) are covered by Hidra. Based on services provided by Hidra, a software assets repository prototype was built. This way, despite some deficiencies observed in the documentation of Cambuci, it was enough to support the performed instantiation. With that, it was possible to evidence the utility of Cambuci to support the development of software assets repositories.

74.5 Conclusion and Future Work

The main contribution of this work is the Cambuci RA that complies with the Software Assets Management Process from ISO/IEC 12207 and aims to support the architectural design during the development of software assets repositories.

Despite the deficiencies in the Cambuci's documentation identified during the performed evaluation, it was observed that Cambuci complies with the software assets repositories domain and can be used to support the development of repositories, as verified by a performed instantiation, hence contributing to the practice of reuse by software developers.

As future works, we have the evolution of Cambuci's documentation to eliminate the identified deficiencies; instantiation and specialization of Cambuci for the development of repositories with specific types of software assets, such as source code, requirements, and test data; and evaluation of Cambuci from the point of view of various *stakeholders* as well.

---

[1] Available at: http://git.ledes.net/glpn/hidra

References

1. Almeida, E., Alvaro, A., Cardoso, V., Mascena, G., Buregio, V., Nascimento, L., Lucredio, D., & Meira, S. (2007). *C.R.U.I.S.E Component Reuse in Software Engineering*. CESAR, e-Book.
2. Bernstein, P. A., & Dayal, U. (1994). An overview of repository technology. In *VLDB 1994*, pages n.
3. Erl, T. (2007). *SOA principles of service design* (1st ed.). Upper Saddle River: Prentice Hall.
4. Frakes, W. B., & Fox, C. J. (1995). Sixteen questions about software reuse. *Communications of the ACM, 38*(6), 75–87.
5. HongMin, R., Jin, L., & JingZhou, Z. (2010). Software asset repository open framework supporting customizable faceted classification. In *ICSESS 2010* (pp. 1–4).
6. Institute of Electrical and Electronics Engineers and Electronics Industry Association. (2008). IEEE/Std 12207 – Systems and software engineering – Software life cycle processes ISO/IEC 12207:2008. Standard.
7. Jamwal, D. (2010). Software reuse: A systematic review. In *INDIACom-010* (pp. 1–7).
8. Meland, P., Ardi, S., Jensen, J., Rios, E., Sanchez, T., Shahmehri, N., & Tø ndel, I. (2009). An architectural foundation for security model sharing and reuse. In *ARES 2009* (pp. 823–828).
9. Morisio, M., Ezran, M., & Tully, C. (2002). Success and failure factors in software reuse. *IEEE Transactions on Software Engineering, 28*(4), 340–357.
10. Nakagawa, E. Y., Guessi, M., Maldonado, J. C., Feitosa, D., & Oquendo, F. (2014). Consolidating a process for the design, representation, and evaluation of reference architectures. In *WICSA 2014* (pp. 143–152).
11. Oliveira, L. B. R. (2011). *Establishing a service-oriented Reference Architecture for Software Testing Tools*. Master's thesis, University of São Paulo (in Portuguese).
12. Osshiro, M. (2014). Establishing a service-oriented reference architecture for business process line repositories. Master's thesis, Federal University of Mato Grosso do Sul (in Portuguese).
13. Santos, J. F. M., Guessi, M., Galster, M., Feitosa, D., & Nakagawa, E. Y. (2013). A checklist for evaluation of reference architectures of embedded systems. In *SEKE 2013* (pp. 451–459).
14. Silva, L., Paiva, D., Barbosa, E., Braga, R., & Cagnin, M. I. (2014). Onto-resasset development: An ontology for reusable assets specification and management. In *SEKE 2014* (pp. 459–462).
15. Singh, S., Singh, S., & Singh, G. (2010). Reusability of the software. *International Journal of Computer Applications, 7*(14), 38–41.
16. Yan, Z., Dijkman, R., & Grefen, P. (2012). Business process model repositories – Framework and survey. *Information and Software Technology, 54*(4), 380–395.

Extending Automotive Legacy Systems with Existing End-to-End Timing Constraints

75

Matthias Becker, Saad Mubeen, Moris Behnam, and Thomas Nolte

Abstract

Developing automotive software is becoming increasingly challenging due to continuous increase in its size and complexity. The development challenge is amplified when the industrial requirements dictate extensions to the legacy (previously developed) automotive software while requiring to meet the existing timing requirements. To cope with these challenges, sufficient techniques and tooling to support the modeling and timing analysis of such systems at earlier development phases is needed. Within this context, we focus on the extension of software component chains in the software architectures of automotive legacy systems. Selecting the sampling frequency, i.e. period, for newly added software components is crucial to meet the timing requirements of the chains. The challenges in selecting periods are identified. It is further shown how to automatically assign periods to software components, such that the end-to-end timing requirements are met while the runtime overhead is minimized. An industrial case study is presented that demonstrates the applicability of the proposed solution to industrial problems.

Keywords

Component-based software engineering · CBSE · End-to-end timing analysis · Automotive embedded systems · End-to-end delays

75.1 Introduction

The majority of innovations in the automotive industry are driven by advances in the Electric/Electronic (E/E) architecture, where the greater number of those is software driven [1]. In addition to the increasing number of implemented software functions, the complexity of the already existing software functions is increasing as well.

M. Becker (✉) · M. Behnam · T. Nolte
MRTC/Mälardalen University, Västerås, Sweden
e-mail: matthias.becker@mdh.se; moris.behnam@mdh.se;
thomas.nolte@mdh.se

S. Mubeen
MRTC/Mälardalen University, Västerås, Sweden

Arcticus Systems AB, Järfälla, Sweden
e-mail: saad.mubeen@mdh.se

One example is the Engine Management System (EMS), where the complexity of implemented control functions is steadily increasing to achieve low fuel consumption on one side, but also to meet the strict emission guidelines imposed by governmental bodies [2].

To tackle this increasing system complexity, several *abstraction levels* were conceived by academia and industry [1, 3], following the principle of separation of concerns. Each abstraction level provides a complete definition of the system for a given concern. EAST-ADL [3] is an Architecture Description Language (ADL) that models the software architecture at four levels of abstraction, starting from feature modeling in the *Vehicle Level*, followed by the *Analysis Level* where the main focus lies on the consistency of requirements. The remaining two levels focus on the functional design and the allocation of Software Components (SWC) to the hardware elements in the *Design Level*, and

© Springer International Publishing AG 2018
S. Latifi (ed.), *Information Technology – New Generations*, Advances in Intelligent
Systems and Computing 558, DOI 10.1007/978-3-319-54978-1_75

the concrete implementation emerges in the *Implementation Level*, with the possibility to analyze system properties. A detailed description of the different abstraction levels can be found in [3, 4].

Correctness of automotive applications depends not only on the computation of correct values but also on their timely delivery. The allocation of SWCs to hardware elements in the design level also allows to specify timing requirements on them. That is, the last two abstraction levels contain fine-grained timing information. While such timing constraints are generally specified on the execution of specific SWCs, additional constraints may exist on the propagation of data through a chain of SWCs. The most prominent timing constraints being the *data age* and the *reaction delay* [5]. These constraint types are defined in EAST-ADL, as well as in AUTOSAR [1] which provides a standardized software architecture for an automotive embedded system. In this work we focus on the data age, as it is most important for control applications [6].

While the complexity of automotive systems is increasing with each product generation, it is important to support the extension of legacy software in order to reduce the development cost and time. This is especially important since development cycles tend to get smaller and smaller while the system complexity increases. Extending existing systems is a non-trivial task if constraints on the data propagation through a chain of SWCs are specified. Adding an additional SWC may or may not violate the end-to-end constraint, depending on the component's period and execution time.

This work presents novel methods to select periods for additionally added SWCs in a chain of SWCs, targeting the design level where *no* information of the concrete execution platform is available. The main contributions of this work are:

- Challenges system designers face when legacy component chains with specified end-to-end constraints need to be extended are identified.
- A heuristic solution is introduced, with linear computational overhead, that automatically selects the periods of new SWCs in order to reduce the maximum data age in a way such that the system is not over-provisioned, while end-to-end timing constraints are met. Two versions are presented, the first version solely applies timing analysis in the design level, whereas the second solution also specifies a partial ordering on the SWC instances.
- An industrial case study is performed to demonstrate the applicability of the approach to industrial problems. The results additionally highlight the benefits of the approach which additionally specifies a partial order on SWC instances, for both the required runtime of the algorithm and the resulting solution quality.

The rest of the paper is organized as follows. In Sect. 75.2, related work is discussed, followed by background information and the component model in Sect. 75.3. The challenges that arise when legacy systems are extended are discussed in Sect. 75.4. In Sect. 75.5 we present the selection process for component periods. A case study is presented in Sect. 75.6, and finally, conclusions and future work are presented in Sect. 75.7.

75.2 Related Work

Model-driven engineering [7] and component-based software engineering [8, 9] have received significant attention in the automotive domain in recent years. In this context, several frameworks have been developed. For instance, EAST-ADL [3] and EAST-ADL-like models are used at the higher abstraction levels, while AUTOSAR [1], Rubus Component Model (RCM) [10], ComDES [11] and several others are used at the implementation level.

The TIMMO-2-USE [12] project was started out of the need for a timing model in AUTOSAR. The TADL2 language [13] is used to describe timing constraints in all abstraction levels of EAST-ADL, with focus on AUTOSAR at the implementation level.

Mubeen et al. [14] provide a method to refine end-to-end timing requirements early during the development of the systems that reuse the complete or partial models of legacy nodes. However, they do not support the extension of internal software architectures of nodes and allocation of periods to new components, which is the main goal of our current work. In short, the work in [14] focuses on the reuse of complete nodes as black boxes, whereas our current work focuses on the reuse of software components and software component chains, thus providing a higher precision in the corresponding timing analysis results.

Zeng et al. [15] define an Integer Linear Programming (ILP) formulation to assign activation models to distributed automotive effect chains. They do not focus on the optimization of periods which makes their work orthogonal to ours.

In [16], Davare et al. focus on the optimization of task periods in distributed automotive systems. Similar to our work the authors assume that legacy systems are extended and the main objective is to find a period setting that satisfies end-to-end delay constraints. In contrast to our work, the work in [16] uses the convex optimization process which relies on information that are available only at the implementation level. Whereas our methods extract the required information at higher levels of abstraction, i.e. at the design level.

To the best of our knowledge all these works require information only accessible at the implementation level. The main reason being that timing analysis is performed which implicitly assumes the underlying scheduling framework. In contrast, our work relies on the high-level timing analysis presented in [17]. No assumptions are made on the scheduling framework used at the implementation level. This makes it equally applicable for dynamically- or offline-scheduled systems. The same authors have later showed in [18] that the timing analysis framework can reduce pessimism from the analysis results by leveraging information from the implementation level.

75.3 Background and Component Model

This section introduces the required background information on the end-to-end timing constraints, as well as the component model which is basis of this work.

75.3.1 Component Model

Each application is modeled using a number of SWCs. The definition of the component model is according to the EAST-ADL specification [3], targeting the design level, where the functionality is modeled using FunctionTypes (we refer to a FunctionType as the Design Level Software Component (DS-SWC)).

A DL-SWC can itself contain a number of DL-SWCs. On the lowest level a DL-SWC is called elementary DL-SWC. In this work we focus on these elementary DL-SWC. This means the model is flattened such that only elementary DL-SWCs exist (Fig. 75.1).

Elementary DL-SWC must follow a synchronous execution model, namely the *read-execute-write* model, see Fig. 75.2. Each execution can thus be described in three steps.

- **Read:** The DL-SWC creates local copies of all input data.
- **Execute:** During the execution the DL-SWC only operates on local copies of the input and output data.

This has the advantage to result in a coherent view of all data without the need for any additional mechanisms.

- **Write:** The DL-SWC writes the values of the local copies of output data to the output data so that the other DL-SWCs can read it.

Each DL-SWC can provide a number of so-called FunctionFlowPorts. Each port has either input or output functionality, a required data type and further specifications for minimum and maximum supported values.

In order to connect two FunctionFlowPorts, a FunctionConnector is used. Only ports of same data characteristic can be connected.

EAST-ADL discusses port functionality for discrete and continuous functions. In this work, however, we only focus on the former function type.

75.3.1.1 Timing Model for DL-SWCs

Timing behavior can be defined for each DL-SWC. The read and write access is performed instantaneously, however the execution phase has a specified *Execution Time Constraint*. This property defines the lower and upper bound for the execution time of the DL-SWC. This does not include preemptions of execution which may be experienced by the SWC at runtime. The upper bound of a DL-SWC's execution time constraint is defined as C. In addition, a constraints can be added to define the trigger behavior of the DL-SWC. In this work we focus only on the *Periodic Constraints*. This means the DL-SWC is triggered with a fixed period of T time units. We assume the implementation level does not affect the periodicity of the tasks. In the automotive industry periods are generally assigned from a set of periods [19]. $\mathcal{P}$ is defined as the set of all possible periods. In this work, a DL-SWC is represented by f and characterized by the tupple $\{C, T\}$.

75.3.1.2 Communication Mechanism

Already at the design level, EAST-ADL defines semantics on how each DL-SWC shall access the data provided by the FunctionFlowPort. Single buffer is used with each port, where read access is non-consuming (i.e. the value stays in the buffer after the DL-SWC reads it), and write access has overwrite semantics (i.e. a last-is-best semantic is used).

Following the read-execute-write semantic, all input ports are sampled at the invocation and all output ports are updated at the completion of a DL-SWC. Since the read and write phases are performed instantaneously, each communication requires only one buffer. This communication scheme can be visualized as shown in Fig. 75.3. Each communication channel is represented by a shared register (i.e. a buffer of size 1). The output port of DL-SWC f_1 writes to the register R_1, and the input port of DL-SWC f_2 reads from the same register.

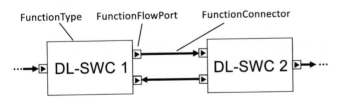

Fig. 75.1 Two design level software components connected by FunctionConnectors

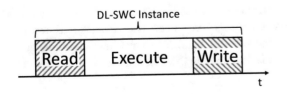

Fig. 75.2 The read-execute-write semantic

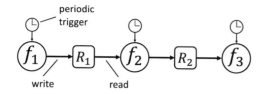

Fig. 75.3 Three periodically triggered DL-SWC and the resulting communication via shared registers

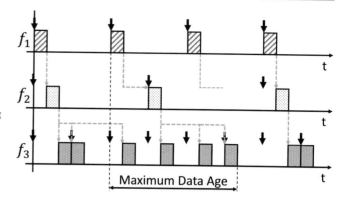

Fig. 75.4 Data propagation in an effect chain of three DL-SWCs (f_1 to f_3) that are activated with different periods. The Data propagation in the schedule is shown by *gray dashed arrows* and the maximum data age is highlighted

75.3.2 End-to-End Delay

In addition to the timing constraints that can be defined for individual DL-SWCs, timing constraints can also be defined on a chain of DL-SWCs. Such a chain is called *Effect Chain* and describes timing constraints from an initial input flow port, through a chain of components, until a final output flow port. This work focuses on the *data age* constraint type. The data age can best be described from the perspective of the last DL-SWC in an effect chain. If the last DL-SWC has finished its execution, i.e. the output is written, how much time has past since the initial input flow port of the effect chain has been sampled, that has effect on this output.

Data age is hence of high importance for each instance of the last DL-SWC in an effect chain. Since runtime scheduling decisions, as well as possibly different periods within the effect chain, have effect on the data age, each instance needs to be examined. The maximum value among all instances needs to be less than the specified data age constraint.

Figure 75.4 presents an example of an effect chain consisting of three DL-SWCs, f_1, f_2, and f_3. A potential schedule of this effect chain is shown to illustrate the example. The periodic releases of several instances of each DL-SWCs are marked with down-pointing arrows. Note that all three components are triggered at different rates. f_1 is the first DL-SWC and is triggered with the middle rate. f_2 is triggered with the slowest rate. It can be seen that the first two instances of f_2 consume the data produced by the respective first and second instance of f_1. Due to the different execution rates, however, the third instance of f_2 consumes the data produced by the fourth instance of f_1. Hence, the third value of f_1 is overwritten and thus never used. Similarly, f_3, which operates at the highest rate in the example, reads the same values more than once. This can, for example, be seen for the first three instances which all consume the data produced by the first instance of f_2. The maximum data age in this example can be found starting from the second instance of f_1, and the finishing time of the sixth

instance of f_3. Note that this is only the maximum data age for this particular schedule.

75.3.3 Implementation Level

At the implementation level, each DL-SWC is translated into a SWC which can be translated to code and executed on the target hardware using a runtime framework. AUTOSAR provides one such framework. In this framework, a SWC includes a collection of smaller entities, so-called runnables [20]. Runnables represent the schedulable entities of AUTOSAR applications and are in turn mapped to tasks, that are then scheduled by the fixed-priority preemptive Operating System (OS) [21]. The RCM provides an alternative execution model and runtime support. The benefits compared to AUTOSAR are that DL-SWCs are not further divided at the implementation level. Thus, one DL-SWC maps to one SWC at the implementation level. In RCM, these SWCs can be thought of as tasks which are scheduled using offline and online scheduling techniques. Since this work focuses on the design level, we restrict ourselves on using the RCM only, since benefits for the analysis arise if a one-to-one mapping between DL-SWCs and SWCs exists.

75.3.4 Calculating the End-to-End Delays

In the implementation level, several methods exist that allow to calculate the maximum data age of effect chains. Most approaches rely on either task response times [22] or recorded scheduling traces, which then allow to apply the methods described in [5]. The drawback of these approaches is that either input format requires scheduling knowledge which is not available at the design level.

Alternatively, [17] presents a scheduling and hardware agnostic method to calculate the maximum data age of effect chains of task sets on a single ECU. In this method, every possible valid data path is examined, where a data path is defined by a possible sequence of DL-SWC instances the data may be propagated through (for example the respective first instances of the DL-SWCs in Fig. 75.4). A data path is valid, if there exists a schedule where the data possibly propagates from the beginning of the effect chain up to its end, while all involved tasks have the chance to be scheduled within their deadlines. Note that this does not guarantee any schedulability of the system and solely focuses on the data age.

While methods based on [5] rely on information that is only available at the implementation level, the method described in [17] can be applied already at the design level. The only prerequisite being that each DL-SWC requires a specified *Execution Time Constraint*, and a *Periodic Constraint*. Thus, this analysis will be applied in the later parts of this paper.

75.3.5 Job-Level Dependencies

Several domains face the problem of having to tackle data propagation through multi-rate effect chains. One approach to increase the predictability of such systems is the usage of *job-level dependencies*. A job-level dependency is specified between two consecutive DL-SWCs of an effect chain. In contrast to DL-SWC-level dependencies, a job-level dependency is specified on instance level. I.e. it is defined which instance of the preceding DL-SWC must terminate before a specific instance of the succeeding DL-SWC is allowed to start (this dependency pattern repeats at the Least Common Multiple (LCM) of the two DL-SWCs). Thus, by specifying job-level dependencies, a partial ordering is introduced on the set of all DL-SWC instances that need to be executed. In [17], a heuristic solution is presented to synthesize job-level dependencies such that a specified data age constraint is met. In [23] it is shown how to schedule such systems using the fixed-priority scheduling.

75.4 Challenges of Extending Legacy Systems

Several challenges exist in the selection of periods for DL-SWCs that are added to a legacy effect chain. This section first outlines general observations and later discusses computational issues that arise.

75.4.1 Implications of the Period Selection

A first intuitive approach is to assign the smallest possible period to all new DL-SWCs. This will result in minimum data age, when compared to the other alternative periods. However, once the transition to the implementation level is taken, this will also result in the highest processor utilization. The required processor utilization of a task can be described as C/T, the relation of execution time to its period. But also more instances of the SWC need to be executed which then additionally result in higher scheduling overheads.

On the other hand, one might assign the largest possible period to all new DL-SWCs. This results in the minimum runtime overhead but also in the largest maximum data age delay.

It can thus be seen that the problem at hand can be described as an optimization problem. The main objective being the selection of a set of periods that yield the least constraint model. We define the least constraint model as the model that conforms to the specified data age constraint but also results in the least runtime overhead once translated to the implementation level.

We can thus describe the objective of the priority assignment: Assign periods out of the set $\mathcal{P}$ to the new DL-SWCs, such that the runtime overhead (i.e. how much processor time is needed) is minimized, while the data age constraint is met.

The required utilization can be described as follows:

$$U_{\text{Chain}} = \sum_{\forall \text{DL}-\text{SWC}_i \in \text{Chain}} \frac{C_i}{T_i} \qquad (75.1)$$

The equation calculates the processor utilization that is required to execute all DL-SWCs instances of the effect chain. If the selected period for a DL-SWC is smaller, U_{Chain} increases.

75.4.2 Exhaustive Search

Performing an exhaustive search over all period combinations for new DL-SWC can be applied. The clear downside of this approach is, however, the computational complexity.

Lets assume an effect chain is extended by a number of n DL-SWCs that each is assigned a period T, where $T \in \mathcal{P}$. The number of different periods p that may be assigned to DL-SWCs is given by the cardinality of $\mathcal{P}$. For an exhaustive search it is then required to examine all possible combinations, i.e. the complexity is $\mathcal{O}(n^p)$.

75.5 Synthesize Periods for Partial Component Chains

This section describes the heuristic that is used to generate the periods for all DL-SWCs that have been added to the effect chain.

75.5.1 Algorithm Description

The proposed heuristic is presented in Algorithm 1. The algorithm assigns periods based on the observation that larger task periods generally lead to a larger end-to-end delay. Since the objective of the approach is to select periods in a way to minimize runtime overheads, larger periods are favorable. Similarly, if tasks need to be assigned a smaller period, they should be selected based on their upper bound of the execution time constraint C, since it will minimize the impact on the total utilization (see Equation 75.1).

Input parameters of the algorithm are the effect chain ζ, the maximum data age constraint D, and the set of all possible periods $\mathcal{P}$.

At the outset, the algorithm creates the set $\mathcal{N}$ that includes all DL-SWCs that have no period specified (line 2). All these DL-SWCs are then assigned an initial period which is equivalent to the maximum possible period out of $\mathcal{P}$ (line 4).

The effect chain created in this way has the minimum possible utilization U_{Chain}.

Algorithm 1: AssignPeriods($\zeta, D, \mathcal{P}$)

1 **begin**
2      $\mathcal{N} = \{DL\_SWC_i | T_i = \emptyset, SWC_i \in \zeta\}$;
3   **for** $\forall DL\_SWC_i \in \mathcal{N}$ **do**
4          $T_i = \max(\mathcal{P})$;
5   **while** checkChain(ζ) $> D$ **do**
6          $\mathcal{H} = \{DL\_SWC_i | \max_{\forall DL\_SWC_i \in \mathcal{N}}(T_i)\}$;
7   **if** $T_{\mathcal{H}} = \min(\mathcal{P})$ **then**
8   **return** unschedulable;
9          $DL_SWC_i = $ getSWCminWCET($\mathcal{H}$);
10          $T_i = $ getNextPeriod($T_i, \mathcal{P}$);
11   **return** success;

In line 5 the algorithm checks if the resulting data age of the effect chain is larger than the specified constraint D. If so, the set of all tasks which have the highest period in $\mathcal{N}$ is created as $\mathcal{H}$ (line 6). If the period of this set is equal to the minimum period in $\mathcal{P}$, all tasks are already reduced to their minimum possible periods while the age constraint is not fulfilled. This means, no successful period assignment could be found and the algorithm returns with *"unschedulable"* (line 8). In all other cases there may still be a period combination that results in the required data age. The algorithm

selects the DL-SWC with the smallest execution time out of all DL-SWCs in $\mathcal{H}$ (line 9). This DL-SWC is then assigned the next smaller period out of $\mathcal{P}$ (line 10). This smallest period is provided by the function getNextPeriod($T_{SWC}, \mathcal{P}$), which returns the next smaller period out of $\mathcal{P}$. The intuition behind this selection is that U_{Chain} is increasing by the minimum possible value compared to all other choices in $\mathcal{H}$ that have larger execution times.

During the execution, the algorithm gradually reduces the periods of new DL-SWCs in order of increasing execution time constraints. During any time, only two different periods are selected for the new DL-SWCs.

The complexity of this algorithm is $\mathcal{O}(n \cdot p)$.

75.5.2 Timing Analysis Variants

The Algorithm 1 can either use the timing analysis directly to verify that the period selection satisfies the data age constraint of the effect chain (in line 5), or job-level dependencies can be synthesized as described in [17]. This decision depends on the concrete development chain and the support for job-level dependencies at the implementation level. The partial order of jobs due to job-level dependencies, however, may allow to specify larger periods which results in smaller system utilization.

75.6 Industrial Case Study

To evaluate the industrial applicability of the proposed method a case study is performed. Each DL-SWC in the effect chain is assigned a period out of the set $\mathcal{P} = 1, 2, 5, 10, 20, 50, 100, 200, 1000$ (all units in ms). This is in line with periods found in automotive control systems [19]. The experiments are performed using two versions of the proposed algorithm. The difference being that version one (No JLD) only performs timing analysis in each iteration, while version two (JLD) also tries to generate job-level dependencies (as described in [17]) while testing the period constellation of the effect chain.

The selected case study is part of an Engine Management System, precisely an effect chain of the Air Intake System (AIS) is shown in Fig. 75.5. The legacy system contains five DL-SWCs, starting with the sampling of the acceleration pedals position. The pedal value is then processed in two stages before it reaches the DL-SWC which is responsible for the control functionality. Finally, the throttle actuation is triggered at the end of the effect chain.

To increase the systems performance, five additional DL-SWCs shall be added to the system. Two DL-SWCs

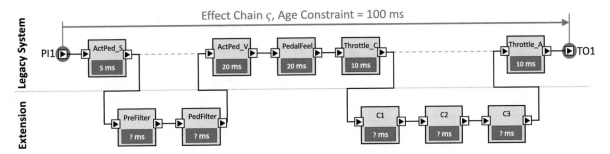

Fig. 75.5 Case study of the effect chain, divided into legacy system and extension

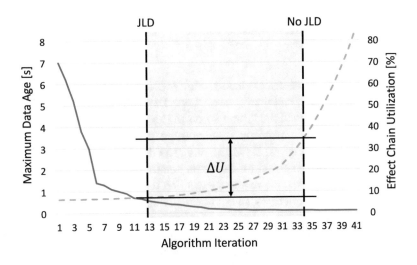

Fig. 75.6 Resulting maximum data age for each algorithm-iteration, shown as *solid curve*. The respective utilization U_{Chain} is shown by the *dashed curve*. In addition, the iteration where the resulting effect chain first satisfies the age constraint is shown for the version with job-level dependencies (JLD) and without (No JLD)

are placed directly after the DL-SWCs that samples the input value. Three DL-SWCs are placed after the control DL-SWC and add additional control functionality to the system. The improved model then consists of ten DL-SWCs, where five of them have no specified period constraint.

The application model is shown in Fig. 75.5. The DL-SWCs of the legacy system have specified execution time constraints of 96 μs, 186 μs, 138 μs, 97 μs, 177 μs, from left to right in Fig. 75.5 respectively. The DL-SWCs which are part of the extension have specified execution time constraints of 133 μs, 158 μs, 136 μs, 199 μs, 161 μs, from left to right in Fig. 75.5 respectively.

The results of the case study are presented in Fig. 75.6. The figure shows the maximum possible data age, as well as the resulting utilization of the effect chain U_{Chain}, plotted for each iteration of the proposed algorithm. It can clearly be seen that the observed data age decreases rapidly. This is due to the variation of periods in $\mathcal{P}$. During the first few iterations, most of the new DL-SWCs are assigned the largest period of 1000 ms. While the algorithm progresses more DL-SWCs are assigned increasingly smaller periods which result in smaller maximum data age. Similarly, it can be observed that the utilization of the effect chain increases

exponentially. Main cause for this behavior is the increasingly smaller periods towards the end of $\mathcal{P}$.

The algorithm which uses only timing analysis to verify the selected periods requires 34 iterations, i.e. 34 different period combinations are examined, where U_{Chain} is 34.8%. In contrast, the algorithm which in addition to the timing analysis tries to synthesize job-level dependencies in a way that the data age constraint is met already finds a valid setting after 11 iterations, where 9 job-level dependencies are specified, and a U_{Chain} is 7.07%.

There is a clear trade-off between the two solutions. The solution with specified job-level dependencies requires much less runtime resources, $\Delta U = 27.76\%$, on the other hand 9 job-level dependencies need to be enforced during runtime. Selecting the better solution heavily depends on the concrete problem and may be subject to the discretion of the system designer.

75.6.1 Required Execution Time

In order to quantify the required execution time of the algorithm, the number of new DL-SWCs in the case study is varied from 0 to 5, where new DL-SWCs are added from

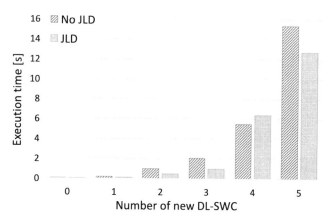

Fig. 75.7 Average execution times of the algorithm without and with job-level dependencies

the beginning of the effect chain in Fig. 75.5 (i.e. one new DL-SWC adds "PreFilter", two new DL-SWC adds "PreFilter" and "PedFilter", and so on.). The experiments are performed on a platform containing an Intel Core i7 processor executing at 2.8 GHz, and 16 GB of RAM. 100 samples are generated for each data point.

It can be seen that the required execution time increases exponentially in both cases. In our concrete implementation of the algorithm, the version which assigns additional job-level dependencies performs better in most scenarios than the basic algorithm that only assigns periods. It performs best in all scenarios except when no new DL-SWC is added (0.59 ms compared to 0.47 ms), and in the case where 4 DL-SWCs are added (6388 ms compared to 5473 ms). It should be noted that the different settings result in placement of new DL-SWCs at different positions in the effect chain. Thus, the results are not general and correspond to the execution times of the algorithm for the software architecture used in the case-study. The better performance of the algorithm with job-level dependencies can be explained by the observation that fewer iterations need to be performed by the algorithm, which increases the industrial applicability. This shows that the larger complexity of the algorithm which assigns job-level dependencies is compensated by the smaller search space that needs to be explored (Fig. 75.7).

75.7 Conclusions and Future Work

Increasing size and complexity of automotive applications calls for model-driven development techniques that allow for faster development cycles. One main concern in today's development cycles is the extension of legacy systems. It is often required to extend the software architectures of legacy systems by adding new SWCs or SWC chains, while the legacy temporal requirements need to be met. Such requirements are typically not only specified on SWC-level but also on a chain of SWCs, where the data propagation through the complete chain is of interest. Thus, software engineers face the challenge to decide the periods of new software components already at early development phases. However, different periods in a chain lead to over- and under-sampling of data which makes it not trivial to determine the end-to-end delay.

The main contributions of this work are the identification of challenges when periods are selected at the design level, and the possible implications at the implementation level. In addition we propose a heuristic algorithm to select periods for new SWCs such that end-to-end constraints are met. This is done in two versions, with and without job-level dependencies that specify a partial execution order of the SWC instances. Finally, an industrial case study is presented to demonstrate the applicability of the proposed algorithm to industrial problems. The evaluations highlight the benefits in low runtime requirements and lower computational times when job-level dependencies are used.

Future work will focus on a proof of concept implementation in an industrial component model and tool chain that supports RCM.

Acknowledgements The work in this paper is supported by the Swedish Knowledge Foundation within the projects PREMISE, PreView, and DPAC. We thank all our industrial partners, especially Arcticus Systems, Volvo Construction Equipment and BAE Systems Hägglunds, Sweden.

References

1. AUTOSAR. Last access Nov 2016. Available at www.autosar.org.
2. Claraz, D., Kuntz, S., Margull, U., Niemetz, M., & Wirrer, G. (2012). Deterministic execution sequence in component based multi-contributor powertrain control systems. *Embedded Real Time Software and Systems Conference (ERTS)* (pp. 1–7).
3. EAST-ADL Association Std. (2014). *EAST-ADL – Domain Model Specification*. V2.1.12.
4. Bucaioni, A., Mubeen, S., Cicchetti, A., & Sjödin, M. (2015). Exploring timing model extractions at EAST-ADL design-level using model transformations. In *12th International Conference on Information Technology: New Generations (ITNG)*.
5. Feiertag, N., Richter, K., Norlander, J., & Jonsson, J. (2008). A compositional framework for end-to-end path delay calculation of automotive systems under different path semantics. In *International Workshop on Compositional Theory and Technology for Real-Time Embedded Systems (CRTS)*.
6. Cervin, A., Henriksson, D., Lincoln, B., Eker, J., & Arzen, K. E. (2003). How does control timing affect performance? Analysis and simulation of timing using jitterbug and truetime. *IEEE Control Systems*, 23(3), 16–30.
7. Schmidt, D. C. (2006) Guest editor's introduction: Model-driven engineering. *Computer*, 39(2), 25–31.
8. Crnkovic, I. & Larsson, M. (2002). *Building reliable component-based software systems*. Norwood, MA: Artech House, Inc.

9. Henzinger, T. A., & Sifakis, J. (2006). The embedded systems design challenge. In *14th International Symposium on Formal Methods (FM)* (Lecture notes in computer science, pp. 1–15). Springer.

10. Hanninen, K., Maki-Turja, J., Nolin, M., Lindberg, M., Lundback, J., & Lundback, K. L. (2008). The rubus component model for resource constrained real-time systems. In *International Symposium on Industrial Embedded Systems (SIES)* (pp. 177–183).

11. Ke, X., Sierszecki, K., & Angelov, C. (2007). COMDES-II: A component-based framework for generative development of distributed real-time control systems. In *13th International Conference on Embedded and Real-Time Computing Systems and Applications (RTAS))*.

12. TIMMO-2-USE. Last access Nov 2016. Available at https://itea3.org/project/timmo-2-use.html.

13. Timing augmented description language (TADL) syntax, semantics, metamodel. (2012). Ver. 2, Deliverable 11, Technical Report.

14. Mubeen, S., Nolte, T., Lundbäck, J., Gålnander, M., & Lundbäck, K.-L. (2016). Refining timing requirements in extended models of legacy vehicular embedded systems using early end-to-end timing analysis. In *13th International Conference on Information Technology: New Generations (ITNG)* (pp. 497–508).

15. Zheng, W., Di Natale, M., Pinello, C., Giusto, P., & Vincentelli, A. S. (2007). Synthesis of task and message activation models in real-time distributed automotive systems. In *Proceedings of the Conference on Design, Automation and Test in Europe (DATE)* (pp. 93–98).

16. Davare, A., Zhu, Q., Di Natale, M., Pinello, C., Kanajan, S., & Sangiovanni-Vincentelli, A. (2007). Period optimization for hard real-time distributed automotive systems. In *44th ACM/IEEE Design Automation Conference* (pp. 278–283).

17. Becker, M., Dasari, D., Mubeen, S., Behnam, M., & Nolte, T. Synthesizing job-level dependencies for automotive multi-rate effect chains. In *Proceedings of the 22th IEEE International Conference on Embedded and Real-Time Computing Systems and Applications (RTCSA)* (pp. 159–169).

18. Becker, M., Dasari, D., Mubeen, S., Behnam, M., & Nolte, T. (2016). Analyzing end-to-end delays in automotive systems at various levels of timing information. In *IEEE 4th International Workshop on Real-Time Computing and Distributed systems in Emerging Applications (REACTION)*.

19. Kramer, S., Ziegenbein, D., & Hamann, A. (2015). Real world automotive benchmarks for free. In *6th International Workshop on Analysis Tools and Methodologies for Embedded and Real-Time Systems (WATERS)*.

20. *AUTOSAR – Software Comp. Template.* (2014). AUTOSAR Std. 4.2.1.

21. *AUTOSAR – Specification of Operating System.* (2014). AUTOSAR Std. 4.2.1.

22. Joseph, M., & Pandya, P. K. (1986). Finding response times in a real-time system. *Computer Journal, 29*(5), 390–395.

23. Forget, J., Boniol, F., Grolleau, E., Lesens, D., & Pagetti, C. (2010). Scheduling dependent periodic tasks without synchronization mechanisms. In *16th IEEE Real-Time and Embedded Technology and Applications Symposium (RTAS)* (pp. 301–310).

Saad Mubeen and Alessio Bucaioni

Abstract

Model- and component-based software development has emerged as an attractive option for the development of vehicle software on single-core platforms. There are many challenges that are encountered when the existing component models, that are originally designed for the software development of vehicular distributed single-core embedded systems, are extended for the software development on multi-core platforms. This paper targets the challenge of extending the structural hierarchies in the existing component models to enable the software development on multi-core platforms. The proposed extensions ensure backward compatibility of the component models to support the software development of legacy single-core systems. Moreover, the proposed extensions also anticipate forward compatibility of the component models to the future many-core platforms.

Keywords

Component-based software engineering • CBSE • Distributed embedded systems • Vehicle software • Component model • Multi-core platforms

76.1 Introduction

Majority of functions in modern vehicles are realized by software that runs on Electronic Control Units (ECUs). The size and complexity of the software is continuously increasing due to the high demand for innovations in the vehicle functionality. Already today, the software in a modern car consists of millions of lines of code that runs on tens of distributed ECUs that can be connected by five or more different types of in-vehicle networks [1]. Moreover, many vehicle functions are required to meet real-time requirements, i.e., logically correct functionality should be provided at the times that are appropriate to the function's environment. Such times are mandated by the timing requirements specified on the functions. Failing to meet the timing requirements results in the system failure in hard real-time systems. In the case of safety-critical systems, violation of a timing requirement can result in catastrophic consequences.

Component-based software engineering [2] and model-driven engineering [3], complemented by the real-time scheduling theory [4, 5], have proven effective in dealing with the software complexity and real-time challenges in single-core distributed embedded systems in the vehicular domain [1, 6–8]. In this regard, several software component models have been developed as a result of the collaboration between academia and the vehicle industry. For example, AUTomotive Open System ARchitecture (AUTOSAR) [7] and the Rubus Component Model (RCM) [8] are the component models that have originated from these collaborations. On the other hand, ProCom [9] and COMDES [10] are the examples of the component models that have originated purely from the academic research.

S. Mubeen (✉) • A. Bucaioni
Mälardalen Real-Time Research Centre (MRTC), Mälardalen University, Västerås, Sweden

Arcticus Systems AB, Järfälla, Sweden
e-mail: saad.mubeen@mdh.se; alessio.bucaioni@mdh.se

© Springer International Publishing AG 2018
S. Latifi (ed.), *Information Technology – New Generations*, Advances in Intelligent Systems and Computing 558, DOI 10.1007/978-3-319-54978-1_76

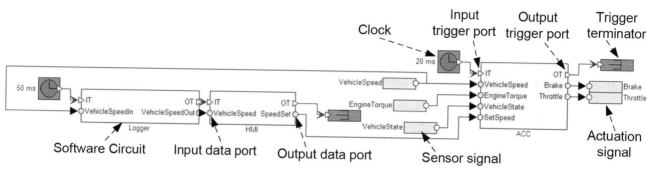

Fig. 76.1 Software architecture of an example system modeled with RCM.

76.1.1 Motivation

The existing single-core platforms fall short of high computational power to support data-intensive sensors and complex coordination among ECUs that is required to support many advanced vehicle features. Recently, multi-core ECUs have been introduced in the vehicular domain to provide such high levels of computational power [11–13]. AUTOSAR has recently introduced guidelines to develop multi-core systems [14]. Using these guidelines, the modeling and runtime support for multicore platforms is discussed in [15]. Other frameworks that are in the scope of this paper include EAST-ADL [16] and AADL [17]. While the existing software development approaches and tools provide good support for single-core platforms, such a support for multi-core platforms in the vehicle industry is yet to mature. This paper takes the first step towards providing an industrial prototype comprising of a component model and a tool chain for the software development of vehicular distributed embedded systems on multi-core platforms.

76.1.2 Paper Layout

The rest of the paper is organized as follows. Section 76.2 presents the research challenges and paper contributions. Section 76.3 discusses the background about the Rubus concept, component model and tool suite. Section 76.4 discusses the structural hierarchy of the existing component models for distributed embedded systems. Section 76.5 presents the proposed extensions to the structural hierarchy. Section 76.6 summarizes the paper and discusses the ongoing work.

76.2 Research Challenges and Paper Contribution

In this paper we identify and target the challenge that is concerned with extension of the existing component models (originally designed for single-core platforms) to support the

vehicle software development on multi-core platforms. In particular, we target the component models that explicitly include the following two features.

(1) A clear separation between the control flow and the data flow among software components is supported. An example of a software architecture with a clear separation between the two flows is shown in Fig. 76.1.
(2) The pipe-and-filter communication style is supported for the interaction among software components.

It is worth mentioning that a large majority of component models with the above-mentioned capabilities facilitate the extraction of timing models from the software architectures at the higher levels of abstraction and at the earlier phases during the development [18]. These timing models are used by the end-to-end timing analysis engines to verify the timing predictability of the systems [19, 20]. Besides, several component models with these features are already used in the vehicle industry, e.g., RCM.

In this paper we aim at answering the following question.

> What extensions are needed in the structural hierarchy of the existing component models for single-core distributed embedded systems to enable them to support the vehicle software development for multi-core distributed embedded systems?

One of the main objectives during the extensions is to ensure backward compatibility with legacy single-core systems, as well as to anticipate the vehicle software development on the future many-core platforms.

76.3 Background: RCM and Rubus-ICE

We consider RCM and its tool suite Rubus-ICE (Integrated software Component development Environment) [21] as a starting point for our work. Currently, the Rubus models and tools support the model- and component-based software development of vehicular distributed embedded systems on single-core platforms. We aim to provide a proof of concept for our approach by extending RCM to support the modeling of the software architectures for multi-core platforms as well.

The Rubus models and tools are developed by Arcticus Systems[1] in collaboration with several academic and industrial partners. RCM and its tool suite have been used in the vehicle industry for over 20 years, e.g., by Volvo Construction Equipment,[2] Haldex,[3] Knorr-Bremse,[4] Mecel,[5] Hoerbiger[6] and BAE Systems Hägglunds.[7] The Rubus concept is based around RCM and Rubus-ICE which is a collection of modeling tools, code generators, analysis tools and runtime infrastructure.

The runtime environment for the Rubus-based applications are supported by the Rubus real-time operating system (RTOS). The Rubus RTOS has been approved as a certifiable RTOS for real-time systems according to the automotive safety standard ISO 26262[8] with respect to the highest safety-integrity level, namely ASIL D. The Rubus RTOS has been ported to many commercial single-core processors including, among others, the Freescale MPC-processors, Texas DSP processors and Infineon processors belonging to the xc167 series. The Rubus RTOS has also been ported to various compiler environments such as Green Hills, WindRiver, Tasking, Microsoft VS and GCC.

RCM and Rubus-ICE allow for the resource-efficient development of predictable, timing analyzable and synthesizable vehicular embedded systems. The predictability of the single-core systems can be verified by the end-to-end response-time and delay analysis supported by Rubus-ICE [20]. The software architecture of a system can be graphically modeled with interconnected software components in RCM. These interconnected components, following a hardware paradigm called the Software Circuits (SWCs), define the structure of the application system that can be analyzed and synthesized entirely within the Rubus environment. Figure 76.1 depicts an example of the software architecture in RCM that is composed of SWCs, interconnections between SWCs and interactions of SWCs with external events and actuators. It should be noted that RCM provides a clear separation between the control flow and the data flow among the software components.

[1] http://www.arcticus-systems.com

[2] https://www.volvoce.com

[3] www.haldex.com

[4] http://www.knorr-bremse.com

[5] http://mecel.se

[6] http://www.hoerbiger.com

[7] http://www.baesystems.com

[8] http://www.iso.org/iso/catalogue_detail?csnumber=43464

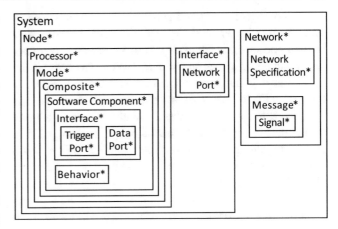

Fig. 76.2 Structural hierarchy in the existing component models for single-core vehicular distributed embedded systems

76.4 Structural Hierarchy in the Component Models Supporting Single-Core ECUs

This section provides a discussion on the structural hierarchy that is found in the majority of existing component models for distributed embedded systems [8–10] as shown in Fig. 76.2. The component models that are focussed in this paper have the following two features. First, these component models use the pipe-and-filter communication for the interaction among the software components. Second, these component models distinguish between the control and data flows among the software components in the software architecture of single-core systems.

In this structural hierarchy, the highest-level hierarchical element is called the *system*. The *system* contains the models of one or more networks and nodes (ECUs). It should be noted that the symbol* in Fig. 76.2 represents one or more elements. A network contains the models of Network Specification (NS) and message objects. The NS is the model representation of a physical network in the vehicle. It is unique for each network communication protocol, e.g., Controller Area Network (CAN), Flexray and switched Ethernet. Each message contains a set of signals that are mapped to it.

The node contains one or more models of processor. Each processor defines a unique runtime environment for the node. Basically, a processor represents the hardware and operating system specific instance of a node. Several processors can be assigned to a single node, e.g., a real hardware target such as the ARM processor or a virtual processor that simulates the instruction set and environment of the hardware target. Note that only one processor can be selected from each node during the deployment. The node also includes an interface that contains one or more network ports that are responsible for sending/receiving messages to/from the network respectively.

Each processor contains one or more models of modes. A mode defines different states of the system. A mode may contain one or more composites. Each composite is a container that encapsulates one or more SWCs. The SWC is the lowest-level hierarchical element that encapsulates basic functions. Each SWC contains only one interface and one or more behaviors. The interface contains only one input trigger port and one or more output trigger ports, input data ports and output data ports. The behavior of an SWC represents a function, e.g., a C function or a simulink block. The SWC has the run-to-completion semantics. This means that upon triggering (activation) it reads data from the input data ports, executes its functionality, writes data at the output data ports and produces an output trigger.

76.5 Proposed Extensions to the Structural Hierarchy

The paper aims at achieving several important goals while addressing the question posed in Sect. 76.2. One goal is to keep the modeling overhead for the user as small as possible. In other words, the extended component model for multi-core platforms should not enforce a lot of burden for the user who already uses the existing component model for developing the software on single-core platforms. Another goal is to support the software development on legacy single-core and contemporary multi-core ECUs, while anticipating further extensions of the component model to support the software development of the future many-core ECUs that will contain several tens of cores connected by on-chip networks.

In order to support the vehicle software development seamlessly on single- and multi-core platforms, we propose extensions to the structural hierarchy of the existing component models by introducing the models of the core, partition and Intra-Processor Communicator (IPC) as shown in Fig. 76.3. The parts of the structure that are above the processor and below the mode in the extended hierarchy remain the same. Note that the symbol* in Fig. 76.3

represents one or more elements. The model of the processor in the extended hierarchy contains one or more cores. A core contains at least one partition. Each partition is assumed to run an instance of the operating system. Each processor also contains one model of the IPC object which handles the inter-core communication as well as the inter-partition communication within each core.

The IPC object can be adapted for any inter-core communication platform. For example, the cores may communicate via direct point-to-point connections, a bus (in the case of contemporary multi-core platforms) or a network (in the case of many-core platforms). The extended structural hierarchy shown in Fig. 76.3 is sufficient to support the software development for multi-core ECUs. Moreover, the extended hierarchy can be used to develop the vehicle software on single-core ECUs by setting the number of cores and partitions each equal to one. The extended hierarchy also allows to reuse the complete models of single-core legacy nodes by allocating the legacy processor to the partition in a single-partition core. The proposed partition model is inline with the partition concept in the ARINC 653 Avionics standard [22]. Hence, the partitions support the development of single- and multi-core systems that have different criticality levels in their software architectures.

76.6 Summary of Ongoing Work

In this paper we have discussed one of the core challenges, that is experienced when an existing component model for the software development of vehicular distributed embedded systems on single-core platforms is extended to support the software development on multi-core platforms. We have addressed the challenge by proposing extensions to the structural hierarchy of the existing component models. The proposed extensions are backward compatible to facilitate the software development on legacy single-core platforms. These extensions also anticipate forward compatibility to support the software development on the future many-core platforms.

Currently, we are developing a hardware model corresponding to the extended structural hierarchy. We also plan to provide a software-to-hardware allocation model. The next step is to provide a formal meta-model definition for the extended structural hierarchy that now supports multi-core platforms. As a proof of concept, we plan to revise and extend the meta-model of Rubus Component Model to incorporate the proposed extensions.

Acknowledgements The work in this paper is supported by the Swedish Knowledge Foundation within the project PreView. We thank all our industrial partners including Arcticus Systems, Sweden, Volvo Construction Equipment (VCE), Sweden and BAE Systems Hägglunds, Sweden.

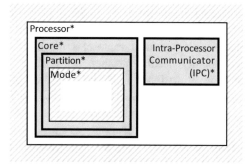

Fig. 76.3 Proposed extensions to the structural hierarchy to support the software development on multi-core platforms

References

1. Broy, M., Kruger, I., Pretschner, A., & Salzmann, C. (2007). Engineering automotive software. *Proceedings of the IEEE, 95*(2), 356–373.
2. Crnkovic, I., & Larsson, M. (2002). *Building reliable component-based software systems*. Norwood, MA: Artech House, Inc.
3. Schmidt, D. C. (2006). Guest editor's introduction: Model-driven engineering. *Computer, 39*(2), 25–31.
4. Audsley, N., Burns, A., Davis, R., Tindell, K., & Wellings, A. (1995). Fixed priority pre-emptive scheduling: an historic perspective. *Real-Time Systems, 8*(2/3), 173–198.
5. Sha, L., Abdelzaher, T., Årzén, K.-E., Cervin, A., Baker, T. P., Burns, A., Buttazzo, G., Caccamo, M., Lehoczky, J. P., & Mok, A. K. (2004). Real time scheduling theory: A historical perspective. *Real-Time Systems, 28*(2/3), 101–155.
6. Henzinger, T. A., & Sifakis, J. (2006). The embedded systems design challenge. In *14th International Symposium on Formal Methods (FM)* (Lecture notes in computer science, pp. 1–15). Springer.
7. AUTOSAR Techincal Overview, Release 4.1, Rev. 2, Ver. 1.1.0., The AUTOSAR Consortium, Oct 2013. http://autosar.org.
8. Hänninen, K. et al. (2008). The rubus component model for resource constrained real-time systems. In *3rd IEEE International Symposium on Industrial Embedded Systems*.
9. Sentilles, S., Vulgarakis, A., Bures, T., Carlson, J., & Crnkovic, I. (2008). A component model for control-intensive distributed embedded systems. In *The 11th International Symposium on Component Based Software* (pp. 310–317).
10. Ke, X., Sierszecki, K., & Angelov, C. (2007). COMDES-II: A component-based framework for generative development of distributed real-time control systems. In *13th International Conference on Embedded and Real-Time Computing Systems and Applications*.
11. SymTA/S for migration from single-core to multi-core ECU-software on infineon microcontrollers. *Electronic Engineering Journal*. 2010.
12. Roland Berger. Consolidation in Vehicle Electronic Architectures, In Think: Aact, July 2015. Available at: https://www.rolandberger. com/en/Publications/pub_-consolidation_-in_-vehicle_-electronic_ -architectures.html. Accessed Oct 2016.
13. Infineon. 32-bit TriCore™Microcontroller. http://www.infineon. com/cms/en/product/microcontroller/32-bit-tricore-tm-microcontro ller/channel.html?channel=ff80808112ab681d0112ab6b64b50805. Accessed Dec 2016.
14. Guide to Multi-Core Systems. Release 4.1, Rev. 3, The AUTOSAR Consortium, Mar 2014. https://www.autosar.org.
15. Moghaddam, A. S. (2013). Performance evaluation and modeling of a multicore AUTOSAR system. Master's thesis, Department of Computer Science and Engineering, Chalmers University of Technology, Sweden.
16. EAST-ADL Domain Model Spec., V2.1.12, Accessed Nov 2016. http://www.east-adl.info/Specification/V2.1.12/EAST-ADL-Specif ication_V2.1.12.pdf.
17. Feiler, P., Lewis, B., Vestal, S., & Colbert, E. (2005). An overview of the SAE Architecture Analysis & Design Language (AADL) Standard: A basis for model-based architecture-driven embedded systems engineering. In *Architecture description languages* (Vol. 176, pp. 3–15). Springer US.
18. Mubeen, S., Mäki-Turja, J., & Sjödin, M. (2014). Towards extraction of interoperable timing models from component-based vehicular distributed embedded systems. In *International Conference on Information Technology: New Generations*. IEEE.
19. Mubeen, S., Nolte, T., Lundbäck, J., Gålnander, M., & Lundbäck, K.-L. (2016). Refining timing requirements in extended models of legacy vehicular embedded systems using early end-to-end timing analysis. In *13th International Conference on Information Technology: New Generations (ITNG)*.
20. Mubeen, S., Mäki-Turja, J., & Sjödin, M. (2013). Support for end-to-end response-time and delay analysis in the industrial tool suite: Issues, experiences and a case study. *Computer Science and Information Systems, 10*(1).
21. Rubus ICE-Integrated Development Environment. http://www. arcticus-systems.com.
22. ARINC Specification 653P1-2. Avionics Application Software Standard Interface Part 1: Required Services. http://www.arinc. com.

On the Impact of Product Quality Attributes on Open Source Project Evolution

77

António César B. Gomes da Silva, Glauco de Figueiredo Carneiro, Miguel Pessoa Monteiro, Fernando Brito e Abreu, Kattiana Constantino, and Eduardo Figueiredo

Abstract

Context: Several Open Source Software (OSS) projects have adopted frequent releases as a strategy to deliver both new features and fixed bugs on time. This cycle begins with express requests from the project's community, registered as issues in bug repositories by active users and developers. Each OSS project has its own priorities established by their respective communities. A a still open question is the set of criteria and priorities that influence the decisions of which issues should be analyzed, implemented/solved and delivered in next releases. In this paper, we present an exploratory study whose goal is to investigate the influence of target product quality attributes in software evolution practices of OSS projects. The goal is to search for evidence of relationships between these target attributes, priorities assigned to the registered issues and the ways they are delivered by product releases. To this end, we asked six participants of an exploratory study to identify these attributes through the data analysis of repositories of three well-known OSS projects: Libre Office, Eclipse and Mozilla Firefox. Evidence indicated by the participants suggest that OSS community developers use criteria/ priorities driven by specific software product quality attributes, to plan and integrate software releases.

Keywords

Software releases • Open Source Software (OSS) projects • Software product quality attributes

77.1 Introduction

The attractiveness of Open Source Software (OSS) projects for both the users' and developers' communities has aroused the interest of Software Engineering researchers [1]. OSS development is a relevant context to study, but it differs from proprietary development in aspects such as development processes, team structure, and developer incentives [2]. Understanding the rationale behind successful OSS initiatives may allow teams of non-OSS projects to reuse best practices on their own projects [3, 4].

According to previous work [5], there are more than 400 active OSS distributions, and each year 26 new projects on average are created. Due to competition and pressure of

A.C.B. Gomes da Silva (✉) • G. de Figueiredo Carneiro
Salvador University (UNIFACS), Salvador, Bahia, Brazil
e-mail: antoniocesar01@gmail.com; glauco.carneiro@unifacs.br

M.P. Monteiro
Universidade Nova de Lisboa (UNL) NOVALINCS/UNL, Lisbon, Portugal
e-mail: mtpm@fct.unl.pt

F.Brito e Abreu
Instituto Universitário de Lisboa (ISCTE-IUL), ISTAR-IUL, Lisbon, Portugal
e-mail: fba@iscte-iul.pt

K. Constantino • E. Figueiredo
Federal University of Minas Gerais (UFMG), Minas Gerais, Brazil
e-mail: kattiana@dcc.ufmg.br; figueiredo@dcc.ufmg.br

© Springer International Publishing AG 2018
S. Latifi (ed.), *Information Technology – New Generations*, Advances in Intelligent Systems and Computing 558, DOI 10.1007/978-3-319-54978-1_77

Fig. 77.1 The ISO/IEC FCD 25010 product quality standard

users and developers, OSS projects need to release new features and bug fixes within increasingly shorter time spans. Releasing software every few weeks is typically referred to as a frequent release cycle, while releasing on larger periods (e.g. quarterly or yearly) is typically referred to as a traditional release cycle [6]. The adoption of frequent releases in OSS projects takes into account that these projects have usually users all over the world, who eagerly download each new version as soon as it is released, and test it as thoroughly as they can. The worldwide dispersion of users often leads to a continuous (24 hours a day) testing process [7]. The process of reporting and resolving issues for a system during its development and/or maintenance is usually supported by an Issue Tracking System(ITS). Typically, an ITS can record the issue type (e.g. defect, enhancement, patch, task), its state (e.g. new, assigned, resolved, closed), the date of submission and of each state change, who was its submitter, was assigned to address it, any comments by others, and indications of severity and/or priority [8].

The community members usually vote for decisions regarding the demands registered in the ITS [9]. However, a still open question is the set of criteria and priorities that influence the decisions of which issues should be analyzed, implemented/solved and delivered in next releases, to the detriment of others. This question is a relevant aspect for the success of such projects. We conducted an exploratory study to investigate the influence of target product quality attributes in software release practices of OSS projects. The goal is to look for evidences on the relationships among these target attributes, the priorities assigned to the registered issues and how they are delivered by product releases.

The rest of this paper is organized as follows: Sect. 77.2 presents some background and related work; Sect. 77.3 outlines our exploratory study; Sect. 77.4 describes the collected data and its analysis; finally, Sect. 77.5 discusses the results of the study, identifies their validity threats and scopes future research on this topic.

77.2 Background and Related Work

The ISO/IEC 25010 international standard for software product quality [10], that replaced the "old" ISO/IEC 9126 standard [11], provides a quality model composed of several software quality characteristics, that are further broken down into many sub-characteristics, as represented in Fig. 77.1. For instance, one of the main characteristics is maintainability, which is further broken down into modularity, reusability, analysability, modifiability and testability. Reusability, for example, is related to Software Reuse, which is defined as the process of building or assembling software applications and systems from previously developed software [12]. Both ISO standard families are based on the model, that the quality of the process influences the internal quality of the software product, which in turn influences its external quality, and then its quality in use [13]. Studies have reported the relevance and influence that software quality attributes have in the software development cycle of a product [14, 15]. This influence is also true for OSS projects. Two surveys in this direction were conducted by Henningsson and Wohlin [16]. The first survey focused on the literature to capture the understanding of the quality attributes in the research community. The second one is an interview survey focused on the perception of the industry regarding the practice and understanding of the quality attributes in an industrial context. The authors concluded that it is clear, from both the literature and the industrial surveys, that relations do exist between many of the software quality attributes.

77.3 Exploratory Study

In this section we present an exploratory study to analyze how software quality attributes influence the prioritization of issues to be implemented/solved and delivered in future releases. Exploratory studies are intended to lay the

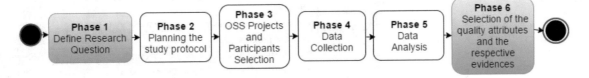

Fig. 77.2 Overview of the exploratory study

groundwork for further empirical work [17]. For this reason, there is no control group to compare to. The adopted strategy consisted in asking developers, outside the community of the selected projects, to search for evidences in data provided by the repositories of the selected OSS projects. This section aims to address the following Research Questions:

RQ1 – *Considering data available in public repositories, is it possible to identify at least three software product quality attributes of the OSS Projects Mozilla Firefox, Libre Office and Eclipse that are prioritized in software release practices?* The identification of software quality attributes in OSS projects is a key to understand their priorities throughout software releases. In practice, depending on the software quality attributes that are prioritized, specific issues are targeted and implemented in the next releases while others will be postponed or even discarded.

RQ2 – *Which evidence from the OSS project repositories support the indication of the tree prioritized software quality attributes?* These evidences are relevant lessons learned that can be referenced and followed by industry practitioners, as well as by other OSS projects.

To answer these questions, we defined the protocol of our study (Sect. 77.3.3) and instantiated it in six study replications (Sects. 77.4.1 to 77.4.6). This protocol is based on the analysis of data that have relationship with bug issues of three OSS projects (Sect. 77.3.2). Prior to the tasks execution, one of the authors conducted a tutorial session (Sect. 77.3.4) to help participants find out evidences through the use of repositories provided by the selected OSS projects.

77.3.1 Data Collection

Data collection. Data collection was carried out directly from answers provided to the questionnaire given to participants. To answer RQ1, we collected information regarding quality attributes and corresponding justifications. To answer RQ2, we collected evidence based on references provided by participants, e.g., print screens and URLs from which relevant information is accessible.

Passing score. After the tutorial, we analyzed data provided by the participants in accordance with the following two criteria. The *Criterion 1* was the indication of at least one quality attribute consistent with data from the analyzed repository. Therefore, inappropriate indications of quality attributes do not meet this criterion. The *Criterion 2* was the indication of evidence from the repositories supporting the choice of the selected quality attribute.

Data analysis. Data from the questionnaires were analyzed in search of the primary strategies employed by participants to answer RQ1 and RQ2 (Fig. 77.2).

77.3.2 Target OSS Projects

This study relies on three highly regarded open source projects presented: *Mozilla Firefox*, *Eclipse*, and *LibreOffice*. We selected those projects due to several reasons. First, they adopted frequent releases implementation [18, 19]. Second, they have a very active developer community comprising 290 (*LibreOffice*), 1087 (*Mozilla Firefox*), and 113 (*Eclipse*) members, respectively. Third, all projects provide a vast repository of documentation, publicly available and readily accessible. Finally, these projects were previously cited in the literature, which enhances their value as the basis for the present study.

77.3.3 The Study Protocol

Six participants took part in this study to answer research questions **RQ1** and **RQ2**. They were also asked to register the collected evidences from the analysis they performed and the time they were identified in the questionnaire form. Moreover, the participants were asked to describe their strategies, as well as their experience while using the OSS project repositories to accomplish the tasks. To answer **RQ1**, we analyzed and compared the attributes indicated by each participant. Additionally, we analyzed the data provided by the participants to identify evidences to justify the attributes so far indicated to answer **RQ2**.

Table 77.1 Pre-study questionnaire

ID	Question	Answers options
OSS-1	My experience as a user of open source software products is?	None/low/ medium/high
OSS-2	My experience as a developer/member of community of open source software products is?	None/low/ medium/high
SRP-1	My theoretical knowledge on software releases is?	None/low/ medium/high
SRP-2	My experience as a user of software releases is?	None/low/ medium/high
SRP-3	My experience as a developer of software releases is?	None/low/ medium/high
QA-1	My theoretical knowledge on software quality product attributes is?	None/low/ medium/high
QA-2	My experience as a user on issues related to software quality product attributes is?	None/low/ medium/high
QA-3	My experience as a developer on issues related to software quality product attributes is?	None/low/ medium/high

Table 77.2 Results of the pre-study questionnaire

ID	P1	P2	P3	P4	P5	P6
OSS-1	None	Medium	Low	High	Medium	Low
OSS-2	None	None	None	None	None	None
SRP-1	Medium	Low	Low	High	Medium	Low
SRP-2	Medium	Low	Low	High	Medium	Medium
SRP-3	Medium	Medium	Low	High	Medium	None
QA-1	Low	Medium	Medium	Medium	Low	Low
QA-2	Low	Medium	Medium	Low	None	Medium
QA-3	Low	Medium	Low	None	None	None

77.3.4 Participants Selection

The study involved six participants recruited from a MSc program in Computer Science. This number of participants offered a reasonable trade-off between the effort to plan and execute the study and detailed qualitative analysis and the generalizability of the results [20, 21]. They were all volunteers and no compensation was provided for their participation in this study.

We were interested in making observations based on a detailed, qualitative analysis of OSS projects repositories, regarding the influence of software product quality attributes on the scheduling of pending issues, rather than testing causality hypotheses using statistical inference. To be eligible for inclusion, participants filled out a pre-study questionnaire (Table 77.1) to describe their profile and experience in software release practices, OSS projects and software quality product attributes. In Table 77.2, we present the answers of each question presented in Table 77.1, using the following ordinal scale: none/low/medium/high.

The Tutorial Session. Prior to the tasks, the participants attended a tutorial session focusing on how to search for

evidences in the *Open Stack IAAS Cloud Platform* project using the data repositories of several tools used during its development: *Bugzilla* (issue tracking system), *Git* (configuration management system), *Gerrit* (code review tool for Git) and a *Wiki*. These tools were also used in the OSS projects selected for this study. We selected a different project in the tutorial session to avoid biasing in the study results. The first part of the tutorial focused on how to understand and search for data within the aforementioned tool repositories for the Open Stack project. The second part focused on using the selected repositories to identify evidences that could establish relationship with corresponding software product attributes prioritized by the Open Stack IAAS Cloud Platform.

One of the first indicators of the connection between the bugs and release planning was found in the interaction between users to request a given fix in the next patch.[1] These bugs also help to identify instances of reuse, through access to the *Gerrit* link provided by one user. Through it, we were able to identify the creation of a method override.

---

[1] https://bugzilla.redhat.com/show_bug.cgi?id=1283721

Use of *Gerrit* also meets the modifiability requirement, as it shows the code before and after the bug fix.[2]

77.4 Data Analysis

This section presents the analysis of data provided by the participants to answer the two research questions stated above.

77.4.1 Participant 1

Data provided by Participant 1 were discarded because criteria 1 and 2 were not met. This decision can be partly explained by the profile of participant 1 presented in Table 77.2. He/she was not familiar with software quality product attributes, neither from the theoretical, nor from the practical perspective. Probably for this reason, this participant could not provide consistent evidences regarding quality attributes, taken from the projects' data repositories.

77.4.2 Participant 2

Participant 2 indicated *Portability* and *Efficiency* as quality attributes for *Mozilla Firefox*. According to Participant 2, the indication of *Portability* was related to the need of the *"the Web browser to be compatible with different technologies and also run in different operating systems"*. The manifesto of *Mozilla Firefox* states that the effectiveness of the Internet as a public resource depends upon interoperability (protocols, data formats, content), innovation and decentralized participation worldwide.[3] And the project seems to follow this principle. Regarding *Efficiency*, this participant found comments registered in Bugzilla issues, comparing *Mozilla Firefox* with *Chrome*, one of its main competitors. The following one, produced by one of the developers, is a good corroborating example: *"significant problem: Firefox is taking an extremely long time to load compared to Chrome because our load time often includes the restoration of a large number of tabs"*.[4]

This participant also identified the following attributes for the *Eclipse* project: *Portability*, *Reliability* and *Efficiency*. The following sentence justified the selection of these three attributes: *"it is an open source tool that provides integration with various other tools. It runs on various versions of*

operating systems (Portability). Reliability is also a concern of Eclipse community during its operation. To deal with this issue, tests are planned and executed to find failures in an attempt to fix them in case they occur"*. Finally, the participant justifies the choice of *Efficiency* with this sentence: *"Response time and resource consumption are also important factors to Efficiency and they are prioritized in corresponding issues that address this subject"*. An example of concern with *Portability* is illustrated in the following issue excerpt, where one developer writes to another developer *"take a look at this problem, which is specific to Linux"*.[5]

The same participant identified the attributes *Functionality* and *Usability* for the *LibreOffice* project. The reason for the indication of these attributes was that *"it is an office suite software, which requires characteristics intrinsic to its application domain and strongly geared to the user needs, i.e. it must provide features that meet the expectations of its users, must be easily understood (easy to use) and have an attractive and modern graphical user interface, contributing to an intuitive and friendly use"*. The participant also identified that issues related to these attributes were marked in the project bug repository as *high priority* for *LibreOffice*. For example, the following issue reports an error related to find and replace functionality. During the developer's interaction it was mentioned that the problem occurred in a specific version: *"this still affects 4.2.0.4, which is now an official release and will be available in LibreOffice 4.2.1"*.[6]

77.4.3 Participant 3

Participant 3 proposed *Security*, *Usability* and *Portability* as attributes for the *Mozilla Firefox* project. The participant observed that *"Firefox is a web browser available for many platforms, including Windows, OS X, Linux, Android and iOS and compatible with the up-to-date Web technologies"*. The participant presented a list conveying the distribution of issues registered in the bug repositories through their several possible status (assigned, closed, new, reopened, resolved, unconfirmed and verified). Considering the status and average bugs solutions of the *Mozilla Firefox* project, the participant concluded that expert users usually mark issues related to *Usability* (Tabbed Browser Toolbar and Customization), *Portability* (Extension compatibility), and *Security* (session restore, disability access) as high priority in the project.

---

[2] https://review.openstack.org/#/c/346796/8/puppet/services/ceph-mon.yaml

[3] https://www.mozilla.org/en-US/about/manifesto/

[4] https://bugzilla.mozilla.org/show_bug.cgi?id=664314

---

[5] https://bugs.eclipse.org/bugs/show_bug.cgi?id=290182

[6] https://bugs.documentfoundation.org/show_bug.cgi?id=74104

For the *Eclipse* project, *Functionality*, *Compatibility* and *Maintainability* were proposed as quality criteria. According to the participant *"Eclipse is a complete multi-platform environment, compatible with different programming languages (C, Java, PHP, etc.) and operating systems (Windows, Linux, iOS, etc.)"*. As evidence, the participant presented a table containing overall statistics on the status and priorities assigned to the solution of *Eclipse* project issues. This table reveals that the highest priorities are given to components of the SWT UI and Core. This prioritizing also takes into account the percentage of issues related to these components.

For the *LibreOffice* project, the selected attributes were *Functionality*, *Usability* and *Maintainability*. The participant mentioned that *"Libre Office is a powerful office suite with a clean interface and rich in productivity tools. It includes several applications, such as Writer (Word processing), Calc (spreadsheet), Impress (presentations), Draw (vector graphics and flowcharts), Base (database) and Math (edit formula). It runs on Linux, iOS, Android and Windows"*. Similarly to the *Eclipse* project, the participant also presented a table containing overall statistics on the status and priorities assigned to the solution of the *LibreOffice* issues. Upon this table, the participant highlighted that *"it was clear that according to data from the last 500 recorded bugs, the terms most frequently used among the highest priority include: crash, merge, update, document, displayed, Calc, Wizard"*.

77.4.4 Participant 4

Portability, *Usability* and *Efficiency* were proposed as quality attributes for *Mozilla Firefox*. The participant justified that the community is concerned with the above quality criteria - *"Firefox for desktop - Reliable, flexible, quick"*; *"Firefox for iOS - Fast, intelligent, Yours"*; *"Committed to you, your privacy and an open Web"*; *"you can modify the Firefox according to your needs"*; *"Save your time and do everything faster"*. The main evidence pointed out by the participant was a bug referring to a version of *Mozilla Firefox*, which occurred when transferring an image saved by the browser to the download folder. This behavior was evidenced only in Linux as stated *"seems to be reproducible only on Linux"*.[7]

For the *Eclipse* project, *Portability*, *Reliability* and *Compatibility* were suggested. The justification betrays a concern, on the part of the community, in meeting the above software quality criteria. *"Eclipse provides IDEs and platforms for almost all languages and architectures"*;

"IDEs built on extensible platforms for creating desktop, web and cloud IDEs"; *"Develop your software wherever you go"*. For this participant, the most prominent bug was evidenced as *"Toolbar does not display custom widgets properly"*. This problem occurred in a newly released version of Eclipse Neon and on the Windows platform, as shown in *"I also tested it on OS X. Seems not a problem there, so I guess this issue is limited to the Windows platform"*.[8]

For *LibreOffice*, *Usability* and *Compatibility* were proposed as relevant quality criteria. As a justification, the participant cites that the system *"includes several applications that make it a powerful free office suite and open source, is compatible with a wide range of document formats such as Microsoft Word, Excel, PowerPoint and Publisher. In addition, it has a modern and open standard, the OpenDocument Format (ODF)"*. From evidences presented, we highlight the concern of the developers with the following bug *"Multiple animated GIFs cause 100% CPU utilization in Impress"*. To fix this bug, a code adjustment was made to improve CPU performance, as presented in *"To reach more with the current approach we would have to re-implement AnimatedGIF import"*.[9, 10]

77.4.5 Participant 5

For the *Mozilla Firefox* project, *Maintainability* and *Reliability* were proposed. The participant provided the following evidence: this project *"provides a platform for presentation of bugs by the developers as well as their status. It uses a version release system in which versions are released after undergoing reliability tests"*. The main evidence presented to justify *Maintainability* was the URL of the project's Bugzilla.

For the *Eclipse* project, *Interoperability* and *Maintainability* were proposed. The following quote was provided as justification: "Eclipse has a high range of compatibility with various development platforms. It boasts a well-designed and well-organized platform for submitting bugs to its developer". The evidence presented to the *Maintainability* was the URL of the Bugzilla project. For *Interoperability*, the following URL from *Eclipse*[11] was presented.

For *LibreOffice*, *Maintainability* and *Usability* were proposed. The choice is justified by pointing that "it features a well-designed and well organized platform for presenting

---

[7] https://bugzilla.mozilla.org/show_bug.cgi?id=1287823

[8] https://bugs.eclipse.org/bugs/show_bug.cgi?id=498196

[9] https://bugs.documentfoundation.org/show_bug.cgi?id=98500

[10] https://cgit.freedesktop.org/libreoffice/core/commit/?id=285744fef87f4ca0278834b97d7f618bdba5f4c0

[11] http://www.eclipse.org/home/newcomers.php

bugs to developers. It encourages users to participate on the improvement the software, through feedbacks. The *LibreOffice* site itself has a Community tab for the community participation". The evidence presented to the *Maintainability* was the URL of the Bugzilla project's. For *Usability*, the participant presented the URL of *LibreOffice*, which highlights the information explaining how new users can participate and interact with the project's products.[12]

77.4.6 Participant 6

For *Mozilla Firefox*, *Reliability* and *Maintainability* were proposed. As justification, "reliability because there is a study specifically for each release version of the software, each release version is kept separate from the others according to the reliability users have the resources belonging to each version, after they have been tested". *Maintainability* because "Mozilla Firefox provides an infrastructure for the management and fixing of bugs according to the versions of the system". The evidence presented to justify *Maintainability* was an issue[13] which a developer making the adjustment removes the comment from the code claiming that the code was self-explanatory *"Because it is no longer relevant. We make the message collapsible if the 'collapsible' proposes truth, and I find the code self-explanatory here"*. However, another developer reports that this is not a good excuse and asks for the re-addition of the comment since *"someone who is unfamiliar with the code can scan it more easily"*.

For *Eclipse*, *Reliability* and *Maintainability* were proposed. As justification, "Compatibility because Eclipse is an open source software with a focus on developing an extensible platform for creating of new projects. *Maintainability* because *Eclipse* provides access to bugs and improvements proposed by its users, and to development code". The evidence presented to justify *Maintainability* in the *Eclipse* project was a high priority bug fix cycle.[14] The fix was very quick during this cycle.

For the *LibreOffice* project, *Maintainability* and *Usability* were suggested as relevant quality attributes. Justification states that *LibreOffice* is "an open source software project that encourages its users, be they developers or not, to test the system so that they contribute to the quality improvement, providing them back information about its use". *Maintainability* is justified because "the official LibreOffice site provides an environment conducive to the identification of system bugs and control of the development of fixes for

those bugs". Evidence cited regarding *Usability* was a sample of the project URLs that encourage users to give their opinion on the *LibreOffice* programs. For *Maintainability*, the URL of the project's Bugzilla is presented, along with the information that this infrastructure describes the bugs raised by developers, their degree of seriousness and status report to help monitoring each bug fix. The site also invites users to participate.[15]

77.5 Conclusions

To point out the selected quality attributes, participants 2, 4 and 6 adopted the strategy to filter issues by their *severity level* together with their respective *bug priority*. These participants justified the use of this strategy considering that projects adopting the frequent release approach need to foster the selection of failures and new features that may have a significant impact on the software product. Participant 5 adopted another strategy. He searched for relevant quality attributes by analyzing each software project profile and scrutinizing data in the software project portals and wikis. However, this strategy revealed as not effective, considering that only selecting quality attributes through portals and wikis, does not necessarily reflect the relationship of these attributes to the release schedules of those projects. Finally, participant 3 performed a quantitative analysis of the issues found in the projects' bug tracking systems, considering the response/solution time for the main components of the projects based on the quality attributes chosen by the participants.

Figure 77.3 depicts the influence of the software product quality attributes on releases according to the participant's perception. The maintainability attribute is the common attribute in the three projects. Moreover in the LibreOffice project, usability was quoted by all the participants, due to the concern of the developer's community with the ease of use of its products by the users, and the strong competition with proprietary and open source products available in the market.

77.5.1 Threats to Validity

Two possible limitations were identified in this exploratory study. The first threat that can be pointed out is the small number of users participating in the study. Since only 6 students were involved, the dimension of the sample may have an influence on the results obtained. To reduce this risk, a tutorial with a test project was presented, in the hope of

---

[12] https://www.libreoffice.org/community/get-involved/

[13] https://bugzilla.mozilla.org/show_bug.cgi?id=1308840

[14] https://bugs.eclipse.org/bugs/show_bug.cgi?id=505535

[15] https://www.libreoffice.org/community/developers/

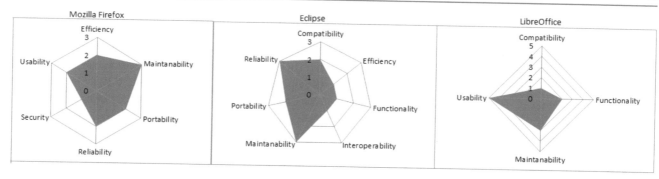

Fig. 77.3 Software product quality attributes prioritized on releases according to participants

filling possible gaps that could possibly occur in the data taken from public repositories. Another potential threat was a possible misinterpretation of the requested activity. To mitigate this risk, the authors were available to answer any doubts that could have arisen during execution of this activity.

77.5.2 Future Works

Considering the attributes selected by the participants of this study and presented in Fig. 77.3, it was possible to identify evidence of the relationship of these target attributes with priorities of registered issues to be fixed and or implemented in next immediate product releases. As ongoing work, we are extending this study with experienced developers in order to contrast with results of this study.

References

1. Michlmayr, M., Fitzgerald, B., & Stol, K.-J. (2015). Why and how should open source projects adopt time-based releases? *Software, IEEE, 32*(2), 55–63.
2. Jonsson, L., Borg, M., Broman, D., Sandahl, K., Eldh, S., & Runeson, P. (2015). Automated bug assignment: Ensemble-based machine learning in large scale industrial contexts. *Empirical Software Engineering*, 1–46.
3. Stol, K.-J., & Fitzgerald, B. (2015). Inner source–adopting open source development practices in organizations: A tutorial. *IEEE Software, 32*(4), 60–67.
4. Rigby, P. C., Cleary, B., Painchaud, F., Storey, M.-A., & German, D. M. (2012). Contemporary peer review in action: Lessons from open source development. *Software, IEEE, 29*(6), 56–61.
5. Adams, B., Kavanagh, R., Hassan, A. E., & German, D. M. (2015). An empirical study of integration activities in distributions of open source software. *Empirical Software Engineering*, 1–42.
6. Mantyla, M. V., Khomh, F., Adams, B., Engstrom, E., & Petersen, K. (2013). On rapid releases and software testing. In *2013 29th IEEE International Conference on Software Maintenance (ICSM)* (pp. 20–29). IEEE.
7. Thomas, L., Schach, S. R., Heller, G. Z., & Offutt, J. (2009). Impact of release intervals on empirical research into software evolution, with application to the maintainability of Linux. *Software, IET, 3*(1), 58–66.
8. Bijlsma, D., Ferreira, M. A., Luijten, B., & Visser, J. (2012). Faster issue resolution with higher technical quality of software. *Software Quality Journal, 20*(2), 265–285.
9. Crowston, K., Annabi, H., & Howison, J. (2003). Defining open source software project success. *ICIS 2003 Proceedings* (p. 28).
10. ISO, I. (2011). Iec 25010: 2011, *Systems and Software Engineering Systems and Software Quality Requirements and Evaluation (SQuaRE) System and Software Quality Models.*
11. ISO/IEC. 2001. *ISO/IEC 9126. Software engineering – Product quality.* ISO/IEC.
12. Shiva, S. G., & Shala, L. A. (2007). Software reuse: Research and practice. In *ITNG* (pp. 603–609).
13. de Souza, G., Mellado, R. P., Montini, D. Á., Dias, L. A. V., & da Cunha, A. M. (2010). Software product measurement and analysis in a continuous integration environment. In *2010 Seventh International Conference on Information Technology: New generations (ITNG)* (pp. 1177–1182). IEEE.
14. Perepletchikov, M., Ryan, C., & Tari, Z. (2005). The impact of software development strategies on project and structural software attributes in SOA. In *OTM Confederated International Conferences On the Move to Meaningful Internet Systems* (pp. 442–451). Springer.
15. Offutt, J. (2002). Quality attributes of web software applications. *IEEE Software, 19*(2), 25.
16. Henningsson, K., & Wohlin, C. (2002). Understanding the relations between software quality attributes-a survey approach. In *Proceedings 12th International Conference for Software Quality.* Citeseer.
17. Wohlin, C., Runeson, P., Höst, M., Ohlsson, M. C., Regnell, B., & Wesslén, A. (2012). *Experimentation in software engineering.* Springer Science & Business Media.
18. da Costa, D. A., Abebe, S. L., McIntosh, S., Kulesza, U., & Hassan, A. E. (2014). An empirical study of delays in the integration of addressed issues. In *ICSME* (pp. 281–290).
19. Gamalielsson, J., & Lundell, B. (2014). Sustainability of open source software communities beyond a fork: How and why has the libreoffice project evolved? *Journal of Systems and Software, 89*, 128–145.
20. Pfleeger, S. L. (1995). Experimental design and analysis in software engineering. *Annals of Software Engineering, 1*(1), 219–253.
21. Yin, R. K., & Campbell, D. (2003). *Case study research: Design and methodsS* (Applied social science research methods series, Vol. 5). Thousand Oaks: SAGE.

AD-Reputation: A Reputation-Based Approach to Support Effort Estimation

Cláudio A.S. Lélis, Marcos A. Miguel, Marco Antônio P. Araújo, José Maria N. David, and Regina Braga

Abstract

Estimating the effort on software maintenance activities is a complex task. When inaccurately accomplished, effort estimation can reduce the quality and hinder software delivery. In a scenario, in which the maintenance and evolution activities are geographically distributed, collaboration is a key issue to estimate and meet deadlines. In this vein, dealing with reputation of developers, as well as establish and promote trust among them, are factors that affect collaboration activities. This paper presents an approach aimed to support effort estimation on collaborative maintenance and evolution activities. It encompasses a model for reputation calculation, visualization elements and the integration with change request repositories. Through an experimental study, quantitative and qualitative data was collected. A statistical analysis was applied and shown that the AD-Reputation is feasible to estimate the effort spent on collaborative maintenance activities.

Keywords

Developer • Model • Reputation • Software Maintenance • Effort

78.1 Introduction

Software systems undergo changes and adaptations over time. To carry out these activities, there is a need of planning by software maintenance teams [1]. Part of this planning is effort estimate, given in time spent on software maintenance activities.

Often time estimates in software are set inaccurately (ad-hoc), causing delays in the delivery of customer requests and software quality. An estimated more correct time and precisely is critical to the viability of the organization's activities [2]. The success or failure of projects depends on the accuracy of the effort and schedule estimates.

Historical information stored throughout the life cycle of a software can help teams to create mechanisms to measure time and effort of the activities to be developed [3]. With historical data, it is assumed that a new project should be run about the same as an earlier and therefore tends to be better than the last, influencing the new estimates [4].

As the software maintenance can be performed collaboratively, reputation is one of the factors that affect collaboration among team members. According to [5], reputation is the perception created by an entity through previous actions, based on their intentions, rules and beliefs.

Considering geographically distributed teams, reputation can be used as a cooperation indicator. In [6], the authors use game theory to understand the emergence of trust in globally

C.A.S. Lélis (✉) • J.M.N. David • R. Braga
Post Graduation Program in Computer Science - Federal University of Juiz de Fora (UFJF), Juiz de Fora, Brazil
e-mail: lelis@ice.ufjf.br; jose.david@ufjf.edu.br; regina.braga@ufjf.edu.br

M.A. Miguel • M.A.P. Araújo
Post Graduation Program in Computer Science - Federal University of Juiz de Fora (UFJF), Juiz de Fora, Brazil

Federal Institute of Education, Science and Technology of Southeast of Minas Gerais –Campus Juiz de Fora (IF Sudeste MG), Juiz de Fora, Brazil
e-mail: marcos.miguel@ice.ufjf.br; marco.araujo@ufjf.edu.br

© Springer International Publishing AG 2018
S. Latifi (ed.), *Information Technology – New Generations*, Advances in Intelligent Systems and Computing 558, DOI 10.1007/978-3-319-54978-1_78

distributed teams. When partners refuse their cooperation, the developer is penalized with the loss of his reputation. This reflects negatively on the implementation of the tasks that the team must play affecting the cohesion among its members. Reputation is a way to establish and promote trust in the team when dealing with maintenance and evolution geographically distributed activities [7].

The reputation also affects the communication. Trainer and Redmiles [8] , for example, investigated the way in which developers and managers use multiple media, or are available via *chat* or are active in the mailing list instead of only via email. As result, they can be well seen by the other members and, consequently, have good reputation and be considered reliable.

In previous work [7, 9], the authors explain that reputation information can be used to estimate effort spent, however It would require further analysis and conduct experimental studies. This work presents an approach consisting of a model calculation reputation developers, visualization and integration with change request repositories to provide evidence of such statement.

The data used in the experiment were obtained as result of a partnership established with a software development company for business management. For confidentiality reasons, the company will be named as partner company. Considering this real context, in order to obtain evidence on the feasibility of the approach, an experimental study was conducted, through which analyzed the change requests (CR) repository of partner company projects.

This paper is structured by this introduction and Sect. 78.2 shows the environmental context that the proposal is inserted and some related work. Section 78.3 details the components of AD-Reputation approach. In Sect. 78.4 the experiment performed and Sect. 78.5 the final considerations.

78.2 Background

The AD-Reputation approach is part of a context (Fig. 78.1) that involves collaboration aspects, visualization and software evolution interacting with other tools. The approach components will be detailed in the next section.

GiveMe Infra [10] is an infrastructure to support collaborative activities in the maintenance and evolution of software made by co-located teams or geographically dispersed presents a set of views that allow the user from the source data repository, obtain visual information about the maintenance and evolution of software.

ArchiRI [9] is an architecture that provides solution based on a logical model and an ontology to represent reputation information and enable interoperability of data. In addition a set of views is adopted to facilitate the analysis of reputation

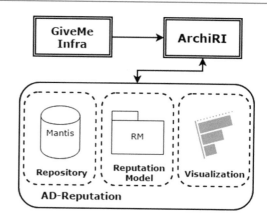

Fig. 78.1 Environment overview

information and assist in decision making by the manager. These resources aim to support managers i) on creation of groups of developers, ii) on exchange of members between groups and iii) assigning maintenance tasks to developers.

Integrating ArchiRI with *GiveMe Infra* fosters collaboration and trust in team. The AD-Reputation approach in this scenario contributes with a mechanism to estimate effort in maintenance tasks, using a reputation calculation model for developers.

The use of visualization to the reputation information is still in its infancy. In [11] the authors intended to enable a visual exploration reputation through a graphical presentation of reference and its context attributes. In online transaction context, Kim et al. [12] use 3D visualization to portray reputation through metrics WOM (*Word of* Mouth) generated from user *feedback*. However, both Sanger and Pernul [11] and Kim et al. [12] are associated with specific contexts without allowing the exchange of information for a software distributed development environment.

The developer's reputation has increasingly aroused the interest of researchers. For example, [13] conducted an experiment to identify how developer's reputation in the OSS context affects the results of his/her code review requests. Developed an approach based on the analysis of the social network formed by developers, on the assumption that the core developers of the network have better reputation than the peripheral developers.

In [14] , have a prototype to find reliable source on the web, associating information reputation of a social network for developers, with the results of a code search engine. With that, established the reputation of source code developed and include a ranking mechanism from the highest scoring code. Our approach also provides a mechanism to rank the developers, but unlike the aforementioned prototype use the source code, the AD-Reputation approach uses CR repositories. Furthermore, the use of historical data is also an important feature because the reputation may change over time [5, 15].

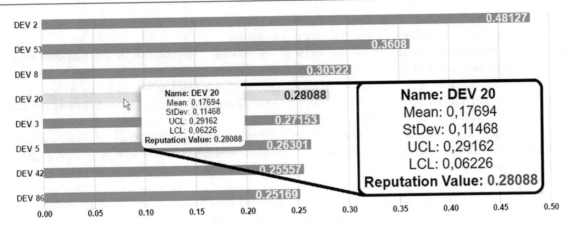

Fig. 78.2 DRR view

Another difference between Ad-Reputation and the two aforementioned work is related to the use of visualization. While addressing the reputation of software developers, do not advance the display elements to support decision making in the estimation of effort spent on development tasks and software maintenance.

78.3   AD-Reputation

The proposed AD-Reputation arose from the need to estimate the effort to deal with a new request, based on historical estimates and effort spent in previous requests. Comprising a reputation model established for developers integrates with the bug tracking system MantisBT,[1] allowing the historical data analysis. A new view to the reputation data is also provided. The components and the process to estimate effort by the approach are detailed below.

78.3.1 Reputation Calculation Model

Based on Bayesian model theory shown in [16] that aims to provide reputation information by calculating an expected value, statistically, for the qualifications that an individual can receive in a specific context. In our case, the probability that the developer be assigned to the next CR, considering the cases dealt with previously. There are two possible situations related to each developer when the team receives a new CR: it is possible that she/he deal with the CR and receives a "yes" qualification to the case, or it is possible that she/he does not deal with the CR and thus receives a "no" qualification. The total CR whose developer received "yes" qualifications $b_{dev}(y)$, represents the number of CR handled

by the developer. Similarly, the sum of all cases ("yes" qualifications plus "no" qualifications) $\sum_{i=1}^{n} b_{dev}(i)$, indicates the total number of CRs received by the team.

This model takes into account the aging of the developer's qualifications, allowing reflect changes of behavior over time. Through a constant W, called a priori, is assigned a weight to previous qualifications in calculating reputation. Usually, its value is assigned by the relationship W = k, where k is the number of possible values that a rating may have, in our context k = 2 ("yes" or "no"). However, this constant can be adjusted to focus more or less recent qualifications, according to necessity [16]. It is also taken into account the initial probability of a rating occurs given by the variable P. Considering k = 2 and a uniform distribution, P = ½.

The formula for calculation of developer's reputation is presented in Eq. (78.1).

$$R_{dev} = \frac{b_{dev}(y) + W \times P}{W + \sum_{i=1}^{n} b_{dev}(i)} \qquad (78.1)$$

The calculated reputation value allows to rank the best developers in each CR and thus select, for the estimation process, only those CRs for which the value is greater, that is, developers with higher reputation have greater trust and therefore a greater assertiveness development time.

78.3.2 Developer Ranking Reputation View

The Developer Ranking Reputation View (Fig. 78.2) uses JavaScript library, d3js,[2] to represent different types of information about developer reputation calculated by model presented above. Through interaction elements is

---

[1] https://www.mantisbt.org/

[2] http://d3js.org

possible to obtain information such as the identity of the developer, the average reputation values, the mean standard deviation, the upper and lower control limits and a color indication about reputation value in relation to control limits. Such information makes possible analyze, graphically, how the reputation of each developer behaves due the others. From this analysis the manager can select only developers with greater reputation and thus better ranked so that requests handled by them are historically considered in effort estimation of new requests.

Through the view depicted in Fig. 78.2 can notice that the five developers with greater reputation in the team are DEV 2, DEV 53, DEV 8, DEV 20 e DEV 3. Therefore, since they are experienced developers and own greater reputation on the team, can have more confidence in the time spent to carry out activities. This analysis helps to rely on the final estimate process.

78.3.3 Estimating Effort

The process adopted by the approach to support effort estimation starts when the user (i) selects the new CR to be analyzed. The approach (ii) collects information (criticity, priority, module and project) associated with the request. Then, (iii) the DRR View is shown as described in the previous subsection. At this stage the user (iv) select developers. Through interaction with the visualization it is possible analyze the reputation information and highlight the selection, so only the requests handled by such developers are aggregated to calculate the estimation. Another possibility is, after interacting with the DRR View, set a limit to the reputation, e.g., average values, to be considered only requests handled by developers with reputation values above the threshold. In the next step, the approach (v) search the repository for CRs that have the same characteristics collected (step ii) of the new CR and was previously handled by the developers selected in the previous step. Finally, (vi) CRs are aggregated by the actual time spent and final estimation is calculated as the arithmetic mean. At this stage the user can request a new selection of developers. The process ends when the user submits the estimation to the repository.

78.4 Experimental Study

This section presents the experimental study conducted. According to the Goal/Question/Metric approach (GQM) [17] the goal can be stated as: "**Analyze** the AD-Reputation approach **in order to** verify the feasibility of use **with respect to** the task of estimating effort spent on maintenance activities **from the point of view** of developers

in the context of collaborative maintenance and software evolution".

In this sense, the experiment was proposed based on a set of maintenance activities to be solved in a 2 week period (Sprint) defined and estimated by the current development team of partner company. A database was considered with the last five years of CRs from partner company.

This experiment included 6 participants. After filling out a form characterization, it was observed that had age between 24 and 40 years, most of them systems analysts and 1 project development manager. The latter had no contact with the CR repository or any knowledge of the internal techniques used by the partner company, to measure effort. The others worked in the company and had knowledge of the project in progress.

Regarding the knowledge of the participant on effort estimation, 50% of respondents said they have a regular knowledge about effort estimation, 2 say they have little knowledge in this area and one of them a high knowledge on the subject. About knowledge of software maintenance activities, two participants said they had high knowledge, other 2 said they had regular knowledge and 2 reported very little knowledge in this area. This diversity shows the possibility of approach help teams with little or no knowledge about effort estimation techniques as well as software maintenance activities.

Participants underwent online training and then the experiment was conducted geographically distributed due to distance between them.

Initially, a Sprint was set (Sprint 182, with 59 CRs) that would be used in the experiment and for which the development team of the partner company needs to plan the effort estimation. In a planning meeting, the estimated effort was manually defined by the Planning Poker technique [18], standard company procedure. The team started the development of the Sprint 182. The starting and ending time for each CR was collected. At this stage there was no involvement of the experiment participants. When the Sprint 182 was concluded, the actual time spent for each CR was obtained by difference between the ending and starting time. The information generated were collected and stored.

During the experiment, each participant received the Sprint 182 to analyze. Participants conducted the analysis by AD-Reputation approach. Reported values for the effort estimation and evaluated qualitatively both the approach and the DRR View.

Through the collected quantitative data (Table 78.1), the statistical analysis was conducted and considered the level of significance 5% ($\alpha = 0.05$). In order to verify the existence of statistically significant differences between the values estimated by Planning Poker (Technique "Estimate PMG" in Fig. 78.3), actual spent and calculated from the approach, the following hypotheses were defined:

Table 78.1 Evaluation scale

1	Does not supports the effort estimation
2	Supports somewhat the effort estimation
3	Indifference
4	Supports the effort estimation
5	Supports very much the effort estimation

Table 78.2 Average ratings

	AD-Reputation	DRR View
DEV-86	3.68	3.00
DEV-20	4.34	4.50
DEV-105	3.07	3.03
DEV-95	3.59	4.46
DEV-94	3.12	3.39
GER-1	4.75	5.00
MEAN	3.76	3.90

```
Kruskal-Wallis Test on Time

Technique      N   Median  Ave Rank      Z
Estimate PMG   6    190,0       6,5   -1,69
Spent          6    264,4      12,5    1,69
AD-Reputation  6    223,2       9,5    0,00
Overall       18                9,5

H = 3,79  DF = 2  P = 0,150
H = 4,08  DF = 2  P = 0,130  (adjusted for ties)
```

Fig. 78.3 Kruskal-Wallis test result

H0: *There is no difference between the average of the estimated predicted time, the actual time spent and the estimated time by AD-Reputation.*

H1: *There is difference between the average of the estimated predicted time, the actual time spent and the estimated time by AD-Reputation.*

Initially, the normality and homoscedasticity of data were tested. For the normality test was used statistical test of Shapiro-Wilk, as the number of samples is less than 30 elements. The result of the samples presented p-value of 0.100, higher than the significance level of 0.05, this fact indicates that the sample has normal distribution. Therefore, the homoscedasticity Levene test was applied. Considering the test for equality of variances, the result of the samples presented a p-value lowers than the significance level of 0.05, this fact indicates that the samples were not homoscedastic, which indicates the need for a nonparametric test to compare the means. The nonparametric Kruskal-wallis test was applied since was a factor (Time) and more than two treatments (Estimate PMG, Spent, AD-Reputation). By applying this test at a 5% significance level, a p-value equal to 0.150 can be seen in Fig. 78.3, thus indicating the acceptance of the null hypothesis that the differences between the means are not statistically significant. So the means are statistically equivalent. Therefore, AD-Reputation is statistically feasible to estimate the effort expended on maintenance activities, since there is no statistical difference between the time estimated by the approach, the values estimated manually by the team and the actual time spent performing the activities .

During the experiment, participants evaluated the degree to which the approach and the DRR View supported the effort estimation. Such assessment followed the scale depicted in Table 78.1 and was repeated for each estimate activity. This evaluation aim to analyze, from the participants point of view, if the approach and visualization are useful for the effort estimating task. Data were collected in a form that also enabled the participants to comment and describe their impressions.

The average ratings in the activities of each participant is arranged in Table 78.2.

From the data in Table 78.2 may be noted that the average of the ratings are higher than 3 (three), so above the indifference area proposed by the scale (Table 78.1). Therefore, in the opinion of the experiment participants approach supports the effort estimation. The same applies to the DRR View. It is noteworthy that the manager who had no knowledge of the processes, either of the company data, evaluated with maximum score DRR View and with 4.75 the approach to support the estimate.

Through textual report, other observations were collected. One of the participants, DEV-105, reported that it would be interesting to offer the DRR View also at time of assigning the task accomplishment. After the experiment, the project manager highlighted assertiveness analysis by reputation.

As threats to validity we can mention the fact that study was applied in a sample of five developers of a company that uses SCRUM methodology, not being extended to other different methodologies. To minimize the effect of this threat, a project manager was involved in the study, who did not know the company, its data and processes.

Throughout the experiment, the database used had well documented CRs, as well as the actual time spent. The application of this study with poorly documented bases and in which the effective time spent in each development is not stored, may achieve different results in the application of the approach.

78.5 Final Considerations

This paper introduced the AD-Reputation approach to support the estimation of effort spent by developers in the collaborative software maintenance activities. An experimental study was conducted and through the statistical analysis of the data, the approach feasibility has been verified. According to participants, both the approach and DRR view support effort estimation. The result of the experiment analysis, reinforce the importance of the approach to estimate effort in software maintenance activities.

As future work we intend to expand the integration to other change request repositories such as Bugzilla.[3] Integration with other visualization tools to diversify the representation of reputation information. In addition, establish new partnerships with other development companies, to supply data.

Acknowledgments To CAPES, FAPEMIG and CNPq for financial support, and partner company for the availability of data and participation in research.

References

1. Rajlich, V. (2014). Software evolution and maintenance. In *Proceedings of the on future of software engineering* (pp. 133–144).
2. Audy, J. L. N., & Prikladnicki, R. (2007). *Desenvolvimento Distribuído de Software*. Rio de Janeiro: Elsevier.
3. Poncin, W., & Serebrenik, A., van den Brand, M. (2011). Process mining software repositories. In *Software maintenance and reengineering (CSMR), 2011 15th european conference on* (pp. 5–14). Oldenburg: Carl von Ossietzky University.
4. McConnell, S. (2006). *Software estimation: Demystifying the black art*. Redmond: Microsoft press.
5. Mui, L., & Mohtashemi, M., Halberstadt, A. (2002). Notions of reputation in multi-agents systems: A review. In *Proceedings of the first international joint conference on Autonomous agents and multiagent systems: Part 1* (pp. 280–287). Bologna.
6. Wang, Yi., & Redmiles, D. (2013). Understanding cheap talk and the emergence of trust in global software engineering: an evolutionary game theory perspective. *CHASE* (pp. 149–152).
7. Lélis, C. A. S., Araújo, M. A. P., David, J. M. N., Carneiro, G. de F. (2016). Investigating reputation in collaborative software maintenance: A study based on systematic mapping. In *Information technology: New generations, 13th international conference on information technology* (pp.615–627). Org. Springer International Publishing.
8. Trainer, E. H., & Redmiles, D. F. (2012). Foundations for the design of visualizations that support trust in distributed teams. *Proceedings of international working confererence on advanced visual interfaces – AVI '12* (pp. 34–41).
9. Lélis, C. A. S., Braga, R., Araújo, M. A. P., David, J. M. N. (2016). ArchiRI - uma arquitetura baseada em ontologias para a troca de informações de reputação Alternative Title : ArchiRI - An ontology-based architecture for the exchange of reputation information. *XII Brazilian Symposium on Information Systerms* (pp. 60–67). Florianopolis.
10. Tavares, J. F., David, J. M. N., Araújo, M. A. P., Braga, R., Campos, F., & Carneiro, G. (2015). Uma Infraestrutura baseada em Múltiplas Visões Interativas para Apoiar Evolução de Software. *iSys-Revista Bras. Sist. Informação, 8*(1), 65–101.
11. Sanger, J., & Pernul, G. (2014). Visualizing transaction context in trust and reputation systems. In *Availability, reliability and security (ARES), 2014 ninth international conference on* (pp. 94–103). Fribourg.
12. Kim, S., Lee, S., Kang, H., & Cho, J. Hybrid WOM collection and visualization method for reputation rating in online community. *Indian Journal of Science and Technology, 8*(18), 1–5.
13. Bosu, A., & Carver, J. C. (2014). Impact of developer reputation on code review outcomes in OSS projects: An empirical investigation. In *Proceedings of the 8th ACM/IEEE international symposium on empirical software engineering and measurement* (p. 33). Torino.
14. Gallardo-Valencia, R. E., Tantikul, P., Sim, S. E. (2010). Searching for reputable source code on the web. In *Proceedings of the 16th ACM international conference on Supporting group work* (pp. 183–186). Sanibel.
15. Jøsang, A., Ismail, R., & Boyd, C. (2007). A survey of trust and reputation systems for online service provision. *Decision Support Systems, 43*(2), 618–644.
16. Jøsang, A., & Quattrociocchi, W. (2009). Advanced features in Bayesian reputation systems. *Lecture Notes in Computer Science, 5695*, 105–114.
17. Basili, V. R., & Weiss, D. M. (1984). A methodology for collecting valid software engineering data. *IEEE Transactions on Software Engineering, SE-10*(6), 728–738.
18. Cohn, M. (2005). *Agile estimating and planning*. Pearson Education. Saddle River: Prentice Hall PTR Upper.

---

[3] https://www.bugzilla.org/

Jisoo Yang and Julian Seymour

Abstract

Modern non-volatile memory storage devices operate significantly faster than traditional rotating disk media. Disk paging, though never intended for use as an active memory displacement scheme, may be viable as a cost-efficient cache between main memory and sufficiently fast secondary storage. However, existing benchmarks are not designed to accurately measure the microsecond-level latencies at which next-generation storage devices are expected to perform. Furthermore, full exploitation of disk paging to fast storage media will require considerations in the design of operating system paging algorithms. This paper presents *pmbench* – a multiplatform synthetic micro-benchmark that profiles system paging characteristics by accurately measuring the latency of paging-related memory access operations. Also presented are sample pmbench results on Linux and Windows using a consumer NAND-based SSD and a prototype low-latency SSD as swap devices. These results implicate operating system-induced software overhead as a major bottleneck for system paging, which intensifies as SSD latencies decrease.

Keywords

Pmbench • Paging performance • SSD Swap

79.1 Introduction

Pmbench is a user-level micro-benchmark designed to profile system paging performance by measuring memory access latency during fault-intensive memory operations. It is intended to serve as a tool for system architects to diagnose and address issues with paging subsystem. Designed with low-latency SSDs in mind, pmbench aims to accurately measure sub-microsecond intervals while incurring minimal overhead.

Pmbench profiles paging performance by collecting memory access latencies from a benchmark run during which a large amount of memory is consumed in a systematic fashion triggering operating system paging activity. It measures the time taken for each memory access and compiles the results by keeping a histogram of the measured latencies. Pmbench is carefully written to minimize unnecessary paging activity by using compact, page-aligned data structures and code segment layout.

Pmbench supports Linux and Windows and could be ported to other POSIX-compliant operating systems. It is launched from the command line, creating one or more threads to conduct paging latency measurement based on user-provided parameters. Upon completion, it generates an XML report, which can be read with a companion GUI tool to visually graph and compare data points between benchmarks and further produce a condensed comma-separated report.

We used pmbench to compare the paging performance of Linux and Windows under a heavy paging load using a NAND-based SSD and an experimental low-latency SSD

J. Yang (✉) • J. Seymour
Department of Computer Science, University of Nevada,
89154 Las Vegas, NV, USA
e-mail: jisoo.yang@unlv.edu; seymou12@unlv.nevada.edu

© Springer International Publishing AG 2018
S. Latifi (ed.), *Information Technology – New Generations*, Advances in Intelligent
Systems and Computing 558, DOI 10.1007/978-3-319-54978-1_79

as swap devices. We confirmed that NAND SSD is entirely unsuitable as DRAM displacement for workloads whose working set exceeds available physical memory. We also found that with low-latency SSD both operating systems suffer from significant software overhead, although Linux paging fares far better than that of Windows.

The rest of this paper is organized as follows: Sect. 79.2 presents background on SSDs and system paging. Section 79.3 motivates pmbench's design and current implementation. Section 79.4 presents pmbench measurements performed on Linux and Windows using low-latency SSDs. Section 79.5 discusses related works. We conclude in Sect. 79.6.

79.2 Background

Solid-State Drives (SSDs) store data electronically in Non-Volatile Memory (NVM) without moving parts, substantially improving bandwidth and latency. SSDs are designed to fill the role occupied by hard drives and thus retain the same storage *block interface*, consisting of a flat address space of fixed-size blocks, whose size typically ranges from 512 to 4K bytes. The host computer reads and writes on a per-block basis, expecting the device to fetch or update the entire block. Almost all storage I/O standards operate under this block interface.

SSDs are equipped with a controller to manage their NVM specifics while retaining the block interface. For example, NAND Flash, used by most of today's SSDs, does not allow in-place writing to a block. Instead, writes must be done on erased blocks, and erasure operations can only be performed in granularities larger than a block. NAND SSD controllers address this issue by implementing an indirection layer that maps a host-visible block address to an internal Flash address. This layer creates the illusion of updating a block in-place, when in fact it writes to a previously erased Flash, and queues the old Flash as invalid and subject to erasure.

Future SSDs equipped with next-generation NVM will outperform current NAND Flash-based SSDs not only by having faster access latencies, but also by supporting in-place update, which significantly reduces overhead from the SSD controller. Therefore, a *low-latency SSD* capable of delivering a 4KiB block within a couple of microseconds in a sustained load is entirely plausible.

SSDs can thus improve the performance of system *paging*, by which an OS transfers the contents of fixed-size memory pages to/from a storage device. Paging implements virtual memory, enabling execution of programs whose collective working set exceeds the size of available physical memory. The storage device used for paging is also known as a *swap device*.

Page fault is the processor protection mechanism allowing operating systems to implement paging. A page fault is triggered when a processor accesses a virtual address that translates to a memory page not physically mapped. Once triggered, the processor invokes the OS's page fault handler, which ensures the page is loaded into memory, updates the corresponding page mapping, then resumes the instruction that triggered the fault.

If the fault address refers to a page which hasn't been mapped, but the content of the page is still in memory, the OS can handle it without triggering storage I/O. This type of fault is known as *soft* or *minor fault* and can be handled very quickly. If the content of the page cannot be found in memory but is instead located on the swap device, the OS must schedule storage I/O to bring in the data. This is known as a *hard* or *major fault* and can take a very long time to resolve.

79.3 Design and Implementation

Pmbench is motivated to profile OS paging performance when a low-latency SSD equipped with next-gen NVM is used as the swap device. We believe, and have confirmed, that current operating systems are inefficient when paging to such devices. As paging was never expected to be fast, its implementation has remained unoptimized, as is apparent when the system swaps to low-latency SSDs. To improve paging performance, detailed quantitative analyses of paging must follow to pinpoint the source of inefficiencies and explore possible solutions.

Pmbench profiles OS paging performance by illustrating the distribution of user-perceived latencies of memory accesses that result in page faults. This requires the systematic consumption of a large quantity of memory in an artificially-generated memory access pattern. Pmbench measures the time taken for each memory access and compiles the results by keeping population counts for latencies measured.

Pmbench's primary technical concern is minimizing overhead from internal memory use. Note that this requirement precludes the use of a trace-based benchmark or recording all measurement data, which would require substantial use of memory and storage I/O, significantly degrading measurement under heavy paging. Although it is impossible to completely prevent interference, efforts were made to minimize memory loads incurred by measurement infrastructure.

The narrow timescale that pmbench tries to accurately measure is another challenge. Events with microsecond-level latency resolution are of interest in profiling operating system paging performance with fast SSDs. In modern systems, a few microseconds translates to tens of thousands

Table 79.1 Command-line options of pmbench. The last argument of the command line specifies the desired duration of the benchmark in seconds. The Initialize -i option was useful in circumventing memory compression newly introduced in Windows 10 [8]

Option (Keyword)	Value	Description
Map size (-m)	Integer	The size of the memory map in mebibytes
Set size (-s)	Integer	The portion of the memory map in mebibytes in which accesses will be performed
Access pattern (-p)	"uniform"	Access pages in a uniform random fashion (default)
	"normal"	Access random pages with normal distribution
	"linear"	Access pages in deterministic fashion with fixed stride
Shape of pattern (-e)	Float	Optional parameter determining the shape of the chosen access pattern's distribution
Read/write ratio (-r)	0–100	Percentage of accesses that will be reads. Default is 50
Number of threads (-j)	Integer	The number of measurement threads
Timestamp method (-t)	"rdtscp"	Use x86's rdtscp instruction (default)
	"rdtsc"	Use x86's rdtsc instruction
	"perfc"	Use OS-provided timestamp method
Delay (-d)	Integer	Delay between accesses in clock cycles. Default is 0
Offset (-o)	Integer	Offset into the page being accessed. Specifying −1 (default) results in uniform random offset
Cold (-c)	Boolean	Skips the warm-up phase of benchmark
Initialize (-i)	Boolean	Randomly initialize the memory map
File name (-f)	String	File name for XML output file

of cycles, necessitating close attention to the measurement methods and delicate instruction-level tuning.

Pmbench generates rich access latency statistics in the form of an XML file, which the companion graphical data analysis tool can import, manage, and use to generate graphs visualizing the latency histograms. This comparison can then be exported to a Comma-Separated Value (CSV) file for further processing/analysis with external applications.

Pmbench is executed from the command line with a variety of optional parameters, summarized in Table 79.1. Pmbench uses a processor timestamp counter to measure the time taken to access an address. The page frame number of each access is determined by a user-selected spatial memory access pattern, each mimicking various statistical distributions; currently the uniform (true random), normal (bell-shaped) and Zipf (long-tail) distributions are supported, along with a deterministic fixed-stride pattern with wraparound. Page offsets are randomly selected with uniform distribution, or can be set to a user-provided fixed value.

Access instructions, hand-written in assembly to ensure precision, are bookended by a pair of timestamps, whose difference in values will measure in clock cycles the time taken for the memory operation to complete. If the accessed address is in main memory and already mapped in its address space, the reference will take only a few clock cycles, typically resulting in less than 0.5 µs. If an access results in either a minor fault or major fault, the timestamp difference will be the fault handling latency. The measured latency is then converted to nanoseconds and used to increment a per-thread occurrence counter. The occurrence counter array is constructed in such a way that it first divides

intervals into buckets of latencies log base-2. Buckets of lower band (between 2^8 and 2^{23} ns) are further divided into 16 sub-bands with linear divisions.[1]

This tiered method of accounting allows pmbench to measure a wide range of latencies, from sub-microsecond to thousands of milliseconds, while preserving a detailed profile of smaller samples within sub-10 microsecond intervals where next-gen SSD devices are expected to operate. Pmbench supports a user-specified number of worker threads and increments per-thread latency counters. Care was taken to construct the in-memory counter so that each measurement thread requires a single page. Pmbench minimizes its code footprint by avoiding library function calls in its critical path.

There are two types of memory access: "read" – aligned 4-byte load and "write" – aligned 4-byte store. The 4K page dedicated to each thread for counting occurrence is divided into two 2K counter arrays, each counting for reads and writes. On operating systems that implement memory compression, the allocated memory map can be optionally initialized with random bits to circumvent its effects. Additionally, pmbench collects OS-reported memory and paging statistics over the course of a benchmark run.

The companion GUI tool produces line graphs from the histograms of pmbench result XML files. The parameter set for graphing/export is specified with dropdown menus; the radio buttons specify which variable will be used as the basis of comparison in the graph as well as exported CSV files.

---

[1] For example, a latency measurement of 9,231 ns will increment the 3rd counter in the 8,192 (2^{13})–16,384(2^{14}) ns bucket.

79.4   Evaluation

To demonstrate its utility, we used pmbench to evaluate the paging characteristics of Linux and Windows using SSDs as swap devices. The devices tested include a conventional NAND-based SSD and a prototype low-latency SSD.

79.4.1 Experimental Setup

All measurements were conducted on a x86 desktop machine with 4.0 GHz quad-core Intel i7-6700K processor and Z170 chipset. The machine is populated with total 16 GiB DRAM truncated to 2 GiB memory by setting OS boot parameters. An SSD is dedicated as the boot drive for both Linux and Windows partitions, each selected upon boot.

Samsung's 850 EVO 500 GB SSD is connected via SATA 6Gb/s and used as the swap device for NAND SSD measurements. The peak performance of the Samsung SSD is rated at 100 μs latency for 4KiB random reads at queue depth of 1, and 25 μs for random writes, but the SSD buffers write in internal DRAM, thus sustained average write latency is expected to be at least 100 μs. For low-latency SSD measurements, we used a prototype SSD which simulates the performance of low-latency NVM media. This low-latency SSD uses NVM Express protocol on 8 lanes of PCIe Gen 2 bus and is capable of sustained latencies of 5 μs for both 4 KiB random reads and writes at queue depth of 1.

On Linux we used a standard installation of Fedora 23 Workstation using stock 64-bit kernel version 4.3.5 with page clustering disabled. For Windows we used 64-bit Windows 10 Professional with a custom-sized pagefile optimized for interactive applications. On both operating systems, only the SSD being tested was enabled as a swap device.

We ran pmbench with normal process priority on four combinations of OS and SSD. The results seen here each represent the average of 25 identical benchmarks run for 5 minutes with two worker threads accessing 8 GiB of virtual memory in a uniformly random fashion. The memory was randomly initialized and the read/write mixture set at 50%.[2] We ran the benchmark after the OS enters quiescent state from fresh boot. We did not disable background processes/services but ensured no significant activity (e.g., antivirus scan) interfered with measurement.

79.4.2 Analyzing Pmbench Result

We illustrate the analysis of pmbench results by using measurement data from Windows 10 with the low-latency SSD as swap device. Figure 79.1 shows the access latency population count (i.e., histogram) plotted over latency as the X-axis in log-scale. The two graphs represent the same data, but the top graph's Y-axis uses a linear scale whereas the bottom graph uses a logarithmic scale. Read counts (black dash) and write counts (gray solid) are plotted separately, but there is no meaningful difference between them for accesses involving major faults.

As shown by the peak of the graph, the most frequent access latency is at 14.1 μs. This implies that the majority of major faults that go through the frequent-path have software overhead of about 9 μs – almost twice the SSD latency.

Left of the peak are a substantial number of accesses not resulting in a page faults. These are the expected 'hits', accessing addresses that just happen to be physically mapped at the time. In our Windows setup around 12% of all accesses result in hits. This is consistent with the fact that around 1 GiB of the 2 GiB system DRAM was available to pmbench after the kernel's pinned memory was accounted for.

The histogram exhibits long-tail, meaning there are a substantial number of extremely long-latency faults that heavily increase average latency and therefore hurt overall paging performance. Not surprisingly, the average latency of all major faults is 36.2 μs, far above the 14.1 μs peak latency. The average pure software overhead – latency from operating system processing – is therefore 31.2 μs, 624% of the 5 μs pure hardware latency.

Pmbench's detailed histogram can guide optimization of paging to low-latency SSDs. Measurement data suggests the frequent-path for major faults is too slow. Frequent-path latency could be reduced by eliminating routines unnecessary for low-latency SSDs, and possibly by employing polled I/O instead of interrupt-driven I/O [22]. Furthermore, there are too many long-latency faults: analysis shows that 33% of total execution time is spent on faults taking longer than 100 μs to handle, and 12.5% on faults taking 1 ms or more. Redesign of process scheduling policy involving paging in low-memory conditions could address these long-latency faults.

79.4.3 Paging Performance Comparison

We performed the same analysis on the data measured from all four combinations of OS and SSD, which is summarized in Table 79.2. Results from NAND SSDs confirm their unsuitability as DRAM displacements for workloads whose working set exceeds available physical memory. Although NAND SSDs can greatly speed up the tasks traditionally

---

[2] Exact command was: pmbench -m 8192 -s 8192 -j 2 -r 50 -d 0 -o -1 -p uniform -t rdtscp -c -i 300

Fig. 79.1 Access latency histogram for Windows 10 with low-latency SSD. The two graphs plot the same data but the bottom graph uses a log scale on Y-axis to highlight the presence of long-latency faults. Both graphs use log scale on X-axis for latencyg

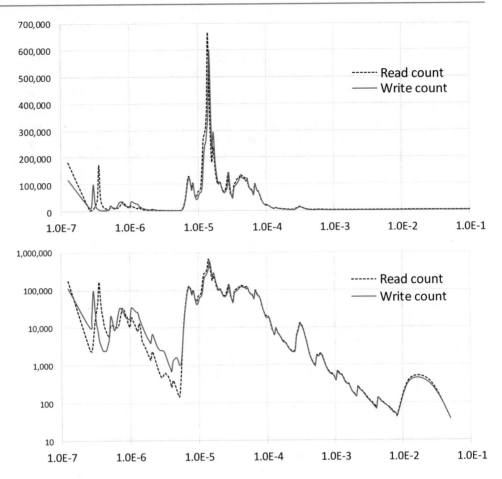

Table 79.2 Comparison of paging overhead. The last row shows the software overhead of paging as percentage of raw media latency, using 5 μs as media latency for the low-latency SSD and 100 μs for NAND SSD. Refer to Sect. 79.4.2 for further explanation of each row

Swap device (SSD) type:		Low-latency SSD		NAND SSD	
Operating system:		Linux	Windows	Linux	Windows
1	Average latency of all accesses including hits:	7.0 μs	32.6 μs	193 μs	369 μs
2	Most frequent major-fault latency:	8.2 μs	14.1 μs	131 μs	385 μs
3	Average latency of all major-faults:	8.6 μs	36.2 μs	236 μs	440 μs
4	Average OS overhead for major-faults:	3.6 μs	31.2 μs	136 μs	340 μs
5	OS overhead as percentage of media latency:	72%	624%	136%	340%

performed by HDDs such as file I/O, its fault service latency of 200 μs or more can easily thrash the system. The maximum sustained bandwidth of swap on NAND SSD is merely 20 MiB/sec, which is three orders of magnitude smaller than DRAM bandwidth.

In contrast, the low-latency SSD looks much more promising for use as DRAM displacement, though further OS optimization seems necessary. Measurements show the low-latency SSD delivering performance far superior to the NAND SSD on both OSes, due primarily to substantially lower media access latency. However, the low-latency media demonstrates that a high percentage of the average

fault latency can be attributed to inefficiencies of the software component.

It also shows that the current version of Linux handles low-latency SSD paging significantly better than Windows 10: Windows is 8.7 times slower than Linux in processing faults in low-memory conditions. Windows suffers from slow frequent-path and inefficient scheduling policy as discussed in Sect. 79.4.2. Linux, though faring better than Windows, still has room for improvement: the average 3.6 μs OS overhead translates to a substantial 14,400 clock cycles. We believe pmbench will prove useful in pinpointing these system bottlenecks.

79.5 Related Works

Low-latency SSDs will use next-generation non-volatile memory (NG-NVM) media. Although the incumbent NAND Flash technology continues to achieve higher bit density by going 3D [17], it suffers from inherent latency and management issues preventing DRAM-like usage [10]. - NG-NVMs include Spin-Transfer-Torque-RAM [13, 18], which is more suited to replace SRAM cache, and Resistive RAM (ReRAM) [19], which is cost-inefficient for mass market.

Phase-Change Memory (PCM) [16, 20] is the most promising NG-NVM technology at present, and thus the technology which our low-latency SSD was chosen to model. 3D Xpoint memory, recently announced for volume production by Intel and Micron, has performance closely resembling that of PCM.

The emergence of SSDs has renewed interest in storage system optimization and differentiation. The idea of using SSDs for DRAM displacement has also been explored for NAND SSDs via a user-level custom memory allocator [5]. NG-NVM media research has focused on its direct integration onto the memory bus [7, 9, 11].

Virtual memory and demand paging are classic topics covered by most OS textbooks and considered matured. Even the ideas of recoverable, persistent virtual memory [15] and compressed virtual memory [21] are not new. However, the rapidly changing platform landscape makes visiting those ideas worthwhile. Failure-atomic `msync()` [14] for example examines the memory persistence issue in the context of widespread Linux [6] and SSD use as a fast-backing persistent store.

The app-based usage pattern in smartphones and tablets led Windows 10 to introduce memory compression, enabling efficient app-swaps [8]. Commercial adoption of memory compression attests to the feasibility of using low-latency SSDs for DRAM displacement via paging. If done efficiently, paging to low-latency SSDs would likely have similar costs to memory compression; pmbench may assist in optimizing operating systems and platforms for this purpose.

Micro-benchmarks are useful for systems engineers to find inefficiencies and fine-tune with surgical precision. LMBench [3, 12] has long been used to measure performance of OS system calls. For storage systems, FIO [1] and IOMeter [2] are well-known micro-benchmarks that produce detailed storage stack statistics. However, existing benchmarks are not designed to accurately measure paging performance with microsecond-level precision. Pmbench aims to fill that void.

79.6 Conclusion

Low-latency next-generation SSDs are already coming to market. Using them as a fast paging space could be a cost-efficient alternative to large pools of DRAM. Using pmbench, we have confirmed that current operating systems are inefficient when paging to such low-latency SSDs. To further improve paging performance, the detailed paging performance data obtained from pmbench can be used by systems developers to locate source of inefficiencies and explore possible solutions.

In the future we intend to improve pmbench by supporting more operating systems, and synthesizing memory access patterns more accurately by analyzing patterns obtained from real workloads.

Acknowledgements We thank the anonymous reviewers for their valuable feedback. This work was supported by an award from Intel Corporation.

References

1. Flexible I/O tester (FIO). https://github.com/axboe/fio. Accessed 28 Oct 2016.
2. Iometer. http://www.iometer.org/. Accessed 28 Oct 2016.
3. Lmbench Source. https://sourceforge.net/projects/lmbench/. Accessed 28 Oct 2016.
4. PMBench. https://bitbucket.org/jisooy/pmbench. Accessed 28 Oct 2016.
5. Badam, A., & Pai, V. S. (2011). SSDAlloc: Hybrid SSD/RAM memory management made easy. In *Proceedings of the 9th USENIX Symposium on Networked Systems Design and Implementation (NSDI)*, Boston, MA.
6. Bovet, D. P., & Cesati, M. (2005). *Understanding the Linux kernel*. Sevastopol: O'Reilly.
7. Caulfield, A. M., Mollov, T. I., Eisner, L. A., De, A., Coburn, J., & Swanson, S. (2012). Providing safe, user space access to fast, solid state disks. In *Proceedings of the 17th International Conference on Architectural Support for Programming Languages and Operating Systems (ASPLOS)*, London, UK.
8. Creeger, E. Windows 10: Memory compression. https://riverar.github.io/insiderhubcontent/memory_compression.html. Accessed 28 Oct 2016.
9. Dulloor, S. R., Kumar, S., Keshavamurthy, A., Lantz, P., Reddy, D., Sankaran, R., & Jackson, J. (2014). System software for persistent memory. In *Proceedings of the 9th European Conference on Computer Systems (EuroSys'14)*, Amsterdam.
10. Grupp, L. M., Davis, J. D., & Swanson, S. (2012). The bleak future of NAND Flash memory. In *Proceedings of the 2012 USENIX/ACM Conference on File and Storage Technologies (FAST)*, San Jose, CA.
11. Lee, B. C., Ipek, E., Mutlu, O., & Burger, D. (2009). Architecting phase change memory as a scalable DRAM alternative. In *Proceedings of the 36th International Symposium on Computer Architecture (ISCA'09)*, Austin, TX.
12. McVoy, L., & Staelin, C. (1996). Lmbench: Portable tools for performance analysis. In *Proceedings of the USENIX 1996 Annual Technical Conference*, San Diego, CA.

13. Mishra, A. K., Dong, X., Sun, G., Xie, Y., Vijaykrishnan, N., & Das, C. R. (2011). Architecting on-chip interconnects for stacked 3D STT-RAM caches in CMPs. In *Proceedings of the 38th International Symposium on Computer Architecture (ISCA'11)*, San Jose, CA.

14. Park, S., Kelly, T., & Shen, K. (2013). Failure-atomic msync(): A simple and efficient mechanism for preserving the integrity of durable data. In *Proceedings of the 8th ACM European Conference on Computer Systems (EuroSys'13)*, Prague, Czech Republic.

15. Satyanarayanan, M., Mashburn, H. H., Kumar, P., Steere, D. C., & Kistler, J. J. (1993). Lightweight recoverable virtual memory. In *Proceedings of the 14th ACM Symposium on Operating System Principles (SOSP'93)*, Asheville, NC.

16. Kau, D., et al. (2009). A stackable cross point phase change memory. In *Proceedings of the 2009 I.E. International Electron Devices Meeting (IEDM)*, Baltimore, MD (pp. 1–4).

17. Im, J.-W., et al. (2015). A 128Gb 3b/cell V-NAND Flash memory with 1gb/s I/O rate. In *Proceedings of the 2015 International Solid-State Circuits Conference (ISSCC)*, San Francisco, CA.

18. Chun, K. C., et al. (2013). A scaling roadmap and performance evaluation of in-plane and perpendicular MTJ based STT-MRAMs for high-density cache memory. *IEEE Journal of Solid-State Circuits, 48*(2), 598–610.

19. Fackenthal, R., et al. (2014). A 16Gb ReRAM with 200MB/s write and 1GB/s read in 27nm technology. In *Proceedings of the 2014 International Solid-State Circuits Conference (ISSCC)*, San Francisco, CA.

20. Choi, Y., et al. (2012). A 20nm 1.8V 8Gb PRAM with 40MB/s program bandwidth. In *Proceedings of the 2012 International Solid-State Circuits Conference (ISSCC)*, San Francisco, CA.

21. Wilson, P. R., Kaplan, S. F., & Smaragdakis, Y. (1999). The case for compressed caching in virtual memory systems. In *Proceedings of the USENIX Annual Technical Conference*, Monterey, CA.

22. Yang, J., Minturn, D., & Hady, F. (2012). When poll is better than interrupt. In *Proceedings of the 2012 USENIX/ACM File and Storage Technology (FAST)*, Santa Clara, CA.

Maen Hammad

Abstract

This paper presents a framework, named DesignObserver, to automatically monitor and track design changes during software evolution. The framework helps in preserving code-design consistency during incremental maintenance activities. The design model is automatically updated based on implemented code changes. Preserving design quality is another important feature of the framework. Design changes are analyzed to determine their violations for pre-defined quality and pattern constraints. Any design change that breaks a design pattern or violates a quality metric is identified and highlighted. The framework also measures the cost of some potential design changes to evaluate their impact on other classes. Finally, designers and their contributions are also identified and reported by the framework to support assigning maintenance tasks. A set of tools that we have previously developed are mainly used to build DesignObserver.

Keywords

Software design • Design quality • Software maintanace and evolution

80.1 Introduction

Software projects are subject to incremental and maintenance activities. Maintenance is necessary to fix bugs, add new functionalities or adapt to environmental changes. Many maintenance tasks result in changing the structural design of software projects. Most of the structural design elements can be modeled by the UML class diagram. For example, adding new functionalities or features to the software results in updating the corresponding UML class diagram of the code by adding new classes and/or methods.

Structural design changes may badly affect the quality of the design. For example, low class coupling and small class size are factors for high quality design. In this case, the addition of a new class or a method needs to be analyzed to check the degree of coupling and the class size. Design changes may increase the complexity of the design that makes it hard to maintain in the future. So, the implemented design changes should be analyzed to check if they violate quality standards set by developers.

High level maintenance tasks need to be implemented by developers who have good design expertise. Identifying these developers is necessary to speed the task assignment process. In open software environments, many developers are involved in the development process. The problem is how to identify specific developers with design expertise. One solution to this problem is to mine historical code changes. Designers can be identified by analyzing historical code changes for developers. Based on historical design changes, designers can be identified and their design contributions can be measured.

Another important issue during maintenance activities is how to keep design consistent with source code. Any code change must be reflected on the design model. Manually updating design models is time consuming process.

M. Hammad (✉)
Department of Software Engineering, The Hashemite University,
Zarqa, Jordan
e-mail: mhammad@hu.edu.jo

© Springer International Publishing AG 2018
S. Latifi (ed.), *Information Technology – New Generations*, Advances in Intelligent
Systems and Computing 558, DOI 10.1007/978-3-319-54978-1_80

Furthermore, developers prefer writing code to complete their tasks rather than updating design documents. Failing in updating design makes design out of date and inconsistent with code.

In this paper, we propose a framework, named DesignObserver, to address all previously discussed issues that face developers during software maintenance. The framework monitors and keeps track on design changes committed by developers. The DesignObserver framework can be used as an integrated environment to handle various design issues for developers during incremental maintenance activities. The framework is mainly built based on a set of tools that we previously developed. These tools are integrated to provide a set of services to developers to monitor design evolution. The proposed framework provides the following three main services:

- Updating design model to keep it consistent with code after each code change.
- Analyzing design changes to identify their impact on design quality.
- Identifying designers with their contributions to the evolved design.

This paper is organized as follows. Section 80.2 discusses the related work in the area. An overview of the proposed framework is presented in Sect. 80.3. The design changes analyzes service is detailed in Sect. 80.4. The second service is presented in Sect. 80.5 following by the code and design consistency service in Sect. 80.6. Section 80.7 presents the preliminary evaluation of DesignObserver. Conclusions and future work are discussed in Sect 80.8.

80.2 Related work

Cazzola et al. [1] proposed an approach to code evolution that supports the automatic co-evolution of the design models. Their approach is based on a set of predefined metadata that the developer should use to annotate the application code and to highlight the refactoring performed on the code. Their approach is not fully automated since it requires developers to annotate their code changes manually.

In [2] D'Hondt et al. discussed their approach to manage the synchronization between design and implementation. The approach uses logic programming called Logic Meta Programming (LMP) to express the design as a set of rules or constraints. Mens et al. [3] proposed the intentional source-code views as a source code modularization mechanism based on crosscutting concerns. The intentional view model is implemented in logic meta-programming language that can reason about and manipulate object-oriented source code directly. Based on intentional views, the IntensiVE tool

suite is presented in [4]. The suite helps a developer in documenting structural source-code regularities, verifying them and offering fine-grained feedback when the code does not satisfy those regularities.

France and Bieman [5] proposed a framework to support managed evolution of OO systems. The framework called multi-view software evolution (MVSE). In MVSE, software evolution is a process in which models are iteratively evaluated and transformed. Aldrich et al. [6] proposed ArchJava, a small, backwards-compatible extension to Java that integrates software architecture specifications into Java implementation code. ArchJava ensures traceability between architecture and code. Ubayashi et al. [7] proposed Archface which is an interface mechanism that takes into account the importance of architecture and integrates an architectural design model with its implementation. Archface realizes the traceability between design and its implementation by enforcing architectural constraints on the program implementation. Aldrich [8] presented a system to statically enforce complete structural conformance between dynamic architectural description and OO implementation code.

Marinescu [9] proposed a mechanism called detection strategy for formulating metrics-based rules that capture deviations from good design principles and heuristics. The approach helps to localize classes or methods affected by a particular design flaw. In our approach, we do not identify design flaws; instead we identify violations of predefined constraints. Gustafsson et al. [10] presented architecture-centric software evolution method. They presented Maisa, a Java-based tool for evaluating software architectures. Maisa is integrated with the reverse engineering tool Columbus [11]. It keeps track of the evolution of the software system by storing the measurements in a database.

The DesignObserver framework does not use logic programming approaches as most of the related work in the area of code-design co-evolution. The framework is also has the advantage of analyzing archived design changes to extract useful information about designers.

80.3 Overview of the Framework

Figure 80.1 shows the structure and the main components of the proposed framework. The DesignObserver framework is applied on the repository of software system managed by subversion systems. The framework provides two major services to developers. It monitors the incremental maintenance tasks that affect design. The goal of monitoring is to keep design consistent with code and to preserve its quality. The other service provided by DesignObserver is the identification of designers and their contributions to design. The goal is to help managers in identifying developers with design expertise.

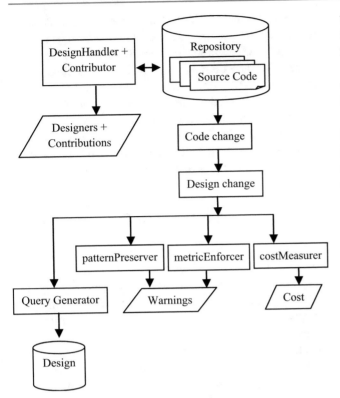

Fig. 80.1 The structure and components of the Design Observer frame work

Any code change is analyzed by the srcTracer [12] tool to identify design changes from the changed code. Any code changes that add/delete a class, a method or a relationship is identified by srcTracer as a design change. The reported design change is processed by three tools. The patternPreserver [13] tool ensures that the design change does not break any predefined design pattern. The second tool, metricEnforcer [14], checks if the change does not violate any predefined object oriented constraints. These two tools warn developers with the identified breaks and violations. The third tool, costMeasurer, calculates the cost of design changes and reports the result.

After the design changes are analyzed, they are transformed by the Query Generator component into XPath queries. These queries are used to update the Design database. This database stores the up to date design model in flexible XML format. So, design is automatically updated after each code change by the transformed queries to keep it consistent with code.

The other service provided by the framework is the identification of designers and their expertise. The DesignHandler [15] and Contributor [16] tools mine the repository to analyze historical code changes to identify designers with their design expertise and contributions. This information helps in recommending a set of developers

to handle high level change requests. It also helps in identifying how developers contributed in shaping the current status of design.

80.4 Checking Design Violations

Code changes, committed by developers, are analyzed to identify their impact on the quality of the design. Each code change is identified as a design change or not. Design change is defined as adding/deleting classes, methods, or relationships (generalizations, associations and dependencies). The identified design changes are analyzed by the following three tools.

80.4.1 PatternPreserver

This tool [13] automatically detects and identifies breaks in predefined design patterns from small and incremental code changes. The tool keeps track on code changes activities committed by developers to determine breaks. Design patterns are represented using XML format in which design elements and their roles in the pattern are saved together. Based on the identified design change, The XML file is parsed to check if the affected design element has a role in a pattern or not. In case a break is detected, a warning message is shown to developers about the design change and the affected pattern.

80.4.2 MetricEnforcer

The metricEnforcer [14] automatically identifies violations of predefined object oriented constraints from code changes. The constraints are; the number of children, the number of methods, the number of inherited methods, the depth of the inheritance tree, the number of called methods, and the number of relationships. The metricEnforcer extracts metrics from the code and view them to developers. Then, they can set the upper limit of each extracted value (i.e. constraints). Then, constraints are saved in XML format. These XML files are parsed to check for violations based on the type of design changes. Violations are reported to developers with all their related information.

80.4.3 CostMeasurer

This tool gives recommendations to developers about the impact of the design change. It helps to measure the effort needed to be applied after committing a major design change. Some design changes require major changes to code in order to cope with the applied design changes.

The tool specifically measures the cost of deleting a class. Deleting a key class may cause major impact on the design. For example, the deletion of a supper class requires rewriting or deleting its subclasses. Another example of a key class is a class that is coupled with many other classes. Deleting such a class requires updating the code of all classes that use the deleted one. Coupling can be measured using the Fan-in value of the class. The Fan-in value of a class A is the number of other classes that use class A. So, deleting a super class or a class that is used by many other classes is a costly design change since it impacts many other classes. The cost of deleting a class A is defined as:

Cost = The number of direct and indirect subclasses of A + The Fan − in value of class A.

The generated cost report includes the names of classes. The cost is calculated by the costMeasurer tool that mainly uses the relFinder [17] tool to find the Fan-in value of the class. The tool transforms the code into the XML representation srcML [18]. In srcML, each code element is tagged with its syntactic information. Then, a set of Xpath queries are used to and parse it and to calculate Fan-in value of a class.

80.5   Identifying Designers

Identifying designers and their contributions to design is another major service provided by DesignObserver. This helps in assigning change requests to developers and in measuring contribution of designers to the evolved design. Two tools are integrated to mine source code repository and identify the design expertise of developers. These tools are DesignHandler and Contributor.

80.5.1 DesignHandler

This tool is developed to automatically handle high level change requests [15]. It prompts users to determine the type of change request. The options are; add/delete feature, add/delete functionality, or design enhancement. Based on the selected option, a list of developers is recommended. The recommendations are based on the number of added/deleted methods, classes or the number of used classes by developers. DesignHandler also builds a knowledge graph for designers. The graph shows the relationship between developers and classes they used during design changes activities.

80.5.2 Contributor

The Contributor tool automatically calculates the design contribution for developers to the evolved structural design [16]. The design contribution of a developer is measured by the total number of added or deleted design elements committed by the developer during specific time duration. Design contributions are categorized into two categories; the addition of new design elements and the deletion of old design elements. So, the tool counts total number of added and deleted design elements by developers. Weight values can also be determined by users to design change. For example, a developer who added one class may be considered to have more design contribution than the one who added one method. Contributor finds the contributions values of a specific developer or set of developers ranked based on their contributions.

80.6   Consistency Between Code and Design

Design should be always kept consistent with source code. Any code change that affects design must be reflected on design documents. It is not easy for developers to go and update design documents after each code change. The framework automatically updates design whenever it is necessary after each code change. In this case, design is kept up-to-date with the code. The design model of the code is stored as XML representation. Each class has a corresponding flexible XML file that stores its methods and relationships with other classes. Figure 80.2 shows a sample class named Subject with its design representation in XML.

The Subject class has the following design elements that are tagged in the XML representation:

```
class Subject public SuperSubject{
    vector <Observer*> views;
    public:
      virtual void attach(Element);
      void notify();
};
```

A) Sample Class

```
<class>
  <name>Subject</name>

  <methods>
    <method><name>attach</name></method>
    <method><name>notify</name></method>
  </methods>

  <Associations>
     <to>Observer</to>
  </Associations>

  <Generalizations>
     <to>SuperSubject</to>
  </Generalizations>

  <Dependencies>
     <to>Element</to>
  </Dependencies>
</class>
```

B) Design Elements in XML

Fig. 80.2 A sample class (**a**) with it design representation in XML (**b**)

- A generalization relationship to class Super Subject.
- Two methods.
- An association relationship to class Observer.
- A dependency relationship to class Element.

The Query Generator component automatically generates the suitable XPath queries that correspond to the identified design changes. These generated XPath queries are executed on the Design database to update the XML representations of changed classes.

We assume that the XML representation of design is already exists since it is build incrementally during the development process. In case the design database is missing, it is automatically generated from the source code. In the first step, the code is transformed into the XML representation srcML [18]. Then, a set of XPath queries are executed to extract the design elements of each class (methods and relationships). Finally, the XML representation for design is generated based on the extracted elements.

80.7 Preliminary Evaluation

A preliminary working version has been implemented for DesignObserver for evaluation and experimental purposes. We designed a set of tasks on selected classes from an open source project. The tasks were designed to violate specific quality and pattern constraints. Two developers were asked to implement the tasks where their code changes are monitored by DesignObserver. We then manually checked DesignObserver results on identifying the violations and updating the design database. We found that Design Observer correctly identified most of the violations in the tasks and correctly updated most of the changed design elements in the database.

80.8 Conclusions and Future work

A framework to automatically monitor the design evolution process has been presented. The framework keeps track on code changes activities that impact design. The main benefits of the proposed framework are; preserving design quality, keeping design consistent with code and identifying designers with their contributions. It also measures the cost of some design changes activities to keep developers aware with the impact of change on other classes. The framework can be used in the development process of open source projects. It can be connected and mine code changes stored in subversion repositories.

We are currently working on a complete working version for DesignObserver followed by comprehensive evaluation.

Our future work aims to extend the framework to include model refactoring. We are working on identifying candidates for model refactoring and preserving design quality standards after applying refactoring. Other extend includes more design changes and quality measures.

References

1. Cazzola, W, Pini, S., Ghoneim, A., Saake, G. (2007). Co-evolving application code and design models by exploiting meta-data. In *Proceedings of the 2007 ACM symposium on applied computing (SAC'07)* (pp. 1275–1279). Republic of Korea.
2. D'Hondt, T., Volder, K. D., Mens, K., Wuyts, R. (2000). Co-evolution of object-oriented software design and Implementation. In *Proceedings of the international symposium software architectures and component technology:The state of the art in research and practice*. Netherlands.
3. Mens K., Mens T., Wermelinger, M. Maintaining software through intentional source-code views. (2002). In *Proceedings of the 14th international conference on software engineering and knowledge engineering (SEKE'02)* (pp. 289–296). Italy.
4. Mens, K., & Kellens, A. (2005). Towards a framework for testing structural source-code regularities. In *Proceedings of the 21st IEEE international conference on software maintenance (ICSM'05)* (pp. 679–682). Hungary.
5. France, R., & Bieman, J. M. (2001). Multi-view software evolution: A UML-based framework for evolving object-oriented software. In *Proceedings of the IEEE international conference on software maintenance (ICSM'01)* (pp. 386–395). Italy.
6. Aldrich, J., Chambers, C., Notkin, D. (2002). ArchJava: Connecting software architecture to implementation. In *Proceedings of the 24th international conference on software engineering (ICSE'02)* (pp. 187–197). USA.
7. Ubayashi, N, Nomura, J., Tamai, T. (2010). Archface: A contract place where architectural design and code meet together. In *Proceedings of the 32nd ACM/IEEE international conference on software engineering (ICSE'10)* (pp. 75–84). South Africa.
8. Aldrich, J. (2008). Using types to enforce architectural structure. In *Proceedings of the 7th working IEEE/IFIP conference on software architecture* (pp. 211–220). Canada.
9. Marinescu, R. (2004). Detection strategies: Metrics-based rules for detecting design flaws. In *Proceedings of the IEEE international conference on software maintenance (ICSM'04)*. USA.
10. Gustafsson, J., Paakki, J., Nenonen, L., Verkamo, A. I. (2002). Architecture-centric software evolution by software metrics and design patterns. In *Proceedings of the 6th european conference on software maintenance and reengineering (CSMR"02)* (pp. 108–115). Spain.
11. Ferenc, R., Magyar, F., Beszedes, A., Kiss, A., Tarkiainen, M. (2002). Columbus - reverse engineering tool and schema for C++. In *Proceedings of the 18th IEEE international conference on software maintenance (ICSM'02)*. Canada.
12. Hammad, M., Collard, M. L., Maletic, J. I. (2009). Automatically identifying changes that impact code-to-design traceability. In *Proceedings of the 17th IEEE international conference on program comprehension (ICPC'09)* (pp. 20–29). Canada.
13. Hammad, M., Hammad, M., Otoom, A. F., & Bsoul, M. (2014). Detecting breaks in design patterns from code changes. *Journal of Software, 9*(6), 1485–1493.
14. Hammad, M., Hammad, M., & Bsoul, M. (2014). An approach to automatically enforce object oriented constraints. *International Journal of Computer Applications in Technology, 49*(1), 50–59.

15. Hammad, M., Hammad, M., & Bani-Salameh, H. (2013). Identifying designers and their design knowledge. *International Journal of Software Engineering and Its Applications, 7*(6), 277–288.

16. Hammad, M., Hammad, M., Bani-Salameh, H., & Fayyoumi, E. (2014). Measuring developers' design contributions in evolved software projects. *Journal of Software, 9*(12), 3005–3011.

17. Hammad, M., Abu-Wandi, R., Aydeh, H. (2016). Automatic reverse engineering of classes' relationships. In *Proceedings of the* 13th *international conference on information technology:New generations (ITNG'16)* (pp. 1267–1272). USA.

18. Collard, M. L., Kagdi, H. H., Maletic, J. I. (2003). An XML based lightweight C++ fact extractor. In *Proceedings of the 11th IEEE international workshop on program comprehension (IWPC'03)* (pp. 134–143). USA.

Generating Sequence Diagram and Call Graph Using Source Code Instrumentation

81

Mustafa Hammad and Muna Al-Hawawreh

Abstract

Understanding the dynamic behavior of the source code is an important key in software comprehension and maintenance. This paper presents a reverse engineering approach to build UML sequence diagram and call graph by monitoring the program execution. The generated models show the dynamic behavior of a set of target methods with time and object creation information. Timing and dynamic behavior details are extracted by instrumenting the target code with a set of calls to a monitoring function in specific instrumentation points in the source code. The proposed approach is applied on a case study to show the effectiveness and the benefits of the generated models.

Keywords

Code instrumentation • Dynamic behaviour • Sequence diagram • Call graph • Dynamic analysis

81.1 Introduction

Software systems are permanently increasing and changing. Generally, software maintainers are not its designers. Therefore, when the maintainers try to modify or change the software, they need to comprehend the implementation of it. This mission is a tedious time consuming and requires a great effort. Moreover, software documentation is insufficient for understanding the software accurately because sometimes the constant changing of the software has not accompanied by modifying of documentation. Hence, it is necessary to develop a technique to extract useful information to understand the program behavior.

Reverse engineering is one of the techniques that provide a solution to increase the overall comprehensibility of the software with little or no additional knowledge about the procedures included in the original program. It is the process of Acquisition a knowledge and the requirement specification of a product from an analysis of its code [1]. Reverse engineering can be done statically or dynamically. The static analysis describes the structure of the program by analyzing the source code, while the dynamic analysis focuses on the behavior of a program during the runtime.

In Object Oriented Program (OOP) the behavioral models (i.e. dynamic models) are of a specific attention, due to the specific characteristics of that system, such as inheritance, polymorphism, and dynamic binding. In fact, it is very difficult to know the dynamic type of object reference, which methods are invoked, and to query the system clock to calculate the time spent in a particular region of the code by just depending on the static analysis of the source code.

Analyzing a dynamic behavior of software, generally, needs code instrumentation where extra code fragment inserts into the source code to capture the behavior of the

M. Hammad (✉)
Department of Information Technology, Mutah University,
Al-Karak, Jordan
e-mail: hammad@mutah.edu.jo

M. Al-Hawawreh
Department of Information Technology, Mutah University,
Al-Karak, Jordan
e-mail: munahawari1@gmail.com

© Springer International Publishing AG 2018
S. Latifi (ed.), *Information Technology – New Generations*, Advances in Intelligent Systems and Computing 558, DOI 10.1007/978-3-319-54978-1_81

program. This paper proposes a new technique that automatically generates two dynamic models; sequence diagram and call graph.

The goal of the proposed approach is provide generic models for software execution to fully understand the dynamic software behavior. We use a source code instrumentation to trace specific methods in a program.

The rest of this paper is organized as follows. Section 81.2 shows the related work. Section 81.3 describes the proposed approach. A case study and discussions are presented in Sect. 81.4. This paper is concluded in Sect. 81.5.

81.2 Related Work

Maintenance of software constitutes a vital point in software life cycle, where between 50% and 80% of maintainers time spent in understanding the programs [2]. To handle this issue, maintainers use reverse engineering techniques to understand complex program.

This discussion outlines several works on reverse engineering techniques of different modeling diagrams. Hammad and Cook [3, 4] have applied a software visualization technique to model the dynamic behavior of wireless sensors. Their work focuses on gathering information about the execution of the sensor program by instrumenting the source code. In addition, they used UML sequence diagram to represent the behavior with timing information.

Sarkar and Chaterjee [5] presented a method to reverse engineer UML sequence diagram from execution traces for Java program; they define how instrumenting the source code to gather information about classes, objects and methods, and it uses this information to generate UML sequence diagrams to describe the dynamic behavior of the program.

Many techniques for the reverse engineering of sequence diagram have been presented [6–8]. All these techniques used one execution trace to support and identify the UML interaction operator. In contrast, *Ziadi et al.*, [9] collected multiple execution traces and used K-tail algorithm to extract labeled transition system, which it translated to a sequence diagram.

The other literature exists [10] wherein a new approach for reverse engineering of UML sequence diagram and state chart based on the dynamic behavior of conformance checking and pattern identification. Alalfi and Cordy [11] used UML 2.1 tool to represent the dynamic behavior of PHP-based web applications. The main goal of this work is to analyze the web application security and understanding it.

A dynamic analysis to generate call graph of java script is presented by Toma and Islam [12]. They used the information gathered early in software life cycles to make or modify the graph that already exist.

Kandala et al., [13] suggested a new tool to generate call graph for C programs. The tool accepts the C program as input, and then analyzes it and generates a call graph starting from Main function. Moreover, it displays some function metrics, such as total number of lines in the function, total number of executable line, number of unreachable line, and the Cyclomatic complexity. This tool provides both dynamic and static view for calling function in the program.

To document software requirements specifications, UML use case diagram is usually adopted to describe the interactions of an actor with the interested system. XMI2UC [14] is an automatic tool to extract a use case from object oriented program based on the program's sequence diagram. This tool is based on filtering process to generate functional sequence diagram. Other works, such as [15–18], applied the reverse engineering techniques to extract the use cases or program classes relationships.

Our work differs from existing approaches in term of providing different dynamic models that visualize the behavior of program during execution. The proposed approach generates both sequence diagrams and modified call graphs with time information.

81.3 Overview of the Framework

The proposed approach facilities the maintainers needs by extracting dynamic models diagrams, which describe the dynamic behavior of the software. Figure 81.1 shows the main steps for the proposed approach. Classes/Objects Finder goal is to identify the instrumentation points in the source code. This step is implemented based on string matching and parsing the main structure of the source code. The second process is set to instrument the code with a set of predefined monitoring function calls, which collect information about the dynamic behavior of program and generate an execution log file. Finally, generate dynamic models after analyzing the generated execution trace file. The generated models include UML sequence diagrams and modified call graphs.

Instrumentation is applied for each target method in two instrumentation points. These points include the starting and ending points of the methods. Figure 81.2 shows an example of a method before and after the instrumentation process. Before the instrumentation process is applied the two instrumentations points are localized in the code, which express the points where a method start and just before finish and return.

As shown in Fig. 81.2, the instrumented code is a call hook for a runtime monitoring function called *MonitFuc*. As a result, the monitoring function is executed two times for each method call. This monitoring function is supported with two parameters to distinguish between invocations

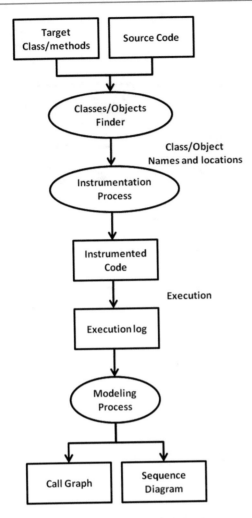

Fig. 81.1 The Architecture of proposed approach

```
Public class C                    Public class C
{                                 {
  Public static foo (){             Public static foo (){
➡ <Instrumentation Point>            MonitFunc("foo", "Start")
  // Method code..                    // Method code..
➡ <Instrumentation Point>            MonitFunc("foo", "End")
  Return ;}                           Return ;}
}                                 }
```

(a) before instrumentation   (b) after instrumentation

Fig. 81.2 Instrumentation process example. (a) before instrumentation (b) after instrumentation

order and locations. The proposed monitoring function simply append the invoked method name, event type (i.e. method starts or ends), and CPU time stamp to an execution log.

When the instrumented source code is executed, an execution log will be created for the target methods in the program. Finally, the generated execution trace information

is processed by the modeling process, which generates a UML sequence diagram and call graph for the target methods.

Sequence diagram is one of dynamic modeling diagrams that describe a dynamic behavior of a program. It shows object interactions coordinated and the message exchanged between them in time sequence. To create a sequence diagram we need to collect target class, methods, and objects name with execution trace. The object and class name are modeled as a rectangle. Each edge connects between two objects or class represents the method name with corresponding time information.

Dynamic call graph is a directed graph that provides calling relations between subroutine in a computer program. Thus, the cycle in the graph refers to the recursive procedure calls. Each method name is presented as a vertex of the graph in a rectangle box. The average execution time and the number of method's invocations in each execution are displayed. Moreover, all object creations during the program execution that happened in the target methods are shown.

The generated models help developers to better understand the dynamic behavior of the source code and capture dynamic metrics and measures, such as Response For Class (RFC) and Coupling Between Objects (CBO), as well as, execution time of a set of interested classes and methods. The details of visualizing process are reported with a case study in the following section.

81.4 Case Study and Descussion

To validate the capabilities of the proposed approach for object oriented programming language, we applied the proposed instrumentation technique to extract dynamic models for JFreeChart open source project (http://www.jfree.org). It is a Java library that is used to create professional charts. As an example, we target the class *BarChartDemo1*, which is part of the package *jfreechart-1.0.16/source/org/jfree/chart/demo*. Timing information and dynamic behavior are captured using instrumentation for a set of target methods. For this case study, the target method is the *CreateDataset* method.

The collected tracing data from the instrumented code showed that the target class has a relationship with *DefaultCategoryDataset* class. Tracing data showed the target method in *BarChartDemo1* class create an object of type *DefaultCategoryDataset*. Figure 81.3 shows the generated sequence diagram that describes the dynamic behavior of one execution of the target class/methods.

As shown Fig. 81.3, *CreateDataset* method uses an object named *dataset* of type *DefaultCategoryDataset*. The target method returns the *dataset* object after invoking *AddValue* method four times. As a result, the instrumentation process

Fig. 81.3 The generated Sequence Diagram for *CreateDataset* method

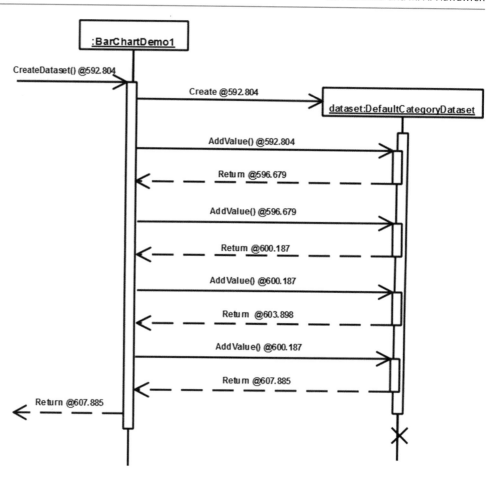

instruments the *AddValue*, as well as, *CreateDataset* methods. The generated sequence diagram shows the dynamic behavior in which those methods are invoked. Timing information, in milliseconds, is displayed on each object message. A global timer is initialized when the target class, *BarChartDemo1*, is used to create a new object. After that the sequence of messages with corresponding time stamped are shown in diagram. For example, as shown in Fig. 81.3, the *dataset* object is created after 592.804 ms from the initialization point. Moreover, the *CreateDataset* returns after 607.885 ms.

Figure 81.4 shows the call graph, which is the second view in the proposed methodology, for the target class/methods in the JFreeChart case study. This view can be seen as a summary of the dynamic behavior for *CreateDataset* and *AddValue* methods. As shown in Fig. 81.4, *CreateDataset* method creates one object named *dataset* from *DefaultCategoryDataset* class and calls *AddValue* method four times for one execution. Furthermore, for each method execution, the average time that is need for a method to return is presented. For instance, the average execution time for method *AddValue* in class *DefaultCategoryDataset* is 3.77 ms.

The generated call graph represents the collected information from execution trace in different ways. The proposed view is based on dynamic analysis, which describes the exact behavior of program execution. On the other hand, generating call graph using static analysis generates overestimation models which may present relationships that never happen.

The generated views are useful for developer or maintainer to better understand the dynamic behavior of their source code. The proposed sequence diagram and call graph present abstraction models that can be used to analyze and compare classes and objects relationships. In addition, the generated models can be used to validate software-requirement properties that are related to execution time.

81.5 Conclusion

This paper proposed a dynamic analysis approach to generate dynamic models. We propose a methodology based on code instrumentation that assists the developers in program comprehension. By using this methodology, it is possible to

Fig. 81.4 The generated Call Graph for *CreateDataset* method

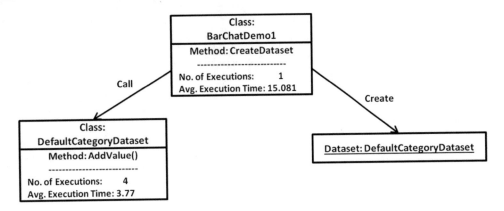

find the interaction between different classes and objects, and capture the execution time for a set of target methods.

The proposed work would be further improved by generating more models, such as use case diagrams. Furthermore, showing more dynamic details of a method execution, such as branches and loops will add a significant platform for software maintainer to study and analyze the dynamic behavior of source code.

References

1. Choudhary, U., & Yadav, M. (2015). Review on reverse engineering techniques of software engineering. *International Journal of Computer Applications, 119*(14), 7–10.
2. Banker, R. D., & Kemerer, F. C. (1987). Factors affecting software maintenance productivity: An exploratory study. *InICIS*.
3. Hammad, M., & Cook, J. (2009). Lightweight monitoring of sensor software. In *Proceedings of the 2009 ACM symposium on applied computing* (pp 2180–2185). USA.
4. Hammadand, M., & Cook, J. (2009). Lightweight deployable software monitoring for sensor networks. *Proceedings of 18th internatonal conference in computer communications and networks(ICCCN)., IEEE* (pp 1–6). USA.
5. Sarkar, K. M., & Chaterjee, T. (2013). Reverse engineering: An analysis of dynamic behavior of object oriented programs by extracting UML interaction diagram. *International Journal of Computer Technology and Applications, 4*(3), 378.
6. Taniguchi, K., Ishio, T., Kamiya, T., Kusumoto, S., Inoue, K. (2005). Extracting sequence diagram from execution trace of Java program. In *Principles of software evolution, eighth international workshop on IEEE* (pp. 148–151). Portugal.
7. Oechsle, R., & Schmitt, T. (2002). Javavis: Automatic program visualization with object and sequence diagrams using the java debug interface (jdi). In *Software visualization* (pp. 176–190). Berlin/Heidelberg:Springer.
8. Lo, D., Maoz, S., Khoo, S.-C.(2007). Mining modal scenariobased specifications from execution traces of reactive systems. In *ASE 02* (pp. 465–468).
9. Ziadi, T., Da Silva, A. M., Hillah, M. L., Ziane, M. (2011). A fully dynamic approach to the reverse engineering of UML sequence diagrams. *IEEE Engineering of Complex Computer Systems (ICECCS)* (pp. 107–116).
10. Guéhéneuc, G. Y., & Ziadi, T. (2005). Automated reverse-engineering of UML v2.0 dynamic models. *Proceedings of the 6th ECOOP workshop on object-oriented reengineering.*
11. Alalfi, H. M., Cordy, R. J., Dean, R. T. (2009). Automated reverse engineering of UML sequence diagrams for dynamic web applications. *IEEE Software testing, verification and validation workshops, (ICSTW'09)* (pp. 287–294). USA.
12. Toma, R. T., & Islam, S. M. (2014). An efficient mechanism of generating call graph for JavaScript using dynamic analysis in web application. *IEEE Informatics, Electronics & Vision (ICIEV)* (pp. 1–6).
13. Kandala, P., Kaur, J., Rao, N. V. T. (2014). Optimizing programs using call graphs. *Compusof An International Journal of Advanced Computer Technology, 3*(2), 561–568.
14. Ebad, S. A., & Ahmed, M. A. (2012). XMI2UC: An automatic tool to extract use cases from object oriented source code. *International Journal of Future Computer and Communication, 1*(2), 193–196.
15. Gomaa, H., & Olimpiew, M. E. (2005). The role of use cases in requirements and analysis modeling. In *Workshop on use cases in model-driven software engineering*. Montego Bay. Jamaica.
16. Gomaa, H., Shin, M. (2003). Variability in multiple-view models of software product lines. In *International workshop on software variability management (SVM)* (pp. 63–70). USA.
17. Khaled, L. (2011). Extracting use case diagram from requirement engineering processes. International Journal of Research and Reviews in Applied Sciences, 9(2), 317–321.
18. Hammad, M., Abu-Wandi, R., Aydeh, Haneen. (2016). Automatic reverse engineering of classes' relationships. *Information Technology: New Generations*. Springer International Publishing (pp. 1267–1272). USA.

Part VIII

High Performance Computing Architectures

Data Retrieval and Parsing of Form 4 from the Edgar System using Multiple CPUs

82

Raja H. Singh, Nolan Burfield, and Frederick Harris Jr.

Abstract

In this paper we present a parallel system that retrieves and parses Form 4 documents from the Securities and Exchange Commission's Electronic Data Gathering, Analysis and Retrieval database (EDGAR). This information is very important for investors looking at insider trading information to make investment decisions. However, the information's usefulness is inversely related to the time it takes to retrieve and analyze the information. A sequential system is slow due to the latency associated with the retrieval and parsing of the Form 4s, which on average exceeds 1000 per day. By making the retrieval and parsing of Form 4s parallel we were able to attain the max speed up of 20x, resulting in parsing of a daily index with 1000 forms in under 30 minutes instead of 9 hours it takes utilizing a single processor.

Keywords

Commodity Cluster • Sun Grid Engine • Investment • Open MPI • Java • MySQL • XML Parsing • Document Object Model.

82.1 Introduction

The Securities and Exchange Commission (SEC) is the regulatory branch of the government that is responsible for "protecting investors, maintaining fair, orderly, and efficient markets, and facilitate capital formation [1]." According to an article on Gallup [2] over 54 percent of Americans own stocks. SEC states that individuals investing in the stock market do so to fund goals such as paying for their homes, sending their children to school, and to secure their futures. Unlike banking, stock investments are not guaranteed by the Federal government and have a high potential of losses. Due to the nature of the investments, the SEC recommends conducting sufficient research about the company prior to investing in their stocks.

The SEC's primary objective is to provide investors with the opportunity to research the company/companies prior to investment. To carry out the said objective, the SEC requires public companies to disclose "meaningful financial and other information to the public [1]." One such form (Form 4) is part of the required disclosure.

82.1.1 Form 4: Insider Trading

Forms 4, 4A, 3, 3A, 5 and 5A are categorized as Insider Trading forms. Contrary to the belief that insider trading is illegal, SEC specifies that it is used for both illegal and legal conduct [3]. In the context of this paper we will refer to insider trading as purchasing or selling of stocks in a company by individuals where they are corporate officers. Any corporate insider (company's officers, directors, or beneficial owners who have more than 10 percent of a company's equity) are required to file a "statement of ownership regarding those securities [3]" using Form 3. Form 4 is used to

R.H. Singh • N. Burfield (✉) • F. Harris Jr.
Department of Computer Science and Engineering, University of Nevada, 89557 Reno, NV, USA
e-mail: nburf@nevada.unr.edu

© Springer International Publishing AG 2018
S. Latifi (ed.), *Information Technology – New Generations*, Advances in Intelligent Systems and Computing 558, DOI 10.1007/978-3-319-54978-1_82

report the changes in ownership i.e. purchasing and selling shares. Form 5 is used to report any transactions that the insider didn't report on form 4; Form 5 is used to report any transactions that should have been reported on Form 4 [3].

According to [4], Form 4 is the most filed form between the years 1994–2011, with more than 4 million filings according to the authors. This also holds true today, for example in the year 2015 there were 397910 Form 4 filings, with 82126 of those Form 4 filings in the fourth quarter of 2015 alone. In order to parse the large number of these forms, a system has to be created to do so automatically at a reasonable speed so that the useful nature of insider trading information is preserved.

82.1.2 Form 4: Potential Uses

We will focus on the retrieval of Form 4. The reason behind this is that several investors look to Form 4 to determine the change of ownership within the company. This is vital information because it indicates the confidence in the company's growth potential. For example, an investor is able to look at Form 4 filed by a high ranking officer and determine if the insider is accumulating or disposing of their holdings in the company. An investor may utilize this information when considering the purchase or sale of shares. For example, if the insider is purchasing shares, it could possibly indicate that they have internal information indicating potential growth of the company. However if an insider disposes shares in the company then it indicates their knowledge of potential downturn in the company's performance in the future. In this paper we will provide an overview of the system, and its optimization through the use of multiple CPUs.

82.2 Prior Works

The financial data vending industry is largely corporate, therefore there have not been many published works on agents downloading and parsing information from the SEC. Up to the point of writing this paper, we were only able to locate two papers that discussed the retrieval of data from the EDGAR system. [4] by Garcia and Norli offer a method to crawl EDGAR for Form 8k, the second most common filed form after Form 4, is used to announce major events to shareholders. However, Garcia and Norli utilize it to analyze CEO turnovers. The writers wrote a Perl program to download all 8k files locally and then parse through and analyze the forms.

The second paper [5], written by D. A. Lyon, discusses parsing the Central Index Keys (CIKs) and correlating them to stock tickers. Currently EDGAR only uses company name

or CIK to present the data on the requested company. Lyon created a graphical user interface that allows the user to enter the company ticker to gather the data from EDGAR; Lyon used Java to write HTML scrapers to gather data from the EDGAR System. We were not able to locate any papers addressing gathering and parsing Form 4, however we did discover companies that charge a monthly subscription fee to disseminate Form 4 information to their subscribers.

82.3 Data Source & Format

SEC maintains all filings on their Electronic Data Gathering, Analysis and Retrieval database (EDGAR) system. The data can be accessed either through their EDGAR search page or through SEC's File Transfer Protocol (FTP) server. The formats offered by the SEC are HTML, "eXtensible Business Reporting Language" (XBRL) and Extensible Markup Language (XML). Our system pulled the Form 4 pages in XML format. Please refer to Fig. 82.1 for the format of the document.

82.3.1 File Transfer Protocol (FTP) Server

The FTP server allows users to log into the system anonymously by utilizing your email address. We utilized this FTP method, making anonymous FTP calls to the server. Refer to [6] for specific instructions on logging into the system. The SEC does request bulk FTP requests be performed between 9:00 p.m. and 6:00 a.m. ET.

82.3.1.1 Indexes
The EDGAR system provides indexes for FTP retrievals. These indexes list Company Name, Form Type, CIK, Date Filed and the File Name; this index also includes the folder path to the document [6]. Please refer to Fig. 82.2 for the image of the EDGAR indexes set-up. The system offers four types of indexes:

- company – Index file sorted by the company name.
- form – Index file sorted by form type.
- master – Index file sorted by the CIK.
- XBRL – "List of submissions containing XBRL financial files, sorted by CIK number; these include Voluntary Filer Program submissions [6]."

The indexes are filed daily on the EDGAR system containing all submissions made to the organization. The system also maintains a quarterly index list containing all filings for the quarter for a given year. These date back to 1994 as shown in Fig. 82.2. Inside each index, as stated above, is the listing of all files based on the type of index

```
                                  ← → C    ⬅ ftp://ftp.sec.gov/edgar/data/1280600/0001179110-16-022918.txt
                              ZIP:                  02139
</SEC-HEADER>
<DOCUMENT>
<TYPE>4
<SEQUENCE>1
<FILENAME>edgar.xml
<DESCRIPTION>FORM 4 -
<TEXT>
<XML>
<?xml version="1.0"?>
<ownershipDocument>

    <schemaVersion>X0306</schemaVersion>

    <documentType>4</documentType>

    <periodOfReport>2016-04-06</periodOfReport>

    <notSubjectToSection16>0</notSubjectToSection16>

    <issuer>
        <issuerCik>0001280600</issuerCik>
        <issuerName>ACCELERON PHARMA INC</issuerName>
        <issuerTradingSymbol>XLRN</issuerTradingSymbol>
    </issuer>

    <reportingOwner>
        <reportingOwnerId>
            <rptOwnerCik>0001325710</rptOwnerCik>
            <rptOwnerName>Sherman Matthew L</rptOwnerName>
        </reportingOwnerId>
        <reportingOwnerAddress>
            <rptOwnerStreet1>128 SIDNEY STREET</rptOwnerStreet1>
            <rptOwnerStreet2></rptOwnerStreet2>
            <rptOwnerCity>CAMBRIDGE</rptOwnerCity>
            <rptOwnerState>MA</rptOwnerState>
            <rptOwnerZipCode>02139</rptOwnerZipCode>
            <rptOwnerStateDescription></rptOwnerStateDescription>
        </reportingOwnerAddress>
        <reportingOwnerRelationship>
            <isDirector>0</isDirector>
            <isOfficer>1</isOfficer>
            <isTenPercentOwner>0</isTenPercentOwner>
            <isOther>0</isOther>
            <officerTitle>EVP &amp; Chief Medical Officer</officerTitle>
            <otherText></otherText>
        </reportingOwnerRelationship>
    </reportingOwner>

    <nonDerivativeTable>
        <nonDerivativeTransaction>
            <securityTitle>
                <value>Common Stock</value>
            </securityTitle>
            <transactionDate>
                <value>2016-04-06</value>
            </transactionDate>
            <transactionCoding>
                <transactionFormType>4</transactionFormType>
                <transactionCode>S</transactionCode>
                <equitySwapInvolved>0</equitySwapInvolved>
                <footnoteId id="F1"/>
            </transactionCoding>
            <transactionAmounts>
                <transactionShares>
                    <value>2080</value>
```

Fig. 82.1 This shows the XML format of the downloaded Form 4 to be parsed by the system

Index of /edgar/daily-index/

Name	Size	Date Modified
🔼 [parent directory]		
📁 1994/		7/10/13, 5:00:00 PM
📁 1995/		7/10/13, 5:00:00 PM
📁 1996/		7/10/13, 5:00:00 PM
📁 1997/		7/10/13, 5:00:00 PM
📁 1998/		7/10/13, 5:00:00 PM
📁 1999/		7/10/13, 5:00:00 PM
📁 2000/		7/10/13, 5:00:00 PM
📁 2001/		7/10/13, 5:00:00 PM
📁 2002/		7/10/13, 5:00:00 PM
📁 2003/		7/10/13, 5:00:00 PM
📁 2004/		7/10/13, 5:00:00 PM
📁 2005/		7/10/13, 5:00:00 PM
📁 2006/		7/10/13, 5:00:00 PM
📁 2007/		7/10/13, 5:00:00 PM
📁 2008/		7/10/13, 5:00:00 PM
📁 2009/		7/10/13, 5:00:00 PM
📁 2010/		7/10/13, 5:00:00 PM
📁 2011/		7/10/13, 5:00:00 PM
📁 2012/		7/10/13, 5:00:00 PM
📁 2013/		10/1/13, 5:00:00 PM
📁 2014/		10/1/14, 5:00:00 PM
📁 2015/		10/1/15, 5:00:00 PM
📁 2016/		4/1/16, 7:01:00 PM
📄 company.20160104.idx	772 kB	1/4/16, 7:01:00 PM

Fig. 82.2 This figure shows the EDGAR file-path with all the daily indexes dating back to 1994

file chosen (i.e. company vs form vs master). Please refer to Fig. 82.3 of the image of a daily index file (type master).

82.4 System Overview

The system is built to pull all data from the SEC and store it into a MySQL database. It relies on multiple classes that can be categorized into specific functionality of the system. Due to the space limitations we will not discuss every class within the system but will instead discuss classes that assist us in giving a general overview the system in the following sections.

82.4.1 FTP

The system relies heavily on the FTP class. The primary function of this class is to open an FTP connection, and connect to a specified destination. The connection stays open until either it is closed or the system exits. This class must be instantiated when used in order to keep a persistent connection and handle connection failures. Running the parallel application several nodes may fail when attempting to pull FTP data, therefore the FTP class must handle this and still grab the necessary data. The approach the system takes is to continue to attempt a connection until the data is

```
Description:          Daily Index of EDGAR Dissemination Feed by Company Name
Last Data Received:   Apr  8, 2016
Comments:             webmaster@sec.gov
Anonymous FTP:        ftp://ftp.sec.gov/edgar/
```

```
Company Name                                              Form Type   CIK
      Date Filed   File Name
-------------------------------------------------------------------------------------------------------
1347 Property Insurance Holdings, Inc.                    8-K         1591890    20160408   edgar/data/1591890/0000914760-16-000397.txt
1st Century Bancshares, Inc.                              PREM14A     1420525    20160408   edgar/data/1420525/0001047469-16-012053.txt
56 Brewing, LLC                                           D           1671634    20160408   edgar/data/1671634/0001671634-16-000001.txt
6D Global Technologies, Inc                               8-K         1382219    20160408   edgar/data/1382219/0001185185-16-004185.txt
86Borders LLC                                             D/A         1622720    20160408   edgar/data/1622720/0001622720-16-000004.txt
A-KNH-33-Fund, a series of AngelList-SDA-Funds, LLC       D           1671392    20160408   edgar/data/1671392/0001671392-16-000001.txt
A-MJQ-13-Fund, a series of AngelList-BaRs-Funds, LLC      D           1669974    20160408   edgar/data/1669974/0001669974-16-000001.txt
A-NHN-34-Fund, a series of AngelList-S2Al-Funds, LLC      D           1671393    20160408   edgar/data/1671393/0001671393-16-000001.txt
A-PKY-15-Fund, a series of AngelList-MaBr-Funds, LLC      D           1671391    20160408   edgar/data/1671391/0001671391-16-000001.txt
A-VRO-11-Fund, a series of AX-IA-Funds, LLC               D           1671390    20160408   edgar/data/1671390/0001671390-16-000001.txt
A10 Networks, Inc.                                        3           1580808    20160408   edgar/data/1580808/0001209191-16-113760.txt
AB CAP FUND, INC.                                         DEF 14C     81443      20160408   edgar/data/81443/0001193125-16-534930.txt
AB CAP FUND, INC.                                         DEFA14C     81443      20160408   edgar/data/81443/0001193125-16-534946.txt
ABERCROMBIE & FITCH CO /DE/                               8-K         1018840    20160408   edgar/data/1018840/0001193125-16-535655.txt
ABIOMED INC                                               8-K         815094     20160408   edgar/data/815094/0001193125-16-535148.txt
ABNER HERRMAN & BROCK LLC                                 13F-HR      1038661    20160408   edgar/data/1038661/0001038661-16-000007.txt
ABRAMSON STEVEN V                                         4           1232980    20160408   edgar/data/1232980/0001005284-16-000182.txt
ACADIA REALTY TRUST                                       8-K         899629     20160408   edgar/data/899629/0001144204-16-093488.txt
ACCELERON PHARMA INC                                      4           1280600    20160408   edgar/data/1280600/0001179110-16-022918.txt
ACCELERON PHARMA INC                                      4           1280600    20160408   edgar/data/1280600/0001179110-16-022919.txt
ACHILLION PHARMACEUTICALS INC                             8-K         1070336    20160408   edgar/data/1070336/0001193125-16-535465.txt
ACORN COMPOSITE Corp                                      4           1519311    20160408   edgar/data/1519311/0001079792-16-002595.txt
```

Fig. 82.3 This figure shows contents of a daily index file. It is an index of all the filings for the specified date, including the name of the filing entity, the type of form filed, the CIK for the filer and the location (file-path) of the form

pulled. To avoid corruption in the data FTP is set to transfer as binary, which is not the default for the Java Apache Commons-net FTPClient [7].

82.4.2 System Updating

The system initializes by checking the last date the MySQL database was updated and computes the number of days that it needs to retrieve and store from the EDGAR system. To achieve this we created a set of nested classes that compute the date difference between the current time and the last known update date for the database. Since the SEC filings are only accepted and updated Monday through Friday, the system does not include the weekends when trying to retrieve data from the system.

82.4.3 Index Retrieval

The EDGAR system has an efficient method to disperse information on filings for specific dates as explained in prior sections that rely on the "index schema." Our system uses the FTP connections to pull the index for a specified time frame (either daily index for a specific date or a quarterly index for a specific quarter of a year). The classes responsible for this retrieve the index and then parse the index storing each filing into a HashMap, where the key is the type of the filing (ie. Form 4) and the value is an ArrayList of a type DailyData. Please refer to [8] for details about HashMap and ArrayList data structures. The Daily Data is a container that stores every component of the index information listed below:

- Company Name
- Form Type
- CIK
- Date Filed
- File Name
- Folder path to the document

Once the data from the daily index file is parsed, and stored into the appropriate data structure, the information is passed to each form parsing class. Meaning the index retrieval class iterates through the HashMap and invokes the appropriate class to handle each type of the form. As of now our system only has parses for the insider trading forms. The system is designed for easily adding in new SEC form parsers without the need to modify other classes.

82.4.4 Data Retrieval

The system currently has a class to retrieve the list of forms passed to it by the index retrieval class. For example when the Form 4 class is invoked it is passed a list of DailyData objects that contain the location of the forms in the EDGAR system. The class iterates through the list and makes FTP requests to pull the specified form one at at time and parses the retrieved data.

82.4.5 Data Parsing Class

To parse the XML data retrieved from Edgar, we created the Form 4 parsing class. The class builds a Document Object Model (DOM) tree object and recursively iterates through all of the nodes within the tree storing the values into a string builder object. In order for us to parse the retrieved documents, we had to know the structure (values) of the XML document prior to computing. We utilized the insider trading form specifications provided by the SEC in [9] to create the parsing class. If the element/node was not found, a default value of null is stored for the field.

For the purpose of testing the efficiency of parallelization of the bottleneck of the system, we modified the code so that the values were dumped into a CSV file instead of storing it in a database. We decided to not store the values into a database for this paper because the parallel execution of the system. In order to store the data into a MySQL database the system requires a pool of Java Database Connectivity (JDBC) connections, unfortunately it was nonsensical on the Sun Grid Engine (SGE) using MPI. If we had chosen to proceed with the storage of the data into MySQL database, it would have resulted in large network overhead.

Refer to Algorithm 82.1 **for the pseudo algorithm of the system.**

Algorithm 82.1 System Overview

Compute the date difference between current date and last database update date initialize retrieval of data for each day
for each date retrieve the daily index **do**
    for each retrieved Daily index file: initialize index parsing **do**
        for each parsed index file **do**
            Store each form information into a Daily Data object
            Store each Daily Data object into a
            Hash Map <Key:Form Type, Value: List<Daily Data>>
            Invoke the appropriate parsing class to parse and
                          store data into database.
        end for
    end for
end for

Table 82.1 The table above shows the most time intensive function calls.

	50	75	100	250	500	750	1000
Main	27.66	41.06	54.52	135.80	269.36	405.79	541.09
RunForms	27.01	40.50	53.98	135.26	268.81	405.24	540.54
FTPRequest	26.12	38.93	51.68	128.43	256.06	384.51	512.10

82.5 Hardware and Language

The code was written in Java programming language due to the cross platform nature of Java and the vast number of libraries available for software development. Apache Ant was used to build/compile the system and was executed on the Oracle Grid Engine (previously known as the Sun Grid Engine). Message Passing Interface (MPI) was used to handle the communication between the CPUs in the cluster.

The grid has 27 Dell Servers with 16 cores each for a total of 432 cores. Each server has the following specification [10]:

- Dual 8 core Intel E2650v2 2.6 GHz processors
- 256 GB RAM
- 10 GBPS Ethernet
- 1.2 TB of local storage

82.6 Profiling: Parallel Justification

Prior to profiling the code we already knew that the system's latency came from the FTP requests, however it was required to prove this in order to justify the parallelization of the system. We did not use the HPROF profiler because of the complications associated with submitting jobs to the Oracle Grid Engine.

In order to profile the sequential execution, we created a profiling class using the `System.nanoTime()`. Refer to [11] for detailed specifications on the Java system timer. Using the profiling class, we were able to time every function call made by the system. Table 82.1 shows the three of the most time intensive functions in the execution of the program. Please note that the main function calls the RunForms function which entails calls the FTPRequests function. Therefore the times for the calling functions exceed the times of the function(s) being called. Based on the timings shown in Table 82.1 we were able to determine the FTPRequest function was the appropriate function to parallelize because it contributes (on average) 94.5% to the total computational time.

82.7 Sequential Execution

The sequential code is described in Sect. 82.4 System Overview and Algorithm 82.1. The sequential code was also executed on the Oracle Grid Engine to ensure consistency. For the parsing we utilized a single daily index with 1025 Form 4 documents in the index list. We did a total of 7 iterations, increasing the number of Form 4 documents parsed during each iteration; we parsed the following number of forms for each iteration 50, 75, 100, 250, 500, 750 and finally 1000. This allowed us to compare the sequential and parallel executions.

82.8 Parallel Execution

For the parallel execution every computer node (CPU) opens the daily master index shown in Fig. 82.3, parses the index list as mentioned in Algorithm 82.1, and stores it into a HashMap. Next each node strides the Form 4 list extracting DailyData objects for each Form 4 that node is responsible for. The nodes stride the ArrayList of DailyData types based on their rank and the total number of nodes initialized on the grid. Refer to Algorithm 82.2 for an overview of the logic for each node in the parallel execution.

Algorithm 82.2 Parallel System Overview

Retrieve the daily index
for each retrieved Daily index file: initialize index parsing **do**
    for each parsed index file **do**
        Store each form information into a Daily Data object
        Store each Daily Data object into a
        Hash Map <Key:Form Type, Value: List<Daily Data>>
        for each Form Type **do**
            Stride the List<Daily Data>
            Build new list based off of rank
            Invoke the appropriate parsing class to parse
            Output parsed data to CSV
        end for
    end for
end for

Once each node creates a sub list of its Form 4 documents, the rest of the code each node executes is the same as sequential. Due to Java's garbage collection feature, utilizing shared memory was not feasible. Therefore each node ran the same code in parallel without communication with any other nodes. The sequential algorithm we parallelized in this manner is the ideal "embarrassingly parallel problem."

82.8.1 Set-up

As with the sequential execution, there were a total of 7 iterations with each iteration increasing the number of forms as follows: 50, 75, 100, 250, 500, 750, and 1000. However, for each one of the iterations where the number of forms was increased, the parallel execution ran each iteration five times starting with 4 nodes and increasing by 4 up to a total of 20 nodes; the number of nodes used for each form iteration were 4, 8, 12, 16, and 20.

To avoid the networking bottleneck when making FTP calls, we utilized a specific round-robin scheduling scheme to choose workers. The system was initialized so that the slots (number of nodes requested) "are filled one per machine until all machines have one slot filled, and then the next slot on each machine is filled, until all requested slots are provided [10]." However since there are 29 servers and the max number of nodes we requested are 20 at any given time, we never ran into a scenario with more than one node per server/box.

82.9   Results

The results were as expected. The sequential execution of the data retrieval and parsing takes longer than reasonable times. An average daily index contains over 1000 Form 4 documents. At the current execution speeds, a sequential execution of the system takes around 9 hours to parse 1000 forms. This is unacceptable in the financial industries where time is very critical factor when determining how to react on the information attained; the benefit of any information is inversely proportionate to the time. The longer the information is out there, the less it is worth. Once the system was parallelized the runtime drops drastically as shown in Table 82.2. For example, parsing 1000 forms with 4 CPUs resulted in a completion time of around 2 hours and 30 minutes. Figure 82.4 visualizes this and can show that by simply adding in 4 nodes will result in an acceptable data collection time (Fig. 82.5 shows this data with normalized values).

Since the bottleneck of the system could be divided into individual parts and executed independently of each other, the speedup increased in conjunction with the number of processors used. This resulted in a relatively constant speedup for each node size regardless of the number of forms parsed. For example using 4 processors the speedup stayed between 3.5X to 4X. This makes sense, because even

Fig. 82.4 The total time in minutes to finish execution of each processor size at each iteration of quantity of parsed Form 4

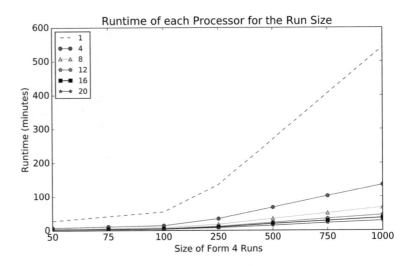

Table 82.2 The table above shows the run times of the program throughout each processor size at each iteration of quantity of parsed Form 4

	1	4	8	12	16	20
50	27.66	7.55	4.33	3.25	2.73	3.26
75	41.06	10.94	5.95	4.33	3.27	2.74
100	54.52	14.05	8.08	5.40	4.53	3.73
250	135.80	34.46	17.97	11.85	9.93	7.97
500	269.36	68.11	34.63	23.16	19.85	14.46
750	405.79	102.05	51.12	35.01	28.56	22.13
1000	541.09	135.54	68.16	45.77	37.15	28.75

Fig. 82.5 This shows that same as Fig. 82.4 with the data scaled to log 10 along the time axis to better show run time differences

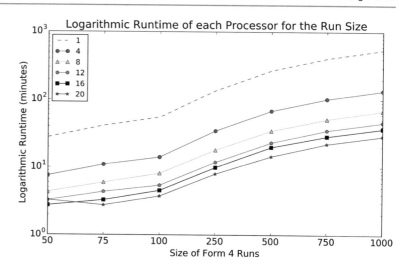

Table 82.3 The table above shows the speedup of the program throughout each processor size at each iteration of quantity of parsed Form 4

	1	4	8	12	16	20
50	1.00	3.66	6.38	8.50	10.11	8.46
75	1.00	3.75	6.89	9.46	12.53	14.97
100	1.00	3.87	6.74	10.09	12.02	14.58
250	1.00	3.94	7.55	11.45	13.67	17.01
500	1.00	3.95	7.77	11.62	13.56	18.61
750	1.00	3.97	7.93	11.58	14.20	18.33
1000	1.00	3.99	7.93	11.81	14.56	18.81

Fig. 82.6 This figure shows the speedup of the program run time, throughout each processor size at each iteration of quantity of parsed Form 4, with the baseline being the sequential execution time (Sequential Run Time/ Parallel Run Time)

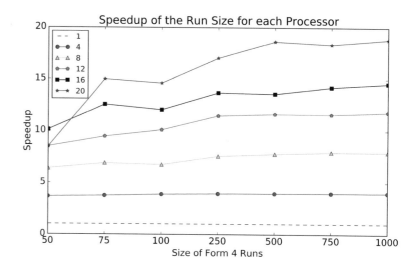

though there was no communication between the processors, they were still vying for the same network resources, resulting in a lag. Table 82.3 shows that values of speedup through all iterations, as the number of forms parsed increases the speedup increases as well. However, speedup of all nodes does not change drastically. With 20 nodes speedup is about 20 when 1000 forms are parsed. Please refer to Fig. 82.6 for the visualization of the speedup values;

it shows the gradual increase of speedup for each number of nodes.

The key function of the system is to retrieve and extract EDGAR data and store it into local database. The speed at which the data is extracted is a very important metric to determine whether the system is feasible for use in the real world scenario. Currently our system (upon parallelization) parses around 34 forms per minute using 20 nodes.

Table 82.4 shows the values for each iteration performed, and these values are visualized in Fig. 82.7. Looking at the trends in the graph it can be seen that the throughput follows the trend of speedup exactly (Fig. 82.8).

82.10 Conclusion and Future Work

As stated in the problem, it is necessary with financial data to receive the information as soon as possible in order to react to the changing markets. In order to do this a delay of 9 hours would not be an acceptable time frame to get information from a Form 4 document filed in the Edgar database. Parallelization of the program successfully dropped that time delay down to only 27 minutes. This time was achieved with running the same algorithm on 20 processors at once, and since the nodes all run the same logic there is no need for communication between them.

Achieving a large speedup was the purpose of the parallelization of the bottleneck within the program. That is to say processing as many forms per minute as possible is most beneficial when talking about the financial data. Since an average daily size of Form 4 submissions is around 1000 and these can be fully processed and stored in under 30 minutes the parallel program accomplishes the goals set out to be done.

Table 82.4 The table above shows the throughput of the program throughout each processor size at each iteration of quantity of parsed Form 4

	1	4	8	12	16	20
50	1.80	6.62	11.53	15.37	18.28	15.29
75	1.82	6.85	12.58	17.28	22.90	27.35
100	1.83	7.11	12.36	18.50	22.05	26.75
250	1.84	7.25	13.91	21.08	25.16	31.33
500	1.85	7.34	14.43	21.58	25.18	34.56
750	1.84	7.34	14.67	21.42	26.25	33.88
1000	1.84	7.37	14.67	21.84	26.91	34.77

There is much work that can be done to make the system a comprehensive tool to gather the data from EDGAR. Please see the list below for future work:

- Add parsing class to parse all key forms such as 10-k, 10-Q, 8-K, 13-F, 13-H, etc.
- Add a View to the system that visualizes the data both in a tabular and graphical manner.
- Add a package to the system that extracts historical data so back testing can be performed to check the validity of any trading strategies.
- Create a class that utilizes Map Reduce to parse complex forms with XBRL formats.
- Write a RSS (Rich Site Summary) reader to update the system live based on the up to date filing information.

Form 4 information offers great a great indicator of company's direction, performance/profit wise. The system can be utilized as stand alone for long term investments or it could be used in conjunction with the historical data. For example the system could be utilized to find insiders with a track record of indicating the future behavior of a company. This can be done by back testing their past insider trading actions and comparing them to the historical movement of the company. If they have a track record of making appropriate trades based on possible future movements that specific insider's actions can be tracked to make investment decisions. The possible uses of the insider trading are too many to list them all in this paper.

Acknowledgements This material is based in part upon work supported by the National Science Foundation under grant no. IIA-1301726. Any opinions, findings, and conclusions or recommendations expressed in this material are those of the authors and do not necessarily reflect the views of the National Science Foundation.

Fig. 82.7 This figure shows the throughput of the Form 4 parsing, the system throughput is calculated on forms parsed per minute

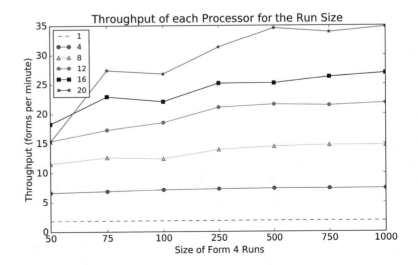

Fig. 82.8 Efficiency

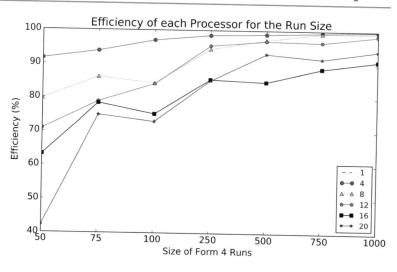

References

1. What We Do. SEC.gov. Web. 05 Apr. 2016.
2. In U.S., 54% Have Stock Market Investments, Lowest Since 1999. Gallup.com. N.p., n.d. Web. 14 Dec. 2015. http://www.gallup.com/poll/147206/stock-market-investments-lowest-1999.aspx.
3. Fast Answers. SEC.gov. Securities and Exchange Commission, n.d. Web. 2 Mar. 2016. https://www.sec.gov/answers/form345. htm. Forms 3, 4, 5
4. García, D., & Norli, Ø. (2012). Crawling EDGAR, *The Spanish Review of Financial Economics, 10*(1), 1–10.
5. Lyon, D. A. (2008). Multi-threaded data mining of EDGAR CIKs (Central Index Keys) from ticker symbols.In *2008 I.E. International Symposium on Parallel and Distributed Processing.*
6. Information for FTP Users. Information for FTP Users. Securities and Exchange Commission. Web. 5 Mar. 2016. https://www.sec.gov/edgar/searchedgar/ftpusers.htm.
7. Apache Commons Net Overview. Apache Commons Net Overview. Apache, n.d. Web. 21 Mar. 2016. https://commons.apache.org/proper/commons-net/.
8. The Java Tutorials. The Java Tutorials. Oracle, n.d. Web. 27 Jan. 2016. https://docs.oracle.com/javase/tutorial/.
9. EDGAR Ownership XML Technical Specification (Version 5.1). EDGAR Ownership XML Technical Specification (Version 5.1).

Securities and Exchange Commission, n.d. Web. 15 Mar. 2016. https://www.sec.gov/info/edgar/ownershipxmltechspec.htm.
10. High Performance Computing (The Grid). The Grid Specifications. Web. 15 Mar. 2016. http://www.unr.edu/it/research-resources/the-grid.
11. Java Platform SE 8. Java Platform SE 8. Oracle. Web. 17 Mar. 2016. http://docs.oracle.com/javase/8/docs/api/.
12. Using EDGAR – Researching Public Companies. Investor.gov. Securities and Exchange Commission, n.d. Web. 1 Feb. 2016. https://www.investor.gov/researching-managing-investments/researching-investments/using-edgar-researching-public-companies.
13. Gupta, Lokesh. "Java XML DOM Parser Example Tutorial." HowToDoInJava. N.p., 31 July 2014. Web. 10 Feb. 2016. http://howtodoinjava.com/xml/java-xml-dom-parser-example-tutorial/.
14. Welcome – Apache Ant. Apache Ant. N.p., n.d. Web. 4 Apr. 2016. http://ant.apache.org/.
15. About JDBC Resources and Connection Pools. (Sun Java System Application Server Platform Edition 8.2 Administration Guide). Oracle, n.d. Web. 24 Feb. 2016. <https://docs.oracle.com/cd/E19830-01/819-4712/ablii/index.html>.
16. MPI Documents. MPI Documents. Web. 12 Mar. 2016. http://www.mpi-forum.org/docs/.
17. Chapter 4 Tutorial. MySQL. N.p., n.d. Web. 29 Jan. 2016. http://dev.mysql.com/doc/refman/5.7/en/tutorial.html.

Fangyang Shen, Janine Roccosalvo, Jun Zhang, Yang Yi, and Yanqing Ji

Abstract

This research introduces an innovative approach for STEM teacher scholarship implementation. The purpose of this research is to introduce a new STEM teacher implementation model. This model will help recruit high-quality Noyce Scholars and retain and train them as effective STEM teachers. In this paper, we share the actual implementation experience of this strategy in the past three years for an existing STEM teacher education project funded by the National Science Foundation (NSF). In addition, we prove that this strategy is effective with two types of evidence. First, we use the survey data collected from over twenty-five STEM teacher candidates. Secondly, we report the actual interview data and student feedback and share the experiences learned from the NSF Noyce project.

Keywords

Noyce • STEM • STEM Teacher Education

83.1 Introduction

There is a significant demand for Science, Technology, Engineering and Mathematics (STEM) teachers throughout the United States, especially in Mathematics and Technology fields. This has become an immense problem for the United States so it is crucial to train STEM teachers. The U.S. government is also aware of this situation. On January 25, 2011, President Obama delivered a State of the Union Address emphasizing the importance of STEM fields in education. The main goal for this plan was to prepare 100,000 new teachers in the fields of science, technology, engineering and math in 10 years [1].

In addition, 28 different organizations joined together to accomplish this goal. One of these organizations was the National Science Foundation which formulated and funded the Robert Noyce Teacher Scholarship Program [2]. This scholarship program aims to recruit extraordinary students with science, technology, engineering and mathematics majors and train them to become STEM teachers in grades K-12, especially in high-need schools and communities. Institutions of higher education then promote and support students with undergraduate STEM majors who also have a genuine interest in becoming future STEM teachers.

New York City College of Technology (City Tech) and Borough of Manhattan Community College (BMCC), neighboring CUNY institutions, implemented a Robert Noyce Teacher Scholarship Project that increases the number of highly qualified STEM teachers in high-needs school districts in Brooklyn and the New York metropolitan area [3]. This will help expand exponentially the numbers of

F. Shen (✉) • J. Roccosalvo
NYC College of Technology, Brooklyn, NY, USA
e-mail: fshen@citytech.cuny.edu; fangyangshen@gmail.com;
jroccosalvo@citytech.cuny.edu

J. Zhang
Univ. of Maryland, Eastern Shore, Princess Anne, MD, USA
e-mail: Jzhang@umes.edu

Y. Yi
The University of Kansas, Lawrence, KS, USA
e-mail: yyi@ku.edu

Y. Ji
Gonzaga University, Spokane, WA, USA
e-mail: ji@gonzaga.edu

© Springer International Publishing AG 2018
S. Latifi (ed.), *Information Technology – New Generations*, Advances in Intelligent
Systems and Computing 558, DOI 10.1007/978-3-319-54978-1_83

students in under-resourced school districts who will be taught by well-qualified STEM professionals. The BMCC-to-City Tech STEM teacher preparation infrastructure will provide a sustainable pathway for the professional teacher preparation of underrepresented students who attend City Tech and BMCC. Through the project website and other mechanisms, the CUNY system, one of the largest urban public higher education systems in the nation, will be able to look to City Tech and BMCC as creators and sustainers of a model pathway with multiple points of entry and an innovative three-tier structure "Explorer, Scholar, Teacher" that maximizes support for STEM students who wish to enter the teaching profession in STEM disciplines and serve their communities.

This three-tiered Noyce partnership is an excellent strategy to recruit and retain high quality STEM teachers. It will recruit students in their first and second years of STEM studies and enroll them as Noyce Explorers (TIER I), lead them through a highly motivating Summer Program that offers a combination of STEM content, STEM pedagogy and provides an opportunity to interact in a pedagogical role with peers and/or younger students through peer tutoring, mentoring and in other quasi-instructional capacities. Noyce Scholars (Tier II) are third and fourth year undergraduates in STEM programs who have committed to becoming STEM teachers in high need schools through generous scholarships. Noyce Scholars will receive $10 K per year as they complete a teacher certificate program in a STEM field. Upon licensure, Noyce Scholars will be designated Noyce Teachers (TIER III) and they will receive support through their initial induction into the profession either from City Tech's Math Education or Career and Technical Teacher Education program. Over the five years' grant period, this program will produce a total of 20 new STEM teachers and create a new STEM teacher preparation pathway for the critically underpowered STEM teaching force in New York City.

One of the main challenges for most of the Noyce Project nationwide is that it is difficult to recruit and retain dedicated, highly qualified Noyce Scholars and turn them into effective STEM teachers. This is no exception for the current City Tech-BMCC Noyce Project. In the past three years, we investigated different approaches to make the Noyce Scholar recruitment and retention mechanism more effective. Eventually, based on three years of implementation and experiences, we proved that the three-tiered Noyce partnership is the most effective structure to recruit and retain Noyce Scholars and produce committed, skilled STEM teachers.

In this paper, we introduce our innovative three-tiered Noyce partnership implementation structure. We also demonstrate the performance of the proposed structure from the independent external program evaluation using the Noyce scholar interview data, student feedback and share the actual implementation experiences learned from the NSF Noyce project. The results indicate that the proposed three-tiered Noyce partnership is an effective mechanism to recruit and retain STEM teachers.

The rest of this paper is organized as follows: Sect. 83.2 reviews the literature for this topic; Sect. 83.3 introduces the three-tiered Noyce partnership structure and implementation in the past three years; Sect. 83.4 presents the survey results, Noyce scholar interview data and student feedback conducted by the independent external program evaluator; Sect. 83.5 summarizes the findings of this study and discusses possible directions for future research.

83.2 Literature Review

Generally, it is well accepted that the U.S. economy should have more workers with high levels of knowledge in Science, Technology, Engineering and Mathematics (STEM). The shortage of these skilled STEM professionals will make U.S. fall behind in the global economic competition. According to [4, 5], the United States might lose its competitive edge in STEM fields while the rest of the world soars ahead, if a nationwide STEM education plan is not proposed and implemented. In reality, American students are falling behind other main competitors such as China, Japan and Germany in the critical STEM subjects. As a result, one of the main challenges facing STEM education today in the United States is to generate and produce more STEM teachers and produce more effective STEM workers. Therefore, it is crucial to develop strong STEM teacher preparation programs nationwide.

In [1], the White House is making an effort to dramatically change this situation in STEM education. The Obama Administration announced an aggressive plan to prepare 100,000 excellent math and science teachers by 2020. Accordingly, the White House announced related steps by the Administration and its partners to help accomplish the President's goal to produce more STEM teachers. One of these steps include a new $22.5 M investment by the Howard Hughes Medical Institute (HHMI) which would approximately double the private-sector investment in the President's initiative.

On the federal level, such as with the National Science Foundation, there is a reputable program called the NSF Noyce Program [2] which will also contribute to the White House's teacher training goal by 2020. The National Science Foundation's Robert Noyce Teacher Scholarship Program seeks to encourage talented science, technology, engineering and mathematics majors and professionals to become K-12 STEM teachers. This program includes creative and innovative proposals that address the critical need for recruiting and preparing highly effective K-12 STEM teachers, especially in high-need local educational agencies. This will ultimately increase the number of students in under-resourced school districts who will be taught by high quality STEM teachers.

There are some existing examples of STEM teacher recruitment and retention programs such as [6–8]. In [6], it describes a STEM teacher recruiting and training practice developed through the Talented Teachers in Training for Texas (T4) program. They discuss implementing three distinctive recruiting experiences—a STEM Master Teacher Job Shadow, a STEM Day and a NASA Aerospace Teachers Program—along with a multiyear scholarship and mentoring program designed to invite preservice teachers into an authentic, sustained academic community of practice supported by high levels of engagement with caring STEM practitioners.

In [7], the paper begins by identifying three main reasons why many STEM-talented students in a university level do not consider enrolling in STEM teacher education programs. Then, based on review literature, a framework for addressing this dilemma is presented and discussed. The proposed framework consists of a set of three principles together with eleven strategies for the operationalization of these principles. During the presentation of the framework, the roles of governments and of universities at the institutional, faculty or division and departmental levels in the operationalization of the framework are examined.

In [8], six rural in-service science teachers were interviewed regarding their perceptions of the benefits and challenges of teaching in rural schools in general and teaching science subjects in particular. Community interactions, professional development and rural school structures emerged as three key factors related to rural teacher retention. Participants viewed each of these factors as having both positive and negative aspects. Findings from [8] confirm the findings of existing literature regarding rural teaching in general, but also provide additional insights into the complexities of rural science teaching in particular. Implications for rural teacher preparation, recruitment and retention are discussed.

In New York City, City Tech and BMCC received a NSF Noyce grant [3]. Both of these neighboring CUNY institutions implemented a Noyce Phase I scholarship program that increases the number of highly qualified STEM teachers in high-needs school districts in Brooklyn and the New York metropolitan area. The BMCC-to-City Tech STEM teacher preparation infrastructure will provide a sustainable pathway for the professional teacher preparation of students. In this project, an innovative strategy/structure called the three-tiered Noyce partnership was implemented. This program aims to recruit students in their first and second years of STEM studies and enroll them as Noyce Explorers (TIER I), lead them through a highly motivating Summer Program at CUNY that offers a combination of STEM content, STEM pedagogy and provides an opportunity to interact in a pedagogical role with peers and/or younger students through peer tutoring, mentoring and in other quasi-instructional capacities. Noyce Scholars (TIER II) are third and fourth year undergraduates in STEM programs who have committed to becoming STEM teachers in high-need schools through generous scholarships ($10 K) per year per Noyce Scholar. This three-tier structure of "Explorer, Scholar, Teacher" will maximize support for STEM students who wish to enter the teaching profession in STEM disciplines and serve their communities. This scholarship program will help supply the rising demand for skilled STEM teachers and professionals.

As a result of the above approaches taken from [6–8], none were able to provide a comprehensive solution to smoothly transfer students interested in STEM education to becoming effective STEM teachers and educators. In [3], we addressed an innovative recruitment strategy and structure to improve STEM education. We also sustained a high retention rate for the recruited STEM teachers based on the results from the independent evaluation report and the experiences and results that we learned from the actual project implementation.

83.3 Three-Tiered Noyce Partnership Structure and Implementation

High-quality STEM teachers can help improve overall student achievement in STEM fields. However, teacher shortages remain a problem for urban schools—especially schools serving minority or low-income students. The Robert Noyce Teacher Scholarship Program at City Tech and BMCC is an advanced three-tier structure that designates potential scholars as Noyce Explorers, continues to support a selected cohort as Noyce Scholars and continues to give professional support as they become Noyce Teachers. As part of this scholarship program, Noyce Scholars participate in annual Noyce teacher training workshops, teaching internships, mentorships and are given the opportunity to practice teaching under the direct supervision of STEM teachers. Noyce Scholars obtain a New York State teaching license as they complete a teacher certificate program and will become STEM teachers. The Noyce Project team includes five faculty members from STEM fields, three faculty members from an education field and a professional external Noyce program evaluator.

To supply the growing demand and reduce turnover, the Noyce Project will specifically foster the development of mathematics and technology teachers and learners in New York City. This program is categorized into three broad-based themes aligning to the Engagement, Capacity, and Continuity (ECC) trilogy [9]. The Noyce Program actively integrates ECC into the design of the program goals and its respective components:

Engagement. It combine evidence-based information with interactive teaching strategies and activities that motivate and develop a strong induction into teaching inquiry and experiential learning, thereby increasing student engagement, interests and motivation.

Capacity. The teacher education programs are situated within the successful infrastructure in the Career and Technical Teacher Education-CTTE and Math Education programs at City Tech to increase our participants' capacity to gain content knowledge and pedagogical skills in their respective discipline. Both of these programs lead to an initial teacher certification in New York State.

Continuity. A dual-discipline specific network system between the Noyce Scholars and the Career and Technical Teacher Education/Math Education faculty was created to support the continuity of material learned both within the discipline and in educational pedagogy. An online and in-person social support system was also created to support student-to-student and faculty-to-student mentoring.

Based on the above theory, we developed the three-tiered Noyce partnership strategy/structure as follows in our project.

TIER I—NOYCE EXPLORER Students enrolled in associate degree programs in a STEM major who have a genuine interest in STEM teaching are encouraged to apply to become Noyce Explorers. Noyce Explorers participate in paid internship placements, The Noyce Explorers Summer Program and various seminars. For the internships, students are placed in New York City High Schools, Middle Schools and/or with City Tech and BMCC faculty. Explorers are also mentored by various STEM faculty members as well. The Noyce Explorers Summer Program incorporates the best evidence-based research as well as high-impact practices effectively used at City Tech and other CUNY colleges. Students are provided with a wide range of experiential-based learning and enrichment activities that include both STEM and Education training. Each workshop covers an umbrella topic interspersed with engaging activities (speakers, site-visits) to expose students to a variety of STEM areas. Topics include Human Computer Interaction, Problem Solving, Web Design, Introduction of Programming, Computer and Data Analysis, Robotics, Geographic Information Systems, Mobile Devices, Alice Training, iPAD for Classrooms, Number and Quantity, Functions and Relations, Statistics, Synthetic and Analytic Geometry, Mathematical Modeling and Mathematical Games.

To promote the trilogy of Engagement, Capacity and Consistency, we focused on three effective techniques to be modeled in the workshops. First, scaffolding was used to provide support for students to teach computer science

Tier I: Noyce Explorer

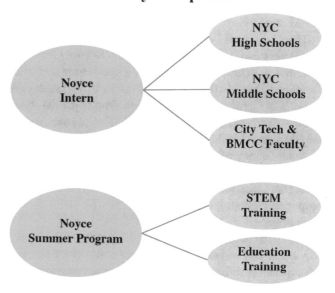

Fig. 83.1 Tier I: Noyce explorer

and math using project- or design-based activities. Second, while using Inquiry Based Learning Cycle, students Engage, Explore, Explain, Elaborate and Evaluate with support from the workshop leaders. Third, participants are organized into groups for team-based projects that will combine collaborative problem-solving and active listening. The infrastructure of Tier I is shown in Fig. 80.1.

TIER II—NOYCE SCHOLAR Noyce Scholars are selected from current Noyce Explorers, transfer students from BMCC and City Tech's STEM associate degree programs and transfer students from other CUNY community colleges. Noyce Scholars are third and fourth year undergraduates enrolled in a STEM baccalaureate program at City Tech and take courses in the Mathematics Education or Career and Technical Teacher Education program. Scholars receive a scholarship of $10 K per year. The main goal in this phase is to train Scholars to become both passionate and knowledgeable STEM teachers. Scholars are engaged in extensive Noyce mentorship and pedagogical and STEM development. They also participate in teaching internships in high schools, middle schools and/or with City Tech and BMCC faculty.

Each Scholar is designated to a mentor. Mentors are faculty members who are experienced teachers and researchers in a STEM discipline. Mentorships have a dual-discipline specific network system: A STEM research component and a pedagogical component. To facilitate a Scholar's development as an expert in a STEM discipline, he or she pursues a teaching-oriented project in a STEM discipline under the supervision of a mentor. Mentors serve as role models for the Scholars. Scholars visit classrooms of their mentors to observe and assist in the development and execution of lesson plans. Scholars receive extensive pedagogical development in their teacher training programs and through their mentorship. Scholars assist in the running of The Noyce Explorers Summer Program, guiding groups of Explorers through group activities and assisting faculty in the running of lessons. They also observe faculty as they model effective pedagogical practices and reflect upon these practices. The outline of this phase is shown in Fig. 80.2.

One important requirement of the scholarship is that the Noyce Scholar will remain in good academic standing and complete the two-year service requirement in a high-needs school for each year the scholarship is given. The project team is always available to advise students regarding coursework and to be sure students are completing requirements on time. If a Scholar is found to be in violation of the scholarship requirements, he or she will be contacted and every effort will be made to be sure the student is able to get back on track in the program. If a student cannot rejoin the program or complete their service agreement, their scholarship will be revoked and the student will be required to repay the monies with interest. After completing the program, Scholars are required to provide the program with annual certification of employment as well as up-to-date contact information. Failure to produce the required information will place a hold on the student's academic record, preventing the request of transcripts in the future. There are two additional steps that are taken to guarantee the previous steps are effective. First, when the students receive an offer to become a Scholar, they are required to sign a two-year service agreement as well as an agreement to provide contact information once they leave the Noyce Program and to allow the Noyce Project team to contact them for future Noyce Program follow-up activities. Second, we

also use a MK database provided by the Office of Assessment and Institutional Research at City Tech to assist the Noyce Project team in monitoring the Scholars' progress after they graduate. The City Tech Financial office will also report students to the National Science Foundation if they do not reply to our notice or repay the funds.

TIER III—NOYCE TEACHERS After graduating from a baccalaureate program, Noyce Scholars are qualified to apply for an initial teacher certificate in New York State which will allow them to teach in a STEM field. After receiving initial certification, Scholars are obligated to teach for two years in a high-needs school district for each year of support in the Noyce Program. Scholars receive guidance and support for their first two years in the classroom through participation in professional development, school-based mentoring and virtual mentoring. Virtual mentoring includes an e-mentoring support network which is an on-line, bi-weekly synchronous meeting among Noyce Scholars facilitated by a STEM faculty mentor. Each cohort of Noyce Teachers presents at a local professional conference in each of the two years after graduation such as the Metro New York meeting of the Mathematical Association of America. Noyce Teachers also serve as role models for the Noyce Explorers and Scholars. The structure of this phase is shown in Fig. 80.3.

The Noyce Program also maintains strong partnerships with City Polytechnic High School of Engineering, Architecture and Technology (City Poly), Pathways in Technology Early College High School (P-TECH), Queens Vocational and Technical High School, Brooklyn Technical High School and Intermediate School 220 located in Brooklyn. Institutional partners include the New York City Department of Education, CUNY Early College High

Tier II: Noyce Scholar

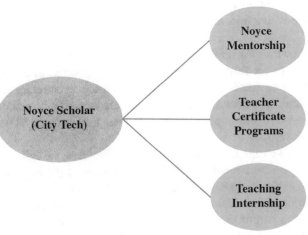

Fig. 83.2 Tier II: Noyce scholar

Tier III: Noyce Teacher

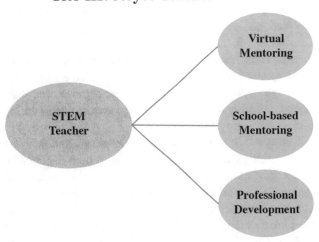

Fig. 83.3 Tier III: Noyce teacher

Fig. 83.4 Instructional experience

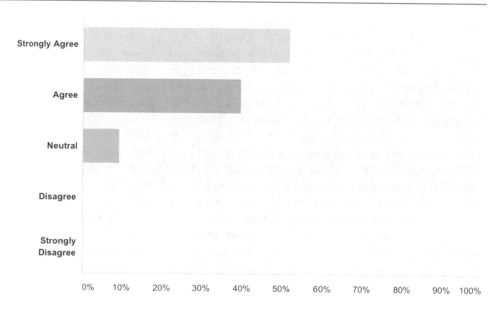

School Initiative, National Academy Foundation (NAF), National Action Council for Minorities in Engineering (NACME) and Project Lead the Way. There are currently 345 students already enrolled into the Noyce Program.

The above description analyzes our three-tiered structure in the Noyce Program that consists of Explorers, Scholars and Teachers. Based on the actual evidence from program implementation and collection of survey data from the program evaluation, we find that our three-tiered structure in the Noyce Program is effective to recruit and retain Noyce Scholars and Teachers. The survey results listed below will further convey how efficient our strategy is to implement STEM teacher scholarships.

83.4 Survey and Interview Results

Both Noyce Explorers and Scholars studying Computer Systems Technology, Mathematics and other STEM majors accepted and retained Noyce Internship positions. They were placed in classroom settings, engaged as tutors in peer-led team learning activities, as well as in-house interns. The Explorers and Scholars were not only able to observe the pedagogies used during the lectures, but also used scaffolding and other teaching strategies to assist students during in-class laboratory exercises.

During the Spring 2016 semester, Explorers and Scholars were placed as interns in college, high school and middle school environments. Those in the college settings were exposed to a variety of pedagogical experiences such as tutoring, laboratory assistance and class presentations in Computer Programming, Mathematics, Computer Technology and Biology. Other interns were placed in middle school and high school Pre-Calculus, Algebra and Chemistry classes.

The Noyce Explorers were surveyed at the end of the Spring 2016 semester. The overall feedback about the Noyce Program and the students' personal experiences with the program were positive. The Explorers were surveyed about 20 questions. Based on the results of the survey, 528 students were in classes that included a Noyce intern. Over 90 percent of the Explorers were interested in becoming STEM educators. Figures 80.4, 80.5 and 80.6 illustrate three important questions of the survey that reflect the content of this paper.

Question 10- After completing my internship, I have greater confidence in my ability to help students learn.

Question 6- During my internship, I gained a significant amount of instructional experience.

From the survey results of Question 10 shown in Fig. 80.4, we can see that the majority of Noyce students earned a lot of experience from the intern programs and have positive impressions on the teaching intern activities.

Question 13- During my internship, I was able to have a positive impact on student learning.

From the survey results of Questions 6 and 13 which is shown in Figs. 80.5 and 80.6, we can see that most of the interns are fully motivated and actively involved in student learning activities. They enjoy STEM teaching activities and are fully encouraged to participate. They have learned skills that they will implement later on when they become STEM educators.

To ensure the quality of the Noyce Program, we collected some direct feedback about our program implementation from the students in the survey. Some of the direct quotes from student surveys were as follows. "I enjoy working and being part of the internship program." "The summer workshop gave me a great interest in teaching math in the future." "This program provides great opportunity for students to

Fig. 83.5 Helping student learning

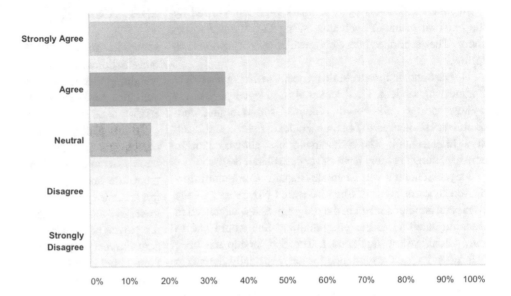

Fig. 83.6 Impact on student learning

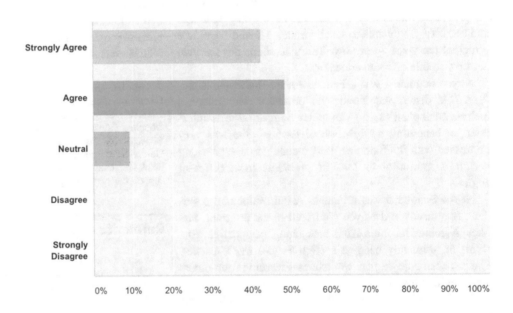

have experience with students." "Learning was fun and easygoing, yet challenging." "What I like most about the workshop is that we were able to stimulate our mind by using critical thinking problems." "I liked the hands-on activities." "It is a great opportunity to know other majors' content specialty and gain new perspectives." "The environment with the students as well as with the instructors and coordinators are excellent".

From the above quotes, it is clear and encouraging that students enjoyed participating in the Noyce Program and described the program as "valuable" and "inspiring." Students expressed that the program was both highly engaging and informative. They were extremely impressed with the knowledge and skills they acquired. They liked the experience of working with others in a collective process while performing hands-on STEM activities. Students benefitted both from solving problems as a team member and independently. Students believed that the program offered various insights from different angles of STEM and motivated them to become successful future STEM educators.

In sum, students strongly agreed that they gained positive experiences and useful information from completing the internship. Students were more comfortable presenting topics to students and were better able to maintain a constructive and consistent rapport with students. They were also better able to make STEM content understandable for students and enjoyed transferring their knowledge and skills to students.

In June 2016, five Noyce Scholars were interviewed by the external program evaluator during the summer workshop. The accurate interview results were reported as follows:

Noyce Scholar 1 was a Male African American Electrical Engineering student. He discovered the Noyce project via college poster. He loves robotics, programming and circuitry. His goal is to obtain a graduate degree and teach at a high school. The Scholarship has allowed him to consider things he may have never considered before.

Noyce Scholar 2 was a Female Hispanic Computer Information Systems student. She discovered Noyce via college email and another student in the program. She is interested in teaching database and/or programming. She would like to start teaching in a high school. The Scholarship has been helpful and a great experience for her, especially the group activities.

Noyce Scholar 3 was a Female African American Computer Information Systems student. She discovered Noyce via professor. She is very interested in networking, database and security. She wants to teach Grades 4-7 and become a principal ten years from now. The workshops have shown her how to think like an educator.

Noyce Scholar 4 was a Female Asian Mathematics student. She discovered Noyce via professor. Her interests include Mathematics and Computer Science. She is interested in becoming a High School teacher. She was very impressed with the Summer Workshops that have provided her the opportunity to look at problems from different angles.

Noyce Scholar 5 was a Female Asian Mathematics student. She discovered Noyce via friend in the program. She likes Algebra the most from a teaching perspective. She wants to ultimately become a High School Math teacher. She has been enjoying the workshops – learning from others and excellent instructors.

All five Noyce Scholars expressed, as a result of the workshop, an increased understanding and interest in becoming STEM educators. Noyce Scholar interviews evaluated STEM knowledge gained as a result of the Noyce internship and scholarship program. The implementation of the STEM teacher scholarship program has proven to improve and stimulate the interests of students in teaching. The significant strategies used in this program have helped guide students to become highly qualified future STEM educators.

The above survey results and interview data demonstrate that the phases of Tiers 1 and 2 in the Noyce project are extremely effective. Further results will be provided as more accurate data is collected over the next few years.

83.5 Conclusion

Recruiting and retaining STEM teachers is of crucial importance nationwide, and it is therefore necessary to give equal weighting to invest large funding and efforts on these related activities to keep the United States' competitiveness in STEM fields in the 21st Century.

In this paper, we proposed the three-tiered Noyce partnership structure and implementation experience in the past three years in an NSF Noyce Scholarship project. We shared the results and experiences that we learned from the project implementation. In addition, we presented survey results, interview data and student feedback and experiences for the Noyce partnership structure conducted by the independent external program evaluator. The results showed that our three-tiered structure was effective and could be applied to many other similar projects nationwide. For future work, we plan to continue to implement our three-tiered Noyce partnership structure for two more years and continue to modify and collect more experiences and meaningful results for the Noyce project which could greatly contribute to future STEM education research.

Acknowledgements This work is supported by the National Science Foundation (Grant Number: NSF 1340007, $1,418,976, Jan. 2014-Dec. 2018, PI: Fangyang Shen; Co-PI: Mete Kok, Annie Han, Andrew Douglas, Estela Rojas, Lieselle Trinidad; Project Manager: Janine Roccosalvo). The Noyce project team would also like to thank Prof. Gordon Snyder for his help on the project's evaluation. We also want to thank all faculty and staffs at both City Tech and BMCC who help and support our Noyce project in the past five years.

References

1. Weiss, R. (2013). *Obama Administration Announces New Steps to Meet President's Goal of Preparing 100,000 STEM Teachers*, Office of Science and Technology Policy Executive Office of the President, Mar 2013.
2. *Robert Noyce Teacher Scholarship Program Solicitation NSF 16–559*, National Science Foundation, 6 Sept 2016.
3. Shen, F., et.al. (2013). *Noyce Explorers, Scholars, Teachers (NEST): Fostering the Creation of Exceptional Mathematics and Technology Teachers in New York City*, NSF Award 1340007, 24 Aug 2013.
4. Rothwell, J. (2014). *Short on STEM Talent*, US NEWS, Sep. 15, 2014.
5. *The STEM Crisis*, https://www.nms.org.
6. Keith, H., et al. (2015). A University approach to improving STEM teacher recruitment and retention. *Kappa Delta Pi Record, 51*(2), 69–74.
7. Lee, K., et al. (2013). The recruitment of STEM-talented students into teacher education programs. *International Journal of Engineering Education, 29*(4), 833–838.
8. Kasey, G., et al. (Spring 2012). Teachers' perceptions of rural STEM teaching: Implications for rural teacher retention. *Rural Educator, 33*, 9–22.
9. Jolly, E., & Campbell, P. (2004). *Engagement, capacity and continuity: A trilogy for student success*. GE Foundation. http://www.campbell-kibler.com/

Improving the Performance of the CamShift Algorithm Using Dynamic Parallelism on GPU

84

Yun Tian, Carol Taylor, and Yanqing Ji

Abstract

The CamShift algorithm is widely used for tracking dynamically sized and positioned objects that appear in a sequence of video pictures captured by a camera. In spite of a great number of literatures regarding CamShift on the platform of CPU, its research on the massively parallel Graphics Processing Unit(GPU) platform is quite limited, where a GPU device is an emerging technology for high-performance computing. In this work, we improve the existing work by utilizing a new strategy – Dynamic Parallelism (DP), which helps to minimize the communication cost between a GPU device and the CPU. As far as we know, our project is the first proposal to utilize DP on a GPU device to further improve the CamShift algorithm. In experiments, we verify that our design is up to three times faster than the existing work due to applying DP, while we achieve the same tracking accuracy. These improvements allow the CamShift algorithm to be used in a more performance-demanding environment, for example, in real-time video processing with high-speed cameras or in processing videos with high resolution.

Keywords

GPU • CUDA • CamShift • Object tracking • Dynamic parallelism

84.1 Introduction

The *Continuously Adaptive Mean Shift (CamShift)* algorithm is widely used for tracking dynamically sized and positioned objects that appear in a sequence of video frames captured by a camera. For example, in transportation surveillance, a video camera constantly records all passing-by vehicles in a section of freeway. Then it attempts to identify at what time a specific vehicle enters and leaves the section of freeway, and to count the total number of vehicles in a real-time manner. In such applications, the CamShift algorithm continuously tracks and locks the same object that appears on multiple video frames.

84.1.1 MeanShift

The CamShift algorithm is based on the MeanShift algorithm. The MeanShift algorithm is used to find the mode or peak in a static probability distribution. It iteratively climbs the gradient of a probability distribution until it reaches a gradient of zero, indicating the zenith of the distribution. It is a non-parametric technique similar to kernel density estimations, however it operates on discrete, rather than continuous, distributions. The MeanShift algorithm proceeds in the following manner.

Y. Tian (✉) • C. Taylor
Department of Computer Science, Eastern Washington University, 99004 Cheney, WA, USA
e-mail: ytian@ewu.edu; ctaylor@ewu.edu

Y. Ji
Department of Electrical and Computer Engineering, Gonzaga University, 99258 Spokane, WA, USA
e-mail: ji@gonzaga.edu

© Springer International Publishing AG 2018
S. Latifi (ed.), *Information Technology – New Generations*, Advances in Intelligent Systems and Computing 558, DOI 10.1007/978-3-319-54978-1_84

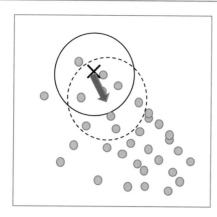

Fig. 84.1 Mean shift

Step (1),   An initial search window size and location is chosen, such as the large solid-line circle in Fig. 84.1, while the small dots are data samples.

Step (2),   It computes a mean location I_l of all data samples enclosed within the search window, which is represented by the ending point of the red arrow (the bottom-right end) in Fig. 84.1, while the starting point of the arrow (the top-left end) denotes the geometrical center of the search window. We will discuss how to calculate the value I_l in the Sect. 84.1.3.

Step (3),   The search window is then re-centered at the mean location I_l. The dotted circle represents the new search window after it moves to the new center location in Fig. 84.1.

Step (4),   It repeats the step 2 and step 3 until the difference between current and previous search window centroids is equal to or less than a predefined minimum threshold, which is referred to as *convergence*.

The step 2 above computes a mean shift vector originating from the center of the window, exemplified by the arrow in Fig. 84.1. The peak of a distribution is reached upon convergence, that is, when the mean shift vector is zero or close enough to zero [1].

84.1.2 CamShift

Suppose that we have a sequence of video frames, $V = \{ v_1, v_2, v_3, \ldots, v_n \}$. An object B appears on a subset of successive frames or all frames, but maybe at different positions on different frames. That is, the object was moving during the video was recorded. Next, we describe how to apply the CamShift algorithm in tracking the object B on different video frames.

Step (1),   A search window W is chosen to enclose the object B on the first video frame v_1.

Step (2),   Then the step 2 and step 3 of the *MeanShift* algorithm is repeatedly applied to the current video frame. Once the MeanShift algorithm converges on the current video frame, it yields the mean location I_l on the current frame for the *last* time, denoted by F_l, which is the position of B on the current frame.

Step (3),   The size of the search window is adjusted according to the information under W.

Step (4),   On the next video frame in the sequence, we place W centered at the location F_l that has been calculated in the step 2, then repeat the step 2 and the step 3 for the next video frame. The iterative nature of CamShift renders the algorithm quite computationally intensive.

84.1.3 Compute I_l and Adjust Search Window Size in CamShift

The CamShift algorithm uses a color histogram to represent the statistical color information of the object that we attempt to track, tightly enclosed by the *initial* search window. A histogram H basically records, for each color value c, the number of image pixels $H(c)$ that have the color c. The histogram is pre-defined and saved for repeated use. The hue component in the HSV color space is used to create the histogram because it is less susceptible to the effects of lighting changes than a combination of RGB values which include the saturation level [2].

We describe in the below how to compute the mean location $I_l(x_c, y_c)$, where x_c and y_c are the x-coordinate and y-coordinate of the mean location respectively. The zeroth moment is defined as:

$$M_{00} = \sum_x \sum_y P(I(x,y)) \tag{84.1}$$

where $I(x, y)$ is a hue value of the video frame at a particular x and y-coordinate under the search window, and $P(I(x, y))$ denotes the ratio of $H(I(x, y))$ to the total number pixels under the search window, meaning the probability of the color I(x,y) being in the object to be tracked. The first-order moments for the x and y-coordinates are defined as:

$$M_{10} = \sum_x \sum_y x P(I(x,y));$$
$$M_{01} = \sum_x \sum_y y P(I(x,y)); \tag{84.2}$$

The x and y-coordinates of the new mean location I_l for the search window are derived from the following equations:

$$x_c = \frac{M_{10}}{M_{00}}; \quad y_c = \frac{M_{01}}{M_{00}}; \qquad (84.3)$$

The 2nd-order moments for the x and y-coordinates are defined as:

$$M_{11} = \sum_x \sum_y xyP(I(x,y));$$
$$M_{20} = \sum_x \sum_y x^2 P(I(x,y)); \qquad (84.4)$$
$$M_{02} = \sum_x \sum_y y^2 P(I(x,y));$$

Calculating a rectangular shaped window size is computable based on the zeroth moment. If the zeroth-order moment is a lower value, then the search window should decrease in size. This assumes extraneous pixels under the search window with low probability values were involved in the zeroth moment calculation. Conversely, if the zeroth-order moment has a higher value, then the search window should increase in size. This assumes the search window was filled with higher probability values of matching the object's color profile and has not expanded to extremities of the object yet. The width of a rectangular shaped search window is defined as:

$$width = 2 * \sqrt{\frac{M_{00}}{max}} \qquad (84.5)$$

The height of the search window is arbitrarily sized by multiplying the width by a constant factor. The constant factor is determined through experiment beforehand. Certain assumptions about the dimensions of the object need to be made to tune the ratio of height and width. Bradski's original goal for CAMSHIFT was to track human faces. He set the ratio at 1.2 to produce an elongated search window proportional to a human face.

The search window can also be elliptically shaped and re-sized using the following equations, as described in the paper [3]:

$$length = \sqrt{\frac{(a+c) + \sqrt{b^2 + (a-c)^2}}{2}};$$
$$width = \sqrt{\frac{(a+c) - \sqrt{b^2 + (a-c)^2}}{2}}; \qquad (84.6)$$

where:

$$a = \frac{M_{20}}{M_{00}}; \quad b = 2(\frac{M_{11}}{M_{00}} - x_c y_c);$$
$$c = \frac{M_{02}}{M_{00}} - y_c^2; \qquad (84.7)$$

84.1.4 Graphics Processing Unit (GPU)

Graphics Processing Unit (GPU) is a massively parallel computing architecture that consists of thousand of processor cores, but with a small physical size. Tasks on a CPU can be offloaded to a GPU device and to be parallelized. A GPU device is able to execute a large number of operations simultaneously, thus is potentially capable of significantly speeding up computations. A *kernel* on a GPU device is a procedure that can be executed on thousand of processor cores simultaneously, and all running instances on these processors in parallel is called a *grid*. The process that creates these running instances and causes them to execute is referred to as *kernel launch*.

84.2 Existing Work and Their Limitations

Research show that applications using a GPU device outperform its counterpart on the CPU by a *factor* of 10 to more than 100, equivalent to a supercomputer on your desktop [4–9]. For example, in our previous work, we parallelized the problem of finding the longest repeat in a DNA sequence on a GPU device [9], while this work focuses on a different algorithm – CamShift that processes video data. Our previous work allows scientists to solve the longest repeat problem in 1/10th of the execution time, compared against existing solutions. In contrary, it costs only as 1/10th much as that of quad-core CPUs which deliver an equivalent supercomputing performance and 1/20th the power consumption [10]. Nevertheless, it is notoriously challenging to use a GPU device efficiently, which requires new strategies to re-organize or re-order computing instructions or input data sets, different from the strategies employed on the traditional CPU.

We identified two research venues that attempt to improve the CamShift algorithm. First, some research work focus on improving tracking accuracy and recovering lost object on a CPU [2, 3, 11–13]. For example, Li et al. not only utilized color information, but also took into account texture information of video frames in calculating the mean location [3]. In the second venue, some researchers emphasize to improve the performance of the CamShift algorithm, making it run *faster*. In this project, we follow the second direction and employ a new strategy to further accelerate the CamShift algorithm.

The CamShift algorithm has been extensively studied on the CPU, yet its research on the massively parallel GPU platform is quite limited. Exner et al. utilized a GPU device to accelerate the CamShift algorithm. They speeded up the mean location I_l calculation by using *OpenGL* framework [14]. In that work, the authors had to repurpose their input as 3-dimensional points in order to use the OpenGL

Fig. 84.2 Exiting CamShift implementation using a GPU device

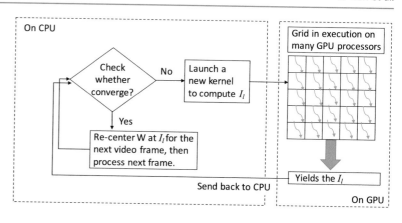

framework, which is quite inconvenient. *OpenGL* is originally designed for 3D graphics rendering on GPU devices, but not suitable for general-purpose computations, thus it is subject to great limitations [15]. In another paper, Jo et al. successfully implemented a CamShift program on a GPU device with real-time performance [16].

We identify one *disadvantage* in both Exner's [14] and Jo's work [16], where their design is shown in Fig. 84.2. The step 2 and step 3 of the MeanShift algorithm is repeatedly performed on a GPU device and produces an intermediate mean location I_l. Then I_l has to be transferred back to the CPU, and let the CPU determine whether MeanShift reaches convergence. If it is not converged, it repeats that step 2 and step 3. Otherwise, it runs the MeanShift algorithm on the next video frame. Large overheads are created by transferring data back and forth between the CPU and the GPU device, which bottlenecks the performance of the entire system.

We explain the motivation of their design in the following. Suppose if the GPU device is responsible to check whether it is converged, then it is not necessary to send I_l back to the CPU. In this case, if a test shows not converged, then on the GPU device it has to compute a new I_l with a new kernel. But previously, GPU devices did not provide such functionality that allows to launch new kernels inside an existing kernel. Without other choices, the I_l value had to be transferred back to the CPU for convergence checking, and let the CPU to launch a new kernel.

84.3 Our Contributions

In this project, we improve the aforementioned existing work by utilizing a new strategy – *Dynamic Parallelism (DP)*, which helps to minimize communications between a GPU device and a CPU. As far as we know, our project is the *first* proposal to utilize DP on a GPU device to further improve the CamShift algorithm. In experiments, we verify that our CamShift with DP implementation is up to *three* times faster than the existing designs without DP implementation, while we achieve the same tracking accuracy. These

improvements allow the CamShift algorithm to be used in a more performance-demanding environment, for example, in real-time video processing with high-speed cameras or in processing videos with high resolution.

84.3.1 Dynamic Parallelism (DP)

DP in a GPU device enables a kernel to spawn and synchronize nested kernels, as shown in Fig. 84.3. The parent *Grid A* is created after the CPU launches a kernel *A*. Then in Grid A, the GPU device is able to launch a child kernel *B* which then creates *Grid B* for execution. After *Grid B* is complete, the control is returned to the parent grid. In this way, applications with iterative steps of kernel launches can shift the control to the parent kernel away from the CPU's responsibilities. The parent kernel can collect the output produced by its child kernel and control the next launch of children accordingly.

84.3.2 Our Design and Advantage

Figure 84.4 shows the overview of the design of our improved CamShift using DP. For each video frame v_i, the CPU launch a parent kernel on the GPU device. Then the parent kernel spawns a child kernel repeatedly for computing the intermediate mean location I_l until a convergence is encountered. Compared with the previous design shown in Fig. 84.2, our approach only sends I_l back *once* to the CPU per each video frame, while the previous work has to send back *each* computed I_l to the CPU for testing if convergence is reached. Our design greatly reduces the communication cost between the CPU and the GPU device.

84.4 Implementation

As we zoom in Fig. 84.4, the detailed operations of our design are shown in Fig. 84.5. It presents our preprocessing operations and the steps performed on a GPU device for

Fig. 84.3 Dynamic Parallelism parent and child grid execution

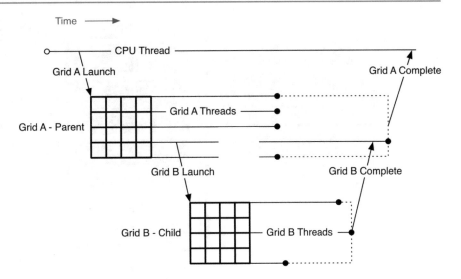

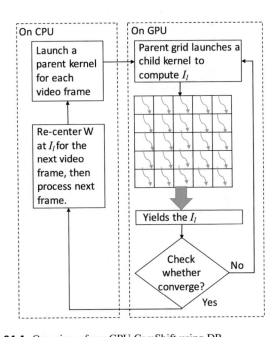

Fig. 84.4 Overview of our GPU CamShift using DP

processing each video frame. The CPU launches a *parent* kernel that runs on the GPU device. Then, with DP, the parent kernel launches two child kernels for calculating the moment values, using the equations that are described in Sect. 84.1.3. In this project, we use the CUDA framework to implement the CamShift algorithm on a GPU device.

84.4.1 Preprocessing

(1) *Build A Histogram for the Object to Track*: After the histogram is constructed on the CPU, it is copied into the GPU *device* constant memory for use in the kernels. The histogram for the target object is statically sized because we attempt to track the same object on multiple video frames. Constant memory is chosen for the histogram's storage because the histogram is a read-only.

(2) *RGB color to HSV color Conversion*: The conversion of the RGB video frame to HSV is parallelized using a kernel on the device as an optimized preprocessing step before loading the video frame hue values into the *global* memory on the GPU device. The hue array is the input to the CamShift algorithm, and is statically allocated in the global memory on the device since the video frame has a fixed size throughout the life of the parent kernel. The array needed to be updated with each subsequent video frame, which requires launching a *new* parent kernel on the device.

84.4.2 Parent Kernel for Processing One Video Frame

The major purpose of the parent kernel is to compute the mean location $I_l(x_c, y_c)$, and to evaluate whether CamShift converges on the current frame, without sending back the I_l value to the CPU. According to the equations presented in Sect. 84.1.3, the I_l depends on the *moment* values that are calculated with summation. Summation can be implemented using a common parallel pattern *Reduction* on a GPU device.

(1) *Parallel Reduction on a GPU device*: A reduction is a simple parallel algorithm used to compute the summation of a large linear array. It takes a divide and conquer approach, which is described in the paper [17]. The complete reduction in GPU has to launch two kernels one after another:

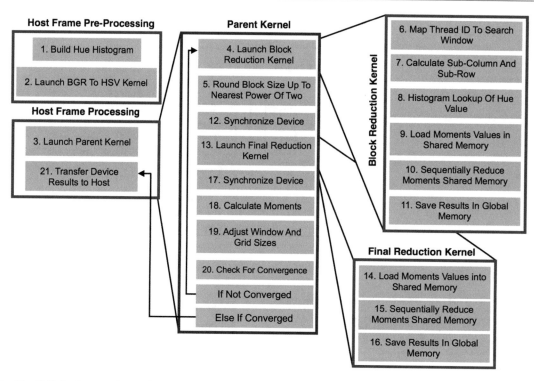

Fig. 84.5 Our CamShift implementation on GPU with DP

Fig. 84.6 Two-kernel reduction overview in [17]

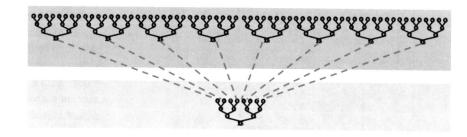

(1) The input array is divided into subsections across multiple thread blocks. Each thread block is able to work on a different subsection in parallel. All thread blocks created by a kernel launch comprise the *grid*.

(2) A first kernel reduces each subsection into an intermediate total by the different thread blocks. This first kernel outputs an array of intermediate totals, which is illustrated by the top box in Fig. 84.6.

(3) A second kernel reduces the intermediate totals into a final total sum of the entire input array, illustrated by the bottom box in Fig. 84.6.

Figure 84.7 illustrates how *one* thread block reduces a specific subsection of the input array into *one* intermediate total.

(2) *Steps in the Parent Kernel*:

In Fig. 84.5, the step 4 and the step 13 launch two child kernels to calculate the moment values using parallel reduction. Explicit device synchronization is required between reduction kernel launches, shown in the step 12 in Fig. 84.5. Then the parent grid calculates the new centroid and search window dimensions based on the second child reduction values in step 18, and updates the variables used in the first reduction kernel representing the search window dimensions and positions in step 19. If it is converged, i.e., the new centroid roughly overlaps with the previous centroid, then the parent kernel terminates. The *host* CPU will then copy the memory from the *global* memory and update its variables storing the new centroid and the top-left and bottom-right corners of the rectangular search window for display. However, if there is no convergence, then the parent kernel calculates the new grid size for the first child kernel and repeats the two-step reduction kernel launches again. The only per-frame memory transfer between host and device is the host-to-device image hue array and the device-to-host new centroid and height and width after convergence.

(3) *Challenge and Solution*: The thread block in our first child kernel is a one-dimensional thread block, same as the settings shown in Fig. 84.7. However, different from the existing work in Fig. 84.7, the input data of our project is an image, which conceptually is a two-dimensional array. But underlying, each video frame is *physically* stored in a one-dimensional memory by linearizing the two-dimensional array with a row-major order. In Fig. 84.8, the conceptual video frame in two dimensions is shown on the top of the

diagram, and its physical storage in the GPU memory is shown on the bottom of the diagram.

As illustrated in Fig. 84.8, our task is to perform the parallel reduction on the pixels under the red rectangular search window. In other words, the data to be summed up is not contiguous in the physical memory. Therefore, we have to adapt the existing parallel reduction algorithm in Fig. 84.7 for our purpose.

The block size is set to 1024 in our first kernel, then the number of blocks is calculated by rounding up to the nearest integer value of the window size divided by the block size. Each thread block has three statically allocated shared memory arrays of floating-point precision, each with the same size as the thread block, representing the zeroth and first-order moments respectively.

Within a one-dimensional grid and thread block, we identify each thread within the grid by assigning it an unique *absolute Thread ID*, based on the following equations:

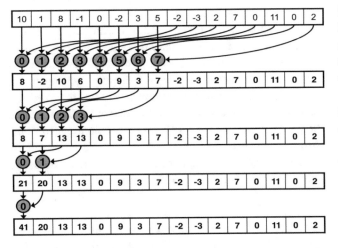

Fig. 84.7 Parallel reduction within one thread block, where thread IDs are shown in the colored *circles* in [17]

(1) threadIdx.x signifies the relative thread ID within one specific block.
(2) absoluteThreadID = blockIdx.x * blockDim.x + threadIdx.x

The *absoluteThreadID* corresponds to a relative position within the search window. Translating this relative index

Fig. 84.8 Example mapping thread ID of 3 to search window index in video frame

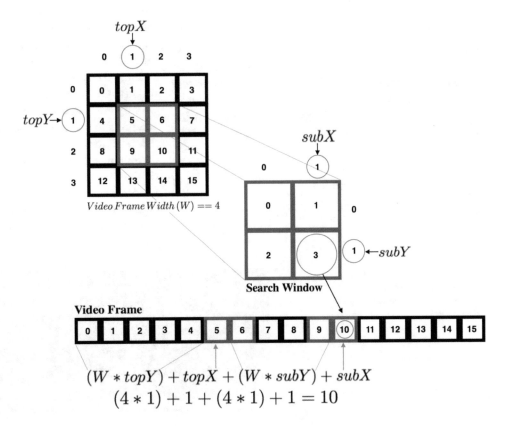

value to its absolute frame index within the entire hue array was computed using the following equation, which is also illustrated in Fig. 84.8.

$$absolute frame index = (W*topY) + topX + \\ (W*subY) + subX \tag{84.8}$$

topX: x-coordinate of the top left corner of the search window in the video frame.

topY: y-coordinate of the top left corner of the search window in the video frame.

W: the width of video frame

subX: x-coordinate relative to within the search window

subY: y-coordinate relative to within the search window

The subX and subY coordinates were calculated from the thread ID using the following equations:

$$subX = i/w \\ subY = i \bmod w \tag{84.9}$$

where:

i: the absolute thread ID within the grid

w: search window width

84.5 Experimental Results

The test results are collected in this project on a server with an Intel i7 CPU with 6 cores and a clock speed of 3.50 GHz. The server runs a Linux operating system with a kernel version of 4.4.0-36-generic. The server has four Nvidia GTX 970 GPUs with 1.2 GHz clock speed, connected to the host CPU with PCIe 4.0 interface. We implement the project and perform tests in CUDA 7.0 package.

Table 84.1 Performance of our design compared with existing work

	200 × 200	400 × 400	1080 × 720
GPU CamShift without DP	0.38 ms	0.61 ms	1.01 ms
GPU CamShift with DP	0.19 ms	0.19 ms	0.4 ms
Total speedups	2.0	3.2	2.5

We implement two different versions of the CamShift algorithm, a GPU CamShift without DP and a GPU CamShift with DP. Reading the video frames into memory and writing the processed video frames to an output video file share the same OpenCV library calls in both versions.

The computational task timed and compared is the CamShift statistical moment computation and the convergence checking. We do not have to compare the RGB-to-HSV conversion, because it is a shared operation in both designs. The time cost of the memory transfer for each video frame and the time cost of transferring I_l values are included in the results for the GPU CamShift without DP version, because this transfer is necessary when the CPU has to check convergence and to launch next kernel. However, the frequency of this memory transfer is greatly reduced in the GPU CamShift with DP version because the hue array of a video frame has been already stored on the GPU device and the parent kernel on the GPU device is able to perform convergence checking.

We execute both implementations for more than two hundred video frames, on which only one object is to be tracked. Each video frame is an 1080-by-720 image. We calculate the *average* time cost for the computational task that is performed in all iterations of the MeanShift algorithm, that is, the step 2, the step 3 and the step 4 presented in Sect. 84.1.1. We present the test results in Table 84.1, which suggests that our design can be up to *three* times faster than the previous work, because DP effectively reduces the communication cost. In other words, by applying DP, we are able to reduce the communication cost by more than 50%.

Each column in Table 84.1 shows the average time cost for different-sized search windows. We also observe that as we vary the search window size, our GPU CamShift with DP continues to outperform the existing GPU CamShift without DP. The reason why the search window size affects the performance gains is that, in each iteration CamShift algorithm uses the color information enclosed within the search window on an input video frame to calculate the new center of the search window, as shown in Equations 84.1, 84.2 and 84.3. Meanwhile, Fig. 84.9 shows our design is able to

Fig. 84.9 In our experiments, the same object is accurately tracked and locked on different video frames

Fig. 84.10 Extension for multi-object tracking

accurately identify and lock the object on different video frames.

In addition, in tests we verified that it is effective to use DP in CamShift to track multiple objects on a GPU device. As shown in Fig. 84.10, with various image size and object size, the proposed algorithm is effective to track multiple objects that appear in a sequence of videos frames. When tracking two objects on GPU, the CamShift algorithm with DP is two times quicker on average than its counterpart without DP.

84.6 Conclusion and Future Work

In this work, we improve the existing CamShift algorithm on GPU by utilizing a new strategy – Dynamic Parallelism (DP), which helps to minimize the communication cost between a GPU device and the CPU. As far as we know, our project is the first proposal to utilize DP on a GPU device to further improve the performance of the CamShift algorithm. In experiments, we verify that our design is up to three times faster than the existing GPU CamShift implementation due to the use of DP, while we achieve the same tracking accuracy. In the future, we will investigate using DP to improve other iterative algorithms.

References

1. Cheng, Y. (1995). Mean shift, mode seeking, and clustering. *IEEE Transactions on Pattern Analysis and Machine Intelligence, 17*(8), 790–799.
2. Bradski, G. R. (1998). Computer vision face tracking for use in a perceptual user interface.
3. Li, J., Zhang, J., Zhou, Z., Guo, W., Wang, B., & Zhao, Q. (2011). Object tracking using improved camshift with surf method. In *2011 International Workshop on Open-Source Software for Scientific Computation (OSSC)* (pp. 136–141). Piscataway: IEEE.
4. Che, S., Boyer, M., Meng, J., Tarjan, D., Sheaffer, J., & Skadron, K. (2008). A performance study of general-purpose applications on graphics processors using CUDA. *Journal of Parallel and Distributed Computing, 68*(10), 1370–1380.
5. Nickolls, J., & Dally, W. (2010). The GPU computing era. *Micro, IEEE, 30*(2), 56–69.
6. Okuyama, T., Ino, F., & Hagihara, K. (2008). A task parallel algorithm for computing the costs of all-pairs shortest paths on the cuda-compatible GPU. In *Proceedings of the 2008 I.E. International Symposium on Parallel and Distributed Processing with Applications* (pp. 284–291). Washington, DC: IEEE Computer Society. [Online]. Available: http://0-portal.acm.org.umiss.lib.olemiss.edu/citation.cfm?id=1493613.1494203
7. Owens, J., Houston, M., Luebke, D., Green, S., Stone, J., & Phillips, J. (2008). GPU computing. *Proceedings of the IEEE, 6*(5), 879–899
8. Satish, N., Harris, M., & Garland, M. (2009). Designing efficient sorting algorithms for manycore GPUs.
9. Tian, Y., & Xu, B. (2015). On longest repeat queries using GPU. In *International Conference on Database Systems for Advanced Applications* (pp. 316–333). Springer.
10. The NVIDIA Tesla C2050 and C2070 GPU. [Online]. Available: http://www.nvidia.com/docs/IO/43395/NV_DS_Tesla_C2050_C2070_jul10_lores.pdf
11. Bay, H., Tuytelaars, T., & Van Gool, L. (2006). Surf: speeded up robust features. In *Computer vision–ECCV 2006* (pp. 404–417). Berlin/Heidelberg: Springer.
12. Fu, M., Cai, C., & Mao, Y. (2015). An improved camshift algorithm for target recognition. In *Ninth International Symposium on Multispectral Image Processing and Pattern Recognition (MIPPR2015)* (pp. 981208–981208). International Society for Optics and Photonics.
13. Sirikuntamat, N., Satoh, S., & Chalidabhongse, T. H. (2015). Vehicle tracking in low hue contrast based on camshift and background subtraction. In *2015 12th International Joint Conference on Computer Science and Software Engineering (JCSSE)* (pp. 58–62). Piscataway: IEEE.
14. Exner, D., Bruns, E., Kurz, D., Grundhöfer, A., & Bimber, O. (2010). Fast and robust camshift tracking. In *2010 I.E. Computer Society Conference on Computer Vision and Pattern Recognition Workshops (CVPRW)* (pp. 9–16). San Francisco: IEEE.
15. Owens, J. D., Luebke, D., Govindaraju, N., Harris, M., Krüger, J., Lefohn, A. E., & Purcell, T. J. (2007). A survey of general-purpose computation on graphics hardware. *Computer Graphics Forum, 26*(1), 80–113. Wiley Online Library.
16. Jo, J. H., & Lee, S. G. (2013). Cuda based camshift algorithm for object tracking systems.
17. Harris, M., et al. (2007). Optimizing parallel reduction in CUDA. *NVIDIA Developer Technology, 2*(4).

Jun Zhang and Fangyang Shen

Abstract

Laplace Transform is among a few of these transforms playing important roles in pure and applied mathematics. Sumudu Transform is relatively new but has many good properties for solving problems in computational science. In this work, the authors introduce Sumudu Transform in computational approach which leads to various interesting and useful applications. Traditionally, the invention of mathematical formulae and proof of mathematical identities belong to the intelligent human beings. But in this work, we shall show that Sumudu Transform can be used to prove existing formulae and generate new mathematical identities automatically without embellishment. To show how it works, a good number of sample formulae and identities are provided for demonstration, including some famous ones such as Euler's Formula, de Moivre's Identity and Pythagorean Identity. Not only the work presented here is straightforward to introduce to students but also can be useful to assist new research works related to computational applications!

Keywords

Sumudu transform · Formula · Identity · proof · Generation

General Terms

Type: Regular Research Paper

85.1 Introduction

The Laplace Transform is one of the most famous integral transforms and has been well studied and widely used in mathematics and engineering for centuries. Laplace transform is very important, for example, under this transform,

J. Zhang (✉)
Department of Math and Computer Science, University of Maryland Eastern Shore, 21853 Princess Anne, MD, USA
e-mail: jzhang@umes.edu

F. Shen
Department of Computer Systems Technology, New York City College of Technology, City University of New York, 11201 Brooklyn, NY, USA
e-mail: fshen@citytech.cuny.edu

the differential operator is converted into multiplication, so differential equations become algebraic equations. As a result, Laplace transform is a useful tool for dealing with linear systems described by ordinary differential equations. Many undergraduate and graduate programs offer courses on the Laplace Transform. The Sumudu Transform is relatively new, but it is as powerful as the Laplace Transform and has some good features [1–5]. Sumudu transform was introduced and studied in a traditional way as other integral transforms by many researchers recently. In this paper, the authors shall introduce the Sumudu Transform through a new approach, a computational approach which can be implemented into a system to solve problems automatically. Traditionally, the invention of mathematical formulae and proof of mathematical identities belong to the intelligent human beings. But in this work, we shall show that Sumudu Transform can be used to prove existing formulae and generate new mathematical identities automatically without embellishment. To show how it works, a good number of

© Springer International Publishing AG 2018
S. Latifi (ed.), *Information Technology – New Generations*, Advances in Intelligent Systems and Computing 558, DOI 10.1007/978-3-319-54978-1_85

sample formulae and identities are provided for demonstration, including some famous ones such as Euler's Formula, de Moivre's Identity and Pythagorean Identity. Not only the work presented here is straightforward to introduce to students but also can be useful to assist related new research works. In this work, identity and formula are used interchangeably without much difference.

Let's assume that f is a function of x. The Sumudu Transform of f is defined as

$$F(z) = S[f(x)] = \int_0^\infty \frac{1}{z} e^{-x/z} f(x) dx. \quad (85.1)$$

We say $f(x)$ as the original function of $F(z)$ and $F(z)$ as the Sumudu Transform of $f(x)$. We also refer to $f(x)$ as the inverse Sumudu Transform of $F(z)$. The symbol S denotes the Sumudu Transform. The function $\frac{1}{z} e^{-x/z}$ is called the kernel of the transform.

Here are some basic properties of the Sumudu Transform:

(1) *Linearity*

$$S[c_1 f(x) + c_2 g(x)] = c_1 S[f(x)] + c_2 S[g(x)]. \quad (85.2)$$

(2) *Convolution*

$$S[(f * g)(x)] = z S[f(x)] * S[g(x)]. \quad (85.3)$$

(3) *Laplace-Sumudu Duality*

$$L[f(x)] = S[f(1/x)]/z, S[f(x)] = L[f(1/x)]/z. \quad (85.4)$$

(4) *Derivative*

$$S[f^m(x)] = S[f(x)]/z^m - f(0)/z^m - \cdots - f^{m-1}(0)/z. \quad (85.5)$$

More properties can be found in [1–5].

A very interesting fact about Sumudu Transform is that the original function and its Sumudu Transform have the same Taylor coefficients except for a factor $n!$. This fact is illustrated by the following theorems:

Theorem 1.1 ([6]). *If*

1. *$f(x)$ is bounded and continuous,*
2. *$F(z) = S[f(x)]$, and*
3. *$F(z) = \sum_{n=0}^\infty a_n z^n$,*

then

$$f(x) = \sum_{n=0}^\infty a_n \frac{x^n}{n!}. \quad (85.6)$$

Theorem 1.2 ([2]). *The Sumudu Transform amplifies the coefficients of the power series function,*

$$f(x) = \sum_{n=0}^\infty a_n x^n, \quad (85.7)$$

by mapping f(x) to the power series function,

$$S[f(z)] = \sum_{n=0}^\infty n! a_n z^n. \quad (85.8)$$

As a matter of fact, generating functions are a very important technique in discrete mathematics and algorithm analysis. For a given sequence a_n, there are two classical generating functions:

$$g(x) = \sum_{n=0}^\infty a_n x^n, \quad h(x) = \sum_{n=0}^\infty \frac{a_n}{n!} x^n. \quad (85.9)$$

$g(x)$ is called the ordinary generating function of the sequence a_n and $h(x)$ is the exponential generating function. Theorem 1.1 and Theorem 1.2 are inverse to each other and give a complete relationship about coefficients under Transform. Based on Theorem 1.1 and Theorem 1.2, there is another interesting fact about Sumudu Transform:

Proposition 1.3. *The Sumudu Transform of an exponential generating function is its ordinary generating function; the inverse Sumudu Transform of an ordinary generating function is its exponential generating function.*

These theorems serve as a base to calculate the general terms of Taylor series expansions, in turn, lead to various applications.

85.2 Application in Coefficient Calculation

Based on the above theory, the coefficients of original functions can be calculated from the coefficients of the Sumudu Transforms by using the following algorithm:

Algorithm 85.1

(1) Input: $f(x)$.
(2) Calculate the Sumudu Transform $S[f(x)]$.
(3) Calculate the nth coefficient $a_n = [z^n] S[f]$.
(4) Output: $a_n / n!$.

In the Maple software, this algorithm can be implemented as the following `coefficient` procedure.

with(inttrans);
with(genfunc);
coefficient := *proc(f, x)*

```
laplace(f, x, s);
subs(s = 1/t, %)/t;
rgf_expand(%, t, n);
simplify((%)/n! );
  return %;
end;
```

The above implementation uses the Laplace Transform function in the Maple software. Alternatively, an implementation without using Laplace Transform would be like the following:

```
with(genfunc);
coefficient: = proc (f, x)
  int( f * exp(−x/t)/t, x =
    0. . infinity) assuming t > 0;
  simplify(%, radical, symbolic);
  simplify(rgf_expand(%, t, n)/n! );
  return %;
end;
```

Theoretically, if the Sumudu Transform of a function is rational, then it can be calculated by the `coefficient` procedure automatically without human interaction. In practice, because of Maples's limitations on parameterized improper integrals, the `coefficient` procedure does not always work. However, it still works for a wide range of functions, including all these functions whose Sumudu Transforms are rational functions.

The *coefficient* procedure can be used to extract the nth coefficients for an infinite number of functions, including but not limited to: $sin(ax)$, $cos(ax)$, e^{ax}, $e^{bx}sin(ax)$, $e^{bx}cos(ax)$, $sinh(ax)$, $cosh(ax)$, $e^{bx}sinh(ax)$, $e^{bx}cosh(ax)$, $sin(ax)cos(bx)$, $sin^{100}(x)$, $cos^{100}(x)$, $sinh^{100}(x)$, $cosh^{100}(x)$, etc. Here we assume a, b are constants.

85.3 Automatic Proof of Identities

Let $f(x) = e^{ix} − cos(x) − isin(x)$, then applying the `coefficient` procedure presented above to extract the nth coefficients of $f(x)$, we have: $a_n = 0$ for all $n = 0, 1, 2, 3,$ …. Since $f(x)$ is analytic, we prove $e^{ix} = cos(x) + isin(x)$.

This is the famous Euler's formula. This result is of fundamental importance in such fields as harmonic analysis. Before Euler's work, mathematicians did not know the relationship between exponential functions and trigonometric functions, so these two sets of functions were studied separately. It took a genius like Leonhard Euler to prove this formula. With the Sumudu Transform, the simple

coefficient procedure can help us to prove this formula straightforwardly.

Now, let $f(x) = cos(nx) + isin(nx) − (cos(x) + Isin(x))^n$, by applying the `coefficient` procedure to extract the nth coefficients of $f(x)$, we have: $a_n = 0$ for all $n = 0, 1, 2, 3,$ …. Since $f(x)$ is analytic, so $f(x) = 0$. We have $cos(nx) + isin(nx) = (cos(x) + Isin(x))^n$, this is the famous de Moivre's Identity.

By the same way, we can automatically prove the following formulae listed in Table 85.1 below.

Table 85.1 lists only some sample examples, much more formulae can be simply proved by the same way automatically. Some examples in the list are very famous and widely used. The list can be extended to infinite in length. More advanced algorithms can be developed for such a purpose.

85.4 Automatic Generation of New Identities

The list showed in Table 85.1 are existed formulae or identities. Now we shall generate some new formulae or identities automatically. The generated formulae or identities may or may not in any reference.

Now we consider the product of series.

Theorem 4.1. : *Given two power series expansions*

$$g(x) = \sum_{n=0}^{\infty} b_n x^n, \qquad x \in B(0, r)$$

and

$$h(x) = \sum_{n=0}^{\infty} c_n x^n, \qquad x \in B(0, R),$$

the power series of their product $g(x)h(x)$ is given by

$$g(x)h(x) = \sum_{n=0}^{\infty} a_n x^n, \qquad x \in B(0, r) \cap B(0, R),$$

where

$$a_n = \sum_{k=0}^{n} b_k c_{n-k}. \qquad for \quad n = 0, 1, 2, \cdots$$

This theorem is very easy to verify and can be found in many textbooks.

Applying Theorem 4.1, if $f(x) = g(x)h(x)$ and we know the nth coefficients of the Sumudu Transform of $g(x)$ and h

Table 85.1 List of mathematical formulae

(Assume x is a variable, and a, b are constants)	
Formula	Name
$e^{Ix} = cos(x) + Isin(x)$	Euler's formula
$cos(nx) + Isin(nx) = (cos(x) + Isin(x))^n$	De Moivre's identity
$cosh(nx) + sinh(nx) = (cosh(x) + sinh(x))^n$	Hyperbolic functions
$sin^2(x) + cos^2(x) = 1$	Pythagorean identity
$sin(a + x) = sin(a)cos(x) + cos(a)sin(x)$	Sine sum identity
$sin(a - x) = sin(a)cos(x) - cos(a)sin(x)$	Sine subtract identity
$cos(a + x) = cos(a)cos(x) - sin(a)sin(x)$	Cosine sum identity
$cos(a - x) = cos(a)cos(x) + sin(a)sin(x)$	Cosine subtract identity
$sin(2x) = 2cos(x)sin(x)$	Double angle sine formula
$cos(2x) = cos^2(x) - sin^2(x)$	Double angle cosine formula
$cos(2x) = 2cos^2(x) - 1$	Double angle cosine formula
$cos(2x) = 1 - 2sin^2(x)$	Double angle cosine formula
$sin(3x) = -sin^3(x) + 3cos^2(x)sin(x)$	Triple angle sine formula
$sin(3x) = -4sin^3(x) + 3sin(x)$	Triple angle sine formula
$cos(3x) = cos^3(x) - 3sin^2(x)cos(x)$	Triple angle cosine formula
$cos(3x) = 4cos^3(x) - 3cos(x)$	Triple angle cosine formula
$sin(nx) = 2cos(x)sin((n - 1)x) - sin((n - 2)x)$	Recursive sine formula
$cos(nx) = 2cos(x)cos((n - 1)x) - cos((n - 2)x)$	Recursive cosine formula
$sin^2(x) = (1 - cos(2x))/2$	Power reduction sine formula
$sin^3(x) = (3sin(x) - sin(3x))/4$	Power reduction sine formula
$sin^4(x) = (3 - 4cos(2x) + cos(4x))/8$	Power reduction sine formula
$sin^5(x) = (10sin(x) - 5sin(3x) + sin(5x))/16$	Power reduction sine formula
$cos^2(x) = (1 + cos(2x))/2$	Power reduction cosine formula
$cos^3(x) = (3cos(x) + cos(3x))/4$	Power reduction cosine formula
$cos^4(x) = (3 + 4cos(2x) + cos(4x))/8$	Power reduction cosine formula
$cos^5(x) = (10cos(x) + 5cos(3x) + cos(5x))/16$	Power reduction cosine formula
$sin^2(x)cos^2(x) = (1 - cos(4x))/8$	Power reduction formula
$sin^3(x)cos^3(x) = (3sin(2x) - sin(6x))/32$	Power reduction formula
$sin^4(x)cos^4(x) = (3 - 4cos(4x) + cos(8x))/128$	Power reduction formula
$sin^5(x)cos^5(x) = (10sin(2x) - 5sin(6x) + sin(10x))/512$	Power reduction formula
$2cos(a)cos(x) = cos(a - x) + cos(a + x)$	Production for sum formula
$2sin(a)sin(x) = cos(a - x) - cos(a + x)$	Production for sum formula
$2sin(a)cos(x) = sin(a + x) + sin(a - x)$	Production for sum formula
$2cos(a)sin(x) = sin(a + x) - sin(a - x)$	Production for sum formula
$sin(x) = (e^{Ix} - e^{-Ix})/2I$	Exponential definition of sine
$cos(x) = (e^{Ix} + e^{-Ix})/2$	Exponential definition of cosine
$sinh(x) = (e^x - e^{-x})/2$	Exponential definition of sinh
$cosh(x) = (e^x + e^{-x})/2$	Exponential definition of cosh
$cosh^2(x) - sinh^2(x) = 1$	Hyperbolic identity

(x), then we can calculate the *nth* coefficient $f(x)$ using the following algorithm.

Algorithm 85.2

---

(1) Input: $g(x)$, $h(x)$.
(2) Calculate the Sumudu Transform $S[g]$.
(3) Calculate the Sumudu Transform $S[h]$.
(4) Initialize: $c_n = 0$.
(5) For $k = 0$ to n do

Calculate the kth coefficient $p_k = [z^k]S[g]$.
Calculate the $n - k$th coefficient
$q_{n-k} = [z^{n-k}]S[h]$.
Add $p_k \ q_{n-k}/k! \ (n - k)!$ to c_n.
End for
(6) Output: c_n.

For example, let $g(x) = e^{ax}$, $h(x) = sinh(bx)$, $f(x) = e^{ax} sinh(bx)$, where a and b are constants, then

$$S[g] = \frac{1}{1 - az},$$

$$S[h] = \frac{bz}{1 - (bz)^2},$$

$$[z^n]S[g] = a^n,$$

$$[z^n]S[h] = \frac{b^n - (-b)^n}{2},$$

$$[x^n]f = \sum_{k=0}^{n} \frac{a^k(b^{n-k} - (-b)^{n-k})}{2k!(n-k)!}.$$

By applying `coefficient` defined above directly to $f(x) = e^{ax}\sinh(bx)$, we have

$$[x^n]f = \frac{(a+b)^n - (a-b)^n}{2n!}.$$

Then we have the following identity:

$$\sum_{k=0}^{n} \frac{a^k(b^{n-k} - (-b)^{n-k})}{2k!(n-k)!} = \frac{(a+b)^n - (a-b)^n}{2n!}.$$

We can even generate new identities as simple as it is showed in the following examples.

Let $g(x) = e^x$, $h(x) = \sin(x)$, applying the `coefficient` procedure defined above directly to $g(x)$, we have:

$$[x^n]g = \frac{1}{n!}$$

Applying the `coefficient` directly to $h(x)$, we have:

$$[x^n]h = \frac{\sin(\frac{n\pi}{2})}{n!}$$

Applying the `coefficient` directly to $f(x) = g(x)h(x)$, we have:

$$[x^n]f = \frac{I((1-I)^n - (1+I)^n)}{2n!}$$

By Theorem 4.1, we have the following identity:

$$\sum_{k=0}^{n} \frac{\sin(k\pi/2)}{k!(n-k)!} = \frac{I((1-I)^n - (1+I)^n)}{2n!}$$

Now, let's look at another example. Let $g(x) = e^{ax}$, $h(x) = \sin(bx) + \cosh(cx)$, $f(x) = g(x)h(x)$, where a, b and c are constants.

Applying the `coefficient` directly to $g(x)$, we have:

$$[x^n]g = \frac{a^n}{n!}$$

Applying the `coefficient` directly to $h(x)$, we have:

$$[x^n]h = \frac{(-bI)^n I - (bI)^n I + c^n + (-c)^n}{2n!}$$

Applying the `coefficient` directly to $f(x)$, we have:

$$[x^n]f = \frac{(-bI + a)^n I - (bI + a)^n I + (a+c)^n + (a-c)^n}{2n!}$$

By theorem 4.1, we have the following identity:

$$\sum_{k=0}^{n} \frac{a^k((-bI)^{n-k}I - (bI)^{n-k}I + c^{n-k} + (-c)^{n-k})}{2k!(n-k)!}$$

$$= \frac{(-bI + a)^n I - (bI + a)^n I + (a+c)^n + (a-c)^n}{2n!}$$

By the same way, we generated a list of identities in Table 85.2. The list is just a sample, and an infinite number of such identities can be generated. We keep the identities the same forms as were generated from the machine with little modification. Advanced methods can be developed to generate more complicated and interesting identities.

For each identity or formula in Tables 85.1 and 85.2, the calculation time is very quick with the answer in seconds by an average laptop. Tests were done on a laptop with Intel (R) Core(TM) i5-5300U CPU @ 2.30 GHz 2.30 GHz, and RAM of 8.00 GB.

85.5 Conclusions

We introduced Sumudu Transform in this work in computational approach. The Sumudu Transform is relatively new but very powerful and has plentiful applications. Sumudu Transform can be used to solve problems in a computational approach. The computational solutions can be implemented in computer algebra systems such as Maple to solve the problems automatically. For demonstration, we implemented a Maple procedure called `coefficient`. This procedure is simple but very powerful, and it can be used to calculate the general term of Taylor's coefficients for a large number of functions. By applying the procedure, a large number of formulae can be proved automatically, and new identities can be generated systematically. In addition to this paper, the same authors also wrote reference [4]; both works are related to Sumudu transform, even the

Table 85.2 Sample list of generation of new identities

(Assume n is a positive integer, a, b, c are constants)

$$\sum_{k=0}^{n} \frac{\sin(k\pi/2)}{k!(n-k)!} = \frac{I((1-I)^n - (1+I)^n)}{2n!}$$

$$\sum_{k=0}^{n} \frac{\cos(k\pi/2)}{k!(n-k)!} = \frac{(1-I)^n + (1+I)^n}{2n!}$$

$$\sum_{k=0}^{n} \frac{1+(-1)^{n-k+1}}{2k!(n-k)!} = \frac{2^{n-1}}{n!}$$

$$\sum_{k=0}^{n} \frac{1+(-1)^{n-k}}{2k!(n-k)!} = \frac{2^{n-1}}{n!}$$

$$\sum_{k=0}^{n} \frac{\sin(k\pi/2)\cos((n-k)\pi/2)}{k!(n-k)!} = \frac{I((-2I)^n - (2I)^n)}{4n!}$$

$$\sum_{k=0}^{n} \frac{\sin(k\pi/2)(1+(-1)^{n-k+1})}{2k!(n-k)!} = \frac{-I((-1-I)^n - (-1+I)^n - (1-I)^n + (1+I)^n)}{4n!}$$

$$\sum_{k=0}^{n} \frac{\sin(k\pi/2)(1+(-1)^{n-k})}{2k!(n-k)!} = \frac{-I((-1+I)^n - (-1-I)^n - (1-I)^n + (1+I)^n)}{4n!}$$

$$\sum_{k=0}^{n} \frac{\cos(k\pi/2)(1+(-1)^{n-k+1})}{2k!(n-k)!} = -\frac{(-1-I)^n + (-1+I)^n - (1-I)^n - (1+I)^n}{4n!}$$

$$\sum_{k=0}^{n} \frac{\cos(k\pi/2)(1+(-1)^{n-k})}{2k!(n-k)!} = \frac{(-1-I)^n + (-1+I)^n + (1-I)^n + (1+I)^n}{4n!}$$

$$\sum_{k=0}^{n} \frac{(1+(-1)^{k+1})(1+(-1)^{n-k})}{4k!(n-k)!} = \frac{2^n(1-(-1)^n)}{4n!}$$

$$\sum_{k=0}^{n} \frac{Ia^k((-Ib)^{n-k} - (bI)^{n-k})}{2k!(n-k)!} = \frac{I((\frac{a^2+b^2}{bI+a})^n - (-\frac{a^2+b^2}{bI-a})^n)}{2n!}$$

$$\sum_{k=0}^{n} \frac{a^k((-Ib)^{n-k} + (bI)^{n-k})}{2k!(n-k)!} = \frac{(-\frac{a^2+b^2}{bI-a})^n + (\frac{a^2+b^2}{bI+a})^n}{2n!}$$

$$\sum_{k=0}^{n} \frac{a^k(b^{n-k} - (-b)^{n-k})}{2k!(n-k)!} = -\frac{(a-b)^n - (a+b)^n}{2n!}$$

$$\sum_{k=0}^{n} \frac{a^k((b^{n-k} + (-b)^{n-k})}{2k!(n-k)!} = \frac{(a-b)^n + (a+b)^n}{2n!}$$

$$I\sum_{k=0}^{n} \frac{((-aI)^k - (aI)^k)((bI)^{n-k} + (-bI)^{n-k})}{4k!(n-k)!} = I\frac{(-I(a-b))^n - ((a-b)I)^n - ((a+b)I)^n + (-I(a+b))^n}{4n!}$$

$$I\sum_{k=0}^{n} \frac{((-aI)^k - (aI)^k)(b^{n-k} - (-b)^{n-k})}{4k!(n-k)!} = I\frac{(\frac{a^2+b^2}{aI+b})^n - (-\frac{a^2+b^2}{aI-b})^n - (\frac{a^2+b^2}{aI-b})^n + (-\frac{a^2+b^2}{aI+b})^n}{4n!}$$

$$I\sum_{k=0}^{n} \frac{((-aI)^k - (aI)^k)(b^{n-k} + (-b)^{n-k})}{4k!(n-k)!} = I\frac{((bI-a)I)^n - ((bI+a)I)^n + (-(bI+a)I)^n - (-I(bI-a))^n}{4n!}$$

$$\sum_{k=0}^{n} \frac{((-aI)^k + (aI)^k)(b^{n-k} - (-b)^{n-k})}{4k!(n-k)!} = \frac{(-\frac{a^2+b^2}{aI-b})^n + (\frac{a^2+b^2}{aI+b})^n - (-\frac{a^2+b^2}{aI+b})^n - (\frac{a^2+b^2}{aI-b})^n}{4n!}$$

$$\sum_{k=0}^{n} \frac{((-aI)^k + (aI)^k)(b^{n-k} + (-b)^{n-k})}{4k!(n-k)!} = \frac{((bI+a)I)^n + ((bI-a)I)^n + (-(bI-a)I)^n + (-I(bI+a))^n}{4n!}$$

$$\sum_{k=0}^{n} \frac{(a^k - (-a)^k)(b^{n-k} + (-b)^{n-k})}{4k!(n-k)!} = \frac{(a-b)^n - (-a-b)^n - (-a+b)^n + (a+b)^n}{4n!}$$

$$\sum_{k=0}^{n} \frac{a^k((-bI)^{n-k}I - (bI)^{n-k}I + (cI)^{n-k} + (-cI)^{n-k})}{2k!(n-k)!} = -\frac{(bI+a)^nI - (-bI+a)^nI - (cI+a)^n - (-cI+a)^n}{2n!}$$

(continued)

Table 85.2 (continued)

(Assume n is a positive integer, a, b, c are constants)

$$\sum_{k=0}^{n} \frac{a^k((bI)^{n-k}I - (-bI)^{n-k}I + (-c)^{n-k} - (c)^{n-k})}{2k!(n-k)!} = \frac{(bI+a)^n I - (-bI+a)^n I - (a+c)^n + (a-c)^n}{2n!}$$

$$\sum_{k=0}^{n} \frac{a^k((bI)^{n-k}I - (-bI)^{n-k}I - (-c)^{n-k} - (c)^{n-k})}{2k!(n-k)!} = \frac{(bI+a)^n I - (-bI+a)^n I - (a+c)^n - (a-c)^n}{2n!}$$

$$\sum_{k=0}^{n} \frac{a^k((bI)^{n-k} + (-bI)^{n-k} - (-c)^{n-k} + (c)^{n-k})}{2k!(n-k)!} = \frac{(a+c)^n + (bI+a)^n - (a-c)^n + (a-bI)^n}{2n!}$$

$$\sum_{k=0}^{n} \frac{a^k((bI)^{n-k} + (-bI)^{n-k} + (-c)^{n-k} + (c)^{n-k})}{2k!(n-k)!} = \frac{(a+c)^n + (bI+a)^n + (a-c)^n + (a-bI)^n}{2n!}$$

$$\sum_{k=0}^{n} \frac{a^k((b)^{n-k} - (-b)^{n-k} + (-c)^{n-k} + (c)^{n-k})}{2k!(n-k)!} = \frac{(a+c)^n - (a-b)^n + (a-c)^n + (a+b)^n}{2n!}$$

$$I\sum_{k=0}^{n} \frac{a^k((-I(cI-b))^{n-k} - ((cI-b)I)^{n-k} - (-I(cI+b))^{n-k} + ((cI+b)I)^{n-k})}{4k!(n-k)!}$$
$$= I\frac{(bI+a-c)^n - (-bI+a+c)^n - (-bI+a-c)^n + (bI+a+c)^n}{4n!}$$

$$\sum_{k=0}^{n} \frac{a^k((-\frac{b^2+c^2}{bI-c})^{n-k} + (\frac{b^2+c^2}{bI+c})^{n-k} - (-\frac{b^2+c^2}{bI+c})^{n-k} - (\frac{b^2+c^2}{bI-c})^{n-k})}{4k!(n-k)!}$$
$$= -\frac{(-bI+a-c)^n - (bI+a+c)^n - (-bI+a+c)^n + (bI+a-c)^n}{4n!}$$

$$\sum_{k=0}^{n} \frac{a^k((-I(cI+b))^{n-k} + (I(cI-b))^{n-k} + (I(cI+b))^{n-k} + (-I(cI-b))^{n-k})}{4k!(n-k)!}$$
$$= \frac{(bI+a+c)^n + (-bI+a+c)^n + (-bI+a-c)^n + (bI+a-c)^n}{4n!}$$

$$\sum_{k=0}^{n} \frac{a^k((b-c)^{n-k} - (-b+c)^{n-k} + (b+c)^{n-k} - (-b-c)^{n-k})}{4k!(n-k)!} = -\frac{(a-b+c)^n + (a-b-c)^n - (a+b+c)^n - (a+b-c)^n}{4n!}$$

$$\sum_{k=0}^{n} \frac{a^k((b+c)^{n-k} - (-b-c)^{n-k})}{2k!(n-k)!} = \frac{(a+b+c)^n - (a-b-c)^n}{2n!}$$

$$I\sum_{k=0}^{n} \frac{a^k((\frac{b^2+c^2}{bI+c})^{n-k} - (-\frac{b^2+c^2}{bI-c})^{n-k} - (\frac{b^2+c^2}{bI-c})^{n-k} + (-\frac{b^2+c^2}{bI+c})^{n-k})}{4k!(n-k)!}$$
$$= -I\frac{(\frac{a^2-2ac+b^2+c^2}{bI+a-c})^n - (-\frac{a^2-2ac+b^2+c^2}{bI-a+c})^n - (\frac{a^2+2ac+b^2+c^2}{bI+a+c})^n + (-\frac{a^2+2ac+b^2+c^2}{bI-a-c})^n}{4n!}$$

introductions for Sumudu transform are not very different, but reference [4] is a general introduction about the transform, and this work is focusing on the automatical proof or generation on identities or formulae. Both papers complement each other and obviously there is significant new material in this work.

References

1. Zhang, J. (2008). Sumudu transform based coefficients calculation. *Nonlinear Studies, 15*(4), 355–372.
2. Belgacem, F. B. M. (2006). Introducing and analysing deeper Sumudu properties. *Nonlinear Studies, 13*(1), 23–41.
3. Kiligman, A., & Altun, O. (2014). Some remarks on the fractional Sumudu transform and applications. *Applied Mathematics and Information Sciences, 8*(6), 2881–2888.
4. Zhang, J., Shen, F., & Liu, C. (2016) A computational approach to introduce Sumudu Transform to students. In *The Proceedings of the 12th International Conference on Frontiers in Education: Computer Science and Computer Engineering*, Las Vegas.
5. Zhang, J. (2007). A Sumudu based algorithm for solving differential equations. *Computer Science Journal of Moldova, 153*(45), 303–313.
6. Widder, D. V. (1971). *An introduction to transform theory*. New York: Academic Press.
7. Petkovsek, M., Wilf, H. S., & Zeilberger, D. (1996). A = B. A K Peters, Ltd.
8. Ravenscroft, R. A. (1994). Rational generating function applications. In R. Lopez (Ed.), *Maple V: Mathematics and its application*. Boston: Birkhaser.
9. Ravenscroft, R. A., & Lamagna, E. A. (1989). Symbolic summation with generating functions. In *Proceedings of the International Symposium on Symbolic and Algebraic Computation* (pp. 228–233). New York: Association for Computing Machinery.
10. Salvy, B., & Zimmermann, P. (1994). GFUN: A maple package for the Manipulation of generating and holonomic functions in one variable. *ACM Transactions on Mathematical Software, 20*, 163–177.

A Multiobjective Optimization Method for the SOC Test Time, TAM, and Power Optimization Using a Strength Pareto Evolutionary Algorithm

Wissam Marrouche, Rana Farah, and Haidar M. Harmanani

Abstract

System-On-Chip (SOCs) test minimization is an important problem that has been receiving considerable attention. The problem is tightly coupled with the number of TAM bits, power, and wrapper design. This paper presents a multiobjective optimization approach for the SOC test scheduling problem. The method uses a *Strength Pareto Evolutionary Algorithm* that minimizes the overall test application time in addition to power, wrapper design and TAM assignment. We present various *experimental results* that demonstrate the effectiveness of our method.

Keywords

SOC test scheduling • Multiobjective optimization • SPEA2 • Metaheuristics.

86.1 Introduction

System-on-Chip is a design methodology that integrates *predesigned* and *preverified* intellectual property blocks or cores are embedded on a single die [1]. A core maybe designed in a hierarchical fashion and thus may embed other cores. A major challenge in the SOC design paradigm involves integrating the IP blocks and developing a test methodology for post-manufacturing tests since cores can only be tested after integration. A generic conceptual test access architecture for an embedded core, introduced by Zorian et al. [2], consists of a *test source* and a *sink*, a *test*

access mechanism and a *core wrapper*. The test source is used for test stimulus generation while the response evaluation is carried out by the test sink. The *source* and the *sink* can be either implemented off-chip or on-chip. Test access mechanism (TAM) serves as a "test data highway" that transports test patterns between the *source* and the *core* as well as between the *core* and the *sink*. Finally, the core is surrounded with test logic, known as the *test wrapper*, that provides switching functionality between *normal* access and *test* access via the TAM [3].

One of the main challenges in core-based designs is test time reduction which aims at maximizing the simultaneous test of all cores. The problem, known as test scheduling, determines the order in which various cores are tested and has been shown to be $\mathcal{NP}$-complete [4]. Hierarchical cores further complicate the SOC test scheduling process especially in the case of hard or legacy cores. Hierarchical cores have multiple levels of test hierarchy with the top level is the SOC itself and consists of several mega-cores that have their own embedded cores. Cores at level n are *parent* cores with respect to the cores at level $n + 1$ [5]. Finally, power dissipation strongly impacts test parallelism since it depends on the switching activity resulting from the application of test vectors to the system [6].

W. Marrouche (✉)
Department of Electrical & Computer Engineering, American University of Beirut, Beirut, Lebanon
e-mail: wissam.marrouche@gmail.com

R. Farah
Department of Computer & Software Engineering, École Polytechnique de Montréal, Quebec, Canada
e-mail: rana.farah@polymtl.ca

H.M. Harmanani
Department of Computer Science, Lebanese American University, Byblos, Lebanon
e-mail: haidar@lau.edu.lb

© Springer International Publishing AG 2018
S. Latifi (ed.), *Information Technology – New Generations*, Advances in Intelligent Systems and Computing 558, DOI 10.1007/978-3-319-54978-1_86

This paper presents a multiobjective approach for test minimization, variable TAM assignment and partitioning, and wrapper design for hierarchical SOCs using a strength Pareto evolutionary algorithm. The method takes into consideration power and precedence constraints. Formally, given a SOC with N_C cores, a total TAM width $\mathcal{W}$, a set of design and test constraints including power and precedence constraints, and a set of parameters for each mega-core, the problem we address in this paper is to minimize the overall test time such that, (i) the test schedule for the entire SOC is efficient, (ii) the TAM wires are optimally partitioned and assigned to cores (iii) the wrapper configuration for each core is determined, (iv) $\mathcal{W}$ is not exceeded, and (v) hierarchical cores receive at least their prespecified TAM widths.

86.1.1 Related Work

Several researchers addressed the test scheduling problem but mostly assumed flat cores with a single level TAM; however, this assumption is only valid if the embedded cores are *mergeable*. Iyengar et al. [7] first formulated the integrated wrapper/TAM optimization problem using ILP and developed later several heuristics for solving the problem. Huang et al. [8] modeled the problem as a restricted 3-D bin packing problem and proposed a heuristic to solve it. Goel et al. [9] proposed an efficient heuristic for fixed-width architecture while Xu et al. [10] proposed a method for multi-level test access mechanism that facilitates test data reuse for hard mega-cores in hierarchical SOCs. Su et al. [11] formulated the problem using graph-based approach and solved it using tabu search. Chakrabarty et al. [12] proposed a combination of integer linear programming, enumeration and efficient heuristics in order to solve the problem for hierarchical cores. Wuu et al. [13] extended the wrapper design method presented by Zou et al. [14] and solved the test scheduling problem by mapping it into a floor planning problem. Wang et al [15] proposed a test scheduling algorithm that is based on the 2-D bin-packing model. Ooi et al. [16] proposed an enhanced rectangle packing test scheduling algorithm efficiently compacts the test scheduling floor plan. SenGupta et al. [17] proposed a new test planning and test access mechanism method for stacked integrated circuits.

86.1.2 Strength Pareto Evolutionary Algorithm

Evolutionary algorithms are effective optimization techniques that are global in scope. Dominance-based evolutionary algorithms have emerged as reliable approaches for generating Pareto optimal solutions to multi-objective optimization problems. SPEA2 [18] is an improved strength Pareto evolutionary algorithm that incorporates a fine-grained fitness assignment strategy, a density estimation technique, and an enhanced archive truncation method. The algorithm uses genetic recombination and mutation in order to locate and maintain a front of non-dominated solutions, ideally a set of Pareto optimal solutions. SPEA2's Pareto measure is the number of solutions that dominates a candidate solution, and its crowding measure is based on the distance to other individuals in the multiobjective space [19]. The algorithm maintains, separate from the population, an *archive* of the non-dominated set which provides a form of elitism.

The remainder of the paper is organized as follows. In Sect. 86.2 we formulate the multiobjective SOC test scheduling problem and describe the genetic encoding, the initial population, the genetic operators, and the fitness function. Section 86.3 presents the annealing test scheduling while Sect. 86.4 presents the hierarchal test scheduling algorithm. We conclude with experimental results in Sect. 86.5.

86.2 Problem Formulation

The test time of a SOC depends on the individual cores test times as well as on the test start times. The cores test times are based on TAM assignments which consequently affect the wrapper design. Test adaptation is performed by serially connecting internal scan-chains, the wrapper input cells and the wrapper output cells in order to form wrapper chains. There is a clear trade-off between test time and TAM capacity. For example, while the length of the wrapper chains directly affects the core's test time, the number of wrapper chains affects the number of TAM bits as each wrapper chain's input and output must be connected to a TAM wire. The test time t_i of core i is determined by the shortest and the longest wrapper chains as follows [20]:

$$t_i = (1 + \max(s_i, s_o)) \times p + \min(s_i, s_o), \qquad (86.1)$$

where, s_i and s_o denote respectively the scan-in and scan-out time for the core and p_i denotes the number of test patterns applied on the core i. Embedding cores adds an additional test conflict problem arising from the fact that it may not be possible to test the parent and the child cores concurrently due, for example, to a conflict in the use of the input wrapper cells. Therefore, an effective test scheduling approach must minimize the test time while addressing test resources allocation and conflicts arising from the use of shared test access mechanism.

In this paper we explore design trade-offs among conflicting objectives by integrating test scheduling, wrapper design and TAM assignment using a multiobjective evolutionary algorithm. In what follows, we describe the multiobjective evolutionary test scheduling algorithm.

86.2.1 Chromosomal Representation

A chromosome is represented using a vector whose length is equal to the number of cores in the SOC. Each chromosome is represented using a vector whose length is equal to the number of cores in the SOC and corresponds to a candidate solution. Every gene corresponds to a block that includes the assigned *TAM bits*, *core test time*, and *power* as shown in Fig. 86.1. During every generation, chromosomes are selected in order to seed the next generation, yielding sub-optimum solutions.

86.2.2 Initial Population

Increasing the number of TAM bits may not guarantee decreasing the test time which may hit a Pareto optimal point. For all points with the same test time, a Pareto-optimal point is the point where the least number of TAM wires is used [21]. The test time does not decrease even if the number of wrapper chains increases. Figure 86.2a shows that the test time decreases as the number of TAM bits

increases for core 5 in the *d695* benchmark, and hits several Pareto optimal points.

At the beginning of each run, the algorithm generates a pool of cores where the test time of each core corresponds to a specific Pareto optimal test point, TAM bits, and power. For example, the pool for the d695 benchmarks includes 82 cores; the pool for core 5 is shown in Fig. 86.2b. The initial population is next generated by randomly selecting cores from the pool, and populating all chromosomes in order to create an initial population of size 70. The wrapper for each core is designed using an improved Best Fit Decreasing (BFD) heuristic [13]. Finally, each individual is test scheduled using a bin packing simulated annealing algorithm.

86.2.3 Archive

Evolutionary algorithms replace individuals with more fit ones using various strategies. However, individuals that are more *fit* than the rest take over the population very quickly leading to premature convergence. This problem is alleviated in *SPEA2* using an archive, A_t, of non-dominated solutions. During each generation, the *best* non-dominated solutions are added to the archive. The archive *size* is set to 70.

86.2.4 Fitness Function

The test scheduling problem has multiple and conflicting objectives that include test time, power, and TAM bits allocation. We optimize all three objectives using a vector-valued fitness function $f = (Test\ Time, Power, TAM\ Width)$, based on the *Pareto dominance concept*. An objective vector f_1 is said to dominate another objective vector f_2 ($f_1 \succ f_2$) if no component of f_1 is smaller than the corresponding component of f_2 and at least one component is greater [18].

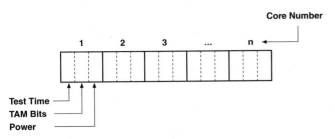

Fig. 86.1 Chromosome representation

Fig. 86.2 (a) Pareto test points for core 5, d695, (b) Cores pool

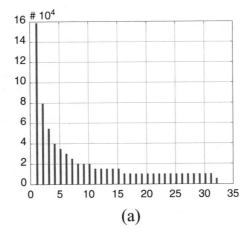

<TAM Bits, Test Time, Power>	
<1, 158507, 690>	<10, 19825, 690>
<2, 79364, 690>	<11, 15072, 690>
<3, 54576, 690>	<12, 15051, 690>
<4, 39847, 690>	<13, 14940, 690>
<5, 34699, 690>	<16, 10210, 690>
<6, 29814, 690>	<17, 10166, 690>
<7, 24930, 690>	<18, 10056, 690>
<8, 20089, 690>	<25, 9946, 690>
<9, 19935, 690>	<26, 9945, 690>
	<32, 5215, 690>

(a)   (b)

Each individual i in the archive A_t as well as in the population P_t is assigned a strength value, $Strength(i)$, representing the number of solutions it dominates:

$$Strength(i) = \sum_{i,j \in P_t \cup A_t,} [i \succ j] \qquad (86.2)$$

The bracketed notation $[S]$ is 1 if S is true and 0 otherwise. The raw fitness $f(i)$ of an individual i is calculated as follows:

$$f(i) = \sum_{j \in P_t \cup A_t, j \succ i} Strength(j) \qquad (86.3)$$

The raw fitness is determined by the strengths of its dominators in the *archive* and in the *population*. Thus, the fitness to be minimized, $f(i) = 0$, corresponds to a non-dominated individual, while a high $f(i)$ value means that i is dominated by many individuals. Ties with the same fitness are broken by a nearest neighbor density estimation function $D(i)$:

$$D(i) = \frac{1}{\sigma_i^k + 2} \qquad (86.4)$$

Where σ_i^k is the Euclidean distance of the objective values between a given solution i and the kth nearest neighbor, $k = \sqrt{(N_P + N_A)}$.

86.2.5 Selection and Reproduction

We use a multi-objective tournament selection of size 2. Thus, during each generation, two individuals are selected randomly and the fittest is preserved for the next generation.

86.2.6 Genetic Operators

We explore the design space using two genetic operators, *mutation* and *crossover*, that are applied iteratively with their corresponding probabilities. Both operators ensure that the generated solutions are feasible, and are followed by the annealing bin packing test scheduling algorithm.

86.2.6.1 Mutation
Mutation is used to explore the search space while inhibiting premature convergence. The operator selects a random chromosome. A random core is next selected and replaced with another random core from the cores pool using a probability

$P_m = 80\%$. The operator results with an increase or decrease in the number of TAM bits.

86.2.6.2 Crossover
We use multiple point uniform crossover which adds more variability than fixed-point crossover with probability $P_c = 40\%$. The crossover operator takes two randomly selected chromosomes, and creates two offspring by selecting each gene from either parent with probability P_c. The uniform crossover generates a diverse new offspring than the traditional one-point or two-point crossover.

86.3 Annealing Bin Packing Algorithm

Once the individuals have been created using the evolutionary operators, the algorithm uses an annealing bin-packing strategy in order to compute the test time for the new population. The annealing algorithm uses a floorplanning sequence-pair representation where each configuration is represented using an ordered pair (S_+, S_-) of block permutations. Together, the two permutations represent geometric relations between every pair of blocks.

Each annealing configuration represents a test schedule but with a different cost. The search space is explored using the following neighborhood functions:

(1) *Test Schedule Exploration:* the algorithm randomly selects a random core i from the current configuration and changes its starting time S_i to the *end time* F_j of a randomly chosen core j by adjusting the corresponding elements in the sequence-pair.
(2) *Test Schedule Swap:* the algorithm randomly selects and swaps two random elements in the sequence-pair.

The neighborhood functions use a constructive approach that leads to incremental feasible test schedules. The variation in the cost functions, Δ_C, is computed and if negative then the transition from C_i to C_{i+1} is accepted. If the cost function increases, the transition is accepted with a probability based on the *Boltzmann* distribution. The temperature is gradually decreased from a high starting value, $T_0 = 1000$, where almost every proposed transition, positive or negative, is accepted to a freezing temperature, $T_f = 0.1$, where no further changes occur. As for the number of iterations, M, and the iteration multiplier, β, they were respectively set to 5 and 1.05. The algorithm stops after a Max_{Time} of 6000 ms.

Algorithm 1 Hierarchical test scheduling algorithm

```
 1: function HIERARCHICAL_TEST_SCHEDULE()
 2:     for all levels starting from level n − 1 to level 0 do
 3:         find all mega cores
 4:         for each mega core do
 5:             assign TAM bits for the children cores
 6:             calculate the test time for each child core
 7:             Schedule all children cores using simulated annealing
 8:             calculate overall test time
 9:         end for
10:         if (level == 0) then
11:             calculate the overall time of the SOC
12:         end if
13:     end for
14:     for (i = 0; i < n; i + +) do
15:         Select a random core i from any level
16:         Move to the next/previous Pareto-optimal point
17:         if the transformation is accepted then
18:             Annealing_Bin_Packing_Algorithm()
19:         end if
20:         Update 𝒯_i and all parents cores that contain this core.
21:     end for
22: end function
```

86.4   Hierarchical Test Scheduling Algorithm

The hierarchical core model is a recursive model where a wrapped parent core has an external TAM that connects externally to the parent core and an internal TAM that consists of the TAM architecture of the children cores. We start with the *mega-cores* that are at level n where each mega-core is considered as a separate SOC. The embedded child cores are initially assigned TAMs based on Algorithm 1 where we define a dominant core as a core such that when allocated the same TAM width as all its peers it returns a larger test time than its peers. For cores that are at level n and that have more than one peer, the algorithm allocates half the TAM bits of the parents to the child cores. During the TAM assignment process, the algorithm ensures that no mega-core is assigned more than the specified TAM bits and that no core is assigned more than the TAM bits of its parent. Once the TAM assignment is performed, the wrapper design is determined using the BFD algorithm. The algorithm next iterates over each mega-core using our simulated annealing in order to determine the *test schedule* as well as the wrapper design for the mega-core itself. The test wrapper is next designed by configuring the scan elements into wrapper chains using the BFD algorithm. On the other hand, test data serialization maybe necessary since a mega-core width maybe assigned less than the TAM width it requires. Thus, a mega-core i with top-level TAM width w_i maybe provided with w_i^* TAM bits such that $w_i^* \leq w_i$. The test time for mega-core i then reduces to $\mathcal{T}_i^* = \left\lceil \frac{w_i}{w_i^*} \right\rceil * \mathcal{T}_i$ where $\mathcal{T}_i$ is the total testing time for the embedded cores on the internal TAM partition for mega-core i. Finally, the testing time for mega-core i must include the test time for the top-level test

time and this reduces to: $\mathcal{T}_i^{W_i^*} = \mathcal{T}_i^* + \mathcal{T}_i^{Mega}$ where $\mathcal{T}_i^{W_i^*}$ is the testing time for mega-core i with an external TAM of width w_i^* and $\mathcal{T}_i^{Mega}$ is the testing time at the system level TAM architecture. Once all cores at level i have been processed, the algorithm recurses back to level $i − 1$ and considers the remaining mega-cores.

Algorithm 2 Multi-Objective Test Scheduling Algorithm()

Input: A set of Cores
Output: Minimum test Schedule

```
 1: N_P = N_A = 70
 2: N_t ← 20
 3: P_0 ← Initial(population)
 4: A_0 ← ∅
 5: t ← 1
 6: while t < N_t do
 7:     Design the cores wrappers using the modified BFD algorithm
 8:     Schedule all S_i ∈ P_t ∪ A_t using simulated annealing
 9:     for all S_i ∈ P_t do
10:         f_{S_i} ← fitness values of individuals in P_t and A_t
11:         Rank all S_i by their fitness value and the k-nearest neighbor
12:     end for
13:     A_{t+1} ← All non-dominated individuals in P_t ∪ A_t
14:     if size(A_{t+1}) > N_A then
15:         Truncate (A_{t+1})
16:     else if size(A_{t+1}) < N_A then
17:         A_{t+1} ← best non-dominated individuals in P_t ∪ A_t
18:     end if
19:     Create N_P offspring using Mutation and crossover
20:     for (i = 0; i < N_P; i + +) do
21:         Selects two random individuals from P_t
22:         Keep the best
23:     end for
24:     t ← t + 1
25: end while
```

86.5   Experimental Results

The proposed algorithm was implemented using `Python`, and tested on the ITC'02 benchmarks. The algorithm starts by creating an initial population of size 70 where each chromosome represents a candidate solution that has a different test time, power, and wrapper design. The algorithm also creates an archive of the same size. The archive is initially empty.

During every generation, chromosomes are selected for reproduction using the *mutation* and *crossover* operators, resulting in a new population that is equal in size to the initial population. The algorithm schedules the newly generated individuals using simulated annealing in order to minimize the test scheduling time and power. Next, all individuals are assigned a fitness function, and the new population is next selected using a tournament selection of size 2. All the best solutions are maintained in the archive using elitism. Figure 86.3 shows the archive cost fitness function quality which monotonically decreases with each evolutionary generation.

Table 86.1 shows the results of our algorithm with power constraints. We compare our hierarchical test scheduling algorithm with Chakrabarty et al. [12], Wang et al. [15],

Fig. 86.3 Archive quality in the case of the D965 benchmark. (**a**) TAM Width. (**b**) Test Time. (**c**) TAM, Test Time, and Power

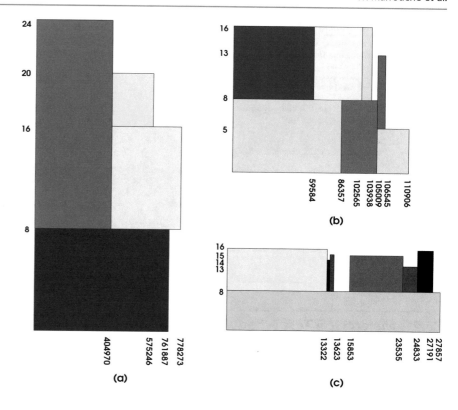

Table 86.1 ITC'02 benchmark test times with power dissipation (in mW)

	TAM bits							
	16	24	32	40	48	56	64	80
d695	412224 (1177)	27822 (2463)	20766 (2364)	17657 (2321)	16612 (1960)	15111 (2587)	3392 (3746)	10634 (3746)
p22810	412224 (2947)	312910 (3265)	247737 (2992)	199110 (4184)	176758 (3920)	159160 (5426)	143760 (3735)	110137 (4727)
p93791	1803764 (9886)	1157365 (15891)	1068967 (10784)	914832 (21440)	772649 (16415)	630287 (19026)	612505 (20720)	502997 (23429)

and Xu et al. [10]. Detailed results comparisons are shown in Table 86.2. The TAM widths supplied to the mega-cores were determined to be 8 bits for SOCs *p22810* and *a586710*, and 16 bits for SOCs *p34392* and *p93791*. Figure 86.4 illustrates the hierarchical test schedule for the *p34392* SOC benchmark using a TAM width of 16-bits. The resulting schedule takes into consideration two megacores, and the system was able to test it in 778,273 cycles. We also compare our method to various researchers [7–9, 11, 14, 14, 21–23] for flat designs based on fixed-width as well as based on flexible width TAM. Detailed results comparisons are shown in Table 86.3.

86.6 Conclusion

We presented a multiobjective approach for test minimization, variable TAM assignment and partitioning, and wrapper design for hierarchical SOCs using a strength Pareto evolutionary algorithm. The algorithm was able to achieve near optimal solutions in a reasonable amount of time for big systems.

Table 86.2 Hierarchical ITC'02 SOC test times comparisons

SOC	W	[10]	[15]	[12]	Ours
a586710	16	5.27×10^7	–	4.44×10^7	**42117546**
	24	3.06×10^7	–	3.06×10^7	**22973206**
	32	2.19×10^7	–	2.28×10^7	**21058772**
	40	1.91×10^7	–	2.50×10^7	**17229905**
	48	1.53×10^7	–	2.14×10^7	**13401037**
	56	–		2.14×10^7	**13031723**
	64	1.41×10^7		2.14×10^7	**13031723**
p93791	16	**1865140**	–	–	1953223
	24	1486628	–	1650880	**1377332**
	32	1486628	1937130	1021320	**991915**
	40	**801271**	–	916852	845391
	48	801271	1272314	681816	**670660**
	56	–	–	632125	**629510**
	64	553775	947554	**521064**	545583
p22810	16	496804	–	505858	**421422**
	24	353619	–	412682	**305936**
	32	**280634**	466044	396473	232531
	40	**280634**	–	366260	280634
	48	**280634**	338374	366260	280634
	56	–	–	–	280634
	64	**280634**	285231	366260	280634
p34392	16	1467705	–	–	**1188916**
	24	1467705	–	1347023	**778273**
	32	776537	–	788873	**713071**
	40	776537	–	728426	**606261**
	48	**60261**	–	618597	**606261**
	56	–		618597	**606261**
	64	**60261**	–	618597	**606261**

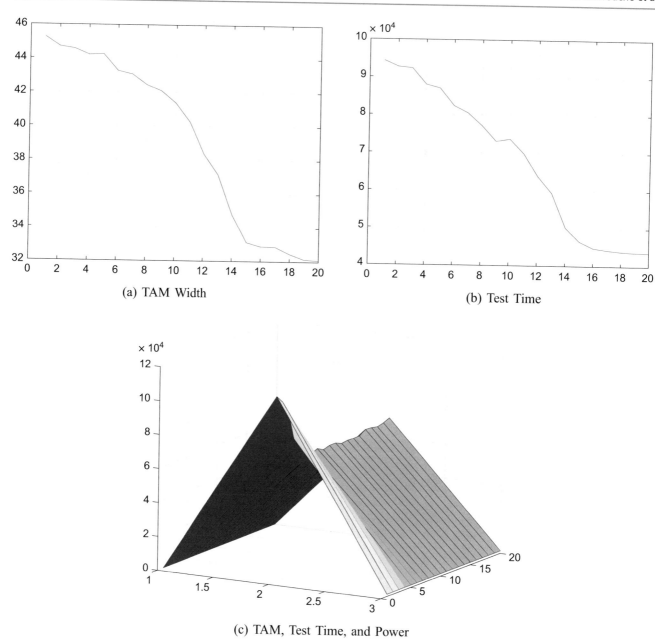

(a) TAM Width

(b) Test Time

(c) TAM, Test Time, and Power

Fig. 86.4 (a) Hierarchical test schedule for p34392; (b) Mega-core 2 test schedule; (c) Mega-core 1 test schedule

Table 86.3 ITC'02 benchmark results

SOC	W_{max}	Fixed-width					Flexible-width				
		[7]	[22]	[9]	[14]	[21]	[8]	[11]	[14]	[23]	Ours
d695	16	42568	42644	44307	–	44545	42716	41905	41604	41847	**36937**
	24	28292	30032	28576	–	31569	28639	28231	28064	29106	**26482**
	32	21566	22268	21518	–	23306	21389	21467	21161	20512	**20089**
	40	17901	18448	17617	–	18837	17366	17308	16993	18691	**16284**
	48	16975	15300	14608	–	16984	15142	14643	14182	17257	**14093**
	56	13207	12491	12462	–	14974	13208	12493	12085	–	**11519**
	64	12941	12941	11033	–	11984	11279	11036	10723	13348	**10643**
g1023	16	–	–	34459	–	–	31444	32602	31139	–	**15063**
	24	–	–	22821	–	–	21409	22005	21024	–	**15063**
	32	–	–	16855	–	–	16489	17422	15890	–	**15063**
	40	–	–	14794	–	–	14794	14794	14794	–	**14794**
	48	–	–	14794	–	–	14794	14794	14794	–	**14794**
	56	–	–	14794	–	–	14794	14794	14794	–	**14794**
	64	–	–	14794	–	–	14794	14794	14794	–	**14794**
p22810	16	462210	468011	458068	434922	489192	446684	465162	438619	473418	**421422**
	24	361571	313607	299718	313607	330016	300723	317761	289287	352834	297596
	32	312659	246332	222471	245622	245718	223462	236796	218855	236186	227938
	40	278359	232049	190995	194193	199558	184951	193696	175946	195733	186788
	48	278359	232049	160221	164755	173705	167858	174491	147944	159994	166169
	56	268472	153990	145417	145417	157159	145087	155730	126947	–	144374
	64	260638	153990	133405	133628	142342	128512	145417	109591	128332	121961
p34392	16	998733	1033210	1010821	1021510	1053491	1016640	995739	944768	–	**873120**
	24	720858	882182	680411	729864	759427	681745	690425	628602	–	**587034**
	32	591027	663193	551778	630934	544579	553713	544579	544579	–	**544579**
	40	544579	544579	544579	544579	544579	544579	544579	544579	–	**544579**
	48	544579	544579	544579	544579	544579	544579	544579	544579	–	**544579**
	56	544579	544579	544579	544579	544579	544579	544579	544579	–	**544579**
	64	544579	544579	544579	544579	544579	544579	544579	544579	–	**544579**
p93791	16	1771720	1786200	1791638	1775586	1932331	1791860	1767248	1757452	1827819	1777840
	24	1187990	1209420	1185434	1198110	1310841	1200157	1178776	1169945	1220469	1204168
	32	887751	894342	912233	936081	988039	900798	906153	878493	945425	943087
	40	698583	741965	718005	734085	794027	719880	737624	718005	787588	760868
	48	599373	599373	601450	599373	669196	607955	608285	594575	639217	638993
	56	514688	514688	528925	514688	568436	521168	539800	509041	–	557890
	64	460328	473997	455738	472388	517958	549233	485031	447974	457862	489591
u226	16	–	–	18663	–	–	13416	18663	13333	–	**8109**
	24	–	–	13331	–	–	10750	14745	8084	–	**4157**
	32	–	–	10665	–	–	6746	10665	6746	–	**4157**

(continued)

Table 86.3 (continued)

SOC	W_{max}	Flexible-width					Fixed-width				
		[7]	[22]	[9]	[14]	[21]	[8]	[11]	[14]	[23]	Ours
	40	–	–	8084	–	–	5332	8084	5332	–	4157
	48	–	–	7999	–	–	5332	7999	5332	–	4157
	56	–	–	7999	–	–	4080	7999	4080	–	4157
	64	–	–	7999	–	–	4080	7999	4080	–	4157
f2126	16	–	–	372125	–	–	357109	357089	357088	–	335666
	24	–	–	335334	–	–	335334	335334	335334	–	335334
	32	–	–	335334	–	–	335334	335334	335334	–	335334
	40	–	–	335334	–	–	335334	335334	335334	–	335334
	48	–	–	335334	–	–	335334	335334	335334	–	335334
	56	–	–	335334	–	–	335334	335334	335334	–	335334
	64	–	–	335334	–	–	335334	335334	335334	–	335334
t512505	16	–	–	10530995	–	–	10531003	11210100	10530995	–	10530995
	24	–	–	10453470	–	–	10453470	10525823	10453470	–	10453470
	32	–	–	5268868	–	–	5268872	6370809	5268868	–	5268868
	40	–	–	5228420	–	–	5228420	5240493	5228420	–	5228420
	48	–	–	5228420	–	–	5228420	5239111	5228420	–	5228420
	56	–	–	5228420	–	–	5228420	5228474	5228420	–	5228420
	64	–	–	5228420	–	–	5228420	5228489	5228420	–	5228420
a586710	16	–	–	41523868	–	–	42198943	42,067,708	32,626,782	–	32417445
	24	–	–	28,716,501	–	–	27,785,885	27,907,180	23,413,604	–	22973206
	32	–	–	22,475,033	–	–	21,735,586	22,70,4821	18,838,663	–	17195389
	40	–	–	19,048,835	–	–	19,041,307	19,041,307	14,260,216	–	14249791
	48	–	–	15,315,476	–	–	15,071,730	15,212,440	12,811,087	–	12811087
	56	–	–	13,415,476	–	–	14,945,057	13,401,034	12,573,448	–	11486603
	64	–	–	12,700,205	–	–	12,754,584	11,567,464	10,65,9014	–	9572169
d281	16	–	–	8444	–	–	7948	8156	7946	–	7347
	24	–	–	6408	–	–	5486	5830	5485	–	4992
	32	–	–	5084	–	–	4070	4640	4070	–	3926
	40	–	–	3964	–	–	3926	3926	3926	–	3926
	48	–	–	3926	–	–	3926	3926	3926	–	3926
	56	–	–	3926	–	–	3926	3926	3926	–	3926
	64	–	–	3926	–	–	3926	3926	3926	–	3926

References

1. Saleh, R., Wilton, S., Mirabbasi, S., Hu, A., Greenstreet, M., lemieux, G., Pande, P., Grecu, C., & Ivanov, A. (2006). System-on-chip: Reuse and integration. In *Proceedings of the IEEE*.
2. Zorian, Y., Marinissen, E., & Dey, S. (1998). Testing embedded core-based system chips. In *Proceedings of ITC*.
3. Bushnell, M., & Agrawal, V. (2000). *Essentials of electronic testing for digital, memory & mixed-signal VLSI circuits*. New York: Kluwer-Academic Publishers.
4. Chakrabarty, K. (2000). Test scheduling for core-based systems using mixed-integer linear programming. *IEEE Transactions on CAD*.
5. Sehgal, A., Goel, S., Marinissen, E., & Chakrabarty, K. (2004). IEEE P1500-Compliant test wrapper design for hierarchical cores. In *Proceedings of ITC*.
6. Chakrabarty, K. (1999). Test scheduling for core-based systems. In *Proceedings of ITC*.
7. Iyengar, V., Chakrabarty, K., & Marinissen, E. (2002). Test wrapper and test access mechanism co-optimization for system-on-a-chip. *JETTA*.
8. Huang, Y., Reddy, S., Cheng W.-T., Reuter, P., Mukherjee, N., Tsai, C., Samman, O., & Zaidan, Y. (2002). Optimal core wrapper width selection and SOC test scheduling on 3-D bin packing algorithm. In *Proceedings of ITC*.
9. Goel, S., & Marinissen, E. (2003). SOC test scheduling design for efficient utilization of bandwidth. In *ACM TODAES*.
10. Xu, Q., & Nicolici, N. (2004). Time/area tardeoffs in testing hierarchical SOCs with hard mega-cores. In *Proceedings of ITC*.
11. Su, C., & Wu, C. (2004). A graph-based approach to power-constrained test scheduling. *JETTA, 20*.
12. Chakrabarty, K., Iyengar, V., & Krasniewski, M. (2005). Test planning for modular testing of hierarchical SOCs. In *IEEE Transactions of CAD*.
13. Wuu, J.-Y., Chen, T.-C., & Chang, Y.-W. (2005). SOC test scheduling using the B*-tree based floorplanning technique. In *Proceedings of ASP-DAC*.
14. Zou, W., Reddy, S. R., Pomeranz, I., & Huang, Y. (2003). SOC test scheduling using simulated annealing. In *Proceedings of VTS*.
15. Wang, T.-P., Tsai, C.-Y., Shieh, M.-D., & Lee, K.-J. (2005). Efficient test scheduling for hierarchical core based designs. In *Proceedings of TSA-DAT*.
16. Ooi, C. Y., Sua, J. P., & Lee, S. C. (2012). Power-aware system-on-chip test scheduling using enhanced rectangle packing algorithm. *CEE*.
17. SenGupta, B., & Larsson, E. (2014). Test planning and test access mechanism design for stacked chips using ILP. In *Proceedings of VTS*.
18. Zitzler, E., Laumanns, M., & Bleuler, S. (2004). A tutorial on evolutionary multiobjective optimization. In X. Gandibleux (Ed.), *Metaheuristics for multiobjective optimisation*. Berlin/Heidelberg/New York: Springer.
19. Luke, S. (2015). *Essentials of metaheuristics*. Available online on http://cs.gmu.edu/~sean/book/metaheuristics/. Lulu.
20. Marinissen, E., Goel, S., & Lousberg, M. (2000). Wrapper design for embedded core test. In *Proceedings of ITC*.
21. Iyengar, V., Chakrabarty, K., & Marinissen, E. J. (2002). On using rectangle packing for SOC wrapper/TAM co-optimization. In *Proceedings of VTS*.
22. Iyengar, V., Chakrabarty, K., & Marinissen, E. (2002). Efficient wrapper/TAM co-optimization for large SOCs. In *Proceedings of DATE*.
23. Pouget, J., Larsson, E., & Peng, Z. (2005). Multiple-constraint driven system-on-chip time optimization. *JETTA*.

Priscila Alves Macanhã, Danilo Medeiros Eler, Rogério Eduardo Garcia, and Wilson Estécio Marcílio Junior

Abstract

Several works have presented distinct ways to compute feature descriptor from different applications and domains. A main issue in Computer Vision systems is how to choose the best descriptor for specific domains. Usually, Computer Vision experts try several combination of descriptor until reach a good result of classification, clustering or retrieving – for instance, the best descriptor is that capable of discriminating the dataset images and reach high correct classification rates. In this paper, we used feature descriptors commonly applied in handwritten images to improve the image classification from fruit datasets. We present distinct combinations of Zoning and Character-Edge Distance methods to generate feature descriptor from fruits. The combination of these two descriptor with Discrete Fourier Transform led us to a new approach for acquire features from fruit images. In the experiments, the new approaches are compared with the main descriptors presented in the literature and our best approach of feature descriptors reaches a correct classification rate of 97.5%. Additionally, we also show how to perform a detailed inspection in feature spaces through an image visualization technique based on a similarity trees known as Neigbor Joining (NJ).

Keywords

Computer vision • Fruit classification • Feature descriptor • Handwritten character • Image visualization

87.1 Introduction

A classification task is commonly based on previous information, features or structures acquired from a specific dataset. In computer vision, image datasets are used in the wholly process and the data information is computed based on feature descriptor methods applied to each image. Thus, a set of features is assigned to each image for generating the feature space from the dataset. Therefore, the feature acquisition step plays an important role in computer vision systems, because the feature space quality can impact the accuracy of classification, clustering and retrieving tasks. Hence, several works have presented efforts to compare several feature descriptor methods applied in different dataset domains.

Fruit classification systems have been focus of computer vision over the years. The main feature descriptor methods presented in the literature are based on texture, shape and color [1–3]. Seng e Mirisaee [4] proposed a system capable to classify fruit with 90% of classification rate. They used feature descriptor methods based on color, shape and size to an automatic fruit recognition. Ganesan et al. [5] proposed a system based on color and texture. Their approach reaches 86% by combining all feature vectors. Zawbaa et al. [6] also

P.A. Macanhã (✉) • D.M. Eler • R.E. Garcia • W.E.M. Junior
Departamento de Matemática e Computação, Faculdade de Ciências e Tecnologia, UNESP – Univ Estadual Paulista, Presidente Prudente, Presidente Prudente, SP, Brazil
e-mail: priscila.macanha@gmail.com; daniloeler@fct.unesp.br; rogerio@fct.unesp.br; wilson_jr@outlook.com

© Springer International Publishing AG 2018
S. Latifi (ed.), *Information Technology – New Generations*, Advances in Intelligent Systems and Computing 558, DOI 10.1007/978-3-319-54978-1_87

proposed the fruit classification based on color and shape. Their best results were gotten with apples and oranges, resulting in 85% to apples and 65% to oranges. An specific fruit recognition system to green citric fruits was proposed by Sengupta and Lee [7], in which they proposed features computation based on shape and texture, resulting in a correct classification rate of 81.7%.

In this work, we propose new approaches of feature descriptor methods to improve the accuracy of fruit classification systems. For that, we employee two well known descriptor methods commonly used in handwritten character recognition: Zoning [8] and Character-Edge Distance [9]. Our methodology is based on the combination of these two methods alone and with Discrete Fourier transform, which is used to decrease the feature space dimensionality [10].

The main contributions of this paper are the new feature descriptor approaches based on Zoning, Character-Edge Distance and Discrete Fourier Transform. We present comparisons of distinct features spaces computed from these combinations and the best descriptor approaches are chosen based on classifiers accuracy. One of the proposed approaches which combines all techniques reaches 97% of correct classification rate. We also present comparisons of the proposed approaches with known descriptor methods presented in the literature of fruit classification. Finally, we used an image visualization approach to perform a detailed inspection of feature spaces. This visual approach reveals the effects of the feature space quality on clustering similar images.

This work is organized as follows: Sect. 87.2 describes the background of the feature descriptor methods employed and proposed in this work; Sect. 87.3 presents the experiments performed with the descriptor methods described in Sect. 87.2 and a comparison with other descriptors presented in the literature; Sect. 87.4 shows an application of image visualization for performing detailed inspections in features spaces; Sect. 87.5 presents the conclusion and further works.

87.2 Feature Descriptor Methods

This work proposed new approaches of feature descriptors based on two main methods Zoning [8] and Character-Edge Distance [9] commonly used in handwritten character classification. We used the original methods and a combination of them, as well as the Discrete Fourier Transform to reduce the feature space dimensionality. These two main descriptor methods are presented in the following.

(a) Zoning: The Zoning method consists of dividing an image into regions and calculating the percentage of active pixels for each region. As shown in Fig. 87.1, this work used

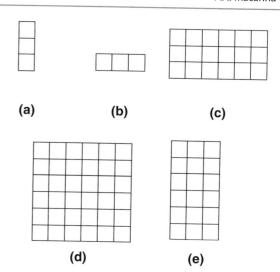

Fig. 87.1 Distinct configurations of Zoning descriptor method: (**a**) zoning 3 × 1; (**b**) zoning 1 × 3; (**c**) zoning 3 × 6; (**d**) zoning 6 × 6; and (**e**) zoning 6 × 3

five different configurations of Zoning technique to determine the regions in which the image is divided. They are: 3 × 1, 1 × 3, 3 × 6, 6 × 3, 6 × 6. These divisions were selected to cover the discriminant features for each fruit.

(b) Character-Edge Distance: The technique called Character-Edge Distance counts the number of background pixels from the border to the character. This counting process is performed in eight directions, as shown in Fig. 87.2. Thus, the technique results in eight attributes corresponding to Character-Edge Distance from eight positions.

Our proposed feature descriptor approaches use these methods alone, combined and with Discrete Fourier Transform. The next section describes all combination of feature descriptors employed in this work.

87.2.1 New Feature Descriptor Approaches

Zoning and Character-Edge Distance methods were combined in new approaches of feature descriptors to fruit classification, as presented next:

- **Zoning:** it corresponds to count the active pixels from five zonings, as previously described. The resulting feature vectors are aggregated to a unique vector composed by 78 descriptors – zoning 3 × 1, 1 × 3, 3 × 6, 6 × 3 and 6 × 6;
- **Zoning and Fourier Transform:** this method apply Discrete Fourier Transform [11] in zoning feature vector to reduce the dimensionality. In this paper, three descriptor are obtained of each extremity from the vector computed with Fourier Transform;

Fig. 87.2 Eight positions to compute the Character-Edge Distance descriptor

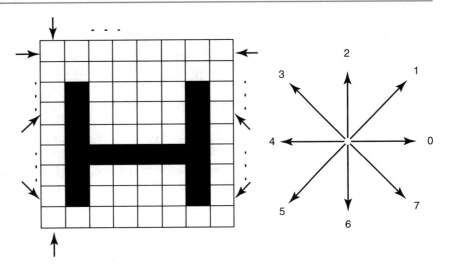

- **Character-Edge Distance with Fourier Transform:** this method apply Character-Edge Distance and aggregates the resulting distance vector to a unique feature vector. After that, the Discrete Fourier Transform is employed to result in a feature vector with less descriptor, by getting three feature from the vector extremities;
- **Character-Edge Distance with Fourier Transform to each distance vector:** in this method the Discrete Fourier Transform is applied to each one of eight distance vector computed with Character-Edge Distance. Thus, the resulting feature vector contains 48 (6 × 8) features;
- **Character-Edge Distance with Zoning:** this method divide the image in similar zones and apply the Character-Edge Distance to each zone. The resulting vector are aggregated in a unique vector;
- **Character-Edge Distance with Fourier Transform in the unique vector from each zone:** this method, as the previous one, divide the image in zones. Following, the Character-Edge Distance is employed to each zone, resulting in a unique aggregated distance vector per zone. Each resulting vector is transformed by Discrete Fourier method, from which three descriptor are gotten from the extremities;
- **Character-Edge Distance with Fourier Transform in each vector from each zone:** this method consist of applying the Fourier Transform to each distance vector computed by Character-Edge Distance in each zone that divide the image.

The next section describes the experiments performed with feature spaces computed with these previously presented feature descriptor approaches.

87.3 Experiments

Previous section described several combinations of Zoning [8] and Character-Edge Distance [9] descriptors, as well as new approaches based on Discrete Fourier Transform (DFT). This section present some experiments with feature spaces of all descriptor methods and proposed approaches. The experiments were performed with a dataset composed by 163 images divided in 15 classes: pineapple, avocado, banana, cherry, apricot, kiwi, orange, lemon, apple, mango, watermelon, melon, strawberry, pear and grape. Those images were acquired from the Internet by *Google Image* searches. All images has a white background, as shown in Fig. 87.3 [12].

All feature spaces were classified with *Multilayer Perceptron* and *k-NN* classifiers from Weka,[1] with default parameters. In the classification tasks, the dataset was divided in two parts: 66% from 163 images (108 images) as training and 34% (55 images) to test.

The Table 87.1 presents the correct classification rate of the classification task with feature spaces computed from distinct descriptor methods. The first column present the feature space description, the second and third columns present, respectively, the correct classification rates for Multi Layer Perceptron (MLP) and k-Nearest Neighbour (k-NN) classifiers. It is worthy to note that reducing the feature space dimensionality, by employing Discrete Fourier Transform, improved a lot the classification rate. Additionally, the best feature space was generated with the proposed approach of combining Zoning, Character-Edge Distance and Discrete Fourier Transform (i.e., Character-Edge Distance with Zoning 3 × 3 with Fourier Transform applied to

---
[1] Weka is a system composed by several data mining algorithms – available in http://www.cs.waikato.ac.nz/ml/weka/.

the unique vector), resulting in 97.54% of correct classification rate.

We also perform some experiments with other methods presented in the literature to compare and confirm that our approach is better than other descriptor methods employed in fruit classification. The main methods are based on color, shape and texture [1–3]. They are:

- **Color:** color methods use the pixel information to compute feature descriptors. Generally, mean, variance and deviation are computed from red, green and blue channels;
- **Shape based on Fourier Transform:** this method computed the Fourier Transform from the fruit contour

and consider only ten descriptors from each vector extremity – 20 descriptors for x and y contour positions;

- **Texture:** the main methods presented in the literature are Gabor filters [13] and Haralick features [14];
- **Fourier Transform from Histogram:** this method employ the Fourier Transform to the histogram computed from the gray scale image;
- **Fourier Transform from Image:** concentric circles are employed as masks in the entire image and the 2D Discrete Fourier Transform is employed in each. After, the power spectrum is computed in the frequency domain.

Table 87.2 presents the classification tasks performed with feature spaces computed from these methods commonly used in the literature. The best descriptor method – color with background and shape (Fourier from contour) – reaches 80% against 97.54% of our proposed descriptor approach (as shown in Table 87.1).

Next section shows a detailed inspection of the fruit dataset and the effects of feature space quality on image similarities. For that, an image visualization technique was employed to compute graphical representations which reflect the similarities from a feature space.

87.4 Detailed Inspection with Image Visualization

Image Visualization approaches based on multidimensional projection or point placement techniques aim to show the image relationship through graphical representations computed from feature spaces. Usually, dimensionality reduction methods are employed to show in the projected space the structures and similarities from the original multidimensional space (i.e., feature space) [15, 16]. Additionally, point placement techniques are also employed in this task, for instance, techniques based on trees can stablish the image similarity with a hierarchical organization of the nodes which represents each image. Thus, a node is associated with other if they have similar image content – this content

Fig. 87.3 Fruit images from the dataset used in the experiments

Table 87.1 Comparison of distinct feature spaces. The correct classification rates were computed with Multilayer Perceptron and k-NN classifiers

Feature descriptor	MLP	k-NN
Zoning 3 × 3	73.62%	63.80%
Zoning 3 × 3 and 6 × 6 with DFT	81.59%	63.80%
Zoning 3 × 3 and 6 × 6 with DFT in each zone	49.08%	47.85%
Char-Edge Dis	16.56%	28.22%
Char-Edge Dis with DFT	92.63%	63.19%
Char-Edge Dis with DFT to each distance vector	59.50%	44.17%
Char-Edge Dis with Zoning 3 × 3	22.69%	36.81%
Char-Edge Dis with DFT in unique vector to each zone 3 × 3	97.54%	70.55%
Char-Edge Dis with DFT in each vector to each zone 3 × 3	77.91%	58.28%

Table 87.2 Other descriptor methods presented in the literature. The correct classification rates were computed with Multilayer Perceptron and k-NN classifiers

Description	MLP	k-NN
Color with background	78.19%	67.28%
Color without background	72.73%	69.10%
Shape (Fourier from contour)	43.64%	43.64%
Color with background and shape (Fourier from contour)	80.00%	67.28%
Color without background and shape (Fourier from contour)	70.91%	65.46%
Texture (Gabor and Haralick)	47.85%	35.58%
Color and Texture (Gabor and Haralick)	63.80%	53.37%
Fourier from histogram and image, mean and deviation	74.55%	65.46%

similarity is based on the computed feature vectors. The resulting graphical representation based on trees presents an useful hierarchical interpretation that reveals both local and global similarity relationships [15] from the feature space. For instance, Fig. 87.4a, b present two results of the point placement technique based on tree known as Neighbor Joining (NJ) [17]. In this graphical representation, each point is assigned to an image and each color indicates a label (class). Note that similar instances (images) are placed in a same branch.

This kind of similarity based visualization technique is highly dependent of the feature space quality [15] – the image visualization only reflects the similarity computed from the feature space. Thus, if the feature space is poor the clusters of similar images on graphical representations are also poor. Figure 87.4 presents two NJ trees computed from two distinct feature spaces. The first, presented in Fig. 87.4a, was computed with the feature space generated from our best descriptor approach – Character-Edge Distance with Fourier Transform in the unique vector from each zone 3×3. The second, presented in Fig. 87.4b, was computed with the worst feature space presented in Table 87.1 – Character-Edge Distance alone. Figure 87.4c, d present, respectively, the same trees of Fig. 87.4a, b, but instead of using points the image was employed as visual marks. So, this kind of graphical representation can be used to explore the wholly dataset similarity and also to evaluate the feature space quality as well as the descriptor methods effects on image similarities. For instance, when comparing Fig. 87.4c, d, it is notable that Fig. 87.4c presents better clusters of similar images, confirming the evaluation based on classifiers presented in previous section.

87.5 Conclusions and Future Works

Computer Vision is a valuable research area composed by several techniques for image recognition, classification, segmentation, as well as applications in distinct domains.

An important step in Computer Vision is the feature descriptor computation, because the better is the feature space quality the higher is the accuracy of a Computer Vision system.

This paper presented distinct approaches of employing feature descriptor methods from handwritten character images to compute feature spaces from fruit images dataset. To compare all feature spaces, we used Multilayer Perceptron (MLP) and k-Nearest Neighbour classifiers to choose the best features based on correct classification rate. The experiments shown that our approach of combining Character-Edge Distance and Zoning with Discrete Fourier transform produced a good feature space capable of discriminating almost all fruits – the best feature space computed with our approach reaches 97.54% of correct classification rates.

The classification rate is a global measure and can hide detailed inspection on how the feature space quality affects the image similarities as well as clusters cohesion and separation. Therefore, image visualization techniques can be employed to compute graphical representations to reveal image similarities, enabling a detailed inspection of the feature space. In this paper, we show the impact of two distinct feature spaces in the image similarities – the worst feature space, computed with Character-Edge Distance descriptor method; and the best feature space, computed based on our proposed approach which combined the following descriptor methods Character-Edge Distance with Fourier Transform in the unique vector from each zone 3×3. The impact of the high quality feature space was presented the graphical representation based on tree, in which groups of similar images was placed together.

In further works, new fruit datasets will be used to evaluate the proposed feature descriptor approaches. Additionally, other image visualization approaches will also be employed to compare which one creates the best graphical representation from feature space computed from fruits.

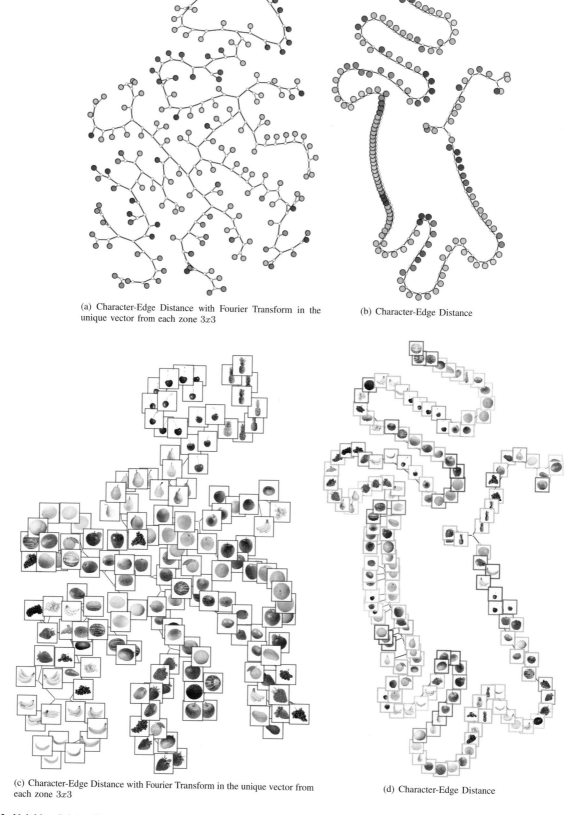

(a) Character-Edge Distance with Fourier Transform in the unique vector from each zone $3x3$

(b) Character-Edge Distance

(c) Character-Edge Distance with Fourier Transform in the unique vector from each zone $3x3$

(d) Character-Edge Distance

Fig. 87.4 Neighbor Joining Tres computed from the best (**a** and **c**) and the worst (**b** and **d**) feature space. Each fruit image were used as visual mark in (**c**) and (**d**)

Acknowledgements The authors acknowledge the financial support of the Brazilian financial agency São Paulo Research Foundation (FAPESP) – grants 2013/03452-0 and 16/11707-6.

References

1. Naskar, S., & Bhattacharya, T. (2015). A fruit recognition technique using multiple features and artificial neural network. *International Journal of Computer Applications, 116*(20), 23–28. Published by Foundation of Computer Science (FCS), NY, USA.

2. Jana, S., & Parekh, R. (2016). Intra-class recognition of fruits using color and texture features with neural classifiers. *International Journal of Computer Applications, 148*(11), 1–6.

3. dos Santos, A. F., Eler, D. M., Artero, A. O., Dias, M. A., & Pola, I. R. V. (2016). Combining feature extraction techniques for fruit recognition. In *Proceedings of the XII Workshop de Visão Computacional* (WVC).

4. Seng, W. C., & Mirisaee, S. H. (2009). A new method for fruits recognition system. In *ICEEI*, Selangor.

5. Arivazhagan, S., Shebiah, R. N., Nidhyan, S. S., & Ganesan, L. (2010). Fruit recognition using color and texture features. *Journal of Emerging Trends in Computing and Information Sciences, 1*(2), 90–94.

6. Zawbaa, H., Abbass, M., Hazman, M., & Hassenian, A. (2014). Automatic fruit image recognition system based on shape and color features. In A. Hassanien, M. Tolba, & A. Taher Azar (Eds.), *Advanced machine learning technologies and applications* (Communications in computer and information science, Vol. 488, pp. 278–290). Cham: Springer.

7. Sengupta, S., & Lee, W. S. (2014). Identification and determination of the number of immature green citrus fruit in a canopy under different ambient light conditions. *Biosystems Engineering, 117*(3), 51–61. Image Analysis in Agriculture.

8. Freitas, C. O. A., Oliveira, L. S., Aires, S. B. K., & Bortolozzi, F. (2008). Metaclasses and zoning mechanism applied to handwriting recognition. *Journal of Universal Computer Science, 14*(2), 211–223.

9. Miranda, R. A. R., Silva, F. A. D., Pazoti, M. A., Artero, A. O., & Piteri, M. A. (2013). Algoritmo para o reconhecimento de caracteres manuscritos. *Colloquium Exactarum, 5*, 109–127.

10. Macanha, P. A., Eler, D. M., & Artero, A. O. (2015). Comparison of computer techniques for handwritten character recognition. In *Proceedings of the XI Workshop de Visão Computacional* (WVC, pp. 418–422).

11. Huang, K., & Aviyente, S. (2006). Rotation invariant texture classification with ridgelet transform and fourier transform. In *2006 I.E. International Conference on Image Processing* (pp. 2141–2144).

12. Nishida, M., Eler, D. M., Artero, A. O., & Dias, M. A. (2015). Comparison of feature spaces in fruit recognition. In *Proceedings of the XI Workshop de Visão Computacional* (WVC, pp. 71–76).

13. Manjunath, B. S., & Ma, W. Y. (1996). Texture features for browsing and retrieval of image data. *IEEE Transactions on Pattern Analysis and Machine Intelligence, 18*(8), 837–842.

14. Tuceryan, M., & Jain, A. K. (1998). Texture analysis. In *The handbook of pattern recognition and computer vision* (2nd ed., pp. 207–248). World Scientific Publishing Co.

15. Eler, D. M., Nakazaki, M. Y., Paulovich, F. V., Santos, D. P., Andery, G. F., Oliveira, M. C. F., Batista Neto, J., & Minghim, R. (2009). Visual analysis of image collections. *The Visual Computer, 25*(10), 923–937.

16. Bodo, L., de Oliveira, H. C., Breve, F. A., & Eler, D. M. (2016). Performance indicators analysis in software processes using semi-supervised learning with information visualization. In *13th International Conference on Information Technology: New Generations (ITNG 2016)* (Advances in intelligent systems and computing, pp. 555–568). Las Vegas: Springer.

17. Cuadros, A. M., Paulovich, F. V., Minhgim, R., & Telles, G. P. (2007). Point placement by phylogenetic trees and its application for visual analysis of document collections. In *IEEE Symposium Visual Analytics Science and Technology 2007 (VAST 2007)*, Sacramento (pp. 99–106).

A No-Reference Quality Assessment Method of Color Image Based on Visual Characteristic

Ling Dong and Xianqiao Chen

Abstract

With the advent of the information age, the image plays an increasingly important role in people's daily work and lives. People's requirements of image quality are increasing. In order to make the assessment result of image quality more in line with person's subjective feeling, a improved no-reference quality assessment method of color image was presented. Firstly, divide the image into several blocks and calculate contrast of each block under the Lab color model, then take each block's effect on the whole visual perception with masking effect into consideration, optimize the whole assessment result of image quality , to achieve the objective and subjective consistency. Using the images of several image databases and the given subjective scores of the images to do the experiment. The experimental results show that the presented method has a good consistency with subjective feelings of image quality, and it's assessment result is more accurately compare to other methods.

Keywords

Color image quality assessment • Visual characteristic • No-reference • Lab color model • Expand region

88.1 Introduction

In the process of image's acquisition, transmission, storage and display, there may be different types and different degrees of distortion, so the content of the image presented to human have differences to the native content [13, 14].

The objective assessment method of image quality can be divided into three kinds: the full-reference, the reduced-reference and the no-reference. Yang Chunling et al. [1] proposed a DWTSSIM method, and the structural similarity was applied to each frequency band after wavelet decomposition. Li Q et al. [2] proposed a DNT algorithm based on the Wavelet-Domain Natural Image Statistic Model, which replaces the GGD Model with a GSM Model, simplifying the process parameters. Lou Bin [3] proposed a no-reference assessment method based on the characteristic that image distortion would change the linear distribution of image energy spectrum to evaluate the image quality.

In practice, it's difficult to obtain the original reference image for many scenes. So the full-reference and reduced-reference assessment methods have lost practical significance. The evaluation accuracy of existed no-reference assessment method still needs to be improved because the knowledge of image characteristics, visual perception system is limited. In this paper, an improved no-reference quality assessment method of color image was proposed to improve the evaluation accuracy.

L. Dong (✉) • X. Chen
College of Computer Science and Technology, Wuhan University
of Technology, Wuhan, Hubei 430063, China
e-mail: 1059675029@qq.com; chenxq1121@qq.com

© Springer International Publishing AG 2018
S. Latifi (ed.), *Information Technology – New Generations*, Advances in Intelligent
Systems and Computing 558, DOI 10.1007/978-3-319-54978-1_88

88.2   Quality Assessment Model of Color Image

In order to make the evaluation result close to the subjective feeling, the color image can be processed in CIE LAB color space. In this paper, we propose a new method which is based on the study of Sen D et al. [6] to evaluate the image quality. We add the visual masking effect model to make the color image quality assessment more in line with people's subjective feeling and improve the accuracy of evaluation result.

(1) Contrast Calculation of Single Channel Gray Signal

The local contrast of gray scale image is described in local band contrast ratio: the ratio of the change of local gray value to local gray mean value. The mathematical form is as follows:

$$C(x, y) = \frac{\beta(x, y)}{\lambda(x, y)} \tag{88.1}$$

$$\beta(x, y) = g(x, y)^* b(x, y) \tag{88.2}$$

$$\lambda(x, y) = g(x, y)^* l(x, y) \tag{88.3}$$

(x,y) is the spatial coordinate of pixel in an image, $g(x,y)$ is the gray-scale representation of image, $b(x,y)$ is a band-pass filter, $l(x,y)$ is a low-pass filter that allows all frequencies below $b(x,y)$ to pass, and * represents the convolution operation.

There are two kinds of receptive fields: one is the center-on peripheral-off type; the other is the center-off peripheral-on type.

The Difference of Gaussian Model can be a good simulation of gray scale image of the center-peripheral receptive field [7], the difference of two Gaussian filters is used to simulate the effects of central-peripheral receptive field, the mathematical expression is as follows:

$$O(x, y) = Ct(x, y) - Sr(x, y) = g(x, y)^* (G_1 - G_2) \tag{88.4}$$

$$Ct(x, y) = g(x, y)^* G_1(x, y) \tag{88.5}$$

$$Sr(x, y) = g(x, y)^* G_2(x, y) \tag{88.6}$$

$O(x,y)$ is the output of center-peripheral receptive field model, $C_t(x,y)$ is the center field, $S_r(x,y)$ is the peripheral field, $G_1(x,y)$ and $G_2(x,y)$ are two Gaussian low-pass filters, the mathematical form is $\exp[-(x^2 + y^2)/(2\sigma^2)]$, the standard deviation of G_2 is greater than the standard deviation of G_1,

so G_1-G_2 is a band-pass filter. In the study of this paper, $\sigma_{G_2} = M \times \sigma_{G_1}, M = 3$.

Therefore, we can define the contrast of gray image by combining local band-limited contrast and center-periphery receptive field:

$$C(x, y) = \frac{\beta(x, y)}{\lambda(x, y)} = \frac{O(x, y)}{Sr(x, y)} \tag{88.7}$$

(2) Contrast Calculation of Single Channel color Signal

In LAB color space, the center-periphery receptive field of color image is composed of red-green and yellow-blue components, and may be one of the followings: center-red periphery-green, center-green peripheral-red, center-yellow periphery-blue, center-blue periphery-yellow.

The retinal center-peripheral receptive field model of color signal(a and b) [8] can be simulated with a low-pass filter $G_{1+}G_2$, and can set the same parameters as gray scale signal. Therefore:

$$O(x, y) = Ct(x, y) + Sr(x, y) = f(x, y)^* (G_1 + G_2) \tag{88.8}$$

$O(x,y)$ is the output of center-peripheral receptive field model, $C_t(x,y)$ is the center field, $S_r(x,y)$ is the peripheral field, $G_1(x,y)$ and $G_2(x,y)$ are two Gaussian low-pass filters, the mathematical form is $\exp[-(x^2 + y^2)/(2\sigma^2)]$, the standard deviation of G_2 is greater than the standard deviation of G_1, so $G_{1+}G_2$ is a low-pass filter. $f(x,y)$ is one of the a, b components of color signal.

With the combination of local band-limited contrast and retinal center-periphery receptive field model, the definition of color signal (a and b) contrast is as follow:

$$C(x, y) = \frac{O(x, y) - \bar{f}(x, y)}{Sr(x, y)} \tag{88.9}$$

$O(x, y) - \bar{f}(x, y)$ simulate a band-pass filter and its output is basically the same as low-pass filter (except for the zero-frequency point). When calculating the contrast of a component, $\bar{f}(x, y)$ is the average of the a component. When calculating the contrast of b component, $\bar{f}(x, y)$ is the average of the b component.

(3) Multi-channel Contrast Fusion

Considering the multi-channel feature of vision system, many low-pass filters, many band-pass filters (G_1-G_2) and many low-pass filters ($G_{1+}G_2$) can be obtained by setting many σ values for $G_1(x,y)$, σ_{G_1} can be set as follows:

$$\sigma_{G_1} = \frac{2\log(1/M^2)}{v^2(1-M^2)}, \quad v = \frac{72\pi}{80}, \quad \frac{69\pi}{80}, \quad ..., \quad \frac{6\pi}{80}, \quad \frac{3\pi}{80}$$

Then the 24 contrast values are combined using sub-threshold and supra-threshold perceived contrast, which can be modeled by Lp-norm.

The mathematical expression of $C(x,y)$ with Lp-norm under the gray signal is as follow:

$$C_P(x,y) = \left(\sum_{\sigma_{G_1}} \left(P^{-1} \times \left| C_{\sigma_{G_1}}(x,y) \right| \right)^P \right)^{\frac{1}{P}} \quad (88.10)$$

Where,

$$P = \exp\left(-\frac{\log(1/M^2)}{1-M^2} \right) - \exp\left(-\frac{\log(1/M^2)}{1-M^2} \times M^2 \right), \quad \text{and}$$

the normalized operation of p ensures that the peak amplitude of each underlying channel is one.

The mathematical expression of $C(x,y)$ with Lp-norm under color signals is as follow:

$$C_P(x,y) = \left(\sum_{\sigma_{G_1}} \left(0.5 \times \left| C_{\sigma_{G_1}}(x,y) \right| \right)^P \right)^{\frac{1}{P}} \quad (88.11)$$

Studies have shown that sub-threshold perceived contrast can be simulated by $P = 1$, and supra-threshold perceived contrast can be simulated by $P = \infty$.

(4) Contrast Calculation of Color Image

The local contrast of color image is defined as:

$$C(x,y) = C_L(x,y) + C_a(x,y) + C_b(x,y) \quad (88.12)$$

Where $C_L(x,y)$ is the contrast of L component and is calculated under gray signal:

$$C_L(x,y) = C_1(x,y)^{0.5} \times C_\infty(x,y)^{0.5} \quad (88.13)$$

$C_a(x,y)$ and $C_b(x,y)$ represent contrasts of a and b component respectively, calculated under color signal:

$$C_{a,b}(x,y) = C_1(x,y)^{0.5} \times C_\infty(x,y)^{0.5} \quad (88.14)$$

(5) Contrast Optimization under Visual Masking Effect

This paper adds visual masking effect to the local contrast of color image to explore the effect of overall image structure on human perception. Distortions of different image regions have different visual characteristics. The distortion of texture region is masked by its own texture, the accurate measurement of the distortion is more complex due to changeable structure. Due to the effect of visual attention mechanism, the smooth region is often less attractive and its structure is simple, so the distortion of it has a little impact on visual experience of the whole image. The expand region (near the texture region) is less affected by its own contrast and structure, and the smooth region near it has a prominent effect on it, therefore, the masking effect of smooth region on expand region is considered in this paper. The following mathematical model is used to simulate the visual masking effect:

$$C(i,j) = \begin{cases} \dfrac{C(i,j)}{1 - (C(i,j) - c)}, & (i,j) \in ER \\ C(i,j), & \text{else} \end{cases} \quad (88.15)$$

(i,j) represents the position of a local block in the global image after segmentation, ER represents the expand region, and c is the contrast threshold that makes smooth region has visual masking effect on expand region.

Therefore, in this study, the general idea of the image quality assessment is as follows:

1) Dividing image into $m \times n$ subblocks;
2) formula (88.8), (88.9), (88.11) and (88.14) are used to obtain the contrast value $C_a(x,y), C_b(x,y)$ of the a and b components at each point(x,y) in the color signal; formula (88.4), (88.7), (88.10) and (88.13) are used to obtain the contrast value $C_L(x,y)$ of the L component at each point(x,y) in the gray signal;
3) Using formula (88.12) to obtain the contrast value $C(x,y)$ at each point(x,y);
4) The average value of $C(x,y)$ in block (i,j) is taken as the contrast value $C(i,j)$of the block,where $i = 1,2,..., m$, $j = 1,2,..., n$;
5) The normalization operation is performed for the contrast value of the respective subblocks:

$$C(i,j) = \frac{C(i,j) - \min}{\max - \min}$$

Where min is the minimum value of all $C(i, j)$, max is the maximum value of all $C(i, j)$;

6) For each block, the normalized contrast value is masked by formula (88.15) to obtain the contrast value based on the image structure features;
7) Calculating $\dfrac{1}{m*n} \sum_{i,j} C(i,j)$ as the contrast value of the whole image, that is, the quality evaluation result of the image.

88.3 Experimental results and analysis

In order to verify the validity of the no-reference image quality assessment method proposed in this paper, we use images and DMOS values used as the standard of subjective quality perception from the LIVE2 [9] image database. The larger the DMOS value, the worse the image quality. This experiment is carried out under Win7, using VS2015, OpenCV3.0, MATLAB R2012b and other software tools.

88.3.1 Quantitative Analysis of Expand Region

To facilitate the calculation of Eq. (88.15), it's necessary to numerically denote the expand region ER and the contrast threshold c of image. Firstly, 100 images with different structures are selected from LIVE2. Using Contourlet transform to extract the contour texture information of these images. The decomposition parameters are set as pfilt = '9–7', dfilt = 'pkva', nlevs= [2,3,4]. And then do morphological dilation operation of texture information, to obtain the expand region information, and the structural element of the dilation operation is set as [0 0 1 0 0;0 1 1 1 0; 1 1 1 1 1;0 1 1 1 0;0 0 1 0 0]. Figure 88.1 is the processed results for lighthouse.bmp ,and Fig. 88.2 is for monarch.bmp.

According to step 1) to step 5) in the general idea, calculate the contrast of expand region block before optimization. The experimental results show that the non-optimized contrast value of expand region block is basically distributed in the range of (0.13, 0.37), so it can be generally believed that the contrast of smooth region is less than 0.13, so the contrast threshold $c = 0.13$. At the same time, in order to avoid the occurrence of the situation in which the contrast of texture region block is also within the range of (0.13, 0.37) because the image is too blurred, it is required that at least one of the four adjacent blocks' contrast of expand region block is less than 0.13, that is to say, there's smooth region block existed in the 4 adjacent blocks. Therefore, the quantized definition of expand region block is that if at least one of the contrast values of the 4 adjacent blocks of one block is smaller than the contrast threshold, and the contrast value of the block is meet the condition $0.13 < C < 0.37$, then the block is an expand region block.

88.3.2 Contrast Optimization Analysis Under Visual Masking Effect

The experiments in this section randomly selected 200 images from the LIVE2. The experimental examples are shown in Figs. 88.3 and 88.4. Table 88.1 shows the experimental results, C_1 means the contrast without visual

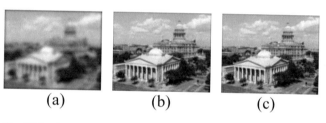

(a)   (b)   (c)

Fig. 88.3 churchandcapitol.bmp (**a**) Distorted version 1, (**b**) Distorted version 2, (**c**) Distorted version 3

(a)   (b)   (c)

Fig. 88.4 carnivaldolls.bmp (**a**) Distorted version 1, (**b**) Distorted version 2, (c) Distorted version 3

(a)   (b)   (c)

Fig. 88.1 lighthouse.bmp (**a**) Original image (**b**) texture information (**c**) expand information

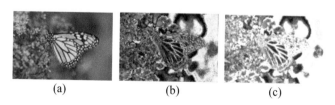

(a)   (b)   (c)

Fig. 88.2 monarch.bmp (**a**) Original image (**b**) texture information (**c**) expand information

Table 88.1 The experimental results of the example image

		DMOS	C_1	C_2
Churchan-dcapitol	(a)	82.2887	0.2365	0.2401
	(b)	46.4481	0.4526	0.5123
	(c)	30.9535	0.5273	0.6189
Carnivald-olls	(a)	63.2726	0.2951	0.3618
	(b)	41.6514	0.5086	0.5498
	(c)	19.7131	0.8084	0.8205

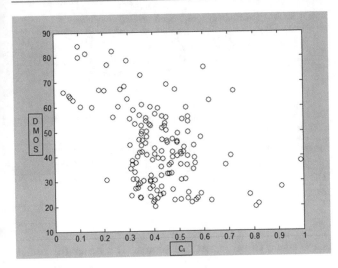

Fig. 88.5 C_1 and DMOS

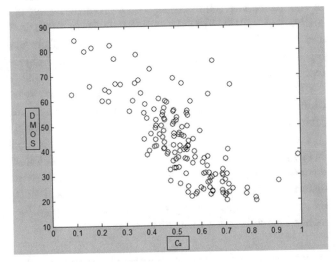

Fig. 88.6 C_2 and DMOS

Table 88.2 Performance comparison before optimization and after optimization

	CC	SROCC	RMSE
before	0.7926	0.7869	12.3758
after	0.8158	0.8194	10.0427

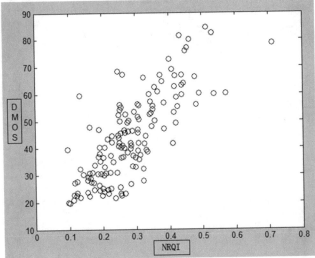

Fig. 88.7 NRQI and DMOS

88.3.3 Comparative Analysis of Related Methods

In order to verify the effectiveness of the method proposed in this paper, some relative methods was tested, and the experiment was carried out using the 200 images selected in Sect. 88.3.2. Experimental results in Ref. [4] use the proposed NRQI as the evaluation results, the better the image quality, the smaller the NRQI value. The results of the general no-reference assessment method in Ref [5] use the proposed Q as the evaluation results, the better the image quality, the smaller the Q value. Ref [10] use NPSNR as the evaluation results in experiments, the better the image quality, the bigger the NPSNR value. The scatter plots of NRQI and DMOS, Q and DMOS, NPSNR and DMOS are shown in Figs. 88.7, 88.8, and 88.9, respectively. Table 88.3 is the performance analysis of the methods in Ref [4], Ref [5], Ref [10].

Combining the data in Tables 88.2 and 88.3, it can be seen that for the various types of distortion in LIVE2, C_2, Q and NRQI have good subjective and objective consistency, and C_2 is better than Q and NRQI.

masking effect, C_2 means the improved contrast. Comparing the experimental results in Figs. 88.3, 88.4 and Table 88.1, we can see that the clearer the image, the smaller the DMOS and the larger the value of C_1 and C_2.

The scatter plots of C_1 and DMOS, C_2 and DMOS are shown in Figs. 88.5 and 88.6. As can be seen from them, C_1 is more dispersed than C_2, C_2 is closer to the human observation than C_1.

Table 88.2 shows the results of performance analysis before optimization and after, where, CC is the Pearson correlation coefficient, SROCC is the Spearman rank order correlation coefficient, RMSE is the root mean square error, the closer the value of CC and SROCC is to "1" and the smaller the RMSE value, the better the performance of the assessment method.

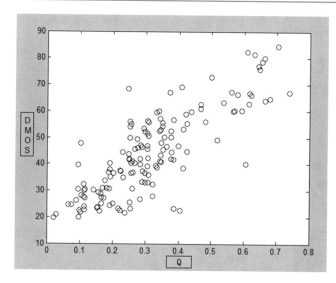

Fig. 88.8 Q and DMOS

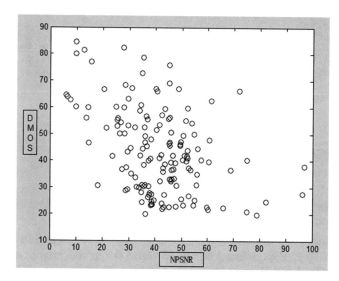

Fig. 88.9 NPSNR and DMOS

Table 88.3 Performance of related methods

	CC	SROCC	RMSE
Ref. [4](NRQI)	0.8103	0.8169	11.0367
Ref. [5](Q)	0.8052	0.8043	11.6853
Ref. [10](NPSNR)	0.6312	0.6375	31.5376

88.3.4 Performance Analysis on Other Image Databases

In order to further evaluate the expansibility of the method in this paper, experiments were carried out on IVC [11] image database and TID2008 [12] image database.

In this section, 185 color distortion images and corresponding MOS values from IVC are used. MOS value

Table 88.4 Performance comparison on IVC database

	This paper	Ref. [4]	Ref. [5]	Ref. [10]
CC	0.8384	0.8204	0.8143	0.6264
SROCC	0.8307	0.8175	0.8089	0.6186

Table 88.5 Performance comparison of "Local block-wise distortions of different intensity"

	This paper	Ref. [4] (NRQI)	Ref. [5] (Q)	Ref. [10] (NPSNR)
CC	0.7483	0.6534	0.7648	0.7213
SROCC	0.7401	0.6361	0.7556	0.7088

is the mean of all subjective scores. The better the image quality, the bigger the MOS value. The experimental results are shown in Table 88.4.

It can be seen from Table 88.4 that the proposed method in this paper has better assessment performance than the methods in Ref [4], Ref [5] and Ref [10]. From Table 88.3 and Table 88.4, it can be found that for common distortion types, the proposed method has good subjective and objective consistency and can represent the latest research results.

100 images with the distortion type of "Local block-wise distortions of different intensity" from TID2008 are selected. The experimental results are shown in Table 88.5.

It can be seen from Table 88.5 that the proposed method in this paper has no absolute advantage in the evaluation of distortion images with distortion type of "Local block-wise distortions of different intensity", and the lack of local block information of high intensity affects the evaluation performance. This will be the focus of future research.

88.4 Conclusion

In this paper, we consider the effect of local perception on the whole visual perception and add the visual masking effect model on the basis of the local band-limited contrast and retinal center-periphery receptive field, to improve the calculation of the contrast, make the color image quality assessment more in line with subjective feelings. The experimental results showed the validity of the no-reference image quality assessment method proposed in this paper, and the method can be used as the representative research results. The next step is to study the effect of distortion type of "Local block-wise distortions of different intensity" on the threshold value in the quantitative analysis of expand region. At the same time, we will study more distortion types and establish a adaptive threshold model to get better consistency of subjective and objective assessment for each distortion type.

Acknowledgements This work was financially supported by National Science and Technology Supporting Project, China (No.2015BAG20B02,No.2015BAG20B00)

References

1. Yang, C., & Gao, W. (2009). Research on image quality assessment in wavelet domain based on structural similarity[J]. *Acta Electronic Sinica, 37*(4), 845–849.
2. Li, Q., & Wang, Z. (2009). Reduced-reference image quality assessment using divisive normalization-based image representation[J]. *IEEE Journal of Selected Topics in Signal Processing, 3*(2), 202–211.
3. Bin, L. (2009). *Research on image quality assessment based on NSS and HVS[D]*. China: Zhejiang University.
4. Li, C., Ju, Y., Bovik, A. C., et al. (2013). No-training, no-reference image quality index using perceptual features[J]. *Optical Engineering, 52*(5), 532–543.
5. Qingbing, S. (2013). *Research on novel methods of reduced reference and no reference image quality assessment[D]*. China: Jiangnan University.
6. Sen, D., & Pal, S K. (2011). *Retinal visual system based contrast measurement in images*[C]// International Conference on Communications and Signal Processing. IEEE, 51–55.
7. Rodieck, R. W., & Stone, J. (1965). Analysis of receptive fields of cat retinal ganglion cells[J]. *Journal of Neurophysiology, 28*(5), 832–849.
8. Rovamo, J. M., Kankaanpää, M. I., & Kukkonen, H. (1999). Modeling spatial contrast sensitivity functions for chromatic luminance-modulated gratings[J]. *Vision Research, 39*(14), 2387–2398.
9. Sheikh, H. R., Wang, Z., Cormack, L., & Bovik, A. C. *LIVE Image Quality Assessment Database Release 2*, http://live.ece.utexas.edu/research/quality.
10. Wang, Z., & Wen, X. (2006). No-reference digital image quality evaluation based on perceptual masking[J]. *Computer Applications, 26*(12), 2838–2840.
11. Ninassi, A., Le Callet, P., & Autrusseau, F. *Subjective Quality Assessment-IVC Database 2005 [DB/OL]*. http://www2.irccyn.ec-nantes.fr/ivcdb
12. Ponomarenko, N., Lukin, V., Zelensky, A., Egiazarian, K., Carli, M., & Battisti, F. (2009). TID2008 – A database for evaluation of full-reference visual quality assessment metrics. *Advances of Modern Radio electronics, 10*, 30–45.
13. Yoneyama, A., & Minamoto, T. (2015). *No-reference image blur assessment in the DWT domain and blurred image classification* [C]//*international conference on information technology – New generations* (pp. 329–334). IEEE.
14. Omura, H., & Minamoto, T. (2015). *Image quality assessment for measuring the degradation by using the dual-tree complex discrete wavelet transform*[C]// International Conference on Information Technology – New Generations. IEEE, 323–328.

Breno Santana Santos, Methanias Colaço Júnior, and Maria Augusta S.N. Nunes

Abstract

Empathy plays an important role in social interactions, such an effective teaching-learning process in a teacher-student relationship, and company-client or employee-customer relationship to retain potential clients and provide them with greater satisfaction. Increasingly, people are using technology to support their interactions, especially when the interlocutors are geographically distant from one another. This has a negative impact on the empathic capacity of individuals. In the Computer Science, there are different approaches, techniques and mechanisms to promote empathy in social or human-computer interactions. Therefore, this article presents a systematic mapping to identify and systematize the approaches, techniques and mechanisms used in computing to promote empathy. As a result, we have identified existing approaches (e.g. collaborative learning environment, virtual and robotics agents, and collaborative/affective games) to promote empathy, the main areas involved (e.g. human-computer interaction, artificial intelligence, robotics, and collaborative systems), the top researchers and their affiliations who are potential contributors to future research and, finally, the growth status of this line of research.

Keywords

Empathy • Rapport • HCI • Systematic mapping study • Secondary study

89.1 Introduction

Empathy can be understood as affective or intellectual reciprocity, that is, a fundamental mutual understanding for the creation of affective ties, be it friendship or love [1]. It can also be considered an emotional response to another person by sharing their affective state as well as being essential for social interactions [2].

For the relationships to be more effective, the presence of empathy in social interactions is paramount, such as the teacher-student relationship for an effective teaching-learning process [3], a doctor-patient relationship for a more humane and consistent treatment [4, 5], and company-client or employee-customer relationship to retain potential clients and provide them with greater satisfaction [6, 7].

Increasingly people tend to interact through technological resources, being a negative factor for the development of their empathic abilities, and there is still little research that seeks to promote empathy in social interactions [8]. There are several approaches in Computer Science to promote empathy in social or human-computer interactions, such as the identification of emotional states [9] and empathic virtual agents [6]. Also, among the various areas of Computer Science, the most outstanding for this line of research are Human-Computer Interaction (HCI), Artificial Intelligence, Robotics, Virtual Reality and Collaborative Systems.

Finally, this systematic mapping aims to identify and systematize the approaches, techniques or mechanisms

B.S. Santos (✉) • M.C. Júnior • M.A.S.N. Nunes
Postgraduate Program in Computer Science (PROCC), Federal University of Sergipe (UFS), São Cristóvão, Sergipe, Brazil
e-mail: breno1005@hotmail.com; mjrse@hotmail.com; gutanunes@gmail.com

© Springer International Publishing AG 2018
S. Latifi (ed.), *Information Technology – New Generations*, Advances in Intelligent
Systems and Computing 558, DOI 10.1007/978-3-319-54978-1_89

used to generate empathy in social or human-computer interactions. For this, the articles of important databases of the Computing area were mapped.

This article is organized as follows: in Sect. 89.2, the method adopted in this mapping is presented; In Sect. 89.3, the results analysis is described; In Sect. 89.4, the threats of validity for this study is discussed; Finally, in Sect. 89.5, the conclusion is presented.

89.2 Method

In order to study and map the state of the art about approaches, techniques, concepts or mechanisms used to promote empathy in social or human-computer interactions, the Systematic Literature Mapping (SLM) method was adopted for this article. According to [10] and [11], it consists of a systematic protocol for searching and selecting relevant studies with the aim of extracting data and mapping results to a specific research problem.

Thus, in order to follow the method of systematic mapping, in this section, it is described how the search and selection process of the primary studies was performed. For this, it was necessary to define the research questions, the search and selection strategy and the criteria for selection.

Finally, the process of searching and selecting primary studies is described in detail below.

89.2.1 Research Questions

In order to reach the objective proposed by this mapping, the following research questions were elaborated:

RQ1: What are the used approaches, techniques, or mechanims in Computer Science to promote empathy?
RQ2: What are the countries that have more researchers who published on this field?
RQ3: What is the most common journal/proceeding where the researchers publish about this area?
RQ4: Who are the researchers that have been publishing in this field?
RQ5: What is the most representative researches affiliation in this field?
RQ6: What are the years of publications that researchers have been publishing most part of papers on this field?

89.2.2 Search and Selection Strategy

For the execution of the search, the following databases of Computer Science were selected: Scopus and Brazilian Digital Library of Computation (BDBComp). For the use without

restrictions of download in both bases, it was used the portal of journals of CAPES (https://www.periodicos.capes.gov.br).

The choice of the Scopus base was based on the fact that it included articles from several databases, such as IEEE, Elsevier, ACM and others [12]. The choice of BDBComp was due to the need to search for national papers.

In the Scopus database, the filtering tools (advanced search) were used to search the abstract, research area and language, thus reducing the number of articles outside the scope of this search line. However, in the BDBComp database, a simple search was performed because there was no advanced search engine available on that base.

The strings used in databases to search for primary studies were:

- **Scopus:** "ABS((empath* OR rapport OR sympath*) AND (approach OR techn* OR concept OR mechanism OR engine)) AND (LIMIT-TO(SUBJAREA,"COMP")) AND (LIMIT-TO(LANGUAGE,"English"))".
- **BDBComp:** Due to the lack of an advanced search engine, searches were performed for paper titles that had the keywords "empatia", "empátic" ("empática" and "empático"), "empath" ("empathy" and "empathic"), "rapport", "simpatia", "simpátic" ("simpática" and "simpático"), and "sympath" ("sympathy" and "sympathic").

Note that only English keywords were used in the search process for primary studies for the Scopus database, whereas for the BDBComp database, keywords were used in English and Portuguese.

The searches with the strings were carried out in May of 2016 using the defined terms. In the Scopus database, 1014 articles were found (in English), while in the database BDBComp database, only one article was found (in Portuguese), totaling 1015 articles found.

Note that no article was found that had the Portuguese terms in the BDBComp database. However, the only article in Portuguese was found using the word "empath", which corresponded to the word "empathy" contained in the title of the work.

With the completion of the search, the filtering process of the articles found was started, based on the selection criteria.

89.2.3 Selection Criteria

In order to filter relevant articles for the purpose of this systematic mapping, the inclusion and exclusion criteria of these articles were defined. The study included the following inclusion criteria:

- Articles that specify the approaches, techniques or mechanisms used to promote empathy in social or

Table 89.1 Results of the searches in the databases and results of the application of the selection criteria

Databases	Results of the searches	Application of the selection criteria
Scopus	1014	45
BDBComp	1	1

human-computer interactions have been included, since they are fundamental to answer our research questions;

- The articles whose year of publication was from 2011 onwards were included;
- Only papers of journals or proceeding/conference were included.

The confirmation of the inclusion criteria was given by analyzing the abstract and introduction of each article.

In parallel, the articles were analyzed in relation to exclusion criteria. The exclusion criteria described below were also applied to them:

- Duplicate articles have been excluded;
- Articles whose publication year was less than 2011 were excluded;
- Articles whose sources were not journals or proceeding/ conferences were excluded;
- Articles that did not demonstrate the approaches, techniques or mechanisms used to promote empathy in social or human-computer interactions were excluded.

After applying the inclusion and exclusion criteria of the found articles, they were evaluated. From 1015 found articles, 46 were selected to compose the primary studies. It is noteworthy that there were two duplicate articles.

In Table 89.1, the search results in the databases are presented using the strings, and application of the selection criteria. Note that, from 1015 found articles in the search process, after reading and applying the selection criteria, there was a high reduction rate in the total number of articles, since only 46 of them were qualified by the selection criteria.

After the selection, the primary studies were referred for reading and analysis. The results can be found in next section.

89.3 Results and Discussion

In this section, we present the results of the analysis of the primary studies, thus responding to the research questions presented in previous section of this mapping.

The answer for research question 01 is presented in Fig. 89.1. We present a characterization of the approaches,

techniques, and mechanisms identified to promote empathy. The most representative approaches among the primary studies were "Empathic virtual agent" (26.09%), "Empathic robotic agent/device" (23.91%), "Collaborative/empathic learning environment" (13.04%), and "Collaborative/affective game" (13.04%). In addition to the characterization, we have identified the different approaches presented in the primary studies, such as the identification of emotional states [9, 13, 14], mimicry/mirroring of verbal and nonverbal behavior [15, 16], virtual and robotics agents [16, 17, 18, 19] and text mining techniques [9, 15, 20].

The answer for research question 02 is presented in Fig. 89.2. We consider only the top 10 countries that had more publications for this line of research. These countries (number of papers in parentheses) were: United Kingdom (13); Germany (07); United States (05); Italy (05); Sweden (05); Japan (04); Portugal (04); Spain (03); Netherlands (03) and Brazil (02).

The answer for research question 03 is presented in Fig. 89.3, which shows the 10 most common journals/ proceedings where the computer scientists have been publishing their papers about approaches to promote empathy. The most representative journals/proceedings were "Autonomous Agents And Multi Agent Systems", "International Journal Of Social Robotics", "Journal On Multimodal User Interfaces", and "Proceedings Of The ACM Conference On Computer Supported Cooperative Work CSCW", which they have 02 papers each. With just one paper, the publication vehicles were: "Computer Animation And Virtual Worlds", "Computers And Education", "Computers In Human Behavior", "Conference On Human Factors In Computing Systems Proceedings", "Cyberpsychology Behavior And Social Networking", and "IEEE Transactions On Affective Computing".

The answer for research question 04 is presented in Fig. 89.4. We present the top 10 authors who appeared in 02 papers or more, and they (number of papers in parentheses) were: Aylett, R. (04); Castellano, G. (04); Barendregt, W. (03); De Rossi, D. (03); Mazzei, D. (03); Paiva, A. (03); Cavazza, M. (02); Hastie, H. (02); Kappas, A. (02) and Lazzeri, N. (02). The authors usually participate also as co-authors of published papers, then the most ranked could be, hypothetically, chief of some lab or research group interested in this research field.

The answer for research question 05 is presented in Fig. 89.5. We present the top 10 universities that published at least 01 paper. They (number of papers in parentheses) were: Heriot-Watt University (06); University of Birmingham (04); Instituto Superior Tecnico (03); Goteborgs Universitet (03); Universita di Pisa (03); University of Glasgow (02); The Royal Institute of Technology KTH (02); Jacobs University Bremen (02); University of Teesside (02) and Latha Mathavan Engineering College (01).

Fig. 89.1 Characterization of
the approaches, techniques, and
mechanisms

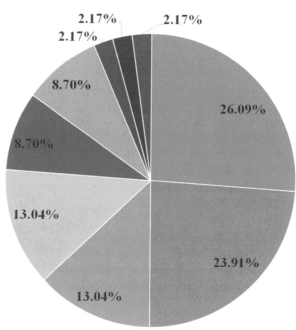

Characterization

- Empathic virtual agent
- Empathic robotic agent/device
- Collaborative/empathic learning environment
- Collaborative/affective game
- Architecture/Component/Framework for analysis of affective characteristics
- Conversation system with analysis of affective characteristics
- Psychometric analysis tool/system
- Empathic SNS
- Web Usage Mining

Fig. 89.2 The most
representative countries in
this area

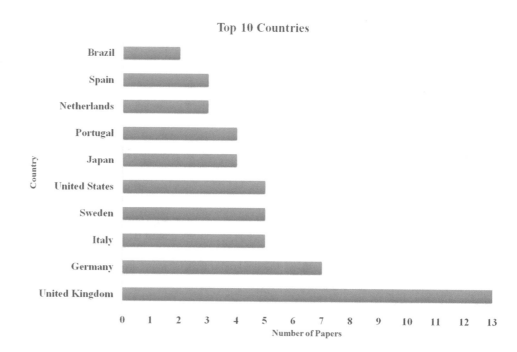

Fig. 89.3 The most representative journals/proceedings in this research field

Fig. 89.4 The most representative authors

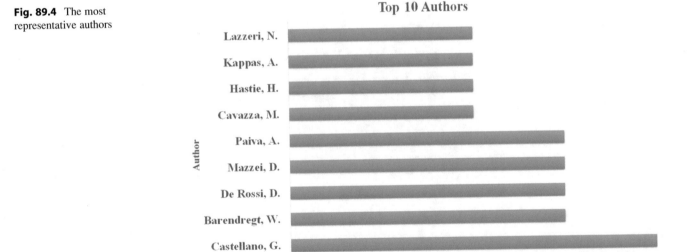

Finally, the answer for research question 06 is presented in Fig. 89.6. The year with the highest number of publications was 2014, with 12 published articles. Still, according to Fig. 89.6, 09 articles were published in 2013, 06 in 2012 and 05 in 2011. There is a noticeable decrease in the number of publications in 2015, with only 9 articles.

As previously mentioned, we only consider the publications that occurred from 2011 in order to verify what is being researched the most recent in the state of the art. We perceived that until May 2016, 05 articles were published, however, these were not counted in Fig. 89.6.

Fig. 89.5 The most representative affiliations in the research field

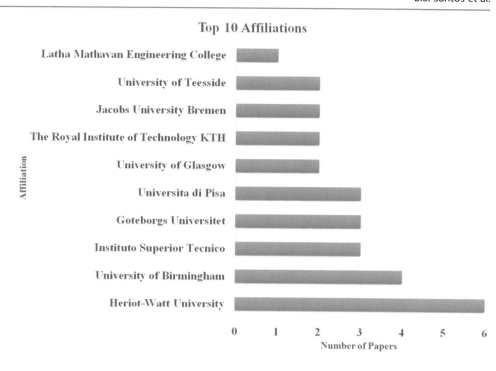

Fig. 89.6 Number of published papers per year in the research field

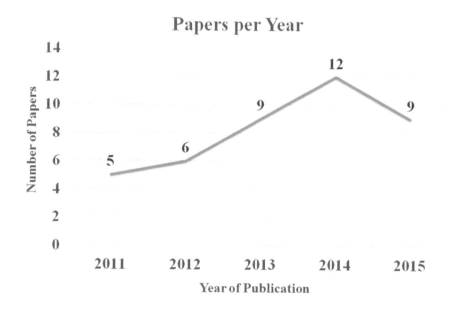

89.4   Threats to Validity

The threats to validity for the present study were:

- **Selection Bias:** At the start of the process, we applied the inclusion and exclusion criteria based on our judgment and the studies were included or excluded into mapping study. This means that some studies could have been categorized incorrectly. With the intention of mitigate this threat, we discussed the study protocol among the researchers to guarantee a common understanding. After

that, two researchers analyzed each paper and third researcher was involved in discussion when needed.

- **Construct validity:** The research questions and strings defined in this study may not completely cover those Computer Science area related to approaches, techniques or mechanisms to promote empathy. In order to reduce this bias, we performed a search for some papers related to empathy in some databases (IEEE, ACM, ScienceDirect, Google Scholar, and PubMed) to extract relevant terms that were associated with empathy. After that, we discussed about selected terms and, finally, defined the research questions and strings.

- **Data extraction:** Bias or problem of extraction can affect the characterization of approaches and the analysis of selected studies. In order to reduce this bias, we discussed deeply the categories and our extraction form. The data extraction was carried out jointly by two researchers. If the researchers disagree about a classification or extracted information, we used a third opinion to resolve differences and reach agreement to make sure that the extracted data are valid and clear for further analysis.
- **External validity:** We performed a systematic mapping study over studies published from 2011 for Scopus and BDBComp databases. This point implies that we may have missed some relevant studies contained in these databases and in others ones no used. Thus we cannot generalize our conclusions for all approaches to promote empathy in the Computer Science area. However, our outcomes allow us to draw insights to guide further investigations.

89.5   Conclusion

In this paper, we performed a systematic mapping that aimed to identify and analyze the approaches, techniques or mechanisms used in Computer Science to promote empathy in social or human-computer interactions.

The systematic mapping process was conducted using a search protocol, and selection of studies that specified the method used in this work. From the specified method, we extracted and analyzed data from 46 papers, which have identified some trends regarding this line of research.

Several approaches have been defined to promote empathy. However, the most relevant ones were virtual agents, robotic agents/devices, collaborative/empathic learning environments and collaborative/affective games (RQ1), which are based on several strategies, for example, identification of emotional states [9, 13, 14] and imitation of verbal and nonverbal behaviors [15, 16]. It was also possible to identify which countries have been dedicated to this line of research, especially the United Kingdom and Germany (RQ2).

There are several journals and proceedings in Computer Science where papers related to the generation of empathy can be published, however, the most common publication vehicles are those related to the areas of HCI, Artificial Agents, Robotics and Collaborative Systems (RQ3). We also identify the authors and universities that most contribute to this line of research (RQ4 and RQ5), thus facilitating the identification of potential collaborators for future research in this area.

Finally, the growth of this area is remarkable, where the year 2014 is the most significant in terms of publications, and in 2015 there has been a decrease (not very representative) in the number of publications (RQ6).

Despite the apparent need to analyze more research work in the area in question, our results are consistent with what exists from current research on empathy generation.

As future work, we will investigate the gaps identified in this mapping study. We believe that this research presents relevant results to the academy, providing support of how empathy has been treated in Computer Science for the realization of more empathic social or human-computer interactions, becoming a source of relevant consultation for this research line. This mapping can be extended by changing search strings, the search questions, or the inclusion and exclusion criteria. A systematic review can also be carried out in this line of research.

References

1. Cabral, Á. (1996). *Dicionário técnico de psicologia*. São Paulo: Editora Cultrix.
2. Ju, J. Y., Yoo, J. S., Lee, J., & Kwon, H. (2015). Breadcrumb SNS: Asynchronous empathy chat for smart city residents. In *2015 8th International Conference on Mobile Computing and Ubiquitous Networking, ICMU 2015*. Hakodate City, Hokkaido, Japan, (pp. 13–18).
3. Rossi, P. G., & Fedeli, L. (2015). Empathy, education and AI. *International Journal of Social Robotics, 7*(1), 103–109.
4. Patel, R. A., Hartzler, A., Pratt, W., Back, A., Czerwinski, M., & Roseway, A. (2013). Visual feedback on nonverbal communication: A design exploration with healthcare professionals. In *Proceedings of the 2013 7th International Conference on Pervasive Computing Technologies for Healthcare and Workshops, PervasiveHealth 2013*. Venice, Italy, (pp. 105–112).
5. Xiao B., Georgiou, P. G., Imel, Z. E., Atkins, D. C., & Narayanan, S. S. (2013). Modeling therapist empathy and vocal entrainment in drug addiction counseling, In *Proceedings of the Annual Conference of the International Speech Communication Association, INTERSPEECH*. Lyon, France, (pp. 2861–2865).
6. Menendez, C., Eciolaza, L., & Trivino, G. (2014). Generating advices with emotional content for promoting efficient consumption of energy. *International Journal of Uncertainty, Fuzziness and Knowledge-Based Systems, 22*(5), 677–697.
7. Wieseke, J., Geigenmüller, A., & Kraus, F. (2012). On the role of empathy in customer-employee interactions. *Journal of Service Research, 15*(3), 316–331.
8. Carrier, L. M., Spradlin, A., Bunce, J. P., & Rosen, L. D. (2015). Virtual empathy: Positive and negative impacts of going online upon empathy in young adults. *Computers in Human Behavior, 52*, 39–48.
9. Dey, L., Asad, M-U., Afroz, N., & Nath, R. P. D. (2014). Emotion extraction from real time chat messenger. In *2014 International Conference on Informatics, Electronics and Vision, ICIEV 2014*. Dhaka, Bangladesh, (pp. 1–5).
10. Petersen, K., Feldt, R., Mujtaba, S., & Mattsson, M. (2008). Systematic mapping studies in software engineering, *EASE'08 Proceedings of the 12th International Conference on Evaluation and Assessment in Software Engineering*. Italy, (pp. 68–77).
11. Kitchenham, B. (2004). Procedures for performing systematic reviews, *Keele, UK, Keele University*, vol. 33, no. TR/SE-0401, p. 28.
12. SCOPUS, Scopus – Elsevier Database. (2016). [Online]. Available: http://www.scopus.com/. Accessed: 03-May-2016.
13. Kummer, N., Kadish, D., Dulic, A., & Najjaran, H. (2012). The empathy machine. In *Conference proceedings – IEEE International*

Conference on Systems, Man and Cybernetics. Seoul, South Korea. (pp. 2265–2271).

14. Acosta, J. C., & Ward, N. G. (2011). Achieving rapport with turn-by-turn, user-responsive emotional coloring. *Speech Communication, 53*(9–10), 1137–1148.

15. Colaco, M., de Souza, J. G., and Goncalves, C. A. (2012). An approach to improve the empathy of text interactions in collaborative systems, *Collaborative Systems (SBSC), 2012 Brazilian Symposium*. São Paulo, Brazil. (pp. 11–15).

16. Hasler, B. S., Hirschberger, G., Shani-Sherman, T., & Friedman, D. A. (2014). Virtual peacemakers: Mimicry increases empathy in simulated contact with virtual outgroup members. *Cyberpsychology, Behavior Social Networling, 17*(12), 766–771.

17. Mazzei, D., Lazzeri, N., Billeci, L., Igliozzi, R., Mancini, A., Ahluwalia, A., Muratori, F., & De Rossi, D. (2011). Development and evaluation of a social robot platform for therapy in autism, In *Proceedings of the Annual International Conference of the IEEE Engineering in Medicine and Biology Society, EMBS*. Boston, Massachusetts, USA. (pp. 4515–4518).

18. Sieber, J. M. & Kistler, R. (2014). Monkey Business, In *TEI 2014 – 8th International Conference on Tangible, Embedded and Embodied Interaction, Proceedings*. Munich, Germany. (pp. 335–336).

19. Malik, N. A., Yussof, H., Hanapiah, F. A., Rahman, R. A. A., & Basri, H. H. (2015). Human-robot interaction for children with cerebral palsy: Reflection and suggestion for interactive scenario design. *Procedia Computer Science, 76*, 388–393.

20. Pandian, P. S., & Srinivasan, S. (2014). An all-inclusive review on various techniques of web usage mining. *Journal of Theoretical and Applied Information Technology, 59*(3), 621–631.

Multi-camera Occlusion and Sudden-Appearance-Change Detection Using Hidden Markovian Chains

Xudong Ma

Abstract

In this paper, a new object tracking algorithm using multiple cameras for surveillance applications is proposed. The proposed algorithm is for detecting sudden-appearance-changes and occlusions. We use a hidden Markovian statistical model, where the random events of sudden-appearance-changes and occlusions are the hidden variables. The tracking algorithm uses both a discriminative model and a generative model for the being-tracked object. The prediction errors in the generative model are used as the observed random variables in the hidden Markovian model. We assume that the prediction errors are exponentially distributed, when no sudden-appearance-changes and occlusion occurs. And the prediction errors are assumed uniformly distributed, when such random events occur. Almost all state-of-the-art discriminative model based object tracking algorithms need to update the discriminative models on-line and thus suffer a so called drifting problem. We show in this paper that the obtained sudden-appearance-changes and occlusion estimations can be used to alleviate such drifting problems. Finally, we show some experimental results that our algorithm detects the sudden-appearance changes and occlusions reliably and can be used for alleviating the drifting problems.

Keywords

Computer vision · Object tracking · Hidden Markovian model · Bayesian estimation · Surveillance

90.1 Introduction

In surveillance applications, the ability of long-term tracking a certain person or object from security cameras is highly desirable. Usually the security personnel can label a suspicious person in the video. The tracking system can then track this suspicious person in the videos without further human inputs for minutes or hours.

One situation that the visual tracking systems for the surveillance applications may deal with frequently is the sudden-appearance-changes of the being tracked person or object. For example, the suspicious person may take off his/her jacket, pull up his/her hood, or abandon some luggage in order to fool the surveillance system. Such sudden-appearance-changes are suspicious activities and usually should be reported to the security personnel in real-time.

Unfortunately, it is usually difficult for computer vision algorithms to distinguish between a sudden-appearance-change to occlusions. Without the ability of distinguishing between sudden-appearance-changes to occlusions, the visual tracking systems may generate a large number of false alarms. Detecting occlusions correctly is also important for enhancing the reliability of visual tracking algorithms.

This paper was originally submitted to Xinova LLC as a response to a Request for Invention (RFI) on new event monitoring methods.

X. Ma (✉)
Pattern Technology Lab, LLC, Apache Junction, AZ, USA
e-mail: xma@ieee.org

© Springer International Publishing AG 2018
S. Latifi (ed.), *Information Technology – New Generations*, Advances in Intelligent Systems and Computing 558, DOI 10.1007/978-3-319-54978-1_90

It is well-known that for visual tracking, there exists a so called "high-adaptability-to-drifting-resistance trade-off" problem. That is, the discriminative model may be wrongly tuned to occluder. And the tracking algorithm may start to track the occluder instead. In Fig. 90.2 of Sect. 90.4, we actually show an example of such drifting phenomena.

In this paper, we propose a new occlusion and appearance-change detection method. The proposed real-time visual tracking system uses multiple surveillance cameras. Initially, the security personnel provides one bounding box of the suspicious person for each video frame sequence. The visual tracking system then tracks the whereabouts of the suspicious person in real-time.

Our method uses both generative and discriminative models for the video frame streams. For each camera, one discriminative model is maintained for discriminating the image patches that contain the being-tracked person to the image patches that do not contain the being-tracked person. Similarly, one generative model is maintained for each camera. In this paper, we use a recently proposed compressive sensing and naive Bayes based classier in [1] as the discriminative model. We use linear sub-space models as the generative models. That is, we assume that the image patches containing the being-tracked person from several adjacent video frames all are vectors within a certain affine sub-space.

A center component of our method is a hidden Markovian model for the prediction errors of the generative models. That is, whenever a new video frame is received, the new image patch containing the being-tracked person is predicted from the previous such image patches. The hidden Markovian model thus contains a visible part and a hidden part. The visible part contains the observed prediction errors. And the hidden part contains random variables $O_1(t)$, $O_2(t)$, ..., $O_N(t)$ and $S(t)$. The binary random variable $O_n(t)$ denotes whether an occlusion has occurred for the n-th camera at time t. And the binary random variable $S(t)$ denotes whether a sudden-appearance-change of the being-tracked person has occurred at time t.

We assume some parametric probability distributions for the hidden Markovian model. The probabilities of $O_n(t)$ and $S(t)$ are estimated from the observed prediction error $z_n(t)$ by using sequential Bayesian estimation. An alarm signal may be raised, if we detect a high probability of $S(t) = 1$, a sudden-appearance-change occurred. The estimated probabilities of $O_n(t)$ and $S(t)$ are also used for adjusting the learning rates of the discriminative and generative models. It should be intuitively clear that any appearance-change of the being tracked objects may result in significant prediction errors at the same time at all the cameras, but occlusions usually result in significant prediction errors only at a few cameras. Their probabilities can thus be estimated accordingly.

Please note that the above hidden Markovian statistical model is the centerpiece of the proposed occlusion and sudden-appearance-change detection methods. The statistical model works also well with other discriminative and generative models.

There is a large literature on visual tracking algorithms, such as [2–4]. The approaches of using multiple cameras for visual tracking have become attractive in the recent years, due to the availability of large quantities of low-cost commodity cameras. In [5, 6], approaches are discussed, where the responsibilities of tracking may be passed from one camera to another camera. In [7], from each camera, a statistical estimation of the location of the being tracked person or object is obtained independently. The independent estimations are then fused into a joint location estimation. In [8], video frames from multiple cameras are projected on a reference frame using homography transforms, such that the signals corresponding to the being tracked person or object may be added constructively.

The rest of the paper is organized as follows. In Sect. 90.2, we discuss the proposed visual tracking system and the hidden Markovian model. In Sect. 90.3, we present the sequential Bayesian estimation methods for the hidden Markovian model. Simulation results of the proposed method is provided in Sect. 90.4. Finally, some concluding remarks are presented in Sect. 90.5.

90.2 Visual Tracking System and Hidden Markovian Model

A block diagram of the proposed visual tracking system is shown in Fig. 90.1. The system uses multiple cameras (only 2 are shown here). The system starts tracking a suspicious person, after the security personnel provides one bounding box of the suspicious person for each camera. For each camera, there is a real-time tracking sub-system as in [1] (shown as classifiers in Fig. 90.1). All such real-time tracking sub-systems work almost independently, except that their learning parameters λ are controlled by the center controller.

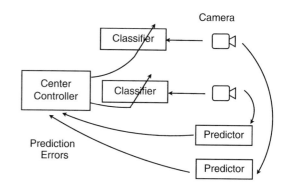

Fig. 90.1 Block diagram of the visual tracking system

Fig. 90.2 Drifting problem of the real-time compressive tracking algorithm with a fixed learning parameter $\lambda = 0.85$. When the being-tracked person is completely occluded by the passenger at frame 577 (sub-figure (**c**)), the discriminative model is updated according to the appearance of the passenger. Thus, a tracked loss occurs (All our figures are best viewed in color)

The learning parameters λ control how much each individual real-time compressive tracking sub-system should update the discriminative model after receiving each new video frame. As the being-tracked person changes his/her pose, orientation etc., the appearance of the person may change smoothly. Thus, each compressive tracking sub-system may update its discriminative model according to each newly observed video frame. The parameter λ is a real number between 0 and 1. If $\lambda = 1$, then the compressive tracking sub-system does not update its model. If $0 < \lambda < 1$, then the compressive tracking sub-system updates the discriminative model using a combination of past observed video frames and the newly observed video frame.

Our proposed method adjusts the learning parameters λ based on the probabilities of sudden-appearance-change $\mathbb{P}[S(t)]$ and the probabilities of occlusions $\mathbb{P}[O_n(t)]$. Let us assume that we use N security cameras. Suppose at each time $t = 1, 2, \ldots$, we receive one video frame $x_n(t)$ at each camera, where $1 \leq n \leq N$. Each tracking sub-system then finds one image patch $Y_n(t)$ that contains the being-tracked person, where $Y_n(t)$ is a column vector. That is, $Y_n(t)$ is the vector obtained by stacking the pixels in the image patch.

For each camera, we use one generative model (shown as predictors in Fig. 90.1). Each predictor maintains an estima-tion of an affine subspace $Space_n(t)$, such that all the past observed $Y_n(t - 1)$, $Y_n(t - 2),\ldots$ roughly lie within this affine space. We can then define a prediction error $Z_n(t)$ as the distance of the newly observed $Y_n(t)$ to this affine space $Space_n(t)$. Note that there exist efficient algorithms for computing the affine space $Space_n(t)$, such as in [9].

We may then estimate the probabilities of sudden-appearance-change $\mathbb{P}[S(t)]$ and the probabilities of occlusions $\mathbb{P}[O_n(t)]$ from $Z_n(t)$ based on the following hidden Markovian model. We assume that the probability density function of $Z_n(t)$

$$\begin{aligned} &\mathbb{P}[Z_n(t)|S(t), O_n(t)] \\ &= \begin{cases} \exp(-x/\mu)/\mu, & \text{if } S(t) = O_n(t) = 0 \\ 1/M, & \text{otherwise} \end{cases} \end{aligned}$$

where, $S(t) = 1$ indicates that a sudden-appearance-change has occurred at time t and $O_n(t) = 1$ indicates that an occlusion has occurred at the n-th camera at time t. In other words, if $S(t) = O_n(t) = 0$, then the prediction error $Z_n(t)$ is exponentially distributed. Otherwise, the prediction error $Z_n(t)$ is uniformly distributed between 0 and M. We further assume that $Z_n(t)$ and $Z_m(t)$ are statistically independent conditioned on $S(t)$ and $O_1(t), \ldots, O_N(t)$, if $n \neq m$. We assume that each random process $O_n(1), O_n(2), \ldots, O_n(t)$ is a Markovian random process. Similarly, we also assume that the random process $S(1)$, $S(2)$, ..., $S(t)$ is Markovian.

We may then use the Bayesian decision methods to detect occlusions at each camera $O_n(t)$ and the sudden appearance change of the being tracked person or object $S(t)$ by computing the probabilities

$$\mathbb{P}[S(t) = 1|Z_{1:N}(1:t)]$$
$$= \mathbb{P}[S(t) = 1|Z_1(1:t), Z_2(1:t), \ldots, Z_N(1:t)]$$
$$\mathbb{P}[O_n(t) = 1|Z_{1:N}(1:t)]$$
$$= \mathbb{P}[O_n(t) = 1|Z_1(1:t), Z_2(1:t), \ldots, Z_N(1:t)]$$

where $Z_n(1:t)$ denotes the collection of observed prediction errors $Z_n(1)$, $Z_n(2)$, ..., $Z_n(t)$, and $Z_{1:N}(1:t)$ denote the collection of variables $Z_1(1:t)$, ..., $Z_N(1:t)$. We show in Sect. 90.3 that these probabilities can be recursively calculated in a very efficient way.

The proposed method raises an alarm signal, whenever the probability $\mathbb{P}[S(t) = 1|Z_{1:N}(1:t)]$ goes over a certain threshold. The method may also adjust the learning parameters λ of the compressive tracking sub-systems according to the probability $\mathbb{P}[O_n(t) = 1|Z_{1:N}(1:t)]$. For example, we may set $\lambda = 1$, whenever the probability $\mathbb{P}[O_n(t) = 1|Z_{1:N}(1:t)]$ is greater than 0.5.

90.3 Recursive Bayesian Estimation

In this section, we derive a recursive formula for calculating $\mathbb{P}[S(t), O_{1:N}(t)|Z_{1:N}(1:t)]$ from $\mathbb{P}[S(t-1), O_{1:N}(t-1)|Z_{1:N}(1:t-1)]$ in Eq. 90.1, where (a) follows from the Markovian properties of the statistical model.

$$\mathbb{P}[S(t), O_{1:N}(t)|Z_{1:N}(1:t)] = \frac{\mathbb{P}[S(t), O_{1:N}(t), Z_{1:N}(1:t)]}{\mathbb{P}[Z_{1:N}(1:t)]} \propto \mathbb{P}[S(t), O_{1:N}(t), Z_{1:N}(1:t)]$$
$$\propto \mathbb{P}[S(t), O_{1:N}(t), Z_{1:N}(t)|Z_{1:N}(1:t-1)]$$
$$= \sum_{S(t-1), O_{1:N}(t-1)} \mathbb{P}[S(t), O_{1:N}(t), S(t-1), O_{1:N}(t-1), Z_{1:N}(1:t)|Z_{1:N}(1:t-1)]$$
$$= \sum_{S(t-1), O_{1:N}(t-1)} \mathbb{P}[S(t-1), O_{1:N}(t-1), |Z_{1:N}(1:t-1)]$$
$$\times \mathbb{P}[S(t), O_{1:N}(t)|S(t-1), O_{1:N}(t-1), Z_{1:N}(1:t-1)]$$
$$\times \mathbb{P}[z_{1:N}(t)|S(t), O_{1:N}(t), S(t-1), O_{1:N}(t-1), Z_{1:N}(1:t-1)]$$
$$\overset{(a)}{=} \sum_{S(t-1), O_{1:N}(t-1)} \mathbb{P}[S(t-1), O_{1:N}(t-1), |Z_{1:N}(1:t-1)]$$
$$\times \mathbb{P}[S(t), O_{1:N}(t)|S(t-1), O_{1:N}(t-1)] \times \mathbb{P}[z_{1:N}(t)|S(t), O_{1:N}(t)] \tag{90.1}$$

90.4 Experimental Result

We use one of the PETS 2007 (Tenth IEEE International Workshop on Performance Evaluation of Tracking and Surveillance) data-sets (available from http://www.cvg.reading.ac.uk/PETS2007/data.html). The data-set consists of 12000 video frames for 4 cameras, 3000 video frames for each camera. A suspicious person enters the scene at roughly frame 500 and drops and leaves behind his backpack at roughly frame 850.

We observe that the real-time compressive tracking systems with a fixed learning parameter in the prior art [1] do have the drifting problem as shown in Fig. 90.2. We

show in Fig. 90.3 that such drifting problem can be avoided by the algorithm proposed in this paper. The occlusion event around frame 577 is detected by our algorithm very clearly with the corresponding probability close to 1.

The proposed algorithm also detects the unusual behavior of the being-tracked person at frame 851 (with the corresponding probability higher than 0.9). The proposed algorithm is able to track the where-about of the suspicious person as shown in Figs. 90.4 and 90.5, where each figure shows the tracking results at one camera.

In all the above experimental results, the tracking results are shown as red bounding boxes. And the frame indexes are labelled in all the images.

Fig. 90.3 Drifting is avoided by the proposed algorithm

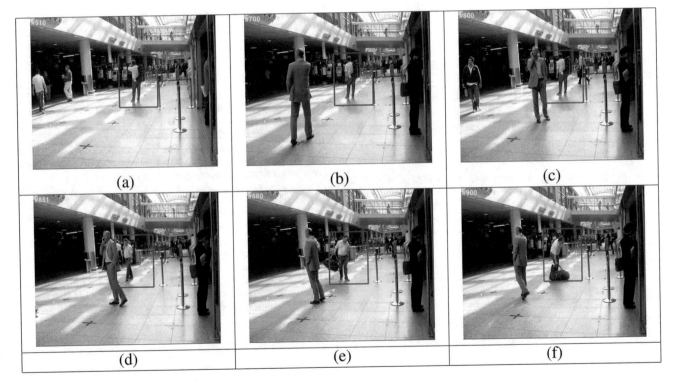

Fig. 90.4 Detecting sudden-appearance change

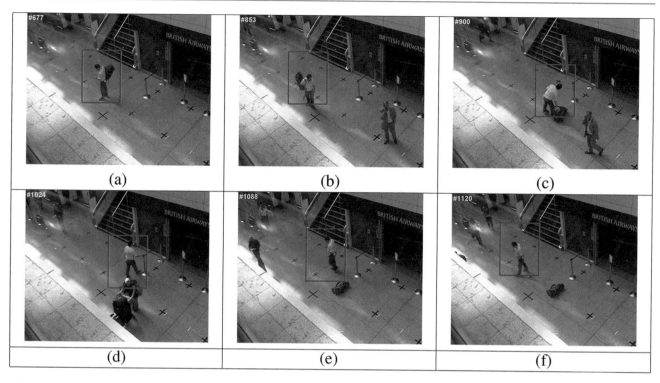

Fig. 90.5 Detecting sudden-appearance change at another camera

90.5 Conclusion

The paper discusses a visual tracking algorithm to detect sudden-appearance-change and occlusions. By experimental results, we show that the proposed algorithm can reliably detect the sudden-appearance-change and occlusion events. Such reliable estimations can also be used to avoid the drifting problems.

Acknowledgements This research was originally submitted to Xinova, LLC by the author in response to a Request for Invention. It is among several submissions that Xinova has chosen to make available to the wider community. The author wishes to thank Xinova, LLC for their funding support of this research. More information about Xinova, LLC is available at www.xinova.com.

References

1. Zhang, K., Zhang, L., & Yang, M.-H. (2012). Real-time compressive tracking. In *12th European Conference on Computer Vision (ECCV)*.

2. Comaniciu, V. R. D., & Meer, P. (2003). Kernel-based object tracking. *IEEE Transactions on Pattern Analysis and Machine Intelligence, 25*(5).

3. Avidan, S. (2004). Support vector tracking. *IEEE Transactions on Pattern Analysis and Machine Intelligence, 26*(8), 1064–1072.

4. Avidan, S. (2007). Ensemble tracking. *IEEE Transactions on Pattern Analysis and Machine Intelligence, 29*(2), 261–271.

5. Quaritsch, M., Kreuzthaler, M., Rinner, B., Bischof, H., & Strobl, B. (2007). Autonomous multicamera tracking on embedded smart cameras. *Journal on Embedded Systems*.

6. Cai, Q., & Aggarwal, J. (1999). Tracking human motion in structured environments using a distributed-camera system. *IEEE Transactions on Pattern Analysis and Machine Intelligence, 21*(12).

7. Bhuyan, M., Lovell, B., & Bigdeli, A. (2007). Tracking with multiple cameras for video surveillance. In *9th Biennial Conference of the Australian Pattern Recognition Society on Digital Image Computing Techniques and Applications*.

8. Eshel, R., & Moses, Y. (2008). Homography based multiple camera detection and tracking of people in a dense crowd. In *IEEE Conference on Computer Vision and Pattern Recognition (CVPR)*.

9. Levy, A., & Lindenbaum, M. (2000). Sequential Karhunen-Loeve basis extraction and its applications to images. *IEEE Transactions on Image Processing, 9*(8), 1371–1374.

Store Separation Analysis Using Image Processing Techniques

Luiz Eduardo Guarino de Vasconcelos, Nelson Paiva O. Leite,
André Yoshimi Kusumoto, and Cristina Moniz Araújo Lopes

Abstract

Store separation flight tests are considered as high-risk activity. These tests are performed to determine the position and attitude of the store after it is deliberately separated or ejected while it is still under the aircraft's area of interference. A process that added value for experimental tests is the photogrammetry. This process extracts data from the cameras, using computational resources that analyze the frames of the videos. Few institutions in the world have an ability to perform this kind of activity. In Brazil, the Flight Tests and Research Institute (IPEV) is one of them. At IPEV, the determination of the store separation trajectory is performed after the flight using a commercial tool. In addition, the process is very inefficient and costly because the activity demands for many weeks of work to analyze the results. Thus, developing a solution to determine the store separation trajectory will increase the efficiency and safety of the flight test campaign. The benefits of such a solution include better use of resources, minimization of workload, and reduced costs and time. In this work, we demonstrate the steps for developing the application that uses a synthetic scenario. The use of a synthetic scenario allows the simulation of different separation scenarios and thus, the application becomes more robust. The experiments performed to validate the application are also demonstrated.

Keywords

Image processing · Store separation · Real time · Synthetic scenario · Camera calibration

L.E.G. de Vasconcelos (✉)
Aeronautics Institute of Technology, ITA, Sao Jose dos Campos, Brazil

National Institute For Space Research, INPE, Sao Jose dos Campos, Brazil
e-mail: du.guarino@gmail.com

N.P.O. Leite · A.Y. Kusumoto
Flight Test & Research Institute, IPEV, Sao Jose dos Campos, Brazil
e-mail: epd@ipev.cta.br

C.M.A. Lopes
Aeronautics Institute of Technology, ITA, Sao Jose dos Campos, Brazil

Aeronautics and Space Institute, IAE, Sao Jose dos Campos, Brazil
e-mail: cmoniz77@gmail.com

91.1 Introduction

The external store separation flight tests (e.g. missiles, fuel tanks) are considered as high risk activity, because during the separation event, the store can perform a different path from the one previously estimated due to non-conformities of the aerodynamic flow, that occurs during the initial phase of store separation. This situation considerably affects flight safety of experimental tests. As a result, there may be delays in product delivery, loss of in-flight test equipment and the aircraft itself.

Simulations in laboratory allow analyzing and estimating the trajectory of the stores in the moment of separation. This assists in determining the risk associated with carrying out store separation flight tests. However, the models and workaround

© Springer International Publishing AG 2018
S. Latifi (ed.), *Information Technology – New Generations*, Advances in Intelligent
Systems and Computing 558, DOI 10.1007/978-3-319-54978-1_91

conditions used in these estimates do not always correspond to the real conditions of the launch, and thus, it becomes necessary the experimental tests. The model of the real separation trajectory, with 6 degrees of freedom (DoF), must be determined during flight test execution. A process that has benefit to the experimental tests is photogrammetry. This process extracts the data from the cameras, using computational resources that analyze the frames of the videos. This analysis requires the identification of multiple marked points of reference on the surface of the store and on the aircraft so that the three-dimensional coordinates can be determined and the store trajectory with 6DoF $(x, y, z,$ roll (ϕ), pitch (θ) e yaw (ψ).

In the world, few institutions have the ability to do this activity. In Brazil, the Flight Test and Research Institute (IPEV) is one of these institutions. At IPEV, the determination of the store separation trajectory is realized after the flight. If the test point is considered safe, a new flight is performed for verification at a new test point, the condition of which is most critical. This process at IPEV is very inefficient and expensive, since it requires many weeks of work for the analysis of the results and the execution of many flight tests, since only one test point can be performed on each flight.

Thus, developing a solution to determine the store separation trajectory will increase the efficiency and safety of this flight test campaign. The benefits of such a solution include the best use of resources (e.g., flights performed and processed in hours rather than weeks), minimizing workload, and reducing costs and time.

The main contribution of this work is the demonstration of the development steps of the application that involves image processing techniques.

The remainder of this work is organized as follows. Section 91.2 has Store Separation overview. Section 91.3 describes in detail the application architecture. Evaluation of the application in a synthetic scenario is described in Sect.

91.4. Section 91.5 describe the results in a pit drop testing. Conclusions are drawn in the end of paper.

91.2 Store Separation

Store separation flight tests can be considered high risk activity because the aerodynamic effect generated between released and/or ejected store and the structure of the aircraft (e.g. airplane wing) is practically unpredictable and can lead to a collision condition with catastrophic consequences [5].

The sequence of images in Fig. 91.1 shows the collision of the store after separation occurred in a real flight test campaign.

Spahr [2] defines store separation analysis as the "determination of the position's position and attitude after it is deliberately separated or ejected while it is still under the aircraft's area of interference".

MIL-STD-1289D [3] establishes requirements and procedures for the installation of all ammunition and store transported in an aircraft and defines that the store is "any device intended for internal or external transportation, affixed to the aircraft and which may be released intentionally or in emergency situations during the flight". In manual MIL-HDBK-1763 [4], the store can be classified as a device that can be jettisoned (e.g. missile, rocket) or not (e.g., fuel tank, photographic POD). This manual further establishes the compatibility requirements between the aircraft, store and its attachment and transportation equipment, which ensure the airworthiness of the store installed in an aircraft. The manual also defines the methodology of tests, instrumentation and data.

In a few years ago, with the evolution of technology, the photogrammetry technique began to be used in store separation tests. The photogrammetry analysis allows analyzing

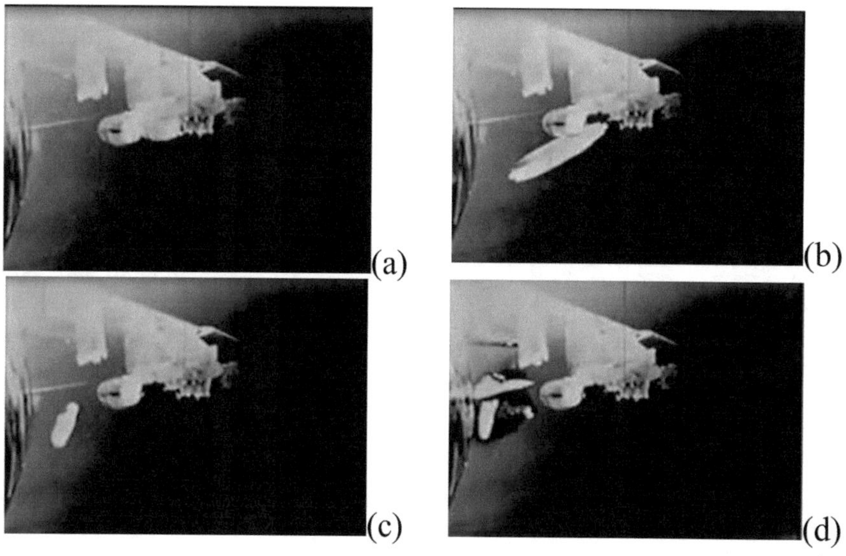

Fig. 91.1 Sequence of photos of an accident occurring during a store separation flight test (**a**) Store on the aircraft (**b**) Frame of 0.2 seconds after store separation (**c**) Frame of 0.5 seconds after store separation (**d**) Moment the store collides with the aircraft

the store trajectories with 6DoF. This is done by capturing images of the store separation using high-speed cameras. These images can be used to determine the position of the store in each frame using image processing techniques and triangulations. By combining the results of individual frames, it is possible to carry out the analysis of the test and define the store trajectory.

The execution of a flight test must necessarily be preceded by a risk analysis. Testing and simulation tools for launch trajectory estimation, such as wind tunnel, simulations and pit drop testing (i.e. weapons delivery and stores separation ground test), are used in this type of test.

The model of the 6DoF separation trajectory can be determined by means of inertial systems, where accelerometers and gyroscopes are installed in the store to obtain information about their angles and acceleration components, or with GNSS Global Navigation Satellite System) receivers installed in the store [6]. However, the use of these devices is limited because the separate load is usually destroyed; retrieving the data collected in the artifact requires the establishment of a channel for transmitting the data with high reliability.

The wind tunnel store separation tests were developed during the 1960s and allow to simulate the store separation trajectory in a controlled environment and to verify the influence of air (i.e. wind) through its flow around the store and the aircraft [7]. In this case, however, in simulations with smaller scale models, the results are generally not similar to the results of the in-flight tests. Another method to estimate the 6DoF trajectory of the store during its separation is the Computational Fluid Dynamics (CFD) simulation; however, its validation requires wind tunnel tests and finally, flight tests.

Pit drop testing allow evaluation under a static condition, in addition to assessing the separation safety requirements. In addition, this test is used to verify the correct operation of all the systems used in the test campaign and to provide reference measurements of the separation trajectory for calibration and validation of the FTI (Flight Test Instrumentation).

The most usual solution is through an optical system that performs the acquisition of images of the store separation. This sequence of images can be used to determine the trajectory of the store during its separation with the use of triangulation techniques [8].

91.3 Application Architecture

The flight test planning is very important. Before any store is ejected from the aircraft, some preparations are required to ensure that the trajectory of the store can be calculated. Preparations include:

91.3.1 Determine the Optical System

The number of cameras, their configurations, position and orientation (i.e. azimuth and elevation) should be determined to optimize the vision of the predicted trajectory.

Analyzes from a single camera result in unstable solutions (have more than one point of convergence), prone to drift (mathematically weak in convergence) and are not redundant (makes troubleshooting difficult). The great advantage is the amount of person-hour for camera setup and data reading. However, viewing an event from the results of a single camera has strong dependence on the judgment of the operator. Thus, the optical system for this study was composed of two cameras installed in a photographic POD.

The high-speed, high-resolution cameras used in pit drop testing were Mikrotron MotionBLITZ EoSens® Cube7. This camera has physical or logical trigger through the network Fast Ethernet or Gigabit (i.e. 100 Mbps/1000 Mbps), Internal memory for recording up to 12 seconds in high resolution and high speed. In the tests, Kowa lenses 6.0 mm were used which produce images with less distortion.

The aircraft used in the tests was the Xavante AT-26. As this aircraft does not have a central station, the photographic POD and the store were installed under the wing in the inboard and outboard stations, respectively. Initially, the proximity of the photographic POD to the store could result in a very small area of capture of the cameras, which would make it difficult to analyze the images. However, the field of view (FOV) of the front and rear cameras was considered satisfactory.

To ensure that the images of the cameras are of the same time, the cameras must be synchronized. A trigger has been developed and adapted in the cockpit of the aircraft so that the pilot can shoot the cameras synchronously. In addition to the trigger being unique, the cameras have an internal sync feature. The start of the recording in the cameras should occur within seconds (between 1 and 5 seconds) before separation of the store, because, cameras have storage limitations.

The higher the acquisition rate of frames per second (fps), the greater the amount of brightness required for recording the image sequence. The tests performed generated videos at 400 fps. The videos captured by the front and back camera were acquired with the resolution of 992 by 1112 pixels and 880 by 1172 pixels, respectively. The difference in resolution did not affect the analyzes. Each recorded video resulted in an average file size of 4 GB.

At the end, the cameras are evaluated before each flight. The evaluation comprises the capture of images in order to provide a basis of comparison for the positioning of the loads because during the flight the conditions can vary. Besides, the setup and the correct operation of the cameras are checked.

Fig. 91.2 Track targets used in tests

91.3.2 Determine the Track Target

The track targets influence image processing algorithms. In this work, in the pit drop testing, four models of track targets on the store surface were used (i.e. circles in black/yellow and black/white, squares in orange/black and black/white (Fig. 91.2)). The size of the target is 10 cm in diameter, if circle, or 10 cm in side, if square. The images are converted to grayscale, thus, black/yellow and black/white combinations are recommended.

91.3.3 Determine the Position and Attitude of Cameras

In order to determine the position of the store in relation to the aircraft, it is necessary to have prior knowledge of the position and the attitude of the cameras in relation to the aircraft. A triangle is placed on the ground visible to the cameras. Then the triangle is captured with the camera, which provides us the 2D vectors from the camera towards the triangle. Then a theodolite is used to measure the position of the plane, the cameras and the triangle. This provides the position/attitude of the triangle in relation to the aircraft, which in turn gives us the position/attitude of the camera relative to the aircraft. The calculation procedure can be performed in CAD or MATLAB software [9].

91.3.4 Measurement of Store and Aircraft Track Targets

In order to be able to determine the position of a store with photogrammetry, a series of points (track targets) with known positions must be visible to the cameras. These points

are adhesives that are glued onto the store. The higher the number of markings visible to any camera, the better the overall solution is. For the calculation of 6DoF, at least 3 non-collinear track targets must be visible in each camera. The track targets are measured with an accuracy theodolite and a steel ruler, and the positions of the track targets relative to the coordinate system are calculated using Matlab. According to [10], great effort is required to position the track targets in the same positions in each load. This is necessary so that new measurements are not taken.

91.3.5 Calibration of the Principal Point of the Cameras and the Focal Length of the Cameras

Each camera with its specific lens requires a calibration to reduce the distortions caused by all the optical elements of the lens. If the optical system changes or if the camera body changes, a new calibration is required. When a photographic POD is used, the calibration should consider the distortion caused by the acrylic. To make the calibration, a plate in chessboard format can be used. The chessboard is captured and a circular correction is calculated (third-order polynomial fit and a compression factor). Poth and Fields [10] used a plate with 202 points to carry out the calibration, while Askmalm e Qwarfort [11] used a plate with 49 points. The errors evaluated in 2D vectors after the optical correction are typically less than 1 mrad (corresponding to a 1 mm error at a distance of 1 meter).

The calibration process is based on the detection of corners of Harris (1988) and the implementation of [12]. The steps used in the algorithm are:

- RGB image is converted to grayscale.
- Contrast adjustment and image brightness based on histogram equalization.
- Applied Sobel filter for edge detection
- Location of corners.
- Refinement of the orientation and location of the corner in subpixel. Lucchese and Mitra [14] show the advantages of working with subpixel.

The algorithm was developed in MATLAB [9].

91.4 Synthetic Scenario Test

Several videos of store separation campaigns carried out by the IPEV were analyzed. All these videos were not enough to develop the tool because they did not have essential information for the analysis, such as: position and orientation of the camera in relation to the store and the aircraft; field of

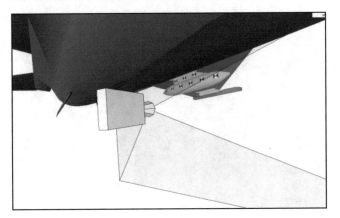

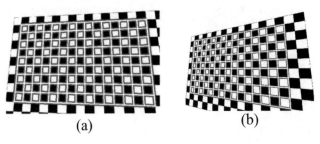

(a) (b)

Fig. 91.4 Plate with 105 calibrated points in both cameras (**a**) front camera (**b**) rear camera

Fig. 91.3 Synthetic scenario of store separation

view of the camera; distance between the cameras; more than one video from the same release (front and back camera); videos synchronized in time; calibration of the cameras and mapping the coordinates of each track target.

Thus, in order to improve the algorithm and reduce the cost of the tests, a synthetic scenario of store separation (Fig. 91.3) was developed to evaluate different launches and to achieve the analysis with the 6 DoF. This scenario was developed in the Sketchup tool [13]. This was possible because the synthetic scenario allows to obtain animations of preprogrammed trajectories that can be exported in video under different points of view (cameras). In this environment, two cameras (front and rear) were introduced. However, others cameras could be added.

Initially, because it was a computational environment, it was considered that the cameras were without distortions, however, it was noticed that the cameras of Sketchup [13] have considerable distortions. Thus, it was necessary to carry out the calibration of the cameras. A plate with squares of 20 cm^2 each was drawn, with an identified area of 105 squares. This plate was captured by the front and rear cameras (Fig. 91.4). In this figure, it is possible to see the area considered in the calibration. It can be seen that the area marked in red represents the area where the store separation occurs.

From the position and orientation of each camera, we can determine the actual azimuth and elevation information for each corner. The field of view (FOV) of each camera is 70°.

For each join, it is necessary to calculate ψ and θ. Equations (91.1) and (91.2) are used to determine azimuth (ψTARGET) and elevation (θTARGET) of each corner, respectively.

$$\psi_{TARGET} = \psi_{CAMERA}$$
$$+ a \tan \left[\frac{x - w/2}{h/2} \tan (FOV/2) \right] \quad (91.1)$$

$$\theta_{TARGET} = \theta_{CAMERA}$$
$$- a \tan \left[\frac{(y - h/2) \tan (FOV/2)}{\sqrt{(h/2)^2 + [(x - w/2) \tan (FOV/2)]^2}} \right]$$
$$(91.2)$$

Where:

- x is the x coordinate of the corner;
- y is the y coordinate of the corner;
- w is the width of the frame;
- h is the height of the frame; and
- FOV is field of view of the camera.

However, it is still necessary to define the actual location of each corner. The chessboard corners are 20 cm apart. Thus, it is necessary to calculate the difference between the position that each corner should be and the position of the corner defined in the image. After this, it is possible to calculate the azimuth (Eq. (91.3)) and the real elevation (Eq. (91.4)).

$$\Psi_{REAL} = atan2 \, (\Delta y - \Delta x) - \pi \quad (91.3)$$

$$\theta_{REAL} = atan \left(\frac{\Delta z}{\sqrt{\sqrt{\Delta x^2 + \Delta y^2}}} \right) \quad (91.4)$$

Where atan2 returns the value close to the interval $[-\pi, \pi]$, known as the 4-quadrant inverse tangent [15].

The errors caused by the distortions are shown in Fig. 91.5. The blue lines shows the errors in x. The red lines shows the errors in y. Errors are calculated by the difference between the actual information and the measured information.

The next step is to define the track targets of the store and the aircraft. After that, targets are identified and tracked in

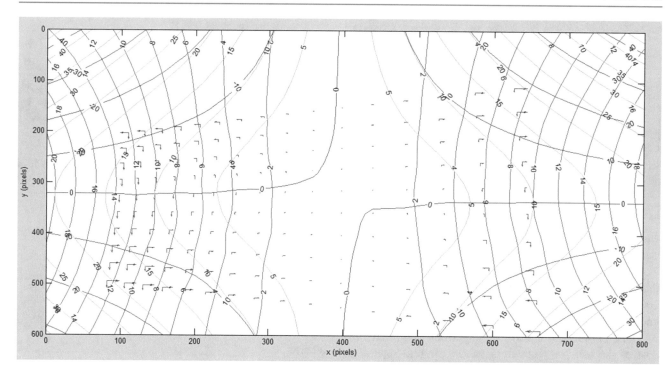

Fig. 91.5 Errors in x and y

Fig. 91.6 Result to front camera

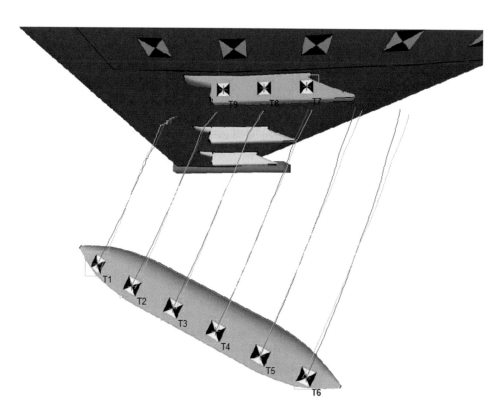

each frame of the video. The track targets were identified using the algorithm described in section III.E. In Fig. 91.6 it is possible to see 3 track targets identified in the aircraft (T9, T8 and T7) and 6 track targets on the store (T1 to T6). At the end of the sequence, we can see the store trajectory trace in red. In green color the corrected tracked information is displayed, i.e. the distortions are corrected from camera calibration.

91.5 Pit Drop Testing

Pit drop testing allow evaluation under a static condition, in addition to assessing the separation safety requirements. In addition, this test is used to verify the correct operation of all systems used in the test campaign and to provide calibration and validation of the store separation trajectory reference measurements.

For a more accurate analysis, the high resolution of the image presents a larger amount of pixels and consequently more information. The possibility of acquiring 400 fps or more allows a better discretization of the trajectory, but the volume of input data for analysis also increases.

Typically, the duration of the store separation trajectory in the aerodynamic influence zone of the aircraft is not greater than two seconds. The best practices, recommended by [1], require the use of cameras with a minimum sampling of 200 fps for separation tests.

The reconstruction of the 6DoF separation trajectory from images through triangulation requires the use of 2 or more high-resolution (i.e. over 720p) and high-speed (i.e. greater than 200 fps) cameras that must be synchronized

and integrated in FTI. This allows the pilot to activate the start of image acquisition when necessary. The cameras are installed inside a photographic POD.

The cameras provide RGB images. The first action is to turn the image into grayscale. After that, the equalization of the histogram of the image is performed in order to improve the brightness and contrast ratio of the image. Figure 91.7 shows the original image of the pit drop testing on the left and the figure on the right shows the image after equalizing the histogram.

The next step is to set the center of each track target so that it can be traced during separation. To do this, the region of interest (ROI) is delimited with the store and the aircraft pylon. This is possible because the position of the store and the pylon relative to the camera will not be changed between the separations tests.

Further, the positions of track targets, aircraft and cameras are previously known and measured from a precision theodolite. An identification algorithm scans the image, looking for corners. It can be seen in Fig. 91.8 that 14 track targets were found (red dots) and other red dots with green circle were found. These last points are incorrect and should be discarded.

Fig. 91.7 Pit drop testing (**a**) original image (**b**) after equalizing histogram

(a)   (b)

Fig. 91.8 Store and pylon with track targets identified

Another algorithm has been implemented so that these incorrect points can be eliminated. The engineer performing the test must select a rectangular area that must comprise the points to be eliminated. Although this process is manual, it is extremely fast (few minutes). Moreover, with the identification algorithm, it is not necessary to perform standard procedures of pattern recognition algorithms (e.g. SIFT, SURF, RNA) [16], such as separating positive and negative samples; carry out training; adjust the algorithm, among others, that can take hours and/or weeks. To carry out the test, at least 3 track targets shall be identified on the pylon (aircraft) and another 3 on the store. In this example, only two track targets have not been identified correctly. In general, 87% of the track targets are found with the algorithm.

In Fig. 91.9, it is possible to observe the center of the highlighted track targets (red color) and a green bound box on each target that serves only visual reference, indicating that the target was identified in the frame. In this type of flight test, environmental conditions vary greatly and may interfere with the performance of the algorithm.

In Fig. 91.10, it is possible to observe to the left the original target and histogram. The image on the right shows the target after equalizing the histogram.

In Fig. 91.11, it can be seen that the aircraft and store track targets have been identified and tracked. The path made by the store, after processing all the frames of the test, can be visualized through the red lines.

91.6 Conclusions

Developing one of a computational solution to determine the trajectory of store separation can increase the efficiency and safety of store separation testing campaigns. The introduction

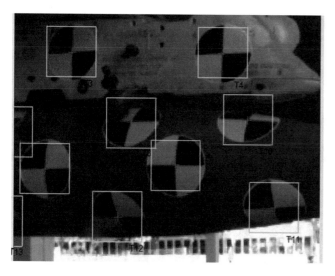

Fig. 91.9 Track targets identified with a bound box

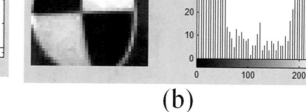

Fig. 91.11 Frame result after pit drop testing

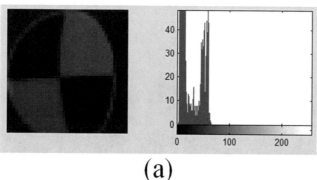

(a)   (b)

Fig. 91.10 Track Target (**a**) original image (**b**) after equalizing histogram

of a photogrammetric application may assist engineers during the flight test to determine whether the flight test point was safe or not.

In this work, the steps for the development of the application and the initial experiments performed to validate the application were demonstrated. The use of a synthetic scenario allows the simulation of different separation scenarios and thus, leave the solution developed, more robust.

Some possibilities of future works are: to develop a 3D solution for the analysis of the store separation; perform pit drop testing and flight tests to improve de tool developed; compare application results with other commercial tools or telemetry data, improve application performance using parallel processing [17].

Acknowledgment The authors would like to thank the unconditional support given by the Instituto de Pesquisas e Ensaios em Voo (IPEV) and Instituto Tecnológico de Aeronáutica (ITA). Also, we'd like to thank FINEP under agreement 01.12.0518.00 that funded the development of this work.

References

1. AGARD. Advisory Group for Aerospace Research and Development. (1986). *Store Separation Flight Testing*. (AGARD Flight Test Technique Series on Store Separation Flight Testing, vol. 5). Research and Technology Organisation (NATO): Neuilly sur Seine.
2. Spahr, H. R. (1975). Theoretical store separation analyses of a prototype store. *Journal of Aircraft, 12*(10), 807–811.
3. United States. (2003). Department of Defense. MIL-STD-1289D:2003: airborne stores, ground fit and compatibility requirements. U.S. Department of Defense: Huntsville.
4. United States. (1998). Department of Defense. Aircraft/Stores compatibility: systems engineering data requirements and test procedures. Wright-Patterson AFB, OH. (MILHDBK-1763): Tuscaloosa.
5. Leite, N. P. O., Vasconcelos, L. E. G., & Kusumoto, A. Y. (2013). Fast on-board tracking system for external stores separation. In *Annual International Telemetering Conference, 49.,* 2013, Las Vegas. Proceedings. Tuscaloosa, AL (US). [S.l: s.n.]. (pp. 1088–1097).
6. Leite, N. P. O. (1997). Sistema de trajetografia GPS diferencial/inercial. 1997. 200 f. Dissertação (Mestrado em Engenharia Eletrônica e Computação) – Instituto Tecnológico de Aeronáutica, São José dos Campos.
7. Denihan, S. G. (2003). *The benefits and risks associated with use of the wind tunnel in safe separation flight test* (74 p). Thesis (M. Sc. in Aviation Systems) – University of Tennessee, Knoxville.
8. Davis, R. E., & Foote, F. C. (1953). *Surveying: Theory and practice*. New York: McGraw-Hill.
9. Mathworks. Disponível em https://www.mathworks.com/. Acessado em 14 Nov 2016.
10. Poth, S.M., & Fields, J.C. (1999). *F/A-18E/F Weapon Separation Photogrammetric Analysis*. 16–20 August 1999. 30th Annual International Symposium.
11. Askmalm, L., & Qwarfort, F. (2001). *Quick Photogrammetric Separation Analysis for the Gripen Test Program*. SFTE 32nd Annual International Symposium. 10–14 September 2001, Seattle.
12. Geiger, A., Moosmann, F., Car, O., & Schuster, B. (2012). *Automatic Calibration of Range and Camera Sensors using a single Shot*. International Conference on Robotics and Automation (ICRA).
13. Sketchup. Disponível em http://www.sketchup.com/. Acessado em 14 Nov 2016.
14. Lucchese, L., & Mitra, S. (2002). Using saddle points for subpixel feature detection in camera calibration targets. In *Asia-Pacific Conference on Circuits and Systems*.
15. Mathworks, Atan2. Disponível em https://www.mathworks.com/help/matlab/ref/atan2.html#buct8h0-4. Acessado em 14 Nov 2016.
16. Kusumoto, A. Y., Vasconcelos, L.E.G., Leite N.P.O., Lopes, C.M. A., & Pirk, R. (2014). Tracking track targets in external store separation using computer vision. In *International Telemetering Conference*, San Diego. Proceedings... [S.l: s.n.].
17. Yazdanpanah, A.P., Mandava, A.K., Regentova, E.E., Muthukumar, V., & Bebis, G. (2014). *A CUDA based implementation of locally-and-feature-adaptive diffusion based image denoising algorithm*, 2014 11th International Conference on Information Technology: New Genrations, Las Vegas, (pp. 388–393). doi:10.1109/ITNG.2014.113.

Ballistic Impact Analysis Using Image Processing Techniques

Luiz Eduardo Guarino de Vasconcelos, Nelson Paiva O. Leite, and Cristina Moniz A. Lopes

Abstract

Materials for ballistic protection are developed with a life-saving purpose and have resistive ability to withstand impacts without damaging the body they are protecting. The development of new materials involves the analysis of deformations occurring after impact. Traditional processes, which use plasticine, strain gages, extensometers, linear variable differential transducers to carry out the measurements, are laborious and inefficient. Few institutions in the world have the ability to develop new ballistic materials. In Brazil, the Aeronautics and Space Institute (IAE), develops new materials using the traditional process through plasticine. Thus, the development of a solution to determine the deformations and their characterizations (diameter, height), can increase the efficiency of this type of test. In this work, the steps for the development of an application that uses image processing techniques to determine the diameter of the deformations are shown. The experiments performed and the results obtained are also shown.

Keywords

Image processing • Ballistic impact • Real time • Deformation • Camera calibration

92.1 Introduction

The materials for ballistic protection are designed to save lives and have resistive ability to withstand high-speed impacts without damaging the body they are protecting. Currently, these materials can be used in different scenarios, such as in the external protection of aircraft, spacecraft, satellites, and so on. The researchers are looking for lighter materials, mainly when applied in vehicles and aircraft, because in this case, the fuel consumption and the center of gravity where the material was applied, are influenced directly by the weight of the material.

Developing a material for ballistic protection that provides greater mobility and safety has been a challenging job for several national and international research centers. In order to develop a new material, it is fundamental to analyze the deformations that occur after high-speed impacts.

In the world, few institutions have the capacity to carry out this activity. In Brazil, the Aeronautics and Space Institute (IAE) is one of these institutions. At IAE, the deformation determination is performed after the end of the tests. The determination of the total deformation (i.e. maximum deformation) is done through a plasticine placed behind the plate and a company outsourced through a mechanical arm determines the plastic deformation (i.e. residual deformation). This process is

L.E.G. de Vasconcelos (✉)
Aeronautics Institute of Technology, ITA, São José dos Campos, Brazil

National Institute For Space Research, INPE, São José dos Campos, Brazil
e-mail: du.guarino@gmail.com

N.P.O. Leite
Flight Test & Research Institute, IPEV, São José dos Campos, Brazil
e-mail: epd@ipev.cta.br

C.M.A. Lopes
Aeronautics Institute of Technology, ITA, São José dos Campos, Brazil

Aeronautics and Space Institute, IAE, São José dos Campos, Brazil
e-mail: cmoniz77@gmail.com

© Springer International Publishing AG 2018
S. Latifi (ed.), *Information Technology – New Generations*, Advances in Intelligent Systems and Computing 558, DOI 10.1007/978-3-319-54978-1_92

laborious, time consuming and costly (i.e. carried out for a few days or weeks).

The process adopted at IAE is very inefficient, because it demands many hours of work. In addition, improvements in the process must be performed in order for quality to be maintained in all tests. The development of a solution to determine in near real time the total and plastic deformations will increase the efficiency of this type of test. In addition to the deformations, deformation characterization (e.g., diameter, area and height of deformations) shall be performed. However, this is complex and long-term work.

In this work, we discuss the first contributions that are to obtain the diameter and the area of the plastic and total deformations through image processing techniques. This paper shows some experiments carried out on hybrid materials composed of UHMWPE (Ultra high molecular weight polyethylene)/polymeric matrix (Dyneema®) that received shots of 9 mm gauge projectiles.

The remainder of this paper is organized as follows. Section 92.2 has ballistic protection overview. Section 92.3 describes in detail the architecture of the proposed algorithm. Evaluation of the application and Results are reported in Sect. 92.4. Conclusions are drawn in the end of paper.

92.2 Ballistic Protection

The first materials for ballistic protection of cast and rolled steel arose in the twentieth century in the First World War. During World War II, it was observed that the tanks and war equipment were very heavy because they were coated with protections of a single material [1]. In this way, it was observed that different materials could be used to carry out the protection, each with a specific objective [2]. According to [3], new technologies have emerged that combine the low weight of the material, the mobility of the user and the capacity of resistance to the ballistic impacts, providing the development of new hybrid structures (i.e., ceramics and polymer).

When a projectile penetrates ballistic protection, a deformation occurs in the material. According to [3], the high tension exerted on the material promotes the breaking of the first fibers, the delamination in the affected region and, finally, the deformation in the back face of the material. According to [4], the problem of excessive deformation in the backsheet should be avoided by means of a material with optimum performance and with an adequate amount of layers of polymer fiber laminates.

The performance of personal ballistic protection and for police and military protection is measured by international standards [5].

Impact speed can be classified into three categories: low speed, high speed and hyper speed [6].

Fig. 92.1 Cone formed in a ballistic panel composed of Dyneema® after impact

The reason for this classification is that, according to [7], there is energy transfer between the projectile and the target, energy dissipation and collateral damages. According to [4], these damages include deformation in the structures of the ballistic panel and the formation of a cone on the posterior face of the target (Fig. 92.1), which is capable of causing serious injury to the user in the case of personal protections.

According to [3], in order to understand the material and the ballistic impact, it is necessary to observe the damages caused after the impact and mechanisms of energy absorption, which may be the formation of "bumps" or cones on the posterior face of the target, the deformation of the secondary wires, the tension in the primary wires / fiber delamination and friction between the projectile and the target.

The cone is formed during ballistic impact. The projectile, when leaving the test piece (equipment that performs the shot) until reaching the target, has kinetic energy that is transformed into mechanical energy as soon as the projectile undergoes the first impact.

According to [8, 9], the thickness and the size of the cone are the parameters that affect the performance of the panel in the ballistic impact. The greater the velocity of the projectile, the greater the probability of drilling the material and the cone be formed. The size of the cone formed at the moment of impact is considered the size of the actual damage caused to the user (i.e. total deformation), since all the kinetic energy of the projectile has been absorbed by the target. After cooling the fibers, only the plastic deformation of the material is observed, since the polymeric material tends to return to the previous stage.

92.3 Application Architecture

A simplified diagram of the experimental setup of the test can be seen in Fig. 92.2.

The cameras used in the tests are from the Mikroton's MotionBLITZ EoSens® Cube7 high-speed data acquisition. For the lighting of the environment were used reflectors that totaled 7500 W. Three notebooks were used for data acquisition of the cameras. A desktop computer was used to control firing and acquisition of projectile velocity data. The cameras were synchronized and the start of the

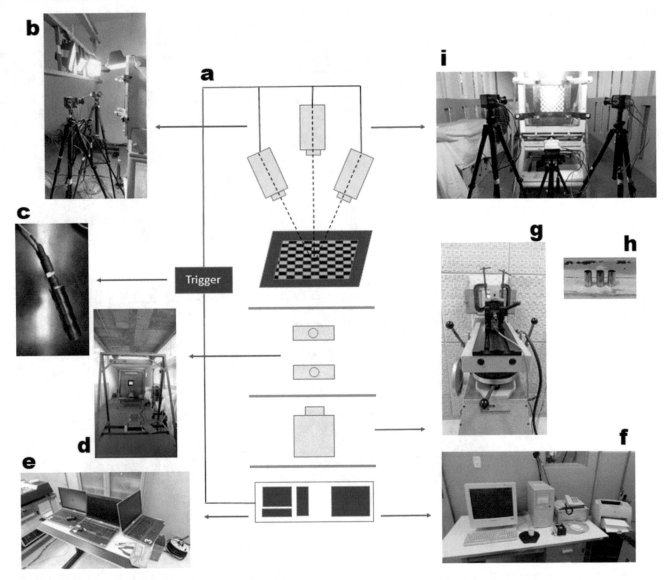

Fig. 92.2 Diagram of the test scenario. (**a**) Diagram. (**b**) Illumination of the environment. (**c**) Device used in the ignition of the cameras. (**d**) Chronographs (**e**) Computers used for camera acquisition and control. (**f**) Computer used to control firing and acquisition of projectile speed data. (**g**) Device used to fire the projectile. (**h**) Used projectiles. (**i**) Positioning of the cameras

acquisition was done through a TTL (Transistor-Transistor Logic) pulse. Notebooks and computer are COTS (Commercial of the shelf). In Fig. 92.3, it is possible to see the cameras positioning.

The diagram shown in Fig. 92.3 is adapted from NIJ 0101.03 Level II [10] and was used in this research. There is a distance of 10 meters between the projectile exit of the gun barrel and the target. In order to measure the velocity of the projectile, speed meters (i.e., chronographs) were positioned at the midpoint of said distance, that is, at 5 meters, the photocells being distant from each other at 4 meters. When the projectile exceeds the first chronograph, the timer starts. When the projectile pass the second chronograph, the stopwatch is finalized, obtaining a measure of

time that is converted into speed. The tests were performed in the IAE ballistic tunnel.

92.3.1 Warm-Up

Test planning is very important. Before any projectile is fired, some preparations are required to ensure that the deformations can be calculated.

The planned activities were: acquisition of the plates. Customizing the plates. The plates were made with light colors (e.g. shades near white or shades of yellow). Then, each plate was painted with a black grid, applying a high contrast pattern, as suggested by [11]; Preparation of the

Fig. 92.3 Cameras positioning

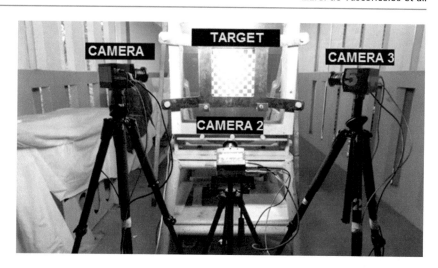

Fig. 92.4 (**a**) plate dimensions
(**b**) identification of impact points
on the plate

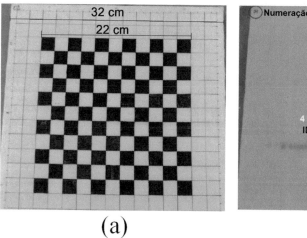

(a)   (b)

device for shooting; positioning of the chronographs, support that will fix each plate, reflectors and cameras; cameras setup. The setup must always be the same; synchronization test of the camera (using the trigger). Measurement of the angles and distances of the cameras relative to the center of the plate using a precision theodolite; Fire (one shot at a time); acquisition of data; extracting information from images; data analysis.

Each plate is approximately 32 cm^2; the working area of the plate used in the test is approximately 22 cm^2 when attached to the firing support. All plates were produced with squares of two cm^2. For the assays, the plates were divided into four quadrants. Each center point, from each quadrant, was demarcated in order to facilitate the identification of the shots. In addition, the center of the plate was demarcated. The standardization adopted for each position of the plate is shown in Fig. 92.4.

A standard sequence for shots was defined in order to allow better analysis of the data. The sequence of shots is SD, IE, SE, ID, CEN and can be seen in Fig. 92.4.

Most plates have never been evaluated. Thus, in order to guarantee the integrity of the cameras used in the test, a shot was performed in the SD (1) position, for most plates, without recording the deformation with the high-speed cameras. For positions IE (2), SE (3) and ID (4), the plates were filmed without any protection in front of the cameras. The CEN position (5) was filmed for the plates that, visually, had not been compromised with the previous shots. At the end, the videos were saved with a standard nomenclature adopted.

92.3.2 Camera Calibration

Every camera has a distortion model for your lenses. Knowing the distortion model is fundamental to determine, with better accuracy, the information of a digital image. According to [12, 13], it is assumed that the model of a pinhole camera has radial and tangential distortion as described in [12, 14, 15]. The Matlab Camera Calibration

Fig. 92.5 Central camera distortion model

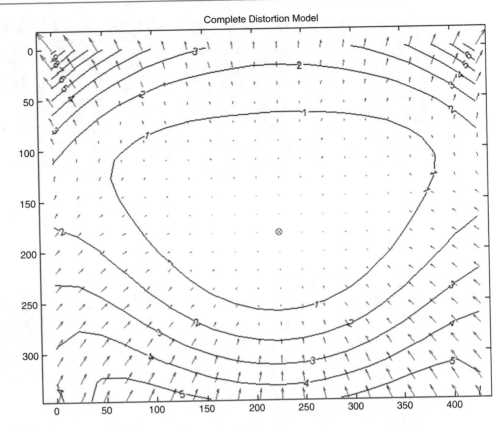

Toolbox [12] is one of the most commonly used tools for camera calibration. This library was also implemented in C++ for OpenCV [13] and uses the distortion model described in [12, 14, 15], which covers a wide range of lenses, and allows to calibrate two video cameras, intrinsically and extrinsically, presenting a chessboard calibration standard in multiple orientations.

Planar plate patterns are generally used as calibration targets because they are inexpensive to make and because they allow location of intersections with subpixel accuracy. The camera calibration method used in this work is performed as follows: the corners of the board are located in the image; the orientations of each location are refined for subpixel accuracy; finally, optimization of distortions is done. In addition, the method used in this work does not require manual intervention.

Figure 92.5 shows the complete distortion model (radial and tangential) of each pixel in the image. Each arrow represents the effective displacement of a pixel induced by lens distortion. It can be seen that the dots at the corners of the image are shifted by approximately six pixels at the top right and nine pixels at the top left of the image. The dot with the cross indicates the center of the image and the circle indicates the location of the main point.

After this, the developed algorithm, adapted from [16], highlights the quadrilaterals considered in the calibration of the camera. Figure 92.6 shows the three cameras with the quadrilateral highlighted.

Then each intersection of the quadrilateral is considered. This intersection is subpixel-accurate. Using the knowledge of the positions of the coordinates of the intersections (corners), it is possible to transform an object from different perspectives into a single coordinate system so that the deformation is measured.

92.4 Experiments and Data Analysis

92.4.1 Tests

Tests were carried out during the months of September and November of 2015. In all, 28 real shots were performed on seven plates and 43 GB of images and videos were produced. All plates were of UHMWPE (Dyneema®) material, varying only the amount of layers of the material. The amounts evaluated had 30, 38, 42 and 48 layers. In general, the greater the amount of layers, the less the deformation. The cameras were configured synchronously, i.e. from a single pulse (trigger), the 3 cameras started recording the images. For this to happen, the cameras were configured using the master-slave sync technique. The master camera was configured with 9000 fps (frames per second) and the slaves

Fig. 92.6 Calibration of the cameras. Identification of each quadrilateral of the plate

17/09/2015 16:34:10 -0346,7[ms]
000014693 EoSens Cube7 [00-11-1c-
f1-74-08] 288x424 8400fps 117µs

17/09/2015 16:34:12 -0346,7[ms] 000011738 EoSens Cube7
[00-50-c2-1d-7f-f1] 448x328 8404fps 116µs

17/09/2015 16:34:15 -0346,7[ms]
000013520 EoSens Cube7 [00-11-1c-
f1-71-39] Mikrotron 304x428 8404fps 116µs
V1.4.0.1

Fig. 92.7 Example of deformation in the 3 cameras

with 9005 fps. According to the manufacturer of the cameras, these five additional frames are used as an interval for synchronization to be possible between the cameras. 7500 W were used for lighting.

In Fig. 92.7, it is possible to observe the variation of the deformations in the three cameras, taking into consideration the lower right quadrant.

Figure 92.8 shows an example of the front and back of a plate after four shots.

Through the metadata of the images, observed at the bottom of the images, it can be seen that the synchronism between the cameras occurred properly. Without this synchronism, it is practically impossible to analyze the data due to the dynamics of the impact.

92.4.2 Data Analysis

All recorded videos must be transformed into images for analysis. This process requires a high computational cost to process the files. The standardization of video file naming

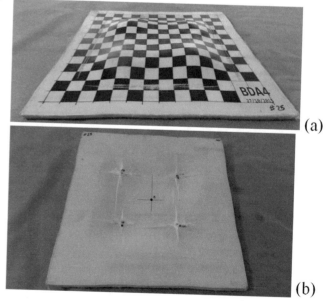

Fig. 92.8 (**a**) Front and (**b**) back of the plate after four shots

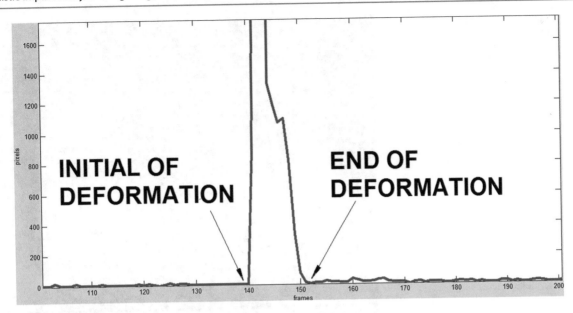

Fig. 92.9   Curve used to detect initial and end of deformation

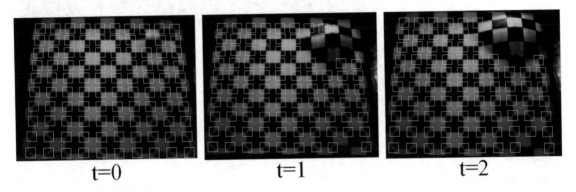

Fig. 92.10   Deformation evolution

allowed the development of an algorithm that performs the reading of all videos on a given test day, creates a "Images" folder for each shot and generates bitmap (bmp) images. Considering the frame rate of 9000 fps, in one second of video, we have 9000 images, and only a few dozen of these images are important for data analysis. Thus, another algorithm was developed, using image processing techniques [17], specifically, image subtraction, that detect the onset of deformation and the post-deformation stabilization (Fig. 92.9). After this, the algorithm considers some frames before the beginning of the deformation and some frames after the beginning of the stabilization of the deformation. The other frames are discarded.

Images are usually preprocessed to enhance the attributes of them to be processed by other algorithms. Preprocessing algorithms are also used to compensate for possible changes

in illumination during image acquisition that may occur in the test environment. In order to determine the area that has been deformed at the impact of the projectile, each intersection of the image is detected and monitored frame by frame. The intersections that undergo deformation are no longer monitored, and represent the deformed area. In Fig. 92.10, the evolution of the deformation area can be visualized.

In order to verify how much each intersection was displaced, the X position, Y position and identifier number for each intersection were printed on the image (Fig. 92.11). In Fig. 92.11 the intersections that were not affected in the impact are in green color and the affected intersections in the impact are in the blue color. Figure 92.11 shows the highlighted deformed area.

To analyze all this collection of images, it was necessary to implement several algorithms to prepare the data for the

Fig. 92.11 Sample of
deformation area

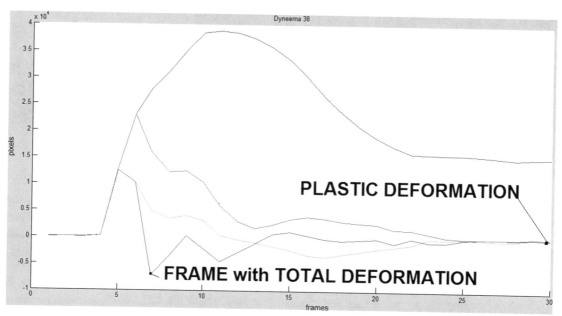

Fig. 92.12 Chart used to detect the image of total and plastic deformation

analysis. One of these algorithms generates a graph that, when analyzed, it was possible to determine the image (frame) that has the total deformation.

In Fig. 92.12, it can be seen that the material undergoes a shock wave, which is propagated over a period that is then stabilized.

The stabilization of material deformation can be seen at the end of the curves (green, red and purple lines), when the values return to zero. The shock wave is the variation of the pixels between the frame of the total deformation until it

returns to zero (plastic deformation). According to [18], the impact speed is, if not the only, one of the most important factors in determining the response at the target. According to [19], in composite materials, the point that receives the high velocity impact, has the highest energy absorption, generating the shock wave, while at low speed impact, the overall deflection of the plate is more important.

The blue line shows the difference in pixels between an image and a base image. The base image is the first image of the sequence (before shooting). The purple line shows the

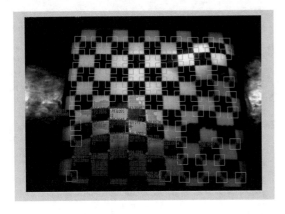
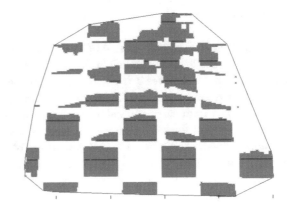

Fig. 92.13 Sample of the deformed area

Fig. 92.14 Area of total and plastic deformation considering IE position

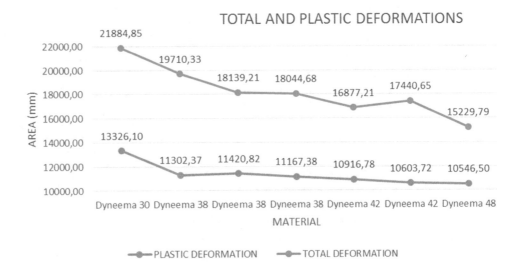

Table 92.1 Information about deformation in IE position

Layers	Total deformation (mm)	Plastic deformation (mm)	Diameter in x axis (mm)	Diameter in y axis (mm)
30	21884,84802	13326,10	134,83871	122,58065
38	19710,32518	11302,37	141,93548	114,83871
38	18139,21005	11420,82	146,45161	107,74194
38	18044,67784	11167,38	136,77419	107,09677
42	16877,21410	10916,78	165,16129	82,58065
42	17440,65323	10603,72	130,96774	98,06452
48	15229,78678	10546,50	123,87097	94,83871

difference in pixels between an image and a base image. The base image changes depending on the number of pixels affected. The red line shows the difference between an image (n) and the previous image (n-1), considering the blue line data. The green line shows the difference between an image (n) and the previous image (n-1), considering the data of the purple line.

In Fig. 92.12, it can be seen that the deformation has a maximum point. In addition, in a few instants (frames) it is possible to visualize the stabilization of the deformation, with values at zero on the y-axis (i.e. zero value - indication that the deformation has been completed).

For each plate, the total and plastic deformation was measured in the five positions (SD, IE, SE, ID, and CEN). The deformed area is drawn from the pixels identified in the image, which have undergone varying luminosity. In Fig. 92.13, an example of the deformed area, in the total deformation, can be visualized on Dyneema plate 42.

Considering only the IE position of the plates, the best plates relative to the deformed area can be seen in Fig. 92.14.

Table 92.1 shows the deformed area in the total and plastic deformations, the deformed diameter in the x and y axes. This table considers only the IE position.

92.5 Conclusions

After performing these tests, more than 43 GB of data was generated in videos and images. Through the application that uses image processing techniques it is possible to determine which materials perform best, analyzing, for the time being, the diameter and the area of the plastic and total deformations. Initially, the materials of the same type were compared, but with the number of layers at 30, 38, 42 and 48. It was possible to notice that the technique developed using image processing maintains the proportion of deformation between these plates. It can be observed that the cameras have distortions and need to be calibrated. The cameras also need to be synchronized so that the analysis can be performed.

A common feature of high-speed cameras is that the higher the acquisition rate, the lower the ROI. The lower the region of interest (ROI) captured by the camera, the number of frames configured may be higher and higher should be the power used in the lighting. The higher the amount of power, the higher the temperature in the test area. The high temperature on the materials can affect the structure of them.

The angles used in the side cameras should be approximately 45° to the center of the plate. The rear camera should be positioned vertically at an angle of approximately 45° to the center of the board. At each new test, angle measurements and distances between the board and the cameras should be performed with precision theodolite. All plates should be marked with squares of 2 cm². The smaller the squares of the boards, the greater the accuracy, however, the power used in the lighting must be greater and the computational cost for the image processing.

In this work, the steps for the development of the application and the initial experiments performed to validate the application were demonstrated. The next steps are: developing a 3D solution for ballistic impact analysis [20]; determine the height of the deformations; compare the results of the application with other methods.

Acknowledgment The authors would like to thank the unconditional support given by the Instituto de Aeronáutica e Espaço (IAE) and Instituto Tecnológico de Aeronáutica (ITA).

References

1. López-Puente, J., Arias, A., Zaera, R., & Navarro, C. (2005). The effect the thickness of the adhesive layer on the ballistic limito of ceramic/metal armours. An experimental and numerical study. *International Journal of Impact Engineering, 32*, 321–336.
2. Sousa, A. N., & Thaumaturgo, C. Geopolímeros para aplicações balísticas. Revista Matéria, v.6, n.2, jan/fev/mar 2002.
3. Noronha, Karina Ferreira. Estruturas híbridas para blindagem balística: estudos da interface da cerâmica/polímero. 2012. 106f. Tese de Mestrado – Instituto Tecnológico de Aeronáutica, São José dos Campos.
4. Mota, J. M. (2010). Desenvolvimento de compósitos híbridos polímero/cerâmica para blindagem balística. Tese de Mestrado do Instituto Tecnológico de Aeronáutica, São José dos Campos.
5. Natinal Institute of Justice. NIJ. Ballistic Resistence of Body Armor. Standard – 0101.06 (2008).
6. Certificação Digital N° 0210646/CA – PUC/Rio. Impacto em Materiais Compósitos.
7. Naik, N. K., Shrirao, P., & Reddy, B. C. K. (2006). Ballistic impact behaviour of woven fabric composites: Formulation. *International Journal of Impact Engineering, 32*, 1521–1552.
8. Cheeseman, B. A., & Bogetti, T. A. (2003). Ballistic impact into fabric and compliant composite laminates. *Composites Structures, 61*, 161–173.
9. Naik, N. K., & Shrirao, P. (2004). Composite structures under ballistic impact. *Composite Structures, 66*, 579–590.
10. National Institute of Justice (NIJ), Standard 0101.03 (1987).
11. Synnergren, P., & Sjodahl, M. (1999). A stereoscopic digital speckle photography system for 3D displacement field measurements. *Optics and Lasers in Engineering, 31*, 425–443.
12. Bouguet, J. Y. (2010). Camera Calibration Toolbox for Matlab, Disponível em http://www.vision.caltech.edu/bouguetj/calib doc
13. Bradski, G. OpenCV, 2011. Disponível em http://opencv.willowgarage.com
14. Heikkila, J., & Silven, O. (1997). A four-step camera calibration procedure with implicit image correction. *Computer Vision and Pattern Recognition*. IEEE. San Juan.
15. Slama, C., Theurer, C., & Henriksen, S. (1980). *Manual of photogrammetry*. Falls Church: American Society of Photogrammetry.
16. Geiger, A., Moosmann, F., Car, O., & Schuster, B. (2012). *Automatic calibration of range and camera sensors using a single shot*. International Conference on Robotics and Automation (ICRA).
17. Gonzalez, R. C., & Woods, R. E. (2010). *Processamento digital de imagens* (3rd ed.). São Paulo: Pearson Prentice Hall.
18. Roylance, D., Wilde, A., & Tocci, G. (1973). Ballistic impact of textile structures. *Textile Research Journal, 43*, 34–41.
19. Cantwell, W. J., & Monton, J. (1989). Comparison of the low and high velocity impact response of CFRP. *Composites, 20*(6), 545–551.
20. Grebennikov, A. (2013). *3D Visualization of Space Distribution of Electrical Characteristics in Compound Structures Reconstructed by General Ray Method*, 2013 10th International Conference on Information Technology:New Generations, Las Vegas, (pp. 743–744). doi:10.1109/ITNG.2013.115.

iHelp HEMOCS Application for Helping Disabled People Communicate by Head Movement Control

Herman Tolle and Kohei Arai

Abstract

The combination of a Head-mounted display (HMDs) with mobile devices, provide an innovation of new low cost of human-computer interaction. Such devices are hands-free systems. In this paper, we introduced a proof of concept of our method on recognizing head movement as the controller of mobile application and proposed a new way of elderly people or disable people communicate with others. The implementation of an *iHelp* application on an iOS devices, shows that the proposed method is appropriate as a real-time human-computer interaction with head movement control only for communicating user need to others through mobile application.

Keywords

Head mounted display • Human-computer interaction • Accelerometer, gyroscope • Assisted technology

93.1 Introduction

The increased popularity of the wide range of applications of which head movement detection is a part, such as assistive technology, teleconferencing, and virtual reality, have increased the size of research aiming to provide robust and effective techniques of real-time head movement detection and tracking [1]. Assistive technology is a technology developed for helping special need people to interact or communicate with others, or for daily activities [2].

Most of the head movement estimation method is based on computer vision approach with image processing [3–5]. Liu et al. [3] introduced a video-based technique for estimating the head pose and used it in an image processing application for a real-world problem; attention recognition for drivers. All computer vision approach require a camera to capture the video or image and to process it which require high computation for the implementation in a real-time implementation.

Another approach for head movement detection is by using sensors such as gyroscopes and accelerometers. King et al. [6] implemented a hands-free head movement classification system which uses pattern recognition techniques with mathematical solutions for enhancement. A dual axis accelerometer mounted inside a hat was used to collect head movement data. A similar method was presented by Nguyen et al. [7]. The method detects the movement of a user's head by analyzing data collected from a dual-axis accelerometer and pattern recognition techniques.

A combination of different techniques can be used in head tracking systems. Satoh et al. [8] proposed a head tracking method that uses a gyroscope mounted on a head-mounted device (HMD) and a fixed bird's-eye view camera responsible for observing the HMD from a third-person viewpoint. Using a fixed camera, customized marker, gyroscope sensor and calibration process makes this proposal

H. Tolle (✉)
Informatics Department of Computer Science Faculty, Brawijaya University, Malang, Indonesia
e-mail: emang@ub.ac.id

K. Arai
Information Science Department, Saga University, Saga, Japan
e-mail: arai@is.saga-u.ac.jp

© Springer International Publishing AG 2018
S. Latifi (ed.), *Information Technology – New Generations*, Advances in Intelligent Systems and Computing 558, DOI 10.1007/978-3-319-54978-1_93

impractical for head tracking tasks. The time complexity of the algorithm has not been investigated which makes it a little far from being used in real-world applications, especially in that the paper mentions no suggested application.

Head-mounted displays (HMDs) embedded in eyeglasses are the next innovation along the path of communication techniques. Such devices are hands-free systems. Although this is not a new idea, currently released and commercially available products (such as the Project Glass by Google) show the immense potential of this technology. They function as stand-alone computers; their light glass frame is equipped with a variety of sensors; a projector displays images and information onto the eye. While wearing these eyeglasses, the user is continuously exposed to the displayed information.

In our previous research work, we propose head movement detection and tracking as a controller for 3D object scene view [9] and the combination of user's head and body movement as a controller for virtual reality labyrinth game [10]. We also proposed a new head pose detection as a controller for a mobile application named as head movement control systems (HEMOCS) [11] that possible for real-time implementation.

In this paper, we introduce a novel implementation of our proposed HEMOCS as an assistive technology for helping disabled and elder people communicate with others. This system is useful in near future because there are more than 1 billion people in the world has some form of disabilities. The proposed system is also applicable for elderly people communicate with the helper. In Japan, there are more than 30 million of elderly people that need help from others. The application is based on the method for recognizing user's

head pose movement in real time basis. The user can easily control the hands-free application by using particular head pose movement only to communicate something to other people who responsible to help them.

93.2 Proposed iHelp HEMOCS System

The proposed iHelp head movement controller system (iHelp HEMOC System) is a system for assist disabled people -who cannot talk and use their hand- to communicate with other people through a hands-free mobile application put on user's head with an HMD as shown in Fig. 93.1. The illustration was taken from Oculus website (Fig. 93.1b), one of the popular virtual reality HMD producer. In this paper, we introduce a new approach for using dummy HMD like *Google Cardboard* with a mobile phone as display and controller (Fig. 93.1a). Besides cheap, this combination gives a flexibility in creating a mobile application for developing communication application and control method using internal sensors data.

The illustration of iHelp HEMOC System for disabled people as shown in Fig. 93.2 has three parts. First is the user wear an HMD, the second part is the ability of the mobile application to talk to the nurse or people near the patient, and the last part is for other people that far from patient. If the patient needs help from others, like asking for someone to bring him/her to the toilet, or want to eat or drink, then the patient can communicate it by moving their head to control what they want to say to others. The system will generate related recorded speech and can also send the request from a patient to other people like nurse or doctor, into another

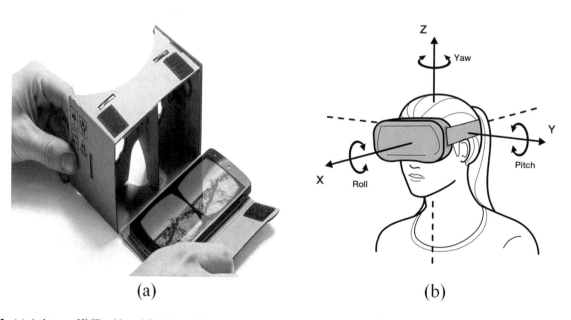

(a) (b)

Fig. 93.1 (a) A dummy HMD with mobile phone [11]; (b) User wear HMD with 3 degrees of freedom of head movement [12]

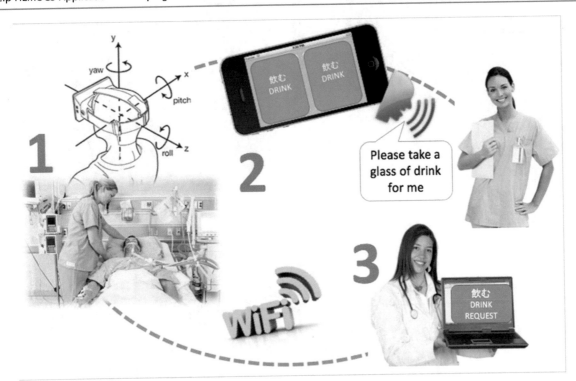

Fig. 93.2 Proposed iHelp HEMOCS system illustration

machine through a wireless network connection. This system work by recognizing user's head movement and interpret it for controlling the iHelp HEMOCS application. The method for identifying head movement is part of our research work on HEMOCS that implemented in a prototype of mobile healthcare application. The word "patient" is changed to "user" from now on for a simplifying explanation of the system.

93.2.1 What Is Head Movement Control Systems

The initial head movement controller system proposed before [10], working with a dummy HMD complementary with a smartphone which has internal inertial sensors like *accelerometer*, *gyroscope*, and *magnetometer*. The user wears an HMD with a smartphone as shown in Figs. 93.1 and 93.2, while system displays a mobile application developed on a smartphone. As an augmented reality application, user can still view something in front of the user while controlling something in the application. The user can control the overlaid text or information by moving the head in a particular movement. The head movement is detected through real-time basis data gathering and analysis from mobile phone's sensors.

The method for detecting user's head pose movement is based on the pattern of data gathered through internal sensors. The human head is limited to 3 degrees of freedom (DOF) of head translation in the pose, which can be characterized by the *pitch*, *roll*, and *yaw* angles as pictured in Fig. 93.2. In the previous research, we proposed a control system using three types of head pose movement with each has two opposite direction as shown in Fig. 93.3. Type of head movement is *Axial rotation left* (H1), *axial rotation right* (H2), *flexion* (H3) and *extension* (H4), *lateral bending left* (H5) and *lateral bending right* (H6). H1 to H6 is used as the code for easily named the head to pose movement. *Axial rotation left* means that user rotates his/her head to the left (around 15–30 degree) from the initial position in the certain speed, and then pose back the head into initial position again. All move should be done in specific head speed movement. This movement feels like user tries to swipe something in the application using their head. All the same process for other 5 types of movement with each direction. Then we have 6 type of head pose movement as the gesture for control something in the mobile application.

93.2.2 Head Pose Gesture Control Function

In HEMOCS, the head movement gesture is act like a swipe type control, or to change the conventional button function. The proposed method for the head movement and each type of control purpose is shown in Table 93.1, where axial rotation left and right is using as selecting control (previous

Fig. 93.3 Six type of head pose movement named as H1 to H6

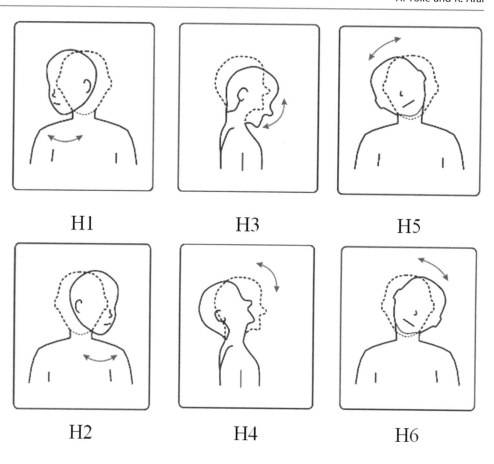

H1   H3   H5

H2   H4   H6

Table 93.1 Design of head pose control function

Code	Head pose type	Control purpose
H1	Axial rotation left (then back)	Move previous
H2	Axial rotation right (then back)	Move next
H3	Flexion (then back)	Choose this item
H4	Extension (then back)	Back to the list
H5	Lateral bending left (then back)	Back to home
H6	Lateral bending right (then back)	Reserved

(H1) or next (H2)), *flexion* for accept (or *tap* or *choose*) control (H3), *extension* for back function (H4), *lateral bending left* or *tilt head toward left shoulder* is for back home (H5) and the opposite direction (H6) is just reserved at this time.

The first thing to investigate is how to detect and recognize the head pose movement. The method in a head movement controlling system is based on four looped process steps as shown in Fig. 93.4 as follows: 1) read sensors data, (2) recognize the data/signal pattern, 3) determine the head movement, and 4) controller response based on determined head movement. First, the system read sensors data using push method, secondly recognized the pattern of sensor's data, then determine which head pose movement type it is, lastly is the response effect for control something

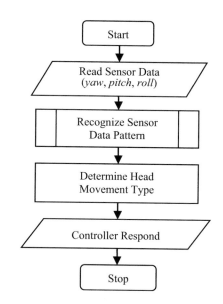

Fig. 93.4 Proposed head movement control system process step

in the mobile application that correlated with the particular detected head movement type. A preliminary investigation has to be done to recognize the pattern of accelerometer and gyro signal.

Fig. 93.5 Signal pattern of coremotion attitude while user turn the head to the right (H2)

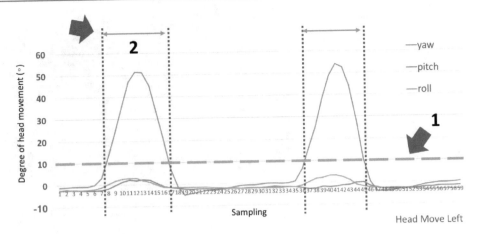

93.2.3 Method for Detecting Head Pose Movement

The method is based on preliminary research work for recognizing the signal pattern from the data gathered from smartphone's internal sensors [10]. The method for detecting and recognizing the head pose is based on the analyzing of *attitude* data including *yaw*, *roll*, and *pitch*. We proposed the general algorithm as shown in Fig. 93.4 and sample of *pseudocode* of detecting head pose algorithm as shown in Fig. 93.7. Attitude data pattern for H1 and H2 is affected only by *yaw* data as shown in Fig. 93.5. The system then starts to count if *yaw* degree is higher than a specific threshold. We use 10 degrees as the base threshold line (point 1) since the amplitude of pitch and roll is below the threshold value when user's H1 or H2 head movements. The number of the counter when user's head move more than 10 degrees is using to determine that user is in the motion of H1 or H2 (point 2). If the counter is in the specific number between *yMin* and *yMax* then system recognized it as user head movement in H1 or H2 pose. Determine the H1 or H2 is based on the peak value of the absolute degree level, if positive then determine as H1, and H2 for negative value. For H3-H4 and H5-H6 we use *roll* and *pitch* respectively (Fig. 93.6).

There are two challenges in this system when using *CoreMotion* attitude data. These are, how to get high accuracy when combine all the head pose movement detection; and the second is about user first head orientation. The first problem is an algorithm problem to combine all processes with high accuracy. For the second problem because *attitude* data is in degree (after conversion from radian to degree), which means that first front position as zero degrees, then we have to consider about user orientation movement. If the user change his/her base front position orientation then the algorithm is not working anymore.

User first front orientation position is a challenging because in practice, user's head can move any direction

```
function DetectPose()
{
    yawData = CoreMotion.Attitude.Yaw
    if |yawData| > yawTreshold
        yawCount++
    else
        allCount++
    if (yawCount > yawMin) && (yawCount < yawMax)
    {
        if yawData < 0
            moveLeft()
        else
            moveRight()
    }
    if allCount > 60
        yawCount = 0
}
```

Fig. 93.6 Pseudocode of head pose detection

and orientation, but only 6 specific head pose should detected, recognized and use as controller. We improved the algorithm to ensure that our proposed system is robust and able to avoid any kind of head movement beside 6 types of head pose defined as H1 to H6. The threshold number is a static number while the user's base front orientation is adaptable depend user's head base front orientation. If the user move the head's orientation more than specific cycle then we change the front base orientation based on current orientation.

93.3   Implementation and Evaluation

93.3.1   Implementation of iHelp HEMOCS

There are plenty of mobile application can implement HEMOCS as new control method. This method is applicable for any kind of mobile application in the area where user should not use their hand for controlling something in the

Fig. 93.7 Implementaion of
HEMOC system

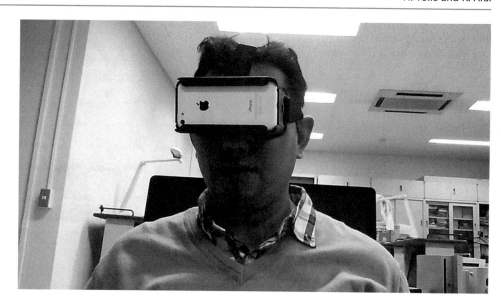

Table 93.2 Design of iHelp HEMOCS menu

Menu	Function	Generated voices
Hello	Greetings	Hello, nice to meet you
Help	Need help	I need help, please help me
Hungry	Want to eat	I'm hungry, please bring me the food
Drink	Want to drink	I'm thirsty, please bring me the drink
Toilet	Want to go to toilet	Please bring me to the toilet
Thank you	Say thank you	Thank you for your help

screen, for example for driver, welder, and etc. In the area of healthcare, HEMOCS can implemented in the mobile application for usage by disabled people where they cannot talk and using their hands for controlling an application. Figure 93.7 show the illustration of implementation of HEMOCS in HMD based mobile application. This time we implement HEMOCS as assisted communication device for disabled people.

We develop *iHelp HEMOCS* application on iOS based mobile device using iPhone 5C and an HMD cardboard. The iPhone was chosen because it has internal sensors with an excellent accuracy and *CoreMotion* library for simplification of reading sensors data. System also required a network communication capabilities while simple implementation for mobile wireless connection with *AirPlay* feature on iOS environment.

The implementation of the application is based on typical patient's need when they cannot do it by themselves directly. These requirement is based on common nursing system in Japan like care workers (*kaigofukushishi*) for elderly people that sometimes has language problem because the workers are from different country which cannot understand clearly what desired by patient. The requirement of the iHelp HEMOCS are as follows: (1) System should exactly recognized particular head movement (pose); (2) Robust

controller system that avoid coincide movement; (3) Customized menu (Request Items to communicate); (4) Customized instruction on chosen menu; (5) Support multi language; and (5) Able to communicate with other devices through wireless connection.

We create six common request for a patient, as shown in Table 93.2. Patient can ask someone for help, request for food or drink, request for bring the patient to toilet, and also can say greetings and thank you. These six menus is modular since all information is put on an XML file and can change externally. Each menu has its own recorded sound. When a menu is chosen by user then system will generate the related recorded voice and also send the request to other connected devices Tables 93.3 and 93.4.

User can control the menu with head pose movement control only as shown in Fig. 93.8. User move the head in H1 or H2 movement for select the menu, and H3 for chose a menu. Then system will generate a recorded voice and send the request to other devices. After a chosen menu finish to playing the recorded voice, user can back to the menu with H4 movement or H5 for back to initial home screen. These head movement control system then can control what user want to say to other through a mobile application. Since this is just a first model, the modification of menu and information is possible depend on what user or patient need.

Table 93.3 Accuracy evaluation of each pose on 3 user

Head pose	Accuracy (%)			Average (%)
	U1	U2	U3	
H1	90	85	90	88.33
H2	90	90	93	91.00
H3	70	73	77	73.33
H4	75	83	73	77.00
H5	95	83	83	87.00
H6	60	68	73	67.00
Average	80	80.33	81.50	80.61

Table 93.4 Accuracy evaluation based on movement duration

No	Movement duration	Max degree	Accuracy
1.	≤0.3	29.11	0%
2.	0.4–0.6	41.63	100%
3.	0.7–0.9	48.40	100%
4.	1–1.2	50.50	100%
5.	>= 1.3	50.33	0%

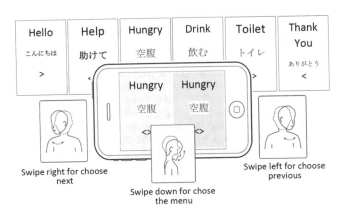

Fig. 93.8 Implementaion of HEMOC system

93.3.2 Accuracy Evaluation

Implementation of iHelp HEMOC System should evaluate to measure the level of success of implementation. Accuracy parameter is use to evaluate the performance of proposed method on detecting user head movement. Accuracy is the overall success rate, when user start to move the head into particular move and system respond it with corresponding control type on the application. We evaluate the accuracy using 3 user act as the patient using iHelp HEMOCS. Experiment results as shown in Table 93.5 shows the accuracy level of each movement type (H1 to H6) on each user (U1 to U3). From the evaluation results we got average accuracy is 80.61% for all head movement type. In particularly, head movement H4 and H6 has lowest accuracy because some errors on determination of movement type. These errors happen because user move their head in various speed.

Table 93.5 Usability evaluation results

No	Usability factors	Average score
1.	Functionality	92.73
2.	Easy to use	80.00
3.	Effectiveness	83.64
4.	Satisfaction	80.00
5.	Understandable	76.36
	Average	87.27

Different user average is almost same but not identical on each movement type. Using the same evaluation process, but categories it into head movement duration as shown in Table 93.3 shows that 100% accuracy is achieved when user move the head in the speed between 400 millisecond and 1.2 second, with average head maximum degree is around 46.84°.

93.3.3 Usability Evaluation

Usability evaluation is conducted to measure the satisfaction of the user while using the new proposed controller systems. Experiments results by 25 users trying to use the *iOS based* iHelp Application with head pose movement control as shown in Table 93.5. We evaluate 5 usability factor as follows: *functionality*, *easy to use*, *effectiveness*, *satisfaction*, and *understandable*. Highest average score 92.73% is achieved in the factor of "*functionality*" which means that user thinks this control type is suits for usage by disabled person. The least average score 76.36% is achieved in the factor of "*understandable*", which means that some user still difficult to use the new control system for the first time even they think that is effective. The average of usability factors results reach 87.27% categorized as 'acceptable' which means that the user satisfied with iHelp HEMOCS mobile application controlling by using their head movement only. Some user's feel that the speed of head movement should calibrating for each user before they start to use as controller. So the system will adaptable with user's head movement speed.

93.4 Conclusion and Future Works

The combination of smartphone, head-mounted display, and head movement control system, is cheap, modularity development, and potential usage in different field of application. Some area of the implementation of these combination are healthcare, entertainment, game application, virtual reality and augmented reality.

Implementation of head movement control system in the mobile healthcare application is successfully implemented in the initial application of iHelp HEMOCS. Experimental results of the prototype system show that this system has an acceptance accuracy level within certain movement speed. It means that this novel system has proven as a new interaction method between disabled people with others through a mobile application by head movement control only. This opens up a broad space for further application for patients with certain special needs.

However, advance experiment should be done with real patient or impaired people to get the best sensitivity level of head pose movement speed. In other hand, the improvement of the method accuracy is also important as well as sensitivity. In the near future, we will continuing to improve the iHelp HEMOC System for real usage as a healthcare application for helping impaired patient. More experiment with real patient on different type of patient, for example for elderly people, stroke patient, or speech impaired people. We hope this technology will have significant role in human life.

Acknowledgement This research work is funded by Dikti SAME 2015 project of Indonesian Ministry of Research, Technology & Higher Education, as collaboration research between Brawijaya University, Indonesia and Saga University, Japan.

References

1. Al-Rahayfeh, A., & Faezipour, M. (2013). Eye tracking and head movement detection: A state-of-art survey. *IEEE Journal of Translational Engineering in Health and Medicine, 1*, 11–22.
2. Manogna, S., Vaishnavi, S., & Geethanjali, B. (2010). Head movement based assist system for physically challenged. In *Proceeding 4th ICBBE*, Chengdu, China, (pp. 1–4).
3. Liu, K., Luo, Y. P., Tei, G., & Yang, S. Y. (2008). Attention recognition of drivers based on head pose estimation. In *Proceeding IEEE VPPC*, Harbin, China, (pp. 1–5).
4. Murphy-Chutorian, E., & Trivedi, M. M. (2010). Head pose estimation and augmented reality tracking: an integrated system and evaluation for monitoring driver awareness. *IEEE Transactions on Intelligent Transportation Systems, 11*(2), 300–311.
5. Kupetz, D. J., Wentzell, S. A., & BuSha, B. F. 2010. Head motion controlled power wheelchair. In *Proceeding IEEE 36th Annual northeast bioengineering conference*, New Jersey, (pp. 1–2).
6. King, L. M., Nguyen, H. T., & Taylor, P. B. (2005). Hands-free head-movement gesture recognition using artificial neural networks and the magnified gradient function. In *Proceeding 27th Annual conference engineering medicine biology*, Shanghai, China, (pp. 2063–2066).
7. Nguyen, S. T., Nguyen, H. T., Taylor, P. B., & Middleton, J. (2006). Improved head direction command classification using an optimised Bayesian neural network. In *Proceeding 28th Annual international conference EMBS*, New York, (pp. 5679–5682).
8. Satoh, K., Uchiyama, S., & Yamamoto, H. (2004). A head tracking method using bird's-eye view camera and gyroscope. In *Proceeding 3rd IEEE/ACM ISMAR*, Arlington, USA, (pp. 202–211).
9. Arai, K., Tolle, H., & Serita, A. (2013). Mobile devices based 3D image display depending on user's actions and movements. *International Journal of Advanced Research in Artificial Intelligence (IJARAI), 2*(6), 71–78.
10. Tolle, H., Pinandito, A., Adams J, E. M., & Arai, K. (2015). Virtual reality game controlled with user's head and body movement detection using smartphone sensors. *ARPN Journal of Engineering and Applied Sciences (JEAS), 10*(20), 9776–9782.
11. Tolle, H., & Arai, K. (2016). Design of head movement controller system (HEMOCS) for control mobile application through head pose movement detection. *International Journal of Interactive Mobile Technologies (IJIM), 10*(3), 24–28.
12. Tolle, H., & Arai, K. Google cardboard SDK. http://developers.google.com/cardboard. Accessed 23 Jan 2015.
13. Tolle, H., & Arai, K. Oculus rift developer documentation. http://developers.oculus.com/. Accessed 12 May 2015.

Evaluation of Audio Denoising Algorithms for Application of Unmanned Aerial Vehicles in Wildlife Monitoring

Yun Long Lan, Ahmed Sony Kamal, Carlo Lopez-Tello, Ali Pour Yazdanpanah, Emma E. Regentova, and Venkatesan Muthukumar

Abstract

Unmanned Aerial Vehicles (UAVs) have become popular alternative for wildlife monitoring and border surveillance applications. Elimination of the UAV's background noise for effective classification of the target audio signal is still a major challenge due to background noise of the vehicles and environments and distances to signal sources. The main goal of this work is to explore acoustic denoising algorithms for effective UAV's background noise removal. Existing denoising algorithms, such as Adaptive Least Mean Square (LMS), Wavelet Denoising, Time-Frequency Block Thresholding, and Wiener Filter, were implemented and their performance evaluated. LMS and DWT algorithms were implemented on a DSP board and their performance compared using software simulations. Experimental results showed that LMS algorithm's performance is robust compared to other denoising algorithms. Also, required SNR gain for effective classification of the denosied audio signal is demonstrated.

Keywords

Audio denoising · Unmanned aerial vehicle (UAV) · Acoustic denoising · Wildlife monitoring

94.1 Introduction

In recent years Unmanned Aerial Vehicles (UAVs) have become very affordable and small and are increasingly used in a wide range of wildlife monitoring, emergency response and broader surveillance applications. Commonly, vision-based sensors and processing are used for the above target detection. However, these techniques require a clear line of view, good illumination, and occlusion-free environments for effective detection. The use of acoustic techniques for specific target detection have been proposed as an alternate solution. However, acoustic techniques have

their own set of problems, such as noisy environments, a mixture of different sound sources, reflection from nearby structures to name a few. Especially UAV captured audio signals for target detection have target audio signal polluted with propeller, motor and wind noises. This necessitates the use of robust acoustic denoising algorithms before applying any detection algorithms. Audio denoising aims at offsetting the noises while retaining the required acoustic target signal. The low Signal to Noise Ratio (SNR) due to the pronounced wind, vibration, motor and propeller noise acoustic denoising in UAV faces number of substantial challenges [1].

In this paper, the feasibility of applying various denoising algorithms to noisy target acoustic data are explored. Typically, denoising are classified into diagonal and non-diagonal estimation techniques [2–5]. Also, denoising algorithms are classified based on the domain they operate: time, frequency, and time-frequency domain. There exist

Y.L. Lan • A.S. Kamal • C. Lopez-Tello • A.P. Yazdanpanah (✉)
E.E. Regentova • V. Muthukumar
Department of Electrical and Computer Engineering,
University of Nevada, Las Vegas, NV, USA
e-mail: pouryazd@unlv.nevada.edu

© Springer International Publishing AG 2018
S. Latifi (ed.), *Information Technology – New Generations*, Advances in Intelligent Systems and Computing 558, DOI 10.1007/978-3-319-54978-1_94

various classical denoising algorithm. In this work the following four denoising techniques: (1) time-frequency block-thresholding algorithm, (2) Wiener filter, (3) Least Mean Square (LMS) algorithm, and (4) Wavelet denoising algorithm. The algorithms were implemented in software and hardware. These denoising algorithms were evaluated for signal to noise ratio (SNR), segmental signal to noise ratio (SSNR), log likelihood ratio (LLR) and log spectral power distance (LSPD). The noise sources considered were Additive White Gaussian Noise (AWGN), brown noise, pink noise, quadcopter propeller, custom mid-size motorized UAV (actual recording), and electrical noise.

The remainder of this paper is organized as follows. In Sect. 94.1, describes in details the denoising algorithms implemented in this work. In Sect. 94.2, the comparison results from these denoising algorithms and some improvements are illustrated. Finally, conclusion in presentation in Sect. 94.3.

94.2 Denoising Algorithms

To circumvent this lack of locality property of Fourier transform, the short-time Fourier transform (STFT) is applied to represent the signal both in time and frequency domain [6]. STFT is a revised version of Fourier transform where the signal is divided into small segments (portions) that are small enough to be assumed as stationary segments of the signal. STFT of a signal $x(n)$ is defined as:

$$X(m, \omega) = \sum_{t=-1}^{W} x(t)w(t-m)e^{-j\omega t/W} \qquad (94.1)$$

where $x(t)$ is the signal at time t, $w(t)$ window function, ω is the frequency index, and m is the time index. It is assumed the noise and clean signal are uncorrelated. The noisy audio signal using the additive noise model is expressed as

$$y(t) = x(t) + n(t) \qquad (94.2)$$

where $y(t)$ is the noisy signal, $x(t)$ is the clean signal, and $n(t)$ the noise signal. And applying STFT

$$Y(m, \omega) = X(m, \omega) + N(m, \omega) \qquad (94.3)$$

The noise reduction function $f()$ when applied to the noisy signal $Y(m, \omega)$ results is an estimate clean signal denoted as $\widehat{X}(m, \omega)$ and is expressed as a filter function $H(m, \omega)$

$$\widehat{X}(m, \omega) = H(m, \omega)Y(m, \omega) \qquad (94.4)$$

94.2.1 Wiener Filter

In the frequency domain, Wiener filter [7, 8] the estimated clean signal $\widehat{X}(\omega)$ is expressed as:

$$\widehat{X}(\omega) = H(\omega)Y(\omega) \qquad (94.5)$$

where $H(\omega)$ is the filter frequency response and $Y(\omega)$ is the noisy input signal. The estimation error signal $E(\omega)$ and the mean square error of the denoising algorithm is expressed as:

$$E(\omega) = X(\omega) - \widehat{X}(\omega) = X(\omega) - H(\omega)Y(\omega) \qquad (94.6)$$

$$\xi[|E(\omega)|^2] = \\ \xi[(X(\omega) - H(\omega)Y(\omega))^*(X(\omega) - H(\omega)Y(\omega))] \qquad (94.7)$$

where $\xi[]$ is the expectation function, the symbol * denotes the complex conjugate. The least mean square error Wiener filter in the frequency domain is give as:

$$H(\omega) = \frac{P_{XY}(\omega)}{P_{YY}(\omega)} \qquad (94.8)$$

where $P_{YY}(f) = E[Y(\omega). Y^*(\omega)]$, and $P_{XY}(\omega) = E(X(\omega)Y^*(\omega)]$ are the power spectrum of $Y(\omega)$, and the cross power spectrum of $Y(\omega)$ and $X(\omega)$ respectively.

94.2.2 Least Mean Square Algorithm

Least Mean Square (LMS) is an adaptive iterative denoising algorithm. LMS adaptive algorithm consist of two steps: (1) filtering process, and (2) adaptive process [9, 10]. The weights of the N-tap filter $\mathbf{w}(n)$ is given as $[w_1(n)w_2(n)...w_N(n)]^T$. The input sequence $x(n)$ is also expressed as a vector $\mathbf{x}(n) = [x(n)x(n-1)... w(n-N+1)]^T$, where N indicates immediate past samples of $x(n)$.

The filter output can be mathematically expressed as:

$$y(n) = \mathbf{w}^T(n)\mathbf{x}(n) \qquad (94.9)$$

Consequently the error signal, the difference of desired and output signal can be expressed as:

$$e(n) = d(n) - y(n) = d(n) - \mathbf{w}^T(n)\mathbf{x}(n) \qquad (94.10)$$

An adaptive filtering algorithm adjusts the filter weights $\mathbf{w}^T(n)$ at each time instant based on the measured value of $e(n)$ and is expressed as:

$$\mathbf{w}(n+1) = \mathbf{w}(n) + 2\mu\, e(n)\mathbf{x}(n) \tag{94.11}$$

where, μ is defined as step factor or convergence factor. This factor determines the rate of convergence of the filter.

94.2.3 Block Thresholding

A clean signal $x(t)$ is contaminated by a noise $n(t)$ is represented as:

$$y(t) = x(t) + n(t), t = 0, 1, \ldots, N-1 \tag{94.12}$$

The time-frequency transform of the audio signal $y(t)$ over the time-frequency block (t, ω) is given as:

$$Y(m, \omega) = \sum_{t=0}^{N-1} y(t)w(n-m)e^{-jt\omega} \tag{94.13}$$

The time frequency plane (m, ω) is segmented in I blocks B_i whose shape is chosen arbitrary [11, 12]. The signal estimator $\widehat{x}$ is calculated from the noisy data y with a constant attenuation factor a_i over each block B_i is represented as:

$$\widehat{x}[n] = \sum_{i=1}^{I} \sum_{(m,\omega)\in B_i} a_i Y(m, \omega) g_{m,\omega}[n] \tag{94.14}$$

where the short-time Fourier $g_{m,\omega}$ can be written as $w[n-lu]ex\,p(i2\pi\,k\,n/K)$, where $w[n]$ is a time window of support size K, which is shifted with a step $u \leq K$. m and ω and are respectively the integer time and frequency indexes with $0 \leq l \leq N/u$ and $0 \leq k \leq K$. To compute the value of each a_i, one determines the risk $r = E \parallel f - \widehat{x} \parallel$ such that

$$r = E \parallel x - \widehat{x} \parallel \leq$$
$$1/A \sum_{i=1}^{I} \sum_{(m,\omega)\in B_K} E|a_i Y(m, \omega) - X(m, \omega)|^2 \tag{94.15}$$

Since $Y(m, \omega) = X(m, \omega) + N(m, \omega)$, the upper bound of the above equation is minimized by choosing

$$a_i = 1 - \frac{1}{\widehat{\xi}_i + 1} \tag{94.16}$$

where $\xi_i = \overline{Y_i^2}/\overline{\sigma_i^2} - 1$ is the average a unbiased priori SNR in B_i, and $B_i^{\#}$ is the number of coefficients $(m, \omega) \in B_i$.

$$\overline{Y_i^2} = 1/B_i \sum_{(m,\omega)\in B_i} |Y(m, \omega)|^2 \tag{94.17}$$

94.2.4 Discrete Wavelet Transform

Wavelet transform compared to Short-Time Fourier Transform (STFT) analyzes the signal at different frequencies with different resolutions [13, 14]. This approach is well suited for multiresolution analysis when a signal has high frequency components for short duration and low frequencies for longer duration.

The noisy audio signal considering the additive noise model with Gaussian probability density function with $\mu=0$ and $\sigma^2=1$ is expressed as

$$y(t) = x(t) + n(t) \tag{94.18}$$

By applying the Wavelet transform Ψ the above equation is transformed to

$$\Psi[y(t)] = \Psi[(x(t) + n(t)] = \Psi[x(t)] + \Psi[n(t)] \tag{94.19}$$

Solving for $x(t)$:

$$x(t) = \Psi^{-1}[\Psi[y(t) - n(t)]] \tag{94.20}$$

Since the value of the clean signal $x(t)$ and the noise $n(t)$ are unknown the value of the clean signal is estimated using the expression

$$\widehat{x(t)} = \Psi^{-1}[\Psi[y(t)] - \Psi[n(t)]] \tag{94.21}$$

The function Ψ is a wavelet coefficient expressed as Ψ_{jk}, where j denotes the decomposition level and k is the index of coefficient in the respective level. Soft thresholding is used in wavelet denoising process and there are various schemes to select the threshold for $\Psi[y(t) - n(t)]$ in (20) to arrive to the estimate of $\widehat{x(t)}$, the one adopted in this work is by VisualShrink [15] shown as:

$$\lambda_j = \sigma_n 2\log_2(N) \tag{94.22}$$

where λ_j ia threshold at level j and N is the number of samples and σ is a standard deviation of coefficients at level j.

94.2.5 Implementation of Denoising Algorithms

The above discussed algorithms are implemented in software using MATLAB simulation software, and the LMS and DWT denosing algorithms in hardware on TI's D5535 eZdsp evaluation board. The D5535 eXdsp board has a TMS320C5535 DSP processor, TLV320AIC3204 stereo codec, and a singe input audio input jack. A dataset consisting of four classes (animal, birds, human, vehicle) of audio data with 100 audio files in each of the classes has been collected. The audio dataset was sampled at 8 KHz and each audio data file is of 3.0 secs in duration. To evaluate the performance of denoising algorithms, three sources of noise (Additive White Gaussian Noise, UAV Propeller and Motor noise, and DJI Quadcopter Propeller noise) have been considered. These noise signals have been added to the clean audio dataset to generate audio data with different input SNRs (0 dB, -3 dB, -6 dB, -9 dB, -12 dB, -15 dB, and -20 dB). The noisy input signals are normalized to avoid clipping and saturation during the denoising process.

In the implementation of the Wiener Filter (WF) denoising algorithm, a linear estimate of the original signal is obtained while minimizing the mean squared error between the original and enhanced signal. The length of the filter considered for implementation is L = 6. The noise estimation is performed throughout the length of the input signal. In the implementation of the BTH denoising algorithm, STFT with a window time of 50 ms, a redundancy factor of one are considered. The block size, $B_i^3 = L_i \times W_i$ is chosen adaptively by using the estimated risk (r). The BTH disjoint rectangular block size $B_i^\#$ of 8×16 and 2×4 have a λ value of 1.5 and 4.7 respectively. In the implementation of the LMS denosing algorithm, a FIR filter of length $N = 30$ and convergence or step size (μ) = 0. 017 are chosen for optimal denoising results. The DWT denoising algorithm implementation employs a filter length of six and the level of decomposition was varied from $L = 1 \sim 10$. Using simulation results the optimal decomposition level was determined as L = 6. The type of wavelet considered was Daubechies' wavelet. Based on the software simulations, LMS and DWT denoising algorithms resulted in the highest SNR gain. Therefore, these two denoising algorithms were considered for hardware implementation. The denoising algorithms parameters considered for hardware are identical to the parameters considered for software implementation.

94.2.6 Results

94.2.6.1 Denoising Results

This section discusses the performance of the denoising algorithm in software simulation and hardware implementation. These denoising algorithms were evaluated for the following performance metrics: Signal to Noise (SNR), Log Likelihood Ratio (LLR), and Log Spectral Power Distance (LSPD). Figure 94.1a, shows the SNR gain comparison of the different denoising algorithms. We can conclude that LMS denoising algorithm results in the highest SNR gain after denoising compared to other denoising

algorithms for all classes and noise sources. Also as shown in Fig. 94.1b, SNR gains are higher for Gaussian White noise and UAV propeller+motor noise compared to the DJI quadcopter noise. The algorithms when evaluated for LLR and LSPD show (Fig. 94.2) that LMS and DWT denoising algorithms provide better results (value closer to zero) compared to other denoising algorithms. The LMS and DWT denoising algorithms were implemented on a DSP board (Fig. 94.3a) and their performance evaluated and compared to software simulation (Fig. 94.3b). The better performance of SW compared to HW is due to constrain of the DSP board to implement non-floating point operations. Since the DSP board has only one audio input we also implemented and compared a single input LMS software simulation as shown in Fig. 94.3.

94.2.6.2 Classification Results

As an application of the denoising algorithms implemented, their performance in classification of audio data has been implemented. We were able to achieve an overall accuracy higher than 55% using both KNN and SVM classifiers on the noisy audio dataset without dimensionality reduction. The following features were used for classification: zero crossing rate, energy, energy entropy, spectral centroid, spectral spread, spectral entropy, spectral flux, spectral rolloff, MFCCs (13 values), harmonic ratio, fundamental frequency, chroma vector (12 values), and TESPAR statistics (4 values). Feature dimensionality reduction was performed using both PCA and LDA methods. The accuracy of classification (4 classes) was 74% for SVM and 77% for KNN based classifiers. In both algorithms there is a significant drop in accuracy when SNR drops below 10 dB (<60%).

94.3 Conclusions

In this work, various classical denoising algorithms were implemented in SW and HW. The algorithms were evaluated for the following performance metrics SNR gain, LLR and LSPD. LMS and DWT denoising algorithms produces the best result in SW simulation. This algorithm was also implemented and tested in real-time using a DSP HW board. Comparisons and discussions of SW simulation and HW implementations are provided. Also, as an application of denoising algorithms, classification of the audio dataset has been investigated. Based on the results, the required SNR for acceptable classification accuracy is higher than 10 dB. Based on simulation results, the maximum SNR gain obtained at lower SNR is approximately close to 10dB and at higher SNR is close to 5 dB, we proposed to train our classifiers with noisy datasets to attain the required classification accuracy.

Fig. 94.1 SNR gain comparison
of denoising algorithms. (**a**) SNR
gain comparison of denoising
algorithms for various classes
implemented in SW. (**b**) SNR
gain comparison of denoising
algorithms for various noise
implemented in SW

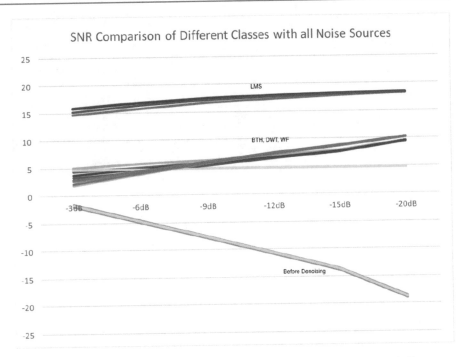

(a) SNR gain comparison of denoising algorithms
for various classes implemented in SW

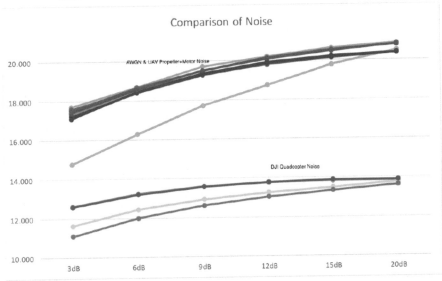

(b) SNR gain comparison of denoising algorithms
for various noise implemented in SW

Fig. 94.2 LLR and LSPD performance comparison of denoising algorithms

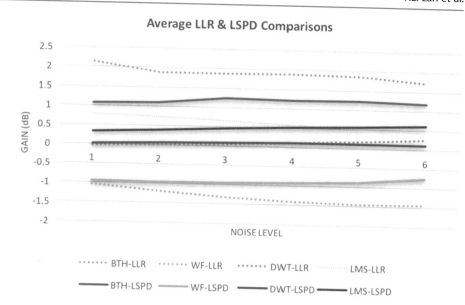

Fig. 94.3 SNR gain comparison of denoising algorithms in SW and HW. (**a**) HW comparison of denoising algorithms (LMS & DWT). (**b**) SW and HW comparison of denoising algorithms (LMS & DWT)

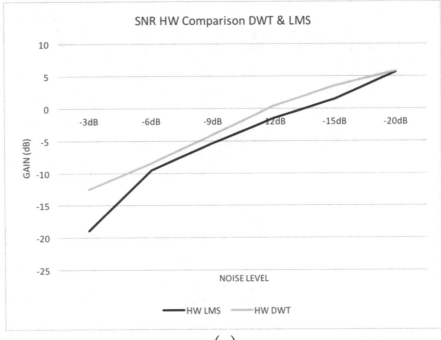

(a)

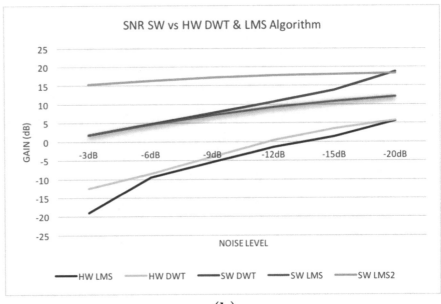

(b)

References

1. Marmaroli, P., & Falourd, X. (2012). A UAV motor denoising technique to improve localization of surrounding noisy aircrafts: Proof of concept for anti-collision systems. *IEEE Acoustics, Speech and Signals*, 333–351.

2. Guoshen, Y., & Mallat, S., & Bacry, E. (2008). Audio denoising by time-frequency block thresholding. *IEEE Transactions on Signal Processing, 56*(5), 1830–1839.

3. Kumari, L., Reddy, K., Krishna, H., & Subash, V. (2011). Time-frequency block thresholding approach for audio denoising. *International Journal of Advances in Science and Technology, 2*(5).

4. Pour Yazdanpanah, A., Mandava, A. K., Regentova, E. E., Muthukumar, V., & Bebis, G. (2014). A CUDA based implementation of locally-and feature-adaptive diffusion based image denoising algorithm. In *11th International Conference on Information Technology: New Generations* (pp. 388–393).

5. Obulesu, K. P., & Uday Kumar, P. (2013). Implementation of time frequency block thresholding algorithm in audio noise reduction. *International Journal of Science, Engineering and Technology Research (IJSETR), 2*(7).

6. Mihov, S.G., Ivanov, R.M., & Popov, A. N. (2009). Denoising speech signals by wavelet transform. *Annual Journal of Electronics*. ISSN:1313–1842.

7. Rox, J., & Vincent, E. (2013). Consistent wiener filtering for audio source separation. *IEEE Signal Processing Letters, 20*(3), 217–220.

8. Mao, R., Zhou, Y., & Liu, H. (2015). An improved iterative wiener filtering algorithm for speech enhancement. In *IEEE International Conference on Computers and Communications (ICCC)* (pp. 436–440).

9. Xu, L., et al. (2010). Research on LMS adaptive filtering algorithm for acoustic emission signal processing. In *Proceeding of the 8th World Congress on Intelligent Control and Automation*, 6–9 July 2010 (pp. 7037–7040).

10. Wu, Y., & Zhao, W. (2013). The improvement of audio noise reduction system based on LMS algorithm. In *International Conference on Computer Science and Application* (pp. 590–594).

11. Cai, T. (1999). Adaptive wavelet estimation: A block thresholding and oracle inequality approach. *Annals Statistics, 27*, 898–924.

12. Cai, T., & Zhouo, H. (2005). A data driven block thresholding approaches to wavelet estimation. Statistics Department, University of Pennsylvania, Technical Report.

13. Gargour, C., Gabrea, M., Ramachandran, V., & Lina, J. (2009). A short introduction to wavelets and their application. *IEEE Circuits and System Magazine, 2*, 57–67. ISSN:1531–636X

14. Donoho, D. L. (1992). Denoising by soft-thresholding. Technical Report no. 409, Stanford University.

15. Wieland, B., Urban, K., & Funken, S. (2009). Speech signal noise reduction with wavelets. Dissertation, Diplomarbeit an der Universitt Ulm.

Development of a Portable, Low-Cost System for Ground Control Station for Drones

Douglas Aparecido Soares, Alexandre Carlos Brandão Ramos, and Roberto Affonso da Costa Junior

Abstract

This paper aims the development of a portable, low-cost software and hardware environment to control the flight of a drone (aircraft up to 5 kg). The project uses the Android operating system of a tablet, what gives the user the benefits of the mobility, ergonomics and ease of field work that are characteristics of this device. The project focuses on portability. To achieve that, the chosen components are generally lighter and easier to transport rather than those traditionally used in a ground control station (GCS). In addition, the components and tools that were used in the project are low-cost, what increases the feasibility of its implementation. The RPA (Remotely Piloted Aircraft) flight control is performed through the communication between the Paparazzi UAV system and the PPRZonDroid application. It is convenient to say that the Paparazzi UAV and the PPRZonDroid application must be previously installed in a Raspberry Pi board and in a tablet, respectively. The use of projected ground control station has a wide area of utilization, since it can be used in various scenarios, both military and civilian.

Keywords

Drone • Raspberry Pi • Paparazzi UAV • Android • Ground Control Station

95.1 Introduction

Although the use of drones seems to be a recent subject, the use of RPAs dates back to the 19th century when the Austrians loaded unmanned balloons with explosives to attack targets in Venice. Even before the First World War, some engineers studied a way to bring an explosive device through the air to a target tens of kilometers away, what later led to the creation of missiles. In 1915, the engineer Nikola Tesla described in a study the military potential of a fleet of unmanned, air combat vehicles [1]. Over the years, there has been ever more research in this area along with the emergence of several companies that are dedicated to develop RPAs as well as equipment and technology to improve their functioning.

Currently, drones are increasingly involved in several applications, that address both military and civilian issues. One of the most important uses for drones is gathering information (photos, videos, etc.) to: perform real-time surveillance of a certain region; verify changes in vegetation, in water resources, etc.; monitor troops advancing; launch bombs and missiles. In [2], there are several examples about the usage of drones in military applications.

Drones can be categorized in various ways, for example regarding size, the maximum altitude reached, the type of fuel used, etc. In all their possible classifications RPAs should always be monitored by a human operator from a ground control station (GCS) [3, 4].

D.A. Soares (✉) • A.C.B. Ramos • R.A. da Costa Junior
Institute of Mathematics and Computer Science, Federal University of Itajuba (UNIFEI), Itajuba, MG, Brazil
e-mail: douglas.asoares@unifei.edu.br; ramos@unifei.edu.br; rcosta62br@unifei.edu.br

© Springer International Publishing AG 2018
S. Latifi (ed.), *Information Technology – New Generations*, Advances in Intelligent Systems and Computing 558, DOI 10.1007/978-3-319-54978-1_95

95.2 Background

To understand the functioning of the developed system some topics and concepts are fundamental, among which can be cited:

- Drone: Also called unmanned aerial vehicle (UAV), or remotely piloted aircraft (RPA), drone is any type of aircraft that does not require onboard pilots to be driven. The vehicle is automatically controlled by electronic and computational means or also can be operated with the aid of a remote control [5].
- Ground control station: A ground control station is a system of radios, and electronic and computing components intended to perform the communication with the RPA, sending and receiving signals and data in order to allow the user to control and monitor the aircraft flight.
- Raspberry Pi: Tiny size and low-cost are the two guidelines of the project called Raspberry Pi. It is a mini-PC that includes: processor, graphics processor, memory card slot, USB interface, HDMI output and their respective controllers. In addition, it also features RAM memory, power input and expansion bus. Though tiny, the Raspberry Pi board is a complete computer [6] (Fig. 95.1).
- Paparazzi UAV: It is an open-source drone hardware and software project encompassing autopilot systems and ground station software for multicopters/multirotors, fixed-wing, helicopters and hybrid aircraft. Paparazzi UAV was designed with autonomous flight as the primary focus and manual flying as the secondary. From the beginning it was designed with portability in mind and the ability to control multiple aircraft within the same system. Paparazzi features a dynamic flight plan system that is defined by mission states and using way points as "variables". This makes it easy to create very complex

fully automated missions without the operators intervention [3] (Fig. 95.2).
- PPRZonDroid: It is an application that is used for controlling a Paparazzi aircraft with an Android device. Server and Android application are connected through Wi-Fi, therefore they must be in the same network. Several clients can be connected at the same time with full or restricted access. This app must work with Android 3.2 and up [7] (Fig. 95.3).

95.3 Materials and Methods

For the development of this project was necessary to pass through some essential steps. First of all, a study concerning the operation of a RPA was carried out, as well as the techniques of its flight control, besides some applications in which it has been used [8, 9, 10].

The system Paparazzi UAV, installed in a Raspberry Pi development board, was used to control the drone flight. The Paparazzi UAV is developed for the Linux operating system and it has a portable version for Android, called PPRZonDroid, which is used in this research.

An important step for the proper functioning of the designed system was the definition of which operating system would be used in Raspberry Pi. Among the several versions of Raspbian and Linux operating systems that were tested, the chosen operating system was Ubuntu 12.04.5 LTS. This choice was made because it was the only OS among the ones tested that allowed the full installation and proper use of the Paparazzi UAV. All the other OSs that were tested presented many errors either during

Fig. 95.1 Raspberry Pi development board Model B+ V1.2 (Available at: <https://www.raspberrypi.org/wp-content/uploads/2014/07/rsz_b-.jpg>. Accessed in September, 2016)

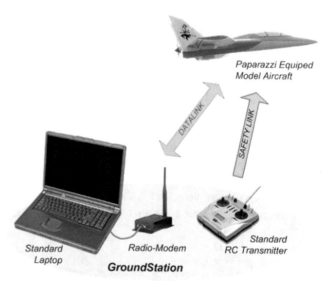

Fig. 95.2 Paparazzi UAV project overview (Available at: <https://wiki.paparazziuav.org/w/images/1/1e/Paparazzi_System_overview.jpg>. Accessed in September, 2016)

Fig. 95.3 Main screen of PPRZonDroid (Available at: <https://wiki.paparazziuav.org/w/images/a/ad/PPRZonDroid_tablet.png>. Accessed in October, 2016)

installation or during the operation of the Paparazzi. The Paparazzi UAV installation process in several operating systems can be found in [11].

After this step, a study was undertaken to define the form of communication between the Paparazzi UAV and the PPRZonDroid to be used. The PPRZonDroid was installed in a HTC Google Nexus 9 tablet. An important observation to be pointed out is that the Paparazzi UAV and the PPRZonDroid need to be on the same network to establish connection, hence both used the Wi-Fi network available on campus of UNIFEI. The tablet directly connects to the network through its own software and hardware systems. On the other hand, the Raspberry Pi had to be connected to the wireless network through the use of a dongle Wi-Fi adapter.

Once Paparazzi UAV and PPRZonDroid are on the same network, certain procedures need to be followed: first of all, the application server must be started in Paparazzi UAV; furthermore, it should be set up a new type of aircraft in Paparazzi Center or another option is to select one of the basic types that the software offers; after starting the simulation in the Paparazzi UAV, it should be opened the PPRZonDroid application in the tablet. In order to enable the communication, it is necessary that both Paparazzi UAV

as PPRZonDroid are configured as described in [7] (Fig. 95.4).

Android was chosen to be used in this research due to some features that are offered by this operating system that are relevant for this project. Among them could be cited: it is an open source system that allows the use of new technologies and it has access to features implemented in the kernel of Linux; it is adopted by a wide variety of devices on the market and it is compatible with several brands of devices; it assures freedom of customization to its users; there is a huge amount of applications already implemented to be ran in it and there are more and more researchers and developers in this area. All these features of Android increase the range of use of the implemented project.

The hardware used in this project is composed of: a) Quadrotor; b) HTC Google Nexus 9 tablet, 2x NVIDIA Tegra K1 @ 2.50 GHz - Android Lollipop operating system (version 5.0.2); c) Raspberry Pi development board Model B + V1.2 (2014); d) USB keyboard; e) USB Mouse; f) HDMI to VGA converter; g) VGA monitor; h) dongle Wi-Fi adapter. Moreover, it is also necessary a pair the antennas that allow the communication between the drone equipped with Paparazzi UAV and the PPRZonDroid. One of the

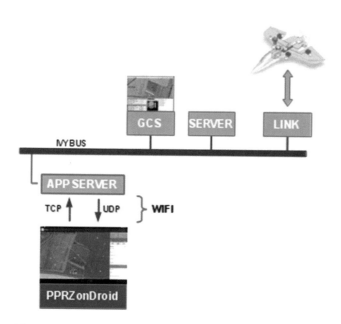

Fig. 95.6 Scheme of the designed system (Source: author)

fast processor and a powerful graphics processor are also desirable features because they allow a better usage of the data gathered during the drone flight by using that data in various applications. Newer versions of Raspberry Pi must be able to run the designed system too.

Fig. 95.4 Scheme of communication between the drone equipped with Paparazzi and PPRZonDroid (Available at: <https://wiki.paparazziuav.org/w/images/2/29/PPRZonDroidOverview.png>. Accessed in October, 2016)

Fig. 95.5 Quadrotor used in the project (Source: author)

antennas is shipped in the drone and the other one is connected to the Raspberry Pi (Fig. 95.5).

The HDMI to VGA adapter was used because the monitor available for this project didn't have a HDMI output. The dongle Wi-Fi adapter is necessary because the Raspberry Pi Model B + V1.2 doesn't have a Wi-Fi board, so it couldn't connect to a wireless network by itself (Fig. 95.6).

To the choice of the tablet that would be used in this project were considered among other characteristics, the storage and processing capacity and the version of the Android operating system. Tablets similar to or with better computational features than that earlier indicated must be able to run the designed system. A large storage capacity, a

95.4 Results and Discussion

From the initial idea to its current stage of development, this project has shown enormous relevance, since the use of RPAs/drones and a large range of researches about them have been widespread throughout the world. Moreover, the financial investment needed to implement the proposed system is relatively low because it consists of a few components and all are low-cost.

This project aims to create a ground-based drone control station more portable than a traditional ground control station. The function performed by the computer/laptop in a traditional GCS is made by a Raspberry Pi board and a tablet in this design.

In addition, the implementation of this project is low-cost if compared to the financial investment required to implement a traditional GCS. The Raspberry Pi board used in the designed system costs approximately $ 27.22 [12] (considering as conversion factor 1 £ = $ 1.2375) and the tablet used costs around $ 399.00 [13], resulting in approximately $ 425.00. This value is relatively smaller than that required to buy a computer/laptop with similar computational characteristics similar than that presented by the tablet and the Raspberry Pi employed in this project.

Although there are several ongoing researches about RPAs, this project deals with an innovative system. In Brazil, similar works are performed within few institutes and universities, among which ITA (Technological Institute of Aeronautics) can be highlighted. Thus, all the solutions to the problems that were faced during the development of this project demanded a considerable effort because there were few technical references for consultation.

As an example of the difficulties met, the definition of the operating system that would be used in the Raspberry Pi was made after several attempts with various versions of Raspbian and Linux operating systems. As can be seen in [14], the process of installation and configuration of the Paparazzi UAV in a Raspberry Pi takes too long (about 4 to 5 hours). Thus, every time that one attempt to use Paparazzi in each one of the tested operating systems ended up on failure, whether by fatal errors during installation or during the execution in the project, a new attempt with another OS demanded a similar amount of time. However, all of these attempts brought benefits to the Paparazzi users community because those failures were reported to the developers and they added information in some pages of the Paparazzi website warning about those errors.

Moreover, as the Paparazzi UAV is an open source system, most of the aid found to resolve errors in its operation comes from its community of users and developers. In some cases, to find a solution to problems met is not always quick or even trivial. For example, it was found an error that is consuming considerable time to be solved: when trying to perform the communication between the PPRZonDroid application and the Paparazzi UAV, a socket failure takes place and the application server stops working, shutting down the communication. Even with the help of Paparazzi system developers, it has not been possible to resolve this error yet. Possible solutions are: to try to use other operating system in the Raspberry Pi or even other versions of Paparazzi UAV and PPRZonDroid; to try to use other model of Raspberry Pi board.

Finally, some suggestions for future works based in this project are: to look for another solution to the error reported above, or to try the possible solutions cited.

95.5 Conclusion

Allowing the RPA's ground control station be deployed in a tablet will guarantee to user a huge range of features that are offered by this type of device, such as mobility, ergonomics, ease to do field services and the possibility of use it in several places and for many purposes. In addition, the projected system has relevance for both civilian and military applications.

This work deals with the development of a portable, low-cost system that allows the user to perform several activities with drones in various environments. After the establishment of communication between the Paparazzi UAV and the PPRZonDroid, all the other settings and the flight control can be carried out using the tablet.

The possible solutions presented for the reported problems can be used as a starting point for future works and researches using this project as a base.

Acknowledgment I thank CAPES (Coordenação de Aperfeiçoamento de Pessoal de Nível Superior) for the scholarship at the Graduate Program in Computer Science and Technology, which provided financial support to this project.

References

1. A origem dos vant – AERO Magazine. Available at: http://aeromagazine.uol.com.br/artigo/origem-dos-vant_1907.html. Accessed Aug 2016.
2. Love, P. (2011). *The UAV Question and Answer Book*. Publisher: createSpace independent publishing platform. ISBN-10: 1463573162; ISBN-13: 978–1463573164.
3. PaparazziUAV. Available at: http://wiki.paparazziuav.org/wiki/Main_Page. Accessed Aug 2016.
4. Brisset, P. (2010). *The Paparazzi UAV System: Solutions Linux*. ENAC. Available at: http://www.recherche.enac.fr/~brisset/PascalBrisset_InfoIndustrielle_Paparazzi.pdf. Accessed Aug 2016.
5. Diferença entre Drone e Robô. Available at: <http://originaleexclusivo.com.br/diferenca-entre-drone-e-robo/. Accessed Sept 2016.
6. Garrett, F. Como funciona o raspberry pi? Entenda a tecnologia e sua aplicabilidade. Available at: http://www.techtudo.com.br/noticias/noticia/2014/11/como-funciona-o-raspberry-pi-entenda-tecnologia-e-sua-aplicabilidade.html. Accessed Sept 2016.
7. PPRZonDroid – PaparazziUAV. Available at: https://wiki.paparazziuav.org/wiki/PPRZonDroid. Accessed Oct 2016.
8. Glover, J. M. (2014). *Drone University*. Publisher: Droneuniversity. ISBN-10: 0692316035; ISBN-13: 978–0692316030.
9. Barnhart, R. K., Hottman, S. B., & Marshall, D. M. (2011) *Introduction to Unmanned Aerial Systems*. Publisher: CRC Press. ISBN-10: 1439835209; ISBN-13: 978–1439835203.
10. Fahlstrom, P. G., & Gleason, T. J. (1993). *Introduction to Uav Systems*. Publisher: Uav Systems Inc. ISBN-10: 9995144328; ISBN-13: 978–9995144326.
11. Installation – PaparazziUAV. Available at: https://wiki.paparazziuav.org/wiki/Installation. Accessed Oct 2016.
12. Raspberry Pi Model B+. Available at: http://uk.rs-online.com/web/p/processor-microcontroller-development-kits/8111284/. Accessed Dec 2016.
13. BUY THE NEXUS 9 – HTC United States. Available at: http://www.htc.com/us/go/buy-now-nexus-9/. Accessed Dec 2016.
14. Installation/RaspberryPi – PaparazziUAV. Available at: https://wiki.paparazziuav.org/wiki/Installation/RaspberryPi. Accessed Oct 2016.

Collision Avoidance Based on Reynolds Rules: A Case Study Using Quadrotors

96

Rafael G. Braga, Roberto C. da Silva, Alexandre C.B. Ramos, and Felix Mora-Camino

Abstract

This work aims to present a collision avoidance algorithm developed to drive a swarm of Unmanned Aerial Vehicles (UAVs) using Reynolds flocking rules. We used small quadrotor aircrafts with 250 mm diameter, equipped with GPS and distance sensors, controlled by a Pixhawk autopilot board and an embedded Linux computer (Raspberry Pi). The control algorithm was implemented in C++ as a package for the ROS platform and runs on the Raspberry Pi, while the Pixhawk is responsible for the low level control of the aircraft. The distance sensors are used to detect nearby robots and based on this information the control algorithm determines the direction the robot should move to avoid collisions with its neighbors. We are able to see that from the individual perception and local interaction of each robot a group behavior emerges and the swarm move together coherently. A simulation environment based on the Gazebo Simulator was prepared to test and evaluate the algorithm in a way as close to the reality as possible.

Keywords

Collision avoidance • Robot Swarm • Unmanned aerial vehicles • Reynolds rules • ROS

96.1 Introduction

UAVs have grown in popularity in the last few years, due to the development of inexpensive, easy to build models such as the quadrotor. This type of vehicle is very maneuverable and can be controlled remotely or even autonomously, making it suitable for many civil and military applications. Examples of civil uses include aerial photography and search and rescue of survivors after a disaster, while military uses include surveillance and monitoring of an area. Since there is no pilot on board they can also be used in dangerous tasks, such as fighting wildfire.

The large range of possible applications has lead to the development of expensive commercial models for professional photography and research, but also cheaper, open-source projects that can be used for robotics research, as demonstrated in [1]. This is the case of the Pixhawk and the Arducopter project, which we used in this work.

Since the control of the quadrotor is already well understood, a new area of interest emerges that is the use of a group of UAVs to perform a mission, instead of only one unit. Many advantages can be obtained using this strategy. A group of robots is able to acquire more data from sensors and cover a bigger area than a single one; the group becomes more aware of its surroundings and is able to more effectively defend itself from threats; and even if some of the robots are lost the mission can be carried on by the remaining ones, thus the system becomes more robust.

R.G. Braga (✉) • R.C. da Silva • A.C.B. Ramos
Institute of Mathematics and Computer Science, Federal University of Itajubá, Itajubá, MG, Brazil
e-mail: fael_gb@yahoo.com.br

F. Mora-Camino
Air Transport Department, Ecole Nationale de l'Aviation Civile, Toulouse, France

© Springer International Publishing AG 2018
S. Latifi (ed.), *Information Technology – New Generations*, Advances in Intelligent Systems and Computing 558, DOI 10.1007/978-3-319-54978-1_96

But while this approach brings us many advantages, many challenges arise as well. When moving as a group, we have to ensure that each robot doesn't collide with its neighbors. To design a control strategy to solve this problem, researchers take inspiration from nature. There are many examples of animals that present group behavior, such as flocks of birds, schools of fishes, swarms of insects and herds of land animals. We observe that these animals are able to move as a group although there is no central control. In fact, each individual perceives the environment around itself and takes its own decisions. From these local interactions, a group behavior emerges.

Our goal is to develop a control algorithm to drive a group of quad-rotor UAVs together while avoiding collisions with each other, using strategies based on behaviors observed in nature. To better simulate the behavior of real groups of animals, our robots use sensors to detect neighbors instead of communicating their positions.

96.2   Background and Related Work

In 1987 Craig Reynolds introduced a set of behavioral rules aiming to simulate the movement of groups of animals, such as flocks of birds or schools of fishes[2]. He created a computer program in which entities called boids move following these rules. To best represent the behavior of real animals, Reynolds didn't create any central control, but instead programmed each boid to sense its own environment and decide where to move. This resulted in a fluid movement very similar to real bird flocks.

The flocking rules are represented in Fig. 96.1 and can be described as the following:

- Separation: boids try to move away from nearby flock mates to avoid collisions;
- Alignment: boids try to match their heading with other boids;
- Cohesion: boids try to move closer to the other boids to form a flock;

These three rules alone are sufficient to make the robots group and move randomly together. But since we want to control the swarm, we incorporated a fourth rule, called

Migration. This rule lets us set a destiny location we call migration point to where the robots try to move.

Since then, many researchers have tried to use Reynolds rules in the area of swarm robotics to drive groups of robots. Some of them are very notable and inspired our work. Hauert et al. used Reynolds rules to create a flock of 10 fixed-wing UAVs both in simulation and reality [4]. However, their UAVs flew at different altitudes, so they didn't have to avoid collisions with each other. In [5] a flock of 3 UAVs flew together, but again, in different altitudes.

Kushleyev et al. developed an impressive flock of 20 miniature quad-rotors capable of assuming many different formations, using a centralized control strategy [6]. Bürkle et al. created an outdoor quadcopter swarm also using central processing at a ground station [7]. However central control is something that do not exist in a real flock of birds and for this reason other works distributed the control, programming each UAV to make its own decisions. A remarkable example is [8], where Vsrhelyi et al. flew a flock of 10 quadcopters in an outdoor environment using Reynolds rules. An application of flocking UAVs can be seen in [9], where a group of simulated UAVs is used to monitor an area.

All these works however rely on communication between robots to create the flocking behavior. This is something that we also do not observe in nature, since animals detect each other through vision or other types of sensing. Moeslinger et al. attacked this issue, creating a swarm of robots that detect each other using infrared sensors [10].

96.3   Materials and Methods

96.3.1 Swarm Robotics

Swarm robotics is a new research field that studies the coordination of a large number of robots, taking inspiration from the behavior of social animals [11]. By understanding and recreating the behavior rules followed by these animals, researchers aim to develop robotic systems that present similar swarm intelligence.

Many animals, such as birds, fishes or insects, although very simple, are able to act as a group and gain many advantages from that. Some of these advantages include

Fig. 96.1 Reynolds flocking rules [3]. (**a**) Separation. (**b**) Alignment. (**c**) Cohesion

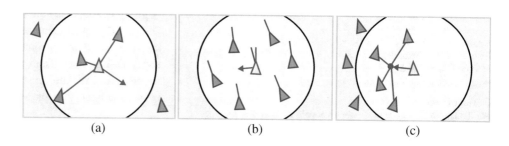

(a)   (b)   (c)

higher chance of finding food, avoiding predators and social interaction [12].

The goal in swarm robotics is usually to design a system composed by a large number of simple agents that move, interact with each other and with the environment following simple rules, such that a collective behavior emerges.

Swarm robotics systems are robust, meaning that the loss of individuals doesn't affect the system, scalable, meaning that they can grow to larger sizes without losing performance, and flexible, meaning that they can perform well in a wide variety of environments and tasks.

96.3.2 Quadrotor Dynamics

The quadrotor is a versatile and very maneuverable VTOL (Vertical Takeoff and Landing) vehicle capable of moving with 6 degrees of freedom (x, y, z, roll, pitch and yaw axis, showed in Fig. 96.2). This movement results from the individual control of each of its four motors, executed by a central controller board.

The aircraft flies by spinning its motors, which generates a vertical force that lifts the quadrotor in the air. However each propeller also creates a rotational torque in the Yaw axis that must be eliminated for stable flight. This is obtained by making each motor spin in the opposite direction of the adjacent motors, canceling the torque. Figure 96.2 also shows the direction of rotation of each motor.

It is possible to rotate the quadrotor in the pitch or roll axis by speeding up two motors on the same side and slowing down the two motors on the opposite side. For example, by speeding up the motors on the right and slowing down the motors on the left, the quadrotor rolls to the left. The same applies to pitch, but controlling the front or back motors.

Changing the yaw angle is obtained by speeding up two motors that are at diagonally opposite sides, and slowing down the other two. This way the torque in the yaw axis is no longer canceled and the quadrotor turns.

To move on the X and Y axis the motors must be controlled in a way that the quadrotor leans itself in the direction of the movement. To change altitude, that is, to move on the Z axis, all motors must be speed up or slowed down at the same time.

96.3.3 Pixhawk

The Pixhawk is an open-source flight controller board[13] responsible for the low-level control of the quadrotor. Originally it is intended to be controlled by a human pilot through a RC (Radio Controller) with 6 or more communication channels. The most common control scheme maps 4 analog channels to the axis of roll, pitch, yaw and throttle (z-axis) and uses 2 or more digital channels to change between different configuration settings or flight modes. The Pixhawk is equipped with gyroscope, accelerometers, magnetometer and barometer, and can also be connected to an external GPS module.

The software that runs on the Pixhawk is called Ardupilot. It is responsible for interpreting the sensors readings and signals received from the RC, generate flight data and control each individual motor, as explained on Sect. 96.3.2, to move the UAV in the intended direction. It also provides several flight modes, adapted to different control types and pilot's skill level.

The three most common flight modes are Stabilize, Altitude Hold and Loiter. In Stabilize mode, the pilot have direct control over the roll, pitch, yaw and throttle axis, and is intended to move the UAV by leaning it to the direction of movement. In Altitude Hold mode the UAV tries to maintain a constant altitude, so the pilot doesn't have to worry about the throttle. Loiter mode also maintains altitude and makes the X and Y position being controlled directly by the pitch and roll channels. As we will see, this mode is very useful for our control strategy.

Another useful feature of the Pixhawk is the ability to communicate with other devices through a protocol called MAVLink, which was developed specifically for UAV applications. Commonly the quadrotor is equipped with a telemetry module that connects wirelessly to a ground station, enabling the Pixhawk to exchange MAVLink messages with it. That way it is possible to see flight data in real time and also edit the Pixhawk configuration in the ground station.

Autonomous control of the UAV can be obtained using a small portable computer, such as a Raspberry Pi, to communicate with the Pixhawk using the MAVLink protocol. This way we are able to read flight data and send control commands that override the RC channels, thus controlling the Pixhawk the same way a human pilot would be able to.

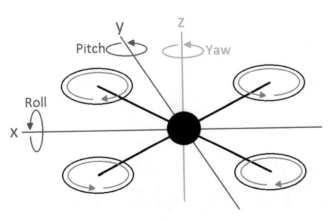

Fig. 96.2 Degrees of freedom achieved by the quadrotor and rotation direction of the motors

96.3.4 Robots

Our flying robots are small quadrotors with 250mm diameter with four 2300KV brushless motors, each one with its own speed controller, connected to a Pixhawk via I2C bus. Each UAV is equipped with a GPS module, which is important for our control strategy and enables the use of loiter flight mode, and a telemetry module, which enables communication with a ground station. Figure 96.3 is a photo of one of our robots.

To obtain autonomous control our UAVs also carry a Raspberry Pi, an embedded Linux computer, where all the control logic is programmed. To be able to detect other robots and obstacles our robots are also equipped with distance sensors, connected to the Raspberry Pi. As we will see, these sensors are crucial for our control algorithm. Figure 96.4 shows a diagram representing all UAV components and their connections.

Fig. 96.3 UAV used in the laboratory

96.3.5 Mission Description

Our system consists of a group of robots that fly over an area and a ground station with which they communicate through a telemetry link. The ground station is responsible for collecting flight data from each robot, but not controlling them. The control algorithm is distributed and runs in each UAV. The objective is to move the swarm to a destination point, using the flocking rules to avoid collision and optimize movement.

We define a global reference frame using latitude, longitude, altitude and heading as coordinates, where each robot is able to localize itself using its embedded sensors. To simplify the control strategy we fix the altitude to a predetermined value, which is valid for most surveillance and search and rescue applications. Given the predetermined altitude, the values for latitude and longitude can be converted to X and Y Cartesian coordinates. Heading is given by the Yaw angle. Thus the pose of each robot i at any time is given by the tuple:

$$Pose_i = (X_i, Y_i, Yaw_i)$$

For our control strategy to work each robot has to determine its own pose and also the pose of its neighbors. The GPS and embedded sensors are sufficient to find the robot's pose. However, GPS localization has very low precision and is unreliable to use for collision avoidance. Luckily, as we will soon see, Reynolds rules require only that a robot knows the relative position of its neighbors to itself, not their global position. This allows us to use distance sensors, which have better precision than GPS, to detect nearby robots. This strategy is closer to how real birds flock and also eliminates the need of communication between robots.

The mission objective is to reach a location called Migration Point defined by X_{mp} and Y_{mp} coordinates. An interface

Fig. 96.4 UAV components

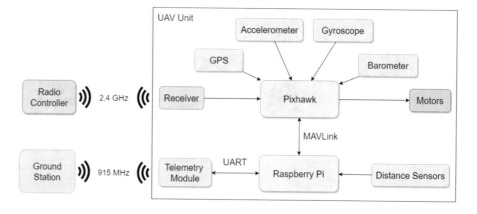

running on the ground station enables an operator to create a migration point during mission. The ground station sends this information to all robots. Again, since GPS precision is poor, we define a tolerance radius R_T around the migration point. The objective is considered achieved by the UAV when it enters this area.

96.4 Results and Discussion

96.4.1 System Architecture

Our algorithm was implemented in C++ and runs on the ROS platform. ROS (Robot Operational System) is an open source technology created to aid researchers in developing robotic applications [14]. ROS provides us with many tools and facilities that were very useful in our work.

A ROS application is a network of nodes that communicate with each other. Each node is an independent program in the system and a Master node also exists to manage the network. Nodes that generate data publish this information in topics in the form of messages, while nodes that need that data subscribe to the corresponding topics and receive the published messages.

In our application the master node runs in the ground station. Each robot runs its own instance of the control node, where our swarming algorithm is implemented. The control node gets the data from the distance sensors and uses the Reynolds rules to calculate the direction the robot should move. It sends this information to another node called mavros that create a MAVLink message and send it over a serial connection to the Pixhawk. Figure 96.5 illustrates the system architecture.

96.4.2 Algorithm Implementation

The control algorithm runs as an independent ROS node on each UAV. In each loop, the algorithm aims to: *i*. Obtain the UAV's own pose from the GPS and the Pixhawk's sensors readings; *ii*. Detect neighbors and obtain their pose relative to the UAV using distance sensors; *iii*. Use Reynolds rules for Separation, Alignment, Cohesion and Migration to determine how to move. The main loop of the algorithm is given in Algorithm 1.

Algorithm 1 Main loop of the swarming algorithm

1: **procedure** SWARM
2:     getPose()
3:     getNeighbors()
4:     *v1* ← separation()
5:     *v2* ← alignment()
6:     *v3* ← cohesion()
7:     *v4* ← migration()
8:     *vres* ← *r1* ∗ *v1* + *r2* ∗ *v2* + *r3* ∗ *v3* + *r4* ∗ *v4*
9:     move(*vres*)
10: **end procedure**

First the program calls the function *getPose()* which calculates the pose of the UAV where the program is running and sets two variables, *this.position* and *this.heading*, representing the UAV's XY position and heading. Then, the program calls another function *getNeighbors()* that reads the data from the distance sensors, calculates the positions and headings of the nearby robots that were detected and stores this data in an array called *neighbors*.

Then the program starts calling the functions where the actual flocking rules are implemented. Each function returns a vector that represent the direction that rule tells the UAV to go. For example, the Separation rule urges the UAV to move away from its neighbors, so the *separation()* function will return a vector pointing to the opposite direction of the other UAVs. Each function will be explained in detail ahead

In line 8 the four vectors are combined trough a weighted sum to generate the final resulting direction the robot should move. The weight applied to each rule determines how much that rule influences in the final result. The values of the weights can be changed to obtain different results. Rules can be even completely removed from the calculation by setting its weight to zero. After some testing we obtained good results using the following values:

$$R1 = 1; R2 = 1.5; R3 = 1; R4 = 1;$$

In line 9 the calculated resulting direction is passed to another function that moves the UAV. This function uses the

Fig. 96.5 System architecture overview

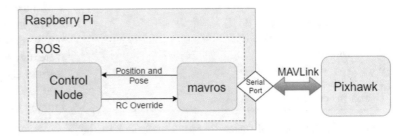

mavros package functions to send MAVLink messages to the autopilot board, overriding the RC channels to drive it the same way a human pilot would do. We put the UAVs in Loiter mode, because in this mode the Pitch, Roll and Yaw channels are directly converted to movement in the X, Y and Yaw axis, thus the control is easier.

Now we will explain each rule and its corresponding function in detail.

Separation Rule: This rule urges the robot to move away from other robots to avoid collisions. This is done creating a vector for each one of the detected neighbors pointing to the exactly opposite direction of that neighbor. These vectors are then combined into a resulting vector and returned to the main program. The implementation is shown in Algorithm 2.

Algorithm 2 Separation Rule

Algorithm 2 Separation Rule
1: **procedure** SEPARATION
2:     Vector $v \leftarrow 0$
3:     **for all** neighbors n **do**
4:         $vn \leftarrow this.$position $- n.$position
5:         $vn.$normalize()
6:         $vn \leftarrow vn * $distance$(this, n)$
7:         $v \leftarrow v - n$
8:     **end for**
9:     **return** v
10: **end procedure**

We also want the robot to move faster the closer it is from its neighbor. We do this adjusting the vector's magnitude. First we normalize it so it becomes a unit vector. Then we divide it by the distance between the two robots. As the distance between two robots gets smaller, the magnitude of the resulting vector becomes bigger.

Alignment Rule: This rule urges the robot to move in the same direction its neighbors are moving. The implementation can be seen in Algorithm 3 We loop trough each detected neighbor and calculate the average direction they are moving using their heading. We then return this vector to the main function.

Algorithm 3 Alignment Rule

Algorithm 3 Alignment Rule
1: **procedure** ALIGNMENT
2:     Vector $v \leftarrow 0$
3:     **for all** neighbors n **do**
4:         $v \leftarrow v + n.$velocity
5:     **end for**
6:     $v \leftarrow v/neighbors.$length
7:     **return** v
8: **end procedure**

Cohesion Rule: This rule tries to move the robot to the center of mass of the other robots, to form a swarm. The implementation is shown in Algorithm 5 and is similar to the Alignment rule, but instead of calculating the average direction we calculate the average XY position of the neighbors. This is the group's center of mass. We create a vector pointing to this location and return it to the main program.

Algorithm 4 Cohesion Rule

Algorithm 4 Cohesion Rule
1: **procedure** COHESION
2:     Vector $v \leftarrow 0$
3:     **for all** neighbors n **do**
4:         $v \leftarrow v + n.$position
5:     **end for**
6:     $v \leftarrow v/neighbors.$length
7:     **return** v
8: **end procedure**

Migration Rule: This is the rule used to make the swarm move to a destination point. Once a migration point is set by the operator, the robots should move to that point. This function is very simple and just returns a vector pointing to the Migration Point. The implementation is seen in Algorithm 5

Algorithm 5 Migration Rule

Algorithm 5 Migration Rule
1: **procedure** MIGRATION
2:     Vector $v \leftarrow 0$
3:     **if** *migrationPoint* **is set then**
4:         $v \leftarrow migrationPoint - this.$position
5:     **end if**
6:     **return** v
7: **end procedure**

96.4.3 Simulation

We decided to run our program in simulation first as a proof-of-concept and also as a way to evaluate its performance as a swarming algorithm. The simulations were run on the Gazebo Simulator [15], a versatile software that is able to simulate entire 3D environments, robots and sensors. All the elements in the simulation can be edited and manipulated through the use of plugins and SDF files, which are based on XML syntax.

An useful feature of the Gazebo simulator is the possibility to integrate it with other platforms. Using software provided by the Pixhawk team and the Gazebo team, we were able to simulate quadrotors running the Pixhawk firmware in Gazebo and integrate it with ROS. This way we are able to test our ROS applications in the simulator as we were running in real UAVs.

Figure 96.6 is a screenshot of the simulator screen showing one quadrotor with one distance sensor in the

Fig. 96.6 Simulated UAV with one distance sensor in the front

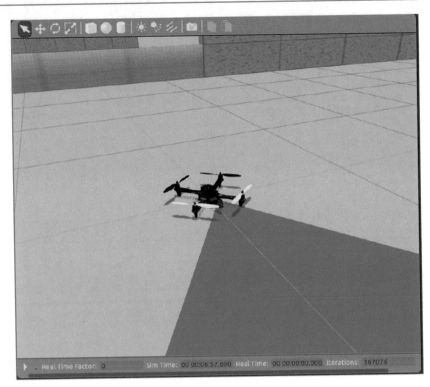

front. Although not used in this work, we are also capable of simulating cameras and other types of sensors.

96.5 Conclusion

This paper presented the design of a swarming algorithm that works on quadrotor UAVs with the Pixhawk autopilot board. Using Craig Reynolds flocking rules the algorithm is able to move a small group of UAVs to a common goal while avoiding collisions with each other. The algorithm was implemented in C++ and runs on the ROS platform. A simulation environment based on the Gazebo simulator was prepared to test and evaluate the algorithm in an use case as close as possible to reality.

This implementation however is still very simple and many other improvements could be done in the future. First, our implementation requires the UAVs to fly at a fix altitude, limiting the flock to a two dimensional movement. This is valid for most monitoring, search and rescue and surveillance applications, but if the swarm must move through a tight space it would be useful to move them in the z axis to concentrate the UAVs.

Another interesting work could be adapting our implementation to avoid collisions with fixed obstacles as well. This would require that the UAVs be able to detect these obstacles and distinguish them from other robots.

Acknowledgements The authors would like to thank the funding institution: CAPES.

References

1. Sa, I., & Corke, P. (2011). Estimation and control for an open-source quadcopter. In *Proceedings of the Australasian Conference on Robotics and Automation.*
2. Reynolds, C. (1987). Flocks, herds and schools: A distributed behavioral model. In *Proceedings of the 14th Annual Conference on Computer Graphics and Interactive Techniques* (pp. 25–34).
3. Boids Homepage. http://www.red3d.com/cwr/boids/. Accessed Sept 2016.
4. Hauert, S., Leven, S., Varga, M., Ruini, F., Cangelosi, A., Zufferey, J. C., & Floreano, D. (2011). Reynolds flocking in reality with fixed-wing robots: Communication range vs. maximum turning rate. *IEEE/RSJ International Conference on Intelligent Robots and Systems* 5015–5020.
5. Quintero, S. A., Collins, G. E., & Hespanha, J. P. (2013). Flocking with fixed-wing UAVs for distributed sensing: A stochastic optimal control approach. *American Control Conference,* 2025–2031.
6. Kushleyev, A., Mellinger, D., Powers, C., & Kumar, V. (2013). Towards a swarm of agile micro quadrotors. *Autonomous Robots, 35,* 287–300.
7. Bürkle, A., Segor, F., & Kollmann, M. (2001). Towards autonomous micro UAV swarms. *Journal of Intelligent & Robotic Systems,* 339–353.
8. Vásárhelyi, G., Virágh, C., Somorjai, G., Tarcai, N., Szörényi, T., Nepusz, T., & Vicsek, T. (2014). Outdoor flocking and formation flight with autonomous aerial robots. *IEEE/RSJ International Conference on Intelligent Robots and Systems,* 3866–3873.

9. De Benedetti, M., D'Urso, F., Messina, F., Pappalardo, G., & Santoro, C. (2015). UAV-based aerial monitoring: A performance evaluation of a self-organising flocking algorithm. In *10th International Conference on P2P, Parallel, Grid, Cloud and Internet Computing* (pp. 248–255).

10. Moeslinger, C., Schmickl, T., & Crailsheim, K. (2009). A minimalist flocking algorithm for swarm robots. In *European Conference on Artificial Life* (pp. 375–382).

11. Brambilla, M., Ferrante, E., Birattari, M., & Dorigo, M. (2013). Swarm robotics: A review from the swarm engineering perspective. *Swarm Intelligence*, 1–14.

12. Tan, Y., & Zheng, Z. Y. (2013). Research advance in swarm robotics. *Defence Technology*, 18–39.

13. Pixhawk Homepage. https://pixhawk.org/. Accessed Sept 2016.

14. ROS Homepage. http://www.ros.org/. Accessed Sept 2016.

15. Gazebo Homepage. http://gazebosim.org/. Accessed Sept 2016.

PID Speed Control of a DC Motor Using Particle Swarm Optimization

Benedicta B. Obeng and Marc Karam

Abstract

This research is based on applying Particle Swarm Optimization (PSO) to the Proportional-Integral-Derivative (PID) speed control of a permanent magnet DC motor. The integration of PID control and PSO optimization technique starts by designing the initial PID control gains so that the DC motor angular speed exhibits a given overshoot (OS) and settling time (t_s). Based on these gains, we designed an initial set of particles that were subsequently modified using PSO algorithm with the goal of reducing OS and t_s. Simulation results led to the desired swarming since the PID gains converged, causing the OS and t_s to get below the desired threshold. Thus, our proposed PID-PSO algorithm was very successful in achieving the optimization goal of reducing OS and t_s by means of PSO-optimized PID control.

Keywords

Particle Swarm Optimization • PID Control

97.1 Introduction

In this research, we apply Particle Swarm Optimization (PSO) to the Proportional-Integral-Derivative (PID) speed control of a permanent magnet DC (PMDC) motor. PSO is a population-based stochastic optimization technique developed by Dr. Eberhart and Dr. Kennedy in 1995 [1] that emulates the behavior of a flock of birds or a school of fish. The system is initialized with a population of random particles and then the search for optima is performed. The potential solutions, called particles, fly through the problem space by following the current optimum particles. Each particle keeps track of its coordinates in the problem space which are associated with the best solution it has achieved so far that is called P_{best}. When a particle takes all the population as its topological neighbors, the best value is a global best and is called G_{best}. The PSO concept consists of

changing the velocity of each particle toward its P_{best} and G_{best} locations.

Our proposed PID-PSO optimization technique starts by designing initial PID control gains that lead to a given overshoot (OS) and settling time (t_s) of the unit-step response of the PMDC motor angular speed. Then, we design an initial set of particles that are to be modified using PSO algorithm in view of reducing OS and t_s.

This paper is organized as follows: In Sect. 97.1, we present the transfer function of the PMDC motor used in this research. In Sect. 97.2, we detail our proposed PID-PSO control technique, with the goal of optimizing the PID control gains using PSO algorithm. Simulation results are presented in Sects. 97.3, and 97.5 is the Conclusion.

97.2 PMDC Motor Model

The permanent magnet DC motor we decided to use in this research has the following transfer function:

B.B. Obeng • M. Karam (✉)
Department of Electrical Engineering, Tuskegee University, Tuskegee, AL, USA
e-mail: bobeng6576@mytu.tuskegee.edu; karam@mytu.tuskegee.edu

© Springer International Publishing AG 2018
S. Latifi (ed.), *Information Technology – New Generations*, Advances in Intelligent
Systems and Computing 558, DOI 10.1007/978-3-319-54978-1_97

Table 97.1 Permanent Magnet DC Motor Parameters

Parameter	Value	Unit
K_t	0.072	N.m/A
J	0.0005	Kg.m^2/rad
b	0.00021	N.m.s/rad
R	0.5	Ω
L	0.002	H
K_b	0.072	V.s/rad

$$G(s) = \frac{\omega(s)}{V(s)} = \frac{K_t}{(sJ + b)(R + sL) + K_b K_t}, \quad (97.1)$$

where ω is the speed of the motor, V is the voltage, K_t is the motor torque constant, J is the moment of inertia of the rotor, b is the damping ratio of the mechanical system, R is the resistance, L is the inductance, and K_b is the electromotive force constant. The values of these parameters are given in Table 97.1.

97.3 PID-PSO Algorithm

In this Section, we present our proposed technique to combine the PID speed control of a PMDC motor and the PSO algorithm overviewed in the previous section in order to optimize the performance of the PID controller.

Our proposed PID-PSO algorithm consists of the following steps:

1. Identify the upper and lower bounds of the PID controller gains K_P, K_I, and K_D and randomly initialize position and velocity of each particle.
2. Using the system's step response, measure the maximum OS, t_s and t_r which are the performance criteria of the system for each particle.
3. Evaluate each particle using the performance criterion given by the following fitness function of the PID controller:

$$W(K) = \left(1 - e^{-\beta}\right)OS + e^{-\beta}(t_s - t_r), \quad (97.2)$$

where $K = [K_P\ K_I\ K_D]$ and β is a weighting factor [2].
4. Calculate the minimum performance criterion of all the particles and update G_{best} and P_{best} of each particle.
5. Update the velocity of each particle using [5]
6. Limit the updated velocity to the interval $[V_{min}, V_{max}]$
7. Update each particle position using [6]
8. If the maximum number of iterations or the desired system performance is not met, then go to step 2, else terminate algorithm.

The particle with the latest G_{best} is the one with the optimal K_P, K_I, and K_D gains. Upon termination of the algorithm, the PID gains are expected to converge to the global optimum of the search space and the fittest vector is taken as a possible solution to the problem. A flow chart summarizing the PID-PSO technique is shown in Fig. 97.1 below.

Fig. 97.1 Flowchart of PID-PSO algorithm

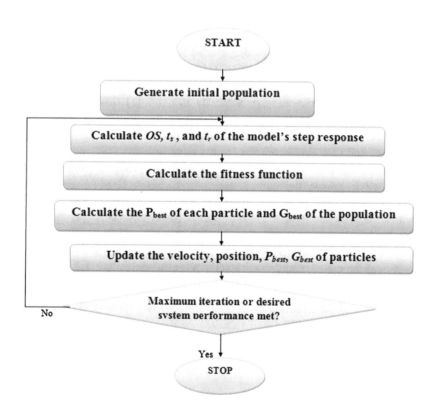

97.4 Application to the PMDC Motor

In this Section, we apply the PID-PSO technique presented in the previous Section to the speed control of a PMDC motor. The goal of the PID-PSO algorithm is to obtain OS less than 2% and t_s less than 1 ms. We started with a manually-tuned PID controller with the following gains $K_P = 100$, $K_I = 7$, and $K_D = 0.03$. The corresponding unit-step response of the DC motor, shown in Fig. 97.2, reveals $OS = 30.23\%$ and $t_s = 2.91$ ms.

We then implemented the PID-PSO algorithm using Matlab software. The particle positions were randomly initialized based on the initial PID gains. After several trials, we decided to choose the values of the PSO parameters listed in Table 97.2 below. Evaluation of the control performance was accomplished by computing the fitness function of the PID controller given by (8) with the objective of reducing OS and t_s as iterations increased.

Effectively, throughout all 30 iterations the K_P, K_I, and K_D particles kept "swarming" (getting closer to each other) as shown in Fig. 97.3, and the best OS and t_s decreased substantially as shown in Figs. 97.4 and 97.5. Table 97.3 shows the global best OS and t_s values at iterations 10, 20, and 30. The corresponding unit-step responses of the PID-PSO controlled DC motor are displayed in Figs. 97.5, 97.6 and 97.7. Both graphical and numerical results show that both global best OS and t_s decrease as iterations increase.

Analyzing these results, we observe that OS varied from the initial 30.23% value to 1.89%, which is below the desired goal of 2%. Similarly can be said of t_s, which varied from 2.91 ms to 0.43 ms, reaching below the desired goal of

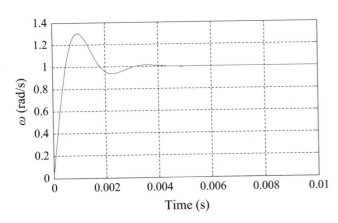

Fig. 97.2 Unit-step response of PID-controlled PMDC motor with $K_P = 100$, $K_I = 7$, and $K_D = 0.03$

Table 97.2 Parameters of the PID-PSO algorithm

Parameter	Value
N	20
ω_{min}	0.9
ω_{max}	0.4
c_1	1
c_2	1
M	30
B	0.3
V_{min}	0.001
V_{max}	2.335

Fig. 97.3 Swarming of the PID-PSO particles

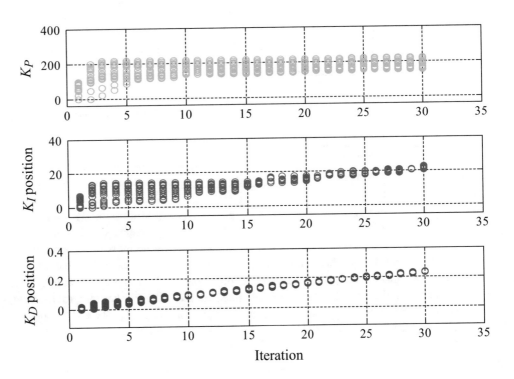

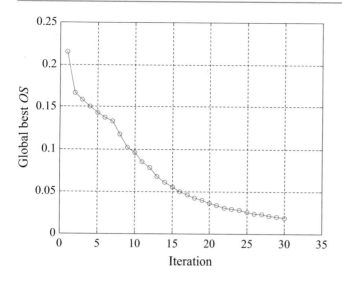

Fig. 97.4 Global best overshoot profile using PID-PSO control

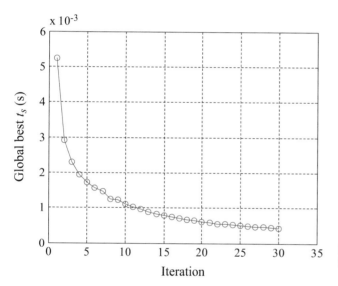

Fig. 97.5 Global best settling time profile using PID-PSO control

Table 97.3 Best *OS* and t_s using PID-PSO control

Iteration	Global best *OS* (%)	Global best t_s (ms)
10	9.62	1.12
20	3.64	0.61
30	1.89	0.43

1 ms. Thus is confirmed the validity of our proposed PID-PSO algorithm as well as the success of its implementation. Moreover, the optimal PID gains obtained at the termination of the PID-PSO optimization process were: $K_P = 155.45$, $K_I = 21.81$ and $K_D = 0.24$.

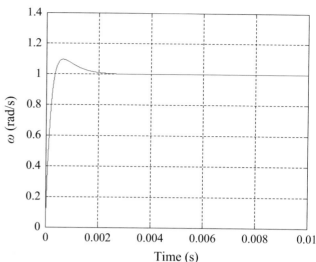

Fig. 97.6 Unit-step response of PID-PSO controlled motor at iteration 10

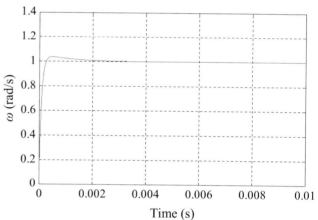

Fig. 97.7 Unit-step response of PID-PSO controlled motor at iteration 20

97.5 Improvement of Pid-Pso Performance

In the previous section, the PID-PSO technique used in controlling the PMDC motor speed led to the successful response shown in Figs. 97.3, 97.4, 97.5, 97.6, 97.7 and 97.8. In this section, we seek to improve the PID-PSO performance by studying the effect of varying the PSO parameters and choosing the ones that lead to the optimal solution with the least number of iterations. The parameters that were varied included the acceleration constants c_1 and c_2, initial and final inertia weight factors ω_{max} and ω_{min} [3], the number of particles N, and the final number of iterations k_{final} [4]. *OS* and t_s values were recorded and analyzed for different cases.

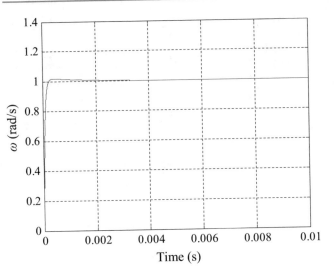

Fig. 97.8 Unit-step response of PID-PSO controlled motor at iteration 30

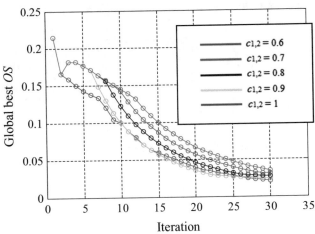

Fig. 97.9 Best *OS* profile for various values of c_1 and c_2

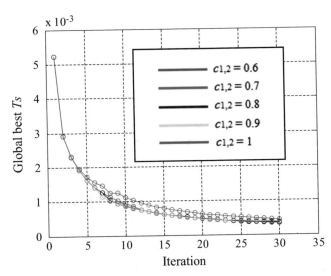

Fig. 97.10 Best *Ts* profile for various values of c_1 and c_2

Table 97.4 Effect of varying c_1 and c_2 on best *OS* and t_s

Case	c_1	c_2	Global best *OS* (%)	Global best t_s (ms)
1	0.6	0.6	3.06	0.34
2	0.7	0.7	2.69	0.36
3	0.8	0.8	2.37	0.39
4	0.9	0.9	2.09	0.44
5	1	1	1.89	0.43

97.5.1 Effect of Varying the Acceleration Constants

Simulations were carried out for five different values of acceleration constants as shown in Table 97.4, where we also recorded the values of the best *OS* and t_s obtained at the final iteration $k = 30$. The corresponding *OS* and t_s profiles are presented in Figs. 97.9 and 97.10.

The simulation results showed that increasing c_1 and c_2 leads to decreasing *OS* but increasing t_s. Consequently, we chose the value 1 for c_1 and c_2 as the outcome of this section leading to a motor speed response with the lowest best *OS* =1.89% but the highest best $t_s = 0.43$ms.

97.5.2 Effect of Varying the Inertia Weight

Four different values of ω_{max} and ω_{min} ranging from 0.5 to 0.8 were adjusted as shown in Tables 97.5 and 97.6 along with the corresponding values of the best *OS* and t_s obtained at the final iteration $k = 30$. The resulting *OS* and t_s profiles with respect to iterations are presented in Figs. 97.11, 97.12, 97.13 and 97.14.

It can be deduced from the simulation results that decreasing initial inertia weight ω_{max} resulted in an increase in *OS* whiles increasing ω_{min} resulted in an increase in the *OS*. In both cases t_s decreased. Thus the best ω_{max} value was found to be 0.7 with an *OS* of 2.68% and t_s of 0.41ms. Furthermore, that of ω_{min} was 0.5 with *OS* of 1.77% and t_s of 0.43ms.

97.5.3 Effect of Varying the Number of Particles

In Table 97.7, we list four different values of N ranging from 30 to 60 and record the corresponding values of the best *OS* and t_s obtained at the final iteration $k = 9$. The resulting *OS* and t_s profiles with respect to iterations are presented in Figs. 97.15 and 97.16.

Table 97.5 Effect of varying maximum weight on Best OS and t_s

Case	ω_{max}	Global Best OS (%)	Global Best t_s (ms)
1	0.5	2.92	0.35
2	0.6	2.77	0.37
3	0.7	2.68	0.41
4	0.8	2.74	0.49

Table 97.6 Effect of varying minimum weight on Best OS and t_s

Case	ω_{min}	Global best OS (%)	Global best t_s (ms)
1	0.5	1.77	0.43
2	0.6	2.14	0.38
3	0.7	2.48	0.35
4	0.8	2.82	0.33

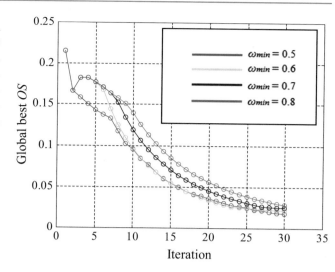

Fig. 97.13 Best OS profile for various values of ω_{min}

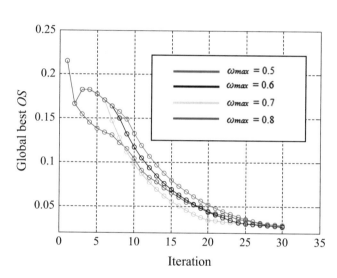

Fig. 97.11 Best OS profile for various values of ω_{max}

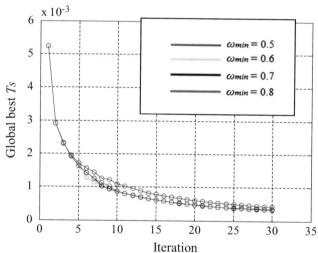

Fig. 97.14 Best Ts profile for various values of ω_{min}

Table 97.7 Effect of varying N on Best OS and t_s

Case	N	Global best OS (%)	Global best t_s (ms)
1	30	8.20	1.6
2	40	3.47	1.1
3	50	3.24	1.2
4	60	3.04	1.1

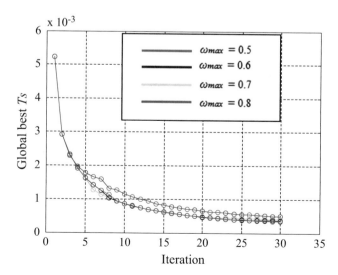

Fig. 97.12 Best Ts profile for various values of ω_{max}

We observed an improvement in the performance of the PID-PSO algorithm when N increases. The OS value decreased with a corresponding decrease of t_s. However, this was achieved with higher computational burden. The lesser the number of swarm particles, the more the number of iterations required for obtaining optimal performance and vice-versa. Large swarm size implies that more particles

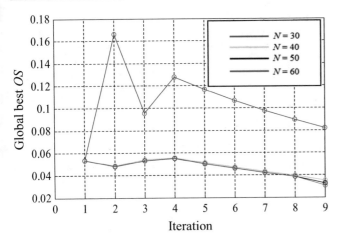

Fig. 97.15 Best *OS* profile for various values of *N*

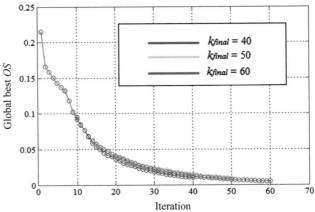

Table 97.8 Effect of varying k_{final} on Best *OS* and t_s

Case	k_{final}	Global best *OS* (%)	Global best t_s (ms)
1	40	0.99	0.33
2	50	0.86	0.28
3	60	0.45	0.22

Fig. 97.17 Best *OS* profile for various values of k_{final}

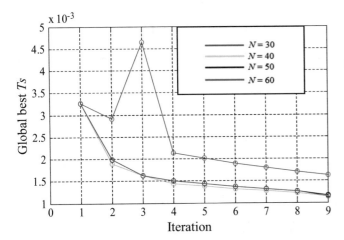

Fig. 97.16 Best *Ts* profile for various values of *N*

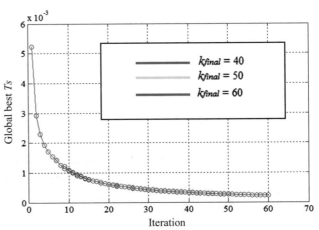

Fig. 97.18 Best *Ts* profile for various values of k_{final}

search the problem space to find optimal solution which results in increased computations. In conclusion, we decided that the best *N* value is 40 with corresponding *OS* and t_s values of 3.47and 1.1 s respectively.

97.5.4 Effect of Varying the Number of Iterations

It is expected that increased iterations lead to better optimization. In Table 97.8, we list three different values of the final number of iterations k_{final} ranging from 40 to 60 and record the corresponding values of the best *OS* and t_s. The

resulting *OS* and t_s profiles with respect to iterations are presented in Figs. 97.17 and 97.18.

From the foregoing analysis, it can be deduced that an increase in the number of iterations produces better results. However, increasing k_{final} also increases the computational burden of the hardware. Therefore, a compromise needs to be reached between these two opposed criteria. Analyzing the results obtained, we decided that, although

Table 97.9 Initial And improved parameters of the pid-pso algorithm

Parameter	Initial	Improved
N	20	40
ω_{min}	0.9	0.5
ω_{max}	0.4	0.7
c_1	1	1
c_2	1	1
k_{final}	30	30

Table 97.10 Best OS and t_s at $k_{final} = 30$ for initial and improved controllers

Iteration	Initial PID-PSO	Improved PID-PSO
Global Best OS (%)	1.89	2.6950×10^{-5}
Global Best t_s (ms)	0.43	0.30

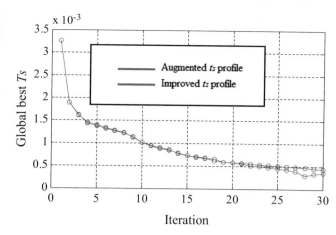

Fig. 97.20 Comparing initial and improved best TS profiles

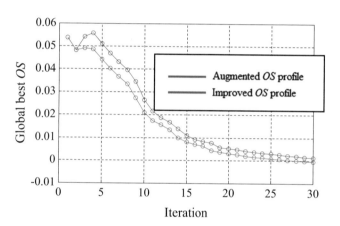

Fig. 97.19 Comparing initial and improved best OS profiles

$k_{final} = 60$ leads to the best performance, $k_{final} = 40$ is the compromise value to choose so that the computations are efficient and practical. The corresponding OS and t_s values are 0.99% and 033ms respectively.

97.5.5 Comparison of Initial and Improved Control System Responses

We carried out the previous analysis by studying the effect of varying the PSO parameters individually. Each analysis resulted in a best performance based on a particular parameter. In this section, we combine all these "best" parameters together and compare the corresponding system responses. Table 97.9 shows the initial parameters as well as the improved parameters.

In Figs. 97.19 and 97.20, we compare both augmented and improved best OS and t_s profiles, while, in Fig. 97.21, we

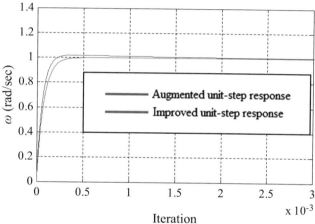

Fig. 97.21 Comparing the initial and improved unit-step responses at $k_{final} = 30$

compare the corresponding motor speed responses. Final results are summarized in Table 97.10 below where it can be seen that the best OS and t_s obtained by using the improved PID-PSO controller are lower than those obtained by using the augmented controller. Thus, our goal of improving the PID-PSO performance by further reducing the best OS and t_s has been achieved.

97.6 Conclusion

In this paper, we have proposed an optimization method based on Particle Swarm Optimization algorithm in order to improve the performance of the PID control of the speed of a PMDC motor. Our simulation results were successful whereas the design goal of reducing the overshoot OS and settling time t_s of the DC motor speed response was met. Practically, our aim was to obtain OS less than 2% and t_s less than 1 ms. Consequently, the PID-PSO-controlled system remained stable and the PSO particles exhibited the swarming effect which led to continually decreasing overshoot and settling time that reached below the design threshold, confirming thus the validity of our proposed optimization method.

Future work includes researching other PSO algorithms that incorporate constriction factors for faster convergence. In addition PSO can be used in conjunction with other tuning methods such as Internal Model Control or Genetic Algorithms. Moreover, we recommend that our work using PSO be compared to other optimization methods in order to evaluate the computational and quantitative efficiency of our proposed PID-PSO algorithm.

References

1. Kennedy, J., & Eberhart, R. (1995). Particle swarm optimization. In *Proceedings of IEEE international conference on neural networks* (pp. 1942–1948). Piscataway.
2. Coelho, L. d. S., & Sierakowski, C. A. (2008). A software tool for teaching of particle swarm optimization fundamentals. *Advances in Engineering Software, 39*, 877–887.
3. Yang, Cheng-Hong., Hsiao, Chih-Jen., & Chuang, Li-Yeh. (2010). Linearly decreasing weight particle swarm optimization with acceleration strategy for data clustering. *IAENG International Journal of Computer Science, 37*(3), IJCS_37_3_05.
4. Shi, Y., & Eberhart, R.C. (1999). Empirical study of particle swarm optimization. In *Proceeding. of congress on volutionary computation* (pp. 1945–1949). Washington, DC.
5. Shi, Y., & Eberhart, R.C. (1998) A modified particle swarm optimizer. In *Proceedings of IEEE congress on evolutionary computation* (pp. 69–73). Anchorage.
6. Allaoua, B., Gasbaoui, B., & Mebarki, B. (2009). Setting up PID DC motor speed control alteration parameters using particle swarm optimization strategy. *Leonardo Electronic Journal of Practices and Technologies, 8*(14), pp. 19–32.

Use of Intelligent Water Drops (IWD) for Intelligent Autonomous Force Deployment

Jeremy Straub and Eunjin Kim

Abstract

This paper presents a decentralized method for autonomously directing the movement of troops or battlefield robots from areas where their capabilities are being underutilized to areas where they are needed. This technique, which relies on limited message passing, does not require a centralized controller and is thus well suited to the battlefield environment where natural or deliberately created conditions may limit communications or render a centralized controller inaccessible. The Colonel Blotto Game (simulation scenario) is extended to provide a testing framework for Intelligent Water Drops (IWD)-derivative methods. The performance of the conventional approach to the Colonel Blotto Game is characterized in application to this extended scenario. Then, an IWD approach is presented and its performance is compared to the conventional method. The IWD approach is shown to outperform the conventional approach, from a gameplay perspective, while having significantly greater processing costs. Finally, the performance of an extended approach, which plays out possibilities for the remainder of the game multiple times before making a decision, is compared with an approach based on making the best decision in the short term without extended network information. The gameplay utility of this extended solver is not demonstrated, despite it having significantly higher computational costs.

Keywords

Autonomous troop deployment • Intelligent water drops • Swarm intelligence • Autonomous control • Autonomous decision making

98.1 Introduction

Troop deployment in an active combat region is an ill posed problem which gains significant complexity from the fact that some communications links may be created on an ad hoc basis while other links may be denied due to environmental factors or deliberate action by the enemy. Troop needs in various areas of the combat zone will change over time, with immediate (and unpredictable) needs being triggered by enemy-initiated engagements and other needs being triggered by planned offensives. Troop movement must be minimized, in the context of providing the required service level, as a cost is incurred in terms of supply consumption and fatigue of the troops.

A very similar problem exists with regards to robotic combat units. These units may be needed for sensing, warfighting and other purposes, at various areas of the battlefield. They are similarly constrained by communications link availability and movement costs. Unlike their human counterparts, however, they cannot act intuitively and must be directed by an algorithmic decision making process.

For both human and robotic warfighters, an autonomous approach to maximizing performance (i.e., winning battles) is desirable. For humans, this can be utilized to direct human

J. Straub (✉) • E. Kim
North Dakota State University, Fargo, ND, USA
e-mail: jeremy.straub@ndsu.edu

© Springer International Publishing AG 2018
S. Latifi (ed.), *Information Technology – New Generations*, Advances in Intelligent
Systems and Computing 558, DOI 10.1007/978-3-319-54978-1_98

troops (either as a command or decision support system); for robots, it can be used for onboard or group decision making. A distributed/decentralized approach where a positive (win-maximizing) outcome is produced as an emergent property of the network facilitates operations in a communications-impaired or communications-denied environment. Swarm-style techniques, such as the Intelligent Water Drop (IWD) algorithm [1] represent one approach to creating this type of system.

This paper presents and assesses an approach to the decentralized allocation of warfighting resources using a modified version of the IWD algorithm. The IWD algorithm uses the concept of how streams and rivers move and distribute sediment to solve problems for which a deterministic algorithmic solution is not possible or not known at present. The application of the IWD algorithm to warfighting resource management allows good decisions to be made based on the latest information available to the deciding-unit. It relies on a probabilistic distribution to ensure that an appropriate number of units are tasked to each area of need.

The paper continues by providing background on troop deployment and the IWD algorithm, in Sect. 98.2. In Sect. 98.3, it then presents a simplified version of Shah-Hosseini's IWD algorithm that is adapted for troop deployment decision making. Next, in Sects. 98.4 and 98.5, the application of the stream/river concepts of bifurcation and distributaries to the IWD algorithm and how to make decisions regarding drop flows is discussed. The application of the bifurcation/distributaries concept to the IWD troop deployment decision making algorithm is considered. Following this, a qualitative evaluation of this approach is performed and presented in Sect. 98.6. Finally, a quantitative evaluation is performed and analyzed (in Sects. 98.7, 98.8 and 98.9) and conclusions are drawn, in Sect. 98.10.

98.2   Background

The use of the IWD alogirthm is considered in this paper for decision making for troop deployment. Unlike other possible approaches which may use expert system-like (e.g., [2]), heuristic (e.g., [3]), or neural network-based techniques (e.g., [4]), IWD is a swarm-style (see, e.g., [5–8] for other examples) technique. This section reviews prior work in areas critical to the understanding and development of the approach presented herein. First, modern concepts related to troop deployment decision making are reviewed. Second, prior work related to the creation and use of the IWD algorithm is discussed.

98.2.1   Troop Deployment

The basic premise of troop deployment in an active combat region is to ensure that a sufficient fighting force is present in each area to have the desired impact (e.g., to win a battle/skirmish or to dominate a battle/skirmish to minimize losses) while not allocating resources beyond what is required to do so (given an acceptable margin for error and considering fog-of-war effects, etc.). This basic concept was represented in the Colonel Blotto Game, introduced initially by Borel [9] in 1921 and studied by others throughout the intervening period [10–12]. This simplified combat simulation presumes that winning a war is based on winning more battles than one's opponent, that all battlefields are equally-valuable and that a side wins a battle by having more troops assigned to the battlefield.

These principles, though, are not typical of modern warfare. Arguably, they may not have ever been more than a significant over abstraction; however, this abstraction has become more inaccurate due to technical and other advances. Recognizing the utility of the concept and the need to update and reform it, a variety of derivative versions and adaptations have been proposed. Gross [13] and Laslier [14] allow different battlefield valuations (but keep the value of the territories the same from a decision-making perspective). Kvasov [15] adds a cost for resource allocation and Roberson and Kvasov [16] consider the case of different sized forces. Hortala-Vallve and Lorente-Saguer [17] propose an updated version of the Colonel Blotto Game where opposing commanders have different valuations for both battlefield wins and the territories potentially won by the battle.

However, the real-world scenario is still more complicated than what is modeled by any of the versions of the Colonel Blotto Game. Specifically, the game neglects the temporal aspects of battles (the fact that fighting forces can be deployed before or during the battle and the intervening time between force assignment and force arrival), the heterogeneous capabilities of units (e.g., different impact levels on battle based on skills and equipment) and restrictions on the paths that troops can travel (e.g., roads, minefields, etc.) that limit where troops can be deployed from.

Chou, Teo and Zheng [18] present work that may contribute partially to a solution. They discuss the deployment of resources in several contexts. In each case, the goal is to minimize the mismatch between need and resources utilized to satisfy that need. They proffer that the introduction of flexibility into the system assists in resolving this mismatch issue; however, it also introduces complexity: a trade-off that must be managed.

Akgün [19] looks at how actual troop (and associated equipment) deployment can be performed. This work reduces a planning process that has generally taken a human commander a week to perform into a computational process taking several hours. The work specifically considers the real-world aspects of the deployment process such as path capabilities, movement speeds and such. The proposed work generated optimal solutions in some, but not all, cases. Further it deals with meeting a pre-existing demand but does not substantively consider the actual demand-creation process.

98.2.2 Need in the Context of Autonomous Warfighters

A decision support system for the deployment of human troops can be used to aid (or even drive) positioning/ repositioning decision making. However, the individual troops can still make their own decisions regarding redeployment, in the event of a communication failure. As these decisions will be lacking updated information (since the beginning of the communications failure) and may be driven by emotional response, they may be sub-optimal. Evans, DeCostanza and Pierce [20], for example, note that, for human soldiers, trust in the electronic system and in others following it is critical to troops' decision making to follow a system's information or recommendations. Knowledge of the lack of current information and a potential belief that others may not be getting commands may lead them to question the system's recommendations. A decision support system (such as that proposed herein) could be similarly impaired by a communications failure (but might be less so, given the peer-to-peer communications model proposed, instead of using the base-to-squad communications approach). It may also be ignored by troops who utilize its lack of updated information as a way to justify following their emotion-driven decision.

For autonomous robots, on the other hand, the system is mission critical, as this type of system (arguably, probably a more complex version that is presented herein) will serve as the primary decision making mechanism for the robot. A distributed or hybrid communications system, such as the version described by Liu et al. [21], seeks to maximize operating potential in a communications-partially-denied environment. A decision making system must, similarly, seek to maximize the correctness of decisions in an environment where communications are partially denied. The latest known information is relied upon in a communications-totally-denied scenario to facilitate communications impairment not being usable as a method of robotic troop disablement.

98.2.3 Intelligent Water Drops

Swarm intelligence techniques produce a system-level behavior as an emergent property of node-level decision making. Techniques such as ant colony optimization [22], the grey wolf optimizer [23], particle swarm optimization [24] and its extension multi-swarm optimization [25] and the artificial bee colony algorithm [26] have been shown to be effective in a wide variety of circumstances. In each case, a large number of agents are created which perform a simplistic decision making process based on local information. The combination of all of these simple agents working in tandem (and seeking to maximize their local metrics) results in a system level outcome. Shah-Hosseini [1] proposed a technique following this swarm concept called the Intelligent Water Drop algorithm, which utilizes a simulation of water flows in a stream/river system as the basis for decision making.

This IWD concept was introduced by Shah-Hosseini [1] in 2007, initially, as an iterative approach to solving the traveling salesman problem. It has been enhanced in several ways (e.g., through the incorporation of a mutation algorithm described in [27]) and has been used for numerous applications including image enhancement [28], the n-queen problem [29], the multidimensional knapsack problem [29], network security [30, 31], supply chain management [32], irrigation system development [33] and trajectory planning [34]. Previous work [35] resulted in the development of a simplified version, but functionally similar, version.

The basic premise of the IWD algorithm is that sediment is picked up by the simulated water drops in areas where the water is fast-moving and deposited by the water drops in areas where the water is slower-moving. Water speed is a function of sediment. This results in areas getting progressively faster and slower as sediment is moved based upon the speed of water flowing across each edge. Note that water flow speed is a function of the initial speed of the water before entering the edge and the effect of the edge on the water's speed.

98.3 Intelligent Water Drops Troop Deployment Algorithm

A simplified version of the IWD algorithm presented in [1] has been developed for the purposes of automating the troop deployment process. This algorithm, referred to herein as the Basic IWD Troop Deployment Algorithm, is depicted in Fig. 98.1. For testing purposes, it begins with the initialization of the network. An initial unit is selected. A run of the simulation will iterate until all units have been processed, so the choice of the initial unit can be arbitrary or

Fig. 98.1 Basic IWD troop deployment algorithm

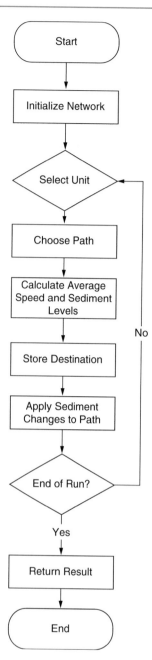

The final destination of the water drop (unit) is stored and sediment changes are applied uniformly to the path (based on the average speed taken and the impact of this average speed on sediment collection/depositing). If additional units remain to be processed, the next unit is selected and the process restarts.

98.4 IWD Bifurcation and Distributaries

In the vernacular of streams and rivers, bifurcation points are locations where a stream or river splits into multiple smaller streams or rivers. Each of these smaller streams or rivers is called a distributary. Bifurcation points and distributaries are depicted in Fig. 98.2.

The key to effectively deploying troops to ideal (or near-ideal) locations is the decision making that occurs at bifurcation points as to what distributary is taken. This should consider the value of the destination achieved and the speed of the path used to get there. A '*goodness*' value is determined for each prospective direction that can be taken from the current node and a probabilistic selection is made based on the relative goodness of the path in question relative to the total goodness of all possible paths. The probability of selection of a given path (x) is thus given by:

$$P_x = \frac{G_x}{\sum_n G_n} \tag{98.1}$$

where P_x is the probability of taking a path x, G_x is the goodness of the path in question and G_n is the goodness of the n-th path. Note that goodness is a function of value obtained divided by the cost of the path that is required to attain it. Thus, goodness is determined by:

$$G_x = \sum_{n=1} \frac{V_n}{\sum_{a=1..n} C_a} \tag{98.2}$$

where Gx is the goodness of a particular direction, Vn is the value of node n and this is divided by the sum of the cost (represented by cost C_a) of all nodes (1..n) traversed to reach the node currently being added to the summation (node n).

To accommodate this form of decision making, the basic IWD grid must, thus be updated with node values. Fig. 98.3 depicts this directional, weighted and path-costed grid. The IWD grid represents the problem and the solution space that that the algorithm is trying to solve. In its most basic form, the grid can be simply be directional with no node values (making all nodes equally valued) and traversal costs. Node labels are included for reference purposes. In the directional,

random; note, however, that because the runs update the sediment levels the choice of the unit may affect the outcome of the simulation. Multiple simulation runs can (and in many cases should) be conducted to identify an optimal or near-optimal solution.

Each unit's run begins with a selection of the entire path that it will take through the grid. This is based on a probabilistic selection based on the levels of sediment that are present in each path connected to the current node. Once the path is selected, the average sediment level across the entire path is computed; from this the average traversal speed is computed.

weighted and path-costed grid, each node is labeled in the format "n:x" where n is the node number and x is the value of traversing this node. The cost of traversing the path between two nodes is indicated on the edge between the nodes. Note two critical features of this grid. First, all edges are directional with a consistent direction of flow

(from upper left to lower right). Because of this, no loops exist (or could exist) in the grid. Second, multiple ways of getting to many nodes exist.

The process of solving a problem using any of the IWD Troop Deployment Algorithms begins with representing the problem and solution space in an appropriate grid. Most will, quite obviously, be more complex than the example shown in Fig. 98.3.

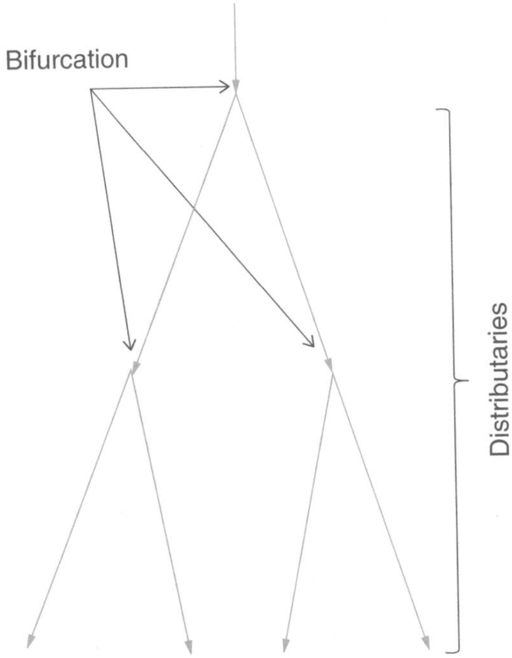

98.5   IWD Bifurcation and Troop Deployment

An updated version of the troop deployment algorithm that considers the impact of bifurcation decision making on troop deployment is now considered. This algorithm, referred to herein as the Bifurcation-Aware IWD Troop Deployment Algorithm, is depicted in Fig. 98.4. The principal change that has been made is that a step for path evaluation has been added prior to the selection of the best path.

The goodness value is computed as described in Sect. 98.4. In the most basic implementation, the percentage of total goodness can be used to determine what percentage of units should be sent down the path. However, this direct correlation may not always be desirable. For example, it may be desirable to send a more-than-proportional amount down the paths deemed particularly good and a less-than-proportional amount down paths deemed less good. For this reason, a revised probability equation is also considered:

Fig. 98.2 Bifurcation points and distributaries

Fig. 98.3 Nodes, edges and associated values

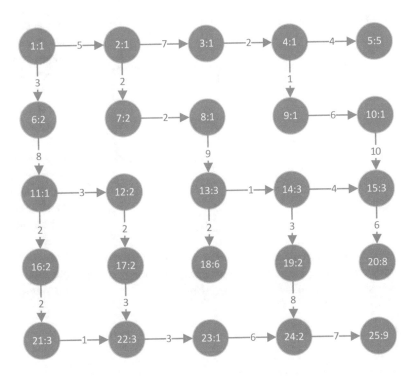

Fig. 98.4 Bifurcation-aware
IWD troop deployment algorithm

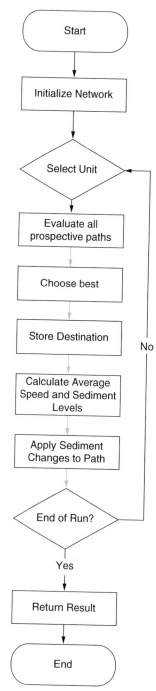

98.6 Qualitative Evaluation

A qualitative evaluation of the proposed algorithm indicates several prospective benefits and drawbacks. The algorithm changes the communications flow. Typically (in a central controller scenario), this would consist of command messages; however using the approach proposed herein, messages now consist of node value, edge cost and connectivity updates. Given that some knowledge of the network being traversed would be expected to be required for the command-based approach (where units are given specific commands to take a particular path and report to a given node), it would seem that the level of communications required for this approach would be less than the level required for the command-based approach. This is, however, by no means certain, as the size of the grid provided as reference (under the centrally commanded approach) may be smaller (perhaps dramatically) than the size used for decision making. Other application-specific elements may also contribute to the relative size of the communications being larger or smaller between the two approaches.

Irrespective of whether size benefits exist or not, the proposed approach is very resilient to communications link failures. Because each unit has a local copy of the current status of the network and this is replicated between units, a unit may lose one link and (thus) lose updates related to one part of the grid network; however, it can continue to use the information regarding this area of the network that it had on-hand at the time of the communication failure. The correctness of this data (and thus its utility for decision making) will obviously decline over time (though some things, like edge costs will likely change infrequently); however, the presence of local network data prevents the loss of a communications channel from having an immediate dramatic (possibly disabling) impact. The unit may continue to receive updates from other units that flow over other links not impacted. It may also receive updates to an area that it has lost a more direct connection to via another, more indirect route.

The approach presented for destination assignment is non-deterministic, but stochastic. Presuming that all units have a suitably good random number generating capability, this should result in an ideal or near-ideal distribution. However, latency of information updating may result in various clustering effects (that have yet to be investigated). The effect of strategically impairing certain communications links (i.e., based on a prospective adversary's tactic) on algorithm performance and results also remains a subject for future work.

$$P_x = c^* \times \frac{G_x}{\sum_n G_n} \qquad (98.3)$$

This approach adds the steering coefficient (c*) term which can be replaced with a constant or equation that will be used to steer units in the desired way.

The uncertainty of the approach may, however, be problematic from the viewpoint of decision makers who may be hesitant to turn over control of high-impact decision making to a system that may not always perform perfectly (or repeatably) and which they may have difficulty understanding. Simulation-based testing (e.g., [36, 37]) may overcome some resistance; however, because of the distributed nature of the system it is not possible to incorporate a centralized override mechanism (which may be the critical impediment for its use for directing humans) directly. Commands or recommendations can, of course, be overridden by the units being given direction by it (in the context of human soldiers); however, units that decide to ignore the system's instructions in deference to their own feeling as to where they should be deployed break the probabilistic distribution and may result in over-responding to areas that appear to have high impact or importance and failing to deploy to areas with less impact or importance.

The proposed approach is applicable to the tasking of all types of units: human or autonomously controlled and units with different movement characteristics. The algorithm's use for heterogeneous units may benefit from an addition to the goodness characterization heuristic calculation to assess the goodness of a particular unit's presence at a given location (based on its suitability for a given task). Presently, this can be taken into account through the creation of unit-type specific grids.

98.7 Research Methods, Evaluation Scenario and Results

The analysis presented in this section begins the process of characterizing the performance of the IWD-based troop deployment algorithm through testing with the Colonel Blotto Game (CBG). The CBG is modified to increase its relevance to modern-day war-fighting. The modifications include the introduction of reinforcement (where troops can be deployed to an ongoing battle that has not yet been won or lost) has been incorporated. Along with reinforcement, the notion of requiring a level of dominance for winning a battle is introduced (i.e., there must be a given percent more troops, instead of allowing a win with a single additional troop compared to the opponent, which would not be realistic). In a case where a battle is not decided by the presence of a dominating force on one side or the other, a portion of the troops (inclusive of autonomous robots, UAVs and other fighting asserts) is lost. This modified approach, referred to as the Colonel Blotto Game with Reinforcement (CBG-R), is still a dramatic oversimplification of a real modern (or any) war-fighting environment; however, the change facilitates initial characterization of the use of the IWD algorithm for this purpose, facilitating an exploratory study.

The work presented herein consists of several parts. First, initial testing focuses on characterizing the performance of a basic (non-IWD) algorithm's performance under the CBG-R. This basic algorithm utilizes a strategy of replacing fallen troops (or autonomous robots, UAVs and other asserts) with a multiple of the number lost. Multiple levels of reinforcement are considered, with side A maintaining a constant level of reinforcement, while the reinforcement level of side B is manipulated. The purpose of this testing is to determine how the different reinforcement strategies impact the performance of the algorithm (in search of an optimal approach). These results are presented in Table 98.1. This non-IWD algorithm will serve as the control to the performance of the IWD-based approaches used in later experiments.

For each test, 10,000 troops are randomly distributed to the 625 locations on a 25 x 25 fully connected grid. At the beginning of each turn, a location is randomly selected to be a prospective battle site. If a battle has not already been won there (by one side or the other), one commences. The troop levels initially present are checked to determine if one side has enough troops present to dominate the other and win. If not, the reinforcement process starts with troops possibly being deployed directly to the battle location by both sides.

Next, the impact of the number of troops available for reinforcement's impact on algorithm selection is characterized. In this experiment, the total level of troops is changed from the 20,000 used in Table 98.1. In Table 98.2, data for a scenario where significantly less reinforcements are available (comparative to the number of troops initially deployed) is presented. In this scenario, only 11,000 total troops are available and 10,000 of these are deployed initially.

Next, the most basic variant of the Bifurcation-Aware IWD Troop Deployment Algorithm is evaluated. Under this approach, a decision as to where to deploy troops to is made at each grid node. Each node can deploy troops only to its adjacent nodes; however, no restriction on movement was incorporated effectively allowing a single troop to move

Table 98.1 Results from Scenario with 20,000 Total Troops and 10,000 Initially Deployed. The reinforcement level is presented in terms of multiples (e.g., 2x) of the number lost

Reinforcement level		Results		
A	B	# A Wins	# B Wins	# Ties
2x	2x	25029	24792	179
2x	2.5x	2788	47154	58
2x	3x	20	49979	1
2x	10x	0	50000	0
2x	100x	10613	39259	128
2x	110x	20695	29119	186
2x	120x	29575	20250	175
2x	150x	45575	4343	82

Table 98.2 Results from Scenario with 11,000 Total Troops and 10,000 Initially Deployed

Reinforcement level		Results		
A	B	# A Wins	# B Wins	# Ties
2x	2x	24904	24917	179
2x	2.2x	25153	24655	192
2x	2.4x	2861	47094	45
2x	2.5x	2901	47055	44
2x	4x	1	49999	0
2x	8x	415	49576	9
2x	10x	10664	39202	134
2x	11x	20469	29355	176
2x	12x	30066	19738	196

Table 98.3 Results from Comparing IWD Approach to the Non-IWD Approach (with 11,000 Total Troops and 10,000 Initially Deployed)

Using IWD flow		Results		
A	B	# A Wins	# B Wins	# Ties
No	No	24933	24895	172
Yes	No	50000	0	0
No	Yes	0	50000	0
Yes	Yes	24824	25150	26

Table 98.4 Comparison of Processing Time (in tics) for IWD versus Non-IWD Approach (Average across 1,000 trials)

	IWD approach	Non-IWD approach
Time - Mean	1,383,063.5	5,021.1
Time - Median	1,434,011.0	4,842.5

Table 98.5 Comparison of Results Using IWD with Single Direction Choice versus Basic IWD (with 11,000 Total Troops and 10,000 Initially Deployed)

Using IWD SDC		Results		
A	B	# A Wins	# B Wins	# Ties
No	No	2461	2535	4
Yes	No	14	4986	0
No	Yes	4992	8	0
Yes	Yes	2487	2508	5

Table 98.6 Comparison of IWD Single Choice versus IWD Projection (1,000 Trials per Condition, 11,000 Total Troops with 10,000 Initially Deployed)

Using long-term solver		Results		
A	B	# A Wins	# B Wins	# Ties
No	No	496	504	0
Yes	No	0	1000	0
No	Yes	1000	0	0
Yes	Yes	501	498	1

across the entire field. Practically, only a portion of the troops in an area are moved (thus making it plausible in many scenarios that troops from A move to B and troops from B move to C, instead of presuming that troops move from A through B to C; however, this is not enforced). The level of movement allowed and the performance of the approaches under different movement conditions will serve as a subject for future work.

Table 98.3 presents a comparison of using the IWD movement approach to the static-deployment approach used for the previous two scenarios tested. Again, this testing was performed with 11,000 total troops of which 10,000 were initially (randomly) deployed. Table 98.4 compares the processing time (mean average based on 1,000 trials) between the local IWD-based movement approach and the non-IWD approach.

Generally, troops don't move individually; they move as units. While the troops under IWD could be taken to represent units, this interpretation limits the analysis possible (making individual soldier/autonomous robot-level analysis impossible). Autonomous robot soldiers will likely utilize different command and control structures to human troops.

In particular, no localized commander/decision maker/supervisor may be required, presuming that each node has sufficient onboard autonomous decision making capabilities. This would make individual troop/robot deployment possible. In Table 98.5, two approaches to the basic Bifurcation-Aware IWD Troop Deployment Algorithm are compared. The approach presented in the previous section (with individual unit-level deployment) is compared to an approach where all individuals must be moved (or not moved) as a unit.

Finally, the full IWD algorithm (i.e., the one presented in Fig. 98.4 utilizing the algorithm's full extrapolation features) is compared to the localized IWD-based decision making one (which was used for the previous two experiments presented in Tables 98.3, 98.4 and 98.5). Under this approach, a full simulation is run five times for each prospective move (e.g., five runs for each location where troops could be moved to). Moving all of the troops as a unit or moving a number that is proportionate to need are compared. The decision that is made is based on whichever approach performs the best across the simulation runs. This data is presented in Table 98.6.

The time cost of the full IWD simulation is now compared with the time required for the localized approximation. These results are presented in Table 98.7, based on the comparison of the average run time of twenty runs with each configuration condition. Note that with only 20 trials, variance exists based on the level of complexity (e.g., how many rounds of reinforcement were required, the number of battles dominated initially, etc.) of the trials.

Table 98.7 Comparison of Time Requirements for IWD Projection versus Not Using

Using long-term solver		
A	B	Time Required
No	No	34274
Yes	No	410597451
No	Yes	106262399
Yes	Yes	1902230202

98.8   Analysis of Results

This section analyzes the data presented in the previous section. The previous section presented data for three different algorithms (direct-to-location/non-IWD, IWD with local consideration only and IWD with projection). The latter two are based on the algorithm presented in Fig. 98.4, with the local consideration approach simply not allowing the algorithm to run to completion (forcing it to make decisions with a very limited subset of data). This data demonstrates several key relationships between the various approaches.

It is clear that, for the direct-to-location/non-IWD replacement deployment approach (the data for which was presented in Tables 98.1 and 98.2), the performance of the algorithm depends on reserve troop depletion. Thus, increasing the (comparative) level of reinforcement deployment increases the potential for victory (percentage of victories), up until a certain point when it causes the side to draw down its troops too quickly and run out of reserves (allowing the other side to win later battles due to still having reinforcements to draw upon).

Table 98.3 shows that the to-location replacement deployment approach is equally matched with itself (allowing for a minor chance-based deviation); however, it does not fare well against the IWD-based localized decision making approach, which deploys all troops to a single depot location and allows them (along with existing troops) to be moved around the grid. Notably the non-IWD approach is significantly faster (see Table 98.4) than the IWD approach, however. In terms of time taken, the non-IWD approach dominates, taking only 0.3% of the time of the localized IWD approach.

The comparison of the individual troop decision making and unit-based movement shows a significant benefit to moving troops individually: the group-based movement is equally paired with itself; however, the individual troop movement approach dominates when the two are compared. The time required was not compared for these two computationally-similar approaches.

The IWD with projection approach (despite taking significantly longer from a calculation perspective) fails to produce superior results, with the localized decision making approach dominating its performance. It (the IWD with projection approach) requires 55,500 times as much processing time as the IWD approach without using projection (comparing the localized IWD-style decision making and simulation-based decision making, for both A and B). When simulation is used for only one side, it requires only 7,540 times as much processing time. The longer time required when pitting the two equally matched approaches, is attributable to the fact that it results in protracted battles (instead of one approach trumping the other quickly). It is important to note that the data in Tables 98.4 and 98.7 cannot be directly compared due to hardware differences between the environments where data was collected.

While the scenarios utilized featured random placement and battle location selection, it is possible that some algorithms may be better suited to specific scenarios than this general (somewhat generic) case. Additional characterization of the relative efficacy (or lack thereof) of the three approaches across additional scenarios will serve as a subject for possible future work.

98.9   Discussion

The initial work presented has demonstrated the efficacy of using an IWD-derived approach over a traditional direct-to-location Colonel Blotto Game-based deployment strategy. Note that in the traditional version deployment is performed only before gameplay starts, not as reinforcements. The ability to deploy reinforcements and the requirement of having a percentage more troops were added to model the complexity of modern warfare more accurately (though even with this change, it is still a dramatic oversimplification).

The work presented also demonstrates the utility of individual troop-based commanding (which would be difficult with human troops, but practical with robotic ones). The benefits, from both a performance and outcome perspective, of using the localized IWD-based decision making approach as opposed to the full multi-game simulation approach are also shown.

For future work and to allow the extrapolation of the results presented herein, several principal areas of consideration exist. The first is validating the results from the Colonel Blotto Game scenarios to real war fighting scenarios. Second, a plethora of permutations of experimental conditions are possible across the numerous variables manipulated in this experiment. It is possible that particular algorithms may exhibit superior performance in alternate (specific) combinations of variable choice selections. This may mean that an implementation would need a mechanism for detecting these special cases and changing the algorithm that it utilizes in applicable circumstances.

98.10 Conclusions and Future Work

This paper has presented the initial work on the development of an algorithm, derived from Shah-Hosseini's [1] Intelligent Water Drop's algorithm, for use in the autonomous deployment of fighting forces. It has shown how a simplified version of the IWD algorithm could potentially be used for this application. It has also presented a version of the IWD algorithm that has been adapted specifically for use in troop deployment which projects downstream information about the edge and node network to determine what path is most desirable to take at each node (and then makes a probabilistic decision as to what path to take, based on its goodness). The simplified algorithm has been shown to provide superior performance in cases where it and the more computationally expensive projection-based algorithm are directly compared. A qualitative evaluation of the proposed algorithms has also been performed and positive and negative aspects of the proposed approach have been enumerated and discussed.

Ongoing work involves the characterization of this approach. Future work will involve further development of the simplified intelligent water drop algorithm for troop deployment, making it more robust and adaptable to more types of scenarios and unit types. This enhanced algorithm will be tested in a simulation environment.

References

1. Shah-Hosseini, H. (2007). Problem solving by intelligent water drops. *Presented at evolutionary computation, 2007. CEC 2007. IEEE congress on*. Singapore.
2. Straub, J., & Reza, H. (2014). The use of the blackboard architecture for a decision making system for the control of craft with various actuator and movement capabilities. *Presented at proceedings of the international conference on information technology: New generations*. Las Vegas.
3. Chang, W., Lo, Y., & Hong, Y. (2009). A heuristic model of network-based group decision making for E-services. *Presented at information technology: New generations, 2009. ITNG'09. sixth international conference on*. Las Vegas.
4. Anyanwu, L. O., Keengwe, J., & Arome, G. A. (2010). Scalable intrusion detection with recurrent neural networks. *Presented at information technology: New generations (ITNG), 2010 seventh international conference on*. Las Vegas.
5. Fister, I., et al. (2013). A comprehensive review of firefly algorithms. *Swarm and Evolutionary Computation, 13*, 34–46.
6. Gosciniak, I. (2015). A new approach to particle swarm optimization algorithm. *Expert Systems Applications, 42(2)*, 844–854.
7. Haack, J. N. et al. (2011). Ant-based cyber security. *Presented at information technology: New generations (ITNG), 2011 eighth international conference on*. Las Vegas.
8. Fink, G. A., et al. (2014). Defense on the move: Ant-based cyber defense. *IEEE Security & Privacy, 12(2)*, 36–43.
9. Borel, E. (1953). The theory of play and integral equations with skew symmetric kernels. *Econometrica: Journal of the Econometric Society, 21*, 97–100.
10. Gross, O. A., & Wagner, R. A. (1950). A continuous colonel blotto game. RAND Corporation Report. Document Number: RM-408.
11. Roberson, B. (2006). The colonel blotto game. *Economic Theory, 29(1)*, 1–24.
12. Hart, S. (2008). Discrete colonel blotto and general lotto games. *International Journal of Game Theory, 36(3-4)*, 441–460.
13. Gross, O. A. (1950). The symmetric blotto game. RAND Corporation Report. Document Number: RM-424.
14. Laslier, J. (2002). How two-party competition treats minorities. *Review of Economic Design, 7(3)*, 297–307.
15. Kvasov, D. (2007). Contests with limited resources. *Journal Economic Theory, 136(1)*, 738–748.
16. Roberson, B., & Kvasov, D. (2012). The non-constant-sum colonel blotto game. *Economic Theory, 51(2)*, 397–433.
17. Hortala-Vallve, R., & Llorente-Saguer, A. (2012). Pure strategy nash equilibria in non-zero sum colonel blotto games. *International Journal of Game Theory, 41(2)*, 331–343.
18. Chou, M. C., Teo, C., & Zheng, H. (2008). Process flexibility: Design, evaluation, and applications. *Flexible Services and Manufacturing Journal, 20(1-2)*, 59–94.
19. Akgün, İ. (2007). Optimization of transportation requirements in the deployment of military units. *Computers & Operations Research, 34(4)*, 1158–1176.
20. Evans, K. M., DeCostanza, A. H., & L. G. Pierce. (2011). Trust in distributed operations. In *Trust in military teams* (p. 89).
21. Liu, C., et al. (2013). Primary research on the collaboration of hybrid unmanned vehicles. 2013 International Conference on Information Science and Computer Technology.
22. Dorigo, M., & Birattari, M. (2010). Ant colony optimization. In *Encyclopedia of machine learning* Anonymous.
23. Mirjalili, S., Mirjalili, S. M., & Lewis, A. (2014). Grey wolf optimizer. *Advances in Engineering Software, 69*, 46–61.
24. Kennedy, J., & Eberhart, R. (1995). Particle swarm optimization. *Presented at proceedings of IEEE international conference on neural networks*. Perth.
25. Blackwell, T., & Branke, J. (2004). Multi-swarm optimization in dynamic environments. In *Applications of evolutionary computing* Anonymous.
26. Karaboga, D., & Basturk, B. (2007). A powerful and efficient algorithm for numerical function optimization: Artificial bee colony (ABC) algorithm. *Journal of Global Optimization, 39(3)*, 459–471.
27. Shah-Hosseini, H. (2012). An approach to continuous optimization by the intelligent water drops algorithm. *Procedia-Social and Behavioral Sciences, 32*, 224–229.
28. Shah–Hosseini, H. (2012). Intelligent water drops algorithm for automatic multilevel thresholding of grey–level images using a modified otsu's criterion. *International Journal of Modelling, Identification and Control, 15(4)*, 241–249.
29. Shah-Hosseini, H. (2009). Optimization with the nature-inspired intelligent water drops algorithm. In: Wellington Pinheiro dos Santos. *Evolutionary Computation*, (pp. 297–320). Rijeka.
30. Qureshi, S., & ul Asar, A. Detection of malicious beacon node based on intelligent water drops algorithm. Proceedings on the International Conference on Artificial Intelligence, 2012. Athens.
31. Qureshi, S., et al. (2011). Swarm intelligence based detection of malicious beacon node for secure localization in wireless sensor networks. *Journal of Emerging Trends in Engineering and Applied Sciences (JETEAS), 2(4)*, 664–672.
32. Moncayo-Martínez, L., & Zhang, D. (2012). Optimisation of safety and in-transit inventory in manufacturing supply chains by intelligent water drop metaheuristic. *Presented at proceedings of the 2012 international working seminar on production economics*. Innsbruck.

33. Hendrawan, Y., & Murase, H. (2011). Neural-intelligent water drops algorithm to select relevant textural features for developing precision irrigation system using machine vision. *Computers Electronics Agriculture, 77(2)*, 214–228.

34. Duan, H., Liu, S., & Wu, J. (2009). Novel intelligent water drops optimization approach to single UCAV smooth trajectory planning. *Aerospace Science and Technology, 13(8)*, 442–449.

35. Straub, J., & Kim, E. (2013). Characterization of extended and simplified intelligent water drop (SIWD) approaches and their comparison to the intelligent water drop (IWD) approach. *Presented at proceedings of the 25th international conference on tools with artificial intelligence*. Herndon.

36. Huber, J., & Straub, J. (2013). A human proximity operations system test case validation approach. *Presented at 2013 I.E. aerospace conference*. Big Sky.

37. Huber, J., & Straub, J. (2013). Validating an artificial intelligence human proximity operations system with test cases. *Presented at proceedings of the SPIE defense, security sensing conference*. Baltimore.

Noshina Tariq and Farrukh Aslam Khan

Abstract

The ubiquity of the Internet has led to develop security measures to protect its services against abusive and malicious attacks. Most of the mercantile websites extensively use CAPTCHAs as a security measure against illegal bot attacks. Their purpose is to distinguish between humans and bot programs in order to defend web services from the bot programs. In this paper, we propose a CAPTCHA scheme that is based on cognitive abilities of human users. We name this scheme as "Match-the-Sound CAPTCHA" or "MS-CAPTCHA". The users are supposed to choose the best-matched object from the rest of the images in order to prove themselves as humans after listening to the sound. A study of 50 candidates is performed to check the performance of our proposed scheme. We also conduct a feedback survey from the candidates to explore the usability features of the proposed MS-CAPTCHA. Our study shows that the users get annoyed with distorted text, audio, and background clutter, whereas they enjoy more image-based and simple audio-based MS-CAPTCHAs.

Keywords

Human Interaction Proof (HIP) · CAPTCHA · Bot attacks · Audio CAPTCHA · Image-based CAPTCHA

99.1 Introduction

Over the past few years, there has been an elevated utilization of the intricate applications on the web in an exceptionally dynamic fashion, raising security issues and concerns. Thus, the inclusion of security pledge against illegal and malicious access is inevitable for organizations to gain user trust and confidence. Therefore, we need to render access to legitimate users only. There are certain conditions where on some open web resources, for example, social networks including Facebook, Twitter etc., and service providers, e.g., Google, Yahoo, Hotmail etc., malicious and spurious bot programs can attack. To fight back such kinds of bot attacks, we use a security mechanism – called Human Interaction Proof (HIP) or Completely Automated Public Turing test to tell Computers and Humans Apart (CAPTCHA). CAPTCHA is a category of HIP tests grounded on artificial intelligence [1]. A user who qualifies this challenge is assumed to be a human and if not then it is classified as a bot [4]. Essentially, there are two main aspects while designing a CAPTCHA. First, it must be robust and strong against attacks. Secondly, a CAPTCHA must be usable and functional. Therefore, researchers are focusing on making it strong as well as usable [3, 6]. The usability feature is somewhat ignored in many researches for designing a CAPTCHA scheme [8]. Yan and El Ahmed [8] discussed

N. Tariq
Department of Computer Science, National University of Computer and Emerging Sciences, A. K. Brohi Road, H-11/4, Islamabad, Pakistan
e-mail: noshtariq@yahoo.com

F.A. Khan (✉)
Center of Excellence in Information Assurance (CoEIA), King Saud University, Riyadh, Saudi Arabia
e-mail: fakhan@ksu.edu.pk

© Springer International Publishing AG 2018
S. Latifi (ed.), *Information Technology – New Generations*, Advances in Intelligent Systems and Computing 558, DOI 10.1007/978-3-319-54978-1_99

the usability issues in a very effective and simple way. They mention that while distorting the text, it is important to contemplate the method and level of distortion because it can make a human user confused in perceiving the right letter. It can eventually become challenging for non-native users to identify the actual text. The authors also suggest the content length i.e., the number of letters in a challenge and the arrangement and presentation of the challenge in terms of fonts, colors, and sizes.

Generally, there are two sets of CAPTCHAs - OCR-based and non-OCR based CAPTCHAs [18]. However, the authors in [19] categorized CAPTCHAs into three sets: OCR-based, Visual Non-OCR-based, and Non-Visual CAPTCHAs. OCR-based challenges are quite vulnerable to attacks as they are dependent on the degree of deformation of the text. The more distorted and deformed these challenges are, the stronger they are against attacks [10, 19]. However, these kinds of CAPTCHAs have a tradeoff between robustness and user-friendliness. If we make a text CAPTCHA more robust by adding more clutter, deformation, and distortion of the text, we make it less user-friendly and less readable. Failure in recognition makes a text CAPTCHA weak and ineffective [9].

In this paper, we propose a novel CAPTCHA scheme called Match-the-Sound CAPTCHA (MS-CAPTCHA) based on perceptual and cognitive abilities of the humans. For a user to prove him/her as a legitimate human user, the user needs to carefully select the best match from a set of images. Our proposed scheme relies on semantics and the relationship between sound and displayed images rather than their low-level features such as shape, color, size, and texture etc. This improves its robustness and strength. To avoid machine-learning attacks, we have shown six different images in the test, which enhance their robustness. As the scheme consists of different images and not a typical text, there is no need to add noise and clutter. This will increase the readability and usability of the test. Similarly, MS-CAPTCHA handles random guess attacks as well. A random guess attack has a probability of 1/6 or 16.7% to be accurate. Nevertheless, it is significantly less than the rate for bot programs to attack a CAPTCHA image [9].

The remainder of the paper is organized as follows: In Sect. 99.2, we discuss some related work about image and sound-based CAPTCHAs. Sect. 99.3 discusses the proposed scheme, whereas Sect. 99.4 shows the experimental details and results. Finally, Sect. 99.5 concludes the paper with possible future directions.

99.2   Related Work

During the past few years, CAPTCHAs have been used as a safety measure against malevolent bot attacks. Text CAPTCHAs are easily broken if they are well segmented,

and if we add more noise, clutter, and deformation, these CAPTCHAs become ineffective. Gao et al. [2] proposed a new audio CAPTCHA in which the user reads the sentence written in the displayed text box. To help the user, the voice waveform diagram is also given as a feedback. They used different books to randomly generate the sentences of 8 to 20 words. The temporally saved sound file is then analyzed to decide the authenticity of the user as a human or a bot. Such kinds of challenges are subject to accent problems. reCAPTCHA was proposed at Carnegie Mellon University by Ahn et al. [17]. Later on, it was sold to Google. It has a sound file in the challenge for human ease. It displays text challenges along with the audio clip, where different numbers and words are spoken in different voices. The advantage of re-CAPTCHA is its scalability, which is why many social networking websites and well-known companies such as Google are using it. On the other hand, the major problem with this kind of challenge is the noise and deformation of the text, which is sometimes very annoying to the users.

Soni and Tiwari [13] proposed an improved method for Collage CAPTCHA in which images are shown on the right and left sides of the screen. The image on the left has the picture of an object and on the right-side image; the label is displayed for the given image. To pass the challenge, the user has to select the right image correlated to the given text image. Then, he/she is allowed to enter the label/name of the image in the text box to complete the test. Gossweiler et al. [11] proposed another CAPTCHA scheme based on image orientation. The user has to set the image in his/her upright orientation. The problem with this scheme is that sometimes users fail to set it at up-right angle due to its actual orientation. The eminence of this CAPTCHA is based on the social feedback mechanism.

Lazar et al. [12] proposed an audio-based HIP in which a series of audios is played with at least one correct option. If the user has entered the correct answer, he gets access to the web page. Agiomyrgiannakis et al. [6] gave the idea of a three-dimensional audio-CAPTCHA. In this scheme, an audio prompt is presented to the user. The proposed three-dimensional audio CAPTCHA encapsulates a target signal having an authentication key and a distraction signal in an auditory environment. The answer is checked against the authentication key and if it matches, the access is granted to the user.

The authors in [18] proposed advanced collage CAPTCHA based on their previous collage CAPTCHA. They ensured the security and resilience in the advanced collage CAPTCHA. In collage CAPTCHA, some random pictures are taken and rotated to some extent and displayed on the screen. The user is then asked to select the desired picture; if the picture is selected correctly, it is presumed to be a legitimate human user. Whereas, in advanced collage CAPTCHA, some pictures are shown on the screen along with some other pictures on the right side of the screen having one object same as the original picture. If the user selects both the objects correctly, he/she is assumed to be a legitimate user.

Yamaguchi et al. [7] proposed an audio CAPTCHA scheme for disabled people. They used a large number of Internet documents for this purpose. Their proposed scheme is highly scalable as they use any number of documents for generating their CAPTCHA scheme. The test is generated from these documents and presented as a verbal/audio sound to the users. They proposed two types of schemes. In the first scheme, the user has to distinguish a different meaning phrase from the other ones. In the second scheme, the user has to identify the most appropriate frequent topic among the given topics. Meutzner et al. [8] proposed an audio CAPTCHA based on basic English core words with pronunciation variety. The desired word is synthesized by the audio signal randomly chosen from the word list. Per CAPTCHA, four to six words are selected. They also added nonsense words in the audio to distract the speech recognizer programs as shown in Fig. 99.1a.

Lazar et al. [16] proposed a new real-time audio-based challenge with fixed time period to be solved, called SoundRight CAPTCHA. A user is asked to recognize an explicit sound among 10 sounds. The study proved that blind users found it more useful with 90% success rate. A visual CAPTCHA along with an audio access button is displayed to the user. On audio alternative selection, an instruction message is popped and the user presses the space bar to initiate the challenge. The user is supposed to click the space bar each time he/she hears a meticulous sound. Figure 99.1b shows the example of an image/audio CAPTCHA.

99.3 Proposed MS-CAPTCHA Scheme

We propose a novel CAPTCHA scheme that relies on perceptual and cognitive abilities of the users. In order to prove a person as a legitimate human user and not as a bot, the user has to carefully select the best match from the given images. This selection entirely depends upon the perception and cognition of a user. We crawled through the Internet and gathered different types of images and their respective sounds, and then cropped them according to our needs and demands. We tried to give diversity in image selection to prevent machine learning and segmentation attacks.

We select six images from the database at runtime randomly and display them to the user. Then, the relevant audio file is also selected and saved as the answer. If the user matches the sound with the correct object image, he/she is assumed to be the legitimate user i.e., a human not a bot. Then, the answer is matched with the already saved answer. The answer is saved on the time of selection of an image from the database randomly. If the answer matches with that, it is concluded that the user is a human, otherwise not. Initially, every image has a weight "50" assigned to it. We have randomly chosen 50 participants to conduct this experimentation. After the experiments, it is decided which image is appropriate and should be kept for use and which one must be discarded. The image having higher weight is meant to

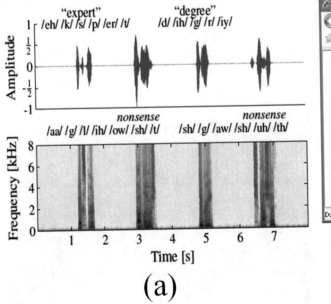

Fig. 99.1 (**a**) Words and nonsense speech, (**b**) An image/audio CAPTCHA

have a higher discernment value and is supposed to be perfect for our scheme. If an image is not correctly identified, its weight is decreased by 5 each time and vice versa. We set a threshold of "20"; if an image gets weight equal to or less than 20, it is discarded from the database automatically. At the end, the results are compared with the results of the re-CAPTCHA scheme to show whether our scheme works better than the text CAPTCHA schemes.

Algorithm 99.1 Match-the-Sound CAPTCHA

Input: *Pictures, their respective sounds, and weights*
Output: *MS-CAPTCHA*
// Gather pictures and their respective sounds from the Internet.
1: *Crawl down pictures from Internet*
2: *Crawl down their sounds from Internet*
3: *Crop the image according to the demand*
//Assign weight and store images in the database.
4: *Assign each Image a weight*
5: *let weight ← 50*
6: *Put them in the database*
 // Making of MS-CAPTCHA
7: *Get 6 Match-the-Sound samples randomly from the database*
8: repeat
9:   *Save respective sound of the sample*
10:   *Display the image*
11:   *Display the sound button*
12:   *Set the answer sound file to audio player*
 // Updating weights of each image.
13:   **if** *answer matches* **then**
14:   *weight = weight + 5*
15:   **else**
16:   *weight = weight - 5*
17:   **end if**
18: until *end of database*

We, very carefully, select only those images that are not typical in nature. If we use the sound of glass breaking, we try to select an image that has an unconventional look. The image has a broken glass, a hand touching the glass, and a face image reflecting from the pieces of the glass. The working of the proposed MS-CAPTCHA is presented in Algorithm 99.1. The examples of MS-CAPTCHA images are shown in Fig. 99.2.

99.4 Experimentation and Results

A high-quality CAPTCHA has two significant traits i.e., security and usability. OCR-based CAPTCHAs are prone to segmentation and recognition attacks as compared to non-OCR based CAPTCHAs. A CAPTCHA is considered good if it is attack-resilient against bot attacks as well as it is user-friendly to allow users to pass it with minimal efforts. Fidas et al. [4] conducted a test on OCR-based CATCHA usability, and the results showed that 61.4% of users find text distortion as a key hindrance. Moreover, 21.4% found background distortion and patterns as a major problem.

reCAPTCHA is a well-known CAPTCHA scheme used by many social networking and commercial websites. In this scheme, two words are selected after performing the OCR test: a control word and an unknown word. The user is verified only by the control word. The main problem for a user is to guess exactly which word is the control word. Sometimes, the distortion level is high, which makes it less user-friendly and the user inputs a wrong text [5]. Lazar et al. [12] discussed that Google's reCAPTCHA has only 46% success rate due to its distortion and difficulty. It is discussed in [14, 15] that reCAPTCHA is not safe and the crack percentage is 52%. It is difficult to segment when played as a non-continuous audio CAPTCHA.

For our experiments, 50 volunteers/candidates were selected for conducting the experiments. 20 different CAPTCHAs were generated dynamically from the database

Fig. 99.2 Examples of Match-the-Sound CAPTCHA

Fig. 99.3 Comparison between MS-CAPTCHA and reCAPTCHA features

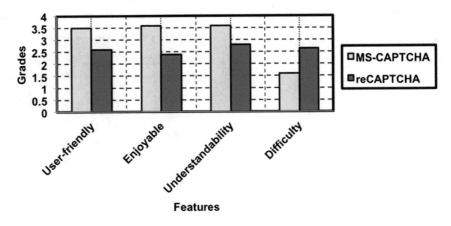

Fig. 99.4 Average time to solve MS-CAPTCHA and reCAPTCHA

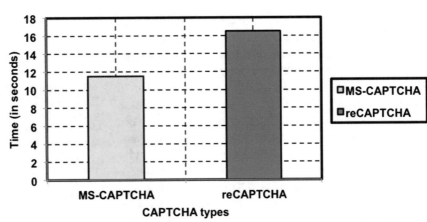

with their respective sounds and shown to the candidates individually. For comparison, we selected reCAPTCHA, which is a type of text CAPTCHA widely used by Google. At the end of each session, a short questionnaire survey was given to the candidates to ask their opinion as a feedback, for example, friendliness, complexity of images, etc. The feedback showed that they were confused with the text and the sound, which was played in reCAPTCHA as it was not very helpful in deciding the appropriate word to input. It has overlapping and distorted target voices with stationary background noise. Whereas, due to the usability feature of MS-CAPTCHA, they found it easy to identify the sound and decided the input image. We also compared our proposed CAPTCHA scheme's feedback results with those of reCAPTCHA, as shown in Fig. 99.3. The figure shows that our scheme is not only more user-friendly, but also more enjoyable and understandable for the users because there is no distortion in the images as it is found in conventional text CAPTCHA schemes. The difficulty level is also low as compared to reCAPTCHA.

According to Fig. 99.3, MS-CAPTCHA scored 3.5, 3.6, 3.6, and 1.6 on the grade sheet for "User-friendly", "Enjoyable", "Understandability", and "Difficulty" features, respectively, whereas reCAPTCHA scored 2.6, 2.4, 2.8, and 2.65 for the same features, respectively. As discussed

earlier, image-based CAPTCHAs are more user-friendly and usable. They are well perceived, recognized, and identified by the human users. Consequently, on average, the users took 11.5 seconds to solve 20 samples of MS-CAPTCHA. On the other hand, they took 16.5 seconds to solve equal number of reCAPTCHAs. The results of average time to solve MS-CAPTCHA and reCAPTCHA are given in Fig. 99.4. The figure clearly shows that MS-CAPTCHA works better than reCAPTCHA and takes much shorter time to solve it. Similarly, the comparison of the hit-rate is also performed for both the schemes. The hit-rate of MS-CAPTCHA is 92.50%, whereas it is only 90.40% for the reCAPTCHA. The results show that the average hit-rate of the proposed MS-CAPTCHA is much higher than reCAPTCHA, as shown in Fig. 99.5.

99.5 Conclusions and Future Work

CAPTCHA provides security against malicious and malevolent bot programs. There are different types of CAPTCHAs used for various web-based applications. Based on the formation and structure of these CAPTCHAs, it is tested whether the user is a human or a machine. However, there is still a question about CAPTCHA's security as it can be

Fig. 99.5 Hit-rate of MS-CAPTCHA and reCAPTCHA

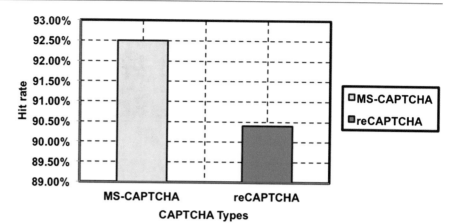

attacked by different CAPTCHA breaking techniques. These techniques are mostly based on machine learning algorithms and segmentation attacks. Non-OCR based CAPTCHAs like audio and image CAPTCHAs are more interesting and usable for human users instead of the distorted text. In designing a strong CAPTCHA, segmentation prevention techniques should be implemented in the text CAPTCHAs and image labeling should be avoided. The source image collection and challenge generation should be fully automated. We designed a novel CAPTCHA called Match-the-Sound CAPTCHA (MS-CAPTCHA) based on the cognitive abilities of human users. The scheme was tested with the help of 50 candidates to check its effectiveness. The results of the study show that MS-CAPTCHA is more user-friendly and efficient, has a higher hit-rate, and takes less time to solve as compared to reCAPTCHA. In the future, we plan to watermark our images in order to make them more robust against machine learning and segmentation attacks. In addition to this, we would also like to watermark the audio file against attacks.

References

1. Banday, M. T., & Shah, N. A. (2011). *A study of CAPTCHAs for securing web services*. arXiv preprint arXiv:1112.5605.
2. Gao, H., Liu, H., Yao, D., Liu, X., & Aickelin, U. (2010). An audio CAPTCHA to distinguish humans from computers. In: *Third international symposium on electronic commerce and security (ISECS 2010)*, July 29–31, Guangzhou, China (pp. 265–269).
3. Yan, J., & El Ahmad, A. S (2008). Usability of CAPTCHAs or usability issues in CAPTCHA design. In *Proceedings of the 4th symposium on usable privacy and security, July 23–25, 2008*, Pittsburgh, Pennsylvania, USA, (pp. 44–52).
4. Fidas, C. A., Voyiatzis, A. G., & Avouris, N. M. (2011). On the necessity of user-friendly CAPTCHA. In *Proceedings of the SIGCHI conference on human factors in computing systems* (pp. 2623–2626). Canada:Vancouver.
5. Saini, B. S., & Bala, A. (2013). A Review of Bot Protection using CAPTCHA for Web Security. *IOSR Journal of Computer Engineering, 8*(6), 36–42.
6. Agiomyrgiannakis, Y., Tan, E., & Abraham, D. J. (2016). *Systems and methods for three-dimensional audio CAPTCHA*. U.S. Patent Application 13/859,979. Feb 16, 2016.
7. Yamaguchi, M., Nakata, T., Okamoto, T., & Kikuchi, H. (2014). An Accessible CAPTCHA System for People with Visual Disability–Generation of Human/Computer Distinguish Test with Documents on the Net. In *Universal access in human-computer interaction. Design for all and accessibility practice.* Springer International Publishing (pp. 119–130).
8. Meutzner, H., Gupta, S., & Kolossa, D. (2015). Constructing Secure Audio CAPTCHAs by Exploiting Differences between Humans and Machines. In *Proceedings of the 33rd annual ACM conference on human factors in computing systems* (pp. 2335–2338). Seoul, Republic of Korea.
9. Dhamija R., & Tygar J. (2005). Phish and Hips: Human interactive proofs to detect phishing attacks. In *Proceedings of Human Interactive Proofs (HIP 2005) LNCS 3517* (pp. 69–83). Bethlehem.
10. Raj, A., Jain, A., Pahwa, T., & Jain, A. (2010). Picture captchas with sequencing: Their types and analysis. *International Journal of Digital Society (IJDS), 1*(3), 208–220.
11. Gossweiler R., Kamvar M., and Baluja S. (2009). What's up captcha?: A captcha based on image orientation. In *Proceedings of the 18th international conference on World Wide Web* (pp. 841–850). Madrid.
12. Lazar, J., Brooks, T. I., Melamed, G., Holman, J. D., & Feng, J. (2014), *Audio based human-interaction proof*. U.S. Patent Application 13/308,011. Mar 4, 2014.
13. Soni, R., & Tiwari, D. (2010). Improved captcha method. *International Journal of Computer Applications, 1*(25), 107–109.
14. Aiswarya, K., & Kuppusamy, K. S. (2015). A study of audio captcha and their limitations. *International Journal of Science and Research, 4*(4), 918–923.
15. Sano, S., Otsuka, T., & Okuno, H. G. (2013). Solving google's continuous audio CAPTCHA with HMM-based automatic speech recognition. In *Advances in information and computer security* (pp. 36–52). Berlin/Heidelberg: Springer.
16. Lazar, J., Feng, J., Brooks, T., Melamed, G., Wentz, B., Holman, J., & Ekedebe, N. (2012). The SoundsRight CAPTCHA: An improved approach to audio human interaction proofs for blind users. In *Proceedings of the SIGCHI conference on human factors in computing systems* (pp. 2267–2276). Austin.
17. Ahn, L., Maurer, B., McMillen, C., Abraham, D., & Blum, M. (2008). reCAPTCHA: Human-based character recognition via web security measures. *Science, 321*(5895), 1465–1468.
18. Shirali-Shahreza, M., & Shirali-Shahreza, S. (2008). Advanced collage captcha. In: *Proceedings of 5th international conference on information technology: New generations (ITNG 2008)*, April 7-9, 2008, Las Vegas, Nevada, USA. (pp. 1234–1235)
19. Shirali-Shahreza, S., & Shirali-Shahreza, M. (2011). Multilingual highlighting CAPTCHA. In: *Proceedings of 8th international conference on information technology: New generations (ITNG 2011)* April 11–13, 2011, Las Vegas, Nevada, USA. (pp. 447–452).

Part XI

Health, Bioinformatics, Pattern Detection and Optimization

Information Technology – Next Generation: The Impact of 5G on the Evolution of Health and Care Services

Christoph Thuemmler, Alois Paulin, Thomas Jell, and Ai Keow Lim

Abstract

As more and more details of 5G technology specifications unveil and standards emerge it becomes clear that 5G will have an enabling effect on many different verticals including automotive, mobility and health. This paper gives an overview about technical, regulatory, business and bandwidth requirements of health care applications including e-connectivity in the pharmaceutical domain, medical device maintenance management, hospital at home, supply chain management, Precision and Personalized medicine, robotics and others based on latest research activity in the field.

Keywords

5G • Healthcare • LPWA • Edge-clouds • Algorithms

100.1 Introduction

Although concrete specifications are still lacking in some areas, 5G technology is approaching with giant footsteps. With vendors having announced trials with fixed mobile access as early as 2017 there can be no doubt that 5G technology has the potential to become a relevant enabler for the Internet of Things and the health vertical. There have been ongoing discussions about the specific requirements of the health sector to establish meaningful service offerings to health care providers, pharmaceutical industry and other relevant stakeholders such as patients, formal and informal carers and social care.

Health care is a relevant market as its overall share in national GDPs is significant with values ranging between 6% in China, 10% in Europe and 18.5% in the United States. eHealth has been going through a typical hype curve starting in the early 2000 fulling high hopes of quick enhancement of quality of care, quality of experience and a reduction in health care costs. However, the uptake of eHealth technology has been protracted and the reason for this protracted uptake is still unclear. The eHealth action plan published by the European Commission in 2012 [1] lists several barriers to deployment of eHealth that could be identified during public consultations and research on the topic:

– lack of awareness of, and confidence in eHealth solutions among patients, citizens and health care professionals
– lack of interoperability between eHealth solutions
– limited large-scale evidence of the cost-effectiveness of eHealth tools and services
– lack of legal clarity for health and wellbeing mobile applications and the lack of

C. Thuemmler (✉) • A. Paulin
Edinburgh Napier University, 10 Colinton Road, Edinburgh EH10 5DT, UK
e-mail: c.thuemmler@napier.ac.uk; alois@apaulin.com

T. Jell
Siemens AG, Infrastructure & Cities Sector, IC MOL TI IMSP, Otto-Hahn-Ring 6, 81739 Munich, Germany
e-mail: thomas.jell@siemens.com

A.K. Lim
Celestor Ltd, 43 Corby Craig Avenue, Bilston, Midlothian EH25 9TL, UK
e-mail: aikeowlim@gmail.com

© Springer International Publishing AG 2018
S. Latifi (ed.), *Information Technology – New Generations*, Advances in Intelligent Systems and Computing 558, DOI 10.1007/978-3-319-54978-1_100

– transparency regarding the utilisation of data collected by such applications
– inadequate or fragmented legal frameworks including the lack of reimbursement
– schemes for eHealth services
– high start-up costs involved in setting up eHealth systems
– regional differences in accessing ICT services, limited access in deprived areas

The World Health Organisation (WHO) concludes in their most recent report on eHealth that there *"is the need for stronger political commitment for eHealth, backed by sustainable funding and for effective implementation of policy"* [2].

Although there is some mentioning of high level issues with regards to interoperability and connectivity both reports say little about the technological readiness of network infrastructures, future applications and business models and the subsequent requirements. This paper will discuss details on the most relevant use case scenarios and estimate their technological requirements based on experimental tests and empiric evidence. Furthermore, the paper will discuss current 5G architecture and business models potentially suitable for the health domain.

This paper is seeking to contribute to the current discussion on 5G specifications in the health domain. In this paper we include a discussion of the empiric evidence for the rapid evolution of eHealth as a 5G application domain, the emergence of low wide area technology and its implication for health care, smart phones, 4G and mm Waves, advanced 5G functionalities, the role of mobile edge clouds, next generation network abilities, and 5G service examples and bandwidth requirements.

100.2 Empiric Evidence for the Rapid Evolution of eHealth as a 5G Application Domain

Several white papers have been created in a bit to establish an outlook onto eHealth use cases and their potential requirements. Thus, in 2014 the European Commission published a "green paper" followed by a public consultation to obtain a better understanding of the strategic requirements for the use of mobile telephony in the health domain [3]; while contributors highlighted the need for regulatory certainty and made a case for more evidence synthesis through large scale implementations there was only limited input with regards to specific applications and concrete requirements [4].

With the emergence of 5G technology and the layout of an aligned international time table several white papers have been published to establish requirements for the application of 5G technologies in the health vertical [5, 6]. Several concrete applications have been highlighted, among them the use of smart pharmaceuticals connected directly to the network utilizing Low Power Wide Area technology, the use of millimetre-waves for quick audio and video connectivity with disease management portals, ultra-low latency to enable robotic assisted surgery and more. As the network capability grows there is strong evidence for the massive use of mobile health applications. The US Food and Drug Administration (FDA) estimated that 500 million smartphone users have been using health care applications by 2015 and that by 2018 50% of the more than 3.4 billion smartphone and tablet users will have downloaded mobile health applications [7]. This is supported by data from the European Commission estimating the value of the global m-Health market at 17.8 Billion Euro by 2017 [8]. In summary, with the growth in network capability and mobile health applications, several white papers have focused on the requirements for the application of 5G technologies in the health domain.

100.3 The Emergence of Low Power Wide Area Technology and Its Implications for Health Care

One of the current challenges in 5G is the lack of crisp technology specifications. There is agreement that 5G will offer more than just a simple performance enhancement compared to previous network generations. Rather than being more of the same but faster and bigger, 5G will present as a set of services, especially addressing the need of the integration of M2M, audio and video and other services spread over a much larger spectrum range than any network generation before.

Low power wide area (LPWA) technology has become increasingly popular mainly through the introduction of smart meters. The roll out of LPWA in Europe is in full swing and mobile telecommunication operators are set to commence to fit their towers with LPWA modules in 2017 or 2018 depending on the geographical area. Gartner forecasts that in general "Low Power short range networks will dominate wireless connectivity through 2025" but predicts a significant role for LPWA devices [9] as an enabler for the Internet of Things (IoT). (See [10] for an overview over security and privacy implications of IoT.)

This is emphasized by predictions of strong growth of LPWA by Cisco in their latest Visual Networking Index update [11].

In fact, with the 3GPP standardization of Narrow Band – Internet of Things (NB-IOT) radio access technology an alternative to Bluetooth for the connection of inhalers and Insulin Pens carrying advanced sensor technologies to remote disease management portal has gained serious momentum. While Bluetooth is dependent on smart phones or tablets as gateway with serious limitations to the number of devices that can be connected and also being affected by the physical limitations of high frequency radio communication (reduced indoor reach and penetration) LPWA modules have superior proprieties due to their operation on a lower frequency band. LPWA technologies seems also promising for indoor use in hospitals for supply chain management and monitoring and maintenance of medical devices such as infusion pumps, hospital beds, wheel chairs, ECGs, etc.

Onduo, a newly announced joint venture between Sanofi and Google, is aiming to provide services for 592 million diabetics globally by 2035 [12]. The number of patients suffering from asthma is on the rise. According to the Global Asthma Network there were 334 million people affected by asthma in 2014 [13]. These figures are likely to climb further. Due to the nature of the treatment of these conditions, which requires the regular use of insulin pens and inhalers it can be assumed that the use of directly connected devices might generate the need for more than 1 billion LPWA connections. There are concerns on the efficiency and cost effectiveness of e-connected devices. However, due to the high costs of hospitalization and the societal burden of inability to work based on the available evidence it must be assumed that devices directly connected to the network by LPWA technology will have a place in the management of chronic, non-communicable conditions. Given the demographic development in almost all industrialized countries and emerging economies it is also safe to say that 20th century hospital care is not a futureproof solution and that in the future more care will be delivered at home and over the Internet. LPWA will play a role in the monitoring of devices at home and in hospitals.

100.4 Smart Phones, 4G and mm Waves

Smart phones and tablets will continue to play an important role in health care as they hold audio and video capability. It will be crucial to provide sufficient bandwidth for audio and video interaction. This is not only relevant for the use of telemedicine and similar services but also for the virtualization of care. In many cases a personal visit from a carer might not be necessary but could be replaced by a video call. Visits to the doctor or the pharmacist to get advice on the use of medical devices might be replaced by video tutorials. They can be replayed at no extra costs. Smart algorithms and autonomous systems might advise people with chronic conditions on best practice beyond their prescriptions to cover non-pharmaceutical aspects of the disease management. Audio and video interfaces will play an important role in the virtualization of care. In this context, real time connectivity and the availability of suitable bandwidth might be required. Indoors this might soon be delivered by mm Waves (fixed mobile access).

100.5 Advanced 5G Functionalities

The integration of LPWA and mm Wave technology are important to the health vertical. However, more complex innovation is required in the access and core network to provide the ability to combine or re-combine data and services in order to add value. This will be achieved through multi domain orchestration and will play a particular huge role in the virtualization of care (Precision Medicine) where different services consumed by a patient need to be integrated in real time to conclude on for example the dose of a particular pharmaceutical required or in the external management of artificial organs such as insulin pumps in real time.

Multi-domain orchestration and multi-tenancy will play a key role in the use of the same physical network infrastructure by different stakeholders simultaneously. While patients will have an interest to combine several individual services other stakeholders using the same infrastructure might have different needs. Healthcare providers might want to see how the prevalence of a particular disease impact their budget. Pharmaceutical companies might want to know how a particular medication is utilized in a region. Self-help groups might want to implement peer to peer communication – all utilizing the same network at the same time.

There are emerging requirements with regards to time synchronization. 5G will see an increased density of access points in the radio access network (RAT) due to the physical proprieties of the new wave formats. On top of this different technologies such as future satellite internet, device-to-device (D2D) communication and terrestrial Internet need to be coordinated. Time synchronization will also play a role in the progressive uptake of parallel processing where centralised public cloud intelligence will be progressively deployed into the network periphery.

100.6 The Role of Mobile Edge Clouds

Especially from the perspective of the health domain the current technology trend towards everything as a service strategies (XaaS) is a difficult one. Patient information is on the one hand a desirable commodity for all kind of right and wrong reasons. On the other hand, health care providers have to obey the existing rules which make them liable for any data loss or data breach. At the same time, Big Data technology seems promising by opening the horizon for new research potentially leading to enhanced quality of care through new knowledge and technology.

The offerings of Google and IBM Watson with regards to smart algorithms are so far based on XaaS, meaning the data has to be delivered to the service provider for processing. This, however, seems incompatible with the duty of the health care provider to keep patient data private and safe. A proxy-server architecture, whereby the service provider positions a server on site is not a real remedy as the service provider is still in full control of the data. From a technological perspective edge clouds significantly reduce the latency, make the service more fluid but does not change the topology. In 2012 the European Commission published a report on the future of cloud computing, suggesting a shift in the existing cloud dogma by sending software to the data rather the other way around [14]. This idea has been developed further under FI-STAR a research project funded by the European Commission [15–17]. As such the edge-cloud could serve as a universal proxy-server in the user domain. This would prevent user lock-in and boost diversity.

100.7 Next Generation Network Abilities

Although 5G will not be fully deployed until at least 2020 there are already speculations on 6G and the next generation network technology ongoing. It seems that 6G will be featured by satellite- and proxy internet. There is a proposal by Elon Musk under the working title Space X to launch 4425 satellites into low earth orbits [18]. While it is unlikely that this will remain the only proposal for space

based internet services it needs to be anticipated that current networks, especially the "core network" will be expanded into space. Proxy-internet will be based on D2D capability and future technologies allowing to create dynamic mash-networks to enhance the reach at the cloud/service edge.

100.8 5G Service Examples and Bandwidth Requirements

Current white papers are not specific about the potential future utilization of 5G technology yet. However, concrete and tangible applications have been discussed within the health care and pharmaceutical industry. Table 100.1 provides an overview about the current suggestions and the potentially required throughput.

We found that typically bandwidth requirements in the health domain are overestimated due a focus on telehealth services and remote diagnostics involving video-streaming. Not enough attention is being paid to M2 M applications. Based on a recent study commissioned by Huawei we had an opportunity to study bandwidth requirement in an urban teaching hospital carefully and found that traffic scenarios are not challenging for the available infrastructure [19, 20]. However, the existing infrastructure does not allow for geographical flexibility. Many hospitals are fitted with Wi-Fi. However, many IT directors perceive Wi-Fi as risky and are not keen to allow devices to access their network.

100.9 Conclusions

5G technology clearly has the potential to provide network capability and connectivity to boost the use of eHealth and mHealth in hospital and the community. Especially LPWA technology may lead to a break-through in the uptake of eHealth technology at home and in hospital. LPWA has the potential to massively grow the number of machine-to-machine (M2M) connections over the coming decade. Overall the need to connect 1000 times more devices via 5G is probably more relevant than providing for huge bandwidth for video streaming. There is the need to formulate new business models describing how different stakeholders can create win-win scenarios.

Table 100.1 5G health vertical use case scenarios, technological, regulatory and business case specifications

Use case	5G technology	Regulatory	Business	Bandwidth/data volume
Smart pharma	• LPWA • mm Waves • Network slicing • Multi domain orchestration • 100 x connections • Intelligent networks • Terminal with network control capacity • 1/10 of energy consumption	• Medical Product Legislation • International standardization • National and EU privacy rules • Safe Harbouring (EU) • National eHealth legislation (for example Germany, Austria (ELGA), Denmark (Sundhed)	• Transition of business models of pharmaceutical industry from manufacturing to service • Massive enhancement of adherence • Prevention of serious episodes, sick days, hospitalization and death • Combination of pharmaceutical and non-pharmaceutical therapy • Patient feedback • Use of authorized data for research purposes • Pharma – Telco joint ventures • Medication incompatibility • GS1 compliance • QoS management • Governance	<1 Mb per unit per day
Precision Medicine/ Personalized Medicine	• LPWA • 4G • mm Waves • Multi domain orchestration • Mobility support at speed	• Medical Product Legislation • International standardization • National and EU privacy rules • ISO 27000 • ISO 80001	• Tagging and tracking of personalized medicines • Time management and synchronization • Telemedicine • VOIP • GS1 compliance	5 Mbps or
Hospital at Home	• mm Waves • 4G • LPWA • Multi domain orchestration • Real-time	• Universal health and care budgets • National "Social Legislation" • National eHealth legislation (for example Germany, Austria, Denmark) • Mandatory conformity marking, such as for example CE • National and EU privacy rules • ISO 27000 • ISO 80001	• Transition of health care delivery to distributed patient centric care • Speed up transition from hospital to home • Improved mental health management • Home diagnostics (for example ultrasound scanner, echocardiogram, Doppler) • Empowering patients to make decisions on care priorities • Releasing "Social Capital" equity • Peer-to-peer • Self-help groups • Monitoring (for example blood glucose, heart rate, blood pressure, regular infusions, feeding pumps, etc) • GS1 compliance • Governance • Billing	5 Mbps
Surgical Robotics	• mm Waves • Low latency	• Medical device regulations • Mandatory conformity marking, such as for example CE • BEREC rules on Net Neutrality	• Specialists to support generalists in peripheral locations • Improvement of quality of care through availability of expert input in real time • Enhancing patient safety • High resolution 3D video imaging (4 k plus) • Augmented reality	>50 Mbps
Social Robotics	• mm Waves • Low latency • Mobility support at speed • Multi-tenancy	• Mandatory conformity marking, such as for example CE • BEREC rules on Net Neutrality	• Wide-spread manual support in activity of daily living (ADL) which might not require a human • HD video • Augmented reality	< 25 Mbps
Antibiotics resistance prophylaxis	• 4G • mm Waves	• Medical product legislation • Mandatory conformity marking, such as for example CE	• Automated screening of samples (swaps, urine, blood)	200 Kb per sample

(*continued*)

Table 100.1 (continued)

Use case	5G technology	Regulatory	Business	Bandwidth/data volume
Medical device management (in hospital and at home)	• LPWA • mm Waves • 4G • Multi-tenancy • Multi-domain orchestration	• Medical product legislation • Warranty • Mandatory conformity marking, such as for example CE • ISO 80001 • ISO 27000	• Obligatory maintenance as defined by medical product legislation • Prevention of nosocomial infections by contamination detection • Malfunction alert • QoS monitoring for vendors and service providers	200 Kb per unit per day
Smart Packaging	• LPWA • Multi domain orchestration • 100 x connections • 1/10 energy consumption	• Medical product legislation ISO 80001 • ISO 27000	• Counterfeiting: 1 in 10 drugs sold globally is fake. For 1000 dollars invested, a criminal can garner 20,000 dollars in profits with trafficking heroin and 400,000 dollars by dealing in counterfeit drugs [21] • Supply chain management of value drugs • Storage and transport conditions (cooling chain, humidity, temperature, etc) Block-chain technology to establish audit-trail GS1-compliance	300 Kb per unit per week
Artificial organs Endo-prosthetics Artificial limbs Body Area Networks	• mm Waves • 4G • LPWA • Multi-tenancy • Mobility support at speed • 1/10 energy consumption	• IEEE 802.15.6 • ISO 80001 • ISO 27000 • BEREC rules on Net Neutrality	• Pacemaker • Insulin pumps • Brain pacemakers • In vivo sensors • Sensors in endo-prosthetic devices etc	200 kb per unit per day
Emergency services	• 4G • mm Waves • LPWA • Terminal with network control capacity • Multi domain orchestration • Multi-tenancy	• BEREC rules on Net Neutrality • Emergency regulations • Mandatory conformity marking, such as for example CE • Medical product legislation	• Tagging and tracking • Telemedicine • Documentation • Modularity • Remote diagnostics (ultrasound)	5 Mbps per unit
Wound-management	• LPWA • 1/10 energy consumption	• Medical product legislation	• Smart wound dressings • Decubitus monitoring	100 kb per unit per day
Wellness and Fitness	• LPWA • 4G • mm Waves • 1/10 energy consumption	• Mandatory conformity marking, such as for example CE • Potentially medical product legislation	• Health insurance discount • Non-pharmaceutical therapy (for example diabetes, high blood pressure and hypercholesterolemia)	<1Mbps

References

1. European Commission. (2012). eHealth Action Plan 2012–2020 – Innovative health care for the 21st Century. http://eur-lex.europa.eu/legal-content/EN/TXT/PDF/?uri=CELEX:52012DC0736&from=EN.
2. WHO. (2016). *From innovation to implementation: eHealth in the WHO European region*. Copenhagen: WHO Regional Office for Europe.
3. European Commission. (2014). Green paper on mobile health ("mHealth"). https://ec.europa.eu/digital-single-market/en/news/green-paper-mobile-health-mhealth.
4. European Commission. (2014). Public consultation on the green paper on mobile health. https://ec.europa.eu/digital-single-market/en/public-consultation-green-paper-mobile-health.
5. 5G PPP. (2015). 5G and e-health. https://5g-ppp.eu/wp-content/uploads/2016/02/5G-PPP-White-Paper-on-eHealth-Vertical-Sector.pdf.
6. WWRF. (2016). A new generation of e-health systems powered by 5G. Wireless World Research Forum. http://www.wwrf.ch/files/wwrf/content/files/publications/outlook/Outlook17.pdf.
7. US FDA. (2015). Mobile medical applications. U.S. Food and Drug Administration. http://www.fda.gov/medicaldevices/digitalhealth/mobilemedicalapplications/.
8. European Commission. (2014). mHealth, what is it? – Infographic. https://ec.europa.eu/digital-single-market/en/news/mhealth-what-it-infographic.

9. Gartner. (2016). Gartner identifies the top 10 internet of things technologies for 2017 and 2018. http://www.gartner.com/newsroom/id/3221818.

10. BITAG. (2016). Internet of Things (IoT) Security and Privacy Recommendations'. Broadband Internet Technical Advisory Group. http://www.bitag.org/documents/BITAG_Report_-_Internet_of_Things_(IoT)_Security_and_Privacy_Recommendations.pdf.

11. Cisco. (2016). Cisco Visual Networking Index: Global Mobile Data Traffic Forecast Update 2015–2020 White Paper. http://www.cisco.com/c/en/us/solutions/collateral/service-provider/visual-networking-index-vni/mobile-white-paper-c11-520862.html.

12. Roland, D., & Lantauro, I. (2016). Google Parent and Sanofi Name Diabetes Joint Venture Onduo. The Wall Street Journal, September 12. http://www.wsj.com/articles/google-parent-and-sanofi-name-diabetes-joint-venture-onduo-1473659627.

13. Global Asthma Network. (2014). The Global Asthma Report 2014. http://www.globalasthmareport.org/.

14. Schubert, L. (2010). The Future of Cloud Computing. Edited by Keith Jeffery and Burkhard Neidecker-Lutz. http://cordis.europa.eu/fp7/ict/ssai/docs/cloud-report-final.pdf.

15. fistar.eu. (2016). FI-STAR. https://www.fi-star.eu/fi-star.html. Accessed 11 Feb.

16. Thuemmler, C., Mueller J., Covaci, S., Magedanz, T., de Panfilis, S., Jell, T., & Gavras, A. (2013). Applying the software-to-data paradigm in next generation e-health hybrid clouds. In *Proceedings of the 2013 10th International Conference on Information Technology: New Generations* (459–463). ITNG '13. Washington, DC: IEEE Computer Society. doi:10.1109/ITNG.2013.77.

17. Paulin, A., & Thuemmler, C. (2016). Dynamic fine-grained access control in E-health using: the secure SQL server system as an enabler of the future internet'. In *Proceedings of the 2016 HealthCom*. IEEE. (pp. 1–4). doi:10.1109/HealthCom.2016.7749462.

18. Caole Villa, A. (2016, November 19). Elon musk to launch super fast internet from space. *Nature World News*. http://www.natureworldnews.com/articles/32332/20161119/elon-musk-launch-super-fast-internet-space.htm.

19. Paulin, A., Thuemmler, C., Lim, A. K., Schneider, A., & Feussner, H. (2016). E-health traffic analysis and 2035 future network requirements – Part I: state of the art service description.

20. Paulin, A., Thuemmler, C., Lim, A. K., Schneider, A., & Feussner, H. (2016). E-health traffic analysis and 2035 future network requirements – Part II: future ecosystem conditions and constraints.

21. Sanofi, Sanofi fights against counterfeit medicines. http://en.sanofi.com/Images/33151_DP_Counterfeit.pdf.

Stelios Sotiriadis, Andrus Lehmets, Euripides G.M. Petrakis, and Nik Bessis

Abstract

This work presents the testing requirements for cloud services including unit and integration testing by identifying services that could communicate with each other according to their APIs. We also present the Elvior TestCast T3 (TTCN-3) testing tool that provides an efficient and easy to use solution for automating functional tests. This allows incremental development where users can test specific systems and features separately as well as the entire system as a whole. We finally demonstrate the empirical results as lessons learned from our experiences when apploying such solution in testing real world cloud health services.

Keywords

Cloud computing • Cloud services • Cloud service testing • TestCast • TTCN-3

101.1 Introduction

Cloud computing offers on demand services over the Internet to users and developers that combine easy access to computing resources with remote data management, elasticity, self service provisioning (allowing users to set-up and launch applications as services) and certain economic benefits. It supports different models to cover a variety of cloud users', such as the Infrastructure, Platform and Software as a Service (IaaS, PaaS and SaaS). The evolution of these, characterizes the so-called Future Internet (FI) concept [1] and allows development of innovative applications from modular services referred to as cloud enablers. In particular, developers build applications by utilizing on-the-self services encapsulating common functionalities (e.g. user authentication, data storage, context data management etc.), instead of re-engineering and implementing services from scratch.

Another important aspect of these services is the modularity that in cloud computing is enabled by deploying SaaS whose specifications are open and are available for utilization using APIs (i.e. as RESTFul interfaces [2]). While these services highlight innovation and promotion of a new easy-going development method, software engineering processes are becoming more and more complex. This is because such enablers have distinct features that impact several different research fields, including software testing [3]. These include execution over virtualized resources [5] that can be highly scalable and elastic, for instance, the users can increase their computational capacities and share

S. Sotiriadis (✉)
Computer Engineering Research Group, University of Toronto, Toronto, Canada

School of Electronic and Computer Engineering, Technical University of Crete, Chania, Greece
e-mail: s.sotiriadis@utoronto.ca

A. Lehmets
Elvior, Tallinn, Estonia
e-mail: andrus.lehtmets@elvior.ee

E.G.M. Petrakis
School of Electronic and Computer Engineering, Technical University of Crete, Chania, Greece
e-mail: petrakis@intelligence.tuc.grl

N. Bessis
Edge Hill Univerity, Ormskirk, UK
e-mail: Nik.Bessis@edgehill.ac.uk

© Springer International Publishing AG 2018
S. Latifi (ed.), *Information Technology – New Generations*, Advances in Intelligent Systems and Computing 558, DOI 10.1007/978-3-319-54978-1_101

common physical resources with other cloud applications or services, thus supporting multi-tenancy. Also such services provide interfaces using APIs that allow their combination to form composite services (e.g. services developed by extending or combining others).

Cloud services' testing introduces a new set of challenges and requirements [4] that have a direct impact on system engineering processes. The testing methodology involves not only validation of a software processes, but also validation of services APIs and of their interactions with other services. This involves testing of the actual service interactions based on their input and output interfaces. The complexity factor is also increased due to the fact that software engineers base their design on abstract and high-level use case models. Thus it becomes a hurdle to design test cases that are based (a) on the application requirements e.g. expecting behaviour of the services according to a use cases and (b) on the services it self (that is related with the functionalities of the cloud enablers).

In this work we present a testing methodology for testing cloud services utilizing the TestCast TTCN3 by Elvior.[1] The tool provides a programming language for testing communication protocols and Web services in an easy and efficient way. Based on this, Sect. 101.2 presents the motivation and background for this work, Sect. 101.3 the proposed testing methodology of cloud services, Sect. 101.4 the analysis of the tool, Sect. 101.5 the lessons learned by applying the tool in real world use cases and in Sect. 101.6 the conclusions and future research directions.

101.2 Motivation and Background

This work is based on the Future Internet Social and Technological Alignment Research (FI-STAR)[2] FP7 project that attempted to establish early trials in the health domain and in the context of FI-PPP EU initiative.[3] FI-STAR purpose was to prepare industry take-up of the developed cloud technology. It also build upon cloud services called Generic Enablers (GEs) provided by FIWARE.[4] GEs are software modules that offer various functionalities along with protocols and interfaces for operation and communication. These include the cloud management for supervision of the underlying infrastructure, the utilization of various IoT devices for data collection, tools for data analytics and communication interfaces for gateways and end-users. All FIWARE GEs are stored in a public catalogue, thus

developers could easily browse and select appropriate APIs to use. As a side-result, FI-STAR aims to create a framework (a software to data approach) to allow GEs to be delivered to different physical locations.

Today, there are new requirements related to the software engineering processes of cloud enablers since there are significant changes taking place as to how to test the new software running on the latest cloud platforms. Especially in the case of software testing, this refers to the gap between software developers and software test engineers. There are three key differentiations among traditional and cloud enablers testing and these are related with (a) scalability of virtual resources, (b) multi-tenancy efficiency that refers concurrent users and (c) dynamic reconfiguration of services. As a result, the cloud testing models have to support different kinds of requirements [14], thus, tend to become more and more complex. In more detail, traditional applications are firstly designed and then tested. This work is motivated by work of [15, 16] where authors suggest that there is a lack of research papers addressing new issues, challenges, and needs in SaaS testing. In [16] we present the details of the unit and integration testing methodology for future Internet cloud services. We focus on cloud services testing and we propose a methodology enabling efficient unit and integration testing of modular services.

101.3 Testing Methodology of Cloud Services

The methodology includes the testing requirements analysis and the test cases analysis of cloud enablers. The testing preparation strategy defines the testing schedule and plan for black box testing of cloud enablers including conceptualization for unit and integration testing. Firstly, the unit testing allows cloud services' testers to execute various tests with different input parameters to the operations and interfaces defined in service specification documents and manuals. The modules together with associated control data, usage procedures, and operating procedures, are tested to determine whether they are properly developed. The integration testing will allow combination and test execution of linked cloud enablers in order to test key functionalities derived by application use-cases.

The testing starts with the unit testing preparation activity where each cloud enabler tester defines the unit test and unit test cases so the unit test defect management process will allow bug tracking and fixing. Then the integration testing includes preparation, execution and defection management process regarding the inter-dependencies of enablers. At this stage, the aim is to exploit the building blocks of the application that map to the key functionalities. In more detail, the unit testing includes the testing activity of the cloud enablers by focusing on the evaluating of their interfaces. Integration

---

[1] http://www.elvior.com/testcast/ttcn-3.

[2] https://www.fi-star.eu.

[3] https://www.fi-ppp.eu.

[4] https://www.fiware.org.

testing includes exploration of the interactions between the services, such as their commincation and their input-output bonds. Next we focus on the TestCast TTCN-3 tool.

101.4 General overview of the TestCast TTCN-3 Tool

The Elvior TestCast is a full featured TTCN-3 tool (a programming language for testing of communication protocols and Web services) for automated testing of cloud systems. TestCast is ideal for incremental development where users can test specific systems and features separately as well as the entire system as a whole. Figure 101.1 demonstrates the Principal architecture of TTCN-3 test environment.

In detail, the System Adapter (SA) is used for communication of test tool (test executable) with System Under Test (SUT). The interface between test tool and SA is standardized and called TRI (TTCN-3 Runtime Interface) – SA is usually a piece of software that can be written in different languages such as C, C++, C#, Java, mainly depending on test tool. The other important interface is TCI (TTCN-3 Control Interface) is essential part of TTCN-3 test environments because its type system is not bind to any binary representation. It is entirely up to the test tool and its codecs to ensure encoding and decoding of data in appropriate format.

The characteristics of the Elvior TTCN-3 tool are as follows. The tool is developed in C# and could be run in Microsoft Windows and Linux platforms. Also, it has a user-friendly graphical interface for test development and management. It has a native compiler that supports all TTCN-3

standards (up to TTCN-3:2015) and includes a native TTCN-3 debugger. The tool supports TTCN-3 test execution and XSD import. It has built in codecs such as textual, binary, XML (via XSD schemas), TCI XML, ASN.1 and supports TRI and TCI mapping for C, C++, C# and Java. Finally, it offers a rich TTCN-3 editor and test suite viewer and a enriched logging (textual and graphical views) and logs analyzing capabilities. The tool provides a test environment for RESTful interfaces testing. The TTCN-3 based test environment architecture for functional Black-Box testing (BBT) includes an HTTP System Adapter, XML and JSON codecs and SUT (implementation using RESTful API). The roles of actors shown in Fig. 101.2 in the TestCast Tool testing environment are presented bellow.

(a) System Under Test (SUT): The implementation using the RESTful API.
(b) TestCast T3 test tool for TTCN-3 test development and execution (it is responsible for execution of TTCN-3 test scripts).
(c) HTTP SA (developed in C#): It completes communication between SUT and test tool. It acts as a server or client depending on test case needs and is controlled by TTCN-3 test script.
(d) XML TestCast T3 built in codec and the external JSON codec: Codecs are responsible for encoding and decoding messages sent/received to/from SUT.

In particular, for RESTful interfaces a generic TTCN-3 framework for sending and receiving HTTP messages was created. The scripts are written and executed in TestCast T3,

Fig. 101.1 Principal architecture of TTCN-3 test environment

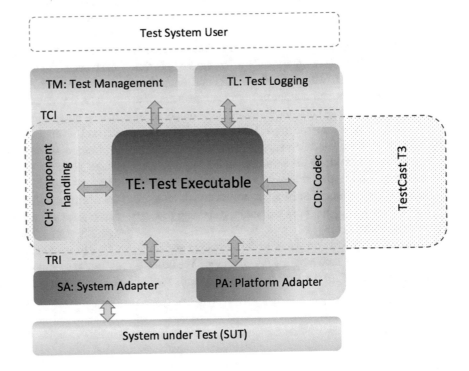

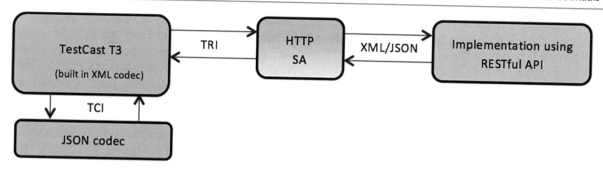

Fig. 101.2 Architecture of BBT environment for RESTful APIs

which sends messages to the HTTP adapter. The System Adapter creates HTTP messages based on the received information from TestCast and sends each of which to the SUT. In practice, the following actions are executed: i) message (hex sequence) is received from the SUT via SA to TestCast, ii) TestCast (TTCN-3 test executable) reads this hex sequence from TTCN-3 port and iii) the message is decoded by TestCast and can be used in TTCN-3 script.

It should be mentioned that TTCN-3 libraries include common types and templates for forming messages for RESTful APIs. These common types and templates are used for creating implementation specific messages and respective test cases.

101.5 Lessons Learned

This section presents the experiences regarding the "lessons learned" from the unit and integration testing of cloud enablers when applied in the FI-STAR project. During this work we focused on the following research contributions.

(a) How to perform testing (regarding unit and integration) of cloud enablers? We presented a methodology that encompasses unit (white and black box) and integration testing (top down and bottom up) of cloud enablers. The cloud enablers tested are decentralized and composed by 3rd party services based on the real world case of FI-STAR.

(b) How to provision decision support to the cloud application developers during the testing process? We provided an integration testing strategy to characterize the functional building blocks of FI applications. In addition the integration testing matrices will assist on testing process management.

(c) How testing can assure standard conformance and automation to ensure extensive testing coverage? We utilized the Elvior TestCast T3 tool to automatize unit testing

and to ensure standards adoption. We presented an example case of NGSI9/10 interfaces testing.

The testing starts at the module level and works outward the integration of the cloud application. Based on the empirical analysis the testing techniques presented could be used at different points of the overall cloud enabler engineering process. The cloud application developer could take advantage of the testing activities of the modular parts in order to conduct testing for the larger group and ensure verification (application works as expected) and validation (compliance to the requirements). In addition, we propose the use of an independent tester in order to remove the conflict of interest that is characteristic when the developer tests its own product.

In general the test execution of cloud enablers could be difficult since it involves a number of actors such as (a) the enabler and (b) the use case application developer. Based on the methodology we executed various test cases from the perspective of both actors. Another vital requirement is the correct depiction of the use case application requirements, since these will play an important role to the identification of the test cases. Here, the assumption is that the tester will have access to the functional building blocks of the application and the chains of enablers' dependencies.

The top down and bottom up approaches will warranty that in integration testing (a) the cloud enabler will be tested according to its specification and (b) according to the requirements of the use case application. We encourage its usage since it provides ready-made unit tests for cloud enablers ensuring adoption of standards. An example was the test cases of the FI-STAR Event Service that demonstrated an error regarding the expected output of a test (that was not NGSI9/10 compliance). This test has revealed a hidden error that the SE owner did not identify at the beginning. Finally, the SE owner considered the results and made appropriate corrections to the beta version of the software. It should be mentioned that test cases, test assets, and test features could be accessible could be open to everyone.

The following is a summary of the innovative features of this work.

(a) We focused on the research gap created, regarding cloud enablers testing, in the area of FI services. Thus we present a methodology to allow efficient testing of cloud enablers belonging to different owners, deployed to various cloud platforms and utilized by different cloud applications.

(b) We presented a methodology that incorporates strategies for unit testing (white and black box approaches) and integration testing (bottom up and top down approaches). The correlation of these will ensure an increased coverage of cloud enablers testing.

(c) We presented the FI-STAR and FIWARE projects as use case scenarios of our methodology by including example cases of real world cloud enablers.

(d) We detailed unit and integration testing matrices to allow efficient capturing of testing requirements. In many instances various FI-STAR SE testers with success have utilized the proposed matrices.

(e) We demonstrated the TestCast tool, an advanced solution for testing according to standards such as compliance with ETSI, OMA, TTCN-3, and NGSI9/10. In many instances the tool surfaced failures and issues that did not identified by the white box testing process.

101.6 Conclusions

In this work we presented the Elvior TestCast T3 tool that includes FIWARE compliance. The proposed solution assists cloud enabler testers to rapidly isolate errors and failures based on the virtualization of services so to identify dependencies with other components. Different from other solutions, the approach provides a complete end-to-end transaction that integrates chains of Applications, SEs and GEs. To demonstrate effectiveness, we presented the real world case scenario of FIWARE and FI-STAR projects. We emphasized the adaption of the proposed methodology and tools to the production of test cases; analysis of data and automation of processes based on real world use case requirements for cloud-based healthcare applications. The environment is highly complex with multi-tenancy, diversity of requirements security and off-premises cloud services that are decentralized in terms of ownership. We anticipate that the methodology will serve cloud testers to ensure proper functionality and test coverage and application developers could utilize results as a reference model for building innovating systems.

References

1. Sotiriadis, S., Zampognaro, P., Petrakis, E., & Bessis, N. (2015). Automatic deployment of cloud services for healthcare application development in FI-STAR FP7. *The 30th IEEE international conference on advanced information networking and applications (AINA-2016)*, Le Régent Congress Centre, Crans-Montana, Switzerland, March 23–25, 2016.
2. Seijas, P. L., Li, H., & Thompson, S.. (2013). Towards property-based testing of RESTful web services. *Proceedings of the twelfth ACM SIGPLAN workshop on Erlang* (pp. 77, 78).
3. Incki, K., Ari, I., & Sozer, H.. (2012). A survey of software testing in the cloud. In *Proceedings of the 2012 I.E. sixth international conference on software security and reliability companion (SERE-C '12)*. IEEE Computer Society, Washington, DC (pp. 18–23).
4. Lima Neto, C. R., & Garcia V. C. (2013). Cloud testing framework. In *Proceedings of the 17th International Conference on Evaluation and Assessment in Software Engineering (EASE '13)*. ACM, New York (pp.252, 255).
5. Vakanas, L., Sotiriadis, S., & Petrakis, E. (2015). Implementing the cloud software to data approach for OpenStack environments, adaptive resource management and scheduling for cloud computing, held in conjunction with PODC-2015, Donostia-San Sebastián, Spain, on 20 July 2015.
6. Ciortea, L., Zamfir, C., Bucur, S., Chipounov, V., & Candea, G. (2010). Cloud9: a software testing service. *ACM SIGOPS Operating Systems Review Archive, 43*(4), 5, 10.
7. Banzai, T., Koizumi, H., Kanbayashi, R., Imada, T., Hanawa, T., & Sato, M. (2010). D-Cloud: Design of a software testing environment for reliable distributed systems using cloud computing technology. *Proceedings of 10th IEEE/ACM international conference on cluster, cloud and grid computing, 2010* (pp. 631, 636). Heraklion.
8. Mathew, R., & Spraetz, R. (2009). Test automation on a SaaS platform. *Proceedings of international conference on software testing verification and validation, 2009* (pp. 317, 325). Seoul, Republic of Korea.
9. Wickremasinghe, B, Calheiros, R. N., & Buyya, R. (2010). CloudAnalyst: a CloudSim-based visual modeller for analysing cloud computing environments and applications. *24th IEEE international conference on advanced information networking and applications (AINA), 2010*. Bethlehem.
10. Sotiriadis, S., Bessis, N., Antonopoulos, N.,& Anjum, A. (2013). SimIC: designing a new inter-cloud simulation platform for integrating large-scale resource management. *27th IEEE international conference on advanced information networking and applications (AINA-2013)*, March 25–28, Barcelona, IEEE Computer Society, Washington, DC (pp. 90–97).
11. Oriol, M., & Ullah, F. (2010). YETI on the cloud, software testing, verification, and validation workshops (ICSTW). *2010 third international conference on* , vol., no., pp.434, 437, 6–10 April 2010. Madrid.
12. Vilkomir, S. (2012). Cloud Testing: A state-of-the-art review. *Information and Security. An international journal, 28*(2), 213, 222.

13. Grabowski, J., Hogrefe, D., Réthy, G., Schieferdecker, I., Wiles, A., & Willcock, C. (2003). An introduction to the testing and test control notation (TTCN-3). *Computer Networks, 42*(3), 375, 403.

14. Heckel, R., & Lohmann, M. (2005). Towards contract-based testing of web services. *Electronic Notes in Theoretical Computer Science, 116*(2005), 145–156.

15. Wang, J., & Meng, F.. (2011). Software testing based on cloud computing. *In Proceedings of the 2011 international conference on internet computing and information services (ICICIS '11)*. IEEE Computer Society, Washington, DC (pp.176, 178).

16. Sotiriadis, S., Lehmets, A., Petrakis, E., & Bessis, N. (2016). Unit and integration testing of future internet cloud services. *The 31st IEEE international conference on advanced information networking and applications (AINA-2017)*, Tamkang University, Taipei, Taiwan, March 27–29, 2017.

Evaluation of High-Fidelity Mannequins in Convulsion Simulation and Pediatric CPR

Paôla de O. Souza, Alexandre C.B. Ramos, Leticia H. Januário,
Ana A.L. Dias, Cristina R. Flôr, Heber P. Pena, Helen C.T.C. Ribeiro,
Júlio C. Veloso, and Milla W. Fiedler

Abstract

This article's objective was evaluating the resources and the fidelity of SimBaby's mannequin for Cardiopulmonary Resuscitation (CPR) and pediatric convulsion training. Education based on training through simulation resorts frequently to high-fidelity mannequins for practicing CPR skills. However, if the simulated patient is not realistic enough, the learning process is compromised. The evaluation of said mannequins' realism is scarce in specialized literature. Methodology: A group of engineers, medics and nurses was chosen for the purposed evaluation. First, the designation of some members of the team as first-aiders for the infant's treatment. The recognition of convulsion signals and the need of CPR was handled by algorithms created for this project. On the second moment of evaluation, the first-aiders executed incorrect CPR maneuvers to evaluate the mannequin's feedback accuracy. Results: the first-aiders recognized the pulse and cyanosis, but did not recognize the convulsion and the inadequate perfusion. The simulator does not distinguish correct or incorrect maneuvers. Conclusion: SimBaby exhibits high technology, but lacks realism simulating convulsions and feedback to CPR.

Keywords

Mannequin • Performance Evaluation • Medical Simulation

102.1 Introduction

Technological advances gave higher esteem and allowed the development of human performance simulators [1] such as software for mimicking real-life scenarios [2] and hi-tech mannequins [3, 4]. As such, the improvement of Simulation Based Medical Education (SBME) gained prominence in medical professionals' education.

However, the relevance of SBME goes beyond didactics, because it assumes quality of assistance, said assistance's safety and the patient's survival. The simulated scenarios also allow the coordination of teams for diagnosis and treatment in emergencies or risk situations, allowing the learning by trial and error without exposing real patients to risks [5].

Experiments applying pedagogic activities on medical education based on the use of simulators show that the most used kind of simulator are high-fidelity mannequins [6]. These mannequins simulate the whole body, are computer-assisted, robotized, and allow the reproduction of diversified clinical situations. They also react to therapeutic interventions and correct or incorrect procedures [7]. In that context, truly "interactive" mannequins have special

P. de O. Souza (✉) • A.C.B. Ramos
Federal University of Itajubá, Itajubá, Brazil
e-mail: paola.souza@unifei.edu.br; ramos@unifei.edu.br

L.H. Januário • A.A.L. Dias • C.R. Flôr • H.P. Pena • H.C.T.C. Ribeiro
J.C. Veloso • M.W. Fiedler
Federal University of São João del Rei, São João Del Rei, Brazil
e-mail: leticiahj@ufsj.edu.br; anaangelica@ufsj.edu.br;
cristinaflor@ufsj.edu.br; heberpaulino@ufsj.edu.br;
helen.cristiny@ufsj.edu.br; juliocesar@ufsj.edu.br;
millawf@ufsj.edu.br

© Springer International Publishing AG 2018
S. Latifi (ed.), *Information Technology – New Generations*, Advances in Intelligent
Systems and Computing 558, DOI 10.1007/978-3-319-54978-1_102

importance in the training for Basic Life Support (BLS), Advanced Life Support (ALS) and Cardiology [8–14].

On the other hand, the training of the population in reanimation maneuvers in adults, as recommended by American Health Association's (AHA) BLS elevated the survival rate of victims. However, this number does not apply to the treatment of children, because the BLS is not as widespread in pediatrics [15]. Data from 2005 to 2007 from Resuscitation Outcomes Consortium – a register of 11 emergency medical systems from USA and Canada – show the survival rates of 3,3% to infants less than 1 year old; 9,1% to children (1 to 11 years old); and 8,9% to teenagers (12 to 19 years old).

Other clinical conditions in infants such as generalized convulsion require immediate care, because, among other reasons, it predisposes the child to a cardiopulmonary arrest. The sudden, involuntary and episodic changes in consciousness, motor activity, behavior, autonomous sensation or functions caused by abnormal electrical discharges in the brain are characteristics that denote convulsions [16].

The recognition of convulsion and cardiac arrest allow the effective intervention. In that sense, technologic tools such as a health-monitoring app will permit medical professionals to monitor their patients in a hospital by portable sensors that are able to collect vital signs in real time [17]. However, this kind of monitoring technology is still not available to most of the population, especially outside of hospitals. In that way, it is the recognition of the cardiac arrest and the immediate and effective cardiopulmonary resuscitation (CPR) that raises the survival rate of children victim of cardiac arrests. However, few children actually receive quality CPR.

AHA [18] directs to the need of better quality CPR given by medical professionals and lifeguards. In that regard, the SBME is an optimization instrument of training and academic learning of health professionals. The training's quality and the acquired skills have direct relation to the quality of the used simulator. However, few studies are available that verify if the high-fidelity mannequins are capable of precisely reproducing the human anatomy and the body's responses to treatment [19].

The research based on international databases utilizing the terms ("mannequin OR manikin) AND (evaluation OR comparison OR improvement) AND (CPR OR cardiopulmonary OR resuscitation)" generated 565 results. Only 13 of the articles had content related to mannequin evaluation, being most of the articles exclusively the evaluation of the mannequin's anatomy. Only two articles did the evaluation of pediatric mannequins. Broadly, the researchers concluded that the studied mannequins' are very different from a real patient and upgrading the simulators was necessary [20–26].

Many mannequins are offered to pediatric reanimation training. Laerdal – the oldest American corporation in mannequin manufacture [1] and one of the worldwide references on the production of medical training simulators – offers the SimBaby, one of the most utilized products for high fidelity simulation in the training of pediatric scenarios. However, no articles were found evaluating the quality of SimBaby on CPR and convulsion simulations.

In this context, the objective of the article was reviewing the resources and the fidelity of the SimBaby mannequin on the simulation of convulsion and of pediatric CPR according to AHA's BLS guide.

102.2 Methodology

The Lab of Simulation and Skills (LAHAS) from Universidade Federal de São João del-Rei (UFSJ) – Campus Centro-Oeste Dona Lindu (CCO) did the SimBaby's review. A team of nine people was created for executing the procedure: one electronics engineer, one computer engineer, one medical doctor specialized in pediatric emergencies, three nurse teachers, two technicians and one nurse from LAHAS.

No specific protocol for evaluating mannequins was found, so the reviewing script was established according to AHA's guide [19] for identifying cardiopulmonary arrest and executing CPR. Algorithms were designed to recognize convulsion signs according to Hockenberry [16].

The script was designed to verify the mannequin's fidelity in reproducing real characteristics of one 6 months-old child with a case of convulsion evolving into bradychardia with an inadequate perfusion, and a respiratory arrest followed by cardiac arrest. The analysis of the mannequin's feedback to the interventions was also planned, as was evaluated the feedback functions that helped verifying intervention procedures during and after the simulation (Chart 102.1).

During the simulation, one part of the team stayed in the lab's control and observation room, where the mannequin's management software was. The other part was designated to assist the infant without being informed on the clinical scenarios that followed, as it was intended to examine the mannequin's realism, and not the scenario itself.

The simulation happened in two moments. On the first, it was expected that the group of medical assistants identified the clinical condition of the simulated patient and made the required interventions. The objective was to find how close to life the mannequin's reproduction of clinical signs was. On the second moment, the team members were asked to make all the intervention procedures with technical errors, as to verify his feedback in said conditions.

1-Turn on the convulsion function and adjust the mannequin's vital signs

Identification of the patient's clinical condition:
- ✓ Is it possible to verify the convulsion?
- ✓ Is it possible to identify the kind of convulsion?
- ✓ Is it possible to find the group of muscles with spasms?

2-Configure a condition of inadequate perfusion, followed by cardiorespiratory arrest.

Verification of the need for CPR:
- ✓ Is there gasping?
- ✓ Is there a lack of breathing?
- ✓ Is the pulse lower than 60 or absent?
- ✓ Are there signs of bad perfusion (paleness, cyanosis)?

3-The team must start the CPR procedure

Feedback during toracic compressions:
- ✓ Are the rates adequate (30:2 with 100/min at 120/min)?
- ✓ Is the depth adequate (about four cm)?
- ✓ Is the place adequate (sternum with two fingers placed right under the intermammary line?)?
- ✓ Is the team being cautious (not pressing the xiphoid and ribs, watching for complete return of the chest after each compression, minimizing compression's interruptions)?

Feedback during ventilations:
- ✓ Is the chest rising?
- ✓ Does the auto inflatable bag provide an air volume of at least 450-500 ml?
- ✓ Are there responses to excessive ventilation?
- ✓ Is there gastric distention?

Chart 102.1 Evaluation script

By the end of the simulations, the team members had debriefing meetings to discuss the simulation experience and conclude the mannequin's evaluation.

102.3 Results and Discussions

The evaluation results are presented on the same sequence as they were executed.

102.3.1 Convulsion

The infant's convulsion presents signs as loss of consciousness, short and sudden contractures of muscle groups, generally in the limbs, and in some cases, apnea [16].

The mannequin was configured to represent a clinical case of convulsion. SimBaby's control software offers two possibilities for adjusting the chest's movement: slow mode and fast mode. It was opt to use the fast mode convulsion, considering these signs more evident. The team did not identify the convulsion, that lasted three minutes.

The SimBaby's convulsions are summed up to chest movements and a discreet head motion. The mannequin does not offer movements on his limbs, which are the most frequent movements on real patients [16].

The SimBaby simulates a convulsion by inflating an air bag located on the mannequin's back of his torso, allowing the adequate movements to the simulation of convulsion. Also, the use of the air bag generates air noises similar to the breathing of the mannequin. Both the torso's movement and the air bag movements can be interpreted as a breathing pattern.

102.3.2 Need of CPR

To evaluate the need of CPR in infants, according to AHA [18], the lifeguard must assume a cardiac arrest if the victim does not respond, does not breathe (or gasps), and has a pulse of <60 beats per minute (or lacks a pulse entirely). The signs of bad perfusion to be taken into consideration are paleness, spots and cyanosis.

Breathing Up to 11 months old, the breathing of infants is mostly abnormal and the respiratory rate varies between 30 and 50 respiratory movements per minute [27]. To decide on the need of CPR on an infant, the lifeguard must observe gasping or respiratory arrest.

During the mannequin's simulation of a convulsion, his respiratory rate, that started adequately, gradually slowed until a respiratory arrest. The team identified the clinical case as a breathing difficulty, and not a respiratory arrest. The recognition of the respiratory arrest only occurred when the team observed the low saturation of 02 on the simulator's monitor.

On the SimBaby the characteristics of the infant's breathing are evident and the rate and depth from the respiratory movements are easily programmable. However, it does not offer a function for gasping. The gasping may be improvised activating groaning sounds and dyspnea manually on the management software.

It is suggested to create in the software an option to simulate gasping where the chest movements and the mannequin's groans can be automatically synchronized.

Pulse The pulse of an infant has a variable rate from 80 to 160, with a medium 120 beats per minute. It can also vary on the intensity that can be felt during palpitation [27]. According to AHA's norms, in a condition of non-responsiveness and lack of breathing, the brachial pulse must be checked on the infant [18].

On the SimBaby it is possible to configure the pulse to thin and full modes and the desired rate. The simulation was configured to a low-frequency thin pulse that evolved to a lack of pulse. It is also possible to identify the cardiac rhythm by the simulator's electrocardiogram monitor (ECG). The team easily detected all the pulse's conditions.

However, the SimBaby's mannequin possesses a palpable brachial pulse only on its left arm, requiring the team to be informed on this condition. In this simulation, one of the team members tried to find pulse on the right arm.

Its suggested that the mannequin should offer a palpable pulse on both arms.

Inadequate perfusion The inadequate perfusion on an infant is identified by brachicardia, in this case a cardiac rate lower than 60 beats per minute, spots and cyanosis. On the SymBaby only labial cyanosis can be configured. The cyanosis is simulated by two blue LEDs that light up inside the mannequin's mouth.

The team easily identified the brachicardia as it is possible to visualize the cardiac frequency on the ECG monitor. The cyanosis was also detected during the simulation, because of the intensity of the lights. However, on debriefing, this function was not considered compatible with the characteristics of a real patient by any member of the team.

102.3.3 Mannequin's Response to the Execution of Pediatric CPR Maneuvers

To execute a quality CPR it must be assured that: thoracic compressions are of an adequate rate and depth, the full chest recoil between compressions, minimization of interruption of compressions and it must be checked to not occur excessive ventilation [19].

Toracic compressions The compressions in infants for a quality CPR must be done on the sternum with two fingers placed right under the intermammary line. The recommended compression-ventilation rate is 30:2, that is, after the initial group of 30 compressions, the airways must be open and proceed with two ventilations. The rate of compressions must be approximately 100 to 120/min. It is also necessary to hit an adequate compression depth (about 1,5 inches or 4 cm) and complete recoil of chest after each compression. The xiphoid and the ribs must not be compressed.

After verification of a CPR by the team, the procedure started by executing thoracic compressions and ventilation on the mannequin.

The adequate spot to executing the compressions was easily found because the bone structure of the mannequin is very close to a real patient. However, the adequate depth could not be hit because of limitations of the mannequin. The SimBaby also offers the resource of monitoring of compressions by a simulated ECG graph. In this case, the adequate compression generates a heartbeat. It is also possible to verify the full chest recoil between compressions.

Ventilation The adequate ventilation in an infant, when done with an auto inflatable bag, must find a volume between 450 and 500 ml. Excessive ventilation must be avoided, and only the necessary force and volume to raise the chest must be used. Each ventilation must be done slowly, over about 1 second [18].

The team executed ventilation on the mannequin, utilizing the auto inflatable bag of 500 ml. It was easily observed that the mannequin's chest expanded according to a real life scenario.

It is also possible to verify the carbon-dioxide graph that shows the ventilation pulses on the simulation monitor.

On a second moment, the team executed the resuscitation maneuvers with an incorrect technique to evaluate the mannequin's feedback in these circumstances.

Incorrect thoracic compressions The execution of compressions in wrong places, such as the abdomen and the infraclavicular region, generated the compression pulses in the same way it would if done on the right spot, as it was shown on the ECG simulation monitor. It was also verified that the heartbeats appear even with compressions that are more superficial. In the same manner, the simulator does not recognize the rate of inadequate compressions.

Incorrect ventilation During the ventilation in a CPR it is possible that gastric distention in an infant occurs. The SimBaby possesses a gastric distention function that was activated during ventilation.

The team executed the hyperventilation and the mannequin did not detect it. The team did not notice the gastric distention.

The results of this article are coherent with the literature on the subject, because the analyzed publications concluded that the mannequins did not possess the physical characteristics and adequate feedback to represent reality.

Extrapolating the evaluation script, the team had certain difficulty using the management software to start, configure and manage the mannequin for the simulation. This happened because numerous errors occurred in the software that required the frequent rebooting of the simulation system. In addition, there was a considerable delay on the mannequin's feedback to the commands executed on the computer happening many times.

102.4 Conclusion

The EBS is considered an efficient method on the formation of the apprentice. That happens because the contact of the apprentice with a realistic scenario where he has the opportunity to fail in a simulated patient, reducing the possibility of risk in a real one. However, if the simulated patient is not realistic enough, the learning process can be compromised.

In that sense, some evaluated clinical situations on the SimBaby (like cyanosis and convulsion) need to improve in realism.

In addition, the SimBaby lacks real time feedback and logs, which could help the apprentice to check his errors, comparing his actions to the parameters. The SimBaby is a high technology and high cost too, but the results point towards a need of further investments in its technology to better his feedback in the simulation of convulsions and CPR-needed clinical cases.

References

1. Cooper, J. B., & Taqueti, V. R. (2008). A brief history of the development of mannequin simulators for clinical education and training. *Postgraduate Medical Journal, 84*(997), 563–570.
2. Botezatu, M., et al. (2010). Virtual patient simulation for learning and assessment: Superior results in comparison with regular course exams. *Medical Teacher, 32*(10), 845–850.
3. Boet, S., Naik, V. N., & Diemunsch, P. A. (2009). Virtual simulation training for fibreoptic intubation. *Canadian Journal of Anaesthesia, 56*(1), 87–88.
4. Beydon, L., et al. (2010). High fidelity simulation in Anesthesia and Intensive Care: context and opinion of performing centres- a survey by the French College of Anesthesiologists and Intensivists. *Annales Françaises d'Anesthèsie et de Rèanimation, 29*(11), 782–786.
5. Gomez, M.V., Vieira, J.E., & Scalabrini, A. (2011). The background of professors in health fields that use simulation as a teaching strategy. *Revista Brasileira de Educação Médica, Rio de Janeiro, 35*(2), 157–162.
6. Simões, A. S. (2014). *A utilização do ensino baseado em simulação na educação continuada de médicos.* Rio de Janeiro: UFRJ/NUTES.
7. Amaral, J. M. V. (2010). Simulação e ensino aprendizagem em pediatria. 1ª Parte: Tópicos essenciais. *Acta Pediátrica Portuguesa, 41*(1), 44–50.
8. Ogden, P. E., et al. (2007). Clinical simulation: importance to the internal medicine educational mission. *Am J Medicine, 120*(1), 820–824.
9. Bradley, P., & Postlethwaite, K. (2003). Simulation in clinical practice. *Medical Education Volume, 37*, 1–5.
10. Halamek, L. P., et al. (2000). Time for a new paradigm in pediatric medical education: teaching neonatal resuscitation in a simulated delivery room environment. Pediatrics. *Volume, 106*, 45–50.
11. McLaughlin, S. A., Doezema, D., & Sklar, D. P. (2009). Human simulation in emergency medicine training: a model curriculum. Acad Emerg Med. *Volume, 9*, 1308–1310.
12. Cooper, J. B., & Taqueti, V. R. (2004). A brief history of the development of mannequin simulators for clinical and education training. *Quality & Safety in Health Care, 3*(Suppl 1), i11–i18.
13. Dalley, P., et al. (2004). The use of high fidelity human patient simulation and the introduction of new anesthesia delivery systems. Anesth Analg. *Volume, 6*, 1737–1741.
14. Issenberg, S. B., McGaghie, W. C., & Petrusa, E. R. (2005). Features and uses of high fidelity medical simulations that lead to effective learning: a BEME systematic review. Med Teach. *Volume, 27*, 10–28.
15. Teixeira, V. C. (2007). Suporte básico de vida em pediatria. *Medicina Preoperatória. 1*(1), 1321–1328.
16. Hockenberry, M. J., Wilson, D., & Wong, D. L. (2013). *Wong's essentials of pediatric nursing.* St. Louis: Elsevier Health Sciences.
17. Messina, M. et al. (2008). Implementing and validating an environmental and health monitoring system, information technology: new generations 2008. *Fifth international conference on ITNG* (pp. 994–999). Washington, D.C.
18. American Heart Association. (2015 November 3). Guidelines for Cardiopulmonary Resuscitation and Emergency Cardiovascular Care. Part 11: Pediatric basic life support. *Circulation. 132* (18 suppl 2)
19. Luca, A. D., et al. (2015). Reliability of manikin-based studies: an evaluation of manikin characteristics and their impact on measurements of ventilatory variables. *Anaesthesia, 70*, 915–921.
20. Pretto, F. (2008). Uso de Realidade Aumentada no Processo de Treinamento em Suporte à Vida. Dissertação (Mestrado) – Pontifícia Universidade Católica do Rio Grande do Sul, Porto Alegre.
21. Schalka, R., et al. (2015). A radiographic comparison of human airway anatomy and airway manikins – implications for manikin-based testing of artificial airways. *Resuscitation, 92*, 129–136.
22. Schebesta, K., et al. (2012). Degrees of reality: Airway anatomy of high-fidelity human patient simulators and airway trainers. *Anesthesiology, 116*(6), 1204–1209.
23. Schebesta, K., et al. (2011). A comparison of paediatric airway anatomy with the Simbaby high-fidelity patient simulator. *Resuscitation, 82*, 468–472.
24. Jackson, K. M., & Cook, T. M. (2007). Evaluation of four airway training manikins as patient simulators for the insertion of eight types of supraglottic airway devices. *Anaesthesia, 62*, 388–393.
25. Madar, J., & Richmond, S. (2002). Improving paediatric and newborn life support training by the use of modified manikins allowing airway occlusion. *Resuscitation, 54*, 265–268.
26. Howells, R., & Madar, J. (2002). Newborn resuscitation training – which manikin. Resuscitation. *Volume, 54*, 175–181.
27. Ministério da Saúde (2012). Saúde da Criança: crescimento e desenvolvimento. Brasilia, DF.

wCReF – A Web Server for the CReF Protein Structure Predictor

103

Vanessa Stangherlin Machado, Michele dos Santos da Silva Tanus,
Walter Ritzel Paixão-Cortes, Osmar Norberto de Souza,
Márcia de Borba Campos, and Milene Selbach Silveira

Abstract

The prediction of protein tertiary structure is a problem of Structural Bioinformatics still unsolved by science. The challenge is to understand the relationship between the amino acid sequence of a protein and its three-dimensional structure, which is related to the function of these macromolecules. Among the methods related to protein structure prediction is CReF (*Central Residue Fragment-based Method*) proposed by Dorn & Norberto Souza. Here we present wCReF, the Web interface for the CReF method developed with a focus on usability. With this tool, users can enter the amino acid sequence of their target protein, and get as a result the approximate 3D structure of a protein without the need to install all the multitude of necessary tools for their use. In order to create an interface that take usability in consideration, we have conducted a study to analyze usability of similar servers (I-TASSER, QUARK and Robetta), guided by experts on both Human-Computer Interaction and Bioinformatics domain areas, using the Nielsens' Heuristic Evaluation method. Evaluation results have served as guiding orientation to design the key features that wCReF must have and then, develop the first version of the interface. As a final product we present the wCReF protein structure prediction server. Furthermore, this study can contribute to improve the usability of existing bioinformatics applications, the prediction servers analyzed and the development of new scientific tools.

Keyword

Prediction server • Usability • Protein Structure Prediction • Bioinformatics

103.1 Introduction

Protein structure prediction servers are automated tools where the user provides as input the sequence of the target protein and the execution parameters, obtaining as a result the three-dimensional model of the protein, which is then used to infer the function of the protein and guide experimental efforts [1]. These servers use different computational approaches to predict the spatial conformation from the amino acid sequence or primary structure. The objective is to understand how the amino acid sequence of a protein and its three-dimensional structure related to each other, and how this relationship helps determine the function of these macromolecules. Anfinsen [2] experiments suggested that the information necessary for the formation of a protein's structure is encoded in its sequence, the question to predict these structures has become one of the biggest research challenges today.

Thus, scientists continue to develop tools for predicting the structure with increasing accuracy [3]. However, scientists have not achieved a universal solution to the problem of

V.S. Machado (✉) • M.dos.S.da.S. Tanus • W.R. Paixão-Cortes • O.N. de Souza • M. de Borba Campos • M.S. Silveira
Faculty of Informatics (FACIN), Pontifical Catholic University of Rio Grande do Sul (PUCRS), Porto Alegre, Brazil
e-mail: vanessa.stangherlin@acad.pucrs.br; michele.silva.004@acad.pucrs.br; walter.paixao-cortes@acad.pucrs.br; osmar.norberto@pucrs.br; marcia.campos@pucrs.br; milene.silveira@pucrs.br

© Springer International Publishing AG 2018
S. Latifi (ed.), *Information Technology – New Generations*, Advances in Intelligent Systems and Computing 558, DOI 10.1007/978-3-319-54978-1_103

protein structure prediction (Protein Structure Prediction Problem – PSP). Therefore, the development of computational methods to predict 3D structures from sequences is a way to solve this problem and fill the gap between the amount of sequences and resolution of these structures [4].

The trend of using prediction servers has been observed when analyzing the previous editions of CASP, a competition started in 1994 [5] where, biannually, the prediction techniques are evaluated and ranked. We can cite as the main protein structure prediction servers in the literature and presented at CASP: Rosetta, now as Robetta server [6], I-TASSER [7] and QUARK [8].

The CReF method [9] has shown good results in the prediction of protein structures [10–12], demonstrating potential for further scientific studies and applications. However, this method does not have a graphical interface for communication with the user, it runs locally on a Linux platform, requiring knowledge by the user for installation, configuration and use.

In order to make the method easier to use for the scientific community, the wCReF server was created. With this more user-friendly interface, we can proceed with further improvements on the method, as more researchers will have access to it.

This paper describes the creation process behind wCReF and discusses usability issues considered in its design and identified during the assessment with Human-Computer Interaction (HCI) and Bioinformatics experts and end users.

103.2 Implementation

103.2.1 Preparing the CReF Environment

Using the instructions provided by CReF documentation, the first step on implementing wCReF was to build an environment where CReF could run with no issues. The difficulties on doing that, depending on your familiarity with this type of task, reinforce the need of an easy to use interface. This step also updated the documentation, describing more thoroughly the server's installation process and usage.

103.2.2 Deciding the Architectural Model

Considering that CReF was an existing method, capable of running standalone, we have decided that wCReF would be a Web interface, serving two purposes. The first being to provide a friendly graphical user interface to execute the method, and the second to serve as a gateway for more information about the CReF method. The Web interface

would be the first element to increase the usability. The last aspect considered on the architecture was that CReF, as it was, could be used by one person at a time, as it did not have any mechanism to allow multiple executions. So, it was designed a service layer, with task scheduling capabilities to allow that multiple executions were queued and executed.

103.2.3 Conducting the Usability Evaluations

Usability seeks to ensure that systems are easy to learn to use, effective and enjoyable [13]. Usability barriers can directly influence user satisfaction, leading to extra time to understand the system and interact with it. These questions become more complex when there is a wider range of user profiles, such as the Bioinformatics area. Despite this, usability seems to not be considered as crucial for many protein structure prediction servers, resulting in hard to use interfaces and users lost to get the results. To avoid that in wCReF, we have performed the following steps:

1) Identify users' needs: to understand the features of a protein structure prediction server, the end user profiles and the commonly performed tasks. wCReF's interface considered both expert and novice users in the Bioinformatics field.
2) Server usability inspection: in order to define wCReF's requirements and features, we have used Nielsen's 10 heuristics from the Heuristic Evaluation [13, 14]: Visibility of system status; Match between system and the real world; User control and freedom; Consistency and standards; Error prevention; Recognition rather than recall; Flexibility and efficiency of use; Aesthetic and minimalist design; Help users Recognize, diagnose, and recover from errors; and Help documentation. The evaluation was conducted in 3D structure prediction servers participants of CASP: QUARK, I-TASSER and Robetta. A total of four experts performed the evaluation: two experts in bioinformatics and two HCI specialists. Based on heuristic evaluation of the servers, we have written the requirements to build the interface.
3) Development of the interface prototype.
4) End users evaluation: we adopted the evaluation by observation and questionnaires to understand the user's feelings regarding the prototype developed. The instrument used was an adaptation of Ssemugabi questionnaire [15, 16]. The ELEVEN (11) study participants are graduate and undergraduate students of courses in the area of Bioinformatics, including Computer Science, Pediatrics and Child Health, Zoology, Medicine and Health Sciences, Molecular and Cell Biology, Pharmaceutical

Biotechnology, and Biological Sciences. Ethical issues were addressed in the study and all have filled the informed consent form.

103.2.4 Algorithm

The Central Residue Fragment method (CReF) applies data mining techniques to the Protein Data Bank [17, 18] in order to predict a protein's backbone torsion angles [9]. The algorithm performs the following steps:

1) Fragments the target sequence using a sliding window with an odd number of residues. For each fragment, goes through steps 2 to 5 to predict its central residue dihedral angles.
2) Performs a BLAST [20] search using the target fragment as query and obtains the top scoring matches to use as templates.
3) Uses the k-means algorithm to cluster the templates' central residue torsion angles (φ and ψ).
4) Selects a representative cluster according to the secondary structure prediction of the target sequence.
5) Uses the selected cluster's centroid as the central residue torsion angles.

By setting the fragment size and BLAST parameters, such as substitution matrix and gap penalty, the user is able to fine tune the prediction process and outcome.

The prediction library has a REST API that interfaces with wCReF. When a new prediction is submitted through wCReF's interface, it is added to a task queue. Workers will then retrieve items from the queue, making the prediction process scalable. The API also allows querying a given task's status and retrieving the prediction results.

103.2.5 User Interface

The wCRef has a common area for all users (Fig. 103.1), with public access and an area that is only accessed when performing login, to make predictions and track the results.

The public pages are composed of the server home page, where general information is presented, as well as external links and information on the research group. In addition to the home page, there are the following pages: user registration, submissions, documentation, staff and talk to us. We emphasize that the submissions page displays all the submissions made in wCReF, but the only available information is the general status of each submission. To follow the result, the user needs to log into the system.

The private pages' content is unique to each user and can be accessed through the wCReF sign in system. It has the main screen for submission, the submissions status page and results page. Users can submit a prediction, access and view the results, and access statistics. It also has a user profile page for the registered user.

To send a prediction of a target protein, the user must enter the name of the protein to be predicted, the query sequence in FASTA format, and optionally modify certain parameters. They can be of two types: the parameters related to wCReF implementation and the parameters used in BLASTp [17] execution, all described in Table 103.1.

In the submissions page, users can track their predictions, almost in real time. As the users send their jobs, they will be added, and presented by the submission number, ID, name of protein, size (number of amino acids) and date of submission. It is also informed the progress status of the prediction on the server (which CReF step is running) and finally the result, with the option to be redirected to the results page or delete it.

The results page summarizes all the information about the prediction, including the input sequence, the prediction of the secondary structure and the predicted three-dimensional structure. The wCReF has a viewer of the molecule directly on the results page. Figures 103.2, 103.3 and 103.4 show sections of wCReF results page. For this example we have used the 1ZDD protein [21]. There is an option to print the results in PDF or download in PDB format.

At the end of wCReF preview page, the approximate 3D structure can be viewed in different ways (Fig. 103.4). The visualization software used in wCReF is iView, an interactive viewer of three-dimensional structures of molecules based on WebGL [22]. Its components are connected to wCReF interface through HTML code.

The wCReF preview page generates the image of the secondary structure with PolyView 2D [23] and displays 3D approximate structure with iView automatically, not requiring, in this way, the installation or configuration of these user components. The prediction of the approximate protein is available in the standard format of the Protein Data Bank, with the extension "pdb" and works on most viewers of biological macromolecules.

It is important to notice that wCReF also makes available extensive help to understand the server's purpose and how to use it, in the form of online help, tutorials and documentation, available to users at any time.

103.3 Discussion

As the wCReF interface has been made regardless of prediction method, any changes in CReF do not change the Web interface. The database, the HTML and CSS and page

Fig. 103.1 wCReF homepage with login area, server usage information, statistics and links to additional resources (All the figures of the article shows the content in Portuguese. wCReF website is being updated to support multiple languages)

Table 103.1 Optional wCReF parameters

	Parameter	Definition
CReF parameters	Fragment size	# of amino acids in a fragment: 5, 7 or 9
	Number of clusters for φ and ψ angles	Define the number of clusters created by the k-means algorithm
	Exclude homologues with similarity over	Threshold to exclude fragments from the template search: 100%, 90%, 80% or 70%
	Exclude homologues based on PDB code	Exclude from the PDB template search the proteins informed with PDB code
BLAST general parameters	Expect threshold	Probability to find this sequence in a random model
	Word size	Sequence length for alignment: The default values are 2 or 3
	Matrix	Associate the score that will be applied to aligned sequences. Select according to sequence type
	Gap Costs	The cost of a gap in the aligned sequence, according to the selected Matrix

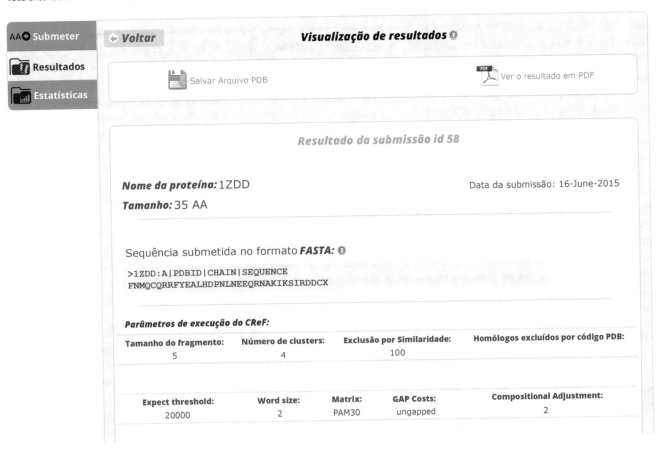

Você encontra-se em: wCReF ❯ Resultados ❯ Visualizar

AA⊙ Submeter
Resultados
Estatísticas

← Voltar   **Visualização de resultados** ⊙

💾 Salvar Arquivo PDB   📄 Ver o resultado em PDF

Resultado da submissão id 58

Nome da proteína: 1ZDD   Data da submissão: 16-June-2015
Tamanho: 35 AA

Sequência submetida no formato **FASTA:** ⊙

```
>1ZDD:A|PDBID|CHAIN|SEQUENCE
FNMQCQRRFYEALHDPNLNEEQRNAKIKSIRDDCX
```

Parâmetros de execução do CReF:

Tamanho do fragmento:	Número de clusters:	Exclusão por Similaridade:	Homólogos excluídos por código PDB:
5	4	100	

Expect threshold:	Word size:	Matrix:	GAP Costs:	Compositional Adjustment:
20000	2	PAM30	ungapped	2

Fig. 103.2 wCReF results page. The first part shows the submission parameters as sent by the user

templates does not change if the method undergoes changes, which facilitates updates and maintenance.

The wCReF interface is a result of the assessments carried out by the evaluations by experts and by users. The experts pointed out the major usability problems found in the interfaces of the evaluated prediction servers (Robetta, QUARK and I-TASSER). With the description of the errors and the observation of these interfaces, it became clear what features the new server should have, and what problems should be avoided during its creation.

The main issues found by experts during the prediction servers' evaluation was navigation and system visibility,

1————————————————35
FNMQCQRRFYEALHDPNLNEEQRNAKIKSIRDDCX
CCHHHHHHHHHHHHHCCCCCHHHHHHHHHHHHHCCC

1————————┬ Numeração dos resíduos de Aminoácidos

H - Hélice α

E - Folha β

C - Estruturas Irregulares - Voltas e Alças

Fig. 103.3 wCReF results page. Secondary structure representation

user login, FASTA format validation, academic email address requirement, information presented as long texts, lack of information on prediction status, such as execution time, and results presentation. Additionally, they also suffered from non-standardized error messages, menus and icons, and non-customizable user interface as it didn't adapt to novices or experts.

The next step was the implementation following the guidance of experts. To the prototype evaluation, we chose to seek end users opinion to the experts' inputs. It was also possible to identify usability problems in wCReF prototype interface.

The evaluation of the prototype, by adapting the Ssemugabi questionnaire [16, 17] was positive. Of the total 972 issues evaluated, approximately 80% were evaluated with positive scores, suggesting that the prototype was well designed considering usability aspects. We would like to highlight that the main positive point, noticed several times in the open questions, was that the interface is "easy to use". We can see some of the positive comments on these issues described by users:

"Very good visualization interface of the predicted structure with all the data we have set for the algorithm execution." (User A, Computer Science)

"Easy to use, very simple." (User A, Computer Science)

"The system is easy for beginners", "the results are presented in a consistent way" and "the help is available at every step." (User B, Biology)

"The system has a simple and clear interface." "The understanding of the method and submissions are easy." (C User, Computer Science)

"The interface is easy to use, beautiful and simple." (User D, Information Systems)

"Interface easy to use." (User E, Mathematics and User G, Pharmacy)

"The wCReF a Web system is used to predict the 3D structure of proteins. The system follows the basic standards

required for usability to new and advanced users responding satisfactorily to the purpose for which it was developed. " (User F, Biological Sciences)

"The interface is simple and lightweight (no overload of colors, images) and has objective menus. "(User H, Biological Sciences)

The interface is simple and offers advanced features without imposing a steep learning curve usually associated with complex systems. For example, data on the prediction are arranged in a single window, without other information that can disrupt the user, being sent directly to the wCReF server and the users do not need to inform their e-mail and password for each new submission.

With our evaluation results, wCReF interface has been improved. The requirements raised in the evaluation include:

- System Visibility: viewing submissions made and their status, updated every 30 seconds.
- Clear information about the need to sign in: displays a visible information that the user is required to register into the server for sending jobs; job queue and sent jobs.
- Minimalist design: reduced menu and priority to the sequence submission area. The menu options are shown only upon request.
- Warning and error messages: displayed in windows on the same page where the error occurred and requiring user confirmation. Attention to colors and symbols to highlight information.
- To prevent mistakes: no indication of what data are required and which are optional. The function of clearing a field is only performed through user confirmation.
- Error handling: if we enter incorrect data, such as an amino acid sequence in incompatible format, it is warned by the server.
- Delete a job or prediction directly: use of a button, without the need for various actions, with user confirmation and the option of sending more than one job at a time to the server.
- Interface consistency: the menus, icons and interface buttons are unique, standardized.
- Prediction result not only sent by email: this makes it difficult to use the server as part of a workflow. There is a page with the results that can be read from time to time to determine whether the prediction was complete.

We suggest that to send the sequence first the user has to perform log in into the system through a previous registration. We found usability problems related to the fact that it was not clear the user needed to login before submitting the prediction. In some of the analyzed servers, login area was displayed after the user sends the prediction, which facilitated the occurrence of the error, or was presented

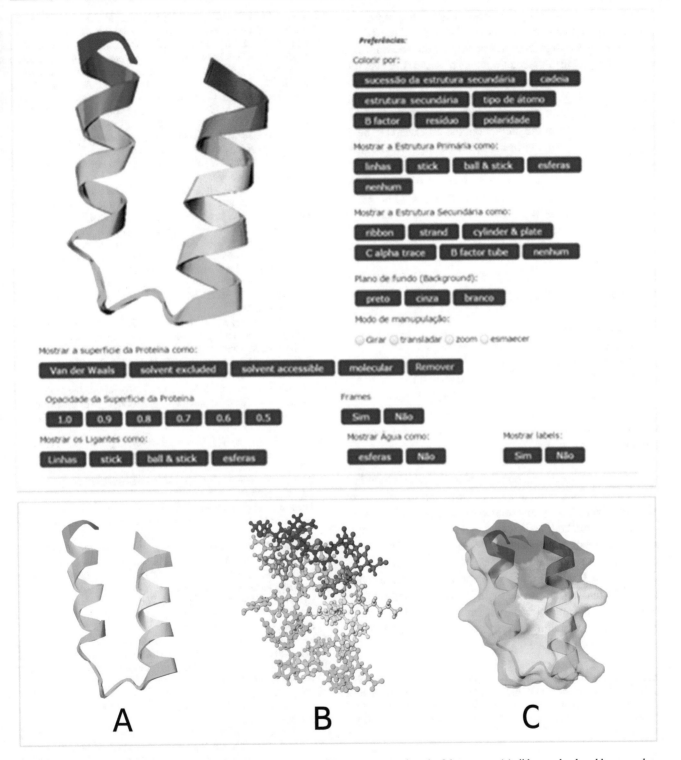

Fig. 103.4   wCReF results page. 3D viewer. Here we can see three different ways to view the 3d structure: (**a**) ribbon and colored by secondary structure succession; (**b**) ball and stick; (**c**) molecular surface

only at the time of submission of the sequence. Therefore, the wCReF sign in area is highlighted on the homepage, and the login button appears on all pages of the interface.

Furthermore, it was found in the evaluated servers that too much text was used when communicating with users, usually differentiated only by titles. There is no standardization on the pages. In wCReF, we have prioritized the use of shortcuts, buttons, messages, display information according to user interest, assembled into blocks or new pages, separated by menus according to the interest or related content.

In the submissions page there is a search field, to perform a search by the date that the prediction was submitted, the name that was assigned to the protein or the identification number (ID). One of the requirements to ensure the visibility was this possibility to inform the user the Jobs list (work) that are running. For ethical reasons, the user can only monitor submissions from this page and view their status. To view the result of the prediction, the user needs to be on his private area. This is important because it ensures that each user can only see her/his results.

The language choice option was also one of the requirements detected by heuristic evaluation. This feature will be added to the wCReF interface, allowing access to researchers and users from other countries, as all the interface will be presented on their primary language.

103.3.1 Availability and Requirements

Project name: wCReF
Project home page: wcref.labio.org
Operating system: platform independent
Technologies: HTML, CSS, Python, MySQL
Other requirements: browser with WebGL support

103.4 Conclusion

wCReF allows the execution of the CReF methodology in a Web environment, without the need for local installation of all packages and programs related to its use, facilitating the use of the method and reducing costs associated with maintaining the computational resources. This also contributes to the new updates being implemented in CReF. We believe this study resulted in an application that is easy to interact, to use and can be used by both experienced users in their scientific research, and for less experienced users, including serving as a support tool for bioinformatics education.

Acknowledgments This work was supported in part by grants ONS (CNPq, 308124/20154; FAPERGS, TO2054-2551/13-0). ONS is a CNPq Research Fellow. VSM and MSST are supported by CAPES/PROSUP PhD scholarships. WRPC is supported by a DELL scholarship.

References

1. Kim, D. E., Chivian, D., & Baker, D. (2004). Protein structure prediction and analysis using the Robetta server. *Nucleic Acids Research, 32*(Web Server issue), W526–W531.
2. Anfinsen, C. B. (1973). Principles that govern the folding of protein chains. *Science, 181*(4096), 223–230.
3. Kozma, D., & Tusnády, G. E. (2015). TMFoldWeb: a web server for predicting transmembrane protein fold class. *Biology Direct, 10*, 54.
4. Lee, J., Wu, S., & Zhang, Y. (2009). Ab Initio protein structure prediction. In D. J. Rigden (Ed.), *From protein structure to function with bioinformatics* (pp. 3–25). Dordrecht: Springer Netherlands.
5. Moult, J., Fidelis, K., Kryshtafovych, A., Schwede, T., & Tramontano, A. (2014). Critical assessment of methods of protein structure prediction (CASP)–round x. *Proteins, 82*(Suppl 2), 1–6.
6. Kim, D. E., Chivian, D., & Baker, D. (2004). Protein structure prediction and analysis using the Robetta server. *Nucleic Acids Research, 32*(Web Server issue), W526–W531.
7. Zhang, Y. (2008). I-TASSER server for protein 3D structure prediction. *BMC Bioinformatics, 9*(1), 40.
8. Zhang, Y. (2014). Interplay of I-TASSER and QUARK for template-based and ab initio protein structure prediction in CASP10. *Proteins Structure Function Bioinformatics, 82*, 175–187.
9. Dorn, M., & Norberto de Souza, O. (2008). *CReF: a central-residue-fragment-based method for predicting approximate 3-D polypeptides structures* (pp. 1261–1267). ACM: New York, NY.
10. Dall'Agno, K. C. M., & Norberto de Souza, O. (2013). An expert protein loop refinement protocol by molecular dynamics simulations with restraints. *Expert Systems with Applications, 40* (7), 2568–2574.
11. Dorn, M., Breda, A., & Norberto de Souza, O. (2008). A hybrid mthod for the protein structure prediction problem. In A. L. C. Bazzan, M. Craven, & N. F. Martins (Eds.), *Advances in bioinformatics and computational biology* (pp. 47–56). Berlin/Heidelberg: Springer.
12. Dorn, M., & Norberto de Souza, O. (2010). Mining the protein data bank with CReF to predict approximate 3-D structures of polypeptides. *International Journal of Data Mining and Bioinformatics, 4*(3), 281–299.
13. Preece, J., Rogers, Y, Afiada, H., Benyon, D., Holland, S., & Carey, T. (1994). *Human-Computer Interaction*. ACM SIGCHI: Essex, UK.
14. Nielsen, J. (1994). *Usability inspection methods* (pp. 413–414). ACM: NewYork.
15. Nielsen, J. (1995). 10 Heuristics for user interface design: Article by Jakob Nielsen, [Online]. Available: http://www.nngroup.com/articles/ten-usability-heuristics/. Accessed 11 Jun 2014.
16. Ssemugabi, S. (2009). Usability evaluation of a web-based e-learning application: a study of two evaluation methods, Thesis.
17. Ssemugabi, S., & de Villiers, R. (2007). *A comparative study of two usability evaluation methods using a web-based e-learning application* (pp. 132–142). ACM: New York.
18. Berman, H. M. (2000). The Protein Data Bank. *Nucleic Acids Research, 28*(1), 235–242.
19. Zardecki, C., Dutta, S., Goodsell, D. S., Voigt, M., & Burley, S. K. (Mar. 2016). RCSB Protein Data Bank: A Resource for Chemical, Biochemical, and Structural Explorations of Large and Small Biomolecules. *Journal of Chemical Education, 93*(3), 569–575.
20. Altschul, S. F., Gish, W., Miller, W., Myers, E. W., & Lipman, D. J. (1990). Basic local alignment search tool. *Journal of Molecular Biology, 215*(3), 403–410.
21. Starovasnik, M. A., Braisted, A. C., & Wells, J. A. (1997). Structural mimicry of a native protein by a minimized binding domain. *Proceedings of the National Academy of Sciences, 94*(19), 10080–10085.
22. Li, H., Leung, K.-S., Nakane, T., & Wong, M.-H. (2014). iview: an interactive WebGL visualizer for protein-ligand complex. *BMC Bioinformatics, 15*(1), 56.
23. Porollo, A. A., Adamczak, R., & Meller, J. (Oct. 2004). POLYVIEW: a flexible visualization tool for structural and functional annotations of proteins. *Bioinformatics, 20*(15), 2460–2462.

Automating Search Strings for Secondary Studies

Francisco Carlos Souza, Alinne Santos, Stevão Andrade, Rafael Durelli, Vinicius Durelli, and Rafael Oliveira

Abstract

Background: secondary studies (SSs), in the form of systematic literature reviews and systematic mappings, have become a widely used evidence-based methodology to to create a classification scheme and structure research fields, thus giving an overview of what has been done in a given research field.

Problem: often, the conduction of high-quality SSs is hampered by the difficulties that stem from creating a proper "search string". Creating sound search strings entails an array of skills and domain knowledge. Search strings are ill-defined because of a number of reasons. Two common reasons are *(i)* insuffient domain knowledge and *(ii)* time and resource constraints. When ill-defined search strings are used to carry out SSs, a potentially high number of pertinent studies is likely to be left out of the analysis.

Method: to overcome this limitation we propose an approach that applies a search-based algorithm called Hill Climbing to automate this key step in the conduction of SSs: search string generation and calibration.

Results: we conducted an experiment to evaluate our approach in terms of sensibility and precision. The results would seem to suggest that the precision and the sensibility our approach are 25.2% and 96.2%, respectively.

Conclusion: The results were promising given that our approach was able to generate and calibrate suitable search strings to support researchers during the conduction of SSs.

Keywords

Secondary studies • Search string • Hill climbing.

104.1 Introduction

Systematic literature reviews (SLRs) and systematic literature mappings (SLMs), also known as secondary studies (SSs), have been widely used in medical research and in the natural sciences since the 1970s and early 1980s. SLR and SLM are considered rigorous methods to map the evidence base in an unbiased way, evaluate the quality of the existing evidence, and synthesize and give an overview of a given research field. Based on the guidelines of these methods, Kitchenham et. al. [11] proposed evidence-based Software Engineering (EBSE) in hopes of fostering the adoption of SSs in Software Engineering (SE). In the context

F.C. Souza (✉) • A. Santos • S. Andrade • V. Durelli
University of São Paulo – USP, São Carlos, São Paulo, SP, Brazil
e-mail: fcarlos@icmc.usp.br; alinne@icmc.usp.br; stevao@icmc.usp.br; durelli@icmc.usp.br

R. Durelli
Federal University of Lavras, Lavras, MG, Brazil
e-mail: rafael.durelli@dcc.ufla.br

R. Oliveira
Federal Technological University of Parana, Dois Vizinhos, PR, Brazil
e-mail: raoliveira@utfpr.edu.br

© Springer International Publishing AG 2018
S. Latifi (ed.), *Information Technology – New Generations*, Advances in Intelligent Systems and Computing 558, DOI 10.1007/978-3-319-54978-1_104

of Software Engineering, SSs rely on the use of an objective, transparent, and rigorous approach for the entire research process in order to minimize bias and ensure future replicability. Rigour, transparency, and replicability are achieved by following a fixed process for all reviews. The fixed process is one of the characteristics that distinguish SSs from traditional literature reviews (TLR).

The conduction of SSs usually include the following steps: first, the research question is deconstructed by considering Population, Intervention, Comparison, and Outcome (PICO) criterion. Terms from this criterion form of the basis of search strings that are used in the literature search. Then a protocol is produced to describe definitions, search strings, search strategy, inclusion and exclusion criteria, and the approach that will be used to synthesize data. This protocol is often peer-reviewed and piloted. This may lead to several revisions in the search strategy. Next, a systematic search is conducted; studies are retrieved from digital scientific databases sources (e.g., IEEE Xplore, ACM, Scopus, Scirus, etc.) [8].

A remarkable problem regarding SSs is the generation of suitable search strings. Ill-defined search string can hinder the search process by returning a significant amount of irrelevant studies. In addition, another problem stems from the different rules employed by digital databases, which may render the search step into a process of trial and error. For instance, to generate a suitable search string researchers need to grasp a set of keywords and synonymous, which is a time and resource consuming task. Also, usually researches need to adjust the same generic search string to several digital databases. According to Kitchenham et. al., [11], four common aspects are associated with having to analyze a high number of irrelevant papers: (*i*) unsatisfactory search string creation; (*ii*) digital databases of studies have different interfaces; (*iii*) digital databases deal with Boolean formulas [5, 8] in a slightly different way from each other; and (*iv*) digital databases have different methods to search the body of the manuscript or even some indexing elements (e.g., title, keywords, and abstract).

In order to overcome these four issues we propose an approach called Search-based String Generation (SBSG) that applies an Artificial Intelligence (AI) technique called Hill Climbing (HC) [12]. The main goals of SBSG is to produce and recommend a suitable search string to be used during the conduction of an SS. In this study we prototype the SBSG approach in a proof-of-concept tool. Specifically, SBSG runs as following: (*i*) researchers need to define a set of parameters: terms, keywords, synonyms, number of iterations (how many times SBSG will run), and a list of control studies[1]; (*ii*) these parameters are used as input to an HC algorithm to create an initial search string, i.e., initial solution; (*iii*) from the initial search string HC generates a set of neighbor search strings, i.e., similar search strings; and

(*iv*) if a neighbor improves the value of the initial solution, i.e., a better suitable search string is generated, then this new search string is selected and becomes the current search string. This process runs interactively until SBSG finds the best suitable search string or reaches the specified number of iterations.

In order to provide some evidence of the applicability of our approach we performed an experiment. More specifically, we used terms, keywords, synonyms, and lists of control studies of five published SSs. Afterwards, we assessed if the search strings generated by our approach were as suitable as the ones used by the published SSs. Two metrics were used to gauge the quality of the generated search strings: precision and sensibility. Experimental results that our approach improved the search strings.

The main contributions of this paper are fourfold: (*i*) a well-defined approach to improve a key step of SSs – search string generation and calibration; (*ii*) an approach to tap into the strengths of a search-based algorithm and improve the applicability of EBSE; (*iii*) a proof-of-concept tool that automatically supports the generation of search strings for SSs; and (*iv*) an experiment to evaluate our approach and implementation thereof in terms of precision and sensibility.

This remainder of this paper is structured as follows: in Sect. 104.2 SSs, HC, and fitness function are described. Section 104.3 details some technical issues concerning the use of the SBSG approach as an effective solution to improve automated searches in SSs. The experiment we carried out is presented in Sect. 104.4. Discussions, general impressions, and threats to validity are presented in Sect. 104.5. Section ?? discusses related work. Finally, Sect. 104.6 presents concluding remarks.

104.2 Background

104.2.1 Secondary Studies

According to Kitchenham et al. [10] "primary studies" are experiments and empirical validations with qualitative or quantitative results to an specific research field. In other hand, "Secondary Studies" (SSs) in SE represent a compilation of several primary studies gathered to find relevant research evidence and answer specific research question. There are two types of SS, they are: SLR and SLM. Both SSs rely upon the use of an objective, transparent and rigorous approach for the entire research process in order to minimize bias and ensure future replicability.

---

[1] Control studies are papers that must be retrieved when search in a given database.

In spite of the growing importance that SSs have been achieving nowadays, they are still new topics to the SE community. Many challenges appear as to how to create a suitable search string. Usually, the search string is generated and calibrated based on a set of terms, keywords, synonyms, etc. Moreover, this whole process is done manually by the researchers, which is time-consuming and error-prone. In addition, if researchers are new in a particular field, then they need to spend more time and dedicate efforts to generate and calibrate search strings. Another challenge is the different rules employed by the digital scientific databases that make the search string calibration almost a process of trial-and-error. Usually, researchers need to rework on setting the same search string in several databases source to identify good studies, avoiding missing papers. We argue that this whole process of creating a suitable search string could be semi-automated by means of HC algorithm.

104.2.2 Hill Climbing and Fitness Function

Search Based Software Engineering (SBSE) is the field of software engineering research and practice that applies search based techniques to solve different optimization problems from diverse SE areas. SBSE approaches allow software engineers to automatically obtain solutions for complex and labor-intensive tasks, contributing to reduce efforts and costs associated to the software development [4, 6]. SBSE consists of search-based algorithms used in SE, such as generic algorithms, generic programming, simulated annealing, and HC [12].

HC is a local search algorithm that combines a general search method with either objective functions or fitness functions for evaluating the states generated by the method [12]. HC aims to identify the best path to be followed in the search. As outcome the technique returns a satisfactory result for a given problem. HC algorithm consists in selecting an initial solution randomly, evaluating and improving it step by step, from the investigation of the neighborhood of the current solution. If a neighbor improves the value of the initial solution, this new solution is then selected and becomes the current solution. This process is performed until it finds an optimal solution or when no neighbor has a better value [12].

The main benefits of HC algorithm are: (*i*) it requires low memory due to fact that only the current state is stored; (*ii*) it is easy to be implemented; and (*iii*) whether the best or optimal solution exists in the search space, HC is able to find that solution in a feasible computational cost. Regarding SE issues, HC algorithm is a simple and powerful SBSE strategy to search optimal solutions in combinatorial search spaces. Then, it is feasible to employ HC algorithm in different SE applications.

HC algorithms must be combined with a proper Fitness Function (FF) that is able to measure the quality of the solutions found [4]. FFs are important components for search and metaheuristics techniques since they predict how close is a solution to be optimal. Metaheuristics are high level strategies to efficiently explore the search space in order to find optimal or near solutions by using different methods, techniques which constitute metaheuristic algorithms range from complex learning processes to simple local search procedures such as a HC algorithm [2]. Thus, in general terms, a FF is an expression that measures the goodness of a candidate solution for solving a given problem.

There are many guidance on setting a FF, however it is not a general task due to the fact that each function depends on the specific problem and its features. The idea is employing heuristics information regarding to the features of an specific problem into a function so that it can be able to assess the adequacy of candidate solutions. Then, there are cases in which the FF is fairly trivial and there are cases in which the designer needs to dedicate efforts to figure which FF is more suitable for a given problem.

104.3 The SBSG Approach

Researchers must generate and calibrate a search string totally manually, which is an error prone activity, time consuming, and labor-intensive. Aiming at applying the concepts of HC algorithm to alleviate the problem of generating suitable search strings for different digital scientific databases, we propose the SBSG approach. Our approach is semi-automatic and combines an HC algorithm with an assess strategy for searching and generating suitable search strings.

Our approach provides a semi-automatic method to produce a good string reducing the aforementioned problems about the manual process. The idea behind the SBSG approach does not replace researchers in the string generation process, on the contrary it assists them. A generic workflow of SBSG is shown in Fig. 104.1. According to the figure, researchers must provide the following parameters: (*i*) a list of keywords, (*ii*) a list of control studies, (*iii*) the set of terms of the string and their respective synonymous, and (*iv*) the number of iterations, i.e, how many times SBSG will run.

Through those parameters, SBSG starts its process with an initial string (S_0) based on PICO criterion. It means that it is necessary at least one keyword to fill out the population, intervention, comparison and outcome. Figure 104.2 shows an illustrative example of PICO criterion.

SBSG employs an HC algorithm that works performing small changes in each part of the string to create a

Fig. 104.1 SBSG's workflow

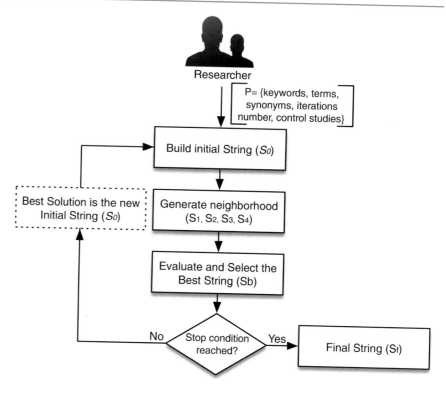

Fig. 104.2 PICO criterion
(Kitchenham et al. [9])

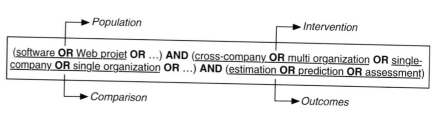

neighborhood of string candidates. First, HC generates the string S_0 though of a set of terms previously defined by the user. Then, a neighborhood (S_1, S_2, S_3, S_4) from S_0 is expanded, which it is based on strategies presented in the Table 104.1.

Each string is assessed through a special FF to search the best one (S_b). This FF is based on an optimal search strategy introduced by Straus & Richardson [13] and Haynes et al. [7] in the context of SSs. The FF proposed is composed of two measures called *sensibility* and *precision*. Figure 104.3 represents the measures aforementioned.

The *sensibility* on search strategy context is a measure to identify all of the relevant studies (**A** – Fig. 104.3) for a specific domain from retrieved studies (**R** – Fig. 104.3) supported by a set of relevant studies previously defined (**C** – Fig. 104.3). The higher **C** is, the higher will be the sensibility score or the opposite. On the other hand, the *precision* is an ability to identify the amount of irrelevant studies (**B** – Fig. 104.3), where **B** = **R**−**A**. When **B** is zero, i.e., no irrelevant study is detected, the precision score is greater. A search string with low precision will lead a lot of

irrelevant studies retrieved. *Sensibility* and *precision* are computed using the Equations 104.1 and 104.2, respectively.

$$S = \frac{A}{A+C} * 100. \qquad (104.1)$$

$$P = \frac{A}{A+B} * 100. \qquad (104.2)$$

The overall process by which a candidate string is evolved also provides an option to reduce researchers' efforts during the study selection stage. This alleviation of human efforts is provided enabling the assess to the quality of each study based on abstract, keywords, and control list. Then, researchers are able to eliminate those studies with zero or a very low fitness value. Therefore, the total fitness (**F**) is computed through Equation 104.3. When **F** is zero, the string returns only irrelevant studies and the higher **F** is, the higher the adequacy of search string will be.

$$F = \frac{(S) + (P)}{2} \tag{104.3}$$

Once no further improvements can be achieved for a search string, the search continues exploring the next neighborhood, starting over with the new current string, until no neighbor leads to improvements. At this point the search restarts at another randomly chosen location in the search space. This is known as a strategy to overcome local optima, enabling the search to explore a wider region of the possible strings for a specific topic.

The quality of the SBSG approach is quantified through search strategies scale using in Dieste & Padua [3], which was inferred from the sensitivity and precision ranges of SLRs in medicine. We have adopted this search strategy because it can qualify how relevant a study is to a particular domain. Table 104.2 provides a scale to measure the quality of search based on the amount of relevant and irrelevant

Table 104.1 Strategies for strings neighborhood

Functions	Changes	Where
Adds or deletes	Synonymous	Population
Adds or deletes	Synonymous	Intervention
Adds or deletes	"-"	In compound words
Adds or deletes	Plural	Of a synonymous
Replaces	The suffix of a Synonymous	Population

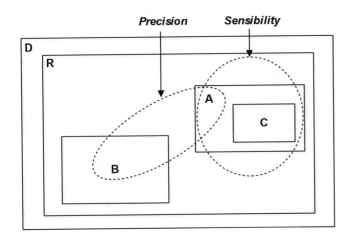

Fig. 104.3 Sensibility and precision search strategies

studies. Assuming the scales for adequate strings, we considered a threshold between 80% and 99%, 20%, and 60% as references for sensitivity and precision, respectively.

104.3.1 Proof-of-Concept Implementation

A tool to fully support SBSG was devised. It owns two modules: (*i*) a Java module; and (*ii*) a Python module. The former is used to generate search Strings following the rules of the IEEE search engine. Based on a parameter defined by hand, this module reads a text file that must follow a pre-defined syntax and then it identifies the PICO's terms in the file. From this starting point, the first module creates a data structure that is able to apply rules from IEEE Xplore[2] to generate real search strings.

The latter module contains a set of scripts devised in Python. These scripts are used to call the first module. As outcome a object in Python is obtained. After, a script is used to request and send the generated search string to the IEEE Xplore digital database. The outcome is an XML (eXtensible Markup Language) file containing all fetched primaries studies's data such as: title, keywords, abstract. Then these data are parsed and analyzed in terms of *sensibility* and *precision*. This second module runs interactively until it finds the best suitable search string or it reaches the specified number of iterations.

104.4 Empirical Evaluation

This section presents the search strings that were used as subject in our evaluation, all of the decision we took when designing our experiment to address three research questions, and the results obtained.

104.4.1 Research Questions

We designed our experiments in a proof-of-concept format aiming to answer the following *Research Questions* (RQ): (i) **RQ₁**: How good is the sensibility of search strings generated when using our approach?; (ii) **RQ₂**: How good is the precision of search strings generated when using our

---

[2] see: http://ieeexplore.ieee.org/

Table 104.2 Search strategy scales

Strategy type	Sensitivity	Precision	Goal
High sensitivity	85–90%	7–15%	Max sensitivity despite low precision
High precision	40–58%	25–60%	Max precision rate despite low sensitivity
Optimum	80–99%	20–25%	Maximize both sensitivity and precision
Acceptable	72–80%	15–25%	Fair sensitivity and precision

approach?; and (iii) **RQ$_3$:** In practice, how efficient is our approach to generate the best search strings?

104.4.2 Goals

The goal of our experimentation can be therefore defined by using the GQM (Goal Question Metric) [1], which can be summarized as: **Analyze** the SBSG approach, **for the purpose of** evaluating it sensibility and effectiveness, **with respect to** improvement of SS's search string, **from the point of view of** researchers, **in the context of** heterogeneous subject secondary studies' search string.

104.4.3 Experimental Design and Execution

To provide empirical evidences to answer our RQs, we have used five SS's search strings that have already been published. During the selection of these SSs we have focused on covering a broad class of SS's search string from different field of research in SE. Summing up, have chosen SS's search string that aimed to identify primary studies of different contexts, such as software testing, software reusability, and model-driven development, etc.

Zhang et. al. [14] identified 11 digital scientific databases used more than once in SS for searching relevant studies in SE. Among them, IEEE Xplore is seen as the main digital source for SSs in SE. The content in IEEE Xplore comprises over 180 journals, over 1,400 conference proceedings, etc. Approximately 20,000 new documents are added to IEEE Xplore each month. Therefore, we have decided to use the IEEE Xplore digital source in our experiments. Furthermore, four reasons contribute to our choice on using IEEE Xplore to automate the SBSG strategy in this study: *(i)* satisfactory search algorithm; *(ii)* bibliographic resources are not limited; *(iii)* recognition of plurals; and *(iv)* IEEE retrieves the largest number of studies with abstract and complete texts.

This experiment was carried out in four steps. Firstly, selected five set of terms, keywords and control list. Secondly, we performed the search with 5, 15, and 30 iterations. Then we performed search for the best string according to three settings: (i) measure the sensibility, precision, and time to the generation; (ii) measure how many iterations needs to find the best one; and (iii) repeat this process 10 times. Finally, the average of the sensibility and precision were computed by the sum obtained in the previous sub-steps.

In order to analyze the gathered data, descriptive statistics have been used. We explored the features of descriptive statistics provided by IBM SPSS Statistics tool.[3] They

Table 104.3 Statistics for sensibility with 5 iterations

Subj.	Minimum	Maximum	Mean	Std. Deviation
String 1	,92	,93	,9316	,00562
String 2	,95	,97	,9677	,00873
String 3	,95	,97	,9656	,00770
String 4	,95	,98	,9800	,01146
String 5	,98	,98	,9815	,00000

summarize data using four numbers: the mean; minimum and maximum values; and standard deviation. Section 104.5 provides the results collected from our proof-of-concept validation.

104.5 Results and Discussion

104.5.1 Sensibility and Precision

First, we performed the descriptive statistics for Sensibility (RQ$_1$) of the generated search strings using our approach. Furthermore, we computed the mean for each Sensibility's iterations ($i = 5$, $i = 15$ and $i = 30$). All search strings had optimum results. The mean sensibility of all evaluated strings with five iterations was high, ranging from 93% to 98%. We noticed from the fifth iteration there were no increases in the sensitivity's mean. These results indicate that search strings with optimum sensibility can be generated through of the first iterations (Table 104.3).

Figure 104.4 shows that, for a small number of iterations the approach obtained a high sensibility for all cases, it means that a string can identify the most of relevant studies. In terms of SLR and SLM, a high sensibility is usually more desired. Therefore, the higher it is, more studies related to the domain must be returned in the search.

We also performed the descriptive statistics for precision (RQ$_2$) according to each iteration ($i = 5$, $i = 15$ and $i = 30$). Most of search strings had good results for five iterations (see Table 104.4). String 1 and String 2 achieved an optimum precision (20%, 21%, respectively), while the String 3 and String 5 had very high precision (41% and 35%, respectively). On the other hand, String 2 reached a low precision (12%). We noticed from the fifth iteration there were no increases in the mean of precision. Therefore, our approach is able to generate through of the first search strings with optimum and high precision.

The precision value for optimal strings must be low compared with the sensibility. Figure 104.5 shows a low value for all cases, this measures gives a proportion of relevant studies retrieved regarding to the number of

---

[3] see: http://www-01.ibm.com/software/analytics/spss/

Fig. 104.4 The best sensitivity by iteration

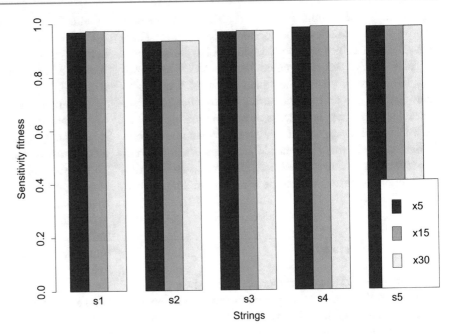

Table 104.4 Statistics for precision with 5 iterations

Subj.	Minimum	Maximum	Mean	Std. Deviation
String 1	,20	,20	,1979	,00232
String 2	,21	,21	,2133	,00000
String 3	,41	,41	,4145	,00000
String 4	,12	,12	,1195	,00000
String 5	,35	,35	,3464	,00000

Fig. 104.5 The best precision by iteration

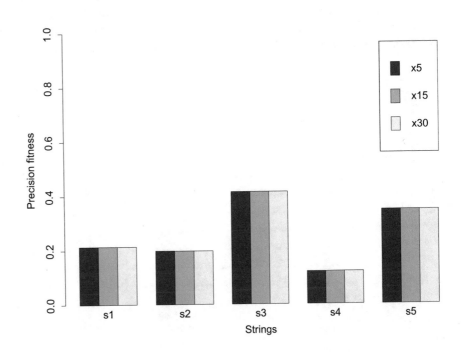

irrelevant studies from the search. Therefore, to our approach these results can be considered satisfactory, since that all strings achieve good precision in a few iterations.

Additionally, we can notice that in the Figs. 104.4 and 104.5 have been obtained similar results. We believe that it occurs due to the initial string be structured to cover the domain as a whole, thus, is almost impossible to start with an initial string with a low fitness. Assuming a good initial strings, the HC needs too less efforts and improvements to find the best one. Therefore, based on this assumption the results tend to be similar from a number of iterations.

104.5.2 Efficiency

The efficiency (RQ₃) of our approach was measured using the time for the search string generation and the number of iterations. The time includes the time of each execution to generate and calibrate the search string and their improvement. Figure 104.6 depicts how sensibility and precision improves in 10 iterations. One can mention that the approach has a fast convergence, since in five or six iterations the optimal solution has been found, i.e it was evaluated only five or six neighborhoods.

In our approach, to avoid results being skewed by the randomness inherent in search techniques, we conducted the experiment using three settings to measure the time and how many iterations were necessary for the search to converge for an optimal solution. We can notice in Fig. 104.7 that the time depends of some important factors, such as, size of string and complexity of it, but even for the worst cases the approach is faster than a string generated by hand. However,

as already mentioned, the HC had a fast convergence, it means that a few iterations were necessary to find a good solution, in general, performing the experiment with the first setting (5 iterations) has been enough to obtain an acceptable string.

Our approach generates an initial string based on the set of parameters provided by the researcher. Based on the first string, our approach generates four additional strings, one in each iteration. These strings are similar to the initial string. This process of deriving more strings from the initial string is analogous to the many steps that researchers usually take to manually fine-tune their search strings. As for our results, we evaluated 25 different strings (in five iterations): they took on average 116 second. Often, when manually fine-tuning search strings, many changes are needed. Therefore, given that this process of improving a given search string by hand might take hours, we conjecture that using our HC-based algorithm is able to significantly speed up the fine-tuning of search strings.

104.5.3 Threats to Validity

Conclusion validity: a possible threat associated to the conclusion of our study is intimately associated with the fact that we have used only IEEE Xplore as the subject database. Generally, researchers use at least three digital libraries as source of information to select primary studies.

Construct validity: regarding the theory behind our experiment and the observations, we believe the main threat on using SBSG is the fact that the approach must be implemented and adapted to the search engine of each digital

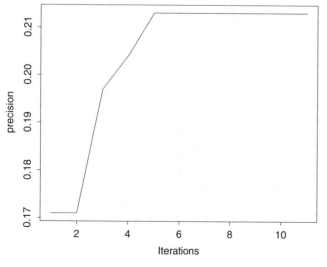
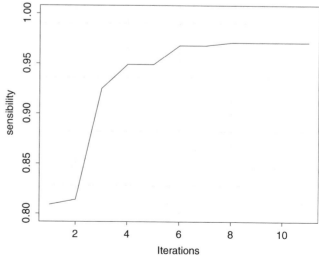

Fig. 104.6 Evolution of the sensibility and precision

Fig. 104.7 Time to generate the search strings

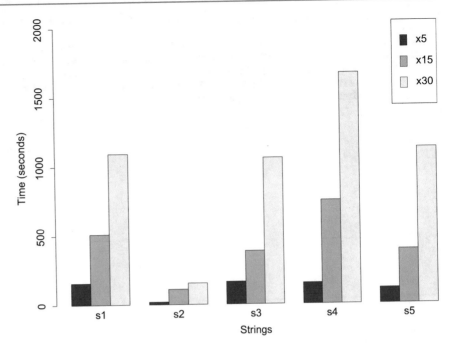

scientific database. This means that if some update or new technology is implemented on the database, our implementation must be adapted to work properly on it.

External validity: considering the generalization of our findings, we believe that the main threat is the fact that the unknown drawbacks on using SBSG would be better observed with subjects conducting their systematic reviews. Moreover, we believe that the fact the we already known the SBSG method could contribute to better results.

104.6 Conclusion and Future Work

In this paper we propose the SBSG approach that applies a search-based algorithm called Hill Climbing to automate thr key step in the conduction of SSs: search string generation and calibration. Using a list of terms, synonyms, keywords, and a set of control studies previously known, SBSG automatically provides a string in the appropriate format according to a pre-determined digital database.

We also carried out an experiment using five real SSs that have already been published. Our study consisted of using SBSG to find search strings with good accuracy in accordance to measures of sensibility, precision, and efficiency. We considered the results satisfactory, once the strings were automatically generated/calibrated and the primary studies identified contained few losses compared to the actual results of the five real SSs. The results shown that the precision and the sensibility our approach are 25,2% and 96,2%, respectively. We believe SBSG is a contribution

once it represents a first step to automatically support the generation of search strings for systematic reviews.

References

1. Basili, V., Caldiera, G., & Rombach, H. (1994). *The goal question metric paradigm* (1st ed.). Wiley.
2. Blum, C., & Roli, A. (2003). Metaheuristics in combinatorial optimization: Overview and conceptual comparison. *ACM Computing Surveys, 35*(3), 268–308.
3. Dieste, O., & Padua, A. (2007). Developing search strategies for detecting relevant experiments for systematic reviews. In *ESEM 2007*, Madrid (pp. 215–224).
4. Gay, G. (2010). A baseline method for search-based software engineering. In *PROMISE 2010*, PROMISE '10, Timisoara (pp. 2:1–2:11). New York, NY: ACM.
5. Hannay, J. E., Dybå, T., Arisholm, E., & Sjøberg, D. I. K. (2009). The effectiveness of pair programming: A meta-analysis. *Information and Software Technology, 51*(7), 1110–1122.
6. Harman, M., McMinn, P., de Souza, J. T., & Yoo, S. (2012). Search based software engineering: Techniques, taxonomy, tutorial. *Empirical software engineering and verification* (pp. 1–59). Berlin/Heidelberg: Springer.
7. Haynes, R. B., Wilczynski, K. A., McKibbon, C. J., & Sinclair, J. C. (1994). Developing optimal search strategies for detecting clinically sound studies in medline. *Journal of the American Medical Informatics Association, 1*, 447–458.
8. Kitchenham, B. (2011). Chapter three – What we can learn from systematic reviews. In *Making software what really works, and why we believe it* (Vol. 1, 1st ed.). Gravenstein Highway North, Sebastopol, CA: O'Reilly Media
9. Kitchenham, B., Mendes, E., & Travassos, G. H. (2007). A systematic review of cross vs. within-company cost estimation studies. *IEEE Transactions on Software Engineering, 33*(5), 361–329.
10. Kitchenham, B. A., Budgen, D., & Pearl Brereton, O. (2011). Using mapping studies as the basis for further research – A participant-

observer case study. *Information and Software Technology, 53*(6), 638–651.

11. Kitchenham, B. A., Dyba, T., & Jorgensen, M. (2004). Evidence-based software engineering. In *ICSE 2004*, ICSE '04, Edinburgh (pp. 273–281). Washington, DC: IEEE Computer Society.

12. Russell, S. J., & Norvig, P. (2003). *Artificial intelligence: A modern approach* (2nd ed.). Upper Saddle River, N.J: Pearson Education.

13. Straus, S., & Richardson, W. (2010). *Evidence-based medicine: How to practice and teach it* (4th ed.). Edinburgh: Churchill Livingstone.

14. Zhang, H., Babar, M. A., & Tell, P. (2011). Identifying relevant studies in software engineering. *Information and Software Technology, 53*(6), 625–637.

Visual Approach to Boundary Detection of Clusters Projected in 2D Space

Lenon Fachiano Silva and Danilo Medeiros Eler

Abstract

Data mining tasks are commonly employed to aid users in both dataset organization and classification. Clustering techniques are important tools among all data mining techniques because no class information is previously necessary – unlabeled datasets can be clustered only based on their attributes or distance matrices. In the last years, visualization techniques have been employed to show graphical representations from datasets. One class of techniques known as multidimensional projection can be employed to project datasets from a high dimensional space to a lower dimensional space (e.g., 2D space). As clustering techniques, multidimensional projection techniques present the datasets relationships based on distance, by grouping or separating cluster of instances in projected space. Usually, it is difficult to detect the boundary among distinct clusters presented in 2D space, once they are projected near or overlapped. Therefore, this work proposes a new visual approach for boundary detection of clusters projected in 2D space. For that, the attributes behavior are mapped to graphical representations based on lines or colors. Thus, images are computed for each instance and the graphical representation is used to discriminate the boundary of distinct clusters. In the experiments, the color mapping presented the best results because it is supported by the user's pre-attentive perception for boundary detection at a glance.

Keywords

Text mining • Document pre-processing • Visualization • Document similarity • Multidimensional projection

105.1 Introduction

Clustering algorithms can be used to divide a dataset in distinct groups (or clusters), enabling users to focus in specific groups or to organize a dataset to facilitate its exploration and understanding. A main issue in most clustering algorithms is how to specify the number of groups to cluster a dataset. Alternatively, some visualization techniques can be employed to present the dataset instances similarities in 2D space, thus, the groups are naturally created according to the feature space that describes the dataset. Usually, multidimensional projection techniques [13] are employed in this visualization process by reducing the dataset dimensionality to two dimensions.

Even though multidimensional projection techniques can represent structures from original multidimensional space in the projected space, groups can be overlapped or share a common boundary. The first case is a consequence of a dimensionality reduction and projection techniques. In the second case, when groups of distinct classes of instances

L.F. Silva (✉) • D.M. Eler
Faculdade de Ciências e Tecnologia, Departamento de Matemática e Computação, UNESP – Universidade Estadual Paulista, Presidente Prudente, SP, Brazil
e-mail: lenon_fachiano@hotmail.com; daniloeler@fct.unesp.br

© Springer International Publishing AG 2018
S. Latifi (ed.), *Information Technology – New Generations*, Advances in Intelligent Systems and Computing 558, DOI 10.1007/978-3-319-54978-1_105

share a common border, the main issue is to identify the boundary of different clusters. In the literature, distinct approaches can be employed to identify boundary among clusters. Usually, they are based on clustering algorithms employed in 2D space [11]; edge connecting based on similarities or triangulations performed in 2D space [10]; approaches for understanding projection and attributes dimensions based on most representative attributes [3]; hybrid visualizations to explain attribute behaviours [1, 5]; and other ways of cluster identification and space division [8, 9]. Even though they can highlight a border among clusters, the presented boundary do not reveal the real boundary from the wholly feature space, but a frontier imposed by 2D space – in the case of clustering algorithms employed in 2D space or space division – or based on some attributes – in the case of highlighting of the most representative attributes.

In this paper, we propose a new approach for boundary detection of clusters projected in 2D space. For that, we use multidimensional projection for cluster formation and instances relationship mapping in 2D space (i.e., projected space). We transform this similarity based exploration in a hybrid exploration by adding other class of visualization techniques based on attributes analysis. Thus, after mapping the instances similarities for 2D space, our hybrid approach generates an image for each instance to highlight the attributes behaviour, enabling the user to differentiate instances from distinct classes that share a common boundary. In this paper we propose two kinds of attributes mapping: one based on lines, in which polylines are computed from attribute values like in Parallel Coordinates technique [7]; and other based on color, in which each attribute value is mapped for a color.

The main contribution of this paper is to aid the dataset exploration based on a new hybrid exploration approach capable of highlight boundary of distinct clusters projected in 2D space. Additionally, this approach uses all dataset attributes to generate the visual representations and can also be used to understand the cluster formation based on the attributes analysis. Thus, the feature space is understood based on the similarities mapping from projection techniques and attributes mapping based on our approach.

This paper is organized as follows. Section 105.2 presents the theoretical foundation of visualization techniques employed in this work: multidimensional projection, parallel coordinates and color mapping. Section 105.3 presents the boundary detection approach proposed in this work. Section 105.4 presents the performed experiments with the proposed approach employed to boundary detection and feature space comprehension. Section 105.5 concludes the paper, summarizing the main achievements and projecting further works.

105.2 Background

Visualization is a research area whose main goal is to enable exploration, understanding and analysis of datasets through interactive visual explorations [2]. Visualization techniques generate graphical representations to facilitate the user comprehension and perception during a dataset exploration [4, 14]. This work is based on three visualization techniques: multidimensional projection, parallel coordinates and color mapping.

The multidimensional projection technique aims to map the instances similarities from the original space in a lower dimensional space. The dataset instances are mapped according to the similarities relationships from the original multidimensional space [11–13]. Thus, groups and neighbourhood from the original space are kept in the lower dimensional space – in this work, they are projected in 2D space. Scatter plots are used as graphical representations, but, instead of using two attributes from the dataset, the X and Y positions are computed based on all attributes that are projected to two dimensions.

The line mapping presented in this work is based on Parallel Coordinates technique [7] which main objective is to show the attributes behaviour and relationship. For that, Parallel Coordinates uses A_k parallel axes representing K attributes from a dataset – each axes assigned to an attribute. The instances are mapped as polylines that intersect each axes at the value of the attribute K_i for the axes A_i. Thus, the user can compare attributes behaviour by looking at the line shapes. In this work, instead of drawing all lines, we only draw one line for each instance.

The last technique employed in this work is a classical color mapping that aims to map data values to color according to a predefined color scale. Thus, each attribute value is mapped to a color, in this case, the user looks to the color information to understand the attribute behaviour.

Next section presents the proposed approach for boundary detection.

105.3 Boundary Detection Approach

This section describes the proposed boundary detection approach based on line and color mapping. This approach can be employed to detect clusters frontier in the projected space. For instance, in Fig. 105.1a is shown a group of instances in 2D space, but the label information is not available. So, detecting the boundary between distinct groups is to laborious, needing other techniques to improve this detection task. As shown in Fig. 105.1b, only with label information it is possible to perfectly detect the boundary between these two clusters.

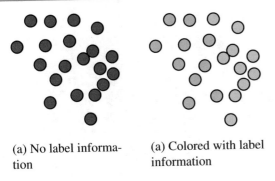

(a) No label informa-tion

(a) Colored with label information

Fig. 105.1 Example of clusters in projected space: (**a**) no label information provided and (**b**) instances are colored based on label information

Our proposed approach aid in boundary detection based on color and line mappings from attributes information of each dataset instance. An example of mapping from three distinct feature vectors is shown in Fig. 105.2. The color mapping is based on a previous defined color scale by means which each attribute value is mapped to a color. Thus, an image is generated for each instance and divided in columns according to the number of attributes. Each column is colored with the color mapping for the respective attribute value. Similarly, for the line mapping, an image is generated with a polyline to represent the attributes behaviour for the respective instance. The polyline is generated based on Parallel Coordinates [7] technique, in which the space is divided in parallel axes – one axis for each attribute – and a polyline intersects these axes in the point correspondent to the attribute value.

Next section show examples of this approach in boundary detection of a dataset.

105.4 Experiments

This section presents experiments to show how the proposed approach can be employed in boundary detection of datasets projected in 2D space. The first experiment used the motivating example presented in Fig. 105.1. For that, the attributes information was used to generate images for each instance based on color and line mappings. Figure 105.3 presents the results of these mappings. The boundary detection based on lines can be performed by looking at the changing in specific attributes behaviour, for instance, we can note in each image that the values of some attributes of the right cluster is lower than the left cluster. On the other hand, the boundary detection based on color is more precise and evident. The left cluster present blue color mapping for some attributes and right cluster present red color mapping for the same attributes.

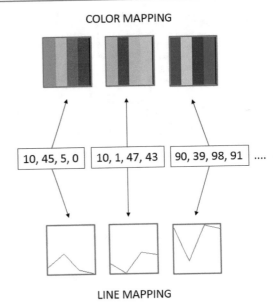

COLOR MAPPING

LINE MAPPING

Fig. 105.2 Boundary detection approach

Other experiment was performed with a dataset composed by three classes – the well known Iris dataset.[1] Figure 105.4a shows the 2D representation of this dataset, in which is not possible to detect the clusters boundary. After applying our approach, color and line mapping were computed. Firstly, by analysing the clusters boundary based on line mapping, as show in Fig. 105.4b, we can detect a left cluster whose boundary was indicated by a blue line. However, the detection of other two clusters transition at right is to laborious when using the line mapping. On the other hand, by using the color mapping, as shown in Fig. 105.4c, the left cluster is evident and the blue line indicates its boundary. The color mapping facilitated the detection of other clusters boundary as indicated by the red line. To validate our approach, in Fig. 105.4d is shown the color information based on instances label (class) as a ground truth.

These experiments show that our approach can be employed to aid in boundary detection of clusters projected in 2D space. Additionally, the perception based on color information enables users to rapidly and accurately detect a boundary among groups of elements.

The proposed approach can also be employed to show bad feature spaces, that is, feature space that cannot discriminate instances from distinct classes. To show this application, we consider an image dataset with MRI images from six distinct classes. As shown in Fig. 105.5a, we can note some groups based on the projected clusters and detect the boundary of some clusters based on color mapping.

---

[1] Iris dataset is available at http://archive.ics.uci.edu/ml/datasets/Iris

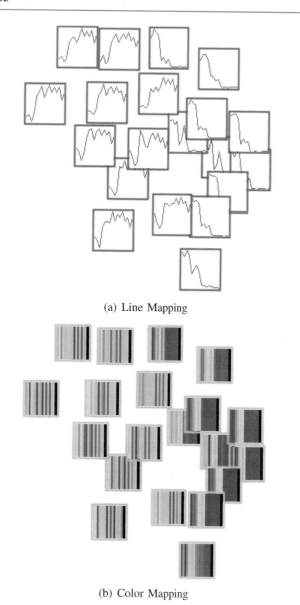

(a) Line Mapping

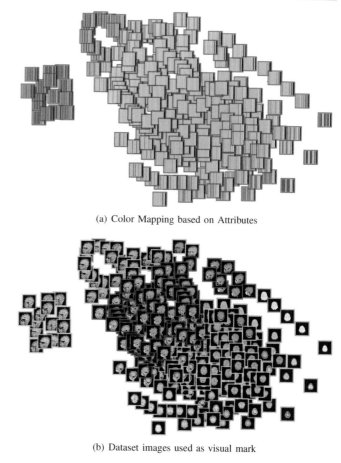

(a) Color Mapping based on Attributes

(b) Dataset images used as visual mark

Fig. 105.4 Boundary detection of a dataset projected in 2D space (**a**). Our approach was employed to show two mappings based on shape (**b**) and color (**c**). In (**d**) is shown the real class information from the dataset

(b) Color Mapping

Fig. 105.3 Boundary detection using our approach by mapping the attributes values for line (**a**) and color (**b**)

However, the boundary detection is too laborious at the centre of the big cluster, where the green color is similar to all attributes of several images. Using the dataset images as visual marks [6] we can perceive that the features (attributes) employed to this dataset is not able to discriminate instances of different classes. There are four classes of head that are too similar. Therefore, these images are considered as a big cluster as shown based on the color mapping. Thus, it is necessary to use other kind of features to separate this classes of images or considering it all as misclassified from the expert.

Next section presents some conclusions and further works.

105.5 Conclusions and Future Works

Boundary detection among distinct cluster is an important step in dataset organizations. In this work, we presented a visual approach based on multidimensional projection technique to reduce the dataset dimensionality and project it in 2D space. Thus, clusters are presented based on instances similarities from the original multidimensional space. To identify the cluster boundaries, we used line and color mappings to show the attributes relationship and highlight the cluster boundaries.

In the experiments, we used the line mapping based on Parallel Coordinates to show the attributes behaviour, by means which the user could identify the transition among clusters by looking at the shape of each line. However, the shape comparison can be too laborious. Therefore, color mapping was also employed as visual characteristic and the boundaries could be rapidly perceived.

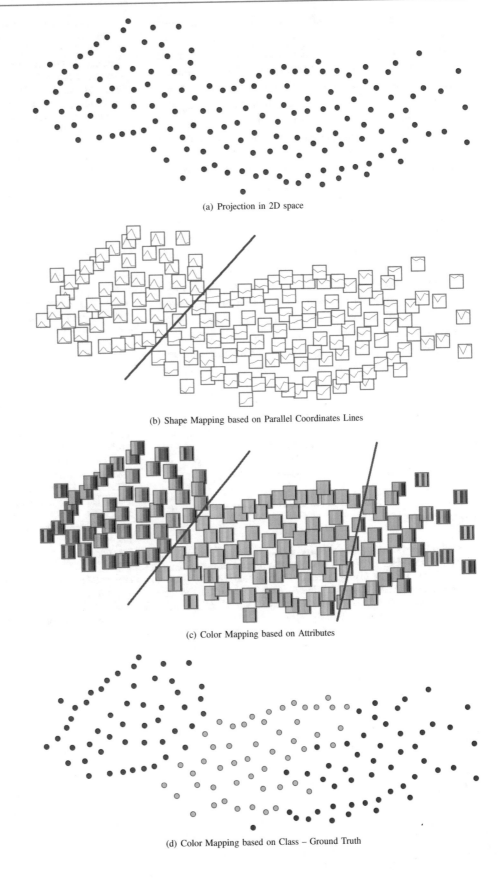

(a) Projection in 2D space

(b) Shape Mapping based on Parallel Coordinates Lines

(c) Color Mapping based on Attributes

(d) Color Mapping based on Class – Ground Truth

Color mapping is more effective in boundary detection tasks because the hue is a pre-attentive visual property. As shown in the results, boundaries could be detected at a glance with color mapping. While the line (shape) requires an attentive user perception, that is, the user needs to focus and search for the boundary detection by comparing line shapes. In further works, other pre-attentive visual properties could be employed as well as combination of shape and color in the same image.

Besides the boundary detection, the last experiment presented in previous section shows that our approach can also be employed to explain the cluster cohesion and separation. Thus, experts can analyse the attributes behaviour in cluster formation and decide the quality of the feature spaces as well as identify some wrong in instances labelling task.

The main limitation of this approach is the size of generated images, because when the dataset have more attributes than the image size in pixels, we cannot generate an image with all attributes – for instance, in text mining, is common a feature space with more than 500 attributes. Therefore, in further works, approaches to feature selection can be employed to show attributes behaviour of datasets with several attributes.

Acknowledgements The authors acknowledge the financial support of the Brazilian financial agency São Paulo Research Foundation (FAPESP) – grant #2013/03452-0, National Counsel of Technological and Scientific Development (CNPq), and Coordenação de Aperfeiçoamento de Pessoal de Nível Superior (CAPES).

References

1. Bodo, L., de Oliveira, H. C., Breve, F. A., & Eler, D. M. (2016). Performance indicators analysis in software processes using semi-supervised learning with information visualization. In *13th International Conference on Information Technology: New Generations (ITNG 2016)* (Advances in intelligent systems and computing, pp. 555–568). Las Vegas, NV: Springer.
2. Card, S., Mackinlay, J., & Shneiderman, B. (1999). *Readings in information visualization: Using vision to think* (Interactive technologies Series). San Francisco, CA: Morgan Kaufmann Publishers.
3. da Silva, R. O., Rauber, P. E., Martins, R. M., Minghim, R., & Telea, A. C. (2015). Attribute-based visual explanation of multidimensional projections. In *EuroVis workshop on visual analytics (EuroVA)*, Cagliari. The Eurographics Association.
4. de Oliveira, M., & Levkowitz, H. (2003). From visual data exploration to visual data mining: a survey. *IEEE Transactions on Visualization and Computer Graphics, 9*(3), 378–394.
5. de Oliveira, R. A. P., Silva, L. F., & Eler, D. M. (2015). Hybrid visualization: A new approach to display instances relationship and attributes behaviour in a single view. In *19th International Conference on Information Visualisation (iV)*, Barcelona (pp. 277–282).
6. Eler, D. M., Nakazaki, M. Y., Paulovich, F. V., Santos, D. P., Andery, G. F., Oliveira, M. C. F., Batista Neto, J., & Minghim, R. (2009). Visual analysis of image collections. *The Visual Computer, 25*(10), 923–937.
7. Inselberg, A. (1985). The plane with parallel coordinates. *The Visual Computer, 1*(2), pp. 69–91.
8. Kandogan, E. (2012). Just-in-time annotation of clusters, outliers, and trends in point-based data visualizations. In *Proceedings of the 2012 I.E. Conference on Visual Analytics Science and Technology VAST'12* (pp. 73–82). Washington, DC: IEEE Computer Society.
9. Nocaj, A., & Brandes, U. (2012). Organizing search results with a reference map. *IEEE Transactions on Visualization and Computer Graphics, 18*(12), 2546–2555.
10. Paulovich, F. V., Nonato, L. G., & Minghim, R. (2006). Visual mapping of text collections through a fast high precision projection technique. In *Proceedings of the Conference on Information Visualization IV'06* (pp. 282–290). Washington, DC: IEEE Computer Society.
11. Paulovich, F. V., Oliveira, M. C. F., & Minghim, R. (2007). The projection explorer: A flexible tool for projection-based multidimensional visualization. In *XX Brazilian Symposium on Computer Graphics and Image Processing, SIBGRAPI 2007* (pp. 27–36). IEEE, Belo Horizonte.
12. Paulovich, F. V., Nonato, L. G., Minghim, R., & Levkowitz, H. (2008). Least square projection: A fast high-precision multidimensional projection technique and its application to document mapping. *IEEE Transactions on Visualization and Computer Graphics, 14*(3), 564–575.
13. Tejada, E., Minghim, R., & Nonato, L. G. (2003). On improved projection techniques to support visual exploration of multidimensional data sets. *Information Visualization, 2*(4), 218–231.
14. Ware, C., (2012). *Information visualization: Perception for design* (Interactive technologies). Amsterdam: Elsevier Science.

Opposition-Based Particle Swarm Optimization Algorithm with Self-adaptive Strategy

106

Xuehan Qin and Yi Xu

Abstract

In view of the problems that the standard Particle Swarm Optimization(PSO) algorithm tend to fall into premature convergence and slow convergence velocity and low precision in the late evolutionary, an Opposition-based Particle Swarm Optimization Algorithm with Self-adaptive Strategy (SAOPSO) is proposed. The algorithm uses an adaptive inertial weighting strategy to balance the ability of global search and local exploration; Meanwhile, a strategy of opposition-based learning is adopted on the elite particle population, which will improve learning ability of particle, expand the search space and enhance the global search capability; In order to avoid falling into the local optimum, which may cause search stagnation, Cauchy mutation strategy with a adaptive probability value is presented to disturb the current global optimal particle. SAOPSO algorithm is compared with other improved PSO on 5 classic benchmark functions, and the experimental results show that ASOPSO algorithm is improved in convergence speed and accuracy of solution.

Keywords

Particle Swarm Optimization algorithm • Adaptive inertia weight • Opposite learning • Cauchy mutation

106.1 Introduction

This particle swarm optimization is an kind of swarm intelligence optimization algorithm based on population search used to find the optimal solution motivated from the behavior of bird flocking. It was introduced in 1995 by Kennedy and Ebenhart [1]. Currently, PSO has become one of the most preferred choices for optimization problems due to its relative ease of operation and capability to quickly arrive at an optimal/near-optimal solution on different benchmark and engineering problems [2, 3]. And it has attracted many domestic and foreign researchers attention and research, and been successfully applied to multi-objective optimization [4], neural network training, pattern recognition, data mining, robot path optimization [5] etc. But the PSO algorithm also has problems such as easy to fall into local optimum, slow convergence speed. In view of these problems, many researchers have made a lot of improvements and achieved a lot of results. The [6] presents an elite reverse learning strategy. In the iterative process, according to the opposition-based learning strategy, the optimal position of the elite individual generates the opposite solution, which can expand the search space, improve population diversity and improve the global search ability. In [7], Particle Swarm Optimization Algorithm with Gaussian Opposition-based Learning is proposed by introducing the Gaussian learning mechanism, random reverse learning and historical best averages.

In this paper, by introducing the nonlinear adaptive inertia weight to balance the global search and local exploitative ability of the algorithm. At the same time, the

X. Qin (✉) • Y. Xu
School of Computer Science and Technology, Wuhan University of Technology, Wuhan, China
e-mail: 1404709790@qq.com; 365220180@qq.com

© Springer International Publishing AG 2018
S. Latifi (ed.), *Information Technology – New Generations*, Advances in Intelligent Systems and Computing 558, DOI 10.1007/978-3-319-54978-1_106

opposition-based learning strategy is adopted on elite particle;. In order to increase the diversity of population, and avoid falling into local optimum causing search stagnation, the current optimal particle adopts Cauchy mutation which will guide the particle to the better solution direction of evolution and improve the global search ability of algorithm.

106.2 A Brief Overview of the PSO Algorithm

The basic idea of PSO is to find the optimal solution through the collaboration and information sharing among the individuals in the group, which is derived from bird's searching food. In PSO algorithm, some particles were initiated, then iterated them until an optimization solution was obtained.

Assume that each swarm has a population of N, having random positions and velocities. The particles are uniformly initialized along the D-dimensional search space. Each particle i consists of a D-dimensional position vector $x_i = (x_{i1}, x_{i2},\ldots,x_{iD})$ and a velocity vector $v_i = (v_{i1}, v_{i2},\ldots,v_{iD})$. The two equations used in the algorithm for updating the velocity and position of the particles are defined as follows.

$$v_{id}^{k+1} = w \cdot v_{id}^k + c_1 r_1 \cdot \left(p_{id}^k - x_{id}^k\right) + c_2 r_2 \cdot \left(p_{gd}^k - x_{id}^k\right)$$

$$(106.1)$$

$$x_{id}^{k+1} = x_{id}^k + v_{id}^{k+1} \tag{106.2}$$

where $i = 1, 2,\ldots,N; d = 1, 2,\ldots,D; x_{id}$ and v_{id} are the position and velocity respectively of the ith particle in the dth dimension. P_{id} is the personal best for particle i in the dth dimension and P_{gd} is the global best in dth dimension. k represents the current iteration. c_1 and c_2 represents self-cognitive parameters and social cognitive parameters. r_1 and r_2 are the random numbers distributed uniformly within the range [0, 1]. w is the inertia weight.

106.3 The SAOPSO Algorithm

106.3.1 Adaptive Inertia Weight

Inertia weight w plays a vital role in controlling the process of exploration and exploitation by maintaining a balance. Using a fixed inertia weight, the algorithm has a faster convergence speed, but it may make the algorithm easily fall into the local optimum. Therefore, Shi Y and Eberhart R C proposed a linear decreasing weight strategy [8], which can improve the performance of the algorithm. However, the searching process of PSO algorithm is very complex and nonlinear, and the variation of inertia weight linear decreasing strategy is too single.

For the particle swarm optimization algorithm, it is expected to have a strong global search ability in the early stage of the algorithm and have a better local search ability in the latter part of the algorithm to improve the convergence precision.

This paper adopts self-regulating inertia weight strategy and it is defined as follows.

$$w_i = \begin{cases} w_{max} & f_i > f_{avg1} \\ w_{min} + \dfrac{(w_{max} - w_{min})\left(f_i - f_{avg2}\right)}{f_{avg} - f_{avg2}} & f_{avg2} \leq f_i \leq f_{avg1} \\ w_{min} & f_i < f_{avg2} \end{cases}$$

$$(106.3)$$

where w_{max}, w_{min} and w_i are the maximum inertia weight, the minimum inertia weight and the current particle inertia weight respectively. f_{avg} is the the mean value of the fitness value of the current population. f_{avg1} is the mean of the fitness value of the particles whose value are greater than f_{avg} in the current population. f_{avg2} is the mean of the fitness value of the particles whose value are less than f_{avg} in the current population.

106.3.2 Opposite-Based Learning

In order to improve the diversity of particles and search for a better optimal solution, this paper uses the elite opposite learning strategy. The concept of Opposition-based Learning (OBL) [9] is first proposed by Tizhoosh. It can improve the convergence speed and precision of the algorithm by applying the opposite learning strategy to PSO.

If x_i ($x_{i1}, x_{i2},\ldots,x_{iD}$) is a point in D-dimensional space, $x_{id} \in (a_d, b_d), d \in [1, 2,\ldots, D]$, the opposite point [10] $ox_{id}(ox_{i1}, ox_{i2},\ldots,ox_{iD})$ of x_i can be obtained according to (106.4).

$$ox_{id} = a_d + b_d - x_{id} \tag{106.4}$$

If the position $p_{id}(t)$ of the tth iteration of the particle x_i is the optimal position, $p_{id}(t)$ is the elite solution of the particle x_i. The solution of ox_i is elite opposite solution of x_i, and it can be obtained according to (106.5).

$$\begin{cases} op_{id}(t) = r_1 a_d(t) + r_2 b_d(t) - p_{id}(t) \\ a_d(t) = \min(x_{id}(t)) \\ b_d(t) = \max(x_{id}(t)) \end{cases} \tag{106.5}$$

where r_1 and r_2 are the random numbers distributed uniformly within the range [0, 1]. $[a_d, b_d]$ is the dynamic boundary of the dth dimension search space. If the opposite solution is out of the range, that is, $op_{id} < a_d$ or $op_{id} > b_d$, ad and b_d, the opposite solution need to be reset. The specific updating formula is shown as follows.

$$op_{id} = a_d + r(b_d - a_d) \qquad (106.6)$$

where r_1 and r_2 are the random numbers distributed uniformly within the range $[0, 1]$.

106.3.3 Dynamic Adjustment of Cauchy Mutation Probability

With the iteration of the PSO algorithm, more and more particles will tend to be consistent, and the particle velocity is also getting smaller and smaller, which may make the algorithm easily fall into local optimum. Therefore, increasing the search space of the best particle p_{id} will effectively reduce the possibility of falling into local optimum and premature convergence.

In this paper, we use the Cauchy mutation strategy to the global optimal position of the particle. The variation probability of particles is dynamically adjusted by the diversity of particles. The diversity of particles can be expressed by the degree of aggregation [11]. The calculation formula of aggregation σ is shown as follows.

$$\sigma = \frac{1}{N} \sum_{i=1}^{N} \left| \frac{f_i - f_{avg}}{f_{max} - f_{min}} \right| \qquad (106.7)$$

where $f_{avg} = \frac{1}{n} \sum_{i=1}^{n} f_i$ represents the average fitness of the current particle swarm.

According to (106.7) can be known: when the fitness of individual particles is close to the average fitness of the current population, the aggregation degree σ is smaller, and the diversity is worse. On the contrary, the aggregation degree σ is bigger, and the diversity is better. In order to ensure the search performance of the algorithm, we hope when the population diversity is poor, the probability of variation is larger, and when the population's diversity is better, the probability of variation is smaller. The mutation probability of the tth *iteration* of the particles is p_m^{t}, and the calculation formula is as follows.

$$p_m^t = \sigma e^{-a\left(1 + t/T\right)} \qquad (106.8)$$

where T is the total number of iterations, and σ is a constant that regulates the speed of mutation probability with the range $[2, 4]$.

The formula of Cauchy mutation is shown as follows.

$$Cauchy = \tan\left(\pi(rand - 0.5)\right) \qquad (106.9)$$

$$p_{gd} = p_{gd}(1 + 0.5cauchy) \qquad (106.10)$$

where, p_{gd} is the global optimal position, and Cauchy is the random number of Gauss distribution and the random number of Cauchy distribution.

During the search process, when a particle flies out of the search space, it is necessary to use the new method of re_initialization. The specific initialization formula is as follows.

$$x_{id} = \begin{cases} x_{id}, & x_{id} \in [x_{min}, x_{max}] \\ x_{min} + r(x_{max} - x_{min}), & \text{others} \end{cases}$$
$$(106.11)$$

106.3.4 The SAOPSO Algorithm

The specific implementation steps are as follows.

Initialization:

for each particle i **do**

Randomly initialize position of each particle x_i in the search range (x_{min}, x_{max}).

Randomly initialize velocity of each particle v_i.

end for

The SAOPSO Loop:

while (not success and $t < = T$)**do**

Calculate the fitness values for each particle.

Calculate the inertia weight w using (106.3).

Find the personal best position of each particle p_{id}.

Find the particle with the best fitness value in the current population.

Assign this position to global best p_{gd}.

for the best particle p_{id} **do**

Calculate the opposite point ox_{id} and elite opposite solution of each particle op_{id} using (106.4), (106.5) and (106.6).

if op_{id} is out of the range

reset the op_{id} using (106.6).

end if

if ($p_{id} > = op_{id}$)

Assign this opposite point to personal best position p_{id}.

end if

end for

Generate the uniform random number r within the range $[0, 1]$.

Calculate the aggregation σ using (106.7).

Calculate the mutation probability using (106.8).

if $(r < p_m^t)$

Adopt the Cauchy mutation strategy to the global optimal position p_{gd} by using (106.9) and (106.10).

end if

Update the velocity of each particle using (106.1).

Update the position of each particle using (106.2).

end while

end

106.4 Simulation Experiment and Result Analysis

106.4.1 Benchmark Function

In this paper, the simulation experiment is carried out under the Matlab environment. In order to verify the validity of the algorithm, 5 typical benchmark functions are used to test this algorithm. The global optimal values of all test functions are 0, and the dimension is 30-dimension. f_1 and f_2 are single peak functions, and f_3, f_4 and f_5 are the multi peak test functions.

(1) Sphere

$$f_1(x) = \sum_{i=1}^{d} x_i^2, x_i \in [-100, 100]$$

(2) Rosenbrock

$$f_2(x) = \sum_{i=1}^{n-1} \left(100\left(x_{i+1} - x_i^2\right)^2 + (x_i - 1)^2\right), x_i \in [-30, 30]$$

(3) Rastrigin

$$f_3(x) = \sum_{i=1}^{n} \left(\left(x_i^2 - 10\cos(2\pi x_i)\right) + 10\right), x_i \in [-5.12, 5.12]$$

(4) Ackley

$$f_4(x) = -20\exp\left(-0.2\sqrt{\frac{1}{n}\sum_{i=1}^{n} x_i^2}\right) - \exp\left(\frac{1}{n}\sum_{i=1}^{n}\cos(2\pi x_i)\right) + 20 + e, x_i \in [-30, 30]$$

(5) Griewank

$$f_5(x) = \sum_{i=1}^{n}\frac{x_i^2}{4000} - \prod_{i=1}^{n}\cos\left(\frac{x_i}{\sqrt{i}}\right) + 1, x_i \in [-600, 600]$$

106.4.2 Parameter Settings

In order to test the performance of the algorithm, the experimental result of SAOPSO algorithm is compared with the results of the basic PSO algorithm, OPSO [12], GOL-PSO [7] and EOPSO [6].The experimental parameters are set as follows: The algorithm parameters of the last four algorithms are consistent with the original document. Population size $N = 40$. The maximum velocity of the particle is half of the search space. The maximum number of iterations $T = 5000$; as for the SAOPSO, the acceleration factor $c_1 = c_2 = 2$, the maximum inertia weight $w_{max} = 0.9$, the minimum inertia weight $w_{min} = 0.4$.

106.4.3 Experimental Results and Performance Comparison

To eliminate the influence of the randomness of the algorithm, all the algorithms are run independently on each benchmark function for 30 times, and the average value is calculated as the final test result. Table 106.1 shows the optimization results of the 5 algorithms on 5 test functions and presents the median, mean and standard deviation of the SAOPSO along with the other algorithms for both

Table 106.1 Experimental results

Func	Test ponit	Algorithm				
		PSO	OPSO	GOL-PSO	EOPSO	SAOPSO
f_1	Mean	1.67E-03	4.49E-36	3.13E-113	0	0
	StD.	1.82E-02	3.25E-35	7.27E-112	0	0
f_2	Mean	2.93E + 01	7.18E + 00	1.81E + 00	3.64E-24	2.18E-36
	StD.	1.14E + 01	3.78E + 01	2.20E + 01	7.38E-23	2.57E-35
f_3	Mean	2.06E + 01	2.21E + 01	2.54E-16	1.09E-12	0
	StD.	1.68E + 01	4.75E + 01	4.26E-16	2.57E-14	0
f_4	Mean	4.45E-01	1.85E-01	7.25E-15	3.79E-15	8.27E-15
	StD.	1.61E-01	5.21E-02	4.31E-16	8.21E-16	5.84E-16
f_5	Mean	1.64E-03	3.81E-01	2.37E-04	1.12E-02	0
	StD.	2.68E-03	6.21E-02	1.27E-03	4.12E-02	0

30-dimensional cases. From the test results in Table 106.1, it can be seen that the SAOPSO algorithm has better convergence and stability compared with the other four algorithms, whether single-peak function or multi-peak function, and the optimal solution of f_1, f_3 and f_5 can be found by using SAOPSO. Compared with PSO and OPSO, SAOPSO achieves a better solution on the test function. In addition to f_4, GOL-PSO is slightly better than SAOPSO, and SAOPSO

achieves better results in the other four test functions; SAOPSO and EOPSO achieve comparable results at f_1, f_4, but SAOPSO is superior to EOPSO on other functions.

In order to analyze the convergence performance of SAOPSO algorithm in the evolutionary process more clearly, Figs. 106.1, 106.2, 106.3, 106.4 and 106.5 shows the adaptive evolution curve of the fitness of each algorithm over the five test functions in the iterative process of the

Fig. 106.1 Sphere

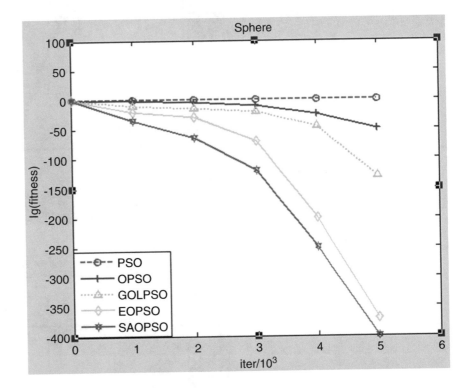

Fig. 106.2 Rosenbrock

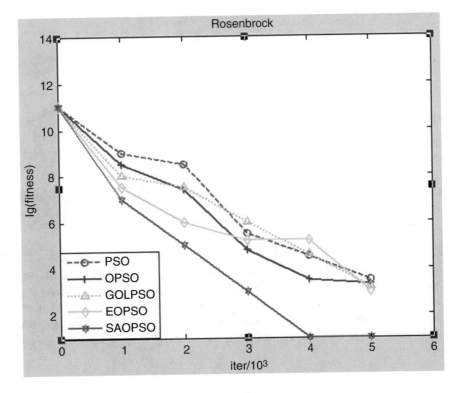

Fig. 106.3 Rastrigin

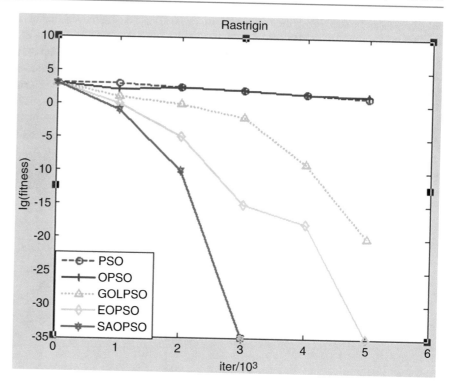

Fig. 106.4 Ackley

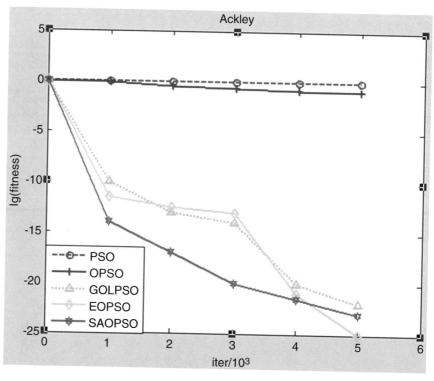

algorithm. In order to facilitate the comparative observation, the vertical axis is the average fitness test function, the average fitness of the test function is taken to the end of the 10 logarithm, abscissa is the number of iterations. As can be seen from Figs. 106.1, 106.2, 106.3, 106.4 and 106.5: Compared with GOL-PSO and EOPSO, the convergence performance of SAOPSO on function f_4 is not much improved, but the convergence performance of the other four functions is obviously improved. But the convergence performance of the other four functions is improved obviously.Overall, SAOPSO has more advantages in convergence accuracy and convergence speed than other algorithm.

Fig. 106.5 Griewank

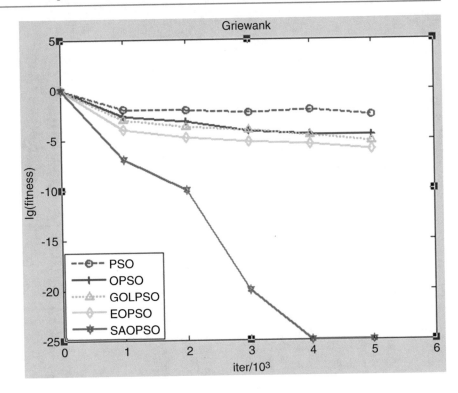

106.5 Conclusion

In this paper, an Opposition-based Particle Swarm Optimization Algorithm with Self-adaptive Strategy is proposed. Adaptive inertia weight strategy automatically adjusts its inertia weight according to the fitness of each particle, which can effectively balance the global search and local detection ability of the algorithm. The elite opposite-based learning strategy is used to generate the opposite solution of the elite particle, which increases the probability of searching the optimal solution and improves the convergence speed. According to the aggregation degree of the particles, the mutation probability is dynamically adjusted, and the Cauchy mutation is performed on the global optimal particles, which can enhance the diversity of the particles and avoid algorithm falling into local optimum, and this guarantees the algorithm converges smoothly and quickly to the global optimal value. The comparison and analysis of experimental results of five typical benchmark functions show that the improved algorithm SAOPSO proposed in this paper improves the convergence speed and convergence precision of the algorithm and shows good stability.

References

1. Kennedy, J., & Eberhart, R. C. (1995). Partical swarm optimization [C], In *Proceedings of the 1995 I.E. international conference on natural networks* (pp. 1942–1948). Piscataway: IEEE.

2. Tanweer, M. R., Suresh, S., & Sundararajan, N. (2015). Self regulating particle swarm optimization algorithm [J]. *Information Sciences, 294*(10), 182–202.

3. Chatterjee, A., & Siarry, P. (2006). Nonlinear inertia weight variation for dynamic adaptation in particle swarm optimization [J]. *Computers & Operations Research, 33*(3), 859–871.

4. Yeh, W. C. (2013). New parameter-free simplified swarm optimization for artificial neural network training and its application in the prediction of time series.[J]. *IEEE Transactions on Neural Networks & Learning Systems, 24*(4), 661–665.

5. Supakar, N., & Senthil, A. (2013). PSO obstacle avoidance algorithm for robot in unknown environment[C]. *International conference on communication and computer vision.* IEEE.

6. Zhou, X., Wu, Z., et al. (2013). [J]. Elite opposition based particle swarm optimization. *Acta Electronica Sinica, 8*, 1647–1652.

7. Zhan, D., Lu, H., et al. (2015). Particle swarm optimization algorithm with Gaussian opposition-based learning [J]. *Journal of Chinese Computer Systems, 36*(5), 1064–1068.

8. Shi, Y., & Eberhart, R. C. (1998). Parameter selection in particle swarm optimization [C]. *International conference on evolutionary programming Vii.* San Diego: Springer (pp. 591–600).

9. Tizhoosh, H. R. (2005). Opposition-based learning: a new scheme for machine intelligence [C]. *International conference on computational intelligence for modelling, control & automation, & international conference on intelligent agents, web technologies & internet commerce.* IEEE Computer Society (pp. 695–701).

10. Liu, H., Wu, Z., Wang, H., et al. (2014). Improved differential evolution with adaptive opposition strategy [C], *IEEE congress on evolutionary computation* (pp. 1776–1783).

11. Huang, S., Tian, N., & Ji, Z. (2016). Study of modified particle swarm optimization algorithm based on adaptive mutation probability [J]. *Journal of System Simulation, 28*(4), 874–879.

12. Wang, H., Li, H., Liu, Y., et al. (2007). Opposition-based particle swarm algorithm with cauchy mutation [C]. *IEEE congress on evolutionary computation, Cec 2007*, 25–28 September 2007, Singapore (pp. 4750–4756).

Raghavendar Cheruku, Doina Bein, Wolfgang Bein, and Vlad Popa

Abstract

Today's recruitment applications are designed not only reduce paperwork but can make a significant contribution to a company's marketing and sales activity. Recruitment websites and software make possible for managers to access information crucial to managing their staff, which they can use for promotion decisions, payroll considerations and hiring. We present a web-based solutions to the recruitment process for small companies in which job seekers (users) can register to a created job position, manage their information, and be informed of the various steps of the hiring process. The project will address three categories of people: a single admin, a number of staff personnel, and a very large number of (about 1000) applicants. This software is very useful to small companies. It reduces the paperwork, takes less time, is very transparent to the users, and is available through Internet.

Keywords

Recruitment • Web-based application • Marketing

107.1 Introduction

Today's recruitment applications are designed to do a whole lot more than just reduce paperwork. They can make a significant contribution to a company's marketing and sales activity. Recruitment websites and software make possible for managers to access information that is crucial to managing their staff, which they can use for promotion decisions, payroll considerations and succession planning. The system makes it friendly to distribute, share and manage the examination entities with higher efficiency and easiness. Many small companies or business are still following the old style process for recruiting people like advertising the vacancy in local and national newspapers, magazines, or posting fliers. Even the selection process itself takes much time even if the company's requirement is urgent. The process starts from application selection, calling applicants for interview providing the results. This process will take nearly a 30–45 days. So because of this small companies it takes more time to recruit applicants. Presently recruitment for small companies is done manually. If a company needs to hire, the company will announce through ads in newspapers, online, or by email. Then interested people will send application to the company. After shortlisting, the company calls some applicants for interviews. It takes about 1–2 weeks to shortlist the application. So that is a long time for a company interested in hiring right away.

We present a web based application in which job seekers (users) can register to an existing job posting (see

R. Cheruku • D. Bein (✉)
Department of Computer Science, California State University, Fullerton, USA
e-mail: raghu@csu.fullerton.edu; dbein@fullerton.edu

W. Bein
Department of Computer Science, University of Nevada, Las Vegas, USA
e-mail: wolfgang.bein@unlv.edu

V. Popa
Liceul Tehnologic Petru Poni, Iasi, Romania
e-mail: pvlad2001@gmail.com

© Springer International Publishing AG 2018
S. Latifi (ed.), *Information Technology – New Generations*, Advances in Intelligent Systems and Computing 558, DOI 10.1007/978-3-319-54978-1_107

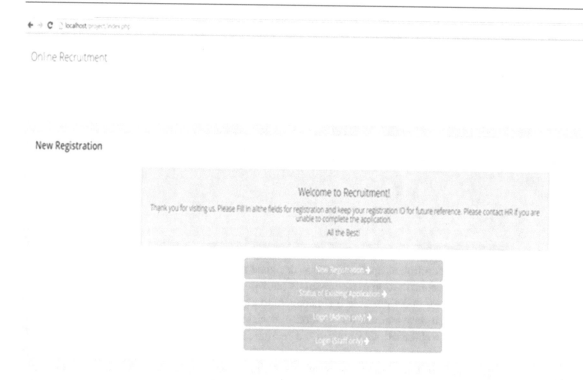

Fig. 107.1 Homepage of recruitment drive

Fig. 107.1). After applying for a job, the job seekers can review all the results and present status, check their success or failure in the hiring process' rounds. The scope of this project is to serve as a common meeting ground for jobseekers and employers, where the job applicants are informed at every step of their status and recruiters find the right applicant to fulfill their needs. This software is very useful to small companies It reduces the paperwork, takes less time, is very transparent to the users, and is available through Internet. So job applicants can apply for a position from anywhere and obtain their applications' status at every stage using some unique ID provided by the software.

The paper is organized as follows. In Sect. 107.2 we present the system analysis and design, followed by the programming languages and software tools used for developing the application in Sect. 107.3. A very general description of the software modules is given in Sect. 107.4. We conclude in Sect. 107.5.

107.2 System Analysis and Design

By definition, system analysis is the detailed study of the various operations performed by the system and their relationships within and outside the system. System analysis is concerned with becoming aware of the problem, identifying the relevant and most decisional variables, analyzing and synthesizing the various factors and determining an optional or at least a satisfactory solution. During this a problem is identified, alternate system solutions are studied and recommendations are made about committing the resources used to the system.

System design provides the procedural details necessary for the logical and physical stages of development. In designing a new system, the system analyst must have a clear understanding of the objectives, which the design is aiming to fulfill. The first step is to determine how the output is to be produced and in what format. Second, input data and master files have to be designed to meet the requirements of the proposed output. The operational phases are handled through program construction and testing.

The system design can be defined as the process of applying various techniques and principles for the purpose of defining a device, a process or a system in sufficient detail to permit its physical realization. Thus system design is a solution to "how to" approach to the creation of a new system. This important phase provides the understanding and the procedural details necessary for implementing the system recommended in the feasibility study. The design step provides a data design, architectural design, and a procedural design.

107.2.1 Output Design

In the output design, the emphasis is on producing a hard copy of the information requested or the screen. Most users now access their reports from either a hard copy or screen display. Computer's output is the most important and direct source of information to the user, efficient, logical, output design should improve the systems relations with the user and help in decision-making. As the outputs are the most important source of information to the user, better design should improve the systems' relations and also should help in decision-making. The output device's capability, print quality, response time requirements etc., should also be considered, form design elaborates the way the output is presented and layout available for capturing information. It is very helpful to produce the clear, accurate and speedy information for end users.

107.2.2 Input Design

In the input design, user-originated inputs are converted into a computer-based system format. It also includes determining the record media, method of input, speed of capture and entry on to the screen. Online data entry accepts commands and data through a keyboard. The major approach to input design is the menu and the prompt design. In each alternative, the user's options are predefined. The data flow diagram indicates logical data flow, data stores, source and destination. Input data are collected and organized into a group of similar data once identified input media are selected for processing.

In this software, the Graphical User Interface (GUI) is an important factor in developing efficient and user friendly software. For inputting user data, attractive forms are designed. User can also select the desired options from the menu, which provides many options. The goal of input design is to make entry as easy, logical and free from errors. Also the input format is designed in such a way that accidental errors are avoided. The user has to input only just the minimum data required, which also helps in avoiding the errors that the users may make. Accurate designing of the input format is very important in developing efficient software.

107.2.3 Logical Design

Logical data design is about the logically implied data. Logical data designing should give a clear understanding and idea about the related data used to construct a form.

A Data Flow Diagram (DFD) is a diagram that describes the flow of data and the processes that change or transform data throughout a system. It's a structured analysis and design tool that can be used for flowcharting in place of, or in association with, information oriented and process oriented system flowcharts. When analysts prepare the Data Flow Diagram, they specify the user needs at a level of detail that virtually determines the information flow into and out of the system and the required data resources. This network is constructed by using a set of symbols that do not imply a physical implementation. The Data Flow Diagram reviews the current physical system, prepares input and output specification, specifies the implementation plan etc. Four basic symbols are used to construct data flow diagrams. They are symbols that represent data source, data flows, data transformations and data storage. The points at which data gets transformed are represented by closed polygons, usually circles, which are called nodes.

107.3 Developing the Application

Programming languages used for this project are PHP programming language, JavaScript, HTML, CSS5. A XAMPP server is used for the PHP programming language.

PHP is a server side scripting language that is embedded in HTML [1]. It is used to manage dynamic content, databases, session tracking, even build entire e-commerce sites. PHP will work with virtually all Web Server software, including Microsoft's Internet Information Server (IIS) but then most often used is freely available Apache Server. PHP will work with virtually all database software, including Oracle and Sybase but most commonly used is freely available MySQL database [1, 2]. In order to process PHP script instructions, a parser must be installed to generate HTML output that can be sent to the web browser. PHP programs can run under various like WAMP, XAMPP etc. WAMP Server is a web development platform which helps in creating dynamic web applications. XAMPP Server is a free open source cross-platform web server package [4].

JavaScript is a dynamic computer programming language. It is lightweight and most commonly used as a part of web pages, whose implementations allow client-side script to interact with the user and make dynamic pages. It is an interpreted programming language with object-oriented capabilities. JavaScript made its first appearance in Netscape 2.0 in 1995 with the name Live Script. The general-purpose core of the language has been embedded in Netscape, Internet Explorer, and other web browsers. The ECMA-262 Specification [5] defined a standard version of the core

JavaScript language. The JavaScript client-side mechanism [6] provides many advantages over traditional CGI server-side scripts.

For example, you might use JavaScript to check if the user has entered a valid email address in a form field. The JavaScript code is executed when the user submits the form, and only if all the entries are valid, they would be submitted to the Web Server. JavaScript can be used to trap user-initiated events such as button clicks, link navigation, and other actions that the user initiates explicitly or implicitly.

HTML stands for Hypertext Markup Language, which is the most widely used language on Web to develop web pages. HTML was created by Berners-Lee in late 1991 but "HTML 2.0" was the first standard HTML specification which was published in 1995. HTML 4.01 was a major version of HTML and it was published in late 1999. Though HTML 4.01 version is widely used but currently we are having HTML5 version which is an extension to HTML 4.01, and this version was published in 2012.

Cascading Style Sheets, referred to as CSS, is a simple design language intended to simplify the process of making web pages presentable [7]. CSS was invented by Hakon Wium Lie on October 10, 1994 and maintained through a group of people within the W3C called the CSS Working Group. The CSS Working Group creates documents called specifications. CSS is easy to learn and understand but it provides powerful control over the presentation of an HTML document. Most commonly, CSS is combined with the markup languages HTML or XHTML.

107.4 Software Modules

The programming languages used for this project are: PHP programming language, JavaScript, HTML, and CSS5. XAMPP a server which is used for the PHP programming language.

There are three types of users. The admin, the staff and a very large number of (about 1000) applicants Admin has the authority to provide login details for staff and to modify the any of recruitment application at any time. The staff are the reviewers of the job applications. The applicant is a user who is applying for the job. He will register to the events created by admin. Using this event link applicant can register for that event. After the completion of registration, applicant will receive the UNIQUE ID. This ID will be used in exam and to check the status of our recruitment process.

Admin has the authority over the entire system. So he can for example modify and delete registered applicant's details. Admin provides login credentials to staff. And he creates an event for recruitment. In Admin Module another module, which is a DASHBOARD where he can view all the shortlisted students for the further process and who were rejected. It is easy for admin to manage the software. Admin can view the all the student details and also the college lists. He can also able to delete the past events/ expired events. Admin can view the all the details of applicants. He can also view all the personnel i.e., applicants, staff members etc. Admin can also include new round if needed for recruiters. Admin provides all the status of applicants who are qualified and who are disqualified. He can UPDATE the status of applicant application. He can access to all the details of applicants, staff members. The main purpose of the admin is to create/ delete Events. He can also view all the past events. He can add/ delete new rounds for interview.

Staff has authority over all applicants. Similar to Admin in Staff module they have a DASHBOARD module where all the shortlisted and rejected applicants' status will be available. Staff members can also see the list of colleges of current applicants, majors, personnel (applicants registered). He can also view all the details in Chart /Table view. A graph is shown and can help the staff to track the number of applicants who are qualified and moved to further rounds, or are disqualified. The staff can also view all the details as same as admin, but they cannot update or delete the data of applicants.

User will get a unique ID when he registers through the created job opening (see Fig. 107.2). Using this ID he can check the status of his recruitment process. With this unique ID he can also edit his details if anything is incorrect.

There are four modules (see Fig. 107.3):

1) New Registration
2) Status of existing Application
3) Login for Admin
4) Login for Staff

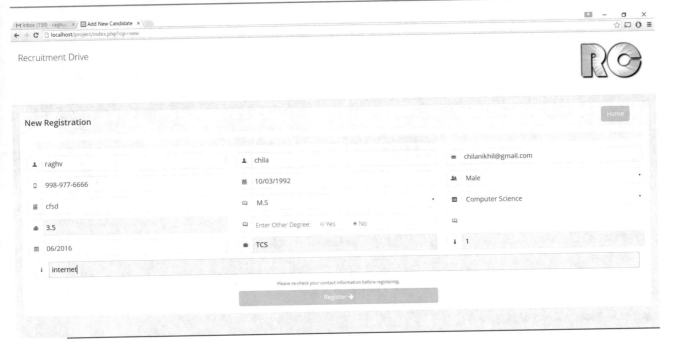

Fig. 107.2 Registration page of recruitment drive

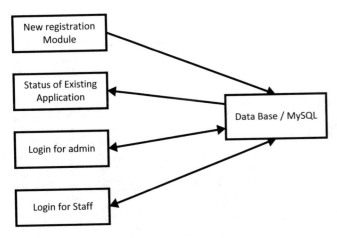

Fig. 107.3 Frontend-backend communication

New Registration In this module applicant has to fill the form to complete the registration.

Status of an Existing Application After registering successfully, a unique ID is created. Using this ID an applicant can check the application status.

Login for Admin The login credential for admin is provided by the software developer. Admin can create login credentials for staff members. His major role is to create events which are job openings. He can also check the status of applicants, details of staff members, and details of all applicants'. He can also edit or update the details of all applicants.

Login for Staff In this module staff members log in with the credentials provided by admin. They can also see the details of applicants. But staff members don't have the access to edit or update the details.

The applicant's details stored in database are shown in Figs. 107.4 and 107.5.

107.5 Conclusions

Recruitment Drive is a web-based application, especially helpful to small companies which require an urgent fill of an open position. This software will reduce paperwork and saves time for small companies to recruit applicants.

One possible future work on this software is to integrate some maintenance examination system. Many small companies use some kind of paper based written examinations.

It will be beneficial to maintain all the records of these examinations. Another feature that will be beneficial will be sending text messages to applicants about their status of application, if they qualified for further or disqualified from the process.

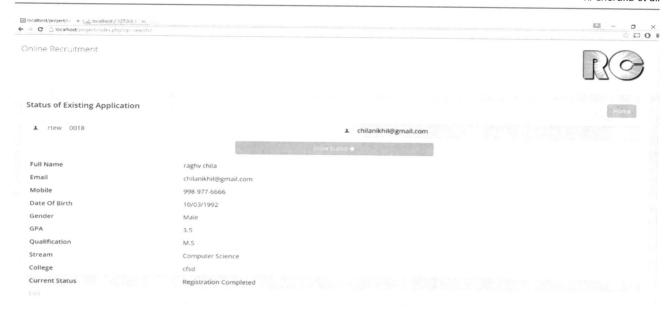

Fig. 107.4 Status of an existing applicant

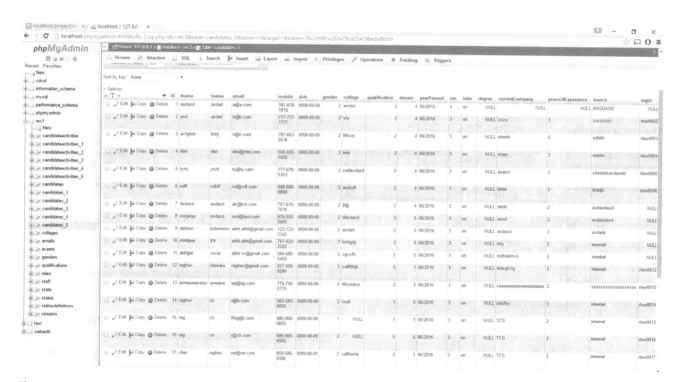

Fig. 107.5 Backend details of applicants

References

1. Welling, L., & Thomson, L. (2008). *PHP and MySQL Web Development* (4th ed.). USA: Addison-Wesley Professional.
2. Nixon, R. (2014). *Learning PHP, MySQL & JavaScript: with jQuery, CSS & HTML5*. Sebastopol, CA: O'Reilly Media.
3. Forbes, A. (2012). *The joy of PHP: a beginner's guide to programming interactive web applications with PHP and MySQL*. Massachusetts: Plum Island Publishing LLC, Rowley.
4. Henderson, J. (2013). *Understanding XAMPP, For Newbies!*. Charleston, SC: CreateSpace Independent Publishing Platform.
5. Standard ECMA-262. Ecma International, [Online]. Available: http://www.ecma-international.org/publications/standards/Ecma-262.htm. Accessed 29 Sept 2016.
6. Duckett, J. (2014). *JavaScript and JQuery: interactive front-end web development*. Indianapolis, IN: Wiley.
7. Robbins, J. N. (2012). *Learning web design: a beginner's guide to HTML, CSS, JavaScript, and web graphics* (4th ed.). Sebastopol, CA: O'Reilly Media.

Nikyle Nguyen and Doina Bein

Abstract

Grading App is a cross-platform desktop application for Windows, Linux, Mac OS X for evaluating student assignments and providing them with a grade and detailed feedback on TITANium. The Grading App can be used for any course in any department. The faculty used the app successfully to grade three sets of assignments (in Fall 2015) and two sets (in Spring 2016) and her grading time was reduced by 80% as compared to manual grading, while providing detailed feedback to the students.

Keywords

Cross-platform application • Web-based application • Evaluating assignments

108.1 Introduction

Evaluating students' work is a meticulous task that requires consistency, accuracy on detecting mistakes, in-depth understanding of the students' thinking mechanism when doing the work, balancing severe versus simple typos when deciding on the number of points to be deducted for each mistake. For a large classroom, with individual projects, that can takes 2–3 days (equivalent of 10–18 hours work) to finish grading student's work for one project, time that can be spend instead on researching the literature for more interesting or challenging projects or just simply keeping up with the topics covered in the class. Due to this, some instructors allow the students to do group projects or just simple reduce the number of projects per class, which is detrimental to individual student learning incremental process.

In some courses, the students have to work individually on developing software projects and they have to submit the project summaries as electronic files, including the source code. While the source code can be easily tested using the appropriate compilers, the project summaries need to be analyzed for correctness by reading them. On reading a project summary, the instructor can pinpoint the mistake a student has made and will need to keep track of all the mistakes. Each mistake will be penalized by a number of points deducted from the maximum number of points assigned to the project.

We have designed and implemented a software product called Grading App designed to shorten the time the faculty spends in grading student assignments submitted as electronic files or handwritten papers. Grading App is a cross-platform desktop application for Windows, Linux, Mac OS X for grading student assignments and providing them with a grade and detailed feedback. The Grading App can be used for any course in any department.

The paper is organized as follows. In Section II we present existing work. In Section III we present the system design and programming environments used for creating the software product. In Section IV we present the capabilities of the software product. We conclude in Section V.

N. Nguyen • D. Bein (✉)
Department of Computer Science, California State University, Fullerton, CA, USA
e-mail: nlknguyen@csu.fullerton.edu; dbein@fullerton.edu

© Springer International Publishing AG 2018
S. Latifi (ed.), *Information Technology – New Generations*, Advances in Intelligent Systems and Computing 558, DOI 10.1007/978-3-319-54978-1_108

108.2 Background

Software was used successfully in evaluating prose of high school [1] and college students. Current advanced software packages permit much more powerful analysis and sometimes do a better job than humans at evaluation and teaching for distance education [2] and support and enhance collaborative learning processes [3].

There have been several software testing products developed over 20 years [4] that test and grade assignments written in some programming languages using some pattern matching techniques for the output of the programs. The assignments are first check for compilation errors, then the output of the program is analyzed for checking the program correctness. Over time, the importance of automated feedback [5] in showing students where they can improve has become a way of encouraging students to improve their code programming skills [6]. Furthermore, a test-driven approach to develop programming assignments can drive the students into writing better code [7].

108.3 Software Design

In order to develop a GUI (Graphical User Interface) desktop application for multiple operating systems (Linux, MacOS, and Windows) with the most code reusability, many GUI technologies were considered before deciding which technology to be used.

Qt [8] is the first cross-platform native application development framework that we considered. Qt is mainly C++ with support for C++ 11 and 14. It also has a custom JavaScript-like language for describing UI components. Although the framework is nice, the packaging process for all operating systems is not easy because of native binary objects required for each platform. Furthermore, since the main language to work with is C++, the overhead of language complexity such as explicit memory management or integrating third-party libraries in the build process will take a large amount of time, which could have been used for developing the application itself. Since our application is not computation intensive, it does not demand high performance where C++ tends to be a viable option. We do not use Qt in favor of more high-level / productive solution.

JavaFX [9] is the second option we considered. JavaFX is the official GUI application development framework for Java. Although there are many alternatives in Java world, JavaFX is latest flagship from Oracle, promising to be the facto standard for building rich client applications with Java. This allows building applications quicker and safer than with C++ in case of Qt. Integrating third-party Java libraries into a Java application build process is easy. Also, packaging a Java application is simple. However, the user needs to install

a Java Virtual Machine (JVM) in order to run the application. We consider that to be too intrusive to the user's machine because of known problems associated with Java on the client side such as update checking or security holes. That is why we do not pick JavaFX.

NW.js [10] (previously known as Node-Webkit) is our choice for cross-platform application development. It is a runtime based on Chromium [11] and Node.js [12] and allows web applications to run as native on multiple operating systems. Chromium is the core of Google Chrome browser. Node.js is a JavaScript runtime built on Chrome's V8 JavaScript engine. NW.js combines the core of Google Chrome browser with complete support for Node.js, which is a server side technology. This is an interesting combination that allows developers to fully utilize web technologies for building desktop applications quickly. Furthermore, with the high productivity of JavaScript/Node.js language and lots of available Node Package Modules [13] as well as many UI (User Interface) frameworks for web application, we can focus more on delivering the application functionality and user experience. The application source code is in plain JavaScript, HTML, and CSS like any web application; therefore, it does not require compilation for each platform. Another thing worth pointing out is that Chromium is a very popular open source project by Google, and it is supported on many platforms. Node.js, on the other hand, is backed by Node.js Foundation, and their agenda is also to bring Node.js to more platforms. That allows this runtime combination to be highly available in many operating systems and architectures. Currently, NW.js is available on Windows (32/64 bit), Linux (32/64 bit), and macOS (64 bit). Packaging the application for each platform is very simple, and the user can click and run the application on their machine in a portable manner without installing any software. NW.js turned out to be the right choice for our application.

JavaScript / Node.js is the main programming language we use for our application. To be more precise, the JavaScript standard that we use is ECMAScript 2015 (often known as ES6). In the build process, the JavaScript source code is transpiled to ES5 version that is more compatible with many runtime environments even though the only target environment for our application is NW.js. In the future, this transpilation will be less needed. Aside from JavaScript, the markup languages that we use are the usual HTML and CSS.

We use AngularJS [14] framework for our application. AngularJS is a JavaScript framework by Google for developing dynamic web applications. It is a browser side technology that provides a well-defined but also opinionated structure to build single-page application that behaves much like a desktop application living in the browser. It has many baked in features such as data binding and dependency injection which eliminate a lot of boilerplate code that we would otherwise need to write to wire up the application. Moreover, its standardized structure allows developers to create and share extensions that other people can plug into their own application easily, which increases code reusability and developer's productivity. Even though the latest AngularJS version is 2, by the time we started our project, version 2 wasn't stable, so we went with version 1.

Material Design [15] is used for the application. Material Design is design guidelines / specifications by Google, and many UI frameworks implement it. Since we use AngularJS, an implementation of Material Design UI components for AngularJS is available as Angular Material [16], and that is what we use. By following the design guidelines and focusing on the user experience, we make the interface simple and straightforward to understand and use.

A JavaScript database called NeDB [17] is used for storing and retrieving data from a local data storage in form of NoSQL database. It is an embedded persistent or in-memory database for Node.js, NW.js, and other environments. This is similar to SQLite for a local SQL database. We use NoSQL because its nature in storing data reflects the data structure of the objects that we deal with in the application source code, and that provides a natural way of thinking when developing. If we use SQL, we will have to map back and forth our object's data structure to tabular structure of a SQL table, which we consider an unnecessary constraint that adds no benefit to our application. In our settings, the data of the grading for a class assignment is stored in a single file, which is the NeDB database file. There is no requirement for a running database management system in order to use the database file. For advance user with programming experience, the database file can be loaded and manipulated programmatically using NeDB module.

108.4 Using GradingApp

The feedback sent back to the student will contain (Figs. 108.1, 108.2, 108.3 and 108.4):

the number of points
if points have been deducted due to some mistake(s), a very detailed explanation of each mistake is given to the student, allowing the student to be able to fix it.

The feedback sent to the students has to be consistent: two students that made the same mistake will have the same amount of points deducted and will receive the same feedback. This can easily be accomplished using some software product, which will maintain a list of mistakes, allowing the user to customize each description of the mistake and the number of points to be deducted. But if the grading is done manually, it will be a tedious task, since the instructor will need to have a sheet with all the mistakes, description for each, the number of points for each, and as s/he grades a project, add to the sheet if a new mistake is found. Additionally, if during the grading process, the instructor decides to modify the number of points for a mistake, then s/he will have to go back, cross out the total points and re-adjust the score.

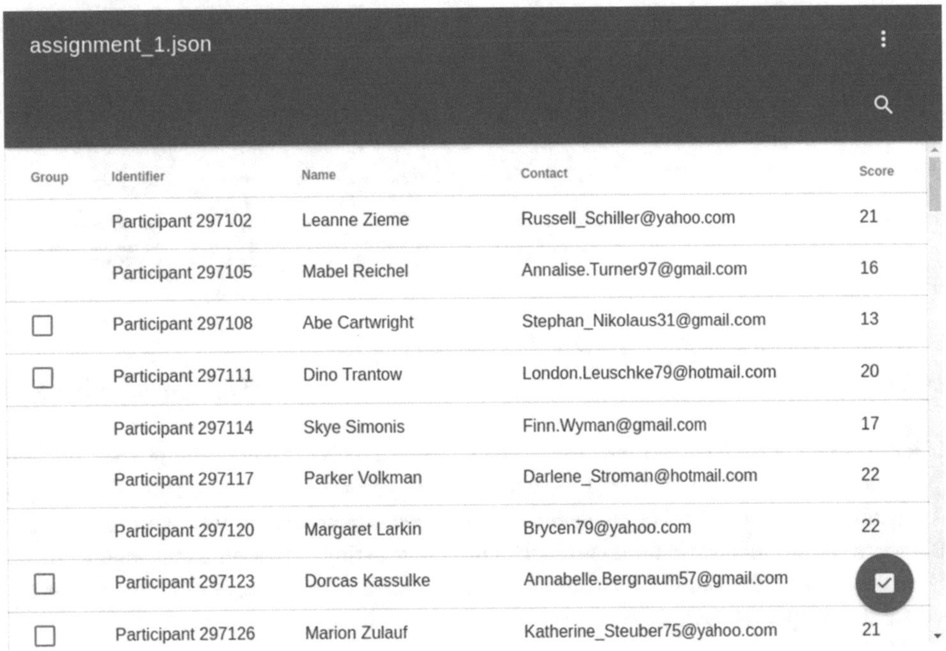

Fig. 108.1 Evaluating students individually or in groups

| | assignment_1.json | | | ⋮ |
| | | | | 🔍 |

Group	Identifier	Name	Contact	Score
	Participant 297102	Leanne Zieme	Russell_Schiller@yahoo.com	21
	Participant 297105	Mabel Reichel	Annalise.Turner97@gmail.com	16
☐	Participant 297108	Abe Cartwright	Stephan_Nikolaus31@gmail.com	13
☐	Participant 297111	Dino Trantow	London.Leuschke79@hotmail.com	20
	Participant 297114	Skye Simonis	Finn.Wyman@gmail.com	17
	Participant 297117	Parker Volkman	Darlene_Stroman@hotmail.com	22
	Participant 297120	Margaret Larkin	Brycen79@yahoo.com	22
☐	Participant 297123	Dorcas Kassulke	Annabelle.Bergnaum57@gmail.com	☑
☐	Participant 297126	Marion Zulauf	Katherine_Steuber75@yahoo.com	21

Fig. 108.2 Importing students' names from some database

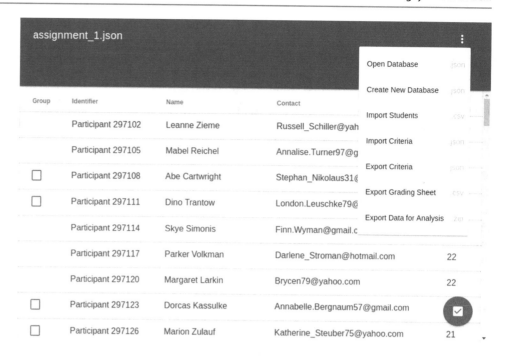

Fig. 108.3 Evaluating mistakes for a group project of two people

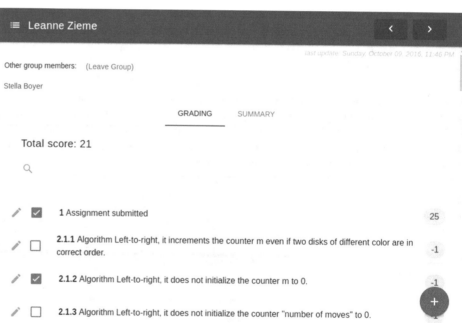

Reasons for re-adjusting the number of points for a mistake could be:

if the mistake is very popular aka over 60% of the students made it, then it may be related to some a general misunderstanding so the number of points deducted may be decreased

if a more serious mistake is found that subsumes and existing mistake, then the simpler mistake may need to have the number of points decreased.

108.5 Conclusions and Future Work

Data mining on the grading table is part of future improvements. Detecting blocks of students that made the same mistake is an NP-complete problem, which translates that an exact algorithm will run in exponential or factorial time, so a heuristics or an approximation algorithm can be used instead.

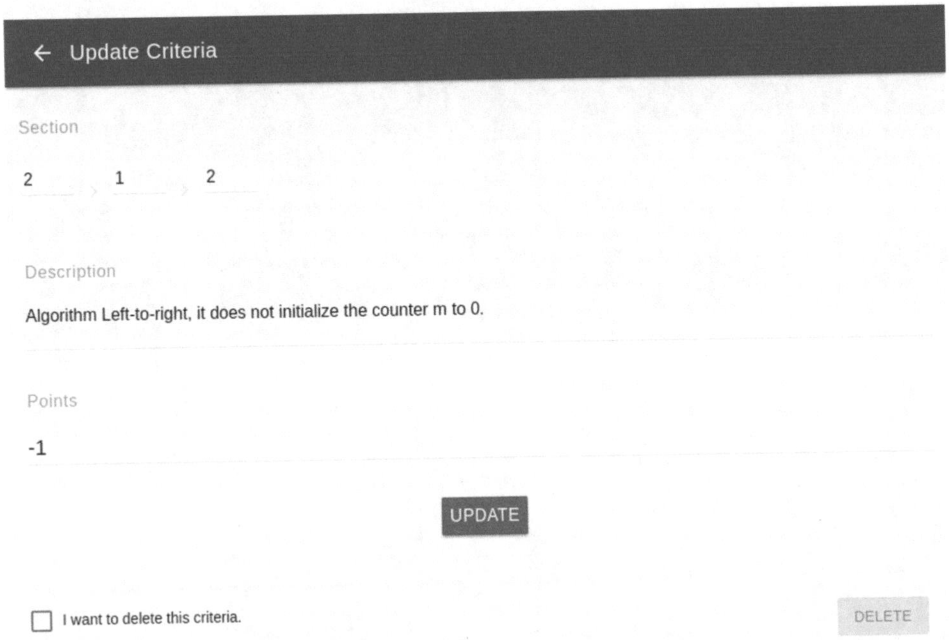

Fig. 108.4 Creating or updating a mistake

← Update Criteria

Section

2   1   2

Description

Algorithm Left-to-right, it does not initialize the counter m to 0.

Points

-1

UPDATE

☐ I want to delete this criteria.   DELETE

References

1. Page, E. B. (1994). Computer Grading of Student Prose, Using Modern Concepts and Software. *The Journal of Experimental Education, 62*(2), 127–142.
2. Jones, A., et al. (1999). Contexts for evaluating educational software. *Interacting with Computers, 11*(5), 499–516.
3. Jonassen, D. H., Peck, K. L., & Wilson, B. G. (1999). *Learning with technology: a constructivist perspective.* Upper Saddle River, NJ: Merrill.
4. Jackson, D., & Usher, M. (1997). Grading student programs using ASSYST. In *Proceedings of the twenty-eighth SIGCSE technical symposium on computer science education*, New York.
5. Jones, E. L. (2001). Grading student programs - a software testing approach. *Journal of Computing Sciences in Colleges, 16*(2), 185–192.
6. Edwards, S. H. (2004).Using software testing to move students from trial-and-error to reflection-in-action. In *Proceedings of the 35th SIGCSE technical symposium on computer science education*, New York City.
7. Edwards, S. H. (2003). Teaching software testing: automatic grading meets test-first coding. In *OOPSLA '03 Companion of the 18th annual ACM SIGPLAN conference on Object-oriented programming, systems, languages, and applications*. Anaheim, CA.
8. The Qt Company. [Online]. Available: https://www.qt.io/. Accessed 18 Oct 2016.
9. Java Platform, Standard Edition (Java SE) 8, Oracle. [Online]. Available: https://docs.oracle.com/javafx/. Accessed 18 Oct 2016.
10. Nw.js community. [Online]. Available: http://nwjs.io/. Accessed 18 Oct 2016.
11. The Chromium projects. [Online]. Available: https://www.chromium.org/. Accessed 18 Oct 2016.
12. Node.js. Node.js Foundation, [Online]. Available: https://nodejs.org. Accessed 18 Oct 2016.
13. npm. npm, Inc., [Online]. Available: https://www.npmjs.com/. Accessed 18 Oct 2016.
14. AngularJS. [Online]. Available: https://angularjs.org. Accessed 18 Oct 2016.
15. Material design. Google, [Online]. Available: https://material.google.com. Accessed 18 Oct 2016.
16. Angular Material. [Online]. Available: https://material.angularjs.org. Accessed 18 Oct 2016.
17. louischatriot/nedb. GitHub, [Online]. Available: https://github.com/louischatriot/nedb. Accessed 18 Oct 2016.

James Andro-Vasko, Wolfgang Bein, Hiro Ito, and Govind Pathak

Abstract

Power-down mechanisms are well known and are widely used to save energy. We consider a device which has states OFF, ON, and a fixed number of intermediate states. The state of the device can be switched at any time. In the OFF state the device consumes zero energy and in ON state it works at its full power consumption. The intermediate states consume only some fraction of energy proportional to the usage time but switching back to the ON state has different constant setup cost depending on the current state. We give a new heuristic to construct optimal power-down systems with few states. The heuristic converges very quickly to an optimal solution.

Keywords

Green computing • Power down problems • Smart grid • Online competitive analysis • Multiple state sytems.

109.1 Introduction

In Information Technology energy consumption is an issue in terms of availability as well as terms of cost. According to Google energy costs are often larger than hardware costs [7]. Ways to minimize energy consumption are crucial and power usage has increasingly become a first order constraint for data centers. A growing body of work on algorithmic approaches for energy efficiency exists, see Albers et al. for a general survey [3].

To manage power usage, power-down mechanisms are widely used: for background on algorithmic approaches to power down see [1, 2, 11, 12]. In power down problems a machine needs to be in an on-state in order to handle

requests but over time it can be wasteful if a machine is on while idling. To increase efficiency, devices are often designed with power saving states such as a "hibernate state", a "suspend state" or various other hybrid states. Power down algorithms exists to control single machines or systems with multiple machines, such as in distributed machine environments.

Power-down is studied for hand-held devices, laptop computers, work stations and data centers. However, recent attention has been on power-down in the context of the smart grid [8]: Electrical energy supplied by sustainable energy sources is more unpredictable due to its dependence on the weather and other factors. When renewables produce a surplus of energy, such surplus generally does not affect the operation of traditional power plants. Instead, renewables are throttled down or the surplus is simply ignored. But in the future where a majority of domestic power would be generated by renewables this is not tenable. Instead it may be the traditional power plant that will need to be throttled down.

Power-down problems are studied in the framework of online competitive analysis, see [4–6, 9, 10]. In online

J. Andro-Vasko • W. Bein (✉) • G. Pathak
Department of Computer Science, University of Nevada, 89154 Las Vegas, NV, USA
e-mail: androvas@unlv.nevada.edu; wolfgang.bein@unlv.edu

H. Ito
School of Informatics and Engineering, The University of Electro-Communications, Tokyo, Japan
e-mail: itohiro@uec.ac.jp

© Springer International Publishing AG 2018
S. Latifi (ed.), *Information Technology – New Generations*, Advances in Intelligent Systems and Computing 558, DOI 10.1007/978-3-319-54978-1_109

computation, an algorithm must make decisions without knowledge of future inputs. Online algorithms can be analyzed in terms of competitiveness, a measure of performance that compares the solution obtained online with the optimal offline solution for the same problem, where the lowest possible competitiveness is best. Online competitive models have the advantage that no statistical insights are needed, instead a worst case view is taken: this is appropriate as request in data centers, or short term gaps in renewable energy supply are hard to predict.

In this paper we study power-down problems with few states. Such systems have not been the focus of extensive research as the algorithm community is generally more interested in general results with an arbitrary (and large) number of states.

Our Contribution: Augustine et al. [6] have given an approximation algorithm to construct a competitive schedule for power-down systems with an arbitrary number of intermediate states. However in practice power-down systems rarely have more than three intermediate states. We present an alternative heuristic to create an optimal power-down mechanism when there are few states. Specifically we describe our heuristic in terms of a quintuple systems, i.e. systems with states "on", "off", as well as three intermediate states.

The paper is organized as follows: In Sect. 109.2 we formally introduce the power-down problem. Section 109.3 gives optimality conditions which are the basis for the heuristic we present in Sect. 109.4. We then validate the heuristic and present simulations for specific quintuple systems. In Sect. 109.5 we give concluding remarks.

109.2 The Power Down Problem

Consider a device with two states, "on"/"off" and a number of intermediate states. In this paper, we consider a five state machine with states denoted by s_0, s_1, s_2, s_3, s_4 in which s_0 is the on-state, s_4 is the off-state, and states $s_1, s_2,$ and s_3 are intermediate states (such as "screen saver", "sleep", and "hibernate".) Each state s_i has an associated power cost a_i – the cost to remain in s_i for time Δt is $\Delta t \cdot a_i$. There is no cost to switch from "on" to "off" or any of the intermediate states. But a fixed cost of 1 occurs to switch to s_0 ("on") from s_4 ("off") and a smaller cost $d_j \in (0, 1)$ occurs to switch to from any of the intermediate states. We assume $a_0 > a_1 > a_2 > a_3 > a_4$ where we normalize $a_0 = 1$ and $a_4 = 0$ since the machine uses no power when it is in the off state. We also have $d_0 < d_1 < d_2 < d_3 < d_4$ where we have normalized $d_4 = 1$ and $d_0 = 0$ since the machine is already in the on state and does not need to power up.

The machine has to serve a sequence of jobs which have start times $t_1{}^s, t_2{}^s, t_3{}^s, \ldots$ and end times $t_1{}^e, t_2{}^e, t_3{}^e, \ldots$ with

$0 < t_1{}^s \le t_1{}^e \le t_2{}^s \le t_2{}^e \le t_3{}^s \le t_3{}^e \le \ldots$ and the device must be "on" during $[t_\ell{}^s, t_\ell{}^e]$. In between requests the device can remain in the on-state or go to the off-state or any of the intermediate states. Note that if $t_{\ell+1}{}^s$ is shortly after $t_\ell{}^e$ it may be wasteful to turn the device off and it could instead be better to leave the machine on or perhaps keep the device in any of the sleep states.

We study the problem in the online model, where an algorithm has to make decisions without knowledge of the entire input, but rather the input arrives a piece at a time. The solution of an online algorithm $\mathcal{A}$ is compared against the solution where the entire request sequence is known in advance. The algorithm which generates this solution is also referred to as the optimal offline algorithm $\mathcal{OPT}$. The ratio between the cost of the online algorithm and the optimal offline algorithm (in the worst case) is called competitive ratio (CR); see [10] for more background on online competitive analysis. For each job ℓ we know that the device must be in the on-state from $t_\ell{}^s$ to $t_\ell{}^e$, as a result in terms of competitive analysis the length of the job is ignored since both $\mathcal{A}$ and $\mathcal{OPT}$ have to be "on": We set $t_\ell := t_\ell{}^s = t_\ell{}^e$ as the point in time when the device has to be in the on-state; the goal is to minimize the power consumption while the machine is idling. An online algorithm for this problem is thus defined by times values x_1, x_2, x_3, x_4 where x_i denotes the switch time from s_{i-1} to s_i.

For time t, the online cost will be $\mathcal{A}(t)$ and the offline cost will be $\mathcal{OPT}(t)$. The competitive ratio $\mathcal{CR}$ at time t is $\mathcal{CR}(t) = \frac{\mathcal{A}(t)}{OPT(t)}$ Given the value for t, the idle time duration, the online cost will be computed in the following way

$$
\mathcal{A}(t) = \begin{cases}
a_0 t & \text{if } t < x_1 \\
a_0 x_1 + d_1 & \text{if } t = x_1 \\
a_0 x_1 + a_1(t - x_1) + d_1 & \text{if } x_1 < t < x_2 \\
a_0 x_1 + a_1(x_2 - x_1) + d_2 & \text{if } t = x_2 \\
\begin{aligned} & a_0 x_1 + a_1(x2 - x1) + \\ & a_2(t - x_2) + d_2 \end{aligned} & \text{if } x_2 < t < x_3 \\
\begin{aligned} & a_0 x_1 + a_1(x2 - x1) + \\ & a_2(x_3 - x_2) + d_3 \end{aligned} & \text{if } t = x_3
\end{cases}
$$

$$
\mathcal{A}(t) = \begin{cases}
\begin{aligned} & a_0 x_1 + a_1(x_2 - x_1) + \\ & a_2(x_3 - x_2) + \\ & a_3(t - x_3) + d_3 \end{aligned} & \text{if } x_3 < t < x_4 \\
\begin{aligned} & a_0 x_1 + a_1(x_2 - x_1) + \\ & a_2(x_3 - x_2) + \\ & a_3(x_4 - x_3) + d_4 \end{aligned} & \text{if } t \ge x_4
\end{cases}
$$

once again we normalize the on and off states so $a_0 = 1$, $d_0 = 0$, $a_4 = 0$, and $d_4 = 1$.

The offline cost will be $OPT(t) = \min\{a_0t, a_1t + d_1, a_2t + d_2, a_3t + d_3, d_4\}$, since the requests are known in advance. Using these costs, we can compute the competitive ratio. The next issue is calculating the values for x_1, x_2, x_3, and x_4 in which we will compute such that we obtain the minimal competitive ratio.

109.3 Optimizing the Online Algorithm

In this section we will consider the worst case costs for $\mathcal{A}(t)$, so we consider when a request arrives it will arrive exactly at times x_1, x_2, x_3, and at or after x_4. So the requests arrive right when the machine switches to a lower power state. We adapt the following proof from [6]:

Theorem 1. *For any system, the worst case competitive ratio occurs at a transition time.*

Proof. Consider $\mathcal{A}(t)$ and $OPT(t)$ and let $\mathcal{A}(t)$ be in state s_i and let $OPT(t)$ be in state s_j. Let t be the moment when the competitive ratio is maximized and t is not a transition time, the competitive ratio at time t will be denoted by ρ.

If $a_i = a_j$, then both $\mathcal{A} = \rho OPT$ will remain the same after t since they are increasing linearly at the same rate. At some point the online algorithm will switch to a lower power state and incur an extra d_{i+1} cost, when a new request arrives, but the offline algorithm will incur a cost of d_i, since the next request is known by the offline algorithm, which is less than d_{i+1} so the competitive ratio will increase which causes a contradiction.

If $a_i > a_j$, then the online algorithm is in a lower power state than the offline algorithm. So in order to have $\mathcal{A}(t) = \rho OPT(t)$, then for some arbitrary value δ, $\mathcal{A}(t - \delta) > \rho OPT(t - \delta)$, where $t - \delta$ will not switch to a lower power state and no request will arrive. This causes a contradiction since the competitive ratio is not maximized at t.

If $a_i < a_j$, then the online algorithm is in a higher power state than the offline algorithm. So in order to have $\mathcal{A}(t) = \rho OPT(t)$, then for some arbitrary value δ, $\mathcal{A}(t + \delta) > \rho OPT(t + \delta)$, where $t + \delta$ will not switch to a lower power state and no request will arrive. This causes a contradiction as well since the competitive ratio is not maximized at t. $\square$

Using Theorem 1, we will consider the worst case competitive ratios at the transition times, and the maximum of all competitive ratios at the transition times will be the competitive ratio so $\max_{i=1}^{n-1}\{CR(x_i)\}$ where $CR(x_i)$ denotes the competitive ratio at time x_i and n will be the number of states. Let $\tau_1 = a_1(x_2 - x_1)$, $\tau_2 = a_2(x_3 - x_2)$, and $\tau_4 = a_3(x_4 - x_3)$. We will compute the x_i values given a competitive ratio.

$$\frac{x_1 + d_1}{x_1} = CR \tag{109.1}$$

$$\frac{x_1 + \tau_1 + d_2}{\min\{x_2, a_1x_2 + d_1\}} = CR \tag{109.2}$$

$$\frac{x_1 + \tau_1 + \tau_2 + d_3}{\min\{x_3, a_1x_3 + d_1, a_2x_3 + d_2\}} = CR \tag{109.3}$$

$$\frac{x_1 + \tau_1 + \tau_3 + \tau_4 + d_4}{d_4} = CR \tag{109.4}$$

In the above equations CR will be the target competitive ratio such that $CR(x_1) = CR(x_2) = CR(x_3) = CR(x_4)$. We solve for x_1, x_2, x_3, and x_4 in Equations 109.1, 109.2, 109.3, and 109.4 respectively. Let $\alpha = x_1(1 - a_1)$, $\beta = x_1(1 - a_1) + x_2(a_1 - a_2)$, $\beta' = x_1(a_1 - 1) + x_2(a_2 - a_1)$, and $\gamma = x_3(a_3 - a_2)$. We obtain:

$$x_1 = \frac{d1}{CR - 1} \tag{109.5}$$

$$x_2 = \max\left\{\frac{\alpha + d_2}{CR - a_1}, \frac{\alpha + d_2 - CR \cdot d_1}{a_1(CR - 1)}\right\}. \tag{109.6}$$

$$x_3 = \max\left\{\frac{\beta + d_3}{CR - a_2}, \frac{\beta' - d_3 + CR \cdot d_1}{a_2 - a_1 \cdot CR}, \frac{\beta + d_3 - CR \cdot d_2}{a_2(CR - 1)}\right\}. \tag{109.7}$$

$$x_4 = \frac{\beta' + \gamma + CR - 1}{a_3} \tag{109.8}$$

Theorem 2. *Given an online algorithm with competitive ration CR for an n state power down problem. Then a necessary condition for CR to be minimal is $CR(x_1) = CR(x_2) = \ldots = CR(x_{n-1}) = CR$.*

Proof. Let us assume that the competitive ratio is not minimized when we have $CR(x_1) = CR(x_2) = \ldots = CR(x_{n-1}) = CR$. We will assume that $CR(x_1) \neq CR(x_2) \neq \ldots \neq CR(x_{n-1})$ will yield the minimal competitive ratio. So the competitive ratio of this system will be $\max_{i=1}^{n-1}\{CR(x_i)\}$, so for some x_i will yield the maximal value and this will be the competitive ratio for our system since we know the competitive ratio is maximized at switch times according to Theorem 1. We will denote this to be CR_{max}. Let us decrease the CR_{max} value to CR'_{max} that is used to compute x_i, we will set CR'_{max} such that CR'_{max} will still remain the maximal competitive ratio for the system. Using CR'_{max} to compute x_i will cause x_i to change and all of the switch times after x_i will also change but the CR values to compute those remaining switch times will remain unchanged, but we have $CR'_{max} < CR_{max}$, so we minimized the competitive ratio for the system so the initial competitive ratio of the system was not minimal which leads to a contradiction. $\square$

Let x_{opt4} be the threshold used that allows $\mathcal{OPT}$ to determine whether to transition into the off state once it starts the idle period. Using the values of the 5 states standby and power up cost,

$$x_{opt4} = \max_t \begin{cases} a_0 t + d_0 = d_4 \\ a_1 t + d_1 = d_4 \\ a_2 t + d_2 = d_4 \\ a_3 t + d_3 = d_4 \end{cases} \tag{109.9}$$

So if the delay time $t < x_{opt4}$, then the offline algorithm will not be in the off state when the request arrives and it will be in the off state otherwise in which its power consumption will simply be d_4 which can be seen in Equation 109.4. In Equation 109.4, we assume that $x_4 \geq x_{opt4}$, since the offline cost was d_4 and not a minimal cost among all the states as we had in Equations 109.1, 109.2, and 109.3.

Theorem 3. *The competitive ratio is minimized when $x_4 = x_{opt4}$.*

Proof. Let $\Phi = x_1 + a_1(x_2 - x_1) + a_2(x_3 - x_2)$, we can rewrite Equation 109.4 to

$$\frac{\Phi + a_3(x_4 - x_3) + d_4}{\min\{x_4, a_1 x_4 + d_1, a_2 x_4 + d_2, a_3 x_4 + d_3, d_4\}} = CR$$

such that we consider all the possible optimal values based when the machine powers down in $\mathcal{A}$. Will we consider $Cost(\mathcal{A}(x_4))$ be the cost of $\mathcal{A}$ at time x_4. If $x_4 < x_{opt4}$ then we have

$$\frac{\Phi + a_3(x_4 - x_3) + d_4}{\min\{x_4, a_1 x_4 + d_1, a_2 x_4 + d_2, a_3 x_4 + d_3\}} = CR'$$

because $\mathcal{OPT}$ will not be in the off state at time x_4. In the above equation, the x_4 term can be factored out of numerator regardless of the minimal cost chosen by $\mathcal{OPT}$. However, that x_4 term will remain in the denominator when after factoring it out of the numerator and if we decrease x_4 by ε such that $x_4 - \varepsilon > x_3$, the competitive ratio grows as long as $\varepsilon > 0$. Now consider the case where $x_4 > x_{opt4}$, then we will have the following

$$\frac{\Phi + a_3((x_4 + \epsilon) - x_3) + d_4}{d_4} = CR'$$

Where $\varepsilon > 0$, we can see that as ε increases the cost of the online algorithm increases and the offline cost remains unchanged so the competitive ratio increases. So in both cases, if $x_4 < x_{opt4}$ or $x_4 > x_{opt4}$, the competitive ratio is greater than when $x_4 = x_{opt4}$, and this is the only value for x_4 which minimizes the competitive ratio. $\square$

109.4 Power Down Heuristic

The previous section implies a natural heuristics for finding the online algorithms with optimal competitiveness. We will now describe this heuristic for a quintuple system; but it can be easily adapted for a system with a different number of states. To begin and chose an initial competitive rations we use the fact from [6] that there exists a schedule that is $(3 + 2\sqrt{2})$–competitive in the worst case. Thus, our heuristic will search competitive ratios in the range $[1, 3 + 2\sqrt{2}]$. Given 5 states with their respective a_i and d_i values, we assign a value for CR and we compute the standby times using Equations 109.5, 109.6, 109.7, and 109.8. The heuristic assigns a value for CR in the range $[1, 3 + 2\sqrt{2}]$; we then compute the standby times and we obtain a schedule for that particular CR value given. This however may not be the optimal competitive ratio for this system. If the value of $x_4 > x_{opt4} + \theta$, where θ is an arbitrarily small constant, then we can choose a new CR that was smaller than the value we assigned earlier since according to Theorem 3, the competitive ratio can only be minimal if $x_4 = x_{opt4}$. However, if $x_4 < x_{opt4}$ then there does not exist a schedule for this five state system that will be CR-competitive. In this case we need to choose a new competitive ratio that is larger than CR. We keep applying the strategy until we find a schedule such that

$x_{opt_4} \geq x_4 \leq x_{opt_4} + \theta$, thus optaining an approximation arbitrarily close the optimal competitive ratio.

Algorithm 106.1 Power-down heuristic

Given values $a_{0 \to 4}$, and $d_{0 \to 4}$
lowerBound = 1, upperBound = $3 + 2\sqrt{2}$;
CR = (lowerBound + upperBound) / 2;
Compute x_1, x_2, x_3, and x_4 using CR;
while $x_4 < x_{opt_4}$ or $x_4 > x_{opt_4} + \theta$ **do**
  if $x_4 < x_{opt_4}$ **then**
    | lowerBound = CR;
  else
    | upperBound = CR;
  end
  CR = (lowerBound + upperBound) / 2;
  Recalculate x_1, x_2, x_3, and x_4;

end

We have implemented Algorithm 106.1 and performed numerous simulations. First we ran a simulation for a quintuple system with the given standby and power up costs, $a_0 = 1\ d_0 = 0\ a_1 = 0.55\ d_1 = 0.225\ a_2 = 0.4\ d_2 = 0.4$ $a_3 = 0.25\ d_3 = 0.60\ a_4 = 0\ d_4 = 1$. The values chosen reflect a typical system with three intermediate states, such as "power save", "suspend", and "hibernate". Using Equation 109.9 the value of $x_{opt_4} = 1.6$. Each iteration of Algorithm 106.1 is tabulated below.

For each iteration, we choose a different value for competitive ratio between the upper bound and the lower bound based on the value of x_4 computed at that iteration. We perform a binary search on the competitive ratio in the range $[1, 3 + 2\sqrt{2}]$, until we find a value such that $x_4 \equiv x_{opt4}$, since we know that $x_4 = x_{opt4}$, according to Theorem 3, yields the optimal competitive ratio and the rest of the x values are used to calculate x_4 and we know that $CR(x_1) = CR(x_2) = CR(x_3) = CR(x_4)$ will yield the competitive ratio according to Theorem 2, and so if both theorems hold, we have the schedule that gives us the optimal competitive ratio for the system. Using the values for x_1, x_2, x_3, and x_4 computed in Table 109.1, we can compute the competitive ratios for any given standby time.

Figure 109.1 displays the competitive ratio for using the optimal transition times we obtained from Table 109.1, we can see that the competitive ratio is always maximized at the transition time, then we can see that the competitive ratio decreases when the standby time diverges away from any transition time, the competitive ratio will continue to decrease until the standby reaches the next transition time. Table 109.2 shows results for different quintuple systems.

Table 109.1 Execution of algorithm 106.1 with sample input

Iteration	x_1	x_2	x_3	x_4
1	0.0932	0.1543	0.2207	9.2632
2	0.1864	0.2920	0.4027	4.0756
3	0.3731	0.6249	1.0401	0.7414
4	0.2486	0.3778	0.5248	2.6309
5	0.2984	0.4438	0.7193	1.7810
6	0.3314	0.4864	0.8488	1.3184
7	0.3142	0.4643	0.7815	1.5509
8	0.3061	0.4538	0.7496	1.6670
9	0.3099	0.4587	0.7645	1.6122
10	0.3121	0.4615	0.7729	1.5816
11	0.3108	0.4598	0.7678	1.6000
Iteration	lowerBound	upperBound	CR	
1	1.000	5.828	3.414	
2	1.000	3.414	2.207	
3	1.000	2.207	1.603	
4	1.603	2.207	1.905	
5	1.603	1.905	1.754	
6	1.603	1.754	1.679	
7	1.679	1.754	1.716	
8	1.716	1.754	1.735	
9	1.716	1.735	1.726	
10	1.716	1.726	1.721	
11	1.721	1.726	1.724	

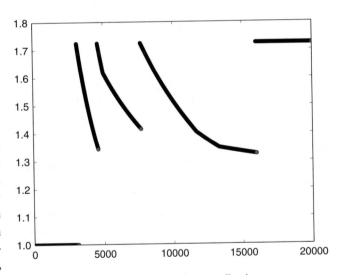

Fig. 109.1 Competitive ratio for various standby times

109.5 Conclusion

Our results show that the heuristic converges very quickly to the optimal solution; in all our simulations fewer than 20 iterations were necessary to get an optimal solution with accuracy of at least two digits. Our approach, though

Table 109.2 Optimal competitive ratio within $\theta = 0.01$ for various a and d costs

i	a	d	x	CR
0	1.0000	0.0000	0.0000	1.701
1	0.7500	0.2500	0.3566	
2	0.5000	0.5000	0.6195	
3	0.2500	0.7500	0.8277	
4	0.0000	1.0000	1.0001	
0	0.0000	1.0000	0.0000	1.739
1	0.6000	0.2000	0.2706	
2	0.4000	0.4000	0.4462	
3	0.2000	0.6000	0.6990	
4	0.0000	1.0000	2.0086	
0	1.0000	0.0000	0.0000	1.7265
1	0.8000	0.1000	0.1376	
2	0.5000	0.4000	0.4614	
3	0.1000	0.8000	0.9003	
4	0.0000	1.0000	2.0043	
0	1.0000	0.0000	0.0000	1.775
1	0.6000	0.1000	0.1290	
2	0.4000	0.3000	0.3744	
3	0.1000	0.6000	0.8256	
4	0.0000	1.0000	4.0083	
0	1.0000	0.0000	0.0000	1.765
1	0.7000	0.2000	0.2614	
2	0.3000	0.4000	0.4492	
3	0.1000	0.8000	1.5343	
4	0.0000	1.0000	2.0001	
0	1.0000	0.0000	0.0000	1.724
1	0.5500	0.2250	0.3108	
2	0.4000	0.4000	0.4598	
3	0.2500	0.6000	0.7678	
4	0.0000	1.0000	1.6000	

presented for the special case of 5 states, can be easily adapted to any system with few states. We chose to present the special case, to simplify the presentation and since 5 state systems are typical in practice. But the theorems in Sect. 109.3 carry over for an arbitrary number of states.

We note that solutions obtained by our heuristic can not only be used to analyse systems but also are useful in their design.

Acknowledgements Discussions with Rüdiger Reischuk of Universität Lübeck during his sabbatical visit are acknowledged. The work of author Wolfgang Bein was supported by National Science Foundation grant IIA 1427584.

References

1. Agarwal, Y., Hodges, S., Chandra, R., Scott, J., Bahl, P., & Gupta, R. (2009). Somniloquy: Augmenting network interfaces to reduce PC energy usage. In *Proceedings of the 6th USENIX Symposium on Networked Systems Design and Implementation*, NSDI'09 (pp. 365–380). Berkeley, CA: USENIX Association.
2. Agarwal, Y., Savage, S., & Gupta, R. (2010). Sleepserver: A software-only approach for reducing the energy consumption of PCs within enterprise environments. In *Proceedings of the 2010 USENIX Conference on USENIX Annual Technical Conference*, USENIXATC'10 (pp. 22–22). Berkeley, CA: USENIX Association.
3. Albers, S. (2010). Energy-efficient algorithms. *Communications of the ACM, 53*, 86–96.
4. Andro-Vasko, J., Bein, W., Nyknahad, D., & Ito, H. (2015). Evaluation of online power-down algorithms. *2015 12th International Conference on Information Technology – New Generations (ITNG)*, Las Vegas (pp. 473–478).
5. Augustine, J., Irani, S., & Swamy, C. (2004). Optimal power-down strategies. In *IEEE Symposium on Foundations of Computer Science* (pp. 530–539). Cambridge University Press.
6. Augustine, J., Irani, S., & Swamy, C. (2008). Optimal power-down strategies. *SIAM Journal on Computing, 37*(5), 1499–1516.
7. Barroso, L. A. (2005). The price of performance. *Queue, 3*(7), 48–53.
8. Bein, W., Bharat, B. M., Bein, D., & Nyknahad, D. (2016). Algorithmic approaches for a dependable smart grid. In S. Latifi (Ed.), *Proceedings of the 13th International Conference on Information Technology*, Las Vegas (pp. 677–687). Springer.
9. Bein, W., Hatta, N., Hernandez-Cons, N., Ito, H., Kasahara, S., & Kawahara, J. (2012). An online algorithm optimally self-tuning to congestion for power management problems. In *Proceedings of the 9th International Conference on Approximation and Online Algorithms*, Warwick (pp. 35–48). Springer.
10. Borodin, A., & El-Yaniv, R. (1998). *Online computation and competitive analysis*. Cambridge: Cambridge University Press.
11. Nedevschi, S., Chandrashekar, J., Liu, J., Nordman, B., Ratnasamy, S., & Taft, N. (2009). Skilled in the art of being idle: Reducing energy waste in networked systems. In *Proceedings of the 6th USENIX Symposium on Networked Systems Design and Implementation*, NSDI'09, Boston (pp. 381–394). Berkeley, CA: USENIX Association.
12. Reich, J., Goraczko, M., Kansal, A., Padhye, J., & Padhye, J. (2010). Sleepless in Seattle no longer. Technical Report MSR-TR-2010-16, Microsoft.

Internet Addiction in Kuwait and Efforts to Control It

Samir N. Hamade

Abstract

Internet Addiction has reached an epidemic level worldwide. Since the 1990's the Internet has exploded to become an important part of our daily live. It was best described as a sword with two edges. On one side, it brought the whole world to our fingertips. On the other side, the excessive use of it can and will lead to a state of mental and psychological disorder, hence the term Internet Addiction Disorder IAD. Kuwait, a tiny nation in the Arab Gulf countries was the first in the area to shed some light on the problem in 2009 by conducting a public awareness campaign in the traditional media. Since then, the government, along with other organizations, started to take some measures to control this disorder without any success because most of the measures were restrictive in nature rather than positive. This paper will revisit the Internet addiction scene among university students in Kuwait after 8 years from publishing the first paper as measured by the Internet Addiction Test (IAT) to measure the level of awareness and percentage of highly addicted students compared to the early results, and describe the efforts taking place to control it at the government, organization, and family level.

Keywords

Internet addiction disorder • Internet addiction test • University students • Preventive measures • Kuwait

110.1 Introduction

Internet Addiction has reached an epidemic level worldwide. Since the 1990's the Internet has exploded to become an important part of our daily live. It was best described as a sword with two edges. On one side, it brought the whole world to our fingertips and become an undisputed opportunity for social connectedness. On the other side, the excessive use of it can lead to a state of mental and psychological disorder, hence the term Internet Addiction Disorder IAD. This paper will shed some light on the Internet addiction problem among university students in the state of Kuwait, measure the level of awareness and the level of addiction among them in two different time periods (2008 and 2016). It will also describe the measures taken to control this disorder or at least stop it from wide-spreading. The outlines of the paper are: review of related literature on the subject followed by the IAD scene in the state of Kuwait; presentation and discussion of the results of the two studies conducted by the author; and explanation of the efforts to control it at the government, organization, and family level.

110.2 Review of Related Literature

Internet addiction disorder made its first significant appearance in the U.S. press in 1995, when an article written by O'Neill [1] was published in the New York Times. O'Neill

S.N. Hamade (✉)
Kuwait University, Kuwait City, Kuwait
e-mail: Samir.hamade@ku.edu.kw

© Springer International Publishing AG 2018
S. Latifi (ed.), *Information Technology – New Generations*, Advances in Intelligent Systems and Computing 558, DOI 10.1007/978-3-319-54978-1_110

quoted addictions specialists and computer industry professionals and likened excessive Internet use to compulsive shopping, exercise, and gambling. The concept did not instantly gain popular interest from journalists, academics, and health professionals until the following year when Kimberly Young presented the results of her research in a paper entitled "Psychology of computer use: XL. Addictive use of the Internet. A case that breaks the stereotype" [2] and later in her famous book *Caught in the net: how to recognize the signs of Internet addiction--and a winning strategy for recovery* [3].

Many researchers such as Kuss, Griffiths & Binder, [4] indicated that the best way to control Internet addiction is to study its relationship with some personality traits that might predispose individuals to Internet addiction. Higher scores on neuroticism [5], agreeableness and emotional stability [6], have been established as potentially important risk factors for Internet addiction. Others such as Chak & Leung, [7] found shyness, loneliness, anxiety and low self-esteem to increase risks of Internet Addiction. Suler [8] discussed the negative effects of Internet addiction, and stated that "people may lose their jobs, or flunk out school, or are divorced by their spouses because they can't resist devoting all of their time to virtual land'. Suler [8] described those people as "pathologically addicted". Engelberg and Sjoberg [9] also discussed the consequences and found that the Internet will cut the users off from real social relationships and ultimately lead to weak participation and involvement in social life.

In terms of treatment and prevention, Van Rooij et al. [10] found that treating therapists agree that a manual-based Cognitive Behavioral Therapy (CBT) and Motivational Interviewing (MI) treatment program, such as the 'Lifestyle Training' program, can be suitable for treating internet addiction. Pontes, Kuss, and Griffits [11] found that both psychological and pharmacological treatments had to be examined in light of existing evidence alongside particular aspects inherent to the patient perspective.

110.3 Internet Addiction in Kuwait

Kuwait, a tiny nation in the Arab Gulf countries was the first in the area to shed some light on the problem in 2009 by conducting a public awareness campaign in the traditional media. Since then, the government, along with other organizations, started to take some measures to control this disorder without any success because most of the measures were restrictive in nature rather than positive.

In 2009 Hamade [12] conducted a study on the use of the Internet among university students across gender at Kuwait University, and measure their awareness and level of addiction to this technology. The results of the study indicated a low level of awareness of Internet addiction among university students. An average of ten percent of students has a high level of addiction that requires treatment, and about 25 percent of them have low level of addiction. Male students were found to be more addicted to the Internet than female students. This is an indication that males in Kuwait enjoy more freedom to spend time outside the house with friends, and visit Internet cafes, game networks, and other places. This freedom enables them to spend more time surfing the Internet, and consequently become more vulnerable to this type of addiction.

In 2016 the author revisited the Internet addiction scene using the Internet Addiction Test (IAT) used by Widyanto & McMurren, [13] and Young & de Adreu, [14] to measure the level of awareness and percentage of highly addicted students compared to the early results. Preliminary results indicated a higher level of awareness and an increase in the highly addictive students, due to the widespread of mobile Internet and the various social media applications.

Table 110.1 shows the definition of Internet addiction provided by the students in the sample. It shows a slight change in the definition of addiction where bad habit has the highest score in 2008 and went down to second place in 2016. At the same time Heavy use went up to the top definition. While both of these definitions indicated a limited understanding of the nature of addiction, it is important to notice that psychological disorder was recognized by more than 27 percent of the students, an increase of 8 percent from the (19%) in the year 2008.

Concerning the solution or treatment of Internet addiction, Table 110.2 indicates that psychological therapy was the top choice for treatment in 2016, while in 2008 was preceded by good advice. This is an indication that among those who recognize Addiction as a disorder they believe that psychological not physical therapy is the solution followed by good advice and medicine.

Table 110.1 Definitions of internet addiction among university students

Internet addiction (IA)	2008 Number Percentage	2016 Number Percentage
Heavy use	060 029.4%	**089** **34.9%**
Bad Habit	**081** **039.7%**	077 31.1%
Psychological disorder	039 019.0%	068 27.4%
Psycho-physical disorder	017 008.3%	011 04.4%
Physical disorder	007 003.4%	003 01.2%
Total	204 100%	248 100%

Table 110.2 Treatment of internet addiction among university students

Internet addiction (IA)	2008 Number Percentage	2016 Number Percentage
Psychological therapy	070 34.2%	**101** **40.7%**
Advice	**088** **43.2%**	074 29.8%
Medicine	013 06.4%	054 21.8%
Physical therapy	025 12.3%	014 05.7%
Other	03.9%	005 02.0%
Total	204 100%	248 100%

Table 110.3 Levels of internet addiction among university students

Internet addiction (IA)	2008 Number Percentage	2016 Number Percentage
No addiction	**133** **65.2%**	**141** **56.9%**
Low addiction	050 24.5%	073 29.4%
High addiction	021 10.3%	034 13.7%
Total	204 100%	248 100%

Table 110.3 represents the level of addiction among students in the sample. Table 110.3, indicates that the number of highly addicted students increased by almost four percent between 2008 and 2016. That is an average of 0.5 percent annually. Also the number of low addicted students went up more than five percent while the students with no addiction symptoms decreased by about 9 percent indicating that the addiction level in general (low and high) had increased to 9 percent.

110.4 Efforts to Control Internet Addiction

The growing concern about the risks of Internet addiction lead to discussions among academics and researchers on the best way or ways to deal with the problem and the highly addicted person and take measures to control it or at least stop it from wide spreading.

Early measures taken to control the increase in Internet Addiction among youth in Kuwait were restrictive in nature. At the government level the ministry of Communication demanded the Internet Service providers (ISP) to install very restrictive filters to control the contents of the web by

Table 110.4 Parental strategies to deal with internet addiction

Strategy	Number Percentage
Order child to avoid certain websites	123 81%
Forbid child from using the Internet unsupervised	117 77%
Restrict time of using the Internet	089 59%
Discuss appropriate us of the Internet	069 45%
Discuss the danger and safety of the Internet	062 41%
Make specific rules to use the Internet	058 38%

blocking most of the websites frequently or heavily used by users such as pornographic sites, online gambling sites, and even online gaming sites. The filters were so restrictive to a degree of over-blocking many medical, health and educational sites because some pictures or texts that were considered inappropriate to the general audiences, especially to children and youths. ISP companies faced a dilemma if they tighten their filters they lose customers who complain about the restrictive filters and if they loosen their filters the government will punish them with fines and closure [15].

Many organizations also took some measures to reduce the heavy use of the Internet among their employees by installing additional filters on their networks blocking many websites and disabling many features and applications such as games and chatting. These organization believe that many employees are wasting company work-hours by doing activities not related to work that lead to low level of performance.

Many families in Kuwait took additional measures for the protection of their children from harmful materials and overuse of Internet activities. Some of them (14.5%) installed Internet filtering software on the home computers or network and took some restrictive measures such restricting the hours for using the Internet or mediating in their children use of the internet by checking websites visited, emails, chats and take counter measures by forbidding their children from spending much time on the Internet, blocking inappropriate websites from porn to gambling to websites inducing crimes, hate, and terrorism even online games and peer-to-peer websites.

Table 110.4 explains the strategies adopted by some families to protect their children from viewing inappropriate materials and becoming Internet addicts [16]. The three highest measures adopted by parents were restrictive in nature forcing children to obey them. They include ordering children to avoid certain websites (81%), forbidding them from using the internet alone (77%) and restricting time spent surfing the Internet (59%). On the other hand, the positive measures such as discussing appropriate Internet use and the danger and safety of the Internet were in the lower percentage of the table (45% and 41% respectively).

110.5 Conclusion and Recommendations

This paper shed some light on the Internet addiction problem in the state of Kuwait, and described some of the measures adopted at the government, organization and parental level. It is clear from this study and other related studies that Internet addiction is a growing problem that is not going to be resolved, and the most that can be done is to alleviate the problem, slow down its progress, and prevent it from spreading, especially among children and youths. As an Information scientist it is my duty to provide all the necessary information available in the literature and let the professionals in the medical, psychological and sociological fields find suitable solutions. However, there is no one solution that fits all. Every country and every society has its own characteristics and sequences and the solutions should be through combined efforts of the medical, sociological, psychological fields in addition to the religious efforts by strengthening the religious belief of the Internet addicts and getting closer to God.

References

1. O'Neill, M. (1995). The lure and addiction of life on line, *The New York Times* Sect. C: 1.
2. Young, K. S. (1996). Psychology of computer use: XL. Addictive use of the internet: A case that breaks the stereotype. *Psychological Reports, 79*(3), 99–902.
3. Young, K. S. (1996). *Caught in the net: how to recognize the signs of Internet addiction–and a winning strategy for recovery.* New York: J. Wiley.
4. Kuss, D. J., Griffiths, M. D., & Binder, J. F. (2013). Internet addiction in students: Prevalence and risk factors. *Computers in Human Behavior, 29*(3), 959–966.
5. Dong, G., Wang, J., Yang, X., & Zhou, H. (2013). Risk personality traits of Internet addiction: a longitudinal study of Internet-addicted Chinese university students. *Asia-Pacific Psychiatry, 5*(4), 316–321.
6. Van der, A.,. N., Overbeek, G., Engels, R. C., Scholte, R. H., Meerkerk, G. J., & Van den Eijnden, R. J. (2009). Daily and compulsive internet use and well-being in adolescence: a diathesis-stress model based on big five personality traits. *Journal of Youth and Adolescence, 38*(6), 765–776.
7. Chack, K., & Leung, L. (2004). Shyness and locus of control as predictors of Internet addiction and Internet use. *Cyber Psychology & Behavior, 7*(5), 559–570.
8. Suler, J. (2004). Computer and Cyberspace Addiction. *International Journal of Applied Psychoanalytic Studies, 1*(4), 395–362.
9. Engelberg, E., & Sjoberg, L. (2004). Internet use, social skills, and adjustment. *Cyber Psychology & Behavior, 7*(1), 41–43.
10. Van Rooij, A. J., Zinn, M. F., Schoenmakers, T. M., & Van de Mheen, D. (2012). Treating internet addiction with cognitive-behavioral therapy: A thematic analysis of the experiences of therapists. *International Journal of Mental Health and Addiction, 10*(1), 69–82.
11. Pontes, M., Kuss, J., & Griffits, M. (2015). Clinical psychology of Internet addiction: a review of its conceptualization, prevalence, neuronal processes, and implications for treatment. *Neuroscience and Neuroeconomics., 4*, 11–23.
12. Hamade, S. (2009). Internet Addiction among University Students in Kuwait. *Digest of Middle East Studies, 18*(2), 4–16.
13. Widyanto, L., & McMurren, M. (2004). The psychometric properties of the Internet addiction test. *Cyberpsychology and Behavior., 7*(4), 445–453.
14. Young, K., & de Adreu, C. N. (Eds.). (2011). *Internet addiction: A handbook and guide to evaluation and treatment.* Hoboken: Wiley.
15. Hamade, S. Internet filtering and censorship. *The fifth international conference of information technology: new generation (ITNG 2008).* Las Vegas, USA. April 7–9, 2008, published in the proceedings by IEEE Computer Society (pp. 1081–1086).
16. Hamade, S. Parental awareness and mediation of children internet use. The 12th *international conference of information technology: new generations (ITNG 2015).* Las Vegas, USA. April 11–14, 2015, published in the proceedings by IEEE Computer Society (pp. 640–645).

Approaches for Clustering Polygonal Obstacles

Laxmi P. Gewali and Sabbir Manandhar

Abstract

Clustering a set of points in Euclidean space is a well-known problem having applications in pattern recognition, document image analysis, big-data analytics, and robotics. While there are a lot of research publications for clustering point objects, only a very few articles have been reported for clustering a distribution of obstacles. In this paper we examine the development of efficient algorithms for clustering a set of convex obstacles in the 2D plane. We present two approaches for developing efficient algorithms for extracting polygonal clusters. While the first method is based on using the nearest neighbor computation, the second one uses reduced visibility graph induced by polygonal obstacles. We also consider the extensions of the proposed algorithms for non-convex polygonal obstacles.

Keywords

Point set clustering · Obstacle clustering · Collision-free paths

111.1 Introduction

In this paper, we investigate the development of efficient algorithms for aggregating a given distribution of polygonal obstacles into clusters, which has application in planning collision free paths and pattern recognition.

Clustering is the process of organizing the given distribution of entities into a few groups such that members in a group are closely related in some measure. The proximity measure to put two entities in the same group depends on the intended applications. One of the most frequently used proximity measures is based on Euclidean distance.

The most commonly used objects for clustering are the distribution of points in Euclidean space. While the problem of clustering points in Euclidean space is a well-investigated problem [6, 11, 13], only a few authors have considered the clustering of polygonal obstacles [9]. A simple instance of

obstacle clustering is shown in Fig. 111.1. This figure shows that obstacles could be clustered in more than one ways. This paper is organized as follows. In Sect. 111.2, we present a critical review of widely used algorithm for clustering points distributed in two dimensions. In particular, we closely examine the quality of solutions obtained by using k-means algorithms and their variations. In Sect. 111.3, we present the main contribution of this paper. We propose two obstacle clustering approaches using Closest Neighbor Computation as an aggregating tool. While the first one is based on the direct computation of shortest separation between all pair of obstacles, the second one uses the visibility graph induced by polygonal obstacles to identify members belonging to the same cluster.

Finally, in Sect. 111.4, we discuss possible extensions and improvement of the pro-posed algorithms.

L.P. Gewali (✉) · S. Manandhar
Department of Computer Science, University of Nevada,
Las Vegas, NV, USA
e-mail: Laxmi.gewali@unlv.edu; manans1@unlv.nevada.edu

© Springer International Publishing AG 2018
S. Latifi (ed.), *Information Technology – New Generations*, Advances in Intelligent
Systems and Computing 558, DOI 10.1007/978-3-319-54978-1_111

Fig. 111.1 Two ways of
clustering obstacles

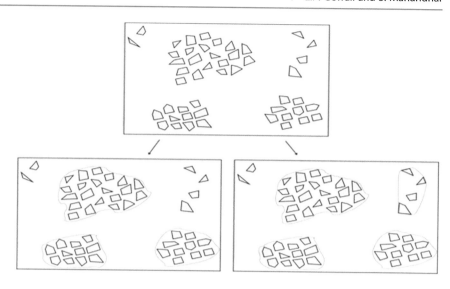

111.2 Preliminaries

Clustering is the process of partitioning a set of objects, usually points, into disjoint groups, such that members close to each other are in the same group. Specifically, any member in a group should have its nearest neighbor in the same group.

Figure 111.2 shows a distribution of points in two dimensions that admit, more or less, distinctly visible clusters. In certain distributions the clusters are not distinctly recognizable. Clustering has important applications in various areas that include pattern recognition, machine learning, image analysis and robotics.

Development of efficient algorithm for identifying clusters in Euclidean space has been considered by several researchers [6, 11, 13]. One of the earliest algorithms for identifying clusters among points distributed in two dimensions is the k-means algorithm proposed by S.P. Lloyd [13]. Several variations of this algorithm have been re-ported [1, 11]. k-means algorithm is an iterative algorithm in which clusters are progressively estimated by a series of refinements. In this algorithm, the number of clusters K is assumed to be given. Furthermore, for each cluster, a representative location is somehow estimated. The algorithm assigns input points to one of the representative points. The points belonging to a particular representative are taken as members of that cluster. Figure 111.3 shows an example of cluster estimation by using the k-means algorithm. The progress of the movement of cluster representatives is shown by directed line segments. It is observed that the extent of the movement of a cluster's representative becomes progressively smaller after each cluster refinement.

The k-means algorithm is fairly simple to understand. However there are some limitations of this approach.

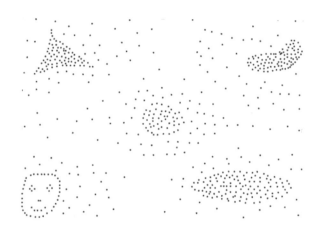

Fig. 111.2 Point distribution admitting distinct clusters

The main drawback of the algorithm is the estimation of number of clusters k and the initial location of their center. For a given input set of points, the output of the algorithm depends upon the initial location of representative points and the value of k. Similarly, the determination of the value of the threshold to stop the iteration of the algorithm is not quite clear. For certain data distributions, using the mean location of the points may not be valid (as observed in [6]).

The location of a representative is updated by computing the mean of the locations of points in that partially determined cluster. The updated point is next used to identify cluster membership. The Euclidean distance between old location and new location of a representative point is taken as the measure of the progress of modification. The updating of cluster identification continues as long as the progress of modification is more than a certain threshold value. The threshold value can be determined in terms of the separation of the closest pair of input points. A formal sketch of k-means algorithm is listed as Algorithm 1.

Fig. 111.3 Illustrating the execution of k-Mean algorithm

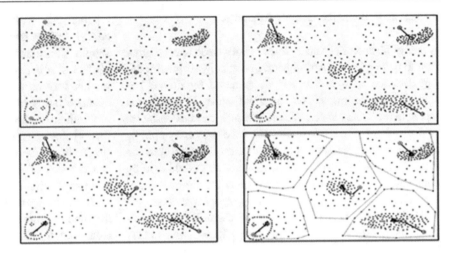

Algorithm 1: (Sketch of k-Mean Algorithm)

Input: A finite set of Points P = {p_1, p_2, . . . , p_n}

Output: Clusters C = {C_1, C_2, . . . , C_k}

1 Let k be the input representing number of clusters in C

2 Let δ be pre-determined threshold value

3 Select epresentative points m_1, m_2, . . . , m_k

4 Let CentroidMovement = δ + 1

5 while CentroidMovement > δ do

    a. For i ← 0 to n-1 do

       Let m_j be the closest representative point to p_i

       Associate the point p_i to the cluster c_j

    b. Compute new representative points new_m_j's as the centroids of the new points associated with c_j's

    c. Set CentroidMovement by comparing m_j's and new_m_j's

    d. Assign new_c_j's to c_j's

6 Output C

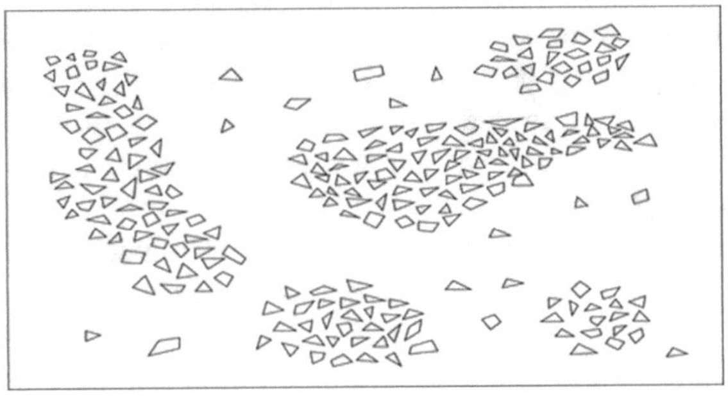

111.3 Clustering Obstacles

111.3.1 Problem Formulation

Consider a collection of polygonal obstacles in two dimensions. For the purpose of clarity of presentation, we consider only convex obstacles. However, at the cost of additional time complexity overhead, the algorithmic techniques presented in this section are applicable even if the obstacles are not convex. A configuration of around 240 polygonal obstacles is shown in Fig. 111.4.

Fig. 111.4 A configuration of 240 obstacles

A visual examination of the obstacles distribution in Fig. 111.4 reveals five clusters. The problem we investigate is the capturing of clusters of obstacles under a certain measure. Broadly speaking, given a parameter δ, we want to group together obstacles whose distance to its nearest neighbor is no more than δ. The problem can be formally stated as follows.

Obstacle Clustering Problem (OCP)

Given: (i) A set of convex obstacles Q_1, Q_2, . . . , Q_m in the plane, (ii) A parameter δ.

Question: Construct obstacle clusters such that any obstacle in a cluster has its nearest neighbor within distance δ.

Remark 1 The distance between two obstacles Q_i and Q_j is the smallest distance between boundary points in Q_i and Q_j. When we use the term *distance*, it is understood to be the Euclidean Distance, i.e. the distance $d(p_i,p_j)$ between two points $p_i(x_i,y_i)$ and $p_j(x_j,y_j)$ is given by $d(p_i,p_j) = ((x_i - x_j)^2 + (y_i - y_j)^2)^{1/2}$. In some applications, distance between two obstacles is measured from their center of gravity. But in our application we measure distance from boundary points.

111.3.2 Proximity Graph Approach

A pair of obstacles O_i and O_j that are very close to each other should belong to the same cluster. Specifically, if the Euclidean distance between O_i and O_j, denoted by $d(O_i,O_j)$ is smaller than the predefined threshold value δ then O_i and O_j are in the same cluster. We refer to such pairs of obstacles as δ-*proximity pairs*. We can connect a δ-*proximity pair* by the edge e that corresponds to the shortest distance between them. So, to identify all obstacle clusters we could begin by connecting proximity pairs by corresponding shortest edges to obtain the δ-*proximity graph* (or δ-*graph* for short).

Each connected component in the δ-graph is a cluster component. Figure 111.5 illustrates these ideas. Figure 111.5 (top part) shows a distribution of convex obstacles with indicated value of predefined parameter δ. The δ-graph for this distribution is shown in the bottom part. A straight forward connection obtained by considering all δ-proximity edges can lead to undesired consequences, which happens when a very small or thin obstacle lies between two δ-proximity pairs.

This is occurring in the top left corner in bottom part of Fig. 111.4, where the min-separation edge between Oi and

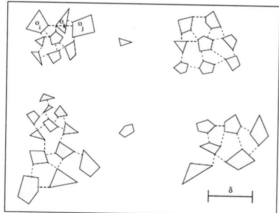

Fig. 111.5 Illustrating δ-proximity graph

Oj intersects with obstacle Ok. We need to discard such edges. Rather than considering all δ-proximity pairs we should consider only those that do not squeeze other obstacles in between them. We refer to the graph obtained in this way as the δ-planar Graph. Computing minimum distance between convex polygons is a well investigated problem in computational geometry [2, 14]. It is known [2, 14] that the minimum distance can be computed in O (log n) time after a pre-processing overhead of O(n). So, we can use one of these two algorithms for computing minimum separation between a pair of convex obstacles.

A direct approach for computing the δ-planar graph is to first find the minimum distance between all O(n2) pairs and then check for intersection of min-separation edges with other polygons. A formal sketch of the algorithm based on this approach is shown as Algorithm 2. Once we have the connecting lines we can run the Breadth First Search (BFS) or Depth First Search (DFS) [3] algorithm to find the connected components and hence identify the obstacle clusters.

Algorithm 2: (Constructing δ-Proximity Graph)

```
Input: i) Obstacles Q1, Q2, . . . . , Qm , ii)
Parameter δ
Output: Connecting Shortest Edges List E
1    E = φ
2    Mark all pairs Unprocessed in M[][]
3    for all unprocessed pair Q[i], Q[j] do
4        let e = distance (Qi , Qj)
5        if |e| < δ and e does not intersect
         with
             other obstacles then
                 E = E U e
7                Mark M[i][j] as Processed
8    Output E
```

The time complexity of Algorithm 2 can be done in straightforward way. Marking the n x n array in step 2 takes $O(n^2)$ time. Step 3 executes in time $O(n^2)$ since there are $O(n^2)$ pairs. Checking the second condition in the if statement (Step 5) takes $O(n)$ time. Hence the total time for the for-loop is $O(n^3)$. Thus the overall time complexity for Algorithm 2 is $O(n^3)$. The time complexity of Algorithm 2 is rather high. Repeatedly checking for intersection of candidate edges with all obstacles is the reason for the high time complexity. With the motivation of developing a faster algorithm, we next examine the feasibility of using a visibility graph for developing a faster algorithm.

111.3.3 Visibility Graph Approach

Let $R = \{O_1,O_2,...O_k\}$ be the given set of polygonal obstacles in the plane. The visibility graph induced by R, denoted as *VG(V, E)* consists of vertices which are precisely the vertices of obstacles in R. That is, the vertex set $V = \{v_i \mid v_i \in O_k\text{'s}\}$. Two vertices $v_i, v_j \in V$ are connected to form a visibility edge (v_i, v_j) if the line segment connecting v_i to v_j does not intersect with any obstacle. Specifically, $E = \{(v_i, v_j) \mid v_i, v_j \in V$ and line segment (v_i, v_j) does not intersect with any obstacles$\}$. Figure 111.6 shows an example of a visibility graph. It is noted that the edges of obstacles are also edges of the visibility graph.

The visibility graph induced by obstacles contains edges corresponding to all visible pairs [4, 5, 10]. Nearer obstacle pair will tend to have shorter visibility edges. Depending on the shape of the obstacles, some closer obstacle pairs can also have longer visibility edges. The edges of the visibility graph can be examined to discard those edges whose 'separation lengths' are longer than δ. Here the term 'separation length' of a visibility edge e_i should be clarified further. Let e_i connects obstacle vertices v_r and v_p and let obstacle edges incident on v_r and v_p be e_1, e_2 and e_3, e_4, respectively, as shown in Fig. 111.6. Then the separation length corresponding to e_i is either projection to one of the obstacle edges (Figs. 111.7 and 111.8) or simply e_i itself.

Let $sep(e_i)$ denote separation length corresponding to visibility edge e_i. The graph obtained by discarding visibility edges whose separation length is greater than δ is called *reduced visibility graph*. The visibility edges in the reduced visibility graph are not necessarily the edges corresponding to minimum distance. However, these edges can be examined locally to construct the corresponding separation length. When separation edges are constructed, some of separation edges could intersect obstacles. In such situations we should be able to find shorter separation edges as stated in the following lemma.

Lemma 1 *The shortest visibility edge e_i from a vertex of an obstacle is such that $sep(e_i)$ can not intersect with another obstacle.*

The connected components of the reduced visibility graph correspond to obstacle clusters. The actual obstacle clusters can be identified by straightforward application of breadth first search or depth first search.

111.4 Discussion

One of the main motivations to write this paper is that the clustering polygonal obstacles can be used to significantly simplify the path planning problem. An extracted cluster can be treated as a single obstacle for planning collision-free paths, which in turn reduces execution time of collision-free path computation. We showed how δ-planer graph induced by polygonal obstacles can be used to identify polygon clusters. This method is simple and easy to understand but the resulting time complexity $O(n^3)$ is rather too high. We explained how the Visibility Graph induced by polygonal obstacles can be used to design a faster version of obstacle clustering. We showed that visibility edges can be examined locally to obtain closest obstacle pairs. This approach exploits faster Visibility Graph construction algorithms to extract obstacle clusters efficiently in $O(n^2)$ time.

Fig. 111.6 Visibility graph induces by polygonal obstacles

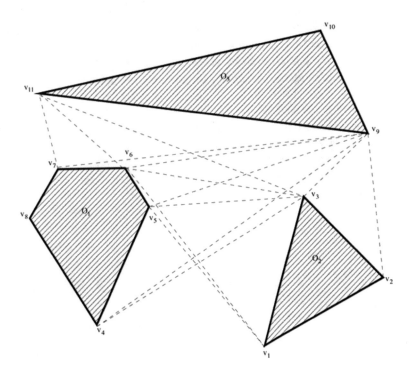

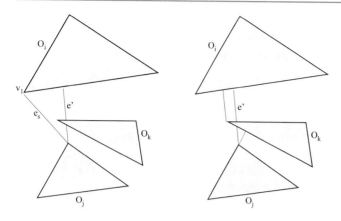

Fig. 111.7 Illustrating 'separation length' of visibility edges

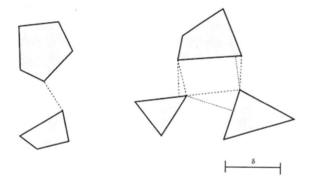

Fig. 111.8 Capturing separation length from reduced visibility edges

We have made some progress in using Voronoi diagram [2, 14] induced by the vertices of the obstacles to identify obstacle clusters. Specifically, the clearance property of the Voronoi diagram can be usedto identify neighboring obstacles efficiently. The detail of this approach is available in [16] and will be reported in future reports. In order to understand the effectiveness of an obstacle clustering algorithm, as a pre-processing step in path planning, it would be necessary to count the number of edges in obstacles before and after the application of clustering. We plan to address this issue in the future. In this paper, we have represented obstacles by convex polygons. In real life, an obstacle can also by nonconvex. In such cases, the algorithms proposed algorithms can still valid with some modifications. One modification could be to convert the non-convex polygonal obstacles into convex by computing their convex hull.

References

1. Alsabti, K., Ranka, S., Singh, V. An efficient K-Means clustering algo-rithm. Available online at https://www.cs.utexas.edu/ kuipers/readings/Alsabti-hpdm-98.pdf
2. De Berg, M., Van Kreveld, M., Overmars, M., Schwarzkoft, O. (1997). Computational Geome-try: algorithms and applications. Berlin Heidelberg: Springer Verlag.
3. Cormen, T. H., Lieserson, C. E., Rivest, R. L., & Stein, C. (2009). *Introduction to Algorithms* (3rd ed.). Cambridge, Massachusetts: The MIT Press.
4. Ghosh, S. K., & Mount, D. M. (1991). An output-sensitive algorithm for computing visibility graphs. *SIAM J. Computation, 20,* 888–910.
5. Goodman, J.E., & O'Rourke, J. (1997). Handbook of discrete and computational geometry. Boca Raton: CRC Press.
6. Clustering Algorithm for Large Data-bases. Available online at http://www.cs.bu.edu/fac/gkollios/ada05/LectNotes/guha98cure.pdf
7. Hershberger, J., & Suri, S. (1997). An Optimal Algorithm for Euclidean Shortest Paths in the Plane. *SIAM Journal on Computing, 28,* 2215–2256.
8. Hershberger, J., Suri, S., Yildiz, H. (2013). A near optimal algorithm for shortest paths among curved obstacles in the plane. SoCG Proceedings of the twenty-ninth annual symposium on computational geometry, (pp. 359–368).
9. Joshi, D. (2011). Polygonal spatial clustering. PhD thesis, University of Nebraska.
10. Kitzinger, J. The visibility graph among polygonal obstacles: a comparison of algo-rithms. Univer-sity of New Mexico. Available online at: http://www.cs.unm.edu/ moore/tr/0305/Kitzingerthesis.pdf
11. Kanungo, T., Mount, D. M., Netanyahu, N. S., Piatko, C. D., & Angela, Y. W. (2002). An efficient k-means clustering algorithm: analysis and implementation. *IEEE Transactions on Pattern Analysis and Machine Intelligence, 24,* 881–892.
12. Lee, D.T. (1978). Proximity and reachability in the plane. PhD thesis, University of Illinois at Ur-bana Champaign, A Doctoral Dissertation.
13. Lloyd, S. P. (1982). Least squares quantization in pcm. *IEEE Transactions of Information Theory, 28,* 129–137.
14. O'Rourke, J. (1998). Computational geometry in C. New York: Cambridge University Press.
15. Sack, J. R., & Urrutia, J. (2000). *Handbook of computational geometry.* Amsterdam: Science Publishers.
16. Manandhar, S. (2016). Efficient algorithms for clustering polygonal obstacles, MS Thesis, Department of Computer Science, University of Nevada, Las Vegas.

Part XIII

Short Papers

Claudio Casado Barragán, Miguel Palma Esquivel,
Cristián Barría Huidobre, Cristián Rusu, Cesar Collazo,
and Clara Burbano

Abstract

Nowadays, where software features such as usability and security are considered ideal, there are currents that at the same time consider these two qualities as adverse, as increasing the standards of one, the other tends to decrease. It is because of the above that it is important to evaluate the usability characteristics in software that focus its operation in the area of information security, through a measurement instrument such as the evaluation heuristic, in order to know the level of usability In this type of software.

Keywords

Usability • Security • Evaluation Heuristics

112.1 Introduction

Currently, one of the parameters for software validation is the degree to which it can be used with effectiveness, efficiency and satisfaction given a specific context of use [5], which definition corresponds to Usability.

C.C. Barragán (✉) • C.B. Huidobre
Universidad Mayor, Computación e Infromática, Santiago, Chile
e-mail: claudio.casadob1@mayor.cl; cristian.barria@mayor.cl

M.P. Esquivel
Universidad Tecnológica de Chile, Computación e Informática, Santiago, Chile
e-mail: miguelpalmae@yahoo.es

C. Rusu
Universidad Católica de Valparaíso, Ingeniería Informática, Santiago, Chile
e-mail: cristian.rusu@ucv.cl

C. Collazo
Universidad del Cauca, Departamento de Sistemas FIET, Popayán, Colombia
e-mail: ccollazo@unicauca.edu.co

C. Burbano
Corporación Universitaria Comfacauca, Departamento de Sistemas, Popayán, Colombia
e-mail: clara_893@hotmail.com

In the case of products oriented to information security, these represent a very high level of difficulty for the average user, which reduces the effectiveness of the Information Security management systems (ISMS), being the factor the greatest obstacle to effective computer security [11].

Considering the usability and security of the factors that can become the ideal characteristics of a system, the problem presented by this research lies in the confrontation between one and the other, where the improvement of one can affect the negative form to the other, The process of designing systems [1, 13].

As he is an expert in the investigation, a test that bases his methodology in heuristic evaluation is realized, on three programs of computer security, that evaluates the usability degree of systems in particular [3], Of acceptance of usability for these software, based In the experience of the authors, in order to obtain a real measure of this characteristic on the different safety programs used in the test. Although there are evaluation heuristics, for this work the J. Nielsen plant is used along with the contributions of D. Pierotti.

The article discusses, in section two, the related works that were contributed to the research. In the section we define the concept of utility and the evaluation heuristic of a job, while in section four, the tools of computer security and are defined three for further evaluation. Finally, in

© Springer International Publishing AG 2018
S. Latifi (ed.), *Information Technology – New Generations*, Advances in Intelligent Systems and Computing 558, DOI 10.1007/978-3-319-54978-1_112

sections five and six, we reviewed the results obtained and the research findings respectively.

112.2 Related Jobs

According to Perurena and Moraguez, despite the efforts made today, software has a poor level of usability, given that they pay more attention to system elements such as performance or reliability, so aspects such as interactive design, adaptability and deployment Of precise and concise information when the user requires it, they pass into the background [8].

Other authors, such as Gómez, focus on the user type and states that some computer applications, from the security area, do not contemplate the user with little experience in its design, presenting usability problems since it does not know the capacity or operation of related options With security, and, therefore, of the product that is being used [9].

Whitten and Tygar say that the aforementioned contradiction between usability and security reaches systems in the latter area, since they are effective only if used correctly. Research shows that 90% of network and computer security flaws are caused by poor antivirus and firewalls, and, given the growing proliferation of mobile devices, notably notebooks and smartphones, networked transactions are increasingly frequent, so the need to make information products intuitive has become a critical issue [11]. Paine and Edwards, based on the importance of good usability for security software, the difference between a good and bad user interface can influence the ability of a user to safely perform tasks on their computer or device [12].

Fierro assures that banking institutions in their attempt to reduce distrust towards electronic banking, focuses their efforts developing solutions with high levels of security, however, this has resulted in very complex applications for the user 10.

According to all of the above and considering that user satisfaction with a product and software intent are affected by features such as usability and security [10], emphasis should be placed on usability in security systems.

112.3 Usability and Heuristic Evaluation

According to ISO, usability corresponds to the degree to which a product can be used by specific users to achieve specific objectives with effectiveness, efficiency and satisfaction in a given context of use. [4].

This is not only about the graphical interface of the system, it also includes the help systems, documentation, installation process and everything related to the interaction between user and product [7].

On the other hand, heuristics is a method of evaluating usability based on previously established principles

[2]. Therefore, the heuristic evaluation aims to measure the quality of any interactive system in relation to its ease of being learned and used by a certain group of users in a given context [4].

In order to carry out the usability evaluation, we analyze the heuristics of several authors, both for web and software in general, using as basis the principles postulated by J. Nielsen next to the adjustments integrated by D. Pierotti, thus proposing a heuristic Of evaluation for information security software that considers principles such as system state visibility, users' language, control and freedom for the user, consistency of standards, assistance to users for recognition, diagnosis and recovery of errors, prevention of errors , Recognition before cancellation, flexibility and efficiency of use, dialog aesthetics and minimalist design, general help and documentation, skills, user interaction pleasant and respectful and privacy.

The evaluation carried out in the present investigation is of qualitative type, so the measurement parameters for each question within the concepts described above are defined as: meets, does not meet or does not apply, only interests to know if the evaluated software complies with it what is being measured.

112.4 Computer Security Tools

The security tools are those that preserve the availability, confidentiality and integrity of information, within an environment in which technological development has made possible the constant Internet connection through multiple devices [6].

In the present study, three of these tools, developed by companies of wide trajectory and of free access to the average user, are analyzed in order to evaluate the usability that they use in their implementationThe selected tools are from Sophos company, its product firewall: UTM, an antivirus, in its free version, AVG company and last PGP encrypter made by Symatec.

Given the evaluation heuristic selected for the test, the criteria of "Compliance", "Not Compliant" and "Not Applicable" were defined for each of the questions in the measurement tool. To perform the measurements, it was defined that the "Compliance" characteristic contributes a greater amount of score to the usability of the system, whereas the criterion "Not Complies" contributes a smaller amount of score. The "Not Applicable" option does not give any score to the usability of the system. At the same time, we defined for each evaluation criterion that provides usability score, a factor with which weights of the results, expressed in a scale where the weighted minimum is 3.8 when all the points of the heuristic The "Not Applicable" parameter and 7.7 when they fully comply with the "Comply" criterion. Independent of the number of questions that each one of the presented heuristics possesses, to later sum the results of each one of them and obtain the final score for each point of the test. Given the above, 75% was defined as the final acceptance parameter for usability in security systems, this being the minimum value that security software can have after the heuristic evaluation to have an acceptable level of usability.

112.5 Results Obtained

The instrument is conformed as shown in Table 112.1, forming a questionnaire of a total of 248 questions to which the user is subjected to be able to carry out the evaluation of the software.

Applying the heuristics to each of the selected tools yielded the results of Table 112.2.

Table 112.1 Structure of the measuring instrument

Heuristics	Questions
Visibility system status	17
User Language	21
User Control and Freedom	21
Awareness and standards	43
Helps users to recognize, diagnose, and recover from errors	16
Preventing errors	14
Recognition before cancellation	36
Flexibility and efficiency of use	14
Aesthetics of Dialogues and Minimalist Design	11
General help and documentation	21
Skills	19
Pleasant and respectful user interaction	12
Privacy	3
TOTAL	248

At the time of obtaining the total usability score of the security software, once the heuristics have been applied, it is represented on a scale of 1 to 100. As mentioned above, a default integrated usability level was stipulated, despite the fact that The system does not meet a specific point. Thus, despite obtaining a total predominance of the parameter "No Compliance" in the evaluation, the system will qualify with 50 points, which indicates, for this investigation, the basic parameter of usability despite being using a basic scale as mentioned above. On the other hand, if the parameter "Complies" obtains a total predominance within the test for a specific system, it will have a score of 100 points.

As a result the PGP encrypter gets 79.9 points out of a total of 100, reflecting that it complies with 80%, the measured heuristic. AVG Antivirus gets 74.9 out of a total of 100 points, reflecting compliance with 75% of the measured heuristic. Finally, the Sophos UTM Firewall obtains 75.9 points out of the 100, thus fulfilling 76% of the heuristic used.

It is evident that the tools meet the 50% required by each heuristic to which they were subjected, this shows that the manufacturers of such software have considered usability elements in their development, taking into consideration the characteristics that the software must meet, or Either, for the function that must comply and a specific user profile. As a final result, it can be said that, in order to obtain a minimum standard of usability, software must generally comply with 50% of the score after the usability test, but since this work seeks to identify if the measured tools comply With security and usability standards, is that a score of at least 75% has been established to consider a large presence of usability standards in the systems. As can be seen in Table 112.2, all the analyzed software comply with this minimum. Even so, it can be seen that PGP software obtains a higher percentage in the evaluation, since it is developed for semi specialized users, since its purpose does not

Table 112.2 Heuristic evaluation results

Heuristics	PGP	AVG	UTM
Visibility system status	5,4	5,7	6,1
User Language	7,1	6,4	6,6
User Control and Freedom	4,6	4,9	3,3
Awareness and standards	6,6	6,8	6,2
Helps users to recognize, diagnose, and recover from errors	6,5	5,3	5,5
Preventing errors	5,2	4,9	3,8
Recognition before cancellation	6,4	5,6	6,4
Flexibility and efficiency of use	3,3	2,5	4,4
Aesthetics of Dialogues and Minimalist Design	7,7	7,7	7,7
General help and documentation	6,4	6,2	6,0
Skills	6,9	6,1	5,5
Pleasant and respectful user interaction	6,1	5,1	6,7
Privacy	7,7	7,7	7,7
TOTAL	79,9	74,9	75,9

represent an information security management of the device on which it works, rather , Represents only a utility or tool for a specific function such as message encryption. Unlike the other two softwares evaluated, which represent programs of protection against threats with functions of management of resources, privileges and security functions. It is because of the above that both AVG Antivirus Free and Sophos UTM Firewall software have a lower usability weight, since its complex functions away from average users, making it more difficult to interact with them.

112.6 Conclusions

According to the problematic that poses the possibility of considering usability and security as inverse or opposite factors of a system, despite being considered ideal characteristics, with the results obtained from the analysis and measurement made to the selected software that have specific functions within Security of information such as encryption of information, protection against virus threats and protection of unauthorized access, it is clear that it is perfectly feasible and convenient to consider usability in the development of computer security tools, both can coexist and strengthen, all Since a tool that is easier for the user, will support the function that it should develop.

The present investigation is based, based on the experience, the score of 75% to indicate that said software, in charge of supporting the management of information security, comply with a good usability level according to its function.

Given the results of the three security software evaluated obtained acceptable scores according to the parameter for minimum acceptance of 75% for usability, in the measurement of this type of tools two important elements must be considered, such as the specific task to be performed with such software and The specific user profile that will carry out the task with this one, since for being computer security tools, they are oriented to a profile and level of specific knowledge.

References

1. Hernández, A., Fernández, R., Hernández, L.A., & Toledano, D.T. (2005). Evaluación de la Usabilidad de los Sistemas de Verificación Biométrica.
2. Juan, M. (2014). Usabilidad de la Seguridad.
3. González, M., Pascual, A., & Lorés, J. (2001). Evaluación heurística, Introducción a la Interacción Persona-Ordenador, AIPO: Asociación Interacción Persona-Ordenador.
4. International Organization for Standarization ISO 9126: Software Engineering-Product quiality. (2001). Geneva.
5. ISO-9241-11 Guidance on Usability.
6. Montesino, R., Baluja W., & Porvén, J. (2013). Gestión automatizada e integrada de controles de seguridad informática, RIELAC, Vol.XXXIV.
7. Ferré, X. (2007). *Principios Básicos de Usabilidad para Ingenieros Software*. Madrid: Universidad Politécnica de Madrid.
8. Perurena, L., & Moráguez, M. (2013). *Usabilidad de los sitios Web, los métodos y las técnicas para la evaluación*. Cuba: Universidad de La Habana.
9. Gómez, A. (2014). *Auditoría de seguridad informática*. Madrid: RA-MA S.A.
10. Fierro, N. (2015). Heurísticas para evaluar la Usabilidad de Aplicaciones Web Bancarias. Pontificia Universidad Católica del Perú.
11. Whitten, A., & Tygar, J. D. (1998). *Usability of security: A case study*. Pittsburgh: Carnegie-Mellon Univ, Dept Of Computer Science.
12. Payne, B. D., & Edwards, W. K. (2008). A Brief Introduction to Usable Security. *IEEE Internet Computing, 12*(3), 13–21.
13. Kainda, R., Flechais, I., & Roscoe, A.W. (2010). Security and usability: Analysis and evaluation, In *ARES'10 International Conference on Availability, Reliability, and Security* (pp. 275–282). IEEE.

Techniques for Detecting, Preventing and Mitigating Distributed Denial of Service (DDoS) Attacks

Judith Clarisse Essome Epoh

Abstract

Even though Internet appears to be one of the successful phenomena of globalization today, web applications, services, and servers are being challenged by multiple vulnerabilities due to multiple penetrations. These security flaws can easily be exploited by malicious actors who will use malware to launch DDoS to damage critical infrastructures in small and large businesses putting their productivity and trust at risk. This paper offers methods that public and private sectors can consider to lessen damages cause by DDoS. The detective techniques will help uncover some early signs of malicious activities in the organization's network. The preventive ones will ensure all methods have been implemented to stop the intrusion from happening. Findings have demonstrated that mitigation mechanism can only be effective with detective and preventive methods. It is vital to keep in mind that attackers are busy developing sophisticated tools to disrupt services and damage systems making traditional security tools ineffective. They need to be replaced by robust security technologies to protect networked systems efficiently as presented in this research. Security awareness as an important network security practice, will educate non-IT professionals, serve as a reminder to IT professionals and result in thwarting insider threats. When all these are successfully implemented, an attacker's chances of launching a successful distributed denial-of-service attack are reduced by 2%.

Keywords

DDoS • Detection • Prevention • Mitigation • Limitations

113.1 Introduction

A distributed denial-of-service (DDoS) attack is known as a threat that causes multiple compromised systems connected to Internet to undergo denial of service for legitimate users of the target system. The target system will be forced to shut down denying services to the system to authorized users. All Internet applications are challenged by many types of penetration attempts in today's digital world. This explains the significant increase of Denial of Service (DoS) and Distributed Denial of Service (DDoS) attacks. Since most consumers are generally connected to the same network, tools used by attackers will pass through ISP network disrupting running services not only to the target infrastructure, but also to consumers. That said, IT security management has to prioritize the detection, prevention and mitigation of DDoS attacks to defend against service loss and disruption effectively. IT managers have to think like attackers by using preattack methods that will help them determine the most appropriate mitigation approaches to follow in order to refrain from losing customers trust. In other words, IT managers will have to assess and select the

J.C.E. Epoh (✉)
Cybersecurity and Information Assurance Department, The Graduate School-University of Maryland University College, 3501 University Blvd. East, Adelphi, MD 20783, USA
e-mail: judithessome2010@yahoo.com

© Springer International Publishing AG 2018
S. Latifi (ed.), *Information Technology – New Generations*, Advances in Intelligent Systems and Computing 558, DOI 10.1007/978-3-319-54978-1_113

security best practices from the set of mechanisms available in the network.

This paper will present the detection, prevention, including mitigation techniques of the DDoS, and some of their limitations.

113.2 DDoS Detection

There are several ways of identifying DDoS every Internet user needs to be aware of.

113.2.1 Around-the-Clock Monitoring

IBM experts advise small and large businesses to put into place a response team that should be available around-the-clock to constantly look for circumstances where the DDoS attack(s) may be a masquerade or diversion to a real attack that may have been occurring into their network [6]. The only drawback of this method could be human error. This means, the response team may be lured by a decoy and fail to act faster. The figure below illustrates how the around-the-clock monitoring can be conducted to detect the DDoS attack in an organization's network (Fig. 113.1).

113.2.2 Pattern and Third-Party Detection

An attack detection strategy follows pattern detection, anomaly detection, and third-party detection mechanisms [10]. However, this section will only focus on the pattern detection and third-party detection mechanisms, because their approach of anomaly detection is similar to NIDS (network-based intrusion detection systems) and will be discussed later.

Pattern detection mechanisms secure the signatures of attacks that have been uncovered in a database. They

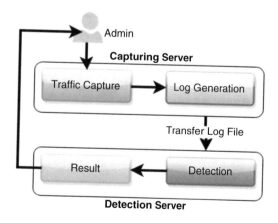

Fig. 113.1 DDoS detection

monitor every interaction and match it with database entries to uncover evidences of DDoS attacks. This detection mechanism is potent in identifying uncovered attacks with no false positives. For instance, Snort is helpful for DDoS defense system in attack detection. The weakness of this detection technique is that it cannot defend against potential and old attacks that do not match the stored signature. Also, a third-party detection technique does not take care of the detection process by itself. It instead depends on an external communication from a vendor that signals the presence of suspicious activities within the system. There was no disadvantage found for this detection tactic.

113.2.3 Dynamic Deterministic Marking and NIDS

IT Vendors may need to consider the approach of dynamic marking and mark-based detection with "Dynamic Deterministic Packet Marking," (DDPM) for fast and effective detection [11]. The dynamic marking technique helps find DDoS operations in large businesses network. In the mark-based method, the detection tool considers the marks of the packets to determine the true origins of DDoS attacks picturing spoofed IP addresses. As much research continues in this space, the only limitation of third-party detection mechanism with DDoS for now is the difficulty of identifying sources of attacks on the Internet-IP traceback, a difficult capability for detecting actual sources of attacks without depending on the IP source field contained in that potentially falsified packet.

Another DDoS detection tactic will be network-based intrusion detection systems (NIDS) very potent in detecting DDoS and other types of malware. NIDS include three beneficial modes, namely, signature detection, anomaly detection, and hybrid.

- A signature-based NIDS analyze the web traffic through its interface, and assess the TCP/IP packets for signatures of threats uncovered in a passive manner. Signature-based NIDS are great in monitoring known threats such as DDoS and preventing them from damaging networked systems, cleaning up resources to investigate potential threats.
- NIDS are built to detect anomalies in an organization web traffic that relies on statistical or baseline models. They monitor the network and send alerts to a network administrator when a user navigates in an environment an attacker can exploit to launch DDoS [2].
- inally, the hybrid system comprises the strengths on both signature-based and anomaly detection NIDS, and associate them to a standalone system to clear the limitations of both models. For instance, commercial NIDS use

signature because they are fast and flexible, combine anomaly detection to reveal malicious activities that will be assessed by response team.

On the other hand, NIDS have some drawbacks. Findings have demonstrated that a signature-based NIDS may fail to detect any attack it does not have a signature for. NIDS generate numerous false alerts due to inappropriate tuning of sensor to match the targeted environment accordingly, inappropriate correlation resulting in multiple alarms for a related incident. NIDS do not have the ability to detect DDoS attack on an encrypted network, cannot deal with high speed networks and cannot intervene during a detected DDoS. Also, NIDS or similar intrusion detection tools cannot prevent and mitigate DDoS or other types of malware. Intrusion detection system works best when complemented with an intrusion prevention system [2]. Finally, detections techniques should be followed by preventive ones for an effective defense.

113.3 Prevention

Prevention methods offer significant security best practices but will never guarantee the complete eradication of DDoS attacks due to new techniques developed by attackers for signatures and patches that are not present in the web application [3].

113.3.1 Kona Site Defender Technology from IBM/Akamai

As DDoS attacks increase in size and sophistication "old-fashioned" defenses of firewalls and intrusion-prevention systems have shown beyond doubt to be obsolete. IBM experts have developed the cloud-based Kona Site Defender technology from Akamai as a multifaceted line of defense against DDoS attacks. IBM and Akamai create a remarkable impact by providing a robust solution to prevent DDoS attacks that meet the bandwidth needs of private and public sectors [6]. Also, resource multiplication techniques provide multiple resources to limit DDoS attacks. For instance, a networked system that employs a pool of machines with a load balancer and installs high bandwidth links between itself and upstream routers. Even though Microsoft has used it as a powerful countermeasure against DDoS attacks in large businesses, the high-cost of Kona Site Defender technology remains one of its main drawbacks. Akamai services for distributed web site hosting redirect all users' requests to an Akamai name server, which will then

redirect the load towards several, geographically distributed web servers hosting replicas of the requested page [6]. However, hackers could launch still Cross-site scripting (XSS) threat and/or Structured Query Language (SQL) injections as most rampant web application vulnerabilities to compromise Akamai.

113.3.2 Security Awareness and Multifactor Authentication

Private and public sectors are recommended to conduct security awareness in the workplace to limit DDoS attacks. By raising awareness of security issues and concerns in a wider context, such as how to better protect their families and personal finances. Employees will be more engaged and their emotional interest will be sparked because they are all being exposed to phishing, password challenges, data theft and other social engineering tactics. Also, Businesses have to subscribe to IT approved vendors to help develop their training program in compliance with NIST SP 800–50, which aims at establishing an Information Technology Security Awareness and Training Program. The implementation of biometric systems, multifactor authentication, regular security audits and vulnerability assessments of systems will be a great addition to current security measures [8]. In addition, the most common indicators every Internet user should know to detect DDoS attacks on web sites, web applications and web services will be, for instance, Apache unusual low performance, inability to accessing Apache, email attachments from unknown sources, a strange rise of spams and more [13]. Finally, IT staff and non-IT staff should be aware that usernames and passwords do not provide adequate authentication. To resolve that issue, Fig. 113.2 explains the raison d'etre of multifactor authentication (something you know, something you have and something you are) as one of the most trusted tactic to avoid unauthorized access into an organization network, today.

Fig. 113.2 Multifactor authentication

113.3.3 Antimalware Software

Antimalware programs such as Nessus (powerful vulnerability scanner) are designed to protect against malware by analyzing the web traffic or by analyzing individual PCs. Antimalware software scans all web applications, services and data for malware prior to damaging a network by stopping any future attacks. The PC scans antimalware software programs could also be used for identifying and fixing security flaws preventing DDoS attacks. This type of DDoS protection is user-friendly and more popular. It provides the tester with a comprehensive list of existing vulnerabilities. This will then help the tester to delete unused files or keep the useful ones, and point out similarities with the current list to a list of existing malware components, eliminating files that match [13]. Individuals as well as organizations should carefully manage information systems risk assessment to impact the business unit the least. The drawback for this method could be an insider threat from a disgruntled network administrator who may decide to install an outdated antimalware program for revenge.

113.3.4 Disabling Unused Services and Other Security Best Practices

Unnecessary services have to be disabled to prevent attackers exploit security flaws found in open ports. It is recommended to install latest security patches from approved IT vendors to eliminate existing vulnerabilities and thwart DDoS attacks take advantage of security weaknesses in the target system. ICMP flood attacks, smurf attacks and more cannot be successful when computers and networks using the same ISP have a disabled IP. Additionally, IP hopping DDoS attacks can be avoided by re- locating IP address of the active host to a pool of similar hosts. On the other hand, black hats can append a domain name service tracing function to the DDoS attacks tools making this technique ineffective [5]. Finally, it appears to be challenging to prevent DDoS attacks without mitigation techniques.

113.4 Mitigation

Small and large businesses should implement a risk analysis and comply with a cybersecurity engineering industry such as the Open Web Application Security Project's (OWASP) recommendations to enhance their abilities to mitigate the risk of a potential DDoS attack by identifying content problems and transport issues (i.e. HTTP and HTML) [1]. The OWASP Foundation is a not-for-profit organization that educates developers, and organizations about the

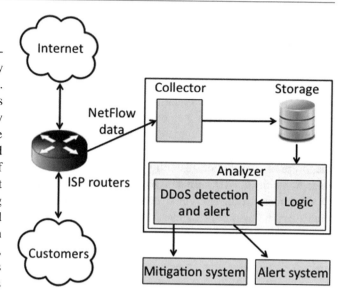

Fig. 113.3 DDoS detection and mitigation

consequences of the most common web applications security vulnerabilities to limit DDoS attacks [9].

The Fig. 113.3 clearly demonstrates that after ISP routers are infected by malware in a company, malware will access the collector. The storage will then communicate with the analyzer which includes the logic. This will then talk to mitigation and alerts systems to stop DDoS attack from happening in its customers' network.

113.4.1 OWASP Guidance in DDoS Mitigations

Following the OWASP framework, to defend against DDoS in a more effective manner, security teams should closely monitor and report the following components for changes that can be addressed by web developers [1]:

1. Address broken authorization, authentication and session management: Access to web servers' has been proven vulnerable like all Internet facing applications. Without a robust authentication mechanism in place, sensitive account information in legitimate users' accounts could be compromised. Therefore, it is important to implement malware detection and eradication software, biometrics systems, multifactor-authentication strong passwords and similar measures would fix any poor access control mechanism.

2. Verify and Validate Insecure Direct Object References— OWASP suggests an attacker can alter authentic data with compromised data if URL is infected or not checked through validation techniques. Hence, it is imperative that organizations identify every security hole in its data to diminish the DDoS threats.

3. Provision Security Misconfiguration—Supported tested configuration, patch management and routine updates of the current database are additional tertiary security measures companies should consider. To keep confidentiality, integrity, and availability of data, security teams should continuously maintain, apply security patches to eliminate security holes [1]. Thus, it is important to closely monitor organizations website/network for security flaws.

4. Encrypt Insecure Cryptographic Storage—Following careful investigation, it is strongly recommended that all personal information such as users' name, date of birth, financial data, and related information should be encrypted. Additionally, the cryptographic keys should be properly secured to lessen DDoS attacks.

5. Harden SOAP and Rest-style Resources to deter failure to restrict URL access—OWASP's standards assert that any URL that can be accessible without authorization should be disabled. Also, legitimate users should be limited to privileges they need to only do their job, namely the least privilege principle and separation of duties for sensitive tasks. They should turnoff auto-execution as well as identifying disgruntled employees.

6. Shore up Insufficient Transport Layer Protection—every company should ensure its data are protected from eavesdropping or snooping. Transport Layer Security (TLS) and cryptographic protocol should include valid Public Key Infrastructure (PKI) certificate to safeguard the confidentiality and authenticity of information [1].

113.4.2 Class Based Queuing Queues and SLA/ IMPERVA and CISCO Mitigations

Class Based Queuing Queues are recommended for different type of packets, and different packets for different type of services are preset while bandwidth is assigned to queues that maintain quality of service during DDoS attack [11]. One of its challenges is that it cannot maintain queues. Moreover, there should be an important service level agreement (SLA) between IT vendors and businesses. These should decide on feasible security measures with the designated IT vendor, including the party responsible for disaster recovery in the event of DDoS threats [12].

In the same manner, Imperva offers Incapsula as a cloud-based security application that mitigates all types of DDoS [7]. Incapsula delivers an incredible defense against all web application, networked systems and TCP/IP-based DDoS attacks. Moreover, DDoS attacks can be mitigated by maintaining reliable and cost-efficient scalability. Its only downfall will be network eavesdropping which aims at compromising the confidentiality and integrity of the data

being read from or written on the company database server [4].

Cisco Systems offer an entire DDoS protection solution based on the principles of detection, diversion, verification, and forwarding to help ensure total protection. However, white hats could still worry about a potential username enumeration or incorrect SQL queries. Finally, to limit DDoS attacks database administrators should ensure that access controls are properly implemented and the use of stored procedures to enable "context-dependent" access controls such as "negative authorizations, separation of duty in the context of role-based access control, and context-, resource-, and data-dependent access control" [4].

113.5 Conclusion

We have presented the detection, the prevention and mitigation techniques to combat DDoS attacks. The detection technique in this paper has shown that DDoS attacks can be detected early if staff members are paying attention to their network; prevention has demonstrated some important mechanisms that can limit a DDoS attack, mitigation includes detection and prevention which appears to be one of the best methods to fight against DDoS attacks in businesses. Additionally, the security requirements of web applications, services, and servers do not provide all the required properties to develop robust security methods. However, a defense from IBM/Akamai, NIDS, OWASP, CISCO, IMPERVA, risk management, multifactor-authentication and more as previously mentioned, have been proven to help limit DDoS efficiently.

As new strategies are being developed every day by black hats, the most important move white hats should consider is working hard towards new security measures to protect data. Information warfare is real in the digital world today. Organizations should invest to secure their assets against DDoS. These have been revealed to be very damaging today. New beneficial measures should be put into place to keep their businesses running, to keep their customers trust and achieve a great productivity. Finally, lawmakers should encourage unity in information security field to help limit insider threats, DDoS attacks, leaks and more.

References

1. Beuchelt, G. (2013). Chapter 8. Securing web applications, services and servers. In J. Vacca (Ed.), *Computer and information security handbook* (2nd ed.). Boston: Morgan Kaufmann Publishers.
2. Day, C. (2013). Chapter 26. Intrusion detection and prevention systems. In J. Vacca (Ed.), *Computer information security handbook*. San Francisco: Morgan Kaufmann Publishers.

3. Douligeris, C., & Mitrokotsa, A. (2004). DDoS attacks and defense mechanisms: Classification and state of the art. *Computer Journal of Networks, 44*(5), 643–666.

4. Gertz, M. (2011). Database security. In H. Bidgoli (Ed.), *Handbook of information security, Volume 3. Database security mechanisms and models*. New York: Wiley.

5. Gupta, B., Joshi, C., & Misra, M. (2010). Distributed denial of service prevention techniques. *International Journal of Computer and Electrical Engineering, 2*, 1793–8163.

6. IBM Global Technology Services (2013, August). Managed distributed denial of service (DDoS) protection. Retrieved from http://www-01.ibm.com/common/ssi/cgi-bin/ssialias?htmlfid=SED0 3135USEN.

7. Imperva (2016). DDoS protection strategies. Retrieved from https:// www.incapsula.com/web-application-ddos-protection-services.html.

8. McMillan, J., & VandenBrink, R. (2009, September 14). GIAC enterprises: Where your fortune is our business. Malware Infection Tiger Team final report. Retrieved from http://www.sans.edu/ studentfiles/projects/200909_.

9. Meyer, R. (2008, Jan 26). Detecting attacks on web applications from Log Files. Retrieved from https://www.sans.org/reading-room/whitepapers/logging/detecting-attacks-web.

10. Mirkovic, J. (n.d.). A taxonomy of DDoS attack and DDoS defense mechanisms. Retrieved from https://www.researchgate.net/profile/ Peter_Reiher/publication/2879658_A_taxonomy_of_DDoS_.

11. Shokri, R., Varshovi, A., H. Mohammadi, Yazdani, N., & Sadeghian, B. (2006, September 13–15). DDPM: Dynamic deterministic packet marking for IP traceback. *IEEE International Conference on Networks* (pp. 1–6). Singapore: IEEE.

12. UMUC (2011). Module 9. Virtualization and cloud computing. Retrieved from https://leoprdws.umuc.edu/cgi-bin/id/FlashSubmit/ fs_link.pl?fs_project_id=385&.

13. Zeltser, L. (2009). Introduction to malware analysis. Retrieved from http://zeltser.com/reversemalware/intro_to_malware_analy sis_201208.pdf.

Next-Generation Firewalls: Cisco ASA with FirePower Services

Taylor J. Transue

Abstract

This paper aims to provide a comprehensive review of the Cisco ASA next-generation firewall with FirePower Services. The product will be introduced as to its purposes and features of why an organization would want to deploy it as a security product in an enterprise or otherwise large scale network. This paper will give insight into the technology behind the Cisco ASA as well as additional features that the FirePower Services adds. I will cover some of the strengths found on the FirePower platform that Cisco offers such as signature-based threat detection and Snort, as well as some of the limitations the platform has compared to other leading vendors in the network security world such as the lack of SSL inspection as the time of this writing.

Keywords

Cisco • Firewall • Security • Technology • Network

114.1 Introduction

Tools such as vulnerability scanners, firewalls, proxies, endpoint protection, and so on all claim to help mitigate attacks and/or respond to them in a timely fashion. Often times, you will see dedicated tools that are proficient in one of those areas. Other times, you will see multiple tools enveloped into one, potentially saving consumers money and time. Cisco, a popular vendor known throughout the world, offers "the industry's first adaptive, threat-focused next-generation firewall designed for a new era of threat and advanced malware protection" [1] to do just that.

114.2 Cisco ASA Technology

The technology behind the Cisco ASA with FirePower services is something that has aged to maturity with time. The Cisco ASA was announced first as the firewall platform for modern day enterprises to secure perimeters using IP and port-based ACLs as well as some additional features. Beyond the basic tasks of a typical firewall by another vendor, the Cisco ASA can also do the following: Antivirus, Antispam, IDS/IPS engine, VPN device, SSL device, and Content inspection [1].

The Cisco ASA was released with multiple platforms, ranging from the 5505 (small business model) to the 5580-40 (large enterprise). However, the intent of this paper is to cover the services provided by the firewall regardless of model.

T.J. Transue (✉)
Robert Morris University, Department of Computer and Information Systems, Moon Township, PA 15108, USA
e-mail: tjtst170@mail.rmu.edu

© Springer International Publishing AG 2018
S. Latifi (ed.), *Information Technology – New Generations*, Advances in Intelligent Systems and Computing 558, DOI 10.1007/978-3-319-54978-1_114

114.3 FirePower Services Technology

The technology behind FirePower services is varied and in depth. There are 4 features to the FirePower services that make this a unique platform. The FirePower services provides AVC, which is application visibility and control. "More than 3000 application-layer and risk-based controls can invoke tailored IPS threat-detection policies to improve security effectiveness" [1]. It also provides an NGIPS solution, or next-generation IPS (intrusion prevention system). Tt also offers a reputation and category-based URL filtering service that provides "comprehensive alerting and control over suspect web traffic. It enforces policies on hundreds of millions of URLs in more than 80 categories" [1]. Lastly, FirePower services offers advanced malware protection which helps administrators discover and stop malware or emerging threats missed by other security layers.

114.4 Strengths of FirePower Services

On the technical side of strengths, Cisco offers a robust signature-based threat detection service. SwishData, a data engineering group, released a Next-Generation firewall capabilities assessment which included 3 top vendors – Cisco, Palo Alto, and Mcafee. Of these vendors, Cisco got rated a "high effectiveness against signature-based threats" [2]. A signature-based threat is a method of detecting an attack by looking for a specific pattern, things like byte sequences in traffic or known malicious instructions seen in malware. This is the core of IDS/IPS devices. This strength can also be credited towards Sourcefire – a company that Cisco acquired in order to build the FirePower services engine. Before being acquired by Cisco, Sourcefire was a renowned network security company most popular for its tool known as Snort – a network intrusion prevention and detection system. This is a solid strength to the Cisco ASA with FirePower services platform. Strong acquisitions like this make for a product that is held to higher standards.

114.5 Limitations of FirePower Services

Like any other product, the Cisco ASA with FirePower Services does have its limitations or weaknesses. The report by SwishData referenced rates the Cisco ASA with Fire-Power a total of 30/50, with other contenders such as McAfee at 46/50 and Palo Alto at 33/50, therefore marking the Cisco ASA the lowest rated out of the 3.

A big weakness found in the Cisco ASA with FirePower services platform is a missing feature that is found in other NGFW platforms – SSL Inspection. SSL Inspection is a technology present in other vendor's platforms, for example, F5 offers SSL inspection devices in their Viprion chassis models. This can be a pretty important selling point to some; SSL traffic is generally not visible to a secure web gateway and therefore cannot be controlled by network or security administrators.

114.6 Conclusion

This paper is meant to be an overview of the product and its strengths/weaknesses into which are most important to a consumer. Though the Cisco ASA with FirePOWER Services does offer many benefits to network or security administrators, it is always best to have multiple vendor platforms and solutions in place with the proper research to what is important for that organization or enterprise. With Cisco's massive background in network devices and now, with a greater focus in security, Cisco is gaining a greater hold in the network security world, making it a viable option to have to protect your company's data.

References

1. Cisco ASA with FirePOWER Services (2016, October). Retrieved from Cisco: https://www.cisco.com/c/dam/en/us/products/collateral/security/asa-firepower-services/at-a-glance-c45-732426.pdf.
2. Next Generation Firewall Capabilites Assessment (2015). Retrieved from SwishData: http://www.swishdata.com/downloads/Next-Generation-Firewall-Capabilities-Assessmet.

DDoS Attacks: Defending Cloud Environments

Michele Cotton

Abstract

The migration of many organizations to cloud-computing environments alters the traditional definition of Distributed Denial of Service (DDoS) targets. This paper presents several techniques to defend, prevent and mitigate Distributed Denial of Service (DDoS) attacks in cloud computing environments. The methods presented in the paper examine the integration of the network as part of the network protection solution with technologies such as shuffling targets, filtering network traffic with BGP and implementing predictive algorithms for future attacks.

Keywords

DDoS · Cloud · Shuffling · SDN · Algorithm

115.1 Introduction

The articles included in this review were selected for the diversity represented in the approach to the ubiquitous network vulnerability Distributed Denial of Service (DDoS) attacks. They also symbolize trends in network defense. As the collective idea of a network architecture migrates to cloud infrastructure [1] the definitions from infrastructure to targets are reinvented.

115.2 Summary: DDoS Defense Mechanisms

This research paper discussed DDoS detection and mitigation methods. It is proposed that the various DDoS attacks require varied responses. Selective blackholing, the proactive mitigation technique is discussed. The reactive defense measures of consecutive packet entropy and the chi-square statistic. The paper is representative of initial change in network topology. Intelligent routers are incorporated as a central element to the protection of the network.

115.2.1 Selective Blackholing

The outer boundaries of a network are managed with the Border Gateway Protocol (BGP). The BGP routers inspect the interior and exterior of packets. Any packets that meet the preset criteria for Denial of Service (DoS) traffic from recognized sources is discarded. The remaining network communication is sent to the original destination without interruption. When selective blackholing is applied the information about packets identified as DoS is sent to the internet service provider (ISP) to update provider that specific criteria were met [3].

Selective blackholing updates ISPs with data for identified distributed denial of service (DDoS) attacks. The method efficiently removes malformed packets from the network, minimizing the potential damage that could be caused. The technique is ineffective for unidentified or modified attack sources. In these scenarios the information necessary to configure the BGP routers to adequately filter packets is not available; limiting the potential use of the practice.

M. Cotton (✉)
Cybersecurity, University of Maryland University College, Adelphi, MD, USA
e-mail: michelecotton9@gmail.com

© Springer International Publishing AG 2018
S. Latifi (ed.), *Information Technology – New Generations*, Advances in Intelligent Systems and Computing 558, DOI 10.1007/978-3-319-54978-1_115

115.2.2 Consecutive Packet Entropy

The essence of entropy in this context is the identification of randomness. The concept of consecutive packet entropy is to inspect packets traversing the network to identify patterns [6]. Packets with header information distinctly different from standard acceptable headers could indicate a DDoS attack. The pro in the scenario is the potential early identification of an attack; potentially decreasing the duration of an attack. The con is if the attacker is aware the implementation of the technique the attack approach can be readily modified to defeat the predictive algorithm [2].

115.2.3 Chi-Square Statistic

This method samples packets on the network in search of entropy patterns. The TCP SYN indicator is used as the baseline indicator of how many packets on average are sent from a single source. The theory is a device under DDoS attack will receive a large number of packets from various sources.

A flaw associated with this theory is globalization. If the website for a small company in New York begins receiving a large number of packets from Germany, there is not enough evidence to confirm the server is under attack. It is just as likely, that someone in Germany heard about the company through social media and decided to share with friends in the same country.

115.3 Summary: Catch Me If You Can: A Cloud-Enabled DDoS Defense

The research paper identifies the proliferation of websites deployed in cloud environment reliance upon open architecture as a driver to examine alternative means of protection form DDoS. The authors [4] indicate the use of packet filters based on documented traffic signatures and revising access control lists have proven ineffective against DDoS attacks. Jia, et al. propose a technique to prevent and mitigate DDoS in a cloud environment.

115.3.1 Shuffling

The procedure involves shuffling the location of an attacked asset to create a quarantine space, protecting other assets on the network. The approach leverages the rapid deployment of machine instances on a network. The challenge is the efforts of a veteran attacker would require more complexity than a single asset move would provide.

115.3.2 Replication

Another action taken by Jia et al. [4] would replicate a machine targeted, move the device to a protected location and deploy a replica in its place to capture information from the attacker. The replica would permit the attacked asset to transform from victim to honeypot. This solution would possibly suffice if the device were under attack from a single source. The technique would not thwart a DDoS with attacks launched from multiple sources simultaneously.

115.3.3 Entropy

This research paper [6] suggesting the use of entropy in two ways with two different algorithms 1) fast heuristic greedy algorithm, as a protective measure to ensure assets are shuffled at the optimal frequency and 2) Maximum Likelihood Estimation (MLE) algorithm to aid in estimating the scale of the attack.

The objective for each of the described algorithms is obtainable. Broad implementation of the concepts presented in this paper is unlikely. Optimal use of the techniques described require use of an intelligent architecture. Network devices capable of deploying resources and hiding others in a rapid fashion. The financial investment necessary to integrate technology with these capabilities would cost much more than the average company could afford to invest. Even in the case where a large enterprise could afford the network infrastructure investment, the technology would present limitations.

Latency is a visible representation of technology limits. In many instances the network would not have enough bandwidth to allocate new device, hide others, capture data and perform analysis real-time. The end-user community and attackers are likely to notice the delayed network performance. Creating undesired attention from both parties.

Positive aspects of the model include low cost for scalability. Compromised assets are cleaned and recycled for reuse on the network. Additionally, the model does allow an organization to implement and manage the solution without the aid of an ISP.

115.4 Summary: DDoS Attack Protection in the Era of Cloud Computing and Software-Defined Networking

The research article was selected for the fusion of the cloud computing environment with Software-Defined Networking (SDN) as a layered solution to DDoS attacks. Cloud-computing is an omnipresent solution regardless of the organization being discussed [9]. SDN is attracting attention because it simplifies network management with centralized device and policy configuration [5]. The inference of a smart network opens the possibility of a new frontier for combating DDoS [7].

Wang, Zheng, Lou, & Hou [9] discuss embedding a DDoS solution within the network architecture via DDoS attack mitigation architecture using software-defined networking (DaMask). This allows SDN to become an active part of DDoS detection. Other aspects of the research paper consider a graphical detection model and predictive model with latent prevention capability.

115.4.1 DaMask

The authors in this research paper assume DaMask implementation in a cloud environment. DaMask is confirmed cloud agnostic consistently defending network assets across all platforms. The benefit SDN offers with centralized network control is a potential liability for network security. The feature leaves all elements of the network connect and therefore vulnerable to a single attack. To resolve this issue, the network operating system (NOS) is divided into virtual segments or sliced. Ownership of the slices is distributed to different parts of the network, establishing partitions between the cloud networks. Traditional forms of DDoS detection algorithms: signature-based, anomaly-based or hybrid can run across the network; isolating any slice deemed compromised.

Two modules of DaMask exist for operational use 1) DaMask-D is a detection system triggered by anomaly-based traffic and 2) DaMask-M is designed to implement a sequence of events in response to a breach and initiate generation of logs. DaMask-D indicates an ongoing attack while DaMask-M takes actions to protect the network.

115.4.2 Graphical Model Detection

The graphical detection model is a type of anomaly-based detection. The distinction is the model is designed with two forward-thinking features a) the functions monitored are pre-selected for the slice based on heuristic analytics and b) data challenges with existing schemas are addressed at the control plane. The visual representation verifies the theory that most attacks follow consistently similar trajectories.

115.4.3 Bayesian Approach

The foundation of the Bayesian model is predictive analytics applied to anomaly-based detection findings. Wang et al. [9] approach the Bayesian network model with a dynamic layer; the inferences are frequently updated and presented in graphical form. This method allows rapid response when changes in the data indicate a new attack approach.

Future architectures where SDN and cloud-computing are implemented could benefit from the use of the DaMask model. The various graphical interfaces could potentially become a dominant DDoS protection and mitigation technique. The resolution is readily implemented, with low overhead cost and willingly scales to accommodate changes in the network.

References

1. Boss, G., Malladi, P., Quan, D., Legregni, L., & Hall, H. (2007). Cloud computing. *IBM White Paper, 321*, 224–231.
2. Chen, P., Qi, Y., & Hou, D. (2015). CHAOS: Accurate and Realtime Detection of Aging-Oriented Failure Using Entropy. https://arxiv.org/abs/1502.00781.
3. Holl, P. (2015). Exploring DDoS Defense Mechanisms. *Future Internet (FI) and Innovative Internet Technologies and Mobile Communications (IITM), 25*, 1.
4. Jia, Q., Wang, H., Fleck, D., Li, F., Stavrou, A., & Powell, W. (2014). Catch me if you can: A cloud-enabled ddos defense. In *2014 44th Annual IEEE/IFIP International Conference on Dependable Systems and Networks* (pp. 264–275). Atlanta: IEEE.
5. Kim, H., & Feamster, N. (2013). Improving network management with software defined networking. *IEEE Communications Magazine, 51*(2), 114–119.
6. Navaz, A. S., Sangeetha, V., & Prabhadevi, C. (2013). Entropy based anomaly detection system to prevent DDoS attacks in cloud. *arXiv preprint arXiv:1308.6745*.
7. Rish, I., & Singh, M. (2000). A tutorial on inference and learning in Bayesian networks. *IBM Watson Research Center*.
8. Sattar, I., Shahid, M., & Abbas, Y. (2015). A review of techniques to detect and prevent distributed denial of service (DDoS) attack in cloud computing environment. *International Journal of Computer Applications, 115*(8), 23–27.
9. Wang, B., Zheng, Y., Lou, W., & Hou, Y. T. (2015). DDoS attack protection in the era of cloud computing and software-defined networking. *Computer Networks, 81*, 308–319.

Cyber Security Policies for Hyperconnectivity and Internet of Things: A Process for Managing Connectivity

Maurice Dawson

Abstract

Hyperconnectivity and Internet of Things are changing the landscape of Information Technology (IT). Architectures are becoming more sophisticated while cyber security is slowing adapting to this shift of Internet-enabled technologies in automotive, industrial, consumer, and networking. This slow adoption of proper security controls, defenses, and aggressive measures is leaving individuals vulnerable. This submission explores how policies can be created that automate the process of device connectivity, and how current frameworks can be used to minimize system risks.

Keywords

Internet of Things • Hyperconnectivity • Complex Systems • Internet of Everything • Connected Devices • Cyber Security Risk Management

116.1 Introduction

For years cyber security has been a concern of the Department of Defense (DoD) allow them to mature their Certification & Accreditation (C&A) process from the Orange Book to the National Institute of Standards and Technology (NIST) Risk Management Framework (RMF) [1]. This has included C&A processes such as Department of Defense Information Technology Certification & Accreditation Process (DITSCAP), Department of Defense Information Assurance Certification & Accreditation Process (DIACAP), Common Criteria, Department of Central Intelligence Directives (DCID), and more. Doing this allowed for the ability to understand how to appropriately apply security controls to systems while learning how to manage risk for complex systems. These cyber security policies are applied to stand-alone systems, enterprise networks, tactical systems, and those entities in a system of systems environment. Even when systems are interconnected, there is a requirement of a certificate of net-worthiness and security testing before continued connection [1]. Data classification and Acquisition Category (ACAT) level drive the need for policy compliance. Through all agencies, this created massive processes for policing the higher levels of government still require an annual auditing known as the Federal Information Security Management Act (FISMA). In the public and private sectors, the cyber security policies are not matured in comparison to the DoD. Thus the problem arises as more companies produce that has Internet-enabled technology without proper cyber security.

116.2 Internet of Things

The Internet of Things (IoT) has spawned a need for seemingly normal devices to have the ability to connect with the Internet. Thus a world is created in which smart technologies enable objects with a network to communicate with each other and interface with human effortlessly [2]. This connected world of convenience and technology does not come without its drawbacks, as interconnectivity implies hackability [3–4]. IoT is found in the automotive, consumer, industrial, and general

M. Dawson (✉)
University of Missouri-St. Louis, 1 University Drive, St. Louis, MO 63121, USA
e-mail: Dawsonmau@umsl.edu; http://www.umsl.edu/

© Springer International Publishing AG 2018
S. Latifi (ed.), *Information Technology – New Generations*, Advances in Intelligent Systems and Computing 558, DOI 10.1007/978-3-319-54978-1_116

computing environments. These environments are incredibly diverse with application in traffic monitors to smart energy grids. Looking beyond IoT to Internet of Everything (IoT) there is a potential market that approximately $14.4 trillion and over 99 percent of physical devices are still unconnected [5].

116.3 Managing Risks

Developers of nongovernmental systems should start using policy, guidance, and directives to formulate baselines [1, 6–8]. NIST Speical Publications (SP) could provide an organization the start they need to start addressing these issues. NIST SP is broken into three sub-series to present cyber security, information systems guidelines, recommendations, and reference materials [1]. The SP 800, Computer Security is the primary outlet for publishing cyber security/information systems guidelines, recommendations, and reference materials. The SP 1800, NIST Cyber Security Practice Guides compliments the SP 800 targeting particular cyber challenges in the public and private sectors. The SP 1800 provide practical, user-friendly guides to facilitate the adoption of standard based approaches to security. The SP 500, Computer Systems Technology is a general Information Technology (IT) sub-series used by NIST's Information Technology Laboratory (ITL) [1].

The RMF is a framework created by the RMF to address risk management that public and private sector can use to baseline security controls in products [1] (NIST, 2012). The RMF uses the risk-based approach to security control selection and specification considers effectiveness, efficiency,

and constraints due to applicable laws, directives, Executive Orders (EO), policies, standards, or regulations. There are six RMF categorization steps that serves as the basis for this NIST guidance [1]. Step 1: Categorize is categorize which the system is assessed and categorized based on an impact analysis. Step 2: Select is chosen which during this period the systems is given a baseline set of security controls that are to be addressed in the design. Step 3: Implement is applied, and during this stage, the controls selected in Step 2 are deployed within the system to included the associated environment of operation. Step 4: Assess is when the controls implemented are assessed to see if they are working as intended and that the desired outcome meets the security requirements for the system. Step 5: Authorize is to get authority for the system to operate based back an acceptable decision upon the acceptable risk for the system. Step 6: Monitor is to continually assess the security control of the system on an ongoing basis. This can include annual security checks to review compliance. This process can not only be implemented but used as a way to measure compliance with federal standards that end up as a performance measurement in the annual FISMA reports.

RMF provides organizations developing Internet-enabled technologies and those that are connected to begin addressing risks associated with their products. Creating a process that allows for updating application software, and system configurations while undergoing an annual compliance helps create a more security hardened product [9–12]. In Fig. 116.1 a process is described that allows devices to be tested according to current standards.

Fig. 116.1 Security testing process for connected devices

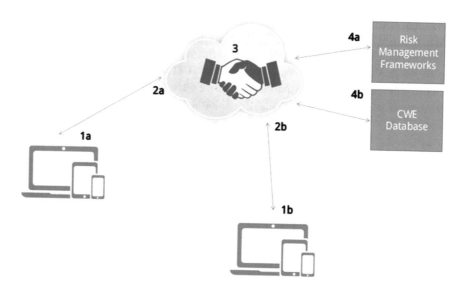

116.4 Hyperconnectivity

The term hyperconnectivity was invented by two Canadian social scientists that discuss the social constructs relating to this term [3–4, 13]. However, other found definitions define this as a to connect oneself to difference information and social stream with a deft facility. Thus IoT, big data, and social media blend allowing vast amounts of information to be shared publicly. As the IoT devices lack the essential security to combat advanced cyber attacks these devices become an unsecured gateway into the system.

Many mobile devices allow for the use of applications to integrate cloud-based applications on the instrument. This coupled with the use of default settings, lack of Anti-Virus (AV) software, and other items that do not allow for defense in depth concept. For example, the default settings on a mobile device may have the Global Positioning System (GPS) enabled. This GPS setting is enabled when needed to use for navigation. If this setting is not changed then when an individual takes photos the latitude and longitude coordinates are created in the Exchangeable Image Format (EXIF) data for the file. This could allow for pattern analysis and know the type of device that took the photo. Knowing the kind of device allows the attacker to research vulnerabilities and exploits specifically targeting that platform. With some open source tools, Open Source Intelligence (OSINT) can be performed that provides data such as behavior analysis, text mining, location analysis, and exploitation of connections [14–15].

Policies be implemented that address the system and components of that system. Once dealing with the system then policies need to be created for the connecting system. As organizations seek to implement Bring Your Device (BYOD), there needs to be a system that addresses the security configuration and a script that runs on the instrument each time to check policy compliance before allowing a connection. However, for household devices, this will be much more complex as there is no central organization responsible for cyber security policy. Devices will need to be outfitted with an automated test before allowing paring or connectivity through a software application.

116.5 Testing

Testing is integral to the software and systems lifecycle for development. However, there is guidance from NIST in the SP 800–15, but there is not truly something that addresses developing tests on commercial devices that provide an analysis that looks at risks [16]. So the need for the development of Built-In Test (BIT) like testing applications that allow

users to set their level of acceptable risk In Fig. 116.1, shown is a testing process for multiple devices. In Steps 1a and 1b devices that decide to pair connect to the web in Steps 2a and 2b. During Step 3 is where a handshake is done Steps 4a and 4b the appropriate security measures are selected to allow the secure connection. In Step 4a an appropriate risk management framework is chosen with security controls being applied to the device. Step 4b looks at the CWE database, uses the appropriate tests for the devices depending up applications discovered. Once tests have been satisfied Step 3 performs a handshake that allows devices to connect. Devices have the ability to perform checks as necessary as possible to remain securely attached. For this process to occur a software-based application will be on the devices that allow connectivity to the Internet. The risk management framework and CWE database get updated daily to ensure that the device owner understands the appropriate risk before deciding to connect or pair device ultimately.

116.6 Conclusion

Creating a software-based application is a method that could significantly reduce risks in a hyperconnected environment. Using an application that can apply a selected framework while providing the end user their risk level if connecting to another device after reviewing CWE compliance has the potential to reduce the number of vulnerabilities drastically. However, there has to be a minimum accepted risk for personal users, and an acceptable associated risk for those in various industries. Future research will look into this concept further while maturing this process to enable implementation at an organization and personal use level.

References

1. NIST, S (2004). 800–37. Guide for the security certification and accreditation of federal information systems.
2. Ashton, K. (2009). That 'internet of things' thing. *RFiD Journal, 22* (7), 97–114.
3. Dawson, M., Eltayeb, M., & Omar, M. (2016). *Security solutions for hyperconnectivity and the internet of things* (pp. 1–347). IGI Global: Hershey.
4. Dawson, M. (2016). Exploring secure computing for the internet of things, internet of wverything, web of things, and hyperconnectivity. In M. Dawson, M. Eltayeb, & M. Omar (Eds.), *Security solutions for hyperconnectivity and the internet of things* (pp. 1–12). IGI Global: Hershey.
5. Bojanova, I., Hurlburt, G., & Voas, J. (2013). Today, the Internet of Things. Tomorrow, the Internet of Everything. Beyond that, perhaps, the Internet of Anything—a radically super-connected ecosystem where questions about security, trust, and control assume entirely new dimensions. information-development, 04.

6. Commanders, C., Defense, U. S. O., & Defense, A. S. O. (2003). Subject: DoD information system certification and accreditation reciprocity. System.

7. Diem, J. W., Smith, J. S., & Butler, L. A. (2009). United States Army Operational Test Command (USAOTC) Integrated Technologies Evolving to Meet New Challenges-A Study in Cross Command Collaboration. ARMY OPERATIONAL TEST COMMAND FORT HOOD TX.

8. Gordon, L. A., Loeb, M. P., & Sohail, T. (2003). A framework for using insurance for cyber-risk management. *Communications of the ACM, 46*(3), 81–85.

9. Tohidi, H. (2011). The Role of Risk Management in IT systems of organizations. *Procedia Computer Science, 3*, 881–887.

10. Saleh, M. S., & Alfantookh, A. (2011). A new comprehensive framework for enterprise information security risk management. *Applied computing and informatics, 9*(2), 107–118.

11. Wood, C. C. (2008). Information security policies made easy, Version 10. Information Shield, Inc.

12. Whitman, M. E., Mattord, H. J. (2011). Principles of information security. Cengage Learning.

13. Fredette, J., Marom, Revital., Steiner, K., Witters, L. (2012). The promise and peril of hyperconnectivity for organizations and societies. The global information technology report, 113–119.

14. Mercado, S. C. (2009). Sailing the sea of OSINT in the information age. Secret Intelligence: A Reader, 78.

15. Glassman, M., & Kang, M. J. (2012). Intelligence in the internet age: The emergence and evolution of Open Source Intelligence (OSINT). *Computers in Human Behavior, 28*(2), 673–682.

16. Scarfone, K. A., Souppaya, M. P., Cody, A., Orebaugh, A. D. (2008). SP 800–115. Technical guide to information security testing and assessment.

DDoS Countermeasures

117

Michael Wisthoff

Abstract

This paper explores three published research articles demonstrating solutions to detect, prevent, and mitigate Distributed Denial of Service (DDoS) attacks. These articles stress the critical severity of DDoS attacks impacting businesses and entities on a worldwide scale and present effective strategies to counter such threats. Cepheli et al. (Journal of Electrical & Computer Engineering 2016: 1–8, 2016) proposes a Hybrid Intrusion Detection System (H-IDS) combining anomaly-based and signature-based intrusion detection mechanisms (Journal of Electrical & Computer Engineering 2016: 1–8, 2016). Kalkan and Alagoz (Computer Networks the International Journal of Computer and Telecommunications Networking 108:199, 11, 2016) presents an active filtering mechanism to stop a DDoS at its origin (Computer Networks the International Journal of Computer and Telecommunications Networking 108:199, 11, 2016). Gupta et al. (International Journal of Computer and Electrical Engineering (IJCEE) 2:268–276, 2010) stress the importance of implementing defense in depth strategies and maintaining fundamental security processes (International Journal of Computer and Electrical Engineering (IJCEE) 2:268–276, 2010). Solutions in these articles are demonstrated in testing and well documented to be highly successful.

Keywords

IDS • IPS • Filtering • Defense in depth • Secure overlay service

117.1 Introduction

Tremendously fast paced development and implementation of digital attacks targeting information systems is continuous and remains a threat to the security of information technology resources. Presently, Distributed Denial-of-service (DDoS) attacks are highly critical threats to various businesses and entities on a global scale. The general strategy of DDoS attacks is to overwhelm target systems with so many requests that it is beyond the maximum amounts these systems can support, which results in the unavailability of system resources. Unlike traditional denial-of-service (DoS) attacks, DDoS' are not deployed from a single host, but rather exploit multiple compromised hosts, zombies, throughout the internet to flood the target systems.

Recent innovative methods to counter DDoS attacks involve combining intrusion detection methods [1, 2, 9], employing active filtering mechanisms [5, 8, 9, 11], and uniquely guarding device identification and authentication across multiple layers [4, 6, 10]. The creation of effective methods to detect, prevent, and mitigate DDoS attacks is essential and various strategies are discussed in a variety of academic journals and research publications [1–11].

M. Wisthoff (✉)
University of Maryland University College, 1616 McCormick Drive, Largo, MD 20774, USA
e-mail: mwisthoff@student.umuc.edu

© Springer International Publishing AG 2018
S. Latifi (ed.), *Information Technology – New Generations*, Advances in Intelligent Systems and Computing 558, DOI 10.1007/978-3-319-54978-1_117

117.2 Hybrid Intrusion Detection Techniques

Firstly, a solution for detection and prevention of DDoS attacks by using a unique implementation of a Hybrid Intrusion Detection System or H-IDS is proposed [1]. This solution incorporates results from using both signature-based and anomaly-based intrusion detection systems (IDS), which analyze the same attack traffic simultaneously [1]. Initially, this solution would develop baseline profiles of normal expected traffic based on packet inter arrival times, packet sizes, and protocol control frequencies [1]. Next, these traffic values are sent to the "anomaly-based detector, which uses multidimensional Gaussian mixture models" or machine learning processes to separate the expected traffic from the unusual [1]. To appropriately differentiate the two traffic types, an Expectation maximization (EM) algorithm is used to approximate parameters of the traffic as compared to predefined limits to identify outliers and generate an alarm [1]. The expectation step (E-step) is determined by iteratively entering approximate values until values meet according to the parameters below [1].

$$\gamma_{ik} = \frac{\omega_k \rho_k(x_i|\theta_k)}{\sum_{k=1}^{K} \omega_k \rho_k(x_i|\theta_k)}, \quad 1 \leq k \leq K, 1 \leq i \leq N$$

The maximizations step (M-step) finds the averages, covariants, and mixing amounts and determines values according to weights using the following equations using variable calculated in the E-step [1].

$$\omega_k = \frac{1}{N} \sum_{i=1}^{N} \gamma_{ik}, \quad 1 \leq k \leq K$$

$$\mu_k = \frac{\sum_{i=1}^{N} \gamma_{ik} x_i}{\sum_{i=1}^{N} \gamma_{ik}}, \quad 1 \leq k \leq K$$

The E and M steps are repeated using the various estimated values and are input to the following equation to calculate their maximum likelihoods [1].

$$\sum_k = \frac{\sum_{i=1}^{N} \gamma_{ik}(x_i - \mu_k)(x_i - \mu_k)^T}{\sum_{i=1}^{N} \gamma_{ik}}, \quad 1 \leq k \leq K$$

This is followed by an analysis of these values by the signature-based IDS to determine if this traffic matches any predefined attack signatures [1]. The outputs from both detectors are then analyzed by the Hybrid Detection Engine (HDE) to determine correlations and the conclusive likelihood that the traffic is an actual attack [1]. In fact, [2] identifies that the collaborated analysis of outputs from both detection methods is a strength of these hybrid systems [2].

Essentially, detection of an attack by the HDE could be used to generate and send an alert to a firewall or Intrusion Prevention System (IPS). These alerts could be used to configure blocks of detected sources of the attacks at the firewall and to take preventative steps using an IPS. Although, [9] notes that due to "spoofing and the large number of hosts used in a DDoS attack, it may be difficult" to block the possibly thousands of source addresses that are participating in the DDoS especially as this may block legitimate sources [9]. However, the accuracy of the described H-IDS solution is impressive. In fact, [1] notes in their testing results that using the "H-IDS solution achieved true positive rates (TPR) of 92.1% when using the DARPA 2000 dataset and 99% using a commercial bank dataset following a penetration test" [1]. As the DARPA 2000 dataset is well known throughout the security community, the highly positive TPR for this dataset adds credibility to the H-IDS solution. Additionally, a considerable benefit of the H-IDS' design is that it can be smoothly integrated into an existing intrusion detection platform or used independently as its own DDoS detection solution.

117.3 Distributed Filtering

Secondly, various filtering techniques can be used for the detection and prevention of DDoS threats [5]. Particularly, [5] notes such filtering solutions include individual reactive filtering like Hop Count filtering, individual proactive filtering like Ingress/Egress filtering, as well as cooperative proactive filtering in their proposed ScoreForCore model [5].

For instance, [11] notes that "hop-count filters are effective in detecting and preventing spoofed source IP addresses" [11]. Hop-Count filters only accept traffic that has traveled the internet using a specific number of router to router hops and will respectively block traffic that does not meet these parameters, which may prevent certain DDoS attacks from entering the network. Such individual reactive filters are only operational post DDoS discovery [5]. Additionally, to prevent DDoS attacks, [8] indicates that "ingress filtering prevents spoofed IP addresses wherein incoming traffic from an external source that has a source address in the private IP scheme is blocked" [8]. Respectively, [8] explains that "egress filtering prevents DDoS attacks by filtering outbound traffic to prevent malicious traffic from getting back to the attacking party" [8]. Both the ingress and egress filters are easy to deploy at the edge router. On the other hand, the proposed ScoreForCore method filters traffic to the packet level and "scores each to determine its legitimacy based on its score value" [5].

Baseline models are determined prior to attack to develop baseline features of normal packets like known IP addresses, ports, protocols, packet sizes, TTL values, and TCP flags [5].

Table 117.1 System terms and parameters

Term	Explanation
SingleNP	A single attribute profile in an attack free period
SingleCP	A single attribute current profile in an attack period
OwnPair	Random attribute pair owned by the router
OwnPairList	A list consists of *OwnPairs*
OPNP	Own pair nominal profile created by considering *OwnPair*
OPNPList	A list of *OPNPs*
ScorePair	The pair that is used for generating current pair profiles
ScorePCP	Current pair profile used for score operations
ScorePNP	Nominal pair profile used for score operations
SuspiciousPair	The attribute pair with the most probable signs for current attack
SPNP	Suspicious pair nominal profile created by considering *SuspiciousPair*
A, B	Attribute A and B
Dev_A	Maximum Deviation for attribute A
$npcp_{an}$	The number of packets with $A = a_n$ in current profile
$npnp_{an}$	The number of packets with $A = a_n$ in nominal profile
dev_{an}	The deviation for attribute value $A = a_n$
$A = a_p$	Attribute A with a value in packet p
$B = b_p$	Attribute B with b value in packet p
$ScorePNP_{(A=ap,B=bp)}$	The number of packets in a nominal profile that have the property of a_p for attribute A and b_p for attribute B
$ScorePCP_{(A=ap,B=bp}$	The number of packets in a current profile that have the property of a_p for attribute A and b_p for attribute B
TPNP	The total number of packets in a nominal profile
TPCP	The total number of packets in a current profile
S_p	Score value of packet p
ScoreList	A list that contains scores of packets
Th	Threshold score value for packet discarding
$\emptyset$	Acceptable traffic
ψ	Total current incoming traffic

Prior to an attack, these baseline features are combined arbitrarily in groups of two "while during an attack, when a congestion is detected, packet based analysis begins" [5]. At this point, baseline feature pairs are compared with features of the incoming traffic to identify values that are outside of the baseline features, which are "probable signs of ongoing attacks and are respectively marked with *SuspiciousPair* values" [5]. Essentially, using the parameters as shown in Table 117.1 [5], the ScoreForCore design communicates the *SuspiciousPair* values to adjacent routers for the purpose of identifying the source and blocking the attack at the source router [5]. For instance, collaborating routers can employ the pseudo code identified in Fig. 117.1 [5] as different routers are denoted by R as well as variations of NR and NR primes to trace a DDoS back to its source where other prevention or mitigation steps could be implemented [5]. Scores are determined by calculating S_p [5].

$$S_p = \frac{ScorePCP_{\left(A=a_p, B=b_p\right)}/TPCP}{ScorePNP_{\left(A=a_p, B=b_p \dots\right)}/TPNP}$$

Ensure: *SuspiciousPair* is determined by R
if *SuspiciousPair* $\in OwnPairList_R$ **then**
    $ScorePair_R=OwnPair_R$ and $ScorePNP_R=OPNP_R$
else if *SuspiciousPair* $\in OwnPairList_{NR}$ for all $NR \in N$ **then**
    $ScorePair_R=SuspiciousPair$ and $ScorePNP_{R'}=OPNP_{NR}$
else if *SuspiciousPair* $\in OwnPairList_{NR'}$ for all $NR' \in N'$
then
    $ScorePair_R=SuspiciousPair$ and $ScorePNP_{R'}=OPNP_{NR'}$
else
    $ScorePair_R = OwnPair_R$ and $ScorePNP_R = OPNP_R$
end if

Fig. 117.1 *SuspiciousPair* pseudo code [5]

Similarly, [9] suggests using such distributed defenses against a DDoS attack, so that upon the detection of an attack, information about the attack is pushed back so that routers close to the origin of the attack can enable filters [9]. This solution appears to achieve this goal. Moreover, as DDoS attacks are largely collaborated DoS attacks among many cooperating zombie machines, [5] points out that to counter such attacks a cooperative defense mechanism is a viable solution [5].

117.4 Prevention Mechanisms

Thirdly, employing defense in depth and fundamental security strategies can guard against DDoS attacks [4]. For instance, [4] suggest a layered defense equipped with "access control lists, firewalls, and intrusion detection systems" [4]. Access control lists and firewalls can certainly be used as a tool in blocking DDoS attacks, but only if they are instructed to block such attack traffic as these tools are not designed to differentiate between legitimate traffic and a possible DDoS [4]. On the other hand, [9] notes that an intrusion detection system (IDS) can observe traffic on a port and attempt to correlate it against known signatures or attack patterns corresponding to malware, port scanning, or DDoS traffic [9]. Basic security techniques like disabling unused services and ports as well as ensuring the prompt installation of the most current security patches on devices throughout the network environment plays a huge role in protecting against DDoS attacks.

These local security maintenance processes are so important because DDoS attacks can be the product of many other types of malware like worms or other viruses that exploit such vulnerabilities. In fact, [7] notes that Botnets are most commonly causes of DDoS attacks [7]. However, [3] points out that while "the largest corporations and entities may be equipped to handle most botnet-based attacks, no one is immune from DDoS attacks launched through botnets" [3]. This fact makes it increasingly vital to ensure basic security techniques are performed to close exploitable gaps in system security. Additionally, [4] suggests using a Secure Overlay Service (SOS) to prevent DDoS attacks [4]. By using such overlay solutions [10] mentions allows a secret node to communicate with an arbitrary other node in such a way that the identity of the secret node is unable to be determined, but is known by and accessible only by previously authorized sources [10]. This process consists of two independent authentications in which one successful authentication will allow transfer of traffic to a secret servlet, which authenticates again and passes only legitimate traffic to the network's edge routers [4].

117.5 DDoS Determinations

Finally, identifying incoming traffic as potential DDoS attacks requires specific parameters and conclusions to be reached as shown in Table 117.2 [1, 5, 6]. For instance, the H-IDS solution can be configured according to the detectors

Table 117.2 Comparison table

Anomaly-based	$isAlarm_a = (0, D_s(P, Q) < \alpha$ is False $isAlarm_a = (1, D_s(P, Q) \geq \alpha$ is True
Signature-based	$isAlarm_r = (0, A(k) = 0$ is False $isAlarm_r = (1, A(k) \geq 0$ is True
ScoreForCore	if $S_p <$ Th then Pass if $S_p >$ Th then Drop
SOS	Must be $n \in N_s$

used whether anomaly-based, signature-based, or the combination. For anomaly-based detectors, the Sibson distance formula below is used to detect a DDoS when compared with the threshold (α), which is determined by the HDE and is factored into the isAlarm$_\alpha$ calculation [1].

$$D_s(P, Q) = \frac{1}{2}\left[D_{kl}\left(P, \frac{1}{2}(P+Q)\right) + D_{kl}\left(Q, \frac{1}{2}(P+Q)\right)\right]$$

For signature-based detectors, outputs are calculated as the time frame index k is a function of the quantity of alerts with an A variable [1]. The HDE can calculate the results and probability of an attack per either an "OR" or "AND" scenario wherein if one or both consider the traffic to be an attack it may be an attack [1].

In ScoreForCore's solution, if the calculated score value (S_p) is less than the threshold (Th) the traffic is passed and otherwise it is blocked [5].

The Secure Overlay Service will only permit traffic from source addresses that prove to be belonging to an overlay node in that $n \in N_s$ is presented to ensure the source is authorized to communicate with the target host [6]. Respectively, all unauthorized sources and all traffic from those sources is denied.

117.6 Conclusion

Moreover, each of the strategies proposed in the research above can be used to effectively detect, prevent, or mitigate DDoS attacks. The solutions described are listed below along with characteristics pertaining to their ease of implementation, overall importance, or management overhead in recognizing or guarding against DDoS attacks. Absolutely, conducting local security maintenance procedures like installing security patches is one of the most important and easiest strategies. On the other hand, Secure Overlay Service networks are hard to deploy but are highly effective preventative mechanisms. Firewalls, access control lists, and intrusion prevention systems require experienced engineers to deploy effectively. Additionally, logging will need to be enabled on these security devices, which are monitored continuously, configured to generate alert notifications, or equipped to perform automated

SmartResponses as necessary. Filtering is simple to implement considering the most appropriate security devices for the environment are in place. Finally, it appears that H-IDS can be implemented with particular ease as the machine learning component would essentially automate updates to rulesets as needed. While these are effective detection, prevention, and mitigation strategies, the ongoing research and development into methodologies to protect against DDoS attacks is imperative.

References

1. Cepheli, Ö., Büyükçorak, S., & Karabulut Kurt, G. (2016). Hybrid intrusion detection system for DDoS attacks. *Journal of Electrical & Computer Engineering, 2016*, 1–8. doi:10.1155/2016/1075648.
2. Dua, S., & Du, X. (2011). *Data mining and machine learning in cybersecurity*. Boca Raton: Taylor & Francis.
3. Dunham, K., & Melnick, J. (2009). *Malicious bots: An inside look into the cyber-criminal underground of the internet*. Boca Raton: CRC Press.
4. Gupta, B. B., Joshi, R. C., & Misra, M. (2010). Distributed denial of service prevention techniques. *International Journal of Computer and Electrical Engineering (IJCEE)*, *2*(2), 268–276. Database: arXiv.
5. Kalkan, K. & Alagoz, F. (2016). A distributed filtering mechanism against DDoS attacks. *Computer Networks the International Journal of Computer and Telecommunications Networking, 108*, p199, 11 p.; Elsevier B.V. Language: English, Database: InfoTrac Computer Database.
6. Keromytis, A. D., Misra, V., & Rubenstein, D. (2002). SOS: Secure Overlay Services, In the Proceedings of. ACM SIGCOMM, pp. 61–72.
7. Kim, D., & Solomon, M. (2014). *Fundamentals of information systems security* (2nd ed.). Burlington: Jones & Bartlett Learning.
8. Oriyano, S. (2014). *CEHv8: Certified Ethical Hacker Version 8 Study Guide*. Indianapolis: Sybex.
9. Prowell, S., Kraus, R., & Borkin, M. (2010). *Seven deadliest network attacks*. Amsterdam/Boston: Syngress.
10. Tarkoma, S. (2010). *Overlay networks: Toward information networking*. Boca Raton: CRC Press.
11. Yu, S. (2014). *Distributed denial of service attack and defense*. New York: Springer.

Lina Pawar, D.G. Khairnar, and Akhil Sureshrao Kulkarni

Abstract

The Long Term Evolution (LTE) handsets operates in the frequency bands which are being affected by the impulsive noise (IN). The double detection method is used to remove IN from the LTE handset. This method uses the conventional threshold method where threshold selection is the major aspect. The LTE handsets consists of two transmitting and receiving antennas, so the output should be considered by the comparison of the two different frequencies received at the receiver. The better signals within the threshold level is mainly considered.

Keyword

LTE • Space time coding • Orthogonal frequency division multiple access (OFDMA) • Multiple input multiple output(MIMO) • IN • Cellular radio

118.1 Introduction

118.1.1 Long Term Evolution

LTE is designed only to support the packet switched services. LTE is mainly used to provide the internet protocol connectivity between the user equipment and the packet data network. LTE works in the frequency range of 700 to 2600MHz and it provides 20 MHz of frequency width. Here to eliminate the IN in double detection method different modulation schemes are being used i.e. single carrier frequency division multiple access (SC-FDMA) for uplink & OFDMA for downlink. Here the work is done on downlink of LTE. The transmitter is base station (eNodeB) and receiver is user equipment. In LTE MIMO is used for transmission of data. Here in this paper only MIMO 2*2 deployment is considered [1, 2].

118.1.2 Impulsive Noise

IN consists of relatively short duration "ON/OFF" noise signals, caused by variety of sources [3]. Mostly IN occurs at very high frequency level up to 7 GHz. But the effect is mainly observed below 3GHz. The IN mostly affects in wideband radio communication. The IN generated due to lightening, transient in transformers, induction motors, sparks produced by manifold, dc electric motors, ignition noise from the spark plug of the petrol engines, arc welders, etc. As the bandwidth of the IN is wider than the communication channel, it is assumed that the impulsive noise pulse duration is shorter than the sampling period i.e. each IN pulse affect only one single sample [4–6].

118.1.3 MIMO

MIMO is mainly used for multiplying the capacity of a radio link using multiple transmit and receive antennas to exploit multipath propagation. This method mainly used to improve the performance of cellular radio networks and enable more aggressive frequency reuse. MIMO technology has been

L. Pawar • D.G. Khairnar (✉) • A.S. Kulkarni
Department of Electronics and Telecommunication, D.Y. Patil College of Engg, Pune, India
e-mail: linadesai.dyp@gmail.com; dgk@ee.iitb.ac.in; Kulkarniakhil108@gmail.com

© Springer International Publishing AG 2018
S. Latifi (ed.), *Information Technology – New Generations*, Advances in Intelligent Systems and Computing 558, DOI 10.1007/978-3-319-54978-1_118

preferred for wireless LAN, 3G mobile phone networks and 4G mobile phone networks. MIMO increases the coverage area and the capacity of the wireless communication system. Capacity of the MIMO can be increased by increasing the number of receiving antennas than the number of transmitting antennas [1].

118.1.4 Alamounti Transmission

For Alamounti transmission channel is considered to be a flat fading Rayleigh multipath channel and modulation is BPSK [1].

The input consideration for Alamounti transmission is

$Input = \{ x_1 , x_2 , \ldots \ldots \ldots , x_n\}$

Here two antennas are present at the transmitter.
In first time slot, x_1 & x_2 transmitted at 1st and 2nd transmitter respectively.
In second time slot, transpose $(-x_2)$ and transpose (x_1) transmitted.
In third time slot, x_3 & x_4 transmitted at 1st and 2nd transmitter respectively.
In fourth time slot, transpose $(-x_4)$ and transpose (x_3) transmitted.

The received signal at the receiver in 1st and 2nd time slot will be

$$Y_1 = [h_1 h_2]^* \begin{bmatrix} x_1 \\ x_2 \end{bmatrix} + n_1$$

And

$$Y_2 = [h_1 h_2]^* \begin{bmatrix} -x^*_2 \\ x^*_1 \end{bmatrix} + n_1$$

Where h1 and h2 represents channel terms.
So combining two equations can be represented as

$$\begin{bmatrix} Y_1 \\ Y_2 \end{bmatrix} = \begin{bmatrix} h_1 & h_2 \\ h^*_2 & -h^*_1 \end{bmatrix}^* \begin{bmatrix} x_1 \\ x_2 \end{bmatrix} + \begin{bmatrix} n_1 \\ n_2 \end{bmatrix}$$

118.2 Background

118.2.1 Addition of IN

MIMO systems combine more technical resources than just spatial diversity which is used in double detection method. Here research was done on the effect of space–time coding when IN enters a MIMO receiver.

118.2.1.1 Space Time Block Coding and Space Frequency Block Coding

Let us consider a MIMO wireless communication system, Alamouti coded, with two transmit antennas and two receive antennas [10]. Let us suppose that we want to transmit two consecutive symbols, i.e. s_1 & s_2

The Alamouti transmission matrix is $\quad c = \begin{pmatrix} s_1 & -s_2^* \\ s_2 & s_1^* \end{pmatrix}$

In the first time slot, Tx_1 transmits s_1 and Tx_2 transmits $-s_2^*$, where * stands for the complex conjugate. In the second time slot, Tx_1 transmits s_2, and Tx_2 transmits s_1^*.

The symbols in time domain with s_1 & s_2 on Tx_1 the original data are transmitted without any modification. Therefore, the subcarrier mapping for Tx_1 results in the inverse discrete Fourier transform as,

$$X_{1,m} = \frac{1}{N}^* \sum_{n=0}^{N-1} s_n^* e^{jm\frac{2\pi n}{N}}$$

Where N is the total number of subcarriers, n is the subcarrier index, and m is the sample index of the time signal.

The subcarrier mapping for Tx_2 is

$$X_{2,m} = \frac{1}{N}^* \sum_{n=0}^{\frac{N}{2}-1} s_{2n}^* e^{jm\frac{2\pi(2n+1)}{N}} - s_{2n+1}^* e^{jm\frac{2\pi(2n)}{N}}$$

118.2.1.2 Space Time Block Coding and IN

At the receiver, the signal in antenna i when two signals at two consecutive time slots ($t = 1, 2$) received is

$$\begin{pmatrix} y_{1,i} \\ y_{2,i} \end{pmatrix} = \begin{pmatrix} s_1 & s_2 \\ -s_2^* & s_1^* \end{pmatrix} \begin{pmatrix} h_{i,1} \\ h_{i,2} \end{pmatrix} + \begin{pmatrix} v_{1,i} \\ v_{2,i} \end{pmatrix}$$

Where $y_{k,i}$ is the whole received signal at time slot k at antenna i. s_k is the subcarrier generated at time slot k. $h_{i,j}$ represents the channel coefficient from transmit antenna j to receive antenna i. Moreover, $y_{k,i}$ is the noise contribution at time slot k, received at antenna I [10].

Above equation can be rewritten as

$$\begin{pmatrix} y_{1,i} \\ y_{2,i} \end{pmatrix} = \begin{pmatrix} h_{i,1} & h_{i,2} \\ h_{i,2}^* & -h_{i,1}^* \end{pmatrix} \begin{pmatrix} s_1 \\ s_2 \end{pmatrix} + \begin{pmatrix} v_{1,i} \\ v^*_{2,i} \end{pmatrix}$$

By multiplying with A^H_i we get which has orthogonal property as,

$$\begin{pmatrix} z_{1,i} \\ z^*_{2,i} \end{pmatrix} = \begin{pmatrix} y_{1,i} \\ y_{2,i} \end{pmatrix} \begin{pmatrix} h_{i,1} & h_{i,2} \\ h_{i,2}^* & -h_{i,1}^* \end{pmatrix}$$

Where i = 1, 2... n.

118.3 Methodology

Methodology mainly consists of the following points.

I.   Threshold detection and blanking method
II. Blanking VS clipping
III. Using receive diversity to eliminate IN
IV. Selection of Threshold

118.3.1 Threshold Detection and Blanking Method

In this method the threshold is set to zero if the value of the signal amplitude exceeding the threshold value. The threshold value is set in terms of the root mean square value of the amplitude of the signal received. Here the threshold value is fixed as a multiple of the RMS value of voltage amplitude.

For ex. $M = 4$ it means that the threshold is fixed at four times the RMS value [7–10].

118.3.2 Blanking Versus Clipping

Blanking and Clipping methods are majorly used when the value of the threshold exceeds than the fixed value. In blanking the samples exceeded than the threshold value are spread out in time domain so as to minimize the effect of the IN. In clipping the samples are made equal to zero exceeding the threshold value. The double detection method supports blanking instead of clipping [7–10].

In Fig. 118.1a, it can be seen that the complex time-domain representation of **s**, which is the true sample that should be received. The IN, i.e., **n**, enters the receiver at the same time. The actual sample that received was **r** = **s** + **n**. The circumference is the amplitude threshold that set to detect IN.

The OFDM receiver obtains the subcarriers after performing a fast Fourier transform (FFT). Therefore, the energy of the IN pulse **n** is spread among the symbol subcarriers, and it may cause decoding errors; the more errors, the more energy it has.

In Fig. 118.1b, the results after the Clipping and Blanking were observed. During clipping, error **c** is added to the signal, and during blanking, the error **b** is added as shown in the figure. We can consider any of these two errors introduced by **n** because both have less energy. From the figure we can easily say that error b will be having less energy than error c. However, blanking has the limits such as if too many samples within the OFDM symbol were affected by IN, the energy of the added contributions of

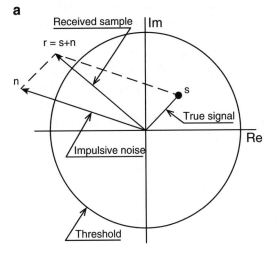

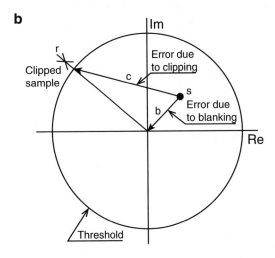

Fig. 118.1 (a) Sample representation (b) Errors due to blanking and clipping

many blanking vectors such as **b** might cause decoding errors in the subcarriers after the FFT.

In Fig. 118.1b, clipping is accomplished at the amplitude of the threshold level. Clipping could also be accomplished at the RMS level, i.e., all samples exceeding the threshold would be clipped to the RMS level (keeping their angle). Clipping at the RMS level offers better results than clipping at the threshold level.

118.3.3 Using Receive Diversity to Eliminate IN

The above two methods were single detection methods. However, the procedure presented by the researchers is to monitor and blank high-amplitude samples detected in both receive antennas at the same time and called as double detection method [7–10].

Fig. 118.2 is a block diagram representing the double-detection procedure. Its description is as follows:

(1) Receive antennas.
(2) Radio frequency tuning, intermediate frequency conversion and all the tasks previous to sampling are used.
(3) Continuous signals $s_1(t)$ and $s_2(t)$ are simultaneously sampled with period T. Discrete signals $s_1[n]$ and $s_2[n]$ are obtained.
(4) The samples of $s_1[n]$ and $s_2[n]$ are checked out to see if their amplitude exceeds the detecting threshold TH.
(5) The logical outputs of the latter stage go through an AND gate. The output is a logical variable called IMP that determines if the sample simultaneously received in $s_1[n]$ and $s_2[n]$ is IN or not.
(6) If high amplitude was detected in $s_1[n]$ and $s_2[n]$ at the same time, the corresponding samples are zeroed in both signals. If not, both signals stay without modification.

LTE systems take advantage of receive diversity. In a MIMO 2×2 deployment, the signal is received on two antennas at the same time.

118.3.4 Selection of Threshold

There are two different types of threshold detection methods.

I. Conservation Threshold
II. Aggressive Threshold

118.3.5 Conservative Threshold

In Conservative threshold the threshold is always set to the higher level. Setting a threshold to higher value, false-alarm probability is very low. So, conservative threshold will hardly be exceeded by signal samples without IN. This constitutes a benefit. From this we can say that if conservative threshold is set, no special advantage will be taken from detecting IN in both receive antennas. The performances of single and double detection will be similar for this method. The reason for this is on the low false-alarm probability: Double detection is good to filter out false-alarm events; if there are no false-alarm events, no improvement will be obtained from it [2].

118.3.6 Aggressive Threshold

Aggressive or demanding thresholds are set at a low amplitude level, i.e., above but not too far away from the signal RMS value. These aggressive thresholds can be used to trigger the blanking procedure on those samples with amplitude slightly above the signal RMS value. Therefore, it is not necessary to set the threshold too low, because the elimination of samples from the signal not required.

On the other hand, aggressive thresholds allow for detecting and blanking those samples that have been affected by moderate-amplitude IN events. These events would have not been detected with thresholds set at higher levels. It is not used in single detection method but in double detection method it is necessarily used as it is very difficult to detect impulsive noise at higher levels [2].

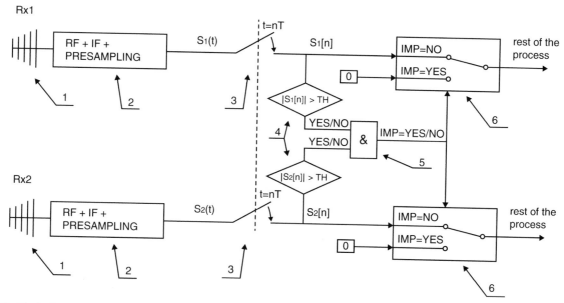

Fig. 118.2 Block diagram of double detection method

Fig. 118.3 Noise observation for sub-carrier 1

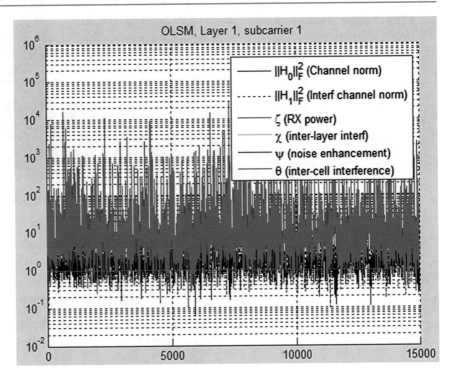

118.4 Results

(1) The Fig. 118.3 is indicating the interference occurred in subcarrier 1 and subcarrier 2 when the signals are transmitted. Here the graph consists of a channel, interference occurred in the channels, the receiver power, and noise enhancement and inter cell interference. The graph gives the basic idea about the noise added in the signals whenever they are being transmitted and received at the receiver.

(2) Figure 118.4 indicating addition of noise in the signal when it is being transmitted through the channel. The figure gives us the information about the channel, interference occurred in the adjacent frequencies, receiver power and about the signals (how the signal looks when noise is added in it) for subcarrier 1, layer 1.

(3) Now the Double Detection Method is being applied for the LTE network and we are calculating the BER vs SNR.

The Fig. 118.5 shows that the value of SNR is increasing as the value of the BER decreasing. Here we are using 1000 number of users. MIMO system is used for Double Detection Method where there are two number of transmitting antennas and one receiving antenna.

The Fig. 118.6 shows that the value of SNR is increasing as the value of the BER decreasing. Here we are using 1000 number of users. MIMO system is used for Double Detection Method where there are two number of transmitting antennas and two receiving antenna.

118.5 Conclusion

(1) From the results obtained we can easily say that the noise in the bit error rate is decreasing with the increase in the signal to noise ratio. As the signal to noise ratio is increasing we can say that the noise content in the signal received at the receiver is minimized.

(2) Also from the observation we can conclude that the method is more effective when we are using more number of receiving antennas.

(3) It has also been shown that the new method is more effective when the signals at both antennas are uncorrelated and when an aggressive threshold is set for detection.

Fig. 118.4 Addition of noise when signal is transmitted in layer 1

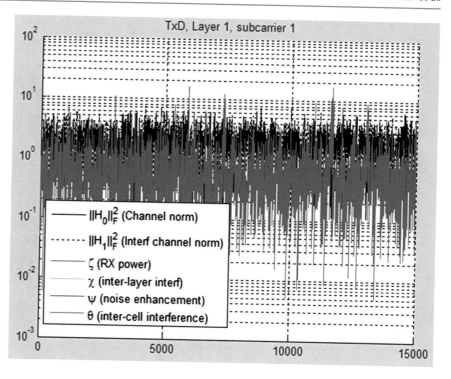

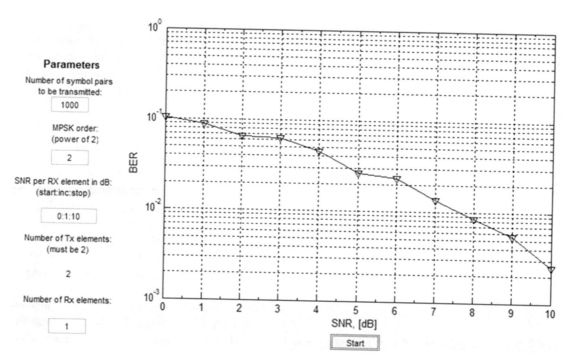

Fig. 118.5 Double detection method with 2 Tx and 1 Rx

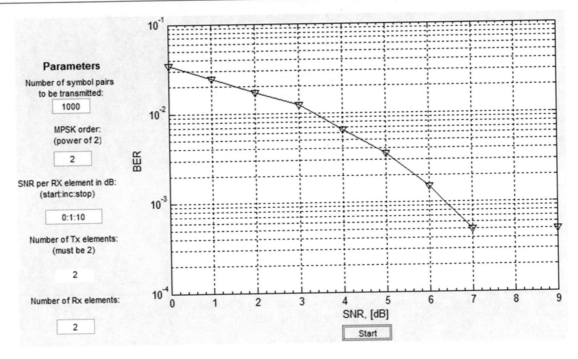

Fig. 118.6 Double detection method with 2 Tx and 2 Rx

References

1. Torío, P., & Sánchez, M. G. (2015). Elimination of impulsive noise by double detection in long-term evolution handsets. *IEEE Transactions on Vehicular Technology, 64*(7), 2875–2882.
2. Nibarger, R., & Teubner, M. (2012). An introduction to long term evolution (LTE). *IEEE Transactions on Vehicular Technology, 60* (10), 1–6.
3. Lin, J., Nassar, M., & Evans, B. L. (2013). Impulsive noise mitigation in power line communications using sparse Bayesian learning. *IEEE Journal on Selected Areas in Communications, 31*(7), 1172–1183.
4. Lin, J., Nassar, M., & Evans, B. L. (2011). Non-parametric impulsive noise mitigation in OFDM systems using sparse Bayesian learning, In *Proceedings IEEE GLOBECOM* (pp. 167–171). Houston.
5. Third Generation Partnership Project. (2011). Technical Specification Group Radio Access Network; Evolved Universal Terrestrial Radio Access (E-UTRA); Physical Channels and Modulation (Release 10), TS 36.211 V10.1.0 (2011–03), pp. 283–290.
6. Nassar, M., & Evans, B. L. (2011). Low complexity EM-based decoding for OFDM systems with impulsive noise, In *Proceedings ASILOMAR 45th Conference on Signals, Systems and Computers* (pp. 1943–1947). Pacific Grove.
7. Zhidkov, S. V. (2008). Analysis and comparison of several simple impulsive noise mitigation schemes for OFDM receivers. *IEEE Transactions on Communications, 56*(1), 5–9.
8. Jiang, T., & Wu, Y. (2008). An overview: Peak-to-average power ratio reduction techniques for OFDM signals. *IEEE Transactions on Broadcasting, 54*(2), 257–268.
9. Torio, P., & Sanchez, M. G. (2007). A study of the correlation between horizontal and vertical polarizations of impulsive noise in UHF. *IEEE Transactions on Vehicular Technology, 56*(5), 2844–2849.
10. Alamouti, S. (1998). A simple transmit diversity technique for wireless communications. IEEE Journal on Selected Areas in Communications. *16*(8), 1451–1458, M. Young, The Technical Writer's Handbook. Mill Valley: University Science, 1989.

A Synchronization Rule Based on Linear Logic for Deadlock Prevention in Interorganizational WorkFlow Nets

Vinícius Ferreira de Oliveira, Stéphane Julia, Lígia Maria Soares Passos, and Kênia Santos de Oliveira

Abstract

This paper presents an approach based on the analysis of Linear Logic proof trees and Interorganizational WorkFlow nets (IOWF-nets) theory to prevent deadlock situations in interorganizational workflow (IOWF) processes. An IOWF can be locally sound but not globally sound. Deadlock situations in IOWF processes come then from message ordering mismatches between several local workflow processes. Scenarios of IOWF-nets can be characterized by sequents of Linear Logic. The analysis of Linear Logic proof trees can be used to detect communication places between distinct WorkFlow nets (WF-nets) that introduce deadlock situations. In this paper, a synchronization rule is proposed in order to prevent deadlock situations caused by asynchronous communication mechanisms used to allow collaboration among individual workflow processes in IOWF processes that are locally sound but not necessarily globally sound.

Keywords

Petri net • Linear logic • Deadlock situation • Interorganizational workflow net • Synchronization rule.

119.1 Introduction

A workflow process corresponds to the automation of a business process during which information or tasks are passed from one participant to another for action, according to a set of procedural rules, representing then the sequences of activities that have to be executed within an organization to treat specific cases to reach well defined goals [1]. Workflow processes that involve several local workflow processes belonging to different organizations and which need to coordinate their actions in order to reach a common goal are known as interorganizational workflow processes [1].

The Soundness property is an important criterion which needs to be satisfied when dealing with workflow processes, once that this property ensures the absence of deadlock and proper termination [2]. But it can be difficult to establish the Soundness correctness of complex interorganizational workflow processes. As a matter of fact, the Soundness correctness of local workflow processes is not a guarantee of the Soundness correctness of the interorganizational workflow model as was shown in [2]. When an interorganizational workflow process is locally sound but not globally sound, a deadlock can be produced from message ordering mismatches as shown in [2]. The only way to prevent deadlock situation caused by message ordering mismatches is then to sincronize some parts of the local processes [2].

A Petri net which models the process aspect of an interorganizational workflow, is called an Interorganizational WorkFlow net (IOWF-net) [2]. In this paper, an approach based on Linear Logic is proposed to deal with deadlock situations in interorganizational workflow processes

V.F. de Oliveira (✉) • S. Julia • L.M.S. Passos • K.S. de Oliveira
Computing Faculty, Federal University of Uberlândia, Uberlândia, MG, Brazil
e-mail: viniciusfdeoliveira@gmail.com; stephane@ufu.br; ligiamaria.soarespassos@gmail.com; keniasoli@gmail.com

© Springer International Publishing AG 2018
S. Latifi (ed.), *Information Technology – New Generations*, Advances in Intelligent Systems and Computing 558, DOI 10.1007/978-3-319-54978-1_119

modeled by IOWF-nets which are locally sound but not necessarily globally sound. The method set in this paper is to replace certain asynchronous communication elements responsible for the deadlock situation by a kind of synchronous communication mechanism, forcing local workflow processes to initiate specific tasks at the same time. In particular, the detection of the asynchronous communication place responsible for the deadlock situation and the non respect of the Soundness property is based on the analysis of the proof tree of a Linear Logic sequent that corresponds to one of the potential scenarios in the IOWF-net.

119.2 Theoretical Background

An IOWF-net which is composed of a number of sound local workflows may be subject to synchronization errors. In addition, it is also possible to have an interorganizational workflow that is globally sound but not locally sound [2]. To define a notion of Soundness suitable for IOWF-nets, Aalst in [2] defined the *unfolding* of an IOWF-net into a WF-net.

In the *unfolded* net, i.e. the U(IOWF-net), all the local WF-nets are connected to each other by a start transition t_i and a termination transition t_o. Moreover, a global source place i and a global sink place o are added in order to respect the basic structure of a simple WF-net. To clarify the concepts defined above, consider the synthetic example presented in [2]. Such U(IOWF-net) presented in Fig. 119.1 has the function of modeling a process that precedes the presentation of a paper at a conference and its detailed description can be found in [2].

An IOWF-net is sound iff it is locally and globally sound. A given IOWF-net is locally sound iff each of its LWF-nets PN_k is sound. An IOWF-net is globally sound iff the U (IOWF-net) is sound [2].

In Linear Logic [3], propositions are considered as resources, which are consumed and produced at each state change [4]. In this paper just two connectives of Linear Logic will be used; the *times* connective, denoted by $\otimes$, represents simultaneous availability of resources and the *linear implies* connective, denoted by $\multimap$, represents a state change.

The translation of a Petri net into formulas of Linear Logic [4] is a relatively simple process: A marking M is a monomial in $\otimes$ and is represented by $M = A1 \otimes A2 \otimes \ldots \otimes Ak$ where Ai are place names; a sequent $M, ti \vdash M'$ represents a scenario where M and M' are respectively the initial and final markings, and ti is a list of non-ordered transitions; a sequent can be proven by applying the rules of the sequent calculus. It was proven in [5] that a proof of the sequent calculus is equivalent to a reachability problem in a Petri net.

A sequent can be proven by applying the rules of the sequent calculus. In this paper, the following rules are used [4]; the $\multimap_L$ rule expresses a transition firing and generates two sequents, such that the right sequent represents the subsequent, which remains to be proven and the left sequent represents the tokens consumed by this firing, the $\otimes_L$ rule transforms a marking in a list of atoms and the $\otimes_R$ rule transforms a sequent such as A, B $\vdash$ A $\otimes$ B into two identity sequents A $\vdash$ A and B $\vdash$ B.

119.3 Synchronization Rule

This approach considers the interorganizational workflow processes modeled by IOWF-nets that are locally sound but not necessarily globally sound. In such systems, deadlock situations come from message ordering mismatches as shown in [6] and are introduced by asynchronous communication elements [2]. According to [2], a deadlock situation can be corrected by replacing an asynchronous communication element (a communication place) by a synchronous communication element (a transition of synchronization).

The approach in this paper consists in replacing some of the asynchronous communication places of the IOWF-net by new communication mechanisms partially synchronous in order to prevent the occurrence of lost messages that, in case of deadlock situations, are generally trapped inside one of the communication places, thus forbidding the unfolded workflow model to respect the Soundness property. Such a substitution can be seen then as a kind of Synchronization Rule.

To apply such a Synchronization Rule, it is necessary first to identify scenarios responsible for deadlock situations and more specifically the asynchronous communication elements that can lead the system to inconsistent states. The detection of inconsistent states in this work will be based on Linear Logic sequent proof trees that correspond to potential scenarios in an U(IOWF-net).

Initially, the elements of the U(IOWF-net) have to be represented through the use of Linear Logic formulas. For each potential scenario of the U(IOWF-net), a Linear Logic sequent is then produced. The detailed method to obtain all Linear Logic sequent candidates to possible collaborations between two or more workflow processes was presented in [7].

A scenario in the context of an U(IOWF-nets) corresponds to a well defined route mapped into the interorganizational workflow process. If the U(IOWF-net) has more than one route (places with two or more output arcs), it is necessary then to build a different Linear Logic sequent for each existing scenario [7]. Each one of these scenarios is then represented by a specific Linear Logic sequent that

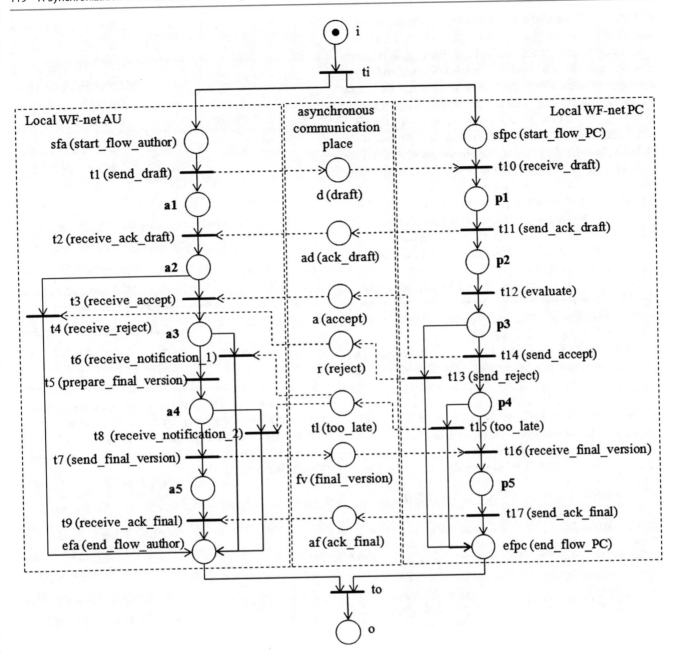

Fig. 119.1 An example of U(IOWF-net)

considers the initial and final markings of the U(IOWF-net) and a non-ordered list of transitions involved in it. After the definition of the Linear Logic sequents that represent all the possible scenarios of the U(IOWF-net), these Linear Logic sequents need to be proven through the building of Linear Logic proof trees.

If the last sequent of a Linear Logic proof tree built for a specific scenario is different from the identity sequent $o \vdash o$, then there is no token in the sink place of the U(IOWF-net), i.e. a deadlock situation occurs before the termination of the process [7].

It is necessary to identify the last transition fired before the deadlock situation. This transition will be called t_{d1}. To identify such a transition t_{d1}, it is necessary to verify in a proof tree leading to a deadlock situation, the transition firing that produces the atom which remains in one of the communication places (a place belonging to P_{AC}) until the conclusion of the Linear Logic proof tree. For each transition of t_{d1} type, a corresponding transition t_{d2} exists. Such a transition, when deadlock situation occurs, is then the output transition of a marked communication place that corresponds to one of the last atoms produced in the proof tree. The deadlock situation corresponds then to the death of transition t_{d2}.

In order to prevent the death of transition t_{d2}, it is necessary to introduce a kind of synchronization rule that corresponds to the following scheduling strategy: each time a transition of type t_{d1} is fired, the corresponding transition of type t_{d2} has to be fired in sequence in order to empty the communication place in which an atom was produced. In practice, such a policy corresponds to the guarantee that for each message sent by a local process, the corresponding answer of the local process will occur with certainty. After the Synchronization Rule application, the cause of the deadlock (a token trapped in a communication place) is removed and the Linear Logic proof can be correctly finalized with, as final state, a single atom produced in the place o (last place of the U(IOWF-net)).

To illustrate the approach, the IOWF-net presented in Fig. 119.1 is considered. As shown in [2], this IOWF-net is locally sound but is not globally sound given that the U(IOWF-net) of Fig. 119.1 is not sound.

For the U(IOWF-net) shown in Fig. 119.1, there exist five different scenarios to be studied. These scenarios are presented in [7]. Each one of these scenarios is then represented by a specific Linear Logic sequent that considers the initial and final markings of the U(IOWF-net) and a non-ordered list of transitions involved in it [7].

The transitions of the U(IOWF-net) shown in Fig. 119.1 are represented by the following formulas of Linear Logic:

$$t_i = i \multimap sfa \otimes sfpc, \qquad t_1 = sfa \multimap a1 \otimes d,$$
$$t_2 = a1 \otimes ad \multimap a2, \qquad t_3 = a2 \otimes a \multimap a3,$$
$$t_4 = a2 \otimes r \multimap efa, \qquad t_5 = a3 \multimap a4,$$
$$t_6 = a3 \otimes tl \multimap efa, \qquad t_7 = a4 \multimap a5 \otimes fv,$$
$$t_8 = a4 \otimes tl \multimap efa, \qquad t_9 = a5 \otimes af \multimap efa,$$
$$t_{10} = sfpc \otimes d \multimap p1, \qquad t_{11} = p1 \multimap ad \otimes p2,$$
$$t_{12} = p2 \multimap p3, \qquad t_{13} = p3 \multimap r \otimes efpc,$$
$$t_{14} = p3 \multimap a \otimes p4, \qquad t_{15} = p4 \multimap tl \otimes efpc,$$
$$t_{16} = fv \otimes p4 \multimap p5, \qquad t_{17} = p5 \multimap af \otimes efpc,$$
$$t_o = efa \otimes efpc \multimap o.$$

The five different scenarios, and consequently Linear Logic sequents, are then the following ones:

$$Sc_1 = i, t_i, t_1, t_2, t_4, t_{10}, t_{11}, t_{12}, t_{13}, t_o \vdash o,$$
$$Sc_2 = i, t_i, t_1, t_2, t_3, t_6, t_{10}, t_{11}, t_{12}, t_{14}, t_{15}, t_o \vdash o,$$
$$Sc_3 = i, t_i, t_1, t_2, t_3, t_5, t_8, t_{10}, t_{11}, t_{12}, t_{14}, t_{15}, t_o \vdash o,$$
$$Sc_4 = i, t_i, t_1, t_2, t_3, t_5, t_7, t_9, t_{10}, t_{11}, t_{12}, t_{14}, t_{16,17,o} \vdash o,$$
$$Sc_5 = i, t_i, t_1, t_2, t_3, t_5, t_7, t_9, t_{10}, t_{11}, t_{12}, t_{14}, t_{15}, t_o \vdash o.$$

The proof trees for scenarios Sc_1, Sc_2 and Sc_3 are deadlock-free scenarios (scenarios that finalized correctly) and can be found in [7]. The proof trees for scenarios Sc_5 and Sc_4 are shown in the following.

The proof tree for scenario Sc_5 is as follows:

$$\cfrac{\cfrac{\cfrac{\cfrac{\cfrac{\cfrac{tl,efpc,\mathbf{a5},\mathbf{fv},t_9,t_o \vdash o}{a4 \vdash a4 \quad tl,efpc,\mathbf{a5} \otimes \mathbf{fv},t_9,t_o \vdash o} \otimes_L}{a3 \vdash a3 \quad tl,efpc,a4,\mathbf{a4} \multimap \mathbf{a5} \otimes \mathbf{fv},t_9,t_o \vdash o} \multimap_L}{a3,tl,efpc,a3 \multimap a4,\mathbf{t_7},t_9,t_o \vdash o} \otimes_L}{\cfrac{p4 \vdash p4 \quad a3,tl \otimes efpc,t_5,t_7,t_9,t_o \vdash o}{p4,a3,t_5,t_7,t_9,p4 \multimap tl \otimes efpc,t_o \vdash o}} \multimap_L}{\cdots}}{\vdots}$$

(proof tree continues)

$$\cfrac{a2 \vdash a2 \quad a \vdash a}{a2,a \vdash a2 \otimes a} \otimes_R$$

For a better analysis of the approach, the proof tree for scenario Sc_4 was divided into two parts, remembering that a Linear Logic proof tree is read from the bottom-up. Part 1 of the proof tree for scenario Sc_4 is in black text, while Part 2 of the proof for scenario Sc_4 is in blue text. The proof tree for scenario Sc_4 is as follows:

$$\cfrac{\cfrac{efa \vdash efa \quad efpc \vdash efpc}{efa,efpc \vdash efa \otimes efpc} \otimes_R \quad o \vdash o}{\cdots} \multimap_L$$

(proof tree continues with steps $\otimes_R$, $\multimap_L$, $\otimes_L$ applied to sequents involving $a5, af, efpc, p5, fv, p4, a4, a3, a2, p3, p2, a1, ad, p1, sfpc, d, sfa, i$ down to)

$$i,t_i,t_1,t_2,t_3,t_5,t_7,t_9,t_{10},t_{11},t_{12},t_{14},t_{16,17,o} \vdash o$$

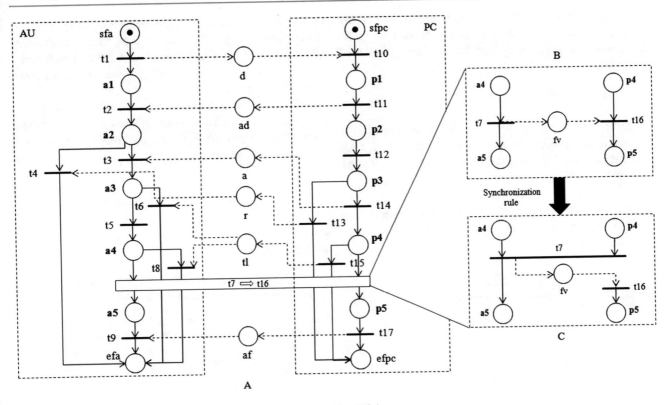

Fig. 119.2 Applying the synchronization rule in IOWF-net shown in Fig. 119.1

According to [7], the last sequent in the proof trees for scenarios Sc_1, Sc_2, Sc_3 and Sc_4 is $o \vdash o$; then these are deadlock-freeness scenarios. The last sequent for scenario Sc_5 is $tl,efp,a5,fv,t_9,to \vdash o$; then scenario Sc_5 is the one where the deadlock situation occurs. It is then necessary to apply the Synchronization Rule in scenario Sc_5 to remove the deadlock.

The last transition that was fired before the deadlock situation is the transition t_7 marked in bold type in the proof tree. In particular, the last atoms $a5$ and fv of the last sequent of the proof tree for scenario Sc_5 are produced when the transition t_7 is fired; however, due to the deadlock situation, these atoms are not consumed at the end of the proof. Transition t_7 will correspond then to the transition t_{d1} of the approach, transition t_{16} to the transition t_{d2}, and the atom fv to the marked asynchronous communication place.

The asynchronous communication place fv present in the last sequent of the proof tree is an output place of transition t_7 (last transition fired in the proof tree). This communication place is the input place of the transition t_{16} of LWF-net PC. The transition t_{16} corresponds then, in scenario Sc_5, to a dead transition.

The Synchronization Rule, shown in Fig. 119.2a, is not a true transition; it is a transition rule that synchronizes part of the communication structure between the AU and PC

LWF-nets. The transformation of the pure asynchronous communication mechanism into a partial synchronous mechanism after the application of the synchronization rule is presented in Fig. 119.2b and c. In Fig. 119.2b, the firing of transition t_7 (transition of type t_{d1}) corresponds only to a necessary condition for the firing of transition t_{16} (transition of type t_{d2}). Such an asynchronous communication protocol is not a guarantee then for the receiving of a sent message; after the firing of transition t_7, a token will remain eventually trapped into the communication place fv. On the contrary, in Fig. 119.2c, the fact to synchronize the execution of the first activity associated to transition t_7 on both local processes AU and PC, corresponds to the guarantee of execution of the called activity t_{16} associated to transition t_{16}. In fact, the firing of a transition of type t_{d1}, after application of the synchronization rule, corresponds to a necessary and sufficient condition for the firing of the transition of type t_{d2}, and, at the end of the associated scenario, no token will remain in intermediary communication places.

After the Synchronization Rule application, the deadlock situation is removed and the scenario Sc_5 will not be able to occur when initiating the execution of the unfolded net U (IOWF-net). In practice, the scenario Sc_5 is then removed from potential scenarios existing when considering the IOWF-net shown in Fig. 119.1.

According to Fig. 119.2c, the transitions t_7 and t_{16} of the U(IOWF-net) are modified after the synchronization rule. They are now represented by the following new formulas of Linear Logic:

$$t_7 = a4 \otimes p4 \multimap a5 \otimes fv, \qquad t_{16} = fv \multimap p5.$$

The transition t_7 and t_{16} do not appear in the Linear Logic sequents of scenario Sc_1, Sc_2 and Sc_3 and the alteration of the model do not modify the proof trees of the corresponding scenarios. On the other hand, both transitions appear in scenario Sc_4. Part 1 of the proof tree for scenario Sc_4 continue unchanged after the application of the synchronization rule. Part 2 of the proof tree needs to be computed again and is the following:

$$
\cfrac{
\cfrac{a3 \vdash a3 \qquad
\cfrac{
\cfrac{a4 \vdash a4 \quad p4 \vdash p4}{a4,p4 \vdash a4 \otimes p4} \otimes_R \quad
\cfrac{
\cfrac{
\cfrac{fv \vdash fv \quad
\cfrac{p5 \vdash p5 \quad
\cfrac{
\cfrac{a5 \vdash a5 \quad af \vdash af}{a5,af \vdash a5 \otimes af} \otimes_R \quad
\cfrac{
\cfrac{efa \vdash efa \quad efpc \vdash efpc}{efa,efpc \vdash efa \otimes efpc} \otimes_R \quad o \vdash o
}{efpc,efa,efa \otimes efpc \multimap o \circ \vdash o} \multimap_L
}{a5,af,efpc,a5 \otimes af \multimap efa,t_o \vdash o} \otimes_L
}{a5,af \otimes efpc,t_9,t_o \vdash o} \multimap_L
}{p5,p5 \multimap af \otimes efpc,t_9,t_o \vdash o} \multimap_L
}{a5,fv,fv \multimap p5,t_9,t_{17},t_o \vdash o} \otimes_L
}{a5 \otimes fv,t_9,t_{16},t_{17},t_o \vdash o} \multimap_L
}{a4,p4,a4 \otimes p4 \multimap a5 \otimes fv,t_9,t_{16},t_{17},t_o \vdash o} \multimap_L
$$

The last sequent in the proof tree for the new scenario Sc_4 is $o \vdash o$; then after the application of the synchronization rule, the scenario Sc_4 is still a deadlock-free scenario.

119.4 Conclusion

This paper presented an approach to prevent deadlock situations for interorganizational workflow processes modeled by IOWF-nets in interorganizational workflow processes that are locally sound but not globally. The synchronization rule presented in this paper implements a kind of local scheduling strategy that guarantee that each time a message is sent, the corresponding answer of the local process will occur with certainty.

As a future work proposal, it will be interesting to propose a kind of quantitative analysis based on symbolic dates, considering in this manner the proof trees of Linear Logic with dates, as presented in [4].

Acknowledgements The authors would like to thank FAPEMIG, FAPERJ, CNPq and CAPES for financial support.

References

1. van der Aalst, W. M. P., & van Hee, K. M. (2004). *Workflow management: Models, methods, and systems.* Cambridge, MA: MIT Press.
2. van der Aalst, W. M. P. (1998). Modeling and analyzing interorganizational workflows. In *International Conference on Application of Concurrency to System Design*, Fukushima (pp. 262–272).
3. Girard, J.-Y. (1987). Linear logic. *Theoretical Computer Science, 50* (1), 1–101.
4. Riviere, N., Pradin-Chezalviel, B., & Valette, R. (2001). Reachability and temporal conflicts in t-time Petri nets. In *Proceedings of 9th International Workshop on Petri Nets and Performance Models*, Aachen (pp. 229–238).
5. Girault, F., Pradier-Chezalviel, B., & Valette, R. (1997). A logic for Petri nets. *Journal Européen des Systèmes Automatisés, 31*(3), 525–542.
6. Xiong, P., Zhou, M., & Pu, C. (2009). A petri net siphon based solution to protocol-level service composition mismatches. In *IEEE International Conference on Web Services, ICWS 2009*, Los Angeles, CA (pp. 952–958).
7. Passos, L. M. S., & Julia, S. (2013). Qualitative analysis of interorganizational workflow nets using linear logic: Soundness verification. In *2013 I.E. 25th International Conference on Tools with Artificial Intelligence*, Herndon, VA (pp. 667–673).

Robert A.N. de Oliveira and Methanias C. Junior

Abstract

Stemming algorithms are commonly used during textual preprocessing phase in order to reduce data dimensionality. However, this reduction presents different efficacy levels depending on the domain it is applied. Hence, this work is an experimental analysis about the dimensionality reduction by stemming a veracious base of judicial jurisprudence formed by four subsets of documents. With such document base, it is necessary to adopt techniques that increase the efficiency of storage and search for such information, otherwise there is a loss of both computing resources and access to justice, as stakeholders may not find the document they need to plead their rights. The results show that, among the stemming algorithms analyzed, the RSLP algorithm was the most effective in terms of dimensionality reduction in the four collections studied.

Keywords

Dimensionality reduction • Experimental analysis • Jurisprudence • Stemming

120.1 Introduction

Every day, the courts, through their magistrates, judge various themes of the Law sphere, generating a large base of legal knowledge that guides new decisions and works as argumentative base to the related parties that plead their interests. Hence, from the corpus formed by uniform set of decisions handed down by the judiciary on a particular subject [1], emerges the concept of jurisprudence, fundamental tool for legal professionals to exercise their role.

Therefore, the objective of this study was to analyze, following an experimental process, described in [2], using quantitative metrics, the effectiveness of stemming on the dimensionality reduction of real jurisprudential bases. The results showed that, depending on the algorithm and the collection, there may be a statistically significant reduction

of these terms in the documents. In fact, the legal universe has its own jargon and we have not found reports in the literature showing this benefit when stemming is applied to jurisprudential bases.

The rest of the paper is structured as follows. Section 120.2 presents the related work. In Sect. 120.3, we present the definition and planning of the experiment. Section 120.4 contains the results of the experiment. Finally, Sect. 120.5 presents the conclusion and future work.

120.2 Related Work

Flores and Moreira [3] measured the impact of stemming on testing collections available in different languages (English, French, Portuguese and Spanish). This way, they collected dimensionality reduction metrics, *overstemming*, *understemming* and also measured the reflection on the application of these algorithms in precision and recall of information retrieval systems. Therefore, due to its scope, the paper did not go into detail on any of the analyzes.

R.A.N. de Oliveira (✉) • M.C. Junior
Universidade Federal de Sergipe UFS, São Cristóvão, Sergipe, Brazil
e-mail: ranomail@gmail.com; mjrse@hotmail.com

© Springer International Publishing AG 2018
S. Latifi (ed.), *Information Technology – New Generations*, Advances in Intelligent
Systems and Computing 558, DOI 10.1007/978-3-319-54978-1_120

It is worth mentioning that, until now, papers that run a detailed analysis of dimensionality reduction per document, like the one presented, were not found. In addition, related work used collections that do not reflect the documents found in the legal universe.

120.3 Definition and Experiment Planning

120.3.1 Goal Definition

The following is the goal formalization, according to GQM model proposed by Basili [4]: **Analyze** stemming algorithms **with the purpose of** evaluating them **with respect to** dimensionality reduction **from the point of view of** data analysts **in the context of** jurisprudential documents.

120.3.2 Planning

Context Selection. The experiment will be in vitro and will use the total base judicial jurisprudence of Supreme Court of the State of Sergipe, formed by four collections: (a) judgments of Appeals Court (181,994 documents); (b) monocratic decisions of Appeals Court (37,142 documents); (c) judgments of Special Courts (37,161 documents); and (d) monocratic decisions of Special Courts (23,151 documents).

Dependent Variable. The average of unique terms per document (UTD).

Independent Variables. Document collection of judgments of Appeals Court (JAC), monocratic decisions of Appeals Court (MAC), judgments of Special Courts (JSC) monocratic decisions of Special Courts (MSC); the stemming algorithms (NoStem, Porter, RSLP, RSLP-S and UniNE).

Hypothesis Formulation.

(a) **Null Hypothesis $H0^{UTD}$:** The stemming algorithms have the same average of unique terms per document.

(b) **Alternative Hypothesis $H1^{UTD}$:** The stemming algorithms have different averages of unique terms per document.

Selection of Participants and Objects. The documents of each collection were chosen randomly, using 95% of trust level in the sample size.

Experiment Project. We will utilize a *randomized complete block design* (RCBD) [2]. This way, NoStem represents the unique terms of the document with no stemming, therefore, it acts as a control group.

Instrumentation. We developed a Java application in order to iterate on each document of the sample, applying stemming algorithms and counting the frequency of unique terms after the execution.

120.4 Results

To answer experimental questions, CSV files generated by Java application were analyzed. The results of stemming impact on the average of unique terms, per document can be seen in Fig. 120.1.

Fig. 120.1 Average unique terms per document obtained by each stemmer

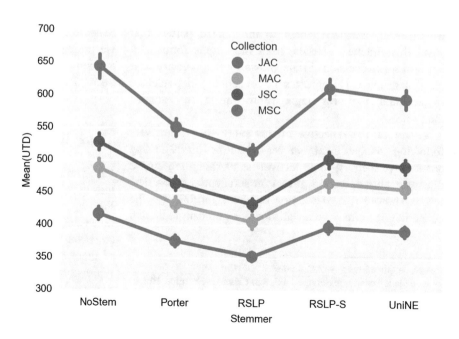

120.4.1 Analysis and Interpretation

Visually, analyzing Fig. 120.1, a stemming application seems to generate differences in the average of reduction of unique terms per document. However, it is not possible to claim that with no statistical evidences that confirm that.

Therefore, we used 95% of trust level ($\alpha = 0.05$), to the entire experiment and, later on, we analyzed if the samples had normal distribution. However, this hypothesis was rejected, since Tests of Shapiro-Wilk obtained p-value below 0.000, lower than significance level adopted, in every collection and algorithm. This way, considering data distribution and RCBD design adopted for the experiment, we performed Test of Friedman to verify Hypothesis 1.

After applying the test, we found a strong evidence for the hypothesis H1^{UTD}, showing that the averages of unique terms per document are not the same among the algorithms, since we verified p-value below 0.000, to every collection, and χ^2 equal to 5,883.84; 5,590.32; 5,863.67 and 5,474.95, referred to collections JAC, MAC, JSC and MSC, respectively. After a post-hoc analysis with Test of Wilcoxon, applying Benferroni correction ($\alpha = \alpha / 10$), we found the following order related to the number of unique terms obtained after stemming: NoStem > RSLP-S > UniNE > Porter > RSLP, to every collection. In other words, RSLP algorithm was the most effective in the reduction of unique terms per document.

Hence, due to the results found, it is possible to say that RSLP algorithm reduces judicial jurisprudence dimensionality more effectively than Porter, UniNE and RSLP-S.

120.4.2 Threats to Validity

Because the data was collected and analyzed by the authors, it happens to be a strong threat to internal and external validities. However, there is not conflict of interest. Thus, there are no reasons to privilege an algorithm rather than another. To mitigate any possible bias, documents were chosen randomly, according to RCBD guidelines.

120.5 Conclusion and Future Work

This paper showed an important contribution related to application of stemming algorithms on jurisprudential bases. Indeed, data dimensionality reduction is used in a variety of text processing techniques, however, we have not found, so far, a qualitative study that analyzes its impact on Brazilian judicial real decisions.

According to experimental results, the use of stemming algorithms reduced the average of unique terms per document. This way, RSLP algorithm was the most effective, among the analyzed algorithms, in terms of dimensionality reduction in the four collections studied.

Finally, for future work, we intend to analyze the reflection of the reduction from the perspective of a judicial information retrieval system, measuring its impact on MAP, R-Precision and Pr@10 metrics.

References

1. Maximiliano, C. (2011). *Hermenêutica e Aplicação do Direito* (20th ed.). Rio de Janeiro: Forense.
2. Wohlin, C., Runeson, P., Höst, M., Ohlsson, M. C., Regnell, B., & Wesslén, A. (2012). *Experimentation in software engineering.* Berlin/Heidelberg: Springer. [Online]. Available: http://link. springer.com/10.1007/978-3-642-29044-2
3. Flores, F. N., & Moreira, V. P.(2016). Assessing the impact of stemming accuracy on information retrieval a multilingual perspective. *Information Processing & Management, 0*, 1–15. [Online]. Available: http://linkinghub.elsevier.com/retrieve/pii/S0306457316 300358
4. Basili, V. R., Caldiera, G., & Rombach, H. D. (1994). The goal question metric approach. *Encyclopedia of Software Engineering, 2*, 528–532. [Online]. Available: http://maisqual.squoring.com/wiki/ index.php/TheGoalQuestionMetricApproach

Applying Collective Intelligence in the Evolution of a Project Architecture Using Agile Methods

Ciro Fernandes Matrigrani, Talita Santos,
Lineu Fernando Stege Mialaret, Adilson Marques da Cunha,
and Luiz Alberto Vieira Dias

Abstract

This paper presents a research using Collective Intelligence combined with the Agile Methods and its best practices to assist in the creation, adaptation, and evolution of design architecture for a data analysis system. In order to achieve this, three courses were integrated, from the graduate program in Electronic and Computer Engineering, at the Brazilian Aeronautics Institute of Technology (Instituto Tecnologico de Aeronautica – ITA): CE-240 Database Systems Project; CE-245 Information Technologies; and CE-229 Software Testing.

Keywords

Collective Intelligence • Agile Methods • Software Architecture • Interdisciplinarity • Scrum

121.1 Introduction

Collective Intelligence is defined as the capability of a group of individuals to improve their ability in specific tasks through information sharing, responding to environmental stimuli to develop jobs [1].

Based on this definition applies Collective Intelligence through the criteria: (i) groups with several individuals overcome and get better results than each individual in a particular task; and (ii) the group gets better performance, when individuals coordinate better their behavior than either individual working in parallel [1, 2].

121.2 The Case Study

The idea was to use social media to extract information referring to potential environmental crises, in order to alert the population and manage the complex situations.

It was used an Interdisciplinary Problem Based Learning (IPBL) technique to provide a scenario where these students can discuss, have a critical sense, and communicate their ideas, to seek a richer learning environment.

Within this IPBL, it was proposed the development of software applied to a social networking system in the cloud. It was developed during three months (Sprints) by students from three different courses.

They were divided into six interdisciplinary developing teams. The project followed an open scope in which the students could create and adapt the architecture as the problem was gaining maturity. The modules goals of each one of the six teams would be integrated over the three Sprints, forming a single real value added deliverable product.

The teams with a superficial view of the problem defined the first project architecture. It was set in the first three weeks of the course in a cycle called Sprint 0. This corresponded to

C.F. Matrigrani • T. Santos (✉) • L.F.S. Mialaret • A.M. da Cunha • L.A.V. Dias
Programa de Pos-graduacao em Engenharia Eletronica e Computacao (PG/EEC-I) Instituto Tecnologico de Aeronautica – ITA, Sao Jose dos Campos, SP, Brazil
e-mail: ciromatrigrani@gmail.com; ssoares.talita@gmail.com; lmialaret@gmail.com; cunha@ita.br; vdias@ita.br

© Springer International Publishing AG 2018
S. Latifi (ed.), *Information Technology – New Generations*, Advances in Intelligent Systems and Computing 558, DOI 10.1007/978-3-319-54978-1_121

Table 121.1 The six Scrum Teams and the number of students from the three courses

Students per Scrum Team/Course	CE-240	CE-245	CE-229
Data collection from social networks (ST01)	2	1	4
Data collection from mobile phones (ST02)	2	2	2
Data sharing among the teams (ST03)	2	1	4
Network infrastructure and safety (ST04)	2	2	3
Data processing (ST05)	2	2	3
Data storage with databases (ST06)	2	1	4

a training phase and leveling knowledge of students of all courses involved.

The Sprint 0 presented the main concepts and the a application of a teaching dynamic called Lego4Scrum [3], in which the students could experience a simulation of these concepts in practice.

At the end of the following three Sprints, the students of each team made a Sprint Retrospective and a Review Meeting [4] in which they discussed the experiences over the Sprint and the possibilities for improvements and adjustments.

Table 121.1 shows the six modules (ST01 to ST06), its implementation, and the number of students involved in each team.

Fig. 121.1 shows the ST01 sending the acquired data crises from a social network using Twitter and Facebook for the ST03 in the first version of the architecture. Technologies and tools were defined that would be used by each team. For example, ST05 has selected Hadoop [5], which would work in the data consolidation.

In the second version, new tools were inserted. The ST05 opted for the addition of the Spark [6] to process and consolidate data acquired from PCDs – Data Collection Platform.

Thus, the third version of the architecture was different from the previous architectures initially proposed by the students, including with the addition of new technologies as Spark and Node.js.

As the solution was gaining maturity, the students were feeling more comfortable in adapting the solution to an also more suitable architecture based on collective opinion.

The main modification in the later Sprint were the addition of a NoSQL database and a relational database for data storage and access. These alterations have enabled the data acquisition by ST01 to be stored again before processing.

121.3 Result Analysis

At the end of three Sprints, the IPBL showed that the architecture was gaining maturity in students' minds from the three different courses. The modifications and the development were democratic and validated by several students' opinions. Thus, the architecture for the problem took advantage of the criteria by applying the Collective Intelligence: (i) by groups with several individuals overcoming and getting better results than each individual in a particular task; and also (ii) by groups getting better performance, when individuals coordinate their behavior, than either individuals working in parallel [1].

This experiment provided the evolution of the software architecture through Collective Intelligence and interaction of six different teams. This interaction has resulted from the evolution of a richer architecture for solving a problem by the students from CE-245, CE-240, and CE-229 courses, based on the collective knowledge acquired during the IPBL.

121.4 Conclusion

The main result of this project was to combine all students' Collective Intelligence for development of a data analysis system. This system was developed using Collective Intelligence and Agile Methods. It has allowed the use of social media aimed to alert the population, due to crises resulted from environmental disasters.

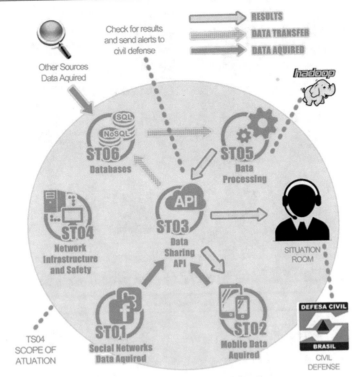

Sprint 0: The IPBL Initial Conceptualization

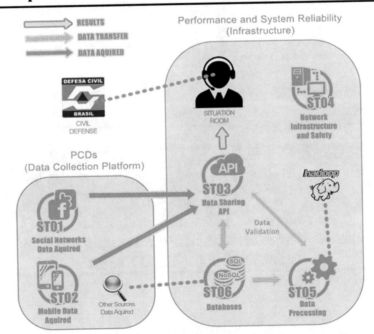

Sprint 1: The adapted IPBL architecture by students after the first Sprint

Fig. 121.1 The evolution of the archtecture

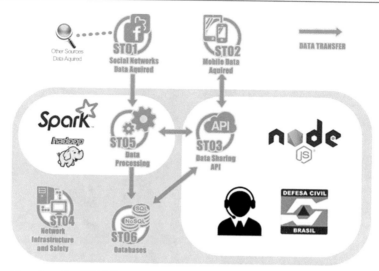

Sprint 2: IPBL architecture proposed by students after the second Sprint

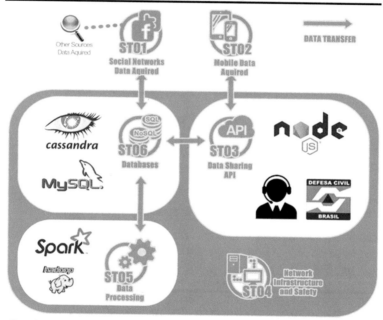

Sprint 3: The evolved IPBL final architecture by students after the third Sprint

Fig. 121.1 (continued)

References

1. Green, B. (2015). Testing and quantifying collective intelligence. In *Proceedings of the Collective Intelligence Conference*. Santa Clara, CA, USA.
2. Lévy, P. (1997). *Collective intelligence: Mankind's emerging world in cyberspace*. Cambridge: Perseus Books.
3. Krivitsky, A. (2011). *Simulação de Scrum com Peças de Lego*. https://www.scrumalliance.org/system/resource_files/0000/3689/ Scrum-Simulation-with-LEGO-Bricks-v2.0.pdf under a Creative Commons Attribution 3.0 Unported License.
4. Kniberg, H., & Sutherland, J., Cohn, M. (2007). Scrum e XP direto das Trincheiras. C4Media of InfoQ.com. https://www.infoq.com/br/minibooks/scrum-xp-from-the-trenches.
5. White, T. (2012). *Hadoop: The definitive guide*. Sebastopol: O'Reilly Media, Inc.
6. Karau, H., Konwinski, A., Wendell, P., & Zaharia, M. (2015). *Learning spark: Lightning-fast big data analysis*. Sebastopol: O'Reilly Media, Inc.

Development of Human Faces Retrieval in a Big Photo Database with SCRUM: A Case Study

Thoris Angelo Pivetta, Carlos Henrique Quartucci Forster,
Luiz Alberto Vieira Dias, and Eduardo Martins Guerra

Abstract

In this paper we present a case study of the use of SCRUM in an applied research scenario, consisting of the detection and identification of human faces in a large database of photos. The study shows how the adoption of agile methods was important to deal with uncertainty and unstable requirements, allowing adaptable development with opportunity to learn from the development process. The implementation of experiments reveals, as our research advances, the applicability of techniques to support the solution for the requirements. All sort of problems with illumination, pose, low quality of image and other type of noise were present as obstacles to accuracy in different approaches for detection and recognition. Fast retrieval of similar faces was also necessary. In the end, the requirements were successfully tackled by developing an efficient and short implementation of face alignment, feature extraction and search.

Keywords

Face detection • Face recognition • Big data • Agile methodology • SCRUM

122.1 Introduction

Agile methods applied in software development are recommended for complex situations when the requirements are not clear or there is an uncertain path towards the final product. But, one very interesting and beneficial characteristic of "agile" is the opportunity to create a learning environment from the development effort [1]. This is the point, we believe, research meets development. And projects involving research should promote experimentation, evaluation of techniques and application of prototypical tools.

In contrast to the acquisition of off-the-shelf products, in an applied research effort, stakeholders are partners of the development team, producing the requirements and learning from the progress of the developers, being able to evolve their initial concepts and beliefs into refined requirements and scope. This close interaction with the development team appears to be costly, but has as aim generating a product that fits the real demands of the final user, with feasible requirements learned in the development process [2].

We describe how a small team (professor and student with cooperation of a scrum master and a product owner), working partial time, guided by the SCRUM methodology, spent the first year of applied research and development of a solution for fast face recognition from a huge database of photos intended for fraud detection in issuing credit for retail purchases. This work involves many sources of uncertainty, including uncertain requirements, lack of understanding of the role of the research team from the stakeholders, uncertain applicability of known techniques and ready-available implementations to address the real requirements, and

T.A. Pivetta (✉) • C.H.Q. Forster • L.A.V. Dias
Instituto Tecnologico de Aeronautica, ITA, Sao Jose dos Campos, Sao Paulo, Brazil
e-mail: thorisangelo@hotmail.com; forster@ita.com.br; vdias@ita.br

E.M. Guerra
Instituto Nacional de Pesquisa Espacial, INPE, Sao Jose dos Campos, Sao Paulo, Brazil
e-mail: guerraem@gmail.com

© Springer International Publishing AG 2018
S. Latifi (ed.), *Information Technology – New Generations*, Advances in Intelligent
Systems and Computing 558, DOI 10.1007/978-3-319-54978-1_122

difficulties in predicting the total effort needed to arrive at a workable solution. By using the agile methodology as a guide and research practice to promote learning during the development, we were able to reach, by the end of the year, an agreement on the requirements and a preliminary working release of the face search engine. In particular, the results of the experiments were used as a feedback to generate relevant information for helping to define the really feasible and desired requirements and the respective most appropriate solutions.

122.2 Implemented Fraud-Prevention Case

The case we describe was to develop an application based on applied research of face recognition intended for fraud detection in issuing credit for retail purchases [3]. We documented our case considering functional and non-functional requirements. Functional requirements include searching the database for similar faces along with their identities. Non-functional requirements include dealing with most of the problematic

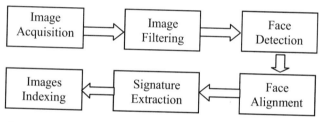

Fig. 122.1 Flowchart of face recoginition process

images of the dataset, responding fast as the number of search requests per minute is quite demanding, and having good accuracy.

For real-world applications, face recognition comprises many steps (Fig. 122.1), which include preparation of image data in pre-processing and normalization steps. Filtering the image is necessary to yield better results [4], and normalization or alignment is necessary for comparison between images.

Beyond this process, an evaluation is necessary to provide statistics for the quality of recognition or recovery, by performing experiments based on data. Results provided are the number of false-positives, false-negatives and accuracy [5].

122.3 The Deliverables

We built a prototype to perform face capture and alignment in real time. Figure 122.2 shows the main screen of the prototype. Only when a face is found and aligned the image on the right is changed. This prototype is useful to capture only the images that can be aligned.

An important type of deliverable in both scientific and applied research is the results of experimental evaluation. Considering the software non-functional requirements, these experiments can also be considered an evaluation of them. We needed to select appropriate methods and tools for the recognition requirements. So, we first tested with ready-made implementations present in OpenCV [6], according to the principle of maximizing value and minimizing first efforts (minimum viable product).

Fig. 122.2 Screenshot of front end application used to align faces

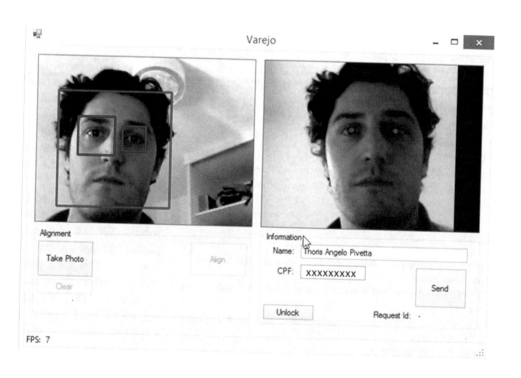

We implemented LBP [7] and ran it with a dataset of 313,643 photos, from which it was possible to detect the eyes and align 219,770, (70%), being about 8% false positive face images. It was able to retrieve 18% of people that have multiple occurrence in the first place and 23% up to the tenth place when fetching the most similar signatures. We performed tests with the problem database (a random selection of 6256 aligned images, from with 1000 with multiple occurrences were used for test) obtaining 28% accuracy on the first image. A new database with more structured images taken from people under some mild control protocol and filtered by software was tested in comparison (2633 aligned images, with 333 multiple occurrences of a person), obtaining 51% accuracy. A search engine has been created considering a biometric signature vector as a chain of bytes. The engine is optimized written in C language. The database is a compact binary file that should fit entirely in RAM. The search time in a 4-core i7 Dell Inspiron 15 7537 laptop with 8 Gbytes of RAM for 200,000 images is 5 seconds. With parallel processing, the average search takes about 1.5 second. These figures happen regardless of the number of similar image indices to be retrieved.

and applied research practice, a productive flow of work was established to mitigate initial uncertainties, foster the understanding of the roles, and reach results that are consistent with the end-user needs instead of just expectations. By promoting "learning from development", our mixed approach allows the progressive refinement of requirements, and discovery of pathways that lead to fruitful solutions. Scrum contributed with: close interaction between stakeholders and development team, quick cycles of producing value, well-defined roles and a mediating product owner, documentation artifacts, effort prediction, visualization of the development status by all the team, and adoption of collaborative tools. Applied research practice contributes with: scientific literature review, hypothesis verification through experiments, evaluation of techniques, development of prototypes, evaluation of prototypical tools and components, and prediction by generalization of experimental results.

Acknowledgments We thank the project SPOT, which is a cooperation of ITA, FCMF (Fundação Casimiro Montenegro Filho) and 2RP, for financing this project, and Cred-System for providing us with the images and the valuable feedback.

122.4 The Agile Methodology

Our development were based on Scrum [8]. As prescribed, we employed smaller time-boxes, and our deliveries and ceremonies were done monthly and we have followed the steps below before each delivery: **Sprint planning** – The first meeting of the month to update the backlog and perform effort estimation with planning poker. **Sprint development** – we had weekly meetings focusing on updating the status of the development using the Kanban and the burndown chart and informing of obstructions to be solved. **Sprint Retrospective** – consolidation meeting with validation of the deliverables by the product owner.

122.5 Conclusion

We presented an applied research project with the aim to develop a solution for face retrieval from large image databases. By mixing concepts from Agile Methodologies

References

1. Beck, K., et al. (2001). *The agile manifesto, 2009*. Snowbird, UT, USA.
2. Leau, Y. B., et al. (2012). Software development life cycle AGILE vs traditional approaches. *International Conference on Information and Network Technology, 37*(1), 162–167.
3. Najarian, K., Ward, K., & Shirani, S. (2013). Biomedical signal and image processing for clinical decision support systems. *Computational and Mathematical Methods in Medicine, 2013*, 1–2.
4. Shrivakshan, G. T., & Chandrasekar, C. (2012). A comparison of various edge detection techniques used in image processing. *International Journal of Computer Science Issues, 9*(5), 272–276.
5. Zhao, W., et al. (2003). Face recognition: A literature survey. *ACM Computing Surveys (CSUR), 35*(4), 399–458.
6. OpenCV. http://opencv.org/. Accessed 11 Nov 2015.
7. Ahonen, T., Hadid, A., & Pietikainen, M. (2006). Face description with local binary patterns: Application to face recognition. *IEEE Transactions on Pattern Analysis and Machine Intelligence, 28*(12), 2037–2041.
8. Boehm, B. (2007). *A survey of agile development methodologies*. Available online at http://agile.csc.ncsu.edu/SEMaterials/AgileMethods.pdf.

Gildarcio Sousa Goncalves, Rafael Shigemura, Paulo Diego da Silva,
Rodrigo Santana, Erlon Silva, Alheri Dakwat, Fernando Miguel,
Paulo Marcelo Tasinaffo, Adilson Marques da Cunha,
and Luiz Alberto Vieira Dias

Abstract

Accidents and crises, whether climatic, economic, or social are undesirably frequent in everyday lives. In such situations, lives are sometimes lost because of inadequate management, lack of qualified and accurate information, besides other factors that prevent full situational awareness. The goal of this work is to report on an academic conceptualization, design, build, test, and demonstration of computer systems, to manage critical information, during hypothetical crises. During the development of an academic system in the second Semester of 2015 at the Brazilian Aeronautics Institute of Technology, the following challenges occurred: strict specifications, agile methods, embedded systems, software testing, and product assessment. Also, some quality, reliability, safety, and testability measurements have been used. At that time, an Interdisciplinary Problem-Based Learning (IPBL) was performed, adding hardware technologies of environment sensors, Radio Frequency Identification (RFID), and Unmanned Aerial Vehicles (UAVs). Software technologies were used for cloud-based web-responsive platform and a mobile application to geographically manage resources at real-time. Finally, the ANSYS® SCADE (Safety-Critical Application Development Environment) was employed to support the embedded and safety-critical portion of this system.

Keywords

Method Scrum • Agile Methods • Embedded Systems • Interdisciplinarity • Problem Based Learning (PBL)

123.1 Introduction

Accidents and crises, whether climatic, economic, or social are undesirably frequent in everyday lives. In such situations, lives are sometimes lost because of inadequate management, lack of qualified and accurate information, besides other factors that prevent full situational awareness. The goal of this work is to report on an academic conceptualization, design, build, test, and demonstration of computer systems, to manage critical information, during hypothetical crises [1].

At that time, an Interdisciplinary Problem-Based Learning (IPBL) was performed, adding hardware technologies of environment sensors, Radio Frequency Identification (RFID), and

G.S. Goncalves • R. Shigemura (✉) • P.D. da Silva •
R. Santana • E. Silva • A. Dakwat • F. Miguel •
P.M. Tasinaffo • A.M. da Cunha • L.A.V. Dias
Computer Science Department Brazilian Aeronautics Institute
of Technology (Instituto Tecnologico de Aeronautica – ITA),
Sao Jose dos Campos, SP, Brazil
e-mail: gildarciosousa@gmail.com; rafael.shigemura@gmail.com;
paulodiego1@gmail.com; rodrigombsantana@gmail.com;
erlonsilva1952@gmail.com; aldakwat@gmail.com; fjmiguel@gmail.
com; tasinaffo@ita.br; cunha@ita.br; vdias@ita.br

© Springer International Publishing AG 2018
S. Latifi (ed.), *Information Technology – New Generations*, Advances in Intelligent
Systems and Computing 558, DOI 10.1007/978-3-319-54978-1_123

Unmanned Aerial Vehicles (UAVs). Software technologies were used for cloud-based web-responsive platform and a mobile application to geographically manage resources at real-time. Finally, the ANSYS® SCADE (Safety-Critical Application Development Environment) was employed to support the embedded and safety-critical portion of this system.

123.2 Accidents and Crises

According to the United Nations Disaster Risk Reduction Office (UNISDR), a disaster is a serious disruption of the functioning of a community or a society involving widespread human, material, economic or environmental losses and impacts, which exceeds the ability of the affected community or society to cope using its own resources, which impacts in hazard exposures, vulnerability conditions, and insufficient capability to reduce or deal with potential negative consequences [1].

Aiming give a proposal solution for this problem, that reduce the consequences of disasters, this work reports the development of a project, during 17 weeks on the second Semester of 2015, at the Brazilian Aeronautics Institute of Technology, involving 80 undergraduate and graduate students from the following courses: CES-65 Embedded Systems Project; CE-230 Software Quality, Reliability, and Safety; CE-235 Real-time Embedded Systems; and CE-237 Advanced Topics in Software Testing.

The MACIS project was developed, using a web platform and the software ANSYS® SCADE Suite and Display, an Integrated-Computer Aided Software Engineering-Environment (I-CASE-E). Its segments were implemented, during four Sprints of four weeks, by 8 different STs, each one comprised of a Product Owner (PO), a Scrum Master (SM), and 8 Team Developers (TDs).

123.3 The Case Study

This section describes the following key concepts, methods, and technologies used in this project development: Software Quality, Reliability, and Safety; Agile Testing; Agile Scrum Method; the ANSYS® SCADE; and also the involved hardware.

The MACIS project was developed, according to quality, reliability, safety, and testability requirements, in line with DO-178C [4] and the DO-278A.

The software quality, reliability, and safety evaluation is a systematic examination of activities throughout all the sprints in an agile development model, using an auditing process to check whether they are aligned with plans and how they were earlier established to verify correct implementations. Quality

never occurs by accident, it is always the result of high intention, sincere effort, intelligent direction, and skillful execution of expected planning [5].

Software testing was used to identify defects in the software and it checks whether the system actually meets its customers' requirements in terms of effectiveness and use. The software product testing basically involves four steps: test planning, test case design, implementation, and evaluation of test results [6].In general, these steps are materialized in four test levels: unit, integration, system, and acceptance [7].

Test Driven Development (TDD) was also used to guide the development-oriented test, because it must be written before implementing the system. Tests are used to facilitate project understanding and to clarify what is expected from the code [8–11].

According to the Scrum Alliance, Scrum is a framework within which people can address complex adaptive problems, while productively and creatively delivering products of the highest possible value. Scrum has three pillars: transparency, inspection, and adaptation, involving teams following roles, ceremonies, and artifacts [2, 3, 12]:

Roles: Product Owner (PO), Scrum Master (SM), and Team Developer (TD);

Ceremonies: Sprint Planning Meetings, Daily Sprints or Weekly Meetings, Sprint Reviews, and Sprint Retrospectives; and

Artifacts: Product Backlog, Sprint Backlogs, Kanban, and Burndown Charts.

The SCADE Suite® I-CASE-E tool, designed for the highest levels of quality and safety, is a Model-Based Design (MBD) environment dedicated to develop safety-critical software. It provides the generation of significant amount of source-code in C and Ada languages through the implementation of native available components standardized according to DO-178C [12].

The MBD paradigm has been widely used for complex projects in various areas of science and engineering. The ability to perform computer simulations in abstract models of reality can provide hundreds of million dollars savings, instead of generating a single physical prototype test [13].

Another highlight worth to mention is the fact that its software development is potentially certifiable by the DO-178C standard and its use significantly reduces the waste of time and other resources [12].

The following hardware were researched, implemented and/or used in the MACIS Project, as Proof of Concept (PoC):Raspberry Pi, Arduino, Noise Sensor (High Sensitive Voice Sensor), RFID (Radio Frequency Identification) Transmitter/Reader, RFID Bracelet, Drone Parrot 2.0 – a drone quadcopter managed via WiFi connection, using an

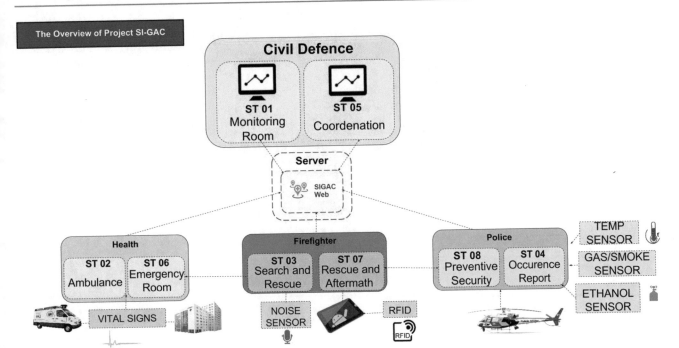

SDK Drone behavior programming with API in C language and GPS; Temperature and Humidity Sensors; and Gas and Smoke Sensor.

123.4 Conclusion

Within a school semester in ITA (seventeen weeks) using agile methods, eight scrum teams coordinated by a General Scrum Master and General Product Owner formed by students from different subjects, both undergraduates and graduates. Despite belonging to the same team and working cooperatively and collaboratively, some team members only met on the final presentation, and yet were able to present a single software product to coordinate efforts and increase efficiency in response to accidents and crises, where several public agencies, (with or without the support of private companies, with independent cultures and management) must act quickly and effectively.

References

1. UNISDR Terminology. Available: https://www.unisdr.org/we/inform/terminology. Access on 18 Mar 2016.
2. Rubin, K. S. (2013). *Essential SCRUM: A practical guide to the most popular agile process*. New York: Addison-Wesley.
3. Sutherland, J., & Schwaber, K. (2013). The definitive guide to scrum: The rules of the game. Available: http://www.scrumguides.org/docs/scrumguide/v1/Scrum-Guide-US.pdf. Access in 18 Mar 2016.
4. RTCA DO-178C. (2011). *Software considerations in airborne systems and equipment certification*. Washington, DC: Radio Technical Commission for Aeronautics (RTCA).
5. Rierson, L. (2013). *Developing safety-critical software: A practical guide for aviation software and DO-178C compliance*. New York: CRC Press.
6. Pressman, R. S. (1997). *Software engineering a practitioners approach* (4th ed.). New York: McGraw-Hill.
7. Copeland, L. (2007). *A practitioner's guide to software test design*. Norwood: Artech House Publishers.
8. Crispin, L., & Gregory, J. (2015). *More agile testing*. New York: Addison-Wesley.
9. Jorgensen, P. C. (2014). *Software testing – A craftsman's approach* (4th ed.). Boca Raton: CRC Press.
10. Astels, D. (2003). *Test-driven development: A pratical guide*. London: Prentice Hall.
11. Beck, K. (2002). *Test-driven development by example*. Boston: Addison Wesley.
12. Esterel Technologies. Available at: http://www.esterel-technologies.com/products/scade-arinc-661/. Accessed on 26 Mar 2016.
13. Esterel Technologies "SCADE Suite". Available at: http://www.esterel-technologies.com/products/scade-suite/. Accessed on 26 Mar 2016.

The Implementation of the Document Management System "DocMan" as an Advantage for the Acceleration of Administrative Work in Macedonia

124

Daut Hajrullahi, Florim Idrizi, and Burhan Rahmani

Abstract

The history of e-government in Macedonia goes back to 2000, with the law for electronic data and that for the electronic signature. From then and until now the Republic of Macedonia has developed significantly in terms of e-government.

Almost all governmental institutions in Macedonia have build applications which will ease the process of dealing with citizens.

In this paper we will analyze the implementation of the "DocMan" application in the Secretariat for European Affairs, which consists of a system for the management with documents.

Keywords

Document Management System • DocMan • Secretariat for European Affairs • Advantages of DMS

124.1 Introduction

More and more governments are using information and communication technology especially Internet or web-based network, to provide services between government agencies and citizens, businesses, employees and other nongovernmental agencies [1].

Aiming towards transparent governance, engaging citizens and citizen-government cooperation, the Macedonian government should have national development strategies of governance.

The primary objective of e-Government should not just be limited to providing information services to its citizens, but it must also include the development of strategic links among various government departments, through the use of communications medium at various levels of government [2].

Document management plays a decisive role in modern e-government applications. As today's authorities have to face the challenge of increasing the efficiency and quality while decreasing the duration of their government processes, a flexible, adaptable document management system is needed for large e-government applications [3]. What we will try to achieve with this paper is to explain the way the application works, the hierarchic way of receiving and delivering documents, as well as to speed up the work through this system.

124.2 System for Document Management "DocMan"

Documents make up one of the most important aspects in the administration process in governmental institutions, as well as their management and their signing, which takes a lot of time in Macedonia.

The hierarchic process of the treatment of the document in governmental institutions works as follows: first, the document is elaborated by the administrator, then, it is sent

D. Hajrullahi • F. Idrizi • B. Rahmani (✉)
State University of Tetova, Computer Science, Tetovo 1200, Macedonia
e-mail: burhanrahmani90@gmail.com

© Springer International Publishing AG 2018
S. Latifi (ed.), *Information Technology – New Generations*, Advances in Intelligent Systems and Computing 558, DOI 10.1007/978-3-319-54978-1_124

to the director of the unit, then it passes through the director of the sector and the state secretary and finally it reaches the due minister.

In the e-Government implementation context, many governments have invested huge amounts of money and manpower to make intergovernmental services both available and user-accepted, although some individuals have no intention of using them. The electronic document management system (EDMS) is the most popular intergovernmental service in the e-Government project [4].

During the analyze conduct in the Secretary for European Affairs of what to expect from a DMS, it showed that its main expectations were to be able to follow at which stage had the document arrived, to have a deadline for its signing or its approval, to centralize the documents in one place, archive them and to have a way for them to be searched and found again.

Along the building of the "DocMan" application it was attempted to take in consideration the needs of the employees of the Secretary.

The implementation of the management system with documents "DocMan" in the Secretary of European Affairs is aimed at shortening the time needed for the approval of he documents. Its purpose is also ecological, as this system allows to avoid needless paper consumption.

124.3 The Advantages of Using the DMS "DocMan"

The advantages of using the document management system "DocMan" in the Secretary were observed during the months long work done with this application. The time needed for the different phases preceding the signing, from the employee to the minister, now is reduced.

Another positive aspect was the start of the gathering of all documents in one place, which also helped in easily finding them. "DocMan" has also impacted in the reduction of paper consumption, as now the documents are sent in electronic form and only at the end, after being approved by all the Secretary instances, it is printed.

124.4 Questionnaire

To make this idea stronger and clear how this software can be important in a working process, we realize to share it through questioners to our collage in administration in Secretariat for European Affairs, where we are part of IT sector.

Questionnaire it was prepared with Google form, online and available for two weeks. The questionnaire includes 6 question completed by 83 participants of Secretariat for European Affairs administration.

First and second questions were "How long they wait for one document to be approved by the head of sector with out "DocMan" and with "DocMan"?

From 83 administrators we get the answer that the head of sectors not delay approval of documents more than one day in standa, and just 3 of the administrators wait for two days, and the similar result it was with "DocMan".

The third question and the fourth question it was related to the state secretary we asked them: "How long they wait for one document to be approved by the state secretary?" (Figs. 124.1 and 124.2)

The fifth and the sixth questions were directed to the highest level: "How long they wait for one document to be approved by Deputy prime Minister without and with "DocMan" ?" (Figs. 124.3 and 124.4)

Fig. 124.1 Answer of question three

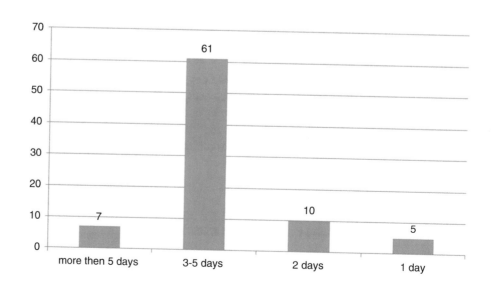

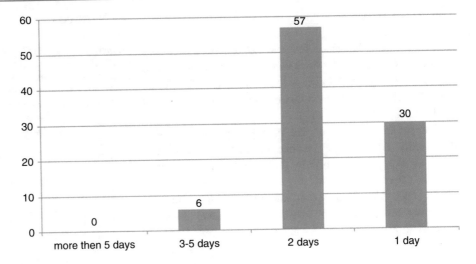

Fig. 124.2 Answer of question four

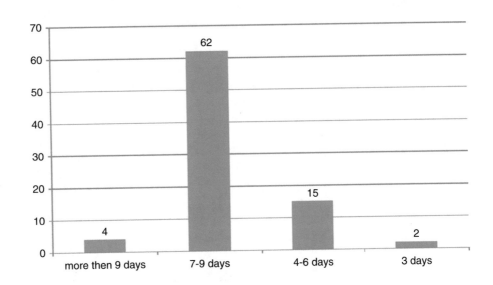

Fig. 124.3 Answer of question five

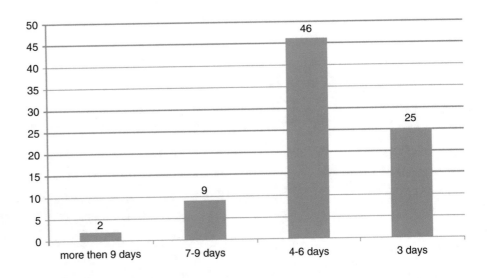

Fig. 124.4 Answer of question six

124.5 Conclusion

Despite the fact that since the year 2000 - more precisely from 2001 to 2015 - a lot of work has been done in order to advance e-governance in Macedonia, there is still a lot of place for improvement in comparison to other countries in the Balkans and in Europe [5].

The implementation of the management system with documents "DocMan" in the Secretary of European Affairs was a challenge of ours of how much could the digitization of data affect the working process in the administration, which, with this paper, we attempted to explain, through clarifying how "DocMan" works and the advantages we have by implementing this or another system for managing documents in the governmental institutions in Macedonia. Also with the questionnaire we wanted to see if administrators agree with us about "DocMan".

References

1. Fang, Z. (2002). E-Government in digital era: Concept, practice, and development. *International Journal of the Computer, the Internet and Management, 10*(2), 1–22.
2. Abdulkadhim, H., Bahari, M., Bakri, A., & Ismail, W. (2015). A research framework of Electronic Document Management Systems (EDMS) implementation process in Government. *Journal of Theoretical and Applied Information Technology, 81*(3), 420–432.
3. Howard, M., & E-Government Across the Globe. (2011, August). How will "e" change Government. *Government Finance Review, 17*(4), 1–9.
4. Hunga, S.-Y., Tanga, K.-Z., Changb, C.-M., & Kea, C.-D. (2009, April). User acceptance of intergovernmental services: An example of electronic document management system. *Government Information Quarterly, 26*(2), 387–397.
5. Rahmani, B., & Idrizi, F. (2015, April–June). An analysis of Macedonia's position in the world ranking in years for e-government and the development of an (online) application for the submission of corruption, *International Journal of Computer Science And Technology (IJCST), 6*(2), 58–61.

Index

© Springer International Publishing AG 2018
S. Latifi (ed.), *Information Technology – New Generations*, Advances in Intelligent
Systems and Computing 558, DOI 10.1007/978-3-319-54978-1